TM 9-2320-272-24-1
5 Ton M939 Series Truck
Direct and General Support
Maintenance Manual
Vol 1 of 4
June 1998

This manual contains maintenance information for the 5 ton M939 US Military Trucks. This is volume 1 of 4 in the Direct / General Support Manual Series. M939 series trucks are a 5 ton heavy duty 6x6 truck. Cargo versions were designed to transport 10,000 pounds of cargo in all terrain and all weather conditions. Originally designed in the 1970's to replace the M39 and M809 series of vehicles. 44,590 units were produced. This manual is printed to help private owners in the maintenance of their vehicles.

Should you have suggestions or feedback on ways to improve this book please send email to Books@OcotilloPress.com

Edited 2021 Ocotillo Press
ISBN 978-1-954285-63-7

Cover Image Credit: LCPL Patrick G. Pressdee, USMC
Image Source: Public Domain work of the US Government

Ocotillo Press
Houston, TX 77017
Books@OcotilloPress.com

Disclaimer: The user of this book is responsible for following safe and lawful practices at all times. The publisher assumes no responsibility for the use of the content of this book. The publisher has made an effort to ensure that the text is complete and properly typeset, however omissions, errors, and other issues may exist that the publisher is unaware of.

ARMY TM 9-2320-272-24-1
AIR FORCE TO 36A12-1C-1155-2-1

This publication supersedes TM 9-2320-272-20-1, October 1985, and changes 1 through 4; TM 9-2320-272-20-2, October 1985, and changes 1 through 3; TM 9-2320-272-34-1, June 1986, and chanaes 1 through 2; TM 9-2320-272-34-2, June 1986, and changes 1 and 2; and TM 9-2320-358-24&P, October 1992

TECHNICAL MANUAL
VOLUME 1 OF 4

UNIT, DIRECT SUPPORT, AND GENERAL SUPPORT MAINTENANCE MANUAL
FOR
TRUCK, 5-TON, 6X6, M939, M939A1, M939A2 SERIES TRUCKS (DIESEL)

HOW TO USE THIS MANUAL	v
INTRODUCTION	1-1
SERVICE AND TROUBLESHOOTING INSTRUCTIONS	2-1
PREVENTIVE MAINTENANCE CHECKS AND SERVICES (PMCS)	2-2
UNIT TROUBLESHOOTING	2-60
UNIT MAINTENANCE	3-1

TRUCK, CARGO: 5-TON, 6X6, DROPSIDE,
M923 (2320-01-050-2084) (EIC: BRY); M923AI (2320-01-206-4087) (EIC: BSS); M923A2 (2320-01-230-0307) (EIC: BS7);
M925 (2320-01-047-8769) (EIC: BRT); M925AI (2320-01-206-4088) (EIC: BST); M925A2 (2320-01-230-0308) (EIC: B58);

TRUCK, CARGO: 5-TON, 6X6 XLWB,
M927 (2320-01-047-8771) (EIC: BRV); M927A1 (2320-01-206-4089) (EIC BSW); M927A2 (2320-01-230-0309) (EIC: BS9);
M928 (2320-01-047-8770) (EIC: BRU); M928AI (2320-01-206-4090) (EIC: BSX); M928A2 (2320-01-230-0310) (EIC: BTM);

TRUCK, DUMP: 5-TON, 6X6,
M929 (2320-01-047-8756) (EIC: BTH); M929AI (2320-01-206-4079) (EIC: BSY); M929A2 (2320-01-230-0305) (EIC: BTN);
M930 (2320-01-047-8755) (EIC: BTG); M930AI (2320-01-206-4080) (EIC: BSZ); M930A2 (2320-01-230-0306) (EIC: BTO);

TRUCK, TRACTOR: 5-TON, 6X6,
M931 (2320-01-047-8753) (EIC: BTE); M931AI (2320-01-206-4077) (EIC: BS2); M931A2 (2320-01-230-0302) (EIC: BTP);
M932 (2320-01-047-8752) (EIC: BTD); M932AI (2320-01-205-2684) (EIC: BS5); M932A2 (2320-01-230-0303) (EIC: BTQ);

TRUCK, VAN, EXPANSIBLE: 5-TON, 6X6,
M934 (2320-01-047-8750) (EIC: BTB); M934A1 (2320-01-205-2682) (EIC: BS4); M934A2 (2320-01-230-0300) (EIC: BTR);

TRUCK, MEDIUM WRECKER: 5-TON, 6X6,
M936 (2320-01-047-8754) (EIC: BTF); M936AI (2320-01-206-4078) (EIC: BS6); M936A2 (2320-01-230-0304) (EIC: BTT).

DEPARTMENTS OF THE ARMY AND THE AIR FORCE

JUNE 1998

WARNING

EXHAUST GASES CAN KILL

1. DO NOT operate vehicle engine in enclosed area.
2   DO NOT idle vehicle engine with windows closed.
3. DO NOT drive vehicle with inspection plates or cover plates removed.
4. BE ALERT at all times for odors.
5. BE ALERT for exhaust poisoning symptoms. They are:
    *   Headache
    *   Dizziness
    *   Sleepiness
    *   Loss of muscular control
6. IF YOU SEE another person with exhaust poisoning symptoms:
    *   Remove person from area
    *   Expose to open air
    *   Keep person warm
    *   Do not permit person to move
    *   Administer artificial respiration or CPR, if necessary*
    *   For artificial respiration, refer to FM 21-11.
7. BE AWARE: The field protective mask for Nuclear, Biological, or Chemical (NBC) protection will not protect you from carbon monoxide poisoning.

    THE BEST DEFENSE AGAINST EXHAUST POISONING IS ADEQUATE VENTILATION.

WARNING SUMMARY

- Hearing protection is required for the driver and passenger. Hearing protection is also required for all personnel working in and around this vehicle while the engine is running (AR-40-5 and TB MED 501).

- If required to remain inside vehicle during extreme heat, occupants should follow the water intake, work/rest cycle, and other stress preventive measures (FM 21-10, Field Hygiene and Sanitation).

- If NBC exposure is suspected, all air filter media should be handled by personnel wearing protective equipment. Consult with your unit NBC officer or NBC NC0 for appropriate handling or disposal instructions.

- This vehicle has been designed to operate safely and efficiently within the limits specified in this TM. Operation beyond these limits is prohibited by IAW AR 70-l without written approval from the commander, U.S. Army Tank-automotive and Armaments Command, ATTN: AMCPEO-CM-S, Warren, MI 48397-5000.

- Never work under dump body unless safety braces are properly positioned. Failure to do this will result in injury to personnel.

- During winching operation, never stand between vehicles. Assistant must remain in secondary vehicle to engage service brake if cable snaps or automatic brake fails while towing vehicle. Failure to do this may result in injury to personnel.

- Accidental or intentional introduction of liquid contaminants into the environment is in violation of state, federal, and military regulations. Refer to Army POL (para. l-7) for information concerning storage, use, and disposal of these liquids. Failure to do so may result in injury or death.

- Cleaning solvents are flammable and toxic. Do not use near open flame and always have a fire extinguisher nearby when solvents are used. Use only in well-ventilated places, wear protective clothing, and dispose of cleaning rags in approved container. Failure to do this will result in injury to personnel and/or damage to equipment.

- Eyeshields must be worn when cleaning with compressed air. Compressed air source will not exceed 30 psi (207 kPa). Failure to do so may result in injury to personnel.

- Extreme care should be taken when removing surge tank filler cap if temperature gauge reads above 175°F (79°C). Steam or hot coolant under pressure will cause injury.

- Alcohol used in the alcohol evaporator is flammable, poisonous, and explosive. Do not smoke when removing alcohol evaporator or adding fluid, and do not drink fluid. Failure to do this will result in injury or death.

- Do not perform electrical circuit testing fuel tank with fill cap or sending unit removed. Fuel may ignite, causing injury to personnel.

- When performing battery maintenance, ensure batteries are seated and clamped down, all rubber boots are installed, clamps are well down on battery posts, and all battery cables lie flat against the top of the batteries. Failure to do this may result in injury to personnel and/or damage to equipment.

- Ensure companion seatbelts are not caught inside battery box. This will cause belts to rot which may lead to iujury of personnel.

- On M938/Al/A2 model vehicles, remove spare tire prior to changing tire and install tire in spare tire carrier after tire change is complete. Operation of crane and/or vehicle engine while vehicle is on jacks may result in injury to personnel or damage to equipment.

- Never assemble or disassemble tire and rim assembly while inflated, use inflation to seat lockring on split rim or tire on two-piece rim, or inflate a tire without a tire inflation cage. Injury to personnel may result.

- Do not disconnect air lines or hoses, remove safety valves or CTIS components, or perform brake chamber repairs before draining air reservoirs. Small parts under pressure may shoot out with high velocity, causing injury to personnel.

WARNING SUMMARY (Contd)

- Remove all jewelry when working on electrical circuits. Jewelry coming in contact with electrical circuits may produce a short circuit, causing extreme heat, explosions, and fling particles of metal. Failure to do so will result in injury or death and damage to equipment.

- Use eyeshields and follow instructions carefully when performing assembling, disassembling, or maintenance on this device. Components of this device are under spring tension and may shoot out at a high velocity. Failure to do so will result in injury to personnel.

- Do not remove hoses with engine running or start engine with hoses removed. High-pressure fluids may cause hoses to whip violently and spray randomly. Failure to do so may result in injury to personnel.

- Keep hands out from between metal surfaces when removing heavy components. Failure to do so may result in injury to personnel.

- Keep personnel out from under equipment and components of equipment when supported by only a lifting device. Sudden loss of lifting power or shift in load may result in injury or death.

- Do not drain engine, transmission, or radiator fluids, or remove lines containing these fluids, when hot. Doing so may result in injury to personnel.

- Vehicle will become charged with electricity if it contacts or breaks high-voltage wires. Do not leave vehicle while high-voltage lines are in contact with vehicle. Failure to do so may result in injury to personnel.

- Wear hand protection when handling lifting and winching cables, hot exhaust components, and parts with sharp edges. Failure to do so may result in injury to personnel.

- Do not perform fuel system procedures while smoking or within 50 ft (15.2 m) of sparks or open flame. Diesel fuel is highly flammable and can explode easily, causing injury or death to personnel and/or damage to equipment.

- Ensure drainvalve on aftercooler is open when filling cooling system. Failure to do so may result in injury to personnel.

- Turbocharger intake fins are extremely sharp and turn at very high rpm. Keep hands and loose items away from intake openings. Failure to do so may result in injury to personnel.

- Do not place hands between frame and radiator when removing screws from trunnion or lifting radiator. Sudden changes in support may cause the radiator to shift, causing injury to personnel.

- Air pressure may create airborne debris. Use eye protection or injury to personnel may result.

- Air system components are subject to high pressure. Always relieve pressure before loosening or removing air system components.

- Wear safety goggles when using a hammer.

- Ether is extremely flammable. Do not perform ether start system procedures near fire. Injury to personnel may result.

ARMY TM 9-2320-272-24-I
AIR FORCE TO 36A12-1C-1155-2-I

TECHNICAL MANUAL
NO. 9-2320-272-24-1

TECHNICAL ORDER
NO. 36A12-1C-1155-2-1

HEADQUARTERS
DEPARTMENTS OF THE ARMY AND THE AIR FORCE
Washington,D.C., 30 JUNE 1998

TECHNICAL MANUAL
VOLUME 1 OF 4
UNIT, DIRECT SUPPORT, AND
GENERAL SUPPORT MAINTENANCE MANUAL
FOR

TRUCK, 5TON, 6X6, M939, M939A1, M939A2 SERIES TRUCKS (DIESEL)

TRUCK	MODEL	EIC	NSN WITHOUT WINCH	NSN WITH WINCH
Cargo, Dropside	M923	BRY	2320-01-050-2084	
Cargo, Dropside	M923A1	BSS	2320-01-206-4087	
Cargo, Dropside	M923A2	BS7	2320-01-230-0307	
Cargo, Dropside	M925	BRT		2320-01-047-8769
Cargo, Dropside	M925A1	BST		2320-01-206-4088
Cargo, Dropside	M925A2	BS8		2320-01-230-0308
Cargo	M927	BRV	2320-01-047-8771	
Cargo	M927A1	BSW	2320-01-206-4089	
Cargo	M927A2	BS9	2320-01-230-0309	
Cargo	M928	BRU		2320-01-047-8770
Cargo	M928A1	BSX		2320-01-206-4090
Cargo	M928A2	BTM		2320-01-230-0310
Dump	M929	BTH	2320-01-047-8756	
Dump	M929A1	BSY	2320-01-206-4079	
Dump	M929A2	BTN	2320-01-230-0305	
Dump	M930	BTG		2320-01-047-8755
Dump	M930A1	BSZ		2320-01-206-4080
Dump	M930A2	BTO		2320-01-230-0306
Tractor	M931	BTE	2320-01-047-8753	
Tractor	M931A1	BS2	2320-01-206-4077	
Tractor	M931A2	BTP	2320-01-230-0302	
Tractor	M932	BTD		2320-01-047-8752
Tractor	M932A1	BS5		2320-01-205-2684
Tractor	M932A2	BTQ		2320-01-230-0303
Van, Expansible	M934	BTB	2320-01-047-8750	
Van, Expansible	M934A1	BS4	2320-01-205-2682	
Van, Expansible	M934A2	BTR	2320-01-230-0300	
Medium Wrecker	M936	BTF		2320-01-047-8754
Medium Wrecker	M936A1	BS6		2320-01-206-4078
Medium Wrecker	M936A2	BTT		2320-01-230-0304

*This publication supersedes TM 9-2320-272-20-1,24 October 1985, and changes 1 through 4; TM 9-2320-272-20-2,25 October 1985, and changes 1 through 3; TM 9-2320-272-34-1, 10 June 1986, and changes 1 through 2; TM 9-2320-272-34-2, 10 June 1986, and changes 1 and 2; and TM 9-2320-358-24&P, 21 October 1992.

This publication is published in four volumes. TM 9-2320-272-24-1 contains chapters 1,2, and 3 (through section IX). TM 9-2320-272-24-2 contains chapters 3 (sections X through XVI) and 4 (sections I through III). TM 9-2320-272-24-3 contains chapter 4 (sections IV through XVI). TM 9-2320-272-24-4 contains chapters 5 and 6 and appendices A through H. Volume 1 contains a table of contents for the entire manual. Volumes 1,2, and 3 contain an alphabetical index covering tasks found in their respective volume. Volume 4 contains an alphabetical index covering all tasks found in the entire manual.

TABLE OF CONTENTS

VOLUME 1 OF 4

TABLE OF CONTENTS (Contd)

HOW TO USE THIS MANUAL

ABOUT YOUR MANUAL

Spend some time looking through this manual. You'll find that it has a new look, different than most TMs you've been using, including its predecessors, TM 9-2320-272-20-1, TM 9-2320-272-20-2, TM 9-2320-272-34-1, TM 9-2320-272-34-2, and TM 9-2320-358-24&P.

New features added to improve the convenience of this manual and increase your efficiency are:

a. **Accessing Information** - These include physical entry features such as the bleed-to-edge indicators on the cover and edge of the manual. Extensive troubleshooting guides for specific systems lead directly to step-by-step directions for problem solving and maintenance tasks.

b. **Illustrations** -A variety of methods are used to make locating and fixing components much easier. Locator illustrations with keyed text, exploded views, and cut-away diagrams make the information in this manual easier to understand and follow.

c. **Commonality** - Only items that are unique to a specific series or vehicle will be given a separate procedure. When only minor differences occur, notes are added to tell you to alter the normal procedure in one way or another to accommodate the differences.

d. **Keying Text with Illustrations** - Instructions/text are located together with illustrations of the specific task you are performing. In most cases, the task steps and illustrations are located on facing pages, with text on the left and illustrations on the right. Continue reading for an example of modular text and illustration layout,

HOW TO USE THIS MANUAL

Here's an example of how to use your manual:

Task: The unit maintenance mechanic of an M939 series vehicle reports that the engine cranks but fails to start.

Troubleshooting steps:

1. Look at the cover of this manual. You'll see subjects listed from top to bottom on the right-hand side.

2. Look at the right edge of the manual. On some of the pages you'll see black bars (edge indicators) that are aligned with the subject bars on the cover. These are the locations of the subjects in the manual.

3. Look for UNIT TROUBLESHOOTING in the subject list on the cover,

4. Turn to the page with the edge indicator matching the subject bar for UNIT TROUBLESHOOTING. Page numbers are also listed next to the subject titles.

5. An edge indicator is placed on the first page of the UNIT TROUBLESHOOTING.

NOTE
If the first page of a subject has an even-numbered page, it will appear on your left, but the edge indicator will still be on the page to your right. So, if the information you are looking for does not start on the page to your right, it will on the page to your left.

6. Look down the list until you find ENGINE. Beneath that heading you will find the symptoms noted by the maintenance mechanic: Engine cranks, fails to start.

7. Turn to the page indicated: page 2-9.

8. On page 2-9, steps/test relating to resolving the malfunction of "Engine cranks, fails to start," are:
Step 1. You inspect the fuel pump shutoff valve and find it is defective and must be replaced. You see para. 4-6 referenced.
Step 2. The rest of the inspection shows no other cause for the problem.

9. Locate para. 4-5, and you will find the detailed procedure for removing the old fuel pump shutoff valve and replacing it with a new one.

v

HOW TO USE THIS MANUAL (Contd)

NOTE
Before starting the maintenance task, look through the entire procedure to familiarize yourself with it.

a. The entire procedure (4-5. FUEL PUMP SHUTOFF REPLACEMENT), according to the THIS TASK COVERS header, includes: a. Removal and b. Installation.

b. The nine basic subheadings listed under INITIAL SETUP outline the task preconditions, materials, tools, manpower requirements, and general safety instructions. They are:

- APPLICABLE MODELS: Any models that require that particular maintenance task.
- TEST EQUIPMENT: Test equipment needed to complete a task.
- SPECIAL TOOLS : Those tools required to complete the task.
- MATERIALS/PARTS: All parts or materials required to complete the task.
- PERSONNEL REQUIRED: The number and type of personnel needed to accomplish a task.

NOTE
If you think you need more help to correctly or safely complete the task (perhaps as a result of an unusual condition, etc.), alert your supervisor and ask for help.

- REFERENCES(TM): Those manuals required to complete the task.
- EQUIPMENT CONDITIQN Notes the conditions that must exist before starting the task. For (4-5. FUEL PUMP SHUTOFF REPLACEMENT), the hood must be raised and secured (TM 9-2320-272-10), and the left splash shield removed (TM 9-2320-272-10).
- SPECIAL ENVIROMENTAL CONDITIONS: Outlines specific environmental conditions necessary to perform the task. For example: Darken an area when adjusting headlight beams.
- GENERAL SAFETY INSTRUCTIONS: Summarizes all safety warnings for the maintenance task.

c. A step-by-step maintenance procedure follows the INITIAL SETUP.

d. At the end of the procedure, FOLLOW-ON TASK(S) will list those additional task(s) that must be performed to complete the procedure. The follow-on tasks for 4-5. FUEL PUMP SHUTOFF REPLACEMENT are: Start engine (TM 9-2320-272-10), check fuel pump shutoff valve for proper operation (TM 9-2320-272-101, install left. splash shield (TM 9-2320-272-10), and lower and secure hood (TM 9-2320-272-10X

10. Refer to our example procedure, 4-5. FUEL PUMP SHUTOFF REPLACEMENT as we review the following points:

a. Modular Text: Both pages of text and illustrations are to be used together. This manual was designed so that the two pages are visible at once, making part identification and procedure sequence easy to follow.

b. Initial Setup: Outlines task conditions.

c. Illustrations: An exploded view of the components removed from the vehicle shows part locations, attachments, and spatial relationships.

11. Refer to TM 9-2320-272-24P, Unit, Direct Support, and General Support Maintenance Repair Parts and Special Tools List for Truck, 5-Ton, 6x6, M939, M939A1, M939A2 Series Trucks, when requisitioning parts and special tools for Unit, Direct Support, and General Support maintenance,

12. Your manual is easier to use once you understand its design. We hope it will encourage you to use it more often.

CHAPTER 1
INTRODUCTION

Section I. GENERAL INFORMATION

1-1. SCOPE

a. This manual contains information and instructions needed to service and maintain 5-ton, 6x6, M939, M939A1, and M939A2 (M939/A1/A2) series vehicles by unit, direct support, and general support maintenance personnel. This manual is intended to be used in conjunction with TM 9-2320-272-10 and TM 9-2320-272-24P for overall completion of tasks found in this manual.

b. The vehicle model numbers and description of vehicles covered in this manual are:

Table 1-1. Vehicle Cross-Reference Table.

VEHICLE DESCRIPTION	M939 SERIES	M939A1 SERIES	M939A2 SERIES
Truck, Cargo, Dropside, Without Winch (WO/W)	M923 (EIC: BRY)	M923A1 (EIC: BSS)	M923A2 (EIC: BS7)
Truck, Cargo, Dropside, With Winch (W/W)	M925 (EIC: BRT)	M925A1 (EIC: BST)	M925A2 (EIC: BS8)
Truck, Cargo, Extra Long Wheelbase (XLWB), Without Winch (WO/W)	M927 (EIC: BRV)	M927A1 (EIC: BSW)	M927A2 (EIC: BS9)
Truck, Cargo, Extra Long Wheelbase (XLWB), With Winch (W/W)	M928 (EIC: BRU)	M928A1 (EIC: BSX)	M928A2 (EIC: BTM)
Truck, Dump, Without Winch (WO/W)	M929 (EIC: BTH)	M929A1 (EIC: BSY)	M929A2 (EIC: BTN)
Truck, Dump, With Winch (W/W)	M930 (EIC: BTG)	M930A1 (EIC: BSZ)	M930A2 (EIC: BTO
Truck, Tractor, Without Winch (WO/W)	M931 (EIC: BTE)	M931A1 (EIC: BS2)	M931A2 (EIC: BTP)
Truck, Tractor, With Winch (W/W)	M932 (EIC: BRD)	M932A1 (EIC: BS5)	M932A2 (EIC: BTQ)
Truck, Van, Expansible	M934 (EIC: BTB)	M934A1 (EIC: BS4)	M934A2 (EIC: BTR)
Truck, Medium Wrecker	M936 (EIC: BTF)	M936A1 (EIC: BS6)	M936A2 (EIC: BTT)

1-2. MAINTENANCE FORMS, RECORDS, AND REPORTS

Department of the Army forms and procedures used for equipment maintenance will be those prescribed by DA Pam 738-750, The Army Maintenance Management System (TAMMS).

1-3. DESTRUCTION OF ARMY MATERIEL TO PREVENT ENEMY USE

Procedures for destruction of Army materiel to prevent enemy use can be found in TM 750-244-6.

1-4. PREPARATION FOR STORAGE OR SHIPMENT

Storage and shipment instructions are found in chapter 6. Additional information can be found in TM 740-90-1, Marking, Packing, and Shipment of Supplies and Equipment: General Packaging Instructions for Field Use.

1-5. OFFICIAL NOMENCLATURE

The nomenclature, names, and designations used in this manual are in accordance with MIL-HDBK-63038-2.

1-6. REPORTING EQUIPMENT IMPROVEMENT RECOMMENDATIONS (EIRs); REPORTING OF ERRORS AND RECOMMENDING IMPROVEMENTS

You can help improve this manual. If you find any mistakes or if you know of a way to improve the procedures, please let us know. Mail your letter, DA Form 2028 (Recommended Changes to Publications and Blank Forms), or DA Form 2028-2 located in back of this manual direct to: Commander, U.S. Army Tank-automotive and Armaments Command, ATTN: AC-NML, Rock Island, IL 61299-7630. A reply will be furnished to you.

1-7. WARRANTY INFORMATION

The 5-ton, 6x6, M939/Al series vehicle Cummins engine (model NHC 250) and Allison transmission (model MT564CR) are warranted in accordance with TB 9-2300-295-15/21. The warranty starts on the date found in block 23, DA Form 2408-9, in logbook. Report all defects in material or workmanship to your supervisor, who will take appropriate action.

1-8. ARMY PETROLEUM, OILS, AND LUBRICANTS (POL)

Proper disposal of hazardous waste material is vital to protecting the environment and providing a safe work environment. Materials such as batteries, oils, and antifreeze must be disposed of in a safe and efficient manner.

The following references are provided as a means to ensure proper disposal methods are followed:
- Technical Guide No. 126 (from the U.S. Army Environmental Hygiene Agency)
- National Environmental Policy Act of 1969 (NEPA)
- Clean Air Act (CAA)
- Resource Conservation and Recovery Act (RCRA)
- Comprehensive Environmental Response, Compensation, and Liability Act
- Emergency Planning and Community Right to Know Act (EPCRA)
- Toxic Substances Control Act (TSCA)
- Occupational Health and Safety Act (OHSA)

The disposal of Army Petroleum, Oils, and Lubricants (POL) products are affected by some of these regulations. State regulations may also be applicable to POL. If you are unsure of which legislation affects you, contact state or local agencies for regulations regarding proper disposal of Army POL.

1-9. USE OF METRIC SYSTEM

The equipment/system described herein contains metric common and special tools; therefore, metric units in addition to U.S. standard units will be used throughout this publication.

1-10. LIST OF ABBREVIATIONS

AOAP - Army Oil Analysis Program
bx- box
CAGEC - Commercial and Government Entity Code
cm - centimeter
CTIS - Central Tire Inflation System
cu-ft - cubic feet
cu-yd - cubic yard
cuM - cubic Meter
ea - each
ft - feet
GM - Grease, Automotive and Artillery
GTW - Gross Towed Weight
gal. - gallon
in. inch
kg- kilograms
km/h - kilometers per hour
km/L - kilometers per liter
kPa - kiloPascal
L - liter
lb-ft - pound-feet
lb - pound
max - maximum
min - minimum

mm - millimeter
mpg - miles per gallon
mph - miles per hour
N•m - Newton meter
NATO - North Atlantic Treaty Organization
OZ - ounce
para. - paragraph
PMCS - Preventive Maintenance Checks and Services
POL - Petroleum, Oils, and Lubricants
psi - pounds per square inch
pt - pint
PTO - Power Takeoff
qt - quart
rpm - revolutions per minute
SEA - Standard, Engineering, and Automotive
TMDE - Test, Measurement, and Diagnostic Equipment
U/M - Unit of Measure
W/W - With Winch
WO/W - Without Winch
XLWB - Extra Long Wheelbase

1-11. GLOSSARY

APPROACH ANGLE - Angle between front tire and front bumper

DEPARTURE ANGLE - Angle between rear tire and rear bumper

FORDING - Crossing through water

GRADE - Steepness of terrain

HYDRAULIC - Operated by oil pressure

OPERATOR - Driver of vehicle

PAULIN - Canvas cover or tarpaulin (tarp)

SLAVING - Jump starting

Section II. EQUIPMENT DESCRIPTION AND DATA

1-12. EQUIPMENT DESCRIPTION AND DATA INDEX

1-13. EQUIPMENT CHARACTERISTICS, CAPABILITIES, AND FEATURES

a. The M939, M939A1, and M939A2 (M939/A1/A2) series of vehicles are varied in design and capabilities. The M939 was a redesign and retrofit of the MS09 series of vehicles, providing enhanced capabilities.

(1) The leading features of the M939 are:

(a) Automatic transmission (Allison MT654)

(b) Hydraulic-assisted power steering system

(c) Complete airbrake system

(d) Improved cooling system

(e) Three-crewmember cab

(f) Tilt hood

(2) Changes were incorporated into later production engines (after engine serial number 11246663) which provided for control of exhaust gas recirculation back to the air intake manifold and the use of top-stop injectors to make up a clean air configuration.

(3) The M939A1 improved on the M939 by adding 14:00xR20 super-sized tires, increasing the minimum road clearance and approach and departure angle. This necessitated a modification to the spare tire rack and lifting device used on all series vehicles.

(4) The M939A2 incorporated a new engine (Cummins 6CTA8.3) and the Central Tire Inflation System (CTIS).

b. The M939/A1/A2 series vehicles can be distinguished from the M809 series by the following features:

(1) On the left side, the exhaust was moved behind the cab and tilted out so exhaust gases clear the side of the vehicle. Hood latches were installed on the sides of the hood near the mirrors. The battery box was incorporated into the companion seat to improve battery life in cold climates. A steering-assist cylinder was installed between the right front frame rail and the right axle hub.

(2) From the front, the hood and fenders are an assembly which tilt forward for access to the engine compartment. A tilt handle and locking device was installed to tilt and hold the hood in a secured open position.

(3) The air filter was moved under the driver's door and the intake stack was brought up behind the cab, even with the cab top.

1-13. EQUIPMENT CHARACTERISTICS, CAPABILITIES, AND FEATURES (Contd)

c. Cargo Trucks With Dropsides: M923/A1/A2 WO/W and M925/A1/A2 W/W.

PURPOSE: These models are used to transport cargo and troops. The vehicles have a payload rating of 10,000 lbs (4,540 kg) and provide 550 cu-ft (15.4 CuM) of cargo space. Removable dropsides and tailgate permit hauling of extra wide loads and easy access for unloading cargo. Troop seats, bows, and canvas are also available. The M925/A1/A2 models have front winches and can be used for recovery operation. The bed of the M923A1/A2 and M925Al/A2 has been shifted back to facilitate a new lifting davit and spare tire mount. When the tire is mounted in its storage location on the M923A1/A2 and M925Al/A2, the top of the tire extends above the minimal reducible height and may need to be removed to obtain the necessary measurement.

M923

M923A1

M923A2

1-13. EQUIPMENT CHARACTERISTICS, CAPABILITIES, AND FEATURES (Contd)

M925

M925A1

M925A2

1-13. EQUIPMENT CHARACTERISTICS, CAPABILITIES, AND FEATURES (Contd)

d. Cargo tick With Extra Long Wheelbase (XLWB): M927/A1/A2 WO/W and M928/A1/1A2 W/W.

PURPOSE: These models are used to transport troops and longer cargo loads. They have the same characteristics as the M923/Al/A2 and M925/A1/A2, but have additional 76 in. (193 cm) of bed space that allows an extra 194 cu-ft (5.4 cuM) of cargo space. Troop seats, bow, and tarpaulin are available. This vehicle has permanent steel-welded sides. The M928/A1/A2 model vehicles have winches and can be used for recovery operations. The bed of the M923A1/A2 and M925A1/A2 has been shifted back to facilitate a new lifting davit and spare tire mount. When the tire is mounted in its storage location on the M923A1/A2 and M925A1/A2, the top of the tire extends above the minimal reducible height and may need to be removed to obtain the necessary measurement.

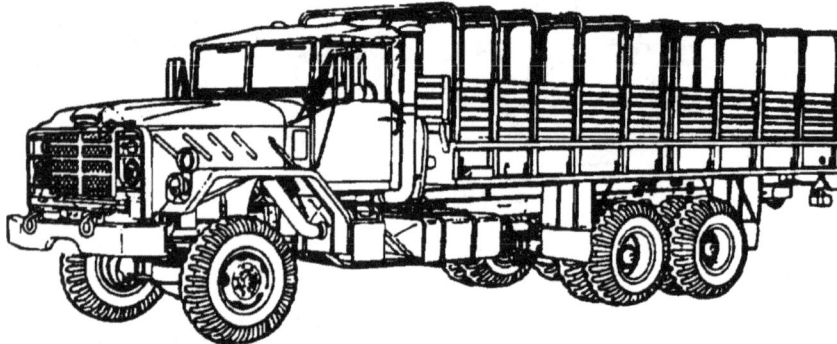

M927

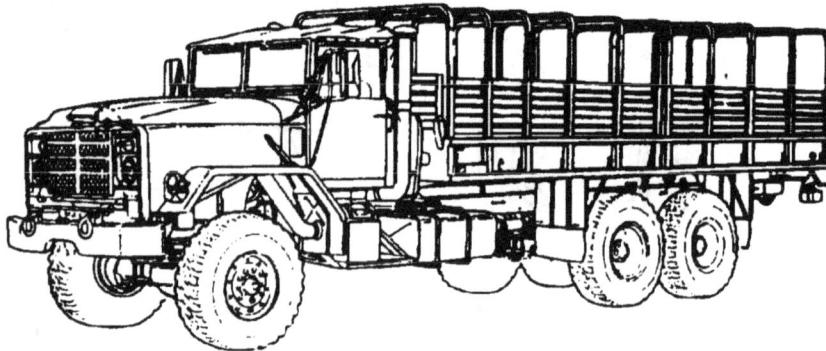

M927A1

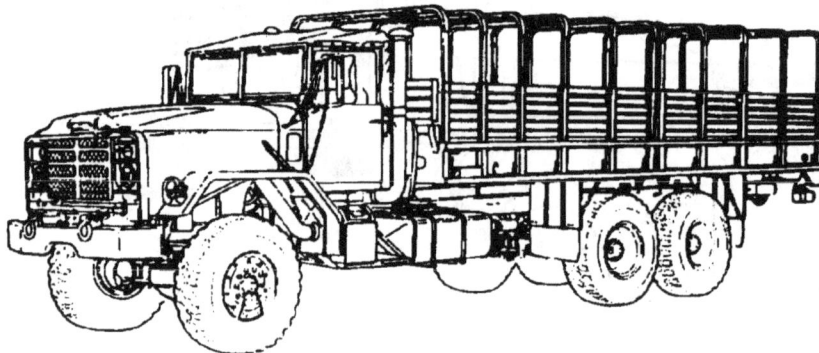

M927A2

1-13. EQUIPMENT CHARACTERISTICS, CAPABILITIES, AND FEATURES (Contd)

M928

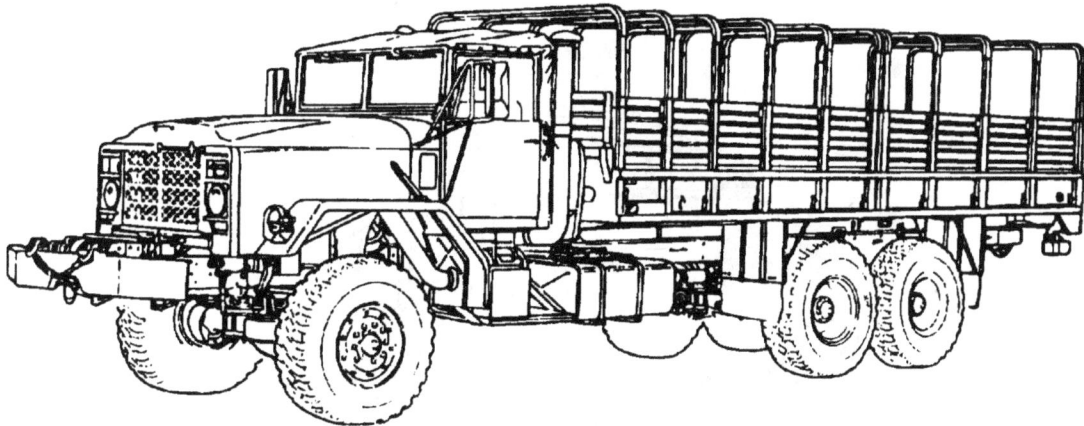

M928A1

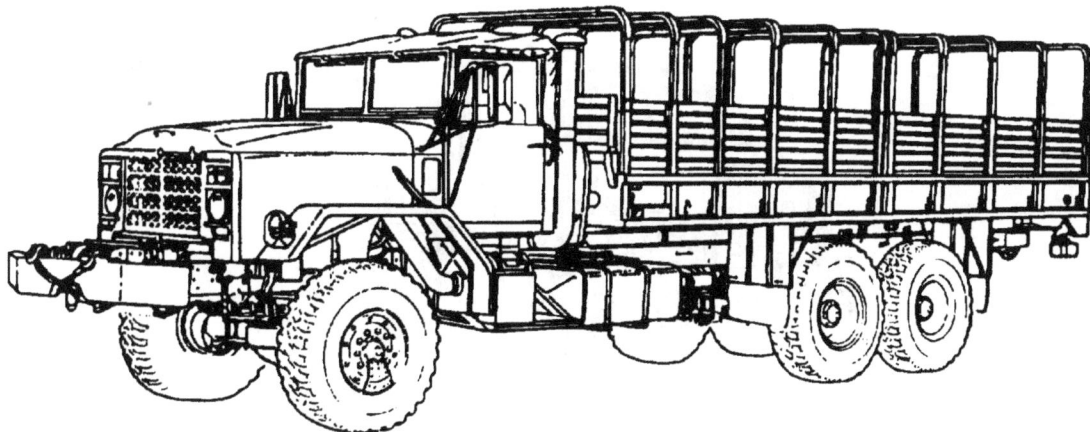

M928A2

1-13. EQUIPMENT CHARACTERISTICS, CAPABILITIES, AND FEATURES (Contd)

e. Dump Truck: M929/A1/A2 WO/W and M930/A1/A2 W/W.

PURPOSE: These models are for hauling and dumping cargo. They have a capacity of 5 cu-yd (3.84 cuM). The bodies have provisions for side racks and troop seats, and bow and tarpaulin for troop transport. The M930/A1/A2 have front winches and can be used for recovery operations. Additional support brackets have been designed on the M929A1/A2 and M930A1/A2 model vehicles and are available to support the bed in a slightly raised position for removal of the 14:00xR20 spare tire.

M929

M929A1

M929A2

1-13. EQUIPMENT CHARACTERISTICS, CAPABILITIES, AND FEATURES (Contd)

M930

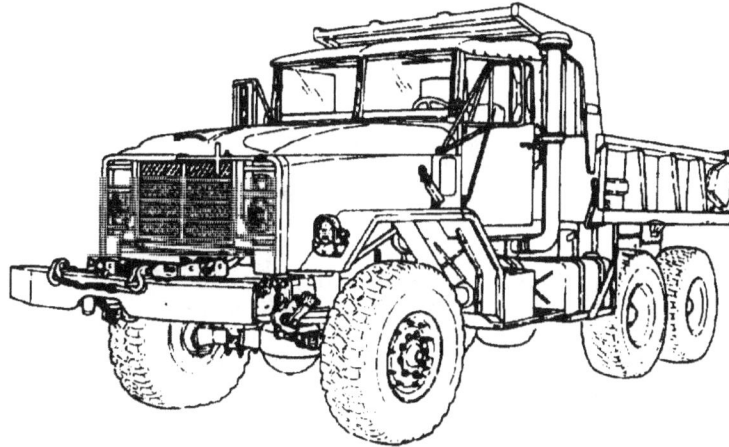

M930A1

M930A2

1-13. EQUIPMENT CHARACTERISTICS, CAPABILITIES, AND FEATURES (Contd)

f. Tractor Trucks: M931/A1/A2 WO/W and M932/A1/1A2 W/W.

PURPOSE: These vehicles are equipped with a fifth wheel and are used for hauling semitrailers. The fifth wheel is capable of pivoting 21 degrees up, 15 degrees down, or 7 degrees sideways. The M932/A1/A2 model vehicles have front winches and can be used for recovery operations.

M931

M931A1

M931A2

l-13. EQUIPMENT CHARACTERISTICS, CAPABILITIES, AND FEATURES (Contd)

M932

M932A1

M932A2

TM 9-2320-272-24-1

1-13. EQUIPMENT CHARACTERISTICS, CAPABILITIES, AND FEATURES (Contd)

g. Expansible Vans: M934/A1/A2.

PURPOSE: These vehicles are used for electronics, maintenance, supply, power, and base station operation. The van body can be expanded when set up in a stationary mode of operation; when mission requires more mobile-type operation, the body is left in the retracted position.

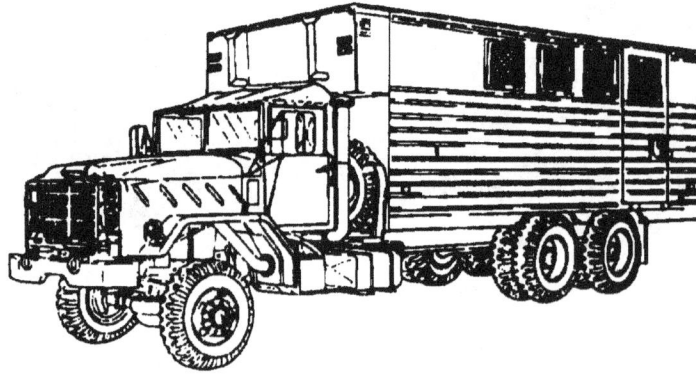

M934

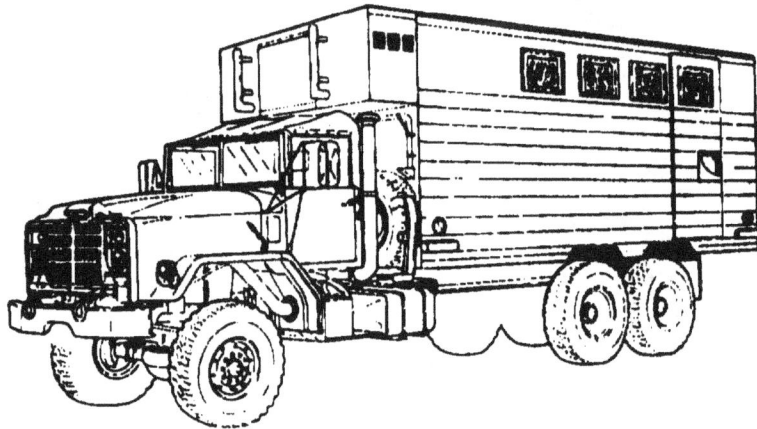

M934A1

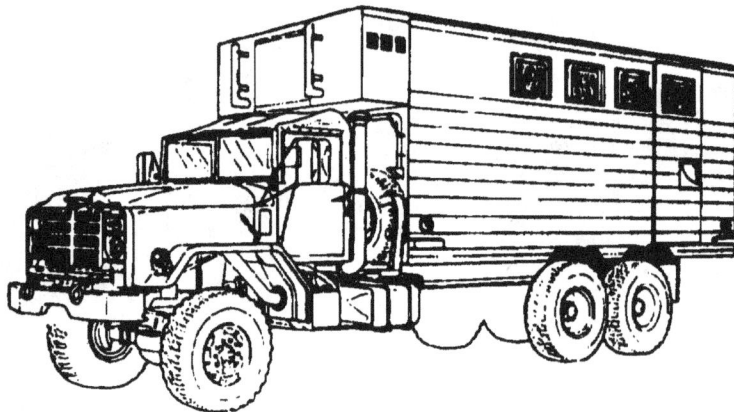

M934A2

1-14

1-13. EQUIPMENT CHARACTERISTICS, CAPABILITIES, AND FEATURES (Contd)

h. Medium Wrecker: M936/Al/A2.

PURPOSE: These vehicles are used for wrecker and salvage operations. They have a revolving hydraulic crane with a self-supported extendable boom. Boom-to-ground supports and outriggers are provided. Crane lifting capacity is 20,000 lb (9,080 kg). The vehicle has a front winch with a 20,000 lb (9,080 kg) capacity. M936/A1 models are equipped with front anchor brackets for heavy straight and side pulls using the front winch. All models are equipped with a 45,000 lb (20,250 kg) capacity rear winch, rear anchor brackets for heavy straight and side pulls, and spring brake override of the Power Takeoff (PTO) air switch for self-recovery operations.

M936

M936A1

M936A2

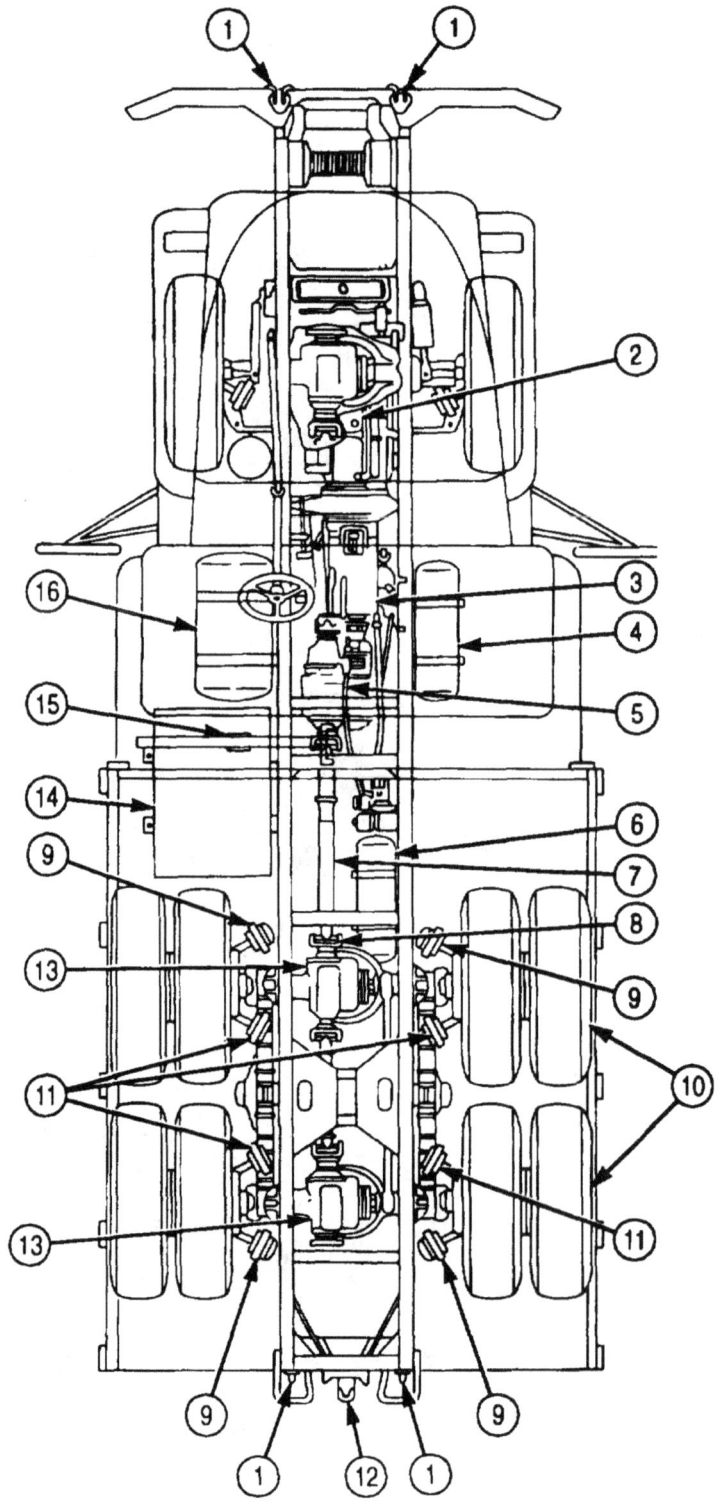

1-14. LOCATION AND DESCRIPTION OF MAJOR COMPONENTS

a. Exterior Components. The components described herein are common to most vehicles covered in this manual. Specific differences can be found in TM 9-2320-272-10 or in table 1-2, Differences Between Models (para. 1-15), of this manual.

(1) LIFTING SHACKLE(S)

(2) ENGINE

(3) TRANSMISSION

(4) PRIMARY AND SECONDARY AIR RESERVOIRS

(5) TRANSFER CASE

(6) WET TANK RESERVOIR

(7) REAR PROPELLER SHAFT(S)

(8) UNIVERSAL JOINT(S)

(9) SERVICE BRAKE CHAMBER

(10) REAR BOGIE

(11) SPRING BRAKE CHAMBER

(12) TOWING PINTLE

(13) REAR DIFFERENTIAL(S)

(14) FUEL TANK(S)

(15) SPARE TIRE CARRIER

(16) AIR CLEANER

1-14. LOCATION AND DESCRIPTION OF MAJOR COMPONENTS (Contd)

Key	Item and Function
1	**LIFTING SHACKLE(S)** - Permit vehicle to be towed by another vehicle or used for tiedown attachment when transporting vehicle.
2	**ENGINE** - There are two different Cummins model engines used, the NHC 250 (M939/A1 series vehicles) and 6CTA8.3 diesel (M939A2 series vehicles), which provide mechanical power to the vehicle and engine-driven subsystems.
3	**TRANSMISSION** - The Allison MT546CR automatic, used on all M939/A1/A2 series vehicles, adapts the engine's output for a varying range of operating speeds.
4	**PRIMARY AND SECONDARY AIR RESERVOIRS** - Provides storage of compressed air and isolation between air subsystems.
5	**TRANSFER CASE** - Directs the engine and transmission output to the specified axles and/or auxiliary equipment.
6	**WET TANK RESERVOIR** - Provides reserve storage of compressed air created by the air compressor or external source when demand is low and releases it when required. This lessens the cycling of the engine-driven air compressor.
7	**REAR PROPELLER SHAFT(S)** - Transmits power between transfer case and forward-rear axle assemblies, and between forward-rear and rear-rear axles assemblies.
8	**UNIVERSAL JOINT(S)** - Provides a flexible point between a component and the propeller shafts. This allows components that cannot maintain precision alignment to transfer power from one point to another, without undue stress **or** breakage.
9	**SERVICE BRAKE CHAMBER** - Mechanical brake actuator that is activated by applying air pressure to an expandable cylinder, causing it to apply braking force to the brakedrum.
10	**REAR BOGIE** - The suspension system comprised of both rear axles, upper and lower torque rods, springs, and seats that support the rear vehicle weight.
11	**SPRING BRAKE CHAMBER** - Mechanical brake actuator that is spring-loaded to apply brakes when air is low or not present. During normal vehicle operation, air piston counteracts the spring tension and allows the brakes to release. Spring brakes can be bypassed by pushing in the spring brake OVERRIDE switch on the instrument panel, or released using procedures found in TM 9-2320-272-10.
12	**TOWING PINTLE** - Provides a secure quick-connect/disconnect for towing vehicles or equipment.
13	**REAR DIFFERENTIAL(S)** - Bi-directionally transfer power from the propeller shaft to the axles, and provide a straight-through connection to power additional propeller shafts.
14	**FUEL TANK(S)** - Provides storage of fuel.
15	**SPARE TIRE CARRIER** - Stores spare tire.

- Davit - Used to load and unload spare tire on M923/A1/A2, M925/A1/A2, M927/A1/A2, M928/A1/A2, M931/A1/A2, and M9321/A1/A2 model vehicles.

- Winch and Swinging Davit - Used to load and unload spare tire on M934A1/A2 model vehicles. Optional kit is available for M934 model vehicle.

- Hoist and Lifting **Eye** - Used to load and unload spare tire on M929/A1/A2, M930/A1/A2, and M934 model vehicles. Optional kit is available to support dump bed in semi-raised position during tire removal.

- Vehicle Crane - Used to load and unload spare tire on M936/A1/A2 model vehicles. **NOTE:** On M936/A1/A2 model vehicles, remove spare tire prior to changing tire, and install tire in spare tire carrier after tire change is complete. Operation of crane and/or vehicle engine while vehicle is on jacks may cause the vehicle to slip off jack.

16	**AIR CLEANER** - Filters air before it enters the intake manifold or turbocharger and collects dust in removable section of filter canister.

1-14. LOCATION AND DESCRIPTION OF MAJOR COMPONENTS (Contd)

b. Interior Cab Components. The components described herein are common to one or more vehicles covered in this manual. Components not found here can be found in TM 9-2320-272-10 or in table 1-2, Differences Between Models (para. 1-15), of this manual.

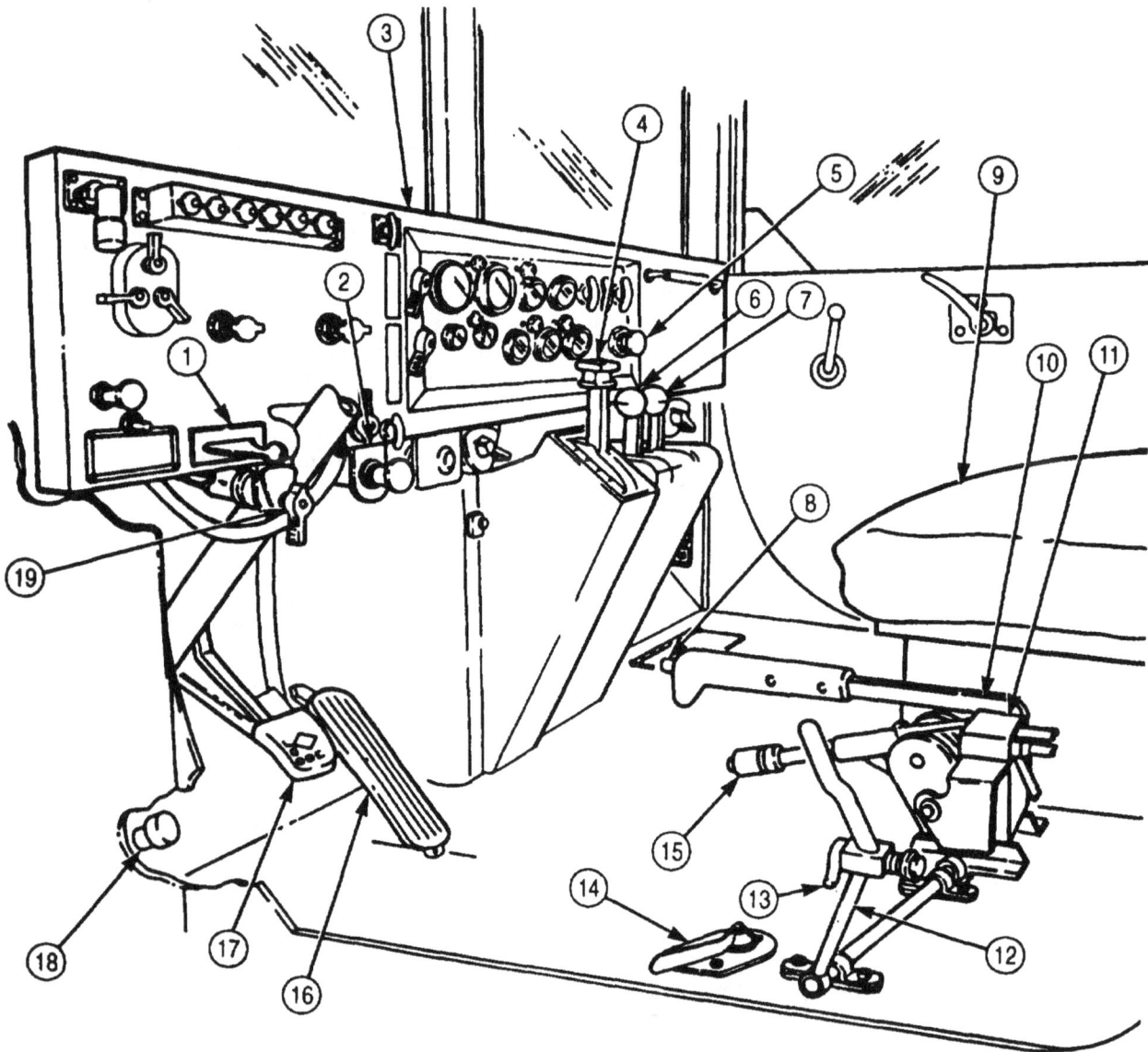

Key Item and Function

1   **FRONT-WHEEL DRIVE LOCK-IN SWITCH** - A pneumatic switch that permits front-wheel drive to be engaged when the transfer case is in HIGH position. Operation of switch is not required when transfer case is in LOW position; system will automatically switch into six-wheel drive.

1-14 LOCATION AND DESCRIPTION OF MAJOR COMPONENTS (Contd)

Key   **Item and Function**

2   **POWER TAKEOFF (PTO) SPRING BRAKE OVERRIDE SWITCH** - Used during rear winch self-recovery operations of M936/A1/A2 model vehicles only. This air control switch must be held in to override spring brake air dump switch on transfer case PTO lever.

3   **INSTRUMENT PANEL** - Houses controls and indicators.

4   **TRANSMISSION SELECTOR LEVER** - Manual control to select driving gear.

5   **SPRING BRAKE OVERRIDE SWITCH** - Pressed in to release spring brakes independent of the mechanical parking brake for tests and adjustments. Can be used to release spring brakes in the event a leak or stoppage occurs in lines between primary and spring brake reservoirs, or primary system air shutoff valve is closed.

6   **TRANSMISSION POWER TAKEOFF (PTO) LEVER** - Used on M925/A1/A2, M928/A1/A2, M929/A1/A2, M930/A1/A2, M932/A1/A2, and M936/A1/A2 model vehicles only. Provides hydraulic power for front winch and/or dump body operation.

7   **WINCH CONTROL LEVER** - Provides control of winch unwinding and winding operations from inside the cab.

8   **TRANSFER CASE SHIFT LEVER SWITCH** - Used in conjunction with transmission selector lever in NEUTRAL position to allow transfer case to be shifted between HIGH and LOW gears.

9   **PASSENGER SEAT** - Combination two-person crew seat, battery box, and storage box. Seats are equipped with two sets of seatbelts, the battery box houses four batteries, and the storage compartment stores technical manuals. NOTE: Ensure companion seatbelts are not caught inside battery box.

10   **TRANSFER CASE SHIFT LEVER** - Pushed down to shift transfer case into HIGH gear, center position for NEUTRAL, and up position for LOW gear. When transfer case is placed in LOW, it will automatically engage six-wheel drive. In HIGH, the instrument panel-mounted front-wheel drive lock-in switch must be used to achieve six-wheel drive operation.

11   **SPRING BRAKE CONTROL SWITCH** - Senses the position of the parking brake control lever and signals the spring brakes to engage when the lever is up and disengage when it is down. Spring brakes can be tested independently from parking brake by raising this switch without pulling up the parking brake control lever (TM 9-2320-272-10).

12   **(a) DUMP BODY CONTROL LEVER** - Used on M929/A1/A2 and M930/A1/A2 to control raising (pulled back) and lowering (pushed forward) of dump body.

        (b) TRANSFER CASE POWER TAKEOFF (PTO) - Used on M936/A1/A2 to control the hydraulic pump that delivers hydraulic pressure to the rear winch and crane. A spring brake air dump switch is installed on the lever when used on these vehicles, which engages the spring brakes. To override this feature during self-recovery operation with the rear winch, the driver must depress and hold the PTO spring brake override air switch on the instrument panel.

13   **DUMP BODY CONTROL LEVER SAFETY LATCH** - Secures dump body control lever or transfer case PTO in NEUTRAL when not in use.

14   **FUEL TANK SELECTOR LEVER** - Used on dual tank model vehicles only. **NOTE:** Tanks must be switched periodically to lessen the buildup of contaminants and fungus.

15   **PARKING BRAKE CONTROL LEVER** - Pulled up to set mechanical brake on output of transfer case and to engage the spring brakes, and down to disengage brakes. Knob on top of handle is turned clockwise to increase braking action on the output of transfer case, and counterclockwise to decrease braking action. Spring brakes engagement is controlled by the spring brake control switch located on the back of the lever.

16   **ACCELERATOR PEDAL** - Foot control to vary the speed of the engine.

17   **BRAKE PEDAL** - Applies the service brakes.

18   **DIMMER SWITCH** - Depressed to change between HIGH and LOW beam setting on headlights.

19   **FUEL GAUGE TANK SELECTOR SWITCH** - Used to switch fuel gauge to read either left or right fuel tank. Installed only on dual tank model vehicles.

1-15. DIFFERENCES BETWEEN MODELS

Table 1-2. Differences Between Models.

EQUIPMENT/FUNCTION	VEHICLE
BODY TYPE	
Cargo Dropside	M923, M923A1, M923A2, M925, M925A1, M925A2
Cargo Fixed Side (XLWB)	M927, M927A1, M927A2, M928, M928A1, M928A2
Crane	M936, M936A1, M936A2
Dump	M929, M929A1, M929A2, M930, M930A1, M930A2
Tractor	M931, M931A1, M931A2, M932, M932A1, M932A2
Van	M934, M934A1, M934A2
CENTRAL TIRE INFLATION SYSTEM	M923A2, M925A.2, M927A2, M928A2, M929A2, M930A2, M931A2, M932A2, M934A2, M936A2
FIELD CHOCKS AND ANCHORS	
Front	M936, M936A1
Rear	M936, M936A1, M936A2
FLOODLIGHTS	M936, M936A1, M936A2
FUEL TANKS	
Dual Tanks 116 gal. (439.1 L)	M929, M929A1, M929A2, M930, M930A1, M930A2, M931, M931A1 M931A2, M932, M932A1, M932A2
Dual Tanks 139 gal. (526.1 L)	M936, M936A1, M936A2
Single Tank 81 gal. (306.6 L)	M923, M923A1, M923A2, M925, M925A1, M925A2, M927, M927A1, M927A2, M928, M928A1, M928A2, M934, M934A1, M934A2
HEAT/AIR CONDITIONED BODY OPERATIONS	
Cargo/Personnel	M923, M923A1, M923A2, M925, M925A1, M925A2, M927, M927A1, M927A2, M928, M928A1, M928A2, M929, M929A1, M929A2, M930, M930A1, M930A2, M934, M934A1, M934A2
Communications/Electronic Repair	M934, M934A1, M934A2
Dump	M929, M929A1, M929A2, M930, M930A1, M930A2
OPERATIONS	
Fifth Wheel	M931, M931A1, M931A2, M932, M932A1, M932A2
Wrecker	M936, M936A1, M936A2

1-15. DIFFERENCES BETWEEN MODELS (Contd)

Table 1-2. Differences Between Models (Contd).

EQUIPMENT/FUNCTION	VEHICLE
TIRES	
ll:00xR20	M923, M925, M927, M928, M929, M930, M931, M932, M934, M936
14:00xR20	M923A1, M923A2, M925A1, M925A2, M927A1, M927A2, M928A1, M928A2, M929A1, M929A2, M930A1, M930A2, M931A1, M93IA2, M932A1, M932A2, M934A1, M934A2, M936A1, M936A2
WHEELBASES	
167 in. (424.2 cm)	M929, M929A1, M929A2, M930, M930A1, M930A2, M931, M93IA1, M931A2, M932, M932A1, M932A2
179 in. (454.7 cm)	M923, M923A1, M923A2, M925, M925A1, M925A2, M936, M936A1, M936A2
215 in. (546.1 cm)	M927, M927A1, M927A2, M928, M928A1, M928A2, M934, M934A1, M934A2
WINCH	
Front	M925, M925A1, M925A2, M928, M928A1, M928A2, M930, M930A1, M930A2, M932, M932A1, M932A2, M936, M936A1, M936A2
Rear	M936, M936A1, M936A2

l-16. EQUIPMENT DATA

Vehicle performance data for the M939/A1/A2 series vehicles are listed in table 1-3. Additional information and equipment service data are in TM 9-2320-272-10.

Table 1-3. Vehicle Performance Data.

	STANDARD	METRIC
1. PAYLOAD		
Carried Weight:		
M923, M923A1, M923A2, M925, M925A1, M925A2, M927, M927A1, M927A2, M928, M928A1, **M928A2**, M929, M929A1, M929A2, M930, M930A1, M930A2.	10,000 lb	4,540 kg
M934, M934A1, M934A2	5,000 lb	2,270 kg
M936, M936A1, M936A2	7,000 lb	3,178 kg
Towed Load on Pintle:		
All (except M936, M936A1, M936A2).	15,000 lb	6,810 kg
M936, M936A1, M936A2	20,000 lb	9,080 kg
On Fifth Wheel:		
M929, M929A1, M929A2, M930, M930A1, M930A2	15,000 lb	6,810 kg
Semitrailer GTW:		
M929, M929A1, M929A2, M930, M930A1, M930A2	37,500 lb	17,025 kg
2. CAPACITIES		
Cooling System:		
All Models	47 qt	44.5 L
Differentials (Each):		
AllModels	12 qt	11.3 L
Engine Crankcase Only:		
M939/A1 Series	23 qt	21.8 L
M939A2 Series	18 qt	17.0 L
Engine Crankcase and Filter:		
M939/A1 Series	27 qt	25.5 L
M939A2 Series	20 qt	18.9 L
Fuel Tank:		
M923, M923A1, M923A2, M925, M925A1, M925A2, M927, M927A1, M927A2, M928, M928A1, M928A2, M934, M934A1, M934A2 (Single Tank)	.81 gal.	306.6 L
M936, M936A1, M936A2 Dual Tanks.	116 gal.	439.1 L
M929, M929A1, M929A2, M930, M930A1, M930A2, M931, M931A1, M931A2, M932, M932A1, M932A2 (DualTanks)	139 gal.	526.1 L
Hydraulic Tank:		
M925, M925A1, M925A2, M928, M928A1, M928A2, M932, M932A1, M932A2	8 gal.	30.3 L
M929, M929A1, M929A2	5 gal	18.9 L
M930, M930A1, M930A2.	6.25 gal.	23.7 L
M936, M936A1, M936A2	100 gal	378.5 L
Steering System:		
M939/A1 Series (Ross)	.5 qt	4.7 L
M939A2 Series (Sheppard).	.3 qt	2.83 L

1-16. EQUIPMENT DATA (Contd)

Table 1-3. Vehicle Performance Data (Contd).

	STANDARD	METRIC
Transmission:		
AllModels(w/oPTO).	17qt	16.1 L
All Models (w/PTO).	19 qt	18.0 L
All Models (w/o PTO and converter dry).	23 qt	22.1 L
All Models (w/PTO and converter dry).	25qt	23.7 L
Transfer Case:		
All Models	6.25 gal.	23.7 L
Winch Gear Case (Front):		
AllModels(W/W)	2.6pt	1.2 L
Winch Gear Case (Rear):		
M936, M936A1, M936A2	7 pt	3.3 L

3. ENGINE

M939/A1 Series Vehicles:

	STANDARD	METRIC
Brake Horsepower.	250 horsepower @2,100 rpm	
Cylinders.	.6 (in-line)	
Fuel Consumption	3-4 mpg	1.3-1.7 km/L
Idle Speed (engine rpm)	600-650 rpm	
Model.	Cummins NHC 250	
Mount(Front).	Trunnion	
Mount (Rear).	Rubber biscuit	
Oil Pressure at Idle.	15 psi	103 kPa
Operating Speed (engine rpm)	1,500-2,100 rpm	
Type	Diesel, normally-aspirated, liquid-cooled	

M939A2 Series Vehicles:

	STANDARD	METRIC
Brake Horsepower.	240 horsepower @ 2,100 rpm	
Cylinders.	.6 (in-line)	
Fuel Consumption	.5.5-6-0 mpg	2.3-2.6 km/L
IdleSpeed.	565-635rpm	
Model	Cummins 6CTA8.3	
Mount(Front).	Trunnion	
Mount (Rear).	Rubber biscuit	
Oil Pressure at Idle.	10 psi	69 kPa
Operating Speed (engine rpm)	2,100 rpm	
Type	Diesel, liquid-cooled, turbocharged, after-cooled	

4. COOLING SYSTEM

M939/A1 Series Vehicles:

	STANDARD	METRIC
Coolant Operating Temperature.	175-195°F	79-91°C
Fan, 6-blade.	.26 in.	660 mm
Thermostat:		
Starts to Open	..175° F	79°C
Fully Open.	185°F	85°C

M939A2 Series Vehicles:

	STANDARD	METRIC
Coolant Operating Temperature.	190-200°F	88-93°C
Fan, 7-blade	.26.5 in.	673 mm
Thermostat:		
Starts to Open.	181°F	83°C
Fully Open.	203°F	95°C

1-16. EQUIPMENT DATA (Contd)

Table 1-3. Vehicle Performance Data (Contd).

Radiator Type .... Crossflow
Surge Tank Cap Pressure .... 14 psi   97 kPa

5. ELECTRICAL SYSTEM

    Alternator:

    Ampere Output (maximum) .... .60 amp
    Model ....
    Voltage Output. .... .28 volts
    Voltage Regulation .... Mounted internal

    Batteries:

    Model 6TN
    Number Required 4
    Plates Per Cell 2 3
    Specific Gravity Full Charge. .... 0 70°F   @ 21°C
    Voltage .... 12 volts
    Protective Control Box. WSU-4001-UT

    Starter:

    Model.. MES6401-CUT
    Voltage.. ....24volts

6. TRANSMISSION (MT654CR)

    Drive Sequence. .... Reverse, Neutral, 1,2,3,4,5
    Drive Range and Shift Control. .... Manual
    Oil Pressure....................................... 26psi   179.3 kPa
    Oil Type OE/HDO-10

    Oil Capacity:

    Drain and Refill. .... .4.25 gal.   16.1 L
    w/o PTO (dry) .... .5.75 gal.   21.8 L
    w/PTO(dry) .... 6.25gal.   23.7 L

    Oil Temperature:

    Maximum ....300°F   149°C
    Normal Operating Temperature. .... 120-220°F   49-104°C

    Power Takeoff:

    Type.. Converter-driven
    Mounting Flange .... One-opening, SAE, 6-bolt

7. TRANSFER CASE

    Model.. T-1138
    Type .... Two-speed synchronized

1-17. DESCRIPTION OF DATA PLATES

The location and contents of caution, data, and warning plates are provided in TM 9-2320-272-10, and a complete list and location of all caution, data, warning, and identification plates is in TM 9-2320-272-24P If any of these plates are worn, broken, painted over, missing, or unreadable, they must be replaced.

Section III. PRINCIPLES OF OPERATION

1-18. GENERAL

This section explains how components of the B-ton M939/Al/A2 series vehicles work together. A functional description of these components and their related parts will be covered in the following paragraphs. To find the operation of a specific system or component, see the principles of operation reference index below.

1-19. PRINCIPLES OF OPERATION REFERENCE INDEX

PARA. NO.	TITLE	PAGE NO.
1-20.	Control System Operation	1-25
1-21.	Power System Operation	1-34
1-22.	Electrical Systems Operation	1-45
1-23.	Compressed Air and Brake System Operation	1-55
1-24.	Hydraulic System Operation	1-71
1-25.	Central Tire Inflation System (CTIS) Operation	1-78

1-20. CONTROL SYSTEM OPERATION

The control system includes those controls and their related parts essential to the operation of the vehicle. These controls are common to all vehicles with the exception of the transmission and transfer case Power Takeoff (PTO) controls. All originate from the cab. Each of these controls and related parts will be described as part of the following systems:

    a. Starting and Ether Starting System Operation (page 1-26).

    b. Accelerator Controls System Operation (page 1-28).

    c. Parking Brake System Operation (page 1-29).

    d. Steering System Operation (page 1-30).

    e. Transmission Control System Operation (page 1-31).

    f. Transfer Case Control System Operation (page 1-32).

1-20. CONTROL SYSTEM OPERATION (Contd)

a. **Starting and Ether Starting System Operation.**

The starting system is identical on all models covered in this manual. It will start the engine in all types of weather and has built-in protection that prevents starting components from reengaging once the engine has been started. Major components of the starting and ether starting system are:

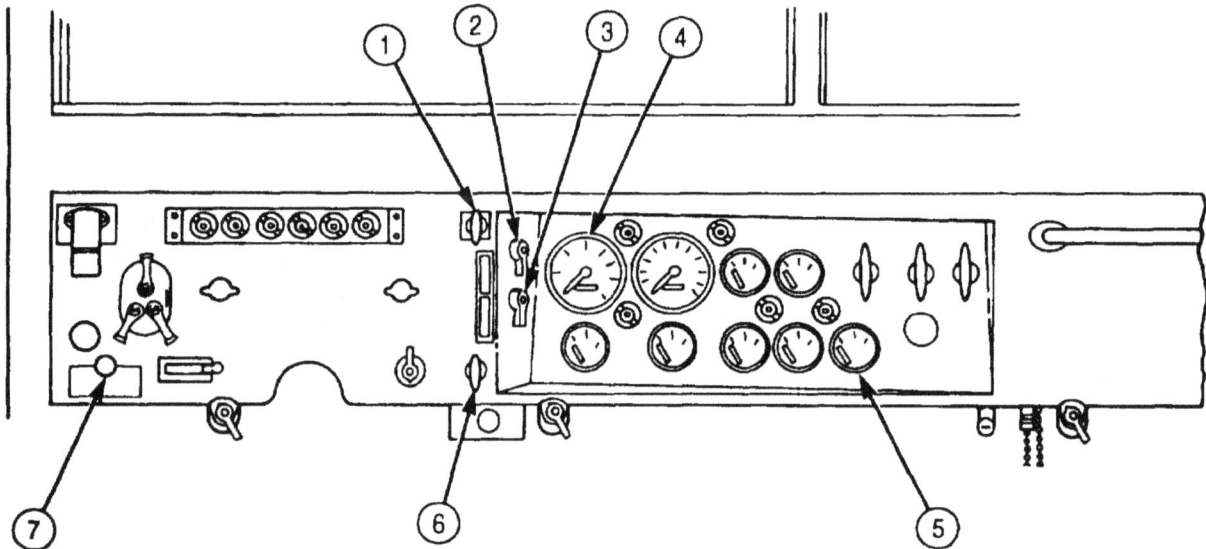

Key	Item and Function
1	**HAND THROTTLE CONTROL** - Used to set engine speed without applying pressure to the accelerator (rotated to lock).
2	**BATTERY SWITCH** -Activates all electrical circuits except arctic heaters.
3	**IGNITION SWITCH** - Has OFF, RUN, and START positions. Switch automatically returns from START to RUN when hand pressure is released.
4	**TACHOMETER** - Indicates speed of engine.
5	**VOLTMETER** - Indicates charging condition of the battery.
6	**EMERGENCY ENGINE STOP** - Control used to shut down engine during emergencies (M934/A1 series vehicles must be reset by unit maintenance).
7	**ETHER START SWITCH** - Injects ether into engine for cold-weather starting.

l-20. CONTROL SYSTEM OPERATION (Contd)

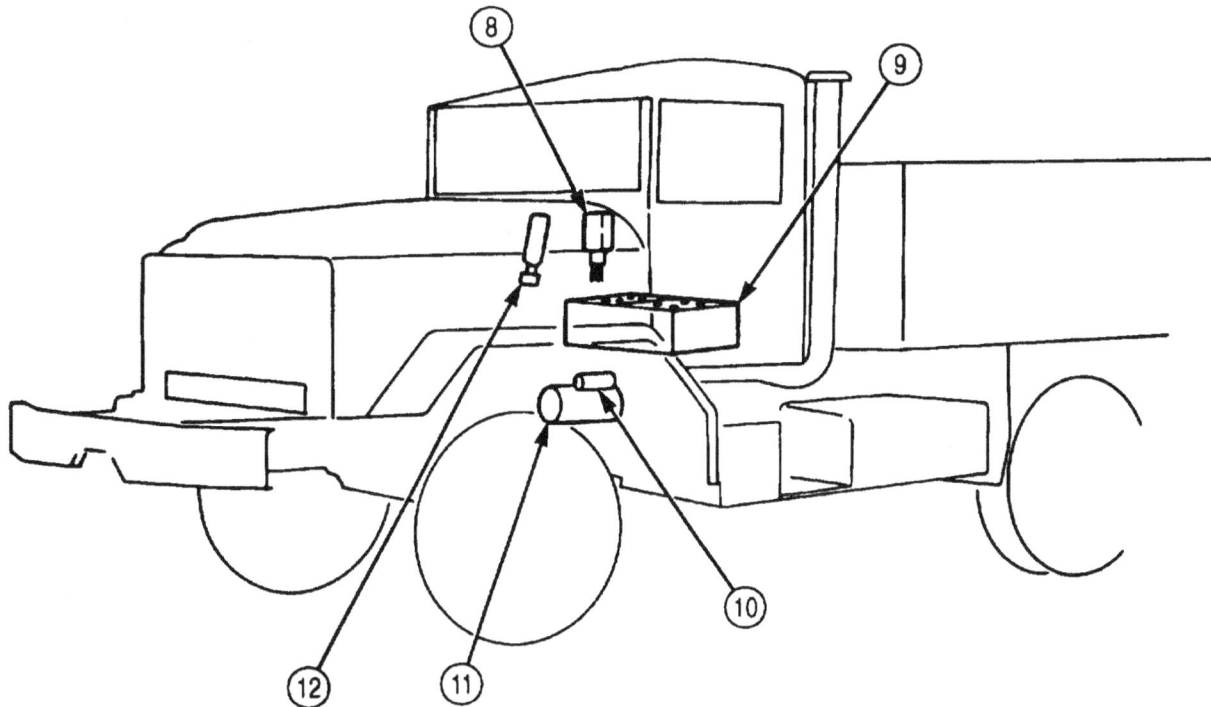

Key Item and Function

8   **PROTECTIVE CONTROL BOX** - Prevents reengagement of starter motor once engine is running.

9   **BATTERIES** - Provide 24-volt electrical current for energizing electrical circuits.

10   **STARTER SOLENOID** - Relays 24-volt battery power to energize starter motor.

11   **STARTER MOTOR** - When energized, converts electrical energy to mechanical power as it engages flywheel to crank engine.

12   **ETHER START CYLINDER** - Stores ether used for cold-weather starting.

l-20. CONTROL SYSTEM OPERATION (Contd)

b. Accelerator Controls System Operation.

The accelerator controls system permits the operator to control vehicle speed and engine power. It is identical on all models covered in this manual. Major components of the accelerator control system are:

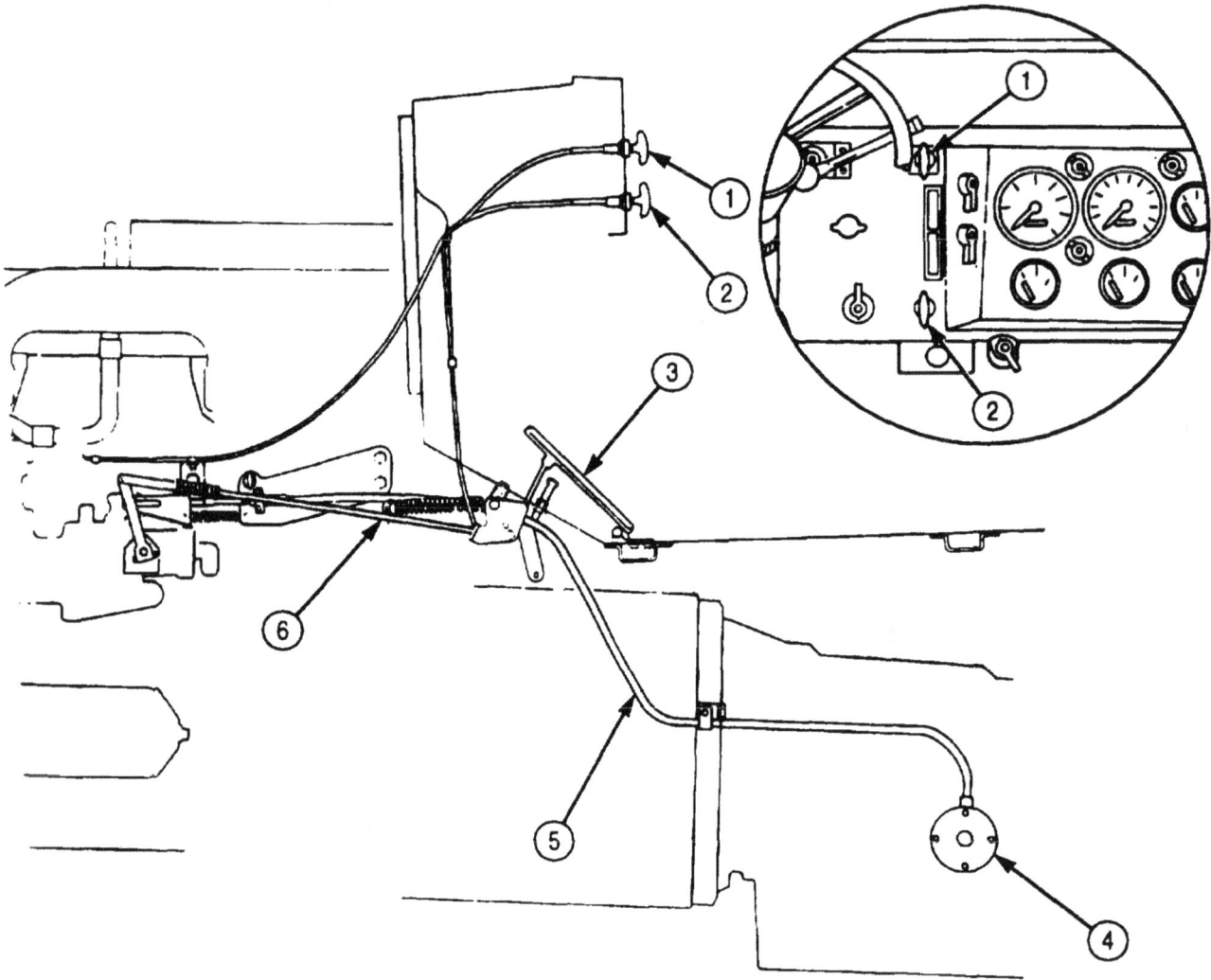

Key Item and Function

1 **HAND THROTTLE CONTROL** - Used to set engine speed without maintaining pressure to the accelerator (rotated to lock).

2 **EMERGENCY** ENGINE STOP CONTROL - Is pulled out to cut off fuel to engine. Used only in an emergency.

3 **ACCELERATOR PEDAL** - Controls engine speed.

4 **MODULATOR** - With transmission selector lever in drive, modulator controls transmission upshifting and downshifting as engine rpm changes.

5 **CABLE** - Connects modulator to fuel pump.

6 **ACCELERATOR LINKAGE** - Links accelerator pedal and throttle control to fuel pump.

1-20. CONTROL SYSTEM OPERATION (Contd)

c. Parking Brake System Operation.

A mechanical and air-actuated brake system performs the following for all vehicles covered in this manual:

(1) Keeps vehicle from rolling once it has stopped.

(2) Slows down or stops vehicle movement.

(3) Provides emergency stopping if there is a complete air system failure.

The mechanical brake system is covered below. The compressed air function of the brake system will be covered in a following paragraph. Major components of the parking brake system are:

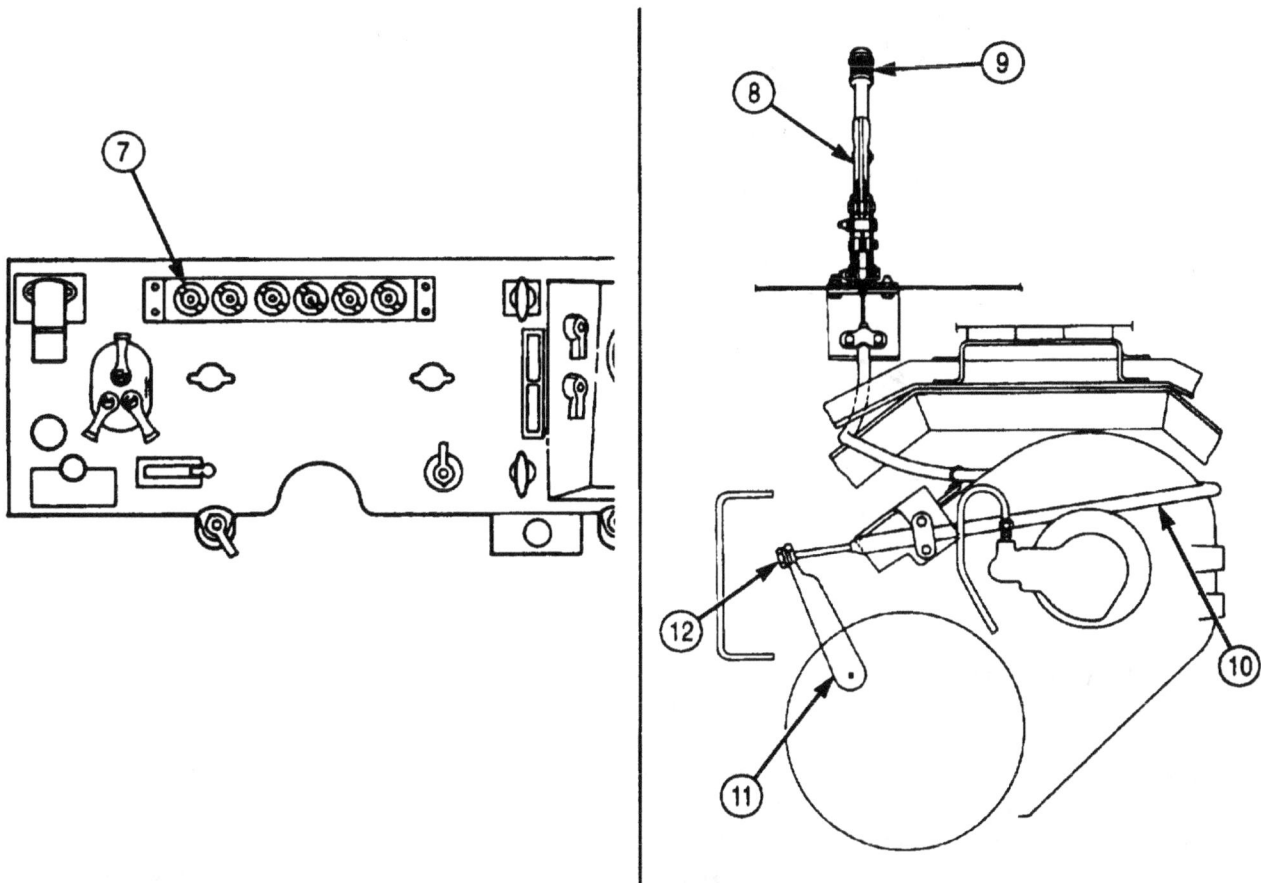

Key Item and Function

7   **PARKING BRAKE WARNING LIGHT** - Illuminates when parking brake is engaged.

8   **PARKING BRAKE CONTROL LEVER** - Is positioned up to engage parking brake; down to disengage parking brake.

9   **PARKING BRAKE CONTROL LEVER ADJUSTING KNOB** - Permits operator to make minor tension adjustment of parking brake.

10   **PARKING BRAKE CABLE** - Links parking brake control lever to brakeshoe lever.

11   **BRAKESHOE LEVER** - Lever turns cam which pushes brakeshoes against drum.

12   **PARKING BRAKE ADJUSTING NUT** - Permits major tension adjustment between parking brake lever and brakeshoes.

I-20. CONTROL SYSTEM OPERATION (Contd)

d. **Steering System Operation.**

The steering system is identical for all models covered in this manual. It is a hydraulically-assisted system that provides ease of turning and control for the operator. Major components of the steering system are:

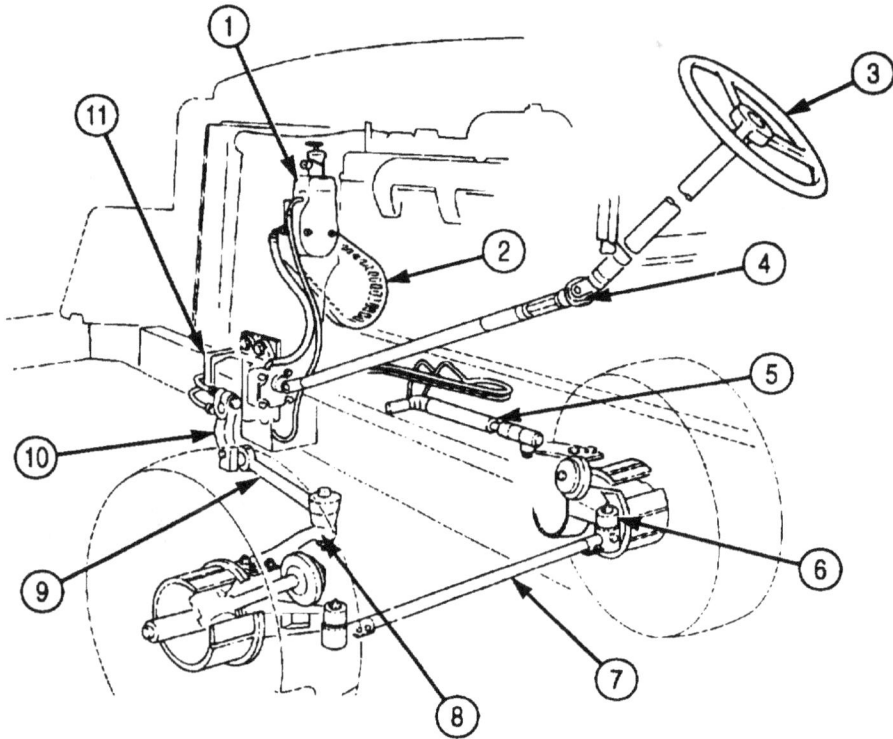

Key	Item and Function
1	**OIL RESERVOIR AND STEERING PUMP** - Combined in one unit, the reservoir serves as an oil fill point and the pump creates pressure.
2	**ACCESSORY DRIVEBELTS** - Transmit mechanical power from accessory drive pulley to steering pump pulley to drive the steering pump.
3	**STEERING WHEEL** - Serves as manual steering control for operator.
4	**STEERING COLUMN UNIVERSAL JOINT** - Connects, at an angle, the steering wheel column and input shaft of power steering gear.
5	**POWER STEERING ASSIST CYLINDER** - Receives hydraulic pressure from steering gear to assist in turning front wheels.
6	**STEERING KNUCKLE** - Serves as pivot point and link for front wheels from tie rod assembly.
7	**TIE ROD ASSEMBLY** - Connects steering knuckles so both wheels turn at the same time.
8	**STEERING ARM** - Connects drag link to steering knuckle.
9	**DRAG LINK** - Transmits movement from steering arm to pitman arm.
10	**PITMAN ARM** - Transfers torque from power steering gear to drag link.
11	**STEERING GEAR** - Converts hydraulic pressure from steering pump to mechanical power at pitman arm.

l-20. CONTROL SYSTEM OPERATION (Contd)

e. Transmission Control System Operation.

The transmission control system permits shifting of transmission, prevents starting of engine with transmission in gear, and prevents shifting of transfer case unless transmission is in neutral. This system also permits engagement of the transmission Power Takeoff (PTO) to provide hydraulic power for auxiliary equipment on M925/A1/A2, M928/A1/A2, M929/A1/A2, M930/Al/A2, M932/A1/A2, and M936/A1/A2 vehicles. Major components of the transmission control system are:

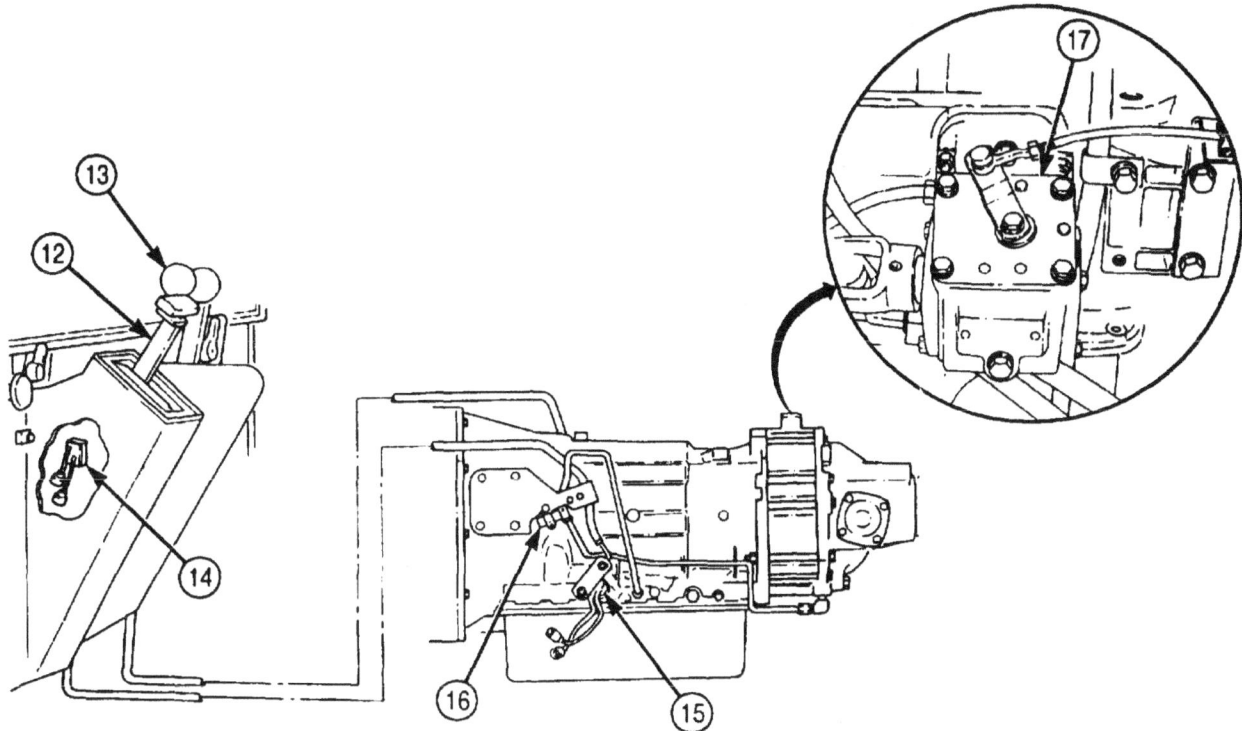

Key	Item and Function
12	**TRANSMISSION SELECTOR LEVER** - Is used to select vehicle driving gear range.
13	**PTO CONTROL LEVER** - Engages transmission power takeoff to provide power for auxiliary equipment.
14	**TRANSMISSION CONTROL SWITCH** - Actuates transmission lockup solenoid valve when transmission selector lever is placed in NEUTRAL and transfer case shift lever lockout switch is pressed.
15	**TRANSMISSION NEUTRAL START SWITCH** - The neutral start switch, wired to the starter switch, prevents the engine from being started with transmission in gear.
16	**TRANSMISSION 5TH-GEAR LOCKUP SOLENOID VALVE** -Activated by transmission **control** switch and transfer case switch, the 5th-gear lockup solenoid valve directs main oil pressure of transmission to the transmission governor system. This puts transmission in 5th-gear, creating less drag on transfer case synchronizer which permits smoother shifting from one transfer case drive range to another. Refer to para. l-20f, Transfer Case Control System Operation, for further details.
17	**TRANSMISSION PTO** - Driven by the transmission, the PTO drives the hydraulic pump which provides hydraulic pressure to power the front winch on M925/A1/A2, M928/Al/A2, M930/Al/A2, M932/Al/A2, and M936/A1/A2 vehicles, and to power the dump body on M929/Al/A2 and M930/Al/A2 vehicles. The PTO is mounted on the right front side of the transmission.

1-20. CONTROL SYSTEM OPERATION (Contd)

f. Transfer Case Control System Operation.

The transfer case control system converts four-wheel driving power into six-wheel driving power, provides smooth shifting of transfer case into high or low driving ranges while vehicle is in motion, prevents transfer case from being shifted with transmission in gear, and provides hydraulic power for auxiliary equipment through PTO.

(1) Six-wheel drive is achieved two different ways, depending on the drive range (high or low) desired. In low range, the transfer case shift linkage automatically moves a cam-actuated valve which dumps air into the front drive cylinder. This forces a piston against the transfer case clutch to engage front-wheel drive. In high range, front-wheel drive is engaged in the same manner except that the front-wheel drive valve is manually actuated by the front-wheel drive lock-in switch on the instrument panel.

(2) In order to shift the transfer case from one driving range to another, an interlock system working in conjunction with the 5th-gear lockup solenoid is used. This system prevents the transfer case from being shifted unless the transmission is in neutral.

(3) With the automatic transmission, several actions must occur in order to shift transfer case from one driving range to another. Because of the interlock system, the transmission must be placed in neutral. The transfer case shift lever switch must also be depressed.

(4) The transfer case control system, through use of a PTO driven by the transfer case, also provides hydraulic power to operate the crane and rear winch on M936/A1/A2 wreckers.

(5) Major components of this system are:

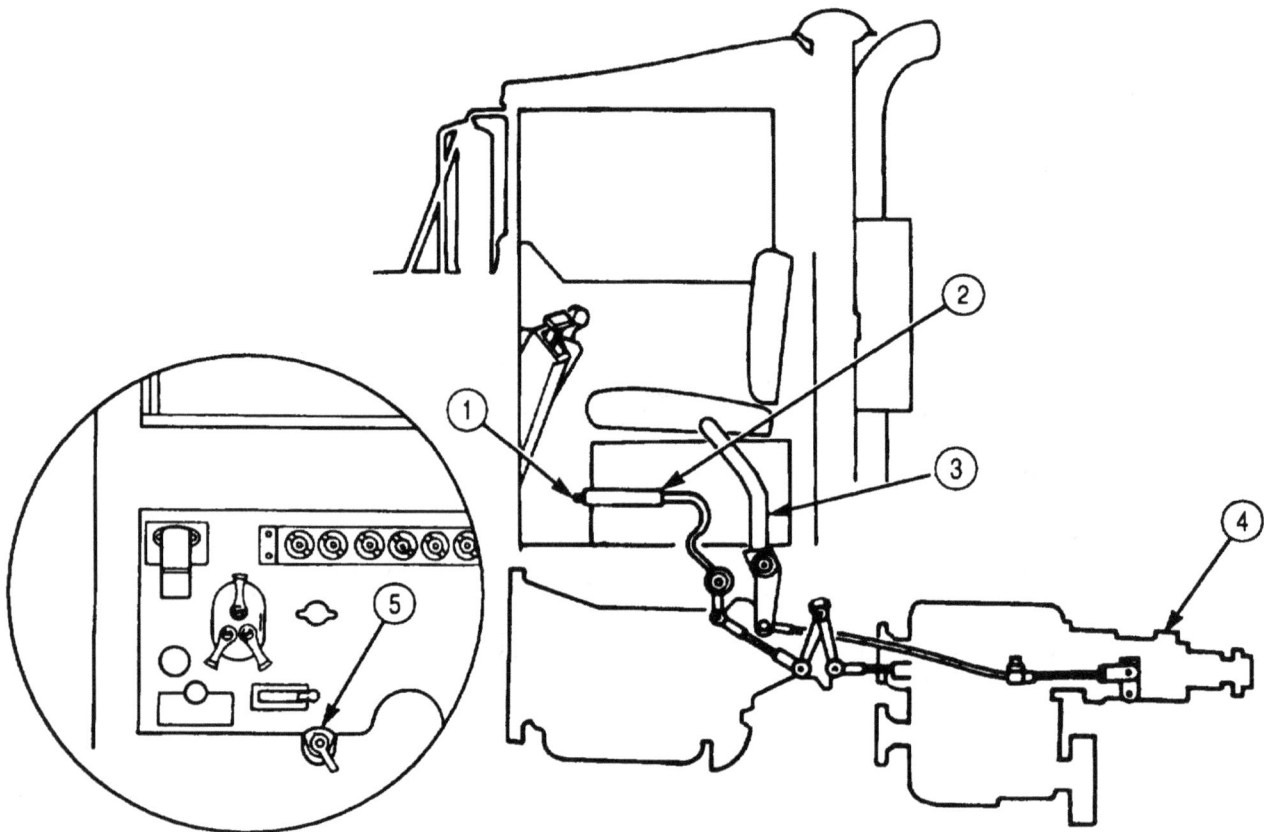

1-20. CONTROL SYSTEM OPERATION (Contd)

Key   Item and Function

1   **TRANSFER CASE SHIFT LEVER SWITCH** - When depressed with transmission in NEUTRAL, signals interlock solenoid valve to exhaust air pressure from interlock air cylinder and actuates lockup solenoid.

2   **TRANSFER CASE SHIFT LEVER** - Is pushed down to HIGH for light load operations, and up to LOW for heavy load operations. Six-wheel drive is achieved automatically when transfer case shift lever is placed in LOW.

3   **TRANSFER CASE PTO CONTROL LEVER** - Manual control for engaging PTO.

4   **TRANSFER CASE PTO** - Mounted and mechanically driven at rear of transfer case, the PTO drives a pump to supply hydraulic pressure to power the rear winch and crane on the M936/A1/A2 wreckers.

5   **FRONT-WHEEL DRIVE LOCK-IN SWITCH** - Manual control for activating front-wheel drive valve to provide front-wheel drive with transfer case in HIGH drive range.

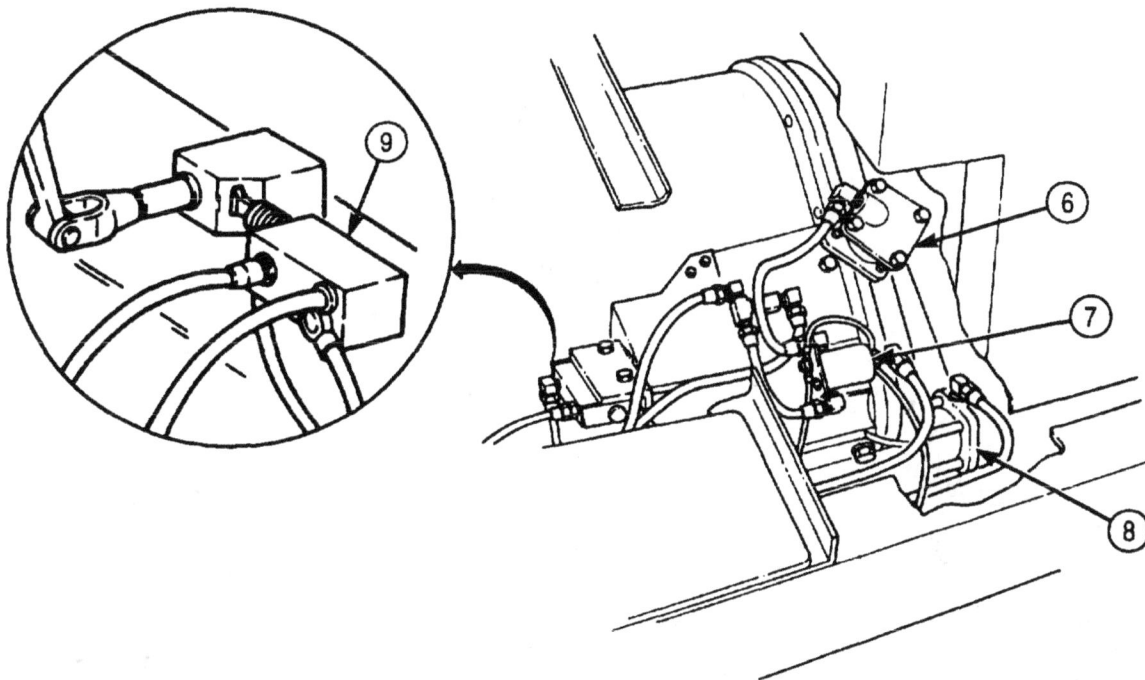

6   **INTERLOCK AIR CYLINDER** - Under air pressure, a piston in the interlock air cylinder forces a shaft against one of three grooves in transfer case shift lever, This prevents transfer case from being shifted with transmission in gear.

7   **INTERLOCK SOLENOID VALVE** - Releases air from interlock air cylinder when transmission is in NEUTRAL and transfer case shift lever switch is depressed. This permits the transfer case high/low shift shaft to move.

8   **FRONT-WHEEL DRIVE AIR CYLINDER** - When under pressure, it moves transfer case clutch forward to engage front-wheel drive.

9   **FRONT-WHEEL DRIVE VALVE** - When tripped by a cam on transfer case shift shaft, the front-wheel drive valve routes air to front-wheel drive air cylinder.

1-21. POWER SYSTEM OPERATION

The power system includes components that supply all vehicles covered in this manual the power to move. Each of these components will be described as part of the following subsystems:

    a. Air Intake System Operation (page 1-34).

    b. Fuel System (Dual Tank) Operation (page 1-36).

    c. Fuel System (Single Tank) Operation (page 1-38).

    d. Exhaust System Operation (page 1-39).

    e. Cooling System Operation (page 1-40).

    f. Engine Oil System Operation (page 1-42).

    g. Powertrain System Operation (page 1-44).

a. Air Intake System Operation.

The air intake system channels and cleans air going to the combustion chamber, where it mixes with fuel from the injectors to provide power for the engine. This system is identical on all models, except where indicated. Major components of the air intake system are:

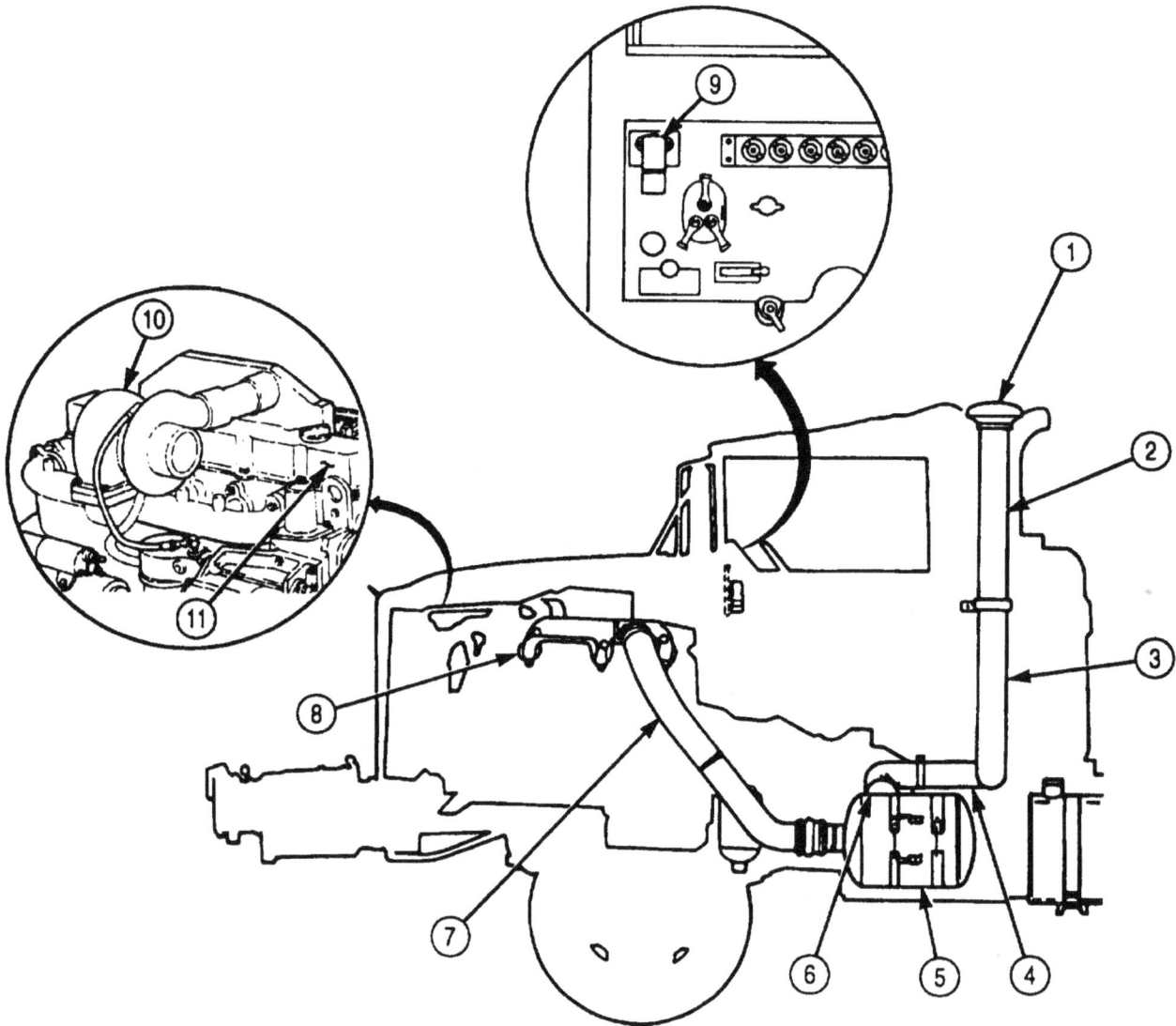

1-21. POWER SYSTEM OPERATION (Contd)

Key	Item and Function
1	**RAIN CAP** - Prevents rain and large objects from entering air intake system.
2	**AIR INTAKE EXTENSION TUBE** - Routes air to air intake system. Can be removed for shipping.
3	**STACK-TO-AIR INTAKE EXTENSION TUBE** - Routes air to air cleaner and is high enough to keep intake opening above fording level.
4	**STACK-TO-AIR CLEANER ELBOW** - Flexible connection between air stack and air cleaner.
5	**AIR CLEANER** - Filters dirt and dust from air.
6	**HUMP HOSE** - Flexible connection between air cleaner and air cleaner outlet tube.
7	**AIR CLEANER OUTLET TUBE** - Routes air from air cleaner to intake manifold.
8	**INTAKE MANIFOLD** - Distributes air to combustion chambers in each cylinder head (M939/A1 series only).
9	**AIR CLEANER INDICATOR** - Shows red when engine air filter needs servicing.
10	**TURBOCHARGER** - Mounts on exhaust manifold and uses spent exhaust gases to drive impeller and pressurize air entering aftercooler (M939A2 series only).
11	**AFTERCOOLER** - Distributes compressed air from turbocharger to combustion chambers while cooling air intake out of turbocharger (M939A2 series only).

1-21. POWER SYSTEM OPERATION (Contd)

b. Fuel System (Dual Tank) Operation.

(1) The fuel system stores, cleans, and supplies fuel to the fuel injectors, where it is mixed with air to initiate engine combustion.

(2) The fuel system is not identical for all models. Vehicles covered in this manual have either one or two tanks. These tanks can also differ in capacity. See table 1-3, Vehicle Performance Data, for these differences.

(3) A typical dual tank fuel system is described below. A single tank fuel system is described in para. 1-21c. Both systems include fuel supply, return, and vent lines to provide fuel flow and release the fumes throughout the system. Major components of the dual tank fuel system are:

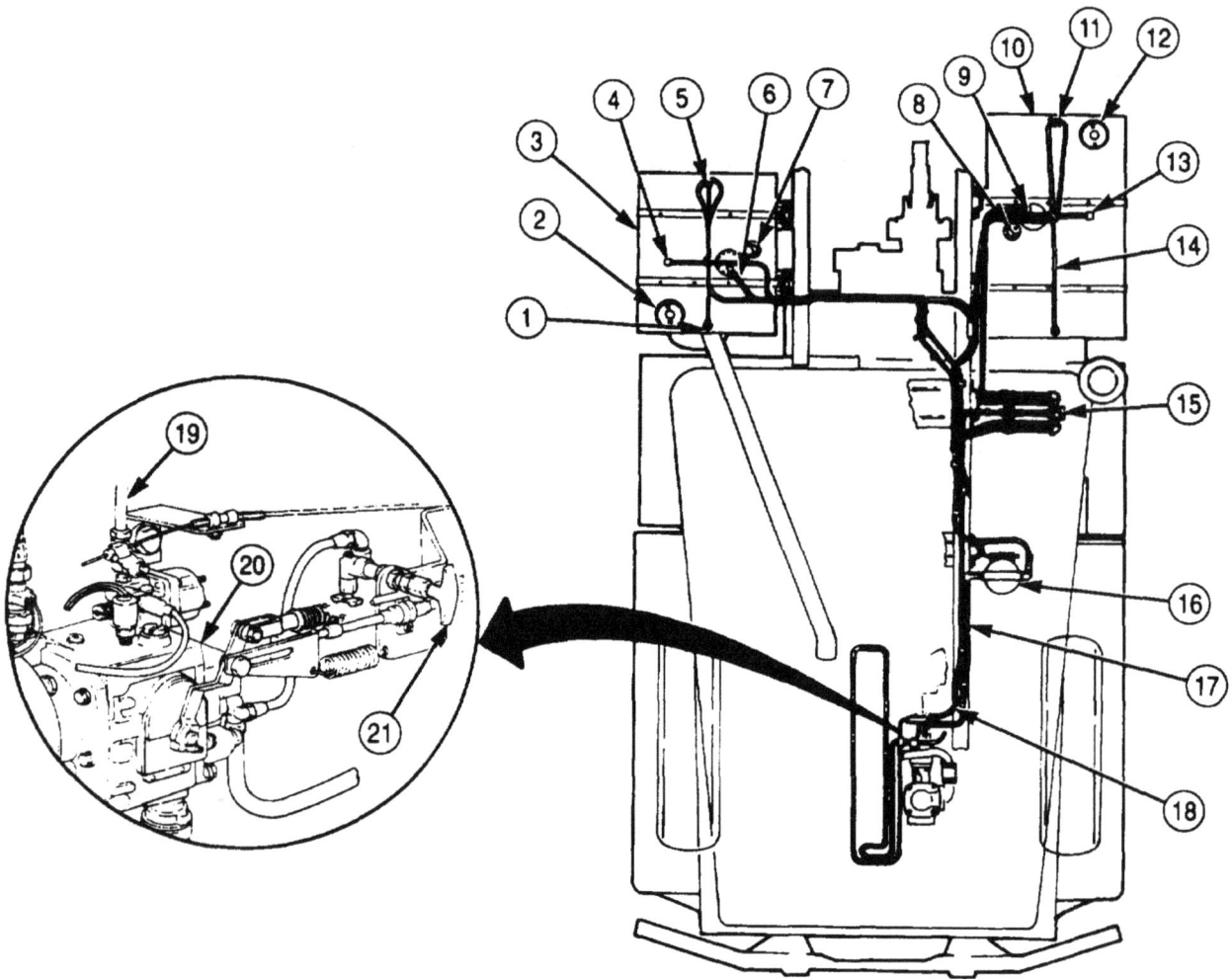

1-21. POWER SYSTEM OPERATION (Contd)

Key Item and Function

1 **RIGHT TANK (FRONT) VENT LINE** - Vents vapors from fuel tank to vent hole in air intake stack.

2 **RIGHT TANK FILLER CAP** - Covers fuel filler opening on right fuel tank.

3 **RIGHT FUEL TANK** - Stores fuel for vehicle use.

4 **RIGHT TANK FUEL RETURN LINE** - Returns unused fuel back to fuel tank.

5 **RIGHT TANK (REAR) VENT LINE** - Vents vapors from fuel tank to vent hole in air intake stack.

6 **RIGHT TANK FUEL SUPPLY LINE** - Directs fuel from tank to fuel filter.

7 **RIGHT TANK FUEL LEVEL SENDING UNIT** - Electrical signal registers fuel level in right tank at gauge on instrument panel.

8 **LEFT TANK FUEL LEVEL SENDING UNIT** - Electrical signal registers fuel level in left tank at gauge on instrument panel.

9 **LEFT TANK FUEL SUPPLY LINE** - Directs fuel from tank to fuel filter.

10 **LEFT FUEL TANK** - Stores fuel for vehicle use.

11 **LEFT TANK (REAR) VENT LINE** - Vents vapors from fuel tank to vent hole in air intake stack.

12 **LEFT TANK FILLER CAP** - Covers fuel filler opening on left fuel tank.

13 **LEFT TANK FUEL RETURN LINE** - Returns unused fuel back to fuel tank.

14 **LEFT TANK (FRONT) VENT LINE** - Vents vapors from fuel tank to vent hole in air intake stack.

15 **FUEL SELECTOR VALVE** - Manual control valve that opens fuel flow to engine from left or right fuel tank.

16 **FUEL FILTER/WATER SEPARATOR** - Filters water and dirt from fuel.

17 **FUEL FILTER-TO-PUMP SUPPLY LINE** - Directs fuel from fuel filter to fuel pump.

18 **FUEL RETURN LINE** - Returns unused fuel back to fuel tanks.

19 **FUEL SUPPLY LINE** - Directs fuel from fuel pump to fuel injectors.

20 **FUEL PUMP** - Draws fuel from tank(s) and pumps it through supply line to fuel injectors.

21 **FUEL PRIMER PUMP** - Purges air from fuel system.

1-21. POWER SYSTEM OPERATION (Contd)

c. Fuel System (Single Tank) Operation.

Major components of the single tank fuel system are:

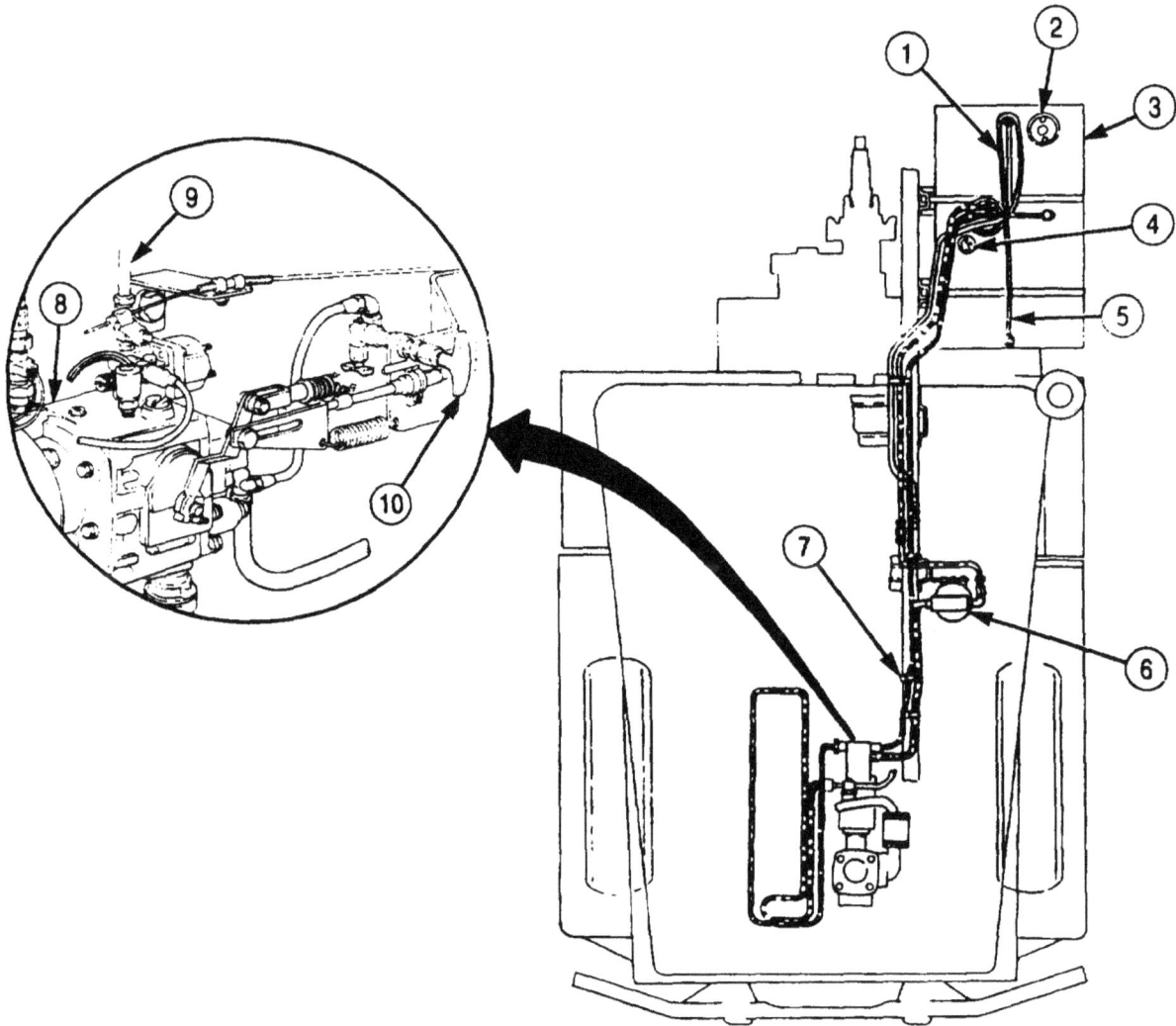

Key   Item and Function

1   **TANK (REAR) VENT LINE** - Vents vapors from fuel tank to vent hole-in air intake stack.

2   **TANK FILLER CAP** - Covers fuel filler opening.

3   **FUEL TANK** - Stores fuel for vehicle use.

4   **FUEL TANK LEVEL SENDING UNIT** - Electrical signal registers fuel level in tank at gauge on instrument panel.

5   **TANK (FRONT) VENT LINE** - Vents vapors from fuel tank to vent hole in air intake stack.

6   **FUEL FILTER/WATER SEPARATOR** - Filters water and dirt from fuel.

7   **FUEL RETURN LINE** - Returns unused fuel back to fuel tank.

8   **FUEL PUMP** - Draws fuel from tank and pumps it through supply line to fuel injectors.

9   **FUEL SUPPLY LINE** - Directs fuel from fuel pump to fuel injectors.

10   **FUEL PRIMER PUMP** - Purges air from fuel system.

1-21. POWER SYSTEM OPERATION (Contd)

d. Exhaust System Operation.

The exhaust system directs exhaust gases away from the vehicle for all models covered in this manual. Major components of the exhaust system are:

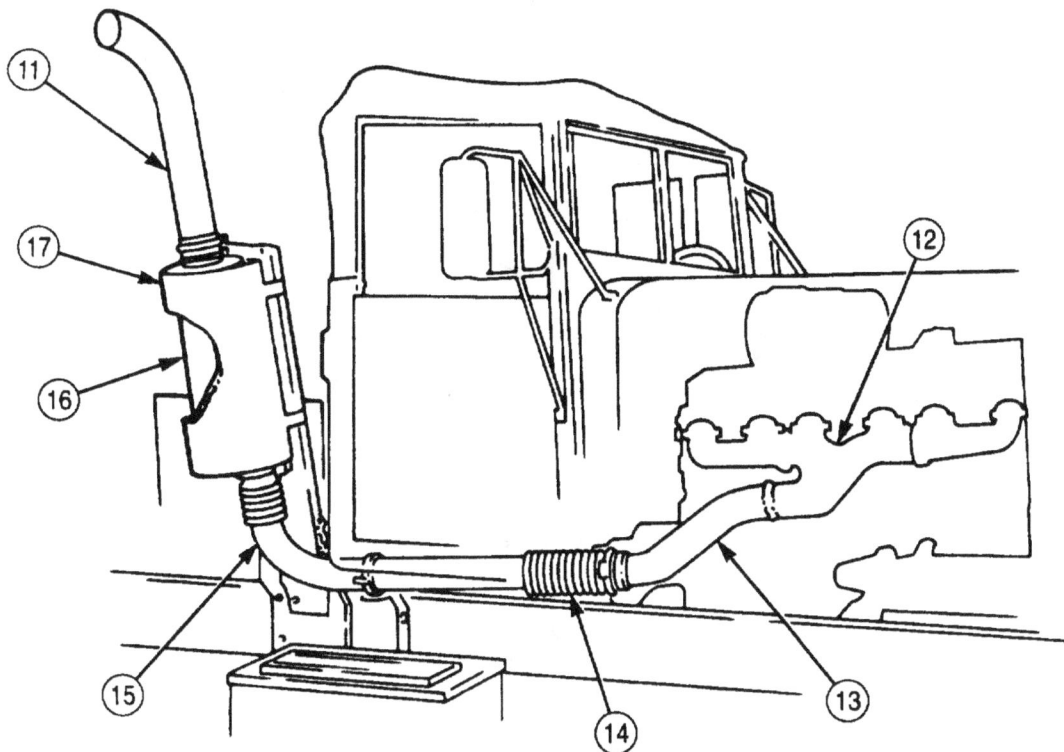

Key Item and Function

11   **EXHAUST STACK** - Directs exhaust from muffler away from vehicle.

12   **EXHAUST MANIFOLD** - Collects exhaust from cylinder head ports and directs it to front exhaust pipe.

13   **FRONT EXHAUST PIPE** - Directs exhaust to rear exhaust pipe.

14   **FLEX PIPE** - Part of rear exhaust pipe; allows flexibility for vibration and expansion in system.

15   **REAR EXHAUST PIPE** - Directs exhaust to muffler.

16   **MUFFLER** - Quiets exhaust noises.

17   **MUFFLER SHIELD** - Protects personnel from muffler heat.

1-21. POWER SYSTEM OPERATION (Contd)

e. Cooling System Operation.

This system provides cooling of the engine and transmission. It differs slightly between the M939/A1 series and M939A2 series vehicles because different engines are used. Major components of the cooling system are:

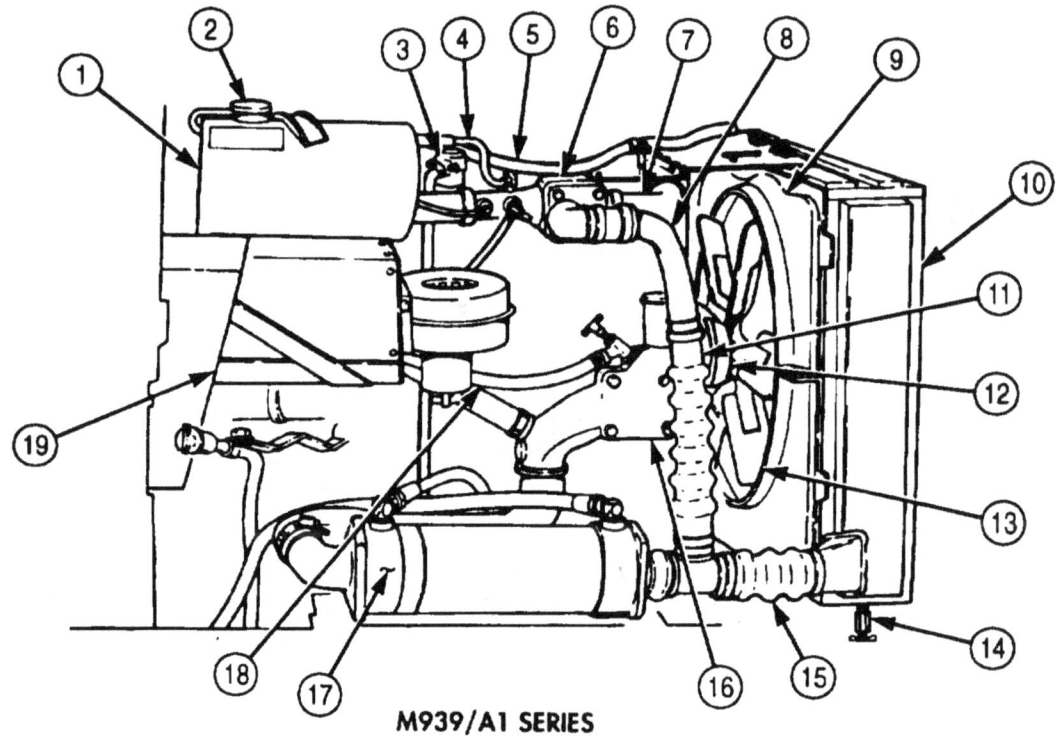

M939/A1 SERIES

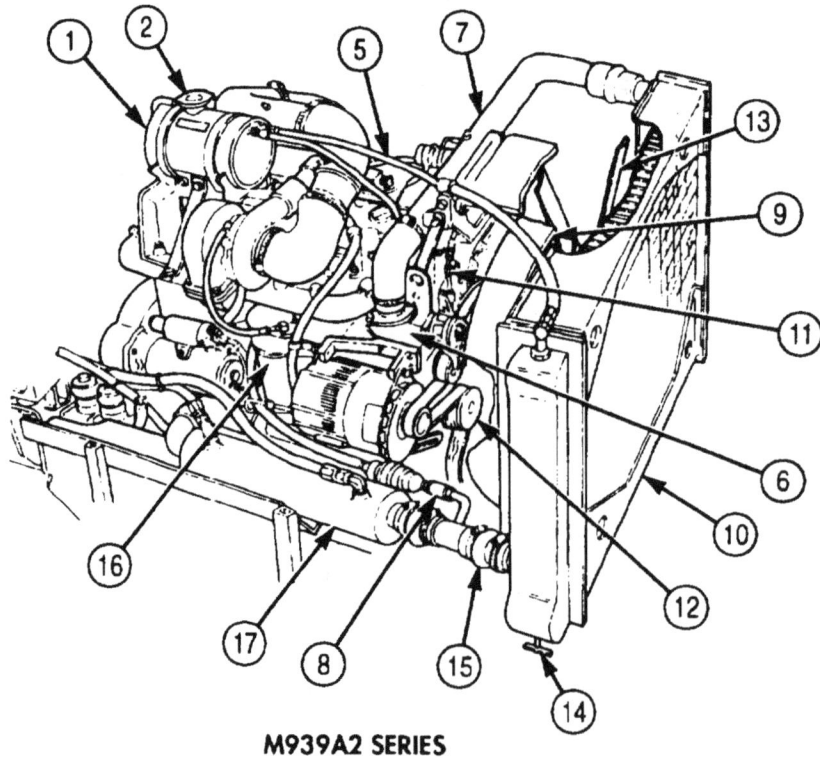

M939A2 SERIES

1-21. POWER SYSTEM OPERATION (Contd)

Key Item and Function

1 **SURGE TANK** - Filling point for cooling system. On M939A2 vehicles, a float sensor monitors water level and illuminates a light on instrument panel.

2 **COOLANT PRESSURE CAP** - Designed to depressurize cooling system and to access cooling system for tilling.

3 **WATER MANIFOLD** - Collects coolant from cylinder heads and directs it to thermostat housing (M939/A1 series only).

4 **SURGE TANK-TO-WATER MANIFOLD VENT** - Vents air trapped in water manifold (M939/A1 series only).

5 **SURGE TANK-TO-RADIATOR VENT** - Vents air in cooling system.

6 **THERMOSTAT** - Shuts off coolant flow to radiator until temperature reaches 175°F (79°C) on M939/A1 series vehicles and 181°F (83°C) on M939A2 series vehicles. Coolant is then directed to the radiator through the radiator inlet hose.

7 **RADIATOR INLET HOSE** - Directs coolant from water manifold to radiator after thermostat has opened.

8 **BYPASS TUBE** - Directs coolant back to transmission oil cooler, where it is then recirculated through the engine block until the thermostat opens.

9 **RADIATOR SHROUD** - Concentrates air flow through the radiator.

10 **RADIATOR** - Directs coolant through a series of fins or baffles so outside air can remove excessive heat from coolant.

11 **FAN CLUTCH** - Regulates use of fan to control engine temperature fan to belt-driven pulley when conditions require additional cooling.

12 **WATER PUMP** - Provides force to move coolant through engine.

13 **FAN** - Provides force to pull air through radiator.

14 **RADIATOR DRAINVALVE** - Permits coolant to be drained from radiator.

15 **TRANSMISSION OIL COOLER HOSE** - Directs coolant to transmission oil cooler.

16 **ENGINE OIL COOLER** - Reduces heat of engine oil (M939/A1 series only).

17 **TRANSMISSION OIL COOLER** - Reduces heat of transmission oil.

18 **ENGINE OIL COOLER-TO-HEATER HOSE** - Directs coolant to personnel water heater when shutoff valve is open (M939/A1 series only).

19 **PERSONNEL WATER HEATER** - Provides heat for cab and personnel (M939/A1 only).

1-21. POWER SYSTEM OPERATION (Contd)

f. Engine Oil System Operation.

The engine oil system provides lubricating oil for internal moving parts. Major components of the engine oil system are:

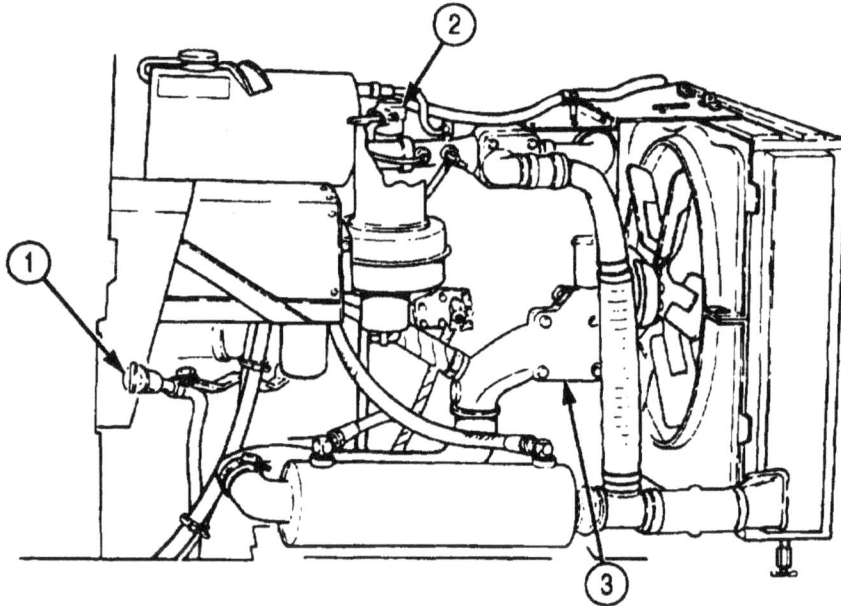

M939/A1 SERIES

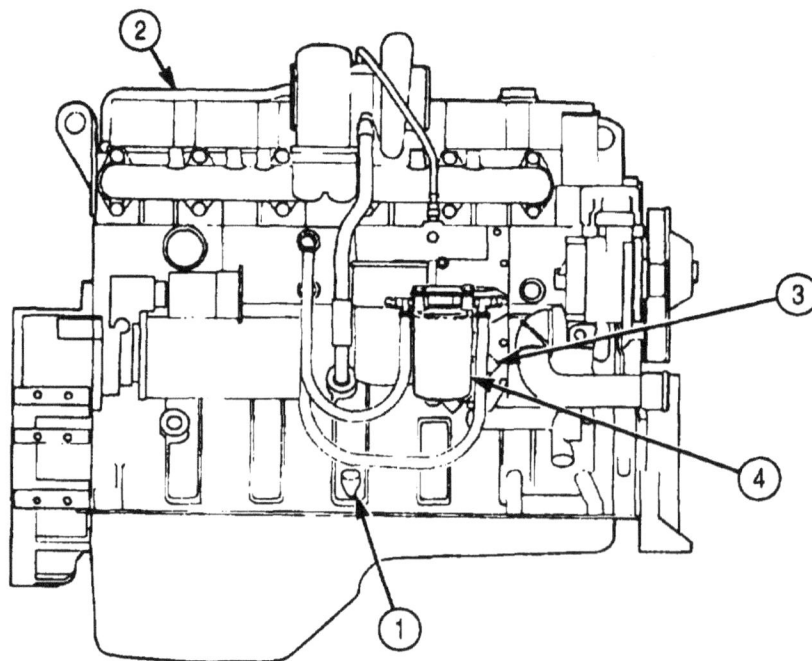

M939A2 SERIES

1-21. POWER SYSTEM OPERATION (Contd)

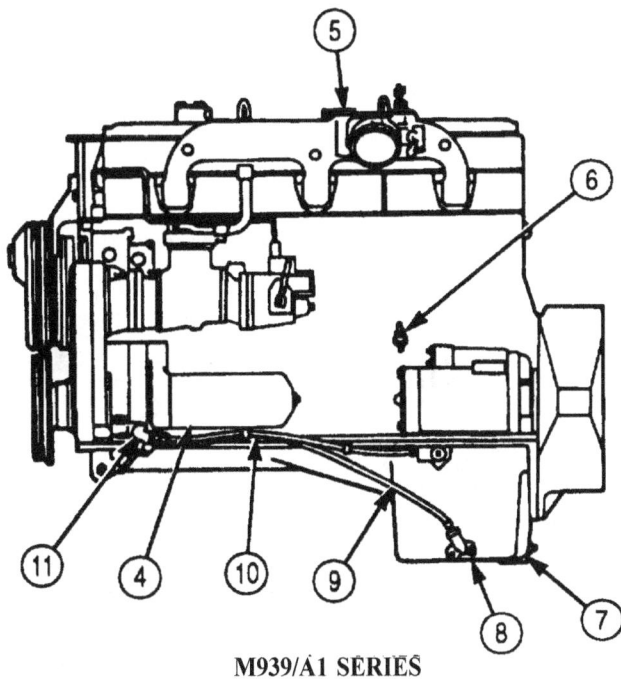

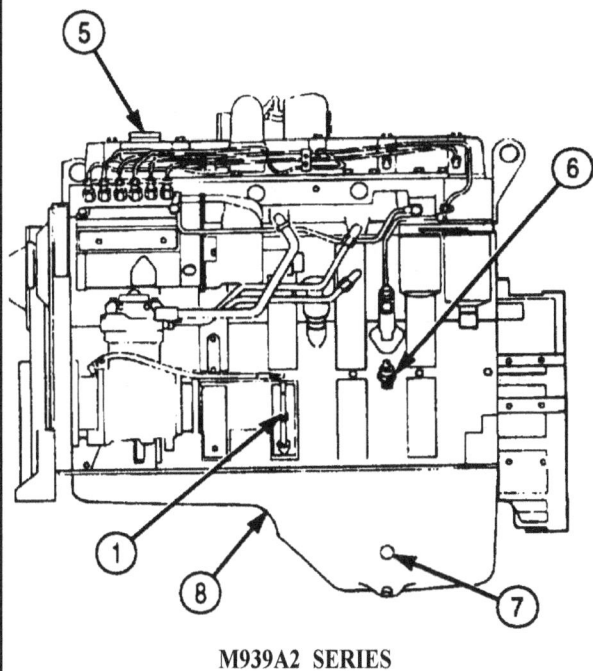

M939/A1 SERIES

M939A2 SERIES

Key Item and Function

1 **OIL DIPSTICK** - Indicates engine oil level.

2 **CRANKCASE BREATHER** - Vents hot engine oil fumes from engine and allows fresh air to enter.

3 **ENGINE OIL COOLER** - Removes heat from engine oil as coolant circulates through internal tubes of oil cooler.

4 **OIL FILTER** - Filters out foreign particles suspended in oil.

5 **OIL FILLER CAP** - Located on rocker lever cover, cap covers engine oil fill opening.

6 **OIL PRESSURE TRANSMITTER** - Sends an electrical signal that indicates engine oil pressure to gauge on instrument panel.

7 **OIL PAN DRAINPLUG** - Plugs engine oil drain opening.

8 **OIL PAN** - Reservoir for engine oil.

9 **OIL SUPPLY LINE** - Carries oil from oil pan to oil pump.

10 **OIL BYPASS RETURN LINE** - Returns oil from oil pump to oil pan.

11 **OIL PUMP** - Provides mechanical pressurization of oil to circulate it through oil system.

1-21. POWER SYSTEM OPERATION (Contd)

g. Powertrain System Operation.

The powertrain system is the same on all models in this manual except the extra-long wheelbase models, which have an additional propeller shaft and center bearing. This system transmits engine power to the axles to put the vehicle in motion. Major components of the powertrain system are:

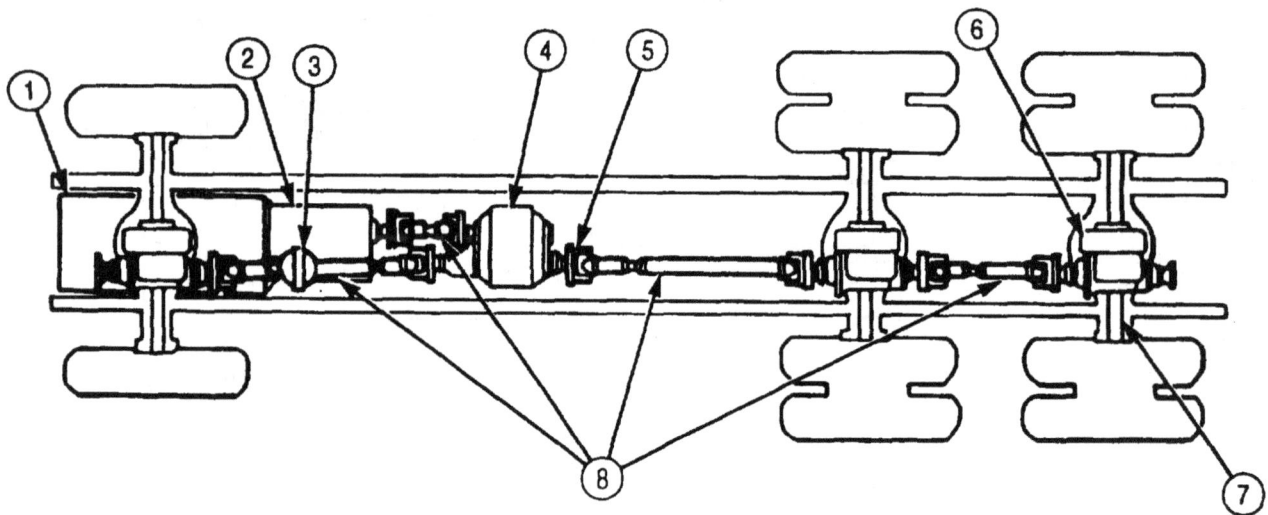

Key Item and Function

1 **ENGINE** - Provides power needed for powertrain component operation.

2 **TRANSMISSION** -Adapts engine power to meet different driving conditions.

3 **CENTER BEARING** - Provides support for propeller shaft to decrease vibration and wear on universal joints (M927/A1/A2, M928/A1/A2, and M934/A1/A2 series only).

4 **TRANSFER CASE** - Distributes power evenly to front and rear axles.

5 **UNIVERSAL JOINTS** - Connections between two propeller shafts that permit one to drive the other even though they may be at different angles.

6 **DIFFERENTIALS** - Distribute power to left and right axle shafts.

7 **AXLES** - Transmit power from differentials to rotate wheels.

8 **PROPELLER SHAFTS** - Serve as driving shafts that connect transmission to transfer case and transfer case to differentials.

1-22. ELECTRICAL SYSTEMS OPERATION

Nearly every component of models covered in this manual is affected by the electrical system. These components and their electrical connections are described as part of the following electrical subsystems:

    a. Battery System Operation (page 1-46).

    b. Starting System Operation (page 1-47).

    c. Ether Starting System Operation (page 1-48).

    d. Generating System Operation (page 1-49).

    e. Directional Signal System Operation (page 1-50).

    f. Heating System Operation (page 1-51).

    g. Indicator, Gauge, and Warning System Operation (page 1-52).

    h. Trailer and Semitrailer Connection System Operation (page 1-54).

Electrical Terms and Definitions.

The following electrical terms and definitions will be frequently referred to throughout this section and should be understood before proceeding:

Alternating Current (AC signal) - Current in a circuit that flows in one direction first, then in the other direction.

Circuit - A complete path for electric current flow between components.

Circuit Breaker - An automatic switch that interrupts current flow in a circuit when the current limit is exceeded.

Direct Current (DC signal) - Current in a circuit that flows in one direction.

Female Connector - One-half of a connector which fits over the other half.

Ground - A common return to complete a path for current flow in a circuit.

Harness - A group of wires connected between devices that are bundled and routed together to prevent damage and make repair and replacement easier.

Male Connector - One-half of a connector which fits the other half.

Polarity - The direction current flows in a circuit (usually positive to negative).

Relay - An electromagnetic device that operates like an automatic switch to control flow of current in the same or different circuit.

Reverse Polarity - The condition that exists when circuit polarity is connected opposite of that which was intended.

Sending Unit - A device that produces an electrical signal and sends this signal to the device which will make use of it.

Sensor - An electrical sensor takes a physical condition (temperature, oil presence or absence) and converts it into an electrical signal.

Splice - A permanent physical connection of two or more wires.

Terminal - Fastener at end of wire used to connect the wire to an electrically-powered device.

1-22. ELECTRICAL SYSTEMS OPERATION (Contd)

a. Battery System Operation.

The battery system is identical for all models covered in this manual and consists of the following major components and circuits:

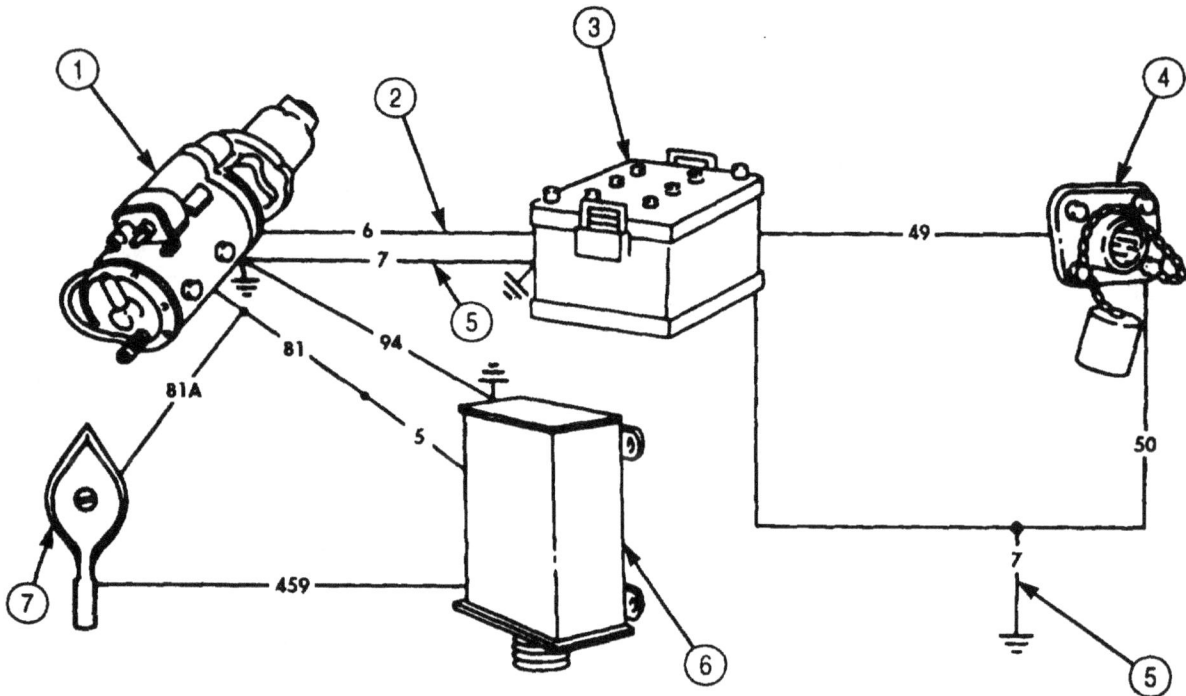

Key Item and Function

1 **STARTER SOLENOID** - Junction point for battery positive lead (circuit 6) and vehicle electrical feed wire (circuit 81).

2 **CIRCUIT 6** - Connects batteries to starting motor and to protective control box through circuit 81.

3 **BATTERIES** - Four 6TN batteries are connected in series parallel to provide 24-volts DC for electrical starter system and 12-volts DC for the heater fan low speed.

4 **SLAVE RECEPTACLE** - Links an external power source directly to the slaved vehicle's batteries to assist in cranking engine when batteries are not sufficiently charged.

5 **CIRCUIT 7** - Provides a ground between starter, battery, and chassis.

6 **PROTECTIVE CONTROL BOX** - Protects the vehicle electrical system in the event battery system polarity is reversed. Connects battery power to vehicle electrical lead through circuit 81 and circuit 5. Connects positive ground through circuit 94 to the starter.

7 **BATTERY SWITCH** - Controls a relay in protective control box through circuit 459 that connects batteries to vehicle electrical load.

1-22. ELECTRICAL SYSTEMS OPERATION (Contd)

b. Starting System Operation.

The starting system is identical for all models covered in this manual and consists of the following major components and circuitry:

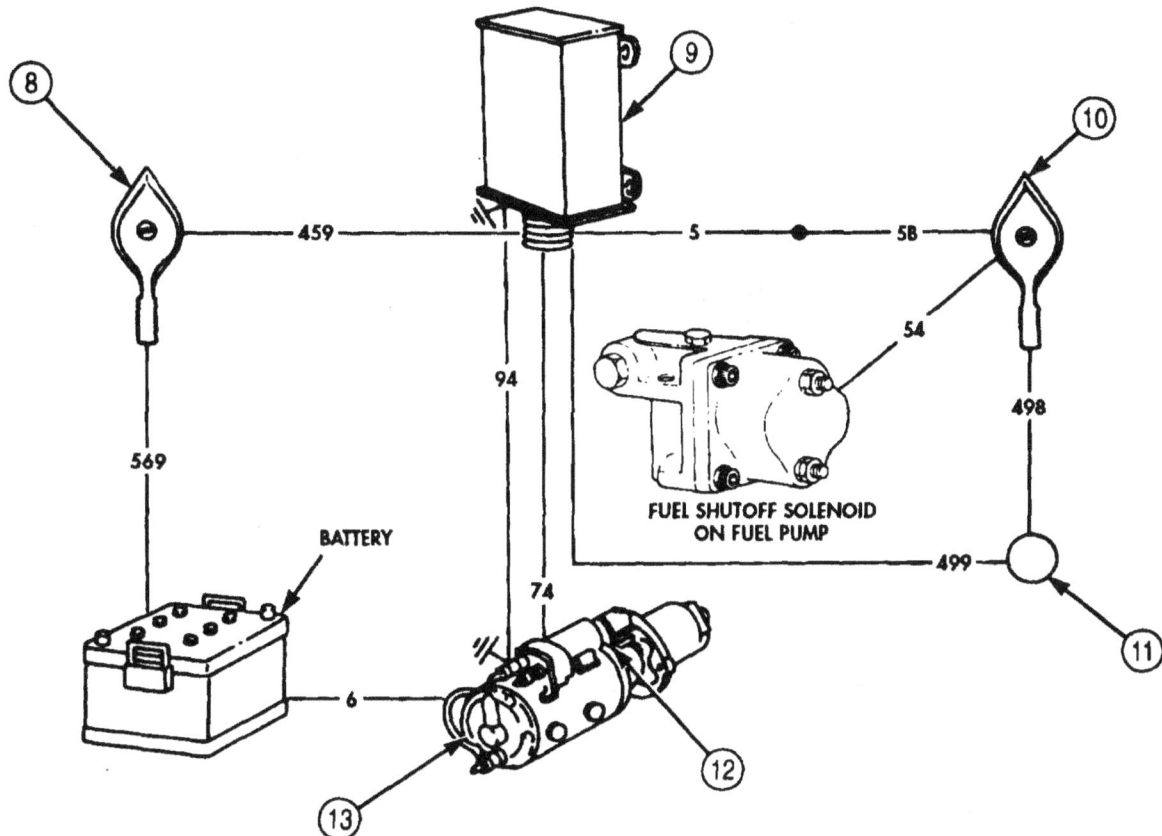

FUEL SHUTOFF SOLENOID
ON FUEL PUMP

BATTERY

Key Item and Function

8 BATTERY SWITCH - Completes circuit 459, closing a relay in the protective control box to supply power to ignition switch through circuits 5 and 5B.

9 PROTECTIVE CONTROL BOX - Locks out starter circuit, which prevents starter from reengaging while engine is running.

10 IGNITION SWITCH - Provides battery power to fuel solenoid through circuit 54 and the neutral start safety switch through circuit 498.

11 NEUTRAL START SAFETY SWITCH - Prevents starter from energizing when vehicle is not in neutral, by deenergizing circuit 499 and a relay in the protective control box, which disconnects power from circuit 74 and the starter solenoid.

12 STARTER SOLENOID - A magnetic relay that is powered by circuit 74 to transmit 24-volt battery power to the starter motor through circuit 6.

13 STARTER MOTOR - Cranks engine for starting. Supplied with 24-volt battery power through circuit 6.

1-22. ELECTRICAL SYSTEMS OPERATION (Contd)

c. Ether Starting System Operation.

The ether starting system is identical for all models covered in this manual and consists of the following major components and circuitry:

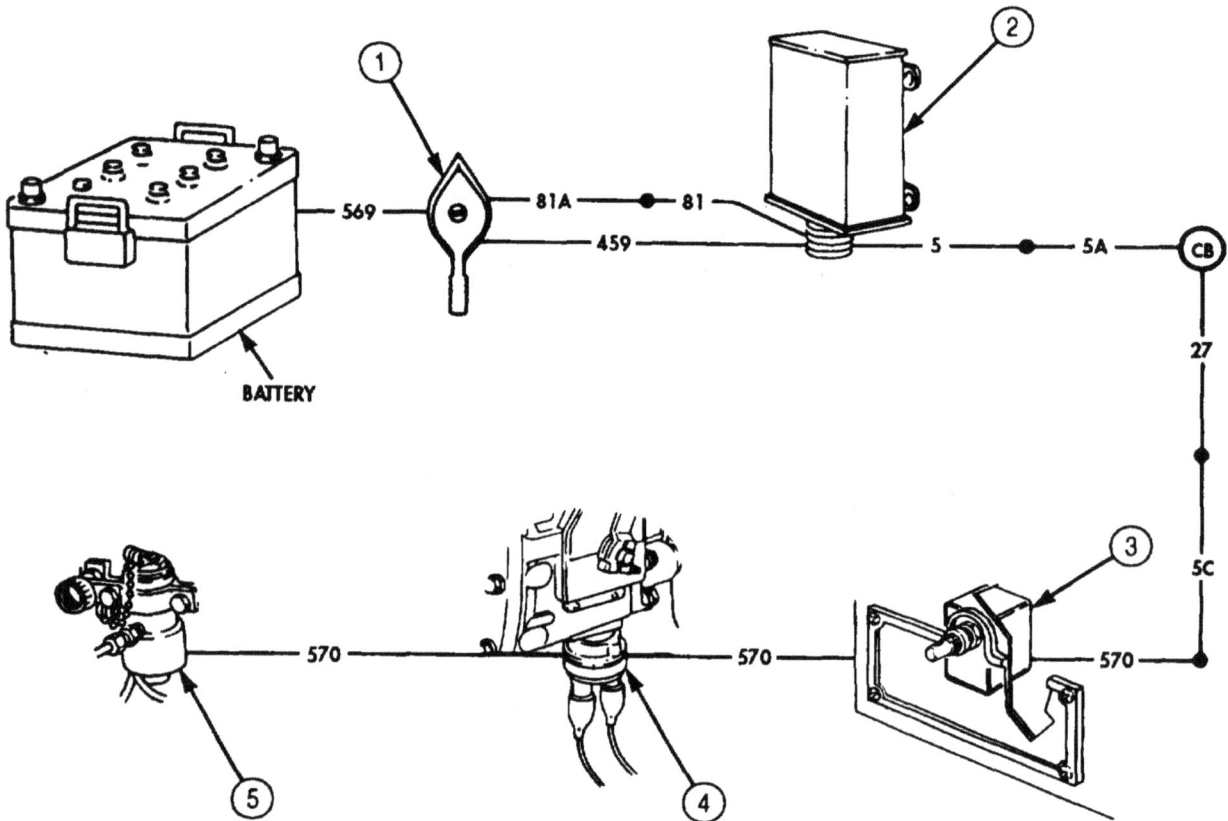

Key Item and Function

1. **BATTERY SWITCH** - Provides 24-volt battery power to protective control box through circuits 459, 81A, and 81.

2. **PROTECTIVE CONTROL BOX** - Energizes ether feed switch through circuits 5, 5A, 27, 5C, and 570.

3. **ETHER FEED SWITCH** - Controls 24-volt power to ether pressure switch through circuit 570.

4. **ETHER PRESSURE SWITCH** - Connects ether feed switch to ether tank valve through circuit 570.

5. **ETHER TANK VALVE** - Is activated through circuit 570 when ether pressure switch is closed and ether feed switch is pressed.

1-22. ELECTRICAL SYSTEMS OPERATION (Contd)

d. Generating System Operation.

The generating system is identical for all vehicles covered in this manual and consists of the following major components and circuitry:

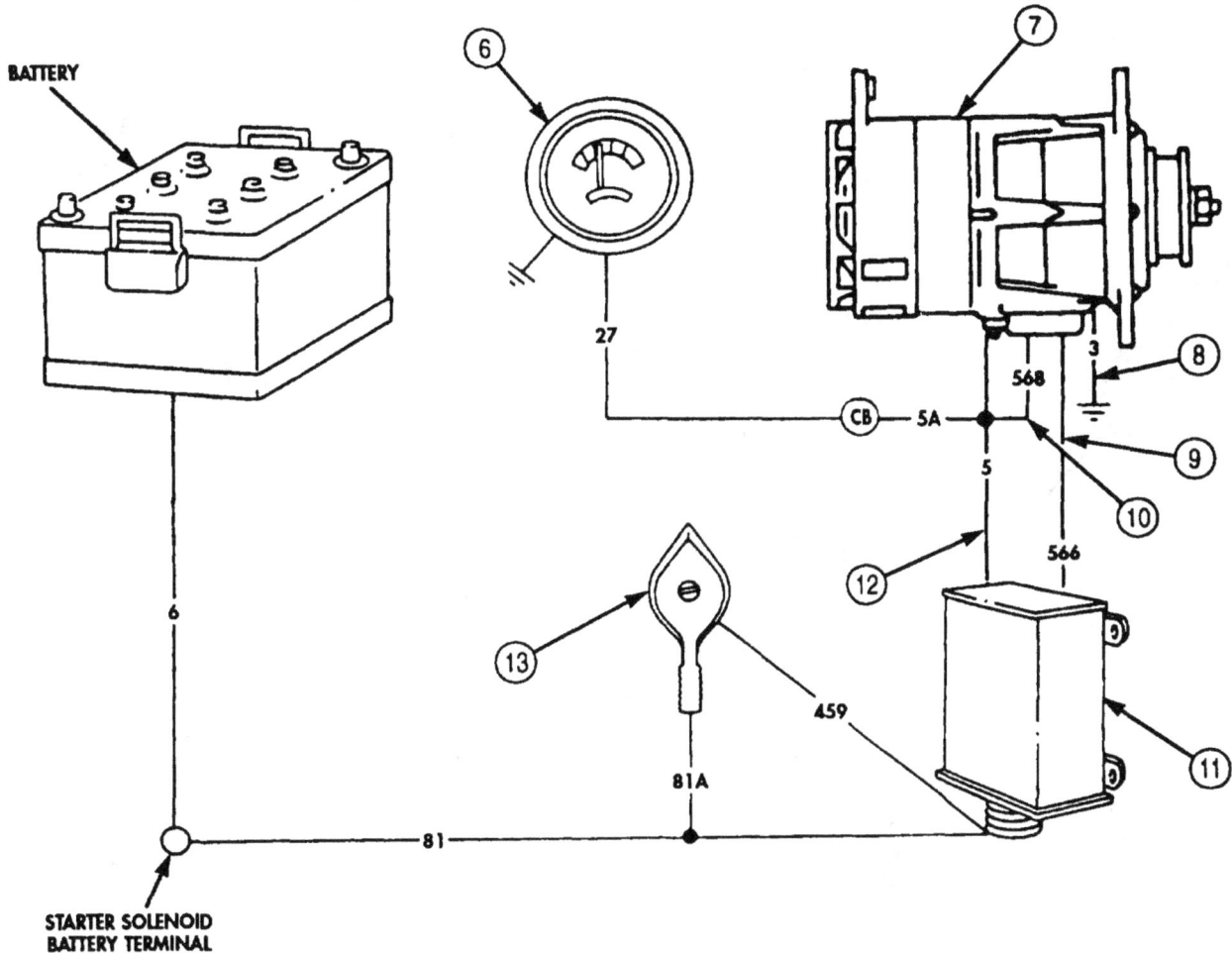

Key Item and Function

6 **VOLTMETER** - Indicates electrical system voltage. It is connected to the electrical system through circuit 27.

7 **ALTERNATOR** - Rated at 26-30 volts, 60 amperes, the alternator assists and recharges batteries during operation. A 100-ampere model is available as a kit.

8 **CIRCUIT 3** - Provides a ground circuit to alternator.

9 **CIRCUIT 566** - Controls a relay in protective control box that prevents starter from reactivating while engine is running.

10 **CIRCUIT 568** - Senses system voltage and excites the alternator field.

11 **PROTECTIVE CONTROL BOX** - Connects circuit 5 to 81 to power the electrical system and charge the batteries.

12 **CIRCUIT 5** - Conducts alternator output to charge batteries and maintain vehicle voltage.

13 **BATTERY SWITCH** - Closes relay in protective control box that connects battery circuits.

1-22. ELECTRICAL SYSTEMS OPERATION (Contd)

e. Directional Signal System Operation.

The directional signal system is identical on all models covered in this manual and consists of the following major components and circuitry:

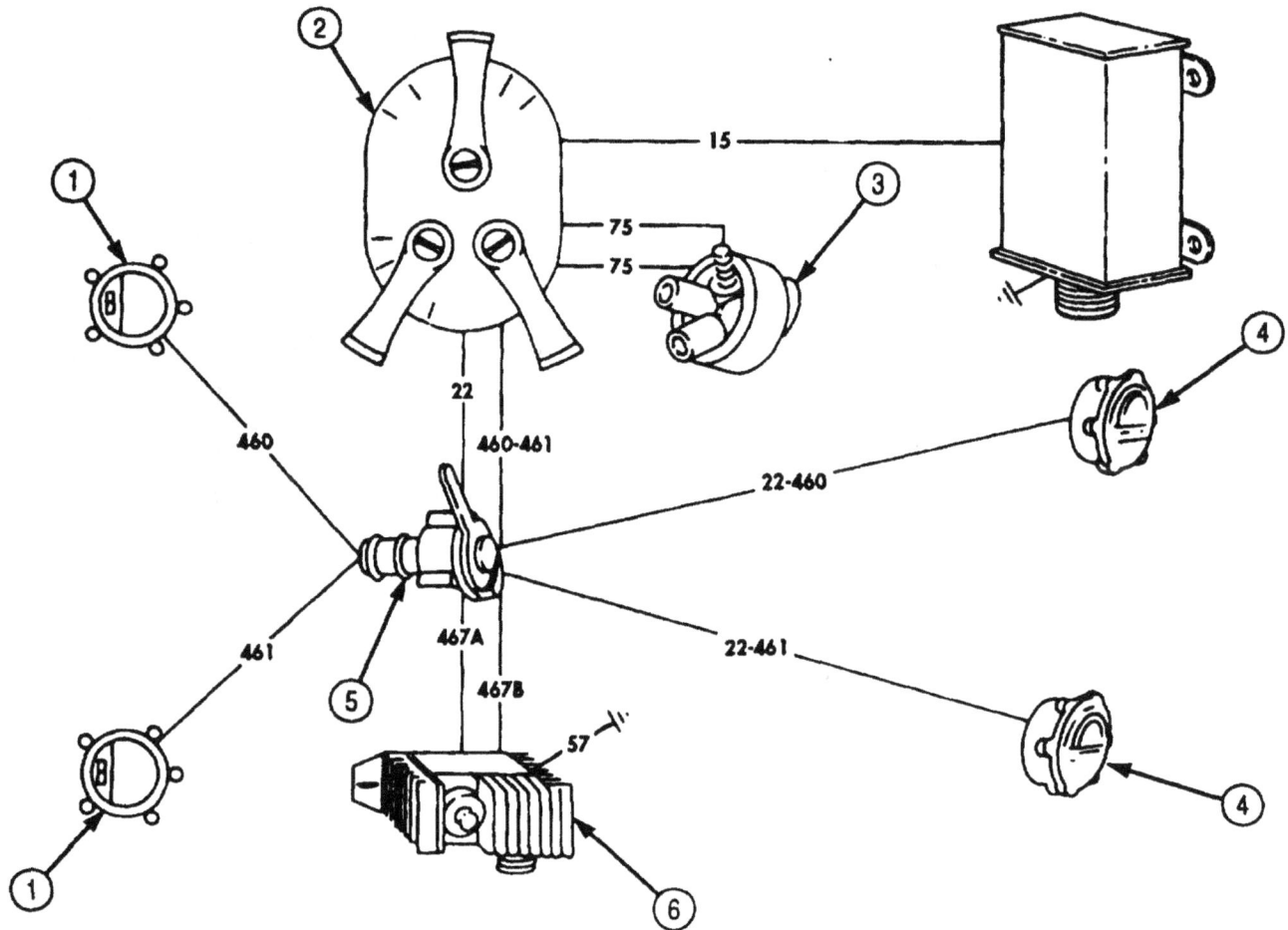

Key Item and Function

1 **FRONT COMPOSITE LAMP** - Receives power from turn signal control through circuits 460 and 461 to indicate turning direction.

2 **LIGHT SWITCH** - Provides battery power to directional signal switch through circuits 460 and 461 and to stoplight switch through circuit 75.

3 **STOPLIGHT SWITCH** - Closing this switch allows power to flow from light switch through circuit 75 to circuit 22 to directional signal switch.

4 **REAR COMPOSITE LAMPS** - Receive power from turn signal control through circuit 22-460 and 22-461 to indicate turning direction.

5 **DIRECTIONAL SIGNAL SWITCH** -A four-position switch that directs power to composite and signal lamps through circuits 460, 461, 22-460, and 22-461 to indicate direction of turn.

6 **TURN SIGNAL FLASHER** - Receives power through circuit 467A and sends intermittent current to the signal lamp through circuit 467B.

1-22. ELECTRICAL SYSTEMS OPERATION (Contd)

f. Heating System Operation.

The electrical portion of the heating system is identical for all models covered in this manual and consists of the following major components and circuitry:

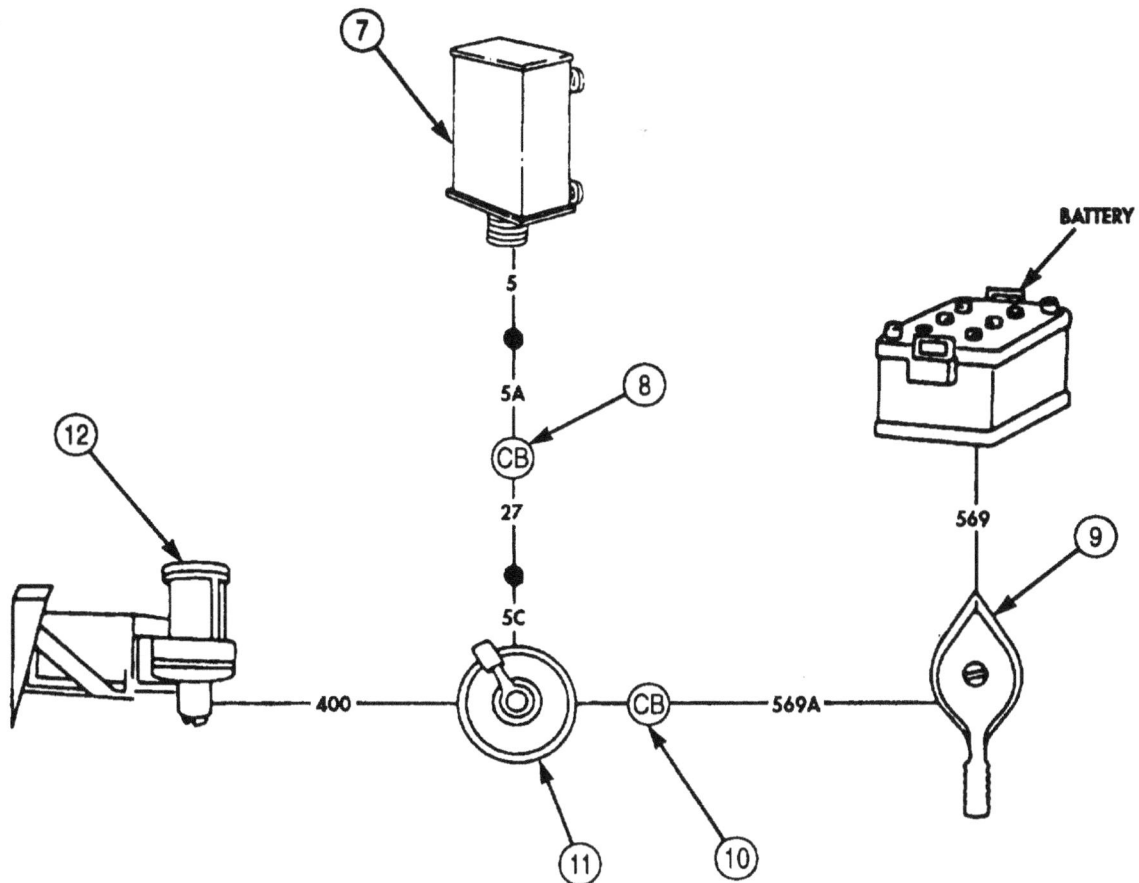

Key Item and Function

7   **PROTECTIVE CONTROL BOX** - Provides 24-volt power to circuit breaker through circuits 5 and 5A and to heater switch through circuits 27 and 5C.

8   **CIRCUIT BREAKER** - Provides overload protection for 24-volt circuits 5, 5A, 27, and 5C leading to the heater switch.

9   **BATTERY SWITCH** - Provides 12-volt battery power from circuit 569 through 569A to the heater.

10   **CIRCUIT BREAKER** - Provides overload protection for 12-volt circuit 569A leading to heater switch.

11   **HEATER SWITCH** - Controls low and high blower motor speed and has two sources of power: 12-volt power is supplied through circuit 569A from battery switch and is used to provide low speed; 24-volt power is supplied through circuit 5C from protective control box and is used to provide high speed.

12   **HEATER BLOWER MOTOR** -A direct current motor controlled by heater switch through circuit 400.

1-22. ELECTRICAL SYSTEMS OPERATION (Contd)

g. **Indicator, Gauge, and Warning System Operation.**

The indicator, gauge, and warning system is comprised of several subsystems:

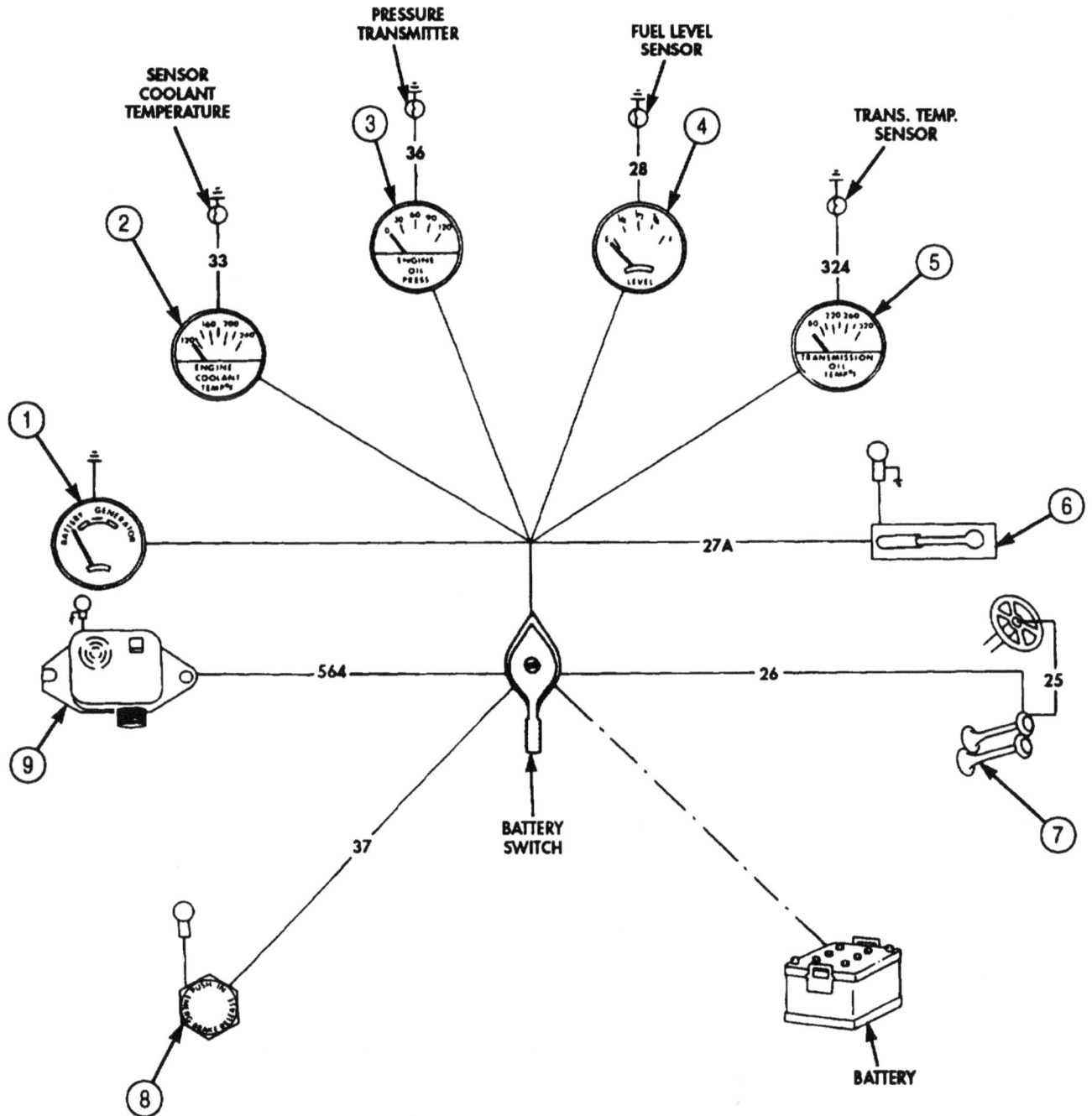

1-22. ELECTRICAL SYSTEMS OPERATION (Contd)

Key Item and Function

1 **VOLTMETER** - Indicates system voltage and is connected to batteries through circuit 27A and to chassis ground through instrument panel.

2 **ENGINE COOLANT TEMPERATURE INDICATOR** - Indicates engine coolant temperature and receives battery power through circuit 27A. Circuit 33 completes the circuit to ground through a coolant temperature sensor that reacts to changes in engine coolant temperature by increasing or decreasing the resistance in ground circuit.

3 **ENGINE OIL PRESSURE INDICATOR** - Indicates engine oil pressure and receives battery power through circuit 27A. Circuit 36 completes the circuit to ground through the oil pressure transmitter located on the engine block.

4 **FUEL INDICATOR** - Indicates fuel level. Receives battery power through circuit 27A. Circuit 28 or 29, depending upon which position fuel selector switch is in, completes the circuit to ground through fuel level sensor.

5 **TRANSMISSION OIL TEMPERATURE INDICATOR** - Indicates transmission oil temperature and receives battery power through circuit 27A. Circuit 324 completes circuit to ground through a temperature sensor located in transmission.

6 **FRONT-WHEEL DRIVE ENGAGEMENT LIGHT** - Informs the operator that front-wheel drive is engaged. The system consists of a normally open pressure switch, which is powered through circuit 27A and an indicator lamp powered through circuit 27A.

7 **HORN SYSTEM** - The horn system consists of an air-operated horn that is controlled by an electric solenoid. The solenoid is powered through circuit 26 and controlled by the horn switch through circuit 25.

8 **SPRING BRAKE WARNING SYSTEM** - Warns the operator that spring brakes are applied. The system consists of normally open pressure switch powered through circuit 37 and an indicator lamp which is powered through circuit 37.

9 **FAILSAFE WARNING SYSTEM** - Intended to give the operator an audible as well as visual signal of a malfunction in one of the primary systems. Power for the system is supplied from the ignition switch through circuit 564. The failsafe module causes an indicator lamp to illuminate and an alarm to sound when air pressure falls below 60 psi (414 kPa) or when parking brake is set.

1-22. ELECTRICAL SYSTEMS OPERATION (Contd)

h. Trailer and Semitrailer Connection System Operation.

The trailer receptacle is identical on all models covered in this manual. The semitrailer receptacle is on the tractor body only.

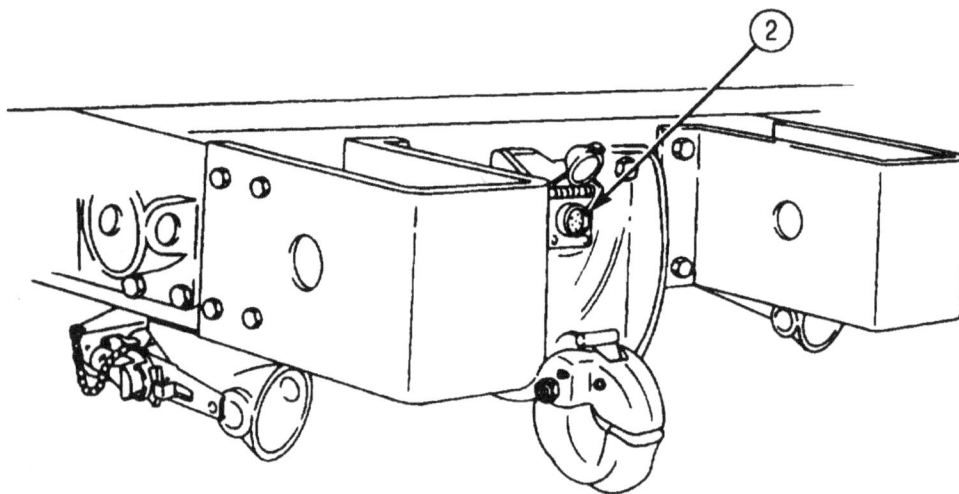

Key Item and Function

1　**TRAILER RECEPTACLE** - Provides vehicle lighting, auxiliary power, and a ground circuit for trailers.

2　**SEMITRAILER RECEPTACLE** - M931/A1/A2 and M932/A1/A2 vehicles equipped with a fifth wheel are provided with a semitrailer receptacle. This receptacle provides vehicle lighting, auxiliary power, and a ground circuit for semitrailers.

1-23. COMPRESSED AIR AND BRAKE SYSTEM OPERATION

The compressed air and brake system takes filtered air, compresses it, and supplies it to various components that enable the operator to slow down or stop the vehicle. This system also supplies compressed air to air-actuated accessories throughout the vehicle. These components and accessories will be described as part of the following systems:

    a. Medium Wrecker Automatic Brake Lock System Operation (page 1-55).

    b. Air Pressure Supply System Operation (page 1-56).

    c. Secondary Service Airbrake System Operation (page 1-60).

    d. Spring Airbrake System Operation (page 1-62).

    e. Primary Service Airbrake System Operation (page 1-63).

    f. Auxiliary Air-Powered System Operation (page 1-66).

    g. Air Venting System Operation (page 1-68).

    h. Central Tire Inflation System (CTIS) (M939A2 series vehicles) (page 1-70).

  a. Medium Wrecker Automatic Brake Lock System Operation.

The M936/A1/A2 Medium Wrecker Automatic Brake Lock System locks the service airbrakes when the transfer case PTO lever is engaged. Major components of the automatic brake lock system are:

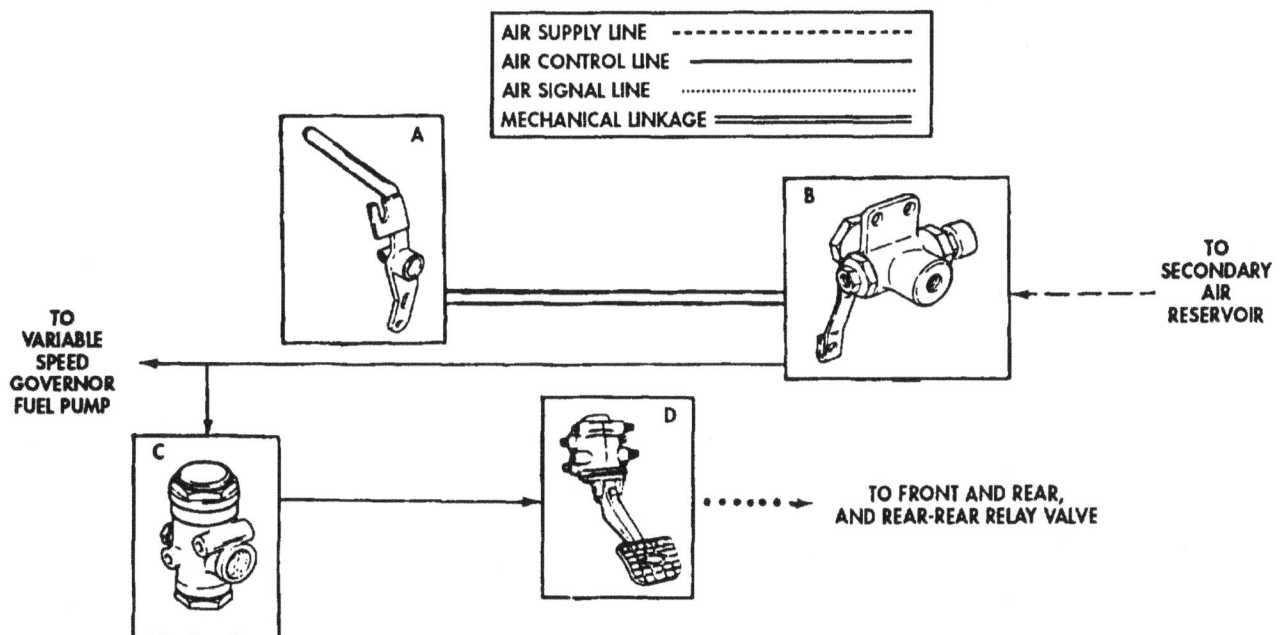

```
AIR SUPPLY LINE    --------------------
AIR CONTROL LINE   ——————————
AIR SIGNAL LINE    ....................
MECHANICAL LINKAGE ================
```

TO SECONDARY AIR RESERVOIR

TO VARIABLE SPEED GOVERNOR FUEL PUMP

TO FRONT AND REAR, AND REAR-REAR RELAY VALVE

Key Item and Function

  A   **TRANSFER CASE PTO LEVER** - Opens the brake lock control valve through mechanical linkage when engaged.

  B   **BRAKE LOCK CONTROL** - Allows air pressure to flow from secondary air reservoir to pressure regulator and activate variable speed governor.

  C   **PRESSURE REGULATOR** - Reduces and regulates system air pressure to 70 psi (483 kPa) for automatic brake lock application.

  D   **TREADLE VALVE** - Connects pressure regulator and service airbrakes.

1-23. COMPRESSED AIR AND BRAKE SYSTEM OPERATION (Contd)

b. Air Pressure Supply System Operation.

(1) A constant air pressure supply is developed by the compressor which is regulated by the governor to maintain 90-120 psi (621-827 kPa) for the airbrake system. Moisture within the system is controlled through the use of either the alcohol evaporator or air dryer. The major components of the system are:

M939A2 ONLY   — — — — —
M939/A1 ONLY   —·—·—·—·—·—
M939/A1 AIR DRYER KIT ·········

BIDIRECTIONAL FEED
(SUBTASK B, STEP 2)

MOISTURE VENT

1-23. COMPRESSED AIR AND BRAKE SYSTEM OPERATION (Contd)

Key Item and Function

A **AIR DRYER** - Installed in supply line to wet tank and removes moisture from inlet air to wet tank (M939/A1 air dryer kit installed only).

B **AIR COMPRESSOR** - Draws in air from the intake manifold and forces it into brake system and wet tank reservoir.

C **SAFETY VALVE** - Located at inlet side of wet reservoir, it prevents pressure buildup by releasing air pressure exceeding 150 psi (1,034 kPa) when the governor fails to regulate air supplied by the compressor.

D **WET TANK RESERVOIR** - Performs two functions:

• Traps water in air reservoir to protect other air systems from freezing or corroding.

• Stores reserve air supply enabling operator to make normal stops when engine stalls or compressor fails.

E **PRESSURE PROTECTION VALVE** - Performs two functions:

• Allows air pressure to build to 60-65 psi (414-448 kPa) before supplying air to auxiliary air-powered equipment.

• Closes off auxiliary air system from other systems if an accessory fails and prevents loss of air from secondary reservoir.

F **ALCOHOL EVAPORATOR (M939/A1 series)** - Helps protect air lines from freezing.

G **WET TANK RESERVOIR DRAINVALVE** - Provides a drain for moisture and air from reservoir.

H **AIR DRYER** - Removes moisture from inlet air to wet tank (M939A2 only).

I **EXPELLO VALVE** - Augments air dryer condensation blowdown by venting moisture when compressor cycles.

J **GOVERNOR** - Trips valve inside compressor to regulate flow of air to the system. When pressure builds to 120-127 psi (827-876 kPa), the governor will close valve.

1-23. COMPRESSED AIR AND BRAKE SYSTEM OPERATION (Contd)

(2) The constant air pressure supply is distributed to the primary service airbrake system (para. e.) and secondary airbrake system (para. c.) through a shutoff and check valve. Air pressure can either be fed from or supplied to another vehicle through the emergency couplings.

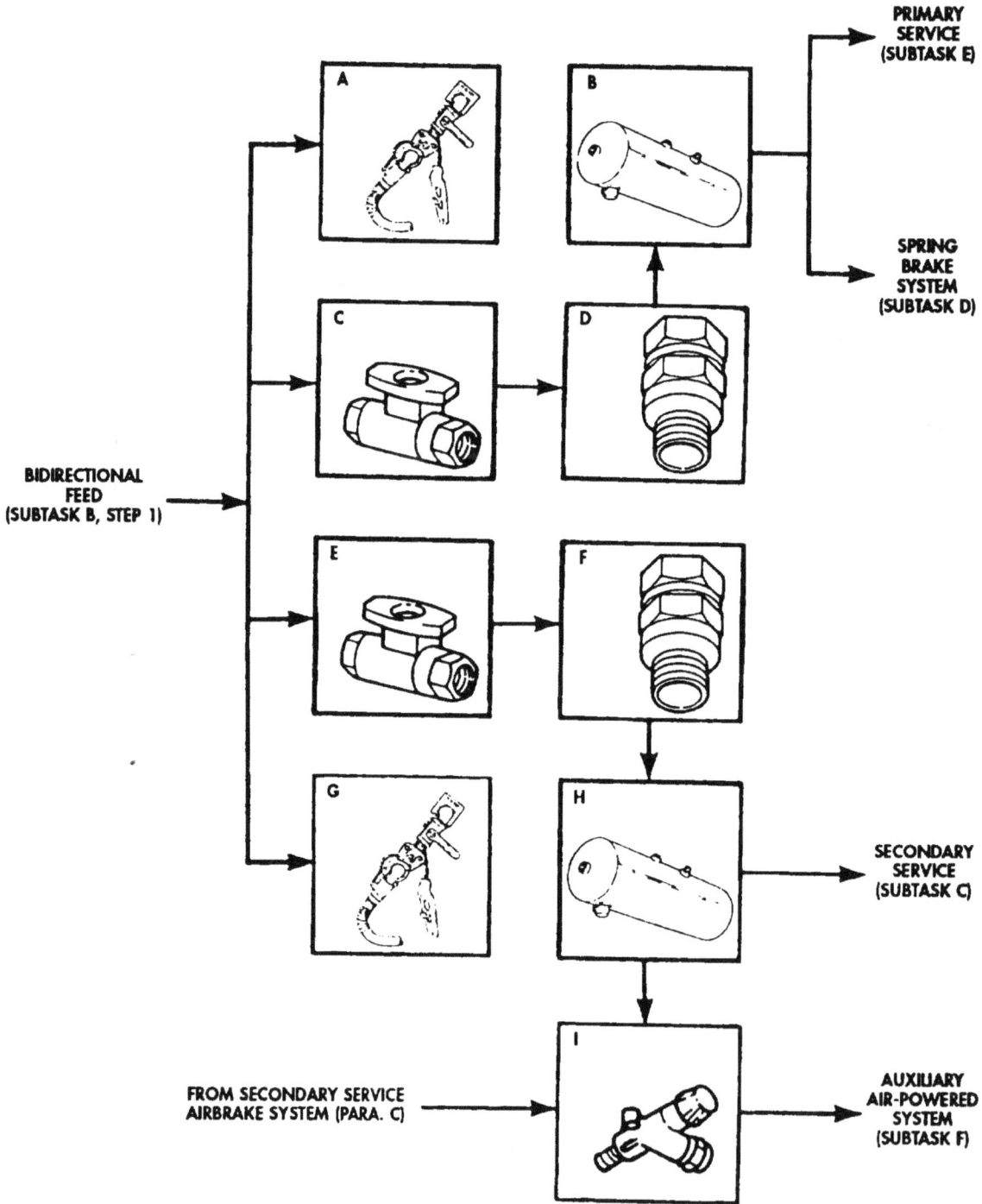

1-23. COMPRESSED AIR AND BRAKE SYSTEM OPERATION (Contd)

Key Item and Function

A **FRONT EMERGENCY COUPLING** -When vehicle is being towed, coupling receives compressed air from towing vehicle's brake system to charge its own brake system.

B **PRIMARY AIR RESERVOIR** - Stores sufficient air pressure to allow operator to make normal brake applications if system pressure fails or engine stalls.

C **PRIMARY FEED CUTOFF** - Manually-operated valve used to isolate pressure leaks in primary air system from draining wet tank (TM 9-2320-272-10).

D **PRIMARY AIR RESERVOIR CHECKVALVE** - Prevents backflow of air from primary tank if wet system develops a leak.

E **SECONDARY FEED CUTOFF** - Manually-operated valve used to isolate pressure leaks in secondary air system from draining wet tank (TM 9-2320-272-10).

F **SECONDARY AIR RESERVOIR CHECKVALVE** - Prevents backflow of air from secondary tank if wet system develops a leak.

G **REAR EMERGENCY COUPLING** - When towing another vehicle, coupling allows pressurized air from wet tank to charge towed vehicle's wet tank.

H **SECONDARY AIR RESERVOIR** - Stores enough air pressure in case constant pressure system fails or engine stalls. The operator can make normal brake application before running out of air.

I **PRESSURE PROTECTION VALVE** - Performs two functions:

(1) Allows air pressure to build to 60-65 psi (414-448 kPa) before supplying air to auxiliary air-powered equipment.

(2) Closes off auxiliary air system from other systems if an accessory fails and prevents loss of air from secondary reservoir.

1-23. COMPRESSED AIR AND BRAKE SYSTEM OPERATION (Contd)

c. **Secondary Service Airbrake System Operation.**

(1) The secondary service airbrake system is made up of two subsystems:

    (a) Secondary constant pressure system provides continuous air pressure to:
- Pedal valve
- Rear relay valve
- Spring brake air reservoir
- Spring parking brake valve

    (b) Secondary signal system serves three functions:
- Contains air pressure only when operator steps on brake pedal.
- Is regulated by various valves to control amount of braking.
- Provides pressure to apply rear two service brakes and intermediate and rear axles stamped with B. Service brakes on rear axle are piggybacked to spring brakes, but operate independently of them.

(2) The secondary constant pressure system is made up of the following components:

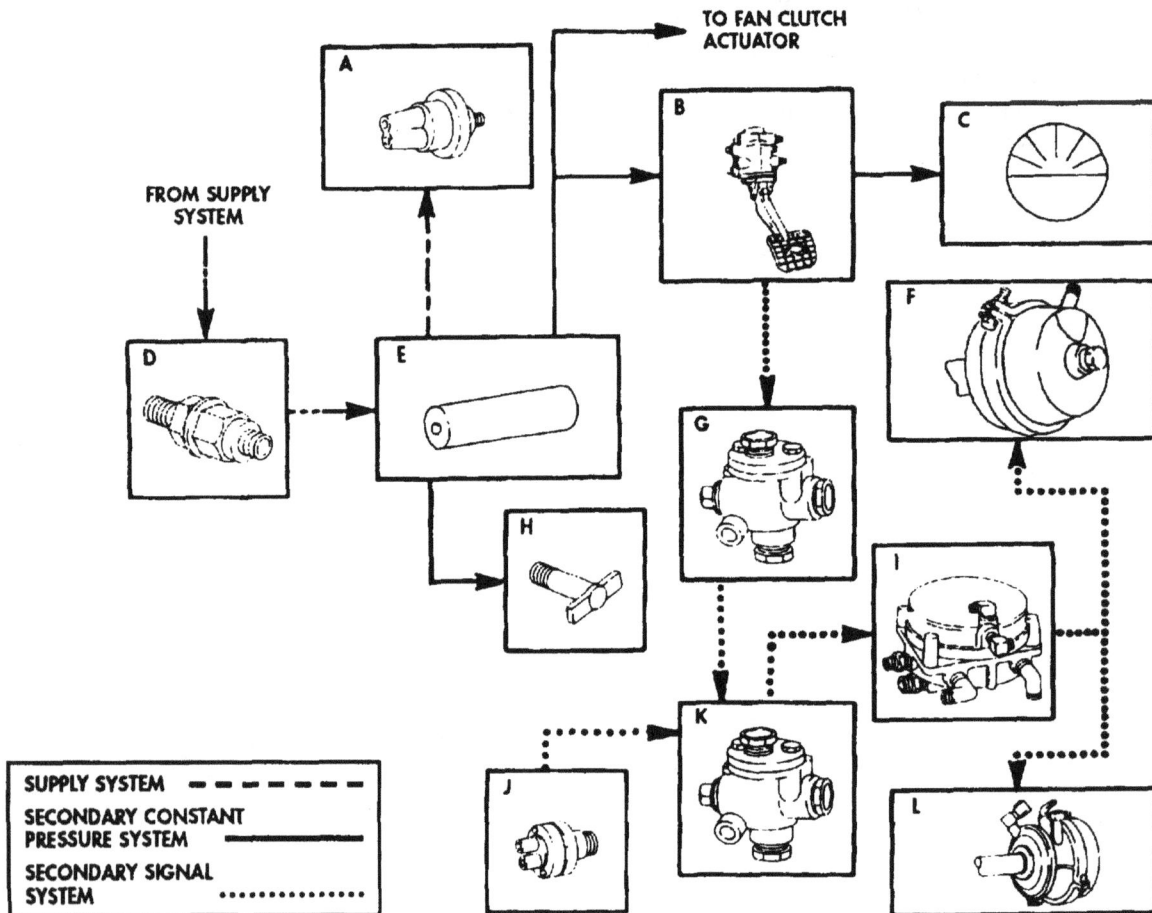

1-60

1-23. COMPRESSED AIR AND BRAKE SYSTEM OPERATION (Contd)

Key Item and Function

A **LOW AIR PRESSURE SWITCH** - Activates warning buzzer and warning lights when air pressure goes below 60 psi (414 kPa).

B **PEDAL VALVE** - Allows air pressure from secondary constant pressure system to flow into secondary signal system when operator depresses brake pedal.

C **SECONDARY AIR PRESSURE GAUGE** - Indicates amount of air pressure in secondary system.

D **ONE-WAY CHECKVALVE** -Allows air pressure to flow into secondary reservoir, but prevents it from coming out if constant pressure system fails or engine stalls.

E **SECONDARY AIR RESERVOIR** - Stores enough air pressure so the operator can make five normal brake applications before running out of air if constant pressure fails or engine stalls.

F **INTERMEDIATE REAR BRAKE CHAMBERS** - Converts air pressure to mechanical force which applies intermediate rear service brake.

G **DOUBLECHECK VALVE NO. 1** - Serves two functions:

- Allows system to receive signal pressure from either pedal valve or, when towed, from brake system of towing vehicle.

- Serves as a tee between front and rear primary signal lines.

H **SECONDARY AIR RESERVOIR DRAINVALVE** - Provides a drain for moisture and air from secondary air reservoir.

I **FRONT RELAY VALVE** - Boosts signal air to rear brake chambers; regulates air pressure to rear brake chambers so operator has control over amount of braking; and releases air pressure to rear brake chambers directly to vent when brake pedal is released.

J **STOPLIGHT SWITCH** - As the brake pedal is depressed, switch receives an air pressure signal at electrical contacts which close to activate circuits to taillights.

K **DOUBLECHECK VALVE NO. 2** - Allows either primary or secondary signal air pressure to activate stoplight switch while keeping the two systems separate.

L **REAR-REAR BRAKE CHAMBERS** - Converts air pressure to mechanical force which applies rear-rear brakes.

1-23. COMPRESSED AIR AND BRAKE SYSTEM OPERATION (Contd)

d. Spring Airbrake System Operation.

The spring airbrake system applies rear brakes when vehicle parking brake is applied or in event of a major brake failure. The spring brake is located on one of the two service brake chambers at each rear wheel. Major components of the spring airbrake system are:

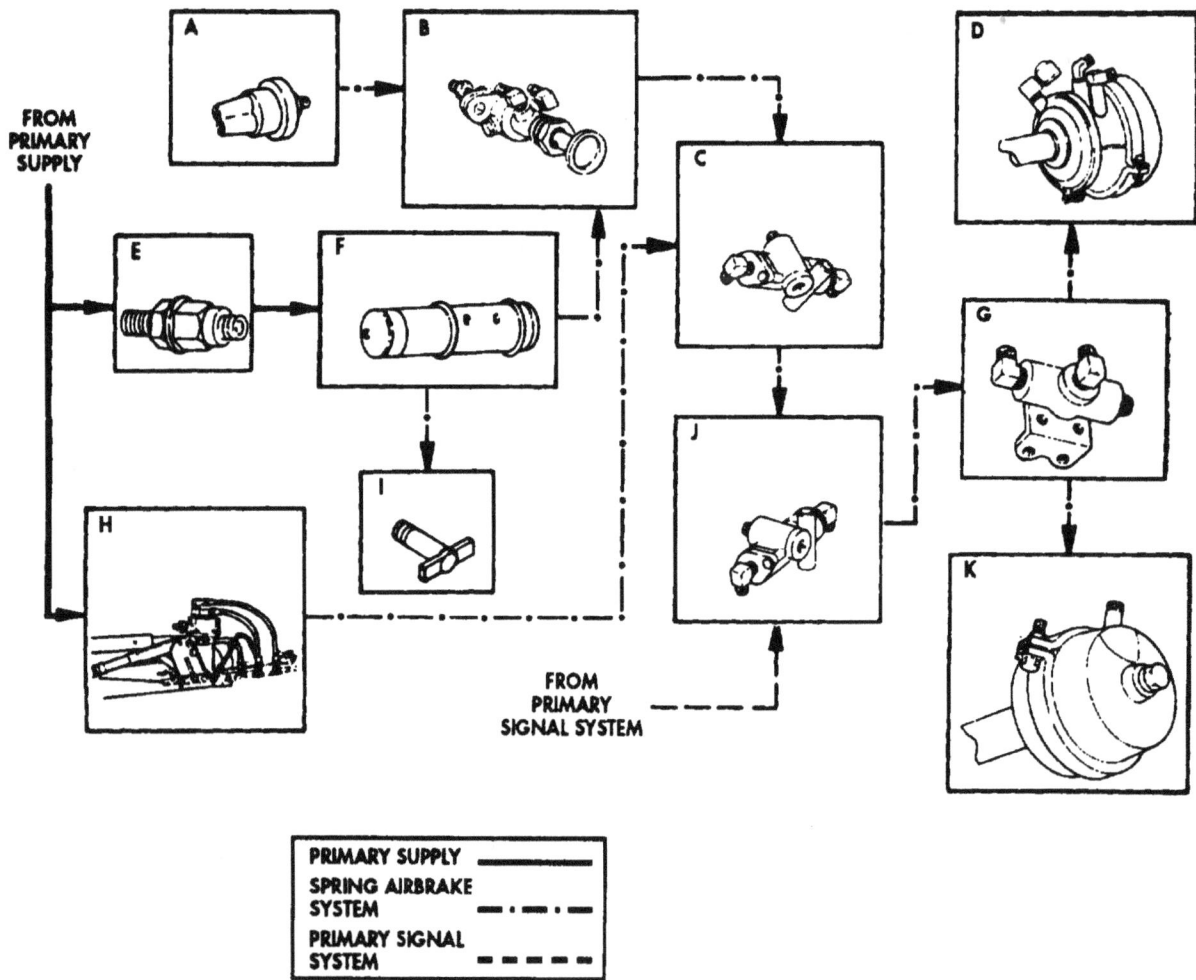

PRIMARY SUPPLY	———
SPRING AIRBRAKE SYSTEM	— · — · —
PRIMARY SIGNAL SYSTEM	— — — —

1-23. COMPRESSED AIR AND BRAKE SYSTEM OPERATION (Contd)

Key Item and Function

A SPRING BRAKE WARNING LIGHT SWITCH - Activates warning light when spring brakes are engaged.

B SPRING BRAKE RELEASE CONTROL VALVE - Pushed in to release spring brakes independently of mechanical parking brake. Control is also used to release spring brakes in order to test and adjust mechanical brake.

C DOUBLECHECK VALVE NO. 4 - Allows spring brake air pressure to come from either release control valve or spring parking brake valve directly to doublecheck valve No. 3.

D INTERMEDIATE FRONT SPRING BRAKE CHAMBER - Contains a large spring which applies rear brakes when spring brake air pressure is released.

E ONE-WAY CHECKVALVE - Allows air pressure to flow into spring brake reservoir, but prevents it from coming out if constant pressure system or primary system fails.

F SPRING BRAKE AIR RESERVOIR - Stores enough air pressure to release spring brakes for emergency operation in event of primary or secondary air system failure.

G QUICK-RELEASE VALVE - Releases spring brake air pressure directly to vent if parking brake has been set or brake system fails.

H SPRING BRAKE VALVE - Automatically sets spring brakes when parking brake is set. Valve can be released independently of parking brake when spring brake control valve is pushed in.

I SPRING BRAKE RESERVOIR DRAINVALVE - Provides a drain for moisture and air from spring brake reservoir.

J DOUBLECHECK VALVE NO. 3 - Allows spring brake air pressure to come from either release control valve or spring parking brake valve directly to doublecheck valve No. 4.

K REAR-BEAR SPRING BRAKE CHAMBER - Contains a large spring which applies rear brakes when spring brake air pressure is released.

e. Primary Service Airbrake System Operation.

(1) The primary service airbrake system is made up of two subsystems:

(a) Primary constant pressure system provides continuous air pressure to:
- Pedal valve
- Rear relay valve
- Spring brake air reservoir
- Spring parking brake valve

(b) Primary signal system serves three functions:
- Contains pressure only when operator steps on brake pedal.
- Is regulated by various valves to give operator control over amount of braking.
- Provides pressure to apply front service brakes and the front two service brakes on intermediate and rear axles stamped with an A. Service brakes on the intermediate axle are piggybacked to spring brakes, but operate independently of them.

1-23. COMPRESSED AIR AND BRAKE SYSTEM OPERATION (Contd)

(2) The primary constant pressure system is made up of the following components:

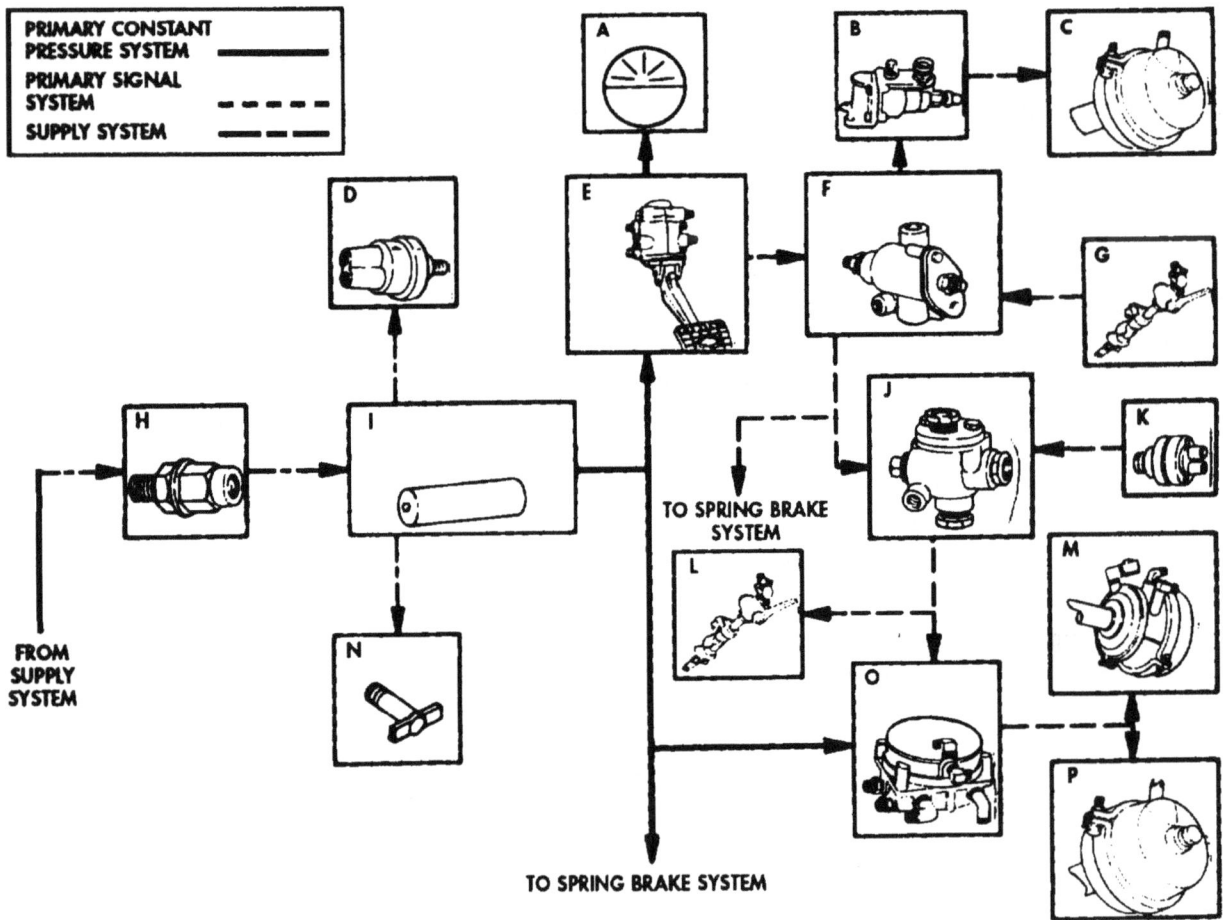

PRIMARY CONSTANT
PRESSURE SYSTEM

PRIMARY SIGNAL
SYSTEM

SUPPLY SYSTEM

A

B

C

D

E

F

G

H

I

J

K

L

M

N

O

P

TO SPRING BRAKE
SYSTEM

FROM
SUPPLY
SYSTEM

TO SPRING BRAKE SYSTEM

1-23. COMPRESSED AIR AND BRAKE SYSTEM OPERATION (Contd)

Key Item and Function

A PRIMARY AIR PRESSURE GAUGE - Indicates amount of air pressure in primary system.

B LIMITING VALVE - Serves three functions:

- Regulates signal air pressure going to front brake chambers so rear brakes are applied first.
- Regulates signal air pressure to front brake chambers so operator has control over amount of braking.
- Releases air pressure in front brake chambers directly to vent in the valve when brake pedal is released.

C FRONT BRAKE CHAMBERS - Convert air pressure to mechanical force which applies front service brakes.

D PRIMARY RESERVOIR LOW AIR PRESSURE SWITCH - Activates warning buzzer and warning light when air pressure goes below 60 psi (414 kPa).

E PEDAL INTO VALVE - Allows air pressure from primary constant pressure system to flow into primary signal system when operator depresses brake pedal.

F DOUBLECHECK VALVE NO. 1 - Serves two functions:

- Allows system to receive signal pressure from either pedal valve or, when towed, from brake system of towing vehicle.
- Serves as a tee between front and rear primary signal lines.

G FRONT SERVICE COUPLING - When vehicle is being towed, coupling is connected to towing vehicle so the brake systems of the two vehicles work together.

H ONE-WAY CHECKVALVE - Allows air pressure to flow into primary reservoir, but prevents it from coming out if constant pressure system fails or engine stalls.

I PRIMARY AIR RESERVOIR - Stores enough air pressure so the operator can make five normal brake applications before running out of air if constant pressure fails or engine stalls.

J DOUBLECHECK VALVE NO. 2 - Allows either primary or secondary signal air pressure to activate stoplight switch while keeping the two systems separate.

K STOPLIGHT SWITCH -As brake pedal is depressed, switch receives an air pressure signal which closes electric contacts turning on stoplight.

L REAR SERVICE COUPLING - When towing another vehicle, coupling is connected to towed vehicle so the brake systems of the two vehicles work together.

M INTERMEDIATE FRONT BRAKE CHAMBERS - Converts air pressure to mechanical force which applies intermediate rear service brake.

N PRIMARY RESERVOIR DRAINVALVE - Provides a drain for moisture and air from primary air reservoir.

O REAR RELAY VALVE - Serves three functions:

- Boosts signal air pressure to rear brake chambers. Air signal from brake pedal opens valve to route constant air pressure to rear brake chambers.
- Regulates signal air pressure from brake pedal to rear brake chambers so operator has control over amount of braking. Regulates amount of constant air pressure going to brake chambers as the operator depresses the brake pedal.
- Releases air pressure in rear brake chamber directly to vent system when brake pedal is released.

P REAR FRONT BRAKE CHAMBERS - Converts air pressure to mechanical force which applies rear service brakes.

1-23. COMPRESSED AIR AND BRAKE SYSTEM OPERATION (Contd)

f. Auxiliary Air-Powered System Operation.

The auxiliary air-powered system consists of air-actuated vehicle accessories. All of these accessories receive air pressure through the accessory manifold and off the pressure protection valve with the exception of the horns. Components of the auxiliary air-powered system are:

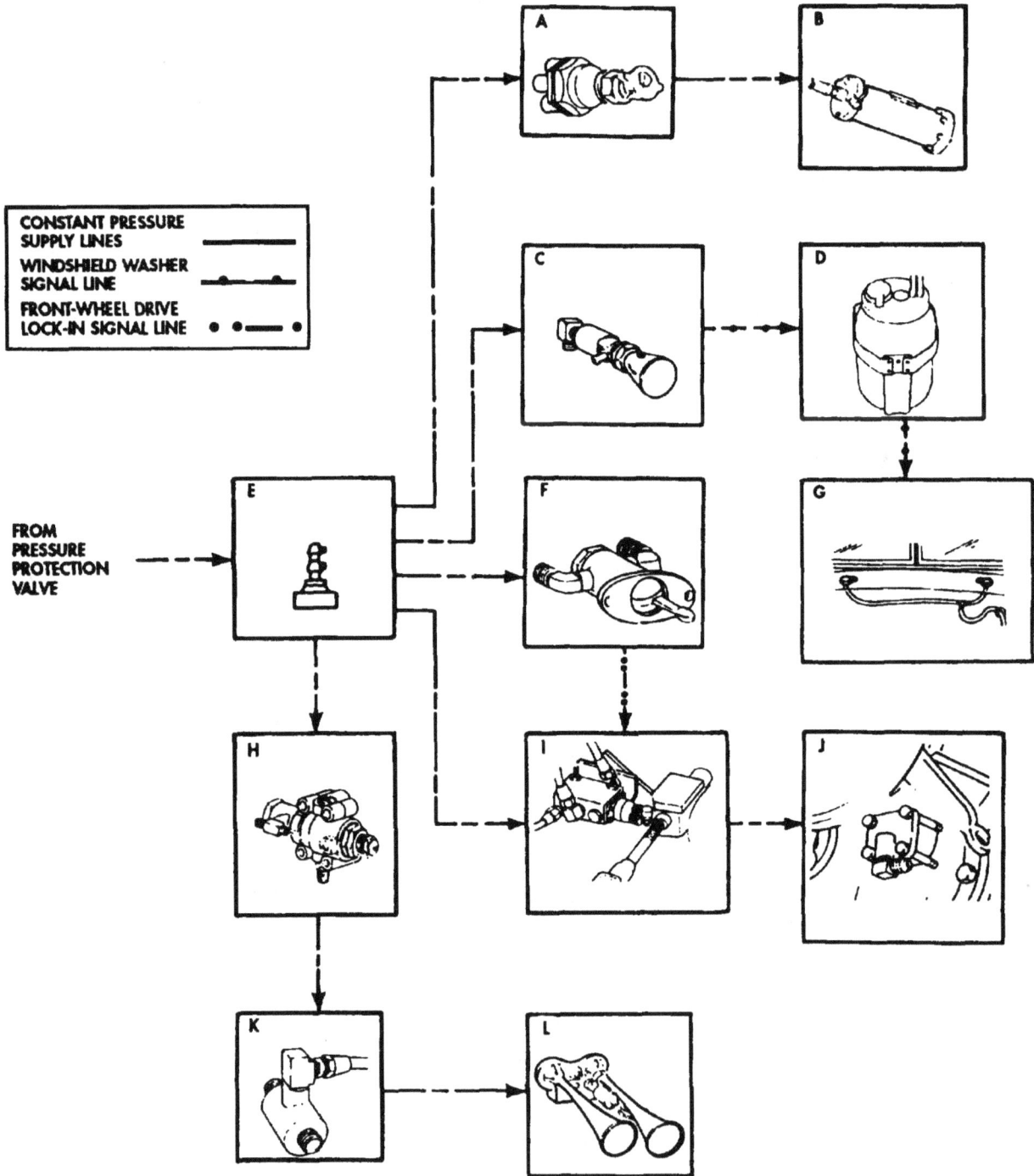

1-23. COMPRESSED AIR AND BRAKE SYSTEM OPERATION (Contd)

Key Item and Function

A WINDSHIELD WIPER CONTROL SWITCH - Opens air pressure valve in wiper motor to operate wipers.

B WINDSHIELD WIPER MOTOR - Air-actuated motor powers windshield wipers.

C WINDSHIELD WASHER CONTROL - Spring-loaded valve that allows air pressure to force washer fluid from washer reservoir to windshield.

D WINDSHIELD WASHER RESERVOIR - Container for windshield washer fluid.

E ACCESSORY MANIFOLD - Receives air pressure from the pressure protection valve and distributes it to the various accessories.

F FRONT-WHEEL DRIVE LOCK-IN SWITCH - Air-actuated switch that engages front-wheel drive when transfer case is in HIGH.

G WINDSHIELD WASHER NOZZLES - Direct washer fluid on windshield.

H GOVERNOR - Serves as a tee between accessory manifold and horn relay valve. It also signals the air compressor to stop compressing air for the supply system when operating pressure has been reached.

I TRANSFER CASE AIR SHIFT CYLINDER - Engages front-wheel drive when it receives air pressure from lock-in switch or engagement control valve.

J FRONT AXLE ENGAGEMENT CONTROL VALVE - Operates off cam, on transfer case shift linkage so front-wheel drive engages automatically when transfer case is put into LOW.

K HORN RELAY VALVE - Electrical signal from horn button on steering wheel opens valve in horn relay, allowing air pressure to sound horns.

L HORNS - Receive air pressure from horn relay valve to sound off.

1-23. COMPRESSED AIR AND BRAKE SYSTEM OPERATION (Contd)

g. Air Venting System Operation.

The air venting system vents air from brake system and powertrain and fuel vapors from fuel system into air intake stack where it is released into the atmosphere. The components of the air venting system are:

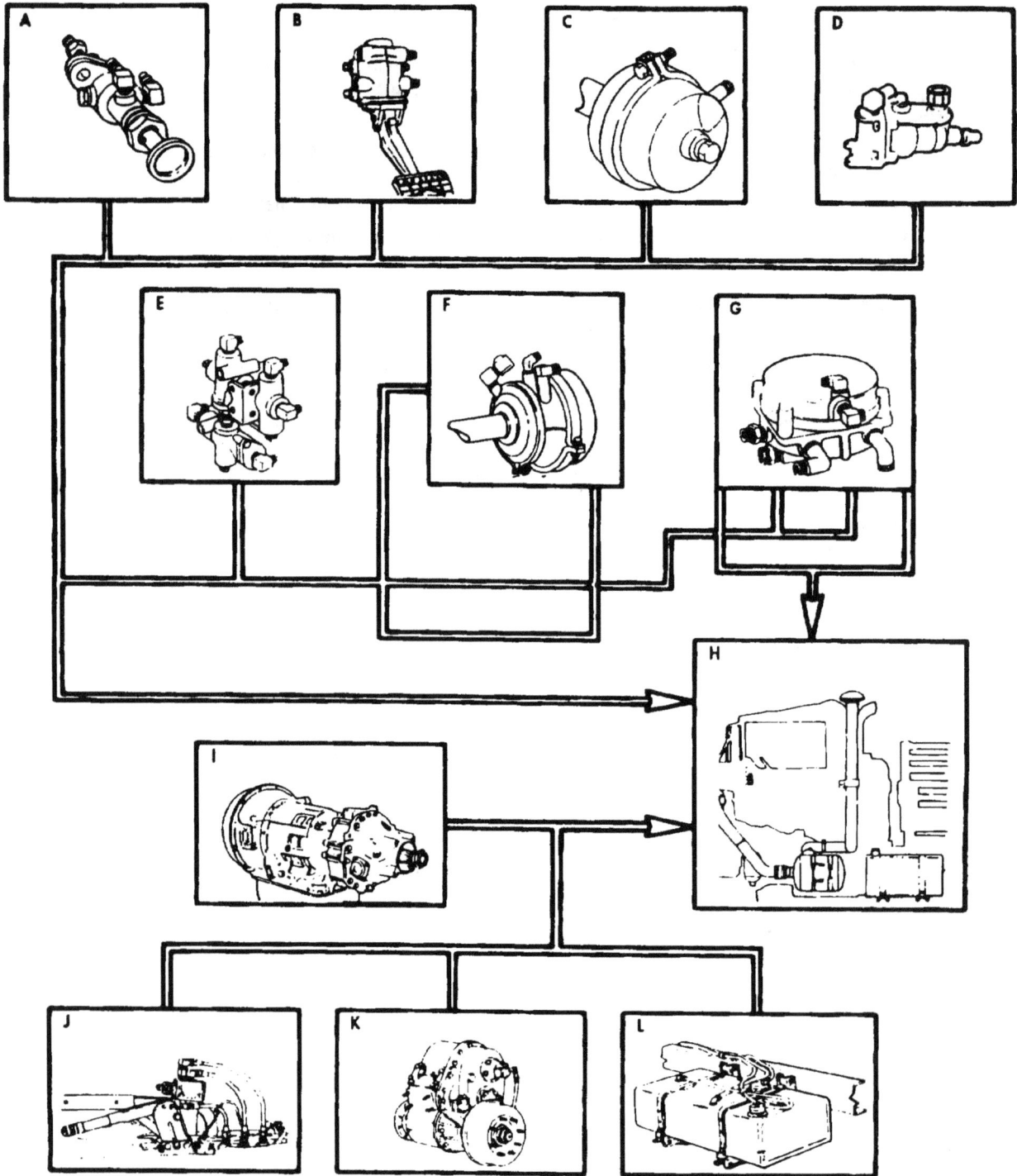

1-23. COMPRESSED AIR AND BRAKE SYSTEM OPERATION (Contd)

Key Item and Function

A   **SPRING BRAKE RELEASE CONTROL VALVE** - This valve functions as an override when a failure in the air supply system (causing spring brakes to engage) occurs. When valve is manually pushed in, emergency air is supplied to the spring brake chambers. This releases the spring brakes, allowing vehicle movement.

B   **PEDAL VALVE** - Vents primary or secondary signal air pressure when pedal is released.

C   **FRONT BRAKE CHAMBER VENT** - Vents air pressure inside chambers when pedal valve is released.

D   **LIMITING VALVE** - Vents signal air pressure going to front brake chambers so rear brakes apply first.

E   **STEP BOX QUICK-RELEASE VALVE** - Vents air pressure from spring brake chambers when parking brake valve has been actuated.

F   **REAR BRAKE CHAMBERS** - Vents ports on chambers to prevent air pressure buildup.

G   **RELAY VALVES** - Vents air pressure in rear brake chambers directly to intake tube when brake pedal is released. Vents signal air pressure through upper port in valve.

H   **AIR INTAKE STACK** - Venting point for the vent system.

I   **TRANSMISSION VENT** - Vents internal air pressure buildup due to internal heat.

J   **SPRING PARKING BRAKE VALVE** - Vents air pressure from air and doublecheck valves No. 3 and No. 4.

K   **TRANSFER CASE VENT** - Vents internal air pressure buildup due to internal heat.

L   **FUEL TANK VENTS** - Vent fuel vapors to prevent partial vacuum from stopping fuel flow.

1-23. COMPRESSED AIR AND BRAKE SYSTEM OPERATION (Contd)

h. Central Tire Inflation System (CTIS) (M939A2 Series Vehicles).

The CTIS is common to all M939A2 series vehicles. This system maintains tire air pressure depending on which road type is selected. If this setting is changed, tires will automatically inflate or deflate to the new setting.

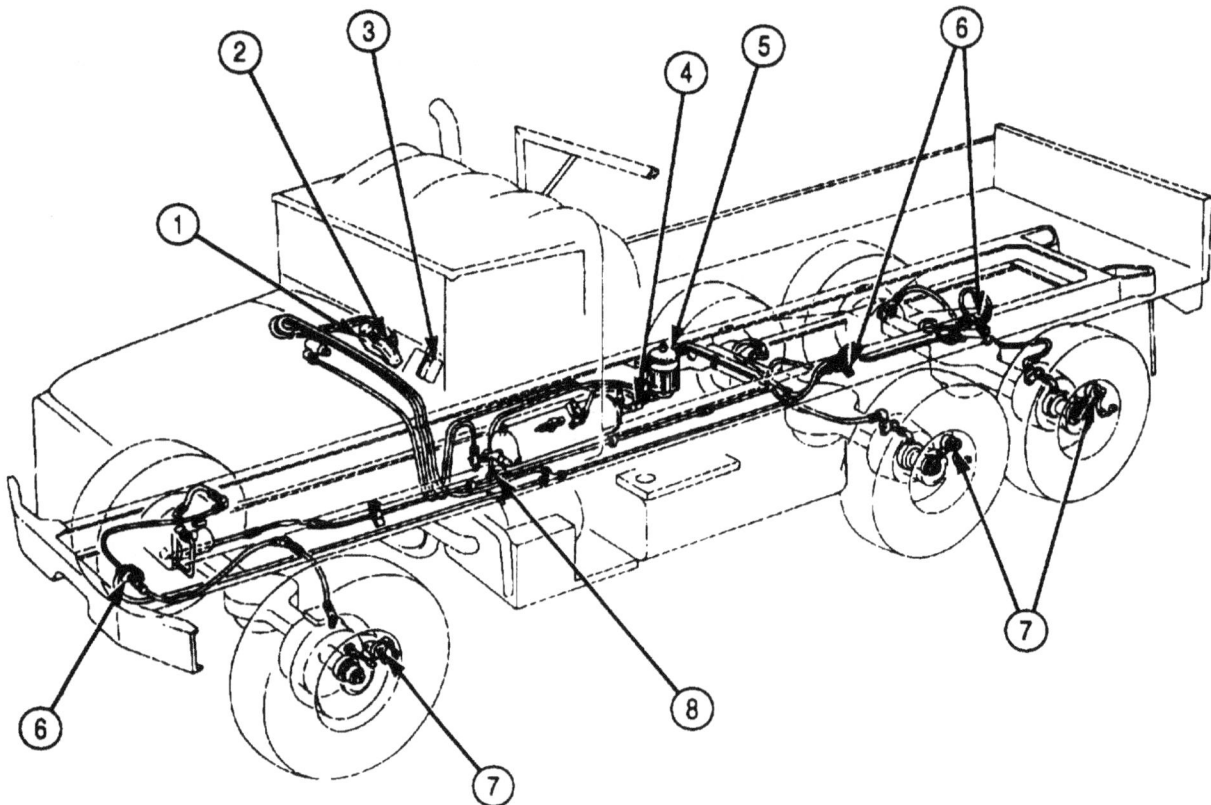

Key Item and Function

1 **PNEUMATIC CONTROLLER** - Directs air pressure according to ECU commands.

2 **PRESSURE TRANSDUCER** - Mounted in pneumatic controller, it measures tire pressure and sends information to ECU.

3 **ELECTRONIC CONTROL UNIT (ECU)** - Contains CTIS selector panel so that operator can change tire inflation during vehicle operation.

4 **AIR PRESSURE SWITCH** - Protects air brake system for a minimum supply of 85 psi (586 kPa) of air.

5 **AIR DRYER AND FILTER** - Separates moisture from compressed air system and filters impurities from compressed air before they enter CTIS.

6 **EXHAUST VALVES** - Exhaust air from tires during deflation.

7 **WHEEL VALVES** - Isolate air pressure in tires during normal operation and for tire removal.

8 **SPEED SIGNAL GENERATOR** - Signals ECU to automatically inflate CTIS when vehicle speed exceeds the top speed setting for the selected mode by 10 mph (16 kmh).

1-24. HYDRAULIC SYSTEM OPERATION

Oil pressure (hydraulics) is used to provide operating power for the auxiliary equipment on the vehicles covered in this manual. The components that provide hydraulic power are discussed in the following order:

    a. Front Winch Hydraulic System Operation (page 1-71).

    b. Rear Winch Hydraulic System Operation (page 1-72).

    c. Dump Body Hydraulic System Operation (page 1-73).

    d. Medium Wrecker Crane Hydraulic System Operation (page 1-74).

a. Front Winch Hydraulic System Operation.

A front winch is installed on M925/A1/A2, M928/A1/A2, M930/A1/A2, M932/A1/A2, and M936/A1/A2 series vehicles. The front winch hydraulic system converts mechanical power at the winch drive motor. The basic operating principles are the same for each model. Major components of this system are:

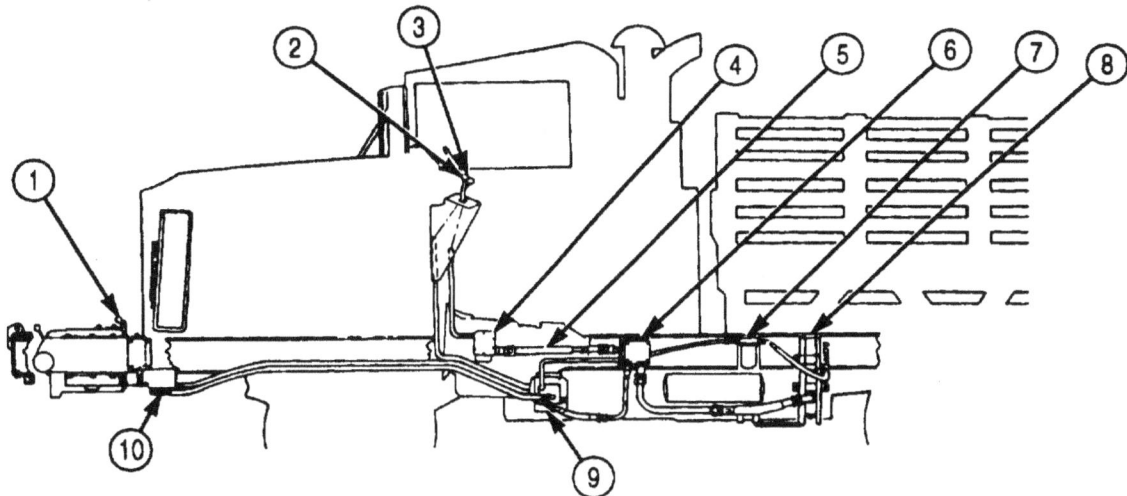

Key Item and Function

1   **CLUTCH LEVER** - Manual control that engages winch drum gear to drive gear of winch motor.

2   **TRANSMISSION PTO CONTROL** - A manually-operated control lever located inside the cab that permits engagement or disengagement of the transmission PTO.

3   **WINCH CONTROL LEVER** -An operator control that determines the hydraulic oil pressure flow from the control valve to the winch motor. The flow of this oil determines the direction the winch drum will turn.

4   **TRANSMISSION PTO** - Uses transmission driving power to provide mechanical driving power for the hydraulic pump.

5   **PTO DRIVESHAFT** - Transmits mechanical power from PTO to the hydraulic pump.

6   **HYDRAULIC PUMP** - Driven by PTO driveshaft, it draws oil from the oil reservoir through hydraulic hoses, then pressurizes and directs this oil to the control valve.

7   **OIL FILTER** - Filters used or bypassed oil from the control valve before it returns to the hydraulic oil reservoir.

8   **HYDRAULIC OIL RESERVOIR** - Storage tank for hydraulic oil.

9   **CONTROL VALVE** - Four-port valve accepts pressurized oil from the hydraulic pump and directs this oil to the winch motor. It also directs oil returning from the winch back to the oil reservoir. The flow of this oil from the valve determines the directional drive of the winch motor.

10   **WINCH MOTOR** - Converts hydraulic power into mechanical power as hydraulic oil is forced through the winch motor.

1-24. HYDRAULIC SYSTEM OPERATION (Contd)

b. Rear Winch Hydraulic System Operation.

A rear winch is installed only on the M936/A1/A2 medium wrecker. It is used primarily to rescue vehicles that have become deeply mired. The rear winch hydraulic system converts mechanical power of the engine into fluid power through use of the hydraulic pump and back into mechanical power at the winch drive motor. The major components of the rear winch hydraulic system are:

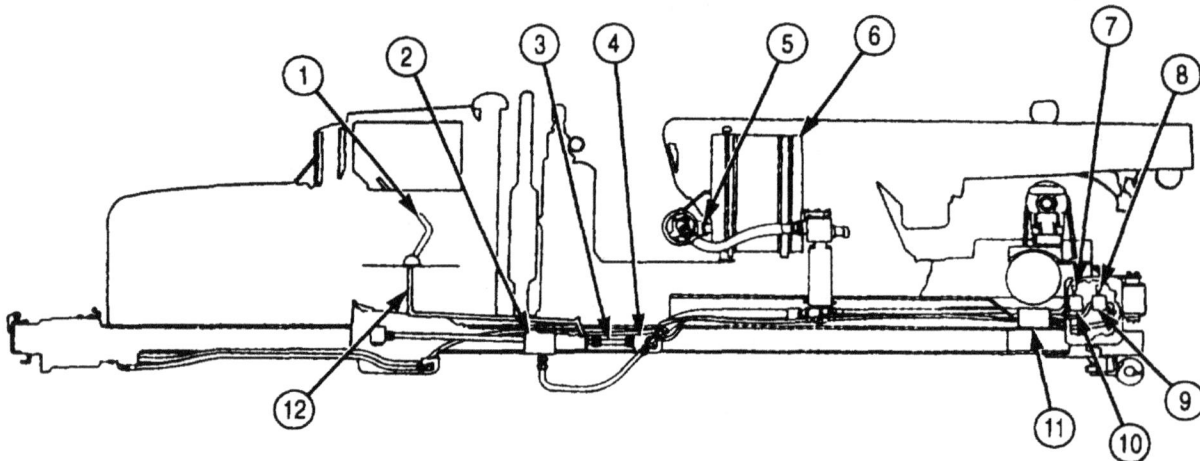

Key Item and Function

1. **TRANSFER CASE PTO CONTROL** - A manually-operated control lever located inside the cab that permits engagement or disengagement of the PTO.

2. **TRANSFER CASE PTO** - Uses driving power of the transfer case to provide mechanical driving power for the hydraulic pump.

3. **PTO DRIVESHAFT** - Transmits mechanical driving power from PTO to the hydraulic pump.

4. **HYDRAULIC PUMP** - Draws oil from hydraulic oil reservoir and directs it to the rear winch control valve and winch drive motor.

5. **OIL FILTER** - Filters used or bypassed oil from the control valve before it returns to the hydraulic oil reservoir.

6. **HYDRAULIC OIL RESERVOIR** - Storage tank for hydraulic oil.

7. **TORQUE CONTROL LEVER** - Controls the operating gear ratio of the winch drive motor. Lever is pulled outward to HIGH for heavy loads or pushed inward to LOW for light loads.

8. **WINCH DIRECTIONAL CONTROL LEVER** - Manually-operated lever that controls the wind and unwind direction of the rear winch drum. Lever does this by opening and closing the directional control valve to the winch motor, and reversing the direction of pressurized hydraulic fluid. Lever is pushed inward to wind and pulled outward to unwind winch cable.

9. **DIRECTIONAL CONTROL VALVE** - Receives pressurized hydraulic oil from the hydraulic pump and directs it to the winch motor. The flow of hydraulic oil to and from this control valve provides forward or reverse driving power to the winch motor. Valve also returns used oil back to the hydraulic oil reservoir from the winch.

10. **TORQUE CONTROL VALVE** - Hydraulically controls the hydraulic oil pressure to engage rear winch drum clutch in high or low gear range.

11. **WINCH MOTOR** - Converts hydraulic power back into mechanical power needed to turn the rear winch drum.

12. **CONTROL LINKAGE** - Connects transfer case PTO control to transfer case PTO.

1-24. HYDRAULIC SYSTEM OPERATION (Contd)

c. Dump Body Hydraulic System Operation.

The dump body is installed on M929/A1/A2 and M930/A1/A2 model vehicles. These models are used to transport and deposit cargo. The dump body hydraulic system converts mechanical power from the engine into fluid power through use of the hydraulic pump. The pump draws fluid from the oil reservoir and then forces it into the control valve. This hydraulic pressure raises and lowers the dump body. Major components of the dump body hydraulic system are:

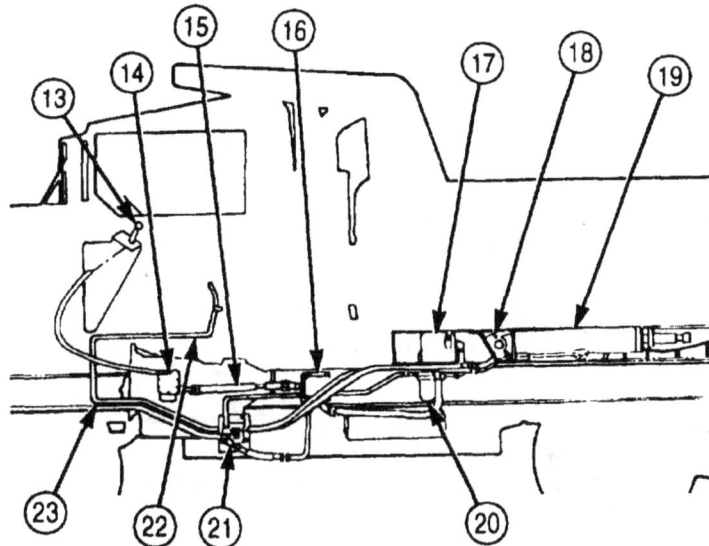

Key Item and Function

13 **TRANSMISSION PTO CONTROL** - A manually-operated control lever located inside the vehicle cab that permits engagement or disengagement of the transmission PTO.

14 **TRANSMISSION PTO** - Uses driving power of the transmission to provide mechanical driving power for the hydraulic pump.

15 **PTO DRIVESHAFT** - Transmits mechanical driving power from the PTO to the hydraulic pump.

16 **HYDRAULIC PUMP** - Driven by the PTO driveshaft, it draws oil from oil reservoir through hydraulic hoses, then pressurizes and directs it to the control valve.

17 **HYDRAULIC OIL RESERVOIR** - Storage tank for hydraulic oil.

18 **DUMP BODY SAFETY LATCH** - Hydraulically-operated in conjunction with the dump body control lever, the safety latch locks dump body in the lowered position and releases it when the control lever is pulled back to the raised position.

19 **DUMP BODY CYLINDER ASSEMBLY** - Consists of two piston-type hydraulic cylinder hoists. Assembly raises and lowers dump body with hydraulic oil, forcing the cylinder upward or downward.

20 **OIL FILTER** - Filters used or bypassed oil from the control valve before it returns to the hydraulic oil reservoir.

21 **CONTROL VALVE** - Four-port valve accepts pressurized oil from the hydraulic pump and directs oil pressure flow from control valve to the hydraulic cylinders. It also directs oil returning from the hydraulic cylinders back to the hydraulic oil reservoir.

22 **DUMP BODY CONTROL LEVER** - An operator control that determines the hydraulic oil pressure flow from control valve to the hydraulic cylinders. The route this oil takes will determine whether the dump will raise or lower.

23 **CONTROL LINKAGE** - Connects dump body control lever inside cab to the control valve.

1-24. HYDRAULIC SYSTEM OPERATION (Contd)

d. Medium Wrecker Crane Hydraulic System Operation.

The M936/A1/A2 medium wrecker is equipped with a hydraulically-operated crane that extends a maximum 18 ft (5 m), elevates 45 degrees, and swings 360 degrees. It is capable of lifting loads up to 20,000 lbs (9,090 kg).

(1) The crane hydraulic system converts power of the engine into fluid power for use by the hydraulic pump. At this pump, oil pressure is supplied to different crane control valves: BOOM, HOIST, CROWD, and SWING. Each of these crane actions are dealt with separately. The major components for raising and lowering the wrecker boom are:

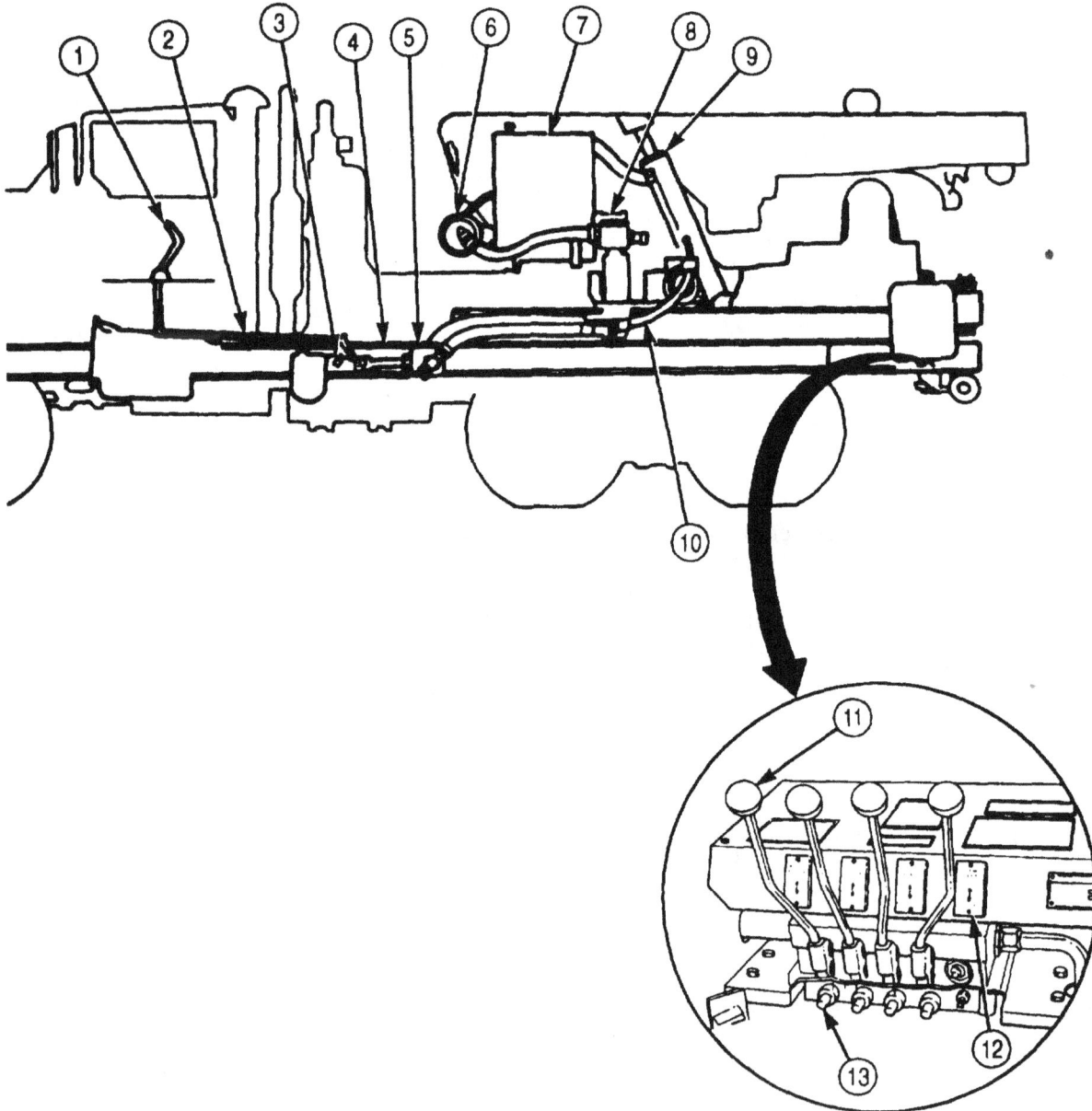

1-24. HYDRAULIC SYSTEM OPERATION (Contd)

Key Item and Function

1   **TRANSFER CASE PTO CONTROL** - A manually-operated control lever located inside the cab that engages and disengages the transfer case PTO.

2   **TRANSFER CASE PTO LINKAGE** - Connects transfer case PTO control to transfer case PTO.

3   **TRANSFER CASE PTO** - Receives driving power from vehicle's engine through the transfer case to provide mechanical driving power for the hydraulic pump.

4   **PTO DRIVESHAFT** - Transmits mechanical driving power from the power takeoff to the hydraulic pump.

5   **HYDRAULIC PUMP** - Draws oil from hydraulic oil reservoir and directs it to valves inside the crane control console.

6   **OIL FILTER** - Filters used or bypassed oil from the control valve before it returns to the hydraulic oil reservoir.

7   **HYDRAULIC OIL RESERVOIR** - Storage tank for hydraulic oil.

8   **SWIVEL VALVE** - Permits oil to channel through pivot post while crane is swinging and eliminates twisting of the hydraulic lines connecting reservoir to the stationary pump.

9   **BOOM LIFT CYLINDER** - A hydraulically-driven piston that extends upward when boom control lever is pulled back to UP position, raising the boom. A check valve located near hydraulic oil inlet hose prevents piston from lowering when control lever is in NEUTRAL. Oil returns through boom control valve back to hydraulic oil reservoir allowing piston to lower when control lever is pushed forward to DOWN position.

10   **BOOM HYDRAULIC LINES** - Carry the hydraulic oil to and from boom lift cylinder. Oil pumped through the bottom lines pushes the lift cylinder piston upward. Oil pumped through the top lines pushes the lift cylinder piston downward. When this downward action occurs, the oil that originally pushes the cylinder upwards is returned to the hydraulic oil reservoir.

11   **BOOM CONTROL LEVER** - Manual control attached to the control valve that determines hydraulic oil flow for raising and lowering action of the boom. Lever is pulled back to raise the boom and pushed forward to lower boom.

12   **CRANE CONTROL CONSOLE** - Houses BOOM, HOIST, CROWD, and SWING levers and their control valves.

13   **BOOM CONTROL VALVE** - Located directly below boom control lever. Valve directs hydraulic oil from the hydraulic pump to the boom lift cylinder for lifting, or out of the lift cylinder and back to the hydraulic oil reservoir for lowering.

1-24. HYDRAULIC SYSTEM OPERATION (Contd)

(2) The major components for raising and lowering the crane cable and hook for the HOIST action are:

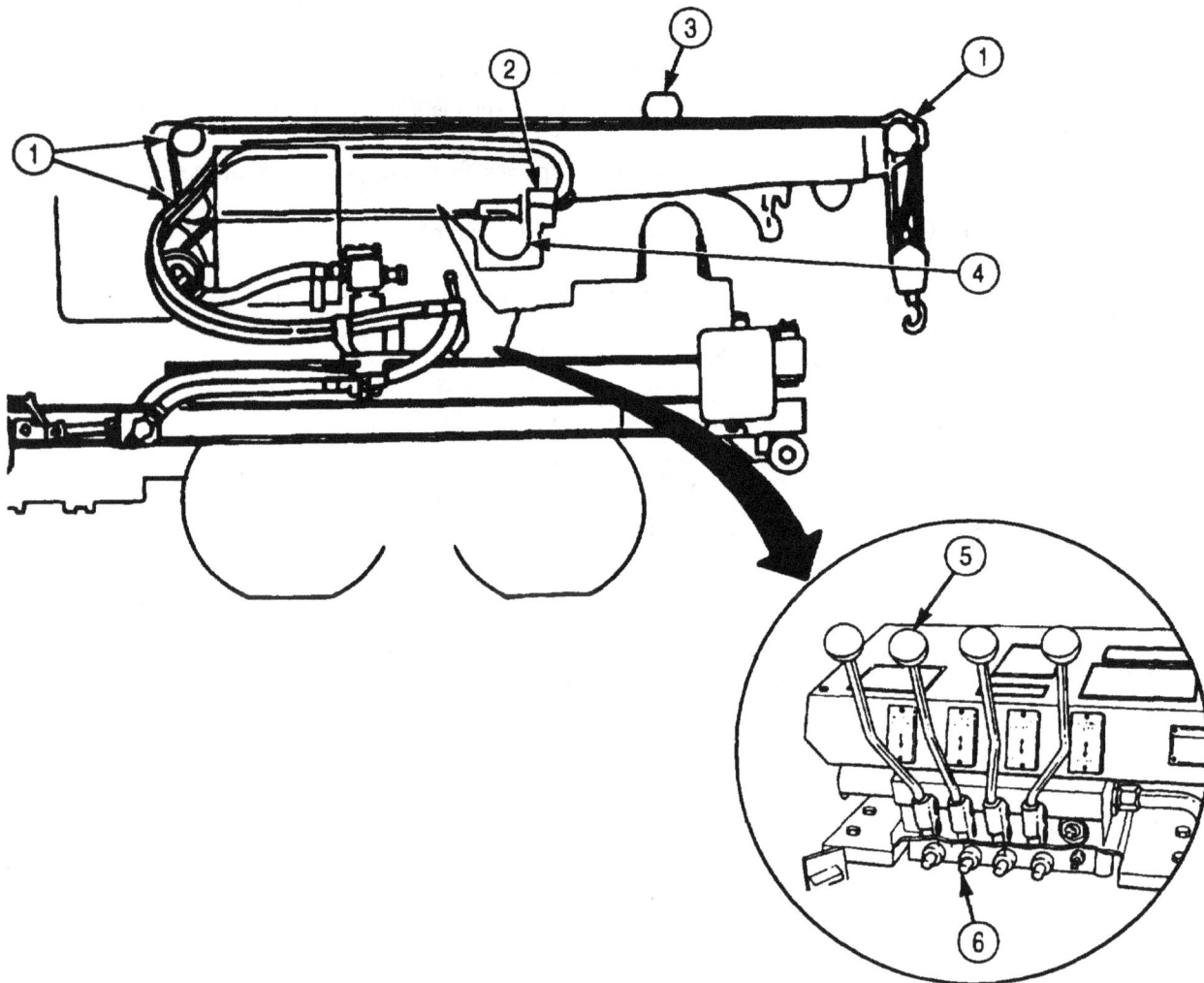

Key Item and Function

  1 **SHEAVES** - Grooved wheels that guide hoist cable through boom.

  2 **HOIST MOTOR ASSEMBLY** - Converts hydraulic power back into mechanical power needed to turn the hoist drum.

  3 **UPPER ROLLER ASSEMBLY** - Prevents cable from contacting inner boom during winding/unwinding.

  4 **CRANE HOIST CABLE DRUM** - Is turned by the worm gear in hoist motor assembly. Drum unwinds cable when turning toward front of vehicle. Drum winds cable when turning toward rear of vehicle.

  5 **HOIST CONTROL LEVER** - Manual control attached to the control valve that determines hydraulic oil flow for the raising and lowering action of the crane hoist cable and hook. Lever is pulled back to raise cable and hook and pushed forward to lower cable and hook.

  6 **HOIST CONTROL VALVE** - Two-way hydraulic valve located under the hoist control lever directs fluid from the hydraulic pump to the hoist motor assembly and back through the valve to the hydraulic oil reservoir.

1-24. HYDRAULIC SYSTEM OPERATION (Contd)

(3) Major components for extending and retracting the boom for the CROWD action are:

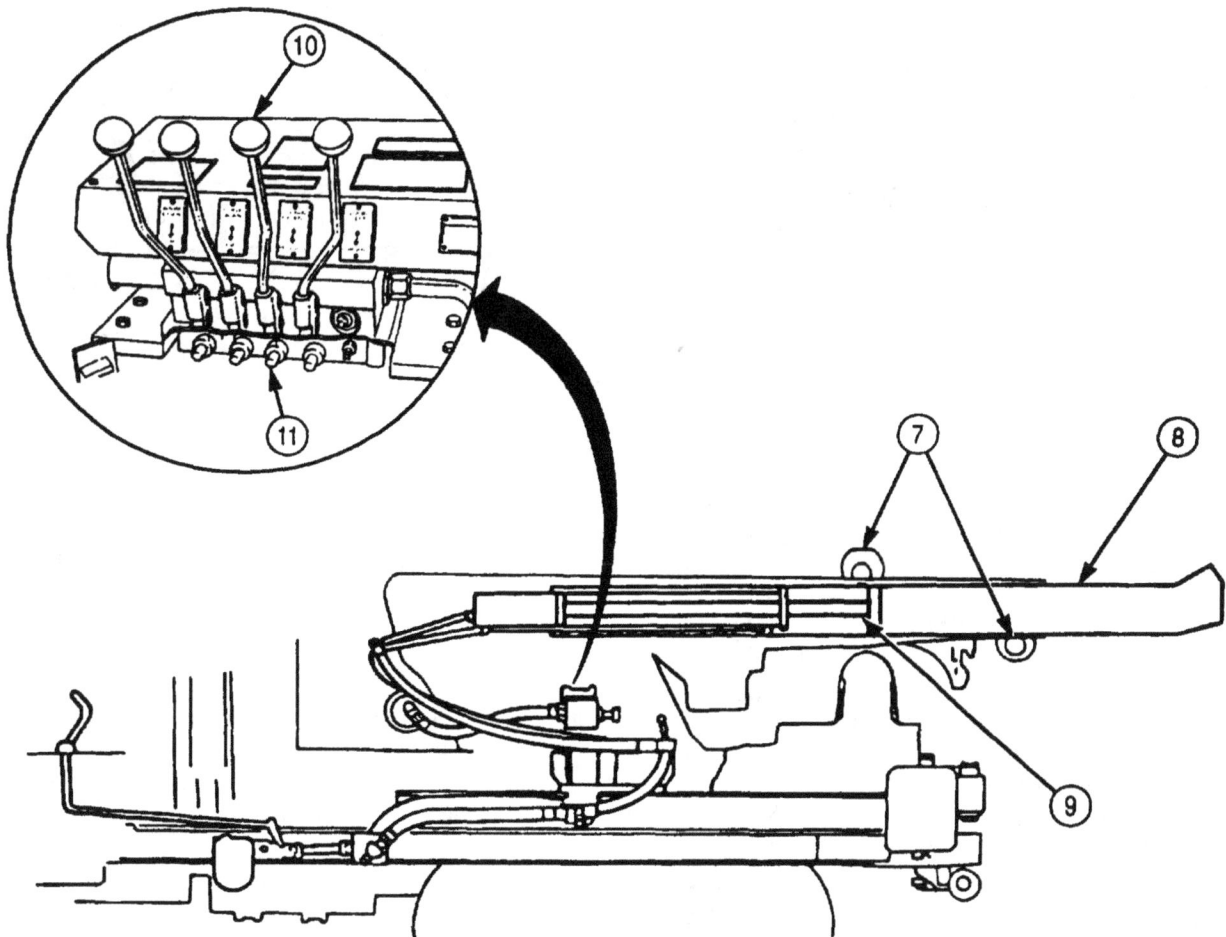

Key Item and Function

7 **ROLLERS** - Guide inner boom assembly and permit smooth extension and retraction of boom.

8 **INNER BOOM ASSEMBLY** - Extends when crowd control lever is pushed forward and retracts when control lever is pulled back.

9 **CROWD CYLINDER** -A hydraulically-driven piston that extends outward when crowd control lever is pushed forward to EXTEND position. Piston is hydraulically-driven back into the cylinder when crowd control lever is pulled back to RETRACT position. This cylinder is contained in the inner boom assembly.

10 **CROWD CONTROL LEVER** - Manual control attached to the control valve that determines oil flow for extending and retracting the crane boom. Lever is pushed forward to extend the boom and pulled back to retract the boom.

11 **CROWD CONTROL VALVE** - Two-way hydraulic valve located directly below crowd control lever. Valve directs hydraulic oil from the hydraulic pump to the crowd cylinder to extend and retract inner boom assembly.

1-24. HYDRAULIC SYSTEM OPERATION (Contd)

(4) The major components for swinging the crane left and right for the SWING action are:

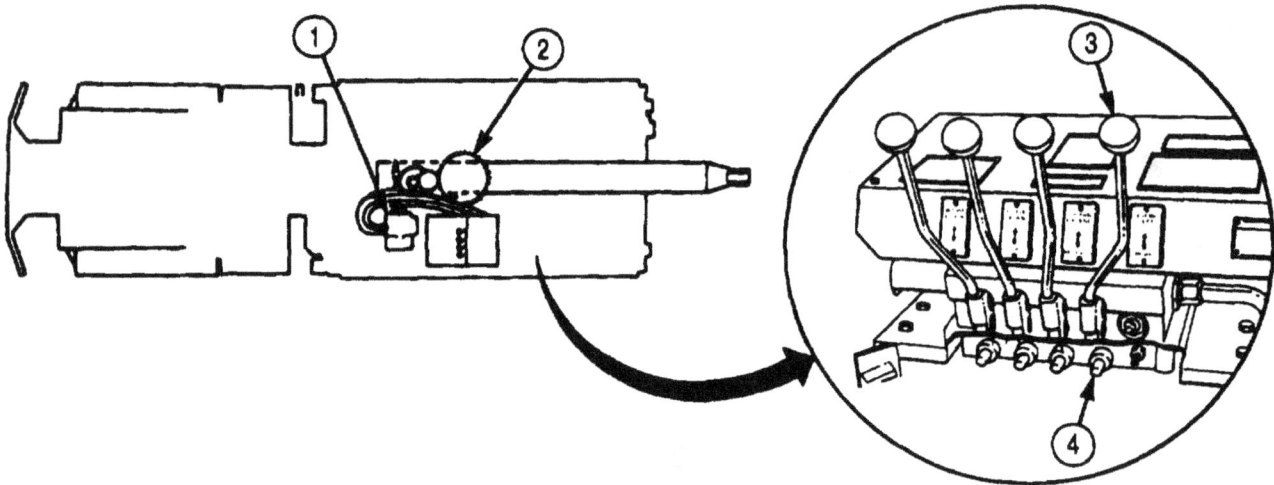

Key Item and Function

1 **SWING MOTOR** - Converts hydraulic power back into mechanical power needed to turn the crane turntable when hydraulic fluid is forced through its worm gear. This gear turns a large gear at the base of the turntable to swing the crane.

2 **TURNTABLE ASSEMBLY** - Driven by the swing motor through a ring gear at the base of the assembly, permits the crane to swing 360 degrees.

3 **SWING CONTROL LEVER** - Manual control attached to the control valve that determines hydraulic oil flow for swinging wrecker boom to the left and to the right. Lever is pushed inward for left boom movement, and pulled outward for right boom movement.

4 **SWING CONTROL VALVE** - Two-way hydraulic valve located directly below swing control lever. Valve directs hydraulic oil from the hydraulic pump to the swing motor assembly and back through the valve to the hydraulic oil reservoir.

1-25. CENTRAL TIRE INFLATION SYSTEM (CTIS) OPERATION

The M939A2 Central Tire Inflation System (CTIS) uses the vehicle's air compressor system, pneumatic valves, and a microprocessor to adjust tire pressure according to mission demands. The system needs clean, dry air to maintain itself. The turbocharger and aftercooler help supply more compressed air when the vehicle is loaded and operating at high rpm.

Compressed air enters air dryer, which filters dirt and ejects moisture from air. System air then enters the wet tank and is available to airbrakes and CTIS. The brake system is protected by an 85 psi (138 kPa) pressure switch in CTIS air line. If wet tank air pressure measures below 85 psi (138 kPa), the CTIS shuts down until pressure switch measures 120 psi (827 kPa).

Air is then routed to pneumatic controller, which consists of three valves and solenoids working in conjunction with the Electronic Control Unit (ECU). The ECU is a microprocessor programmed to issue commands based on information it receives from pressure switch, pressure transducer (which tests tire pressure), and speed signal generator (which monitors vehicle speed). The commands are electrically received by the pneumatic controller, which affects inflation or deflation of tires through quick-exhaust valves.

Six wheel valves isolate tire pressure during normal operation. They also open to allow tire inflation or tire deflation through three quick-exhaust valves.

The system is automatic. The operator only needs to select the mode that best accomplishes the mission.

CHAPTER 2
SERVICE AND TROUBLESHOOTING INSTRUCTIONS

Section I. REPAIR PARTS; TOOLS; SPECIAL TOOLS; TEST, MEASUREMENT, AND DIAGNOSTIC EQUIPMENT (TMDE); AND SUPPORT EQUIPMENT

2-1. SCOPE

This chapter provides mechanics with Preventive Maintenance Checks and Services (PMCS). Refer to TM 9-2320-272-10 for additional PMCS procedures.

2-2. COMMON TOOLS AND EQUIPMENT

For authorized common tools and equipment, refer to the Modified Table of Organization and Equipment (MTOE) applicable to your unit.

2-3. SPECIAL TOOLS, TMDE, AND SUPPORT EQUIPMENT

Special tools; Test, Measurement, and Diagnostic Equipment (TMDE); and support equipment used to maintain the components covered in this manual can be found in Appendix B, Maintenance Allocation Chart (MAC) and TM 9-2320-272-24P.

2-4. REPAIR PARTS

Repair parts are listed and illustrated in TM 9-2320-272-24P.

Section II. SERVICE UPON RECEIPT

2-5. GENERAL

a. Upon receipt of a new, used, or reconditioned vehicle, you must determine if the vehicle has been properly prepared for service. The following steps should be performed:

(1) Inspect all assemblies, subassemblies, and accessories to ensure they are in proper working order.

(2) Secure, clean, lubricate, or adjust as needed.

(3) Check all Basic Issue Items (BII) (TM 9-2320-272-10) to ensure every item is present, in good condition, and is properly mounted or stowed.

(4) Follow general procedures for all services and inspections given in TM 9-2320-272-10.

b. The operator will assist when performing service upon receipt inspections.

c. Refer to TM 9-2320-272-10 when testing equipment for proper operation.

2-6. GENERAL INSPECTION AND SERVICING INSTRUCTIONS

The following steps should be taken while performing general inspections and services:

a. Refer to TM 9-2320-272-10 and LO 9-2320-272-12, as well as other sections of this manual, when servicing, inspecting, and lubricating equipment.

WARNING

Drycleaning solvent is flammable and will not be used near open flame. Keep fire extinguisher nearby. Use only in well-ventilated places. Failure to do so may result in injury to personnel.

b. Using drycleaning solvent, clean all exterior surfaces coated with rustpreventive compounds.

c. Read Processing and Deprocessing Record of Shipment, Storage, and Issue of Vehicles and Spare Engines tag (DD Form 1397), and follow all precautions listed. This tag should be attached to steering wheel, shift lever, or battery switch.

NOTE

If vehicle has been driven to using unit, all the above work should have been completed.

2-7. SPECIFIC INSPECTION AND SERVICING INSTRUCTIONS

The following steps should be taken while performing specific inspections and services:

a. Perform the semiannual, six months or 6,000 miles (9,654 kilometers), unit Preventive Maintenance Checks and Services (PMCS) listed in section III of this chapter.

b. Lubricate the vehicle according to the instructions found in LO 9-2320-272-12. Do not lubricate gearcases or engine unless processing tag states that the oil is unsuitable for 500 miles (805 kilometers) of operation. If oil is suitable, just check level.

c. Schedule semiannual service on DD Form 314 (Preventive Maintenance Schedule and Record).

d. If vehicle is delivered with a dry-charged battery, activate it according to TM 9-6140-200-14.

e. Check vehicle coolant level and determine if solution is proper for climate. (Refer to TB 750-651 for preparation of antifreeze solutions.)

Section III. UNIT PREVENTIVE MAINTENANCE CHECKS AND SERVICES (PMCS)

2-8. GENERAL

The best way to maintain vehicles covered by this manual is to inspect them on a regular basis so minor faults can be discovered and corrected before they result in serious damage, failure of vehicle and equipment, or injury to personnel. This section contains systematic instructions for inspection, adjustment, and correction of vehicle components to avoid costly repairs or major breakdowns. This is referred to as Preventive Maintenance Checks and Services (PMCS).

2-9. INTERVALS

NOTE

Designated intervals are performed under usual operating conditions. PMCS must be performed more frequently when operating under unusual conditions.

2-9. INTERVALS (Contd)

a. Unit maintenance, assisted by operator/crew, will perform the checks and services contained in table 2-1 at the following intervals:

(1) **Semiannual.** Every 6 months or 6,000 miles (9,654 kilometers), whichever occurs first.

(2) **Annual.** Every 12 months or 12,000 miles (19,308 kilometers), whichever occurs first.

(3) **Biennial.** Every 24 months or 24,000 miles (38,616 kilometers), whichever occurs first.

b. Perform all semiannual inspections in addition to annual inspections at the time of the annual inspection. Perform all annual and semiannual inspections in addition to biennial inspections at the time of the biennial inspection.

2-10. REPORTING REPAIRS

All uncorrected defects will be recorded on Equipment Inspection and Maintenance Worksheet, DA Form 2404, in accordance with DA Pam 738-750.

2-11. GENERAL SERVICE AND INSPECTION PROCEDURES

a. While performing specific PMCS procedures, ensure items are correctly assembled, secure, serviceable, not worn, not leaking, and adequately lubricated as defined below:

(1) An item is CORRECTLY ASSEMBLED when all parts are present and in proper position.

(2) When wires, nuts, washers, hoses, or attaching hardware cannot be moved by hand, wrench, or prybar, they are secure.

(3) An item is UNSERVICEABLE if it is worn beyond established wear limits or is likely to fail before the next scheduled inspection.

(4) An item is WORN if there is play between joining parts, or warning and caution plates are not readable.

(5) LEAKS. TM 9-2320-272-10 contains definitions of class I, II, and III leaks and their effect on vehicle operation.

(6) If an item meets the requirements specified by lubrication instructions, LO 9-2320-272-12, then it is ADEQUATELY LUBRICATED.

b. Where the instruction "tighten" appears in a procedure, you must tighten with a wrench to the given torque value even when the item appears to be secure.

WARNING

Drycleaning solvent is flammable and will not be used near open flame. Keep fire extinguisher nearby. Use only in well-ventilated places. Failure to do so may result in injury to personnel.

c. Where the instruction "clean" appears in a procedure, you must use drycleaning solvent, Type II (SD-2) (PD-680) biodegradable, to clean grease or oil from metal parts. After the item is cleaned, rinsed, and dried, apply a light grade of oil to unprotected surfaces to prevent rusting. To clean rubber and plastic materials, use soap and water.

2-12. SPECIFIC PMCS PROCEDURES

a. The preventive maintenance checks and services for which you are responsible are provided in table 2-1. The checks and services listed are arranged in logical order.

b. The following columns are left to right on the PMCS schedule:

(1) **Item Number.** Provides logical order for PMCS performance and is used as a source number for DA Form 2404, on which your PMCS results will be recorded.

2-12. SPECIFIC PMCS PROCEDURES (Contd)

(2) **Interval.** Indicates when check or service is to be performed (para. 2-9).

(3) **Item To Check/Service.**

(a) Lists the system, common name, or location of the item to be inspected.

(b) The letters RPL in this column indicate replacement parts are required to complete the task or procedure. Check mandatory replacement parts list immediately following table 2-1.

(4) **Procedure.** Provides instructions for servicing, inspection, replacement, or adjustment and, in some cases, having an item repaired at a higher level. If a defect is found, repair, fill, replace, or adjust as needed.

(5) **Not Fully Mission Capable If:.** Provides information for deadlining a vehicle when checks or services reveal a defect or deficiency of a component(s) of the vehicle.

Table 2-1. Unit Preventive Maintenance Checks and Services.

ITEM NO.	INTERVAL	ITEM TO CHECK/ SERVICE	PROCEDURE	NOT FULLY MISSION CAPABLE IF:
			SEMIANNUAL INSPECTION **PRIOR TO ROAD TEST** Perform all Before operation and Weekly checks listed in TM 9-2320-272-10 PMCS in addition to those that follow.	
1	Semi-annual	Starter	Start engine (TM 9-2320-272-10). While starting engine, listen for unusual noises and difficult cranking.	Starter is inoperative or makes excessive grinding noise.
2	Semi-annual	Accelerator pedal and engine	**a.** Observe response to accelerator pedal. **1.** Listen for unusual noises. **2.** Observe for hesitation or varying idle speed. **3.** Check for sticking or binding of pedal. If accelerator pedal binds or sticks, troubleshoot (para. 2-21). b. Be alert for excessive vibration, fuel odor, oil, coolant or exhaust dripping, and any indication of malfunction.	**a.** Accelerator pedal is sticking or binding. **b.** Engine knocks, rattles, or smokes excessively.
3	Semi-annual	Throttle	Check throttle control travel and freedom of movement. Pull throttle control all the way out, rotate, and release. Throttle control should return to original position. If control sticks or binds, lubricate (LO 9-2320-272-12) or replace (para. 3-45).	Throttle control is sticking or binding.

Table 2-1. Unit Preventive Maintenance Checks and Services (Contd).

ITEM NO.	INTERVAL	ITEM TO CHECK/ SERVICE	PROCEDURE	NOT FULLY MISSION CAPABLE IF:
3	Semi-annual	Throttle (Contd)	**ROAD TEST** Perform all During operation checks listed in TM 9-2320-272-10 PMCS in addition to those that follow. Drive vehicle at least 5 mi. (8 km) over varied terrain, both on and off road. This will provide ample time to check reported malfunctions and to locate unreported malfunctions. Lubrication intervals of every 1,000 mi. (1,600 km) or monthly, and 3,000 mi. (4,800 km) or 3 months, will be performed with maintenance or, when practical, lubrication services will be made to coincide with the semiannual preventive maintenance services. For this purpose, a 10 percent tolerance (variation) in specified lubrication point mileage is permissible.	
4	Semi-annual	Brakes	**a.** Check brake pedal for free travel. Adjust brake pedal if required (para. 3-196). **b.** Reach a desired speed and lightly apply brake pedal with steady force. Vehicle should slow down immediately and stop smoothly, without side pull or chatter. **c.** After stopping vehicle, and with transmission select lever in 1-5 (drive), release brake pedal. All wheel brakes should release immediately and without difficulty.	
5	Semi-annual	Engine	**NOTE** Ensure engine tachometer does not exceed maximum governed speed of 2,400 rpm with no load. **a.** Check engine throughout the range of operating speeds. Refer to TM 9-2320-272-10. **b.** Check engine instruments. Refer to TM 9-2320-272-10 for proper readings.	
6	Semi-annual	Transmission	**a.** Check transmission oil temperature gauge. Normal range is 120°-220°F (49°-104°C). **b.** Check for response to shifting and smoothness of operation in all speed ranges. Refer to TM 9-2320-272-10 for proper shifting speed ranges.	**a.** Oil temperature exceeds 300°F (149°C).
7	Semi-annual	Transfer case	Shift transfer case lever between HI and LOW positions to ensure proper operation. Observe for smoothness of engagements. Refer to TM 9-2320-272-10 for proper shifting speed ranges and operation.	
8	Semi-annual	Suspension	Observe how vehicle responds to road shock. Constant bouncing or swaying from side-to-side is an indication of a malfunction.	

Table 2-1. Unit Preventive Maintenance Checks and Services (Contd).

ITEM NO.	INTERVAL	ITEM TO CHECK/ SERVICE	PROCEDURE	NOT FULLY MISSION CAPABLE IF:
9	Semi-annual	Emergency engine stop control	**NOTE** • Before M939/A1 series vehicles can be restarted, fuel shutoff valve control lever under the hood must be reset. • This check will be performed with engine running, transmission in N (neutral), and parking brake applied. Check emergency engine stop control: 1. Pull emergency engine stop control out until engine stops. If engine does not stop before emergency engine stop control is pulled completely out, place battery switch in the OFF position, and troubleshoot (para. 2-21). 2. Push emergency engine stop control in until original position is obtained. **AFTER ROAD TEST** Perform all After operation, Weekly, and Monthly checks in TM 9-2320-272-10, then make the following inspections in the order given, including kit items on vehicles so equipped.	Engine fails to shut off.
10	Semi-annual	Air filter	**WARNING** If NBC exposure is suspected, all filter media should be handled by personnel wearing protective equipment. Consult your unit NBC officer or NBC NCO for appropriate handling or disposal instructions. **a.** Inspect filter element for tears and presence of dirt and oil. 1. If dirt is present, clean filter element (TM 9-2320-272-10). 2. If oil is present, replace filter element. b. Check for oil in air cleaner manifold. If present: 1. Check oil level in transmission and transfer case. If oil levels are excessive, drain excess (para. 3-133). 2. Check for fuel diluting the oil. If necessary, change oil and filters (para. 3-133). 3. If oil levels are low, check transfer case interlock air cylinder for leaks (para. 3-143).	

Table 2-1. Unit Preventive Maintenance Checks and Services (Contd).

ITEM NO.	INTERVAL	ITEM TO CHECK/ SERVICE	PROCEDURE	NOT FULLY MISSION CAPABLE IF:
11	Semi-annual	Batteries [RPL]	**WARNING** • Wear safety glasses or goggles when checking batteries. Always check electrolyte level with engine stopped. Do not smoke or use exposed flame when checking batteries; explosive gases are present, and severe injury to personnel can result. • Remove all jewelry such as rings, dog tags, bracelets, etc. If jewelry contacts battery terminal, a direct short may result in instant heating of tools, damage to equipment, and injury or death to personnel. • Remove or disconnect batteries, and turn master battery disconnect switch off prior to performing maintenance in immediate area or working on electrical system. Such disconnections prevent electrical shock to personnel and equipment. • Ensure seatbelts do not come in contact with electrolyte. Damage to strapping will result, leading to injury or death. • Ensure seatbelts are not caught inside battery box when closing cover. Failure to do so will result in injury or death. **a.** Clean and inspect batteries (para. 3-125). Replace if required. **b.** Inspect battery box for security of mounting and completeness of assembly (para. 3-128). **c.** Inspect battery cables and terminals for frays, splits, and security Repair battery cables or terminals (para. 3-131), or replace as necessary (para. 3-124). **e.** Inspect slave receptacle and wiring for security of mounting and damage. Repair (para. 3-131), or replace if damaged (para. 3-127). **f.** Lightly coat battery terminals and slave receptacle contacts with grease.	

LUBRICANTS			EXPECTED TEMPERATURES		
			ABOVE 15°F (ABOVE -9°C)	+40°F TO -15°F (+4°C TO -26°C)	+40°F TO -65°F (+4°C TO -54°C)
GAA Grease, automotive and artillery (MIL-G-10924)			All temperatures		

g. Inspect seatbelts for serviceability and security of mounting. Replace seatbelts if unserviceable or show signs of contact with electrolyte (para. 3-289).

Table 2-1. Unit Preventive Maintenance Checks and Services (Contd).

ITEM NO.	INTERVAL	ITEM TO CHECK/ SERVICE	PROCEDURE	NOT FULLY MISSION CAPABLE IF:
11	Semi-annual	Batteries (Contd)	h. Lubricate hinges on battery box cover and map compartment every 1,000 mi. (1,600 km) or monthly, whichever occurs first.	

LUBRICANTS		EXPECTED TEMPERATURES		
		ABOVE 15°F (ABOVE -9°C)	+40°F TO - 15°F (+4°C TO -26°C)	+40°F TO -65°F (+4°C TO -54°C)
OE/HDO Lubricating oil, internal combustion engine (MIL-L-2104)		OE/HDO 30	OE/HDO 10	
OEA Lubricating oil, internal combustion engine (arctic) (MIL-L-46167)				OEA

ITEM NO.	INTERVAL	ITEM TO CHECK/ SERVICE	PROCEDURE	NOT FULLY MISSION CAPABLE IF:
12	Semi-annual	Cab components	i. Inspect battery box cover and map compartment cover and seals on both for condition and security of installation. Replace seal(s) on battery box cover and/or map compartment cover if damaged (para. 3-123).	

WARNING

If transmission oil temperature is above 220°F (104°C), allow transmission oil to cool before removing dipstick.

CAUTION

- Do not remove transmission dipstick before cleaning dirt away from access plate, filler tube, and dipstick. Dirt may enter and damage transmission.

- Do not overfill transmission. Internal transmission component damage will result.

- Change transmission oil when contamination by fuel, water, and other foreign materials is evident. Failure to do so may result in failure of internal transmission components.

- Shut off engine if transmission oil temperature exceeds 300°F (149°C). Continuing to operate transmission under these conditions may result in failure of internal transmission components.

NOTE

Steps a. through c. apply to M939/A1 series vehicles only. Transmission dipstick and temperature sending unit cannot be accessed through cab floor on M939A2 series vehicles.

Table 2-1. Unit Preventive Maintenance Checks and Services (Contd).

ITEM NO.	INTERVAL	ITEM TO CHECK/ SERVICE	PROCEDURE	NOT FULLY MISSION CAPABLE IF:
12	Semi-annual	Cab components (Contd)	**a.** Open access door in floor of cab. Lubricate access door hinge every 1,000 mi. (1,600 km) or monthly, whichever occurs first.	

LUBRICANTS		EXPECTED TEMPERATURES		
		ABOVE 15°F (ABOVE -9°C)	+40°F TO -15°F (+4°C TO -26°C)	+40°F TO -65°F (+4°C TO -54°C)
OE/HDO Lubricating oil, internal combustion engine (MIL-L-2104)		OE/HDO 30	OE/HDO 10	
OEA Lubricating oil, internal combustion engine (arctic) (MIL-L-46167)				OEA

b. Inspect transmission dipstick and oil:

1. Check for evidence of metal particles. Notify your supervisor if metal particles are found.

2. Check for evidence of dilution from coolant. If oil is diluted by coolant, refer to para. 2-21, mechanical troubleshooting.

3. Check oil level in transmission. If oil levels are excessive, drain the excess (para. 3-133).

c. Inspect transmission temperature sending unit for security of mounting and wiring for frays, splits, breaks, and missing insulation.

d. On vehicles with winch, lubricate winch and Power Takeoff (PTO) control levers every 1,000 mi. (1,600 km) or monthly, whichever occurs first.

e. Lubricate power brake treadle with 3-4 drops of oil every 1,000 mi. (1,600 km) or monthly, whichever occurs first.

LUBRICANTS		EXPECTED TEMPERATURES		
		ABOVE 15°F (ABOVE -9°C)	+40°F TO -15°F (+4°C TO -26°C)	+40°F TO -65°F (+4°C TO -54°C)
OE/HDO Lubricating oil, internal combustion engine (MIL-L-2104)		OE/HDO 30	OE/HDO 10	
OEA Lubricating oil, internal combustion engine (arctic) (MIL-L-46167)				OEA

Table 2-1. Unit Preventive Maintenance Checks and Services (Contd).

ITEM NO.	INTERVAL	ITEM TO CHECK/ SERVICE	PROCEDURE	NOT FULLY MISSION CAPABLE IF:
12	Semi-annual	Cab components (Contd)	**f.** Lubricate transfer case shift linkage pins every 1,000 mi. (1,600 km) or monthly, whichever occurs first. **g.** On M929/A1/A2 and M930/A1/A2 vehicles, lubricate dump lever and crossshaft every 1,000 mi. (1,600 km) or monthly, whichever occurs first (LO 9-2320-272-12).	

LUBRICANTS	EXPECTED TEMPERATURES		
	ABOVE 15°F (ABOVE -9°C)	+40°F TO -15°F (+4°C TO -26°C)	+40°F TO -65°F (+4°C TO -54°C)
GAA Grease, automotive and artillery (MIL-G-10924)	All temperatures		
OE/HDO Lubricating oil, internal combustion engine (MIL-L-2104)	OE/HDO 30	OE/HDO 10	

ITEM NO.	INTERVAL	ITEM TO CHECK/ SERVICE	PROCEDURE	NOT FULLY MISSION CAPABLE IF:
13	Semi-annual	Front winch	**h.** Inspect instrument panel cluster and cab portion of front wiring harness for frays, splits, missing or damaged insulation, or poor connections. Repair or replace affected wiring (para. 3-131). **a.** Inspect winch for security of mounting, loose or missing mounting bolts, and broken or missing parts (para. 3-329). **WARNING** • Wire rope can become frayed or contain broken wires. Wear heavy leather-palmed work gloves when handling wire rope. Frayed or broken wires can injure hands. • Never let moving wire rope slide through hands, even when wearing gloves. A broken wire could cut through glove and cut hand. **b.** Unwind entire cable, soak and clean with new oil, and inspect for kinks, frays, and wear. Refer to TM 9-2320-272-10 for operation.	

LUBRICANTS	EXPECTED TEMPERATURES			
	ABOVE +80°F(+27°C)	+80°F TO +30°F (27°C TO -1°C)	+30°F TO -30°F (-1°C TO -34°C)	-30' TO -65° (-34° TO -54°C)
CW Lubricating oil (VV-L-751)	CW-11C	CW-11B	CW-11A	
Oil lubricating, multipurpose (MIL-L-2105) GO-75				GO-75

			PROCEDURE	
			c. On M936/A1 vehicles, lubricate sheave bearing (1), swivel fitting (3), trolley wheel (4), and level wind frame (2) every 3,000 mi. (4,800 km) or 6 months, whichever comes first. If operation is frequent, continuous, or under severe conditions, service weekly (LO 9-2320-272-12).	

Table 2-1. Unit Preventive Maintenance Checks and Services (Contd).

ITEM NO.	INTERVAL	ITEM TO CHECK/ SERVICE	PROCEDURE	NOT FULLY MISSION CAPABLE IF:
13	Semi-annual	Front winch (Contd)	**d.** Lubricate tensioner sheave pins (5), vertical cable rollers (6), and horizontal cable rollers (7) every 3,000 mi. (4,800 km) or 6 months, whichever occurs first. If operation is frequent, continuous, or under severe conditions, service weekly (LO 9-2320-272-12).	

LUBRICANTS	EXPECTED TEMPERATURES		
	ABOVE 15°F (ABOVE -9°C)	+40°F TO -15°F (+4°C TO -26°C)	+40°F TO -65°F (+4°C TO -54°C)
GAA Grease, automotive and	All temperatures		

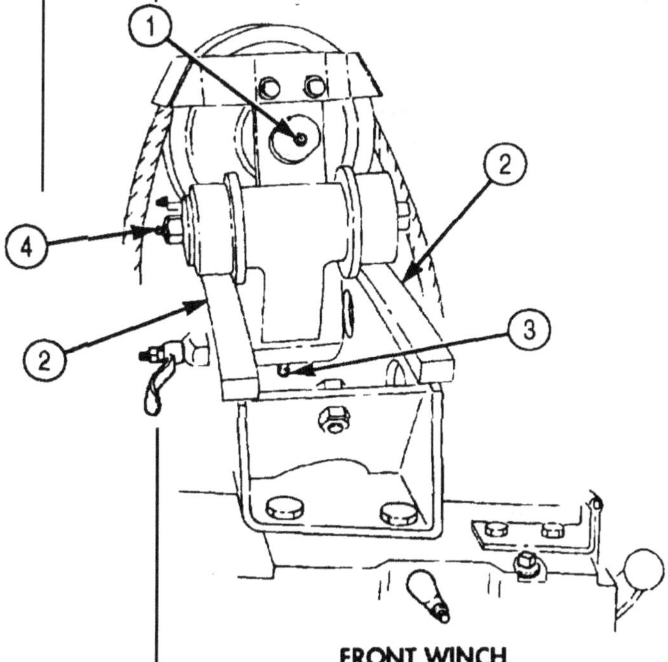

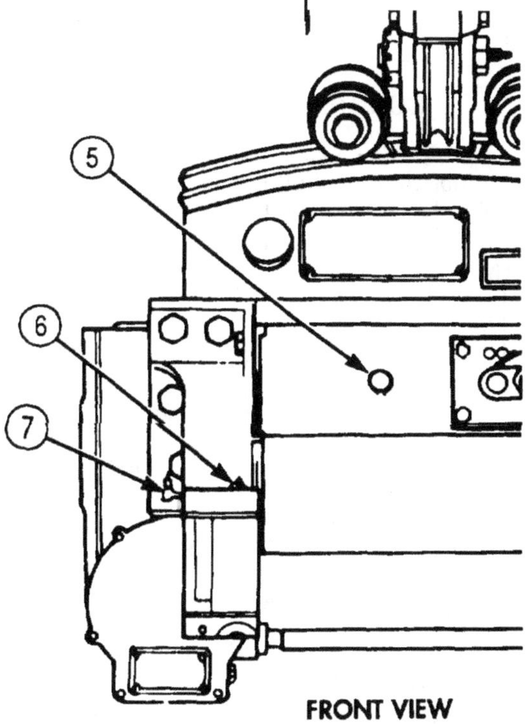

FRONT WINCH (LEFT SIDE)

FRONT VIEW

2-11

Table 2-1. Unit Preventive Maintenance Checks and Services (Contd).

ITEM NO.	INTERVAL	ITEM TO CHECK/ SERVICE	PROCEDURE	NOT FULLY MISSION CAPABLE IF:
13	Semi-annual	Front winch (Contd)	**e.** Check gearcase level. **1.** Clean around and remove oil level sight plug (6), and check oil level. **NOTE** Perform steps 2 through 4 only if oil level is low. **2.** Clean around and remove gearcase fill plug (2). If necessary to drain, remove drain plug (7) and reinstall. **3.** Fill to bottom opening of oil level sight plug (6). **4.** Install filler plug (2) and oil level sight plug (6). f. Check end frame oil level. **1.** Clean around and remove end frame level plug (8) and fill plug (1) and check end frame gearcase oil level. **NOTE** Perform steps 2 and 3 only if oil level is low. **2.** Fill to opening of level plug (8) and install level plug (8). **3.** Install filler plug (1). **g.** Check sight plug (6) and filler plug (2) on gearcase and filler plug (1), level plug (8), and drain plug (9) on end frame for tightness.	

LUBRICANTS	EXPECTED TEMPERATURES		
	ABOVE 15°F (ABOVE -9°C)	+40°F TO -15°F (+4°C TO -26°C)	+40°F TO -65°F (+4°C TO -54°C)
Oil, lubricating, multipurpose (MIL-L-2105) GO-80/90	GO-80/90	GO-80/90	
Oil, lubricating, multipurpose (MIL-L-2105) GO-75			GO-75

h. On M936/A1 model vehicles, lubricate trolley lock (3) every 1,000 mi. (1,600 km) or monthly, whichever occurs first.

i. Lubricate drum lock (4), and tensioner control lever lock (5) every 1,000 mi. (1,600 km) or monthly, whichever occurs first.

LUBRICANTS	EXPECTED TEMPERATURES		
	ABOVE 15°F (ABOVE -9°C)	+40°F TO -15°F (+4°C TO -26°C)	+40°F TO -65°F (+4°C TO -54°C)
OE/HDO Lubricating oil, internal combustion engine (MIL-L-2104)	OE/HDO 30	OE/HDO 10	
OEA Lubricating oil, internal combustion engine (arctic) (MIL-L-46167)			OEA

Table 2-1. Unit Preventive Maintenance Checks and Services (Contd).

ITEM NO.	INTERVAL	ITEM TO CHECK/ SERVICE	PROCEDURE	NOT FULLY MISSION CAPABLE IF:

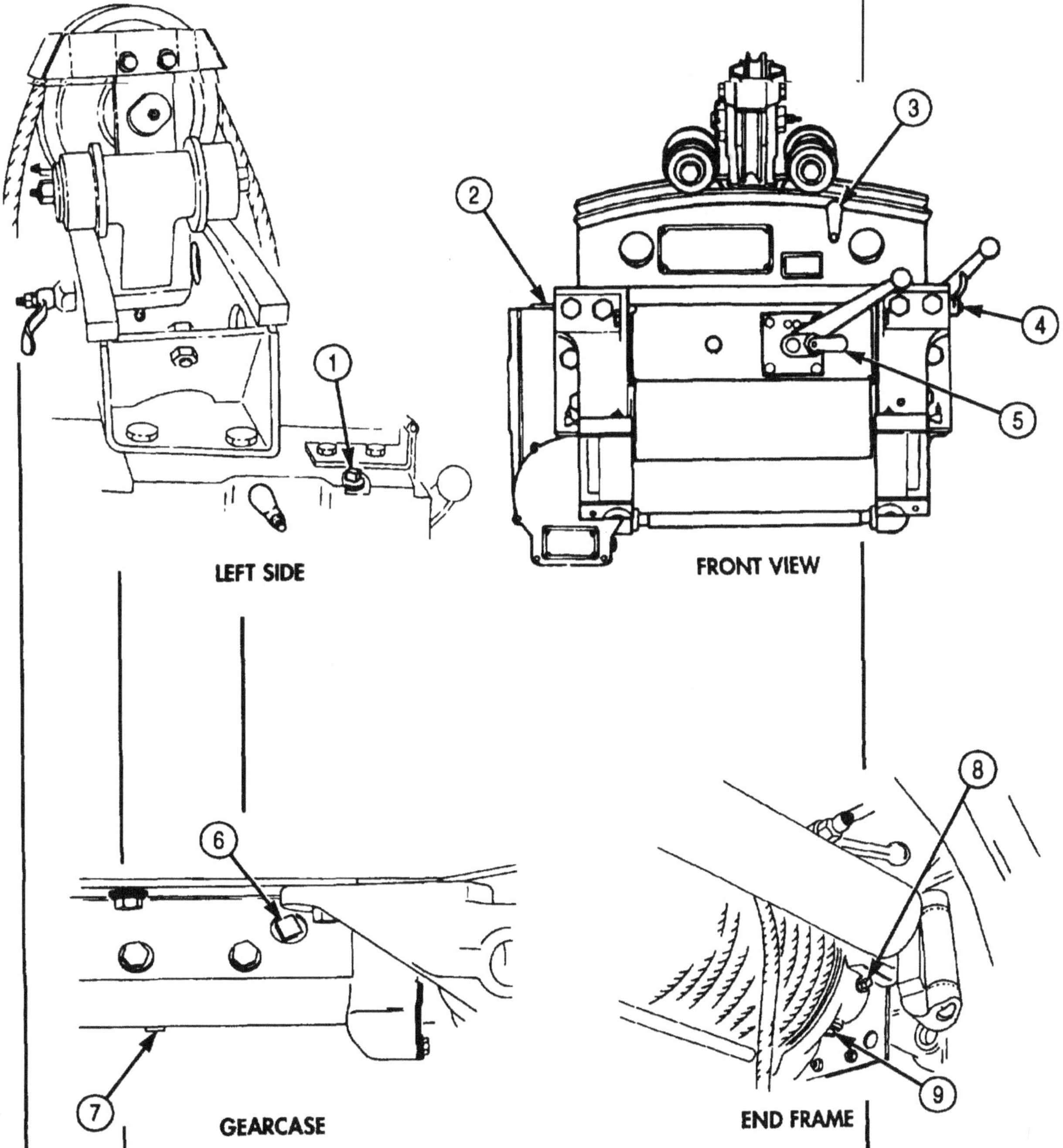
1
2
3
4
5
LEFT SIDE
FRONT VIEW
6
7
8
9
GEARCASE
END FRAME

Table 2-1. Unit Preventive Maintenance Checks and Services (Contd).

ITEM NO.	INTERVAL	ITEM TO CHECK/ SERVICE	PROCEDURE	NOT FULLY MISSION CAPABLE IF:
13	Semi-annual	Front winch (Contd)	**j.** Test drag brake for proper operation (para. 3-324). **k.** Test winch automatic brake for proper operation (para. 3-323).	
14	Semi-annual	Hood	**a.** Inspect hood support bar and locking pins for condition and security of mounting. If support bar or locking pins are defective, repair or replace as required (para. 3-269). **b.** Inspect hood stop cables for condition and security of mounting. If stop cables are defective, repair or replace as required (para. 3-270). **c.** Lubricate hood trunnion every 3,000 mi. (4,800 km) or 3 months, whichever occurs first.	

LUBRICANTS	EXPECTED TEMPERATURES		
	ABOVE 15°F (ABOVE -9°C)	+40°F TO -15°F (+4°C TO -26°C)	+40°F TO -65°F (+4°C TO -54°C)
GAA Grease, automotive and artillery (MIL-G-10924)	All temperatures		

d. Lubricate hood hinges every 1,000 mi. (1,600 km) or monthly, whichever occurs first.

LUBRICANTS	EXPECTED TEMPERATURES		
	ABOVE 15°F (ABOVE -9°C)	+40°F TO -15°F (+4°C TO -26°C)	+40°F TO -65°F (+4°C TO -54°C)
OE/HDO Lubricating oil, internal combustion engine (MIL-L-2104)	OE/HDO 30	OE/HDO 10	
OEA Lubricating oil, internal combustion engine (arctic) (MIL-L-46167)			OEA

ITEM NO.	INTERVAL	ITEM TO CHECK/ SERVICE	PROCEDURE	NOT FULLY MISSION CAPABLE IF:
15	Semi-annual	Front lights and cable assembly	Inspect front light cable assembly wiring for frays, splits, missing or damaged insulation, or poor connections. Repair affected wiring (para. 3-131).	
16	Semi-annual	Motor mount trunnions	**a.** Check two front engine mounting trunnion screws for tightness. If engine mounting trunnion screws are loose, notify your supervisor. **b.** Tighten five lower trunnion mount screws 65-75 lb-ft (88-102 N·m). **c.** On M939A2 series vehicles, check two front mounting locknuts for tightness. Tighten mounting locknuts 75-85 lb-ft (102-115 N·m).	

Table 2-1. Unit Preventive Maintenance Checks and Services (Contd).

ITEM NO.	INTERVAL	ITEM TO CHECK/ SERVICE	PROCEDURE	NOT FULLY CAPABLE IF:
17	Semi-annual	Emergency and service compressed air systems	**a.** Inspect front emergency and service air and dummy couplings for serviceability and seals.	
			NOTE	
			Inspection of emergency and service air lines and fittings will be accomplished over complete vehicle. Tighten, repair, and/or replace components of these compressed air systems as required. If maintenance is required at a higher level, records should reflect closest point of reference to ensure proper identification of components requiring service.	
			b. Inspect air lines, fittings, and emergency shutoff valve for security of mounting, tightness of connections, and damage that could cause air leaks (para. 2-15).	
			c. Ensure front emergency shutoff valve is closed and emergency and service air and dummy couplings are securely fastened (TM 9-2320-272-10).	
			NOTE	
			Inspection of crossmembers, bolts, and rivets will be accomplished over complete vehicle. Tighten, repair, and/or replace components of the frame as required. If maintenance is required at a higher level, records should reflect closest point of reference to ensure proper identification of components requiring service.	
18	Semi-annual	Frame and cross-members	Inspect crossmembers for missing rivets, screws, obstructions to other components, and breaks (TM 9-2320-247-40).	Screws or rivets are loose or missing.
			1. Using a 0.001 in. feeler gauge, check for space between rivet head and the riveted frame members. Penetration of the feeler gauge between the rivet head and riveted member is reason to suspect the riveted connection and/or rivet should be replaced.	
			2. Thoroughly clean, grease, and oil rivet. Using an oil can, apply lubricating oil around the connection. Allow approximately 10-20 seconds for oil to penetrate. Wipe rivet and riveted connection free of oil. Tap rivet with an 8-pound hammer. Any indication of oil around the rivet indicates a loose rivet. Replace all loose rivets (TM 9-2320-247-40). Check all riveted connections for signs of movement, such as bare or shiny spots or other indications of movement between rivet and frame. If movement is indicated, rivet and connection are loose.	

Table 2-1. Unit Preventive Maintenance Checks and Services (Contd).

ITEM NO.	INTERVAL	ITEM TO CHECK/ SERVICE	PROCEDURE	NOT FULLY MISSION CAPABLE IF:
19	Semi- annual	Front wheels, hubs, and drums	**FRONT WHEELS, HUBS, DRUMS, BRAKES, SPRINGS, AXLE, AND STEERING** **WARNING** · Completely deflate tire before removing from axle if there is obvious damage to wheel components. Injury or death to personnel may result from exploding wheel components. · Use caution when inflating tires. Ensure tire is in a tire cage and properly seated on rim before inflating. An improperly seated tire can burst with explosive force. Failure to comply can cause death or serious injury to personnel. · Do not work on any component supported only by lift jacks or hoist. Always use blocks or proper stands to support the component prior to any work. Equipment may fall and cause injury or death to personnel. · Air in system is under pressure. Ensure engine is shut down and all air reservoirs are drained before disconnecting CTIS components to prevent serious injury to personnel. **NOTE** · Similar left and right side components are inspected in the same manner and will be accomplished simultaneously. Procedures cover left side only. · When documenting discrepancies for similar left and right side components, indicate which side is affected. · To simplify maintenance, perform items 19a through 19g simultaneously with items 20a through 20c. **a.** On M939 series vehicles, inspect split locking ring (2) for dents or breaks that could cause them to pop off. If ring is loose, break wheel (4) down and inspect for damage to rim (1) or tire (3). Replace any damaged wheel components (para. 3-221). **b.** On M939A1/A2 series vehicles, inspect wheels (4) for loose or missing rim nuts (5). Tighten rim nuts (5) 210-240 lb-ft (285-325 N·m). Replace any rim stud (6) that is broken or has stripped threads (para. 3-222). **c.** Inspect axle drive companion flange for oil leaks. If oil appears to be leaking from expansion plug (8) on axle drive companion flange (7), notify your supervisor.	

Table 2-1. Unit Preventive Maintenance Checks and Services (Contd).

ITEM NO.	INTERVAL	ITEM TO CHECK/ SERVICE	PROCEDURE	NOT FULLY MISSION CAPABLE IF:

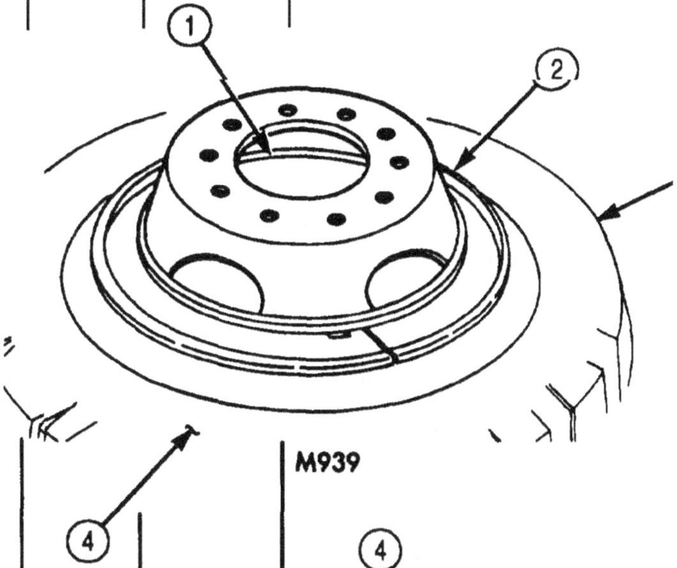

M939

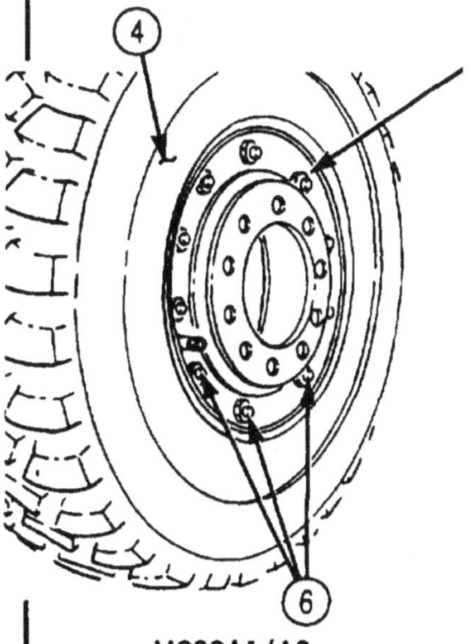

M939A1/A2

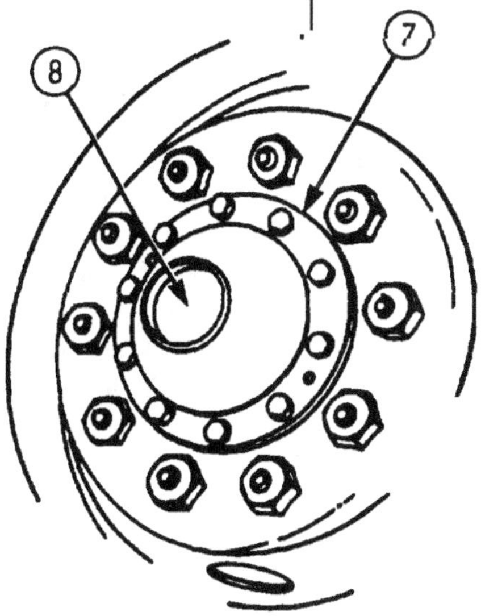

Table 2-1. Unit Preventive Maintenance Checks and Services (Contd).

ITEM NO.	INTERVAL	ITEM TO CHECK/ SERVICE	PROCEDURE	NOT FULLY MISSION CAPABLE IF:
19	Semi-annual	Front wheels, hubs, and drums (Contd) [RPL]	**CAUTION** On M939A2 series vehicles, use care not to damage CTIS air seals when removing axle components. Failure to do so may result in further damage to axle and wheel bearings, in addition to improper operation of CTIS system. **NOTE** Lugnuts on left side have left-hand threads and can be identified by an L, while lugnuts on right side have right-hand threads, and can be identified by an R. **d.** Clean, inspect, and lubricate axle shaft and universal joints (para. 3-154).	

LUBRICANTS			EXPECTED TEMPERATURES		
			ABOVE 15°F (ABOVE -9°C)	+40°F TO -15°F (+4°C TO -26°C)	+40°F TO -65°F (+4°C TO -54°C)
GAA Grease, automotive and artillery (MIL-G-10924)			All temperatures		

| | | [RPL] | **1.** Check bearing sleeve inside spindle for signs of damage. If bearing sleeve is damaged, notify DS maintenance.

2. Replace drive flange gasket with RTV sealant.

e. On M939A2 series vehicles, replace filter in wheel valves (para. 3-448).

WARNING

Do not allow grease or oil to contact brake linings. Linings can absorb grease and oil, causing early glazing and very poor braking action. Failure to comply could cause serious injury or death to personnel.

f. Check brakeshoes for condition and brakeshoe-to-drum clearance (para. 3-179). Replace brakeshoes if worn beyond chamfer on linings (para. 3-180).

g. Inspect CV boots for rips, cracking, punctures, and other damage that could cause a loss of lubrication around CV joint. Replace CV boots if ripped, cracked, punctured, or damaged (para. 3-223). | |

Table 2-1. Unit Preventive Maintenance Checks and Services (Contd).

ITEM NO.	INTERVAL	ITEM TO CHECK SERVICE	PROCEDURE	NOT FULLY MISSION CAPABLE IF:
20	Semi-annual	Air lines and brake chambers	**a.** Inspect front service brake air lines and fittings for loose connections, cracks, splits, or damage that could cause potential air leaks. Tighten loose air lines and fittings connections, and replace any air line or fitting that has cracks, splits, or damage that could cause potential air leaks. **b.** Inspect front service brake chambers for condition and security of mounting. Replace any service brake chamber that is damaged, defective, or inoperative (para. 3-181). **NOTE** Inspection of CTIS air lines and fittings will be accomplished over complete vehicle. Tighten, repair, and/or replace components of the compressed air system as required. If maintenance is required at a higher level, records should reflect closest point of reference to ensure proper identification of components requiring service. **c.** On M939A2 series vehicles, inspect CTIS air lines and fittings for loose connections, cracks, splits, or damage that could cause potential air leaks.	
21	Semi-annual	Steering system	**a.** Inspect steering knuckles, steering gear, tie rod assembly, steering arms, drag link, pitman arm, lower steering gear shaft, and power steering cylinder for breaks, cracks, rust, wear, and signs of unserviceable condition. **1.** Treat for corrosion of steering knuckles, tie rod assembly, steering arms, drag link, pitman arm, or steering gear shaft. **2.** Replace tie rod assembly, drag link, pitman arm, lower steering gear shaft, or power steering cylinder if broken, cracked, worn, or signs of an unserviceable condition are present. **3.** If steering knuckles, steering arms, or steering gear are broken, cracked, worn, or have other signs of an unserviceable condition, notify DS maintenance.	

Table 2-1. Unit Preventive Maintenance Checks and Services (Contd).

ITEM NO.	INTERVAL	ITEM TO CHECK/ SERVICE	PROCEDURE	NOT FULLY MISSION CAPABLE IF:
21	Semi-annual	Steering system (Contd)	**NOTE** Lubricate grease fittings every 3,000 mi. (4,800 km) or 3 months, whichever occurs first. When practical, lubrication services will be made to coincide with the semiannual preventive maintenance service. For this purpose, a 10 percent tolerance (variation) in specified lubrication point mileage is permissible. **b.** Lubricate at steering knuckle grease fittings (1), tie rod assembly grease fittings (4), steering shaft grease fittings (7) and (8), drag link grease fittings (2), and power steering assist cylinder grease fittings (6).	

LUBRICANTS			EXPECTED TEMPERATURES		
			ABOVE 15°F (ABOVE -9°C)	+40°F TO -15°F (+4°C TO -26°C)	+40°F TO -65°F (+4°C TO -54°C)
G A A Grease, automotive and artillery (MID-G-10924)			All temperatures		

ITEM NO.	INTERVAL	ITEM TO CHECK/ SERVICE	PROCEDURE	NOT FULLY MISSION CAPABLE IF:

			NOTE
			An assistant is required to perform the following step.
			
			c. Inspect steering stops for presence and security. If any stop is missing or has broken welds, notify DS maintenance.
			
			d. With engine shut off, turn steering wheel slowly right and left.
			
			    1. Inspect steering gear U-joints while steering wheel is rotated.
			
			    2. Check for free play between steering knuckles and tie rod ends, power assist cylinder, and drag link and pitman arm to drag link. If free play is present, tighten four steering knuckle nuts (5) and drag link-to-pitman arm nut (3) 140-180 lb-ft (190-244 N·m). Tighten to minimum torque and continue to tighten as needed to align slot in nut and cotter pin hole. Do not tighten over 275 lb-ft (373 N·m).
			
			    3. Repeat step 2. If free play is still present, notify DS maintenance.
			
			e. Inspect power steering assist assembly.
			
			    1. Remove stone shield (para. 3-231).
			
			    2. Inspect power steering assist cylinder for condition and security of mounting (para. 3-233).
				d. Movement between cross and cap is present.

Table 2-1. Unit Preventive Maintenance Checks and Services (Contd).

ITEM NO.	INTERVAL	ITEM TO CHECK/ SERVICE	PROCEDURE	NOT FULLY MISSION CAPABLE IF:
21	Semi-annual	Steering system (Contd)	**3.** Inspect hydraulic hoses and fittings between frame and cylinder for loose connections, cracks, splits, or damage that could cause hydraulic leaks. Tighten hydraulic lines and fittings connections. Replace ant hydraulic line or fitting that has cracks, splits, or damage that could cause hydraulic leaks (para. 3-232). **4.** Install stone shield (para. 3-231).	

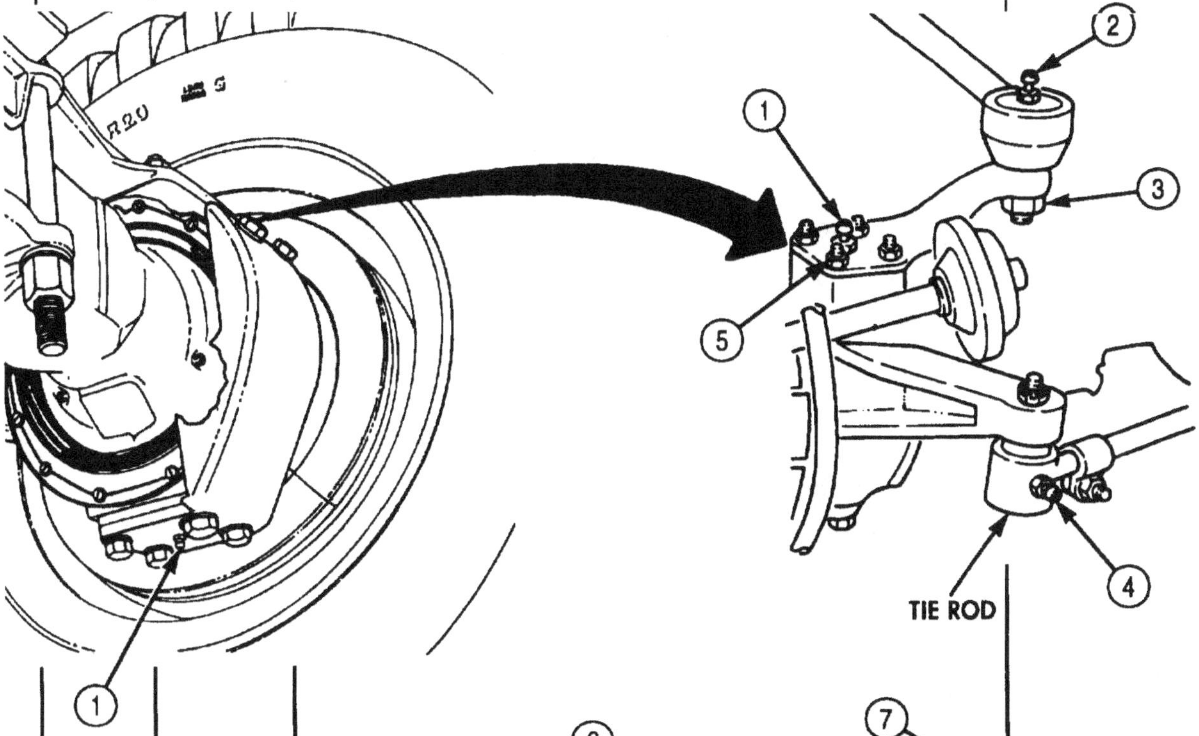

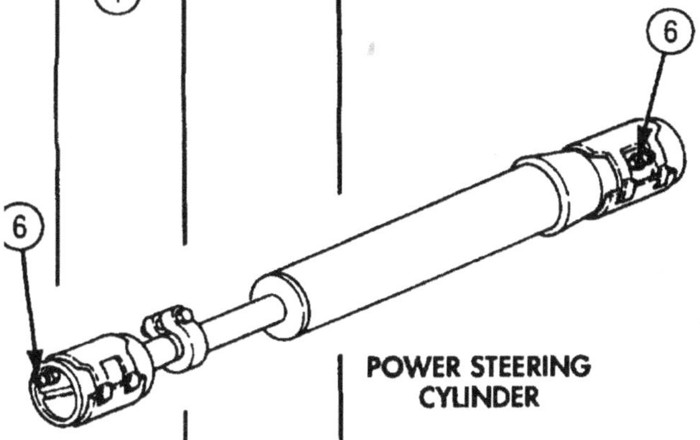

TIE ROD

POWER STEERING CYLINDER

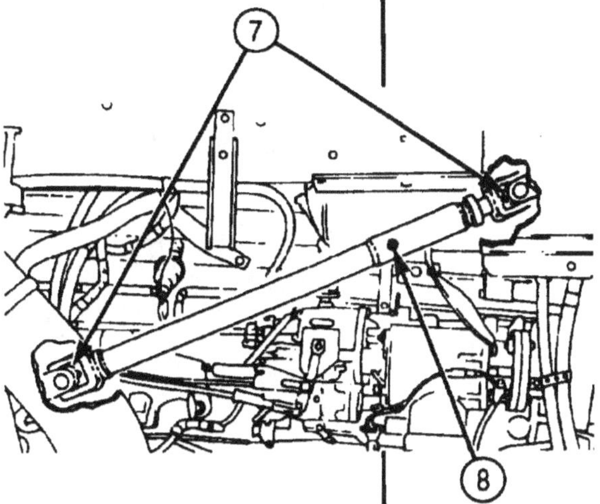

Table 2-1. Unit Preventive Maintenance Checks and Services (Contd).

ITEM NO.	INTERVAL	ITEM TO CHECK/ SERVICE	PROCEDURE	NOT FULLY MISSION CAPABLE IF:
21	Semi-annual	Steering system (Contd)	**f.** Inspect power steering pump for security of mounting, leaks, and signs of damage. Tighten loose mounting hardware. **g.** On M939/A1 series vehicles, inspect power steering pump belts for tension and serviceability (para. 3-230). **h.** Inspect power steering pressure and return hoses and fittings for loose connections, cracks, splits, or damage that could cause potential hydraulic leaks. Tighten loose hydraulic lines and fittings, and replace any hydraulic line or fitting that has cracks, splits, or damage that could cause potential hydraulic leaks. Refer to para. 3-234 for Ross steering gear, or para. 3-235 for Sheppard steering gear. **i.** Inspect power steering hydraulic lines between steering gear and steering assist cylinder and fittings for security of mounting, loose connections, cracks, splits, or damage that could cause hydraulic leaks. Tighten loose hydraulic lines and fittings, and replace any hydraulic line or fitting that has cracks, splits, or damage that could cause hydraulic leaks (para. 3-239). **j.** Inspect steering gear for security of mounting and signs of leaks. **1.** Loosen bracket mounting screws (1). **2.** Tighten steering gear mounting locknuts (2) 260-280 lb-ft (353-380 N·m). **3.** Tighten locknuts (3) 60-70 lb-ft (81-95 N·m). 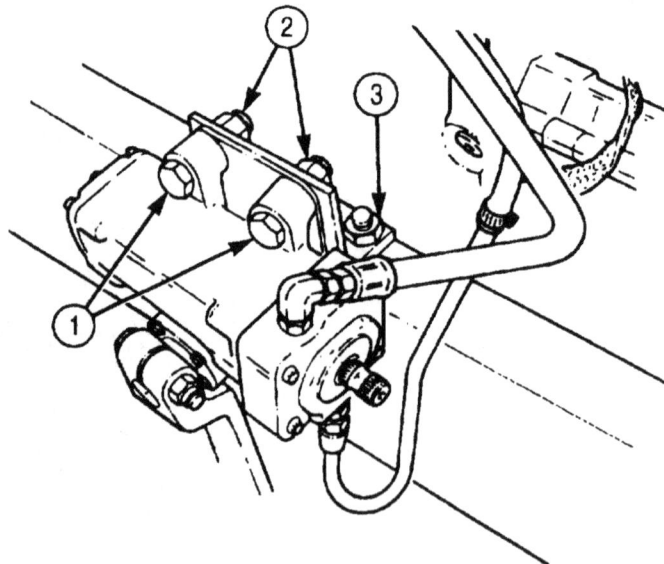	

Table 2-1. Unit Preventive Maintenance Checks and Services (Contd).

ITEM NO.	INTERVAL	ITEM TO CHECK/ SERVICE	PROCEDURE	NOT FULLY MISSION CAPABLE IF:
22	Semi-annual	Front springs, propeller shaft, universal and slip joints, and axle	**a.** Inspect axle (4) for security of mounting on springs (7). Tighten nuts (5) on spring U-bolts (3) 350-400 lb-ft (475-542 N·m). **b.** Inspect springs (7) and shackles (8) for cracks, breaks, and security of mounting. Tighten spring shackle mounting nuts (6) 70 lb-ft (95 N·m). **CAUTION** Wipe fittings clean before servicing to prevent damage to shackle pins and bushings. **c.** Lubricate spring U-bolts (3) and shackles (8) every 3,000 mi. (4,800 km) or 6 months, whichever occurs first.	

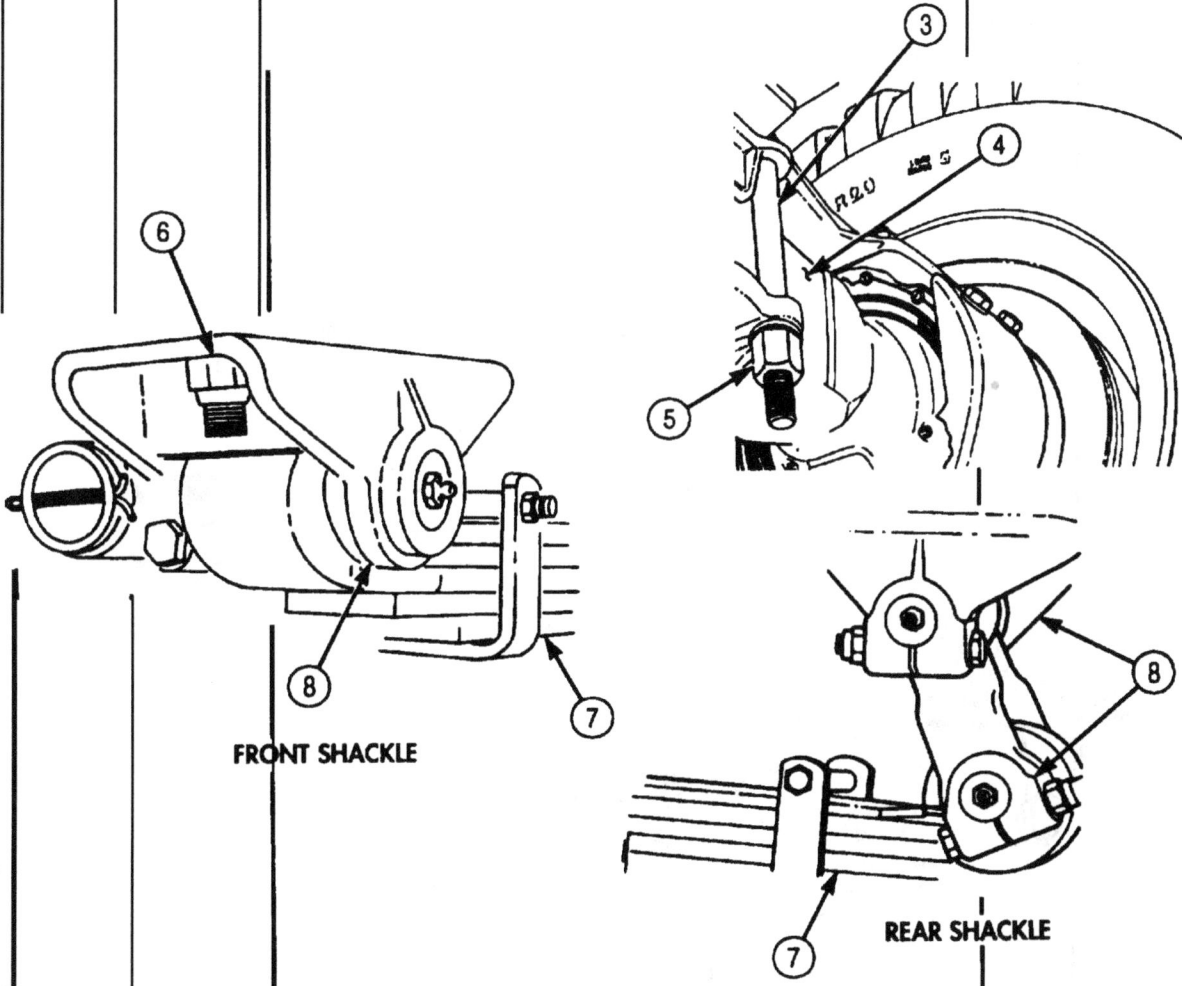

FRONT SHACKLE

REAR SHACKLE

Table 2-1. Unit Preventive Maintenance Checks and Services (Contd).

ITEM NO.	INTERVAL	ITEM TO CHECK/ SERVICE	PROCEDURE	NOT FULLY MISSION CAPABLE IF:
22	Semi-annual	Front springs, propeller shaft, universal and slip joints, and axle (Contd)	1. Apply lubricant to spring pins (1) until it appears between front spring pin (1) and bushing (3) at both ends of shackle (2). If spring pin (1) does not accept lubricant, remove spring pin (1).	

LUBRICANTS		EXPECTED TEMPERATURES		
		ABOVE 15°F (ABOVE -9°C)	+40°F TO -15°F (+4°C TO -26°C)	+40°F TO -65°F (+4°C TO -54°C)
GAA Grease, automotive and artillery MIL-G-10924)		All temperatures		

2. Inspect front spring pin (1) and bushing (3) at both ends of shackle (2) (para. 3-163).

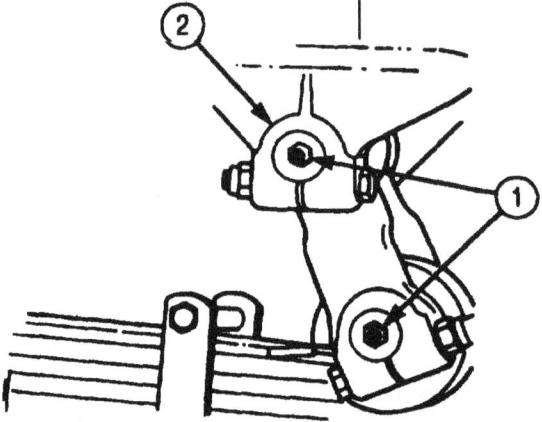

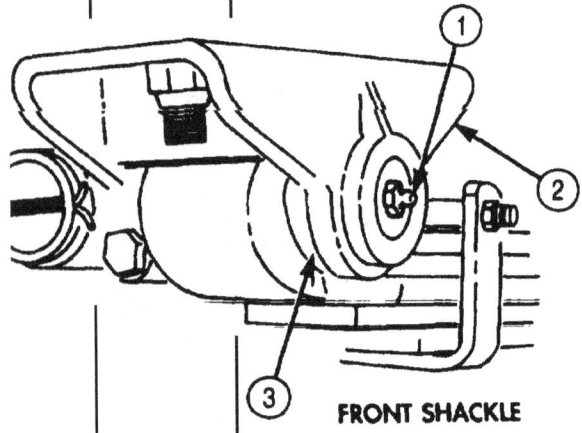

FRONT SHACKLE

Table 2-1. Unit Preventive Maintenance Checks and Services (Contd).

ITEM NO.	INTERVAL	ITEM TO CHECK/ SERVICE	PROCEDURE	NOT FULLY MISSION CAPABLE IF:
12	Semi-annual	Front springs, propeller shaft, universal and slip joints, and axle (Contd)	**NOTE** Lubrication of universal and slip joints will be accomplished while performing other inspection tasks in that same area. Tighten, repair, and/or replace components of universal, slip joints, and propeller shafts when found to be damaged or worn, as required. If maintenance is required at a higher level, records should reflect closest point of reference to ensure proper identification of components requiring service. d. Lubricate universal and slip joints on transfer case-to-front axle propeller shaft adapter every 3,000 mi. (4,800 km) or 3 months, whichever occurs first. e. Inspect universal and slip joints on transfer case-to-front axle propeller shaft for damage or worn components. Replace worn components (para. 3-149). **CAUTION** Breathers and axle around breathers must be wiped clean before servicing to prevent damage to axle from contamination. f. Remove, clean, and lubricate axle breathers every 1,000 mi. (1,600 km) or monthly, whichever occurs first (para. 3-154).	

LUBRICANTS		EXPECTED TEMPERATURES		
		ABOVE 15°F (ABOVE -9°C)	+40°F TO -15°F (+4°C TO -26°C)	+40°F TO -65°F (+4°C TO -54°C)
OE/HDO Lubricating oil, internal combustion engine (MIL-L-2104)		OE/HDO 30	OE/HDO 10	
OEA Lubricating oil, internal combustion engine (arctic) (MIL-L-46167)				OEA

Table 2-1. Unit Preventive Maintenance Checks and Services (Contd).

ITEM NO.	INTERVAL	ITEM TO CHECK/ SERVICE	PROCEDURE	NOT FULLY MISSION CAPABLE IF:
22	Semi-annual	Front springs, propeller shaft, universal and slip joints, and axle (Contd)	**g.** Remove differential fill plug (1) and check oil level in differential (4) every 3,000 mi. (4,800 km) or 3 months, whichever occurs first. Fill if necessary. Level should be within 1/2 in. (12.7 mm) from hole of fill plug (1) when oil is cold, and to the hole of fill plug when hot.	

LUBRICANTS	EXPECTED TEMPERATURES		
	ABOVE 15°F (ABOVE -9°C)	+40°F TO -15°F (+4°C TO -26°C)	+40°F TO -65°F (+4°C TO -54°C)
Oil, lubricating, multipurpose, (MIL-L-2105): GO-80/90	GO-80/90	GO-80/90	
Oil, lubricating, multipurpose, (MIL-L-2105): GO-75			GO-75

h. Inspect differential drainplug (3) and fill plug (1) for tightness and signs of leakage. Tighten drainplug (3) 35-60 lb-ft (47-81 N·m) and fill plug (1) 80-135 lb-ft (108-183 N·m).

i. Inspect differential seals (2) for leaks.

ITEM NO.	INTERVAL	ITEM TO CHECK/ SERVICE	PROCEDURE	NOT FULLY MISSION CAPABLE IF:
23	Semi-annual	Underside of engine and trans-mission	**a.** Inspect underside of engine for fuel, water, and oil leaks. **b.** Inspect oil pan and drainplug for leaks. If oil pan is loose or if leaks are present, tighten oil pan screws 35-40 lb-ft (47-54 N•m). If drainplug is loose or if leaks are present, tighten drainplug 100 lb-ft (136 N•m) (M939/A1 vehicles), 60 lb-ft (80 N•m) (M939A2 vehicles). Notify your supervisor if leaks still occur. **c.** Inspect transmission body for cracks or loose bolts that could cause leaks. **d.** Inspect transmission shift linkage for bends, cracks, and wear that could cause failure. **e.** Tighten transmission oil pan mounting bolts 5 lb-ft (7 N•m). Tighten oil pan drainplug 15-20 lb-ft (20-27 N•m).	

Table 2-1. Unit Preventive Maintenance Checks and Services (Contd).

ITEM NO.	INTERVAL	ITEM TO CHECK/ SERVICE	PROCEDURE	NOT FULLY MISSION CAPABLE IF:

24	Semi-annual	Cooling system	**a.** Inspect coolant lines, hoses, and fittings for loose connections, cracks, frays, wear, and damage that could cause leaks. Tighten loose connections. Replace any oil line, hose, or fitting that is cracked, frayed, worn, or damaged and could cause leaks.
			
			b. Inspect fan actuator and sensor for security of mounting. Inspect air hoses and fittings for loose connections, cracks, frays, wear, and damage that could cause leaks. Tighten loose connections. Replace any air line, hose, or fitting that is cracked, frayed, worn, or damaged and could cause leaks.
			
			c. Inspect radiator core for clogged or bent fins, leaks, and protruding debris, Clean clogged core and remove debris (para. 2-15).
			
			d. On M939A2 series vehicles, inspect power steering cooler core for clogged or bent fins, leaks, and protruding debris. Clean clogged core and remove debris (para. 2-15).
			
			e. Inspect water pump vent/drain for obstructions.
			
			f. Inspect water pump pulley and fan for play.
			
			g. Check drivebelt(s) for proper tension.
			
			    1. For M939A2 series vehicles, check belt tensioner for security of mounting 30 lb-ft (41 N•m) and condition of fan drivebelt (para. 3-71).
			
			    2. For M939/A1 series vehicles, check fan drivebelt tension and adjust as required (para. 3-70).
			
			    3. For M939/A1 series vehicles, check power steering pump drivebelt tension (para. 3-230).
			
			    4. For M939/A1 series vehicles, check water pump drivebelt tension and adjust as required (para. 3-67).
			
			    5. For M939/A1 series vehicles, check alternator drivebelt tension (para. 3-78).
			
			h. Inspect fan blade for security, breaks, missing or loose screws and rivets, and damage which could cause an out-of-balance condition. For M939/A1 series vehicles, refer to para. 3-72, and for M939A2 series vehicles, refer to para. 3-64. Replace fan blade as required.
			
			i. Inspect surge tank, water manifold, thermostat housing, radiator, engine oil cooler, transmission oil cooler, and hoses for leaks, condition, and security of mounting(s).
			
			j. Inspect temperature sending unit for security of mounting. Inspect sending unit wiring for frays, splits, breaks, and worn or missing insulation.
			
			k. Check antifreeze protection temperature and adjust if necessary to meet environmental conditions (para. 3-53).	

Table 2-1. Unit Preventive Maintenance Checks and Services (Contd).

ITEM TO	INTERVAL	ITEM TO CHECK/ SERVICE	PROCEDURE	NOT FULLY MISSION CAPABLE IF:
25	Semi-annual	Compressor	**a.** Check compressor for security of mounting and leaks. 1. If M939/A1 series engine-mounted compressor is loose, notify DS maintenance. 2. If M939A2 series engine-mounted compressor is loose, replace locknut and tighten to 55 lb-ft (77 N•m). **b.** Check condition and security of cooling lines to air compressor. Tighten cooling lines, if loose. Replace cooling lines if split, cracked, or damaged in such a manner as to cause leaks (para. 3-55). **c.** On M939A2 series engines, check condition of compressor oil line and fittings. Tighten fittings and oil line if loose. Replace oil line if split, cracked, worn, or damaged in such a manner as to cause leaks (para. 3-206). **d.** Check condition and security of input tube and hoses, output air line, and governor control air line. Tighten output and governor control air lines, if loose. Replace input tube or hoses if split, cracked, collapsed, distorted, or damaged in such a manner as to prevent a tight seal or restrict incoming air (para. 3-55).	
26	Semi-annual	Engine lubrication system and oil lines	**a.** Inspect all engine oil lines, hoses, and fittings for loose connections, cracks, frays, wear, and damage that could cause leaks. Tighten loose connections, and replace any oil lines, hoses, and fittings that are cracked, frayed, worn, or damaged and could cause leaks. **NOTE** • Oil filter is located on lower front left side of engine on M939/A1 series vehicles and lower front right side of engine on M939A2 series vehicles. • M939/A1 series vehicles utilize an element-type engine oil filter, while M939A2 series vehicles use one of two styles of cartridge-type engine oil filters. **b.** Inspect M939/A1 series vehicles for security of oil filter housing. Ensure filter center bolt is tight. If center bolt is loose, tighten 25-35 lb-ft (34-48 N•m). **c.** Inspect M939A2 series vehicles for security of oil filter head. Ensure spin-on oil filter is tight. If spin-on oil filter is loose, hand-tighten, then tighten an additional 3/4 turn with a wrench. **d.** Check engine oil and dipstick for metal particles. Notify DS maintenance if metal particles are found.	

Table 2-1. Unit Preventive Maintenance Checks and Services (Contd).

ITEM NO.	INTERVAL	ITEM TO CHECK/ SERVICE	PROCEDURE	NOT FULLY MISSION CAPABLE IF:
16	Semi-annual	Engine lubrication system and oil lines (Contd)	**WARNING** Accidental or intentional introduction of liquid contaminants into the environment is in violation of state, federal, and military regulations. Refer to Army POL (para. 1-8) for information concerning storage, use, and disposal of these liquids. Failure to do so may result in injury or death. **NOTE** • Sample oil every 60 days or 1,000 mi. (1,600 km), whichever occurs first. Army Reserve units will sample oil every 120 days or 1,000 mi. (1,600 km). • When AOAP laboratory support is not available, change engine oil every 6,000 mi. (9,600 km) or 6 months, whichever occurs first. • Engine AOAP sampling valve on M939/A1 series vehicles is located on right side of engine attached to oil cooler, On M939A2 series vehicles, valve is located near starter on left side of engine.	
		[RPL]	e. Collect engine oil sample (para. 3-10) or change oil (para. 3-5) as required.	

LUBRICANTS		EXPECTED TEMPERATURES		
		ABOVE 15°F (ABOVE -9°C)	+40°F TO -15°F (+4°C TO -26°C)	+40°F TO -65°F (+4°C TO -54°C)
OE/HDO Lubricating oil, internal combustion engine (MIL-L-2104)		OE/HDO 30	OE/HDO 10	
OEA Lubricating oil, internal combustion engine (arctic) (MIL-L-46167)				OEA

| | | | f. Inspect valve cover(s) and gasket(s) for evidence of leaks. Notify DS maintenance if leak(s) are detected. | |

Table 2-1. Unit Preventive Maintenance Checks and Services (Contd).

ITEM NO.	INTERVAL	ITEM TO CHECK/ SERVICE	PROCEDURE	NOT FULLY MISSION CAPABLE IF:
1 7	Semi-annual	Fuel system engine	**NOTE** • Steps a. and b. apply to M939A1series vehicles and step e. applies to M939A2 series vehicles. All other steps are common to both vehicles. • Replace fuel filter every 3,090 mi. (4,800 km) or 3 months, whichever occurs first.	
		[RPL]	**a.** Replace filter in standard AFC fuel pump on all M939/A1 series vehicles except M936/A1 models (para. 3-31) or in VS fuel pump on M936/A1 models (para. 3-32).	
			b. Inspect fuel filter/water separator mounting and housing for dents and cracks, and damage to inlet, outlet, and bleeder fittings that could cause leaks (para. 2-15).	
		[RPL]	**c.** Install new fuel filter element on M939/A1 series vehicles, or new fuel filter on M939A2 series vehicles (para. 3-30).	
			d. Inspect fuel lines for tightness of connections (para. 3-29).	
			e. Check fuel system components of M939A2 series vehicles.	
			1. Inspect fuel injector lines, injector pump lines, and manifold line and screws for leaks and damage. Tighten fuel injector lines, injector pump lines, and manifold line and screws if leaking, and replace if damaged (para. 3-18).	
			2. Inspect injector line holddown screws for security of mounting. If loose, tighten (para. 3-18).	
		[RPL]	**3.** Replace in-line fuel filter (para. 3-30).	
			f. Check fuel priming pump on M939/A1 series vehicles or fuel priming/transfer pump on M939A2 series vehicles for security of mounting and proper operation.	
			g. Lubricate throttle control and modulator linkages.	

LUBRICANTS	EXPECTED TEMPERATURES		
	ABOVE 15°F (ABOVE -9°C)	**+40°F TO -15°F (+4°C TO -26°C)**	**+40°F TO -65°F (+4°C TO -54°C)**
OE/HDO Lubricating oil, internal combustion engine (MIL-L-2104)	OE/HDO 30	OE/HDO 10	
OEA Lubricating oil, internal combustion engine (arctic) (MIL-L-46167)			OEA

Table 2-1. Unit Preventive Maintenance Checks and Services (Contd).

ITEM NO.	INTERVAL	ITEM TO CHECK/ SERVICE	PROCEDURE	NOT FULLY MISSION CAPABLE IF:
28	Semi-annual	Protective control box	Inspect control box and quick-disconnect for security of mounting and connection. Tighten control box mounting screws and/or quick-disconnect, if loose (para. 3-115).	
29	Semi-annual	Engine compart-ment electrical wiring	Inspect all engine compartment wiring for frays, splits, missing or damaged insulation, or poor connections. Repair or replace affected wiring (para. 3-131).	
30	Semi-annual	Fuel lines and tank	**NOTE** On dual tank fuel systems, inspect left side of tank when performing other tasks conducted in that area of the vehicle. Ensure any problem found while inspecting right and left side component systems is noted in maintenance forms. **a.** Inspect fuel tank(s) for dents, cracks, and broken welds that could cause leaks. **b.** Visually inspect fuel sending unit(s) and wiring for loose connections, presence, frays, splits, and missing insulation (para. 3-131). Repair or replace wiring that is missing or shows signs of frayed, split, or missing insulation. Tighten loose connections (para. 3-92). **c.** Visually inspect tube(s) and hose(s) at fuel tank(s) for loose connections, cracks, and splits. Replace fuel tank tube(s) and hose(s) that are cracked or split (para. 3-26). Tighten loose fuel tank tube(s) and hoses connections. **d.** Inspect fuel tank selector valve on dual tank model vehicles for freedom of operation and security of mounting, and check flex lines for leaks (para. 3-24). **e.** Check fuel tank selector valves for proper operation (TM 9-2320-272-10).	
31	Semi-annual	Transmis-sion and transfer case	**WARNING** • If transmission oil temperature is above 220°F (104°C), allow transmission oil to cool before removing dipstick. • Accidental or intentional introduction of liquid contaminants into the environment is in violation of state, federal, and military regulations. Refer to Army POL (para. 1-8) for information concerning storage, use, and disposal of these liquids. Failure to do so may result in injury or death.	

Table 2-1. Unit Preventive Maintenance Checks and Services (Contd).

ITEM NO.	INTERVAL	ITEM TO CHECK/ SERVICE	PROCEDURE	NOT FULLY MISSION CAPABLE IF:
31	Semi-annual	Transmission and transfer case (Contd)	**CAUTION** • Do not remove transmission dipstick before cleaning dirt away from access plate, filler tube, and dipstick. Dirt may enter and damage transmission. • Do not overfill transmission. Internal transmission component damage will result. • Change transmission oil when contamination by fuel, water, or other foreign materials is evident. Failure to do so may result in failure of internal transmission components. **NOTE** When AOAP laboratory support is not available, change transmission oil every 24,000 mi. (38,000 km) or 24 months, whichever occurs first. **a.** On M939A2 series vehicles, inspect transmission dipstick and oil. **1.** Check for evidence of metal particles. Notify your supervisor if metal particles are found. **2.** Check for evidence of dilution by coolant. If oil is diluted by coolant, refer to para. 2-21, mechanical troubleshooting. **3.** Check oil level in transmission. If oil levels are excessive, drain excess (para. 3-133).	

LUBRICANTS	EXPECTED TEMPERATURES		
	ABOVE 15°F (ABOVE -9°C)	+40°F TO -15°F (+4°C TO -26°C)	+40°F TO -65°F (+4°C TO -54°C)
OE/HDO Lubricating oil, internal combustion engine (MIL-L-2104)	OE/HDO 30	OE/HDO 10	
OEA Lubricating oil, internal combustion engine (arctic) (MIL-L-46167)			OEA

			PROCEDURE	
			b. Inspect transmission oil lines, hoses, and fittings for loose connections, cracks, frays, wear, and damage that could cause leaks. Tighten loose connections, or notify DS maintenance if any oil line, hose, or fitting is cracked, frayed, worn, or damaged and could cause leaks. **c.** Inspect transmission oil filter housing for security of mounting. Ensure spin-on oil filter is tight. If spin-on oil filter is loose, hand-tighten, then tighten an additional 3/4 turn by hand. If leak continues, replace filter (para. 3-139).	

Table 2-1. Unit Preventive Maintenance Checks and Services (Contd).

ITEM NO.	INTERVAL	ITEM TO CHECK/ SERVICE	PROCEDURE	NOT FULLY MISSION CAPABLE IF:
31	Semi-annual	Transmission and transfer case (Contd)	**d.** For M939A2 series vehicles, inspect transmission temperature sending unit for security and signs of leaks, and wiring for frays, splits, breaks, and missing insulation. Tighten temperature sending unit if loose or leaking. Replace if necessary (para. 3-97). Repair or replace wiring that is missing, or shows signs of frayed, split, or missing insulation. Tighten loose connections (para. 3-97). **e.** Inspect transmission PTO unit (winch models and dump trucks only). **1** Inspect for security of mounting and leaks. Notify DS maintenance if mounting is loose or leakage is present. **2.** Inspect oil line and fittings between transmission and PTO unit for loose connections, cracks, frays, wear, and damage that could cause leaks. Notify DS maintenance if mounting is loose or leakage is present. **3.** Inspect hydraulic pump for security of mounting, leaks, and signs of damage that could cause leaks. Notify DS maintenance if mounting is loose or leakage is present. **4.** Inspect hydraulic lines and fittings for loose connections, cracks, frays, wear, and damage that could cause leaks. Notify DS maintenance if mounting is loose or leakage is present. f. Remove transfer case fill plug and check oil level in transfer case every 3,000 mi. (4,800 km) or 3 months, whichever comes first. Fill if necessary. Level should be within 1/2 in. (12.7 mm) from fill hole when oil is cold, or to the fill hole when hot. g. Inspect transfer case drainplug and fill plug for tightness and signs of leakage. h. Inspect transfer case for leaks. Notify DS maintenance if leaks are detected. i. Inspect shift linkage for cracks, bends, and wear. j. Inspect transfer case PTO unit (M936/A1/A2 only). **1.** Inspect for security of mounting and leaks. **2.** Inspect oil and air lines and fittings on transfer case and PTO for loose connections, cracks, frays, wear, and damage that could cause leaks. Tighten loose connections. Replace any oil or air line, hose, or fitting that is cracked, frayed, worn, or damaged and could cause leaks. **3.** Inspect hydraulic pump for security of mounting, leaks, and signs of damage that could cause leaks. Tighten mounting of hydraulic pump (para. 3-393). **4.** Inspect hydraulic lines and fittings for loose connections, cracks, frays, wear, and damage that could cause leaks. Tighten loose connections. Replace any oil line, hose, or fitting that is cracked, frayed, worn, or damaged and could cause leaks (para. 3-387).	

Table 2-1. Unit Preventive Maintenance Checks and Services (Contd).

ITEM NO.	INTERVAL	ITEM TO CHECK/ SERVICE	PROCEDURE	NOT FULLY MISSION CAPABLE IF:
12	Semi-annual	Propeller shafts, and universal and slip joints	a. On front winch and M929/A1/A2 model vehicles only, lubricate transmission PT0 universal joint every 3,006 mi. (4,800 km) or 3 months, whichever occurs first.	

LUBRICANTS		EXPECTED TEMPERATURES		
		ABOVE 15°F (ABOVE -9°C)	+40°F TO -15°F (+4°C TO -26°C)	+40°F TO -65°F (+4°C TO -54°C)
GAA Grease, automotive and artillery (MIL-G-10924)		All temperatures		

			b. On M936/A1/A2 series vehicles, lubricate transfer case PTO universal joint every 3,000 mi. (4,800 km) or 3 months, whichever occurs first.	

NOTE

Lubrication of universal and slip joints will be accomplished while performing other inspection tasks in that same area. Tighten, repair, and/or replace components of universal, slip joints, and propeller shafts when found to be damaged or worn, as required. If maintenance is required at a higher level, records should reflect closest point of reference to ensure proper identification of components requiring service.

c. Lubricate universal and slip joints on transmission-to-transfer case, transfer case-to-differential, differential-to-differential propeller shafts, and speedometer adapter every 3,000 mi. (4,800 km) or 3 months, whichever occurs first.

d. Inspect universal and slip joints on transmission PTO, transfer case FTO, and transfer case-to-front axle, transmission-to-transfer case (para. 3-148), transfer case-to-differential (para. 3-150 or 3-151), and differential-to-differential (para. 3-144) propeller shafts for damaged or worn components.

| 33 | Semi-annual | Air intake tubes | **a.** Inspect vent lines for serviceability and security of connections (para. 2-15). **b.** Inspect air intake piping and air cleaner assembly for condition and security of mounting. Replace defective components and secure loose mountings (para 3-14). **c.** Test air cleaner indicator for proper operation (para. 3-12). | |

Table 2-1. Unit Preventive Maintenance Checks and Services (Contd).

ITEM NO.	INTERVAL	ITEM TO CHECK/ SERVICE	PROCEDURE	NOT FULLY MISSION CAPABLE IF:
34	Semi-annual	Air reservoir and doublecheck valves	**a.** Inspect reservoir tanks, doublecheck valves, and fittings for leaks (para. 2-15). **b.** Inspect condition of reservoir tank for damage that could cause leaks (para. 2-15).	
35	Semi-annual	Winch reservoir	Check hydraulic oil level in tank. If low, fill to top notch on dipstick.	

LUBRICANTS		EXPECTED TEMPERATURES		
		ABOVE 15°F (ABOVE -9°C)	+40°F TO -15°F (+4°C TO -26°C)	+40°F TO -65°F (+4°C TO -54°C)
OE/HDO Lubricating oil, internal combustion engine (MIL-L-2104)		OE/HDO 30	OE/HDO 10	
OEA Lubricating oil, internal combustion engine (arctic) (MIL-L-46167)				OEA

| 16 | Semi-annual | Rear wheels and hubs, and suspension | **WARNING** • Completely deflate tire before removing from axle if there is obvious damage to wheel components. Injury or death to personnel may result from exploding wheel components. • Use caution when inflating tires. Ensure tire is in a tire cage and properly seated on rim before inflating. An improperly seated tire can burst with explosive force. Failure to comply can result in death or serious injury to personnel. • Do not work on any component supported only by lift jacks or hoist. Always use blocks or proper stands to support the component prior to any work. Equipment may fall and cause injury or death to personnel. • Air in system is under pressure. Ensure engine is shut down and all air reservoirs are drained before disconnecting CTIS components to prevent serious injury to personnel. • Do not allow grease or oil to contact brake linings. Linings can absorb grease and oil, causing early glazing and very poor braking action. Failure to do so could cause serious injury or death to personnel. | |

Table 2-1. Unit Preventive Maintenance Checks and Services (Contd).

ITEM NO.	INTERVAL	ITEM TO CHECK/ SERVICE	PROCEDURE	NOT FULLY MISSION CAPABLE IF:
36	Semi-annual	Rear wheels and hubs, and suspension (Contd) [RPL]	**NOTE** • Similar left and right side, and forward-rear and rear-rear components are inspected in the same manner and will be accomplished simultaneously, although procedures are written for left side only. • When documenting discrepancies for similar left and right side, and forward-rear and rear-rear components, indicate which area is affected. • Lugnuts on left side have left-hand threads and can be identified by an L, while lugnuts on right side have right-hand threads and can be identified by an R. **a.** On M939 series vehicles, inspect split locking ring for dents or breaks that could cause them to pop off. If ring is loose, break tire down, and inspect for damage to rim or tire. Replace any damaged wheel components (para. 3-221). **b.** On M939A1/A2 series vehicles, inspect wheels for any loose or missing rim nut. Tighten rim nuts 210-240 lb-ft (285-325 N•m). Replace any rim stud that is broken or has stripped threads (para. 3-222). **c.** Clean, inspect, and lubricate inner and outer rear wheel bearings (3-225). **d.** Inspect service and service/spring brake air lines and fittings for loose connections, cracks, splits, or damage that could cause air leaks. Tighten loose air lines and fittings and replace any air line or fitting that has cracks, splits, or damage that could cause air leaks (para. 3-174). **e.** Check brakeshoe-to-drum clearance and condition of brakeshoes (para. 3-179). If clearance is more than 0.06 in. (1.5 mm), manually adjust brakes or notify your supervisor of inoperative adjusters. Replace brakeshoes if worn beyond chamfer on linings (para. 3-180). **f.** Inspect service brake chambers for condition and security of mounting. Replace service brake chambers and/or component parts if condition could impair operation of brakes (para. 3-181). **g.** Inspect service/spring brake chambers for condition and security of mounting. Replace service/spring brake chambers and/or component parts if condition could impair operation of brakes (para. 3-182). **h.** Inspect rear relay valve for condition and security of mounting and air lines and fittings for loose connections. Replace rear relay valve and/or component parts if condition could impair operation of brakes (para. 3-188). Tighten loose air lines and fittings.	

Table 2-1. Unit Preventive Maintenance Checks and Services (Contd).

ITEM NO.	INTERVAL	ITEM TO CHECK/ SERVICE	PROCEDURE	NOT FULLY MISSION CAPABLE IF:
36	Semi-annual	Rear wheels and hubs, and suspension (Contd)	**i.** Inspect air manifold tee for condition and security of mounting and air lines and fittings for loose connections. Replace air manifold tee and/or component parts if condition could impair operation of brakes (para. 3-189). Tighten loose air lines and fittings. **j.** Inspect M939A2 series vehicles CTIS air lines and fittings for loose connections, cracks, splits, or damage that could cause air leaks. Tighten loose air lines and fittings. **k.** Lubricate spring trunnion bearings every 3,000 mi. (4,800 km) or 3 months, whichever occurs first.	

LUBRICANTS		EXPECTED TEMPERATURES		
		ABOVE 15°F (ABOVE -9°C)	+40°F TO -15°F (+4°C TO -26°C)	+40°F TO -65°F (+4°C TO -54°C)
G A A Grease, automotive and artillery (MIL-G-10924)		All temperatures		

| | | [RPL] | l. Visually inspect spring leaves, retaining clips, and center screws for cracks, breaks, and security of mounting. Replace spring leaves, retaining clips, or center screws if cracked or broken (para. 3-167). Secure mounting if loose.
 m. Inspect torque rods.
    l. Visually inspect rubber for cracks, splits, or out-of-round condition. Replace torque rods if rubber is cracked or out-of-round (para. 3-170).
    2. Check torque rod nuts for tightness. For inspection purposes, torque rod nuts should be 350-700 lb-ft (475-949 N•m). If loose, tighten torque rod nuts 350-400 lb-ft (475-542 N•m). Continue to tighten until cotter pins can be installed.
 n. Inspect both front and rear wearpads for wear. Replace if wearpads are rubbing the spring bracket (para. 3-168).
 o. Tighten nuts on spring U-bolts 300-400 lb-ft (407-542 N•m).
 CAUTION
 Clean breathers and axle around breathers before servicing to prevent damage to axle from contamination. | |

Table 2-1. Unit Preventive Maintenance Checks and Services (Contd).

ITEM NO.	INTERVAL	ITEM TO CHECK/ SERVICE	PROCEDURE	NOT FULLY MISSION CAPABLE IF:
36	Semi-annual	Rear wheels and hubs, and suspension (Contd)	**p.** Remove, clean, and lubricate axle breathers every 1,000 mi. (1,600 km) or monthly, whichever occurs first.	

LUBRICANTS	EXPECTED TEMPERATURES		
	ABOVE 15°F (ABOVE -9°C)	+40°F TO -15°F (+4°C TO -26°C)	+40°F TO -65°F (+4°C TO -54°C)
OE/HDO Lubricating oil, internal combustion engine (MIL-L-2104)	OE/HDO 30	OE/HDO 10	
OEA Lubricating oil, internal combustion engine (arctic) (MIL-L-46167)			OEA

	[RPL]	**q.** Remove fill plug (3) on forward-rear and rear-rear differential (1) and check oil level in differential (1) every 3,000 mi. (4,800 km) or 3 months, whichever occurs first. If necessary, fill differential (1). Level in differential (1) should be within 1/2 in. (12.7 mm) from hole of fill plug (3) when oil is cold, and to hole of fill plug (3) when hot.	

LUBRICANTS	EXPECTED TEMPERATURES		
	ABOVE 15°F (ABOVE -9°C)	+40°F TO -15°F (+4°C TO -26°C)	+40°F TO -65°F (+4°C TO -54°C)
Oil, lubricating, multipurpose (MIL-L-2105) GO-80/90	GO-80/90	GO-80/90	
Oil, lubricating, multipurpose (MIL-L-2105) GO-75			GO-75

	[RPL]	**r.** Inspect differential drainplug (2) and fill plug (3) for tightness and signs of leakage. Tighten drainplug (2) 35-60 lb-ft (47-81 N•m) and fill plug (3) 80-135 lb-ft (108-183 N•m). **s.** Inspect differential seals for leaks. **t.** On M939A2 series vehicles, replace filter and install wheel valves (para. 3-458).	

Table 2-1. Unit Preventive Maintenance Checks and Services (Contd).

ITEM NO.	INTERVAL	ITEM TO CHECK/ SERVICE	PROCEDURE	NOT FULLY MISSION CAPABLE IF:
37	Semi-annual	Towing pintle and glad hand connections	Lubricate towing pintle every 3,000 mi. (4,800 km) or 3 months, whichever occurs first.	

LUBRICANTS		EXPECTED TEMPERATURES		
		ABOVE 15°F (ABOVE -9°C)	+40°F TO -15°F (+4°C TO -26°C)	+40°F TO -65°F (+4°C TO -54°C)
GAA Grease, automotive and artillery (MIL-G-10924)		All temperatures		

ITEM NO.	INTERVAL	ITEM TO CHECK/ SERVICE	PROCEDURE	NOT FULLY MISSION CAPABLE IF:
38	Semi-annual	Glad hand and air lines	**a.** Inspect emergency and service air and dummy couplings for serviceability and tightness of seal. Replace emergency or service air or dummy coupling(s) that are broken, bent, cracked, or have seals that leak (para. 3-208). **b.** Inspect emergency and service air lines, fittings, and shutoff valves for security of mounting, tightness of connections, and damage that could cause air leaks (para. 3-208).	
39	Semi-annual	Rear lighting and trailer receptacle	Inspect rear lights, trailer receptacle, and wiring for damage. Repair if damaged (para. 3-131).	
40	Semi-annual	Special body items	**a.** For M927/A1/A2 and M928/A1/A2 vehicles, lubricate propeller shaft center bearing every 3,000 mi. (4,800 km) or 3 months, whichever occurs first.	

LUBRICANTS		EXPECTED TEMPERATURES		
		ABOVE 15°F (ABOVE -9°C)	+40°F TO -15°F (+4°C TO -26°C)	+40°F TO -65°F (+4°C TO -54°C)
GAA Grease, automotive and artillery (MIL-G-10924)		All temperatures		

WARNING

Support weight of dump body on safety braces when performing maintenance on hoist mechanism with dump body in raised position.

b. For M929/A1/A2 and M930/A1/A2 model vehicles:

l. Lubricate safety latch, lifting arm, cylinder crosshead, body hinge pins, lifting arm rollers, and trunnion pins every 3,000 mi. (4,800 km) or 6 months, whichever occurs first.

Table 2-1. Unit Preventive Maintenance Checks and Services (Contd).

ITEM NO.	INTERVAL	ITEM TO CHECK/ SERVICE	PROCEDURE	NOT FULLY MISSION CAPABLE IF:
40	semi-annual	Special body items (Contd)	2. Lubricate tailgate bearings every 1,000 mi. (1,600 km) or monthly, whichever occurs first.	

LUBRICANTS		EXPECTED TEMPERATURES		
		ABOVE 15°F (ABOVE -9°C)	+40°F TO -15°F (+4°C TO -26°C)	+40°F TO -65°F (+4°C TO -54°C)
GAA Grease, automotive and artillery (MIL-G-10924)		All temperatures		

c. For M931/A1/A2 and M932/A1/A2 vehicles:

    1. Lubricate bushing pins, coupler jaw pin, and walking beam every 1,000 mi. (1,600 km) or monthly, whichever occurs first.

    2. Lubricate lock plunger shaft every 1,000 mi. (1,600 km) or monthly, whichever occurs first.

LUBRICANTS		EXPECTED TEMPERATURES		
		ABOVE 15°F (ABOVE -9°C)	+40°F TO -15°F (+4°C TO -26°C)	+40°F TO -65°F (+4°C TO -54°C)
OE/HDO Lubricating oil, internal combustion engine (MIL-L-2104)		OE/HDO 30	OE/HDO 10	
OEA Lubricating oil, internal combustion engine (arctic) (MIL-L-46167)				OEA

d. For M934/A1/A2 model vehicles:

    1. Perform a general inspection of van body (para. 2-15).

    2. Inspect and operate heater, air conditioner, and ventilators to ensure proper operation (TM 9-2320-272-10).

    3. Lubricate propeller shaft center bearing every 3,000 mi. (4,800 km) or 3 months, whichever occurs first.

    4. Lubricate ratchets and pawl plungers every 6,000 mi. (9,600 km) or 6 months, whichever occurs first.

LUBRICANTS		EXPECTED TEMPERATURES		
		ABOVE 15°F (ABOVE -9°C)	+40°F TO -15°F (+4°C TO -26°C)	+40°F TO -65°F (+4°C TO -54°C)
GAA Grease, automotive and artillery (MIL-G-10924)		All temperatures		

Table 2-1. Unit Preventive Maintenance Checks and Services (Contd).

ITEM NO.	INTERVAL	ITEM TO CHECK/ SERVICE	PROCEDURE	NOT FULLY MISSION CAPABLE IF:
40	semi-annual	Special body items (Contd)	**e.** For M936/A1/A2 model vehicles: 1. Test rear winch automatic brake for proper operation (para. 3-331). **2.** Unwind cable completely and inspect for kinks, frays, and wear (TM 9-2320-272-10). **3.** Lubricate upper, lower, and vertical cable guide rollers every 3,000 mi. (4,800 km) or 6 months, whichever occurs first. If operation is frequent, continuous, or under severe conditions, service weekly.	

LUBRICANTS	EXPECTED TEMPERATURES		
	ABOVE 15°F (ABOVE -9°C)	+40°F TO -15°F (+4°C TO -26°C)	+40°F TO -65°F (+4°C TO -54°C)
GAA Grease, automotive and artillery (MIL-G-10924)	All temperatures		

4. Lubricate winch and hoist cables every 6,000 mi. (9,600 km) or 6 months, whichever occurs first. Unwind entire cable; soak and clean and soak with new OE/HDO 30. Wipe off excess and coat cable and drum with lubricating oil before rewinding (TM 9-2320-272-10). If operation is frequent, continuous, or under severe conditions, service weekly.

LUBRICANTS	EXPECTED TEMPERATURES			
	ABOVE +8O°F(+27°C)	+80°F TO +30°F (27°C TO -1°C)	+30°F TO -30°F (-1°C TO -34°C)	-30" TO -65" (-34" TO -54°C)
C W Lubricating oil (VV-L-751)	CW-11C	CW-11B	CW-11A	
Oil lubricating, multipurpose (MIL-L-2105) GO-75				GO-75

5. Lubricate hoist cable sheaves, boom hinge pins, boom cylinder pins, boom sheave pins, block sheave pins, block hook, cable guide rollers, boom winch drum shaft bearings, and tumable gear every 3,000 mi. (4,800 km) or 6 months, whichever occurs first. If operation is frequent, continuous, or under severe conditions, service weekly. Rotate turntable through full range of travel while lubricating turnable gear.

LUBRICANTS	EXPECTED TEMPERATURES		
	ABOVE 15°F (ABOVE -9°C)	+40°F TO -15°F (+4°C TO -26°C)	+40°F TO -65°F (+4°C TO -54°C)
Oil, lubricating, multipurpose (MIL-L-2105) GO-80/90	GO-80/90	GO-80/90	
Oil, lubricating, multipurpose (MIL-L-2105) GO-75			GO-75

Table 2-1. Unit Preventive Maintenance Checks and Services (Contd).

ITEM NO.	INTERVAL	ITEM TO CHECK/ SERVICE	PROCEDURE	NOT FULLY MISSION CAPABLE IF:
40	Semi-annual	Special body items (Contd)	**6.** Lubricate level wind frame, end frame bearing, and sheave frame pin bearing every 3,000 mi. (4,800 km) or 6 months, whichever occurs first. If operation is frequent, continuous, or under severe conditions, service weekly. **7.** Lubricate tensioner sheave pins, level wind sheave bearing, and level wind trolley wheels every 1,000 mi. (1,600 km) or monthly, whichever occurs first.	

LUBRICANTS	EXPECTED TEMPERATURES		
	ABOVE 15°F (ABOVE -9°C)	+40°F TO -15°F (+4°C TO -26°C)	+40°F TO -65°F (+4°C TO -54°C)
GAA Grease, automotive and artillery (MIL-G-10924)	All temperatures		

8. Lubricate level wind trolley lock and tensioner rocker lever pins every 1,006 mi. (1,600 km) or monthly, whichever occurs first.

LUBRICANTS	EXPECTED TEMPERATURES		
	ABOVE 15°F (ABOVE -9°C)	+40°F TO -15°F (+4°C TO -26°C)	+40°F TO -65°F (+4°C TO -54°C)
OE/HDO Lubricating oil, internal combustion engine (MIL-L-2104)	OE/HDO 30	OE/HDO 10	
OEA Lubricating oil, internal combustion engine (arctic) (MIL-L-46167)			OEA

9. Lubricate turntable bearings every 6,000 mi. (4,800 km) or 6 months, whichever occurs first. Rotate turntable through full range of travel while lubricating turntable bearings.

LUBRICANTS	EXPECTED TEMPERATURES		
	ABOVE 15°F (ABOVE -9°C)	+40°F TO -15°F (+4°C TO -26°C)	+40°F TO -65°F (+4°C TO -54°C)
GAA Grease, automotive and artillery (MIL-G-10924)	All temperatures		

Table 2-1. Unit Preventive Maintenance Checks and Services (Contd).

ITEM NO.	INTERVAL	ITEM TO CHECK/ SERVICE	PROCEDURE	NOT FULLY MISSION CAPABLE IF:
40	Semi-annual	Special body items (Contd)	**10.** Drain and fill hoist gearcase with new oil every 6,000 mi. (9,600 km) or 6 months, whichever occurs first. Remove level plugs, fill plugs, drainplugs, and drain gearcases. Install drainplugs and fill to level plugs. Install level plugs and fill plug.	

LUBRICANTS		EXPECTED TEMPERATURES		
		ABOVE 15°F (ABOVE -9°C)	+40°F TO -15°F (+4°C TO -26°C)	+40°F TO -65°F (+4°C TO -54°C)
Oil, lubricating, multipurpose (MIL-L-2105) GO-80/90		GO-80/90	GO-80/90	
Oil, lubricating, multipurpose (MIL-L-2105) GO-75				GO-75

			11. Check hydraulic oil reservoir for proper level every 1,000 mi. (1,600 km) or monthly, whichever occurs first. Fill as required.	

LUBRICANTS		EXPECTED TEMPERATURES		
		ABOVE 15°F (ABOVE -9°C)	+40°F TO - 15°F (+4°C TO -26°C)	+40°F TO -65°F (+4°C TO -54°C)
OE/HDO Lubricating oil, internal combustion engine (MIL-L-2104)		OE/HDO 30	OE/HDO 10	
OEA Lubricating oil, internal combustion engine (arctic) (MIL-L-46167)				OEA

ITEM NO.	INTERVAL	ITEM TO CHECK/ SERVICE	PROCEDURE	NOT FULLY MISSION CAPABLE IF:
41	Semi-annual	Rifle mounting kit	a. Check rifle top and lower mounts for looseness, binding, and damage. b. Check handle for excessive looseness and damage.	
42	Semi-annual	Machine gun mount	Refer to TM 9-1005-245-14 for Preventive Maintenance Checks and Services (PMCS).	
43	Semi-annual	M-8 chemical alarm	Refer to TM 3-6665-225-12 for PMCS.	
44	Semi-annual	M-11 decontamination unit	Refer to TM 3-4230-204-12&P for PMCS.	

Table 2-1. Unit Preventive Maintenance Checks and Services (Contd).

ITEM NO.	INTERVAL	ITEM TO CHECK/ SERVICE	PROCEDURE	NOT FULLY MISSION CAPABLE IF:
			FINAL ROAD TEST	
			After all services and inspections have been completed, perform a short road test to ensure all corrections have been implemented. Correct any defects or malfunctions that may occur during this test.	
			ANNUAL INSPECTION	
			NOTE	
			• Lubricate vehicle in accordance with LO 9-2320-272-12.	
			• Perform all semiannual checks listed in this table.	
45	Annual	Front end	a. Check front end alignment with toe-in gauge. Correct toe-in is 1/16-3/16 in. (1.588-4.763 mm). When toe-in is correct, tighten crossshaft screws and nuts 35-55 lb-ft (47-75 N•m) (para. 3-152).	
			b. Inspect axle housings and differentials for cracks.	
			c. Inspect shock absorbers and mounting brackets for looseness, wear, cracks, serviceability, and leaks. Replace leaking shock absorbers if more than a class I leak is detected (para. 3-166).	
			d. Check each tire using tire depth gauge. Tread depth should be at least 1/8 in. (3 mm) or as indicated on tire depth gauge.	
46	Annual	Engine compart- ment and cab	**a.** Inspect front cab mounting brackets for security, wear, cracks, splits, broken welds, loose bushings, and missing screws. Replace front cab mounting brackets if worn, cracked, split, or welds are broken. Replace loose insulators and missing screws (para. 3-304).	
			b. Inspect rear cab mounting brackets for security, wear, cracks, splits, broken welds, loose bushings, and missing screws. Replace front cab mounting brackets if worn, cracked, split, or welds are broken. Replace loose bushings and missing screws (para. 3-305).	
			c. Tighten front cab mounting screws.	
			d. Check the following items on M939A2 vehicles:	
			1. Inspect front and rear engine mounting brackets for security, wear, cracks, splits, broken welds, loose bushings, and missing bolts (para. 2-15).	

Table 2-1. Unit Preventive Maintenance Checks and Services (Contd).

ITEM NO.	INTERVAL	ITEM TO CHECK/ SERVICE	PROCEDURE	NOT FULLY MISSION CAPABLE IF:
46	Annual	Engine compartment and cab (Contd)	**2.** Check left-side engine mounting locknuts (1) and right-side engine mounting screws (6) for looseness. If loose, tighten 120-140 lb-ft (163-190 N•m). **3.** Check two upper locknuts (4) securing left engine mount (2). If loose, tighten 90 lb-ft (122 N•m). Check two lower locknuts (3) for looseness. If loose, tighten 60 lb-ft (82 N•m). **4.** Check two upper locknuts (7) securing right engine mount (5) for looseness. If loose, tighten 90 lb-ft (122 N•m). Check two lower locknuts (8) for looseness. If loose, tighten 60 lb-ft (81 N•m).	
47	Annual	Rear suspension	a, Inspect forward-rear and rear-rear axle housings and differentials for cracks.	

WARNING

Do not allow grease or oil to contact brake linings. Linings can absorb grease and oil, causing early glazing and very poor braking action. Failure to comply could cause serious injury or death to personnel.

b. Tighten forward-rear and rear-rear axle drive flange screws 80-105 lb-ft (108-142 N•m).

c. Ensure all differential companion flange mounting screws and U-joint mounting screws are tight. Tighten companion flange mounting screws 30-40 lb-ft (41-54 N•m) and U-joint mounting screws 90-110 lb-ft (122-149 N•m).

d. Test spring seat bearing free play by placing jack under spring seat bracket. Raise vehicle so spring moves freely up and down in guide brackets. Put prybar between U-bolt saddle and lift pin. Pull up on prybar. If there is free play, adjust spring seat bearing (para. 3-169).

Table 2-1. Unit Preventive Maintenance Checks and Services (Contd).

ITEM NO.	INTERVAL	ITEM TO CHECK/ SERVICE	PROCEDURE	NOT FULLY MISSION CAPABLE IF:
48	Annual	Com-pressed air system	Check for faulty check valves at the primary and secondary air tanks and crossed hoses. Crossed hoses occur from A treadle valve air supply hoses at the primary and secondary air tanks, and B primary and secondary air gauge lines at the treadle valve.	

WARNING

- Chock vehicle wheels. Failure to do so may result in injury or death to personnel.
- Wear safety goggles while performing inspection. Line is under pressure. Failure to do so may result in injury to personnel from flying parts or debris.

Build up air pressure in primary and secondary air systems to 50-60 psi (344.75-413.70 kPa). Trace the air line from wet tank (3) to drainvalve (5). Open the drainvalve (5).

Air pressure drops to zero in either primary (6) or secondary (2) air tanks. Replace faulty check valve(s) (1) at affected tank(s).

Air systems maintain pressure.

Build up pressure in the primary and secondary air system to 50-60 psi (344.75-413.70 kPa). Remove air gauge line at the primary (lower) air gauge port of the treadle valve (7). Do not allow air to bleed out slowly as this may affect results. Observe air gauges.

Secondary air gauge drops to zero.

Primary air gauge drops to zero.

Follow air line from primary air tank (6) to drainvalve (4). Open drainvalve (4).

Follow air line from primary air tank (6) to drainvalve (4). Open drainvalve (4).

Air pressure released when drainvalve (4) is opened.

No air pressure released when drain-valve (4) is opened.

Air pressure released when drainvalve (4) is opened.

No air pressure released when drain-valve (4) is opened. Air lines are installed properly. Inspection is complete. Reconnect lower air gauge line at treadle valve (7). See **View D** below.

Switch and connect the primary and secondary air gauge lines at the treadle valve (7). See **View B** below.

Switch and connect treadle supply hoses at the primary (6) and secondary (2) air tanks; and the primary and secondary air gauge lines at the treadle valve (7). See **View A** below.

Switch and connect treadle valve supply hoses at primary (6) and secondary (2) air tanks. Reconnect lower air gauge line at treadle valve (7). See **View C** below.

VIEW A

VIEW B

VIEW C

VIEW D (NORMAL)

Table 2-1. Unit Preventive Maintenance Checks and Services (Contd).

ITEM NO.	INTERVAL	ITEM TO CHECK/ SERVICE	PROCEDURE	NOT FULLY MISSION CAPABLE IF:

Table 2-1. Unit Preventive Maintenance Checks and Services (Contd).

ITEM NO.	INTERVAL	ITEM TO CHECK/ SERVICE	PROCEDURE	NOT FULLY MISSION CAPABLE IF:
49	Annual	Wiring harnesses and compressed air lines	Inspect spare tire carrier for security, completeness of assembly, and proper operation.	
50	Annual	Spare tire carrier	Inspect starter mounting bolts and starter wiring for corrosion and loose connections. Tighten starter mounting bolts 55 lb-ft (75 N•m).	
51	Annual	Parking brake	Inspect brake lever travel (para. 3-172). Replace brakeshoes if brake lever travel is greater than 2 in. (5.1 cm) (para. 3-180).	
52	Annual	Air dryer system	**a.** Inspect air dryer, two purge valves, and check valve for security of mounting and signs of damage that could cause leaks.	
			NOTE	
			Air dryer will whistle when filter needs to be replaced.	
		[RPL]	**b.** Replace filter in air dryer (3-465).	
			c. Inspect filter in water separator (para. 3-466).	
			d. Inspect all tubes and fittings for damage or cracks that could cause leaks.	
53	Annual	Towing pintle	Check operation of towing pintle hook. Inspect pintle and bracket for cracks, breaks, wear, and play of 0.003-0.017 in. (0.07-0.43 mm) (para. 3-241 or 3-242).	

Table 2-1. Unit Preventive Maintenance Checks and Services (Contd).

ITEM NO.	INTERVAL	ITEM TO CHECK/ SERVICE	PROCEDURE	NOT FULLY MISSION CAPABLE IF:
54	Annual	Special body items	a. M929/A1/A2 and M930/A1/A2.	

a. M929/A1/A2 and M930/A1/A2.

WARNING

- Support weight of dump body on safety braces when performing maintenance on hoist mechanism with dump body in raised position.
- Do not drain oil from dump hydraulic pump, reservoir, or cylinder or remove lines containing oil when dump body is supported by hydraulic cylinder. Doing so may result in injury to personnel.
- Do not drain oil from dump hydraulic pump, reservoir, or cylinder or remove lines containing oil when oil is hot. Doing so may result in injury to personnel.
- Do not remove hoses while engine is running or start engine with hoses removed. High-pressure fluids may cause hoses to whip violently and spray randomly. Failure to do so may result in injury to personnel.
- Accidental or intentional introduction of liquid contaminants into the environment is in violation of state, federal, and military regulations. Refer to Army POL (para. 1-8) for information concerning storage, use, and disposal of these liquids. Failure to do so may result in injury or death.

1. Inspect dump body for completeness of assembly. Ensure dump body is aligned with frame.

2. Inspect dump hydraulic lines, hoses, and fittings for leaks and splits. Check hydraulic pump for tightness of mounting.

3. Inspect transmission PTO, hydraulic pump propeller shaft, and hydraulic pump for tight mounting and leaks.

4. Ensure tailgate control rod hand lever locks and unlocks tailgate lower latch. Inspect control linkage for security, bends, and binding.

5. Operate dump body and observe for smooth raising and lowering of dump body.

6. With dump body raised, inspect cylinder piston rods for scoring and wear.

Table 2-1. Unit Preventive Maintenance Checks and Services (Contd).

ITEM NO.	INTERVAL	ITEM TO CHECK/ SERVICE	PROCEDURE	NOT FULLY MISSION CAPABLE IF:
4	Annual	Special body items (Contd) [RPL]	**CAUTION** Remove filler plug slowly to release pressure. Do not overfill. **7.** Drain, clean, and fill winch hydraulic oil reservoir and replace filter (para. 3-336) every 12 months.	

LUBRICANTS	EXPECTED TEMPERATURES		
	ABOVE 15°F (ABOVE -9°C)	+40°F TO -15°F (+4°C TO -26°C)	+40°F TO -65°F (+4°C TO -54°C)
OE/HDO Lubricating oil, internal combustion engine (MIL-L-2104)	OE/HDO 30	OE/HDO 10	
OEA Lubricating oil, internal combustion engine (arctic) (MIL-L-46167)			OEA

| | | | **b.** On M934/A1/A2 model vehicles:

    1. Expand van body and inspect for damaged or broken hinges, fasteners, and latches (para. 2-15).

    2. Check for bent or binding of components or damaged seals or panels that would allow light to shine through or prevent a tight seal. Repair or replace defective components (para. 2-15).

    3. Lubricate worm gear, worm, winch handle shaft, wire rope, winch barrel shaft, and swing davit base every 12,000 mi. (19,000 km) or 12 months, whichever occurs first. | |

LUBRICANTS	EXPECTED TEMPERATURES			
	ABOVE +80°F(+27°C)	+80°F TO +30°F (27°C TO -1 "C)	+30°F TO -30°F (-1°C TO -34°C)	-30° TO -65' (-34° TO -54°C)
CW Lubricating oil (VV-L-751) Oil lubricating, multipurpose (MIL-L-2105): GO-75	CW-11C	CW-11B	CW-11A	GO-75

| | | | **4.** Lubricate support rollers, end rollers, retractor beams, sprocket, and sprocket bushings every 12,000 mi. (19,606 km) or 12 months, whichever occurs first. | |

LUBRICANTS	EXPECTED TEMPERATURES		
	ABOVE 15°F (ABOVE -9°C)	+40°F TO -15°F (+4°C TO -26°C)	+40°F TO -65°F (+4°C TO -54°C)
GM Grease, automotive and artillery (MIL-G-10924)	All temperatures		

Table 2-1. Unit Preventive Maintenance Checks and Services (Contd).

ITEM NO.	INTERVAL	ITEM TO CHECK/ SERVICE	PROCEDURE	NOT FULLY MISSION CAPABLE IF:
54	Annual	Special body items (Contd)	**c.** On M936/A1/A2 model vehicles: **WARNING** • Do not drain oil from wrecker hydraulic pump, reservoir, cylinders, or motors, or remove lines containing oil when oil is hot. Doing so may result in injury to personnel. • Do not remove hoist or winch hydraulic hoses when engine is running or start engine with hoist or winch hydraulic hoses removed. High pressure fluids may cause hoses to whip violently and spray randomly. Failure to do so may result in injury to personnel. • Do not drain oil from wrecker hydraulic pump, reservoir, or cylinders or remove lines containing oil when crane boom is supported by hydraulic cylinder, or hoist or winch cables are supporting a load. Doing so may result in injury to personnel. • Accidental or intentional introduction of liquid contaminants into the environment is in violation of state, federal, and military regulations. Refer to Army POL (para. 1-8) for information concerning storage, use, and disposal of these liquids. Failure to do so may result in injury or death. **1.** Inspect rear winch and winch controls for loose mounting and broken or missing parts. If mounting is loose, tighten. If parts are missing or broken, replace (para. 3-332). **2.** Inspect hydraulic pump for leaks. If leaks are present, notify DS maintenance. **3.** Test automatic hoist drum brake for proper operation (para. 3-323). **4.** With boom raised, inspect crane cylinder piston rods for bends and scoring. If crane cylinder piston rods are bent or scored, notify DS maintenance. **5.** While operating crane, observe that fuel pump governor is maintaining 1,250-1,300 rpm during hoisting operation. Notify DS maintenance if engine is surging erratically (TM 9-2320-272-10). **6.** Inspect and clean hydraulic tank swing motor and hoist crane motor breather caps (para. 2-15). **7.** Inspect reservoir filter case for leaks. Replace filter element (para. 3-336). **8.** Drain, clean, and fill hydraulic oil reservoir every 12 months (LO 9-2320-272-12).	

Table 2-1. Unit Preventive Maintenance Checks and Services (Contd).

ITEM NO.	INTERVAL	ITEM TO CHECK/ SERVICE	PROCEDURE	NOT FULLY MISSION CAPABLE IF:
55	Biennial	Trans- mission	**NOTE** Perform item 55 only if AOAP laboratory support is not available. Change transmission oil, and internal, external, and governor oil filters every 24,000 mi. (38,000 km) or 24 months, whichever occurs first (para. 3-133).	

LUBRICANTS	EXPECTED TEMPERATURES		
	ABOVE 15°F (ABOVE -9°C)	**+40°F TO -15°F (+4°C TO -26°C)**	**+40°F TO -65°F (+4°C TO -54°C)**
OE/HDO Lubricating oil, internal combustion engine (MIL-L-2104)	OE/HDO 30	OE/HDO 10	
O E A Lubricating oil, internal combustion engine (arctic) (MIL-L-46167)			OEA

ITEM NO.	INTERVAL	ITEM TO CHECK/ SERVICE	PROCEDURE	NOT FULLY MISSION CAPABLE IF:
56	Biennial	Transfer case	Inspect tightness of transfer case mounting bolts and mounting brackets. Tighten three transfer case mounting bracket-to-frame bolts 50-60 lb-ft (68-81 N•m) Tighten seven transfer case-to-mounting bracket bolts 125-135 lb-ft (170-183 N•m).	
57	Biennial	Special body items	**a.** On M929/A1/A2 and M930/A1/A2 model vehicles, tighten all dump body mounting bolts 240 lb-ft (325 N•m). **b.** On M931/A1/A2 and M932/A1/A2 model vehicles, inspect fifth wheel for completeness of assembly. Ensure mounting screws are tightened 160-170 lb-ft (217-231 N•m).	

PMCS MANDATORY REPLACEMENT PARTS LIST

The following is a list of parts required when performing semiannual PMCS. The semiannual parts list contains the mandatory replacement parts for one semiannual PMCS. The annual parts list contains the mandatory replacement parts for one semiannual PMCS and the peculiar replacement parts for one annual PMCS.

SEMIANNUAL (6,000 MILE) PMCS MANDATORY REPLACEMENT PARTS LIST

	ITEM NO.	PART NUMBER	NSN	NOMENCLATURE	QTY	STEP NUMBER
	1	MS35338-46	5310-00-722-5658	Lockwasher	10	11a
*	2	MS35338-48	5310-00-003-4094	Lockwasher	20	19d
*	3	7979374		Seal, plain, encased	2	19d
	4	MS35335-33	5310-00-209-0786	Lockwasher	20	19d
	5	10935405	5340-00-450-57 18	Protective caps	A/R	19d
	6	1229-S-513-C	5310-01-062-3384	Lockwasher	8	19d
* **	7	5139123	5310-00-700-7069	Keyway insert	2	19d
**	8	1229-U-1009	5310-00-321-9974	Keyway insert	2	19d
**	9	MS28775-206	5330-01-133-5858	O-ring	2	19d
**	10	A-1205-Z-2132	5330-01-271-9410	Seal, plain, encased	2	19d
**	11	2297-N-5630	4730-01-272-0582	Seal, conical, flared	2	19d
**	12	2208-S-1033	5330-01-272-1148	Gasket	2	19d
**	13	A-1205-N-2120	5330-01-272-1147	Retainer, oil seal (air seals)	4	19d
**	14	599791	4460-01-284-2344	Filter element, fluid	2	19e
	15	MS35769-21	5330-01-352-7768	Gasket	1	22g
	16	C2AZ9324C	4720-00-845-0630	Hose, nonmetallic	1	26e
***	17	154088	5330-00-013-7806	Seal, cap	1	27a
*	18	146483	5310-00-291-9490	Filter element	1	27a
****	19	145504	5330-01-051-4243	Packing, preformed	1	27a
*	20	256476	2910-00-152-2033	Kit, fluid pressure	1	27c
**	21	3313281	2940-01-157-6309	Filter element, fluid	1	27c
**	22	FF-5079	4330-01-309-6189	Filter, fluid, in-line	1	27e3
**	23			O-ring (wheel valve-to-axle)	4	36c
**	24	A3994-1	4720-01-279-3034	O-ring (wheel valve filter)	4	36c
**	25	131245		Locknut	8	36c
**	26	10947447-2		Locknut	40	36c
**	27	A-1205-Z-2132	5330-01-271-9410	Oil seal, axle	4	36c
**	28	1229-U-1009	5310-00-321-9974	Keyway insert	4	36c
**	29	A-1205-N-2120	5330-01-272-1147	Retainer, oil seal (rear air seals)	4	36c
**	30	A-1205-D-2162	5330-01-308-0175	Retainer, oil seal (front air seals)	4	36c
**	31	3286-P-1056	2530-01-285-3563	Slinger ring	4	36c
*	32	10947447		Locknut	40	36c
*	33	5139123	5310-00-700-7089	Keyway insert	4	36c
*	34	741344		Seal assembly	4	36c
*	35	12375801	5330-01-444-8350	Seal assembly	4	36c
*	36	7409553	2590-00-740-9553	Wiper	4	36c
	37	MS24665-500	5315-00-187-9567	Cotter pin	12	36m2
	38	MS35769-21	5330-01-352-7768	Gasket	4	36q
**	39	599791	4460-01-284-2344	Filter element, fluid	4	36t

ANNUAL (12,000 MILE) PMCS MANDATORY REPLACEMENT PARTS LIST
M929/A1/A2 and M930/A1/A2

	ITEM NO.	PART NUMBER	NSN	NOMENCLATURE	QTY	STEP NUMBER
*****	40	PF297		Filter element	1	7

* M939/A1   ** M939A2   *** M936/A1   **** M939/A1 only (except M936/A1)

2-13. GENERAL MAINTENANCE PROCEDURES INDEX

2-14. CLEANING

a. **General Instructions.** Cleaning procedures are the same for the majority of parts and components which compose the vehicle subassemblies. General cleaning procedures are detailed in steps b. through o.

WARNING

Improper cleaning methods and use of unauthorized cleaning solvents may result in injury to personnel.

CAUTION

When cleaning any CTIS components or parts, special care must be taken not to contaminate the system's mating surfaces, tubes, hoses, or passages. Damage to components may result.

b. **The Importance of Cleaning.** Great care and effort are required in all cleaning operations. The presence of dirt and foreign material is a constant threat to satisfactory vehicle operation and maintenance. The following instructions will apply to all cleaning operations.

CAUTION

Keep all related parts and components together. Do not mix parts. Failure to comply may result in damage to parts.

(1) Clean all parts before inspection, after repair, and before assembly.

(2) Hands must be kept free of any accumulation of grease which can collect dust and grit.

(3) After cleaning, all parts must be covered or wrapped in plastic or paper to protect them from dust and/or dirt.

c. **Disassembled Parts Cleaning.** Place all disassembled parts in wire baskets for cleaning.

(1) Dry and cover all cleaned parts.

(2) Place on or in racks and hold for inspection or repair.

(3) All parts subject to rusting must be lightly oiled and wrapped.

d. **Castings.**

WARNING

Drycleaning solvent is flammable and will not be used near open flame. Use only in well-ventilated places and keep tire extinguisher nearby. Failure to do so may result in injury to personnel.

(1) Clean all inner and outer areas subject to grease and oil with drycleaning solvent (TM 9-247).

(2) Use a stiff brush to remove sludge and gum deposits.

2-14. CLEANING (Contd)

WARNING

Compressed air source will not exceed 30 psi (207 kPa). When cleaning with compressed air, eyeshields must be worn. Failure to do so may result in injury to personnel.

(3) Use compressed air to blow out all tapped screw holes and to dry castings after cleaning.

e. External Engine Cleaning. All electrical equipment and other parts that could be damaged by steam-cleaning or moisture must be removed and all openings must be covered before cleaning. Dry with compressed air.

f. Oil Passages. Particular attention must be given to all oil passages in castings and machined parts. Oil passages must be clean and free of any obstructions.

(1) Clean passages with wire probes to break up any sludge or gum deposits.

(2) Wash passages by flushing with solvent (TM 9-247).

WARNING

Compressed air source will not exceed 30 psi (207 kPa). When cleaning with compressed air, eyeshields must be worn. Failure to do so may result in injury to personnel.

(3) Dry passages with compressed air.

g. Seals, Electrical Cables, and Flexible Hoses. Clean with soap and water.

h. Bearings.

(1) Bearings require special cleaning. After removing surface oil and gum deposits, wipe bearings dry; do not use compressed air. After cleaning, coat bearings with oil, wrap, and hold for inspection.

(2) Refer to TM 9-214 for more information on care of bearings.

WARNING

• Drycleaning solvent is flammable and will not be used near open flame. Use only in well-ventilated places and keep fire extinguisher nearby. Failure to do so may result in injury to personnel.

• Compressed air source will not exceed 30 psi (207 kPa). When cleaning with compressed air, eyeshields must be worn. Failure to do so may result in injury to personnel.

i. Machined Tooled Parts. Clean with drycleaning solvent and dry with compressed air.

j. Machined Surfaces. Clean with drycleaning solvent and dry with lint-free cloth.

WARNING

Eyeshields must be worn when cleaning with a wire brush. Flying rust and metal particles may cause injury to personnel using a wire brush.

k. Mated Surfaces. Remove old gasket and/or sealing compound using wire brush and drycleaning solvent.

NOTE

All parts subject to rusting must be lightly oiled and wrapped before being stored.

l. Rusted Surfaces. Clean all rusted surfaces using wire brush and crocus cloth.

m. Oil-bathed Internal Parts. Wipe clean with lint-free cloth.

n. Air-actuated Internal Parts. Wipe clean with lint-free cloth.

o. External Exposed Parts. Wash with soap and water. Rinse thoroughly and air-dry.

2-15. INSPECTION

a. **General Instructions.** Procedures for inspections will be the same for many parts and components which make up the vehicle subassemblies. General procedures are detailed in steps b. through q. Dimensional standards for parts have been fixed at extremely close tolerances, so use specification tables. Use specified inspection equipment for inspection where cracks and other damage cannot be spotted visually. Exercise extreme care in all phases of inspection.

b. **Castings.**

(1) Inspect all ferrous and nonferrous castings for cracks using a magnifying glass and strong light.

(2) Refer to MIL-L-6866, Inspection, Liquid Penetrant Methods, and MIL-L-6868, Inspection Process, Magnetic Particles.

(3) Particularly inspect areas around studs, pipe plugs, threaded inserts, and sharp corners. Replace all cracked castings.

(4) Inspect machined surfaces for nicks, burrs, or raised metal. Mark damaged areas for repair or replacement.

(5) Inspect all pipe plugs, pipe plug openings, screws, and screw openings for damaged or stripped threads. Replace or repair damaged or stripped threads.

(6) Using a straightedge or surface plate, check all gasket mating surfaces, flanges on housings, and supports for warpage. Inspect mating flanges for discolorations which may indicate persistent oil leakage. Replace damaged parts.

(7) Check all castings for conformance to applicable repair standards. Replace damaged castings.

c. **Bearings.** Refer to TM 9-214 for inspection of bearings. Check all bearings for conformance to applicable standards.

d. **Studs, Bolts, and Screws.** Replace if threads are damaged, bent, or stripped.

e. **Gears.**

(1) In strong light, inspect all gears for cracks. No cracks are allowed.

(2) Inspect gear teeth for wear, sharp edges, burrs, and galled or pitted surfaces.

(3) Check keyway slots for wear and/or damage.

f. **Oil Seals.** Oil seals are mandatory replacement items.

g. **Engine Bearings.**

NOTE

- Engine connecting rods and main bearings are serviced in sets. If one bearing fails, all bearings must be replaced.

- Old and new style engine components must be used in sets and not intermixed.

h. **Bushings and Bushing-Type Bearings.**

(1) Check all bushings and bushing-type bearings for secure fit, evidence of overheating, wear, burrs, nicks, and out-of-round condition.

(2) Check for dirt in lubrication holes or grooves. Holes and grooves must be clean and free from damage.

l. **Expansion Plugs.** Inspect for leakage. Replace plugs when leakage is present.

j. **Machined Tooled Parts.** Inspect for cracks, breaks, elongated holes, wear, and chips.

k. **Machined Surfaces.** Inspect for cracks, evidence of wear, galled or pitted surface, burrs, nicks, and scratches.

leakage.

m. **Rusted Surfaces.** Inspect for pitting, holes, and severe damage.

n. **Oil-bathed Internal Parts.** Inspect for cracks, nicks, burrs, evidence of overheating, and wear.

2-15. INSPECTION (Contd)

o. **Air-actuated Internal Parts.** Inspect for cracks, nicks, burrs, evidence of overheating, and wear.

p. **Externally Exposed Parts.** Inspect for breaks, cracks, rust damage, and wear.

q. **Rivets.** Inspect for loose, broken, and missing rivets in accordance with TM 9-450.

2-16. REPAIR

a. General Instructions. Repair of most parts and components is limited to general procedures outlined in applicable maintenance instructions and the following detailed procedures, steps b. through h.

CAUTION

Repaired items must be thoroughly cleaned to remove metal chips and abrasives to prevent them from entering working parts of vehicle. Special care must be taken with CTIS parts, or damage to components may result.

b. **Castings.**

(1) All cracked castings will be replaced.

(2) Only minor repairs to machined surfaces, flanges, and gasket mating surfaces are permitted. Remove minor nicks, burrs, and/or scratches with:

(a) Fine mill file.

(b) Crocus cloth dipped in cleaning solvent.

(c) Lapping across a surface plate.

(3) Remachining of machined surfaces to repair damage, warpage, or uneven surfaces is not permitted. Replace castings.

(4) Repair damaged threaded pipe plug and/or screw holes with a tap. Repair oversize holes with threaded inserts.

c. **Bearings.** Refer to TM 9-214.

d. **Studs.** Replace all bent and stretched studs. Repair minor thread damage with a thread die. Replace studs having stripped or damaged threads as outlined below:

(1) Remove studs using a stud remover. Back studs out slowly to avoid heat buildup and seizure which can cause stud to break off.

NOTE

If welding method is used, refer to TM 9-237 for proper instructions.

(2) If studs break off too short to use a stud remover or a stud extractor, use welding method.

(3) Broken studs can be removed by welding bar stock or a nut to stud and removing with wrench.

(4) Standard studs may have a coarse thread on one end and a fine thread on the other end. The coarse thread end is installed in aluminum casting. Studs having coarse threads on both ends are used in some applications. The shorter threaded end goes into the casting. Refer to appendix D for correct part number.

(5) Replacement studs have a special coating and must have a small amount of antiseize compound (Appendix C, Item 10) applied on the threads before stud is installed. Install replacement stud slowly to prevent heat buildup and snapping off.

e. **Gears.**

(1) Remove gears using pullers.

(2) Use the same method described in step b. for castings to remove minor nicks, burrs, or scratches on gear teeth.

2-16. REPAIR (Contd)

    (3) If keyways are worn or enlarged, replace gear.

    f. Bushings and Bushing-type Bearings. When bushings and bushing-type bearings seize to a shaft and spin in the bore, the associated part must also be replaced.

    g. Oil Sears.

      (1) Using proper oil seal removal tool, remove oil seals without damaging casting or adapter bore.

      (2) Always install new seal in bore using proper seal replacement tool.

    h. Rivets. Replace rivets in accordance with TM 9-450.

2-17. DISASSEMBLY

    a. The work area for disassembly of any item must be kept as clean as possible. This will prevent contamination of internal parts. This is especially true when working with fuel and air systems or CTIS components.

    b. All gaskets, O-rings, and seals removed during repair will be discarded and replaced with new parts. These items are usually damaged during removal. Lockwire, lockwashers, cotter pins, and like items should be discarded during disassembly.

    c. When removing gaskets, preformed packings, or seals, do not use any metal tools that will scratch the sealing surfaces next to these items.

    d. Before disassembling any item, study the illustration carefully, noting the relationship of internal parts. Knowing the details of construction will speed up disassembly and help avoid mistakes.

2-18. ASSEMBLY

    a. Cleanliness is essential in all assembly operations. Dirt and dust, even in small quantities, are abrasive. Parts must be cleaned as specified and kept clean. Wrap or cover parts and components when assembly procedures are not completed immediately.

    b. Lubricate all metal parts with lubricant used during operation (LO 9-2320-272-12).

    c. Installation of cotter pins and lockwire shall be accomplished as specified in assembly procedures.

    d. Critical torque values are specified in the assembly procedure. When not specified, tighten bolts, screws, and nuts in accordance with standard dry torque values (Appendix G).

    e. All fuel, air, and hydraulic components must be kept thoroughly clean at all times. Plug all open ports until the component is installed in the vehicle.

    f. All pressing operations should be accomplished using a suitable press and adapters unless otherwise specified.

Section IV. UNIT TROUBLESHOOTING PROCEDURES

2-19. GENERAL

WARNING

Operation of a deadlined vehicle without preliminary inspection
may cause injury to personnel and/or damage to equipment.

NOTE

If malfunction corrective action does not correct malfunction,
notify direct support maintenance.

a. This section provides information needed to diagnose and correct malfunctions of mechanical, electrical, CTIS, and compressed air and brake system at the unit level of maintenance.

b. The troubleshooting procedures in this section cannot give all the answers or correct all vehicle malfunctions encountered. However, these procedures are a step-by-step approach to a problem that directs tests and inspections toward the source of a problem and a successful solution.

c. Each malfunction symptom given for an individual component or system is followed by step(s) to determine the cause and corrective action you must take to remedy the problem.

d. Before taking any corrective action for a possible malfunction, the following rules should be followed:

(1) Question operator to obtain any additional information that might help you to determine the cause of the problem.

(2) Never overlook the chance the problem could be of a simple origin. The problem could require only a minor adjustment.

(3) Use all senses to observe and locate troubles.

(4) Use test instruments and gauges to help you determine and isolate problems.

(5) Always isolate the system where the malfunction occurs and locate the defective component.

(6) Use standard automotive theories and principles when troubleshooting the vehicles covered in this manual.

e. This section cannot list all malfunctions that may occur. If a malfunction occurs that is not listed in the tables, notify Direct Support CDS) maintenance.

2-20. UNIT TROUBLESHOOTING INDEX

2-21. MECHANICAL TROUBLESHOOTING

NOTE

If malfunction corrective action does not correct malfunction, notify direct support maintenance.

a. This section provides information to diagnose and correct malfunctions of the mechanical system at the unit maintenance level.

b. This section cannot list all mechanical malfunctions that may occur. If a malfunction occurs that is not listed in table 2-2, notify direct support maintenance.

MECHANICAL TROUBLESHOOTING SYMPTOM INDEX

MALFUNCTION NO.	MALFUNCTION	TROUBLESHOOTING PROCEDURE PAGE
	ENGINE	
1.	Engine will not crank	2-64
2.	Engine cranks but will not start.	2-64
3.	Starter cranks engine slowly, hard to start.	2-66
4.	Engine stops during normal operation	2-66
5.	Engine stops when accelerator is returned to idle position	2-66
6.	Engine fails to stop	2-66
7.	Engine misfires during operation	2-67
8.	Engine idle rough, erratic (M939A2 only).	2-67
9.	Poor acceleration and/or lack of power	2-67
10.	Engine surges	2-68
11.	Excessive engine oil loss or consumption	2-68
12.	Engine oil pressure too low or too high at normal operating temperature	2-69
13.	Engine oil contaminated	2-70
14.	Excessive vibrating or clunking	2-70
15.	Excessive fuel consumption.	2-70
	ETHER START SYSTEM	
16.	Engine cranks but will not start in cold weather (fuel system operating properly)	2-71
	EXHAUST SYSTEM	
17.	Exhaust color blue during normal operation	2-71
18.	Exhaust color white during normal operation and idle	2-72
19.	Exhaust color black during normal operation and idle.	2-72
20.	Excessive exhaust noise	2-72
21.	Exhaust fumes in cab	2-73

MECHANICAL TROUBLESHOOTING SYMPTOM INDEX

MALFUNCTION NO.	MALFUNCTION	TROUBLESHOOTING PROCEDURE PAGE

MECHANICAL TROUBLESHOOTING SYMPTOM INDEX

MALFUNCTION NO.	MALFUNCTION	TROUBLESHOOTING PROCEDURE PAGE
	WINCH	
59.	Winch does not operate.	2-84
60.	Winch operates in one direction only	2-85
61.	Drag brake does not operate.	2-86
62.	Winch will not hold load	2-86
63.	Automatic brake overheats	2-86
64.	Front winch operates at one speed only	2-86
65.	Rear winch cable tensioner will not operate.	2-86
66.	Vehicle rolls while operating rear winch.	2-87
	POWER TAKEOFF (PTO)	
67.	Excessive noise at power takeoff.	2-87
68.	Hard shifting of power takeoff	2-87
69.	Leaking lubricant at power takeoff.	2-87
	INSTRUMENTS AND GAUGES	
70.	Speedometer or tachometer noisy or erratic.	2-87
71.	Air pressure gauge inoperative.	2-88
72.	Oil pressure gauge inoperative (oil level is correct)	2-88
	RADIO INTERFERENCE SUPPRESSION	
73.	Interference while vehicle in motion.	2-88
74.	Interference only when engine is running	2-88
	FIFTH WHEEL	
75.	Trailer will not hitch to fifth wheel.	2-88
	MEDIUM WRECKER (M936)	
76.	Hydraulic system does not operate.	2-89
77.	Hydraulic pump noisy.	2-89
	DUMP BODY (M929, M930)	
78.	Dump body will not raise	2-89
79.	Dump body will not lower.	2-90
80.	Dump body will not hold in raised position	2-90
81.	Hydraulic pump noisy.	2-90
82.	Tailgate will not open	2-90
	EXPANSIBLE VAN (M934, M935)	
83.	Side panels hard to retract or expand.	2-90
84.	Side panel cannot be locked in retracted position.	2-90
85.	Van body not waterproof or lighttight.	2-91
86.	Door lock will not operate.	2-91
87.	Heater will not ignite	2-91

Table 2-2. Mechanical Troubleshooting.

MALFUNCTION
TEST OR INSPECTION
CORRECTIVE ACTION

ENGINE

1. ENGINE WILL NOT CRANK

NOTE

If STE/ICE is available, perform NG20 - No Crank - No Start Test (para. 2-24).

Step 1. Check battery and starting systems (Electrical Troubleshooting, para. 2-22).

Step 2. Remove starter (para. 2-82) and visually check starter drive and flexplate ring gear for broken and missing teeth.

    a. Replace starter if teeth are missing (para. 3-82).

    b. If flexplate ring gear is damaged, notify DS maintenance.

    c. Install starter (para. 3-82) if vehicle evacuation is required.

Step 3. Check belt-driven engine accessories for seizure.

    a. Remove alternator, fan, water pump, and power steering pump drivebelts (para. 3-78, 3-70, 3-71, 3-67, and 3-230).

    b. Manually turn drive pulley of each accessory.

    c. Replace component if any accessory drive pulley will not turn (para. 3-80, 3-81, 3-71, 3-68, 3-67, and 3-236).

NOTE

Perform step 4 for M939A2 series vehicles.

Step 4. Check air compressor for seizure.

    a. Remove air compressor (para. 3-206).

    b. Rotate drive gear of air compressor. If drive gear does not turn, replace air compressor (para. 3-206).

    c. Attempt to crank engine. If engine will not crank, notify DS maintenance.

Step 5. Check engine for fluid-locked pistons.

Remove turbocharger exhaust and air intake tubes (para. 3-21). If water is present, notify DS maintenance.

END OF TESTING !

2. ENGINE CRANKS BUT WILL NOT START

NOTE

If STE/ICE is available, perform NG30 Engine Cranks No Start Test (para. 2-24).

Step 1. Ensure emergency engine stop control is in all the way.

Step 2. Press accelerator pedal all the way down to floor and hold, then restart engine.

Step 3. Check air cleaner indicator for air restriction indications.

    a. If red appears at indicator window, inspect air intake stack for restrictions and, if necessary, replace air cleaner element (TM 9-2320-272-10).

    b. Reset air cleaner indicator by pressing button down.

NOTE

Perform steps 4 through 6 for M939A2 series vehicles.

Table 2-2. Mechanical Troubleshooting (Contd).

MALFUNCTION
TEST OR INSPECTION
CORRECTIVE ACTION

Step 4. Check for white smoke during cranking.

    a, If white smoke is seen, air may be in fuel system.

    Prime fuel system (para. 2-22),

    b. If white smoke is seen, coolant may be in combustion chambers.

    Remove dipstick. If coolant is present, notify DS maintenance.

WARNING

- Eyeshields must be worn when cleaning with compressed air. Compressed air source will not exceed 30 psi (207 kPa). Failure to do so may result in injury to personnel.
- Diesel fuel is highly flammable. Do not perform fuel system procedures near open flame. Injury to personnel may result.

Step 5. Check fuel transfer pump for proper operation.

    a. Check for loose fuel supply hose at fuel filter head.

    b. Operate hand primer on the fuel transfer pump and check for fuel.

    If fuel is present, replace fuel transfer pump (para. 3-20).

Step 6. Check for fuel leaking from fuel pressure transducer. If fuel is present:

    a. Check fuel transfer pump for proper operation (refer to step 5).

    b. Check for clogged fuel injector supply line.

    If clogged, clear with compressed air or sturdy wire.

Step 7. Inspect fuel filter for dirty and/or clogged condition (para. 3-30).

    Replace fuel filter if clogged or dirty (para. 3-30).

Step 8. Inspect fuel lines and connections for leaks, breaks, obstructions, or damage.

    a. Visually check for leaks. If leaks are present at connection, tighten.

    b. If leaks are result of cracked, split, or damaged tubing, replace fuel line(s) (TM 9-243).

    C. Disconnect fuel line at both ends. If fuel line is clogged, clear with compressed air or sturdy wire.

    d. Reconnect fuel line.

    e. Replace fuel filter (para. 3-30).

    f. Prime fuel system (para. 3-22).

    g. If not corrected, notify DS maintenance.

Step 9. Check for fuel contamination.

    a. Open drainvalve on fuel filter/water separator and drain approximately 1 pt (0.478 L) of fuel into a glass container.

    b. If contamination is present, continue to drain fuel up to 1 qt (0.946 L).

    c. If contamination is still present, drain fuel tank(s) (para. 3-27).

    d. Clean and flush fuel system. Dry with compressed air.

    e. Replace fuel filter (para. 3-30).

    f. Refill tank(s) with fuel (TM 9-2320-272-10).

    g. Prime fuel system (para. 3-22).

Table 2-2. Mechanical Troubleshooting (Contd).

MALFUNCTION
TEST OR INSPECTION
CORRECTIVE ACTION

WARNING

- Ether is extremely flammable. Do not perform ether start system testing procedures near fire. Injury to personnel may result.
- Eyeshields must be worn when cleaning with compressed air. Compressed air source will not exceed 30 psi (207 kPa). Failure to do so may result in injury to personnel.

Step 10. Check ether start system.

Perform malfunction 16.

END OF TESTING!

3. STARTER CRANKS ENGINE SLOWLY, HARD TO START

Step 1. Press accelerator pedal all the way to floor and hold, then restart engine.

Step 2. In cold weather, ensure proper engine oil is being used.

Add or replace oil (LO 9-2320-272-12).

NOTE

If STE/ICE is available, perform NG80-Starter Circuit Tests (para. 2-24).

Step 3. Check starting system (Electrical Troubleshooting, para. 2-22).

END OF TESTING!

4. ENGINE STOPS DURING NORMAL OPERATION

Step 1. Perform malfunction 2, steps 2, 8, and 9.

Step 2. Check for restrictions in exhaust system.

Ensure exhaust system is not bent, restricted, or damaged.

Replace exhaust system component(s) if bent, restricted, or damaged (paras. 3-48, 3-49, 3-50, or 3-51).

END OF TESTING!

5. ENGINE STOPS WHEN ACCELERATOR IS RETURNED TO IDLE POSITION

Step 1. Check accelerator linkage adjustment.

If necessary, adjust accelerator linkage (para. 3-42 or 3-43).

Step 2. Check if engine idle is too low.

If idle speed is below 600 rpm, notify DS maintenance.

END OF TESTING!

6. ENGINE FAILS TO STOP

Step 1. Inspect emergency stop control cable for damage and proper adjustment.

Adjust or replace emergency stop control cable (para. 3-44).

NOTE

Perform step 2 for M939A2 series vehicles.

Step 2. Check for proper operation of throttle control solenoid and linkage.

Replace throttle control solenoid or linkage if not operating properly (para. 3-46).

END OF TESTING!

Table 2-2. Mechanical Troubleshooting (Contd).

> **MALFUNCTION**
> > **TEST OR INSPECTION**
> > > **CORRECTIVE ACTION**

---

7. ENGINE MISFIRES DURING OPERATION

NOTE

If STE/ICE is available, perform NG90 - Governor/Power Test Fault Isolation (para. 2-24).

Step 1. Perform malfunction 4, steps 1 and 2.

Step 2. Check accelerator linkage and modulator cables for proper operation and adjustment.

    a. If accelerator linkage does not operate properly, adjust or replace as necessary (para. 3-42 or 3-43).

    b. Replace modulator cable if binding, broken, or sticking (para. 3-145).

    c. If not corrected, notify DS maintenance.

END OF TESTING!

8. ENGINE IDLE ROUGH, ERRATIC (M939A2 ONLY)

Check for loose, plugged, or cracked injector tubes.

    a. If loose, tighten injector tubes.

    b. If cracked, replace (para. 3-19) (M939A2 only).

WARNING

Eyeshields must be worn when cleaning with compressed air. Compressed air source will not exceed 30 psi (207 kPa). Failure to do so may result in injury to personnel.

---

    c. If plugged, remove and clear tubes with compressed air .

    d. If not corrected, notify DS maintenance.

END OF TESTING!

9. POOR ACCELERATION AND/OR LACK OF POWER

Step 1. Perform malfunction 4.

Step 2. Inspect accelerator pedal and throttle lever for full travel.

Adjust accelerator pedal and throttle lever travel (para. 3-42 or 3-43).

NOTE

Perform steps 3 through 7 for M939A2 series vehicles.

Step 3. Check Air Fuel Control (AFC) tube for leaks and/or restrictions.

    a. If loose, tighten fittings.

    b. If restricted, replace AFC tube if necessary (para. 3-18).

Step 4. Check for air leaks.

    a. Check crossover tube and aftercooler for leaks.

    b. Tighten all hose clamps.

    c. Replace aftercooler gasket (para. 3-76).

    d. Check for exhaust leaks at turbocharger and exhaust manifold. If exhaust leaks are present, notify DS maintenance.

    e. Replace turbocharger gasket (para. 3-21).

Table 2-2. Mechanical Troubleshooting (Contd).

MALFUNCTION
TEST OR INSPECTION
CORRECTIVE ACTION

Step 5. Check fuel transfer pump for proper operation.

Refer to malfunction 2, step 5.

Step 6. Check for fuel leaking from fuel pressure transducer.

Refer to malfunction 2, step 6.

WARNING

Turbocharger intake fans are extremely sharp and turn at very high rpms. Keep hands and loose items away from intake opening. Failure to do so may cause injury to personnel.

Step 7. Check turbocharger for proper operation.

a. Remove air intake tube from turbocharger inlet (para. 3-14).

b. Start engine (TM 9-2320-272-10) and check turbocharger intake fan for free rotation.

c. If turbocharger intake fan does not rotate freely, replace turbocharger (para. 3-21).

d. If poor acceleration or lack of power exists, notify DS maintenance.

NOTE

If STE/ICE is available, perform NC90 - Governor/Power Test Fault Isolation (para. 2-24).

END OF TESTING!

10. ENGINE SURGES

Step 1. Perform malfunction 2, step 8.

Step 2. Check accelerator linkage and modulator cables for proper operation and adjustment.

a. If accelerator linkage does not operate properly, adjust or replace as necessary (para. 3-42 or 3-43).

b. Replace modulator cable if binding, broken, or sticking (para. 3-145).

Step 3. If engine surge continues, notify DS maintenance.

END OF TESTING!

11. EXCESSIVE ENGINE OIL LOSS OR CONSUMPTION

NOTE

If STE/ICE is available, perform NG90 - Governor/Power Test Fault Isolation (para. 2-24).

Step 1. Check engine oil for overfilling (TM 9-2320-272-10).

If dipstick has been read correctly and indicates excessive oil, drain crankcase to safe operating level (LO 9-2320-272-12). Recheck oil level (TM 9-2320-272-10).

Step 2. Check for external oil leaks.

a. Wipe off edge of rocker arm cover, oil pan, oil filter, and other external surfaces.

b. Start engine (TM 9-2320-272-10) and observe for leaks.

c. Tighten any loose connections or loose screws. If leaking continues, replace defective component or notify DS maintenance.

Table 2-2. Mechanical Troubleshooting (Contd).

MALFUNCTION
TEST OR INSPECTION
CORRECTIVE ACTION

NOTE
Perform step 3 for M939A2 series vehicles.

Step 3. Check turbocharger for oil leaking into air intake or exhaust.

  a. Remove air intake tube from turbocharger inlet.

  b. If oil is present in air intake, or excessive smoke is seen, replace turbocharger (para. 3-21).

END OF TESTING!

12. ENGINE OIL PRESSURE TOO LOW OR TOO HIGH AT NORMAL OPERATING TEMPERATURE

Step 1. Check engine oil level (TM 9-2320-272-10).

If dipstick has been read correctly and indicates low or excessive oil, refill or drain crankcase to proper operating level (TM 9-2320-272-10).

Step 2. Check oil supply lines for cracks, splits, leaks, damage, or obstructions.

  a. Tighten loose fittings and connections.

  b. Replace cracked, split, or damaged oil lines.

WARNING

Eyeshields must be worn when cleaning with compressed air. Compressed air source will not exceed 30 psi (207 kPa). Failure to do so may result in injury to personnel.

  c. Clear clogged or obstructed lines with compressed air or sturdy wire.

Step 3. Inspect oil filter for leaks.

  a. Tighten oil filter.

  b. If leaking persists, replace oil filter (para. 3-5).

NOTE
If STE/ICE is available, perform NG05 - Low Oil Pressure Check (para. 2-24).

Step 4. Check oil pressure gauge, sending unit, and electrical lead for malfunction.

  a. Refer to electrical troubleshooting (para. 2-22).

  b. If defective, replace oil pressure sending unit (para. 3-91).

  c. If defective, replace oil pressure gauge (para. 3-86).

NOTE
Perform step 5 for M939A2 series vehicles.

Step 5. Check oil pressure regulator plunger and ensure it moves freely.

  a. Remove oil filter head (para. 3-4).

  b. Remove oil pressure regulator plunger and spring from oil filter head (para. 3-4).

  c. Clean oil pressure regulator plunger and ensure it moves freely in bore.

  d. If restricted, replace oil pressure regulator plunger.

  e. If low oil pressure is not corrected, notify DS maintenance.

END OF TESTING!

Table 2-2. Mechanical Troubleshooting (Contd).

```
MALFUNCTION
    TEST OR INSPECTION
        CORRECTIVE ACTION
```

13. ENGINE OIL CONTAMINATED

Step 1. Take oil sample to determine type of contamination(s) (para. 3-10).

Step 2. If excessive sludge is present in engine oil:

    a. Review oil and filter change intervals (LO 9-2320-272-12).

    b. Drain and refill engine crankcase (LO 9-2320-272-12).

NOTE

Perform step 3 for M939A2 series vehicles.

Step 3. If fuel is present in engine oil (thin, black oil is an indication):

    a. Check transfer pump for leaking seal.

    b. If leaking, replace transfer pump seal (para. 3-20).

Step 4. If coolant is present in engine oil (milky color):

NOTE

• During engine operation, oil pressure will be higher than coolant pressure. A leak in oil cooler will indicate oil in coolant. But, after engine is shut down, residual pressure in cooling system will cause coolant to leak into oil.

• Perform step b. for M939A2 series vehicles.

    a. Repair or replace oil cooler if leaking (para. 3-4, or notify DS maintenance).

    b. Repair or replace aftercooler if leaking (para. 3-76).

Step 5. Check if metal particles are present in engine oil.

If metal particles are present, notify DS maintenance.

END OF TESTING!

14. EXCESSIVE VIBRATING OR CLUNKING

Step 1. Check engine mounting brackets and pads for looseness or damage.

    a. If loose, tighten.

    b. If engine mounting pad(s) are damaged, notify DS maintenance.

Step 2. Check vibration damper for looseness.

If vibration damper is loose, notify DS maintenance.

END OF TESTING!

15. EXCESSIVE FUEL CONSUMPTION

Step 1. Check air cleaner indicator for indication of air restriction.

    a. If red appears at indicator window, inspect air intake stack for restrictions and, if necessary, replace air cleaner element (TM 9-2320-272-10).

    b. Beset air cleaner indicator by pressing button down.

Step 2. Inspect fuel lines and connections for leaks, breaks, obstructions, or damage.

    a. Visually check for leaks. If leaks are present at connection, tighten.

    b. If leaks are result of cracked, split, or damaged tubing, replace fuel line(s) (TM 9-243).

Step 3. Check fuel return line for bends and kinks.

Straighten or replace fuel return lines if bent or kinked.

END OF TESTING!

Table 2-2. Mechanical Troubleshooting (Contd).

MALFUNCTION
TEST OR INSPECTION
CORRECTIVE ACTION

ETHER START SYSTEM

WARNING

Ether is extremely flammable. Do not perform ether start system testing procedures near fire. Injury to personnel may result.

16. ENGINE CRANKS BUT WILL NOT START IN COLD WEATHER (FUEL SYSTEM OPERATING PROPERLY)

Step 1. Check ether cylinder.

  a. Remove ether cylinder from valve, shake cylinder, and listen for liquid splashing inside cylinder (para. 3-35).

  b. If empty, replace ether cylinder (para. 3-35).

Step 2. Check ether valve for proper operation.

  a. Disconnect tubing at ether valve (para. 3-35).

  b. Press ether start switch. A small amount of ether should be released at ether valve.

  c. If ether is not evident, check electrical system (Electrical Troubleshooting, para. 2-22).

  d. If electrical system operates properly, replace ether valve (para. 3-35).

Step 3. Check thermal close valve and thermal close valve tubing for restrictions.

  a. Disconnect ether valve to thermal close valve tubing at ether valve. Disconnect thermal close valve-to-atomizer tubing at thermal close valve.

  b. With thermal close valve cold, blow compressed air into tubing at ether valve end to determine if system is restricted.

  c. If restricted, disconnect tubing from thermal close valve and check for restrictions in tubing.

  d. If tubing is not restricted, replace thermal close valve (para. 3-405).

  e. If tubing is restricted, replace tubing (para. 3-405).

Step 4. Check atomizer and atomizer tubing for restrictions,

  a. Disconnect tubing between thermal close valve and atomizer at thermal close valve.

  b. Blow compressed air into tubing at thermal close valve to determine if system is restricted.

  C. If restricted, disconnect tubing from atomizer and check for restrictions in tubing.

  d. If tubing is not restricted, replace ether atomizer (para. 3-37 or 3-38).

  e. If tubing is restricted, replace tubing (para. 3-39).

END OF TESTING!

EXHAUST SYSTEM

17. EXHAUST COLOR BLUE DURING NORMAL OPERATION

NOTE

Blue exhaust indicates presence of excess engine oil in cylinder combustion chamber.

Step 1. Check that engine oil grade is correct for vehicle use and climatic conditions (LO 9-2320-272-121.

If oil grade is incorrect, replace engine oil and oil filter (LO 9-2320-272-12).

Table 2-2. Mechanical Troubleshooting (Contd).

MALFUNCTION
TEST OR INSPECTION
CORRECTIVE ACTION

Step 2. Check that engine fuel grade is correct for vehicle use and climatic conditions (TM 9-2320-272-10).

If fuel grade is incorrect, drain complete fuel system and replace with correct grade of fuel (TM 9-2320-272-10).

END OF TESTING!

18. EXHAUST COLOR WHITE DURING NORMAL OPERATION AND IDLE

CAUTION

Thick white smoke indicates coolant is present in engine combustion chambers during operation. When this condition is evident, shut engine down immediately and determine cause. Continued engine operations may result in permanent damage to engine.

Step 1. Check engine temperature. Ensure engine temperature is within normal operating range 175°-200°F (79°-93°C).

If temperature is above normal operating range, perform malfunction 23.

Step 2. Check for presence of water in fuel.

If water is present, drain fuel system, replace fuel filter (para. 3-30), and refill fuel tanks (TM 9-2320-272-10).

Step 3. If problem still exists, notify DS maintenance.

END OF TESTING!

19. EXHAUST COLOR BLACK DURING NORMAL OPERATION AND IDLE

Step 1. Check air cleaner indicator for indication of air restriction,

    a. If red appears at indicator window, inspect air intake stack for restrictions and, if necessary, replace air cleaner element (TM 9-2320-272-10).

    b. Reset air cleaner indicator by pressing button down.

Step 2. Check for fuel contamination.

    a. Open drainvalve on fuel filter/water separator and drain approximately 1 pt (0.473 L) of fuel into a glass container.

    b. If contamination is present, continue to drain fuel up to 1 qt (0.946 L).

    c. If contamination is still present, drain fuel tank(s) (para. 3-26).

    d. Clean and flush fuel system. Dry with compressed air.

    e. Replace fuel filter (para. 3-30).

    f. Refill tank(s) with fuel (TM 9-2320-272-10).

    g. Prime fuel system (para. 3-22).

END OF TESTING!

20. EXCESSIVE EXHAUST NOISE

WARNING

Do not touch hot exhaust system components with bare hands. Severe injury to personnel may result.

Table 2-2. Mechanical Troubleshooting (Contd).

MALFUNCTION
  TEST OR INSPECTION
    CORRECTIVE ACTION

Step 1. Inspect exhaust piper, stack, and muffler for secure connections, cracks, breaks, and excessive rust.

   a. If loose, tighten.

   b. If leaking at pipe connection, replace gasket(s) (para. 3-48).

   c. If cracked, broken, or rusted, replace (para, 3-48). If loose, tighten.

Step 2. Check exhaust manifold for leaks. If leaking, notify DS maintenance.

NOTE
Perform step 8 for M989A2 series vehicles,

Step 3. Inspect turbocharger for secure mounting and exhaust leaks.

   If turbocharger mounting is loose or leaks exhaust, notify DS maintenance.

21. EXHAUST FUMES IN CAB

Inspect exhaust pipes, stack, and muffler for secure connections, cracks, breaks, and excessive rust.

   a. If loose, tighten.

   a. If leaking at pipe connection, replace gasket(s) (para. 3-48).

   b. If cracked, broken, or rusted, replace (para. 3-48).

END OF TESTING!

COOLING SYSTEM

22. ENGINE DOES NOT REACH NORMAL OPERATING TEMPERATURE

WARNING

Extreme care should be taken when removing surge tank filler cap if temperature gauge reads above 175°F (79°C). Steam or hot coolant under pressure may cause injury.

NOTE
If STE/ICE is available, perform NG31 - Gauge Test (para. 2-24).

Step 1. Start engine (TM 9-2320-272-10). Remove surge tank cap and visually check coolant for proper circulation.

   Stop engine. If coolant is seen circulating and temperature is below 175°F (79°C), replace thermostat (para. 3-65 or 3-66).

Step 2. Test coolant temperature gauge, sending unit, and electrical leads for malfunction (Electrical Troubleshooting, para. 2-22).

   a. Replace coolant temperature gauge if defective (para. 3-86).

   b. Replace coolant temperature sending unit if defective (para. 3-93).

NOTE
Perform step 3 for M939A2 series vehicles.

Step 3. Observe fan clutch and actuator for proper operation.

   Replace fan clutch and/or actuator if defective (para. 3-74).

END OF TESTING!

Table 2-2. Mechanical Troubleshooting (Contd).

MALFUNCTION
TEST OR INSPECTION
CORRECTIVE ACTION

23. ENGINE OVERHEATS AS INDICATED BY ENGINE COOLANT TEMPERATURE GAUGE

WARNING

Extreme care should be taken when removing surge tank filler cap if temperature gauge reads above 175°F (79°C). Steam or hot coolant under pressure may cause injury.

Step 1. Check coolant level (TM 9-2320-272-10).

CAUTION

- Do not add coolant to engine when hot. Internal damage to engine could result.
- On M939A2 series vehicles, ensure drainvalve on aftercooler is open. Failure to do so may result in damage to equipment.

      If coolant level is low, fill to proper level (para. 3-53).

Step 2. Check engine oil level (TM 9-2320-272-10).

      If oil level is low, fill to correct level (LO 9-2320-272-12).

Step 3. Inspect drivebelts and tensioner (M939A2) for looseness, absence, and worn conditions.

    a. If loose, tighten.

    b. Replace drivebelt if worn or damaged (paras. 3-67, 3-68, 3-71, 3-80, 3-81, or 3-236).

Step 4. Inspect radiator, hoses, hose connections, and drainvalves for leaks.

    a. Tighten loose hose clamps and fittings.

    b. Tighten or close drainvalves.

    c. Replace defective cooling system components (chapter 3, section V).

Step 5. Check cooling system for clogging.

      If clogged, flush and clean cooling system (para. 3-53).

Step 6. Inspect fan for cracked or missing blades.

      If blades are cracked or missing, replace fan (para. 3-64 or 3-72).

Step 7. Check radiator for airflow obstructions.

      Remove obstructions from front of radiator (TM 750-254).

Step 8. Observe fan clutch and actuator for proper operation.

      If defective, replace fan clutch and actuator (para. 3-73 or para. 3-74).

Step 9. Check water pump for leaks and proper operation.

    a. Start engine (TM 9-2320-272-10). Remove surge tank cap and visually check coolant for proper circulation.

    b. Stop engine. If coolant is seen circulating and temperature is below 175°F (79°C) replace thermostat (para. 3-65 or 3-66).

NOTE

If STE/ICE is available, perform NG31 - Gauge Test (para. 2-24).

Step 10. Test temperature gauge for proper operation (TM 9-2320-272-10).

      If not correct, notify DS maintenance.

END OF TESTING!

Table 2-2. Mechanical Troubleshooting (Contd).

| MALFUNCTION |
| TEST OR INSPECTION |
| CORRECTIVE ACTION |

24. LOSS OF COOLANT

Step 1. Check for leaks in cooling system.

    a. Inspect radiator, hoses, hose connections, and drainvalves for leaks.

    b. Tighten loose hose clamps and fittings.

    c. Tighten or close drainvalves.

    d. Replace defective cooling system components (chapter 3, section V).

Step 2. Check for coolant in transmission oil.

    a. Check transmission oil for presence of coolant.

    b. If coolant is present in transmission oil, check transmission oil cooler for damage and leaks.

    c. If damaged or leaking, replace transmission oil cooler (para. 3-140 or 3-141).

    d. Drain and flush cooling system (para. 3-53).

END OF TESTING!

25. CONTAMINATED COOLANT

Step 1. If coolant appears rusty, drain and flush cooling system (para. 3-53).

Step 2. If transmission oil is present in coolant, perform malfunction 24, step 2.

Step 3. If engine oil is present in coolant, notify DS maintenance.

END OF TESTING!

TRANSMISSION

26. EXCESSIVE NOISE DURING SHIFTING

Step 1. Check transmission fluid level (TM 9-2320-272-10).

Drain or fill fluid as necessary (LO 9-2320-272-12).

Step 2. Check propeller shaft flanges for loose mounting bolts.

If loose, tighten mounting bolts 30-40 lb-ft (41-54 N-m).

Step 3. Inspect propeller shaft universal joints for looseness, wear, and damage.

    a. If loose, tighten universal joint screws 90-110 lb-ft (122-149 N•m).

    b. Replace worn or damaged universal joints (para. 3-154).

END OF TESTING!

27. LOW TRANSMISSION OIL PRESSURE

Step 1. Check transmission fluid level (TM 9-2320-272-10).

Drain or fill fluid as necessary (LO 9-2320-272-12).

Step 2. Replace transmission oil filter (para. 3-133).

END OF TESTING!

28. TRANSMISSION OIL LEAKING

Step 1. Inspect for leak at output shaft.

If output shaft oil seal is leaking, notify DS maintenance.

    a. If loose, tighten drainplug.

    b. If leak continues, notify DS maintenance.

Table 2-2. Mechanical Troubleshooting (Contd).

MALFUNCTION
TEST OR INSPECTION
CORRECTIVE ACTION

Step 2. Inspect transmission oil pan gasket for leaks.
  a. If loose, tighten mounting screws 10-15 lb-ft (14-20 N•m).
  b. If leak continues, replace oil pan gasket (para. 3-133).
Step 3. Inspect transmission housing for leaks.
  If leaking, notify DS maintenance.

                    END OF TESTING!

29. NO RESPONSE TO SHIFT LEVER MOVEMENT
Step 1. Check transmission fluid level (TM 9-2320-272-10).
  Drain or fill fluid as necessary (LO 9-2320-272-12).
Step 2. Check for broken or disconnected shift cable.
  If broken or disconnected, notify DS maintenance.

                    END OF TESTING!

30. ROUGH SHIFTING
Step 1. Check transmission fluid level (TM 9-2320-272-10).
  Drain or fill fluid as necessary (para. 3-133).
Step 2. Check transmission select lever and shift cable for proper operation.
  a. If select lever is damaged, notify DS maintenance.
  b. If shift cable is bent, kinked, broken, or frayed, notify DS maintenance.

                    END OF TESTING!

31. TRANSMISSION OVERHEATS AS INDICATED BY TRANSMISSION TEMPERATURE GAUGE
Step 1. Check transmission fluid level (TM 9-2320-272-10).
  Drain or fill fluid as necessary (para. 3-1331.
Step 2. Check for clogged or restricted transmission oil cooler.
  If clogged or restricted, replace transmission oil cooler (para. 3-140 or 3-141).
Step 3. Check transmission temperature sending unit (Electrical Troubleshooting, para. 2-24).
Step 4. Check for clogged or restricted transmission oil cooler hoses.
  If clogged or restricted, replace transmission oil cooler (para. 3-140 or 3-141).
Step 5. Test transmission temperature gauge for proper operation using a gauge known to be good.
  If test gauge does not indicate overheating, replace transmission temperature gauge (para. 3-86).

                    END OF TESTING!

32. DIRT OR METAL PARTICLES IN TRANSMISSION OIL
Take transmission oil sample. Submit oil sample in accordance with TB 43-0210.

                    END OF TESTING!

33. OIL THROWN FROM TRANSMISSION DIPSTICK TUBE
Step 1. Check transmission fluid level for overfilling (TM 9-2320-272-10).
  Drain fluid to proper level (LO 9-2320-272-12). Recheck transmission for proper fluid level (TM 9-2320-272-10).

Table 2-2. Mechanical Troubleshooting (Contd).

MALFUNCTION
TEST OR INSPECTION
CORRECTIVE ACTION

Step 2. Check for loose dipstick tube.

         If loose, tighten.

NOTE

Perform step 3 if deep water was forded.

Step 3. Remove vent line from transmission. Start vehicle and build up air pressure. If air is coming from line, notify DS maintenance.

END OF TESTING!

34. TRANSMISSION OIL DIRTY, FOAMY, AND/OR MILKY

NOTE

Dirt/grit in transmission oil indicates oil needs to be changed (step 2). Foaminess indicates contamination by air or water (step 3). Milkiness indicates contamination by coolant (step 4).

Step 1. Check oil filter tube seal for damage and leaks.

    a. If damaged, replace seal (para. 3-133).

    b. If leaking, tighten.

Step 2. Check for dirt/grit.

    a. Replace transmission internal oil filter (para. 3-133).

    b. Replace transmission external oil filter (para. 3-133).

    c. Notify DS maintenance for transmission replacement,

Step 3. Check for excessive foaming.

         Ensure transmission fluid is at proper level (TM 9-2320-272-10). Drain or fill fluid as necessary (LO 9-2320-272-12).

Step 4. Check for coolant in transmission fluid.

         If coolant is present in transmission oil, perform malfunction 24, step 2.

END OF TESTING!

TRANSFER CASE

35. HARD SHIFTING OF TRANSFER CASE

Step 1. Check transfer case fluid level (LO 9-2320-272-12).

         Fill or drain to proper level as necessary (LO 9-2320-272-12).

Step 2. Inspect transfer case shift linkage for proper lubrication.

         Lubricate linkage as necessary (LO 9-2320-272-12).

Step 3. Inspect shift linkage for bends, breaks, and missing parts.

         If shift linkage is bent, broken, or has missing parts, notify DS maintenance.

Step 4. Check wire 586 for good ground.

         Repair or replace wire (para. 3-131).

END OF TESTING!

Table 2-2. Mechanical Troubleshooting (Contd).

MALFUNCTION
TEST OR INSPECTION
CORRECTIVE ACTION

36. TRANSFER CASE LEAKING OIL

Step 1. Inspect drainplugs for leaks and looseness.

        If loose and leaking, tighten drainplugs 35 lb-ft (48 N•m).

Step 2. Inspect transfer case housing for leaks.

        If leaking, notify DS maintenance.

                END OF TESTING!

37. EXCESSIVE NOISE

Check transfer case fluid level (LO 9-2320-272-12).

      a. Fill to proper fluid level as necessary (LO 9-2320-272-12).

      b. Tighten filler plug 35 lb-ft (48 N-m).

      c. If problem still exists, notify DS maintenance.

                END OF TESTING!

38. EXCESSIVE VIBRATION

Step 1. Check transfer case fluid level (LO 9-2320-272-12).

      a. Fill to proper fluid level as necessary (LO 9-2320-272-12).

      b. Tighten filler plug 35 lb-ft (48 N•m).

Step 2. Check yoke companion flange screws for proper torque.

        Tighten screws 32-40 lb-ft (43-54 N•m).

                END OF TESTING!

PROPELLER AND DRIVESHAFTS

39. EXCESSIVE NOISE OR VIBRATION

Step 1. Check universal joints for proper lubrication.

        Lubricate universal joints as necessary (LO 9-2320-272-12).

Step 2. Check torque of all propeller shaft flange yoke screws.

        Tighten screws 32-40 lb-ft (43-54 N•m).

Step 3. Inspect propeller shafts for wear and damage.

        Replace any worn or damaged propeller shafts (chapter 3, section VIII).

NOTE

Perform step 4 for M927 and M928 model vehicles.

Step 4. Inspect center bearing for looseness and damage.

      a. Tighten center bearing-to-forward rear axle propeller shaft yoke to transfer-to-center bearing rear flange screws 32-40 lb-ft (43-54 N•m).

      b. If damaged, replace center bearing assembly (para. 3-150).

                END OF TESTING!

Table 2-2. Mechanical Troubleshooting (Contd).

MALFUNCTION
   TEST OR INSPECTION
      CORRECTIVE ACTION

FRONT AND REAR AXLES

40. CONTINUOUS AXLE OR WHEEL NOISE

Step 1. Check to see if front wheel drive is engaged.

      Disengage front wheel drive when traveling on hard surfaces.

Step 2. Check for loose or missing wheel stud nuts.

    a. Tighten stud nuts 400-425 lb-ft (542-576 N•m).

    b. Replace missing stud nuts (para. 3-224).

Step 3. Check lubrication level in axle housing and differential (LO 9-2320-272-12).

      Fill axle housing or differential to proper level if necessary (LO 9-2320-272-12).

Step 4. Check for loose or damaged wheel bearings.

    a. Raise vehicle wheels off ground and support with jack stands.

    b. Use prybar to lift up on bottom of tire. Excessive play indicates improperly adjusted or damaged bearings.

    c. If excessive play is present, adjust or replace wheel bearings (para. 3-225).

              END OF TESTING!

41. OIL LEAKING FROM FRONT WHEEL DRIVE LOCK-IN SWITCH

Step 1. Disconnect air line from cylinder and check for oil in line.

Step 2. If oil is present in line, notify DS maintenance.

              END OF TESTING!

42. DIFFERENTIAL NOISY

Step 1. Perform malfunction 40.

Step 2. Check differential operation. Remove differential propeller shaft(s) (para. 3-148, 3-149, 3-150, or 3-151). Raise wheels (TM 9-2320-272-10), manually turn wheels, and observe differential operation.

    a. If tires will not rotate, check brakeshoe condition and operation (para. 3-176).

    b. Remove brakedrums (para. 3-177). If brake system components are defective, repair or replace (chapter 3, section IX), or notify DS maintenance.

    c. If tires will still not rotate, notify DS maintenance.

    d. If tire rotation drags at some points during full rotation, remove and inspect axle shafts (para. 3-154 or 3-157).

    e. If defective, replace axle shafts (para. 3-154 or 3-157).

              END OF TESTING!

43. DIFFERENTIAL CLUNKS DURING TURNS OR INITIAL TAKEOFF

Step 1. Check condition of differential propeller shaft(s) and universal joint(s) (para. 3-154, 3-148, 3-149, 3-150, or 3-151).

      Repair or replace defective components (para. 3-154, 3-148, 3-149, 3-150, or 3-151).

Step 2. Check front axle shafts and universal joints for defects (para. 3-154).

      Repair or replace defective components (para. 3-154).

Step 3. If internal problem exists in differential, notify DS maintenance.

              END OF TESTING!

Table 2-2. Mechanical Troubleshooting (Contd).

MALFUNCTION
TEST OR INSPECTION
CORRECTIVE ACTION

44. DIFFERENTIAL VIBRATES

Step 1. Check condition of tires and rims. Repair or replace defective components (para. 3-219 or 3-220).

Step 2. Perform malfunction 43, steps 1 and 2.

Step 3. If internal problems in differential exist, notify DS maintenance.

END OF TESTING!

45. DIFFERENTIAL LEAKS OIL

Step 1. Check condition of axle seals. Inspect brakedrum for presence of gear oil.

If gear oil is present in or around brakedrums, replace axle seals (para. 3-160).

Step 2. Check through-shaft for free play.

Place end of prybar between differential housing and companion flange. Apply pressure to prybar.

Step 3. If companion flange moves, notify DS maintenance.

END OF TESTING!

WHEELS AND TIRES

NOTE

A broken or separated radial belt will cause symptoms as outlined in malfunctions 46, 47, 49, and 51.

46. UNEVEN TIRE WEAR

Step 1. Check for loose or missing wheel stud nuts and wheel studs.

   a. Tighten loose wheel stud nuts 400-425 lb-ft (542-576 N•m).

   b. Replace missing wheel stud nuts and/or wheel studs (para. 3-218, 3-219, or 3-222).

Step 2. Check for improper wheel alignment adjustment if wear is on front tires.

Adjust alignment if necessary (para. 3-152).

Step 3. Check wheel bearings for proper adjustment and damage. Raise wheels off ground. Use prybar to lift up on bottom of tire. Excessive play indicates improperly adjusted or damaged bearings.

   a. Adjust wheel bearings (para. 3-225).

   b. Replace damaged bearings (para. 3-223, 3-224, 3-460, or 3-461).

   c. Check that tires are properly inflated (TM 9-2320-272-10).

END OF TESTING!

47. WHEEL SHIMMY OR WOBBLE

Step 1. Inspect wheels for bends and damage.

Replace bent or damaged wheels (para. 3-218 or 3-219).

Step 2. Check wheel bearings for proper adjustment and damage. Raise wheels off ground. Use prybar to lift up on bottom of tire. Excessive play indicates improper adjustment or damaged bearings,

   a. Adjust wheel bearings (para. 3-225).

   b. Replace damaged bearings (para. 3-223, 3-224, 3-460, or 3-461).

   c. Check that tires are properly inflated (TM 9-2320-272-10).

END OF TESTING!

Table 2-2. Mechanical Troubleshooting (Contd).

MALFUNCTION
   TEST OR INSPECTION
      CORRECTIVE ACTION

STEERING

48. HARD STEERING

step 1. Inspect all hydraulic lines and hoses for leaks.

Tighten loose fittings and replace leaking lines (chapter 3, section X).

step 2. Inspect steering linkage for binding, damage, and improper lubrication.

a. Repair or replace binding or damaged linkage (chapter 3, section X).

b. Lubricate linkage (LO 9-2320-272-12).

step 3. Inspect steering knuckles for binding. Raise front wheels off ground. Disconnect drag link at pitman steering arm (para. 3-227 or 3-228). Turn wheels from side to side to determine binding.

Notify DS maintenance if wheels do not turn from side to side without binding.

step 4. Check spring U-bolts for tightness.

Tighten U-bolt(s) 350-400 lb-ft (475-542 N•m).

step 6. Check front wheel alignment.

a. Check front tires for underinflation and uneven tire pressure. Inflate tires to proper pressure (TM 9-2320-272-10).

b. Adjust alignment to specifications (para. 3-152).

END OF TESTING!

49. VEHICLE WANDERS OR PULLS TO ONE SIDE

Step 1. Check front tires for underinflation and uneven tire pressure.

Inflate tires to proper pressure (TM 9-2320-272-10).

Step 2. Check tires for uneven wear (indicates improper wheel alignment).

Adjust front wheel alignment (para. 3-152).

Step 3. Check for dragging brakes. Raise wheels off ground. Spin wheels by hand. Wheels should turn with slight drag when properly adjusted.

Refer to compressed air and brake system troubleshooting (table 2-4, malfunction 4).

Step 4. Check wheel bearings for proper adjustment and damage. Raise wheels off ground. Use prybar to lift up on bottom of tire. Excessive play indicates improper adjusted or damaged bearings.

a. Adjust wheel bearings (para. 3-225).

b. Replace damaged bearings (para. 3-223, 3-224, 3-225, 3-460, or 3-461).

c. Check that tires are properly inflated (TM 9-2320-272-10).

Step 5. Inspect steering assist cylinder for damage and improper adjustment.

Replace or adjust steering assist cylinder (para. 3-233).

Step 6. Inspect tie rod for looseness and damage.

Tighten or replace tie rod (para. 3-163).

Step 7. Check for loose steering gear mounting bolts.

Tighten mounting bolts 260-280 lb-ft (353-380 N-m).

Table 2-2. Mechanical Troubleshooting (Contd).

MALFUNCTION
TEST OR INSPECTION
CORRECTIVE ACTION

Step 8.   Check for worn pitman arm.

      If worn, replace pitman arm (para. 3-227 or 3-228).

Step 9.   Check front spring shackle pins for damage and breaks.

      If damaged or broken, replace front spring shackle pins (para. 3-162).

Step 10. Check wheel alignment for proper adjustment.

      Adjust wheel alignment if necessary (para. 3-152).

                END OF TESTING!

50. EXCESSIVE PLAY IN STEERING WHEEL

Step 1.   Check steering wheel free play. With engine running, place stiff wire long enough to touch steering wheel rim against dash. Turn steering wheel left, then right, until there is resistance. Mark the points on steering wheel where the travel ends. Then measure the distance between these points. If there is more than 2-1/2 in. (6.3 cm) of play, proceed to step 2.

Step 2.   Check drag link for looseness and damage.

      Tighten or replace drag link (para. 3-229).

Step 3.   Inspect tie rod for damage and loose ends. No free play is allowable.

      Tighten or replace tie rod (para. 3-153).

Step 4.   Inspect pitman arm for damage.

      Replace pitman arm if damaged (para. 3-227 or 3-228).

Step 5.   Inspect steering assist cylinder for improper adjustment and damage.

      Replace or adjust steering assist cylinder (para. 3-233).

Step 6.   If 2-1/2 in. (6.4 cm) of play still exists, notify DS maintenance.

                END OF TESTING!

51. SHIMMY

Step 1.   Check for loose or missing wheel stud nuts and wheel studs.

    a.   Tighten loose wheel stud nuts 400-425 lb-ft (542-576 N•m).

    b.   Replace missing wheel stud nuts and/or wheel studs (para. 3-218, 3-219, or 3-222).

Step 2.   Inspect wheels for bends and damage.

      Replace bent or damaged wheels (para. 3-218 or 3-219).

Step 3.   Check wheel bearings for proper adjustment and damage. Raise wheels off ground. Use prybar to lift up on bottom of tire. Excessive play indicates improperly adjusted or damaged bearings.

    a.   Adjust wheel bearings (para. 3-225).

    b.   Replace damaged bearings (para. 3-223, 3-224, 3-225, 3-460, or 3-461).

    c.   Check that tires are properly inflated (TM 9-2320-272-10).

Step 4.   Check front wheel alignment.

    a.   Check front tires for underinflation and uneven tire pressure. Inflate tires to proper pressure (TM 9-2320-272-10).

    b.   Adjust alignment to specifications (para. 3-152).

Table 2-2. Mechanical Troubleshooting (Contd).

```
MALFUNCTION
    TEST OR INSPECTION
        CORRECTIVE ACTION
```

Step 5.   Check for loose front axle steering knuckle(s). Raise wheels off ground. Turn wheels from side to side to observe for loose steering knuckle.

If steering knuckle is loose, notify DS maintenance.

END OF TESTING!

FRAME AND BRACKETS

52. TOWING PINTLE DOES NOT LATCH OR LOCK

Step 1.   Inspect pintle hook for lubrication.

Lubricate pintle hook (LO 9-2320-272-12).

Step 2.   Check pintle hook lock for proper operation.

Replace defective pintle hook (para. 3-242).

END OF TESTING!

53. PINTLE HOOK DOES NOT TURN

Step 1.   Inspect pintle hook for lubrication.

Lubricate pintle hook (LO 9-2320-272-12).

Step 2.   Check clearance between thrust washer and mounting bracket housing with feeler gauge. Clearance should be 0.003 in. - 0.017 in. (0.07 mm - 0.43 mm).

   a.   If clearance is the same completely around, but not 0.003 in. - 0.017 in. (0.07 mm - 0.43 mm), adjust clearance (para. 3-242).

   b.   If clearance is the same completely around, go to step 3.

Step 3.   Inspect pintle hook shaft for bends.

Replace pintle hook if shaft is bent (para. 3-242).

END OF TESTING!

5-54. EXCESSIVELY LOOSE LIFTING SHACKLE

Inspect shackle and shackle pin for bends, wear, and proper size.

If worn, bent, or size is improper, replace shackle and/or shackle pin (para. 3-241).

END OF TESTING!

55. LOOSE SPARE TIRE CARRIER

Step 1.   Check for missing and worn mounting bolts.

If worn or missing, replace mounting bolts.

Step 2.   Inspect carrier frame for damage.

If damaged, replace any unserviceable carrier component (chapter 3, section XI).

END OF TESTING!

SPRINGS AND SHOCK ABSORBERS

56. CONTINUOUS WANDERING OR SWAYING (POOR CONTROL)

Step 1.   Inspect front leaf springs for breaks.

If broken, replace main leaves (para. 3-161).

Step 2.   Inspect shock absorbers for leaks and damage.

If leaking or damaged, replace shock absorbers (para. 3-166).

Table 2-2. Mechanical Troubleshooting (Contd).

MALFUNCTION
  TEST OR INSPECTION
    CORRECTIVE ACTION

Step 3. Inspect spring U-bolts for looseness and damage.
    a. Tighten loose spring U-bolts 350-400 lb-ft (475-542 N•m).
    b. If damaged, replace U-bolts (para. 3-161).
Step 4. Check steering system (malfunction 48).
END OF TESTING!

57. HARSH OR HARD RIDE
Step 1. Check lubrication of springs, pivots, and shackle pins.
    Lubricate springs, pivots, and shackle pins (LO 9-2320-272-12).
Step 2. Check spring shackles for frozen condition.
    Free shackles and lubricate (LO 9-2320-272-12).
Step 3. Check for defective shock absorbers.
    a. Remove shock absorber from top mounting.
    b. Pull up and down on shock absorber to test resistance.
    c. If there is little or no resistance, replace shock absorber (para. 3-166).
END OF TESTING!

58. SPRING LEAF DEFECT
Step 1. Inspect for loose and damaged front spring shackles.
    If loose or damaged, replace shackles (para. 3-162).
Step 2. Inspect spring U-bolts for looseness and damage.
    a. Tighten loose spring U-bolts 350-400 lb-ft (475-542 N•m).
    b. If damaged, replace U-bolts (para. 3-161).
Step 3. Inspect front leaf springs for breaks.
    If broken, replace main leaves (para. 3-161).
END OF TESTING!

WINCH

59. WINCH DOES NOT OPERATE
Step 1. Check hydraulic oil reservoir for proper fluid level.
    a. Fill hydraulic oil reservoir to proper oil level (LO 9-2320-272-12).
    b. Check for evidence of oil leaks and damage to reservoir.
    c. If leaks are found, tighten any loose connections or fittings.
    d. If damaged, replace reservoir (para. 3-337 or 3-338).
Step 2. Check PTO and winch controls for kinks, damage, breaks, and improper lubrication (front winch).
    a. Lubricate if necessary (LO 9-2320-272-12).
    b. If winch or PTO cable is kinked, damaged, or broken, notify DS maintenance.

Table 2-2. Mechanical Troubleshooting (Contd).

MALFUNCTION
TEST OR INSPECTION
CORRECTIVE ACTION

Step 3. Check all oil lines for damage and leaks.

    a. Tighten loose fittings.

    b. Replace leaking or damaged hoses.

Step 4. Check winch oil filter for leaks and clogs (front winch).

    a. Remove center bolt, filter housing, filter, and gasket from filter head.

    b. Check filter and filter head for clogs. Replace filter if clogged; clean filter head if clogged.

    c. Install filter, gasket, and filter housing on filter head with center bolt. Tighten center bolt 30-35 lb-ft (41-48 N•m).

Step 5. Check winch pump for leaks and overheating.

    a. If loose, tighten winch pump connections.

    b. If winch pump is defective, notify DS maintenance.

Step 6. Check winch motor for leaks.

    a. If loose, tighten winch motor connections.

    b. If winch motor is defective, notify DS maintenance.

NOTE

Assistant will help with step 7.

Step 7. Check control valve for proper operation and leaks (front winch).

    a. While assistant operates winch control lever, observe control valve for proper operation.

    b. If control valve is defective, notify DS maintenance.

    c. Tighten any loose fittings.

Step 8. Check if drum clutch is engaged (front winch).

    a. Engage drum clutch (TM 9-2320-272-10).

    b. If drum clutch does not engage, replace front winch (para. 3-329).

Step 9. Check if drum lock is pulled out (front winch).

    a. Pull out drum lock (TM 9-2320-272-10).

    b. If drum lock does not pull out, replace front winch (para. 3-329).

END OF TESTING!

60. WINCH OPERATES IN ONE DIRECTION ONLY

Step 1. Check PTO and winch controls for kinks, damage, breaks, and improper lubrication (front winch).

    a. Lubricate if necessary (LO 9-2320-272-12).

    b. If winch or PTO cable is kinked, damaged, or broken, notify DS maintenance.

NOTE

Assistant will help with step 2.

Step 2. Check control valve for proper operation and leaks (front winch).

    a. While assistant operates winch control lever, observe control valve for proper operation.

    b. If control valve is defective, notify DS maintenance.

    c. Tighten loose fittings.

Table 2-2. Mechanical Troubleshooting (Contd).

MALFUNCTION
  TEST OR INSPECTION
    CORRECTIVE ACTION

  Step 3. Check if air-operated tensioner is released (rear winch).
       Release tensioner (TM 9-2320-272-10).
  Step 4. Check if torque control lever is engaged (rear winch).
       Engage torque control lever (TM 9-2320-272-10).
                    END OF TESTING!

61. DRAG BRAKE DOES NOT OPERATE
  Check drag brake adjustment.
         a. Adjust drag brake (para. 3-324 or 3-330).
         b. If drag brake is defective, replace winch (para. 3-329 or 3-332).
                    END OF TESTING!

62. WINCH WILL NOT HOLD LOAD
  Step 1. Check torque control lever position (rear winch).
       Reposition torque control valve lever to HIGH or LOW position.
  Step 2. Check automatic brake adjustment (para. 3-323).
       Adjust automatic brake screw (para. 3-323 or 3-330).
                    END OF TESTING!

63. AUTOMATIC BRAKE OVERHEATS
  Step 1. Check weight limit of winch. Adjust size of load.
  Step 2. Check automatic brake adjustment (para. 3-323).
       Turn automatic brake screw counterclockwise one-quarter turn. Recheck winch for overheating.
                    END OF TESTING!

64. FRONT WINCH OPERATES AT ONE SPEED ONLY
  Step 1. Check control valve for proper operation and leaks (front winch).
         a. While assistant operates winch control lever, observe control valve for proper operation.
         b. If control valve is defective, notify DS maintenance.
  Step 2. Check winch controls for kinks, damage, breaks, and improper lubrication (front winch).
         a. Lubricate if necessary (LO 9-2320-272-12).
         b. If winch cable is kinked, damaged, or broken, notify DS maintenance.
  Step 3. Check throttle control cable for damage (front winch).
       If throttle control cable is damaged, notify DS maintenance.
                    END OF TESTING!

65. REAR WINCH CABLE TENSIONER WILL NOT OPERATE
  Check air supply lines for leaks.
         a. Tighten loose fittings.
         b. If leaks persist, notify DS maintenance.
                    END OF TESTING!

Table 2-2. Mechanical Troubleshooting (Contd).

MALFUNCTION
TEST OR INSPECTION
CORRECTIVE ACTION

66. VEHICLE ROLLS WHILE OPERATING REAR WINCH

Step 1. Check parking brake for proper adjustment.
Adjust parking brake (TM 9-2320-272-10).

Step 2. Check spring brake caging.
Release spring brake cage (TM 9-2320-272-10).

Step 3. Check field chock positioning.
Position chocks facing load for direct pulls (TM 9-2320-272-10).
END OF TESTING!

POWER TAKEOFF (PTO)

67. EXCESSIVE NOISE AT PTO

Step 1. Inspect PTO universal joint for insufficient lubrication.
If necessary, lubricate universal joint (LO 9-2320-272-12).

Step 2. Inspect propeller shaft for bends.
If bent, replace propeller shaft (para. 3-334 or 3-393).
END OF TESTING!

68. HARD SHIFTING OF PTO

Inspect PTO shift linkage for bends, cracks, and improper lubrication.
a. Lubricate shift linkage if necessary (LO 9-2320-272-12).
b. If bent or cracked, notify DS maintenance.
END OF TESTING!

69. LEAKING LUBRICANT AT PTO

Inspect mounting screws for security.
a. Tighten loose mounting screws.
b. If leak continues, notify DS maintenance.
END OF TESTING!

INSTRUMENTS AND GAUGES

70. SPEEDOMETER OR TACHOMETER NOISY OR ERRATIC

Step 1. Inspect tachometer or speedometer flexible shaft for binding and kinks.
If binding or kinked, replace flexible shaft(s) (para. 3-89 or 3-90).

Step 2. Inspect tachometer or speedometer for proper operation.
a. Test operation of tachometer or speedometer by replacing them with a speedometer or tachometer known to be good (para. 3-88).
b. If test speedometer or tachometer operate properly, replace defective speedometer or tachometer (para. 3-88).
END OF TESTING!

Table 2-2. Mechanical Troubleshooting (Contd).

MALFUNCTION
   TEST OR INSPECTION
      CORRECTIVE ACTION

---

71. AIR PRESSURE GAUGE INOPERATIVE

Replace air gauge with one known to be good (para. 3-87).

    If test gauge works properly, replace defective air gauge (para. 3-87).

        END OF TESTING!

72. OIL PRESSURE GAUGE INOPERATIVE (OIL LEVEL IS CORRECT)

Check oil pressure sending unit for proper operation.

    a. Replace oil pressure sending unit with one known to be good (para. 3-86).

    b. If oil pressure gauge does not work properly, replace oil pressure gauge (para. 3-86).

        END OF TESTING!

RADIO INTERFERENCE SUPPRESSION

73. INTERFERENCE WHILE VEHICLE IN MOTION

Inspect wiring for loose connections and frayed or broken wiring.

    a. Tighten all loose connections,

    b. Replace or repair frayed or broken wiring (para. 3-131).

        END OF TESTING!

74. INTERFERENCE ONLY WHEN ENGINE IS RUNNING

Inspect alternator for defects.

    If defective, replace alternator (para. 3-79 or 3-80).

        END OF TESTING!

FIFTH WHEEL

75. TRAILER WILL NOT HITCH TO FIFTH WHEEL

Inspect coupling jaws for bends and breaks.

    If coupling jaws are bent or broken, replace fifth wheel (para. 3-248).

        END OF TESTING!

MEDIUM WRECKER (M936)

76. HYDRAULIC SYSTEM DOES NOT OPERATE

Step 1. Check hydraulic oil reservoir for proper fluid level (LO 9-2320-272-12).

    Fill oil reservoir to proper fluid level if necessary (LO 9-2320-272-12).

Step 2. Check if PTO is engaged.

    a. Engage PTO control (TM 9-2320-272-10).

    b. If PTO does not engage, perform malfunction 67,

Step 3. Check all hydraulic lines for damage and leaks.

    a. Tighten loose fittings.

    b. If hoses are leaking or damaged, notify DS maintenance.

Table 2-2. Mechanical Troubleshooting (Contd).

MALFUNCTION
TEST OR INSPECTION
CORRECTIVE ACTION

Step 4. Check reservoir drainplug and valve for leaks.
   a. Close valve.
   b. Tighten drainplug.
                    END OF TESTING!

77. HYDRAULIC PUMP NOISY

Check hydraulic oil reservoir for proper fluid level (LO 9-2320-272-12),
   Fill oil reservoir to proper fluid level if necessary (LO 9-2320-272-12).
                    END OF TESTING!

DUMP BODY (M929, M930)

WARNING

When dump body is in raised position, oil in filter system is under pressure. Any movement of the control valve or leakage at the hydraulic cylinder line or hose connection will cause dump body to drop to subframe. Never work under dump body unless safety braces are properly positioned.

78. DUMP BODY WILL NOT RAISE

Step 1. Check if PTO is engaged.
   a. Engage PTO control (TM 9-2320-272-10).
   b. If PTO does not engage, perform malfunction 68.

Step 2. Check all hydraulic lines for damage and leaks.
   a. Tighten loose fittings.
   b. Replace leaking or damaged hoses (para. 3-131).

                    NOTE
            Assistant will help with step 3.

Step 3. Check control valve for proper operation and leaks (front winch).
   a. While assistant operates winch control lever, observe control valve for proper operation.
   b. Tighten loose fittings.
   c. If control valve is defective, notify DS maintenance.

Step 4. Check control valve shaft movement.
   If shaft does not move freely, notify DS maintenance.

Step 6. Inspect pump housing for leaks and signs of overheating when PTO is engaged.
   a. Tighten loose fittings.
   b. Notify DS maintenance.
                    END OF TESTING!

Table 2-2. Mechanical Troubleshooting (Contd).

```
MALFUNCTION
   TEST OR INSPECTION
      CORRECTIVE ACTION
```

79. DUMP BODY WILL NOT LOWER

Step 1. Check for raised safety braces under dump body.

Lower and stow safety braces (TM 9-2320-272-10).

NOTE

Assistant will help with step 2.

Step 2. Check control valve for proper operation and leaks (front winch).

   a. While assistant operates winch control lever, observe control valve for proper operation.

   b. Tighten loose fittings.

   c. If control valve is defective, notify DS maintenance.

END OF TESTING!

80. DUMP BODY WILL NOT HOLD IN RAISED POSITION

Step 1. Check all hydraulic lines for damage and leaks.

   a. Tighten loose fittings.

   b. If hoses are cracked or leaking, notify DS maintenance.

Step 2. Check control valve shaft movement.

If shaft does not move freely, notify DS maintenance.

END OF TESTING!

81. HYDRAULIC PUMP NOISY

Check hydraulic oil reservoir for proper fluid level (LO 9-2320-272-12).

Fill oil reservoir to proper fluid level if necessary (LO 9-2320-272-12).

END OF TESTING!

82. TAILGATE WILL NOT OPEN

Check for bent or broken linkage.

If bent or broken, replace linkage (para. 3-349).

END OF TESTING!

EXPANSIBLE VAN (M934, M935)

83. SIDE PANELS HARD TO RETRACT OR EXPAND

Step 1. Check for dirt and other foreign material in sprocket assembly.

   a. Clean and lubricate sprocket assembly (LO 9-2320-272-12).

   b. If sprocket will not turn or engage properly, notify DS maintenance.

Step 2. Check for corroded, dirty, and damaged rollers.

Clean corroded or dirty rollers. If damaged, notify DS maintenance.

END OF TESTING!

84. SIDE PANEL CANNOT BE LOCKED IN RETRACTED POSITION

Step 1. Visually inspect front edge of side panel to see if it is retracted.

Place heavy block of wood (2 in. x 4 in., or 4 in. x 4 in.) against rub rail at front of panel, Strike block with heavy hammer.

Table 2-2. Mechanical Troubleshooting (Contd).

MALFUNCTION
TEST OR INSPECTION
CORRECTIVE ACTION

    Step 2. Visually check to see if top side panel is too far out to engage edge of roof.

          Place heavy block of wood against flat surface of seal retainer, opposite locking bar at top of side panel. Strike block with heavy hammer.

END OF TESTING!

85. VAN BODY NOT WATERPROOF OR LIGHTTIGHT

    step 1. Visually inspect lower part of side panel for tightness against van body.

          Place heavy block of wood (2 in. x 4 in., or 4 in. x 4 in.) against rub rail at end of side panel where leaks occurs. Strike block with heavy hammer.

    step 2. Check for sagging end panel.

          Add rubber seal material (appendix D, item 632) to seal on outer edge of hinged roof until seal meets top edge of panel door.

    step 3. Check lip of block seal at inner rear corner of hinge roof to see if it is out of position.

          Move side panel out to disengage corner block seal. Push seal lip up into correct position so end panel door properly engages seal when side panel is retracted.

    step 4. Check for loose or worn seal at top of rear doors.

          If worn, recover worn or loose area with sealing material (appendix D, item 632).

END OF TESTING!

86. DOOR LOCK WILL NOT OPERATE

    Step 1. Check for jammed lock bolt.

          If lock bolt is jammed, repair lock assembly (para. 3-364).

    Step 2. Check alignment of lock bolt to striker plate.

          Add or remove shims behind lock until bolt properly engages striker plate (para. 3-364).

END OF TESTING!

87. HEATER WILL NOT IGNITE

    Check electrical system (Electrical Troubleshooting, para. 2-22).

END OF TESTING!

2-22. ELECTRICAL SYSTEM TROUBLESHOOTING

a.   This section provides information to diagnose and correct malfunctions of the electrical system. Because of the complexity, the electrical system is divided into the following functional systems:

- Engine and Vehicle Electrical System (page 2-95)
- Battery System (page 2-97)
- Starting System (page 2-101)
- Generating System (page 2-118)
- Lighting System (page 2-124)
- Directional Signal System (page 2-135)
- Ether Start System (page 2-141)
- Indicator, Gauge, and Warning System (page 2-145)
- Trailer Connection System (page 2-163)
- Heater System (page 2-165)
- 100-Amp Alternator Kit (page 2-169)
- Transfer Case System (page 2-174)
- Protective Control Box Assembly (page 2-176)

b.   Principles of operation showing wiring diagrams for each system can be found in chapter 1. The wiring schematic (appendix H) shows the interrelationship of these systems. Both should be utilized as references when performing electrical troubleshooting (table 2-3).

c.   Each malfunction symptom given for an individual component or system is followed by step(s) that should be taken to determine the cause and then corrective action that must be taken to remedy the problem.

d.   Before taking any action to correct a possible malfunction, the following rules should be followed:

(1)   Question the vehicle operator to obtain any information that might help determine the cause of the problem.

(2)   Never overlook the chance that the problem could be of a simple origin. The problem could be corrected with minor adjustment.

(3)   Use all senses to observe and locate troubles.

(4)   Use test instrument or gauges to help determine and isolate problems.

(5)   Always isolate the system where the malfunction occurs and then locate the defective component.

(6)   Use Principles of Automotive Vehicles, TM 9-8000, when troubleshooting vehicles covered in this manual.

e.   Table 2-3 lists electrical malfunctions that may occur in individual units or systems of the vehicle. This table covers electrical troubleshooting only. Troubleshooting procedures for the mechanical systems can be found in table 2-1, para. 2-21.

f.   This section cannot list all the electrical malfunctions that may occur. If a malfunction occurs that is not listed in table 2-3, notify DS maintenance.

ELECTRICAL TROUBLESHOOTING SYMPTOM INDEX

MALFUNCTION NO.	MALFUNCTION	TROUBLESHOOTING PROCEDURE PAGE

ELECTRICAL TROUBLESHOOTING SYMPTOM INDEX (Contd)

Table 2-3. Electrical Troubleshooting.

MALFUNCTION
TEST OR INSPECTION
CORRECTIVE ACTION

ENGINE AND VEHICLE ELECTRICAL SYSTEM

1. ENGINE AND VEHICLE ELECTRICAL SYSTEM MALFUNCTIONING

    Test 1. Check alternator output.

        Step 1. Remove two screws (2), lockwashers (3), and terminal cover (1) from alternator (4).

        Step 2. Connect black test lead to ground and red test lead to lead 5.

        Step 3. Check alternator output (malfunction 9). Multimeter should indicate 26.5-29.5 volts.

        Step 4. If voltage is within range, go to step 6.

        Step 5. If voltage is not within range, adjust alternator (4). Perform malfunction 14.

        Step 6. Install terminal cover (1) on alternator (4) with two lockwashers (3) and screws (2).

    Test 2. Check vehicle ground straps.

        a. Ensure ground straps are securely fastened and no paint is under starwasher.

        b. Remove paint as necessary.

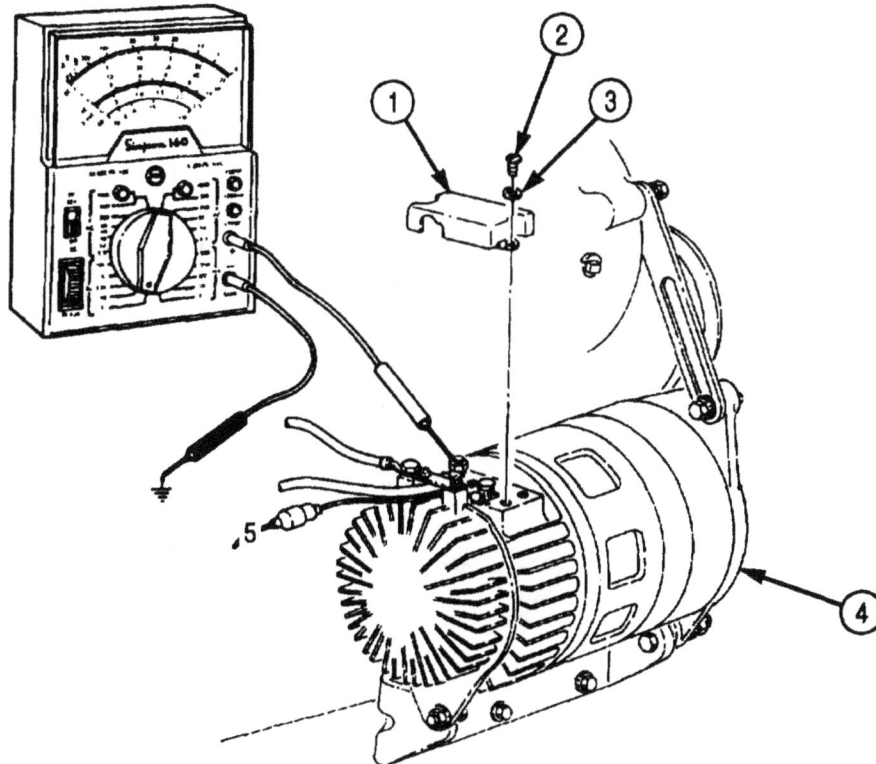

Table 2-3. Electrical Troubleshooting (Contd).

MALFUNCTION
TEST OR INSPECTION
CORRECTIVE ACTION

Test 3. Check protective control box,
    Step 1.   Unplug harness (2) from bottom of protective control box (1).
    Step 2.   Listen for protective control box relay to reset.
    Step 3.   If reset is heard, replace protective control box (1) (para. 3-116).

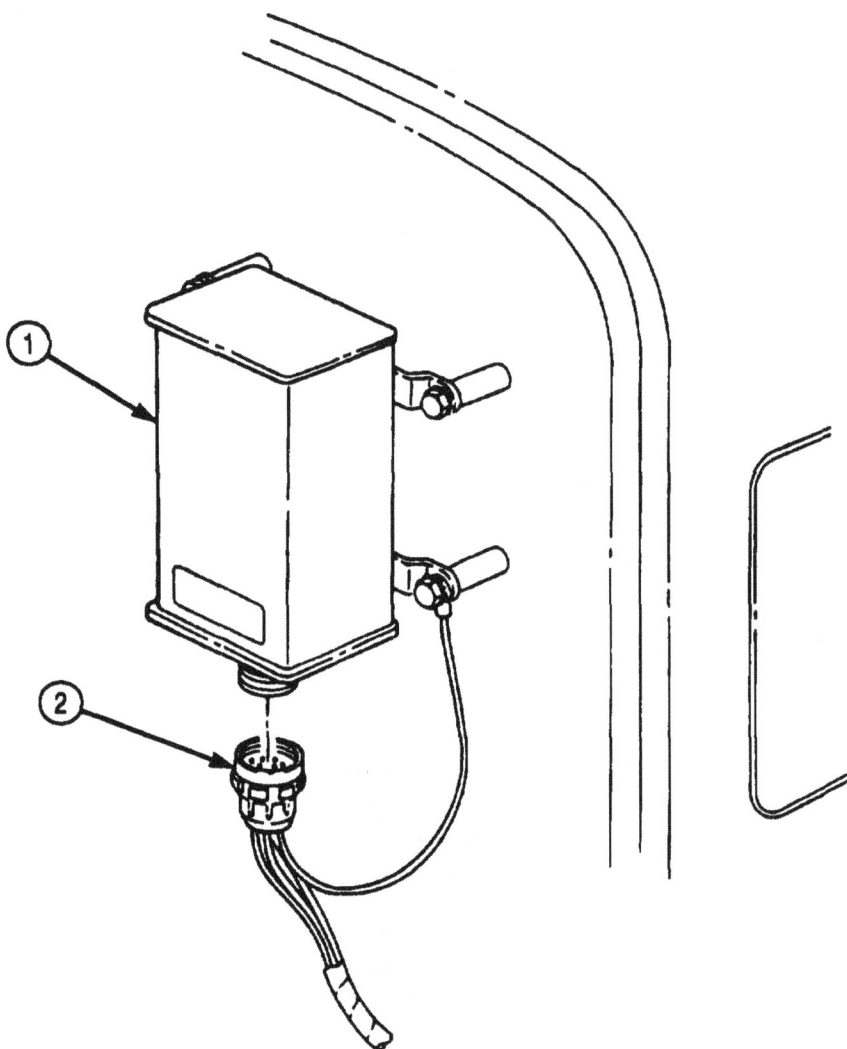

END OF TESTING!

Table 2-3. Electrical Troubleshooting (Contd).

MALFUNCTION
    TEST OR INSPECTION
        CORRECTIVE ACTION

BATTERY SYSTEM

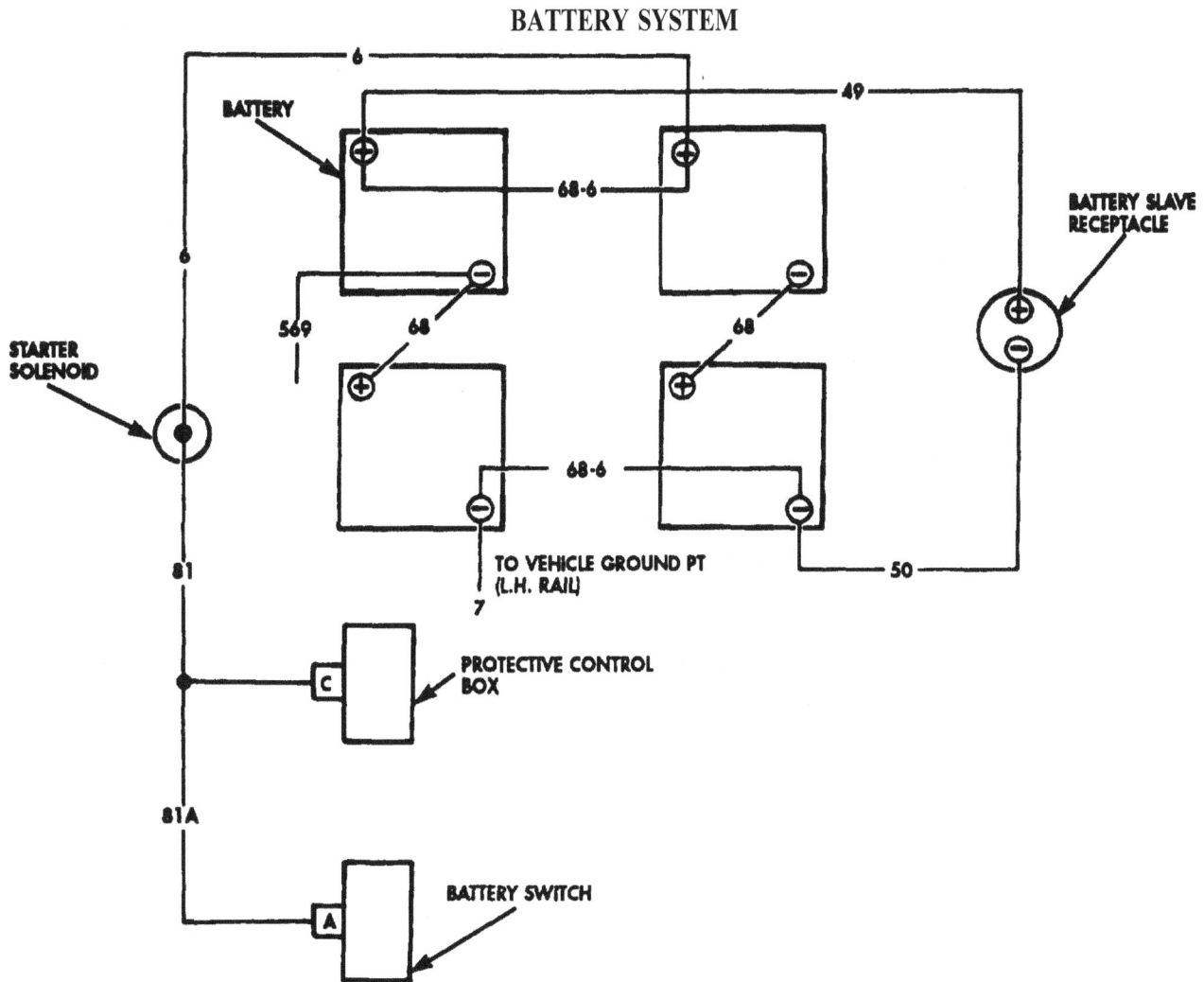

2. BATTERIES ARE HOT, ELECTROLYTE IS BOILING, OR EXCESSIVE USE OF WATER

NOTE

If STE/ICE is available, perform NG50 - Charging Circuit Tests (para. 2-24).

Check electrolyte temperature and specific gravity and note reading (TM 9-6140-200-14).

    a. If temperature is over 120°F (49°C) and specific gravity is 1.280 or greater, batteries are being overcharged. Go to generating system troubleshooting (malfunction 9, test 1).

    b. If battery temperature is over 120°F (49°C), but specific gravity is 1.236-1.260, recharge batteries (TM 9-6 140-200-14).

END OF TESTING!

Table 2-3. Electrical Troubleshooting (Contd).

MALFUNCTION
TEST OR INSPECTION
CORRECTIVE ACTION

3. SPECIFIC GRAVITY WILL NOT INCREASE TO 1.280 UNDER CHARGE

NOTE

If STE/ICE is available, perform NG50 - Charging Circuit Tests (para. 2-24).

Check rate of charging:

Place battery on charge, assuring that cells are gassing freely (TM 9-6140-200-14). Maintain charge rate slightly below heavy gassing.

If specific gravity does not recover to 1.280 in 25 hours of charging, replace battery (para. 3-125).

END OF TESTING!

4. ENGINE WILL NOT CRANK; SOME ELECTRICAL SYSTEMS INOPERATIVE OR WEAK

Test 1. Perform the following steps to inspect batteries:

Step 1. Visually check batteries for cracks, leaks, and corroded or broken terminal posts.

    a. Replace any cracked, leaking, corroded, or broken batteries, or batteries with loose or broken terminal posts (para. 3-125).

    b. Clean corroded terminal posts to bright metal.

Step 2. Check for loose, broken, or worn terminals and cables.

    a. Tighten any loose terminal or cable (para. 3-124 or 3-126).

    b. If any terminal or cable is broken or worn, replace (para. 3-124 or 3-126).

Step 3. Check electrolyte level in each cell (TM 9-6140-200-14).

Fill each cell to fill ring with distilled water (TM 9-6140-200-14).

Step 4. Perform a specific gravity test (TM 9-6140-200-14). Batteries must be 1.225 or greater, temperature corrected, and each cell in a battery must test within 25 points of the others.

    a. Charge all batteries not meeting requirements, and check specific gravity again (TM 9-6140-200-14).

    b. If 25-point variation still exists, replace battery(ies) (para. 3-125).

Step 5. Attempt to crank engine for 15 seconds, then turn battery switch to OFF position. Check batteries for overheating by feeling terminal connections. If battery terminal(s) is hot, loose or corroded connections are indicated.

    a. Tighten all loose connections at batteries.

    b. Tighten battery ground wire (lead 7) at vehicle chassis ground. Tighten battery positive wire (lead 6) at starter solenoid.

NOTE

If STE/ICE is available, perform NG20 - No Crank - No Start (para. 2-24).

Test 2. Test batteries under load to determine adequate current capability and voltage drop during a 15-second amperage load.

Table 2-3. Electrical Troubleshooting (Contd).

MALFUNCTION
    TEST OR INSPECTION
        CORRECTIVE ACTION

Step 1.   Set multimeter to 50-volt range.

Step 2.   Connect multimeter positive lead to starter solenoid terminal 6 and negative lead to ground strap. Multimeter should indicate voltage.

Step 3.   Place headlight and battery switch on ON position for 15 seconds (headlights on high beam). Multimeter should register 1-volt drop from step 2 reading.

Recharge batteries when voltage reading is low (TM 9-6140-200-14).

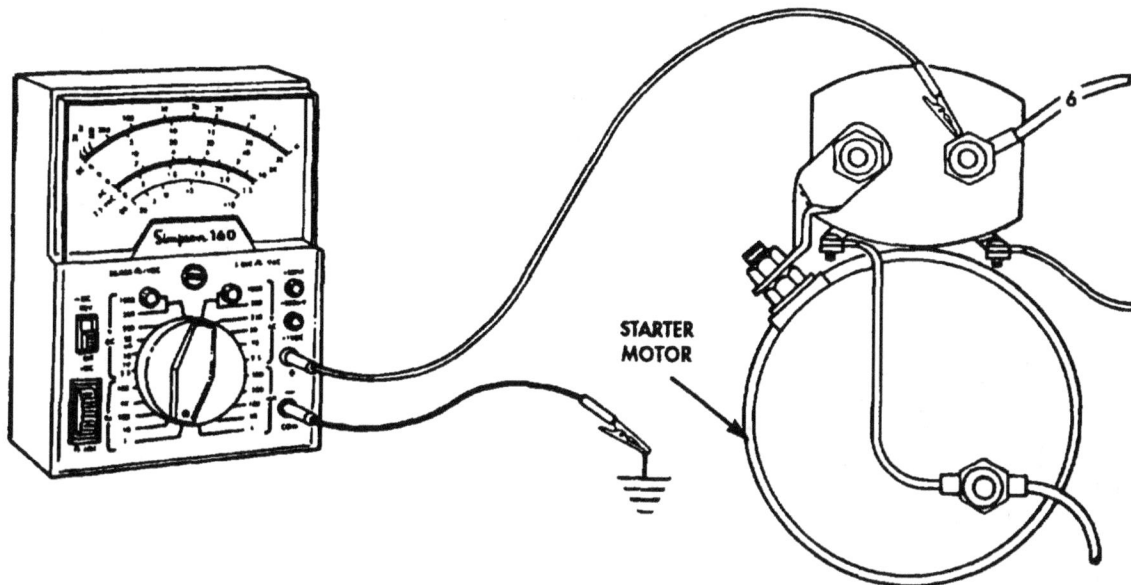

END OF TESTING!

5. ALL VEHICLE ELECTRICAL SYSTEMS INOPERATIVE

Test 1. Check connection of battery cables.

Ensure battery cables are correctly connected to batteries. (See wiring diagram in Appendix H.)

If not correctly connected, secure battery cables (para. 3-124 or 3-126).

Test 2. Inspect batteries (malfunction 4, test 1).

Table 2-3. Electrical Troubleshooting (Contd).

MALFUNCTION
TEST OR INSPECTION
CORRECTIVE ACTION

NOTE

If STE/ICE is available, perform NG81 - Battery Tests (para. 2-24).

Test 3. Check protective control box continuity from pin C to pin D.

Step 1. Place battery switch to OFF position.

Step 2. Disconnect harness connector from protective control box.

Step 3. Set multimeter to RX1 for continuity reading.

Step 4. Apply battery voltage to pin A of control box.

Step 5. Touch negative multimeter lead to pin C and positive lead to pin D on control box.

  a. Continuity should be indicated between pins C and D.

  b. If continuity is not indicated, replace protective control box (para. 3-115).

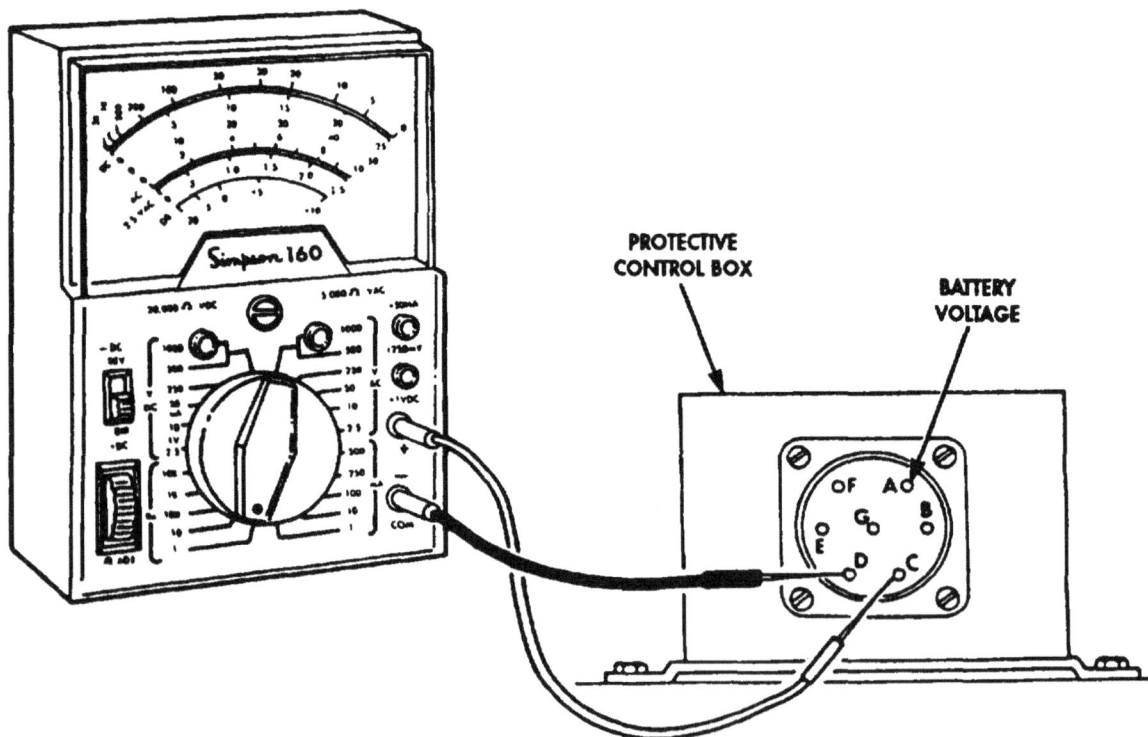

END OF TESTING!

Table 2-3. Electrical Troubleshooting (Contd).

MALFUNCTION
TEST OR INSPECTION
CORRECTIVE ACTION

STARTING SYSTEM

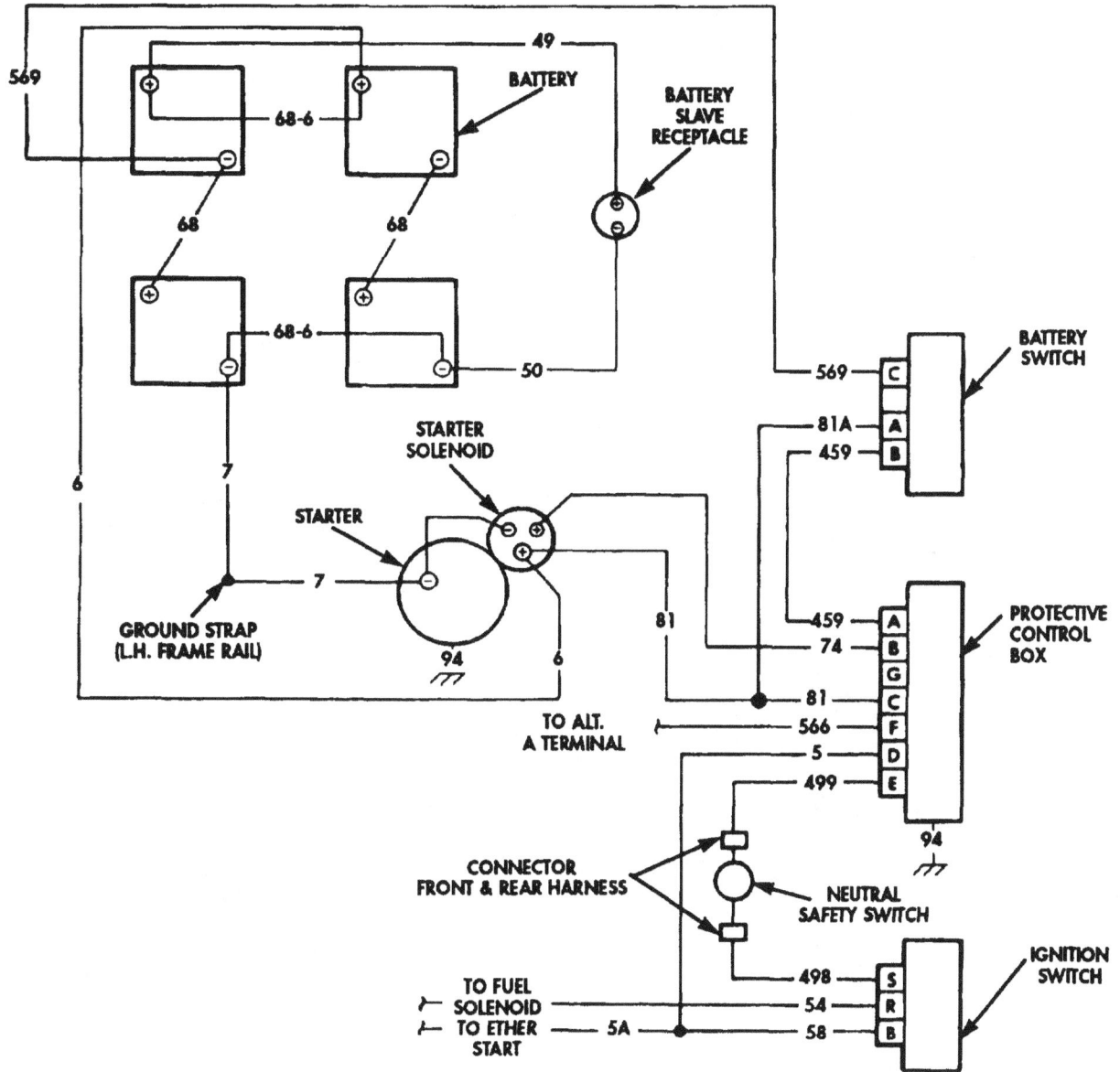

6. STARTER MOTOR INOPERATIVE

NOTE

If STE/ICE is available, perform NG80 - Starter Circuit Tests (para. 2-24).

Table 2-3. Electrical Troubleshooting (Contd).

MALFUNCTION
TEST OR INSPECTION
CORRECTIVE ACTION

Check starter solenoid operation.

    Step 1. Turn starter switch to START position and listen for starter solenoid to engage.

          a. If thump of starter solenoid energizing is heard, perform malfunction 7.

          b. If thump of starter solenoid is not heard, proceed to step 2.

    Step 2. Set multimeter to 50-volt range.

    Step 3. Connect multimeter positive lead to starter solenoid terminal lead 74 and negative lead to starter ground as shown.

    Step 4. Turn starter switch to START position and observe multimeter for a 24-volt reading.

          a. If battery voltage is not indicated, perform malfunction 7.

          b. If battery voltage is indicated, replace starter (para. 3-82).

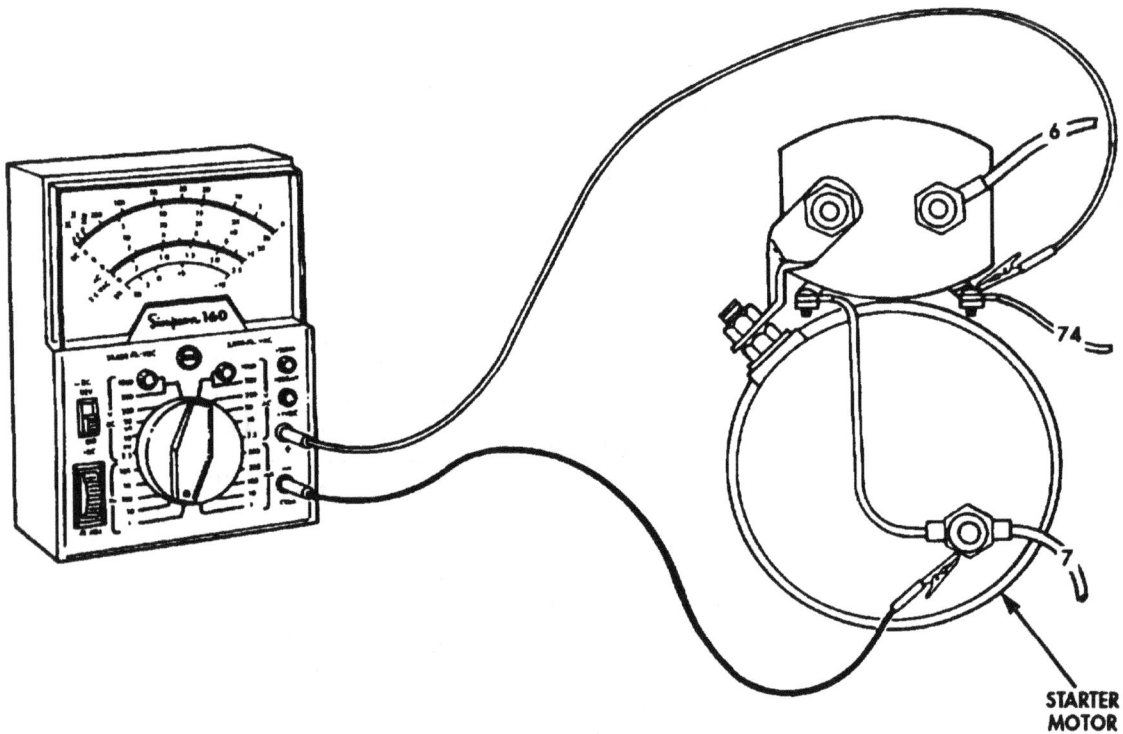

STARTER MOTOR

END OF TESTING!

Table 2-3. Electrical Troubleshooting (Contd).

MALFUNCTION
TEST OR INSPECTION
CORRECTIVE ACTION

7. SOLENOID OPERATES, STARTER OPERATES, BUT ENGINE CRANKS SLOWLY

NOTE

If STE/ICE is available, perform NG80 - Starter Circuit Tests (para. 2-24).

Test 1. Check batteries for overheating. Crank engine for 15 seconds. Check batteries for overheating by feeling battery terminal connections. If battery terminal is hot, a loose or corroded connection is indicated.

    a. Clean corroded connection to bright metal.

    b. Tighten all loose connections at batteries, ground, and starter.

Test 2. Perform a specific gravity test (TM 9-6140-200-14). Batteries must be 1.225 or greater, temperature corrected, and each cell in a battery must test within 25 points of the others.

    a. Charge all batteries not meeting requirements, and check specific gravity again (TM 9-6140-200-14).

    b. If 25-point variation still exists, replace battery(ies) (para. 3-125).

Test 3. Test starter voltage.

    Step 1.   Set multimeter to 50-volt range.

    Step 2.   Connect multimeter positive lead to positive terminal stud of starter and negative lead to terminal stud on end plate of starter.

    Step 3.   Crank engine and observe voltage reading on multimeter. Voltage reading should exceed 22 volts.

    If voltage reading is low, clean and tighten connections.

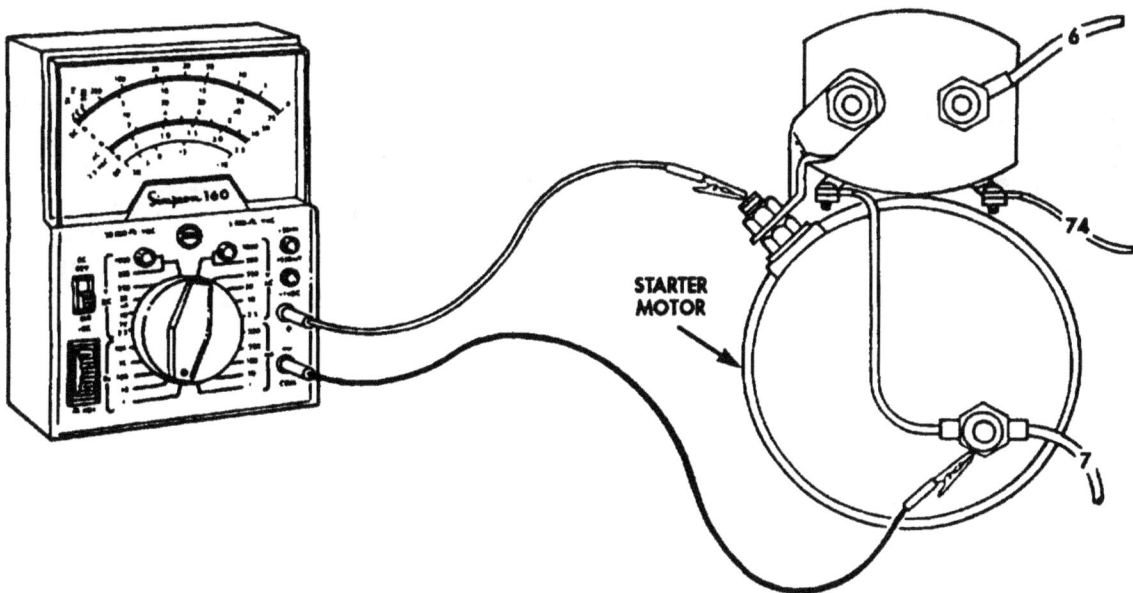

Table 2-3. Electrical Troubleshooting (Contd).

MALFUNCTION
TEST OR INSPECTION
CORRECTIVE ACTION

Test 4. Test starter-to-solenoid strap.

    Step 1. Set multimeter to 10-volt range.

    Step 2. Connect multimeter negative lead to positive terminal stud of starter and positive lead to terminal stud of solenoid.

    Step 3. Crank engine and observe multimeter. A voltage reading exceeding 0.1 volt indicates bad connection at starter terminal stud and terminal stud of solenoid.

            Clean and tighten connections.

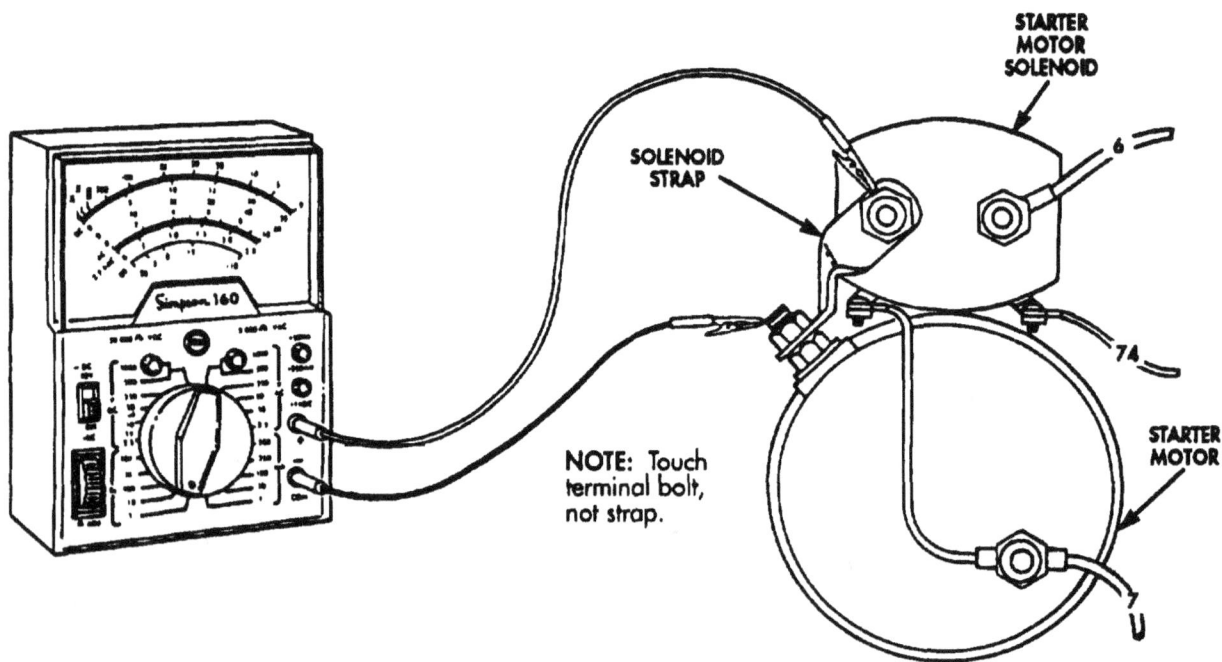

STARTER
MOTOR
SOLENOID

SOLENOID
STRAP

6

74

STARTER
MOTOR

NOTE: Touch
terminal bolt,
not strap.

Table 2-3. Electrical Troubleshooting (Contd).

MALFUNCTION
TEST OR INSPECTION
CORRECTIVE ACTION

Test 5. Test starter solenoid contractors.

    Step 1. Set multimeter to 10-volt range.

    Step 2. Connect multimeter between starter solenoid terminals.

    Step 3. Crank engine and observe multimeter. A voltage reading exceeding 0.4 volt indicates defective starter solenoid.

        a. Replace starter (para. 3-82).

        b. If malfunction still exists, go to tests 6, 7, and 8.

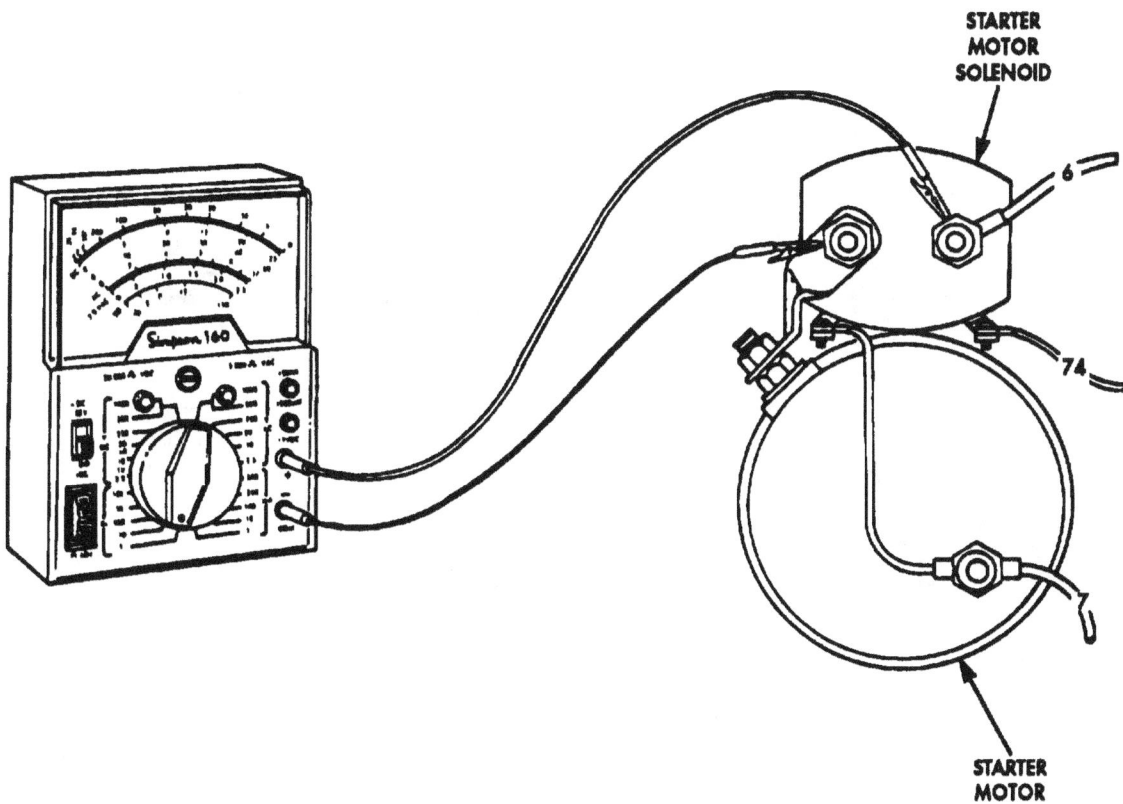

STARTER
MOTOR
SOLENOID

STARTER
MOTOR

Table 2-3. Electrical Troubleshooting (Contd).

MALFUNCTION
TEST OR INSPECTION
CORRECTIVE ACTION

Test 6. Test negative cable 7 voltage drop from batteries to starter.

  Step 1.   Set multimeter to lo-volt range.

  Step 2.   Connect multimeter positive lead to terminal stud on end plate of starter and negative lead to grounding point at batteries.

  Step 3.   Crank engine and observe multimeter. A voltage reading exceeding 0.4 volt indicates loose or corroded connection(s).

           Clean and tighten connections at batteries, starter, and chassis (para. 3-124).

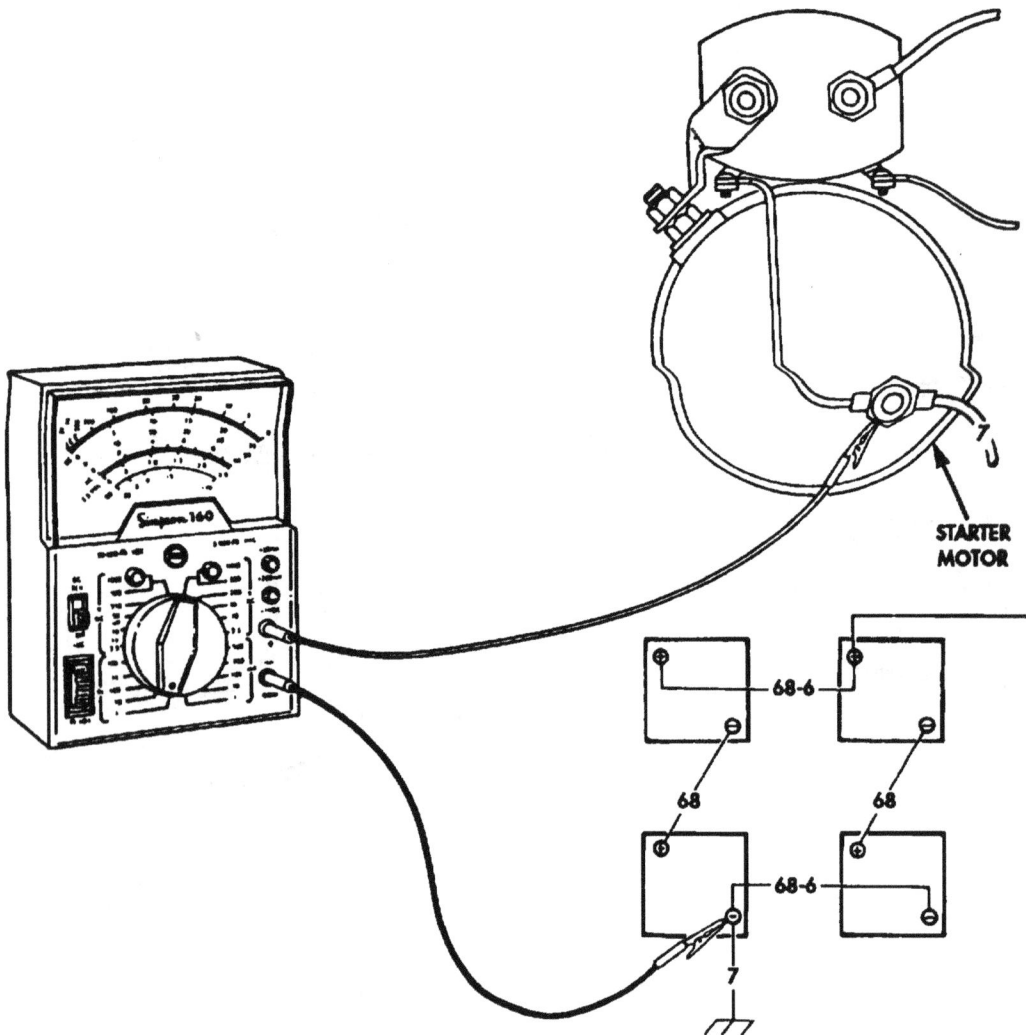

Table 2-3. Electrical Troubleshooting (Contd).

MALFUNCTION
> **TEST OR INSPECTION**
> > **CORRECTIVE ACTION**

Test 7. Test positive cable 6 from batteries to starter solenoid.

    Step 1.   Set multimeter to lo-volt range.

    Step 2.   Connect multimeter positive lead to positive terminal point on batteries and negative lead to positive terminal on starter solenoid.

    Step 3.   Crank engine and observe multimeter. A voltage reading exceeding 0.4 volt indicates loose or corroded connection.

              Clean and tighten connections at batteries, starter, and chassis (para. 3-124).

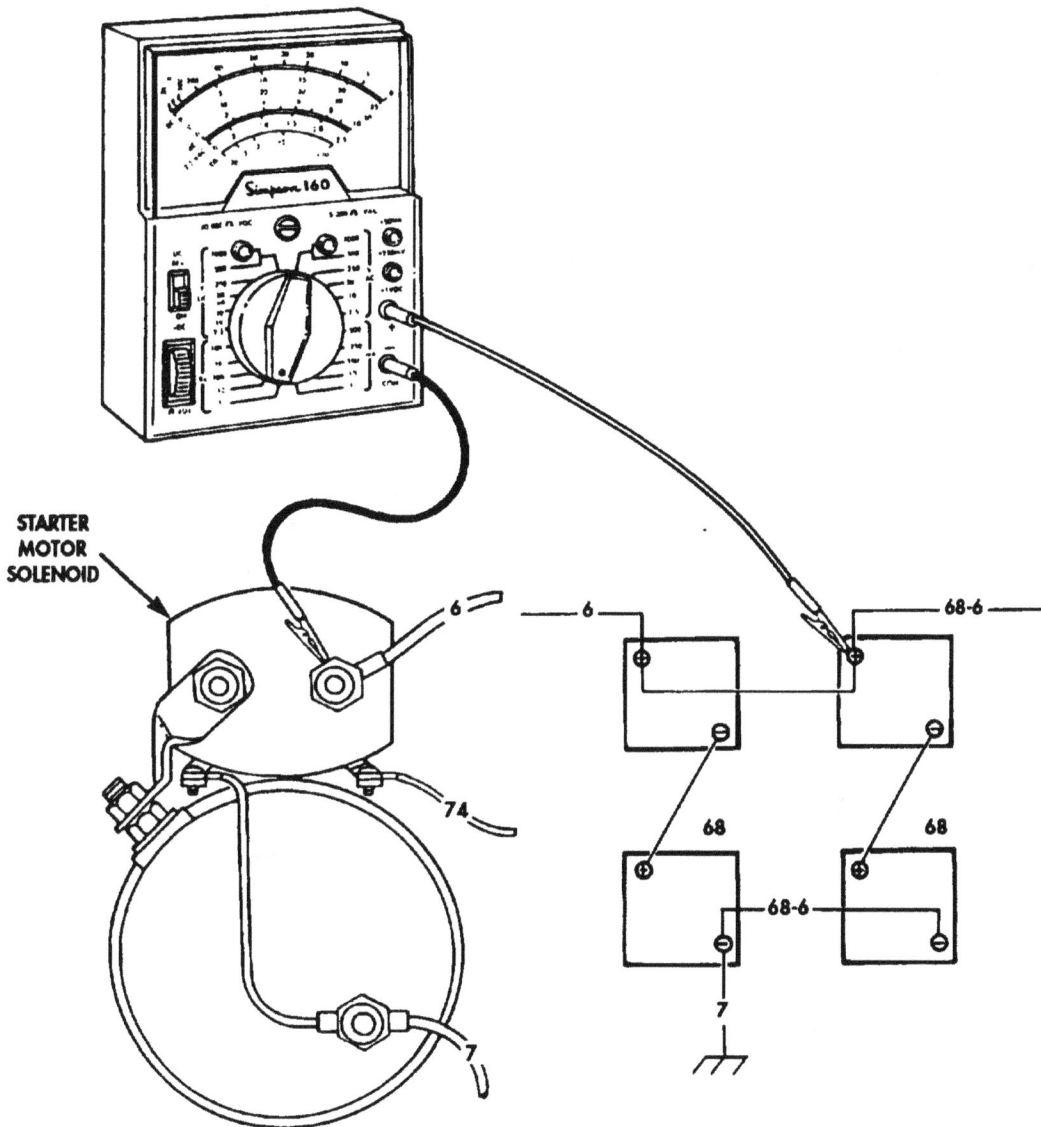

Table 2-3. Electrical Troubleshooting (Contd).

| MALFUNCTION |
| TEST OR INSPECTION |
| CORRECTIVE ACTION |

Test 8. Test battery voltage after cranking load is applied.

    Step 1.   Set multimeter to 50-volt range.

    Step 2.   Connect multimeter directly across battery terminal posts: positive lead to positive post, and negative lead to negative post.

    Step 3.   Engage emergency stop control and crank engine for 15 seconds (TM 9-2320-272-10). Voltage reading should be 20 volts or more after cranking has stopped.

          a. If battery voltage is less than 20 volts, perform malfunction 2.

          b. If battery voltage exceeds 20 volts, replace starter (para. 3-82).

    Step 4.   Disengage emergency stop control and crank engine (TM 9-2320-272-10).

             If engine still cranks slowly, notify DS maintenance.

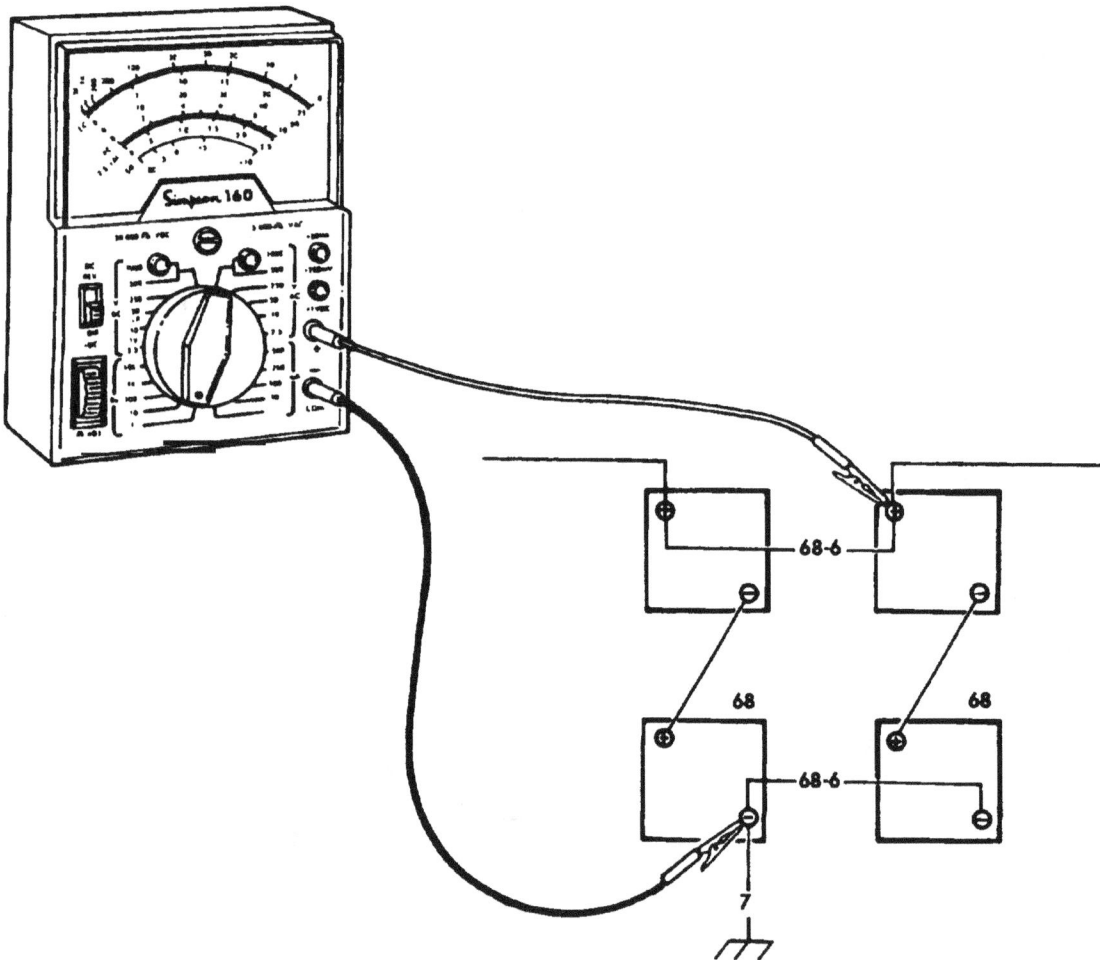

END OF TESTING!

2-108

Table 2-3. Electrical Troubleshooting (Contd).

MALFUNCTION
TEST OR INSPECTION
CORRECTIVE ACTION

8. STARTER MOTOR INOPERATIVE; NO SOLENOID THUMP

Test 1. Test battery and starter systems (malfunctions 2, 3, 4, 6, and 7).

Test 2. Test starter lockout reset.

   Step 1.   Disconnect harness connector from protective control box.

   Step 2.   Connect harness connector to protective control box to reset starter lockout switch.

   Step 3.   Start engine (TM 9-2320-272-10).

      a. If engine fails to start, go to test 3.

      b. If engine starts, replace alternator (para. 3-79 or 3-80).

Test 3. Test battery switch for voltage.

   Step 1.   Set multimeter to 50-volt range.

   Step 2.   Disconnect lead 81A (pin A) from battery switch.

   Step 3.   Connect multimeter positive lead to contact end of lead 81A and negative lead to ground.

      a. Multimeter should indicate battery voltage. If voltage is not indicated, repair broken lead 81A/81 (para. 3-131).

      b. If voltage is indicated, go to step 4.

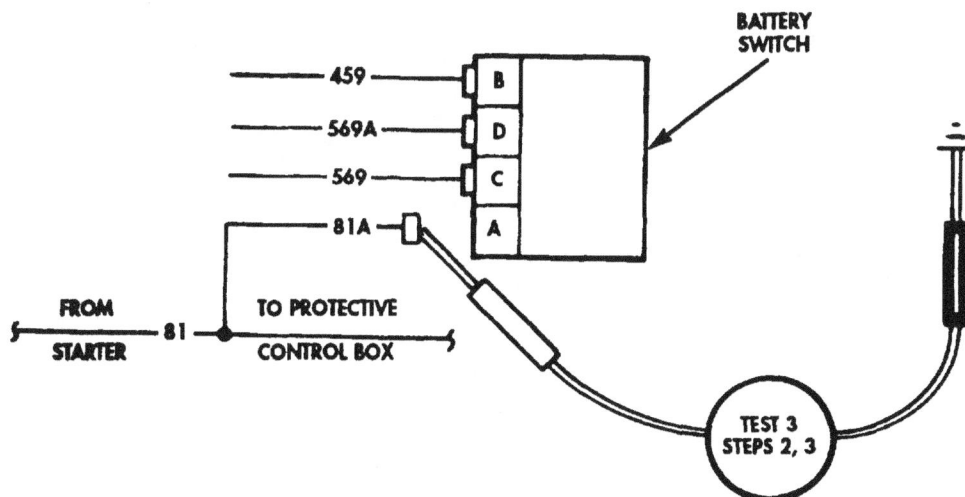

Table 2-3. Electrical Troubleshooting (Contd).

MALFUNCTION
TEST OR INSPECTION
CORRECTIVE ACTION

Step 4. Connect lead 81A to battery switch pin A.

Step 5. Disconnect lead 459 (pin B) at battery switch.

Step 6. Place battery switch to ON position.

Step 7. Connect multimeter positive lead to contact end of pin B.

Step 8. Connect multimeter negative lead to ground.

    a. Multimeter should indicate battery voltage. If voltage is not indicated, replace battery switch (para. 3-107).

    b. If voltage is indicated, connect lead 459 to battery switch (pin B), place battery switch to OFF position, and go to test 4.

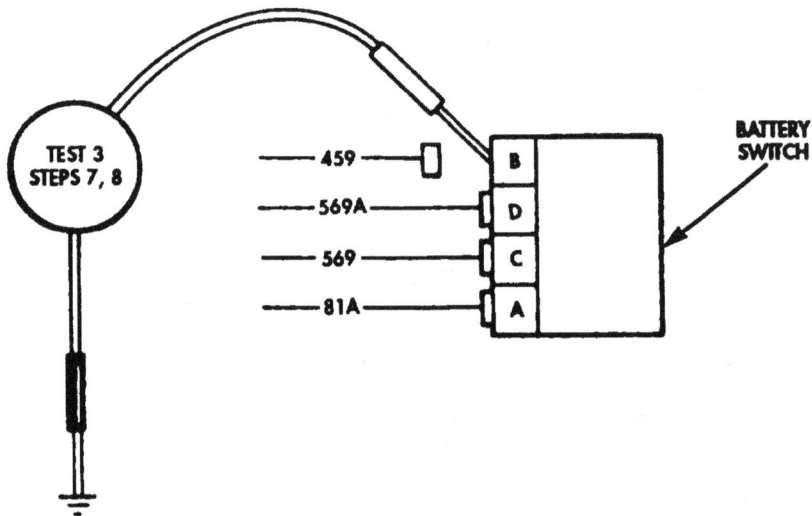

Table 2-3. Electrical Troubleshooting (Contd).

MALFUNCTION
TEST OR INSPECTION
CORRECTIVE ACTION

Test 4. Test voltage input to protective control box.

Step 1. Set multimeter to 50-volt range.

Step 2. Disconnect harness connector at protective control box.

Step 3. Place battery switch to ON.

Step 4. Connect multimeter positive lead to pin A (lead 459) at harness connector and negative lead to ground.

   a. Multimeter should indicate voltage. If voltage is not indicated, repair broken lead 459 (para. 3-131).

   b. If voltage is indicated, go to step 5.

Step 5. Place battery switch to OFF position.

Step 6. Connect multimeter positive lead to pin C (lead 81) at harness connector and negative lead to ground.

   a. Multimeter should indicate voltage. If voltage is not indicated, repair broken lead 81 (para. 3-131).

   b. If voltage is indicated, go to test 5.

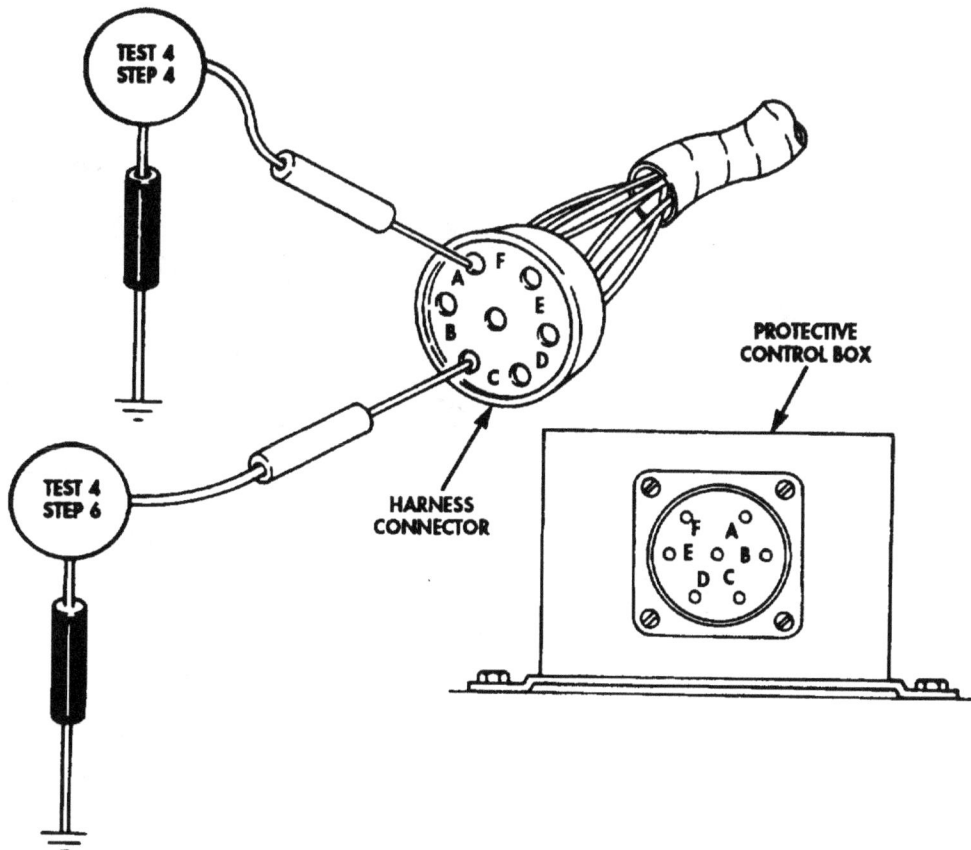

2-111

Table 2-3. Electrical Troubleshooting (Cantd).

MALFUNCTION
TEST OR INSPECTION
CORRECTIVE ACTION

Test 5. Test continuity of protective control box harness.

    Step 1. Dirconnect battery ground cables (para. 3-126).

    Step 2. Set multimeter to RX1 for continuity reading.

    Step 3. Disconnect lead 74 at starter solenoid.

    Step 4. Connect multimeter positive lead to pin B (lead 74) at harness connector and negative lead to lead 74 at starter solenoid.

        a. Multimeter should indicate continuity. If continuity is not indicated, repair broken lead 74 (para. 3-131).

        b. If continuity is indicated, go to step 5.

    Step 5. Connect lead 74 at starter solenoid.

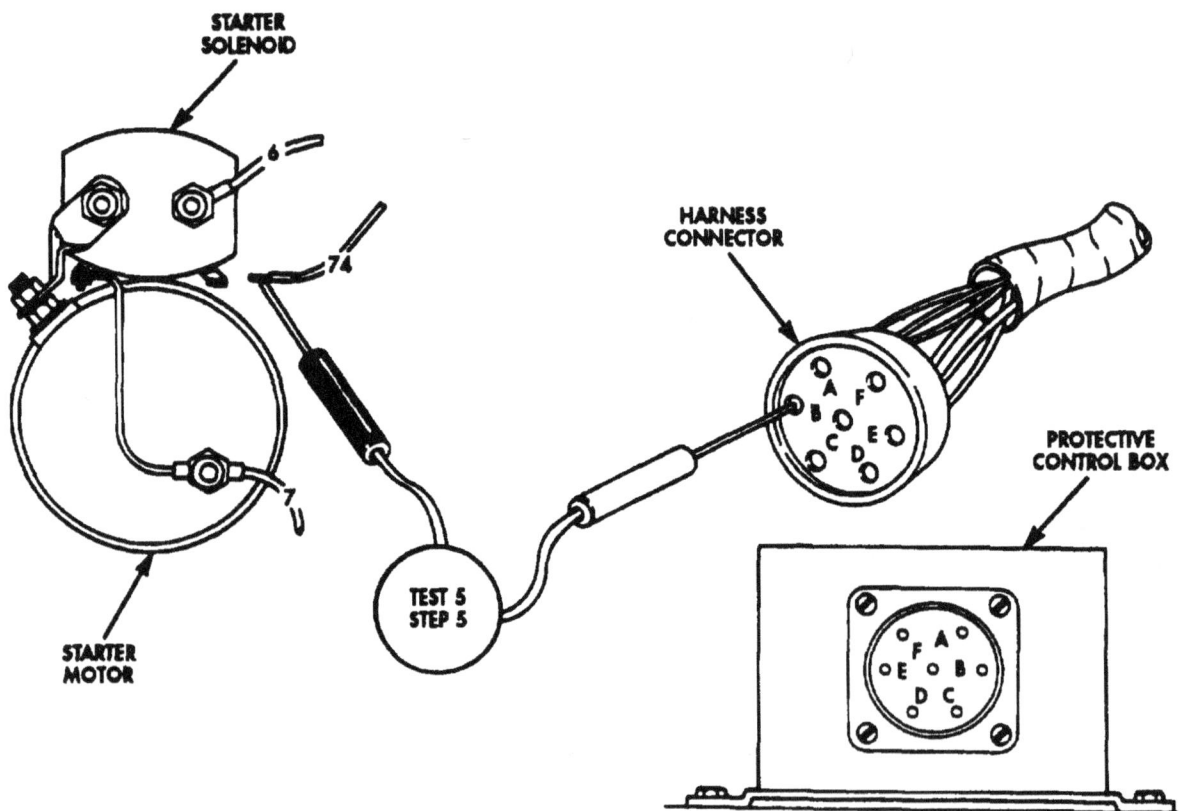

Table 2-3. Electrical Troubleshooting (Contd).

MALFUNCTION
TEST OR INSPECTION
CORRECTIVE ACTION

Step 6. Disconnect lead 5B (pin B) at starter switch.

Step 7. Connect multimeter positive lead to pin D at harness connector and negative lead to contact end of lead 5B at starter switch.

    a. Multimeter should indicate continuity. If continuity is not indicated, repair broken lead 5A/5B (para. 3-131).

    b. If continuity is indicated, go to test 6.

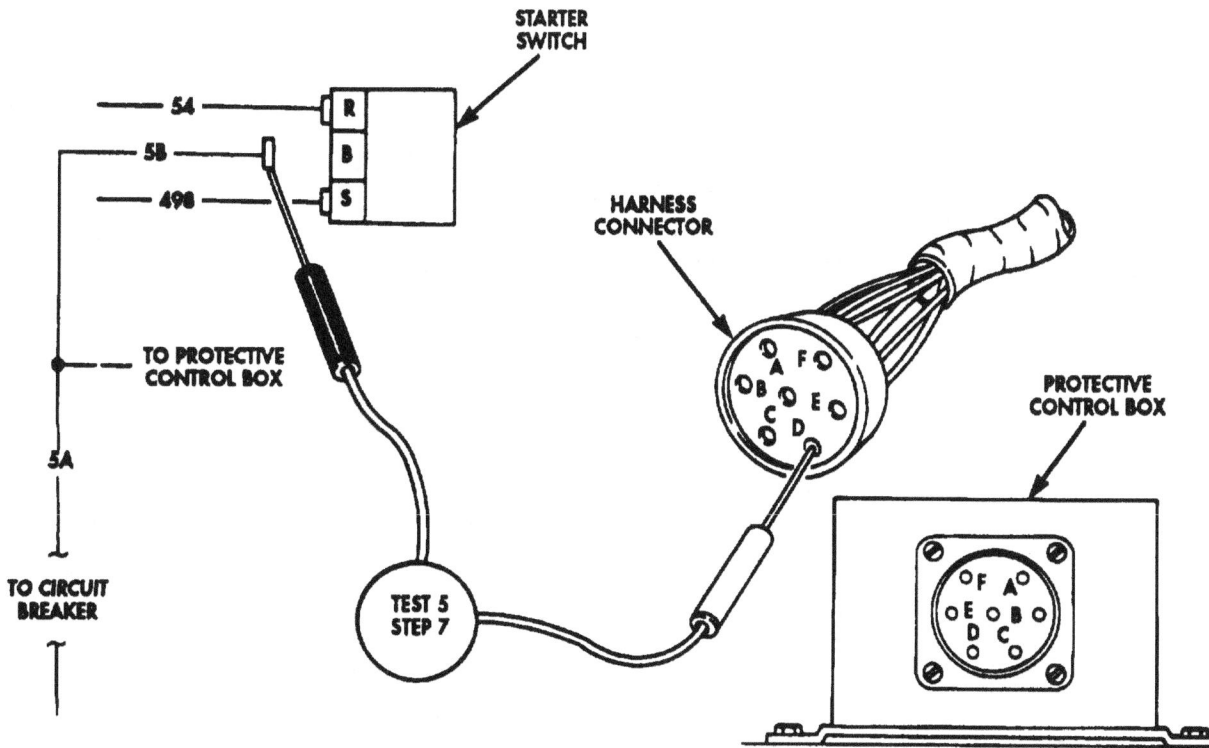

Table 2-3. Electrical Troubleshooting (Contd).

MALFUNCTION
    TEST OR INSPECTION
        CORRECTIVE ACTION

    Test 6. Test starter switch for continuity.

        Step 1. Disconnect lead 498 (pin S) at starter switch.

        Step 2. Connect multimeter positive lead to pin B and negative lead to pin S.

        Step 3. Place stark switch to START position and hold.

            a. Multimeter should indicate continuity. If continuity is not indicated, replace starter switch (para. 3-107).

            b. If continuity is indicated, connect lead 5B to starter switch (pin B) and go to test 7.

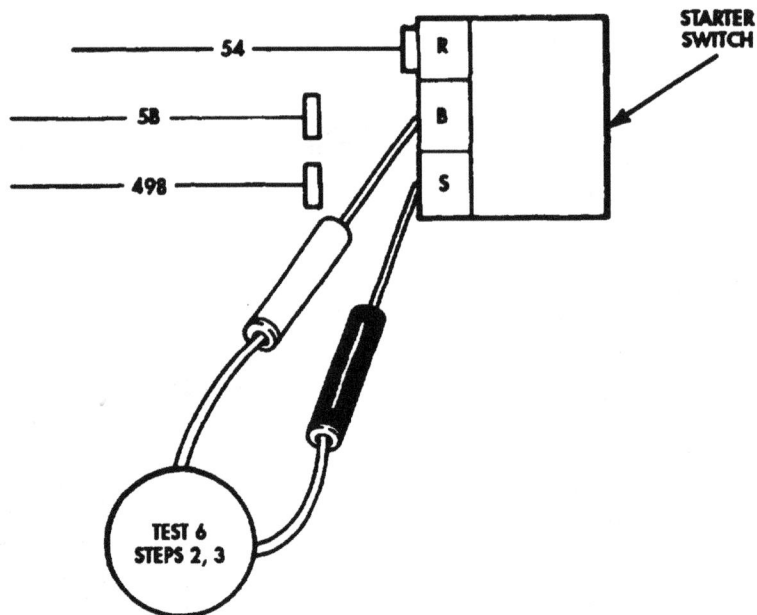

Table 2-3. Electrical Troubleshooting (Contd).

```
┌─────────────────────────────────────────────────────────────────────────────┐
│ MALFUNCTION                                                                   │
│     TEST OR INSPECTION                                                        │
│           CORRECTIVE ACTION                                                   │
└─────────────────────────────────────────────────────────────────────────────┘
```

Test 7. Test continuity of protective control box harness.

   Step 1.   Set parking brake (TM 9-2320-272-10).

   Step 2.   Place transmission in neutral.

   Step 3.   Disconnect lead 498 at neutral start switch.

   Step 4.   Connect multimeter positive lead to contact end of lead 498 and negative lead to contact end of lead 498 at neutral start switch.

       a. Multimeter should indicate continuity. If continuity is not indicated, repair broken lead 498 (para. 3-131).

       b. If continuity is indicated, connect lead 498 to start switch (pin S) and go to step 5.

   Step 5.   Disconnect lead 499 at neutral start switch.

   Step 6.   Connect multimeter positive lead to neutral start switch contact end of lead 499 and negative lead to neutral start contact end of lead 498.

       a. Multimeter should indicate continuity. If continuity is not indicated, replace neutral start switch (para. 3-98).

       b. If continuity is indicated, connect lead 498 to neutral start switch and go to step 7.

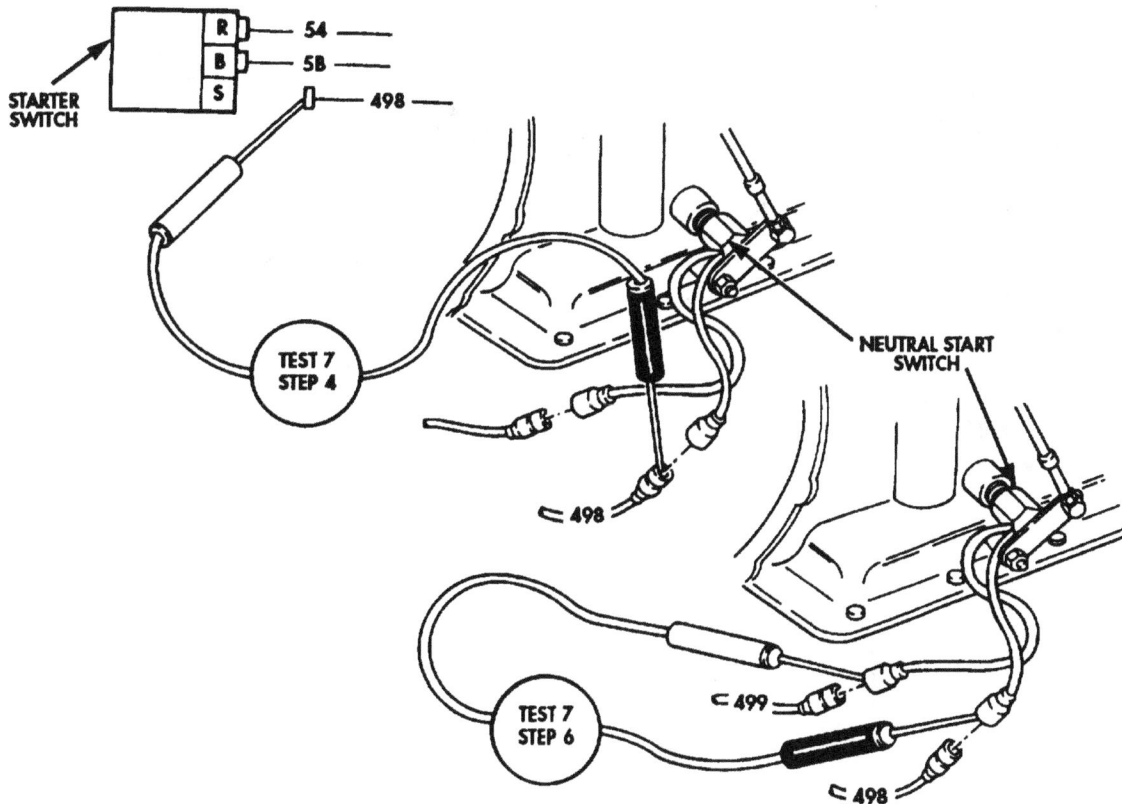

Table 2-3. Electrical Troubleshooting (Contd).

MALFUNCTION
    TEST OR INSPECTION
        CORRECTIVE ACTION

Step 7. Disconnect harness connector at protective control box.

Step 8. Connect multimeter positive lead to contact end of lead 499 and negative lead to pin E at harness connector.

    a. Multimeter should indicate continuity. If continuity is not indicated, repair broken lead 499 (para. 3-131).

    b. If continuity is indicated, connect leads 499 and 498 to neutral start switch and go to test 8.

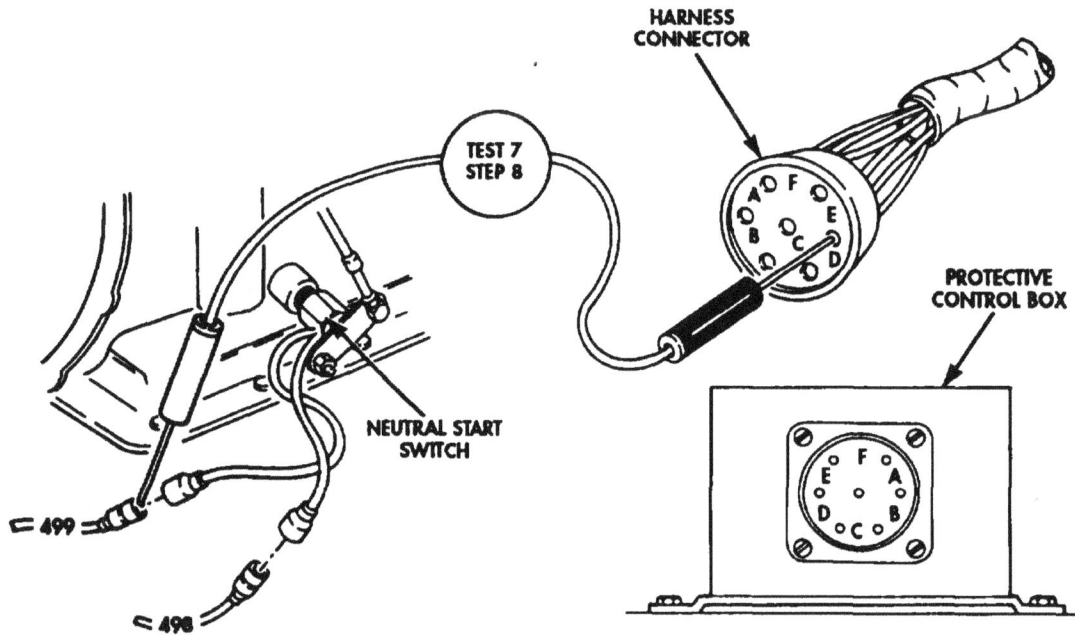

Table 2-3. Electrical Troubleshooting (Contd).

MALFUNCTION
TEST OR INSPECTION
CORRECTIVE ACTION

Test 8. Test continuity of protective control box harness.

NOTE

Sealant must be removed before removing leads.

Step 1. Disconnect lead 566 at alternator.

     a. Remove two screws (2), lockwashers (3), and terminal cover (1) from alternator (4).

     b. Remove nut (7), washer (6), lockwasher (5), and lead 566 from alternator (4).

Step 2. Connect multimeter positive lead to pin F at harness connector and negative lead to contact end of lead 566 at alternator.

     a. Multimeter should indicate continuity. If continuity is not indicated, repair broken lead 566 (para. 3-131).

     b. If continuity is indicated, connect harness connector to protective control box. DO NOT connect lead 566 to alternator. Go to test 9.

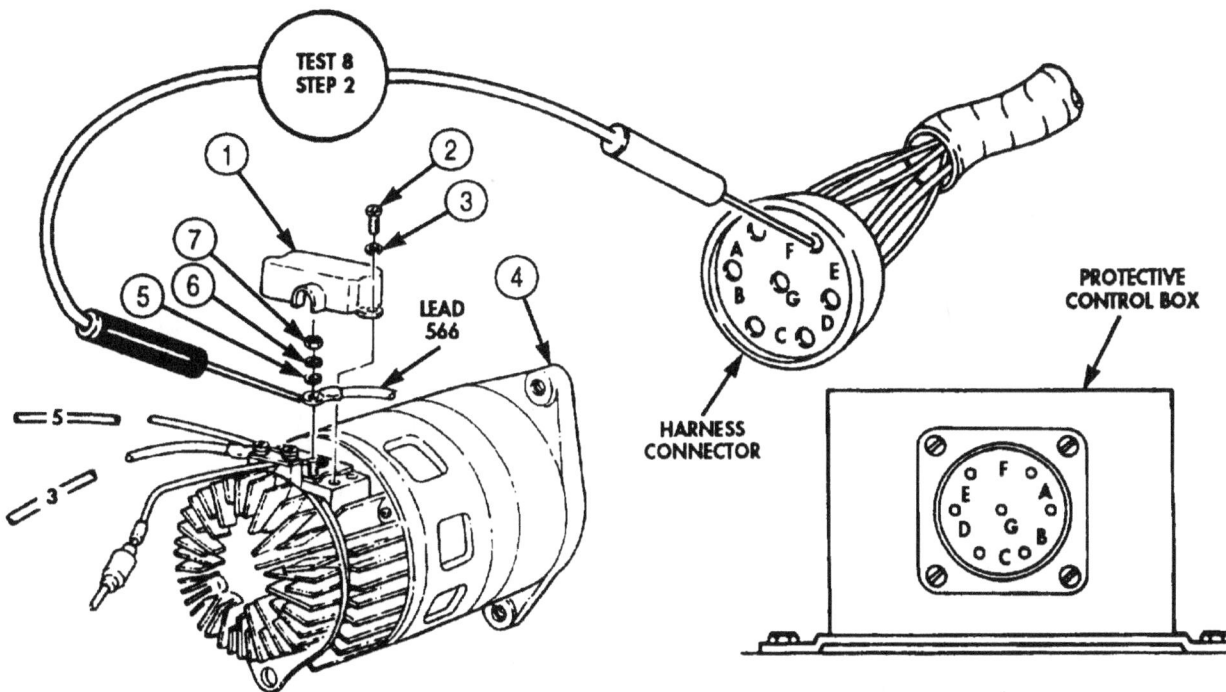

Test 9. Test starter lockout signal.

    Attempt to start engine with lead 566 disconnected from alternator.

       a. If engine fails to start, replace protective control box (para. 3-115).

       b. If engine starts, replace alternator (para. 3-79 or 3-80).

END OF TESTING!

Table 2-3. Electrical Troubleshooting (Contd).

MALFUNCTION
TEST OR INSPECTION
CORRECTIVE ACTION

GENERATING SYSTEM

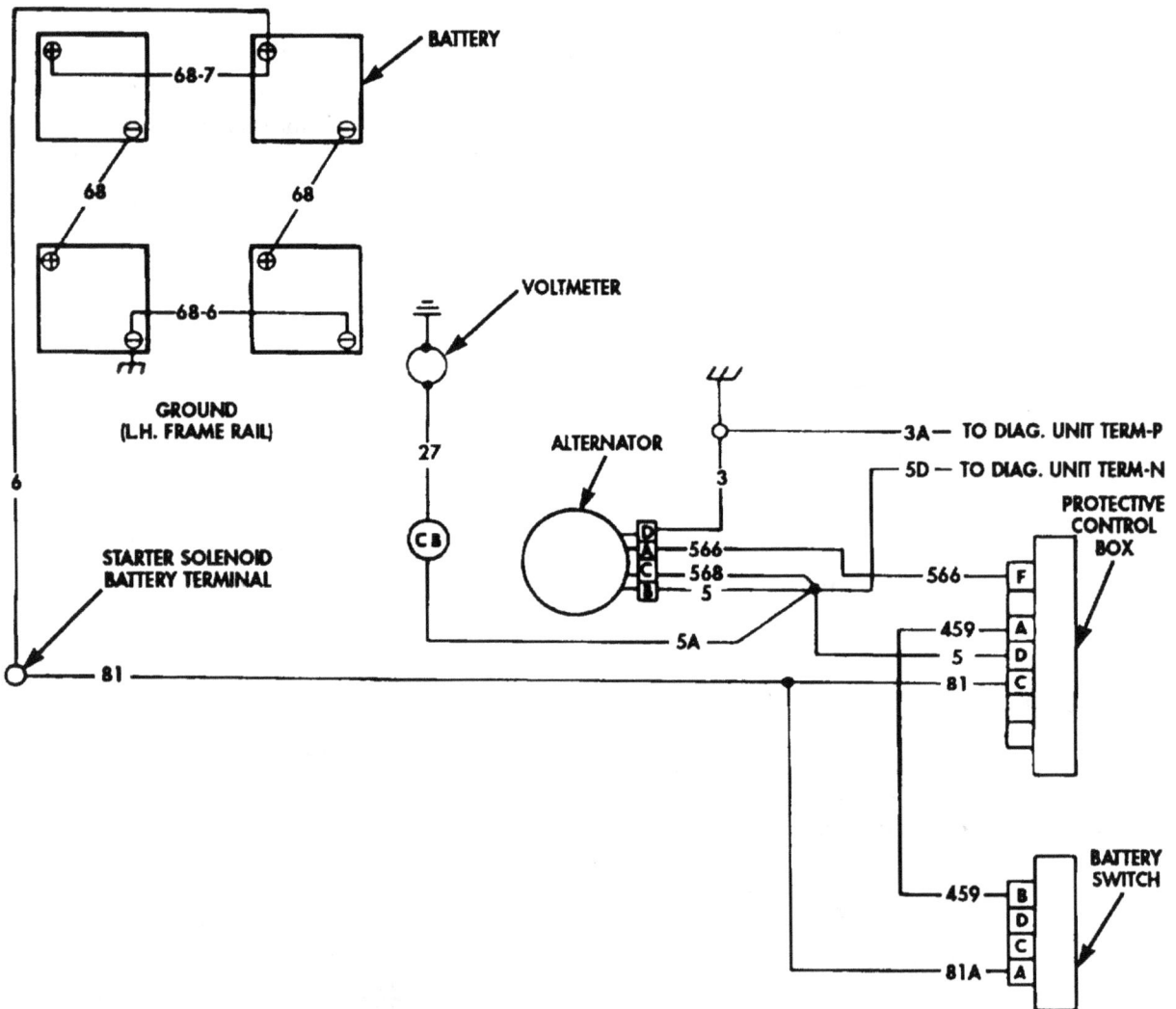

9. BATTERIES HOT OR BOILING; CORRECT SPECIFIC GRAVTY Of ALL CELLS IS 1.280

NOTE

If STE/ICE is available, perform NG50 - Charging Circuit Tests (para. 2-24).

Table 2-3. Electrical Troubleshooting (Contd).

| MALFUNCTION |
| TEST OR INSPECTION |
| CORRECTIVE ACTION |

Test charging voltage.

    Step 1.   Set multimeter to 50-volt range.

    Step 2.   Connect multimeter directly across battery terminal posts: positive lead to positive post and negative lead to negative post.

    Step 3.   Start engine (TM 9-2320-272-10) and allow engine to stabilize at 700-800 rpm.

                Multimeter should indicate 26.5-29.5 volts. If not, replace alternator (para. 3-79 or 3-80).

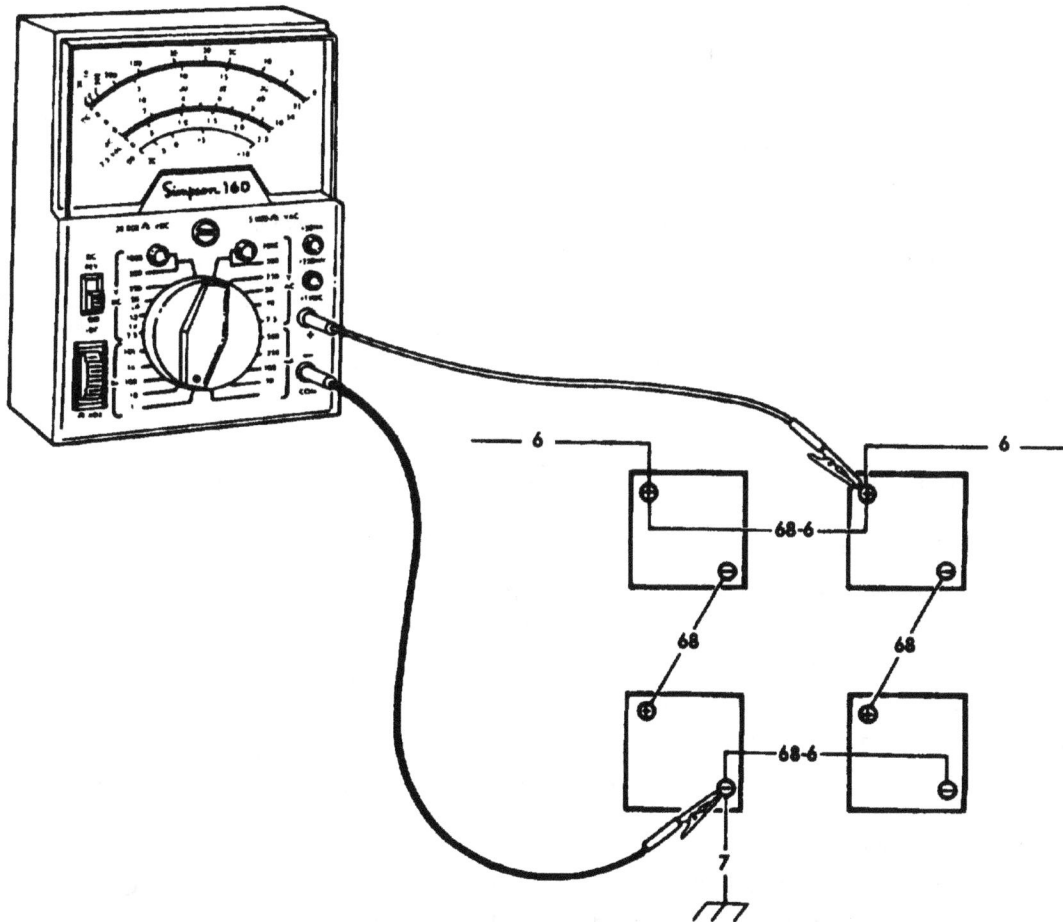

END OF TESTING!

Table 2-3. Electrical Troubleshooting (Contd).

MALFUNCTION
    TEST OR INSPECTION
        CORRECTIVE ACTION

10. BATTERIES USE EXCESSIVE WATER

NOTE

If STE/ICE is available, perform NG81 - Battery Tests or NG50 - Charging Circuit Tests (para. 2-24).

Test charging voltage (malfunction 9).

END OF TESTING !

11. BATTERIES RUN DOWN IN SERVICE

NOTE

If STE/ICE is available, perform NG50 - Charging Circuit Tests (para. 2-24).

Test 1. Check for loose, broken, or missing alternator or fan drivebelts.

    a. Adjust loose drivebelts (M939/A1) (para. 3-78).

    b. Replace broken or missing drivebelts (para. 3-78 or 3-71).

Test 2. Test charging voltage (malfunction 9).

If proper voltage is indicated, problem is not in generating system. Perform battery system troubleshooting, malfunctions 2 through 5.

END OF TESTING!

12. NO ALTERNATOR OUTPUT

NOTE

If STE/ICE is available, perform NG50 - Charging Circuit Teats (para. 2-24).

Test 1. Check for loose, broken, or missing alternator or fan drivebelts.

    a. Adjust loose drivebelts (M939/A1) (para. 3-78).

    b. Replace broken or missing drivebelts (para. 3-78 or 3-71).

Test 2. Test alternator lead 568 for voltage.

Step 1. Disconnect lead 568 at alternator.

Step 2. Set multimeter to 50-volt range.

Step 3. Connect multimeter positive lead to lead 568 and negative lead to vehicle chassis ground.

Step 4. Place battery switch to ON position.

    a. Multimeter should indicate battery voltage.

    b. If battery voltage is indicated, place battery switch to OFF position and replace alternator (para. 3-79 or 3-80).

    c. If battery voltage is not indicated, go to step 5.

Table 2-3. Electrical Troubleshooting (Contd).

MALFUNCTION
TEST OR INSPECTION
CORRECTIVE ACTION

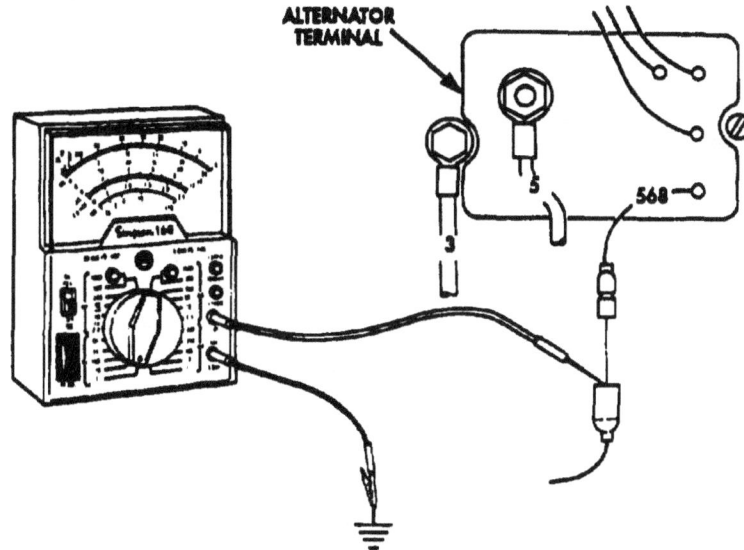

Step 5. Touch multimeter positive lead to lead 5 of alternator terminal.
a. Multimeter should indicate battery voltage.
b. If battery voltage is indicated, repair broken lead 568 (para. 3-131).
c. If battery voltage is not indicated, go to step 6.

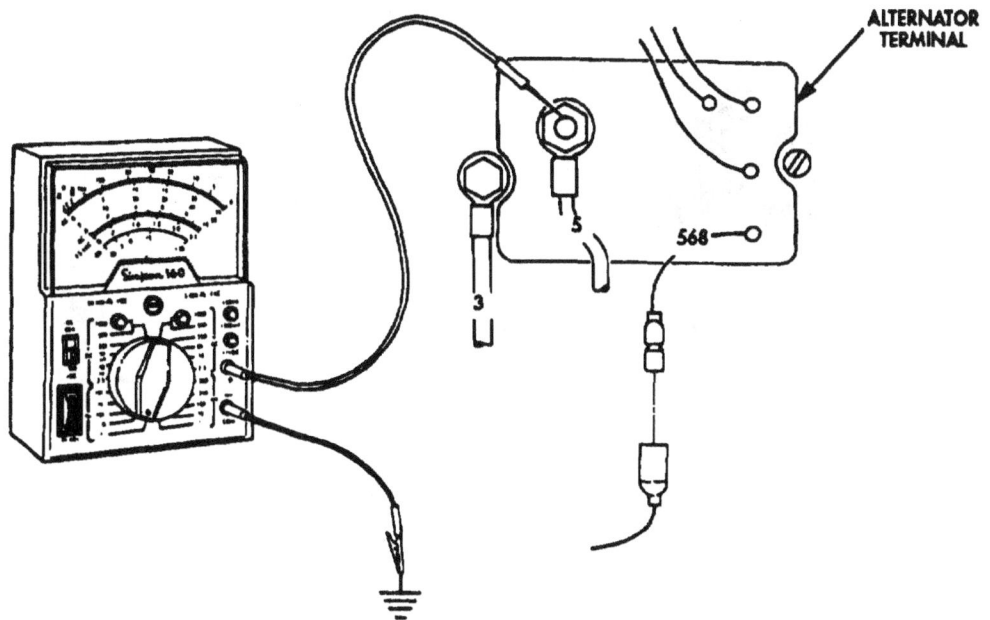

Table 2-3. Electrical Troubleshooting (Contd).

MALFUNCTION
    TEST OR INSPECTION
        CORRECTIVE ACTION

Step 6.   Set multimeter to RX1 for continuity reading.

Step 7.   Disconnect battery ground cable (para. 3-126).

Step 8.   Connect multimeter positive lead to lead 568 and negative lead to lead 5 of alternator terminal.

Step 9.   Multimeter should indicate continuity. If continuity is not indicated, there is an open lead.

         Repair broken lead 568 (para. 3-131).

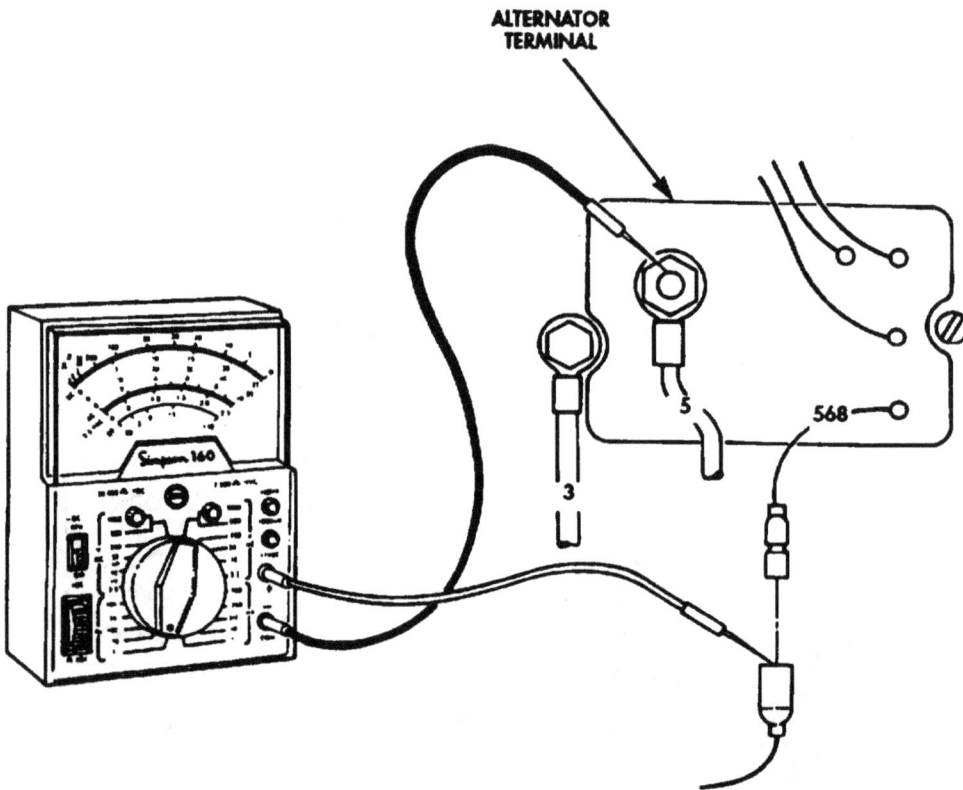

END OF TESTING!

Table 2-3. Electrical Troubleshooting (Contd).

MALFUNCTION
TEST OR INSPECTION
CORRECTIVE ACTION

13. ALTERNATOR OUTPUT VOLTAGE LOW

N O T E

If STE/ICE is available, perform NG50 - Charging Circuit Tests (para. 2-24).

Test 1. Check for loose, broken, or missing alternator or fan drivebelts.

    a. Adjust loose drivebelts (M939/A1) (para. 3-78).

    b. Replace broken or missing drivebelts (para. 3-78 or 3-71).

Test 2. Test battery voltage (malfunction 9).

END OF TESTING!

14. ALTERNATOR ADJUSTMENT

N O T E

Sealant must be removed from terminals before testing alternator output.

Step 1. Remove two screw-assembled lockwashers (2) and terminal cover (1) from alternator (4).

Step 2. Start engine (TM 9-2320-272-10) and raise engine speed above normal idle 600-650 rpm (M939/A1), 565-635 rpm (M939A2).

Step 3. Put a load on alternator (4) by operating driving lights (TM 9-2320-272-10).

Step 4. Using multimeter, check alternator (4) output voltage. Connect black test lead to ground 3. Connect red test lead to lead 5. Output voltage should be 26.5 to 29.5 volts. If adjustment is required, go to step 5. If no adjustment is required, go to step 8.

Step 5. Remove pipe plug (3) from alternator (4).

Step 6. Turn adjusting screw counterclockwise to increase voltage; clockwise to decrease voltage.

Step 7. Apply sealing compound (Appendix C, Item 65) to threads of pipe plug (3). Using hex-head driver, install pipe plug (3) on alternator (4) and tighten to 30-40 lb-in. (3-4 N•m).

Step 8. Turn off driving lights and stop engine (TM 9-2320-272-10).

Step 9. Seal terminal connections with adhesive sealant (Appendix C, Item 4).

Step 10. Install terminal cover (1) on alternator (4) with two screw-assembled lockwashers (2).

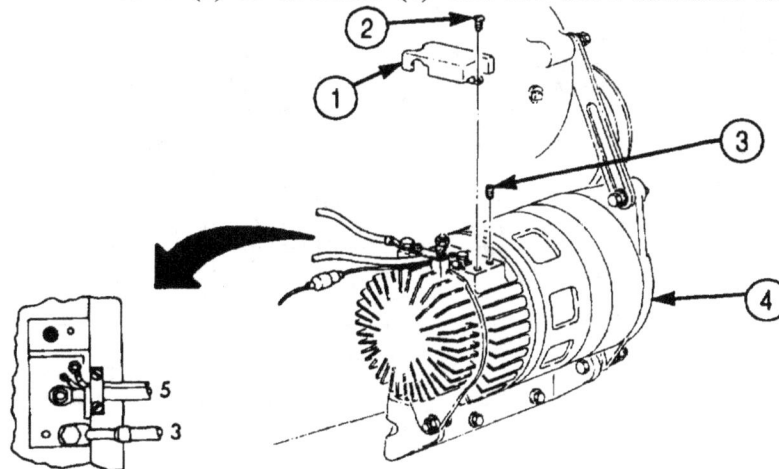

END OF TESTING!

Table 2-3. Electrical Troubleshooting (Contd).

MALFUNCTION
TEST OR INSPECTION
CORRECTIVE ACTION

15. BATTERY INDICATOR GAUGE IN RED POSITION

NOTE

If STE/ICE is available, perform NG50 - Charging Circuit Tests
(para. 2-24).

Test 1. Test battery voltage (malfunction 9).

Test 2. Check alternator for overheating.

   Step 1.   Run engine for approximately 10 minutes.

   Step 2.   With engine off, check alternator for high temperature by holding hand near alternator.

   Step 3.   If alternator is hot, disconnect lead 568 at alternator and allow alternator to cool.

   Step 4.   Connect lead 568 to alternator and start engine. If battery indicator returns to high red position and alternator heats up again, replace alternator (para. 3-79 or 3-80).

END OF TESTING!

LIGHTING SYSTEM

NOTE: Broken lines indicate turn signal circuits.

Table 2-3. Electrical Troubleshooting (Contd).

MALFUNCTION
TEST OR INSPECTION
CORRECTIVE ACTION

16. LAMPS WILL NOT LIGHT

    Test 1. Check for defective bulb.

            a. Replace bulb with one known to be good.

            b. If bulb does not light, go to test 2.

    Test 2. Check for corrosion or dirt in sockets or on terminals.

            a. Clean corroded connections.

            b. Clean dirt from sockets and terminals.

    Test 3. Check lamp holders for loose connections and broken wire terminals.

            a. Tighten all loose connections.

            b. Repair or replace broken wire terminals (para. 3-131).

    Test 4. Test lighting system harness connector voltage.

      Step 1. Place battery switch to OFF position.

      Step 2. Remove light switch from instrument panel (para. 3-108).

      Step 3. Place battery switch to ON position.

      Step 4. Set multimeter to 50-volt range.

      Step 5. Connect multimeter negative lead to vehicle chassis ground and positive lead to connector socket pin F.

            a. If battery voltage is not indicated, go to step 6.

            b. If battery voltage is indicated at socket pin F, connect a jumper wire from socket pin F to socket of faulty lead as shown.

            c. If lamps lights with jumper wire connected, replace light switch (para. 3-108).

      Step 6. Check wiring harness for loose connections.

            a. Tighten all loose connections.

            b. Repair broken wiring harness (para. 3-131).

CAUTION

Lead 15 or wire 15 is hot whenever the batteries are connected.

Table 2-3. Electrical Troubleshooting (Contd).

MALFUNCTION
TEST OR INSPECTION
CORRECTIVE ACTION

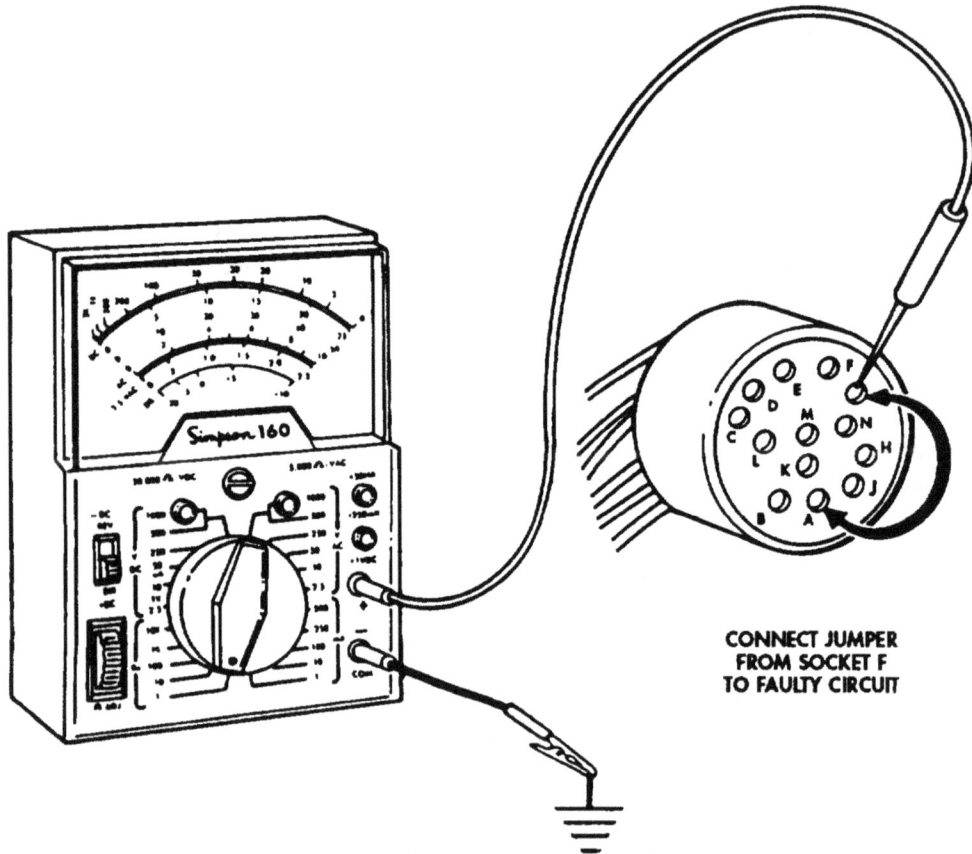

CONNECT JUMPER
FROM SOCKET F
TO FAULTY CIRCUIT

SOCKET	WIRE NO.	CIRCUIT
A	75	STOPLIGHT SWITCH
B	40	PANEL LIGHTS
C	22	DIRECTIONAL CONTROL
D	19	B.O DRIVING LIGHT
E	20-24	B.O. MARKER LIGHTS
F	15	BATTERY POS. 24 VOLTS
H	21	SERVICE REAR LIGHTS
J	460-461	DIRECTIONAL INDICATOR
K	75	STOPLIGHT SWITCH
L	491	SERVICE PARKING LIGHTS
M	16	SERVICE HEADLIGHTS
N	23	B.O. STOPLIGHT

END OF TESTING !

Table 2-3. Electrical Troubleshooting (Contd).

MALFUNCTION
TEST OR INSPECTION
CORRECTIVE ACTION

17. HEADLAMP (ONE SIDE) INOPERATIVE

Test 1. Test headlamp connector voltage.

Step 1. Place battery switch to OFF position.

CAUTION

Lead 15 or wire 15 is hot whenever the batteries are connected.

Step 2. Disconnect leads 18 and 91 at headlights.

Step 3. Place battery switch to ON position.

Step 4. Place light switch to SERVICE LIGHT position and depress high-beam selector switch to LOW BEAM position.

Step 5. Set multimeter to 50-volt range.

Step 6. Connect multimeter positive lead to lead 18 and negative lead to lead 91.

    a. Multimeter should indicate battery voltage. If battery voltage is not indicated, reposition selector lever switch from SERVICE DRIVE to OFF and back to SERVICE DRIVE.

    b. If malfunction is not cleared, wiring harness from headlamp to selector switch or the switch itself is defective. Perform malfunction 18.

    c. If voltage is present and malfunction is not cleared, perform malfunction 18, test 2.

Step 7. Visually inspect headlamp body pin connector for corrosion and loose connections.

    a. Clean corroded parts.

    b. Tighten loose connections.

    c. If malfunction is not cleared, perform malfunction 18, test 2.

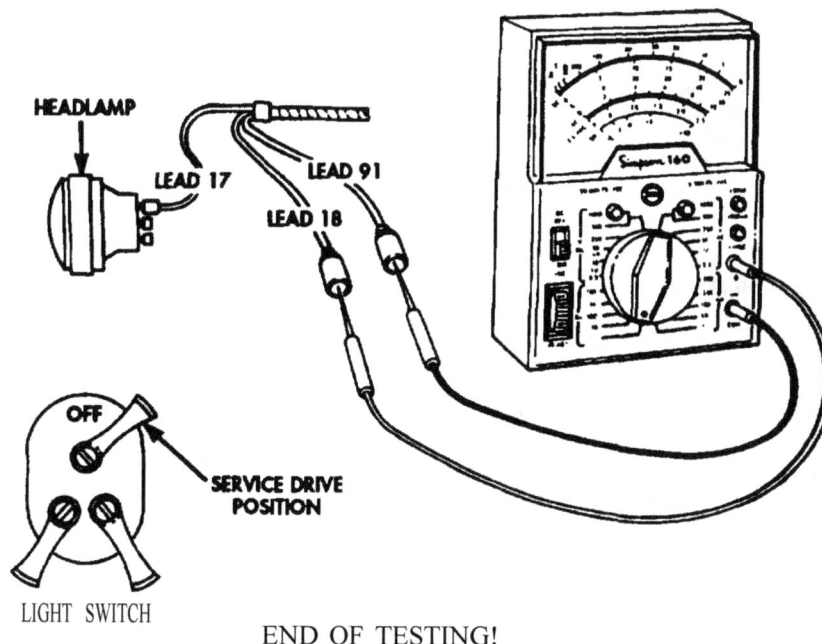

END OF TESTING!

Table 2-3. Electrical Troubleshooting (Contd).

MALFUNCTION
TEST OR INSPECTION
CORRECTIVE ACTION

---

18. HEADLAMP (BOTH SIDES) INOPERATIVE

Test 1. Test headlamps connector voltage for both sides (malfunction 17, test 1).

Test 2. Test headlamp high-beam selector switch.

Step 1. Place battery and light switch to OFF' position.

CAUTION

Lead 15 or wire 16 is hot whenever batteries are connected.

Step 2. Disconnect leads 17 and 18 at high-beam selector switch.

Step 3. Place battery and light switch to ON position.

Step 4. Connect multimeter positive lead to terminal 17 on high-beam selector switch and negative lead to vehicle chassis ground.

Step 5. Note multimeter voltage reading.

Step 6. With multimeter negative lead connected to ground, connect positive lead to exposed terminal 18 on high-beam selector switch.

Step 7. Note multimeter voltage reading,

    a. If battery voltage was indicated at terminals 17 or 18 on switch, but not at headlamps, the wiring harness is defective. Notify DS maintenance.

    b. If battery voltage was not indicated at terminals 17 or 18 on switch, go to step 8.

Step 8. Operate high-beam selector switch. Repeat steps 4 through 7.

    a. If battery voltage was indicated at terminals 17 and 18 alternately as switch was operated, but not at the headlamps, wiring harness is defective. Notify DS maintenance.

    b. If battery voltage was not indicated at terminals 17 or 18 alternately as switch was operated, go to step 9.

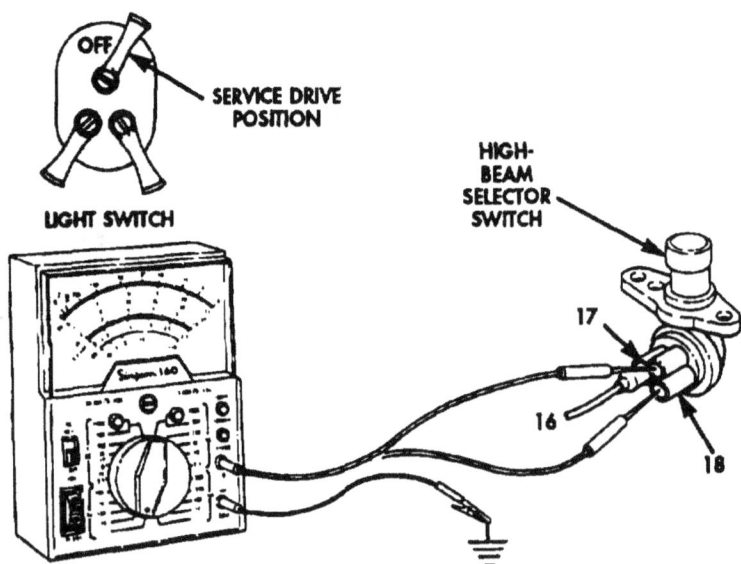

Table 2-3. Electrical Troubleshooting (Contd).

MALFUNCTION
TEST OR INSPECTION
CORRECTIVE ACTION

Step 9. Place battery switch and light switch to OFF position.

Step 10. Disconnect lead 16 from high-beam selector switch,

Step 11. Place battery switch to ON position and light switch to SERVICE DRIVE position.

Step 12. With negative lead of multimeter connected to ground, connect positive lead of multimeter to lead 16 (not switch terminals), Multimeter should indicate battery voltage.

    a, If battery voltage is indicated, replace high-beam selector switch (para. 3-112).

    b. If battery voltage is not indicated, perform malfunction 16, test 4.

CAUTION

Lead 15 or wire 15 is hot whenever the batteries are connected.

WIRE NO.	CIRCUIT
16	LIGHT SWITCH TO SELECT OR SWITCH
17	HIGH BEAM
18	LOW BEAM

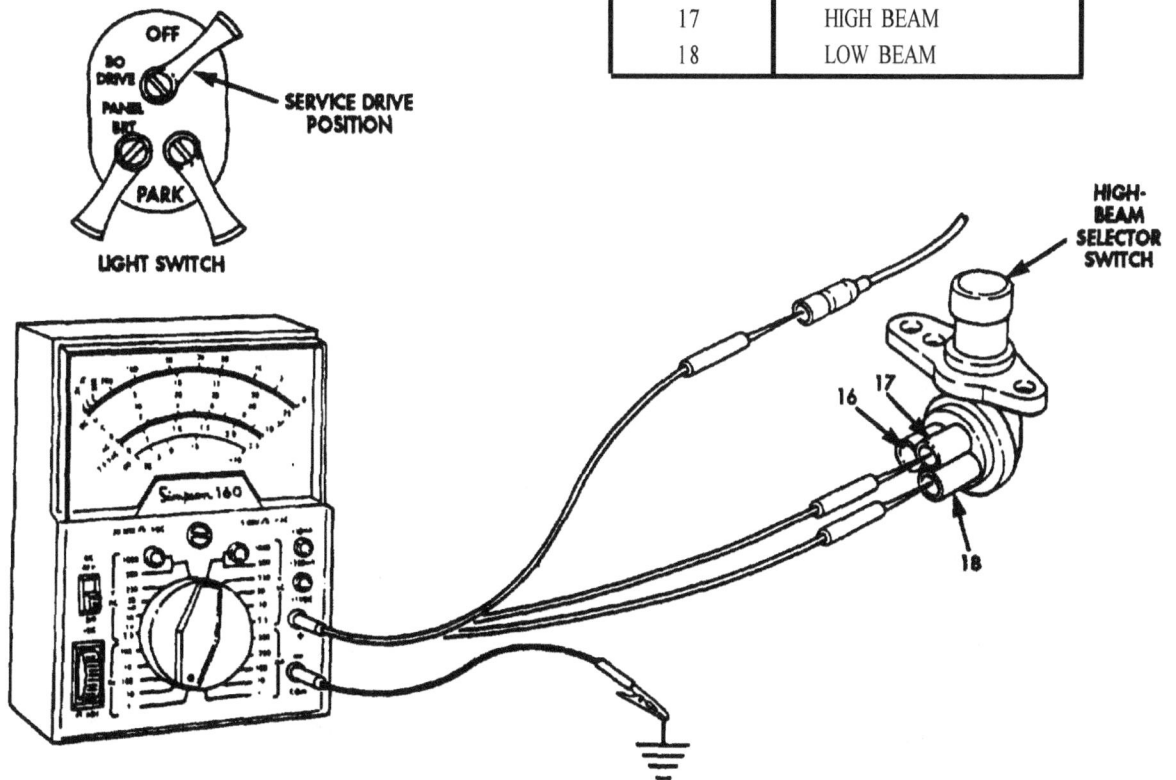

END OF TESTING!

Table 2-3. Electrical Troubleshooting (Contd).

MALFUNCTION
    TEST OR INSPECTION
        CORRECTIVE ACTION

---

19. BIACKOUT DRIVE OR BLACKOUT MARKER LIGHT INOPERATIVE

Test front blackout lamp connector voltage.

    Step 1.   Place battery switch to OFF position,

<u>CAUTION</u>

           Lead 15 or wire 15 is hot whenever the batteries are connected.

    Step 2.   Disconnect lead 19 at blackout drive light. Disconnect lead 20 at both left and right front composite lights.

    Step 3.   Place battery switch to ON position.

    Step 4.   Place light switch to B.O. DRIVE position.

    Step 5.   Set multimeter to 50-volt range.

    Step 6.   Connect multimeter negative lead to vehicle chassis ground. Touch positive lead to each of disconnected lead wires and note voltage indication through each wire.

           a. If voltage was indicated, replace blackout light or lamp (para. 3-117).

           b. If battery voltage is not indicated, perform malfunction 16, test 4.

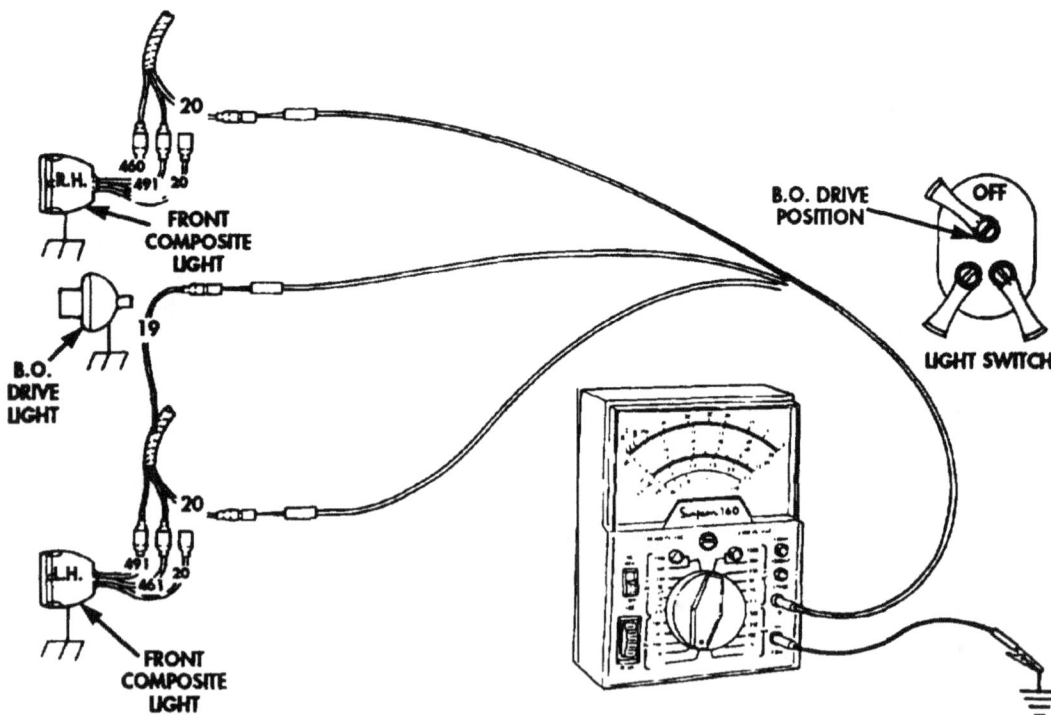

END OF TESTING!

Table 2-3. Electrical Troubleshooting (Contd).

MALFUNCTION
TEST OR INSPECTION
CORRECTIVE ACTION

20. STOPLIGHT INOPERATIVE

    Test stoplight switch voltage.

      Step 1. Place battery switch to OFF position.

CAUTION

          Lead 15 or wire 15 is hot whenever the batteries are connected.

      Step 2. Disconnect both leads 75 at stoplight switch.

      Step 3. Place battery switch to ON position and light switch to STOPLIGHT position.

NOTE

      For stoplight lead test, brake pedal must be depressed and air
      pressure maintained.

      Step 4. Set multimeter to 50-volt range.

      Step 5. Connect multimeter negative lead to good ground and positive lead to each exposed lead 75 and note voltage indicated.

          a. Meter should indicate battery voltage through one of the leads 75. If no voltage is present, perform malfunction 16, test 4. If voltage is indicated, connect that wire to one terminal of stoplight switch.

          b. With brake pedal depressed, connect multimeter positive lead to exposed stoplight switch terminal. Multimeter should indicate battery voltage.

          c. If battery voltage is indicated, stoplight switch is operational. If no voltage is indicated, replace stoplight switch (para. 3-111).

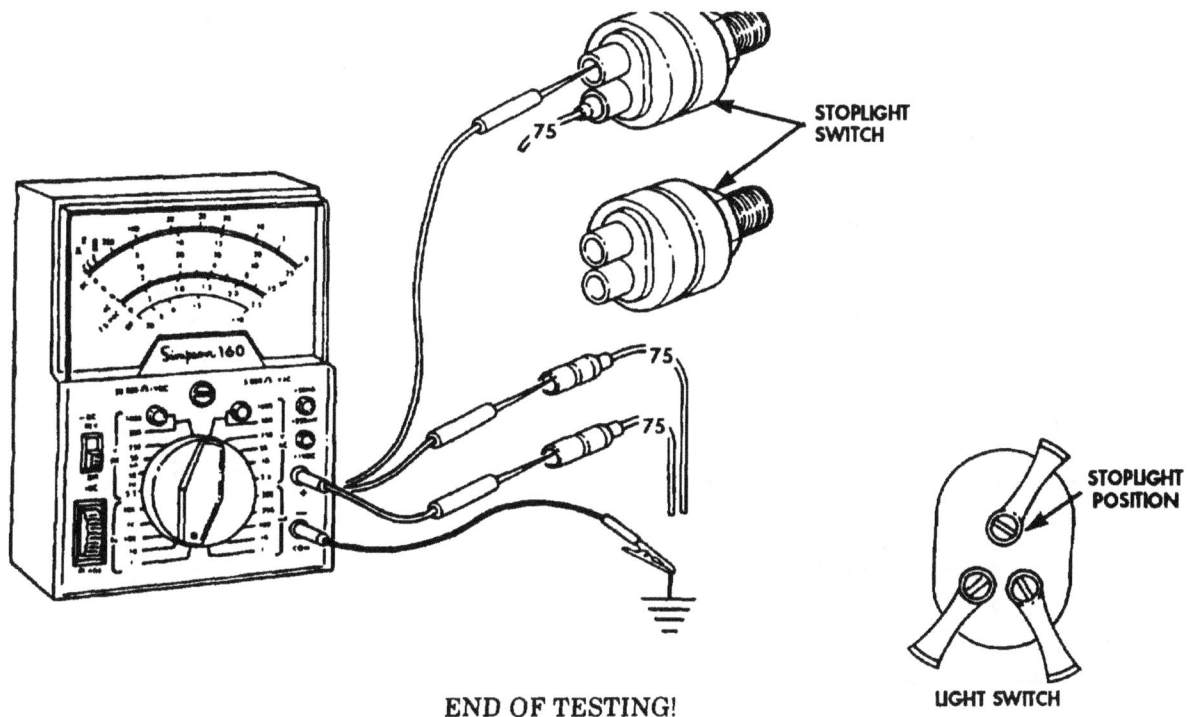

STOPLIGHT SWITCH

STOPLIGHT POSITION

LIGHT SWITCH

END OF TESTING!

Table 2-3. Electrical Troubleshooting (Contd).

```
MALFUNCTION
     TEST OR INSPECTION
               CORRECTIVE ACTION
```

21. REAR LIGHTS INOPERATIVE

Test rear lamp connector voltage.

Step 1. Place battery switch to OFF position.

CAUTION

Lead 15 or wire 15 is hot whenever the batteries are connected.

Step 2. Disconnect lead wire corresponding to inoperative rear lamp.

Step 3. Place battery switch to ON position.

Step 4. Place light switch to position corresponding to disconnected lead wire.

NOTE

For stoplight lead test, brake pedal must be depressed and air pressure maintained.

Step 5. Set multimeter to 50-volt range.

Step 6. Connect multimeter negative lead to vehicle chassis ground and positive lead to disconnected lead wire. Light switch must be in the corresponding position. Note voltage indication.

  a. If voltage was indicated, replace bulb (para. 3-118). If new bulb fails to operate, perform malfunction 16, test 2.

  b. If battery voltage is not indicated, perform malfunction 16, test 4.

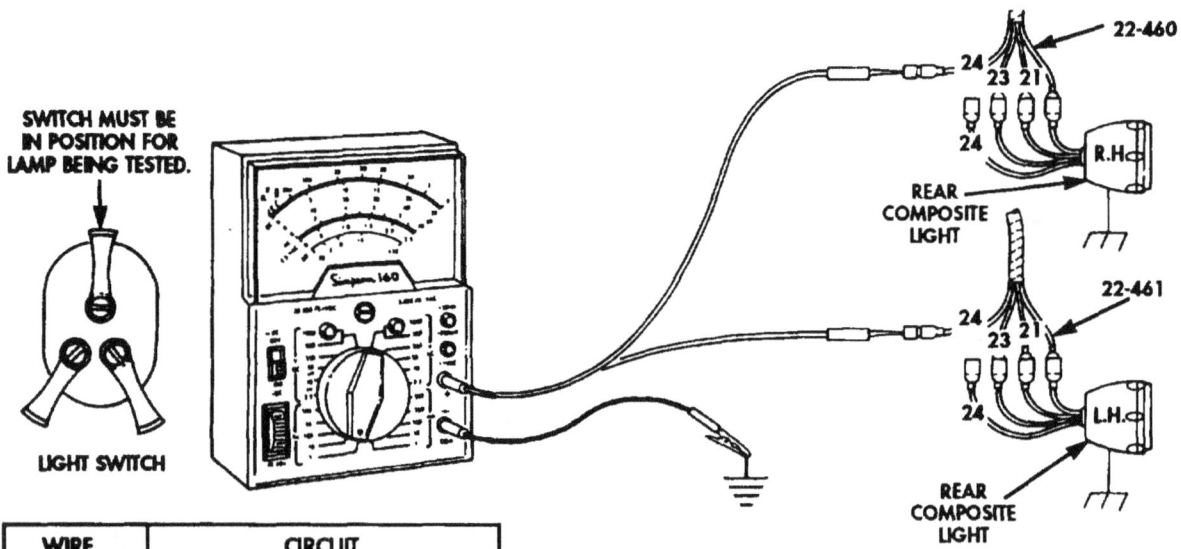

WIRE	CIRCUIT
21	SERVICE TAILLIGHT
22-460	SERVICE STOPLIGHT - RIGHT
22-461	SERVICE STOPLIGHT - LEFT
23	B.O. STOPLIGHT
24	B.O. REAR MARKER

END OF TESTING!

Table 2-3. Electrical Troubleshooting (Contd).

MALFUNCTION
TEST OR INSPECTION
CORRECTIVE ACTION

22. ONE OR MORE TRAILER LIGHTS INOPERATIVE

Test 1. Test trailer receptacle voltage.

Step 1. Place battery switch to ON position.

CAUTION

Lead 15 or wire 15 is hot whenever the batteries are connected.

Step 2. Place light switch to position corresponding to inoperative lamp.

NOTE

For stoplight lead test, brake pedal must be depressed and air pressure maintained.

Step 3. Set multimeter to 50-volt range.

Step 4. Connect multimeter negative lead to socket terminal pin L on receptacle. Touch positive lead to appropriate connector socket of lead being tested. Light switch must be in the corresponding position.

    a. Battery voltage should be indicated at connector socket being tested.

    b. If voltage was not indicated at one or more of pins being tested, go to test 2.

    c. If voltage is indicated, disconnect and reconnect male connector to ensure positive connection. If trailer lamps still do not light, check male connection for corrosion.

    d. If trailer lamps still do not light, check trailer lighting system.

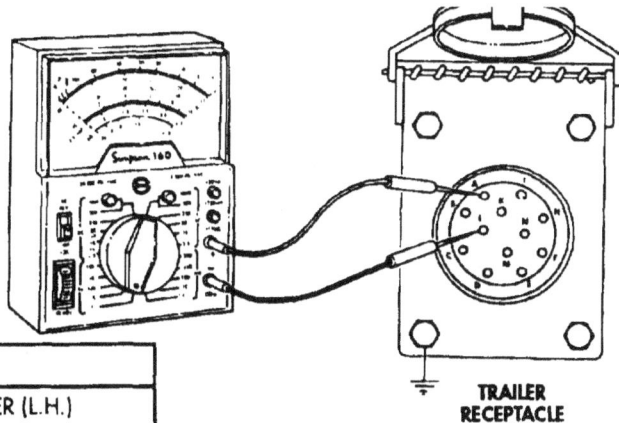

TRAILER RECEPTACLE

SWITCH MUST BE IN POSITION FOR LAMP BEING TESTED.

LIGHT SWITCH

PIN	WIRE NO.	CIRCUIT
A	24	REAR B.O. MARKER (L.H.)
B	22-461	SERVICE STOPLIGHT (L.H.)
C	24	REAR B.O. MARKER (R.H.)
D	90	GROUND TO FRAME
E	21	SERVICE TAILLIGHT
F	23	B.O. STOPLIGHT
H	490	B.O. MARKER LIGHTS
J	22-460	SERVICE STOPLIGHT (R.H.)
K	37	AUXILIARY POWER
L	90	GROUND TO FRAME
M	NONE	NOT USED
N	NONE	NOT USED

NOTE: Refer to trailer manual for type and location of lights on trailer.

2-133

Table 2-3. Electrical Troubleshooting (Contd).

MALFUNCTION
TEST OR INSPECTION
CORRECTIVE ACTION

Test 2. Check trailer receptacle ground.

  Step 1. Set multimeter on Rx1 for continuity reading.

CAUTION

Lead 15 or wire 15 is hot whenever the batteries are connected.

  Step 2. Connect multimeter positive lead to lead 90 at socket (receptacle terminal L and then D) and negative lead to good ground. Multimeter should indicate continuity from both terminal L and D to ground.

      a. If continuity was indicated, perform malfunction 16, test 4.

      b. If continuity was not indicated, repair broken lead 90 (para. 3-131).

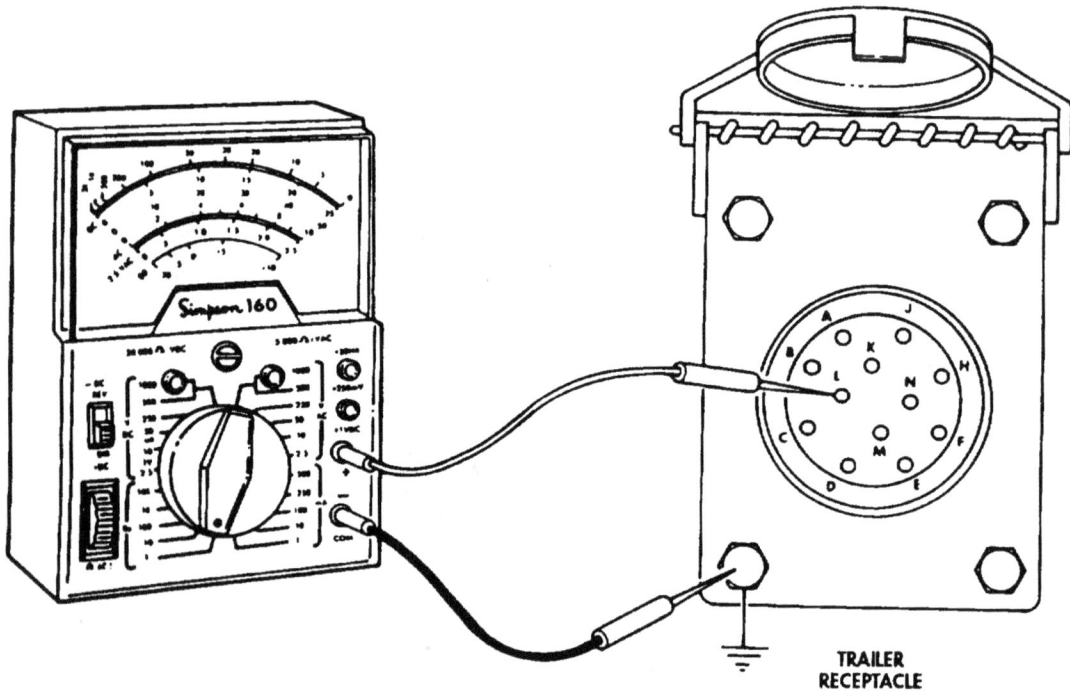

TRAILER
RECEPTACLE

END OF TESTING!

Table 2-3. Electrical Troubleshooting (Contd).

MALFUNCTION
TEST OR INSPECTION
CORRECTIVE ACTION

DIRECTIONAL SIGNAL SYSTEM

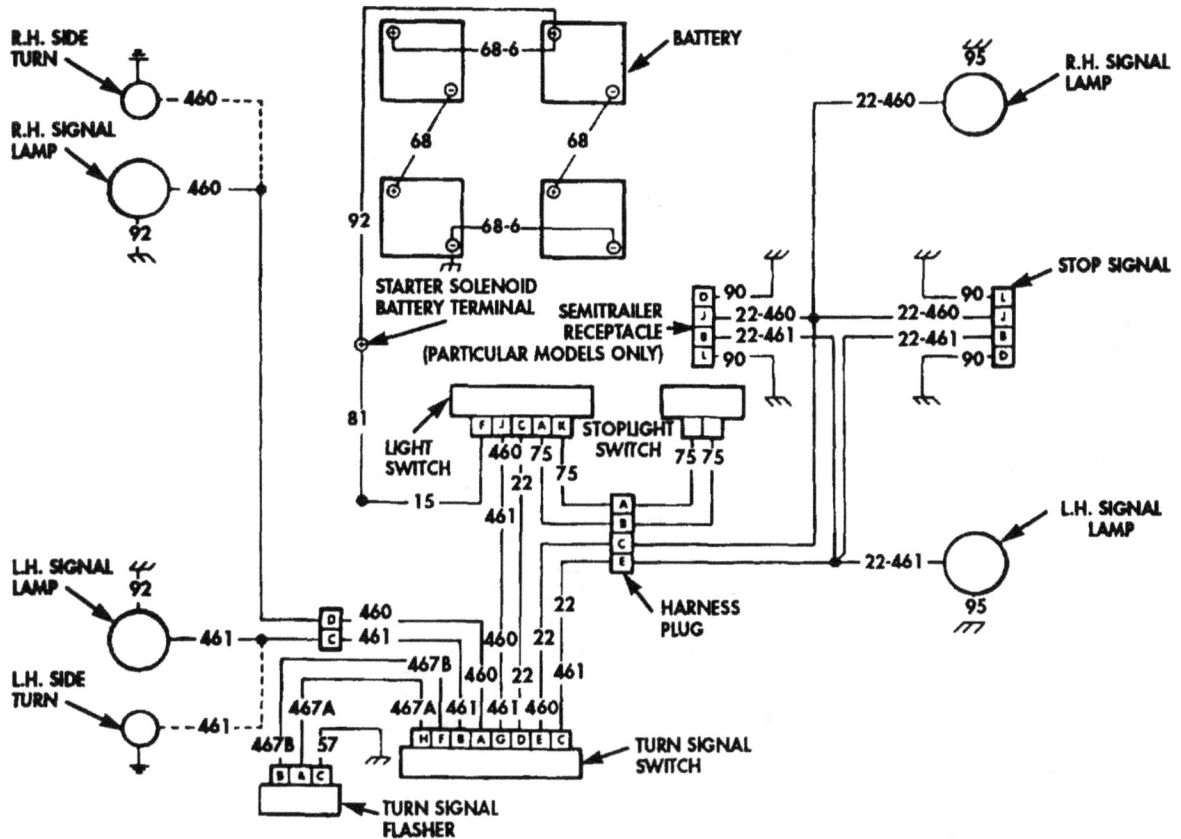

23. INDIVIDUAL LIGHTS DO NOT LIGHT WITH DIRECTIONAL SIGNAL LEVER IN ANY POSITION

    Test 1. Check for defective lamp.

        Step 1.   Place battery switch to ON position and light switch to STOP TURN position.

        Step 2.   With air pressure maintained, depress brake pedal.

            a. If stop signal lamp lights, go to test 2.

            b. If stop signal lamp does not light, replace defective lamp (para. 3-118).

    Test 2. Test wiring harness voltage.

        Step 1.   Place battery switch to OFF position.

        Step 2.   Disconnect leads 460, 461, 22-460, and 22-461 from lights.

        Step 3.   Place battery switch to ON position.

        Step 4.   Place light switch to STOP TURN position.

        Step 5.   Place directional signal control lever to HAZARD warning position.

Table 2-3. Electrical Troubleshooting (Contd).

MALFUNCTION
TEST OR INSPECTION
CORRECTIVE ACTION

Step 6. Set multimeter to 50.volt range.

Step 7. Connect multimeter negative lead to vehicle chassis ground. Touch positive lead to center contact of leads 460,461, 22-460, or 22-461.

    a. Multimeter needle should deflect at a rate of 1 to 2 cycles per second. If multimeter deflects, clean light socket.

    b. If multimeter does not deflect, go to test 3.

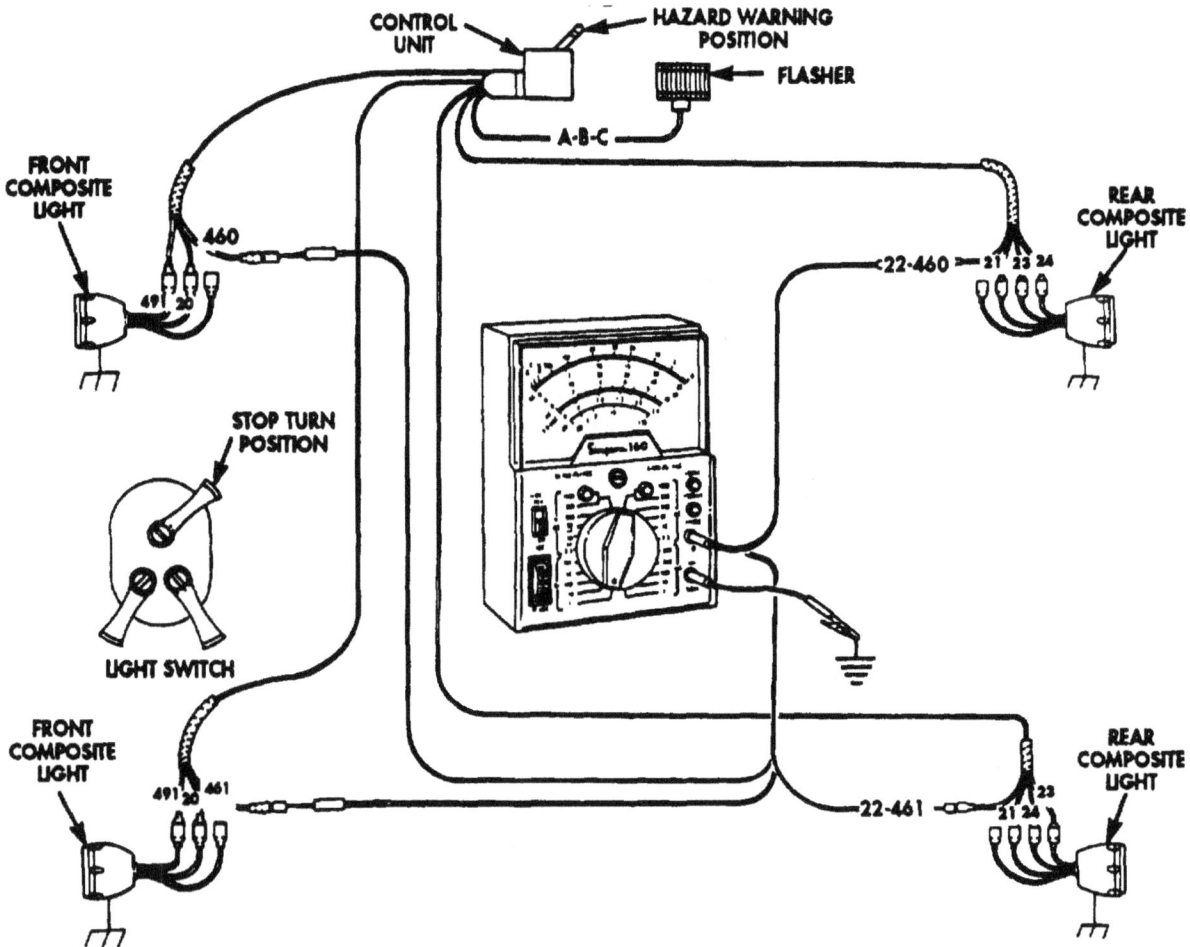

Test 3. Test wiring harness continuity.

Step 1. Place battery switch to OFF position.

Step 2. Disconnect wire connector from control unit and disconnect wire from defective light at lamp base.

Step 3. Place jumper wire from disconnected light wire at light to ground.

Step 4. Set multimeter on RX1 for continuity reading.

Step 5. Connect multimeter negative lead to vehicle chassis ground. Touch positive lead to control unit harness connector socket point for wire that was jumped to ground.

Table 2-3. Electrical Troubleshooting (Contd).

MALFUNCTION
TEST OR INSPECTION
CORRECTIVE ACTION

Step 6. Multimeter should read continuity. If continuity is not read, check wiring harness for broken, frayed, or pinched wires.

    a. Repair broken, frayed, or pinched wires (para. 3-131).

    b. If system is still not operational, go to test 4.

Test 4. Test wiring harness for short.

Step 1. Remove jumper wire from light wire to ground that was connected in test 3, step 3.

Step 2. With multimeter set for continuity, and wires disconnected as in step 1, connect multimeter negative to vehicle chassis ground. Touch positive lead to wire corresponding to defective lead.

Step 3. Multimeter should not read continuity (infinite reading). If there is a reading, the wiring harness has a shorted lead.

Step 4. Check harness for short lead by visually locating frayed or pinched wires.

    a. Repair frayed or pinched wires (para. 3-131).

    b. If system is still not operational, perform malfunction 24.

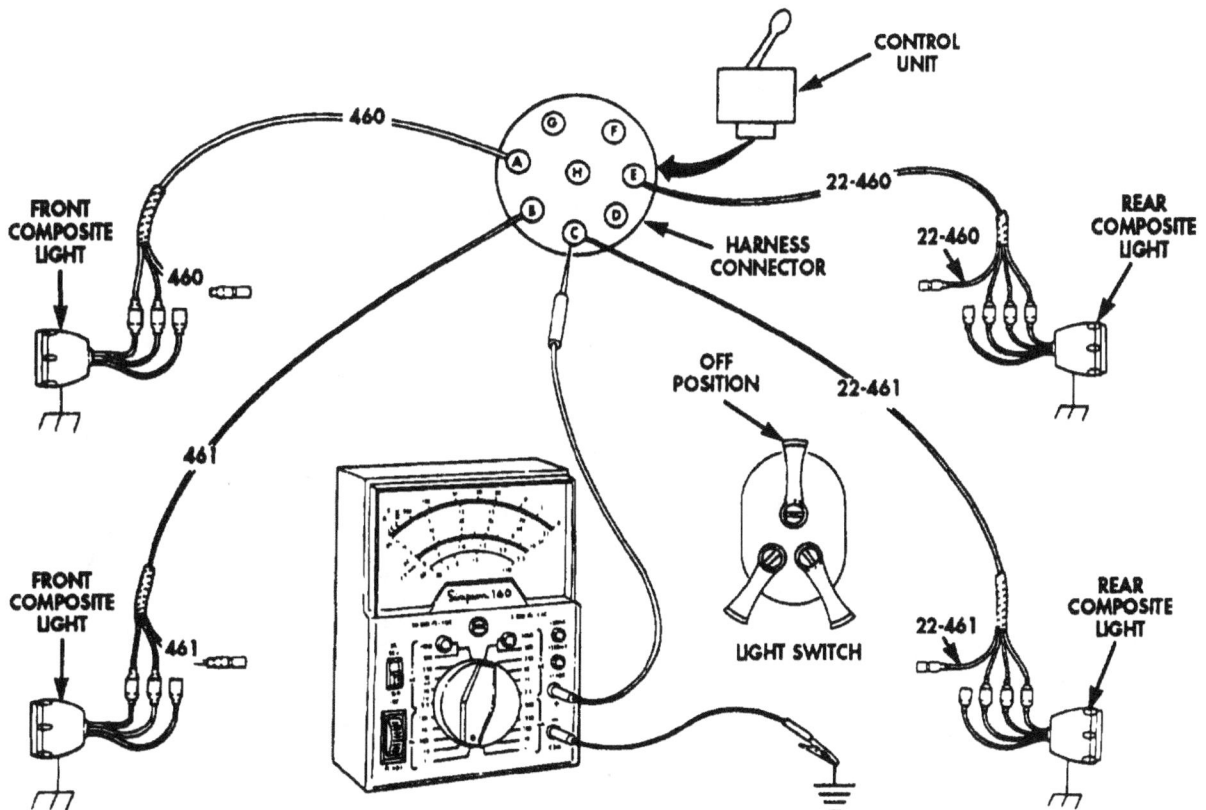

END OF TESTING!

Table 2-3. Electrical Troubleshooting (Contd).

MALFUNCTION
TEST OR INSPECTION
CORRECTIVE ACTION

24. NO LIGHTS OPERATE WITH DIRECTIONAL SIGNAL CONTROL LEVER IN ANY POSITION

Test 1. Test control unit voltage feed.

Step 1. Place battery switch to OFF position.

Step 2. Disconnect connector control unit harness from directional signal control unit.

Step 3. Place battery switch to ON position.

Step 4. Place light switch to STOP TURN position.

Step 5. Set multimeter to 50-volt range.

Step 6. Connect multimeter negative lead to vehicle chassis ground and positive lead to connector terminal point G.

a. Multimeter should indicate battery voltage. If battery voltage is not indicated, perform malfunction 16, test 4.

b. If battery voltage is indicated, go to test 2.

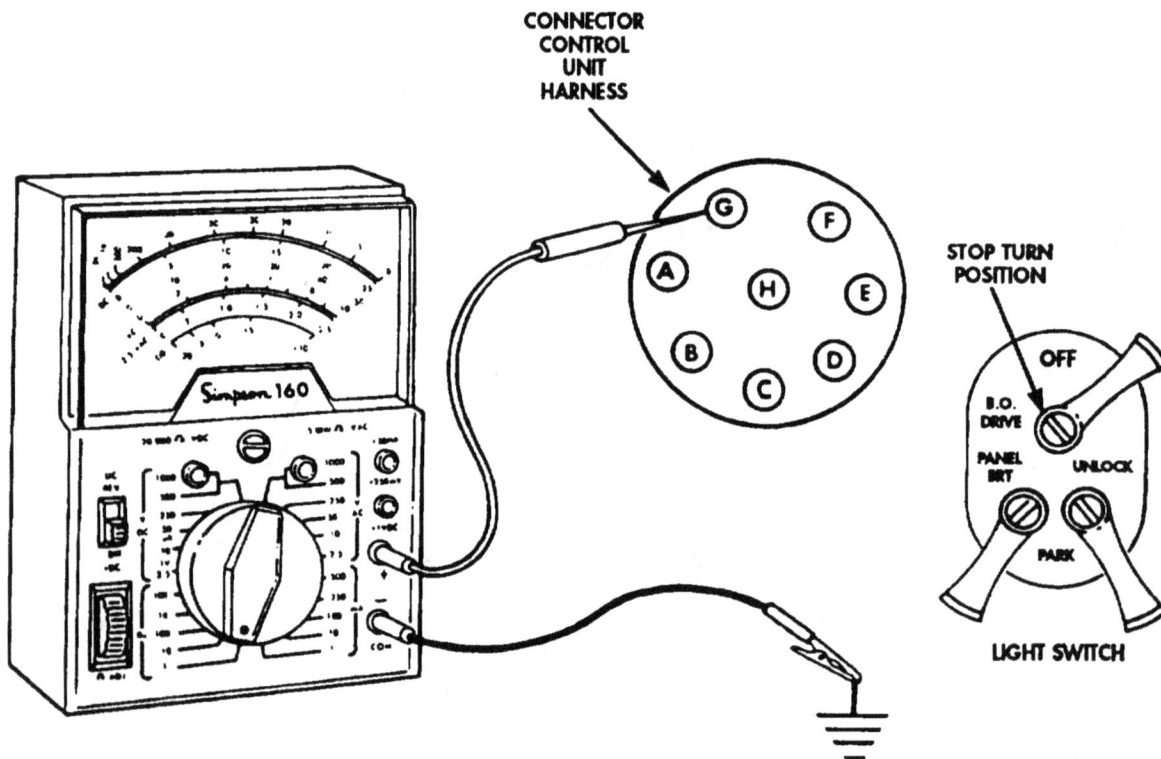

Table 2-3. Electrical Troubleshooting (Contd).

MALFUNCTION
TEST OR INSPECTION
CORRECTIVE ACTION

Test 2. Test flasher harness continuity.

    Step 1.   Set multimeter to RX1 for continuity reading, and connect positive lead to terminal A of flasher unit. Connect negative lead to terminal H of control unit cable, and observe multimeter for continuity.

    Step 2.   Connect multimeter negative lead to terminal F of control unit socket and positive lead to terminal B of flasher unit. Observe multimeter for continuity.

    Step 3.   Connect multimeter positive lead to terminal C of flasher unit connector. Connect negative lead to ground, and observe multimeter for continuity.

        a. If any lead does not have continuity, repair (para. 3-131).

        b. If all leads have continuity and turn signals still do not work, replace flasher (para. 3-114).

        c. If turn signals still do not work, perform malfunction 25.

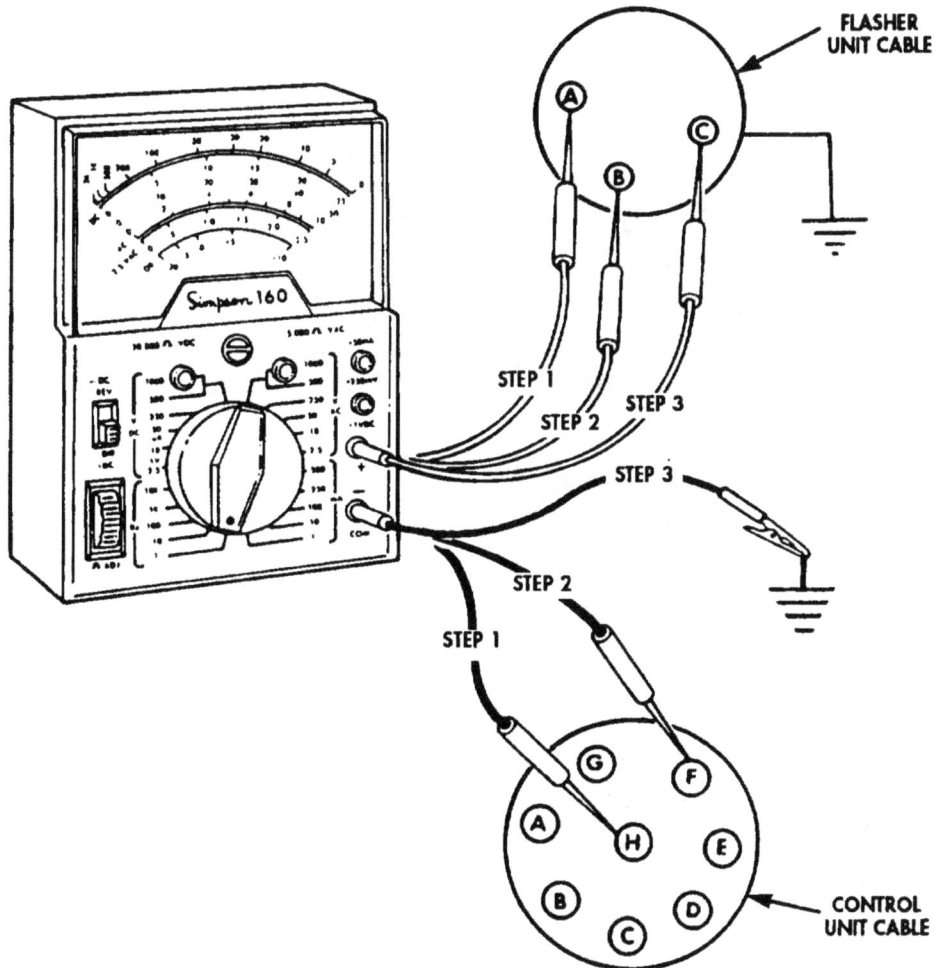

END OF TESTING!

Table 2-3. Electrical Troubleshooting (Contd).

MALFUNCTION
    TEST OR INSPECTION
        CORRECTIVE ACTION

25. SYSTEM OPERATES INCORRECTLY IN ONE OR MORE POSITIONS OF DIRECTIONAL SIGNAL CONTROL LEVER

Test directional signal control unit continuity.

Step 1.   Place battery switch to OFF position.

Step 2.   Remove harness connector from control unit.

NOTE

Remove indicator lamp before making tests.

Step 3.   Remove indicator lamp from directional signal control (para. 3-113).

Step 4.   Set multimeter to RX1 for continuity reading.

Step 5.   Set control lever in each of four operating positions and test.

If any lead does not test as shown in tables below, replace directional signal control unit (para. 3-113).

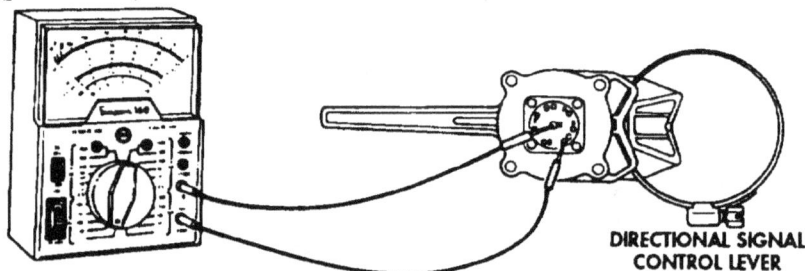

DIRECTIONAL SIGNAL
CONTROL LEVER

CONTROL UNIT TEST CHART

A. DIRECTIONAL SIGNAL CONTROL LEVER IN NEUTRAL POSITION			C. DIRECTIONAL SIGNAL CONTROL LEVER IN RIGHT TURN POSITION		
FROM PIN	TO PIN	CONTINUITY INDICATED	FROM PIN	TO PIN	CONTINUITY INDICATED
H	A	Open	F	G	Shorted
H	B	Open	H	A	Shorted
H	C	Open	H	E	Shorted
H	E	Open	H	B	Open
D	C	Shorted	H	C	Open
D	E	Shorted	D	C	Shorted
F	G	Open	D	E	Open
B. DIRECTIONAL SIGNAL CONTROL LEVER IN LEFT TURN POSITION			**D. DIRECTIONAL SIGNAL CONTROL LEVER IN HAZARD WARNING POSITION**		
FROM PIN	**TO PIN**	**CONTINUITY INDICATED**	**FROM PIN**	**TO PIN**	**CONTINUITY INDICATED**
H	B	Shorted	H	A	Shorted
H	C	Shorted	H	B	Shorted
H	A	Open	H	C	Shorted
H	E	Open	H	E	Shorted
F	G	Shorted	D	E	Open
D	E	Shorted	D	C	Open
D	C	Open	F	G	Shorted

END OF TESTING!

Table 2-3. Electrical Troubleshooting (Contd).

MALFUNCTION
TEST OR INSPECTION
CORRECTIVE ACTION

ETHER START SYSTEM

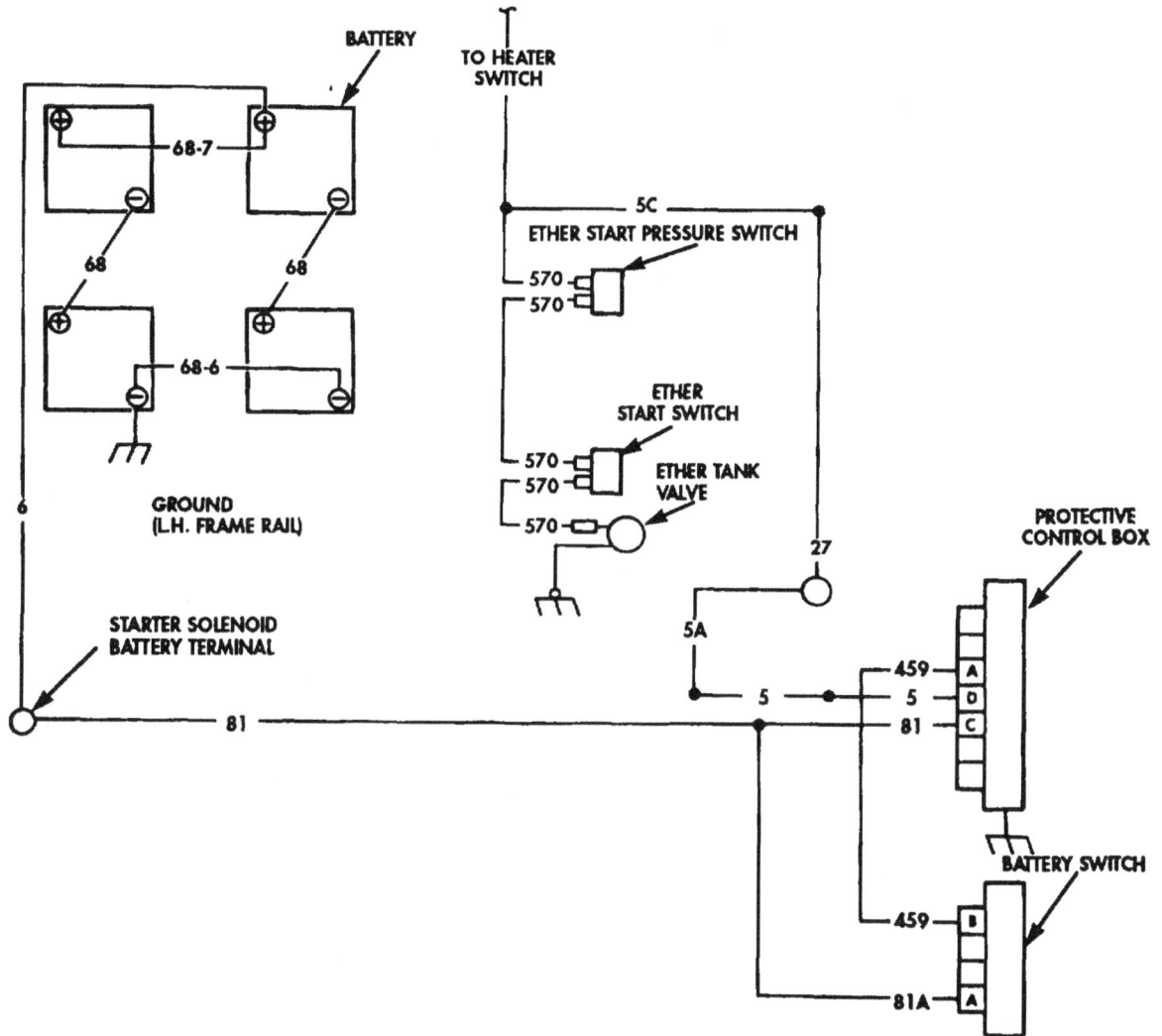

26. ENGINE CRANKS BUT WILL NOT START (FUEL AVAILABLE)

Test 1. Check ether cylinder.

Remove ether cylinder from ether valve (para. 3-35); shake and listen for liquid splashing inside cylinder.

   a. Replace cylinder if empty (para. 3-35) and try to restart engine.

   b. If cylinder is full, reinstall (para. 3-35) and go to test 2.

Table 2-3. Electrical Troubleshooting (Contd).

MALFUNCTION
TEST OR INSPECTION
CORRECTIVE ACTION

Test 2. Check ether start system electrical source.

   Operate personnel heater blower motor with switch in HIGH position (TM 9-2320-272-10).

      a. If personnel heater blower motor does not operate, perform malfunction 39.

      b. If personnel heater blower motor operates, go to test 3.

Test 3. Test lead 570 for voltage at ether valve.

   Step 1. Disconnect lead 570 at ether valve.

   Step 2. Set multimeter to 50-volt range. Connect positive lead to disconnected lead 570 and multimeter negative lead to ground.

   Step 3. Turn battery switch to ON position.

   Step 4. Crank engine, depress ether start switch, and observe multimeter.

      a. If meter indicates battery voltage and ether valve does not function, replace ether valve (para. 3-35).

      b. If no voltage is indicated, go to test 4.

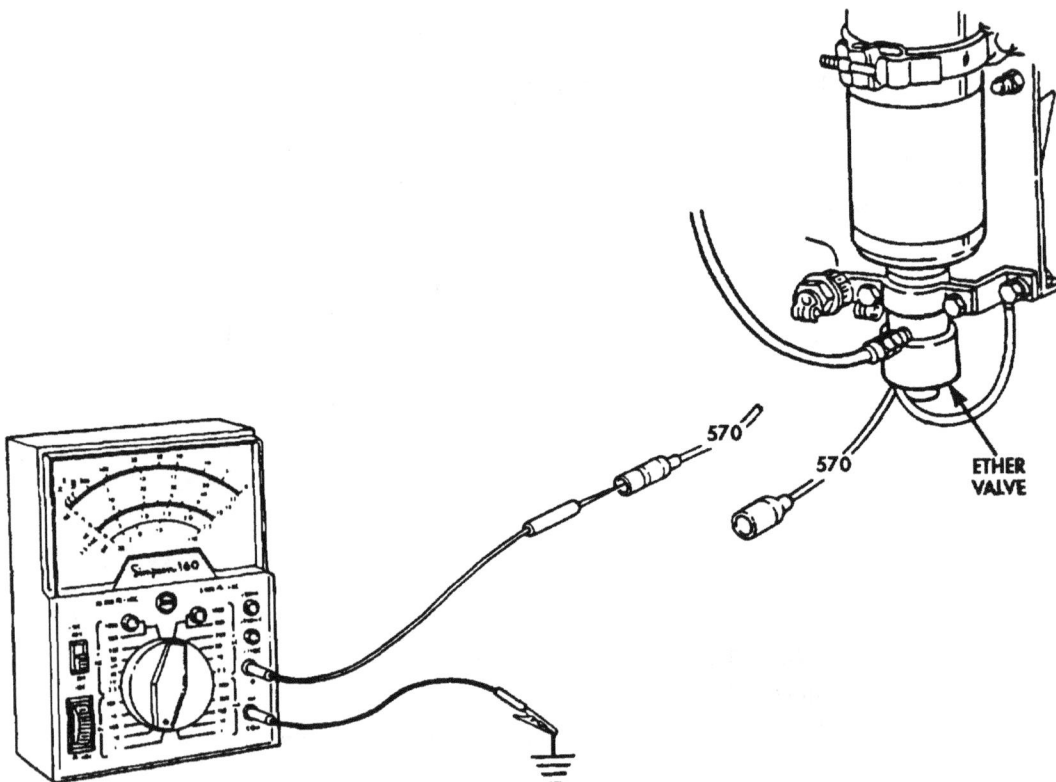

2-142

Table 2-3. Electrical Troubleshooting (Contd).

MALFUNCTION
TEST OR INSPECTION
CORRECTIVE ACTION

Test 4. Test lead 570 for voltage at ether start pressure switch.

   Step 1. Disconnect lead 570 to ether valve from ether start pressure switch.

   Step 2. Connect positive lead to lead 570 terminal on ether start pressure switch and negative lead to chassis ground.

   Step 3. Crank engine, depress ether start switch, and observe multimeter.

       a. If meter indicates battery voltage, repair lead 570 to ether valve (para. 3-131).

       b. If no voltage is indicated, go to step 4.

   Step 4. Disconnect lead 570 at ether start switch.

   Step 5. Connect multimeter positive lead to lead 570 from ether start switch and negative lead to chassis ground.

   Step 6. Crank engine, depress ether start switch, and observe multimeter.

       a. If multimeter indicates battery voltage, replace ether start pressure switch (para. 3-33).

       b. If no voltage is indicated, go to test 5.

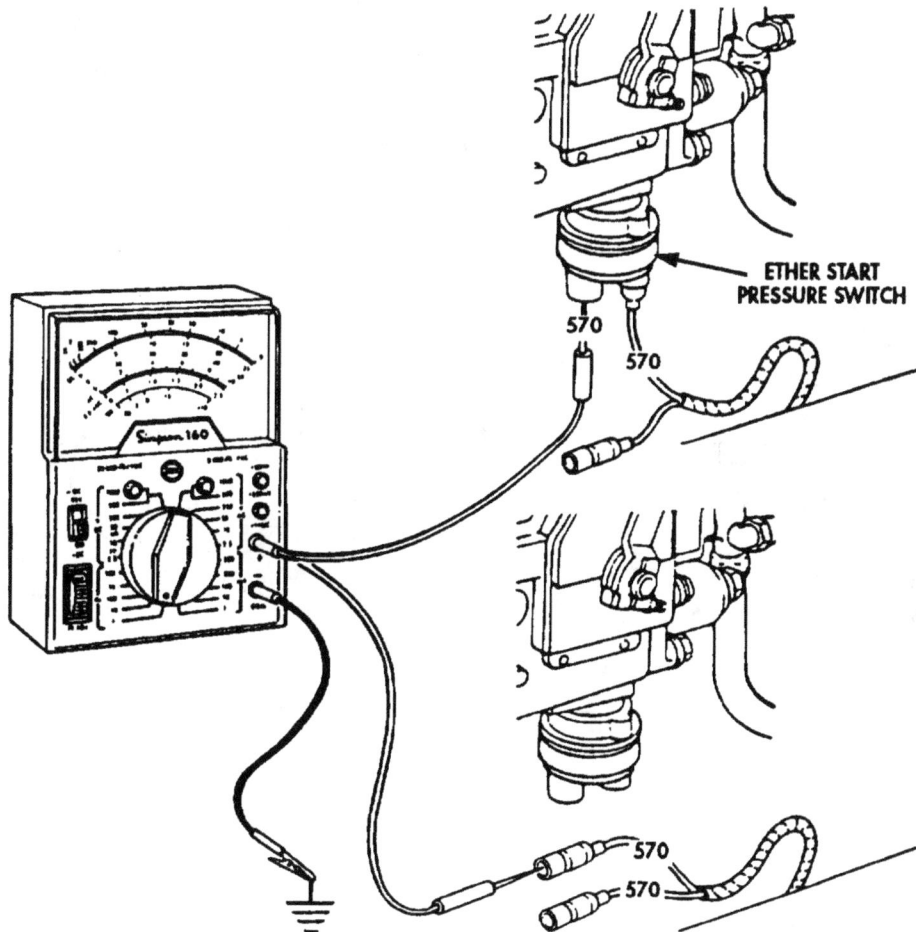

Table 2-3. Electrical Troubleshooting (Contd).

MALFUNCTION
TEST OR INSPECTION
CORRECT ACTION

Test 5. Test lead 570 for voltage at ether start switch.

  Step 1. Dieconned lead 570 to ether start pressure switch from ether start switch.

  Step 2. Connect multimeter positive lead to lead 570 ether start switch connector and negative lead to chassis ground.

  Step 3. Depress ether start switch and observe multimeter.

      a. If meter indicates battery voltage, replace lead 570 to ether start pressure switch (para. 3-131).

      b. If no voltage is indicated, proceed to step 4.

  Step 4. Dieconnect lead 570 to lead 5C from ether start switch.

  Step 6. Connect multimeter positive lead to lead 570 and negative lead to chassis ground.

      a. If meter indicates battery voltage, replace ether start switch (para. 3-33).

      b. If no voltage is indicated, replace lead 570 to lead 5C connector.

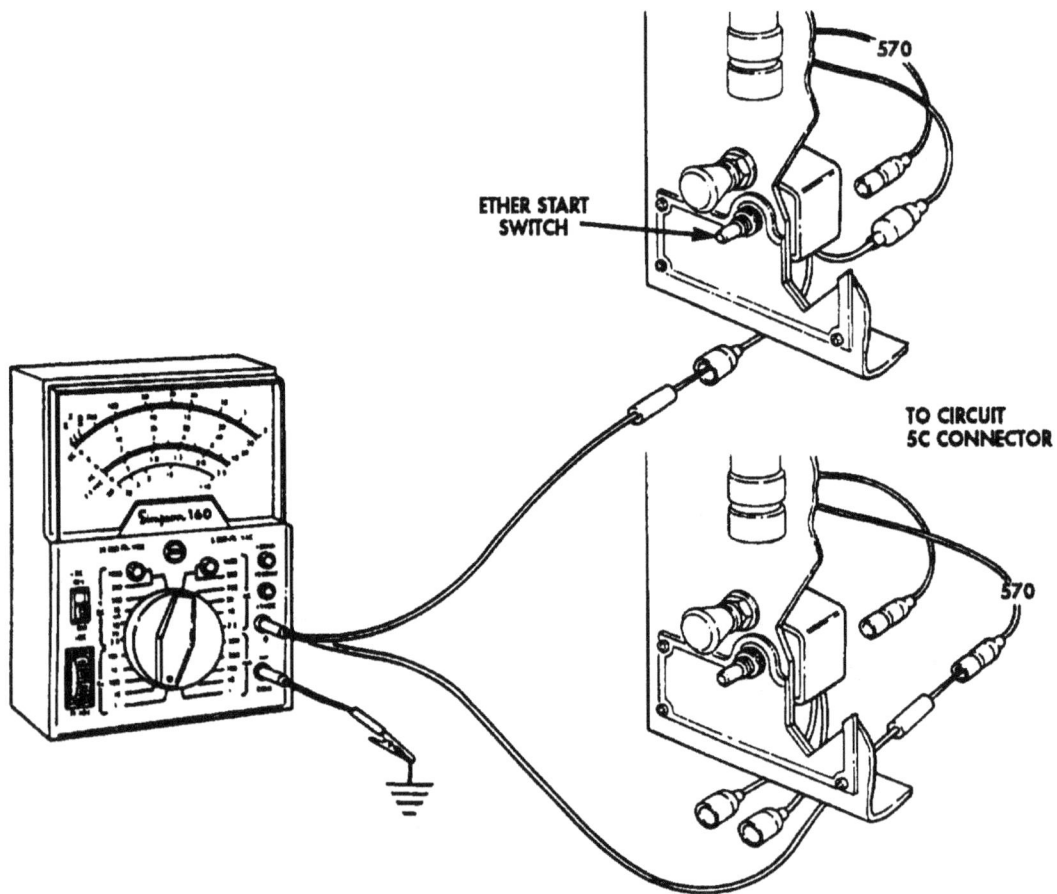

END OF TESTING!

Table 2-3. Electrical Troubleshooting (Contd).

MALFUNCTION
TEST OR INSPECTION
CORRECTIVE ACTION

INDICATOR, GAUGE, AND WARNING SYSTEM

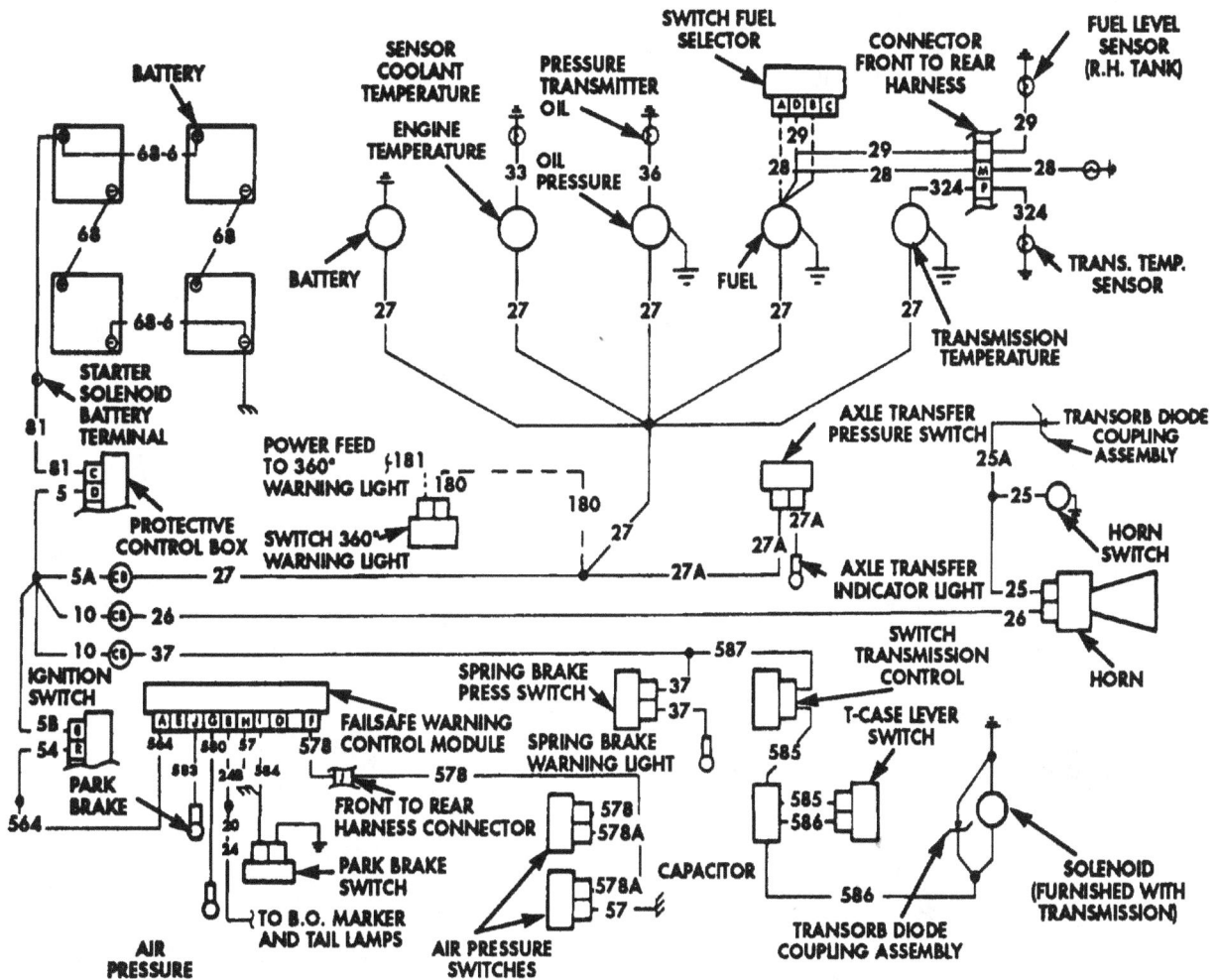

27. ALL GAUGES INOPERATIVE

NOTE
If STE/ICE is available, perform NG31 - Gauge Test (para. 2-24).

Test 1. Test instrument cluster voltage feed.

   Step 1. Place battery switch to OFF position.

   Step 2. Disconnect lead 5A at circuit breaker.

   Step 3. Place battery switch to ON position.

   Step 4. Set multimeter to 50-volt range.

Table 2-3. Electrical Doubleshooting (Contd).

MALFUNCTION
TEST OR INSPECTION
CORRECTIVE ACTION

Step 5. Connect multimeter negative lead to vehicle chassis ground and positive lead to lead 5A Multimeter should indicate battery voltage.

    a. If battery voltage is indicated, go to step 6.

    b. If battery voltage is not indicated, perform malfunctions 1, 2, 3, 5, and 6.

Step 6. Place battery switch to OFF position.

Step 7. Connect lead 5A and disconnect lead 27 at circuit breaker.

Step 8. Place battery switch to ON position.

Step 9. With multimeter negative lead connected to ground, touch positive lead to circuit 27 terminal of circuit breaker. Multimeter should indicate battery voltage.

    a. If battery voltage is not indicated, replace defective circuit breaker (para. 3-105).

    b. If battery voltage is indicated, go to test 2.

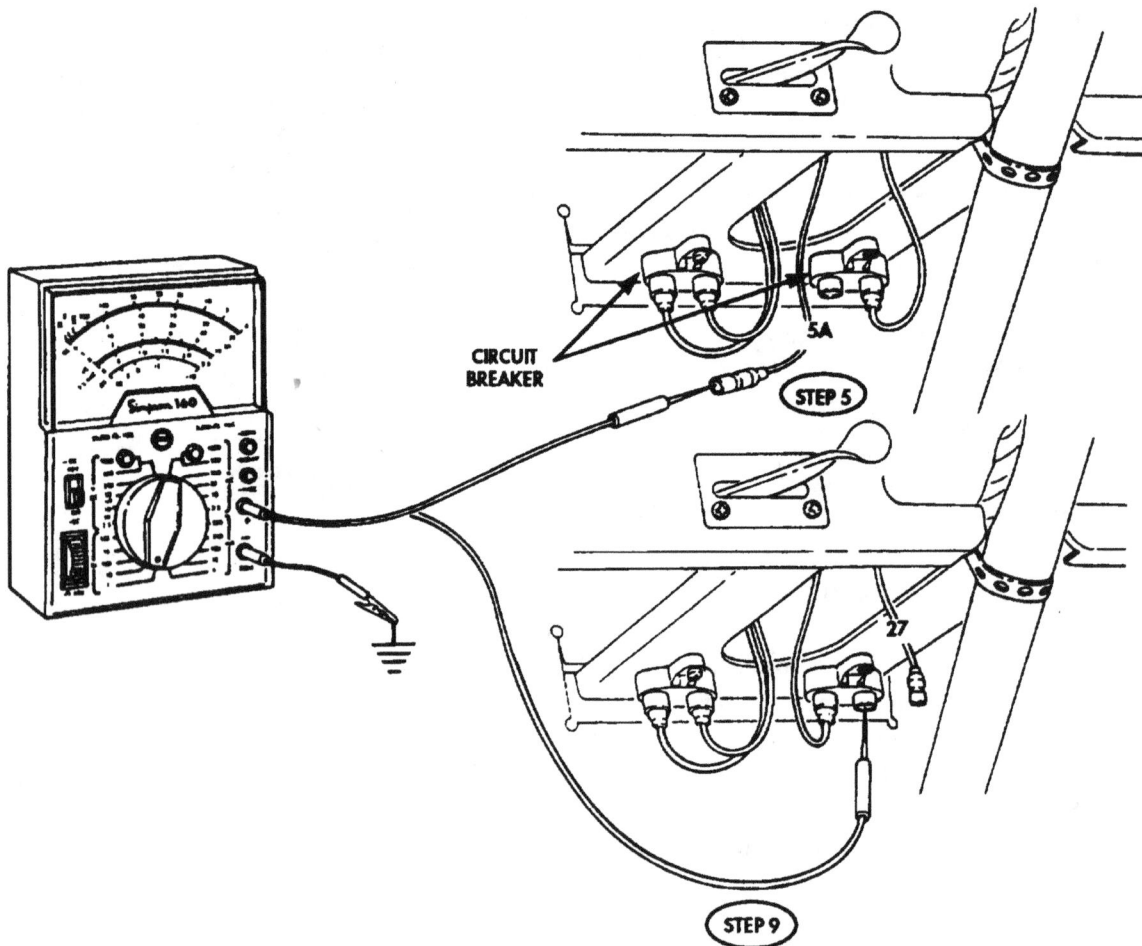

Table 2-3. Electrical Troubleshooting Contd).

MALFUNCTION
  TEST OR INSPECTION
    CORRECTIVE ACTION

Test 2. Test instrument cluster for short circuit.
  Step 1.   Place battery switch to OFF position.
  Step 2.   Remove instrument cluster panel (para. 3-83).
  Step 3.   Disconnect lead 27 from each gauge.
  Step 4.   Place battery switch to ON position.
  Step 5.   Connect one lead of a 24-volt test lamp to lead 27 disconnected from engine temperature gauge and other lead to ground. Reconnect each lead 27 to its appropriate gauge, one at a time, and note lamp indication. Test lamp should remain lit each time.
  Step 6.   Disconnect lead 27 at oil pressure gauge. Connect one lead of test lamp to lead 27 disconnected from oil pressure gauge and other lead to ground. Reconnect lead 27 to engine temperature gauge. Test lamp should light.

  If test lamp is dim, blinks on and off, or goes out when lead is connected to gauge, replace oil pressure gauge (para. 3-86).

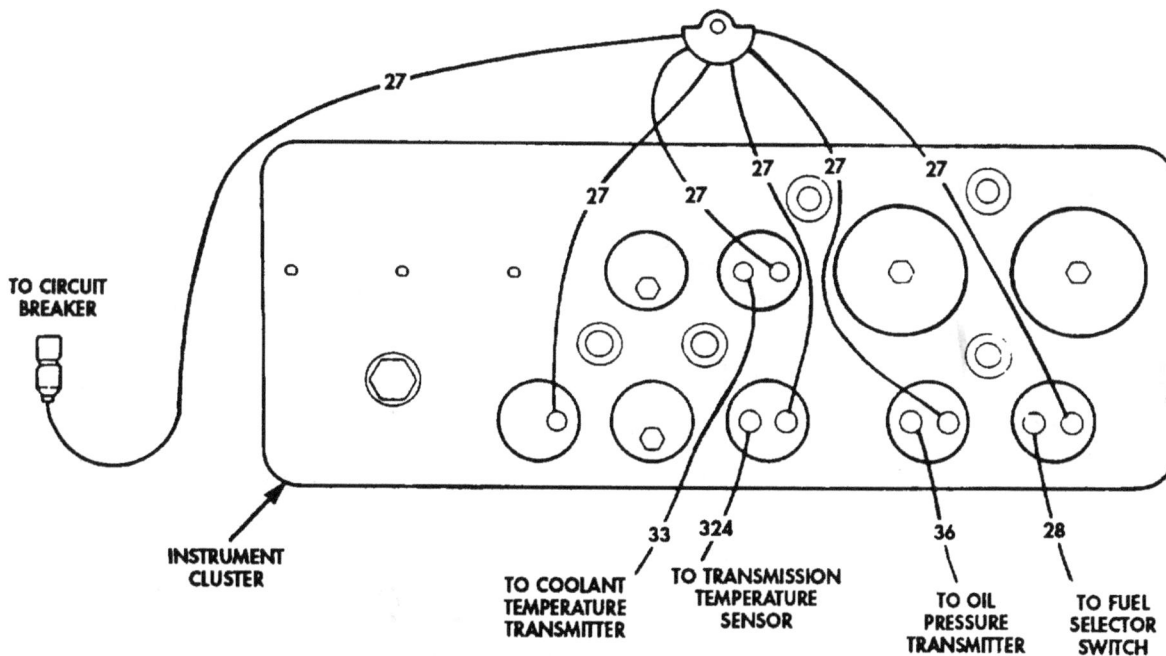

Table 2-3. Electrical Troubleshooting (Contd).

MALFUNCTION
TEST OR INSPECTION
CORRECTIVE ACTION

Test 3. Test instrument cluster voltage feed.

    Step 1.   Place battery switch to OFF position.

    Step 2.   Remove instrument cluster panel (para. 3-83).

    Step 3.   Disconnect lead 27 from each gauge.

    Step 4.   Place battery switch to ON position.

    Step 5.   Set multimeter to 50-volt range.

    Step 6.   Connect multimeter negative lead to vehicle chassis ground and positive lead to lead 27. Multimeter should indicate battery voltage.

            If battery voltage is not indicated, disconnect instrument cluster wiring harness lead 27 from power source lead 27 and go to step 7.

    Step 7.   With multimeter negative lead connected to ground, touch positive lead to circuit 27 connector at instrument cluster, Multimeter should indicate battery voltage.

            a. If battery voltage is indicated, replace instrument cluster wiring harness (para. 3-130).

            b. If battery voltage is not indicated, and voltage was correct in test 1, step 9, repair power source lead 27 (para. 3-131).

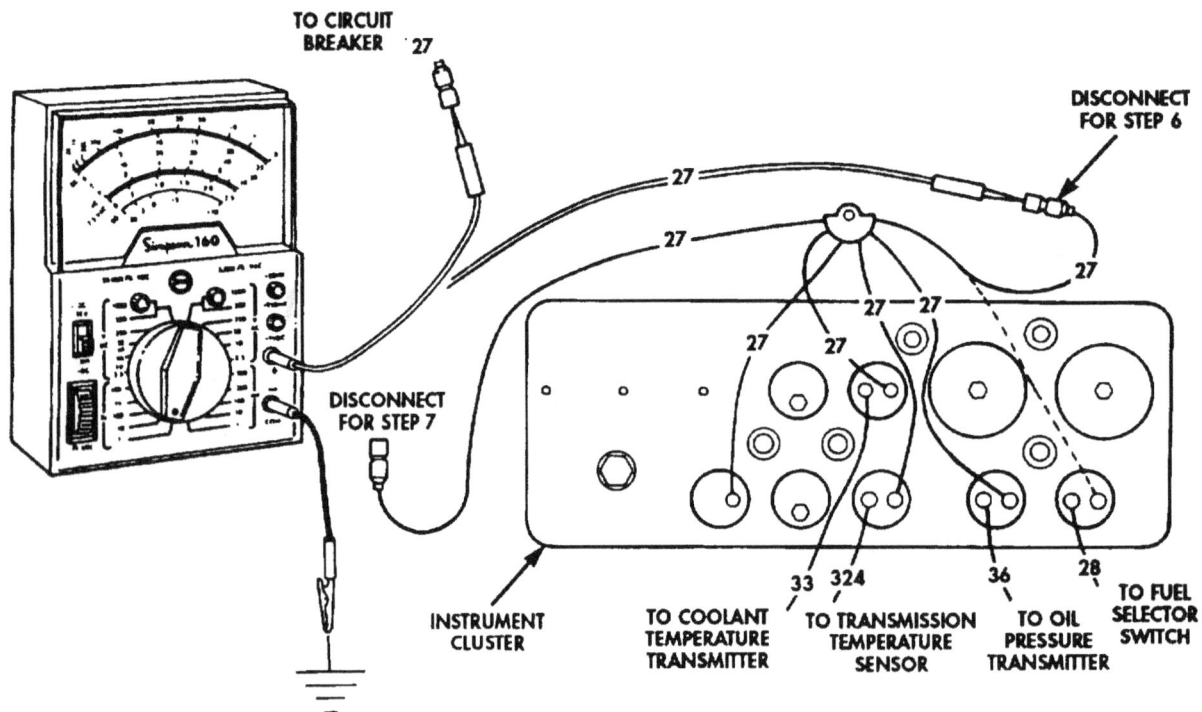

END OF TESTING!

Table 2-3. Electrical Troubleshooting (Contd).

MALFUNCTION
TEST OR INSPECTION
CORRECTIVE ACTION

28. ONE GAUGE INOPERATIVE

NOTE

If STE/ICE is available, perform NG31 - Gauge Test (para. 2-24).

Test individual gauge voltage.

    Step 1.   Place battery switch to OFF position.

    Step 2.   Remove instrument cluster panel (para. 3-83).

    Step 3.   Disconnect lead 27 from inoperative gauge.

    Step 4.   Place battery switch to ON position.

    Step 5.   Set multimeter to 50-volt range.

    Step 6.   Connect multimeter negative lead to vehicle chassis ground and positive lead to disconnected lead 27. Multimeter should indicate battery voltage.

        a. If battery voltage is not indicated, replace instrument cluster wiring harness (para. 3-130).

        b. If battery voltage is indicated, replace defective gauge (para. 3-86).

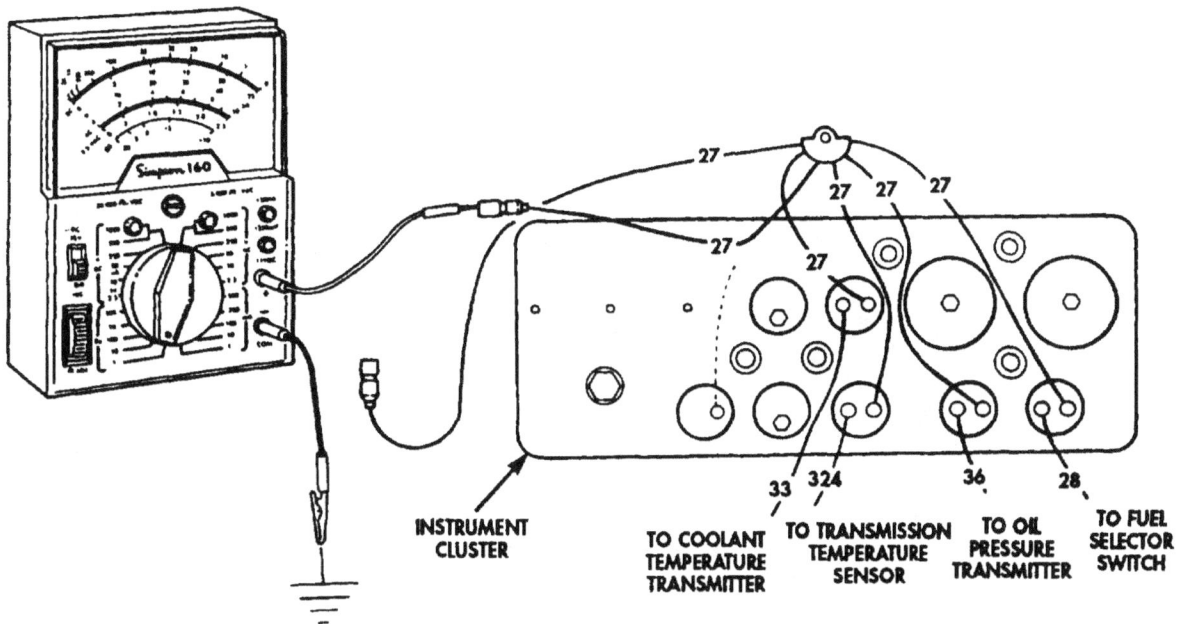

END OF TESTING!

Table 2-3. Electrical Troubleshooting (Contd).

MALFUNCTION
    TEST OR INSPECTION
        CORRECTIVE ACTION

---

29. TEMPERATURE GAUGE INOPERATIVE (COOLANT)

NOTE
If STE/ICE is available, perform NG31 - Gauge Test (para. 2-24).

Test 1. Test temperature gauge voltage. Perform malfunction 27.

Test 2. Test coolant temperature gauge sending unit.

  Step 1. Allow engine to cool.

  Step 2. Set multimeter to RX4 for resistance reading.

  Step 3. Disconnect lead 33 from temperature sending unit.

  Step 4. Start engine (TM 9-2320-272-10).

  Step 5. Connect multimeter negative lead to vehicle engine ground and connect positive lead to sending unit terminal 33. The multimeter reading should decrease as engine temperature increases.

    If resistance does not show any decrease as temperature increases, replace temperature sending unit (para. 3-93).

Test 3. Test lead 33 continuity.

  Step 1. Disconnect lead 33 from temperature gauge and temperature sending unit.

  Step 2. Attach temperature gauge end of lead 33 to chassis ground,

  Step 3. Set multimeter to RX1 for continuity reading.

  Step 4. Connect multimeter negative lead to vehicle chassis ground and positive lead to sending unit end of lead 33. Multimeter should read continuity.

    If continuity is not present, repair lead 33 (para. 3-131).

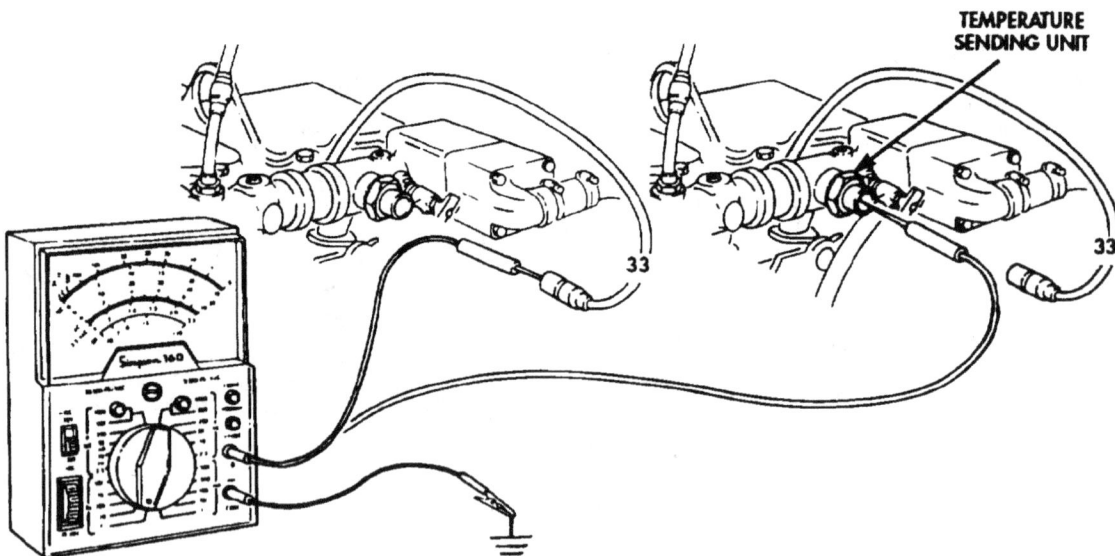

END OF TESTING!

Table 2-3. Electrical Troubleshooting (Contd).

MALFUNCTION
TEST OR INSPECTION
CORRECTIVE ACTION

30. FUEL GAUGE INOPERATIVE

WARNING

Do not perform testing near fuel tank with fill cap or sending unit removed. Fuel may ignite, causing injury to personnel.

NOTE

If STE/ICE is available, perform NG31- Gauge Test (para. 2-24).

Test 1. Test fuel level gauge voltage (malfunction 28).

Test 2. Test fuel level sending unit.

    Step 1.   Disconnect lead 28 (left-hand tank) or lead 29 (right-hand tank) from fuel level sending unit.

    Step 2.   Place battery switch to ON position.

    Step 3.   Place fuel selector switch (if equipped) to affected fuel tank.

    Step 4.   Set multimeter on 50-volt range.

    Step 5.   Connect multimeter negative lead to chassis ground and positive lead to disconnected lead. Multimeter should indicate battery voltage.

              If battery voltage is indicated, replace fuel level sending unit (para. 3-92).

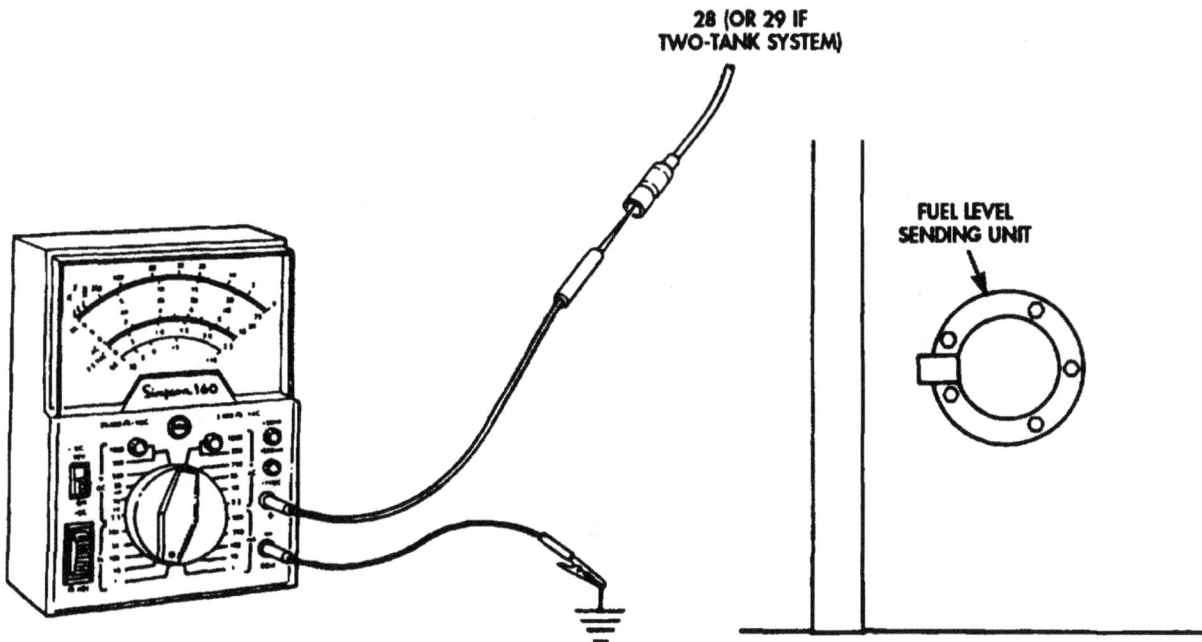

28 (OR 29 IF TWO-TANK SYSTEM)

FUEL LEVEL SENDING UNIT

Table 2-3. Electrical Troubleshooting (Contd).

MALFUNCTION
TEST OR INSPECTION
CORRECTIVE ACTION

Test 3. Test circuit wire of affected tank.

   Step 1. Dieconnect lead 28 (left-hand tank, pin B) or lead 29 (right-hand tank, pin D) from fuel selector switch.

   Step 2. Place battery switch to ON position.

   Step 3. Place fuel selector switch (if equipped) to affected fuel tank.

   Step 4. Set multimeter on 50-volt range.

   Step 5. Connect multimeter negative lead to chassis ground and positive lead to pin B or pin D on selector switch. Multimeter should indicate battery voltage.

      a. If battery voltage is indicated, repair lead from switch to sending unit (para. 3-131).

      b. If battery voltage is not indicated, go to test 4.

Test 4. Test fuel selector switch.

   Step 1. Dieconnect lead 28 from pin A of fuel selector switch.

   Step 2. Place battery switch to ON position.

   Step 3. Set multimeter to 50-volt range.

   Step 4. Connect multimeter negative lead to vehicle chassis ground and positive lead to connector of disconnected lead 28. Multimeter should indicate battery voltage.

      a. If battery voltage is indicated, replace fuel selector switch (para. 3-109).

      b. If battery voltage is not indicated, repair lead 28 (para. 3-131).

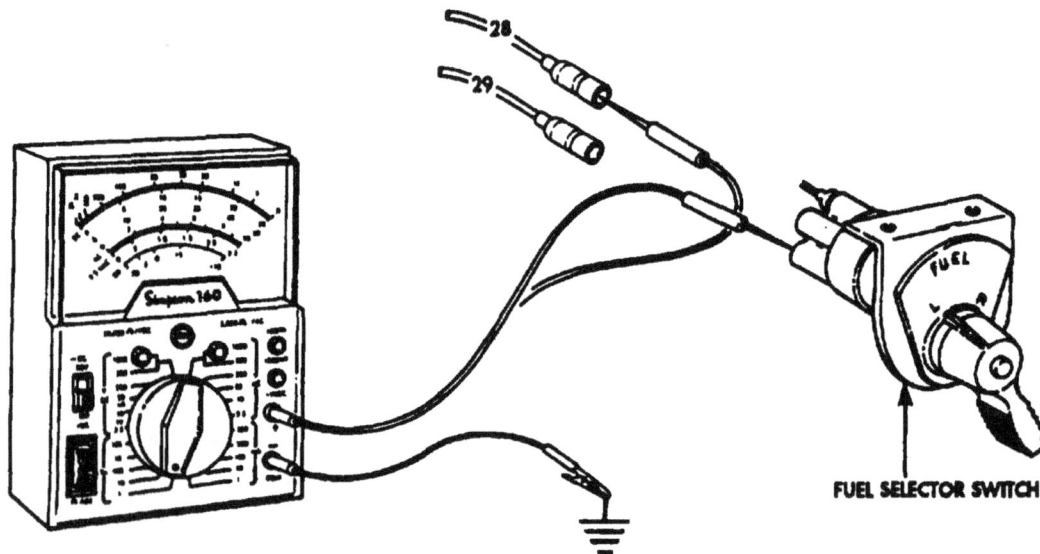

FUEL SELECTOR SWITCH

END OF TESTING!

Table 2-3. Electrical Troubleshooting (Contd).

MALFUNCTION
TEST OR INSPECTION
CORRECTIVE ACTION

31. BATTERY/ALTERNATOR GAUGE INOPERATIVE

NOTE

If STE/ICE is available, perform NG31 - Gauge Test (para. 2-24).

Test 1. Test battery/altamator gauge voltage. Perform malfunction 28.

Test 2. Check battery/alternator gauge for continuity.

CAUTION

Place battery switch to OFF position. If ground is open and battery voltage is present, the multimeter may be damaged.

Step 1. Set multimeter to RX1 for continuity reading.

Step 2. Connect multimeter negative lead to vehicle chassis ground and positive lead to indicator case. Multimeter should read continuity.

   a. If multimeter doer not read continuity, clean and tighten indicator mounting points.

   b. If multimeter doer read continuity, replace battery/alternator gauge (para. 3-86).

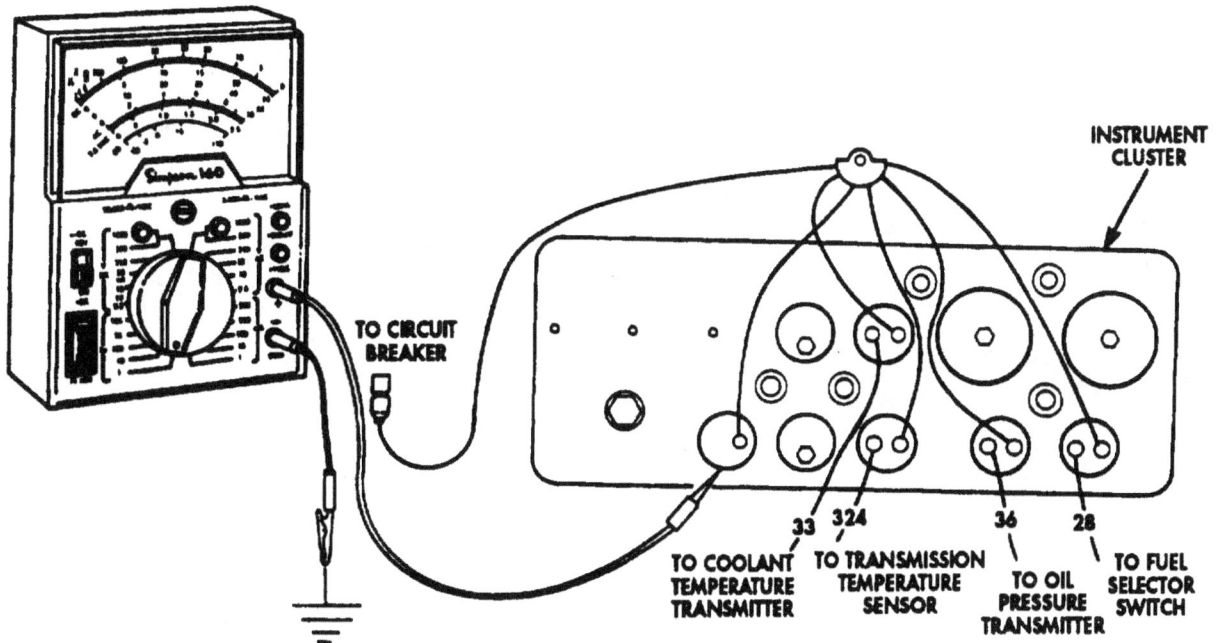

END OF TESTING!

Table 2-3. Electrical Troubleshooting (Contd).

MALFUNCTION
TEST OR INSPECTION
CORRECTIVE ACTION

32. HORN INOPERATIVE

    Test 1. Test horn circuit breaker voltage.

        Step 1. Place battery switch to OFF position.

        Step 2. Disconnect lead 10 from circuit breaker.

        Step 3. Place battery switch to ON position.

        Step 4. Set multimeter to 50-volt range.

        Step 5. Connect multimeter negative lead to vehicle chassis ground and positive lead to lead 10. Multimeter should indicate battery voltage.

            a. If battery voltage is not indicated, perform malfunctions 2, 3,4, 6, and 7.

            b. If battery voltage is indicated, go to step 6.

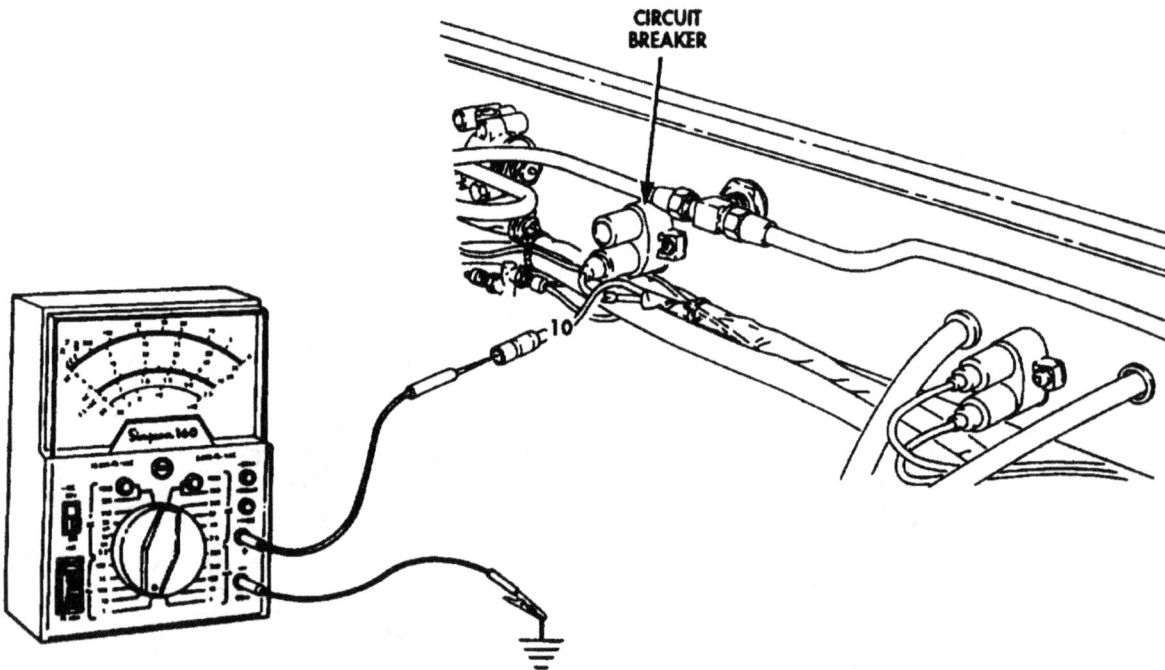

Table 2-3. Electrical Troubleshooting (Contd).

MALFUNCTION
TEST OR INSPECTION
CORRECTIVE ACTION

Step 6. Place battery switch to OFF position.

Step 7. Connect lead 10 and disconnect lead 26 from circuit breaker.

Step 8. Place battery switch to ON position.

Step 9. Connect multimeter negative lead to vehicle chassis ground and positive lead to lead 26. Multimeter should indicate battery voltage.

If battery voltage is not indicated, replace circuit breaker (para. 3-105).

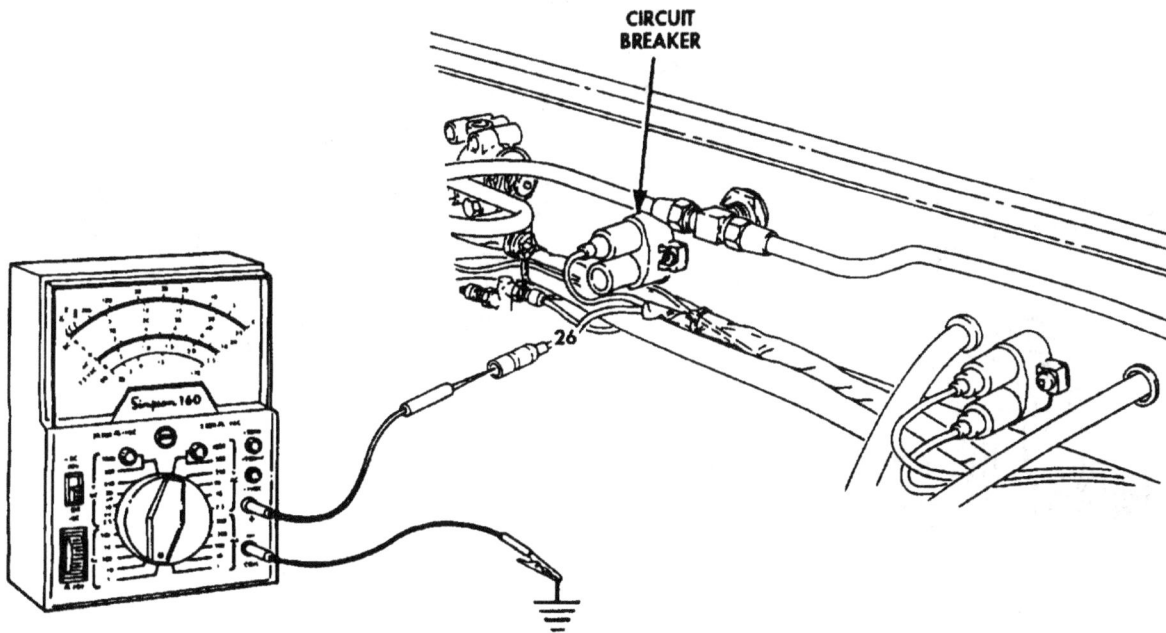

Table 2-3. Electrical Troubleshooting (Contd).

MALFUNCTION
TEST OR INSPECTION
CORRECTIVE ACTION

Test 2. Test horn circuit voltage.

    Step 1.   Disconnect leads 25 and 26 from horn solenoid.

    Step 2.   Set multimeter to 50-volt range.

    Step 3.   Connect multimeter negative lead to vehicle chassis ground and positive lead to lead 26. Multimeter should indicate battery voltage.

        a. If battery voltage is not indicated, repair or replace lead 26 (para. 3-131).

        b. If battery voltage is indicated, proceed to step 4.

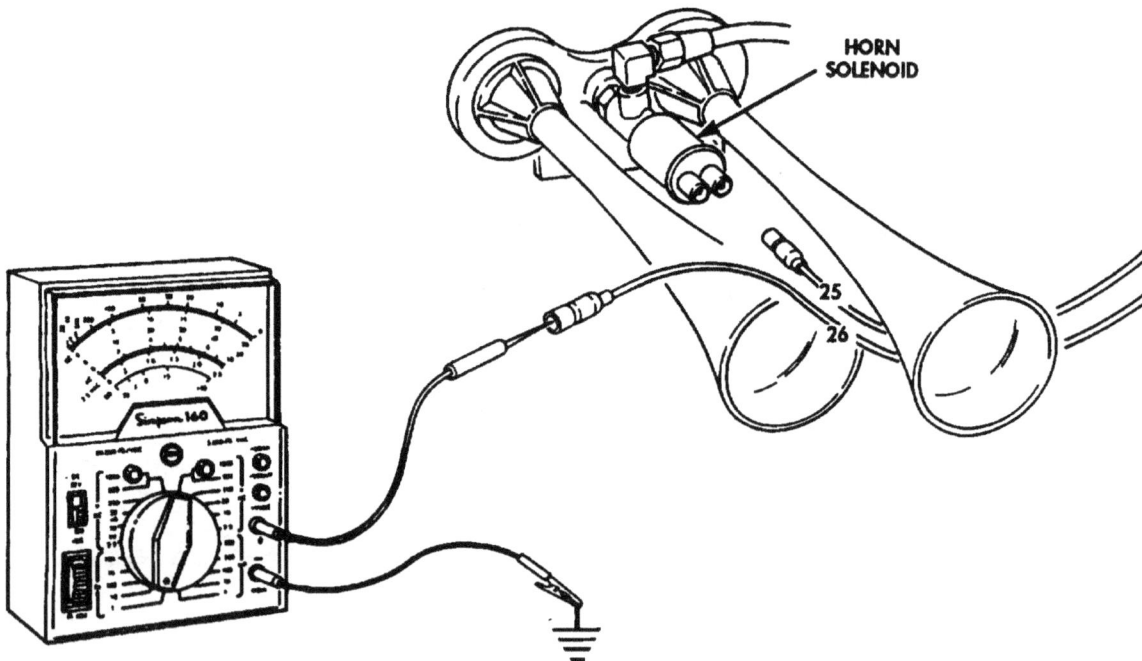

Table 2-3. Electrical Troubleshooting (Contd).

MALFUNCTION
TEST OR INSPECTION
CORRECTIVE ACTION

Step 4. Connect lead 26 to horn solenoid, Connect a jumper wire to vehicle chassis ground and the other end to lead 25 terminal on horn solenoid. Horn should sound.

    a. If horn does not sound, replace horn solenoid (para. 3-103).

    b. If horn sounds, connect lead 26 and go to test 3.

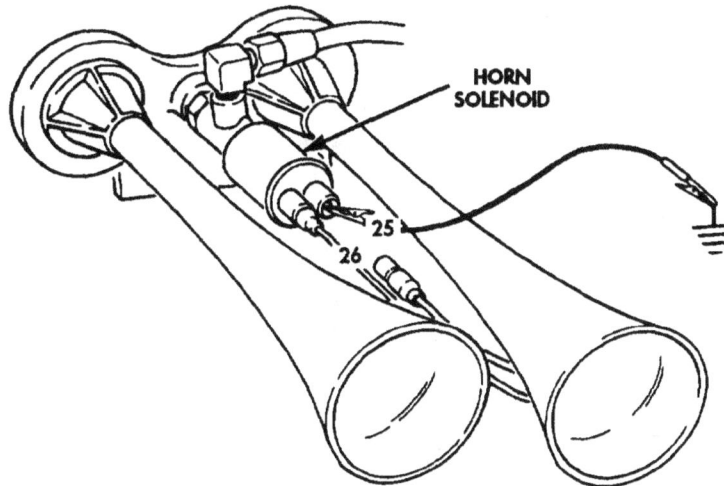

Test 3. Test horn switch.

  Step 1. Disconnect lead 25 from horn switch.

  Step 2. Connect one end of a jumper wire to vehicle chassis ground and touch opposite end to lead 25 connector. Horn should sound.

    If horn sounds, replace horn switch (para. 3-104).

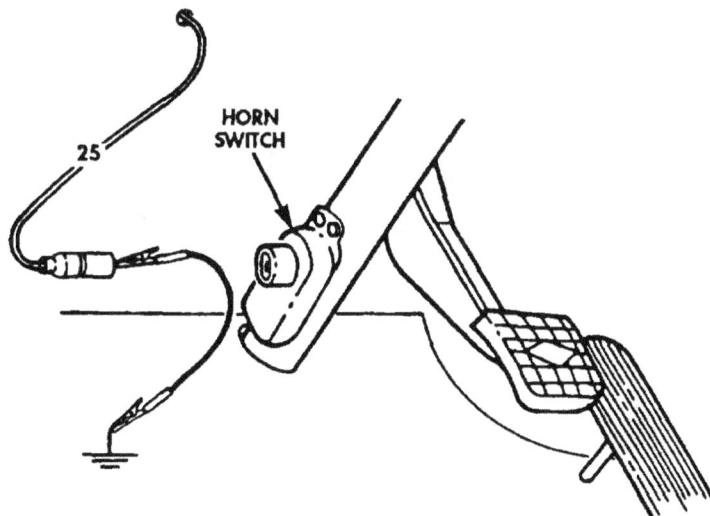

END OF TESTING!

Table 2-3. Electrical Troubleshooting (Contd).

MALFUNCTION
TEST OR INSPECTION
CORRECTIVE ACTION

33. LOW AIR PRESSURE WARNING BUZZER WILL NOT SHUT OFF (AIR PRESSURE GAUGES AT NORMAL SYSTEM PRESSURE)

Test 1. Test resistance at low air pressure switches.

Step 1. Start engine (TM 9-2320-272-10) and allow air pressure to reach normal operating pressure.

Step 2. Disconnect leads 578 and 578A from low air pressure switch A.

Step 3. Set multimeter to RX4 for resistance reading.

Step 4. Touch positive lead of multimeter to contact end of pressure switch A.

Step 5. Touch negative lead of multimeter to other contact end of pressure switch A.

  a. If resistance is 300 ohms or less, go to step 6.

  b. If resistance is more than 300 ohms, replace low air pressure switch (para. 3-86).

Step 6. Connect leads 578 and 578A to low air pressure switch A.

Step 7. Disconnect leads 57 and 578A from low air pressure switch B.

Step 8. Touch positive lead of multimeter to contact end of pressure switch B.

Step 9. Touch negative lead of multimeter to other contact end of pressure switch B.

  a. If resistance is 300 ohms or less, stop engine, connect leads, and go to test 2.

  b. If resistance is more than 300 ohms, replace low air pressure switch (para. 3-86).

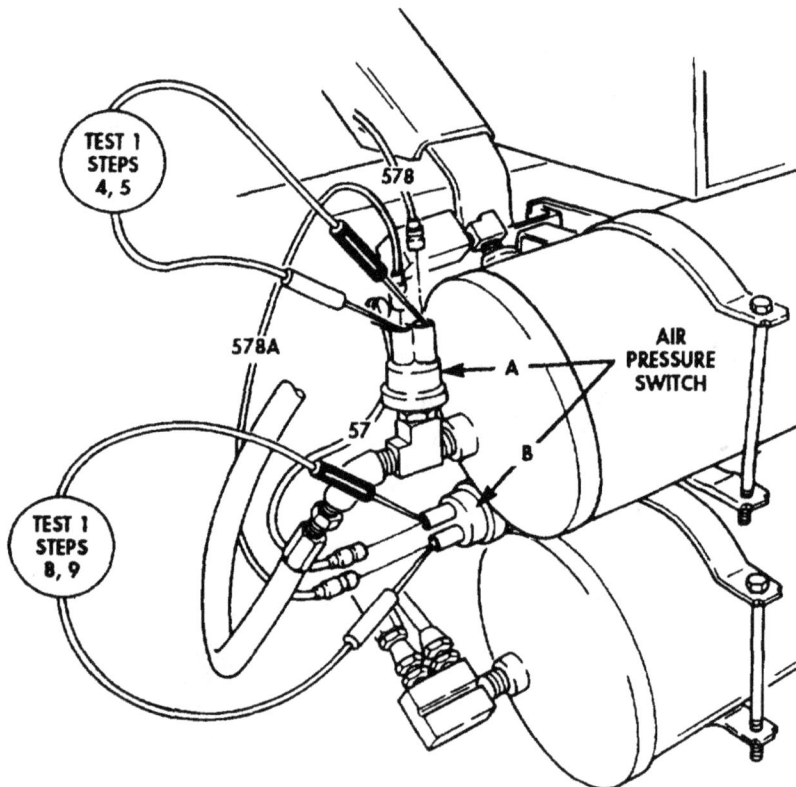

Table 2-3. Electrical Troubleshooting (Contd).

MALFUNCTION
    TEST OR INSPECTION
        CORRECTIVE ACTION

Test 2. Test continuity at air pressure switches.

    Step 1.   Start engine (TM 9-2320-272-10) and allow air pressure to reach normal operating pressure.

    Step 2.   Set multimeter to RX1 for continuity reading.

    Step 3.   Disconnect lead 57 at air pressure switch B.

    Step 4.   Touch positive lead of multimeter to contact end of lead 57 and negative lead of multimeter to frame ground.

        a. If continuity is present, go to step 5.

        b. If continuity is not present, replace lead 57 (para. 3-131).

    Step 5.   Connect lead 57 to air pressure switch B.

    Step 6.   Disconnect lead 578 from air pressure switch A.

    Step 7.   Touch positive lead of multimeter to contact end of lead 578 and negative lead of multimeter to frame ground.

        a. If continuity is present, go to step 8.

        b. If continuity is not present, replace lead 578 (para. 3-131).

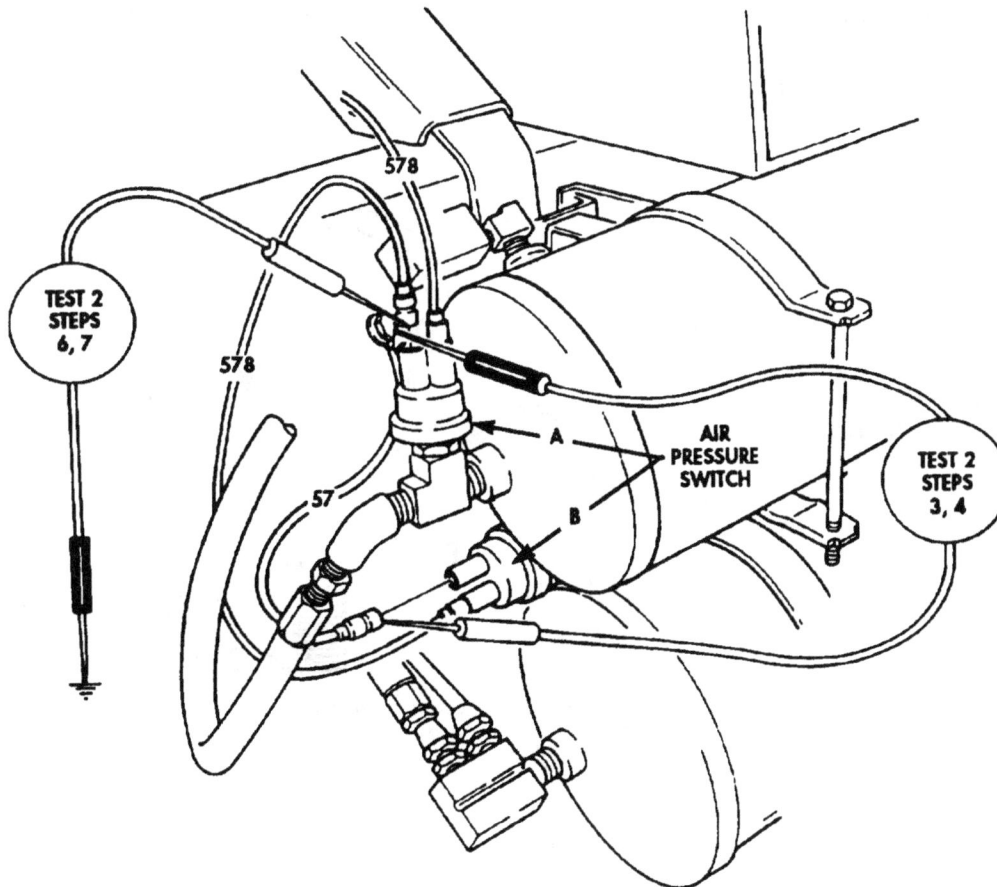

Table 2-3. Electrical Troubleshooting (Contd).

MALFUNCTION
    TEST OR INSPECTION
        CORRECTIVE ACTION

Step 8.   Connect lead 578 to air pressure switch A.

Step 9.   Disconnect wiring harness at failsafe warning control module.

Step 10. Touch positive lead of multimeter to pin F (lead 578) and negative lead of multimeter to frame ground.

      a. If continuity is present, go to step 11.

      b. If continuity is not present, replace lead 578 (para. 3-131).

Step 11. Touch positive lead of multimeter to pin H (lead 57) and negative lead of multimeter to frame ground.

      a. If continuity is present, go to step 12.

      b. If continuity is not present, replace lead 57 (para. 3-131).

Step 12. Touch positive lead of multimeter to pin D (lead 350) and negative lead of multimeter to frame ground.

      a. If continuity is present, go to step 13.

      b. If continuity is not present, replace lead 350. (Refer to Digest Article, Engine Overheating Warning System.)

Step 13. Ensure parking brake is engaged (TM 9-2320-272-10).

Step 14. Touch positive lead of multimeter to pin I (lead 584) and negative lead of multimeter to frame ground.

      If low air warning buzzer will not shut off, replace.

      a. If continuity is not present, replace lead 584 (para. 3-131).

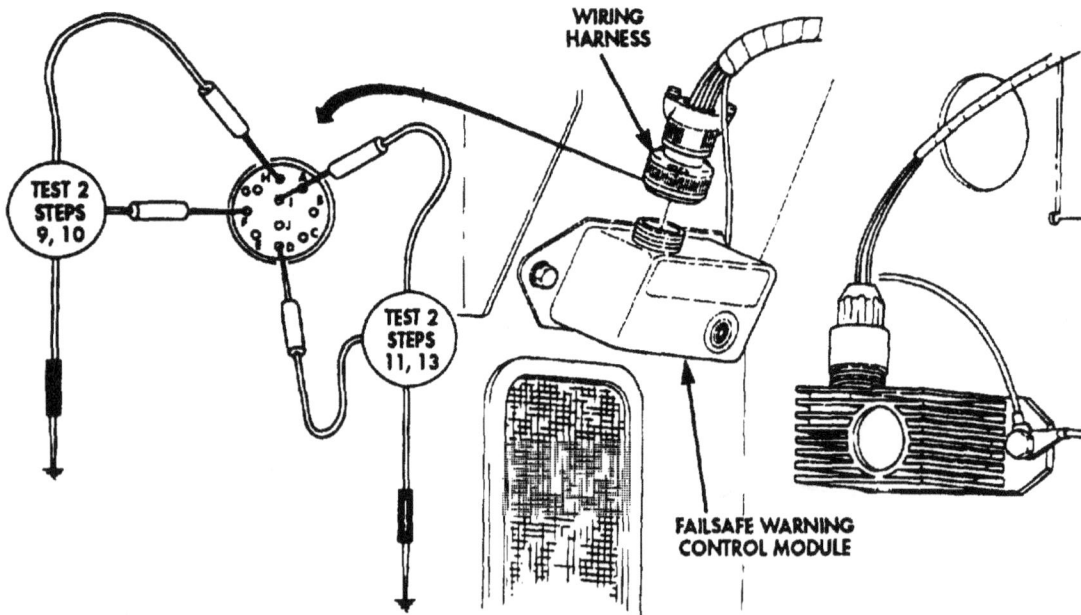

END OF TESTING!

Table 2-3. Electrical Troubleshooting (Contd).

MALFUNCTION
    TEST OR INSPECTION
        CORRECTIVE ACTION

34. LOW AIR PRESSURE WARNING BUZZER WILL NOT SHUT OFF AND PARKING BRAKE WARNING LIGHT FLASHES WITH HAND/PARKING BRAKE IN ANY POSITION (AIR PRESSURE NORMAL)

Step 1.   Set multimeter to RX1 for continuity reading.

Step 2.   Disconnect lead 584 from parking brake switch (under cab).

Step 3.   Check for continuity through parking brake switch to ground lead 584 (with parking brake released).

Step 4.   If continuity is not present, replace parking brake switch (para. 3-107).

END OF TESTING!

35. SPRING BRAKE WARNING LIGHT INOPERATIVE WITH SPRING BRAKE OVERRIDE ENGAGED

Test 1. Check spring brake override warning light.

Replace lamp with one known to be good (para. 3-84).

Test 2. Test spring brake pressure switch.

Step 1.   Place battery switch to ON position.

Step 2.   Disconnect two leads 37 from spring brake pressure switch.

Step 3.   Connect a jumper wire to both leads 37. Warning light should operate.

If warning light is on, replace spring brake pressure switch (para. 3-100).

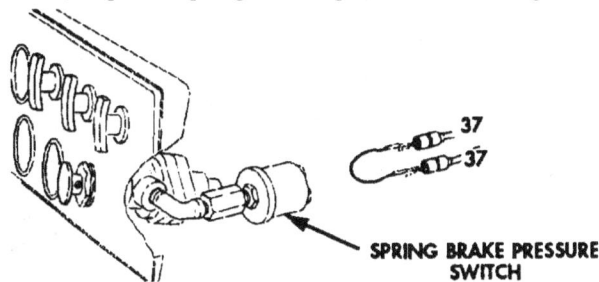

Test 3. Test spring brake override warning light circuit breaker voltage.

Step 1.   Place battery switch to OFF position.

Step 2.   Disconnect lead 10 from circuit breaker.

Step 3.   Place battery switch to ON position.

Step 4.   Set multimeter to 50-volt range.

Step 5.   Connect multimeter negative lead to vehicle chassis ground and positive lead to lead 10. Multimeter should indicate battery voltage.

If battery voltage is not indicated, perform malfunctions 2, 3, 4, 6, and 7.

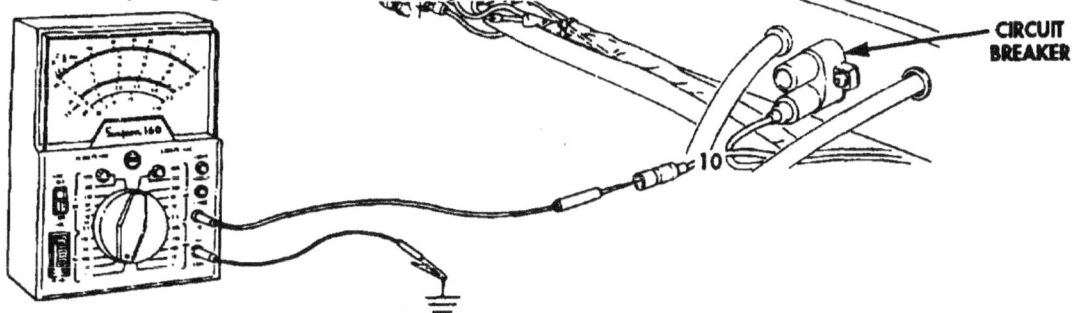

Table 2-3. Electrical Troubleshooting (Contd).

MALFUNCTION
TEST OR INSPECTION
CORRECTIVE ACTION

Step 6. Place battery switch to OFF position.

Step 7. Connect lead 10 and disconnect lead 37 from circuit breaker.

Step 8. Place battery switch to ON position.

Step 9. Connect multimeter negative lead to vehicle chassis ground and positive lead to lead 37 terminal of circuit breaker. Multimeter should indicate battery voltage.

      If battery voltage is not indicated, replace defective circuit breaker (para. 3-105).

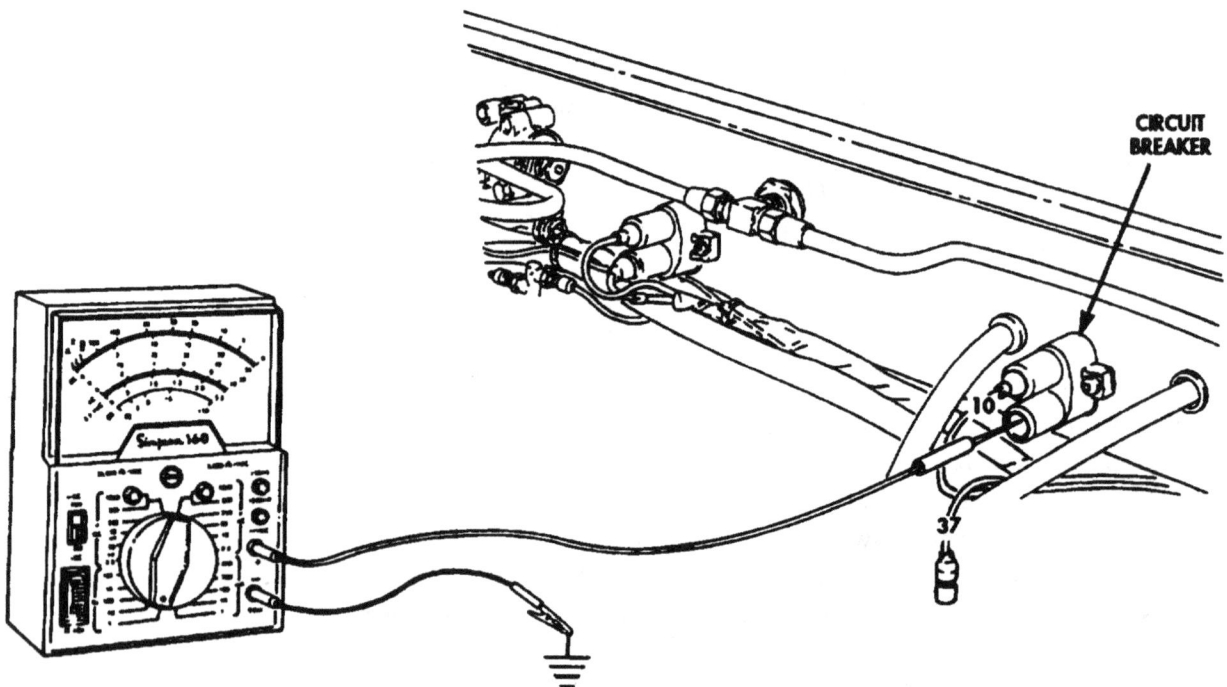

END OF TESTING!

Table 2-3. Electrical Troubleshooting (Contd).

MALFUNCTION
TEST OR INSPECTION
CORRECTIVE ACTION

TRAILER CONNECTION SYSTEM

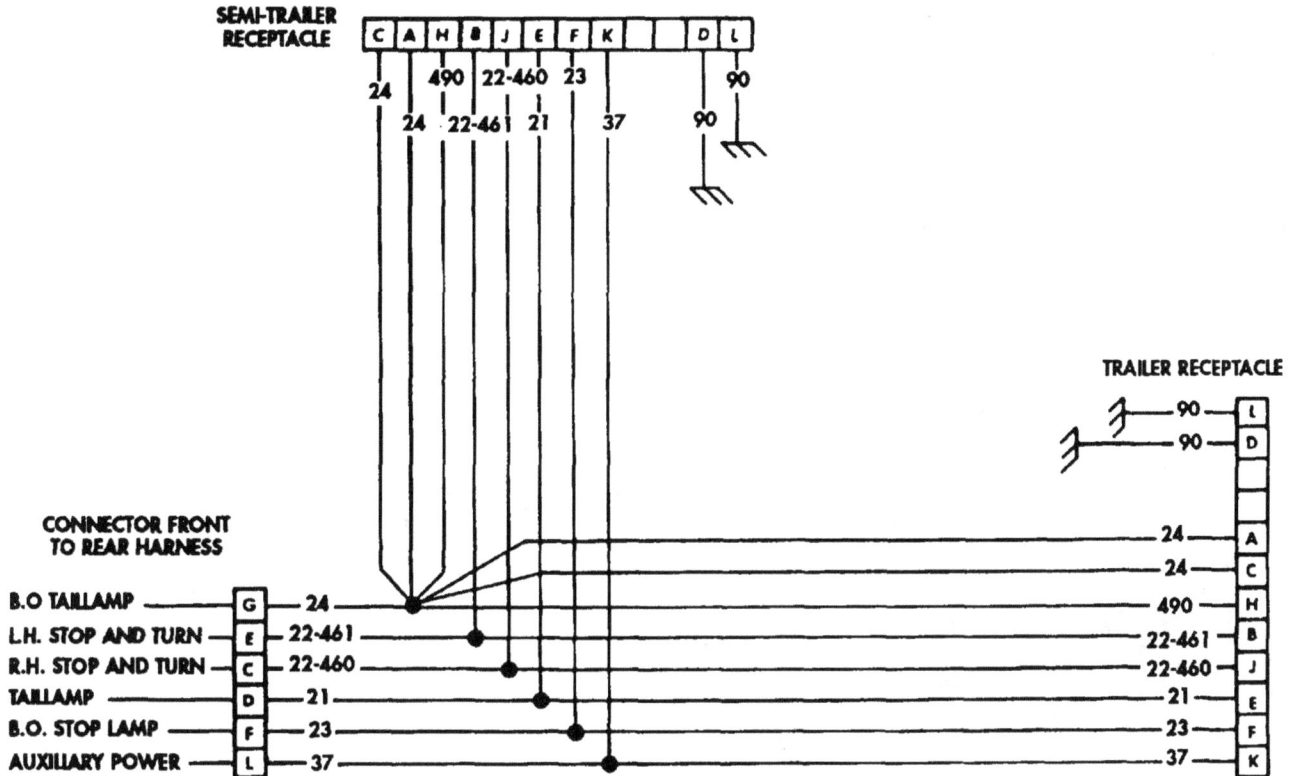

36. ONE OR MORE LIGHTING SYSTEMS DO NOT LIGHT ON TRAILER

Test 1. Test trailer connecting receptacle voltage (malfunction 22).

Test 2. Test trailer connecting cable continuity.

  Step 1. Remove connecting cable from vehicle.

  Step 2. Place both ends of cable at a convenient point near each other. Open and lock hinged covers on each end.

  Step 3. Set multimeter to RX1 for continuity reading.

  Step 4. Connect multimeter leads from socket letter A on one connector to socket letter A on other connector. Note multimeter reading.

NOTE

Wires M and N are not used and continuity is not required.

Table 2-3. Electrical Troubleshooting (Contd).

MALFUNCTION
TEST OR INSPECTION
CORRECTIVE ACTION

Test 3. Test each wire the same until all 10 wires have been tested. Continuity should be read on each wire.

    a. If continuity is not read on any wire, replace cable.

    b. If continuity is read, plug cable into vehicle receptacle and test receptacle voltage at trailer end of cable. Perform malfunction 22.

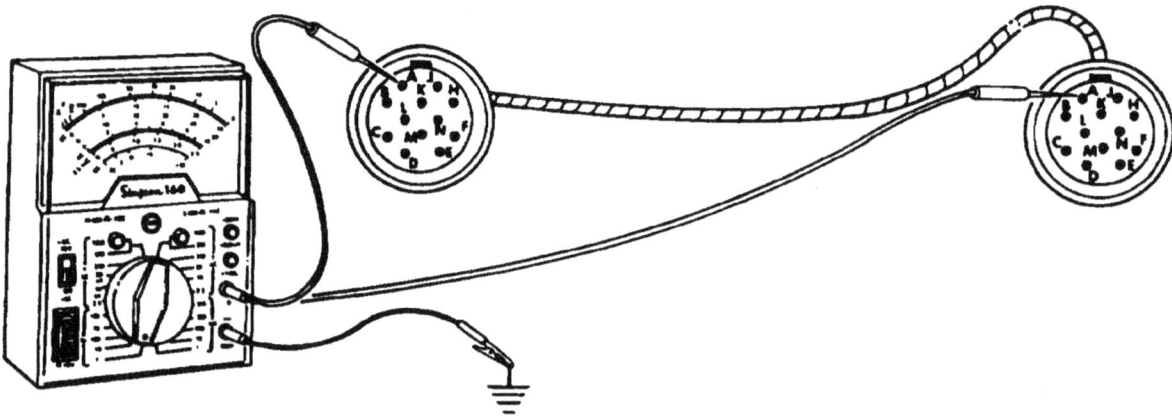

END OF TESTING!

Table 2-3. Electrical Troubleshooting (Contd).

MALFUNCTION
TEST OR INSPECTION
CORRECTIVE ACTION

HEATER SYSTEM

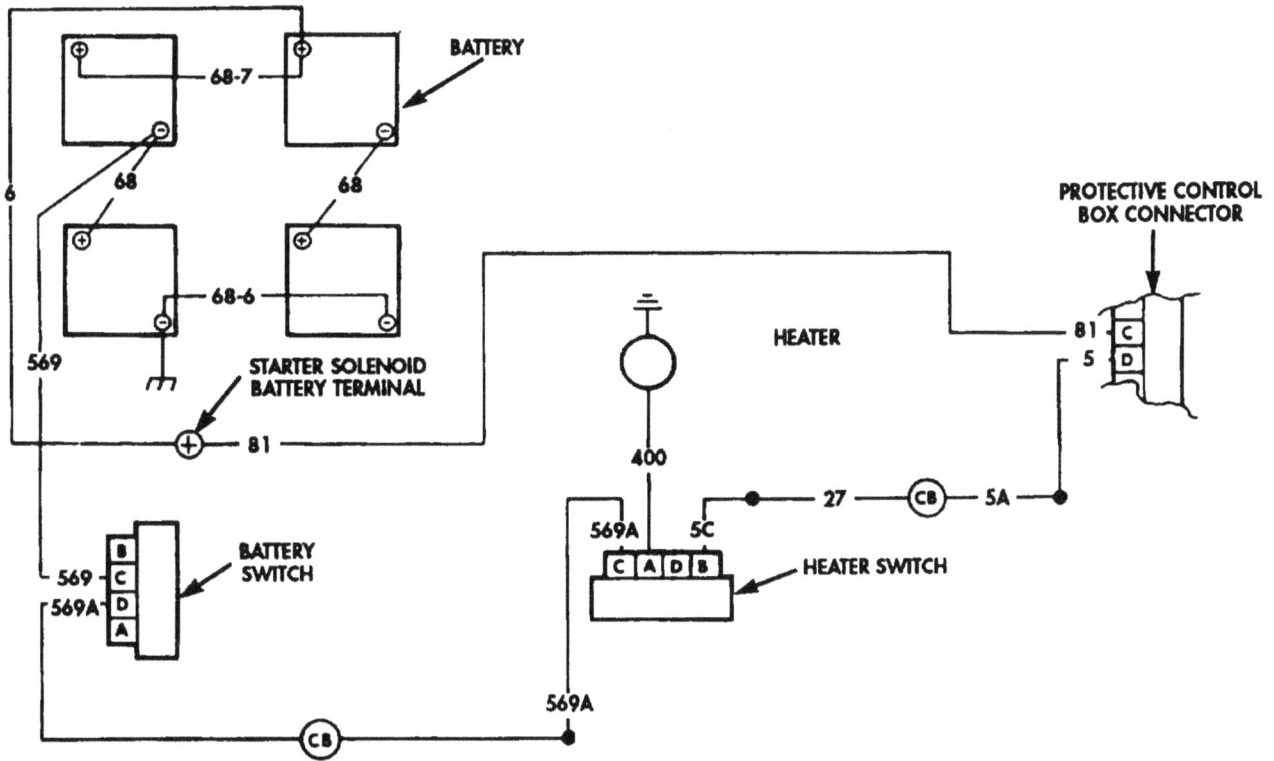

37. HEATER WILL NOT OPERATE WITH SWITCH IN LOW POSITION (HIGH POSITION OPERATION NORMAL)

    Test 1. Test lead 569A for voltage.

       Step 1.   Place battery switch to OFF position.

       Step 2.   Disconnect lead 569A from heater switch.

       Step 3.   Place battery switch to ON position.

       Step 4.   Set multimeter to 50-volt range.

       Step 5.   Connect multimeter negative lead to vehicle chassis ground and positive lead to lead 569A. Multimeter should indicate 12 to 13-volts.

            a. If battery voltage is not indicated, go to test 2.

            b. If battery voltage is indicated, replace defective heater switch (para. 3-107).

Table 2-3. Electrical Troubleshooting (Contd).

MALFUNCTION
TEST OR INSPECTION
CORRECTIVE ACTION

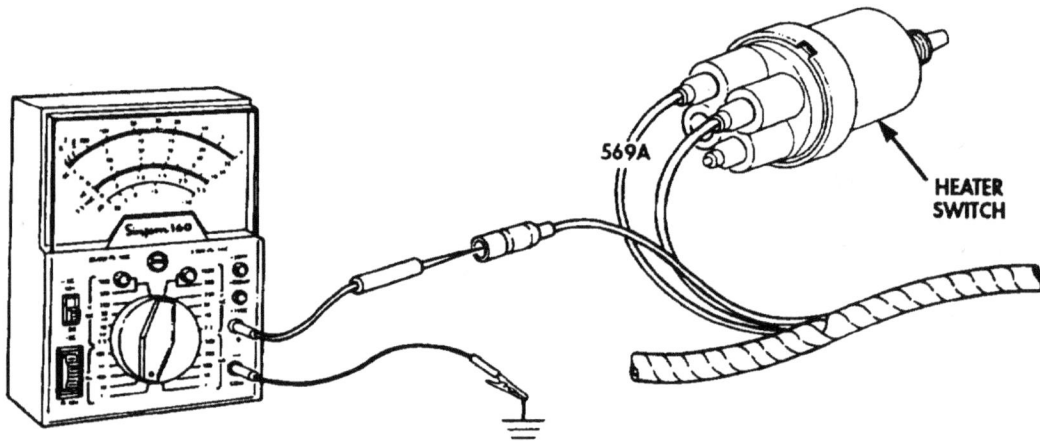

Test. 2. Test circuit breaker.

    Step 1. Disconnect lead 569A to heater switch from circuit breaker.

    Step 2. Set multimeter to 50-volt range, connect negative lead to vehicle chassis ground, and positive lead to exposed circuit breaker terminal.

    Step 3. Place battery switch to ON position and observe multimeter.

    Step 4. Disconnect lead 569A to battery switch from circuit breaker. Connect multimeter positive lead to wire and observe multimeter.

        a. If multimeter indicated 12-13 volts in step 3, repair or replace lead 569A between circuit breaker and heater switch (para. 3-131).

        b. If multimeter indicated 12-13 volts in step 4, replace circuit breaker (para. 3-105).

END OF TESTING!

Table 2-3. Electrical Troubleshooting (Contd).

MALFUNCTION
TEST OR INSPECTION
CORRECTIVE ACTION

38. HEATER WILL NOT OPERATE WITH SWITCH IN HIGH POSITION (LOW POSITION OPERATION NORMAL)

Step 1. Place battery to OFF position.

Step 2. Disconnect lead 5C from heater switch.

Step 3. Place battery switch to ON position.

Step 4. Set multimeter to 50-volt range.

Step 5. Connect multimeter negative lead to vehicle chassis ground and positive lead to lead 5C. Multimeter should indicate 12-13 volts.

    a. If battery voltage is not indicated, perform malfunction 27, test 1.

    b. If battery voltage is indicated, replace defective heater switch (para. 3-107).

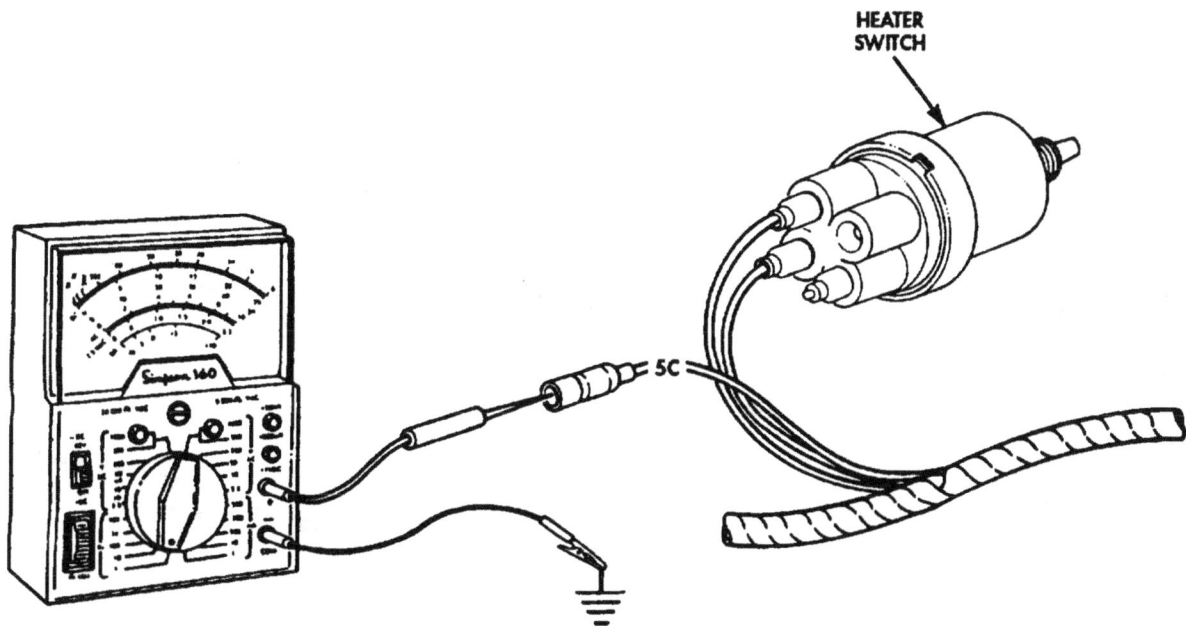

END OF TESTING!

Table 2-3. Electrical Troubleshooting (Contd).

MALFUNCTION
TEST OR INSPECTION
CORRECTIVE ACTION

39. HEATER WILL NOT OPERATE IN LOW OR HIGH POSITION

Step 1.   Place battery to OFF position.

Step 2.   Disconnect lead 400 from heater switch.

Step 3.   Place battery switch to ON position.

Step 4.   Set multimeter to 50-volt range.

Step 5.   Connect multimeter negative lead to vehicle chassis ground. Touch positive lead to lead 400 terminal on heater switch and note voltage reading. Multimeter should indicate 12-13 volts with heater switch in LOW or HIGH position.

    a.   If battery voltage is not indicated, replace heater switch (para. 3-107).

    b.   If battery voltage is indicated, clean and tighten heater ground point connection.

    c.   If heater still does not operate, replace heater (para. 3-292).

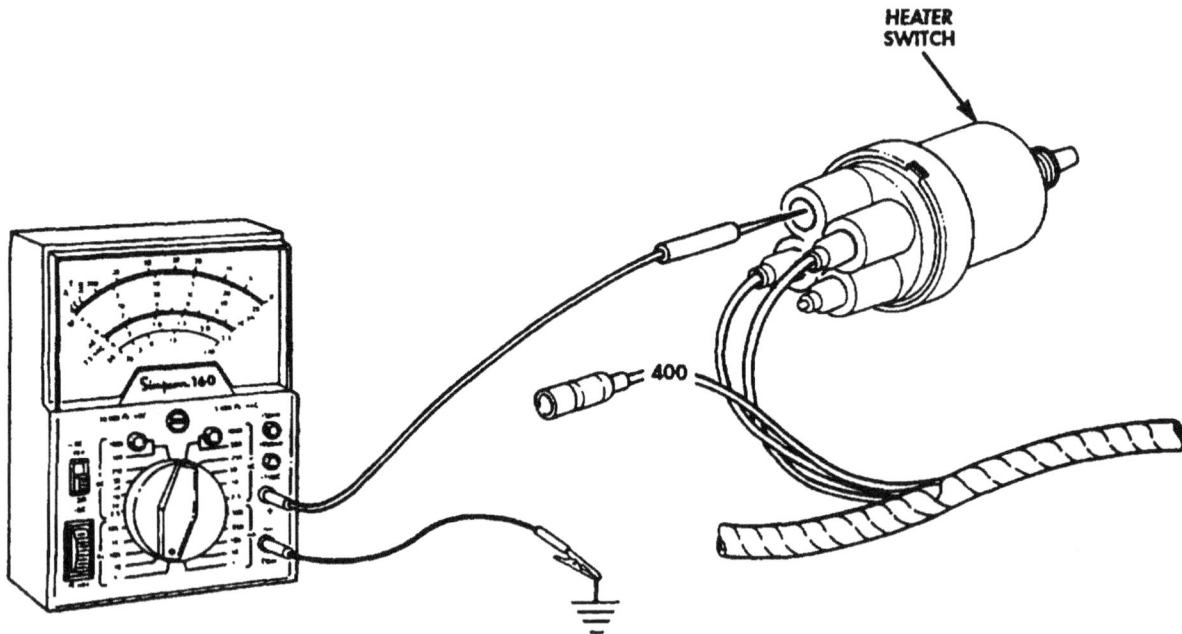

HEATER SWITCH

400

END OF TESTING!

Table 2-3. Electrical Troubleshooting (Contd).

MALFUNCTION
TEST OR INSPECTION
CORRECTIVE ACTION

100-AMP ALTERNATOR KIT

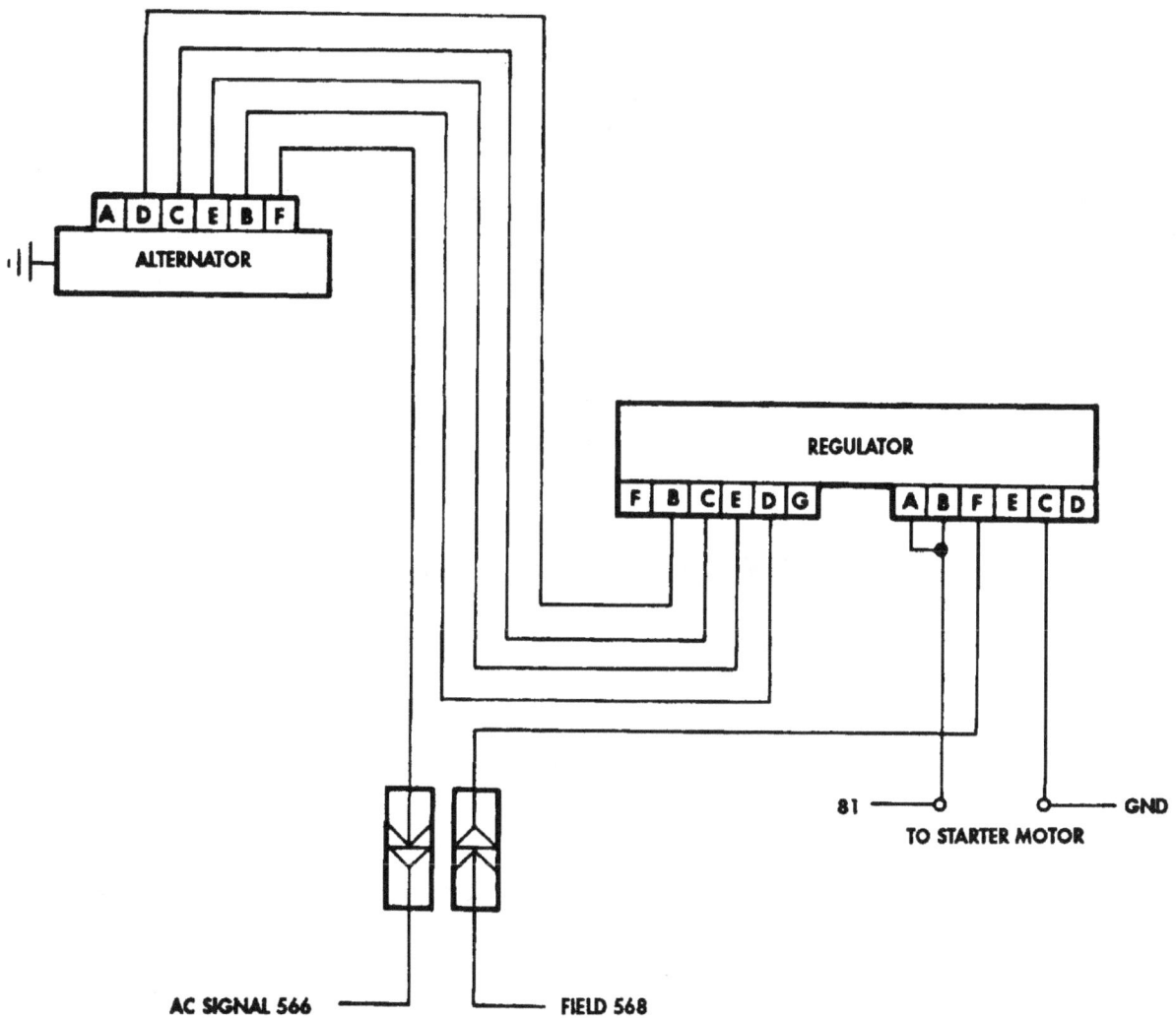

40. BATTERIES HOT OR BOILING; CORRECT SPECIFIC GRAVITY OF ALL CELLS IS 1.280

NOTE

If STE/ICE is available, perform NG50 - Charging Circuit Tests (para. 2-24).

Test charging voltage (malfunction 9).

END OF TESTING!

Table 2-3. Electrical Troubleshooting (Contd).

> MALFUNCTION
>> TEST OR INSPECTION
>>> CORRECTIVE ACTION

41. BATTERIES USE EXCESSIVE WATER

NOTE

If STE/ICE is available, perform NG81 - Battery Tests or NG50 - Charging Circuit Tests (para. 2-24).

Test charging voltage (malfunction 9).

END OF TESTING!

42. BATTERIES RUN DOWN IN SERVICE

NOTE

If STE/ICE is available, perform NG50 - Charging Circuit Tests (para. 2-24).

Test 1. Check for loose, broken, or missing alternator or fan drivebelts.

    a. Adjust loose drivebelts (M939/A1) (para. 3-78).

    b. Replace broken or missing drivebelts (para. 3-79 or 3-80).

Test 2. Test charging voltage (malfunction 9).

    a. If proper voltage is indicated, problem is not in generating system. Refer to battery system troubleshooting.

    b. If proper voltage is not indicated, perform malfunction 44.

END OF TESTING!

43. NO ALTERNATOR OUTPUT

NOTE

If STE/ICE is available, perform NG50 - Charging Circuit Tests (para. 2-24).

Test 1. Check for loose, broken, or missing alternator or fan drivebelts.

    a. Adjust loose drivebelts (M939/A1) (para. 3-78).

    b. Replace broken or missing drivebelts (para. 3-79 or 3-80).

Test 2. Test continuity of alternator harness.

  Step 1. Disconnect quick-disconnect for lead 566.

  Step 2. Disconnect voltage regulator harness connector at alternator and voltage regulator.

  Step 3. Set multimeter to RX1 for continuity reading.

  Step 4. Connect multimeter negative lead to contact end of lead 566 and positive lead to pin F. Continuity should be present.

    a. If continuity is present, go to step 5.

    b. If continuity is not present, replace voltage regulator harness (para. 3-131).

  Step 5. Connect multimeter positive lead to contact end of pin D at voltage regulator harness connector and negative lead to contact end of pin B at alternator harness connector. Continuity should be present.

    a. If continuity is present, go to step 6.

    b. If continuity is not present, replace voltage regulator harness (para. 3-131).

Table 2-3. Electrical Troubleshooting (Contd).

MALFUNCTION
TEST OR INSPECTION
CORRECTIVE ACTION

Step 6.   Connect multimeter positive lead to contact end of pin E at voltage regulator harness connector and negative lead to contact end of pin E at alternator harness connector. Continuity should be present.

        a. If continuity is present, go to step 7.

        b. If continuity is not present, replace voltage regulator harness (para. 3-131).

Step 7.   Connect multimeter positive lead to contact end of pin C at voltage regulator harness connector and negative lead to contact end of pin C at alternator harness connector. Continuity should be present.

        a. If continuity is present, go to step 8.

        b. If continuity is not present, replace voltage regulator harness (para. 3-131).

Step 8.   Connect multimeter positive lead to contact end of pin B at voltage regulator harness connector and negative lead to contact end of pin D at alternator harness connector. Continuity should be present.

        a. If continuity is present, go to test 3.

        b. If continuity is not present, replace voltage regulator harness (para. 3-131).

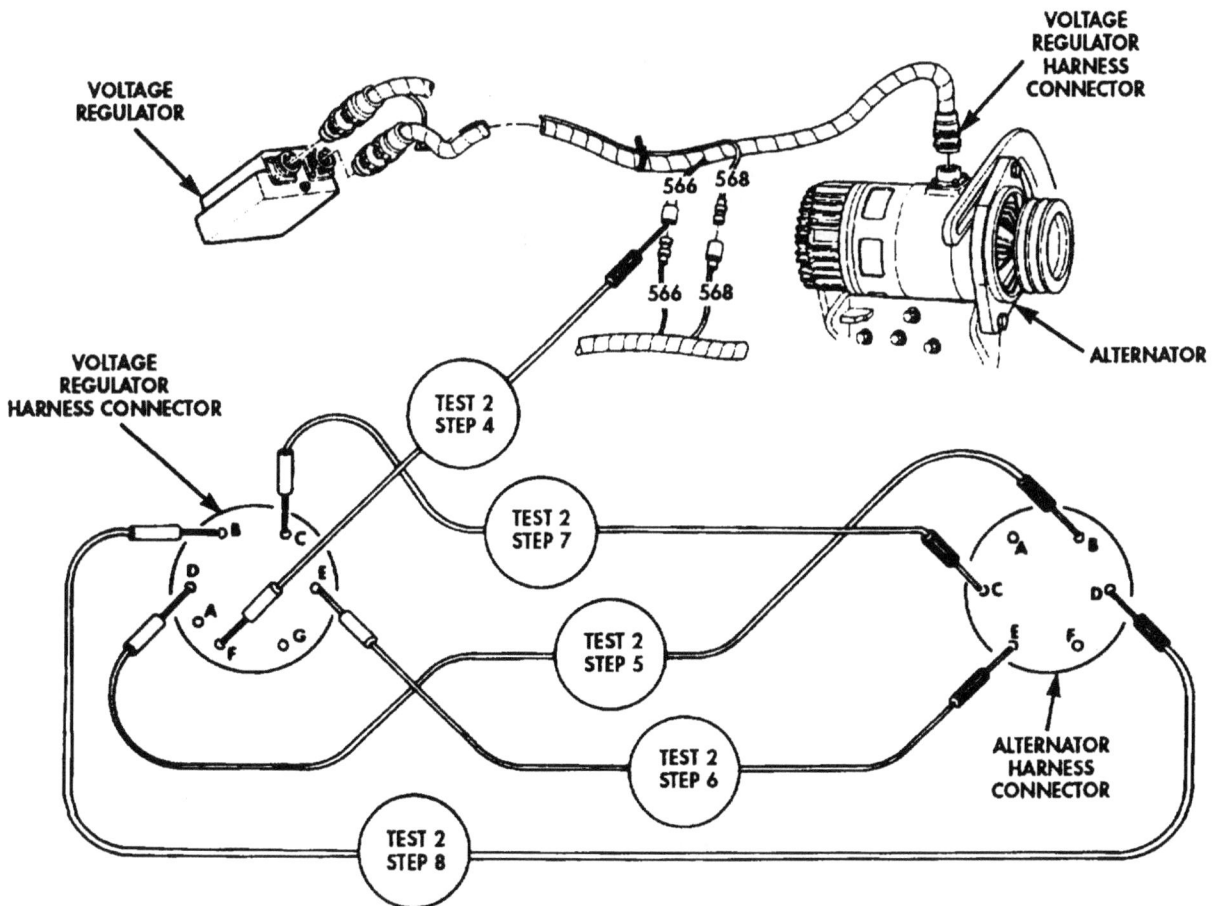

Table 2-3. Electrical Troubleshooting (Contd).

```
MALFUNCTION
    TEST OR INSPECTION
        CORRECTIVE ACTION
```

Test 3. Test continuity of starter harness.

Step 1. Disconnect quick-disconnect from lead 568.

Step 2. Disconnect voltage regulator harness connector at voltage regulator.

Step 3. Disconnect voltage regulator harness at starter motor and starter solenoid.

Step 4. Connect multimeter positive lead to contact end of lead 81 at starter solenoid and negative lead to pin B at voltage regulator harness connector. Continuity should be present.

    a. If continuity is present, go to step 5.

    b. If continuity is not present, replace voltage regulator harness (para. 3-131).

Step 5. Connect multimeter positive lead to contact end of lead 81 at starter solenoid and negative lead to pin A at voltage regulator harness connector. Continuity should be present.

    a. If continuity is present, go to step 6.

    b. If continuity is not present, replace voltage regulator harness (para. 3-131).

Step 6. Connect multimeter negative lead to contact end of pin C at voltage regulator harness connector and positive lead to contact end of lead at starter motor. Continuity should be present.

    a. If continuity is present, go to step 7.

    b. If continuity is not present, replace voltage regulator harness (para. 3-131).

Step 7. Connect multimeter positive lead to contact end of lead 568 and negative lead to contact end of pin F at voltage regulator harness connector. Continuity should be present.

    a. If continuity is present, notify DS maintenance.

    b. If continuity is not present, replace voltage regulator harness (para. 3-131).

Table 2-3. Electrical Troubleshooting (Contd).

MALFUNCTION
TEST OR INSPECTION
CORRECTIVE ACTION

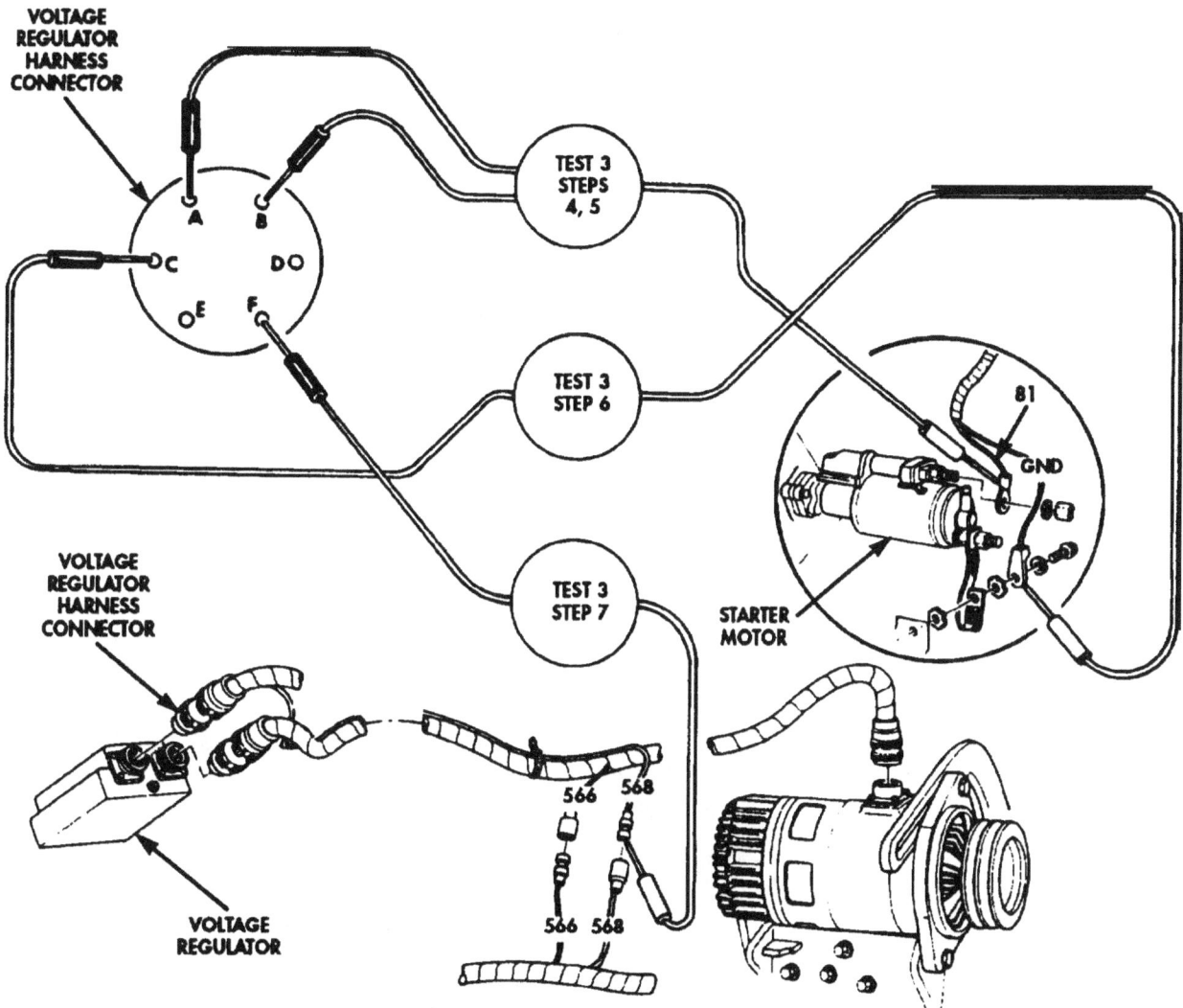

END OF TESTING!

Table 2-3. Electrical Troubleshooting (Contd).

MALFUNCTION
TEST OR INSPECTION
CORRECTIVE ACTION

TRANSFER CASE SYSTEM

44. TRANSFER CASE CONTROL LEVER WILL NOT SHIFT FROM LOW TO HIGH OR HIGH TO LOW WHEN VEHICLE IS IN MOTION

Test 1. Test transfer case control circuit breaker (malfunction 35, test 3).

Test 2. Test transfer case control switch.

Step 1. Place battery switch to OFF position.

Step 2. Disconnect lead 585 from transfer case control switch.

Step 3. Place battery switch to ON position.

Step 4. Set multimeter to 50-volt range.

Step 5. Connect multimeter negative lead to vehicle chassis ground and positive lead to lead 585. Multimeter should indicate battery voltage.

    a. If battery voltage is not indicated, repair or replace lead 585 (para. 131).

    b. If battery voltage is indicated, connect lead 585 and disconnect lead 586 from control switch.

Step 6. Place transmission control lever to NEUTRAL position. With negative lead of meter connected to ground and, with transfer case switch button depressed, touch positive lead to lead 586 terminal on control switch. Multimeter should indicate battery voltage.

If battery voltage is not indicated, notify DS maintenance.

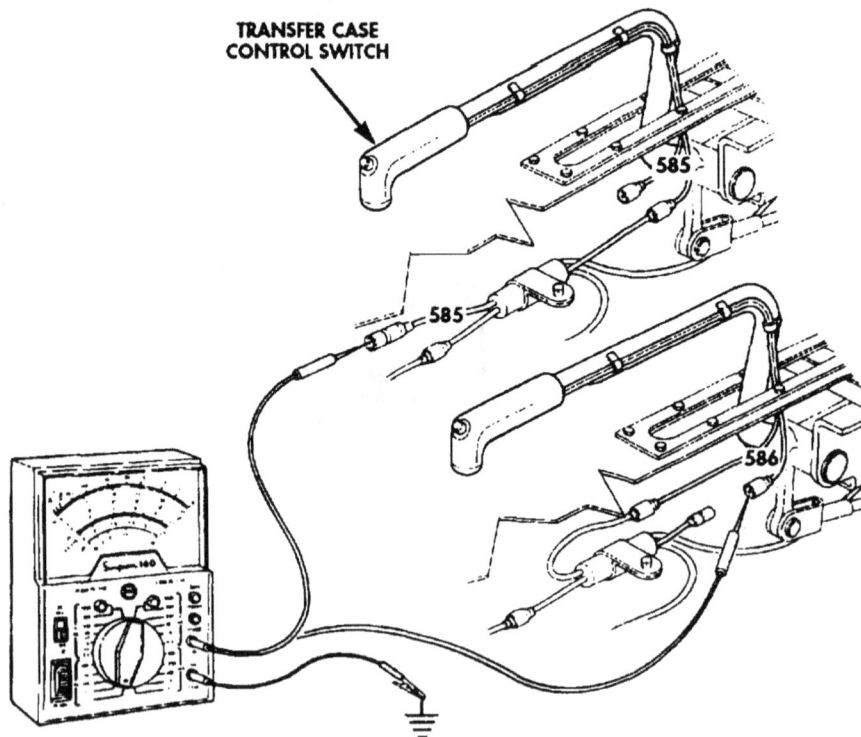

END OF TESTING!

Table 2-3. Electrical Troubleshooting (Contd).

MALFUNCTION
TEST OR INSPECTION
CORRECTIVE ACTION

45. TRANSFER CASE SHIFTS HARD WHILE ENGINE IS RUNNING

Test transfer case switch lever for voltage.

Step 1. Set multimeter to 24-volt range.

Step 2. Disconnect lead 586 from transfer case shift lever switch.

Step 3. Touch positive lead of multimeter to contact end of transfer case shift lever switch and negative lead to ground.

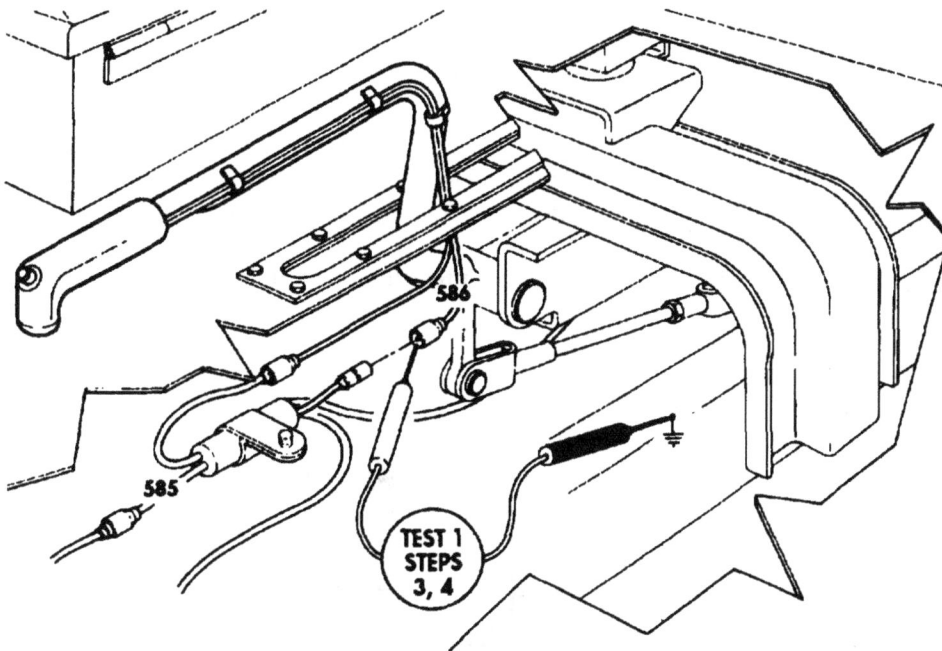

END OF TESTING!

Table 2-3. Electrical Troubleshooting (Contd).

MALFUNCTION
TEST OR INSPECTION
CORRECTIVE ACTION

NOTE

Assistant will help with step 4.

Step 4. Have assistant position ignition switch to ON and push transfer case shift lever button. Voltage should be present.

    a. If voltage is present, linear valve is defective. Notify DS maintenance.

    b. If voltage is not present, transfer case shift lever switch or wiring is defective. Notify DS maintenance.

PROTECTIVE CONTROL BOX ASSEMBLY

46. PROTECTIVE CONTROL BOX ASSEMBLY TEST

Initial check - front wiring harness disconnected at box.

Step 1. Confirm that continuity exists between pin E and ground.

Step 2. Confirm that continuity exists between pins B and D, pins C and D, pin A and ground, pin B and ground, pin C and ground, and pin D and ground.

Step 3. Apply 24 volts to pin A. Confirm that continuity exists between pins C and D, but not between pins B and D.

Step 4. If findings are other than as indicated, protective control box is malfunctioning. Replace protective control box (para. 3-115).

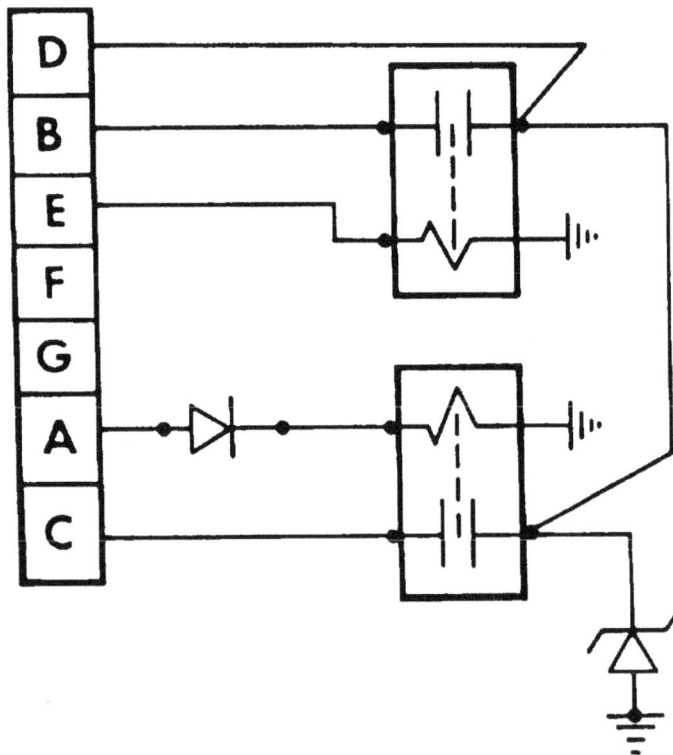

END OF TESTING!

2-23. COMPRESSED AIR AND BRAKE SYSTEM TROUBLESHOOTING

a. This section contains troubleshooting information and tests for locating and correcting malfunctions which may develop in the compressed air and brake system on vehicles covered in this manual. Each symptom or malfunction given for an individual component or system is followed by step(s) you should take to determine the cause and corrective action needed to remedy the problem. Compressed air and brake system is divided into the following functional systems:

- Parking Brake (page 2-179)
- Service Brakes (page 2-180)
- Compressed Air Supply (page 2-197)
- Air-operated Accessories (page 2-213)

b. This troubleshooting guide is used to keep your vehicle operating and trouble-free as much as possible in the quickest, easiest way.

(1) When a malfunction occurs, start at the source of the trouble and keep working back to eliminate elements that are not defective to isolate the parts of the system that are. When found, this part, line, or element is repaired, replaced, or otherwise corrected to get your vehicle running again.

(2) In the compressed air and brake system, emphasis in the troubleshooting guide is placed on leaks, blockages in tubing and hoses, and correct pressure for operating various air-operated devices. Leaks and blockages are easily determined using methods indicated in the tests.

(3) Correct air pressure requires a gauge of known accuracy, which is a special tool, to test whether an element of the system is functioning or operating properly when supplied with the correct air pressure.

c. Before taking any corrective action to correct a possible malfunction, the following rules should be followed:

(1) Question operator to obtain any additional information that might help you to determine the cause of the problem.

(2) Never overlook the chance the problem could be of a simple origin. The problem could require only a minor adjustment.

(3) Use all senses to observe and locate troubles.

(4) Use test instruments and gauges to help you determine and isolate problems.

(5) Always isolate the system where the malfunction occurs and locate the defective component.

(6) Use standard automotive theories and principles when troubleshooting the vehicles covered in this manual

(7) Operate the vehicle yourself to ensure the operator's description of the problem is correct.

(8) Always wear eyeshields when troubleshooting the compressed air system.

d. The compressed air schematic (Appendix H) shows the interrelationship of the compressed air and brake system, and should be used as a reference when performing troubleshooting procedures. Table 2-4 lists malfunctions of the compressed air and brake system that may occur.

e. This section cannot list all compressed air and brake system malfunctions that may occur. If a malfunction occurs that is not listed in table 2-4, notify DS maintenance.

WARNING

- Do not disconnect air lines before draining air reservoirs. Small parts under pressure may shoot out with high velocity, causing injury to personnel.
- Hearing protection is required for the driver and passenger. Hearing protection is also required for all personnel working in and around the vehicle while the engine is running. (Refer to AR 40-5 and TB MED 501.)

COMPRESSED AIR AND BRAKE SYSTEM TROUBLESHOOTING INDEX

MALFUNCTION NO.	MALFUNCTION	TROUBLESHOOTING PROCEDURE PAGE

PARKING BRAKE

1. Parking brake does not hold vehicle on grade 2-179
2. Parking brake drags, as indicated by smoking or burning smell. 2-179

SERVICE BRAKES

3. Insufficient brakes (vehicle stopping distance too long, no apparent air system failure with gauges at normal operating pressure, warning buzzer not sounding) 2-180
4. Vehicle pulls right or left when applying brakes 2-194
5. Vehicle rear brakes grab or drag 2-196
6. Vehicle vibrates, chatters, or bounces when brakes are applied 2-196
7. Brakes squeal 2-196
8. Warning buzzer sounds when brakes are applied (primary and secondary gauge pressure drops below 55-65 psi (379-448 kPa). 2-196

COMPRESSED AIR SUPPLY

9. No air pressure (warning buzzer sounding, air pressure not building to normal operating range as indicated by gauges) 2-197
10. Air pressure does not build to normal operating pressure (above 80 psi (552 kPa)) according to gauges. , ... 2-200
11. Air pressure builds slowly (takes excessive amount of time to build to 100 psi (690kPa) 2-200
12. Air pressure exceeds maximum (gauges show over 130 psi (896 kPa)), safety valve opens to release pressure. 2-200
13. Primary pressure gauge reads no pressure, low pressure, or builds to normal operating pressure slowly. Secondary pressure gauge reads normal (engine idling, brake pedal not applied). 2-202
14. Secondary pressure gauge reads no pressure, low pressure, or builds to normal operating pressure slowly. Primary pressure gauge reads normal (engine idling, brake pedal not applied). . . . . , . . . 2-206
15. Primary air system fails to maintain pressure (no major leaks, air can be heard escaping into air intake stack, parking brake applied) 2-208
16. Secondary air system fails to maintain pressure (no major leaks, air can be heard escaping into air intake stack, parking brake applied) 2-208
17. Warning buzzer fails to sound or fails to shut off on low pressure (below 55-65 psi (379-448 kPa)), air pressure system operating normally. . . . 2-210
18. Spring brakes do not release (vehicle brakes grab or drag). 2-210
19. Spring brakes do not set (gauges at normal operating pressures, air exhausting not heard when parking brake applied). 2-212

AIR-OPERATED ACCESSORIES

20. All air-operated accessories do not operate (horn, windshield wipers, windshield washers, transfer case controls), gauges at normal operating pressure 2-213
21. Cooling fan does not operate, engine temperature above 200°F (93°C) as indicated by temperature gauge. 2-213
22. Cooling fan does not stop running, engine temperature below normal operating range (override bolt not installed) 2-214
23. Front wheel drive does not engage (front wheel drive lock-in switch engaged and transfer case shift lever in HIGH position). 2-214
24. Horn does not work (gauges at normal operating pressure, warning buzzer not sounding) 2-218

Table 2-4. Compressed Air and Brake System Troubleshooting.

```
MALFUNCTION
    TEST OR INSPECTION
        CORRECTIVE ACTION
```

PARKING BRAKE

1. PARKING BRAKE DOES NOT HOLD VEHICLE ON GRADE

Test 1. Check parking brake adjustment.

    a. Release spring brakes (TM 9-2320-272-10).

    b. Turn knob (1) on parking brake lever (2) clockwise to increase braking action.

    c. If adjustment does not correct malfunction, perform test 2.

Test 2. Inspect parking brake cable (3) for binding or breaks.

    a. Replace cable (3) if binding or broken (para. 3-175).

    b. If cable (3) is not damaged, perform test 3.

Test 3. Inspect parking brakeshoes (6) for wear.

    a. Push brakeshoe lever (4) clockwise by hand and measure brakeshoe lever travel.

    b. If travel is more than 2 in. (5) mm), replace parking brakeshoes (6) (para. 3-176).

Test 4. Inspect for broken or faulty actuating plate (7).

    a. Move brakeshoe lever (4) back and forth. Observe operation of brakeshoe lever (4), spring (5), and actuating plate (7).

    b. If actuating mechanism is defective, notify DS maintenance.

    c. If actuating mechanism is serviceable, perform malfunction 19.

END OF TESTING!

2. PARKING BRAKE DRAGS, AS INDICATED BY SMOKING OR BURNING SMELL

Ensure parking brake lever (2) is not partially engaged.

    a. Fully release parking brake.

    b. If parking brake still drags, perform malfunction 1, tests 1, 2, and 4.

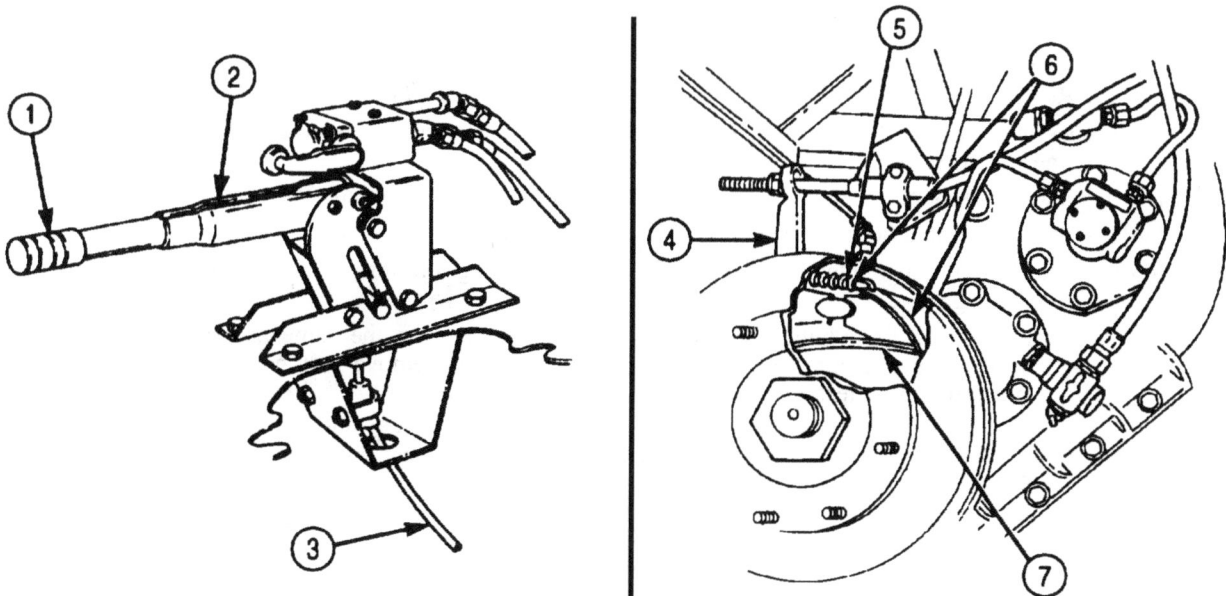

END OF TESTING!

Table 2-4. Compressed Air and Brake System Troubleshooting (Contd).

MALFUNCTION
  TEST OR INSPECTION
    CORRECTIVE ACTION

SERVICE BRAKES

3. INSUFFICIENT BRAKES (VEHICLE STOPPING DISTANCE TOO LONG, NO APPARENT AIR SYSTEM FAILURE WITH GAUGES AT NORMAL OPERATING PRESSURE, WARNING BUZZER NOT SOUNDING)

NOTE

Assistant will help with this procedure.

Test 1. Check lines and hoses for leaks.

Direct assistant to fully apply service brakes.

    a. If air leaks are found, repair as necessary.

    b. If no air leaks in lines and hoses are found, perform test 2.

Test 2. Check service brake chambers (1) for leaks through vent lines (2).

WARNING

Do not look in service chamber vent port when performing test.
Injury to personnel may result.

Step 1.   Disconnect vent line (2) from service brake chamber (1).

Step 2.   Direct assistant to apply service brakes and feel for evidence of escaping air at vent port (3).

    a. If air is escaping, air is present. Notify DS maintenance.

    b. If no air is escaping, install vent line (2) on service brake chamber (1).

    c. Check all remaining vent lines (2) on service brake chambers (1) before proceeding to test 3.

Test 3. Check front service brake chambers (1) for proper pressure.

Step 1.   Stop engine and open all drainvalves (4), in sequence shown, until brake system air pressure is vented.

Step 2.   Close drainvalves (4).

Step 3.   Disconnect delivery line (5) from service brake chamber (1).

Step 4.   Remove adapter fitting (7) from service brake chamber (1).

Step 5.   Connect tee (6) to service brake chamber (1) and connect test gauge and delivery line (5) to tee (6).

Step 6.   Direct assistant to start engine and build air supply to normal operating pressure.

Step 7.   Direct assistant to fully apply and hold service brakes.

Step 8.   Check test gauge reading and compare with reading on instrument panel primary air pressure gauge.

    a. If readings from both front service brake chambers are the same as reading from primary air pressure gauge, stop engine, drain air, remove test gauge, connect lines, and proceed to test 4.

    b. If test gauge readings are nonexistent or far less than primary air pressure gauge, stop engine, drain air, remove test gauge, connect lines, and perform tests 6 through 10.

Table 2-4. Compressed Air and Brake System Troubleshooting (Contd).

MALFUNCTION
TEST OR INSPECTION
CORRECTIVE ACTION

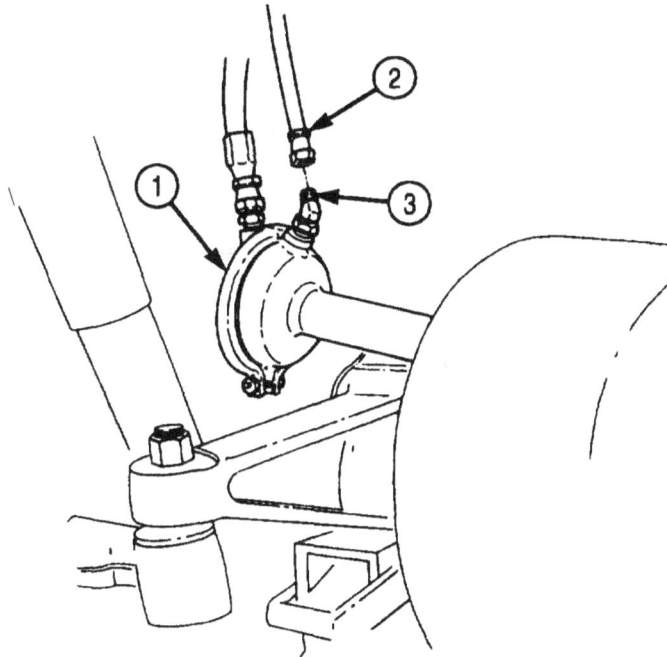

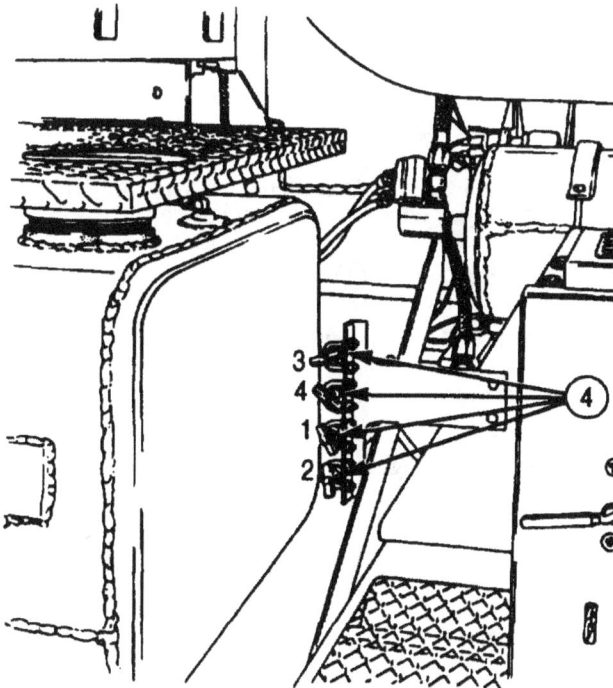

DRAINVALVE SEQUENCE

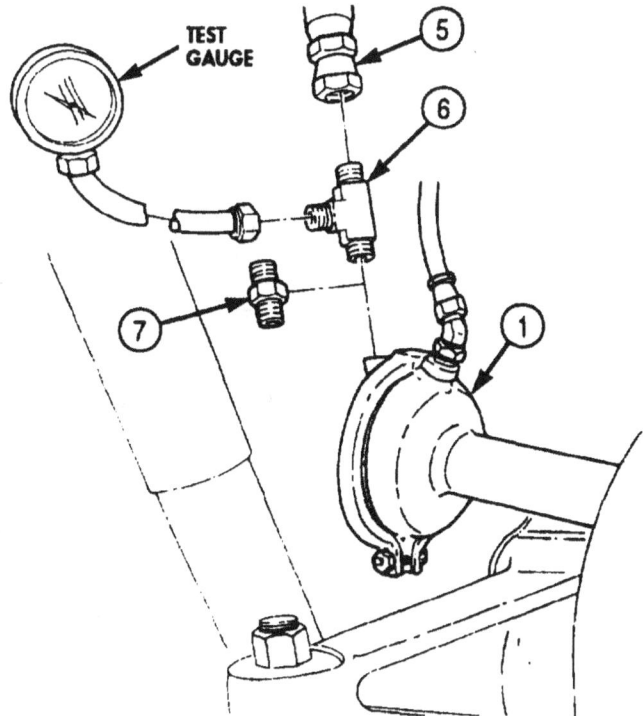

Table 2-4. Compressed Air and Brake System Troubleshooting (Contd).

MALFUNCTION
    TEST OR INSPECTION
        CORRECTIVE ACTION

Test 4. Check service brake chambers (2) and (3) for proper pressure.

NOTE

Air pressure to service brake chambers must be compared with
reading on primary air pressure gauge. Air pressure to spring and
service brake chambers must be compared with reading on
secondary air pressure gauge.

Step 1. Install test gauge on service brake chambers (2) and (3) and check for proper pressure as
outlined in test 3, steps 1 through 7.

Step 2. Check test gauge reading and compare with reading on instrument panel primary air
pressure gauge.

    a. If reading on all four service brake chambers (2) and (3) are the same as reading
from primary air pressure gauge, stop engine, drain air, remove test gauge, connect
lines, and proceed to test 5.

    b. If reading from two forward-rear rear service brake chambers (2) are not the same, stop
engine, drain air, remove test gauge, connect lines, and perform test 11.

    c. If reading from two rear-rear forward service brake chambers (3) are not the same, stop
engine, drain air, remove test gauge, connect lines and perform test 12.

    d. If reading from all service brake chambers (2) and (3) are not the same, stop engine,
drain air, remove test gauge, connect lines, and perform test 13.

Test 5. Check spring and service brake chambers (1) and (4) for proper pressure.

Step 1. Install test gauge on spring and service brake chambers (1) and (4) and check for proper
pressure as outlined in test 3, steps 1 through 7.

Step 2. Check test gauge reading and compare with reading on instrument panel secondary air
pressure gauge.

    a. If reading on all four spring and service brake chambers (1) and (4) are the same as
reading on secondary air pressure gauge, stop engine, drain air, remove test gauge,
connect lines, and proceed to test 17.

    b. If reading from two forward-rear forward spring and service brake chambers (1) are
not the same, stop engine, drain air, remove test gauge, connect lines, and perform
test 14.

    c. If reading from only two rear-rear rear spring and service brake chambers (4) are not
the same, stop engine, drain air, remove test gauge, connect lines, and perform
test 15.

    d. If reading from all four spring and service brake chambers (1) and (4) are not the
same, stop engine, drain air, remove test gauge, connect lines, and perform test 16.

Test 6. Check limiting valve (7) for venting (front service brake chambers test low or nonexistent
air pressure).

WARNING

Do not look into limiting valve vent port when performing test.
Injury to personnel may result.

Step 1. Direct assistant to apply and hold service brake pedal.

Step 2. Disconnect vent line (5) from limiting valve vent port (6).

Step 3. Feel for escaping air at limiting valve vent port (6).

    a. If air is felt, replace limiting valve (7) (para. 3-190).

    b. If no air is felt, reconnect vent line (5) and proceed to test 7.

Table 2-4. Compressed Air and Brake System Troubleshooting (Contd).

MALFUNCTION
TEST OR INSPECTION
CORRECTIVE ACTION

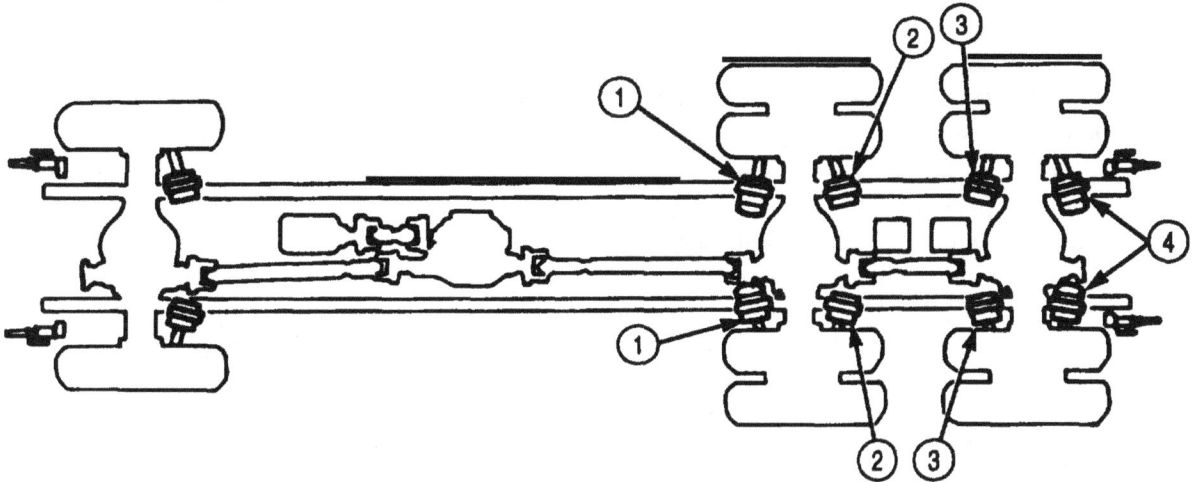

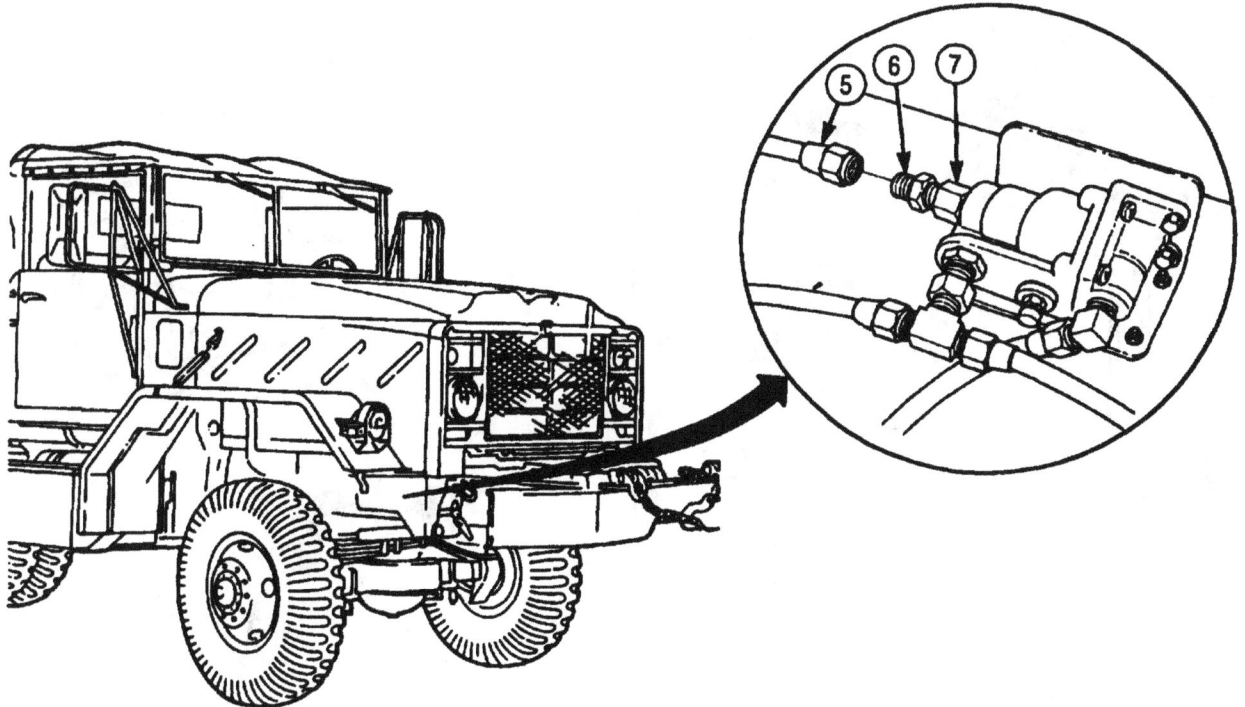

Table 2-4. Compressed Air and Brake System Troubleshooting (Contd).

MALFUNCTION
TEST OR INSPECTION
CORRECTIVE ACTION

Test 7. Check limiting valve (3) for proper air delivery pressure (front service brake chambers test low or nonexistent air pressure).

Step 1. Stop engine and open all drainvalves (1), in sequence shown, until brake system air pressure is vented.

Step 2. Close drainvalves (1).

Step 3. Disconnect air delivery lines (6) and (4) and remove tee (5) from limiting valve (3).

Step 4. Connect test gauge to delivery port (2) of limiting valve (3).

Step 5. Direct assistant to start engine and allow air pressure to build to normal operating pressure.

Step 6. Direct assistant to fully apply and hold service brakes.

Step 7. Check reading on test gauge.

   a. If test gauge reads normal operating pressure, check air delivery lines (6) and (4) for blockage. Repair or replace lines (6) and (4) as necessary (TM 9-243).

   b. If pressure is below normal or nonexistent, stop engine, drain air, connect lines to limiting valve (3), and proceed to test 8.

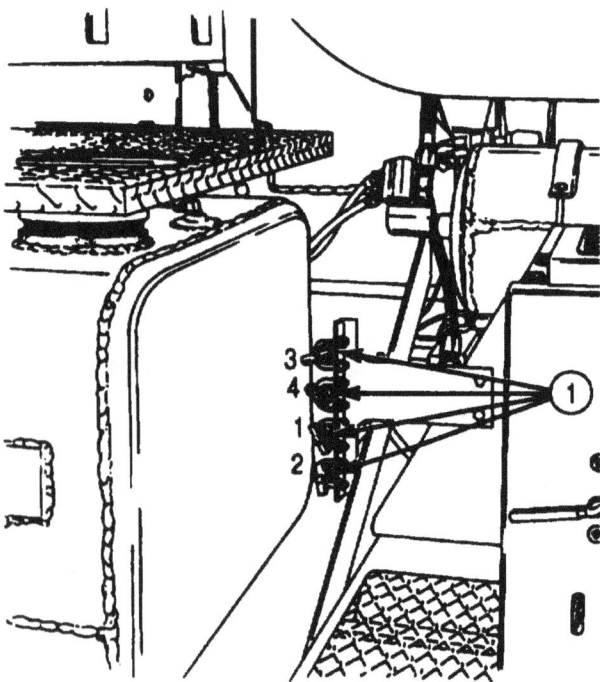

DRAINVALVE SEQUENCE

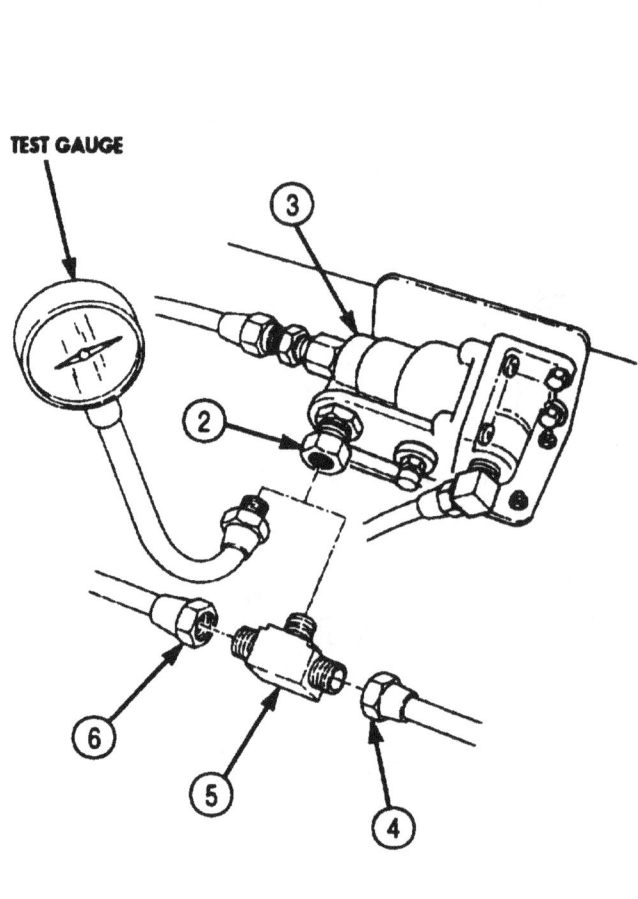

Table 2-4. Compressed Air and Brake System Troubleshooting (Contd).

MALFUNCTION
TEST OR INSPECTION
CORRECTIVE ACTION

Test 8. Check limiting valve (3) for proper air supply pressure (front service brake chambers test low or nonexistent air pressure).

Step 1. Stop engine and open all drainvalves (1), in sequence shown, until brake system air pressure is vented.

Step 2. Close drainvalves (1).

Step 3. Disconnect air supply line (7) from limiting valve (3).

Step 4. Connect test gauge to air supply line (7).

Step 5. Direct assistant to start engine and allow air pressure to built to normal operating pressure.

Step 6. Check reading on test gauge.

    a. If test gauge reads normal operating pressure, replace defective limiting valve (3) (para. 3-190).

    b. If pressure is below normal or nonexistent, check air supply line (7) for blockage. Repair or replace as necessary. If no blockage is present, stop engine, drain air, remove test gauge, connect air supply line (7), and proceed to test 9.

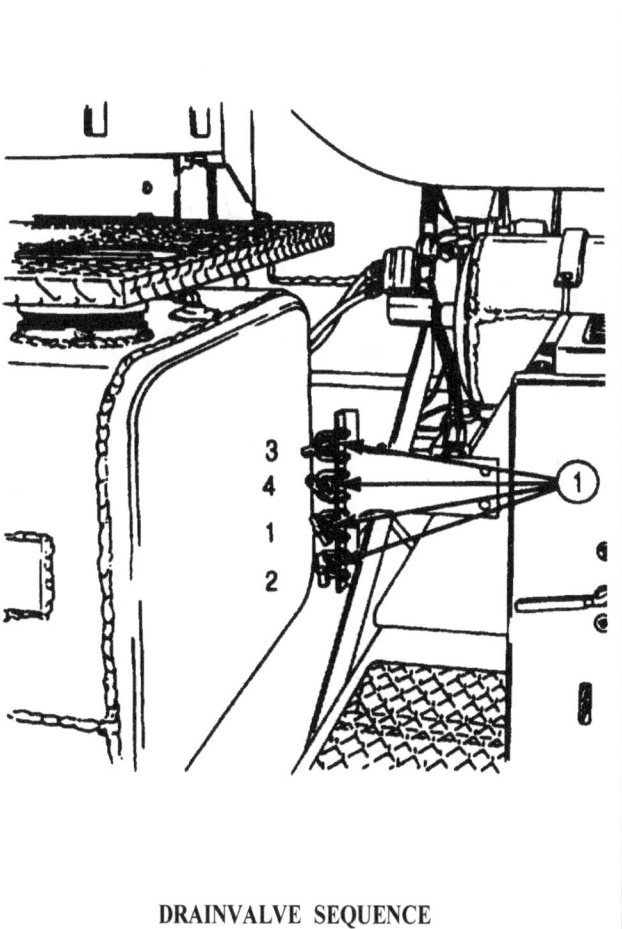

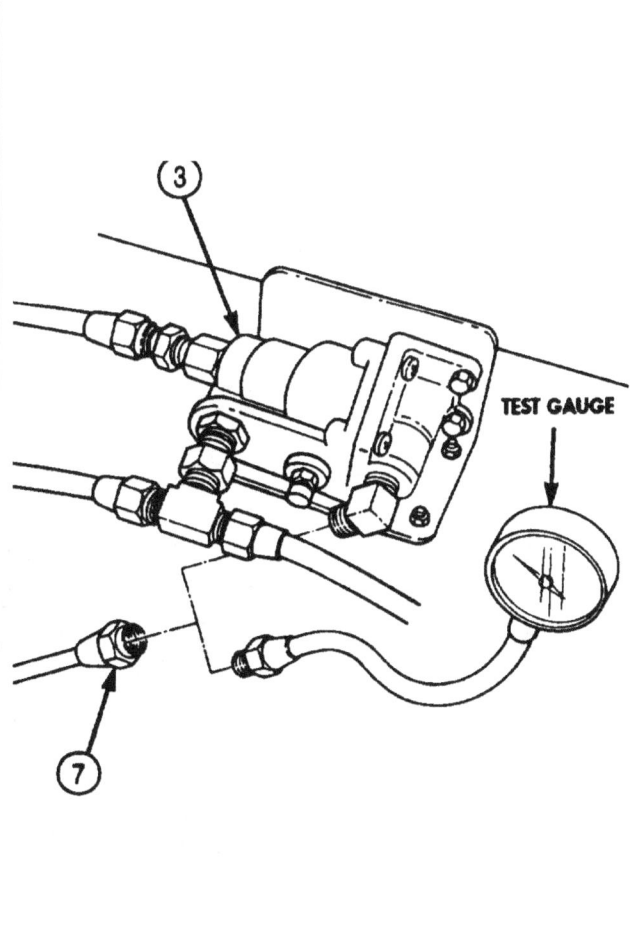

DRAINVALVE SEQUENCE

Table 2-4. Compressed Air and Brake System Troubleshooting (Contd).

MALFUNCTION
    TEST OR INSPECTION
        CORRECTIVE ACTION

Test 9. Test doublecheck valve No. 1(2) supply pressure (front service brake chambers test low or nonexistent air pressure).

WARNING

Do not depress service brake pedal during installation of test gauge. This may result in injury to personnel.

Step 1. Direct assistant to keep foot off service brake pedal (6).

Step 2. Disconnect air supply line (1) from doublecheck valve No. 1 (2).

Step 3. Connect test gauge to air supply line (1). Direct assistant to start engine and allow air pressure to build to normal operating pressure.

Step 4. Depress service brake pedal (6).

Step 5. Check reading on test gauge.

    a. If pressure reads normal and delivery line (3) is not blocked (as determined by test 8), replace defective doublecheck valve No. 1 (2) (para. 3-212).

    b. If pressure is below normal or nonexistent, stop engine, drain air, remove test gauge, connect delivery line (3), and perform test 10.

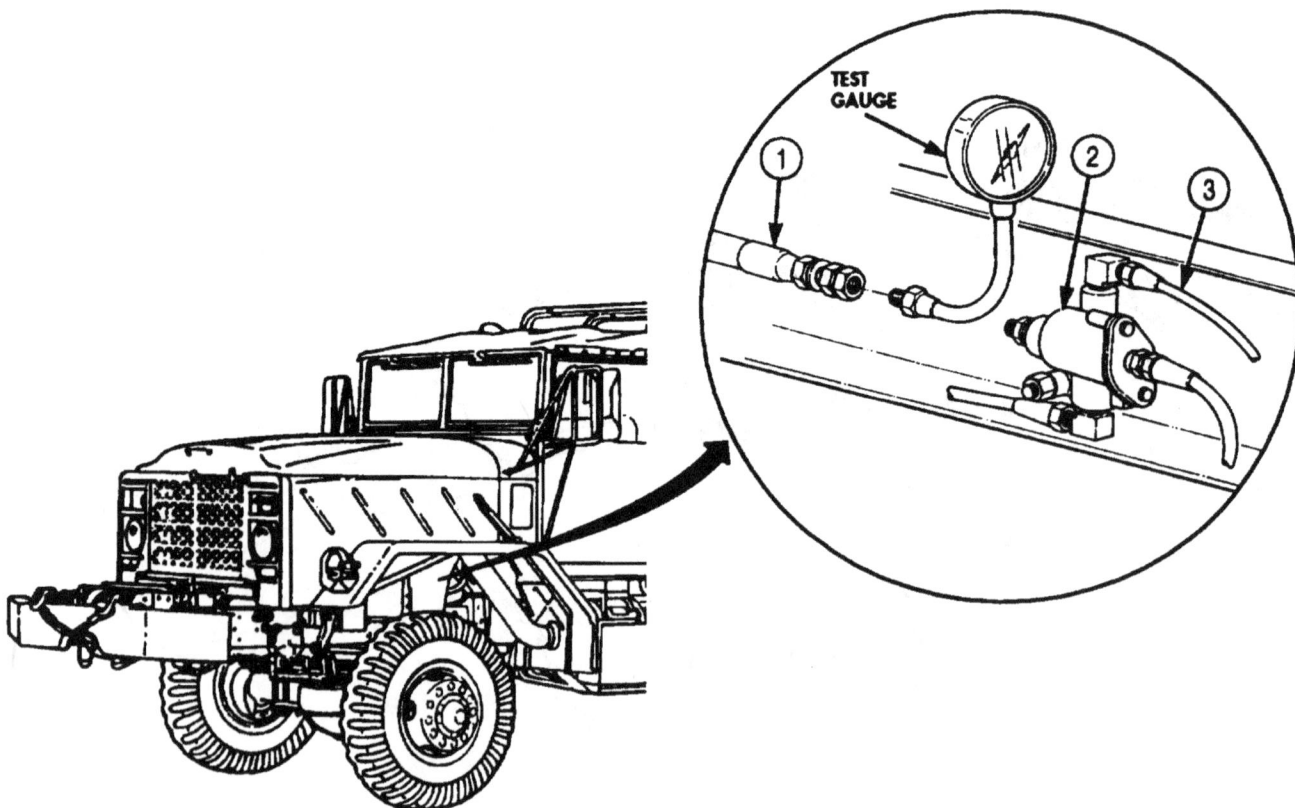

Table 2-4. Compressed Air and Brake System Troubleshooting (Contd).

MALFUNCTION
   TEST OR INSPECTION
      CORRECTIVE ACTION

Test 10. Check service brake pedal primary air system delivery port (4) (front service brake chambers test low or nonexistent air pressure).

WARNING

- Do not depress service brake pedal during removal of plug and/or installation of test gauge. Doing so may result in injury to personnel.
- Remove correct plug only. Removal of incorrect plug may result in injury to personnel.

Step 1. Remove plug (7) from service brake pedal primary air system delivery port (4) on brake pedal valve (5).

Step 2. Connect test gauge to delivery port (4). Direct assistant to start engine and allow air pressure to build to normal operating pressure.

Step 3. Depress service brake pedal (6).

Step 4. Check test gauge.

  a. If air test gauge reads normal, check delivery line (1) to doublecheck valve No. 1 (2) for blockage. Clear blockage as necessary.

  b. If air pressure is below normal or nonexistent, remove test gauge and replace service brake pedal valve (5) (para. 3-197 or 3-198).

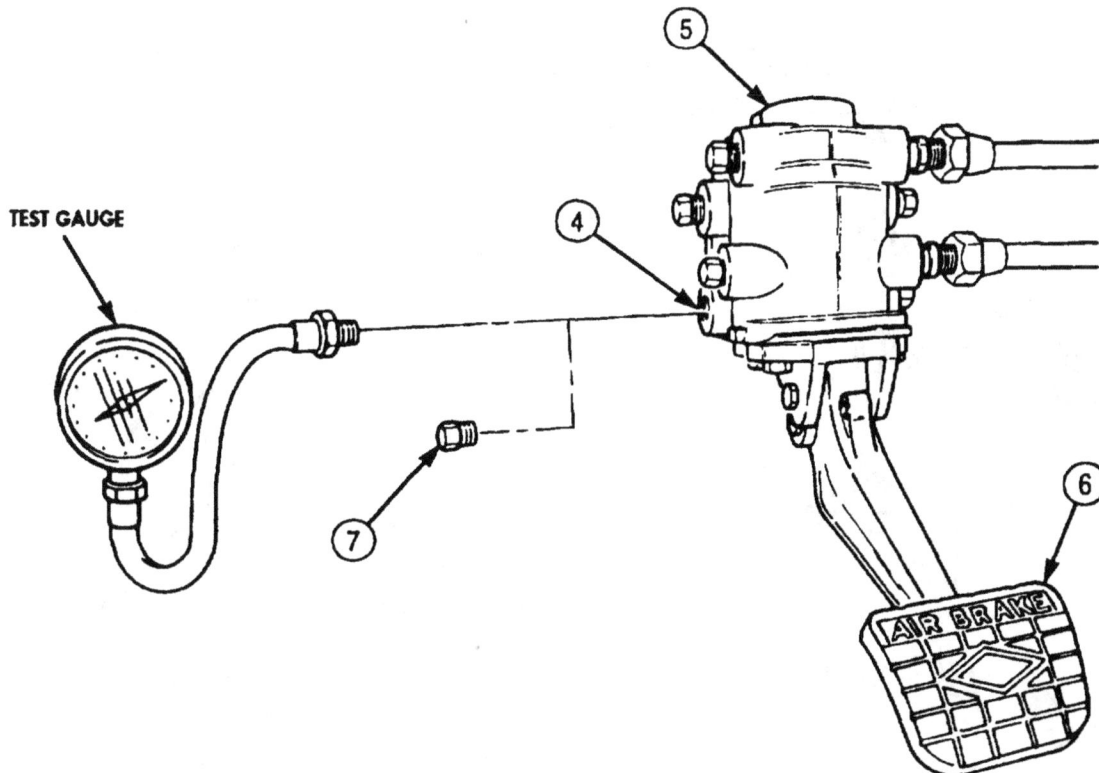

TEST GAUGE

Table 2-4. Compressed Air and Brake System Troubleshooting (Contd).

MALFUNCTION
    TEST OR INSPECTION
        CORRECTIVE ACTION

Test 11. Check primary air system relay valve (3) for proper pressure (forward-rear rear service brake chambers (1) test low or nonexistent air pressure).

Step 1. Stop engine and open all drainvalves (2), in sequence shown, until brake system air pressure is vented.

Step 2. Close drainvalves (2).

Step 3. Disconnect delivery line (4) from forward-rear rear service brake chambers (1) and remove elbow (5) from delivery port (6) on relay valve (3).

Step 4. Connect test gauge to delivery port (6).

Step 5. Direct assistant to start engine and allow air system to build to normal operating pressure.

Step 6. Direct assistant to fully apply and hold service brakes.

Step 7. Check reading on test gauge.

    a. If pressure is below normal, stop engine, drain air, remove test gauge, connect air line, and proceed to test 12.

    b. If pressure is normal, inspect relay valve to service brake chamber delivery lines (4) for blockage. Repair or replace as necessary (TM 9-243).

    c. Stop engine, drain air, remove test gauge, and connect delivery lines (4).

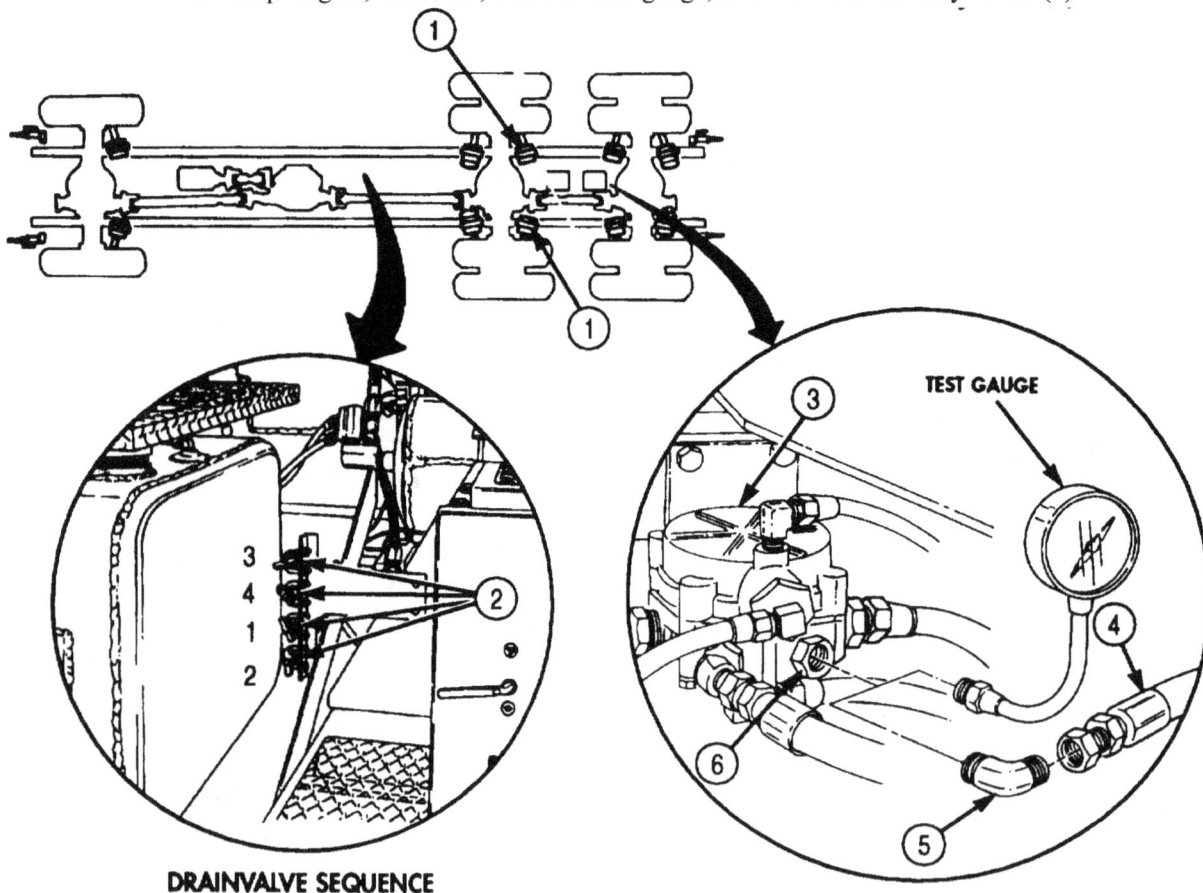

DRAINVALVE SEQUENCE

TEST GAUGE

Table 2-4. Compressed Air and Brake System Troubleshooting (Contd).

MALFUNCTION
　　TEST OR INSPECTION
　　　　CORRECTIVE ACTION

　Test 12. Check primary air system relay valve (3) for proper pressure (rear-rear forward service brake chambers (7) test low or nonexistent air pressure).

　　Step 1. Stop engine and open all drainvalves (2), in sequence shown, until brake system air pressure is vented.

　　Step 2. Close drainvalves (2).

　　Step 3. Disconnect delivery line (9) from rear-rear forward service brake chambers (7) and remove elbow (10) from relay delivery valve port (11) on relay valve (3).

　　Step 4. Connect test gauge to delivery valve port (11).

　　Step 5. Direct assistant to start engine and allow air system to build to normal operating pressure.

　　Step 6. Direct assistant to fully apply and hold service brakes.

　　Step 7. Check reading on test gauge.

　　　　a. If pressure is below normal, proceed to test 13.

　　　　b. If pressure is normal, inspect relay valve to service brake chamber delivery lines (9) for blockage. Repair or replace as necessary (TM 9-243).

　　　　c. Stop engine, drain air, remove test gauge, and connect delivery lines (9).

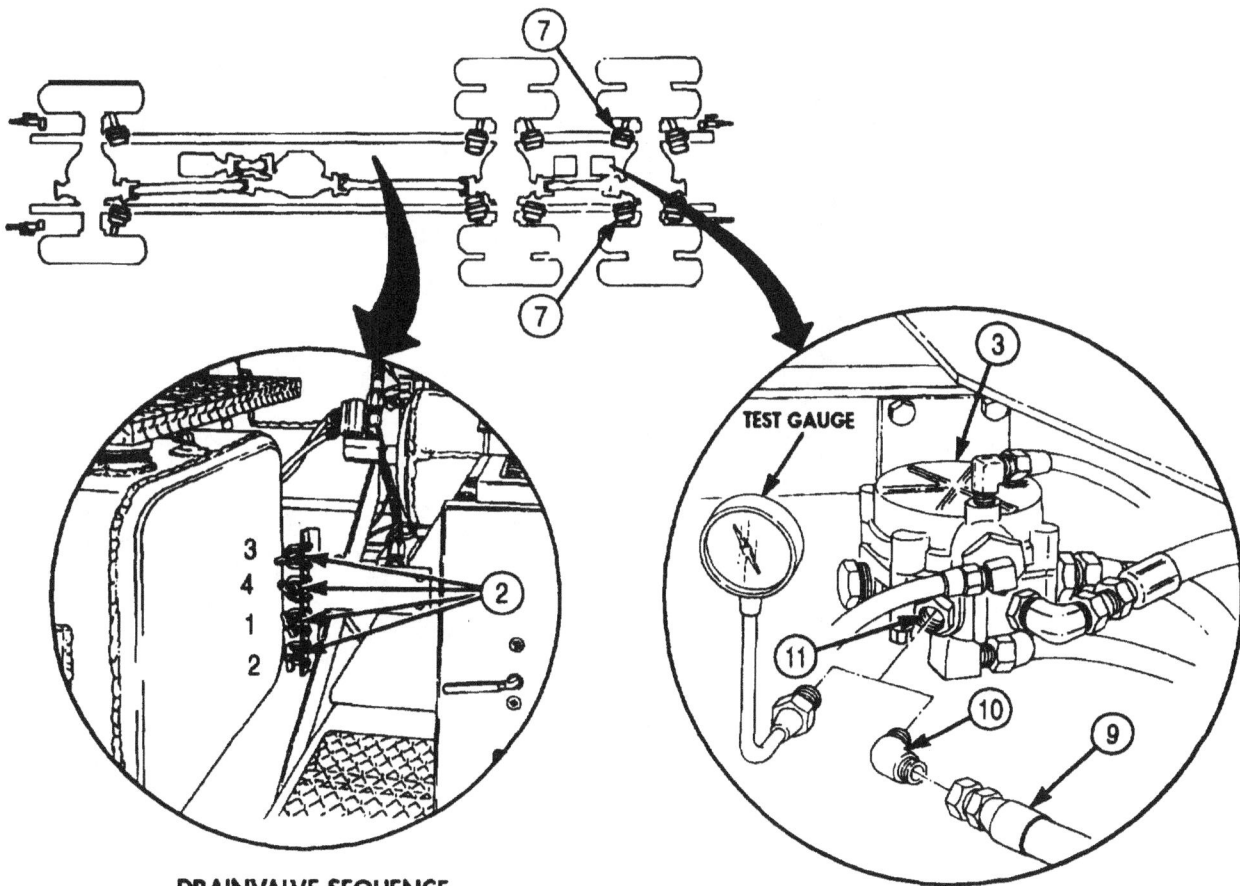

DRAINVALVE SEQUENCE

Table 2-4. Compressed Air and Brake System Troubleshooting (Contd).

MALFUNCTION
  TEST OR INSPECTION
    CORRECTIVE ACTION

Test 13. Check primary air tank to relay valve supply line (3) (all four primary air system service brake chambers test low or nonexistent air pressure).

Step 1. Stop engine and open all drainvalves (1), in sequence shown, until brake system air pressure is vented.

Step 2. Close drainvalves (1).

Step 3. Disconnect supply line (3) from primary air system relay valve (2).

Step 4. Connect test gauge to supply line (3).

Step 5. Direct assistant to start engine and allow air system to build to normal operating pressure.

Step 6. Direct assistant to fully apply and hold service brakes.

Step 7. Check reading on test gauge.

    a. If pressure is below normal, inspect supply line (3) for blockage. Repair or replace as necessary.

    b. If pressure is normal, replace defective primary air system relay valve (2) (para. 3-187 or 3-188).

    c. Stop engine, drain air, remove test gauge, and connect supply line (3).

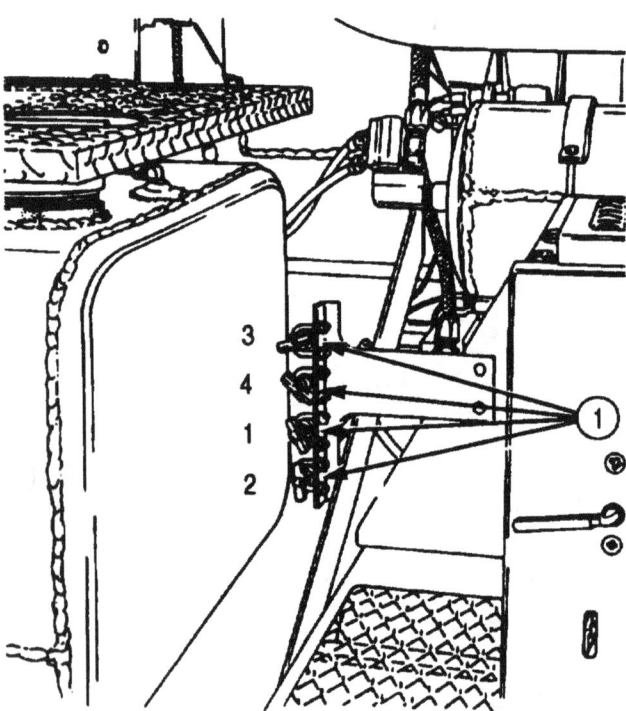

DRAINVALVE SEQUENCE

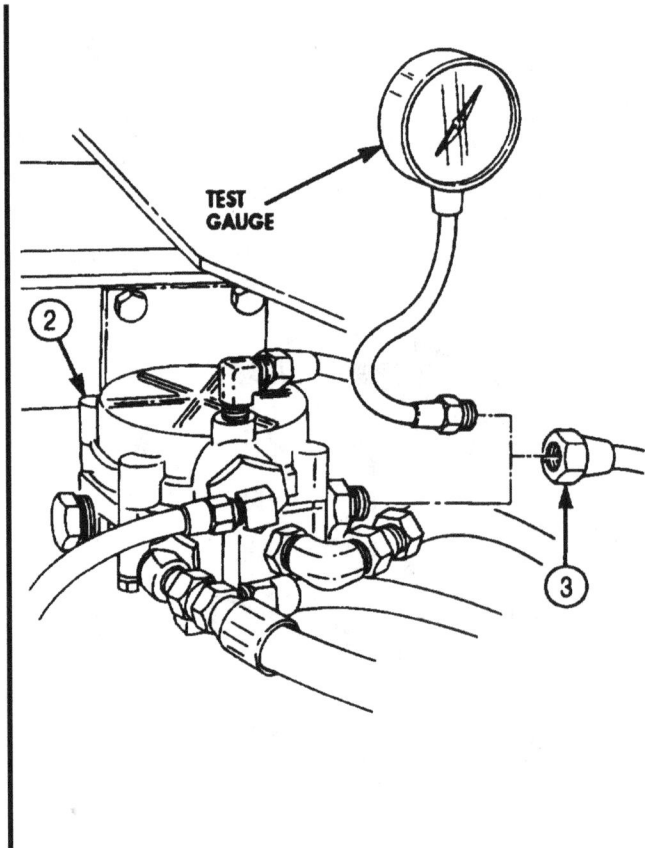

Table 2-4. Compressed Air and Brake System Troubleshooting (Contd).

| MALFUNCTION |
| TEST OR INSPECTION |
| CORRECTIVE ACTION |

Test 14. Check secondary air system relay valve (5) for proper pressure (only spring and service brake chambers (4) test low or nonexistent air pressure).

Step 1. Stop engine and open all drainvalves (1), in sequence shown, until brake system air pressure is vented.

Step 2. Close drainvalves (1).

Step 3. Disconnect delivery line (8) from service brake chambers (4) and remove elbow (7) from relay valve (6).

Step 4. Connect test gauge to delivery port (6).

Step 5. Direct assistant to start engine and allow air system to build to normal operating pressure.

Step 6. Direct assistant to fully apply and hold service brakes.

Step 7. Check reading on test gauge.

    a. If pressure is below normal, proceed to test 16.

    b. If pressure is normal, repair or replace blocked delivery line (8).

    c. Stop engine, drain air, remove test gauge, and connect delivery line (8).

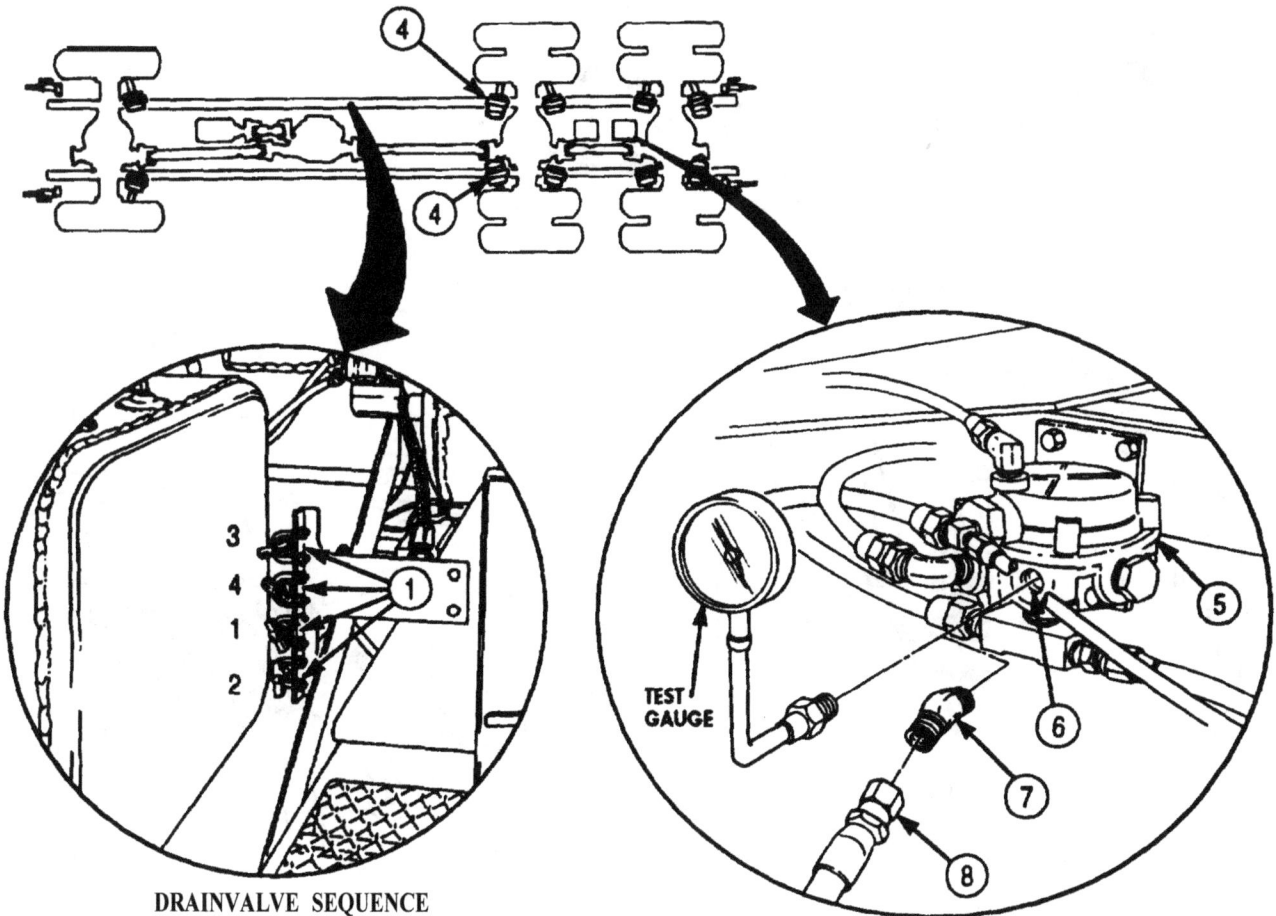

DRAINVALVE SEQUENCE

Table 2-4. Compressed Air and Brake System Troubleshooting (Contd).

MALFUNCTION
TEST OR INSPECTION
CORRECTIVE ACTION

Test 16. Check secondary air system relay valve (3) for proper pressure (only spring and service brake chambers (1) test low or nonexistent air pressure).

  Step 1. Stop engine and open all drainvalves (2), in sequence shown, until brake system air pressure is vented.

  Step 2. Close drainvalves (2).

  Step 3. Disconnect delivery line (6) from service brake chambers (1) and remove elbow (5) from relay valve (3).

  Step 4. Connect test gauge to delivery port (4).

  Step 6. Direct assistant to start engine and allow air system to build to normal operating pressure.

  Step 6. Direct assistant to fully apply and hold service brakes.

  Step 7. Check reading on test gauge.

    a. If pressure is below normal, proceed to test 16.

    b. If pressure is normal, repair or replace blocked delivery line (6).

    c. Stop engine, drain air, remove test gauge, and connect delivery line (6).

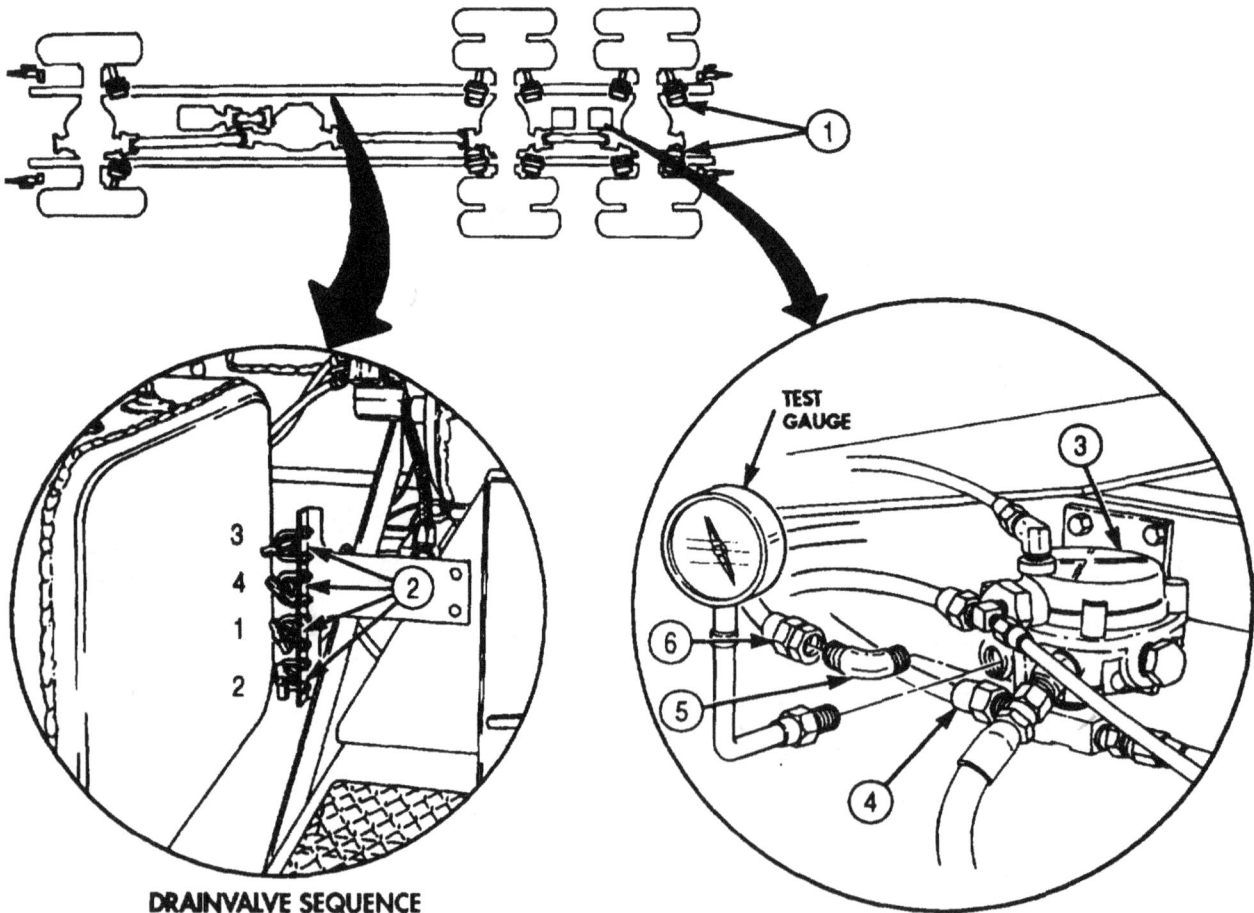

DRAINVALVE SEQUENCE

Table 2-4. Compressed Air and Brake System Troubleshooting (Contd).

MALFUNCTION
    TEST OR INSPECTION
        CORRECTIVE ACTION

Test 16. Check secondary air tank to relay valve supply line (7) (all four secondary air system spring and service brake chambers (1) test low or nonexistent air pressure).

Step 1.   Stop engine and open all drainvalves (2), in sequence shown, until brake system air pressure is vented.

Step 2.   Close drainvalves (2).

Step 3.   Disconnect supply line (7) from secondary air system relay valve (3).

Step 4.   Connect test gauge to supply line (7).

Step 6.   Direct assistant to start engine and allow air system to build to normal operating pressure.

Step 6.   Direct assistant to fully apply and hold service brakes.

Step 7.   Check reading on test gauge.

    a. If pressure is below normal, inspect supply line (7) for blockage. Repair or replace as necessary.

    b. If pressure is normal, replace secondary air system relay valve (3) (para. 3-187 or 3-188).

    c. Stop engine, drain air, remove test gauge, and connect supply line (7).

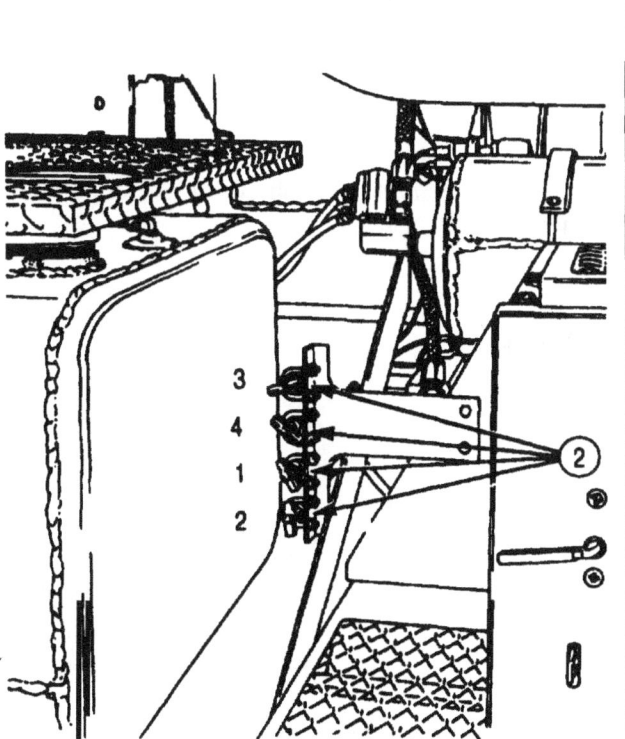

DRAINVALVE SEQUENCE

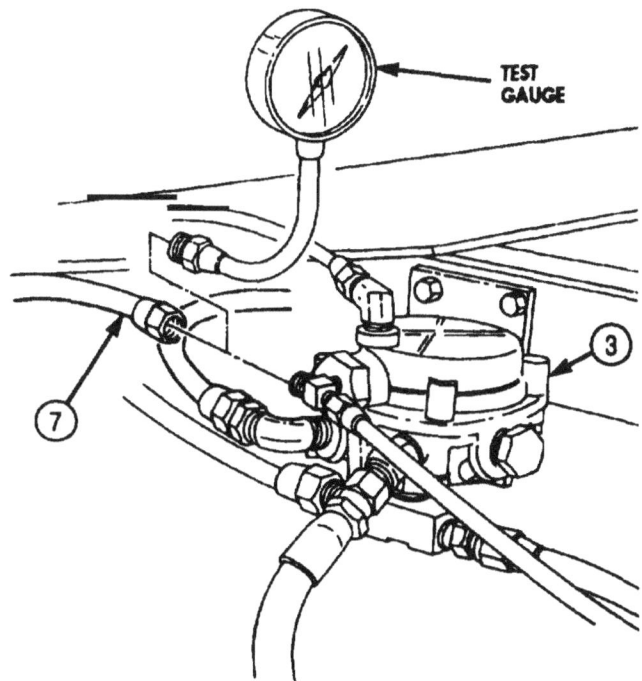

Table 2-4. Compressed Air and Brake System Troubleshooting (Contd).

MALFUNCTION
  TEST OR INSPECTION
    CORRECTIVE ACTION

Test 17. Inspect service brakes.

  Step 1.   Remove brake dustcovers (para. 3-178).

  Step 2.   Inspect chamfer (2) on lining of brakeshoes (1) for wear.

  If brakeshoes (1) are worn to depth of chamfer (2) or less than 5/16 in. (7.94 mm), replace brakeshoe (para. 3-180).

  Step 3.   Inspect for oil on brakeshoes (1) or drum.

  a. If oil is found, replace brakeshoes (1) (para. 3-180).

  b. Replace front axle oil seal (para. 3-160) and clean hub and drum.

  Step 4.   Inspect actuator seals (3) for rotted, tom, or worn condition. If defective, notify DS maintenance.

END OF TESTING!

4. VEHICLE PULLS RIGHT OR LEFT WHEN APPLYING BRAKES

NOTE

  Vehicle pulling to right or left indicates a malfunction in one of two front wheel service brakes.

Test 1. Determine which brake is grabbing.

  Step 1.   In a safe test area, bring vehicle to a hard, sudden stop.

  Step 2.   Inspect front wheels for excessive heat on wheel and drum. Inspect for smoke or skid marks.

  a. If grabbing is found, go to test 3.

  b. If no signs of grabbing are present, perform test 2.

Test 2. Check front service brake chambers for proper air pressure. Perform malfunction 3, test 3.

Test 3. Remove front brake dustcovers (para. 3-178) and inspect service brake.

  Step 1.   Perform malfunction 3, test 17.

  Step 2.   Inspect brakeshoe lining to drum clearance (para. 3-179).

Test, 4. Perform mechanical troubleshooting (para. 2-21, malfunction 47).

Table 2-4. Compressed Air and Brake System Troubleshooting (Contd).

MALFUNCTION
TEST OR INSPECTION
CORRECTIVE ACTION

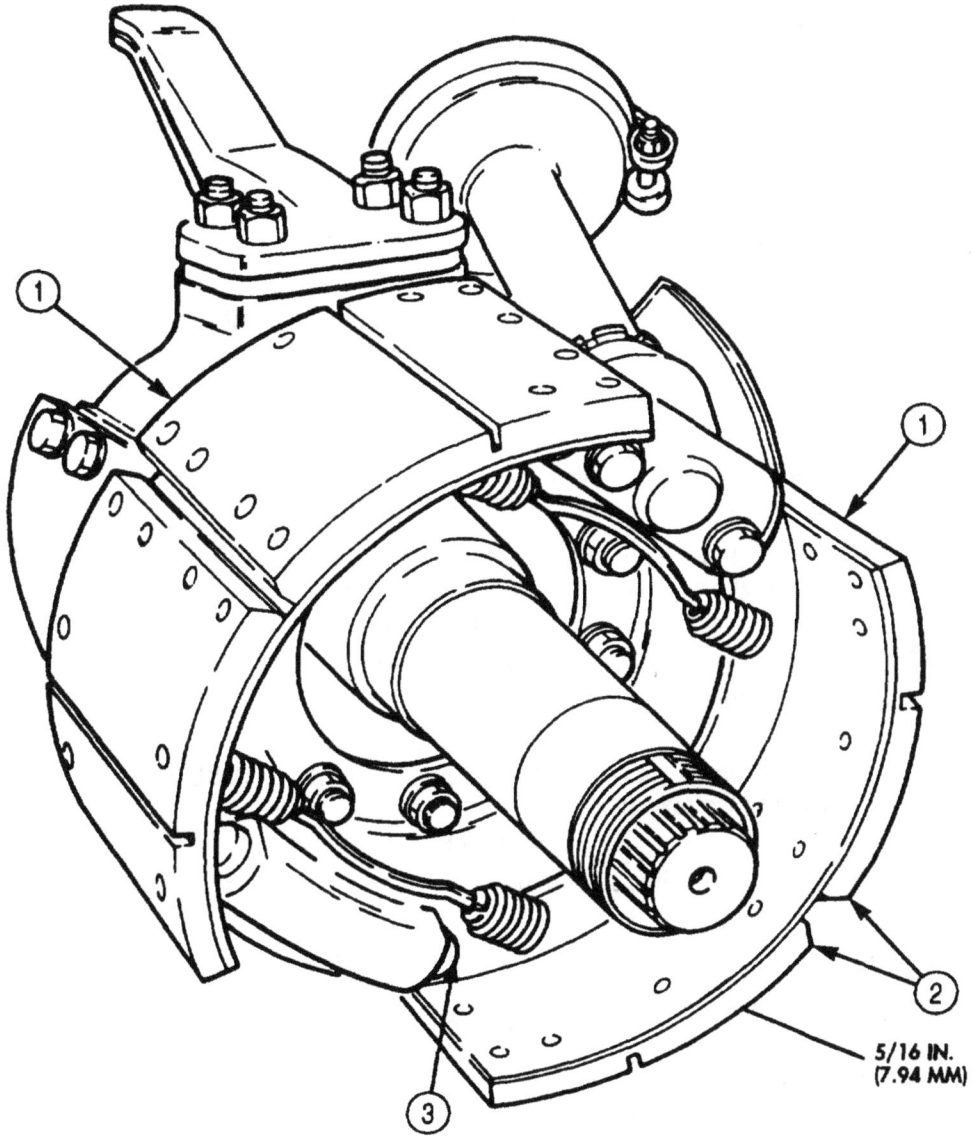

5/16 IN.
(7.94 MM)

END OF TESTING!

Table 2-4. Compressed Air and Brake System Troubleshooting (Contd).

MALFUNCTION
 TEST OR INSPECTION
  CORRECTION ACTION

---

5. VEHICLE REAR BRAKES GRAB OR DRAG

Test 1. Inspect rear brakes to isolate malfunction.

Step 1. Inspect for signs of locking, smoke, skid marks, and excessive heat on wheels and drums.

a. If locking is found, perform malfunction 4, test 3 on rear brakes.

b. If no sign of locking is found, go to step 2.

Step 2. Inspect for leaks. Apply brakes, listen, and locate any defective lines or fittings.

a. If leakage is indicated, repair or replace defective lines or fittings as necessary (TM 9-243).

b. If no leakage is found, go to test 2.

Test 2. Check rear brake chambers for proper air pressure. Perform malfunction 3, tests 4 and 5.

END OF TESTING!

6. VEHICLE VIBRATES, CHATTERS, OR BOUNCES WHEN BRAKES ARE APPLIED

Perform malfunction 5.

END OF TESTING!

7. BRAKES SQUEAL

Inspect for glazed and worn brake linings.

Step 1. Remove brakedrums (paras. 3-223 and 3-224).

Step 2. Perform malfunction 4, test 3.

Step 3. Inspect for glazed lining. Lining should appear dull.

If lining is shiny, remove glaze with wire brush.

Step 4. Inspect for dirt or metal trapped in brakeshoes.

Clean out with wire brush.

END OF TESTING!

8. WARNING BUZZER SOUNDS WHEN BRAKES ARE APPLIED (PRIMARY AND SECONDARY GAUGE PRESSURE DROPS BELOW 55-65 PSI (379-448 kPa)

NOTE

This malfunction indicates a major leak in air delivery system.

Listen and locate leaks.

Step 1. Direct assistant to start engine and allow air system to build to normal operating pressure.

Step 2. Direct assistant to fully apply and hold service brakes.

Step 3. Inspect for leaks.

Locate and repair leaks as necessary.

END OF TESTING!

Table 2-4. Compressed Air and Brake System Troubleshooting (Contd).

MALFUNCTION
TEST OR INSPECTION
CORRECTIVE ACTION

COMPRESSED AIR SUPPLY

9. NO AIR PRESSURE (WARNING BUZZER SOUNDING, AIR PRESSURE NOT BUILDING TO NORMAL OPERATING RANGE AS INDICATED BY GAUGES)

Test 1. Check for leaks.

    Step 1.   Direct assistant to start engine.

    Step 2.   Listen and locate leaks.

        a. If leakage is found, repair lines or fittings as necessary.

        b. If no leakage is found, go to test 2.

Test 2. Check air compressor (1) with engine running.

    Step 1.   Loosen and bleed air from governor signal line (2) to air compressor (1).

    Step 2.   Feel if air compressor outlet line (3) is hot.

        a. If air compressor (1) is operating normally, outlet line (3) will be hot (under great pressure). Proceed to test 3 if outlet line (3) is hot.

        b. If outlet line (3) is not hot, proceed to step 3.

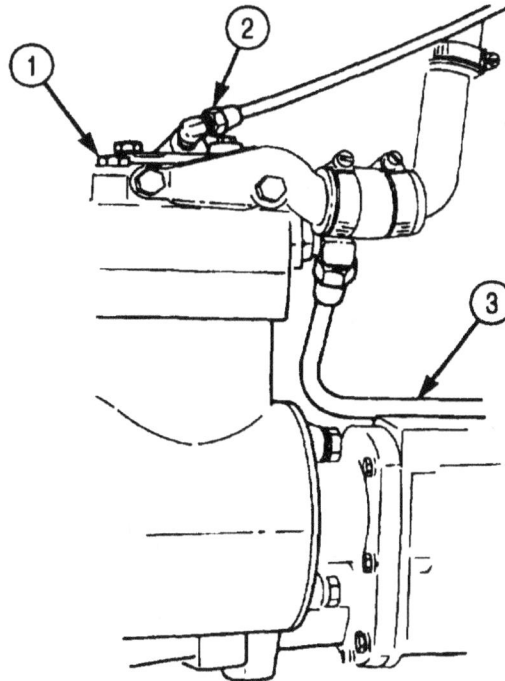

Table 2-4. Compressed Air and Brake System Troubleshooting (Contd).

MALFUNCTION
   TEST OR INSPECTION
      CORRECTIVE ACTION

WARNING

Loosen outlet line at air compressor very slowly. Stop procedure and tighten fitting the moment air begins to escape. Injury to personnel may result if line is accidentally disconnected from a serviceable operating compressor.

Step 3. Carefully loosen fitting (2) until air is heard escaping.

   a. If air is heard, tighten fitting and proceed to test 3.

   b. If air is not present, notify DS maintenance of defective air compressor (1).

Test 3. Inspect outlet line (3) for damage that could restrict airflow.

   a. If damage is present, repair or replace outlet line (3) (TM 9-243).

   b. If no damage is present, perform test 4.

Test 4. Check outlet line for blockage.

Step 1. Stop engine and open all drainvalves (6), in sequence shown, until brake system air pressure is vented.

Step 2. Close drainvalves (6).

Step 3. Disconnect outlet line (3) at wet supply tank (5) and install tee (4) between tank (5) and outlet line (3).

Step 4. Connect test gauge to tee (4).

Step 5. Direct assistant to start engine and allow air system to build to normal operating pressure.

Step 6. Compare test gauge reading with gauge readings on instrument panel.

   a. If readings compare (below 80 psi (552 kPa)), perform test 5.

   b. If test gauge reads normal operating pressure, test instrument panel gauges for proper operation. Perform malfunctions 13 and 14.

   c. Stop engine, drain air, remove test gauge, and connect outlet line (3).

Test 5. Check air governor.

Step 1. Replace governor with one known to be good (para. 3-207).

Step 2. Start engine and allow air pressure to reach normal operating pressure.

   a. If air pressure is normal, governor was defective.

   b. If air pressure remains below 80 psi (552 kPa), air compressor (1) is defective. Notify DS maintenance.

Step 3. Check primary air system relay valve; perform test 12.

Table 2-4. Compressed Air and Brake System Troubleshooting (Contd).

MALFUNCTION
TEST OR INSPECTION
CORRECTIVE ACTION

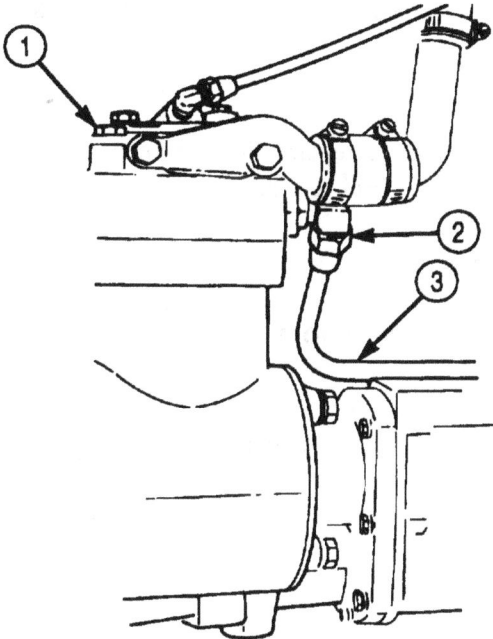

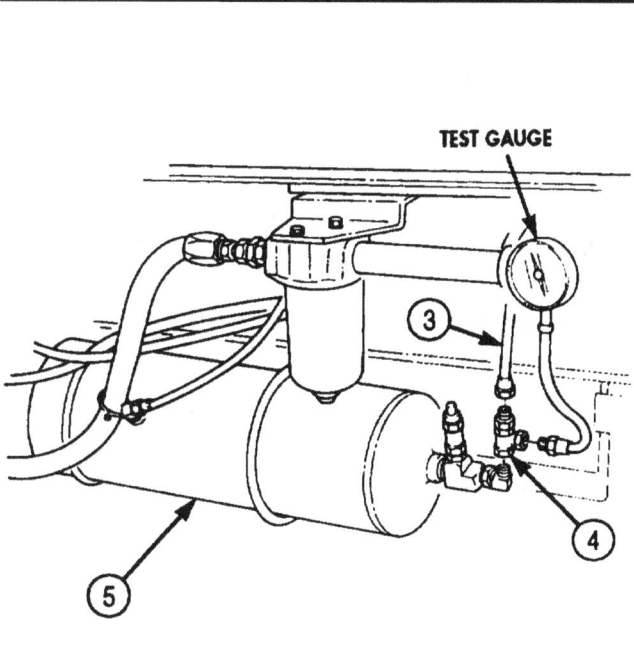

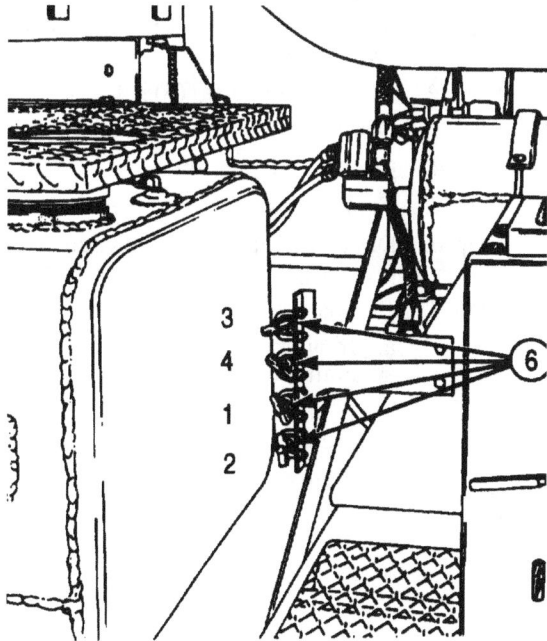

TEST GAUGE

DRAINVALVE SEQUENCE

END OF TESTING!

Table 2-4. Compressed Air and Brake System Troubleshooting (Contd).

```
MALFUNCTION
    TEST OR INSPECTION
        CORRECTIVE ACTION
```

10. AIR PRESSURE DOES NOT BUILD TO NORMAL OPERATING PRESSURE (ABOVE 80 PSI (552 kPa)) ACCORDING TO GAUGES

Test 1. Inspect for leaks.

Listen and locate leakage.

        a. If leakage is indicated, repair or replace lines or fittings as necessary.

        b. If no leakage is indicated, perform test 2.

Test 2. Adjust air governor (para. 3-207).

Test 3. Check air compressor for proper operation (malfunction 9, tests 2, 3, and 4).

END OF TESTING!

11. AIR PRESSURE BUILDS SLOWLY (TAKES EXCESSIVE AMOUNT OF TIME TO BUILD TO 100 PSI (690 kPa))

Perform malfunction 10.

END OF TESTING!

12. AIR PRESSURE EXCEEDS MAXIMUM (GAUGES SHOW OVER 130 PSI (896 kPa)), SAFETY VALVE OPENS TO RELEASE PRESSURE

Test 1. Check for air loss through air accessories. Perform malfunction 20.

Test 2. Check governor signal line (1) for leaks, bends, and clogs.

    Step 1. If leaking, bent, or clogged, repair or replace governor signal line (1) (TM 9-243).

    Step 2. If not leaking, bent, or clogged, proceed to test 3.

Test 3. Check governor signal line (1) pressure.

    Step 1. Stop engine and open all drainvalves (5), in sequence shown, until brake system air pressure is vented.

    Step 2. Close drainvalves (5).

    Step 3. Disconnect governor signal line (1) from air governor (2).

    Step 4. Connect adapter fitting (4) to air governor (2) and tee (3) to adapter (4).

    Step 5. Connect test gauge to tee (3).

    Step 6. Direct assistant to start engine and allow air system to build to normal operating pressure.

    Step 7. Compare test gauge reading with gauge readings on instrument panel.

        a. If air pressure readings are the same on test and instrument panel gauges, notify DS maintenance.

        b. If air pressure reading is below 80 psi (552 kPa), replace air governor (2) (para. 3-207).

        c. Stop engine, drain air, remove test gauge, and connect signal line (1).

Table 2-4. Compressed Air and Brake System Troubleshooting (Contd).

MALFUNCTION
    TEST OR INSPECTION
        CORRECTIVE ACTION

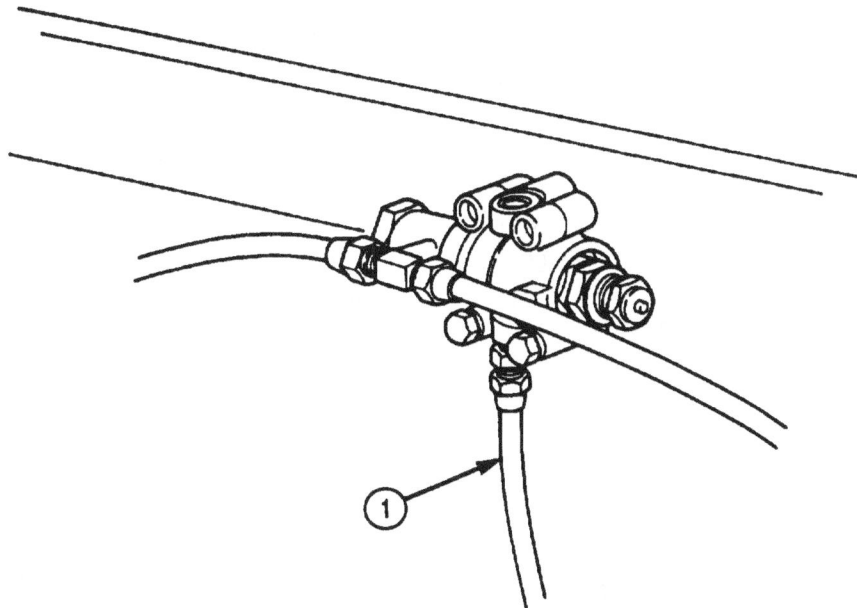

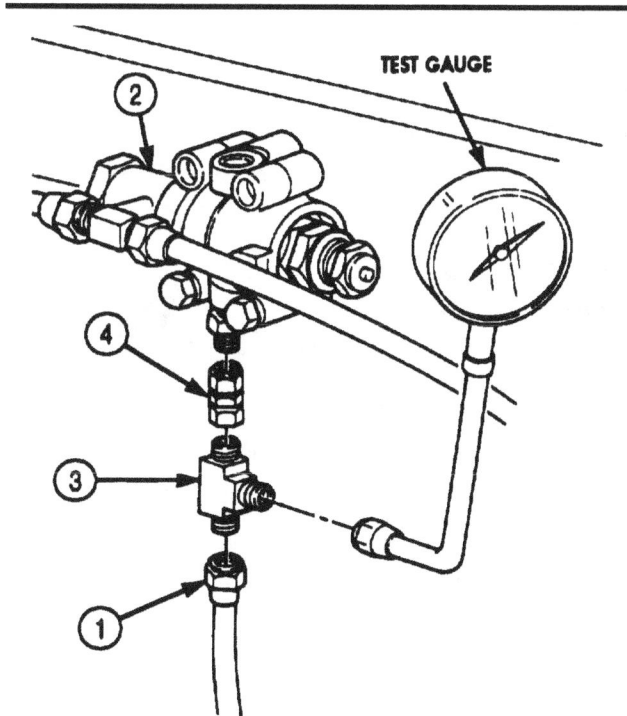

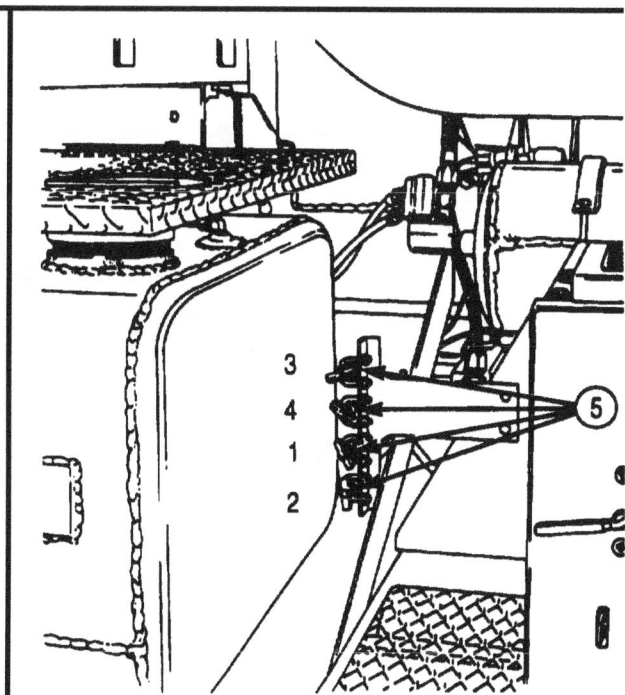

TEST GAUGE

DRAINVALVE SEQUENCE

3
4
1
2

END OF TESTING!

Table 2-4. Compressed Air and Brake System Troubleshooting (Contd).

```
MALFUNCTION
    TEST OR INSPECTION
        CORRECTIVE ACTION
```

13. **PRIMARY PRESSURE GAUGE READS NO PRESSURE, LOW PRESSURE, OR BUILDS TO NORMAL OPERATING PRESSURE SLOWLY. SECONDARY PRESSURE GAUGE READS NORMAL (ENGINE IDLING, BRAKE PEDAL NOT APPLIED)**

Test 1. Check for leaks.

Listen and locate leakage.

a. If leakage is indicated, repair or replace air lines or fittings as necessary.

b. If no leakage is indicated, perform test 2.

Test 2. Check primary air pressure gauge (3).

Step 1. Stop engine and open all drainvalves (1), in sequence shown, until brake system air pressure is vented.

Step 2. Close drainvalves (1).

Step 3. Disconnect air line (2) from primary air pressure gauge (3).

Step 4. Connect test gauge to air line (2).

Step 5. Direct assistant to start engine and allow air system to build to normal operating pressure.

Step 6. Check test gauge reading.

a. If test gauge reads normal operating pressure, replace primary air pressure gauge (3) (para. 3-207).

b. If test gauge reads no pressure, low pressure, or pressure is building slowly, perform test 3.

c. Stop engine, drain air, remove test gauge, and connect air line (2).

Test 3. Check air pressure at brake pedal valve supply port (7).

Step 1. Stop engine and open all drainvalves (1), in sequence shown, until brake system air pressure is vented.

Step 2. Close drainvalves (1).

Step 3. Disconnect primary pressure gauge air line (5) from pedal valve (4).

Step 4. Connect test gauge to supply port (6) of pedal valve (4).

Step 5. Direct assistant to start engine and allow air system to build to normal operating pressure.

Step 6. Check test gauge reading.

a. If test gauge reads normal operating pressure, repair or replace clogged primary pressure gauge air line (5).

b. If test gauge reads no pressure, low pressure, or pressure is building slowly, perform test 4.

c. Stop engine, drain air, remove test gauge, and connect air line (5).

Table 2-4, Compressed Air and Brake System Troubleshooting (Contd).

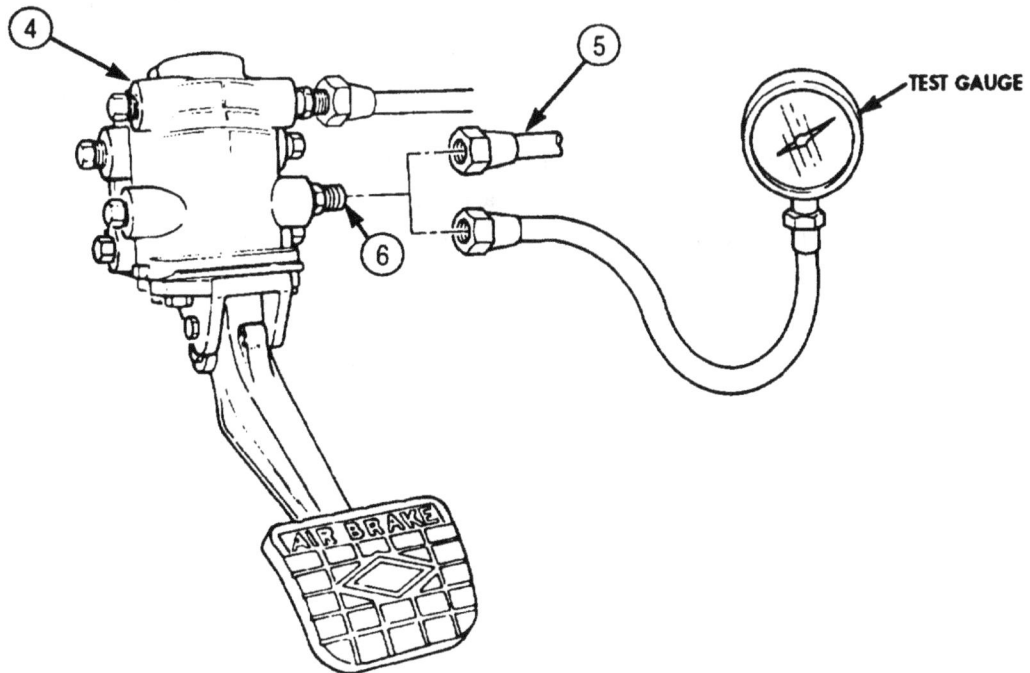

MALFUNCTION
TEST OR INSPECTION
CORRECTIVE ACTION

DRAINVALVE SEQUENCE

TEST GAUGE

TEST GAUGE

Table 2-4. Compressed Air and Brake System Troubleshooting (Contd).

MALFUNCTION
TEST OR INSPECTION
CORRECTIVE ACTION

Test 4. Check primary reservoir air pressure.

   Step 1. Stop engine and drain primary air reservoir at drainvalve (6).

   Step 2. Remove drainvalve (6) from primary drain line (5).

   Step 4. Connect test gauge to primary drain line (5).

   Step 5. Direct assistant to start engine and allow air system to build to normal operating pressure.

   Step 6. Check test gauge reading.

      a. If test gauge reads normal operating pressure, stop engine, drain air, remove test gauge, connect drain line (5), and perform test 5.

      b. If test gauge reads no pressure, low pressure, or pressure is building slowly, perform test 6. Remove test gauge and install drainvalve (6) in primary drain line (5).

Test 5. Check primary air system supply line (3) from primary air reservoir to brake pedal valve (1).

   Step 1. Stop engine and open remaining drainvalves (7), in sequence shown, until brake system air pressure is vented.

   Step 2. Disconnect primary air system supply line (3) from brake pedal valve fitting (4) and connect test gauge to supply line (3).

   Step 3. Direct assistant to start engine and allow air system to build to normal operating pressure.

   Step 4. Check test gauge reading.

      a. If test gauge reads normal operating pressure, replace brake pedal valve (1) (para. 3-197 or 3-198).

      b. If test gauge reads no pressure, low pressure, or pressure is building slowly, repeat steps 1 and 2, reinstall primary air system supply line (3), and proceed to test 6.

Test 6. Check one-way check valve (2).

   Step 1. Stop engine and open all drainvalves (6) and (7) until brake system air pressure is vented.

   Step 2. Remove one-way check valve (2) (para. 3-186).

   Step 3. Inspect check valve (2) for clogging and damage.

      a. If clogging or damage is indicated, replace defective one-way check valve (2) (para. 3-186).

      b. If no clogging or damage is present, repair or replace clogged primary air supply line (3).

Table 2-4. Compressed Air and Brake System Troubleshooting (Contd).

MALFUNCTION
TEST OR INSPECTION
CORRECTIVE ACTION

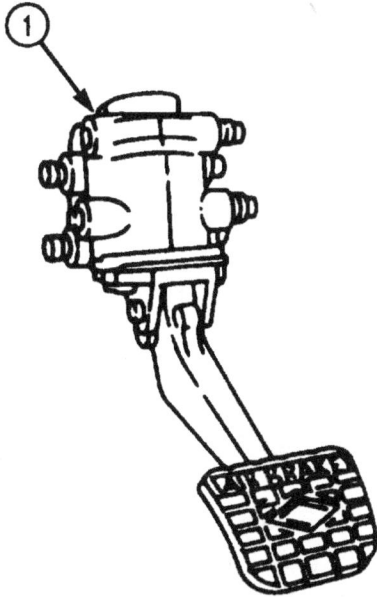

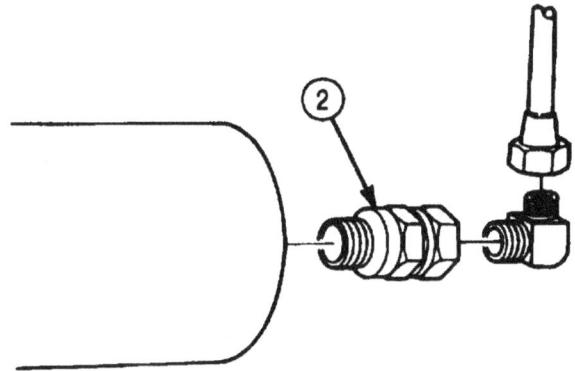

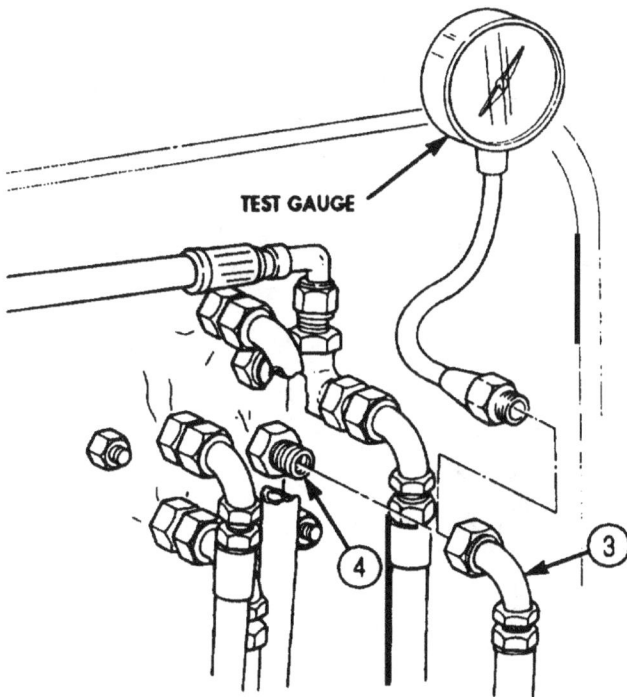

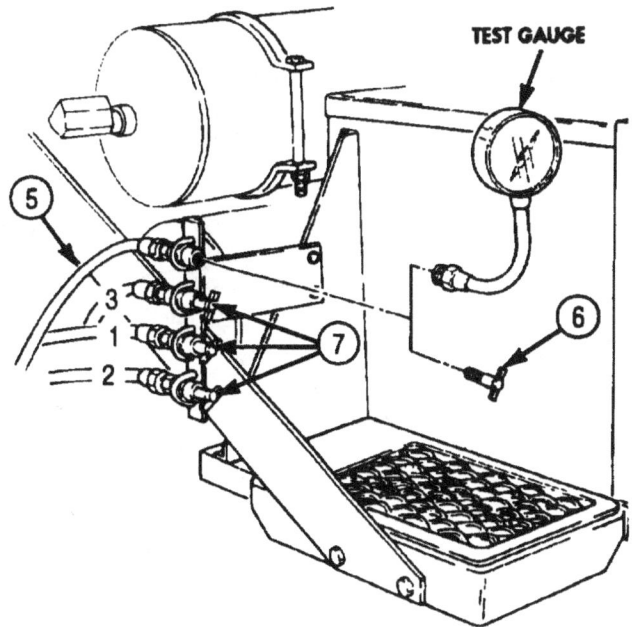

END OF TESTING!

Table 2-4. Compressed Air and Brake System Troubleshooting (Contd).

```
MALFUNCTION
    TEST OR INSPECTION
        CORRECTIVE ACTION
```

14. SECONDARY PRESSURE GAUGE READS NO PRESSURE, LOW PRESSURE, OR BUILDS TO NORMAL OPERATING PRESSURE SLOWLY. PRIMARY PRESSURE GAUGE READS NORMAL (ENGINE IDLING, BRAKE PEDAL NOT APPLIED)

Test 1. Perform malfunction 13, test 1.

Test 2. Check secondary air pressure gauge (3).

   Step 1. Stop engine and open all drainvalves (1), in sequence shown, until brake system air pressure is vented.

   Step 2. Close drainvalves (1).

   Step 3. Disconnect air line (2) from secondary air pressure gauge (3).

   Step 4. Connect test gauge to air line (2).

   Step 5. Direct assistant to start engine and allow air system to build to normal operating pressure.

   Step 6. Check test gauge reading.

      a. If test gauge reads normal operating pressure, replace secondary air pressure gauge (3) (para. 3-87).

      b. If test gauge reads no pressure, low pressure, or pressure is building slowly, perform test 3.

      c. Stop engine, drain air, remove test gauge, and connect air line (2).

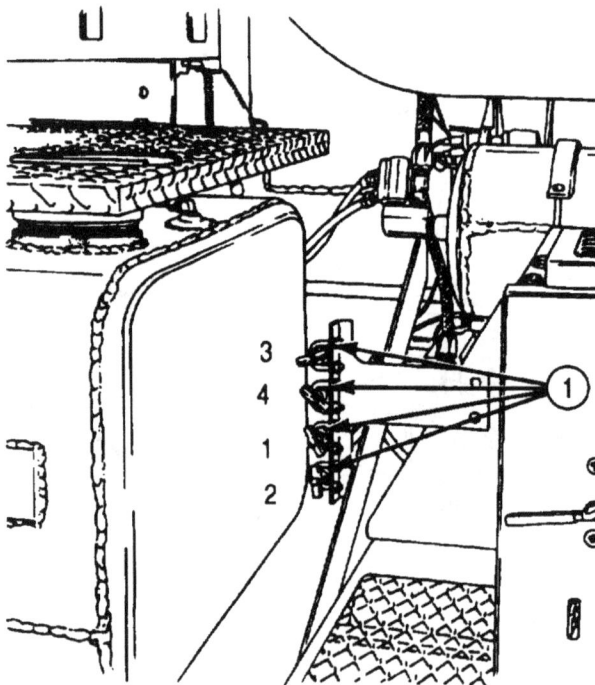

DRAINVALVE SEQUENCE

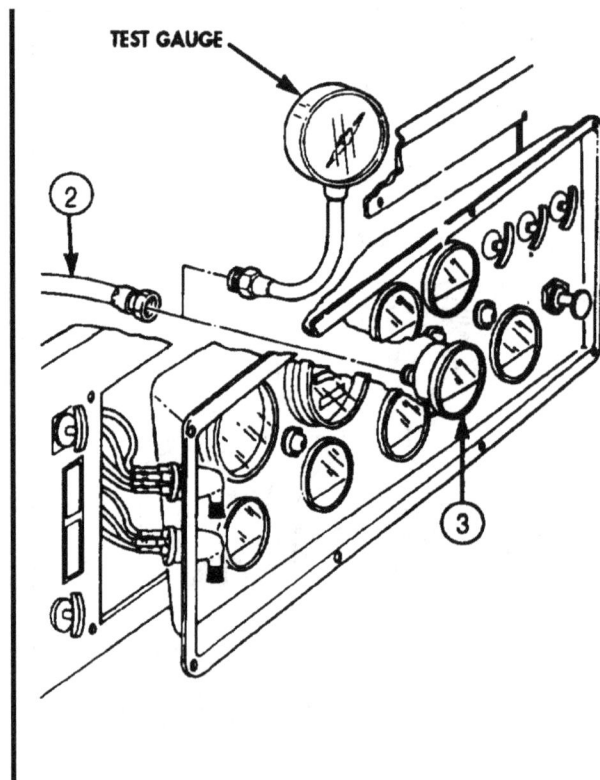

Table 2-4. Compressed Air and Brake System Troubleshooting (Contd).

MALFUNCTION
   TEST OR INSPECTION
       CORRECTIVE ACTION

Test 3. Check air pressure at brake pedal valve supply port (5).

   Step 1.   Stop engine and open all drainvalves (1), in sequence shown, until brake system air pressure is vented.

   Step 2.   Close drainvalves (1).

   Step 3.   Disconnect secondary pressure gauge air line (6) from pedal valve supply port (5).

   Step 4.   Connect test gauge to supply port (5) of pedal valve (4).

   Step 5.   Direct assistant to start engine and allow air system to build to normal operating pressure.

   Step 6.   Check test gauge reading.

       a.   If test gauge reads normal operating pressure, repair or replace clogged air line (6) to secondary air pressure gauge (3) on instrument panel.

       b.   If test gauge reads no pressure, low pressure, or pressure is building slowly, perform test 4.

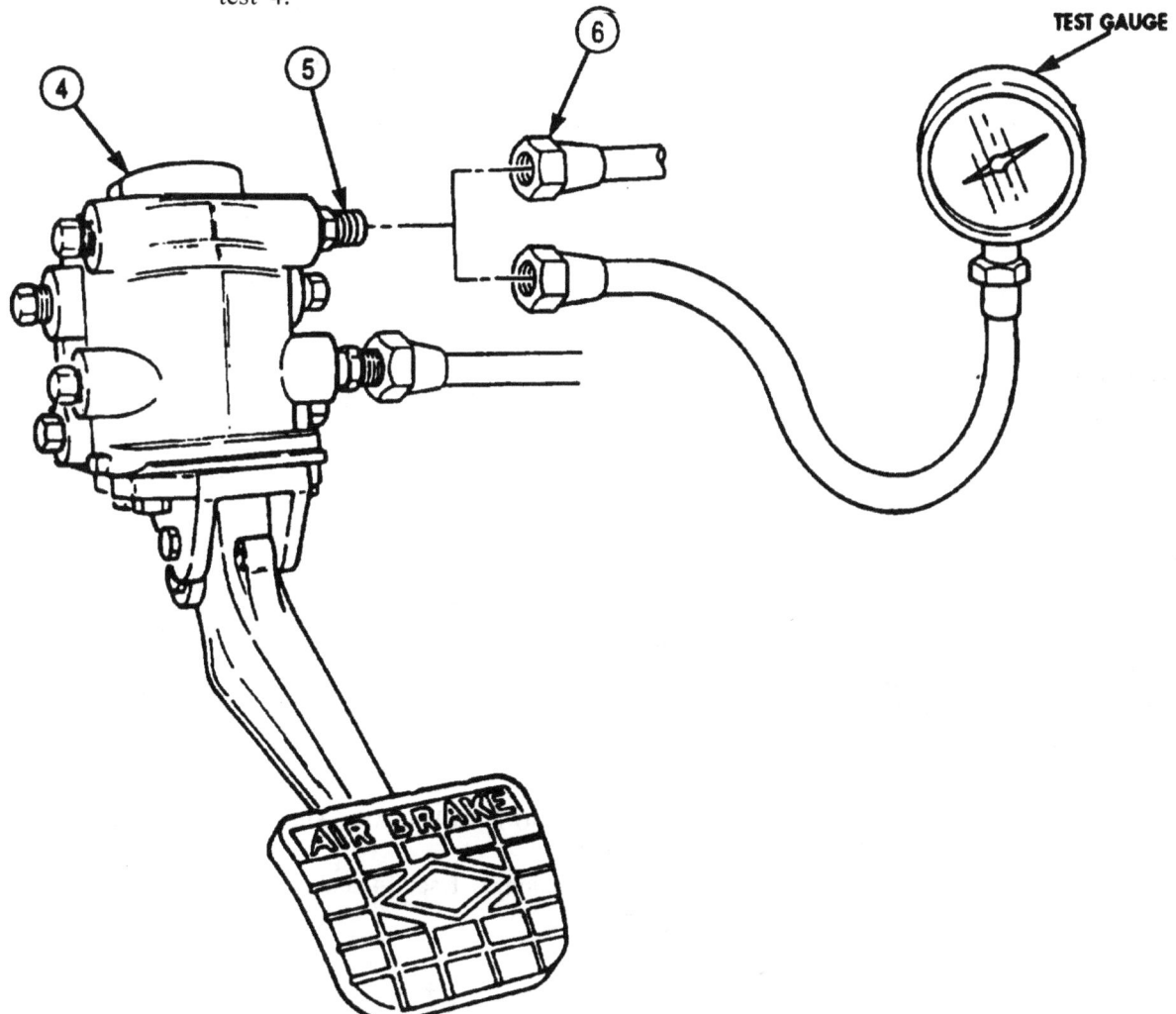

Table 2-4. Compressed Air and Brake System Troubleshooting (Contd).

MALFUNCTION
  TEST OR INSPECTION
    CORRECTIVE ACTION

Test 4. Check secondary reservoir air pressure.

Step 1. Stop engine and drain secondary air reservoir at drainvalve (4).

Step 2. Remove drainvalve (4) from secondary drain line (6).

Step 3. Connect test gauge to secondary drain line (6).

Step 4. Direct assistant to start engine and allow air system to build to normal operating pressure.

Step 6. Check test gauge reading.

   a. If test gauge reads normal operating pressure, perform test 5.

   b. If test gauge reads no pressure, low pressure, or pressure is building slowly, perform test 6.

   c. Stop engine, drain air, remove test gauge, and install drainvalve (4).

Test 6. Check supply line (1) from secondary air reservoir to brake pedal valve.

Step 1. Stop engine and open all drainvalves (3), (4), and (5), in sequence shown, until brake system air pressure is vented.

Step 2. Disconnect secondary air system supply line (1) from brake pedal valve fitting (2) and connect test gauge to supply line (1).

Step 3. Direct assistant to start engine and allow air system to build to normal operating pressure.

Step 4. Check test gauge reading.

   a. If test gauge reads normal operating pressure, replace brake pedal valve (para. 3-197 or 3-198).

   b. If test gauge reads no pressure, low pressure, or pressure is building slowly, repeat steps 1 and 2, install secondary air system supply line (1), and proceed to test 6.

Test 6. Check one-way check valve.

Step 1. Stop engine and open all drainvalves (3), (4), and (5) in sequence shown, until brake system air pressure is vented.

Step 2. Remove one-way check valve (para. 3-186).

Step 3. Inspect check valve for clogging and damage.

   a. If clogging or damage is indicated, replace defective one-way check valve (para. 3-186).

   b. If no clogging or damage is present, repair or replace clogged secondary air supply line (1).

END OF TESTING!

15. PRIMARY AIR SYSTEM FAILS TO MAINTAIN PRESSURE (NO MAJOR LEAKS, AIR CAN BE HEARD ESCAPING INTO AIR INTAKE STACK, PARKING BRAKE APPLIED)

Perform malfunction 3, test 2.

END OF TESTING!

16. SECONDARY AIR SYSTEM FAILS TO MAINTAIN PRESSURE (NO MAJOR LEAKS, AIR CAN BE HEARD ESCAPING INTO AIR INTAKE STACK, PARKING BRAKE APPLIED)

Test 1. Check secondary relay valve. Perform malfunction 3, tests 14 and 15.

Test 2. Check brake pedal valve. Perform malfunction 14, test 3.

Table 2-4. Compressed Air and Brake System Troubleshooting (Contd).

MALFUNCTION
    TEST OR INSPECTION
        CORRECTIVE ACTION

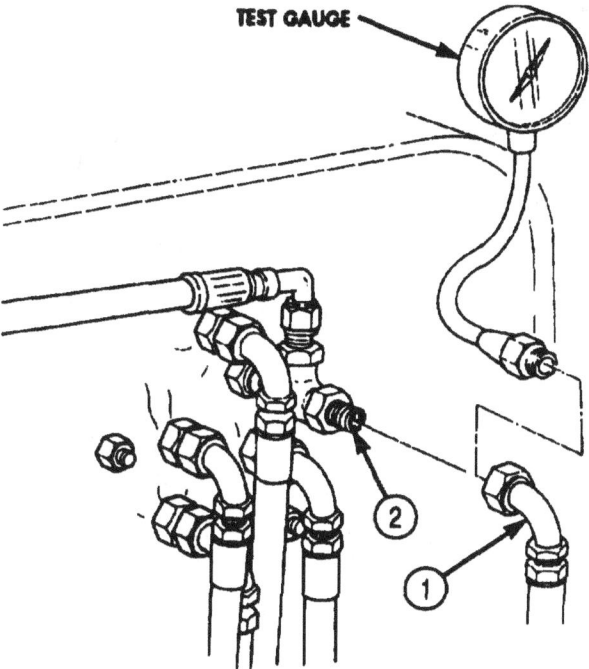

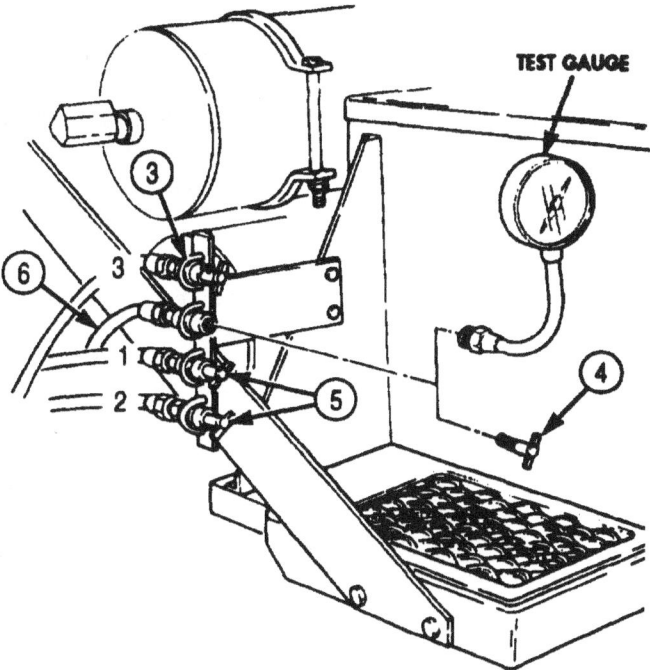

DRAINVALVE SEQUENCE

END OF TESTING!

Table 2-4. Compressed Air and Brake System Troubleshooting (Contd).

```
MALFUNCTION
    TEST OR INSPECTION
        CORRECTIVE ACTION
```

17. WARNING BUZZER FAILS TO SOUND OR FAILS TO SHUT OFF ON LOW PRESSURE (BELOW 55-65 PSI (379-448 kPa)), AIR PRESSURE SYSTEM OPERATING NORMALLY

See Electrical Troubleshooting, para. 2-22.

END OF TESTING!

18. SPRING BRAKES DO NOT RELEASE (VEHICLE BRAKES GRAB OR DRAG)

NOTE

If only one spring brake chamber fails to release, perform malfunction 5.

Test 1. Isolate malfunction.

Full out emergency spring brake release control (1) and move vehicle forward to determine if spring brakes release.

    a. If spring brakes release, perform test 2.

    b. If spring brakes do not release, perform tests 2, 3, and 4.

Test 2. Check supply pressure to parking brake valve (4).

    Step 1. Inspect parking brake valve (4) to determine if mechanical actuator (2) is stuck in engage position.

        a. If actuator (2) is sticking, pull back rubber boot (3) and apply a few drops of light machine oil (Appendix C, Item 47) to actuator (2).

        b. If malfunction continues, go to step 2.

    Step 2. Start engine and allow air system to build to normal operating pressure.

WARNING

Loosen supply line at valve very slowly. Stop procedure and tighten fitting of supply line the moment air begins to escape. Injury to personnel may result if line is accidentally disconnected from valve.

    a. Slowly loosen supply line (5) to check for air pressure.

    b. If pressure is not present, proceed to test 3.

    c. If pressure is present, proceed to test 4.

Table 2-4. Compressed Air and Brake System Troubleshooting (Contd).

MALFUNCTION
TEST OR INSPECTION
CORRECTIVE ACTION

Test 3. Check quick-release valve (10).

  With engine idling and parking brake released, listen for sounds of air escaping through air intake stack (7).

  a. If escaping air is present, replace quick-release valve (10) (para. 3-216).

  b. If escaping air is not present, inspect line (9) between quick-release valve (10) and doublecheck valve (8). If clogged, repair or replace line (9). If not clogged, replace doublecheck valve (8) (para. 3-216).

WARNING

  Loosen delivery line at valve very slowly. Stop procedure and tighten fitting of delivery line the moment air begins to escape. Injury to personnel may result if delivery line is accidentally disconnected from valve.

Test 4. Check parking brake valve (4) delivery pressure (supply pressure confirmed in test 2). With engine running and parking brake released, slowly loosen delivery line (6).

  a. If escaping air is not present, adjust or replace parking brake valve (4) (para. 3-194).

  b. If escaping air is present, spring brake diaphragms are defective. Notify DS maintenance.

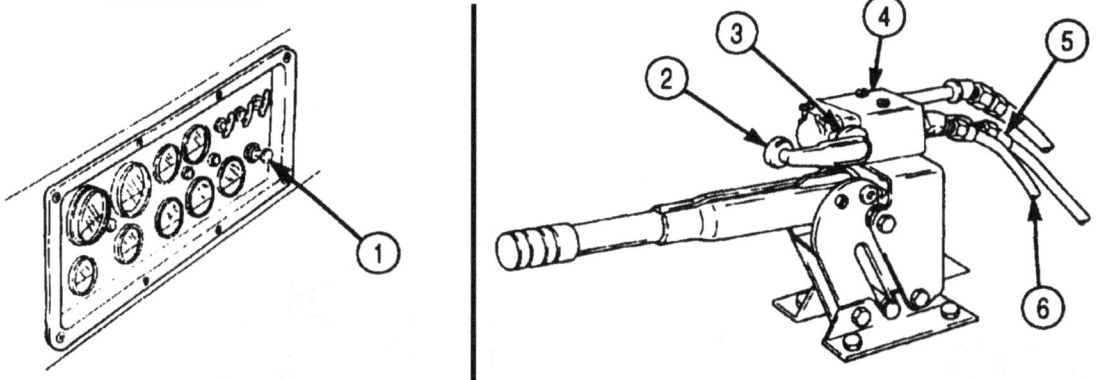

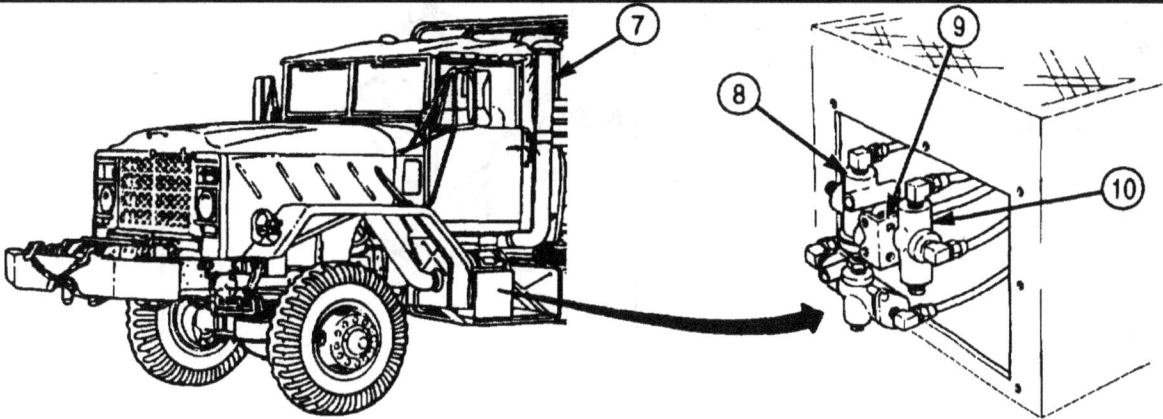

END OF TESTING!

Table 2-4. Compressed Air and Brake System Troubleshooting (Contd).

MALFUNCTION
TEST OR INSPECTION
CORRECTIVE ACTION

19. SPRING BRAKES DO NOT SET (GAUGES AT NORMAL OPERATING PRESSURES, AIR EXHAUSTING NOT HEARD WHEN PARKING BRAKE APPLIED)

Test 1. Ensure spring brake release control (1) is not out.

Test 2. Ensure parking brake is fully applied.

Test 3. Inspect for crimping in air lines (3) on quick-release valve (2).

      a. Replace air line(s) (3) if crimped.

      b. If no damage to lines (3) is apparent, perform test 4.

Test 4. Check parking brake valve (4).

   Step 1. Disconnect vent line (6) from parking brake valve vent port (5).

   Step 2. With engine running and parking brake released, apply parking brake.

   Step 3. Feel for air venting through parking brake valve vent port (5).

      a. If escaping air is present, replace parking brake valve (4) (para. 3-194).

      b. If escaping air is not present, replace quick-release valve (2) (para. 3-216).

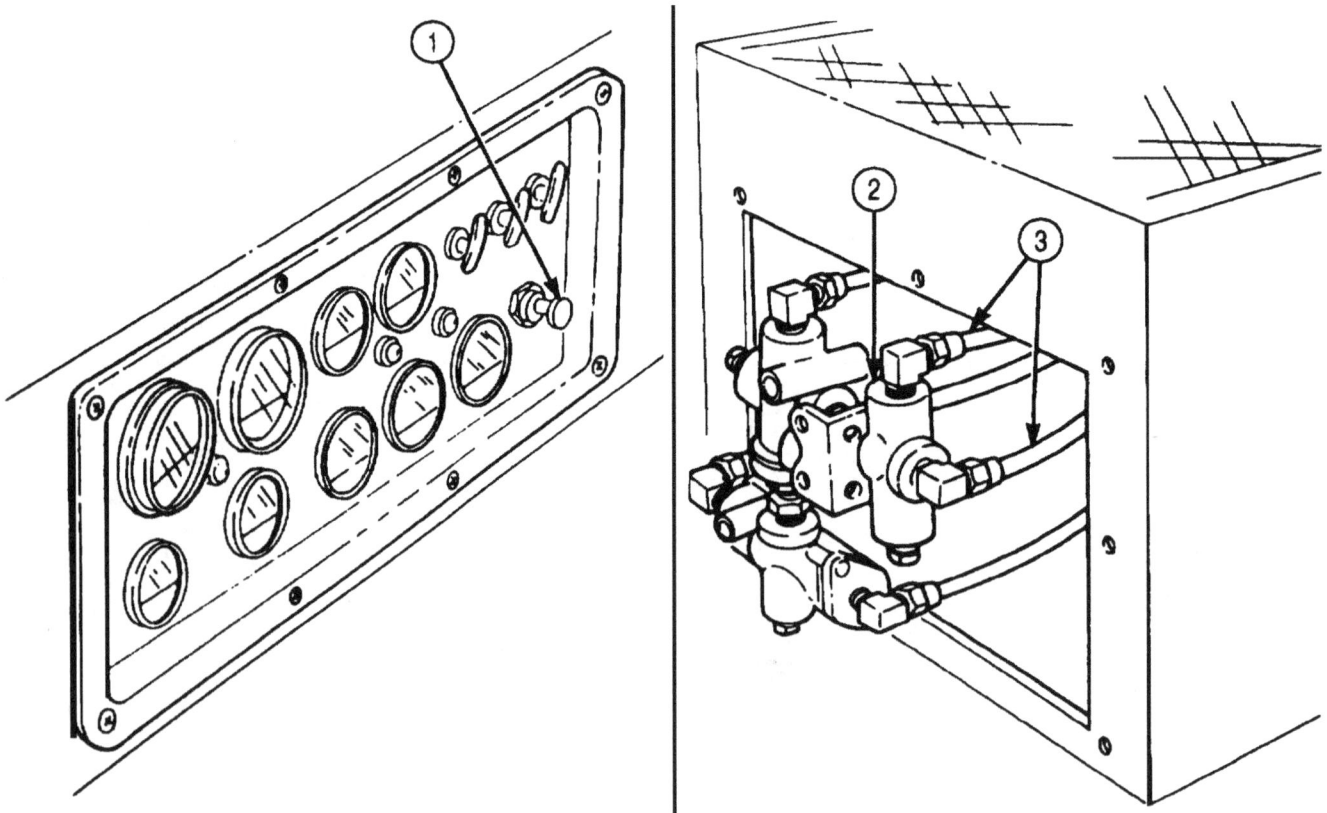

Table 2-4. Compressed Air and Brake System Troubleshooting (Contd).

MALFUNCTION
TEST OR INSPECTION
CORRECTIVE ACTION

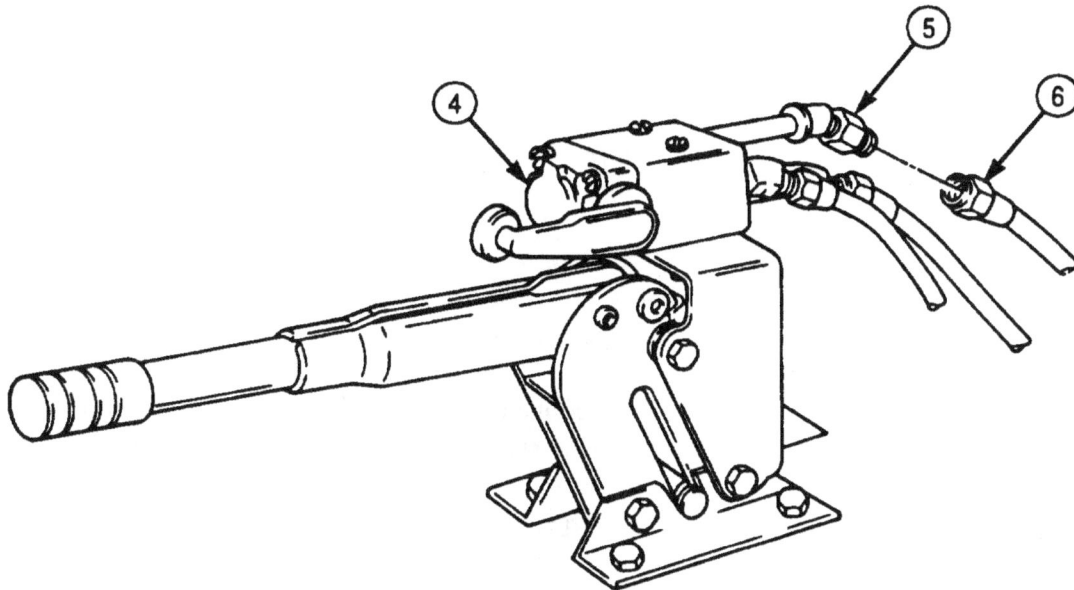

END OF TESTING!

AIR-OPERATED ACCESSORIES

20. ALL AIR-OPERATED ACCESSORIES DO NOT OPERATE (HORN, WINDSHIELD WIPERS, WINDSHIELD WASHERS, TRANSFER CASE CONTROLS), GAUGES AT NORMAL OPERATING PRESSURE

Check air pressure gauges on instrument panel.

    a. If air gauges indicate normal operating pressure, defect is in air lines between the governor and accessories or in accessories themselves.

    b. If gauges indicate excessive air pressure, replace pressure protection valve (para. 3-187).

END OF TESTING!

21. COOLING FAN DOES NOT OPERATE, ENGINE TEMPERATURE ABOVE 200°F (93°C) AS INDICATED BY TEMPERATURE GAUGE

Test 1. Check availability of air as indicated on secondary air system pressure gauge.

If air pressure is below normal operating pressure, perform malfunction 14.

Test 2. Check temperature gauge for proper operation. See Electrical Troubleshooting, para. 2-22.

Test 3. Check fan drive clutch actuator for proper operation.

    a. Remove and replace fan drive clutch actuator with one known to be good (para. 3-73 or 3-74).

    b. If fan remains inoperative, notify DS maintenance.

END OF TESTING!

Table 2-4. Compressed Air and Brake System Troubleshooting (Contd).

MALFUNCTION
    TEST OR INSPECTION
        CORRECTIVE ACTION

---

22. COOLING FAN DOES NOT STOP RUNNING, ENGINE TEMPERATURE BELOW NORMAL OPERATING RANGE (OVERRIDE BOLT NOT INSTALLED)

Replace fan drive clutch actuator (para. 3-73 or 3-74).

END OF TESTING!

23. FRONT WHEEL DRIVE DOES NOT ENGAGE (FRONT WHEEL DRIVE LOCK-IN SWITCH ENGAGED AND TRANSFER CASE SHIFT LEVER IN HIGH POSITION)

Test 1. Isolate malfunction.

  Shift transfer case shift lever in LOW position.

      a. If front wheel drive continues to malfunction, perform test 2.

      b. If front wheel drive operates, perform test 5 and 6.

Test 2. Inspect for air leakage in air system.

  Step 1. Shift transfer case lever in HIGH position and engage front wheel drive lock-in switch.

  Step 2. Start engine and allow air pressure to build to normal operating pressure.

  Step 3. Stop engine.

  Step 4. Observe secondary air pressure gauge.

      a. If gauge indicates a steady loss of air, perform malfunction 14.

      b. If gauge indicates no loss of air, proceed to step 5.

  Step 5. Shift transfer case lever in LOW position and observe secondary air pressure gauge.

      a. If gauge indicates a steady loss of air, perform tests 3 and 4.

      b. If gauge indicates no loss of air, perform tests 7 and 8.

Test 3. Inspect air line (2) from actuator valve (1) to transfer case (transfer case shift lever in LOW position, secondary air pressure gauge indicates steady loss of air).

      a. If leakage is present, repair or replace air line (2) (TM 9-243).

      b. If no leakage is present at air line (2), perform test 4.

Test 4. Inspect supply line (3) for leakage (transfer case shift lever in LOW position, secondary air pressure gauge indicates steady loss of air).

      a. If leakage is present, repair or replace air line (3) (TM 9-243).

      b. If no leakage is present at air line (3), notify DS maintenance of defective actuator valve (1).

Test 5. Check for air pressure at air line (4) from actuator valve (1) to front wheel drive lock-in switch (front wheel drive functions with transfer case shift lever in LOW position, but does not function with transfer case lever in HIGH position and front wheel drive lock-in switch engaged).

  Step 1. Disconnect air line (4) from actuator valve (1) and connect test gauge to air line (4).

  Step 2. Start engine, place transfer case shift lever in HIGH position, and engage front wheel drive lock-in switch.

  Step 3. Observe test gauge.

      a. If air pressure is not indicated, perform test 6.

      b. If air pressure is present, notify DS maintenance of defective actuator valve (1).

      c. Stop engine, drain air, remove test gauge, and connect air line (4).

Table 2-4. Compressed Air and Brake System Troubleshooting (Contd).

MALFUNCTION
TEST OR INSPECTION
CORRECTIVE ACTION

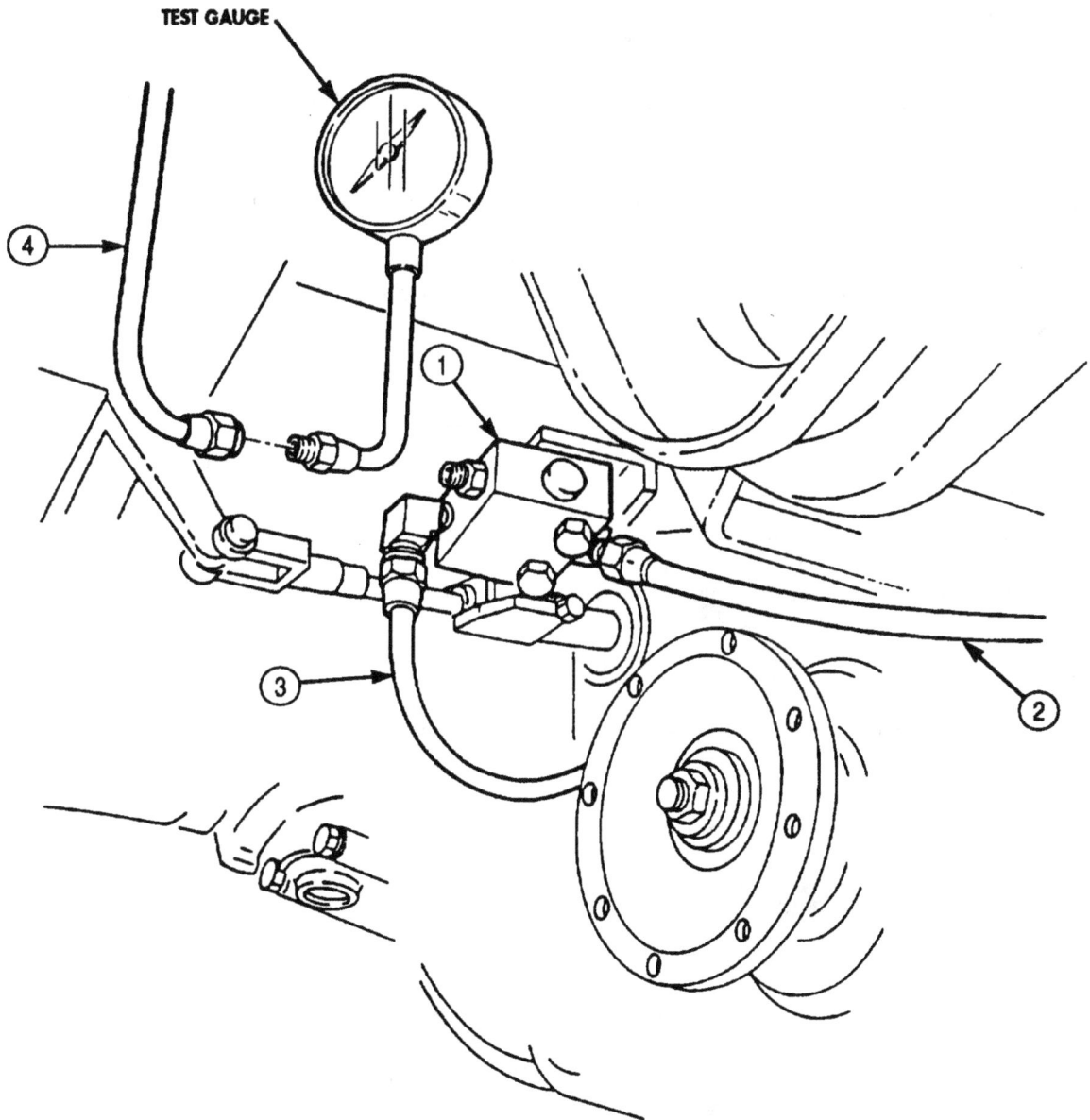

Table 2-4. Compressed Air and Brake System Troubleshooting (Contd).

MALFUNCTION
TEST OR INSPECTION
CORRECTIVE ACTION

Test 6. Check for air pressure at front wheel drive lock-in switch (2).

 Step 1. Disconnect air line (1) from front wheel drive lock-in switch (2) and connect test gauge to front wheel drive lock-in switch (2).

 Step 2. With engine at idle, ensure transfer case shift lever is in HIGH position, and engage front wheel drive lock-in switch.

 Step 3. Observe test gauge.

 a. If air pressure is not indicated, replace front wheel drive lock-in switch (2) (para. 3-101).

 b. If air pressure is present, repair or replace air line (1).

Test 7. Check for air pressure from actuator valve air delivery line (4) at transfer case (supply pressure normal).

 Step 1. Disconnect air delivery line (4) and connect test gauge to air line (4).

 Step 2. With engine at idle, place transfer case shift lever in LOW position.

 Step 3. Observe test gauge.

 a. If air delivery pressure is normal, notify DS maintenance that transfer case does not engage front wheel drive.

 b. If air delivery pressure is below normal, perform test 8.

 c. Stop engine, drain air, remove test gauge, and connect delivery line (4).

Test 8. Check air pressure at actuator valve (3) (supply pressure normal).

 Step 1. Disconnect air delivery line (4) at actuator valve (3) and connect test gauge to air delivery port (5).

 Step 2. With engine at idle, place transfer case shift lever in LOW position.

 Step 3. Observe test gauge.

 a. If air delivery pressure is normal, repair or replace blocked air delivery line (4). (TM 9-243).

 b. If air delivery pressure is below normal, actuator valve (3) is defective. Notify DS maintenance.

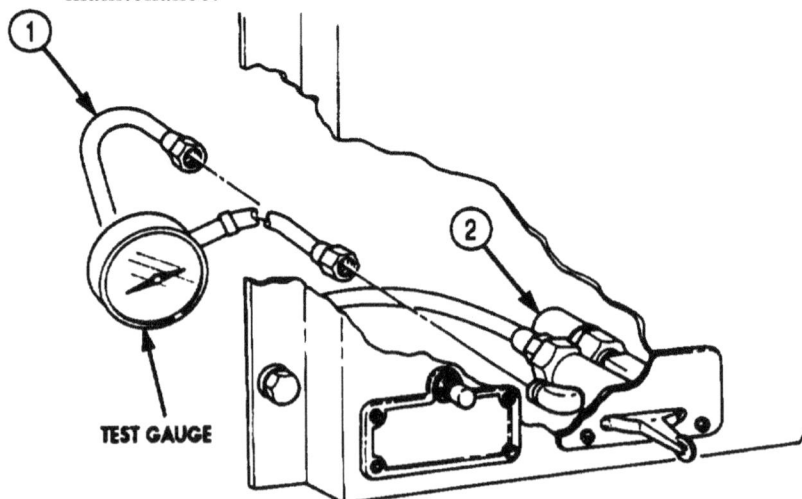

Table 2-4. Compressed Air and Brake System Troubleshooting (Contd).

MALFUNCTION
TEST OR INSPECTION
CORRECTIVE ACTION

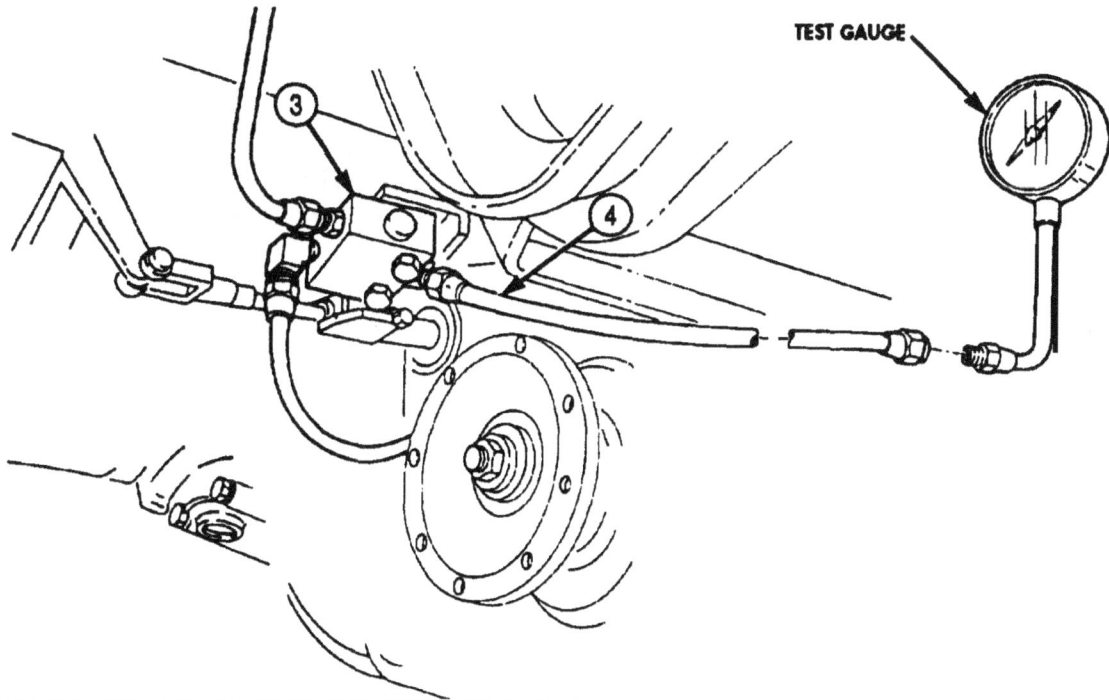

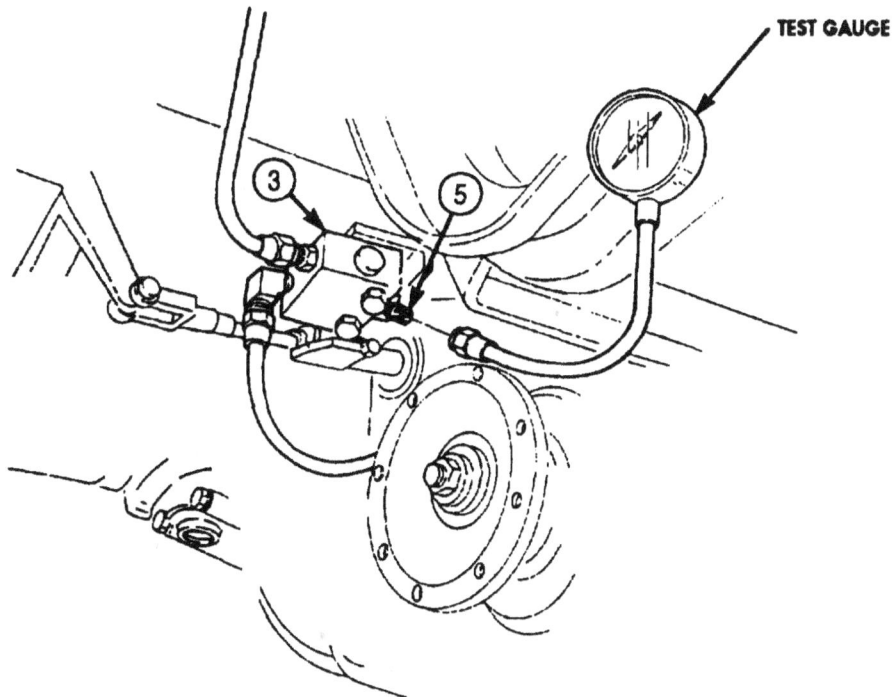

END OF TESTING!

Table 2-4. Compressed Air and Brake System Troubleshooting (Contd).

MALFUNCTION
TEST OR INSPECTION
CORRECTIVE ACTION

24. HORN DOES NOT WORK (GAUGES AT NORMAL OPERATING PRESSURE, WARNING BUZZER NOT SOUNDING)

    Step 1. Check air supply at air electric valve (1).

    Step 2. Loosen supply line (2) slowly.

          a. If air is not heard escaping, repair or replace supply line (2) (TM 9-243).

          b. If air is heard escaping, perform Electrical Troubleshooting, para. 2-22.

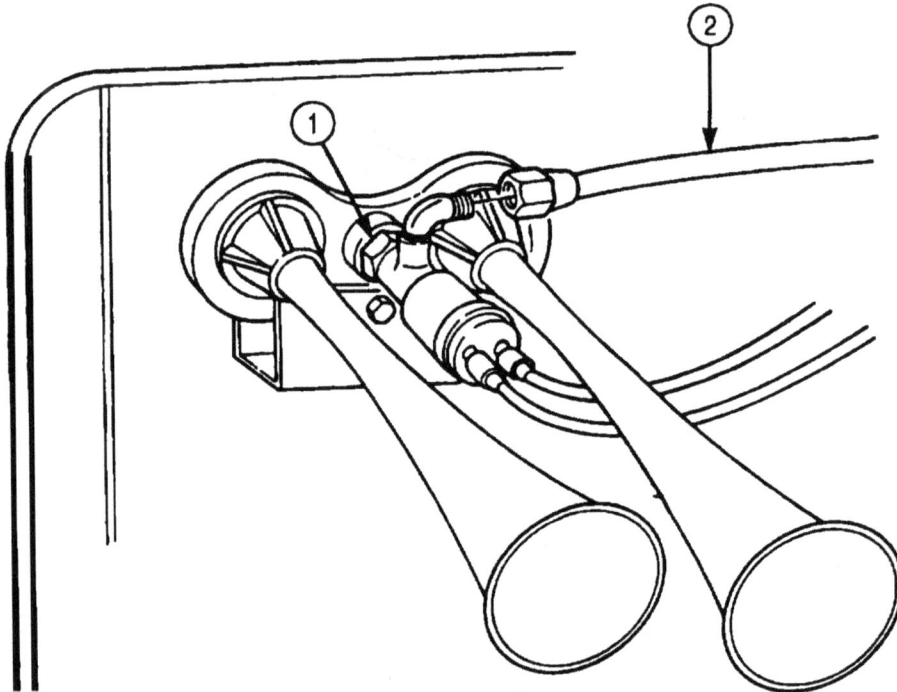

END OF TESTING!

2-24. STE/ICE TROUBLESHOOTING (SIMPLIFIED TEST EQUIPMENT FOR INTERNAL COMBUSTION ENGINES)

a. General. This section is applicable only if STE/ICE is available. This section contains information and tests which may be used with STE/ICE to locate malfunctions that may develop in the vehicle. The tests can be used during troubleshooting, Preventive Maintenance Checks and Service (PMCS), or after replacing parts to isolate malfunctions, anticipate failures, and ensure that proper repairs have been made.

STE/ICE is used primarily with the vehicle electrical system. These tests cannot cover all possible troubles which may occur. If a particular malfunction is not covered, refer to Electrical Troubleshooting, para. 2-22, and locate the troubleshooting procedure for the malfunction observed. To obtain the maximum number of observed symptoms of the malfunction, question the operator.

b. STE/ICE Chain Index. PMCS (para. 2-12) contains a list of various troubles which may occur during operation or inspection of the vehicle. When one of the malfunctions listed occurs, the mechanic proceeds to the associated STE/ICE Chain Index.

c. Engine Chain Indexes. Tables 2-5 and 2-6 provide indexes for GO and NO-GO test chains for M939/A1/A2 series vehicles. The test chains are presented on the pages which follow.

NOTE

Use vehicle entry 06 on test select switches for M939/A1 series vehicles; 31 for M939A2 series vehicles.

Table 2-5. Go-Chain Test Index, Combined Mode Chain.

GO TEST NUMBER	MODE	TEST TITLE	PAGE NUMBER
GO1	DCA	VTM Connections and Checkout	2-220
GO2	TK	First Peak Test - Starter Current	2-225
GO3	TK	Engine Start - Lubrication Check	2-227
GO4	DCA	Charging Circuit and Battery Voltage Test	2-230
GO5	DCA	Engine Warmup/Coolant Check/Oil Pressure Test	2-232
GO6	DCA	Governor Check/Power Test	2-234
GO7	DCA	Idle Speed/Governor Check	2-236
GO8	DCA	Compression Unbalance Test	2-237
—	—	DC Voltage Test	2-239

Table 2-6. No-Go-Chain Test Index, Combined Mode Chain.

NO-GO TEST NUMBER	MODE	TEST TITLE	PAGE NUMBER
NG05	TK	Low Oil Pressure Check	2-241
NG20	DCA	No Crank - No Start	2-243
NG30	DCA	Engine Crank - No Start	2-244
NG31	DCA	Gauge Test	2-247
NG50	DCA	Charging Circuit Tests	2-249
NG80	DCA-TK	Starter Circuit Tests	2-250
NG81	DCA	Battery Tests	2-255
NG90	DCA	Governor/Power Test Fault Isolation	2-261
NG130	DCA	Engine Tightness Test	2-266

Table 2-5. STE/ICE GO-Chain Tests.

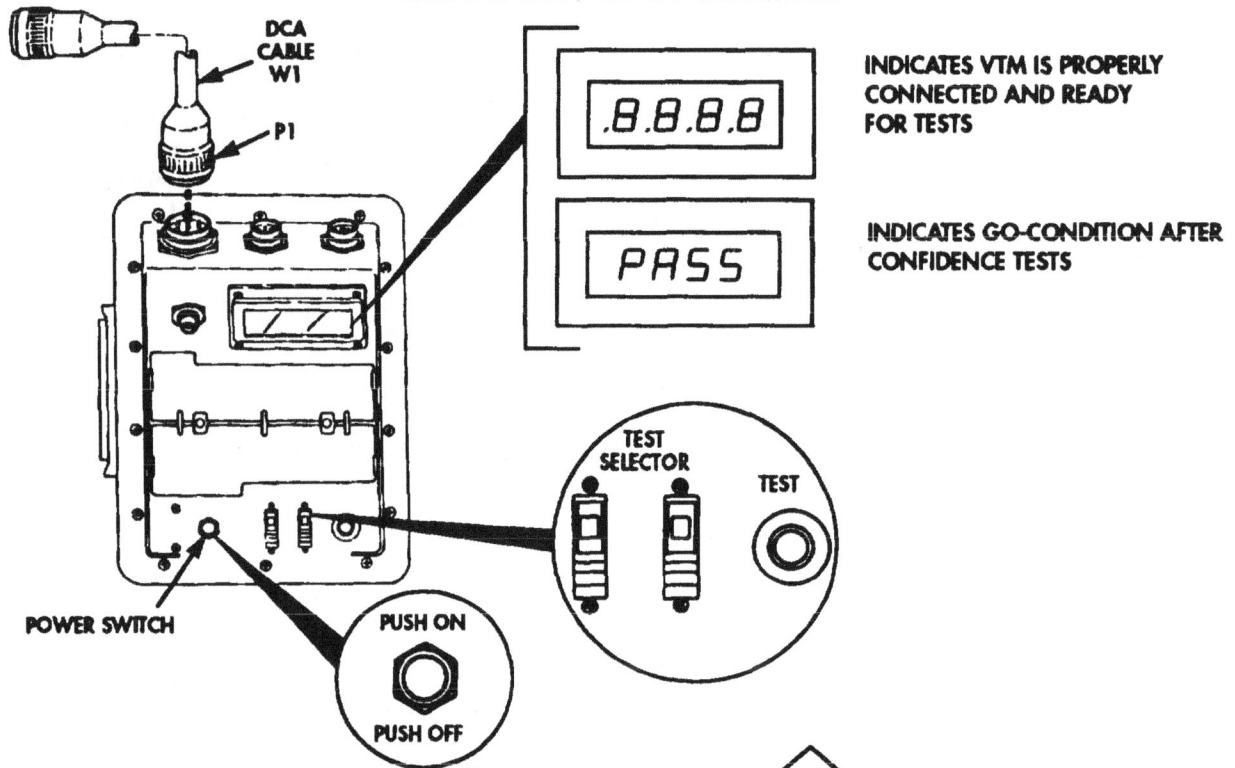

INDICATES VTM IS PROPERLY CONNECTED AND READY FOR TESTS

INDICATES GO-CONDITION AFTER CONFIDENCE TESTS

.8.8.8.8

PASS

DCA CABLE W1

P1

TEST SELECTOR

TEST

POWER SWITCH

PUSH ON

PUSH OFF

| G01 | VTM CONNECTIONS AND CHECKOUT | DCA MODE |

CAUTION

- Do not connect or disconnect VTM while vehicle is running.
- Connect DCA cable W1 to VTM before connecting to diagnostic connector.

1 CONNECT VTM TO VEHICLE DIAGNOSTIC CONNECTOR:

- PULL OFF the VTM power switch.
- Connect DCA cable W1 to VTM.
- Connect DCA cable to vehicle.
- Turn vehicle battery switch to ON position.
- PUSH ON the VTM power switch.
- Verify the display indicates .8.8.8.8 for approximately 2 seconds and then changes to - - - -.

- Does VTM display .8.8.8.8 and then change to - - - -?

.8.8.8.8

- - - -

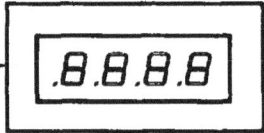

AFTER 2 SECONDS

YES

NO

- Proceed to step 3.

- Proceed to following page.

3

Table 2-5. STE/ICE GO-Chain Tests (Contd).

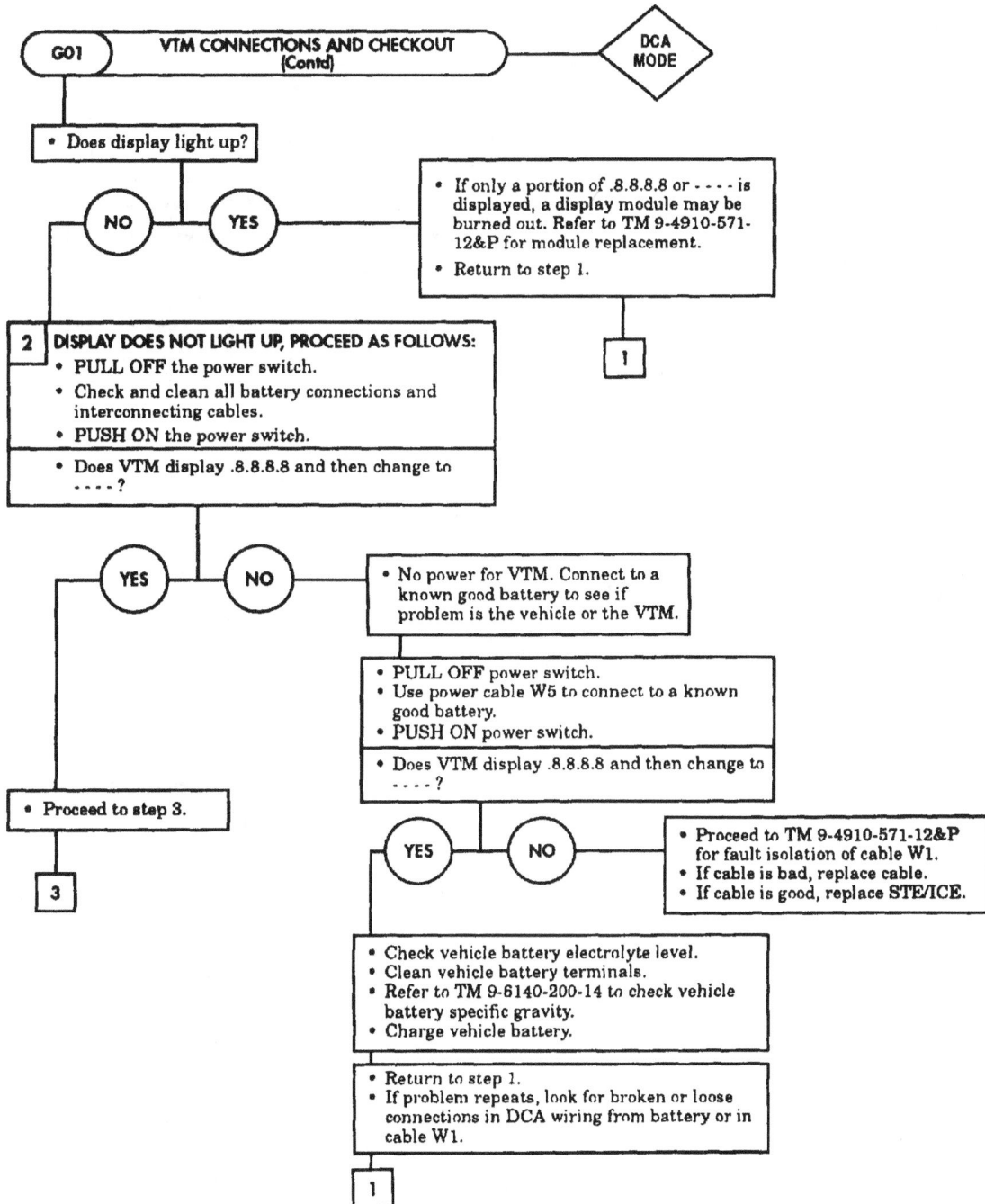

```
( GO1 )──── VTM CONNECTIONS AND CHECKOUT ────◇ DCA
                    (Contd)                     MODE ◇
```

• Does display light up?

```
      (NO)    (YES)
```

• If only a portion of .8.8.8.8 or - - - - is displayed, a display module may be burned out. Refer to TM 9-4910-571-12&P for module replacement.

• Return to step 1.

[1]

| 2 | DISPLAY DOES NOT LIGHT UP, PROCEED AS FOLLOWS: |

• PULL OFF the power switch.
• Check and clean all battery connections and interconnecting cables.
• PUSH ON the power switch.

• Does VTM display .8.8.8.8 and then change to
 - - - - ?

```
      (YES)    (NO)
```

• No power for VTM. Connect to a known good battery to see if problem is the vehicle or the VTM.

• PULL OFF power switch.
• Use power cable W5 to connect to a known good battery.
• PUSH ON power switch.

• Does VTM display .8.8.8.8 and then change to
 - - - - ?

• Proceed to step 3.

[3]

```
      (YES)    (NO)
```

• Proceed to TM 9-4910-571-12&P for fault isolation of cable W1.
• If cable is bad, replace cable.
• If cable is good, replace STE/ICE.

• Check vehicle battery electrolyte level.
• Clean vehicle battery terminals.
• Refer to TM 9-6140-200-14 to check vehicle battery specific gravity.
• Charge vehicle battery.

• Return to step 1.
• If problem repeats, look for broken or loose connections in DCA wiring from battery or in cable W1.

[1]

Table 2-5. STE/ICE GO-Chain Tests (Contd).

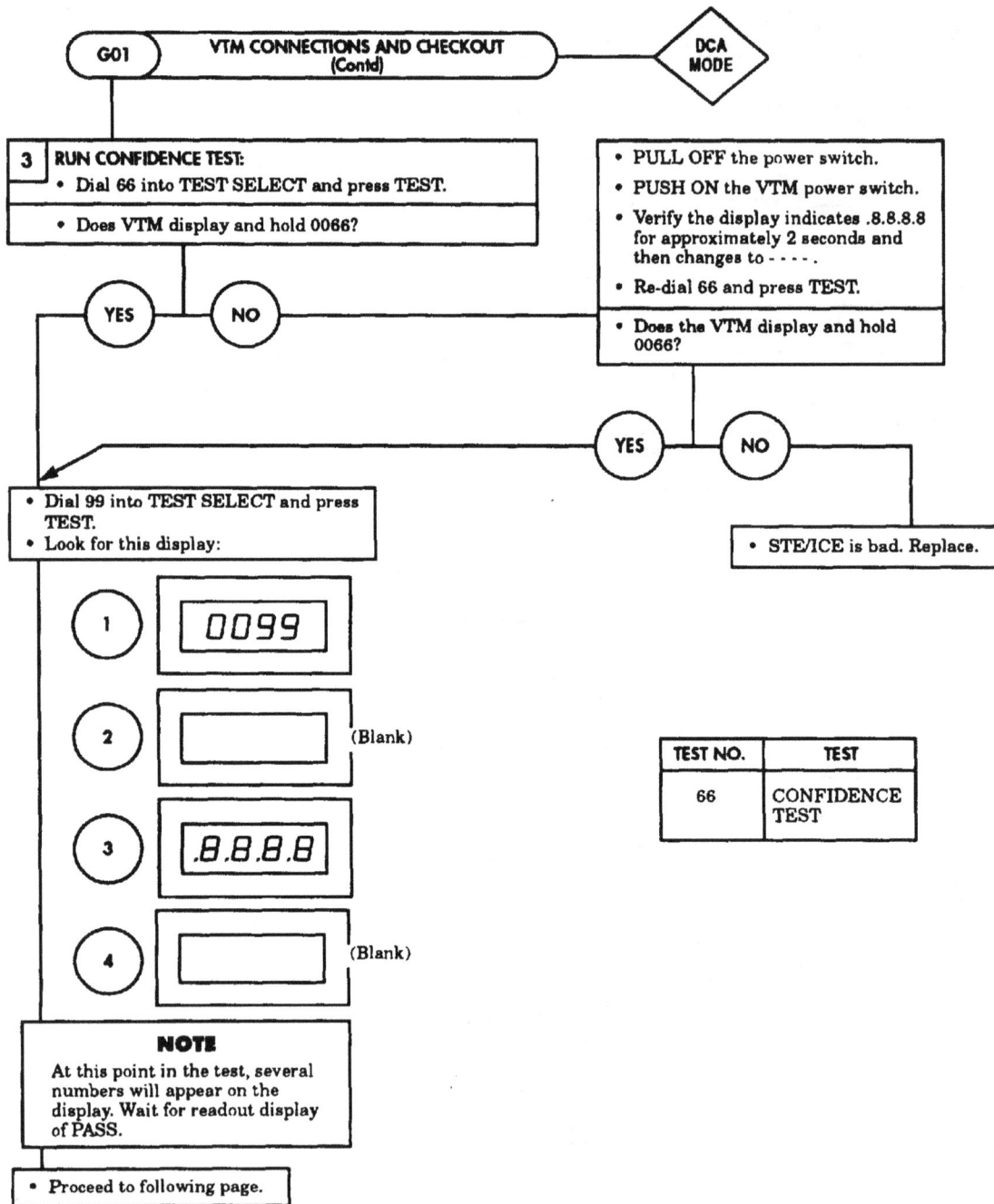

```
( G01 )──┤ VTM CONNECTIONS AND CHECKOUT ├────◇ DCA
              (Contd)                            MODE ◇
```

```
┌───┬──────────────────────────────────────┐   ┌──────────────────────────────────┐
│ 3 │ RUN CONFIDENCE TEST:                  │   │ • PULL OFF the power switch.      │
│   │  • Dial 66 into TEST SELECT and press │   │ • PUSH ON the VTM power switch.   │
│   │    TEST.                              │   │ • Verify the display indicates   │
│   ├──────────────────────────────────────┤   │   .8.8.8.8 for approximately 2   │
│   │  • Does VTM display and hold 0066?    │   │   seconds and then changes       │
│   └──────────────────────────────────────┘   │   to - - - - .                   │
│                                               │ • Re-dial 66 and press TEST.     │
│          ( YES )   ( NO )                     ├──────────────────────────────────┤
│                                               │ • Does the VTM display and hold  │
│                                               │   0066?                          │
│                                               └──────────────────────────────────┘
│
│                                                       ( YES )   ( NO )
```

```
┌──────────────────────────────────────┐
│ • Dial 99 into TEST SELECT and press   │
│   TEST.                                │         ┌──────────────────────────┐
│ • Look for this display:               │         │ • STE/ICE is bad. Replace.│
└──────────────────────────────────────┘         └──────────────────────────┘
```

(1)   ┌───────────┐
         │   0099   │
         └───────────┘

(2)   ┌───────────┐
         │   │   (Blank)
         └───────────┘

TEST NO.	TEST
66	CONFIDENCE TEST

(3)   ┌───────────┐
         │ .8.8.8.8 │
         └───────────┘

(4)   ┌───────────┐
         │   │   (Blank)
         └───────────┘

```
┌──────────────────────────────────────┐
│              NOTE                      │
│ At this point in the test, several     │
│ numbers will appear on the             │
│ display. Wait for readout display      │
│ of PASS.                               │
└──────────────────────────────────────┘
```

```
┌──────────────────────────────────────┐
│ • Proceed to following page.           │
└──────────────────────────────────────┘
```

Table 2-5. STE/ICE GO-Chain Tests (Contd).

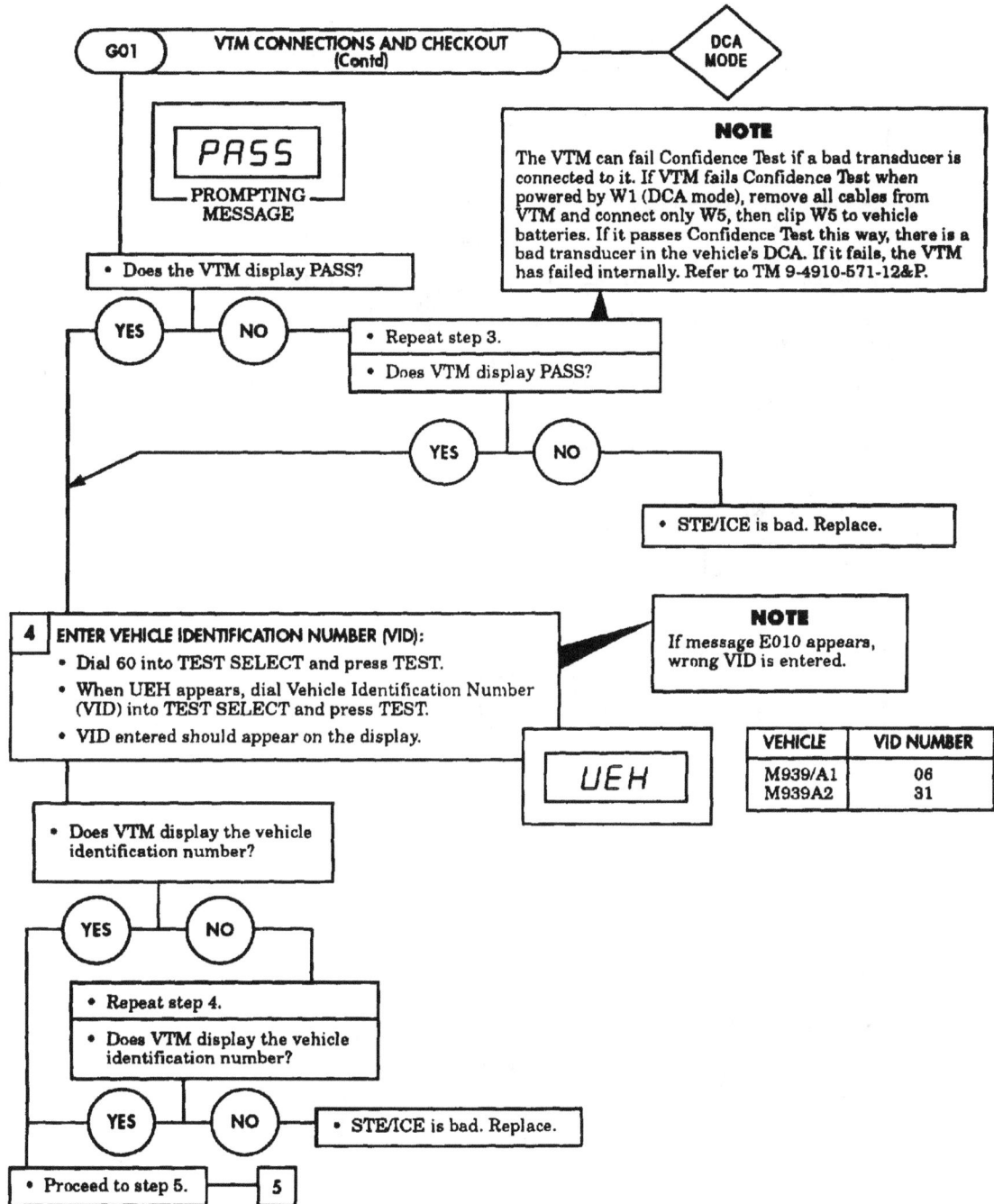

(G01) **VTM CONNECTIONS AND CHECKOUT**
         (Contd)   ◇ DCA MODE

```
   PASS
```
PROMPTING
MESSAGE

NOTE

The VTM can fail Confidence Test if a bad transducer is connected to it. If VTM fails Confidence Test when powered by W1 (DCA mode), remove all cables from VTM and connect only W5, then clip W5 to vehicle batteries. If it passes Confidence Test this way, there is a bad transducer in the vehicle's DCA. If it fails, the VTM has failed internally. Refer to TM 9-4910-571-12&P.

• Does the VTM display PASS?

(YES)   (NO)

• Repeat step 3.
• Does VTM display PASS?

(YES)   (NO)

• STE/ICE is bad. Replace.

NOTE

If message E010 appears, wrong VID is entered.

4 **ENTER VEHICLE IDENTIFICATION NUMBER (VID):**
• Dial 60 into TEST SELECT and press TEST.
• When UEH appears, dial Vehicle Identification Number (VID) into TEST SELECT and press TEST.
• VID entered should appear on the display.

```
   UEH
```

VEHICLE	VID NUMBER
M939/A1	06
M939A2	31

• Does VTM display the vehicle identification number?

(YES)   (NO)

• Repeat step 4.
• Does VTM display the vehicle identification number?

(YES)   (NO)   • STE/ICE is bad. Replace.

• Proceed to step 5.   — [5]

Table 2-5. STE/ICE GO-Chain Tests (Contd).

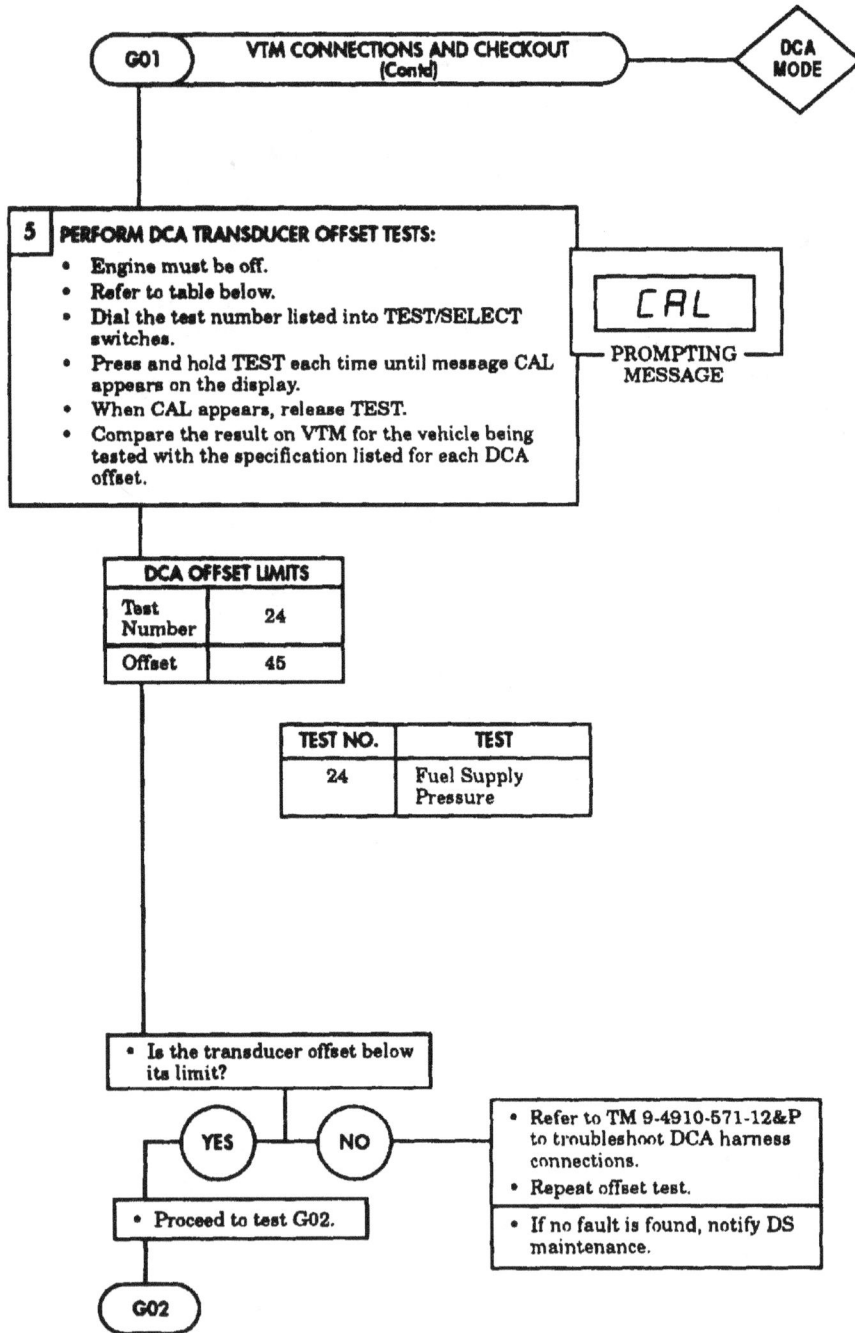

G01 — VTM CONNECTIONS AND CHECKOUT (Contd) — ◇ DCA MODE

5 | PERFORM DCA TRANSDUCER OFFSET TESTS:
- Engine must be off.
- Refer to table below.
- Dial the test number listed into TEST/SELECT switches.
- Press and hold TEST each time until message CAL appears on the display.
- When CAL appears, release TEST.
- Compare the result on VTM for the vehicle being tested with the specification listed for each DCA offset.

`CAL` — PROMPTING MESSAGE

DCA OFFSET LIMITS

Test Number	24
Offset	45

TEST NO.	TEST
24	Fuel Supply Pressure

- Is the transducer offset below its limit?

YES → Proceed to test G02.

NO →
- Refer to TM 9-4910-571-12&P to troubleshoot DCA harness connections.
- Repeat offset test.
- If no fault is found, notify DS maintenance.

G02

Table 2-5. STE/ICE GO-Chain Tests (Contd).

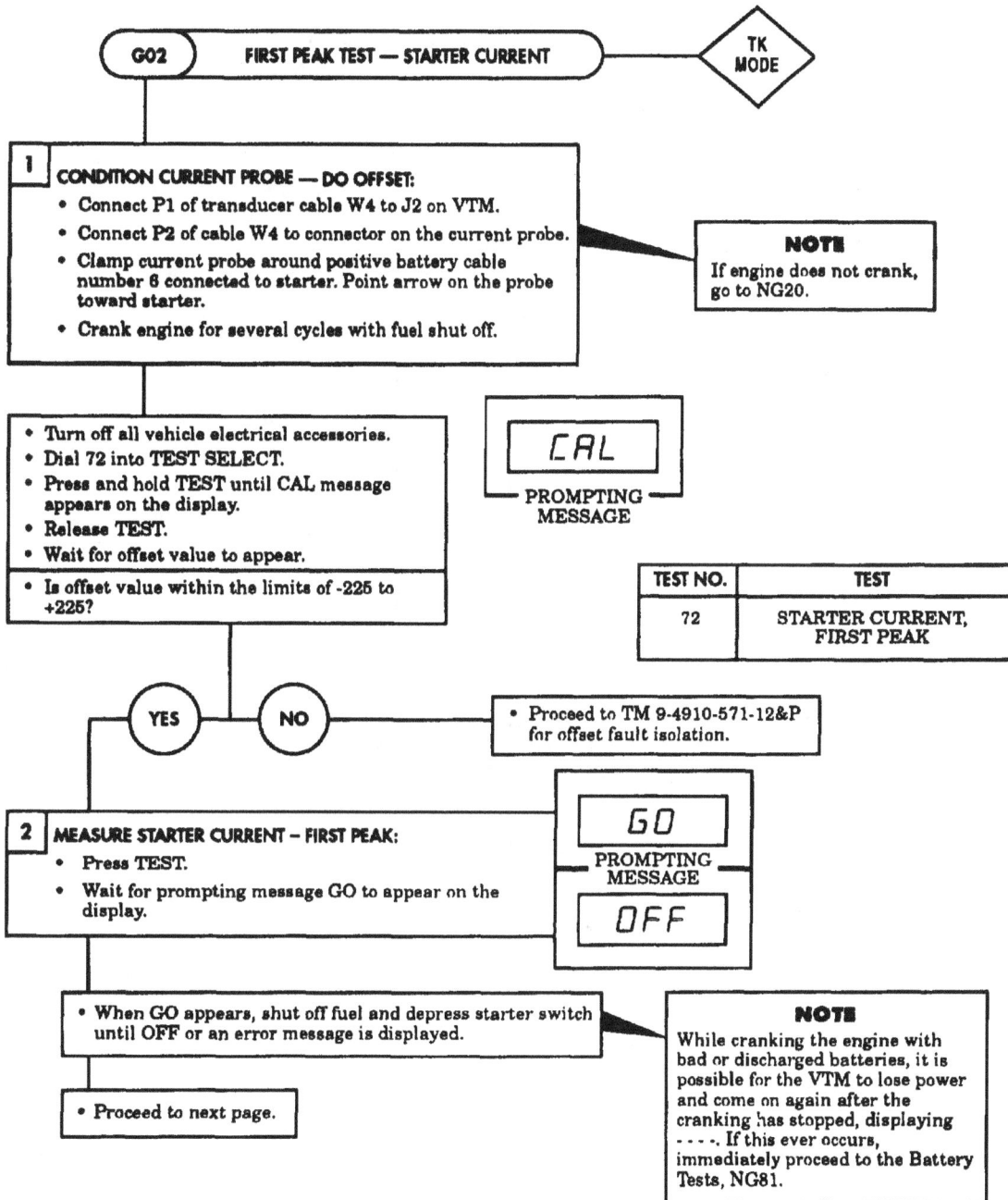

```
( GO2 )    FIRST PEAK TEST — STARTER CURRENT           < TK MODE >
```

1 CONDITION CURRENT PROBE — DO OFFSET:
- Connect P1 of transducer cable W4 to J2 on VTM.
- Connect P2 of cable W4 to connector on the current probe.
- Clamp current probe around positive battery cable number 6 connected to starter. Point arrow on the probe toward starter.
- Crank engine for several cycles with fuel shut off.

NOTE
If engine does not crank, go to NG20.

- Turn off all vehicle electrical accessories.
- Dial 72 into TEST SELECT.
- Press and hold TEST until CAL message appears on the display.
- Release TEST.
- Wait for offset value to appear.

- Is offset value within the limits of -225 to +225?

```
[ CAL ]
— PROMPTING —
MESSAGE
```

TEST NO.	TEST
72	STARTER CURRENT, FIRST PEAK

(YES)   (NO)   - Proceed to TM 9-4910-571-12&P for offset fault isolation.

2 MEASURE STARTER CURRENT – FIRST PEAK:
- Press TEST.
- Wait for prompting message GO to appear on the display.

```
[ GO ]
PROMPTING
MESSAGE
[ OFF ]
```

- When GO appears, shut off fuel and depress starter switch until OFF or an error message is displayed.

NOTE
While cranking the engine with bad or discharged batteries, it is possible for the VTM to lose power and come on again after the cranking has stopped, displaying - - - -. If this ever occurs, immediately proceed to the Battery Tests, NG81.

- Proceed to next page.

Table 2-5. STE/ICE GO-Chain Tests (Contd).

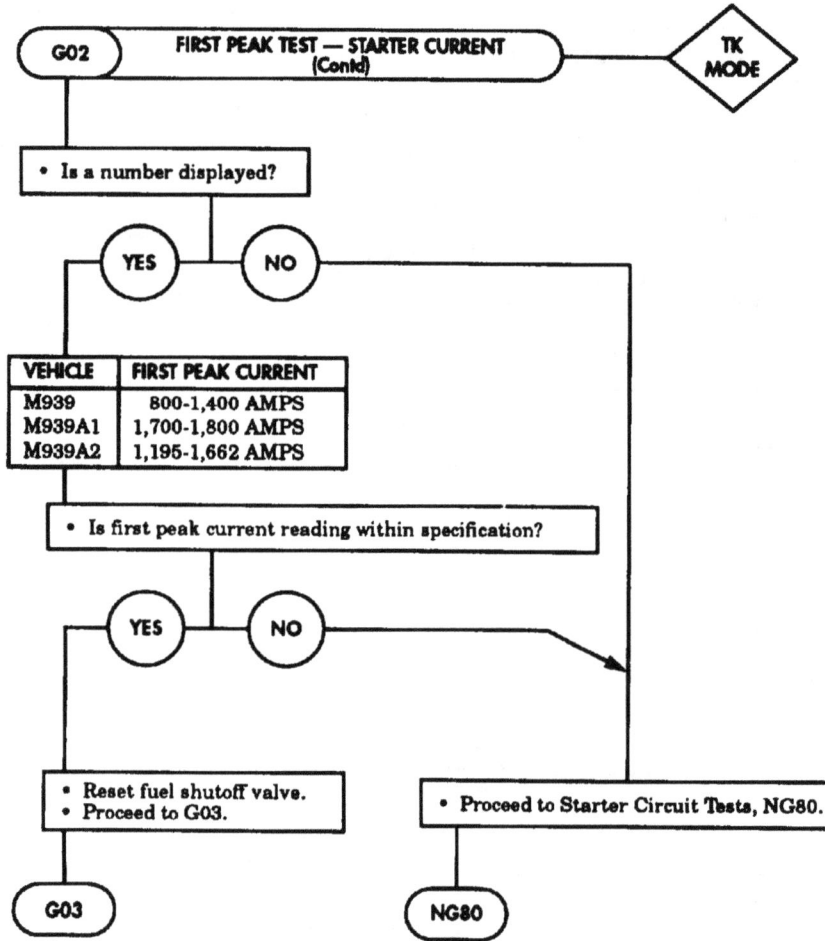

G02 — FIRST PEAK TEST — STARTER CURRENT (Contd) — TK MODE

- Is a number displayed?

YES / NO

VEHICLE	FIRST PEAK CURRENT
M939	800-1,400 AMPS
M939A1	1,700-1,800 AMPS
M939A2	1,195-1,662 AMPS

- Is first peak current reading within specification?

YES / NO

- Reset fuel shutoff valve.
- Proceed to G03.

- Proceed to Starter Circuit Tests, NG80.

G03

NG80

Table 2-5. STE/ICE GO-Chain Tests (Contd).

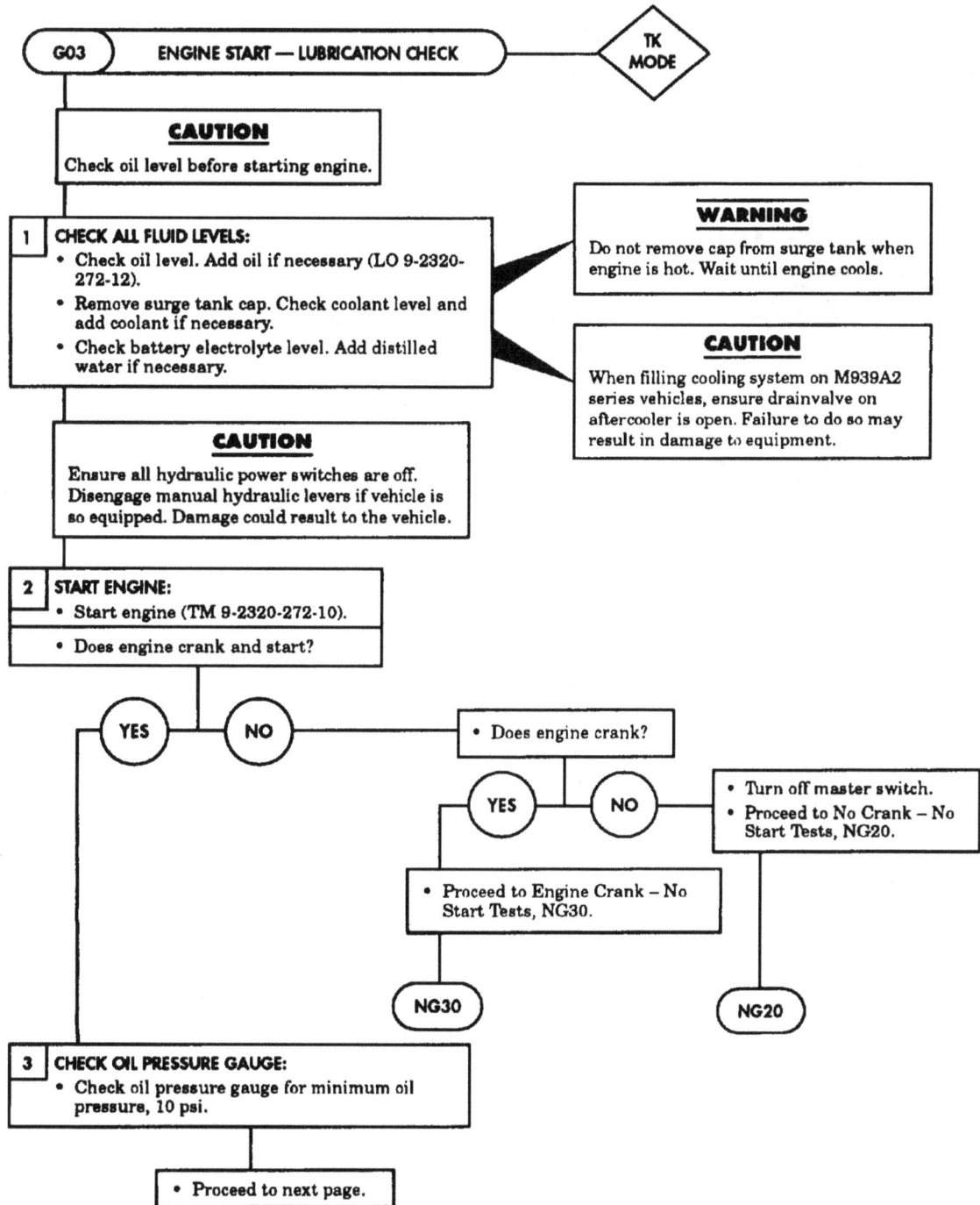

| G03 | ENGINE START — LUBRICATION CHECK | TK MODE |

CAUTION
Check oil level before starting engine.

1 CHECK ALL FLUID LEVELS:
- Check oil level. Add oil if necessary (LO 9-2320-272-12).
- Remove surge tank cap. Check coolant level and add coolant if necessary.
- Check battery electrolyte level. Add distilled water if necessary.

WARNING
Do not remove cap from surge tank when engine is hot. Wait until engine cools.

CAUTION
When filling cooling system on M939A2 series vehicles, ensure drainvalve on aftercooler is open. Failure to do so may result in damage to equipment.

CAUTION
Ensure all hydraulic power switches are off. Disengage manual hydraulic levers if vehicle is so equipped. Damage could result to the vehicle.

2 START ENGINE:
- Start engine (TM 9-2320-272-10).
- Does engine crank and start?

YES   NO

- Does engine crank?

YES   NO

- Turn off master switch.
- Proceed to No Crank – No Start Tests, NG20.

- Proceed to Engine Crank – No Start Tests, NG30.

NG30   NG20

3 CHECK OIL PRESSURE GAUGE:
- Check oil pressure gauge for minimum oil pressure, 10 psi.

- Proceed to next page.

2-227

Table 2-5. STE/ICE GO-Chain Tests (Contd).

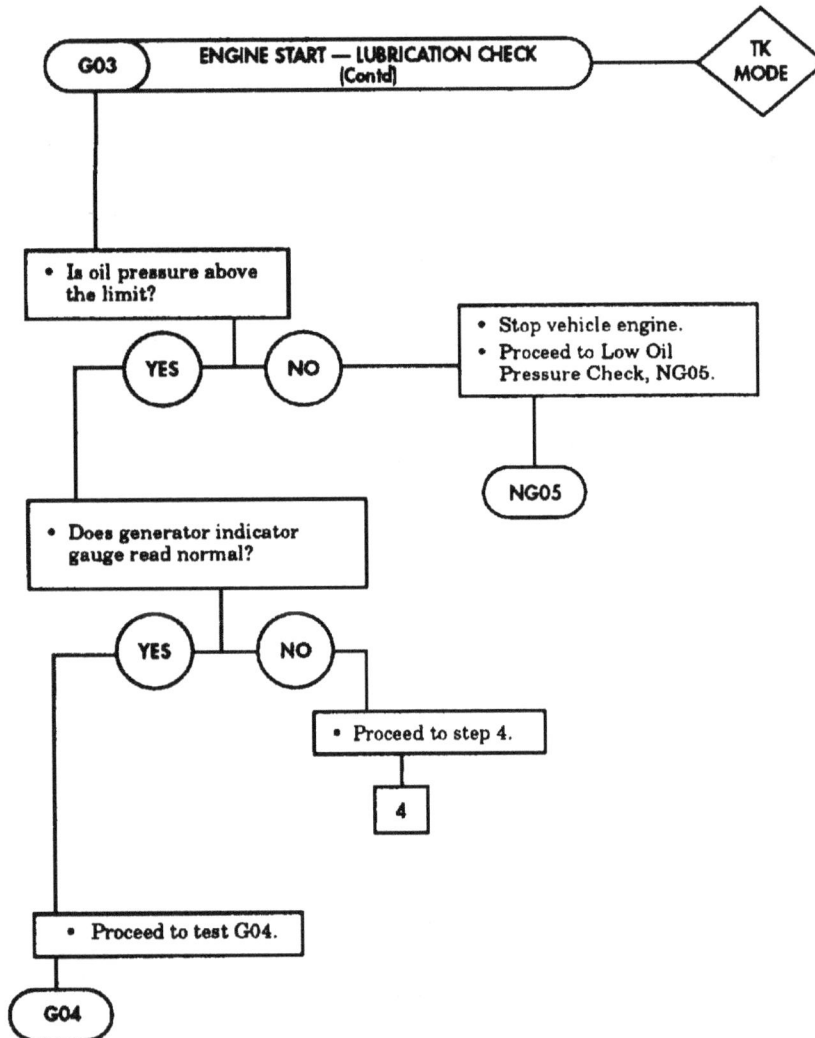

```
┌──────────────────────────────────────────┐        ╱╲
( G03   ENGINE START — LUBRICATION CHECK    )──────╱ TK ╲
└──────────────────────────────(Contd)──────┘      ╲MODE╱
      │                                              ╲╱
      │
      │
  ┌──────────────────────┐
  │ • Is oil pressure above │
  │   the limit?          │
  └──────────────────────┘
         │        │
       ┌───┐   ┌───┐         ┌─────────────────────────┐
       │YES│   │NO │─────────│ • Stop vehicle engine.   │
       └───┘   └───┘         │ • Proceed to Low Oil     │
         │                   │   Pressure Check, NG05.  │
         │                   └─────────────────────────┘
         │                            │
         │                         ( NG05 )
  ┌──────────────────────┐
  │ • Does generator indicator │
  │   gauge read normal?  │
  └──────────────────────┘
         │        │
       ┌───┐   ┌───┐
       │YES│   │NO │
       └───┘   └───┘
         │        │
         │     ┌─────────────────────┐
         │     │ • Proceed to step 4. │
         │     └─────────────────────┘
         │              │
         │            ┌───┐
         │            │ 4 │
         │            └───┘
  ┌─────────────────────┐
  │ • Proceed to test G04. │
  └─────────────────────┘
         │
      ( G04 )
```

Table 2-5. STE/ICE GO-Chain Tests (Contd).

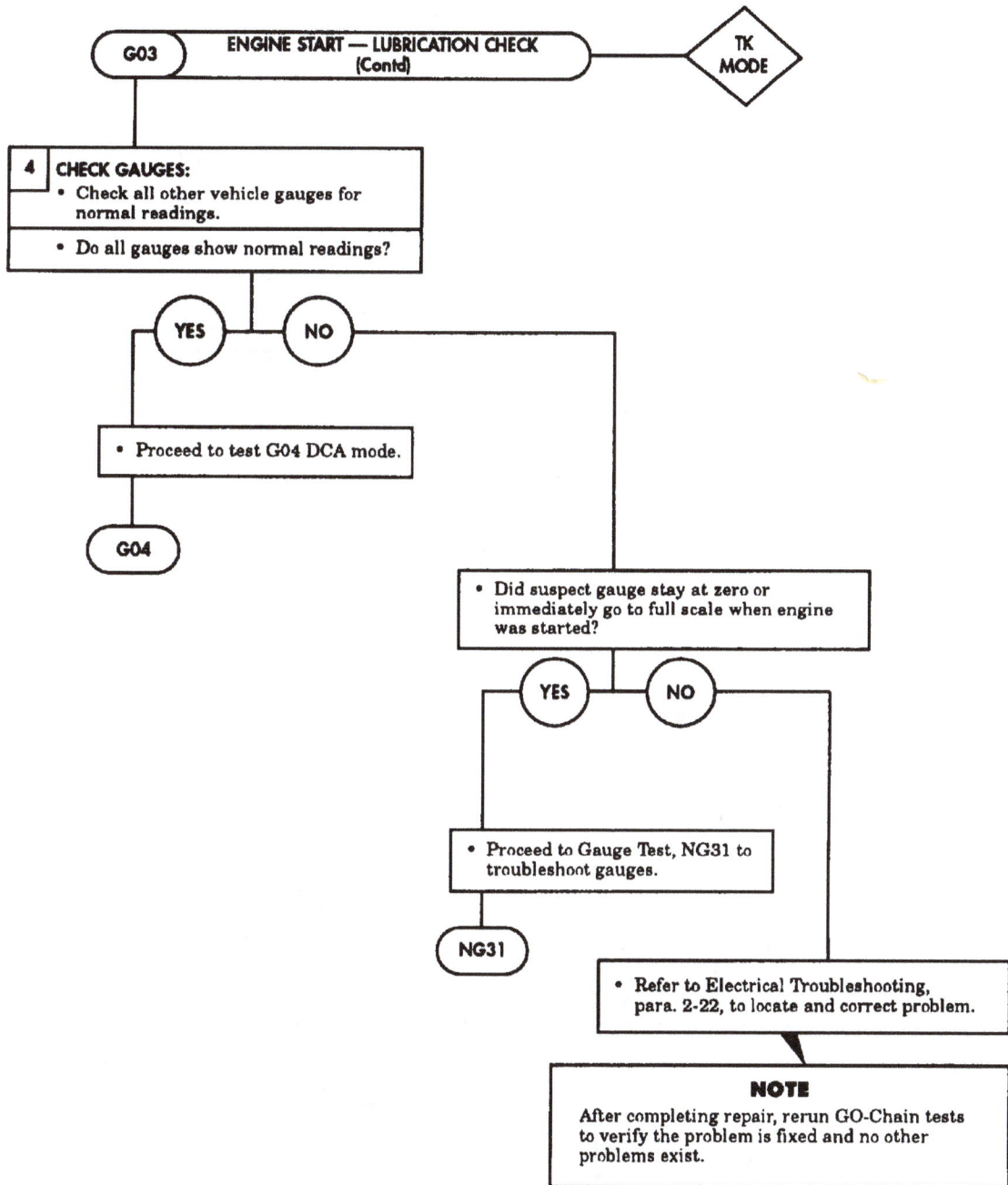

G03 — ENGINE START — LUBRICATION CHECK (Contd) — **TK MODE**

4 CHECK GAUGES:
- Check all other vehicle gauges for normal readings.
- Do all gauges show normal readings?

YES **NO**

- Proceed to test G04 DCA mode.

G04

- Did suspect gauge stay at zero or immediately go to full scale when engine was started?

YES **NO**

- Proceed to Gauge Test, NG31 to troubleshoot gauges.

NG31

- Refer to Electrical Troubleshooting, para. 2-22, to locate and correct problem.

NOTE
After completing repair, rerun GO-Chain tests to verify the problem is fixed and no other problems exist.

Table 2-5. STE/ICE GO-Chain Tests (Contd).

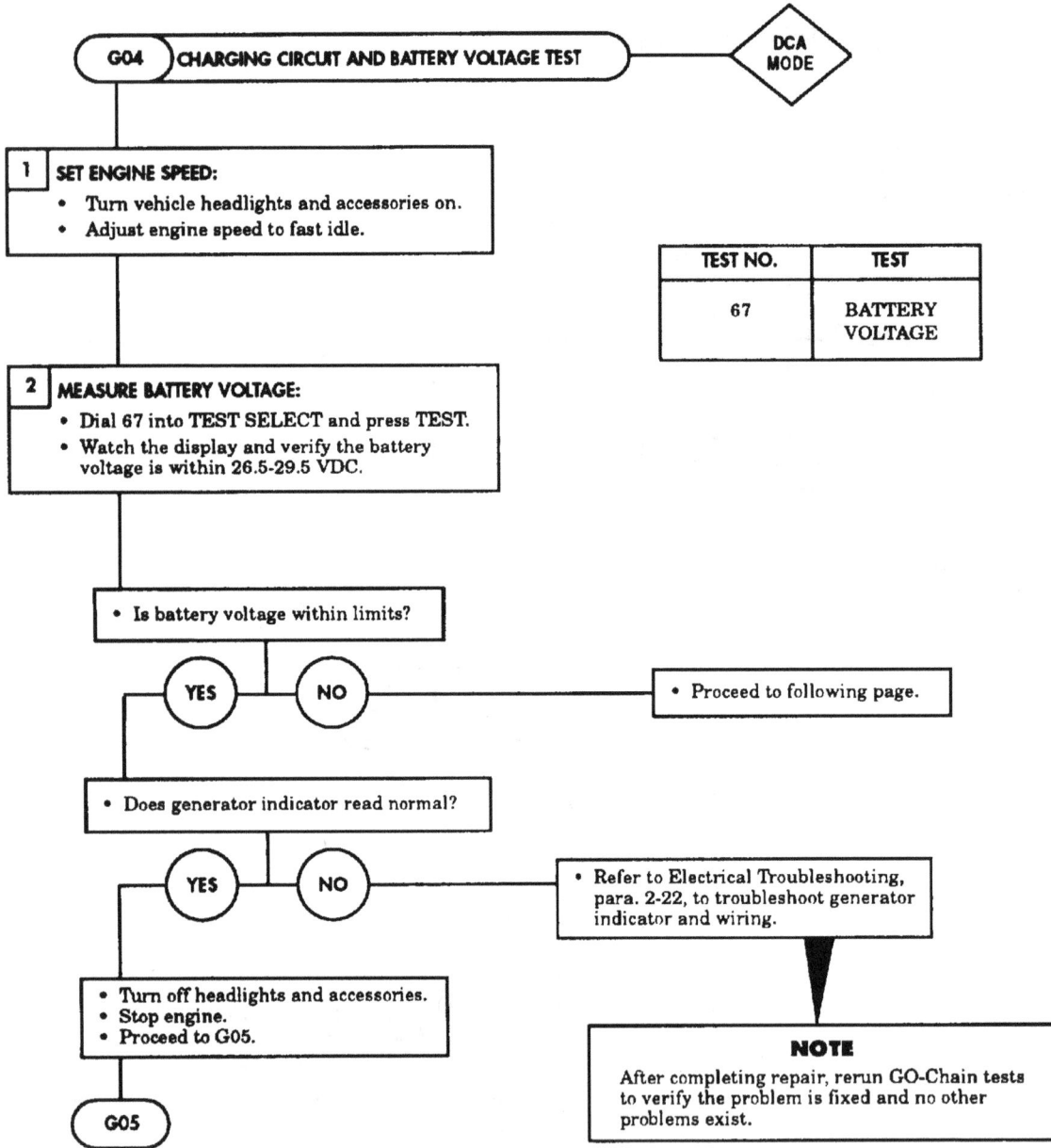

GO4) CHARGING CIRCUIT AND BATTERY VOLTAGE TEST — DCA MODE

1 SET ENGINE SPEED:
- Turn vehicle headlights and accessories on.
- Adjust engine speed to fast idle.

TEST NO.	TEST
67	BATTERY VOLTAGE

2 MEASURE BATTERY VOLTAGE:
- Dial 67 into TEST SELECT and press TEST.
- Watch the display and verify the battery voltage is within 26.5-29.5 VDC.

- Is battery voltage within limits?

YES   NO → • Proceed to following page.

- Does generator indicator read normal?

YES   NO → • Refer to Electrical Troubleshooting, para. 2-22, to troubleshoot generator indicator and wiring.

- Turn off headlights and accessories.
- Stop engine.
- Proceed to G05.

G05

NOTE

After completing repair, rerun GO-Chain tests to verify the problem is fixed and no other problems exist.

Table 2-5. STE/ICE GO-Chain Tests (Contd).

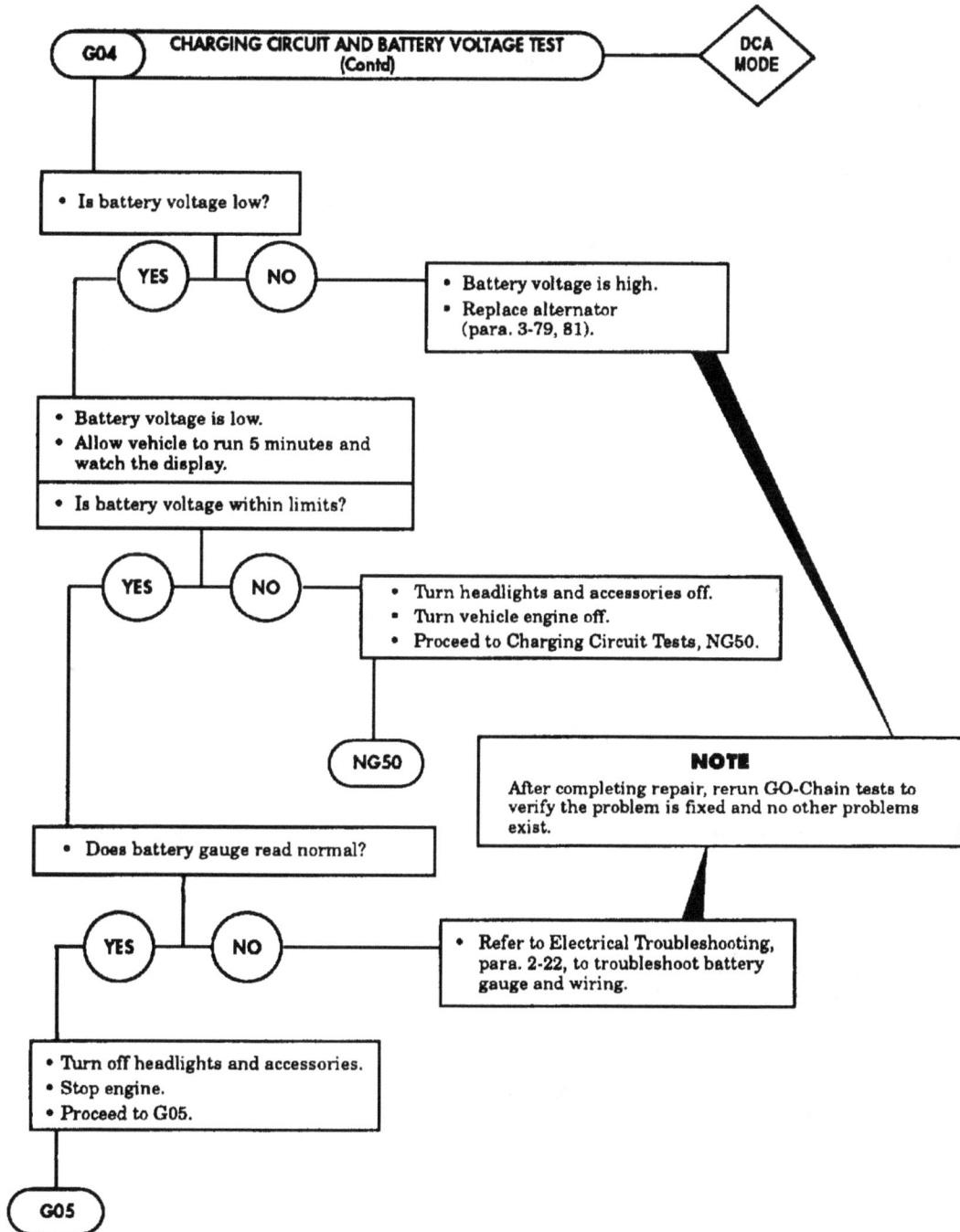

G04   CHARGING CIRCUIT AND BATTERY VOLTAGE TEST
(Contd)   DCA
                                                     MODE

• Is battery voltage low?

YES   NO

• Battery voltage is high.
• Replace alternator
  (para. 3-79, 81).

• Battery voltage is low.
• Allow vehicle to run 5 minutes and
  watch the display.

• Is battery voltage within limits?

YES   NO

• Turn headlights and accessories off.
• Turn vehicle engine off.
• Proceed to Charging Circuit Tests, NG50.

NG50

NOTE
After completing repair, rerun GO-Chain tests to
verify the problem is fixed and no other problems
exist.

• Does battery gauge read normal?

YES   NO

• Refer to Electrical Troubleshooting,
  para. 2-22, to troubleshoot battery
  gauge and wiring.

• Turn off headlights and accessories.
• Stop engine.
• Proceed to G05.

G05

Table 2-5. STE/ICE GO-Chain Tests (Contd).

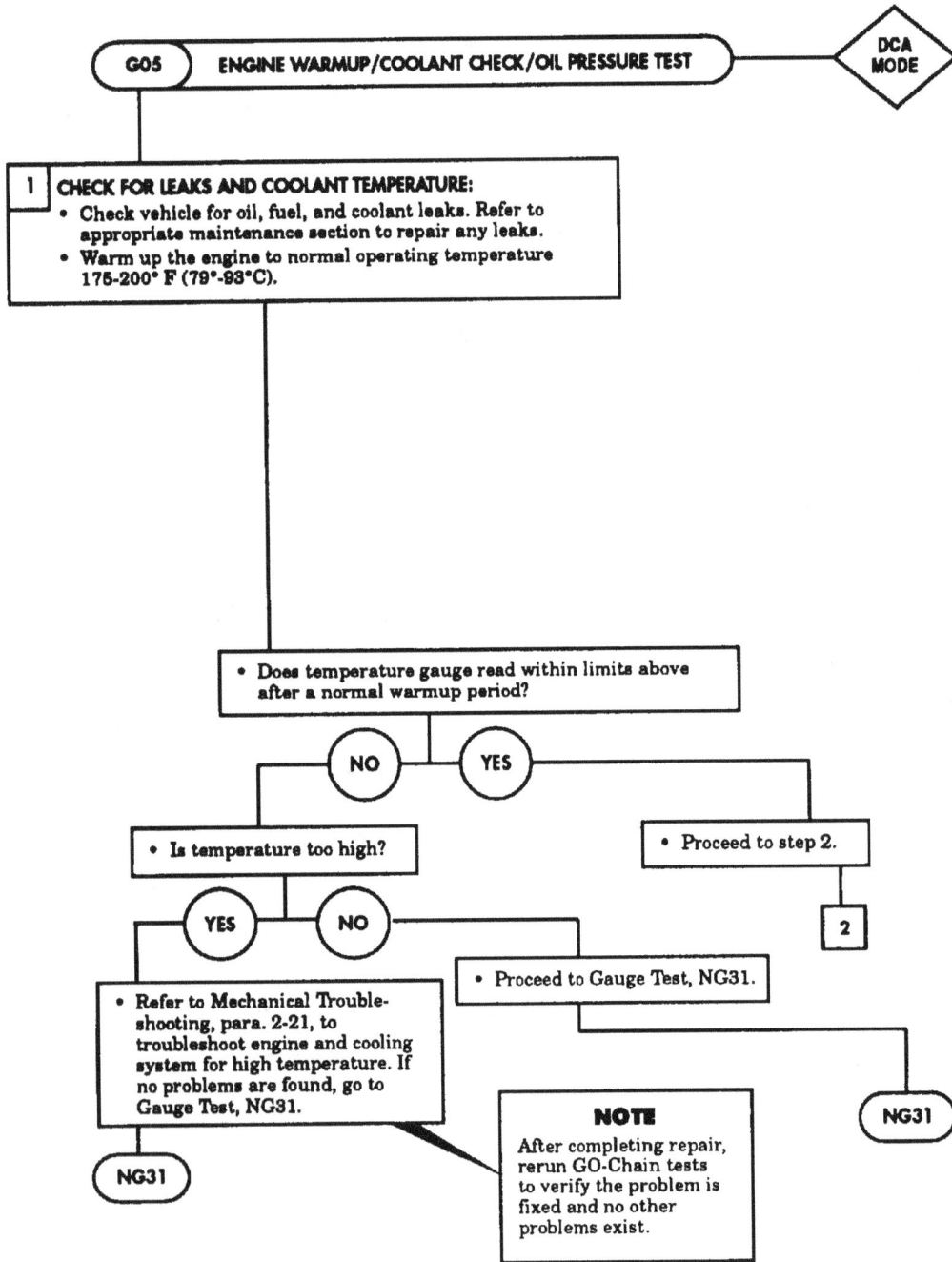

```
┌─────────┐   ┌──────────────────────────────────────────────────┐      ◇ DCA
( G05 )   │ ENGINE WARMUP/COOLANT CHECK/OIL PRESSURE TEST │        MODE
└─────────┘   └──────────────────────────────────────────────────┘
```

1 | **CHECK FOR LEAKS AND COOLANT TEMPERATURE:**
- Check vehicle for oil, fuel, and coolant leaks. Refer to appropriate maintenance section to repair any leaks.
- Warm up the engine to normal operating temperature 175-200° F (79°-93°C).

- Does temperature gauge read within limits above after a normal warmup period?

(NO)   (YES)

- Is temperature too high?

- Proceed to step 2.

2

(YES)   (NO)

- Refer to Mechanical Trouble-shooting, para. 2-21, to troubleshoot engine and cooling system for high temperature. If no problems are found, go to Gauge Test, NG31.

- Proceed to Gauge Test, NG31.

NOTE

After completing repair, rerun GO-Chain tests to verify the problem is fixed and no other problems exist.

(NG31)

(NG31)

Table 2-5. STE/ICE GO-Chain Tests (Contd).

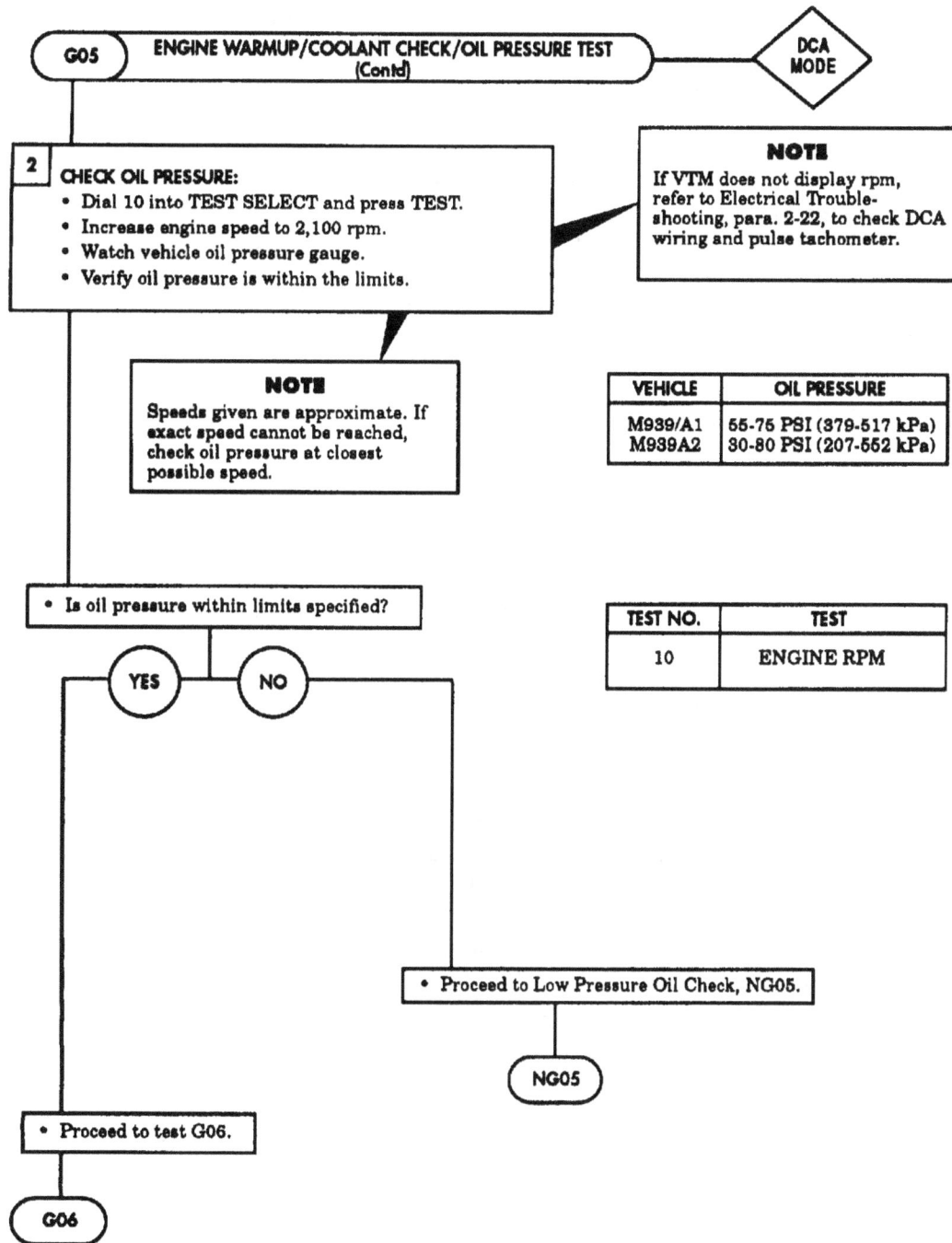

```
( G05 )──[ ENGINE WARMUP/COOLANT CHECK/OIL PRESSURE TEST ]───◇ DCA
                              (Contd)                            MODE
```

2 CHECK OIL PRESSURE:
  - Dial 10 into TEST SELECT and press TEST.
  - Increase engine speed to 2,100 rpm.
  - Watch vehicle oil pressure gauge.
  - Verify oil pressure is within the limits.

NOTE

If VTM does not display rpm, refer to Electrical Trouble-shooting, para. 2-22, to check DCA wiring and pulse tachometer.

NOTE

Speeds given are approximate. If exact speed cannot be reached, check oil pressure at closest possible speed.

VEHICLE	OIL PRESSURE
M939/A1	65-75 PSI (379-517 kPa)
M939A2	30-80 PSI (207-552 kPa)

- Is oil pressure within limits specified?

(YES)   (NO)

TEST NO.	TEST
10	ENGINE RPM

- Proceed to Low Pressure Oil Check, NG05.

(NG05)

- Proceed to test G06.

(G06)

Table 2-5. STE/ICE GO-Chain Tests (Contd).

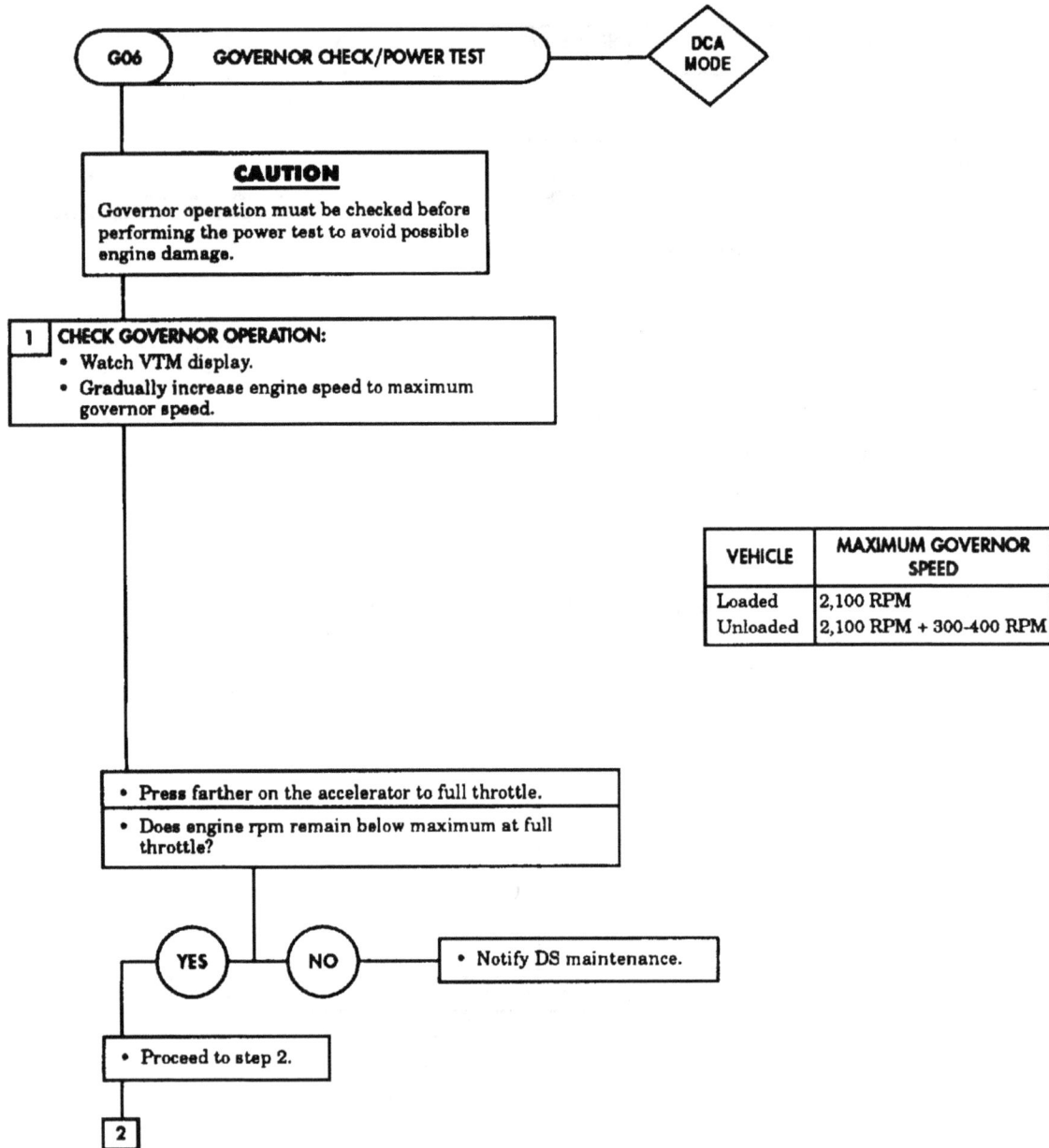

```
( G06 )   GOVERNOR CHECK/POWER TEST          < DCA
                                               MODE >
```

CAUTION

Governor operation must be checked before
performing the power test to avoid possible
engine damage.

1 CHECK GOVERNOR OPERATION:
  • Watch VTM display.
  • Gradually increase engine speed to maximum
    governor speed.

VEHICLE	MAXIMUM GOVERNOR SPEED
Loaded	2,100 RPM
Unloaded	2,100 RPM + 300-400 RPM

  • Press farther on the accelerator to full throttle.
  • Does engine rpm remain below maximum at full throttle?

(YES)   (NO)   • Notify DS maintenance.

  • Proceed to step 2.

2

Table 2-5. STE/ICE GO-Chain Tests (Contd).

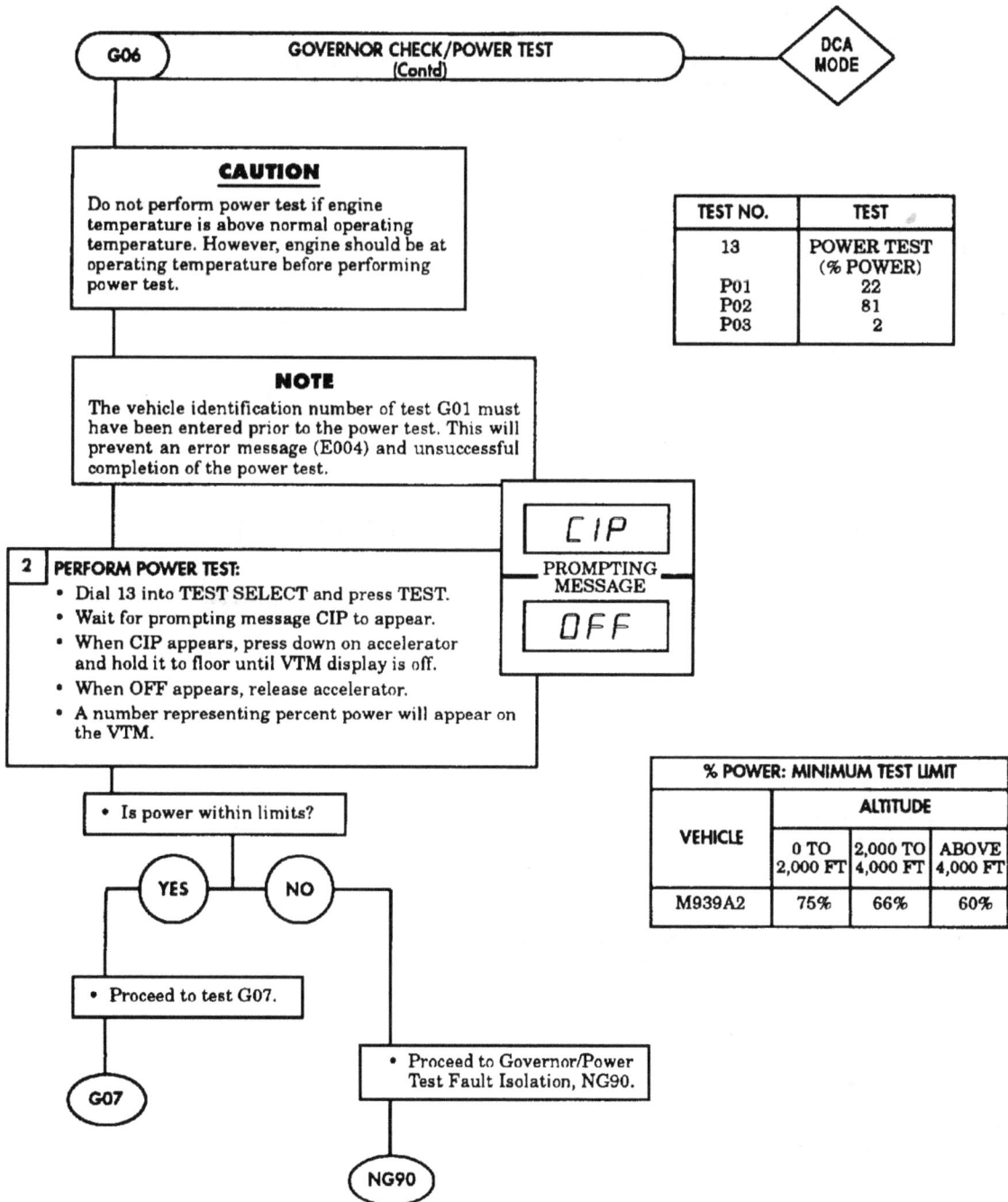

G06 — GOVERNOR CHECK/POWER TEST (Contd) — **DCA MODE**

CAUTION

Do not perform power test if engine temperature is above normal operating temperature. However, engine should be at operating temperature before performing power test.

TEST NO.	TEST
13	POWER TEST (% POWER)
P01	22
P02	81
P03	2

NOTE

The vehicle identification number of test G01 must have been entered prior to the power test. This will prevent an error message (E004) and unsuccessful completion of the power test.

```
CIP
```
PROMPTING MESSAGE
```
OFF
```

2 PERFORM POWER TEST:
- Dial 13 into TEST SELECT and press TEST.
- Wait for prompting message CIP to appear.
- When CIP appears, press down on accelerator and hold it to floor until VTM display is off.
- When OFF appears, release accelerator.
- A number representing percent power will appear on the VTM.

- Is power within limits?

YES → • Proceed to test G07. → **G07**

NO → • Proceed to Governor/Power Test Fault Isolation, NG90. → **NG90**

% POWER: MINIMUM TEST LIMIT			
VEHICLE	ALTITUDE		
	0 TO 2,000 FT	2,000 TO 4,000 FT	ABOVE 4,000 FT
M939A2	75%	66%	60%

Table 2-5. STE/ICE GO-Chain Tests (Contd).

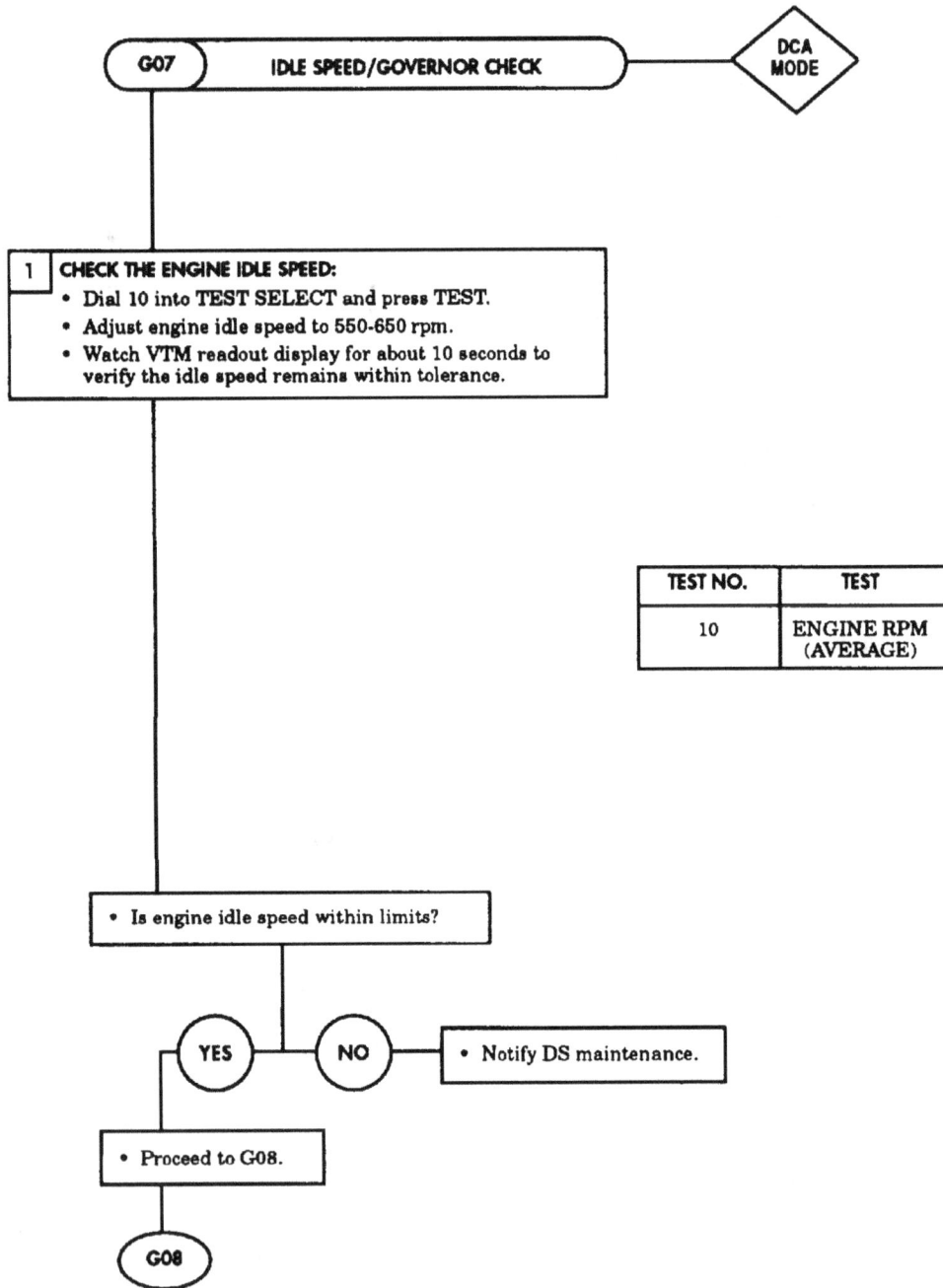

G07 — IDLE SPEED/GOVERNOR CHECK — DCA MODE

1 CHECK THE ENGINE IDLE SPEED:
- Dial 10 into TEST SELECT and press TEST.
- Adjust engine idle speed to 550-650 rpm.
- Watch VTM readout display for about 10 seconds to verify the idle speed remains within tolerance.

TEST NO.	TEST
10	ENGINE RPM (AVERAGE)

- Is engine idle speed within limits?

YES   NO → • Notify DS maintenance.

- Proceed to G08.

G08

Table 2-5. STE/ICE GO-Chain Tests (Contd).

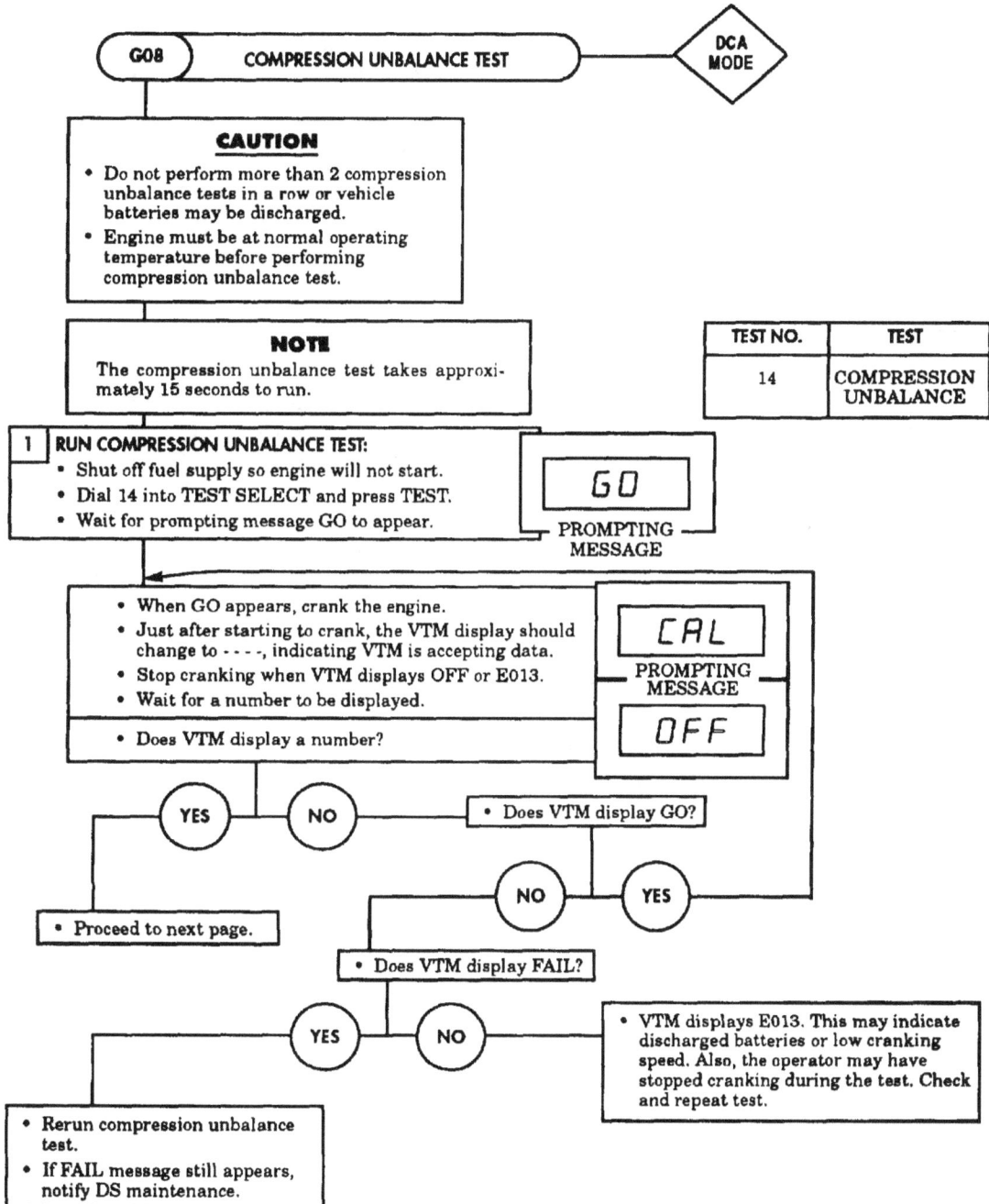

Table 2-5. STE/ICE GO-Chain Tests (Contd).

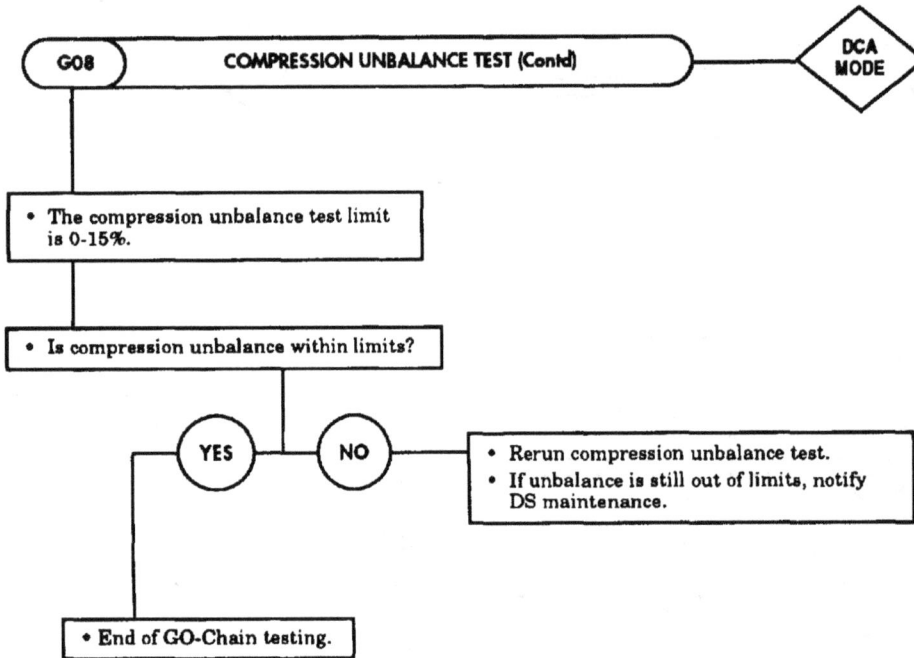

```
┌──────────────────────────────────────────────────┐        ◇
│ GO8    COMPRESSION UNBALANCE TEST (Contd)          │───── DCA
└──────────────────────────────────────────────────┘      MODE
   │
┌──────────────────────────────────────┐
│ • The compression unbalance test limit │
│   is 0-15%.                             │
└──────────────────────────────────────┘
   │
┌──────────────────────────────────────┐
│ • Is compression unbalance within limits? │
└──────────────────────────────────────┘
        │
    (YES)   (NO) ──┌──────────────────────────────────────┐
                   │ • Rerun compression unbalance test.    │
                   │ • If unbalance is still out of limits, │
                   │   notify DS maintenance.               │
                   └──────────────────────────────────────┘
   │
┌──────────────────────────┐
│ • End of GO-Chain testing. │
└──────────────────────────┘
```

Table 2-5. STE/ICE GO-Chain Tests (Contd).

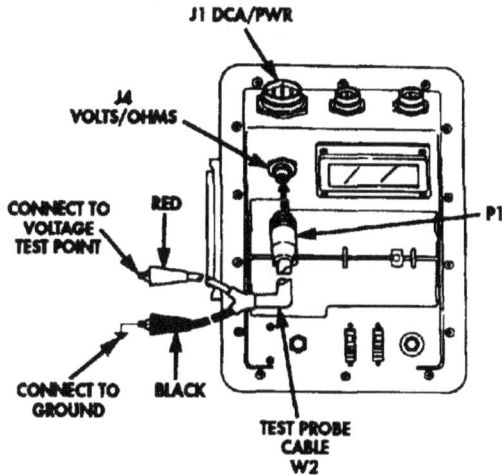

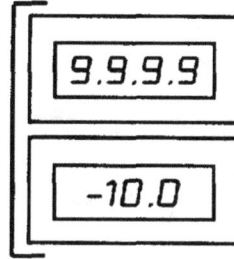

THIS DISPLAY INDICATES VOLTAGE MEASURED IS GREATER THAN 45 VOLTS.

IF TESTING A POSITIVE (+) VOLTAGE, NEGATIVE (-) DISPLAY INDICATES TEST LEADS ARE REVERSED. REVERSE LEADS FOR (+) READING.

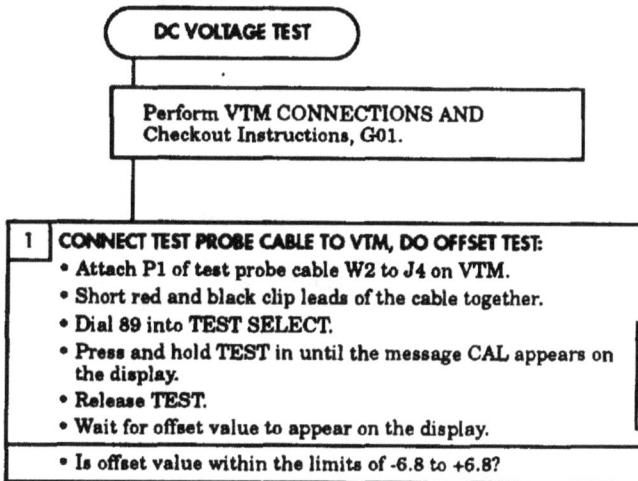

TEST NO.	TEST
89	0-45 VOLTS DC
27	FUEL SOLENOID VOLTAGE
82	ALT/GEN OUTPUT VOLTAGE
84	ALT/GEN NEGATIVE CABLE VOLTAGE DROP

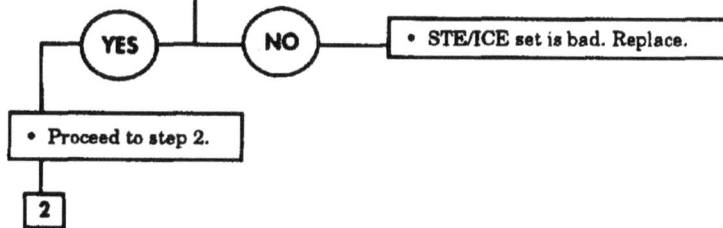

DC VOLTAGE TEST

Perform VTM CONNECTIONS AND Checkout Instructions, G01.

1 CONNECT TEST PROBE CABLE TO VTM, DO OFFSET TEST:
• Attach P1 of test probe cable W2 to J4 on VTM.
• Short red and black clip leads of the cable together.
• Dial 89 into TEST SELECT.
• Press and hold TEST in until the message CAL appears on the display.
• Release TEST.
• Wait for offset value to appear on the display.

• Is offset value within the limits of -6.8 to +6.8?

CAL

PROMPTING MESSAGE

YES

NO → • STE/ICE set is bad. Replace.

• Proceed to step 2.

2

Table 2-5. STE/ICE GO-Chain Tests (Contd).

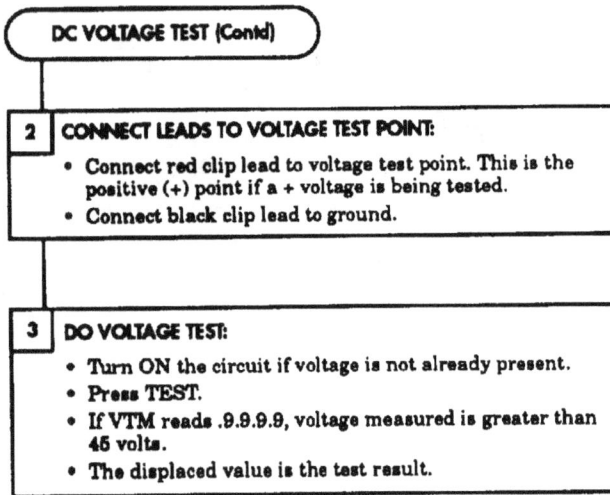

┌─────────────────────────────────────┐
│ DC VOLTAGE TEST (Contd)   │
└─────────────────────────────────────┘

┌───┬──┐
│ 2 │ CONNECT LEADS TO VOLTAGE TEST POINT: │
│   │ • Connect red clip lead to voltage test point. This is the │
│   │   positive (+) point if a + voltage is being tested. │
│   │ • Connect black clip lead to ground.   │
└───┴──┘

┌───┬──┐
│ 3 │ DO VOLTAGE TEST: │
│   │ • Turn ON the circuit if voltage is not already present. │
│   │ • Press TEST.   │
│   │ • If VTM reads .9.9.9.9, voltage measured is greater than │
│   │   45 volts.   │
│   │ • The displaced value is the test result.   │
└───┴──┘

DC VOLTAGE (TEST 89). This test allows VTM to be used as a voltmeter. The voltmeter is automatically ranged (autoranged) through three voltage ranges: 0-0.5 volt, 0-4.5 volts, and 0-45 volts DC. The decimal point will automatically move to the correct position. Each time you want to make a measurement, connect red lead to positive (+) side and black lead to negative (-) side of cable or item being tested. If polarity is reversed, you will get a minus (-) sign in the readout; however, the numerical value is correct.

Table 2-6. STE/ICE NO-GO-Chain Tests (Contd).

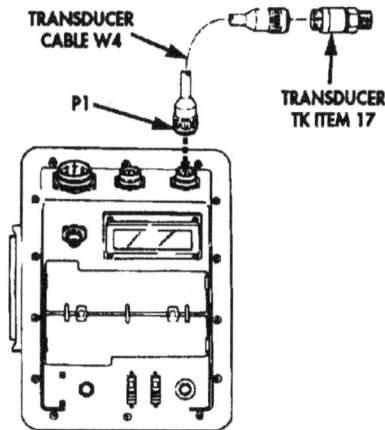

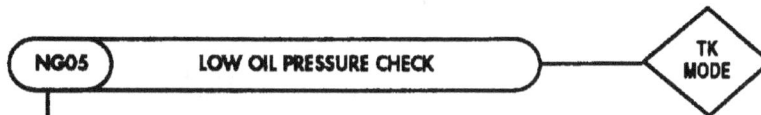

TEST NO.	TEST
01	INTERLEAVE
50	0-1,000 PSIG PRESSURE

NG05 — LOW OIL PRESSURE CHECK — TK MODE

1 INSTALL TRANSDUCER — DO OFFSET:
- Stop vehicle engine.
- Remove oil pressure sending unit.
- Install pressure transducer TK item 17 (blue stripe) in place of sending unit on engine.
- Connect P1 of transducer cable W4 to J1 or J2 on VTM.
- Connect P2 of transducer cable to connector on pressure transducer.

- Dial 50 into TEST SELECT.
- Press and hold TEST until CAL message appears on display.
- Release TEST.
- Wait for offset value to appear on the display.
- Is offset value within the limits -150 to +150?

CAL
PROMPTING MESSAGE

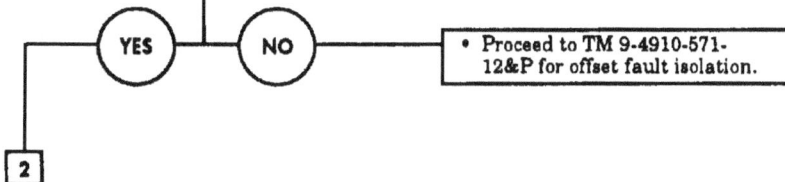

YES   NO

- Proceed to TM 9-4910-571-12&P for offset fault isolation.

2

Table 2-6. STE/ICE NO-GO-Chain Tests (Contd).

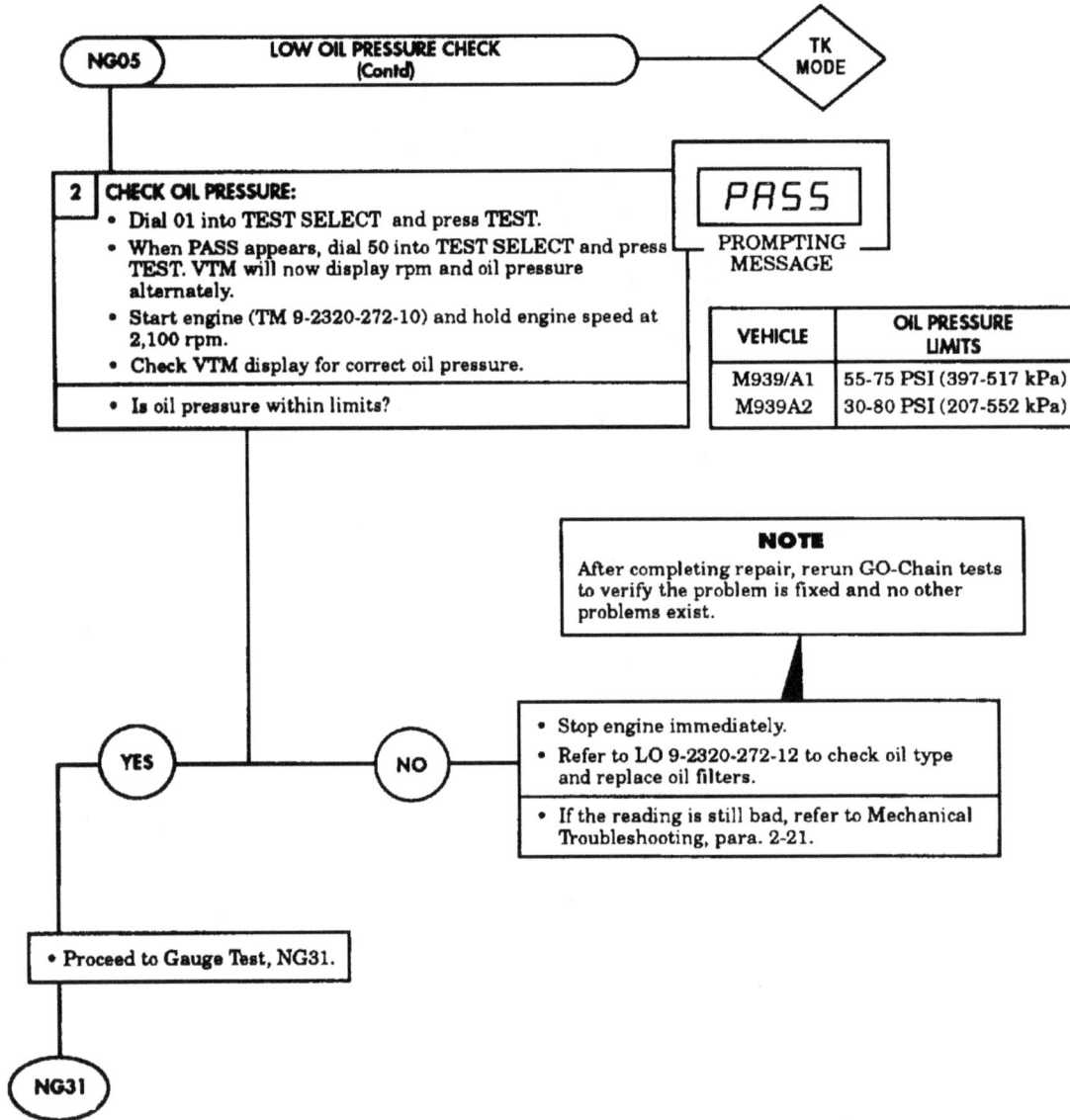

NG05 — LOW OIL PRESSURE CHECK (Contd) — TK MODE

2 CHECK OIL PRESSURE:
- Dial 01 into **TEST SELECT** and press **TEST**.
- When PASS appears, dial 50 into **TEST SELECT** and press **TEST**. VTM will now display rpm and oil pressure alternately.
- Start engine (TM 9-2320-272-10) and hold engine speed at 2,100 rpm.
- Check VTM display for correct oil pressure.

- Is oil pressure within limits?

```
PASS
```
PROMPTING MESSAGE

VEHICLE	OIL PRESSURE LIMITS
M939/A1	55-75 PSI (397-517 kPa)
M939A2	30-80 PSI (207-552 kPa)

NOTE
After completing repair, rerun GO-Chain tests to verify the problem is fixed and no other problems exist.

YES

NO
- Stop engine immediately.
- Refer to LO 9-2320-272-12 to check oil type and replace oil filters.
- If the reading is still bad, refer to Mechanical Troubleshooting, para. 2-21.

- Proceed to Gauge Test, NG31.

NG31

Table 2-6. STE/ICE NO-GO-Chain Tests (Contd).

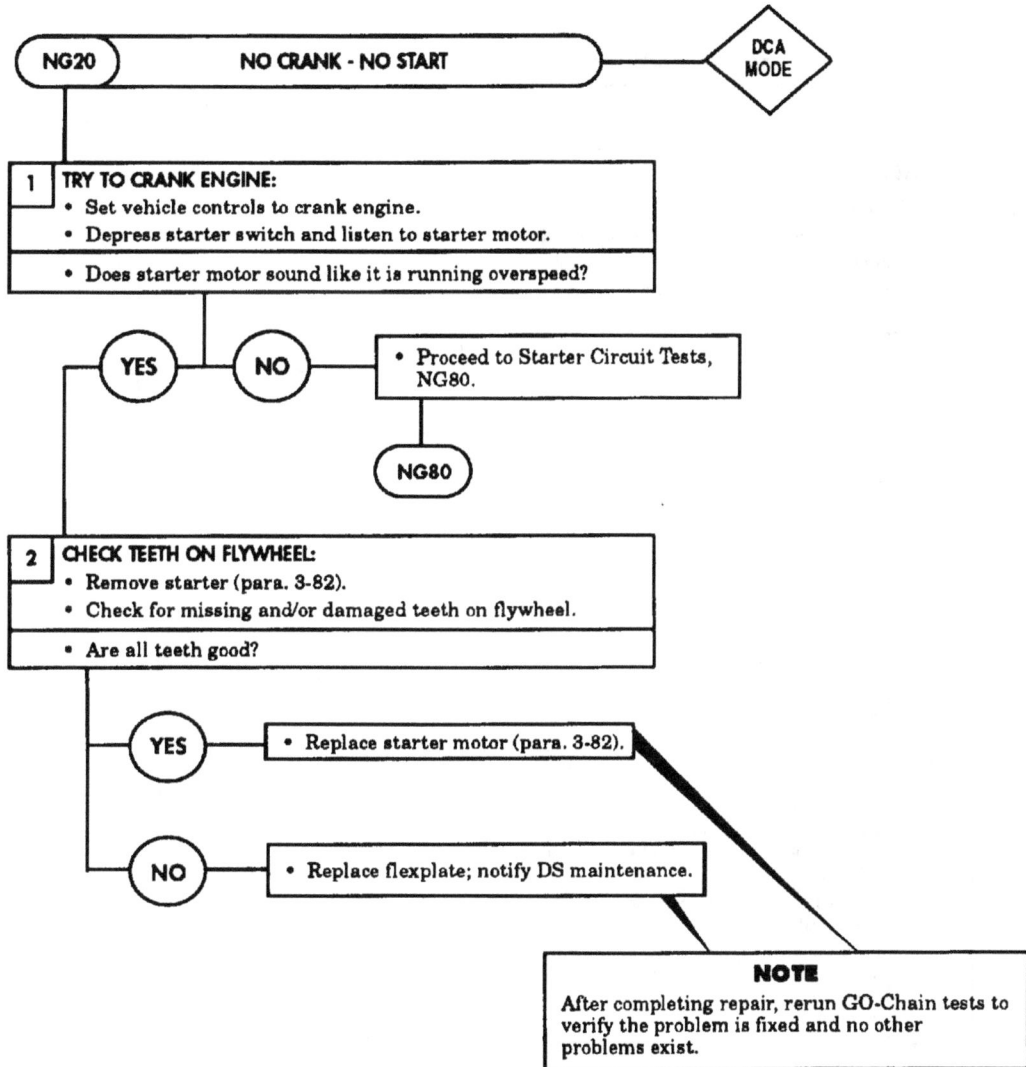

NG20 — NO CRANK - NO START — DCA MODE

1 TRY TO CRANK ENGINE:
- Set vehicle controls to crank engine.
- Depress starter switch and listen to starter motor.
- Does starter motor sound like it is running overspeed?

YES

NO — • Proceed to Starter Circuit Tests, NG80.

NG80

2 CHECK TEETH ON FLYWHEEL:
- Remove starter (para. 3-82).
- Check for missing and/or damaged teeth on flywheel.
- Are all teeth good?

YES — • Replace starter motor (para. 3-82).

NO — • Replace flexplate; notify DS maintenance.

NOTE
After completing repair, rerun GO-Chain tests to verify the problem is fixed and no other problems exist.

Table 2-6. STE/ICE NO-GO-Chain Tests (Contd).

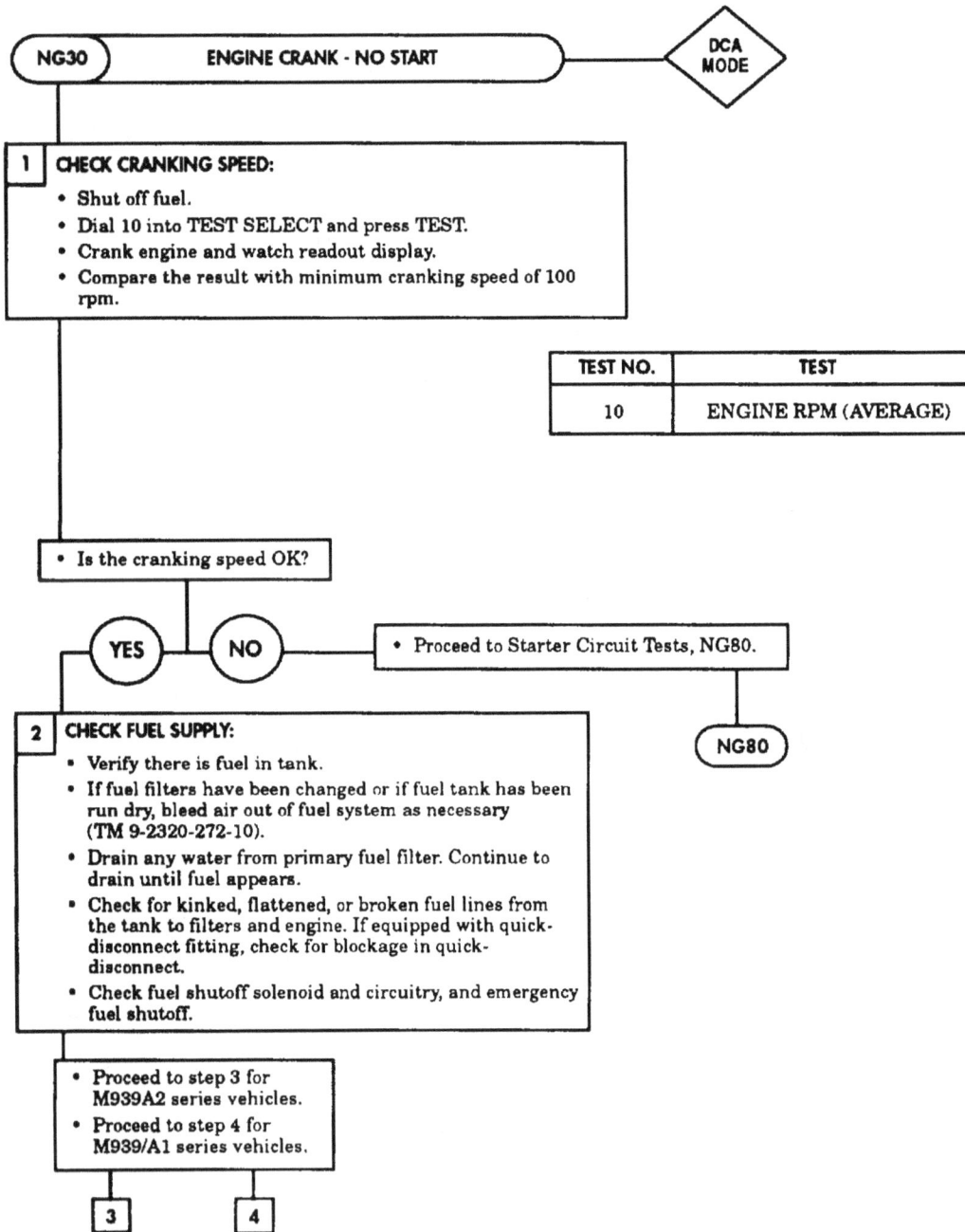

(NG30)   **ENGINE CRANK - NO START**   ◇ DCA MODE ◇

1 **CHECK CRANKING SPEED:**
- Shut off fuel.
- Dial 10 into TEST SELECT and press TEST.
- Crank engine and watch readout display.
- Compare the result with minimum cranking speed of 100 rpm.

TEST NO.	TEST
10	ENGINE RPM (AVERAGE)

- Is the cranking speed OK?

(YES)   (NO)   • Proceed to Starter Circuit Tests, NG80.

(NG80)

2 **CHECK FUEL SUPPLY:**
- Verify there is fuel in tank.
- If fuel filters have been changed or if fuel tank has been run dry, bleed air out of fuel system as necessary (TM 9-2320-272-10).
- Drain any water from primary fuel filter. Continue to drain until fuel appears.
- Check for kinked, flattened, or broken fuel lines from the tank to filters and engine. If equipped with quick-disconnect fitting, check for blockage in quick-disconnect.
- Check fuel shutoff solenoid and circuitry, and emergency fuel shutoff.

- Proceed to step 3 for M939A2 series vehicles.
- Proceed to step 4 for M939/A1 series vehicles.

[3]   [4]

Table 2-6. STE/ICE NO-GO-Chain Tests (Contd).

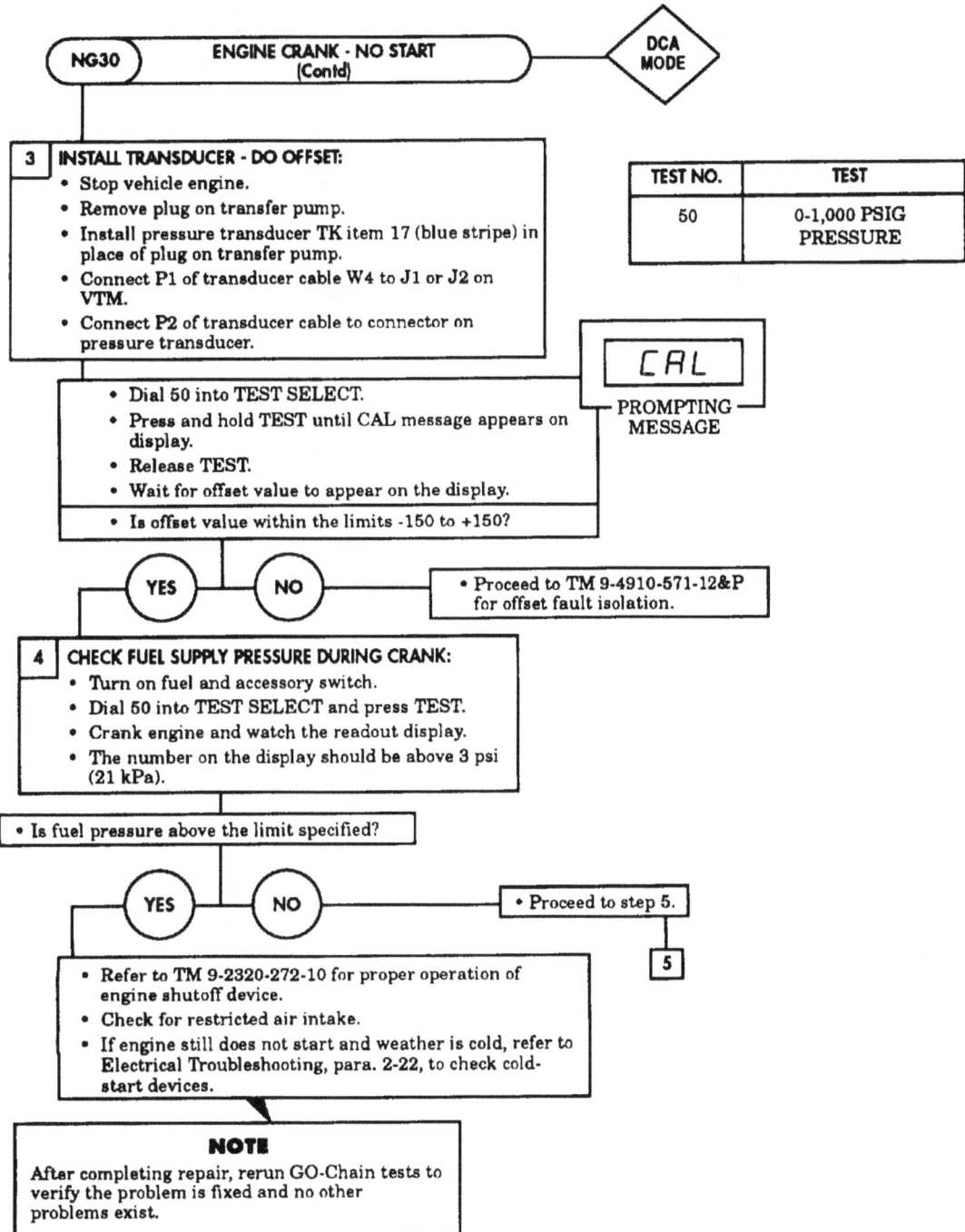

NG30 — ENGINE CRANK - NO START (Contd) — **DCA MODE**

TEST NO.	TEST
50	0-1,000 PSIG PRESSURE

3 | INSTALL TRANSDUCER - DO OFFSET:
- Stop vehicle engine.
- Remove plug on transfer pump.
- Install pressure transducer TK item 17 (blue stripe) in place of plug on transfer pump.
- Connect P1 of transducer cable W4 to J1 or J2 on VTM.
- Connect P2 of transducer cable to connector on pressure transducer.

[CAL] — PROMPTING MESSAGE

- Dial 50 into TEST SELECT.
- Press and hold TEST until CAL message appears on display.
- Release TEST.
- Wait for offset value to appear on the display.
- Is offset value within the limits -150 to +150?

YES / **NO**

- Proceed to TM 9-4910-571-12&P for offset fault isolation.

4 | CHECK FUEL SUPPLY PRESSURE DURING CRANK:
- Turn on fuel and accessory switch.
- Dial 50 into TEST SELECT and press TEST.
- Crank engine and watch the readout display.
- The number on the display should be above 3 psi (21 kPa).

- Is fuel pressure above the limit specified?

YES / **NO**

- Proceed to step 5.

5

- Refer to TM 9-2320-272-10 for proper operation of engine shutoff device.
- Check for restricted air intake.
- If engine still does not start and weather is cold, refer to Electrical Troubleshooting, para. 2-22, to check cold-start devices.

NOTE

After completing repair, rerun GO-Chain tests to verify the problem is fixed and no other problems exist.

Table 2-6. STE/ICE NO-GO-Chain Tests (Contd).

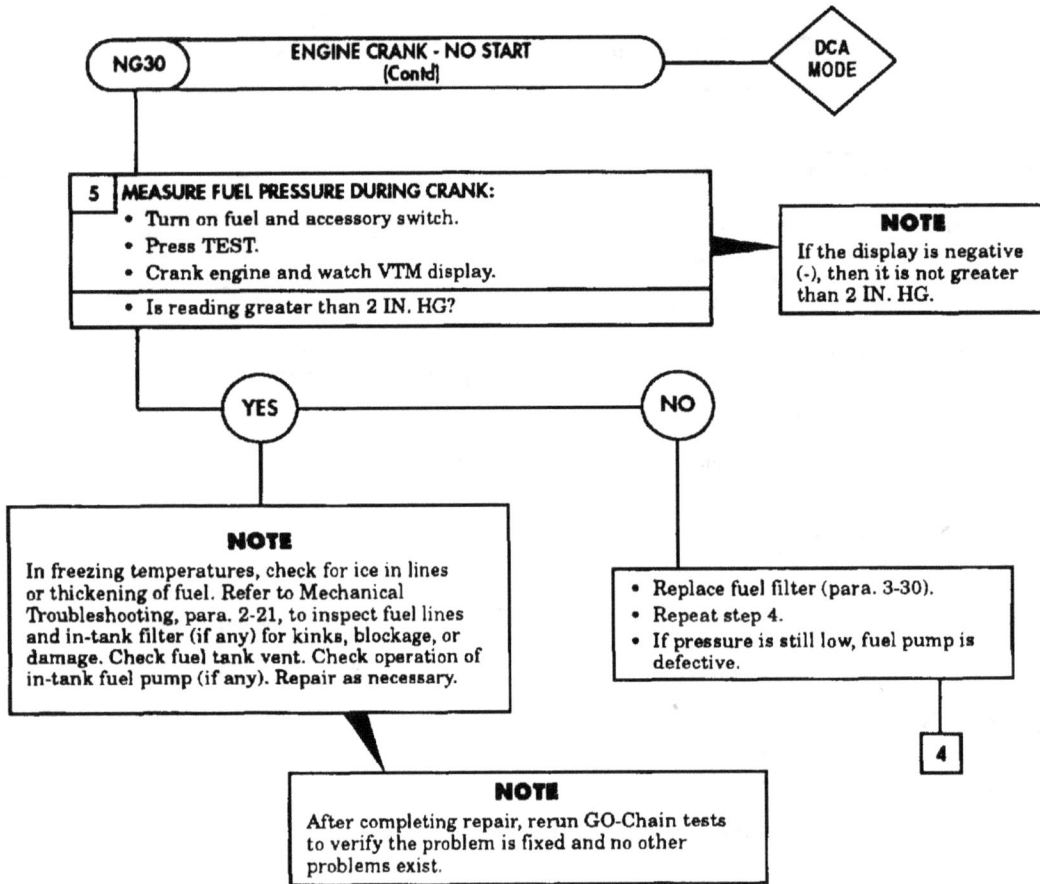

NG30 — ENGINE CRANK - NO START (Contd) — DCA MODE

5 **MEASURE FUEL PRESSURE DURING CRANK:**
- Turn on fuel and accessory switch.
- Press TEST.
- Crank engine and watch VTM display.
- Is reading greater than 2 IN. HG?

NOTE
If the display is negative (-), then it is not greater than 2 IN. HG.

YES

NO

NOTE
In freezing temperatures, check for ice in lines or thickening of fuel. Refer to Mechanical Troubleshooting, para. 2-21, to inspect fuel lines and in-tank filter (if any) for kinks, blockage, or damage. Check fuel tank vent. Check operation of in-tank fuel pump (if any). Repair as necessary.

- Replace fuel filter (para. 3-30).
- Repeat step 4.
- If pressure is still low, fuel pump is defective.

4

NOTE
After completing repair, rerun GO-Chain tests to verify the problem is fixed and no other problems exist.

Table 2-6. STE/ICE NO-GO-Chain Tests (Contd).

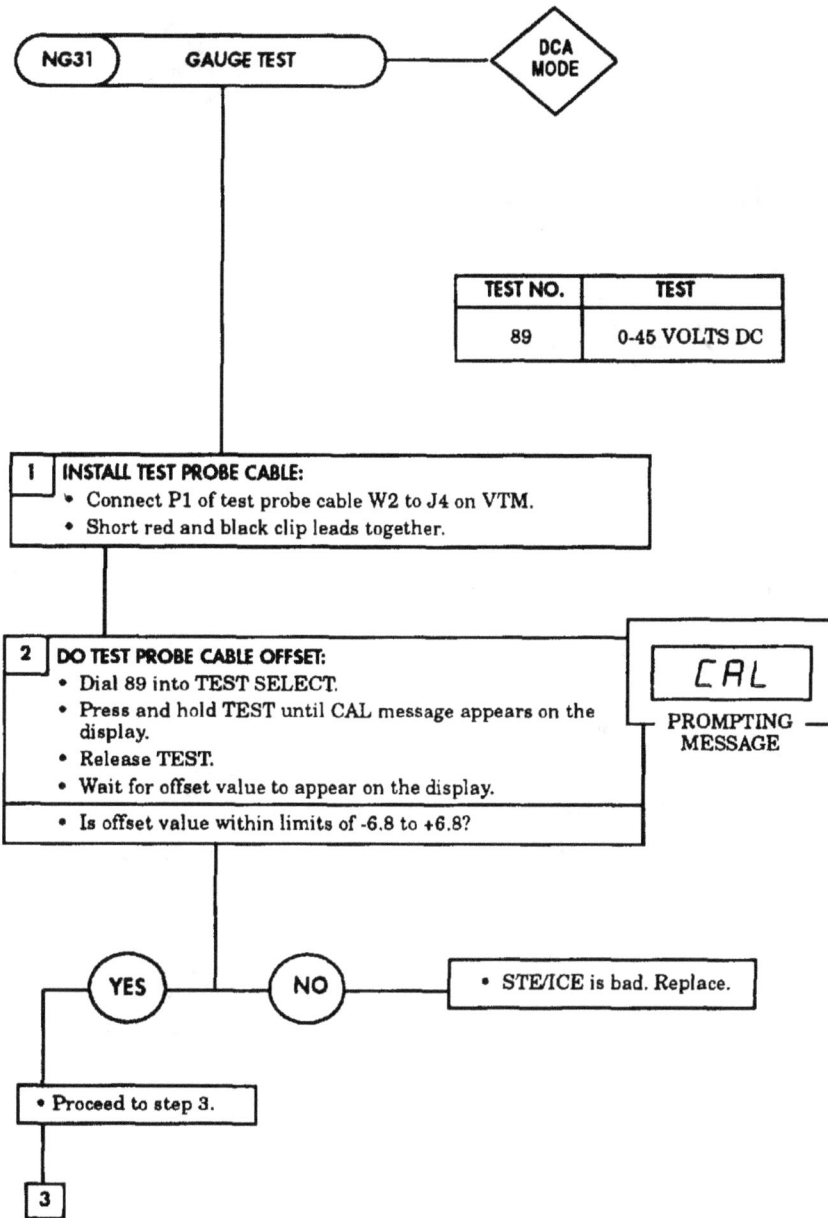

TEST NO.	TEST
89	0-45 VOLTS DC

NG31   GAUGE TEST   DCA MODE

1 | INSTALL TEST PROBE CABLE:
- Connect P1 of test probe cable W2 to J4 on VTM.
- Short red and black clip leads together.

2 | DO TEST PROBE CABLE OFFSET:
- Dial 89 into TEST SELECT.
- Press and hold TEST until CAL message appears on the display.
- Release TEST.
- Wait for offset value to appear on the display.

- Is offset value within limits of -6.8 to +6.8?

CAL

PROMPTING MESSAGE

YES   NO   • STE/ICE is bad. Replace.

• Proceed to step 3.

3

Table 2-6. STE/ICE NO-GO-Chain Tests (Contd).

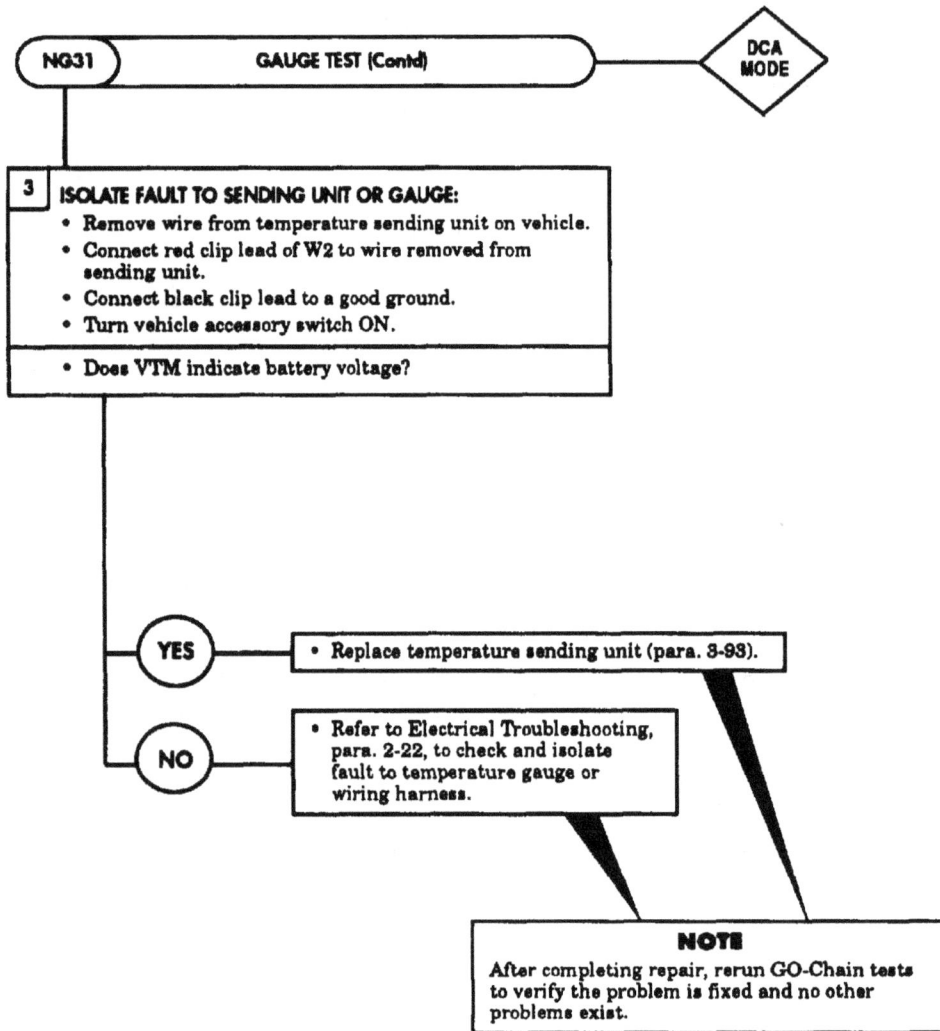

NG31   GAUGE TEST (Contd)   DCA MODE

3 | **ISOLATE FAULT TO SENDING UNIT OR GAUGE:**
- Remove wire from temperature sending unit on vehicle.
- Connect red clip lead of W2 to wire removed from sending unit.
- Connect black clip lead to a good ground.
- Turn vehicle accessory switch ON.

- Does VTM indicate battery voltage?

YES
- Replace temperature sending unit (para. 3-93).

NO
- Refer to Electrical Troubleshooting, para. 2-22, to check and isolate fault to temperature gauge or wiring harness.

NOTE

After completing repair, rerun GO-Chain tests to verify the problem is fixed and no other problems exist.

Table 2-6. STE/ICE NO-GO-Chain Tests (Contd).

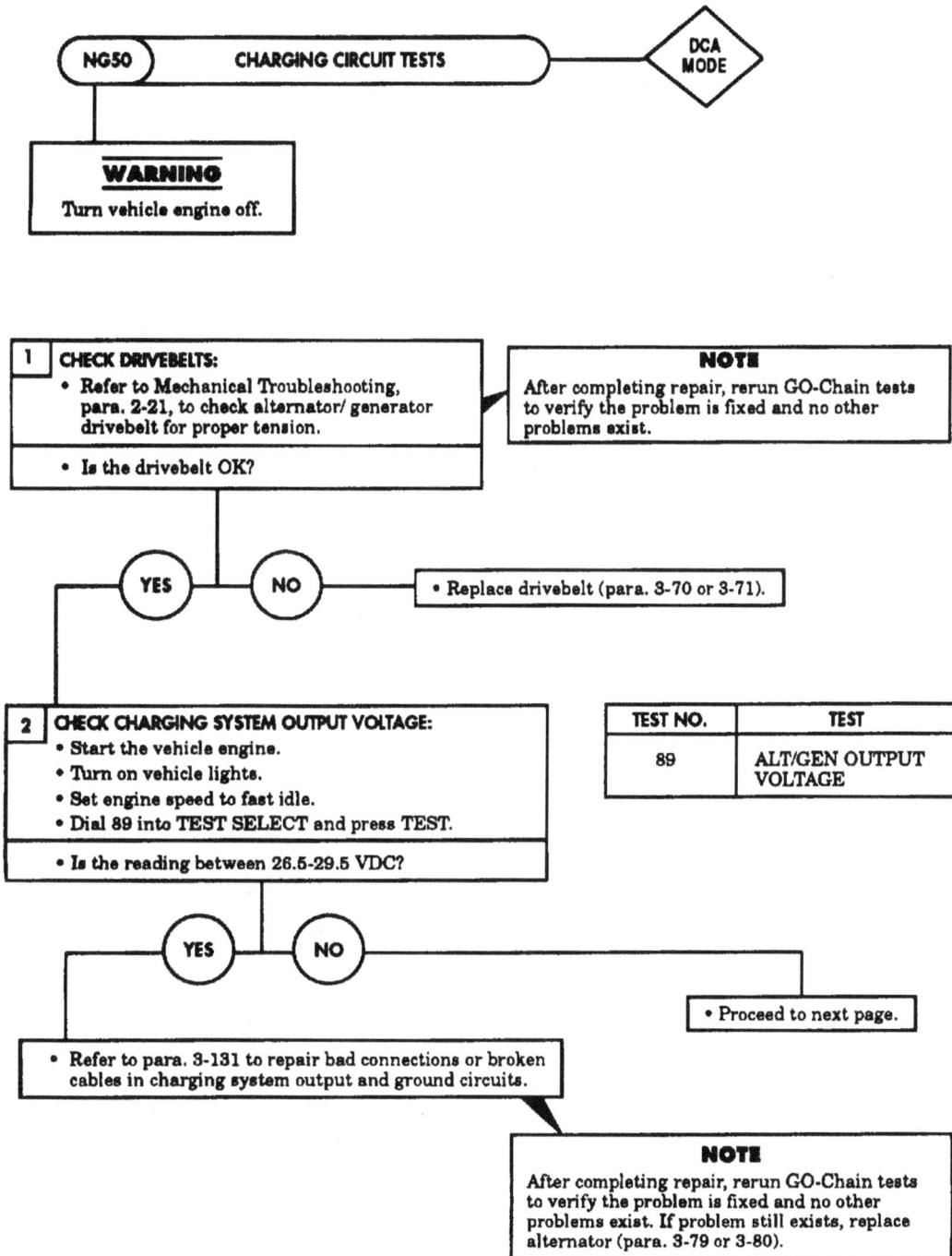

(NG50) CHARGING CIRCUIT TESTS ◇ DCA MODE

WARNING

Turn vehicle engine off.

1 CHECK DRIVEBELTS:
* Refer to Mechanical Troubleshooting, para. 2-21, to check alternator/generator drivebelt for proper tension.
* Is the drivebelt OK?

NOTE
After completing repair, rerun GO-Chain tests to verify the problem is fixed and no other problems exist.

(YES) (NO) • Replace drivebelt (para. 3-70 or 3-71).

2 CHECK CHARGING SYSTEM OUTPUT VOLTAGE:
* Start the vehicle engine.
* Turn on vehicle lights.
* Set engine speed to fast idle.
* Dial 89 into TEST SELECT and press TEST.
* Is the reading between 26.5-29.5 VDC?

TEST NO.	TEST
89	ALT/GEN OUTPUT VOLTAGE

(YES) (NO)

• Proceed to next page.

• Refer to para. 3-131 to repair bad connections or broken cables in charging system output and ground circuits.

NOTE
After completing repair, rerun GO-Chain tests to verify the problem is fixed and no other problems exist. If problem still exists, replace alternator (para. 3-79 or 3-80).

2-249

Table 2-6. STE/ICE NO-GO-Chain Tests (Contd).

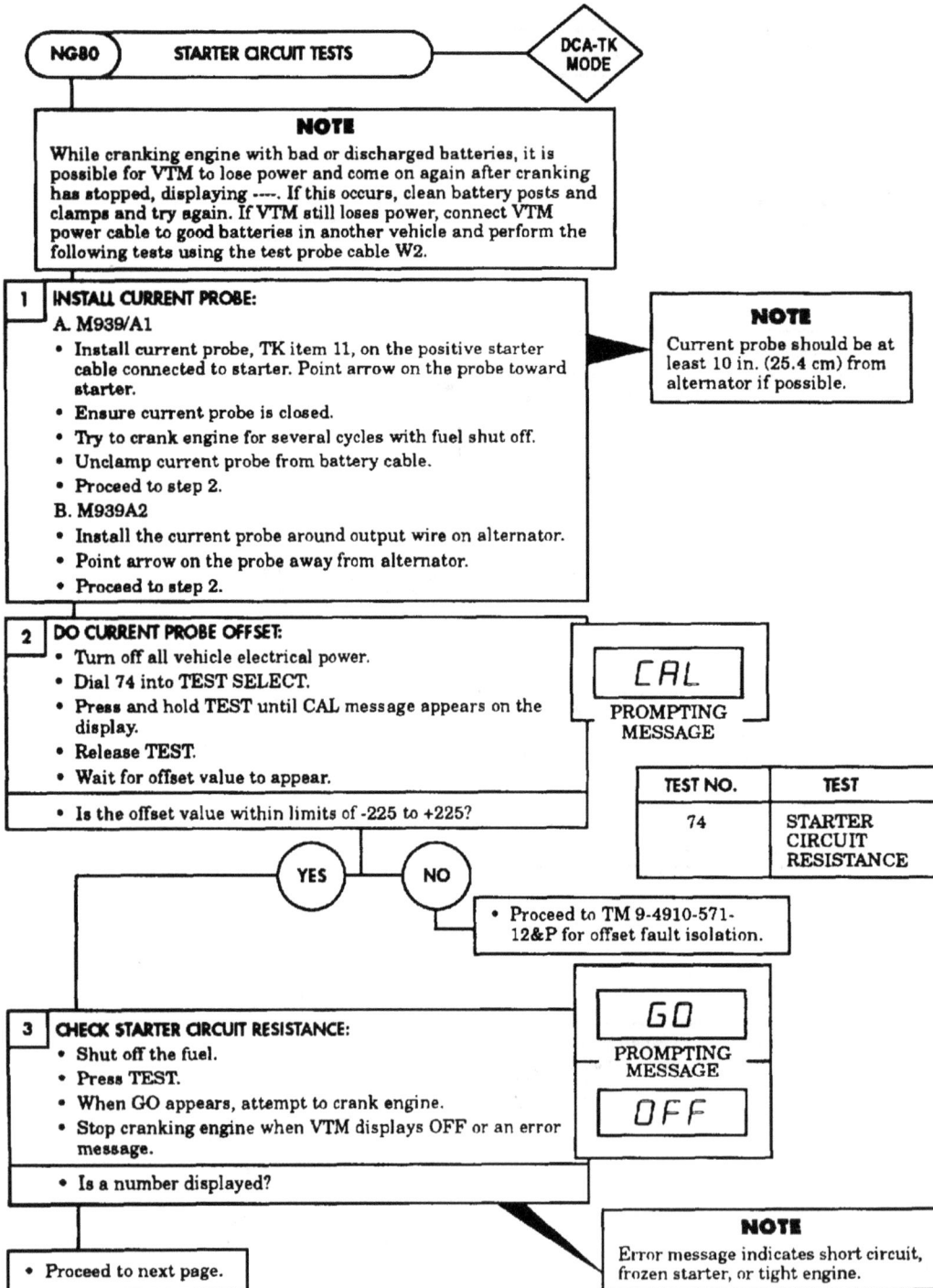

NG80   STARTER CIRCUIT TESTS   ◇ DCA-TK MODE

NOTE

While cranking engine with bad or discharged batteries, it is possible for VTM to lose power and come on again after cranking has stopped, displaying ----. If this occurs, clean battery posts and clamps and try again. If VTM still loses power, connect VTM power cable to good batteries in another vehicle and perform the following tests using the test probe cable W2.

1 INSTALL CURRENT PROBE:

A. M939/A1

- Install current probe, TK item 11, on the positive starter cable connected to starter. Point arrow on the probe toward starter.
- Ensure current probe is closed.
- Try to crank engine for several cycles with fuel shut off.
- Unclamp current probe from battery cable.
- Proceed to step 2.

B. M939A2

- Install the current probe around output wire on alternator.
- Point arrow on the probe away from alternator.
- Proceed to step 2.

NOTE

Current probe should be at least 10 in. (25.4 cm) from alternator if possible.

2 DO CURRENT PROBE OFFSET:

- Turn off all vehicle electrical power.
- Dial 74 into TEST SELECT.
- Press and hold TEST until CAL message appears on the display.
- Release TEST.
- Wait for offset value to appear.

- Is the offset value within limits of -225 to +225?

```
 CAL
```
PROMPTING MESSAGE

TEST NO.	TEST
74	STARTER CIRCUIT RESISTANCE

YES   NO

- Proceed to TM 9-4910-571-12&P for offset fault isolation.

3 CHECK STARTER CIRCUIT RESISTANCE:

- Shut off the fuel.
- Press TEST.
- When GO appears, attempt to crank engine.
- Stop cranking engine when VTM displays OFF or an error message.

- Is a number displayed?

```
 GO
```
PROMPTING MESSAGE

```
 OFF
```

NOTE

Error message indicates short circuit, frozen starter, or tight engine.

- Proceed to next page.

Table 2-6. STE/ICE NO-GO-Chain Tests (Contd).

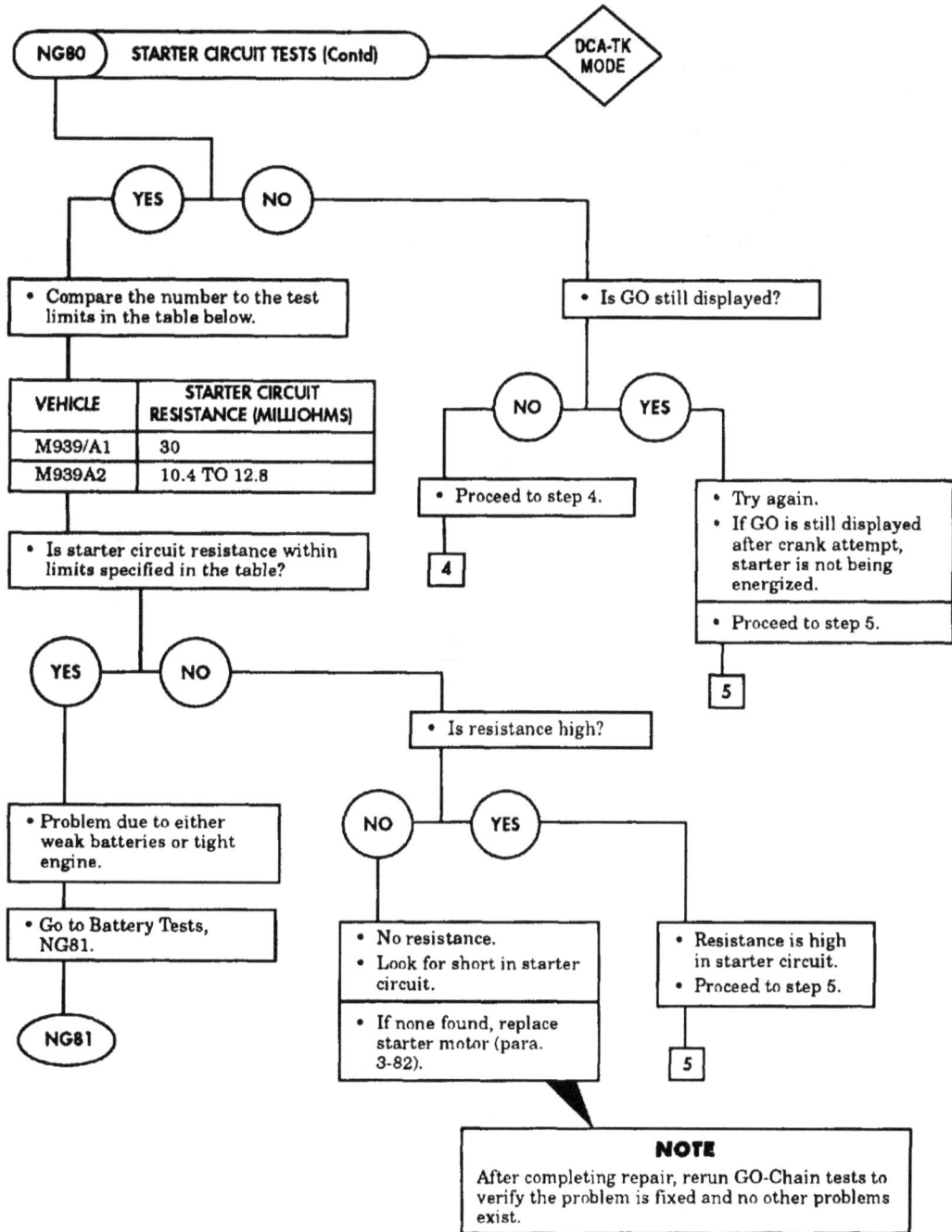

(NG80) STARTER CIRCUIT TESTS (Contd) ◇ DCA-TK MODE

YES / NO

- Compare the number to the test limits in the table below.

VEHICLE	STARTER CIRCUIT RESISTANCE (MILLIOHMS)
M939/A1	30
M939A2	10.4 TO 12.8

- Is starter circuit resistance within limits specified in the table?

YES / NO

- Problem due to either weak batteries or tight engine.

- Go to Battery Tests, NG81.

(NG81)

- Is GO still displayed?

NO / YES

- Proceed to step 4.

[4]

- Try again.
- If GO is still displayed after crank attempt, starter is not being energized.
- Proceed to step 5.

[5]

- Is resistance high?

NO / YES

- No resistance.
- Look for short in starter circuit.
- If none found, replace starter motor (para. 3-82).

- Resistance is high in starter circuit.
- Proceed to step 5.

[5]

NOTE

After completing repair, rerun GO-Chain tests to verify the problem is fixed and no other problems exist.

Table 2-6. STE/ICE NO-GO-Chain Tests (Contd).

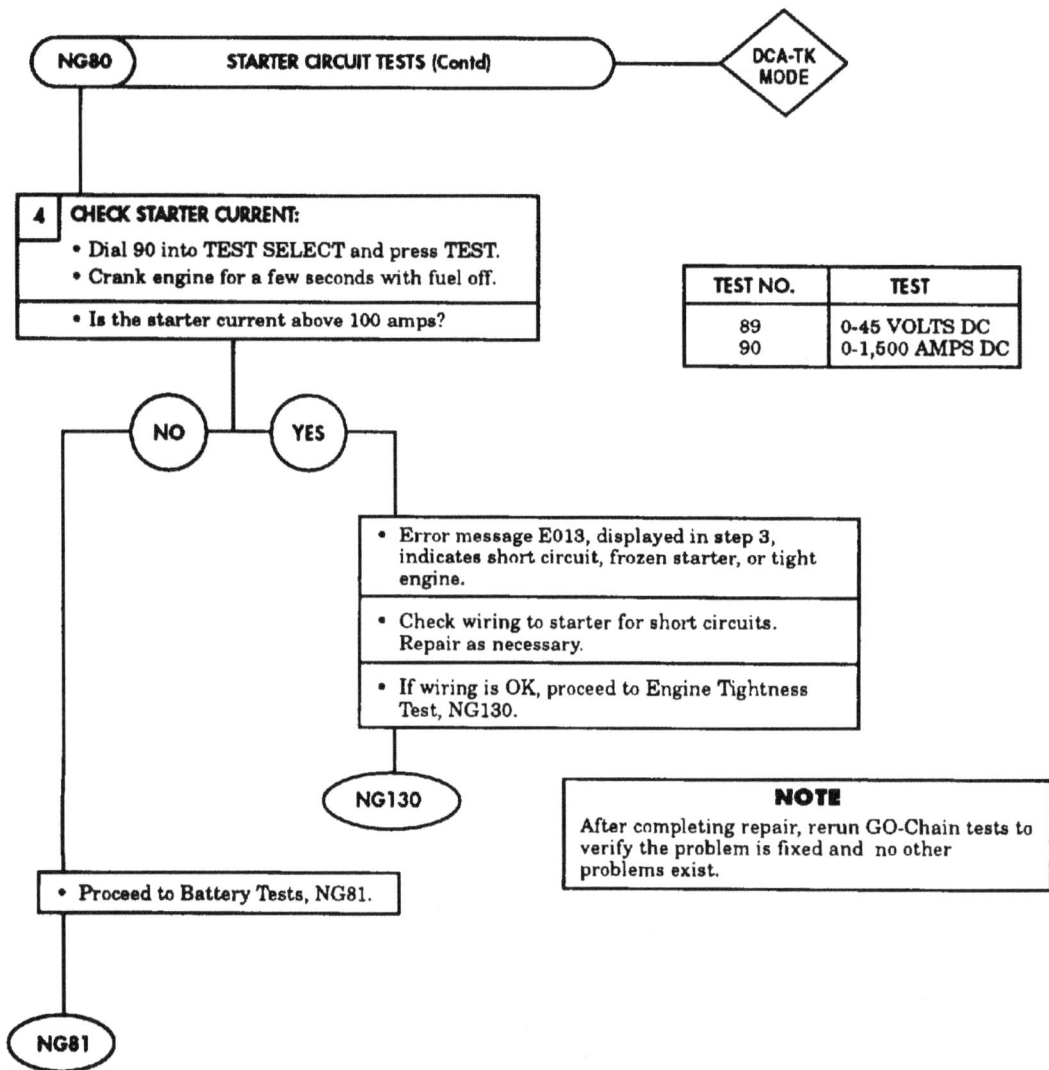

NG80 — STARTER CIRCUIT TESTS (Contd) — DCA-TK MODE

4 **CHECK STARTER CURRENT:**

- Dial 90 into TEST SELECT and press TEST.
- Crank engine for a few seconds with fuel off.

- Is the starter current above 100 amps?

TEST NO.	TEST
89	0-45 VOLTS DC
90	0-1,500 AMPS DC

NO   YES

- Error message E013, displayed in step 3, indicates short circuit, frozen starter, or tight engine.

- Check wiring to starter for short circuits. Repair as necessary.

- If wiring is OK, proceed to Engine Tightness Test, NG130.

NG130

NOTE

After completing repair, rerun GO-Chain tests to verify the problem is fixed and no other problems exist.

- Proceed to Battery Tests, NG81.

NG81

Table 2-6. STE/ICE NO-GO-Chain Tests (Contd).

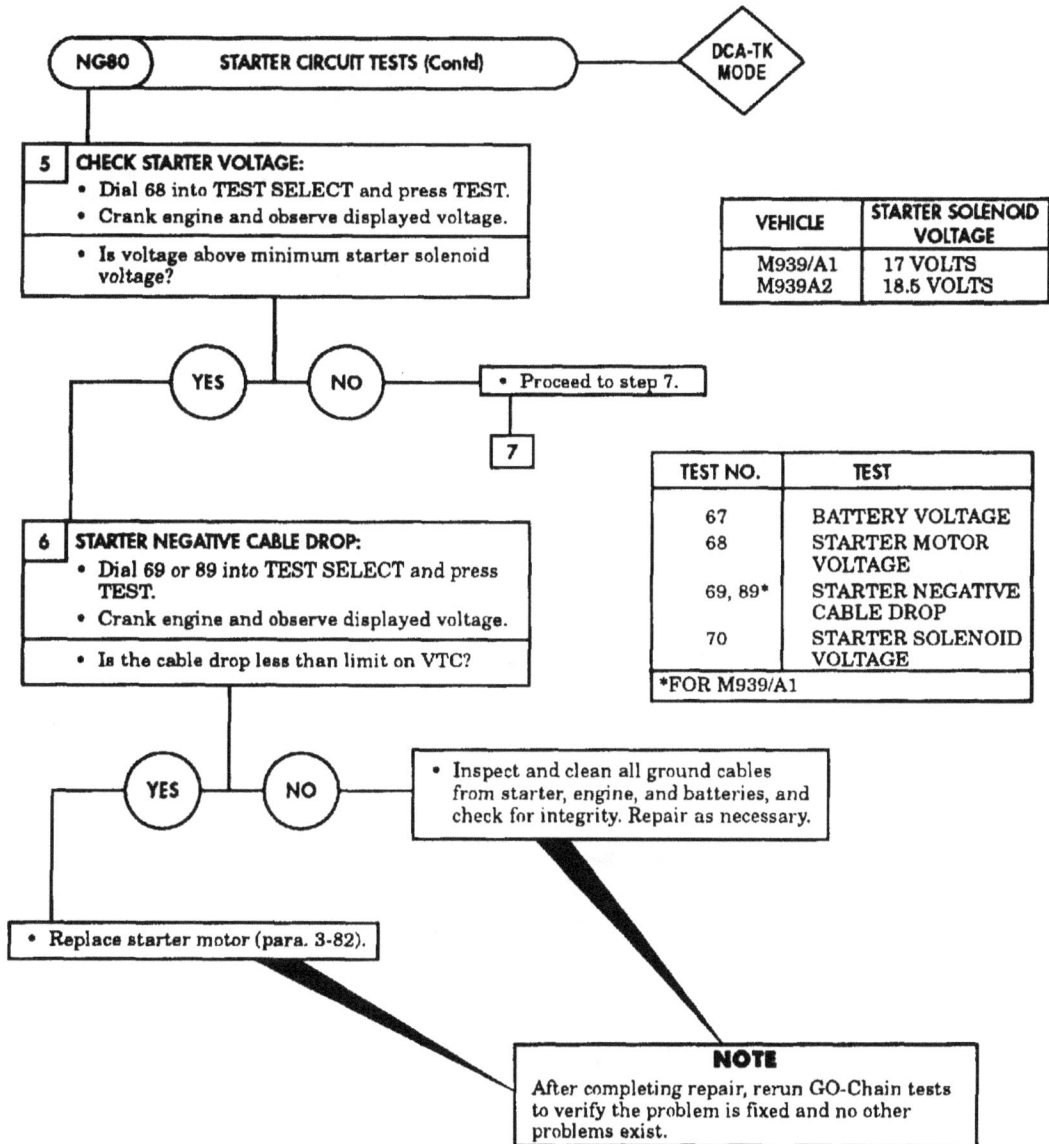

```
┌──────────────────────────────────────────────────────┐        ◇ DCA-TK
( NG80 )     STARTER CIRCUIT TESTS (Contd)              │          MODE
└──────────────────────────────────────────────────────┘
```

5 CHECK STARTER VOLTAGE:
- Dial 68 into TEST SELECT and press TEST.
- Crank engine and observe displayed voltage.

- Is voltage above minimum starter solenoid voltage?

VEHICLE	STARTER SOLENOID VOLTAGE
M939/A1	17 VOLTS
M939A2	18.5 VOLTS

(YES)   (NO)   → • Proceed to step 7.

[7]

TEST NO.	TEST
67	BATTERY VOLTAGE
68	STARTER MOTOR VOLTAGE
69, 89*	STARTER NEGATIVE CABLE DROP
70	STARTER SOLENOID VOLTAGE
*FOR M939/A1	

6 STARTER NEGATIVE CABLE DROP:
- Dial 69 or 89 into TEST SELECT and press TEST.
- Crank engine and observe displayed voltage.

- Is the cable drop less than limit on VTC?

(YES)   (NO)   → • Inspect and clean all ground cables from starter, engine, and batteries, and check for integrity. Repair as necessary.

• Replace starter motor (para. 3-82).

NOTE

After completing repair, rerun GO-Chain tests to verify the problem is fixed and no other problems exist.

Table 2-6. STE/ICE NO-GO-Chain Tests (Contd).

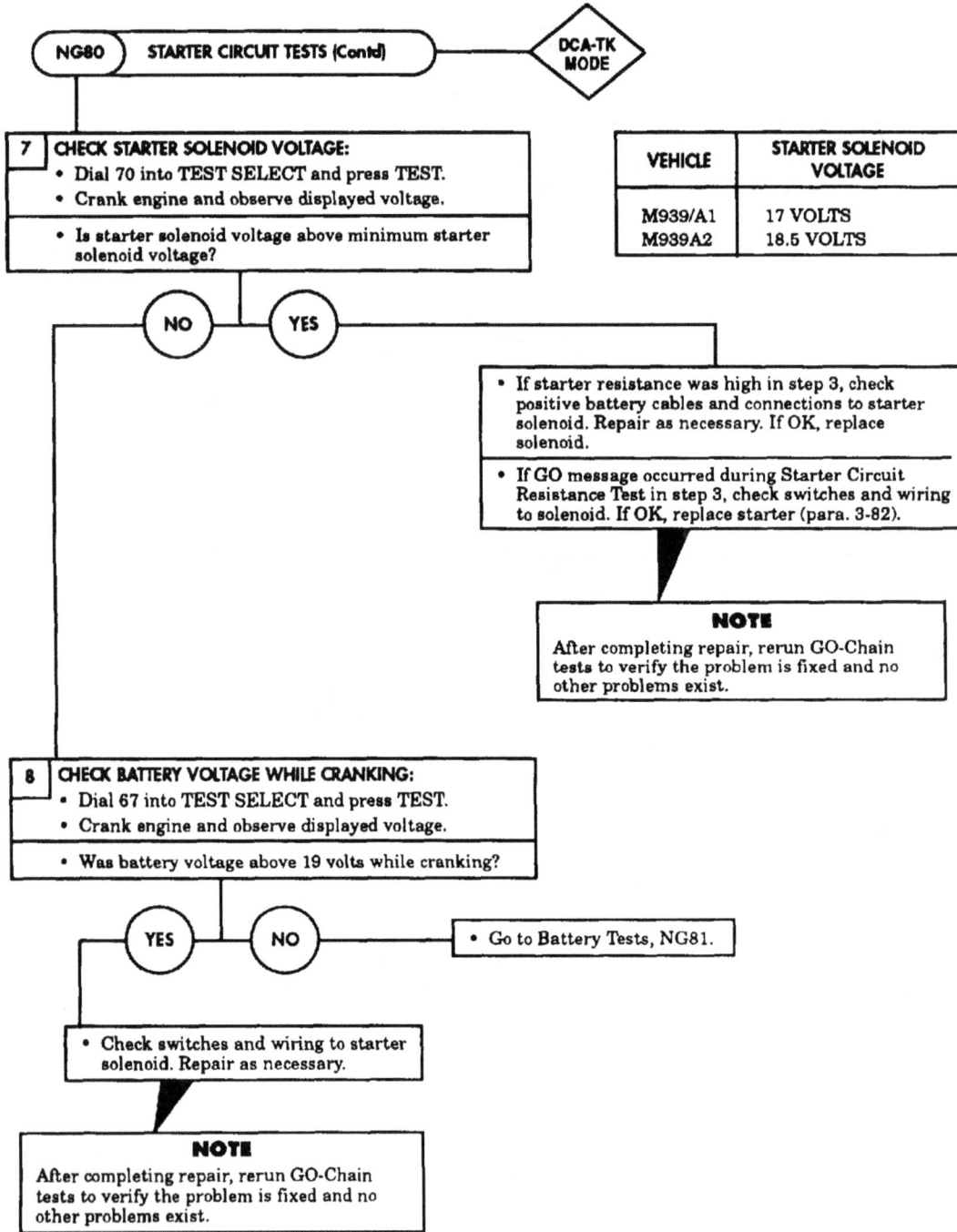

(NG80) STARTER CIRCUIT TESTS (Contd) ◇ DCA-TK MODE

7 | CHECK STARTER SOLENOID VOLTAGE:
- Dial 70 into TEST SELECT and press TEST.
- Crank engine and observe displayed voltage.

- Is starter solenoid voltage above minimum starter solenoid voltage?

VEHICLE	STARTER SOLENOID VOLTAGE
M939/A1	17 VOLTS
M939A2	18.5 VOLTS

(NO)   (YES)

- If starter resistance was high in step 3, check positive battery cables and connections to starter solenoid. Repair as necessary. If OK, replace solenoid.

- If GO message occurred during Starter Circuit Resistance Test in step 3, check switches and wiring to solenoid. If OK, replace starter (para. 3-82).

NOTE

After completing repair, rerun GO-Chain tests to verify the problem is fixed and no other problems exist.

8 | CHECK BATTERY VOLTAGE WHILE CRANKING:
- Dial 67 into TEST SELECT and press TEST.
- Crank engine and observe displayed voltage.

- Was battery voltage above 19 volts while cranking?

(YES)   (NO)   - Go to Battery Tests, NG81.

- Check switches and wiring to starter solenoid. Repair as necessary.

NOTE

After completing repair, rerun GO-Chain tests to verify the problem is fixed and no other problems exist.

Table 2-6. STE/ICE NO-GO-Chain Tests (Contd).

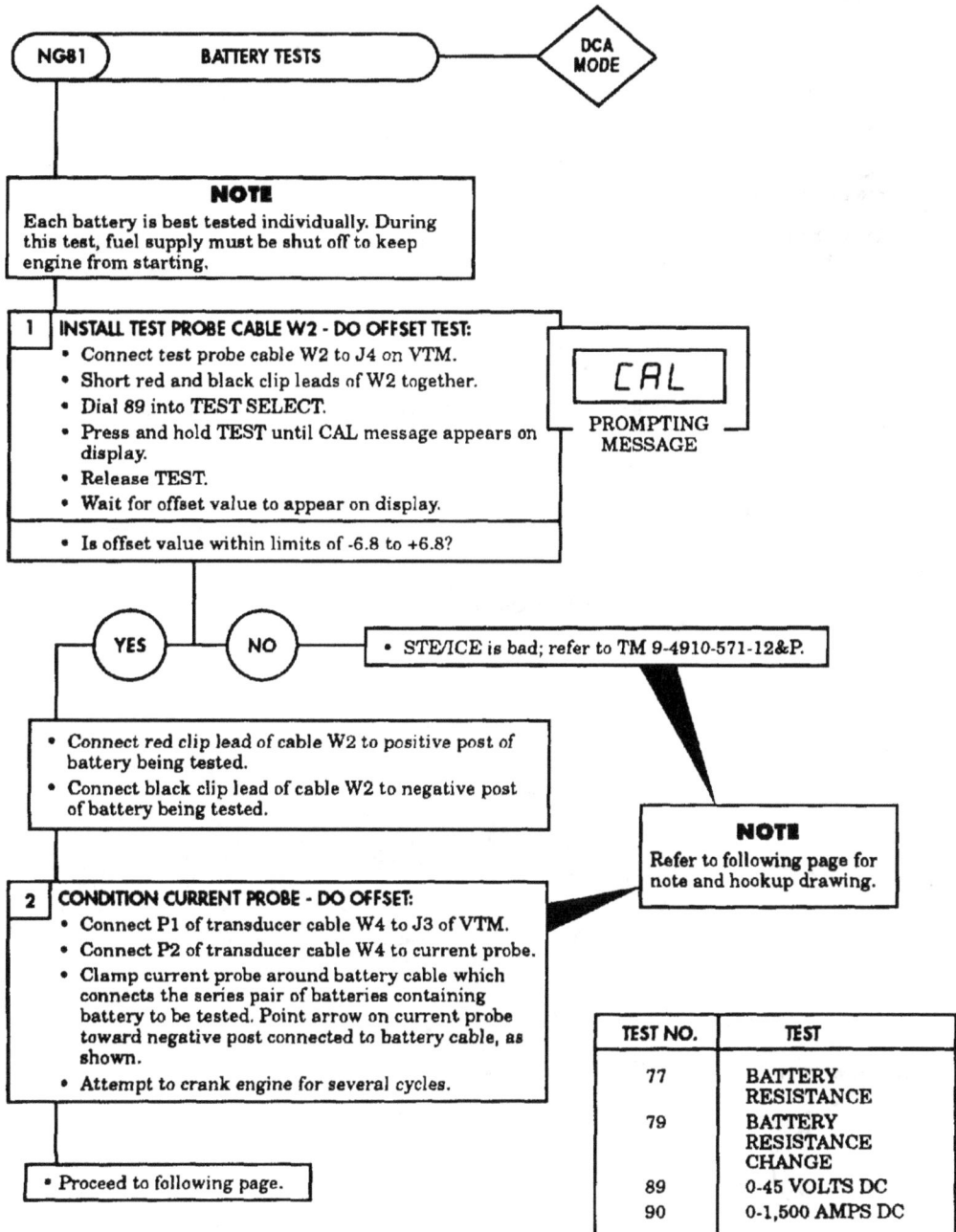

NG81 — BATTERY TESTS — ⬦ DCA MODE

NOTE
Each battery is best tested individually. During this test, fuel supply must be shut off to keep engine from starting.

1 INSTALL TEST PROBE CABLE W2 - DO OFFSET TEST:
- Connect test probe cable W2 to J4 on VTM.
- Short red and black clip leads of W2 together.
- Dial 89 into TEST SELECT.
- Press and hold TEST until CAL message appears on display.
- Release TEST.
- Wait for offset value to appear on display.

CAL

PROMPTING MESSAGE

- Is offset value within limits of -6.8 to +6.8?

YES / **NO** — • STE/ICE is bad; refer to TM 9-4910-571-12&P.

- Connect red clip lead of cable W2 to positive post of battery being tested.
- Connect black clip lead of cable W2 to negative post of battery being tested.

NOTE
Refer to following page for note and hookup drawing.

2 CONDITION CURRENT PROBE - DO OFFSET:
- Connect P1 of transducer cable W4 to J3 of VTM.
- Connect P2 of transducer cable W4 to current probe.
- Clamp current probe around battery cable which connects the series pair of batteries containing battery to be tested. Point arrow on current probe toward negative post connected to battery cable, as shown.
- Attempt to crank engine for several cycles.

- Proceed to following page.

TEST NO.	TEST
77	BATTERY RESISTANCE
79	BATTERY RESISTANCE CHANGE
89	0-45 VOLTS DC
90	0-1,500 AMPS DC

Table 2-6. STE/ICE NO-GO-Chain Tests (Contd).

NOTE

TEST PROCEDURE:

1. Test each battery of a series pair, then proceed to batteries of next series pair.

2. To find the series pair of batteries, find pairs for which negative terminal of one battery is connected by a cable to positive terminal of another battery. This makes the two batteries a series pair. For example, in the figure below, batteries A and B are a series pair, and batteries C and D are also a series pair.

3. To test battery A or B, clamp current probe around cable connecting battery A and battery B. Point arrow on current probe in the direction of negative post connected to the cable.

4. The test probe cable W2 is first connected to battery A for testing battery A.

5. The test probe cable W2 is then connected to battery B for testing battery B. (Current probe in same place as for testing battery A.)

6. To test battery C or D, clamp current probe around cable connecting battery C and battery D. Point arrow on current probe in the direction of negative post connected to the cable.

7. Test probe cable W2 is then connected to battery C for testing battery C.

8. Test probe cable W2 is then connected to battery D for testing battery D.

**BATTERY CABLE
INSTALLATION SEQUENCE**

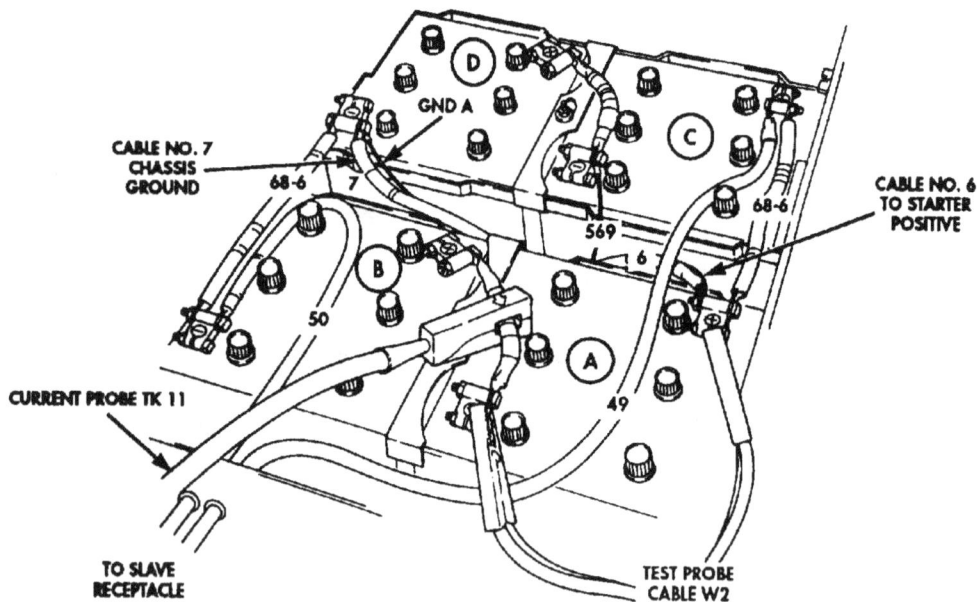

Table 2-6. STE/ICE NO-GO-Chain Tests (Contd).

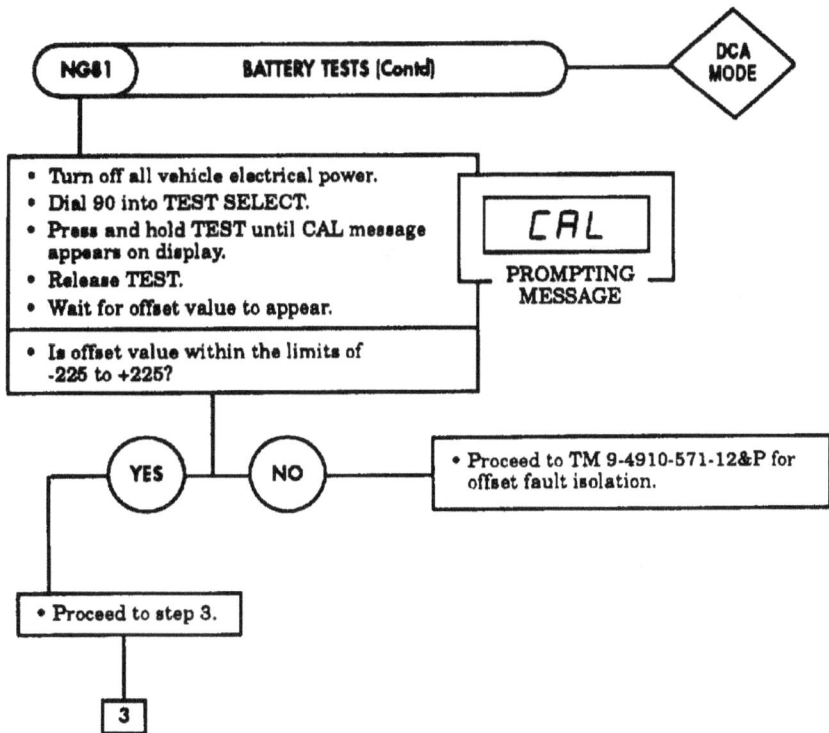

(NG51) BATTERY TESTS (Contd) ◇ DCA MODE ◇

- Turn off all vehicle electrical power.
- Dial 90 into TEST SELECT.
- Press and hold TEST until CAL message appears on display.
- Release TEST.
- Wait for offset value to appear.

┌──────────┐
│ CAL   │
└──────────┘
PROMPTING
MESSAGE

- Is offset value within the limits of -225 to +225?

(YES)   (NO)   • Proceed to TM 9-4910-571-12&P for offset fault isolation.

• Proceed to step 3.

[3]

Table 2-6. STE/ICE NO-GO-Chain Tests (Contd).

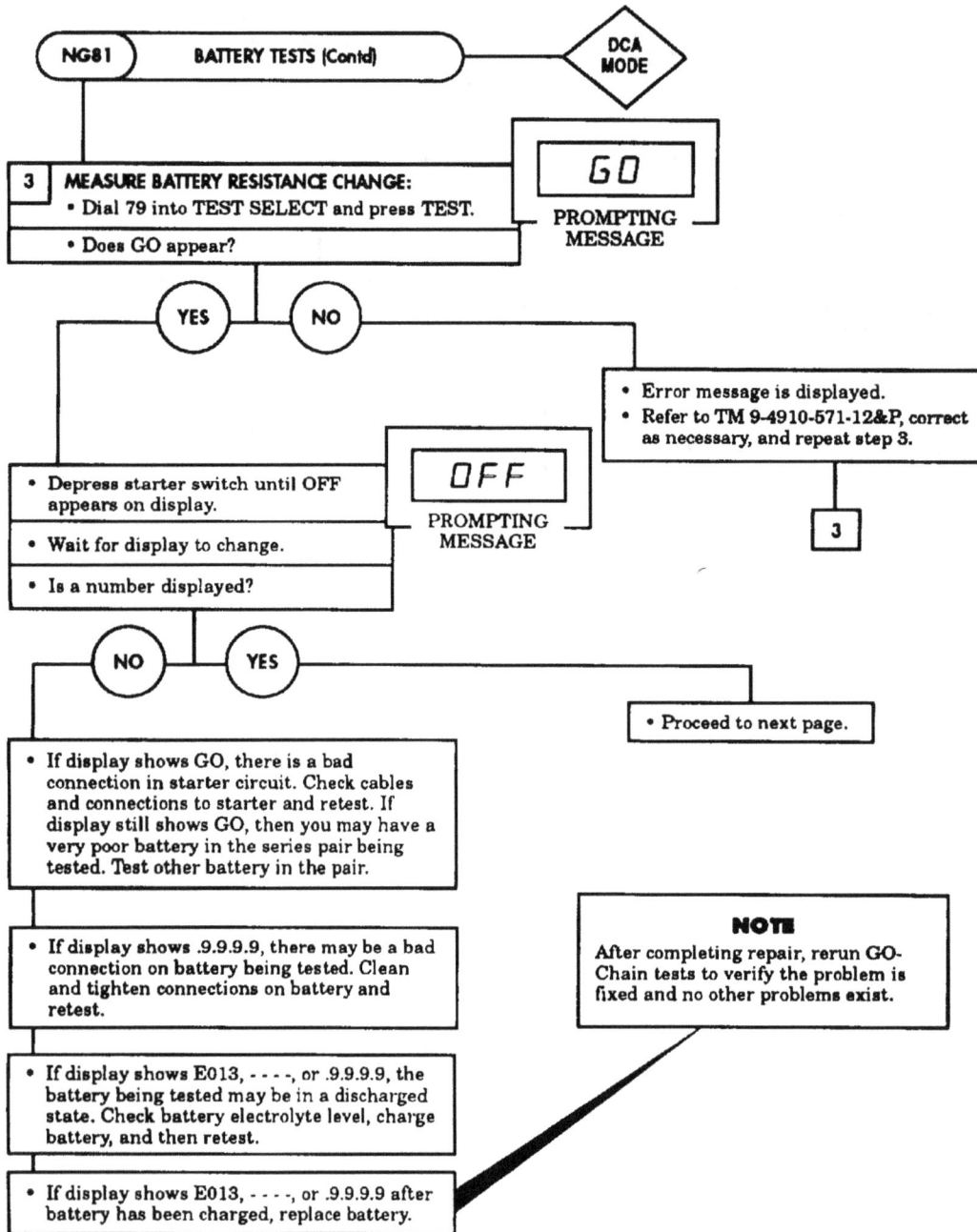

NG81 — BATTERY TESTS (Contd) — DCA MODE

3 MEASURE BATTERY RESISTANCE CHANGE:
• Dial 79 into TEST SELECT and press TEST.

```
GO
```
PROMPTING MESSAGE

• Does GO appear?

YES   NO

• Error message is displayed.
• Refer to TM 9-4910-571-12&P, correct as necessary, and repeat step 3.

3

• Depress starter switch until OFF appears on display.

```
OFF
```
PROMPTING MESSAGE

• Wait for display to change.
• Is a number displayed?

NO   YES

• Proceed to next page.

• If display shows GO, there is a bad connection in starter circuit. Check cables and connections to starter and retest. If display still shows GO, then you may have a very poor battery in the series pair being tested. Test other battery in the pair.

• If display shows .9.9.9.9, there may be a bad connection on battery being tested. Clean and tighten connections on battery and retest.

NOTE
After completing repair, rerun GO-Chain tests to verify the problem is fixed and no other problems exist.

• If display shows E013, - - - -, or .9.9.9.9, the battery being tested may be in a discharged state. Check battery electrolyte level, charge battery, and then retest.

• If display shows E013, - - - -, or .9.9.9.9 after battery has been charged, replace battery.

Table 2-6. STE/ICE NO-GO-Chain Tests (Contd).

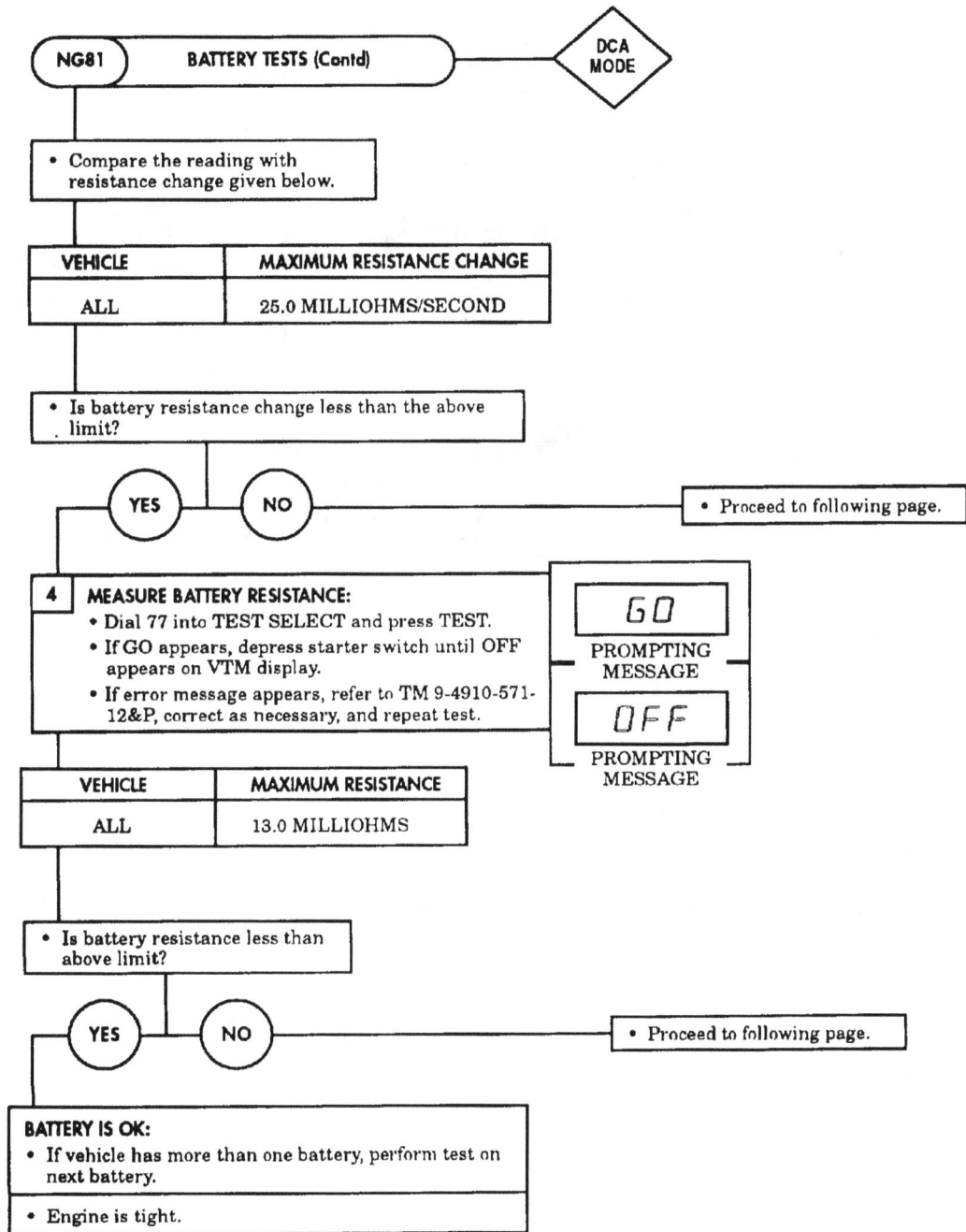

NG81 — BATTERY TESTS (Contd) — DCA MODE

- Compare the reading with resistance change given below.

VEHICLE	MAXIMUM RESISTANCE CHANGE
ALL	25.0 MILLIOHMS/SECOND

- Is battery resistance change less than the above limit?

YES   NO — • Proceed to following page.

4 MEASURE BATTERY RESISTANCE:
- Dial 77 into TEST SELECT and press TEST.
- If GO appears, depress starter switch until OFF appears on VTM display.
- If error message appears, refer to TM 9-4910-571-12&P, correct as necessary, and repeat test.

GO
PROMPTING MESSAGE

OFF
PROMPTING MESSAGE

VEHICLE	MAXIMUM RESISTANCE
ALL	13.0 MILLIOHMS

- Is battery resistance less than above limit?

YES   NO — • Proceed to following page.

BATTERY IS OK:
- If vehicle has more than one battery, perform test on next battery.
- Engine is tight.

2-259

Table 2-6. STE/ICE NO-GO-Chain Tests (Contd).

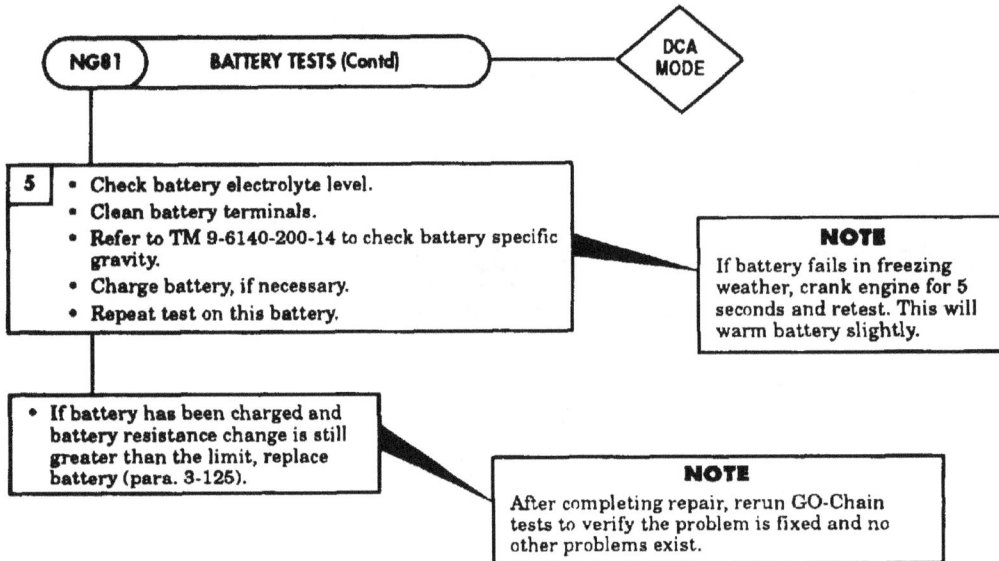

```
┌──────────┐                                          ╱╲
│ NG81  BATTERY TESTS (Contd)                    ╱ DCA ╲
└──────────┘                                    ╲ MODE ╱
                                                  ╲╱
```

5
- Check battery electrolyte level.
- Clean battery terminals.
- Refer to TM 9-6140-200-14 to check battery specific gravity.
- Charge battery, if necessary.
- Repeat test on this battery.

NOTE

If battery fails in freezing weather, crank engine for 5 seconds and retest. This will warm battery slightly.

- If battery has been charged and battery resistance change is still greater than the limit, replace battery (para. 3-125).

NOTE

After completing repair, rerun GO-Chain tests to verify the problem is fixed and no other problems exist.

Table 2-6. STE/ICE NO-GO-Chain Tests (Contd).

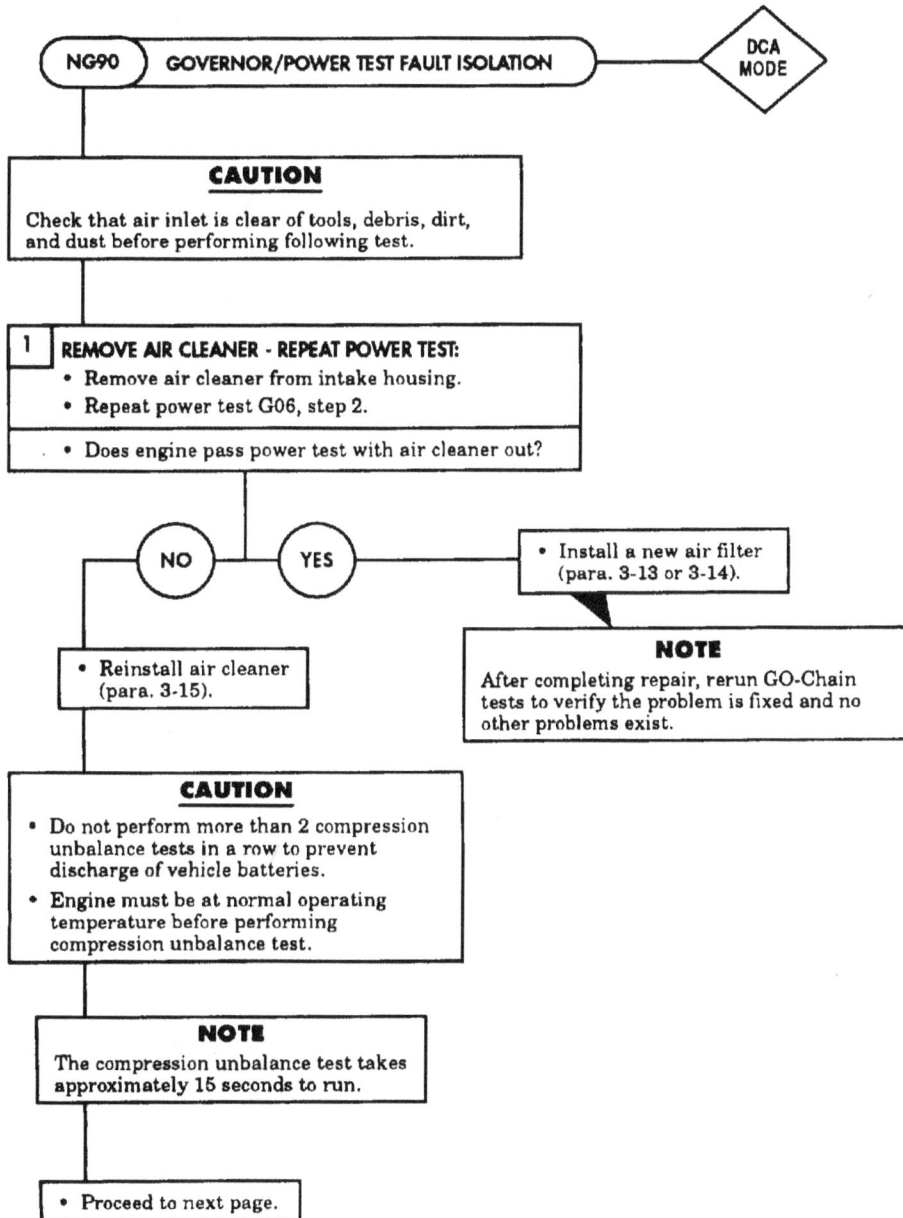

NG90) GOVERNOR/POWER TEST FAULT ISOLATION ◇ DCA MODE

CAUTION

Check that air inlet is clear of tools, debris, dirt, and dust before performing following test.

1 REMOVE AIR CLEANER - REPEAT POWER TEST:
- Remove air cleaner from intake housing.
- Repeat power test G06, step 2.
- Does engine pass power test with air cleaner out?

NO   YES

- Install a new air filter (para. 3-13 or 3-14).

NOTE

After completing repair, rerun GO-Chain tests to verify the problem is fixed and no other problems exist.

- Reinstall air cleaner (para. 3-15).

CAUTION
- Do not perform more than 2 compression unbalance tests in a row to prevent discharge of vehicle batteries.
- Engine must be at normal operating temperature before performing compression unbalance test.

NOTE

The compression unbalance test takes approximately 15 seconds to run.

- Proceed to next page.

Table 2-6. STE/ICE NO-GO-Chain Tests (Contd).

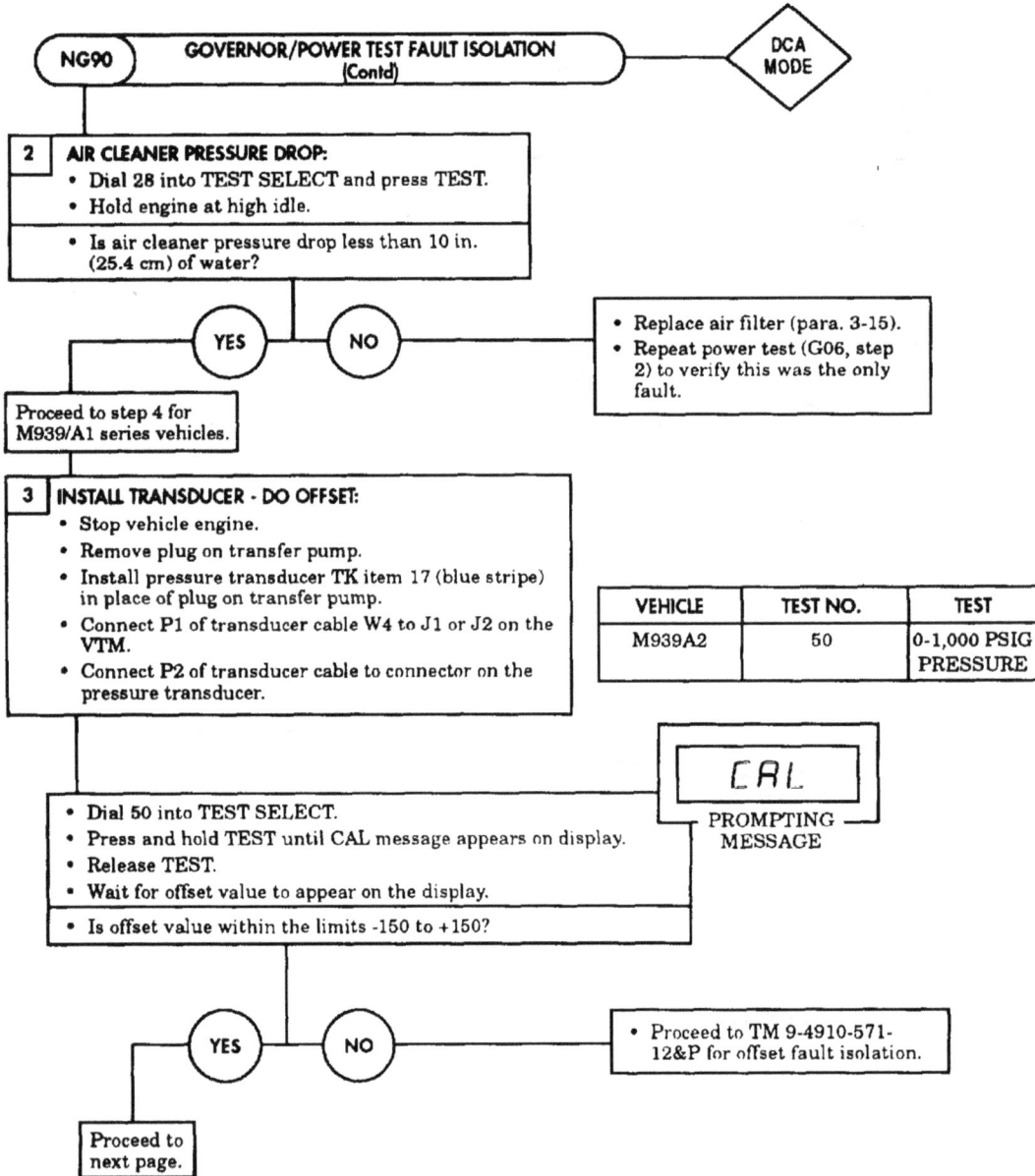

```
┌────┐  GOVERNOR/POWER TEST FAULT ISOLATION       ◇ DCA
│NG90│            (Contd)                           MODE
└────┘
```

2 AIR CLEANER PRESSURE DROP:
- Dial 28 into TEST SELECT and press TEST.
- Hold engine at high idle.

- Is air cleaner pressure drop less than 10 in. (25.4 cm) of water?

(YES)   (NO)

- Replace air filter (para. 3-15).
- Repeat power test (G06, step 2) to verify this was the only fault.

Proceed to step 4 for M939/A1 series vehicles.

3 INSTALL TRANSDUCER - DO OFFSET:
- Stop vehicle engine.
- Remove plug on transfer pump.
- Install pressure transducer TK item 17 (blue stripe) in place of plug on transfer pump.
- Connect P1 of transducer cable W4 to J1 or J2 on the VTM.
- Connect P2 of transducer cable to connector on the pressure transducer.

VEHICLE	TEST NO.	TEST
M939A2	50	0-1,000 PSIG PRESSURE

```
 CAL
```
PROMPTING MESSAGE

- Dial 50 into TEST SELECT.
- Press and hold TEST until CAL message appears on display.
- Release TEST.
- Wait for offset value to appear on the display.

- Is offset value within the limits -150 to +150?

(YES)   (NO)

- Proceed to TM 9-4910-571-12&P for offset fault isolation.

Proceed to next page.

Table 2-6. STE/ICE NO-GO-Chain Tests (Contd).

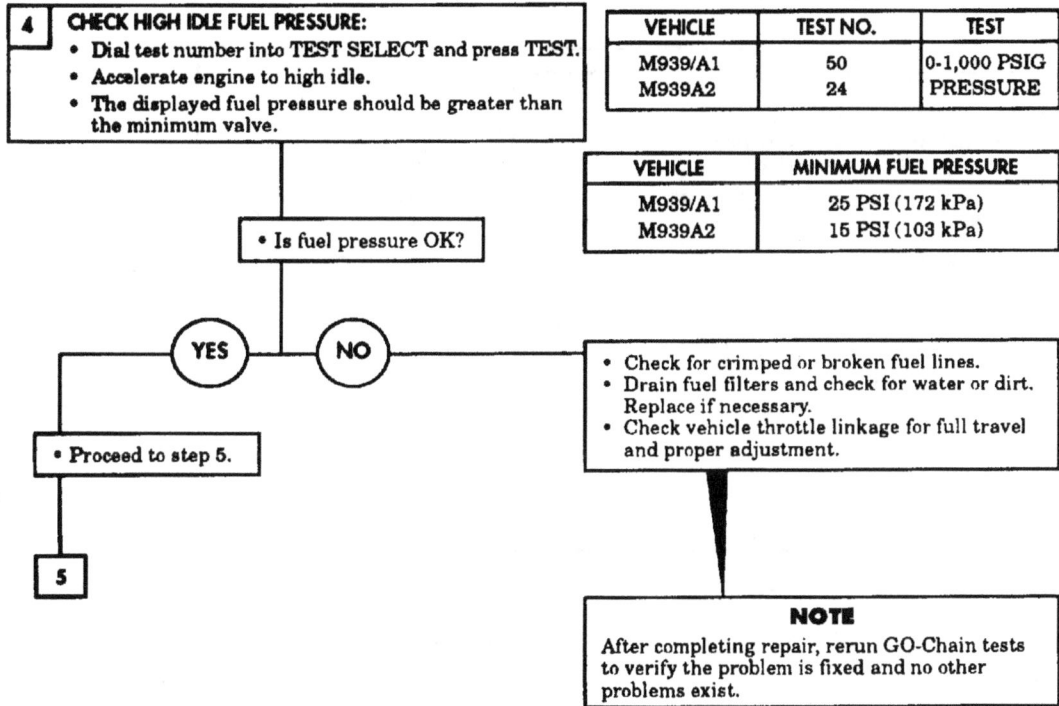

4 **CHECK HIGH IDLE FUEL PRESSURE:**
- Dial test number into TEST SELECT and press TEST.
- Accelerate engine to high idle.
- The displayed fuel pressure should be greater than the minimum valve.

VEHICLE	TEST NO.	TEST
M939/A1	50	0-1,000 PSIG PRESSURE
M939A2	24	

VEHICLE	MINIMUM FUEL PRESSURE
M939/A1	25 PSI (172 kPa)
M939A2	15 PSI (103 kPa)

- Is fuel pressure OK?

YES **NO**

- Check for crimped or broken fuel lines.
- Drain fuel filters and check for water or dirt. Replace if necessary.
- Check vehicle throttle linkage for full travel and proper adjustment.

- Proceed to step 5.

5

NOTE
After completing repair, rerun GO-Chain tests to verify the problem is fixed and no other problems exist.

Table 2-6. STE/ICE NO-GO-Chain Tests (Contd).

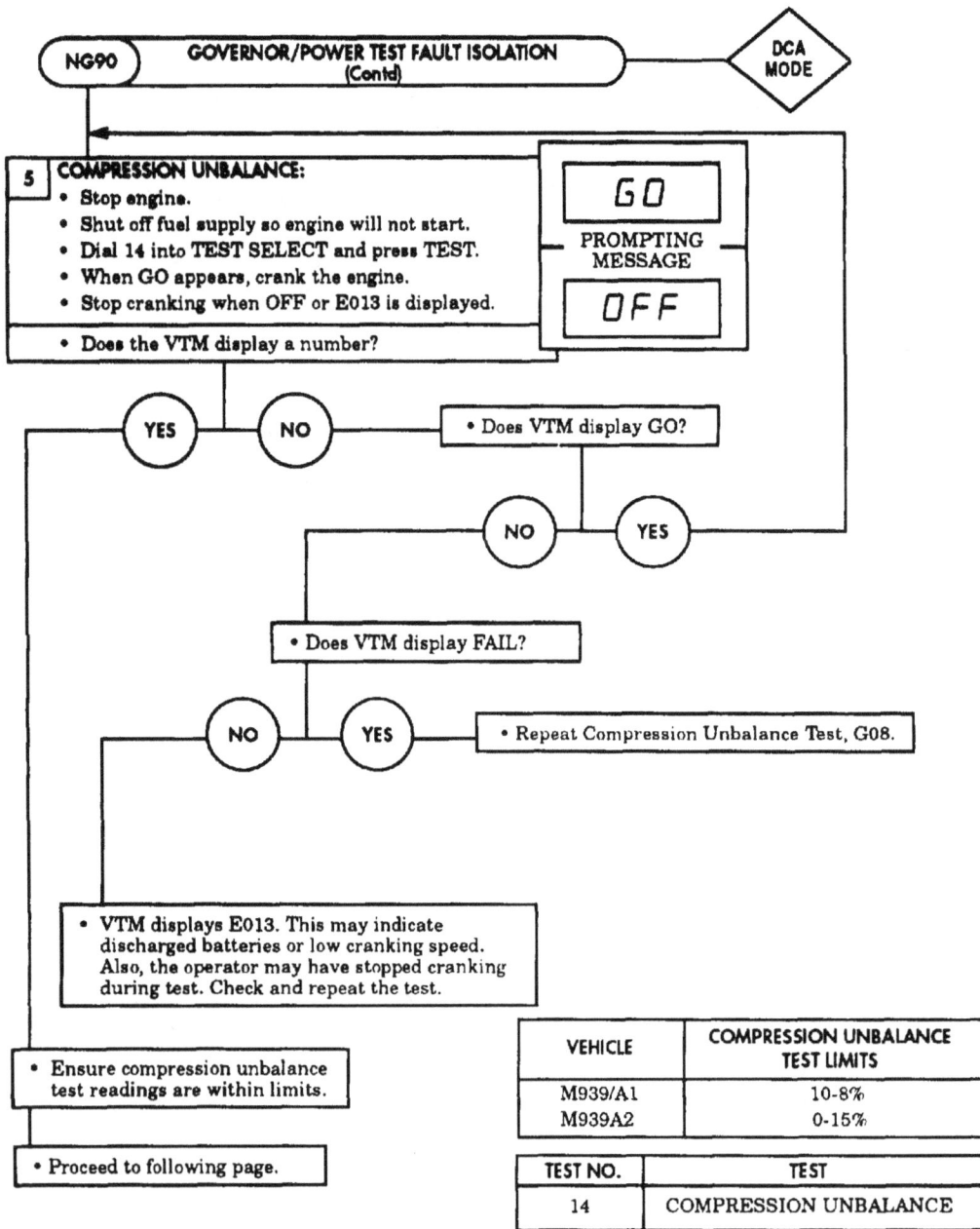

(NG90) GOVERNOR/POWER TEST FAULT ISOLATION (Contd)   ◇ DCA MODE

```
5  COMPRESSION UNBALANCE:
   • Stop engine.
   • Shut off fuel supply so engine will not start.
   • Dial 14 into TEST SELECT and press TEST.
   • When GO appears, crank the engine.
   • Stop cranking when OFF or E013 is displayed.
```

```
      GO
   PROMPTING
    MESSAGE
     OFF
```

• Does the VTM display a number?

(YES)   (NO)   • Does VTM display GO?

(NO)   (YES)

• Does VTM display FAIL?

(NO)   (YES)   • Repeat Compression Unbalance Test, G08.

• VTM displays E013. This may indicate discharged batteries or low cranking speed. Also, the operator may have stopped cranking during test. Check and repeat the test.

• Ensure compression unbalance test readings are within limits.

• Proceed to following page.

VEHICLE	COMPRESSION UNBALANCE TEST LIMITS
M939/A1	10-8%
M939A2	0-15%

TEST NO.	TEST
14	COMPRESSION UNBALANCE

Table 2-6. STE/ICE NO-GO-Chain Tests (Contd).

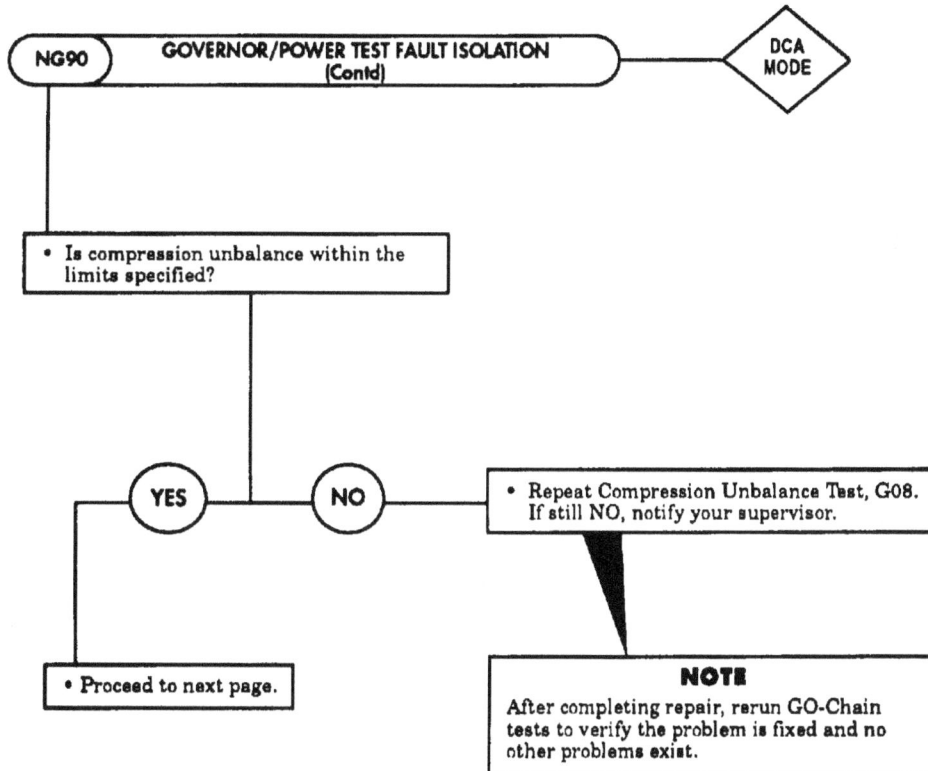

NG90 GOVERNOR/POWER TEST FAULT ISOLATION (Contd) → DCA MODE

- Is compression unbalance within the limits specified?

YES → • Proceed to next page.

NO → • Repeat Compression Unbalance Test, G08. If still NO, notify your supervisor.

NOTE

After completing repair, rerun GO-Chain tests to verify the problem is fixed and no other problems exist.

Table 2-6. STE/ICE NO-GO-Chain Tests (Contd).

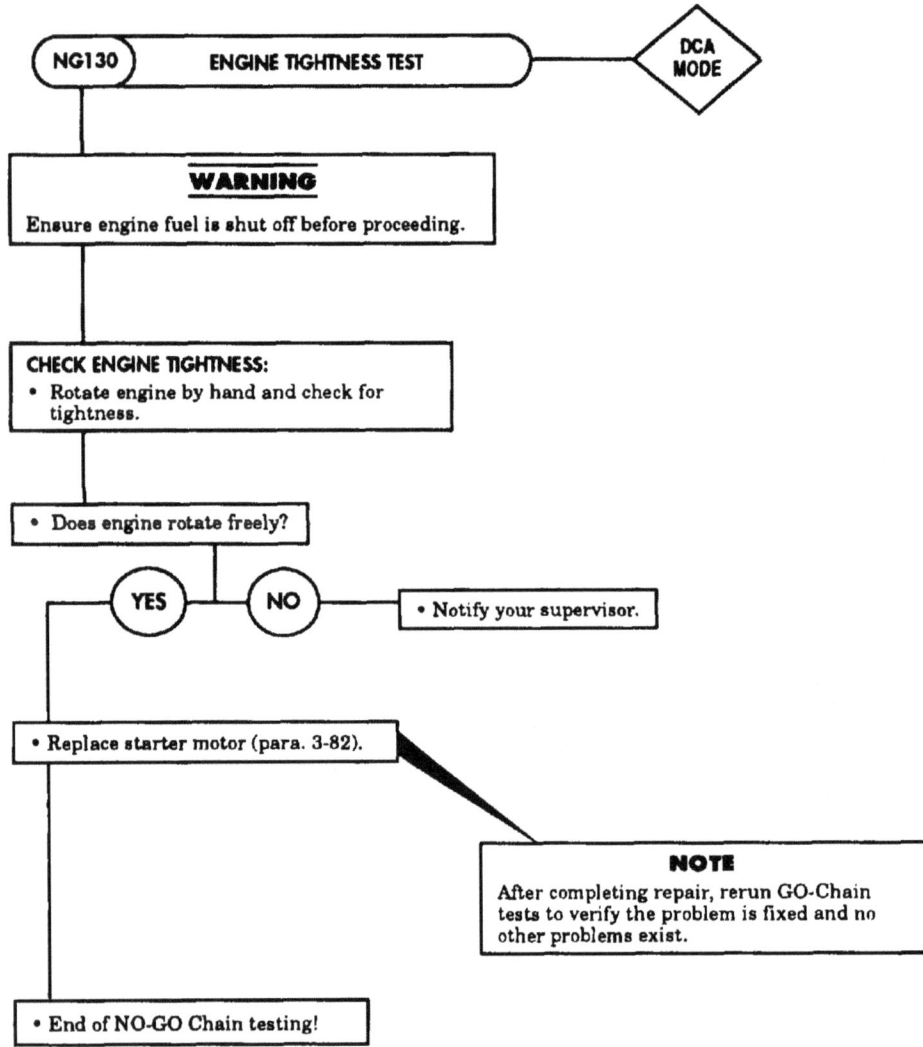

NG130 — ENGINE TIGHTNESS TEST — DCA MODE

WARNING

Ensure engine fuel is shut off before proceeding.

CHECK ENGINE TIGHTNESS:
- Rotate engine by hand and check for tightness.

- Does engine rotate freely?

YES   NO — • Notify your supervisor.

- Replace starter motor (para. 3-82).

NOTE

After completing repair, rerun GO-Chain tests to verify the problem is fixed and no other problems exist.

- End of NO-GO Chain testing!

2-25. CENTRAL TIRE INFLATION SYSTEM (CTIS) TROUBLESHOOTING

WARNING

Operation of a deadlined vehicle, without preliminary inspection, may cause injury to personnel and/or damage to equipment.

NOTE

• Prior to troubleshooting tests, vehicle air pressure must be at operating level.

• Notify DS maintenance if corrective action does not correct the malfunction.

a. Troubleshooting procedures in this section are organized in a step-by-step format that directs tests and inspections toward the source of a problem and a successful solution.

b. The Symptom Index, used in conjunction with the Functional Troubleshooting Diagram (Figure 2-1), is designed to direct the technician to the appropriate troubleshooting section where specific problems and solutions are listed.

NOTE

Repair of CTIS wiring harness is not recommended. However, connectors are available for repair. Refer to para. 3-131 for wiring harness repair procedures.

c. For most system components connected by wiring or pneumatic tubing, intermittent malfunctions are the most difficult to correct. In many instances, erratic operation can be traced to faulty electrical or pneumatic connections. If a symptom description of the malfunction cannot be found in the Symptom Index, checking the continuity and integrity of all connections is the first step to finding the problem.

d. Before taking any action to correct a possible malfunction, the following rules should be followed:

(1) Question operator to obtain any information that might help determine the cause of the problem.

(2) Never overlook the chance that the problem could be of simple origin and could be corrected with minor adjustment.

(3) Use all senses to observe and locate troubles.

(4) Use test instruments or gages to help determine and isolate the problem.

(5) Always isolate the system where mal-function occurs and then locate the defective component.

e. This section cannot list all mechanical malfunctions that may occur. If a malfunction occurs that is not listed in table Central Tire Inflation System (CTIS) Troubleshooting Symptom Index, notify direct support maintenance.

CENTRAL TIRE INFLATION SYSTEM (CTIS) TROUBLESHOOTING SYMPTOM INDEX

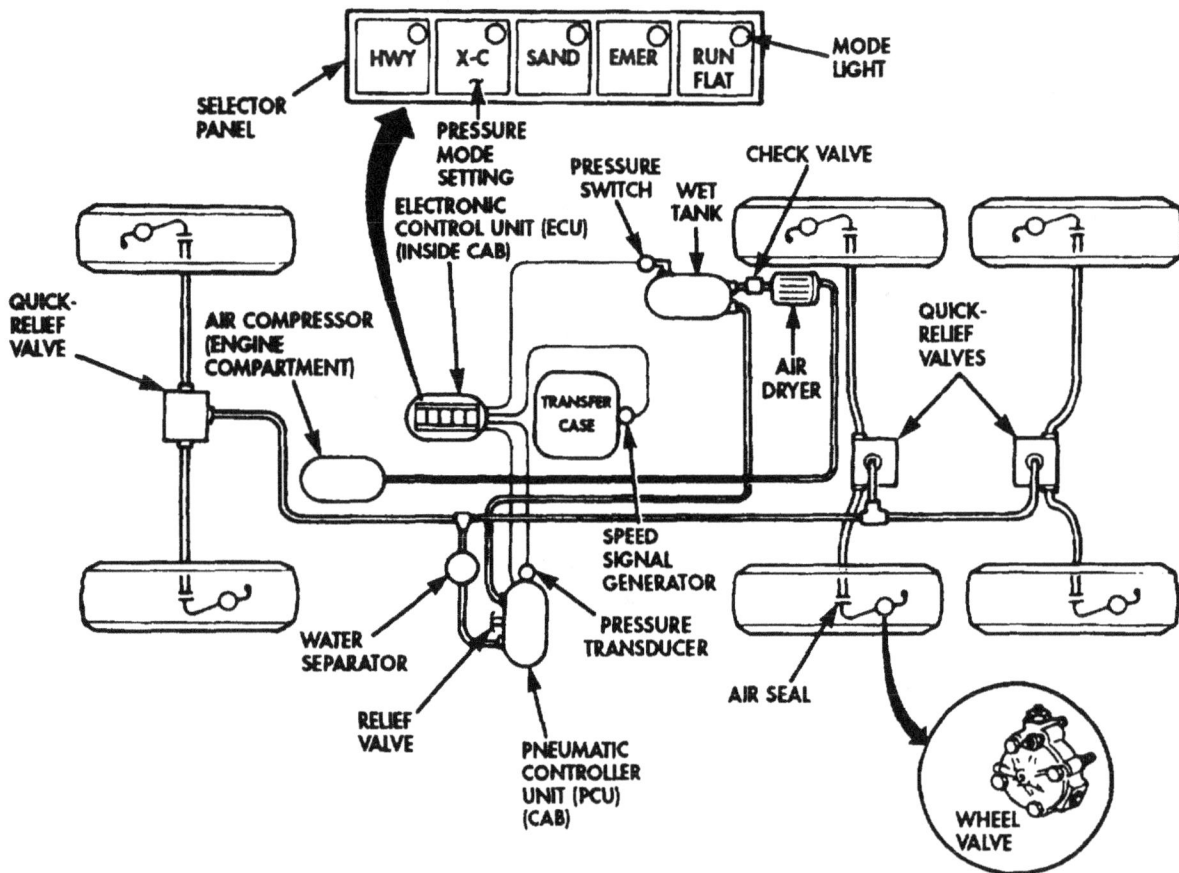

Figure 2-1. Functional Troubleshooting Diagram - Central Tire Inflation System (CTIS).

Table 2-7. CTIS Troubleshooting.

```
MALFUNCTION
   TEST OR INSPECTION
      CORRECTIVE ACTION
```

WARNING

Release all air pressure before loosening or removing air system component(s). Failure to do so may result in injury to personnel.

1. **TWO PRESSURE MODES INDICATED WITH STEADY MODE LIGHTS**

NOTE

- The Electronic Control Unit (ECU) has five pressure mode settings, with each setting having a mode light. If, during programmed operation or after manually selecting a pressure mode setting, two mode lights remain on, the CTIS will discontinue operation. Two steady mode lights indicate that a particular inflate or deflate sequence has taken longer than the programmed limits allow, and has shut the system off with air pressure between the modes, indicated by lights.

- If system was inflating when it stopped between modes, perform tests 1 through 3.

- If system was deflating when it stopped between modes, perform tests 4 and 5.

Test 1. Check air supply to CTIS.

WARNING

Compressed air source will not exceed 30 psi (207 kPa). When cleaning with compressed air, eyeshields must be worn. Failure to do so may result in injury to personnel.

Step 1. Inspect air lines between wet tank and pneumatic controller for leaks, cracks, damage, or restrictions.

 a. Check for leaks at connections using soapy water. If leaks are present at connections, tighten.

 b. If leaks are result of cracked, split, damaged, or restricted tubing, replace tubing (para. 3-452).

 c. Disconnect air line at both ends. If air line is clogged, clear with compressed air or sturdy wire.

 d. Connect air line.

Step 2. Determine adequacy of air supply from vehicle air compressor.

Repeat inflation sequence with engine running at 1000 rpm for three to five minutes. If two steady mode lights persist, refer to Compressed Air and Brake Systems Troubleshooting (para. 2-23).

Step 3. Check water separator and filter element for clogs and damage. Replace filter element or water separator if clogged or damaged (para. 3-465 or 3-466).

Test 2. Check air supply to tires.

Step 1. Inspect air lines between pneumatic controller and quick exhaust valves for leaks, cracks, damage, or restrictions.

Refer to malfunction 1, test 1, step 1.

Step 2. Inspect air lines between quick exhaust valves and wheels for leaks, cracks, damage, or restrictions.

Table 2-7. CTIS Doubleshooting (Contd).

MALFUNCTION
    TEST OR INSPECTION
        CORRECTIVE ACTION

Refer to malfunction 1, test 1, step 1.

Step 3. Perform leak test on hub air seals (para. 3-462). Replace faulty air seals (para. 3-460 or 3-461).

Test 3. Check condition of pneumatic controller.

Listen for clicking noise when selecting mode change. If click is not heard, replace or repair pneumatic controller (para. 3-454).

Test 4. Determine ability of system to maintain correct deflate control pressure.

WARNING

Compressed air source will not exceed 30 psi (207 kPa). When cleaning with compressed air, eyeshields must be worn. Failure to do so may result in injury to personnel.

Step 1. Check quick exhaust valves for obstructions. If clogged, clear with compressed air or sturdy wire.

Step 2. Check for continuous and audible air flow from relief valve on pneumatic controller during deflation. If air flow is continuous and audible, replace faulty relief safety valve(s) (para. 3-455).

Test 5. Determine reason for false indication.

Step 1. Perform leak test on hub air seals (para. 3-462). Replace faulty air seals (para. 3-460 or 3-461).

CAUTION

Turn battery switch to "OFF" position before disconnecting or connecting ECU wiring harness. Failure to do so may result in damage to ECU.

Step 2. Disconnect CTIS wiring harness cannon at ECU and measure voltage between pins X and V on cannon. Turn battery switch to "ON" position and check voltage.

a. If voltage is below 23 volts, replace CTIS wiring harness (para. 3-470).

b. Troubleshoot vehicle electrical system (para. 2-22).

Step 3. Check connection of electrical lead on pressure transducer at pneumatic controller. Clean and tighten.

Step 4. Replace pressure transducer (para. 3-453).

Step 5. Replace pneumatic controller (para. 3-454).

Step 6. Replace ECU (para. 3-468).

END OF TESTING!

2. FOUR PRESSURE MODE LIGHTS FLASHING

Test 1. Determine which tire(s) is low.

Step 1. Measure and record pressure of each tire (TM 9-2320-272-10). If any tire(s) is significantly outside limits, determine the cause.

Step 2. Inspect air lines between wheel valve and wheel rim for leaks, cracks, damage, and restrictions.

Refer to malfunction 1, test 1, step 1.

Table 2-7. CTIS Troubleshooting (Contd).

```
MALFUNCTION
    TEST OR INSPECTION
        CORRECTIVE ACTION
```

Step 3. With wheel valve still connected to tire, disconnect wheel valve and air manifold from hub, and check for air loss from air manifold. Place air manifold in container of water and look for air bubbles. Persistent air bubbles from air manifold indicates faulty wheel valve.

    a. If wheel valve does not leak, dry thoroughly and install.

    b. If leaking, replace or repair wheel valve (para. 3-456 or 3-457).

Step 4. After correcting malfunction, start engine (TM 9-2320-272-10) and press RUN FLAT to override ECU warning. System will then equalize all tires and inflate to selected tire pressure mode.

Test 2. Determine any non-tire related cause of four mode light indication.

Step 1. Inspect air supply lines between pneumatic controller, quick exhaust valves, and each wheel for leaks, cracks, damage, or restrictions.

    Refer to malfunction 1, test 1, step 1.

Step 2. Perform leak test on hub air seals (para. 3-462). Replace faulty air seals (para. 3-460 or 3-461).

Step 3. Check operation of pneumatic controller by listening for clicking noise when selecting mode change. If click is not heard, replace or repair pneumatic controller (para. 3-454).

CAUTION

Turn battery switch to "OFF" position before disconnecting or connecting ECU wiring harness. Failure to do so may result in damage to ECU.

Step 4. Replace ECU with known good ECU and operate CTIS (para. 3-468).

END OF TESTING!

3. FIVE INDICATOR LIGHTS FLASHING

NOTE

The ECU has a self-diagnostic capability which prevents operation of the CTIS when key elements or components of the system malfunction. When the ECU shuts down due to such system malfunction, the condition will be indicated by the flashing of all five mode lights on ECU selector panel. Since it is unlikely that the CTIS would operate properly, there is no option to override the ECU until the problem is corrected. An indication of five flashing lights can be caused by any of four conditions:

Test 1. Check air lines to tires.

Step 1. Inspect air lines between pneumatic controller and quick exhaust valves for leaks, cracks, damage, and restrictions.

    Refer to malfunction 1, test 1, step 1.

Step 2. Inspect air lines between quick exhaust valves and wheels for leaks, cracks, damage, and restrictions.

    Refer to malfunction 1, test 1, step 1.

Step 3. Perform leak test on hub air seals (para. 3-462). Replace faulty air seals (para. 3-460 or 3-461).

Table 2-7. CTIS Troubleshooting (Contd).

```
MALFUNCTION
     TEST OR INSPECTION
          CORRECTIVE ACTION
```

Test 2. Check for operation of pneumatic controller.

        Select a different mode and listen for clicking noise. If click is not heard, repair or replace pneumatic controller (para. 3-454).

Test 3. Determine if pressure switch circuit is functional.

    Step 1.   Check connection of electrical lead on pressure switch on wet tank. Clean and tighten.

CAUTION

Turn battery switch to "OFF position before disconnecting or connecting ECU wiring harness. Failure to do so may result in damage to ECU.

    Step 2.   Disconnect CTIS wiring harness cannon at ECU. Disconnect electrical lead at pressure transducer and check for continuity to ECU cannon. If harness is faulty, repair connections (para. 3-131) or replace CTIS wiring harness (para. 3-470).

    Step 3.   Connect multimeter to pressure switch at pin locations A and B. Start engine (TM 9-2320-272-10) and operate until air pressure gauge, located on instrument cluster, indicates more than 120 psi. Shut off engine and drain secondary tank (TM 9-2320-272-10) until air pressure gauge indicates less than 80 psi. Pressure switch should open (no continuity). If pressure switch remains closed (continuity) with secondary tank less than 80 psi, replace pressure switch (para. 3-467).

    Step 4.   With engine running, drain secondary air tank (TM 9-2320-272-10) until air pressure gage indicates less than 80 psi. If gauge does not reach 120 psi or greater, adjust air governor (para. 3-207).

    Step 5.   Connect CTIS wiring harness cannon to ECU. With battery and CTIS switches in "ON" position, check voltage from ECU to pressure switch. If voltage is less than 23 volts, replace ECU (para. 3-468).

    Step 6.   Connect CTIS wiring harness electrical lead to pressure switch.

Test 4. Determine if pressure transducer circuit is functional.

    Step 1.   Check connection of electrical lead on pressure transducer, located on pneumatic controller. Clean and tighten.

CAUTION

Turn battery switch to "OFF" position before disconnecting or connecting ECU wiring harness. Failure to do so may result in damage to ECU.

    Step 2.   Disconnect CTIS wiring harness cannon at ECU. Disconnect electrical lead at pressure gage and check for continuity to ECU cannon. If harness is faulty, repair connections (para. 8-131) or replace CTIS wiring harness (para. 3-470).

    Step 8.   Check voltage to ECU from vehicle's electrical supply.

            Refer to malfunction 1, test 5, step 2.

    Step 4.   Disconnect CTIS wiring harness electrical lead from pressure transducer.

    Step 5.   Check voltage from ECU to pressure transducer. Voltage should be 4.75-5.25 volts. If voltage is within limits, replace pressure transducer (para. 3-453).

    Step 6.   If voltage from ECU to pressure transducer is outside 4.75-5.25 volts, and voltage from vehicle's electrical supply to ECU is 23 volts or more, replace ECU (para. 3-468).

    Step 7.   Connect CTIS wiring harness electrical lead to pressure transducer.

END OF TESTING!

Table 2-7. CTIS Troubleshooting (Contd).

MALFUNCTION
    TEST OR INSPECTION
        CORRECTIVE ACTION

---

4. **SYSTEM REPEATEDLY RESUMES CYCLING 30 SECONDS AFTER MODE LIGHT STOPS FLASHING**

NOTE

The ECU is programmed to check tire pressures 30 seconds after completing a pressure changing sequence. The ECU does this to verify that all wheel valves are properly closed before making pressure checks at fifteen minute intervals. If the ECU repeatedly checks pressure 30 seconds after a mode light stops flashing, it is indicative of one or more tires losing air.

Test 1. Determine if tire(s) is losing air.

    Shut off engine (TM 9-2320-272-10) and listen for air leaks.

    Refer to malfunction 2, test 1, steps 1 and 2.

Test 2. Determine if wheel valve(s) is closing correctly.

  Step 1. If individual tire(s) is low and leaks cannot be traced to faulty air lines, determine if wheel valve(s) is leaking.

      Refer to malfunction 2, test 1, steps 3 and 4.

  Step 2. If both tires on same axle are low, check for restricted air line(s) to quick exhaust valve.

      Refer to malfunction 1, test 1, step 1.

  Step 3. Check quick exhaust valve for restriction.

      Refer to malfunction 1, test 4, steps 1 and 2.

  Step 4. Run engine at 1000 rpm and select HWY mode. Upon reaching air pressure indicated by steady mode light reading, shut off engine. Wait for audible air flow from pneumatic controller to stop.

  Step 5. Disconnect and plug air line connected to top of each quick exhaust valve. Check shut-off function of pneumatic controller by repeating step 4. If shut-off function does not occur, replace or repair pneumatic controller (pare. 3-454).

  Step 6. Connect air lines to top of each quick exhaust valve.

END OF TESTING!

5. **SYSTEM SHUTS OFF WRING INFLATION, SINGLE MODE LIGHT CONTINUES TO FLASH**

NOTE

If CTIS continuously shuts off air to wheel valves during air compressor recovery, perform the following teat.

Check for air line leaks.

Inspect air lines from pneumatic controller and quick exhaust valves for leaks, cracks, damage, or restrictions.

Refer to malfunction 1, test 1, step 1.

END OF TESTING!

6. **SYSTEM FAILS TO DEFLATE, PARTIALLY DEFLATES, OR TIRE PRESSURES ARE IMBALANCED**

NOTE

During deflation, all tires respond independently, Generally, failure of the entire system to deflate indicates a control problem. Failure of an individual tire to deflate indicates a problem at the wheel.

Test 1. Determine control pressure during deflation.

Table 2-7. CTIS Troubleshooting (Contd).

MALFUNCTION
TEST OR INSPECTION
CORRECTIVE ACTION

      Step 1. Inspect relief valve on pneumatic controller. If poppet is stuck out, replace relief valve (para. 3-455).

      Step 2. Determine if quick exhaust valve(s) is malfunctioning.

              Refer to malfunction 1, test 4, steps 1 and 2.

      Step 8. With CTIS at Hwy air pressure, turn battery switch to "OFF" position. Check for air leaks from pneumatic controller. If air is leaking, replace or repair pneumatic controller (para. 3-454).

      Step 4. Inspect air lines between pneumatic controller, quick exhaust valves, and wheels for leaks, cracks, damage, and restrictions.

              Refer to malfunction 1, test 1, step 1.

      Step 5. Perform leak test on hub air seals (para. 3-462). Replace faulty air seals (para. 3-460 or 3-461).

      Step 6. While operating CTIS, select a mode to cause deflation. As tire deflation occurs, a small amount of air should escape from relief valve. If no air escapes from relief valve, pneumatic controller is not functioning properly Replace or repair pneumatic controller (para. 3-454).

  Test 2. Determine if wheel valve(s) is closing properly.

WARNING

Air pressure in tire is released during step 1. Eyeshields must be worn. Failure to do so may result in injury to personnel.

CAUTION

Do not inflate any tire on the M939A2, including the spare, unless external air source is known to be as clean and dry as air supplied by vehicle. If in doubt, use vehicle's air supply. Contaminated air may disable the CTIS.

      Step 1. Check for obstructed air flow between wheel valve and wheel.

           a. Manually inflate any tire that does not deflate properly to 80 psi (552 kPa).

           b. Disconnect hose assembly from wheel valve. Note consistency of air flow from hose connected to wheel. If air flow varies, or is intermittent, sufficient water has collected in tire to cause icing of air passage. (This can happen even at temperatures above freezing.)

           c. Repeatedly till tire with dry air and deflate until flow from hose connected to wheel is consistent.

           d. Connect hose assembly to wheel valve.

      Step 2. Remove filter from wheel valve and replace filter if obstructed (refer to Chapter 2, PMCS).

      Step 3. Retest system by performing step 1. If not operational, replace wheel valve (para. 3-456 or 3-457).

END OF TESTING!

7. CONTROL PANEL LIGHTS WORK, SYSTEM FAILS TO INFLATE OR DEFLATE

  Test 1. Determine adequacy of air supply to CTIS.

              Refer to malfunction 1, test 1, steps 1 and 2.

Table 2-7. CTIS Troubleshooting (Contd).

MALFUNCTION
TEST OR INSPECTION
CORRECTIVE ACTION

CAUTION

Turn battery switch to "OFF" position before disconnecting or connecting ECU wiring harness. Failure to do so may result in damage to ECU.

Test 2. Check connections to pneumatic controller solenoids.

Disconnect CTIS wiring harness cannon from ECU, and disconnect connector from solenoid. Check for continuity. If harness is faulty, repair connections (para. 3-131) or replace CTIS wiring harness (para. 3-470).

Test 3. Check for operation of pneumatic controller.

Refer to malfunction 3, test 2.

Test 4. Determine if pressure switch is functional.

Refer to malfunction 3, test 3, steps 1 through 3.

END OF TESTING!

8. LOSS OF AMBER WARNING LIGHT AND/OR OVERSPEED PRESSURE CHANGE

NOTE

The CTIS incorporates an overspeed warning and an automatic pressure increase to prevent operation of the vehicle with insufficient tire pressure. This system incorporates a sensor to measure vehicle speed, a panel mounted flashing light, and a connection to the vehicle blackout lights to prevent flashing of the amber warning light when the vehicle is operating under blackout conditions. The ECU periodically measures the speed of the vehicle and compares the speed to programmed limits. If these limits are exceeded, the warning light and automatic pressure increase functions are actuated.

Test 1. Determine condition of warning light.

Step 1. Ensure vehicle blackout light switch is "OFF." If warning light does not flash, replace lamp with known good lamp.

Step 2. Select EMERGENCY setting on ECU to cause CTIS to deflate. Amber warning light should flash as soon as emergency pressure is obtained (steady emergency mode light). Amber warning light should go out when blackout switch is turned on.

NOTE

The ECU limits time spent in the EMERGENCY mode to ten minutes. Emergency mode must be re-selected to maintain correct test conditions.

Step 3. If warning light did not flash in step 2, disconnect blackout wire (wire #463) from CTIS wiring harness. If this causes amber warning light to flash, check vehicle wiring harness from source of blackout signal. Repair vehicle wiring harness connections as necessary (para. 3-131).

Table 2-7. CTIS Troubleshooting (Contd).

MALFUNCTION
   TEST OR INSPECTION
      CORRECTIVE ACTION

CAUTION

Turn battery switch to "OFF" position before disconnecting or connecting ECU wiring harness. Failure to do so may result in damage to ECU.

Step 4. If disconnecting blackout wire in step 3 did not cause amber warning light to flash, check CTIS wiring harness for shorts between blackout connection and ECU. If CTIS wiring harness is faulty, repair connections (para. 3-131) or replace CTIS wiring harness (para. 3-470).

Step 5. Replace ECU with known good ECU (para. 3-468).

Test 2. Determine condition of speed transducer.

Step 1. Check for continuity between ECU and speed signal generator connector. If CTIS wiring harness is faulty, repair connections (para. 3-131) or replace CTIS wiring harness (para. 3-460).

Step 2. Replace speed signal generator (para. 3-471).

Step 3. Replace ECU with known good ECU (para. 3-468).

END OF TESTING!

9. SYSTEM OVER-INFLATES TIRES

CAUTION

• When vehicle is connected to an external air source, ensure that air is clean and dry. Dirty and wet air may result in CTIS malfunctions.

• Do not operate CTIS while vehicle is connected to an external air source. This may result in over-inflation or damage to tires.

NOTE

The ECU is programmed in conjunction with vehicle air compressor output. This procedure requires air source supplied by vehicle for accurate diagnosis.

Determine accuracy of system pressure reading.

Manually balance tire pressures with external air source. Remove external air source and run CTIS. Check operation of pressure transducer.

Refer to malfunction 3, test 4, steps 1 through 7.

END OF TESTING!

10. SLOW AIR RECOVERY OR OCCASIONAL LOW AIR WARNING DURING BRAKING

Determine if pressure switch has failed in closed position.

Step 1. Check CTIS wiring harness for continuity between pressure switch and ECU cannon.

Refer to malfunction 3, test 3, steps 1 and 2.

Step 2. Check pressure switch for continuity.

Refer to malfunction 3, test 3, steps 3 through 6.

END OF TESTING!

CHAPTER 3

UNIT MAINTENANCE

Section I. ENGINE SYSTEM MAINTENANCE

3-1. ENGINE SYSTEM MAINTENANCE INDEX

PARA. NO.	TITLE	PAGE NO.
3-2.	Crankcase Breather and Tube Maintenance	3-2
3-3.	Oil Dipstick Tube Maintenance	3-8
3-4.	Engine Oil Cooler (M939A2) Maintenance	3-12
3-5.	Engine Oil Filter Maintenance	3-18
3-6.	Valve Cover (M939A2) Maintenance	3-22
3-7.	Oil Pump Pickup Hose Maintenance	3-32
3-8.	Oil Pump Return Hose Maintenance	3-34
3-9.	Front Sump Tube Maintenance	3-36
3-10.	Engine AOAP Valve Maintenance	3-38

3-2. CRANKCASE BREATHER AND TUBE MAINTENANCE

THIS TASK COVERS:

a. Breather and Tube Removal (M939/A1)
b. Breather Tube Removal (M939A2)
c. Cleaning and Inspection

d. Breather Tube Installation (M939A2)
e. Breather and Tube Installation (M939/A1)

INITIAL SETUP:

APPLICABLE MODELS

All

TOOLS

General mechanic's tool kit (Appendix E, Item 1)

MATERIALS/PARTS

Locknut (M939/A1 early model engines)
  (Appendix D, Item 272)
Lockwasher (M939/A1 early model engines)
  (Appendix D, Item 354)
Drycleaning solvent (Appendix C, Item 71)
Rags (Appendix C, Item 58)
Adhesive sealant (M939/A1)
  (Appendix C, Item 4)
Antiseize compound (Appendix C, Item 10)

REFERENCES (TM)

TM 9-2320-272-10
TM 9-2320-272-24P

EQUIPMENT CONDITION

• Parking brake set (TM 9-2320-272-10).
• Hood raised and secured (TM 9-2320-272-10).
• Right splash shield removed (M939/A1)
  (TM 9-2320-272-10).
• Coolant drained (M939A2) (para. 3-53).

GENERAL SAFETY INSTRUCTIONS

Drycleaning solvent is flammable and toxic. Do not
use near an open flame.

NOTE

References to early model engines refer to engines with a serial
number before 11246664 used in M939/A1 series vehicles.

a. Breather and Tube Removal (M939/A1)

NOTE

Perform steps 1 and 2 on M939/A1 vehicles with late model
engines and steps 3 through 5 on M939/A1 vehicles early model
engines.

1. Remove four clamps (2), two breather hoses (3), and breather tube (4) from crankcase breather (1)
    and elbow (5).

2. Remove elbow (5) from air connector (6).

3. Remove locknut (17), screw (13), and clamp (16) from mounting bracket (12) and breather tube (8).
    Discard locknut (17).

4. Remove screw (15), lockwasher (14), and mounting bracket (12) from engine (11). Discard
    lockwasher (14).

5. Remove two clamps (9), breather hose (10), and breather tube (8) from crankcase breather (1).

CAUTION

Perform step 6 only if crankcase breather requires replacement.
Removal will permanently damage breather, and it must be
replaced.

6. Mark position of crankcase breather (1) on valve cover (7) and remove breather (1).

3-2. CRANKCASE BREATHER AND TUBE MAINTENANCE (Contd)

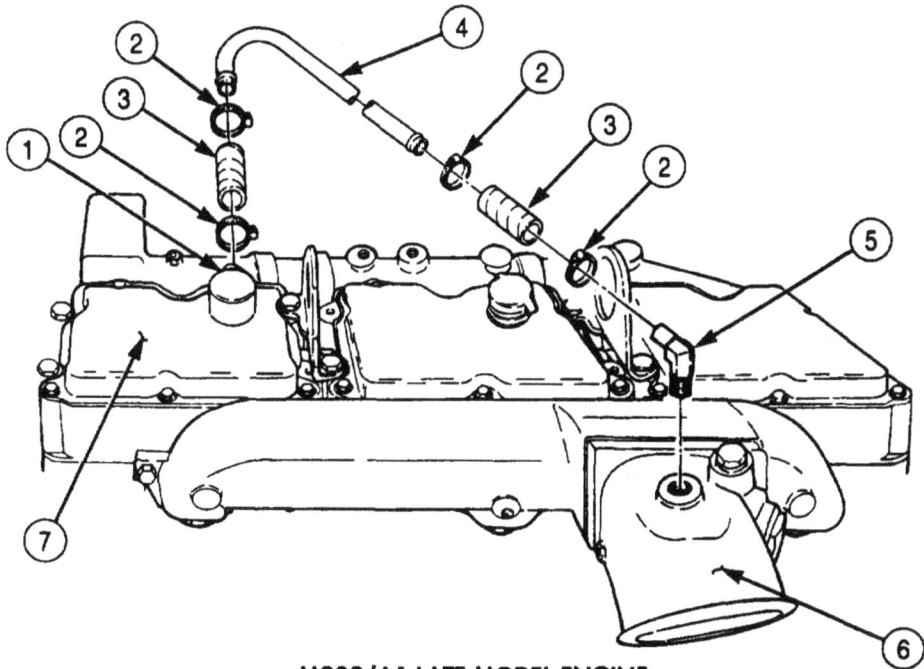

M939/A1 LATE MODEL ENGINE

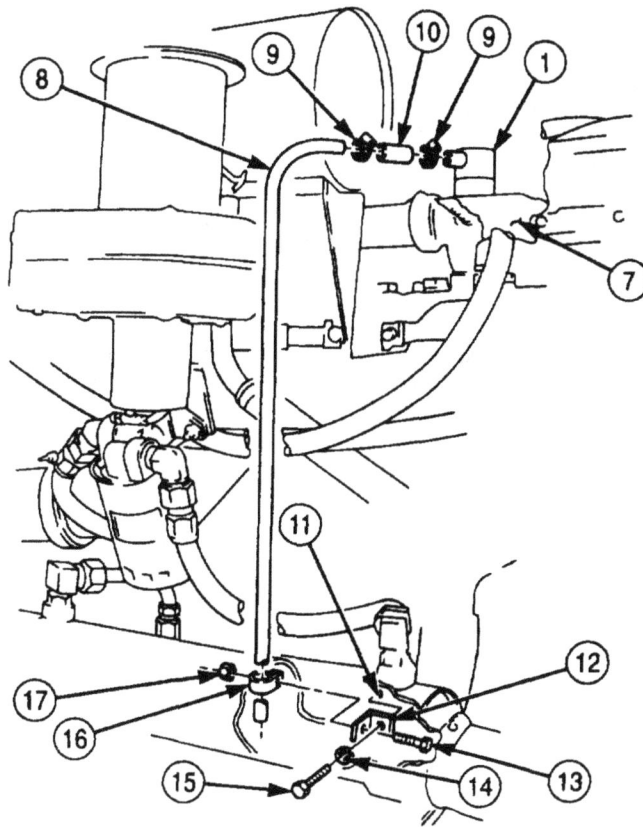

M939/A1 EARLY MODEL ENGINE

3-2. CRANKCASE BREATHER AND TUBE MAINTENANCE (Contd)

b. Breather Tube Removal (M939A2)

1. Remove four hose clamps (11) and two hoses (12) from aftercooler tubes (8) and aftercooler (13).
2. Remove two screws (4), clamps (5), and breather tube (6) from valve cover (7).
3. Remove two hose clamps (3) and hose (2) from vent connector (1) and breather tube (6).
4. Remove hose clamp (9) and hose (10) from breather tube (6).

c. Cleaning and Inspection

WARNING

Drycleaning solvent is flammable and toxic. Do not use near an open flame and always have a fire extinguisher nearby when solvents are used. Use only in well-ventilated places, wear protective clothing, and dispose of cleaning rags in approved container. Failure to do this may result in injury to personnel and/or damage to equipment.

1. Clean breather tube (6) with drycleaning solvent and dry with clean rag.

NOTE

Perform step 2 for M939/A1 series vehicles and only if previously removed.

2. Clean crankcase breather (14) and mating surface on valve cover (16).
3. Inspect breather tube (6) for obstructions and kinks that could cause restrictions. Remove obstructions; replace if kinked.
4. Inspect hose (10) for obstructions, flat spots, or other deformities that could cause restrictions. Remove obstructions; replace hose (10) if bent, deformed, or torn.

d. Breather Tube Installation (M939A2)

1. Install hose (10) on breather tube (6) with clamp (9).
2. Install hose (2) on breather tube (6) and vent connector (1) with two clamps (3).
3. Install two clamps (5) on valve cover (7) with two screws (4).
4. Install two hoses (12) on aftercooler (13) and aftercooler tubes (8) with four clamps (11).

CAUTION

For M939A2 series vehicles, ensure drainvalve on aftercooler is open when filling cooling system. Failure to do so may result in damage to equipment.

3-2. CRANKCASE BREATHER AND TUBE MAINTENANCE (Contd)

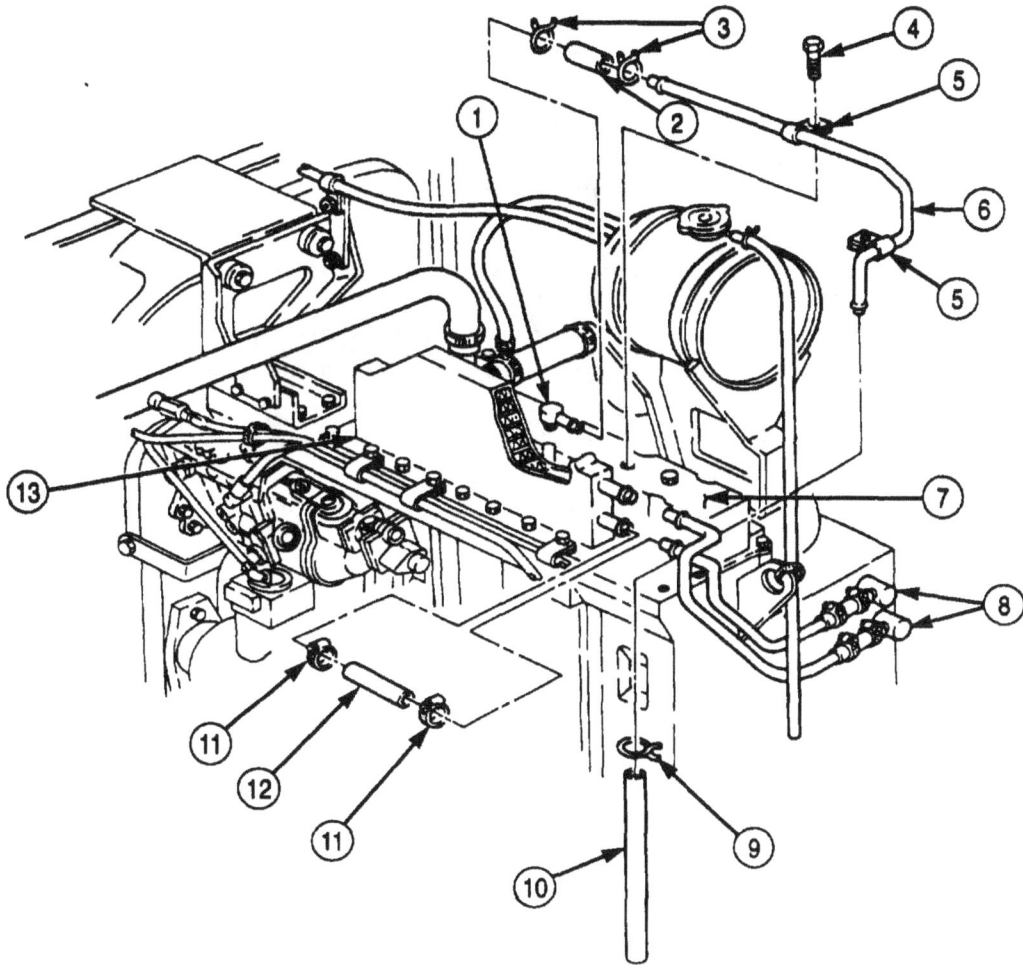

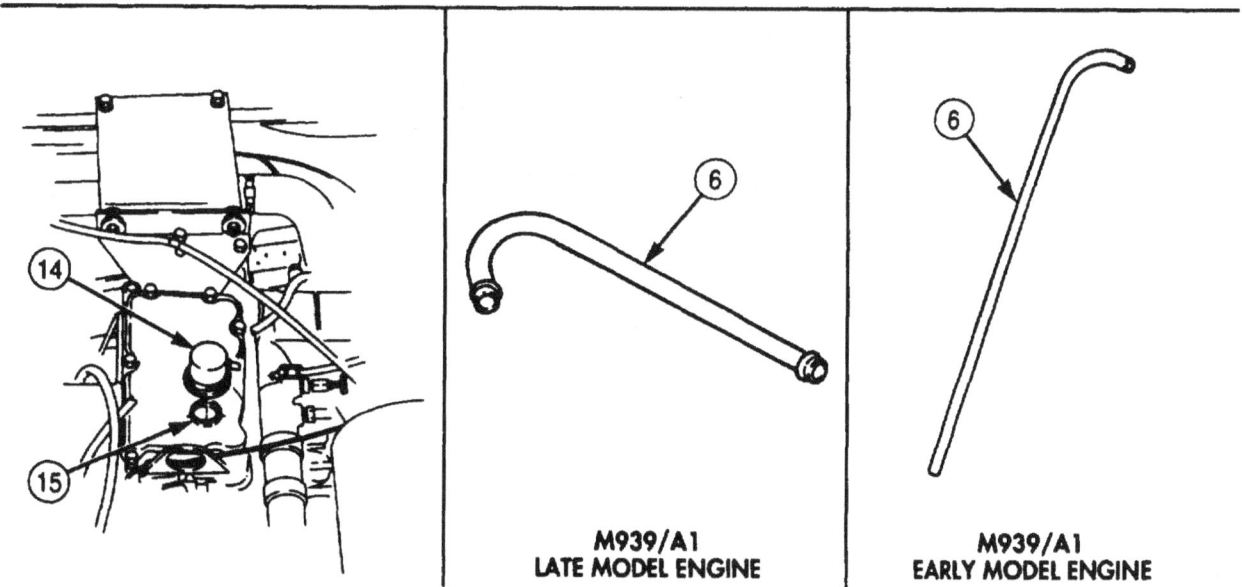

M939/A1
LATE MODEL ENGINE

M939/A1
EARLY MODEL ENGINE

3-2. CRANKCASE BREATHER AND TUBE MAINTENANCE (Contd)

e. Breather and Tube Installation (M939/A1)

NOTE

Perform steps 1 and 2 if breather was removed.

1. Apply adhesive sealant to mating surfaces of crankcase breather (1) and valve cover (7).

2. Install crankcase breather (1) on valve cover (7) and align with mark.

NOTE

Perform steps 3 and 4 on vehicles with late model engines and
steps 5 through 7 on vehicles with early model engines.

3. Apply antiseize compound to threads of elbow (5) and install on air connector (6).

4. Install two breather hoses (3) and breather tube (4) on elbow (5) and crankcase breather (1) with four hose clamps (2).

5. Install breather hose (10) and breather tube (8) on crankcase breather (1) with two clamps (9).

6. Install mounting bracket (12) on engine (11) with new lockwasher (14) and screw (15).

7. Install clamp (16) on mounting bracket (12) and breather tube (8) with screw (13) and new locknut (17).

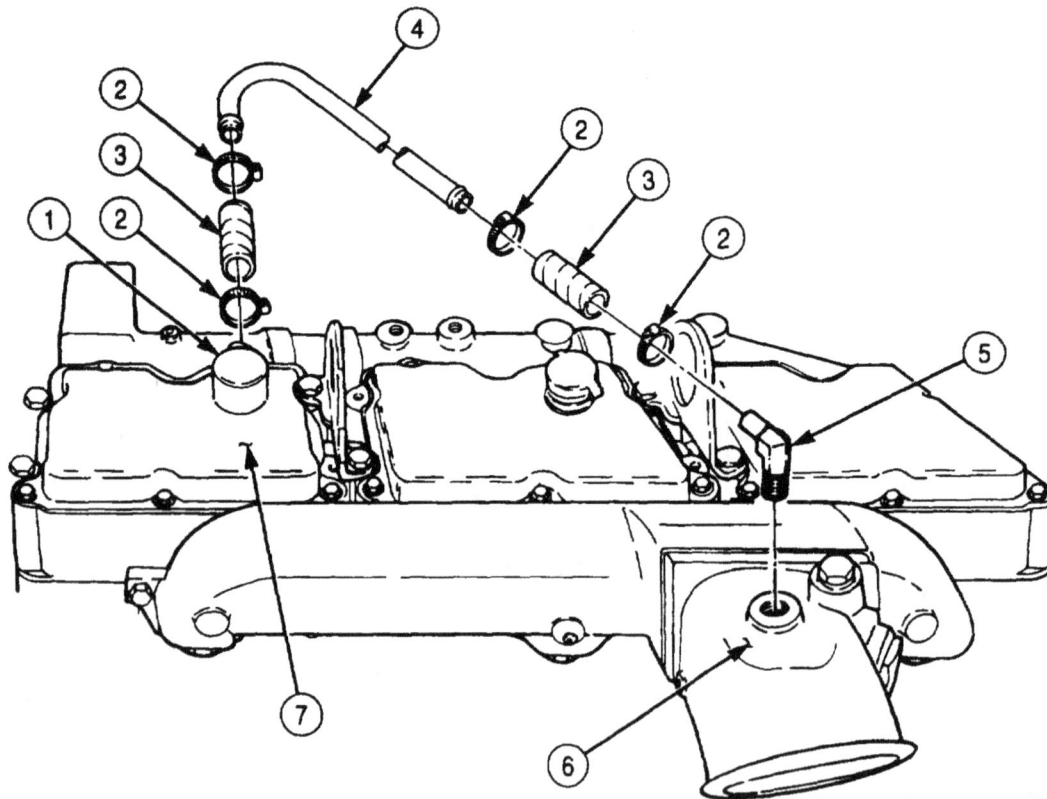

M939/A1 LATE MODEL ENGINE

3-2. CRANKCASE BREATHER AND TUBE MAINTENANCE (Contd)

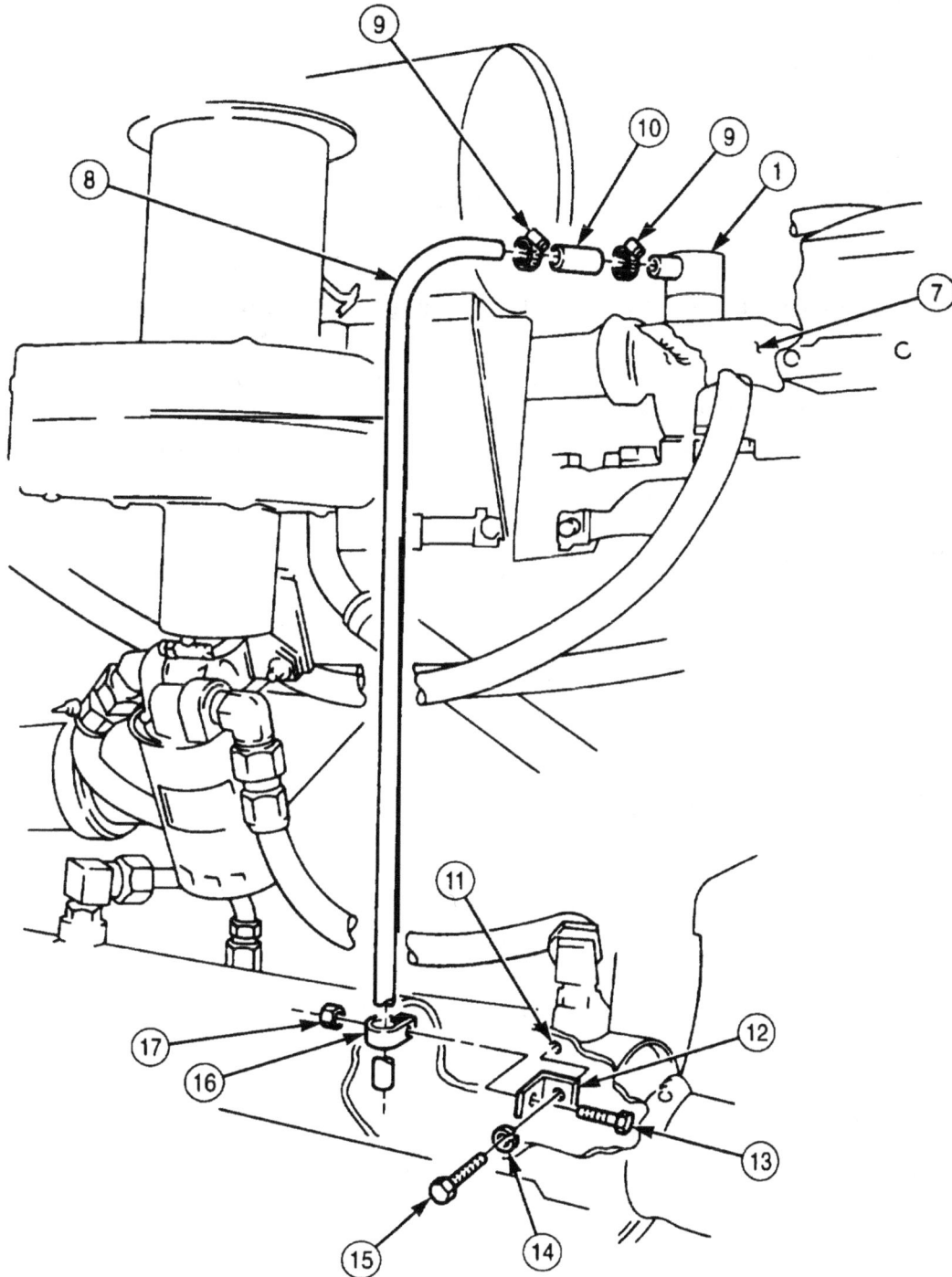

M939/A1 EARLY MODEL ENGINE

FOLLOW-ON TASKS: • Fill coolant to proper level (M939A2) (para. 3-53).
                   • Install right splash shield (M939/A1) (TM 9-2320-272-10).

3-3. OIL DIPSTICK TUBE MAINTENANCE

THIS TASK COVERS:

a. Removal (M939/A1)   d. Installation (M939/A1)
b. Removal (M939A2)   e. Installation (M939A2)
c. Inspection

INITIAL SETUP:

APPLICABLE MODELS   REFERENCES (TM)
All   TM 9-2320-272-10
                                     TM 9-2320-272-24P
TOOLS
General mechanic's tool kit (Appendix E, Item 1)   EQUIPMENT CONDITION
                                     • Hoed raised and secured (TM 9-2320-272-10).
MATERIALS/PARTS   • Parking brake set (TM 9-2320-272-10).
Lockwasher (M939/A1) (Appendix D, Item 354)   • Right splash shield removed (TM 9-2320-272-10).
Lockwasher (M939/A1) (Appendix D, Item 358)
Adhesive sealant (M939A2) (Appendix C. Item 5)

---

a. Removal (M939/A1)

1. Remove dipstick (1) from dipstick tube (9)

2. Remove nut (12), lockwasher (11), washer (10), screw (7), and clamp (8) from dipstick tube (9) and bracket (4). Discard lockwasher (11).

3. Remove dipstick tube (9) from oil pan (6).

4. Remove screw (2), lockwasher (3), and bracket (4) from engine block (5). Discard lockwasher (3).

b. Removal (M939A2)

1. Remove dipstick (14) from dipstick tube (15).

2. Remove dipstick tube (15) from engine block (5).

c. Inspection

1. Inspect dipstick (1) or (14). Replace if cracked or broken.

NOTE

Perform step 2 for M939/A1 series vehicles.

2. Inspect rubber compression seal (13) on dipstick (1). Replace dipstick (1) if compression seal (13) is damaged.

3-3. OIL DIPSTICK TUBE MAINTENANCE (Contd)

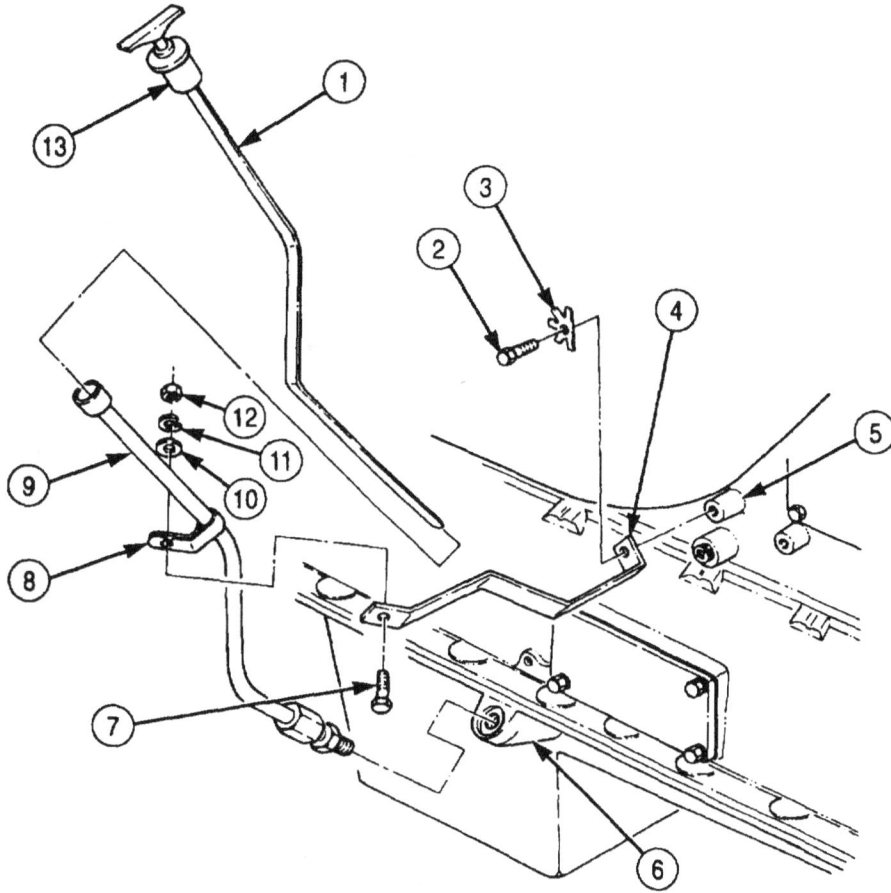

M939/A1

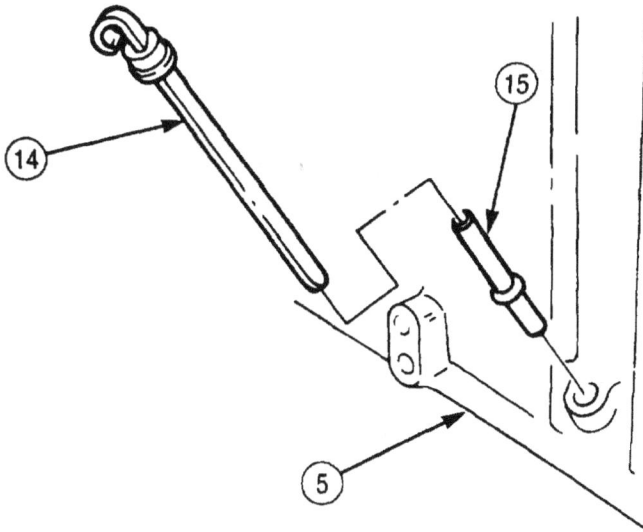

M939A2

3-3. OIL DIPSTICK TUBE MAINTENANCE (Contd)

d. Installation (M939/A1)

1. Install bracket (8) on engine block (9) with new lockwasher (7) and screw (6).
2. Install dipstick tube (13) on oil pan (10).
3. Install clamp (12) on dipstick tube (13) and bracket (8) with screw (11), washer (14), new lockwasher (15), and nut (16).
5. Install dipstick (5) in dipstick tube (13).

e. Installation (M939A2)

1. Apply adhesive sealant to mating surface of oil dipstick tube (2).
2. Align oil dipstick tube (2) with bore (3) and, using soft-headed hammer, drive oil dipstick tube (2) until it seats in the bore (3) of engine block (4).
3. Install dipstick (1) in dipstick tube (2).

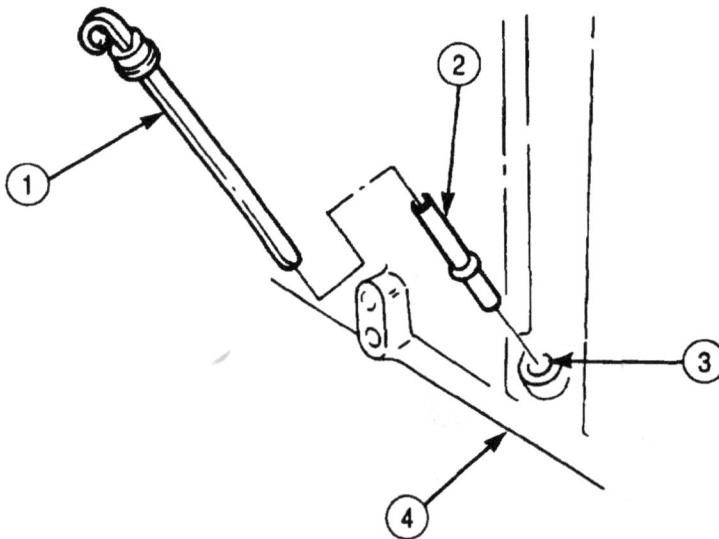

M939A2

3-3. OIL DIPSTICK TUBE MAINTENANCE (Contd)

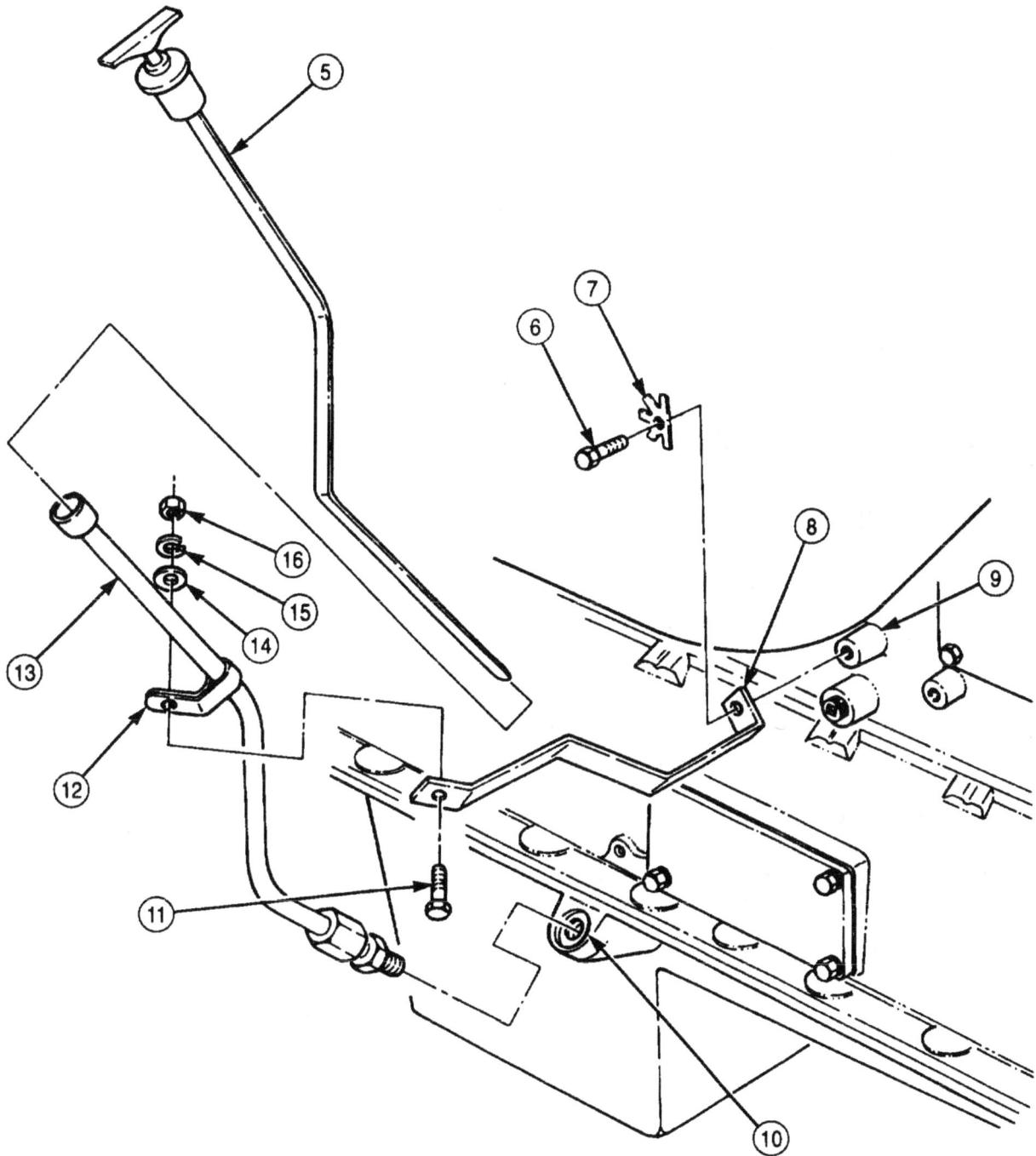

M939/A1

FOLLOW-ON TASKS: • Install right splash shield (TM 9-2320-272-10).
• Start engine (TM 9-2320-272-10) and check for oil leaks.

3-4. ENGINE OIL COOLER (M939A2) MAINTENANCE

THIS TASK COVERS:

a. Removal
b. Disassembly
c. Cleaning and Inspection

d. Assembly
e. Installation

INITIAL SETUP:

APPLICABLE MODELS

M939A2

TOOLS

General mechanic's tool kit (Appendix E, Item 1)
Torque wrench (Appendix E, Item 144)
Soft-jawed vise

MATERIALS/PARTS

Bypass valve (Appendix D, Item 31)
Gasket (Appendix D, Item 215)
Gasket (Appendix D, Item 230)
O-ring (early model engines only)
  (Appendix D, Item 476)
Washer (Appendix D, Item 716)
Antiseize tape (Appendix C, Item 72)

REFERENCES (TM)

TM 9-2320-272-10
TM 9-2320-272-24P

EQUIPMENT CONDITION

• Parking brake set (TM 9-2320-272-10).
• Hood raised and secured (TM 9-2320-272-10).
• Right splash shield removed (TM 9-2320-272-10).
• Coolant drained (para. 3-53).
• Engine oil drained and filter removed (para. 3-5).
• Alternator removed (para. 3-80).

GENERAL SAFETY INSTRUCTIONS

Pressure regulator is under spring tension. Use care when removing plug.

NOTE

References to early model engines refer to engines with a serial number before 44487830 used in M939A2 series vehicles.

a. Removal

1. Loosen clamp (12) and remove hose (13) from nipple (11).

NOTE

Perform step 2 on vehicles with late model engines and step 3 on early model engines.

2. Remove oil supply line (1), adapter (2), and O-ring (3) from oil filter head (4). Discard O-ring (3).

3. Remove oil supply line (1) and elbow (15) from oil filter head (4).

4. Remove eleven screws (6), clamp (5), wiring harness (7), oil filter head (4), gasket (8), oil cooler core (9), and gasket (10) from engine block (14). Discard gaskets (8) and (10).

3-4. ENGINE OIL COOLER (M939A2) MAINTENANCE (Contd)

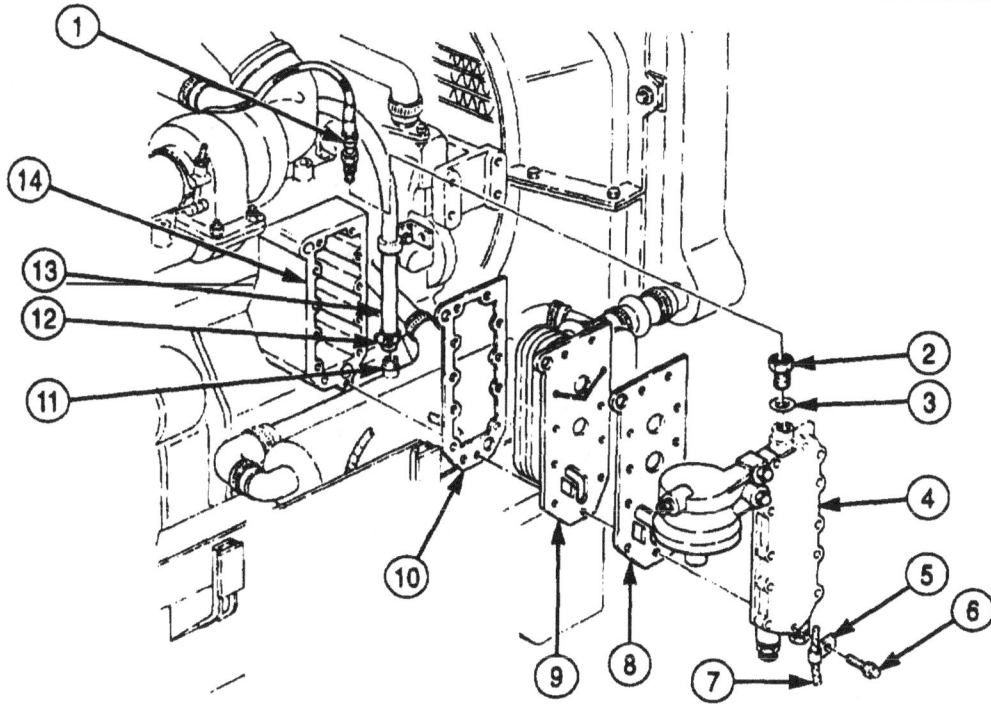

LATE MODEL ENGINE

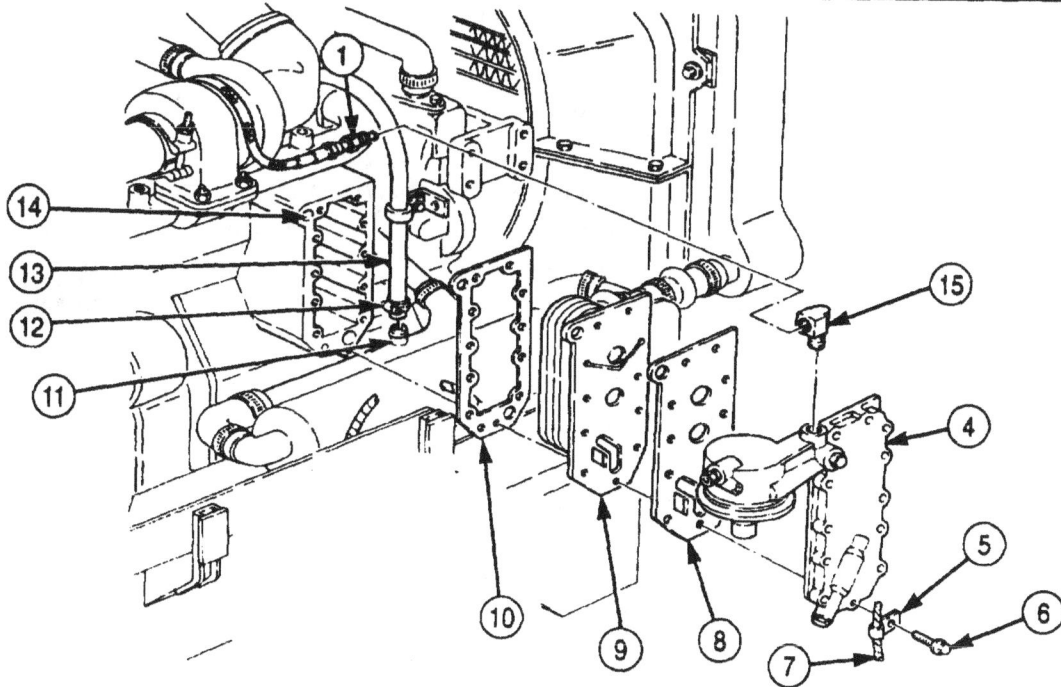

EARLY MODEL ENGINE

3-4. ENGINE OIL COOLER (M939A2) MAINTENANCE (Contd)

b. Disassembly

1. Install oil filter head (3) in soft-jawed vise.

WARNING

Pressure regulator is under spring tension. Use care when
removing plug. Failure to do so may result in injury to personnel,

2. Remove plug (7), washer (6), spring (8), and pressure regulator plunger (9) from port (10) and bore (11). Discard washer (6).
3. Remove oil filter adapter (1) from oil filter head (3).
4. Remove plugs (2) and (4) from oil filter head (3).

CAUTION

When removing bypass valve, be careful not to score oil filter
head. Failure to do so may cause an improper seal resulting in
damage to engine.

5. Using screwdriver, position blade of screwdriver behind bypass valve (5) and pry bypass valve (5) out of oil filter head (3). Discard bypass valve (5).
6. Remove oil filter head (3) from soft-jawed vise.

c. Cleaning and Inspection

1. For general cleaning instructions, refer to para. 2-14.
2. For general inspection instructions, refer to para. 2-15.
3. Inspect pressure regulator plunger (9) for freedom of movement in bore (11). Replace plunger (9) if sticking or binding in bore (11).
4. Inspect spring (8) for pits and cracks. Replace spring (8) if damaged.

d. Assembly

NOTE

Ensure bypass valve is completely seated in oil filter head.

1. Install oil filter head (3) in soft-jawed vise. Install new bypass valve (5) on oil filter head (3).
2. Wrap male threads of oil filter adapter (1) and plugs (2), (4), and (7) with antiseize tape.
3. Install oil filter adapter (1) on oil filter head (3). Tighten oil filter adapter (1) 60 lb-ft (81 N•m).
4. Install pressure regulator plunger (9), spring (8), new washer (6), and plug (7) in bore (11) at port (10). Tighten plug (7) 60 lb-ft (81 N•m).
5. Install plugs (2) and (4) on oil filter head (3).
6. Remove oil filter head (3) from soft-jawed vise.

3-4. ENGINE OIL COOLER (M939A2) MAINTENANCE (Contd)

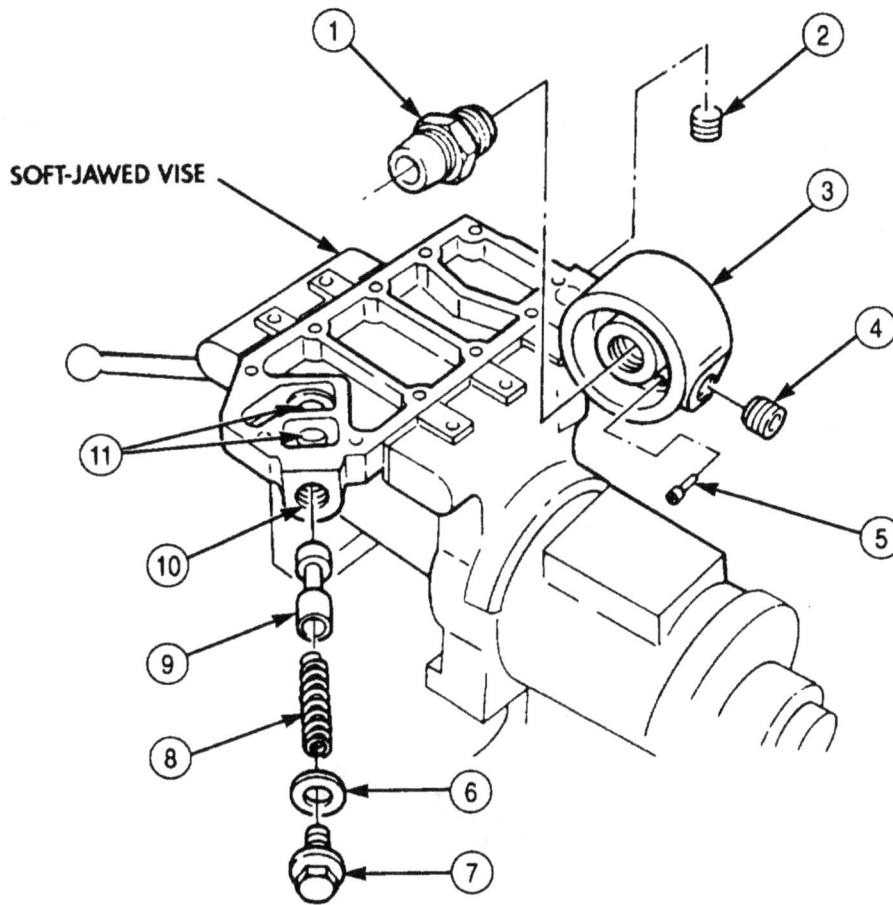

SOFT-JAWED VISE

3-4. ENGINE OIL COOLER (M939A2) MAINTENANCE (Contd)

e. Installation

NOTE

- If a late model engine oil filter head is to be installed on an early model engine, the oil supply system must be replaced, including oil pump, pressure regulator valve, oil supply line. and elbow.

- Perform step 1 if new oil cooler core is being installed.

- Perform step 3 on vehicles with late model engines and step 4 on vehicles with early model engines.

1. Remove two shipping plugs (16) from new oil cooler core (8). Discard shipping plugs (16).

2. Install new gasket (9), oil cooler core (8), new gasket (7), oil filter head (3), wiring harness (6), and clamp (4) on engine block (13) with eleven screws (5).

3. Install new O-ring (15) and adapter (14) on oil filter head (3) and connect oil supply line (1) to adapter (14).

4. Install elbow (2) on oil filter head (3) and connect oil supply line (1) to elbow (2).

5. Install hose (12) on nipple (10) with clamp (11).

CAUTION

When filling cooling system on M939A2 series vehicles, ensure drainvalve on aftercooler is open. Failure to do so may result in damage to equipment.

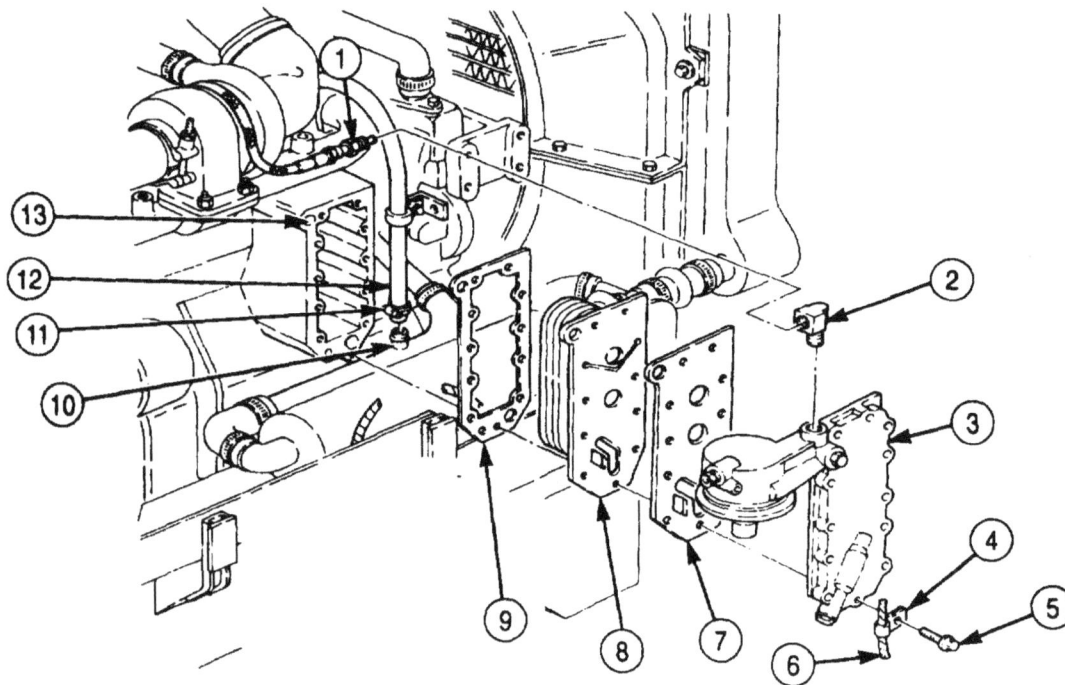

EARLY MODEL ENGINE

3-4. ENGINE OIL COOLER (M939A2) MAINTENANCE (Contd)

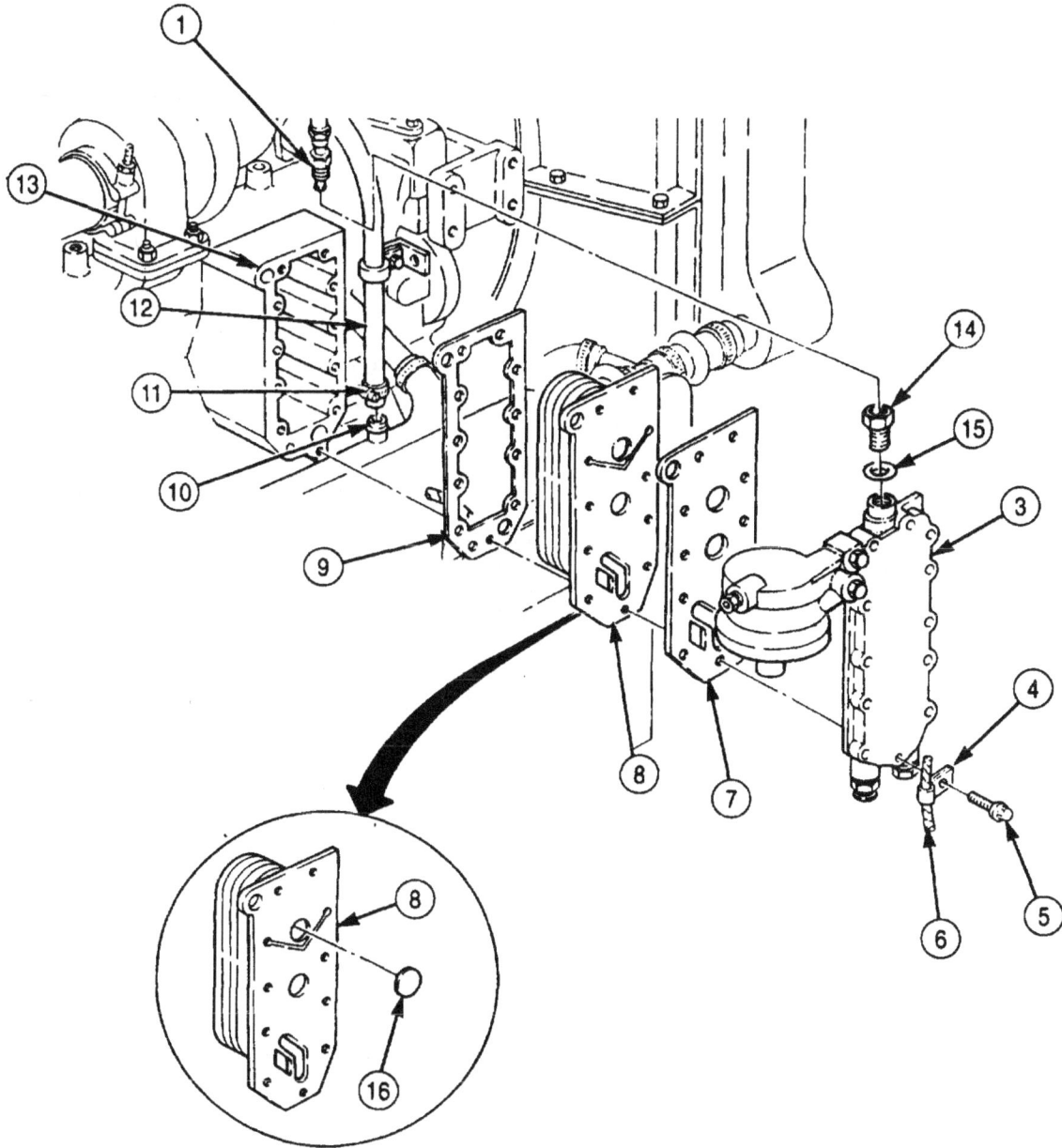

LATE MODEL ENGINE

FOLLOW-ON TASKS: • Fill cooling system to proper level (para. 3-53).
• Install alternator (para. 3-80).
• Install filter and fill engine oil (para. 3-5).
• Start engine (TM 9-2320-272-10) and check for leaks.
• Install right splash shield (TM 9-2320-272-10).

3-5. ENGINE OIL FILTER MAINTENANCE

THIS TASK COVERS:

a. Draining Oil
b. Oil Filter Removal
c. Filter Shell and Bolt Disassembly (M939/A1)

d. Cleaning and Inspection
e. Filter Shell and Bolt Assembly (M939/A1)
f. Oil Filter Installation

INITIAL SETUP:

APPLICABLE MODELS
All

TOOLS
General mechanic's tool kit (Appendix E, Item 1)
Torque wrench (Appendix E, Item 144)

MATERIALS/PARTS
Seal (Appendix D, Item 596)
Filter (M939A2) (Appendix D, Item 120)
Gasket (M939/A1) (Appendix D, Item 145)
Kit, filter (M939/A1) (Appendix D, Item 125)
Seal (M939/A1) (Appendix D, Item 594)
Drycleaning solvent (Appendix C, Item 71)
Oil, diesel (Appendix C, Item 41)
Rags (Appendix C, Item 58)

REFERENCES (TM)
LO 9-2320-272-12
TM 9-2320-272-10
TM 9-2320-272-24P

EQUIPMENT CONDITION
• Parking brake set (TM 9-2320-272-10)
• Hood raised and secured (TM 9-2320-272-10).
• Right splash shield removed (M939A2) (TM 9-2320-272-10).
• Left splash shield removed (M939/A1) (TM 9-2320-272-10).

GENERAL SAFETY INSTRUCTIONS
• Store and dispose of used oil properly.
• Do not perform task when engine is hot or drain oil when hot.
• Drycleaning solvent is flammable and toxic.
• When cleaning with compressed air, wear eyeshield and ensure source pressure does not exceed 30 psi (207 kPa).

WARNING

• Accidental or intentional introduction of liquid contaminants into the environment is in violation of state, federal, and military regulations. Refer to Army POL (para. 1-8) for information concerning storage, use, and disposal of these liquids. Failure to do so may result in injury or death.

• Do not drain oil from engine or remove lines containing oil when engine is hot. Doing so may result in injury to personnel.

NOTE

References to early model engines refer to engines with a serial number before 444487830 used in M939A2 series vehicles.

a. Draining Oil

NOTE

• Have adequate drainage container ready to catch oil.

• Perform steps 1 and 2 for M939/A1 series vehicles.

1. Remove drainplug (2) from filter shell (1) and drain oil completely from filter shell (1).

2. Install drainplug (2) in filter shell (1) and tighten drainplug (2) 25-35 lb-ft (34-47 N•m).

3. Remove drainplug (4) and seal (5) from oil pan (3) and drain oil. Discard seal (5).

3-5. ENGINE OIL FILTER MAINTENANCE (Contd)

WARNING

- Drycleaning solvent is flammable and toxic. Do not use near an open flame and always have a fire extinguisher nearby when solvents are used. Use only in well-ventilated places, wear protective clothing, and dispose of cleaning rags in approved container. Failure to do this may result in injury to personnel and/or damage to equipment.

- Eyeshields must be worn when cleaning with compressed air. Compressed air source will not exceed 30 psi (207 kPa). Failure to do so may result in injury to personnel.

4. Clean drainplug (4) with drycleaning solvent and wipe area around drain hole (6).

5. Install new seal (5) and drainplug (4) on oil pan (3). Tighten drainplug (4) 100 lb-ft (136 N•m) on M939/A1 series vehicles and 60 lb-ft (81 N•m) on M939A2 series vehicles.

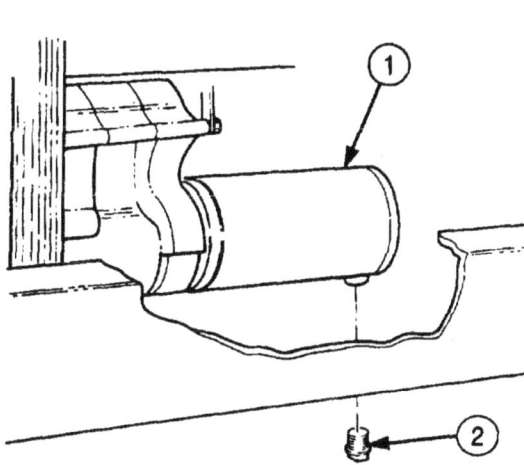

M939/A1

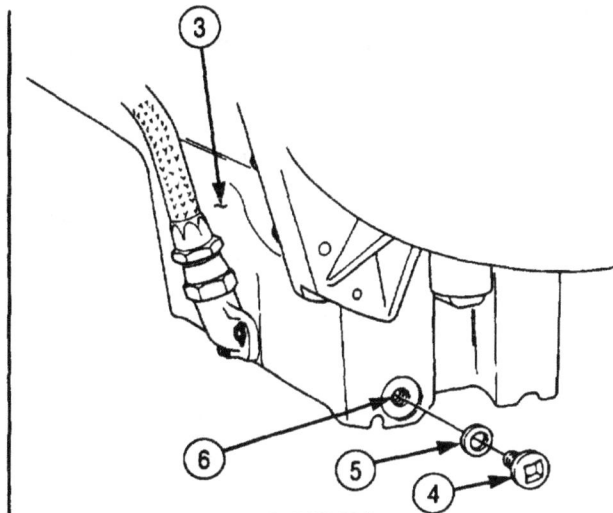

M939/A1

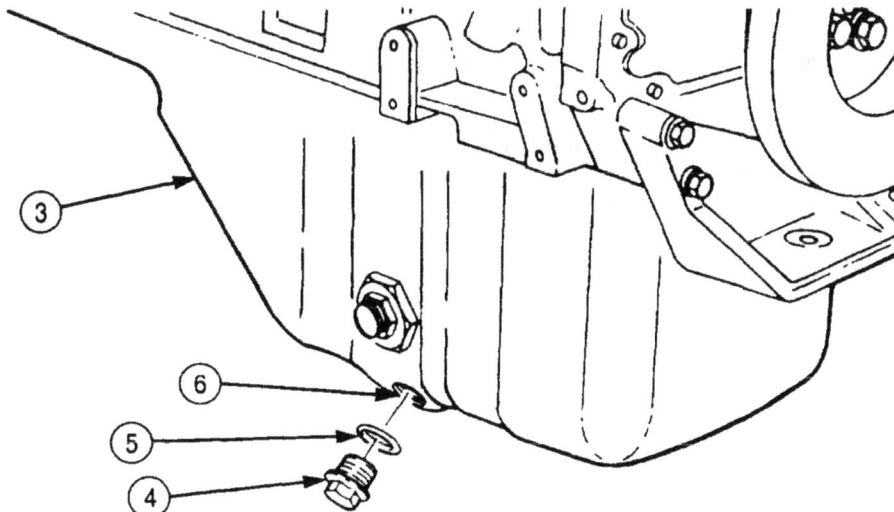

M939A2

3-5. ENGINE OIL FILTER MAINTENANCE (Contd)

b. Oil Filter Removal

NOTE

- Perform steps 1 and 2 for oil filter on M939/A1 series vehicles and step 3 for oil filter on M939A2 series vehicles.
- Support oil filter shell while loosening center bolt. Lightly tap filter shell with palm of hand to loosen from filter base.

1. Loosen center bolt (9) until oil filter shell (11) can be removed from filter base (1), and remove oil filter shell (11) from filter base (1).

2. Remove oil filter element (3) and seal (2) from oil filter shell (11). Discard oil filter element (3) and seal (2).

3. Remove oil filter (13) from adapter (12). Discard oil filter (13).

c. Filter Shell and Bolt Disassembly (M939/A1)

1. Remove center bolt retaining pin (4), filter element support (5), gasket (6), washer (7), and spring (8) from center bolt (9) and oil filter shell (11). Discard gasket (6).

2. Remove center bolt seal (10) and center bolt (9) from oil filter shell (11). Discard center bolt seal (10).

d. Cleaning and Inspection

WARNING

- Drycleaning solvent is flammable and toxic. Do not use near an open flame and always have a fire extinguisher nearby when solvents are used. Use only in well-ventilated places, wear protective clothing, and dispose of cleaning rags in approved container. Failure to do this may result in injury to personnel and/or damage to equipment.
- Eyeshields must be worn when cleaning with compressed air. Compressed air source will not exceed 30 psi (207 kPa). Failure to do 80 may result in injury to personnel.

1. Clean center bolt (9) and oil filter shell (11) with drycleaning solvent and dry with compressed air.

2. Inspect center bolt (9). Replace if threads are stripped.

3. Inspect filter base (1). If cracked, nicked, or threads are stripped, notify DS maintenance.

4. Inspect filter element support (5). Replace if cracked or grooved.

5. Inspect spring (8). Replace if cracked or broken.

e. Filter Shell and Bolt Assembly (M939/A1)

1. Install new center bolt seal (10) and center bolt (9) in oil filter shell (11).

2. Install spring (8), washer (7), new gasket (6), filter element support (5), and center bolt retaining pin (4) on center bolt (9) and oil filter shell (11).

3-5. ENGINE OIL FILTER MAINTENANCE (Contd)

f. Oil Filter Installation

NOTE

- Perform steps 1 through 3 for oil filters on M939/A1 series vehicles and steps 4 and 5 for oil filters on M939A2 series vehicles.
- Ensure correct filter is used during installation. Filter element and oil filter shell seal used on the M939/A1 series vehicle engines are supplied as a kit. Early and late model engines in M939A2 series vehicles use different filters.

1. Install new oil filter element (3) in oil filter shell (11).
2. Install new seal (2) and oil filter shell (11) on filter base (1). Hand-tighten center bolt (9) until oil filter shell (11) seats in filter base (1).
3. Tighten center bolt (9) 25-35 lb-ft (34-47 N•m).
4. Fill new oil filter (13) with clean engine oil and install on adapter (12). Hand-tighten.
5. Tighten oil filter (13) an additional 3/4 turn.

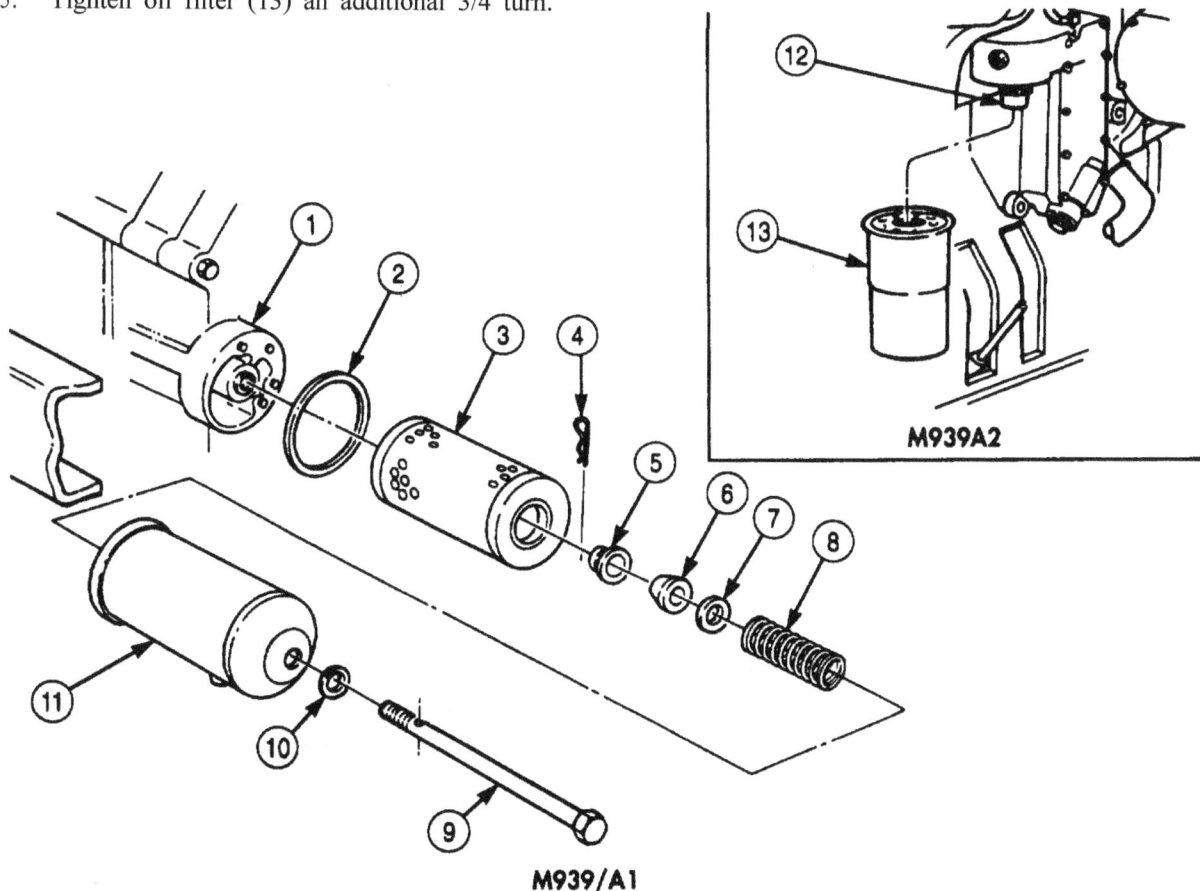

M939A2

M939/A1

FOLLOW-ON TASKS:
- Fill engine with oil to proper level (LO 9-2320-272-12).
- Start engine (TM 9-2320-272-10) and check for leaks.
- Check oil level and add oil as necessary (TM 9-2320-272-10).
- Install right splash shield (M939A2 only) (TM 9-2320-272-10).
- Install left splash shield (M939/A1 only) (TM 9-2320-272-10).

3-6. VALVE COVER (M939A2) MAINTENANCE

THIS TASK COVERS:

a. Removal
b. Cleaning and Inspection   c. Installation

INITIAL SETUP:

APPLICABLE MODELS

M939A2

TOOLS

General mechanic's tool kit (Appendix E, Item 1)
Torque wrench (Appendix E, Item 144)

MATERIALS/PARTS

Gasket (Appendix D, Item 160)
Two locknuts (Appendix D, Item 280)
Six O-rings (Appendix D, Item 427)
Seal (Appendix D, Item 600)
O-ring (Appendix D, Item 431)
Cap and plug set (Appendix C, Item 14)

REFERENCES (TM)

TM 9-2320-272-10
TM 9-2320-272-24P

EQUIPMENT CONDITION

• Parking brake set (TM 9-2320-272-10).
• Hood raised and secured (TM 9-2320-272-10).
• Reservoirs drained (TM 9-2320-272-10).
• Coolant drained (para. 3-53).
• Air cleaner hose removed (para. 3-13).

GENERAL SAFETY INSTRUCTIONS

• Drycleaning solvent is flammable and toxic.
• Ensure drainvalve on aftercooler is open when filling cooling system.

---

a. Removal

CAUTION

Cover turbocharger ports immediately after removing attaching parts to prevent foreign objects from lodging in blades of turbocharger.

NOTE

• Tag all lines and connections for later installation.
• Plug all lines when removed to prevent dripping fluids and contaminated lines.

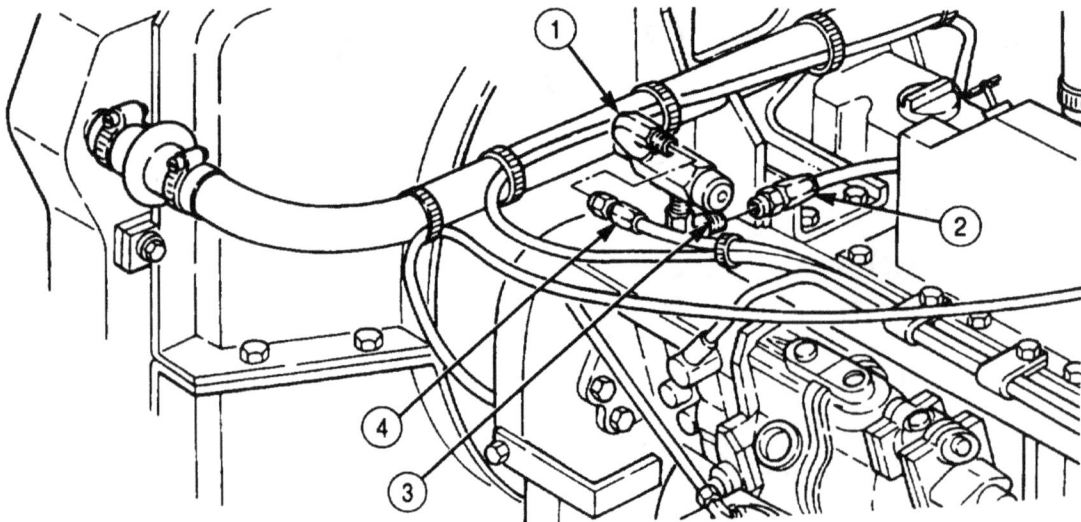

3-6. VALVE COVER (M939A2) MAINTENANCE (Contd)

1. Disconnect air lines (2) and (4) from elbows (3) and (1).
2. Remove four clamps (14) and two hoses (15) from aftercooler tubes (11) and aftercooler (16).
3. Remove two screws (8) and clamps (9) from breather tube (7) and valve cover (10).
4. Remove clamps (5) and hose (6) from vent connector (17) and breather tube (7).
5. Remove clamp (12) and hose (13) from breather tube (7).

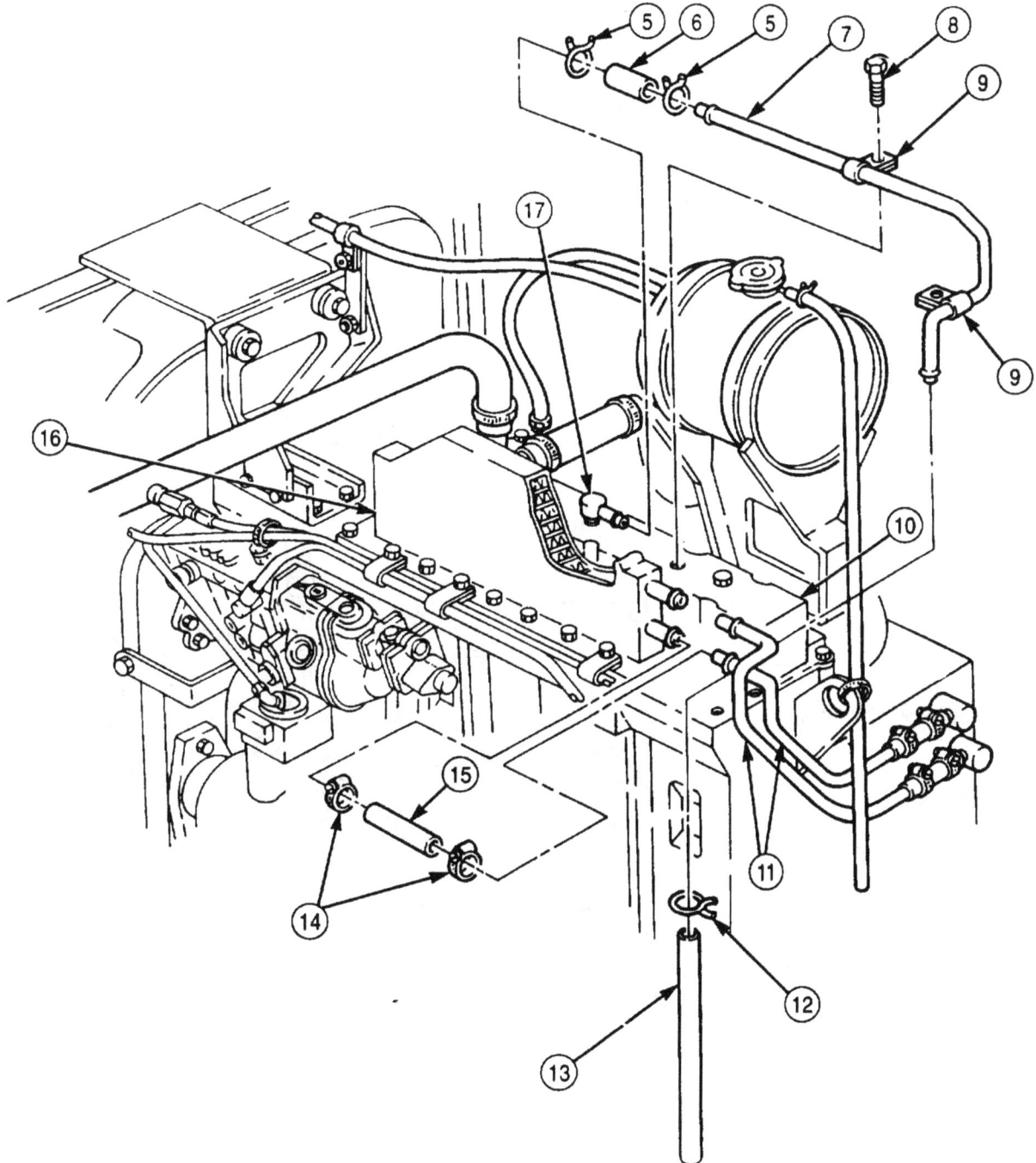

3-6. VALVE COVER (M939A2) MAINTENANCE (Contd)

6. Remove clamps (1) from radiator inlet tube (2), hose (10), and cable (11).
7. Remove two locknuts (6), washers (5), screws (3), and washers (4) from bracket (9) and inlet tube (2). Discard locknuts (6).
8. Loosen clamps (7) and (12) on elbow (8) and hose (13).
9. Remove radiator inlet tube (2) from elbow (8) and hose (13).

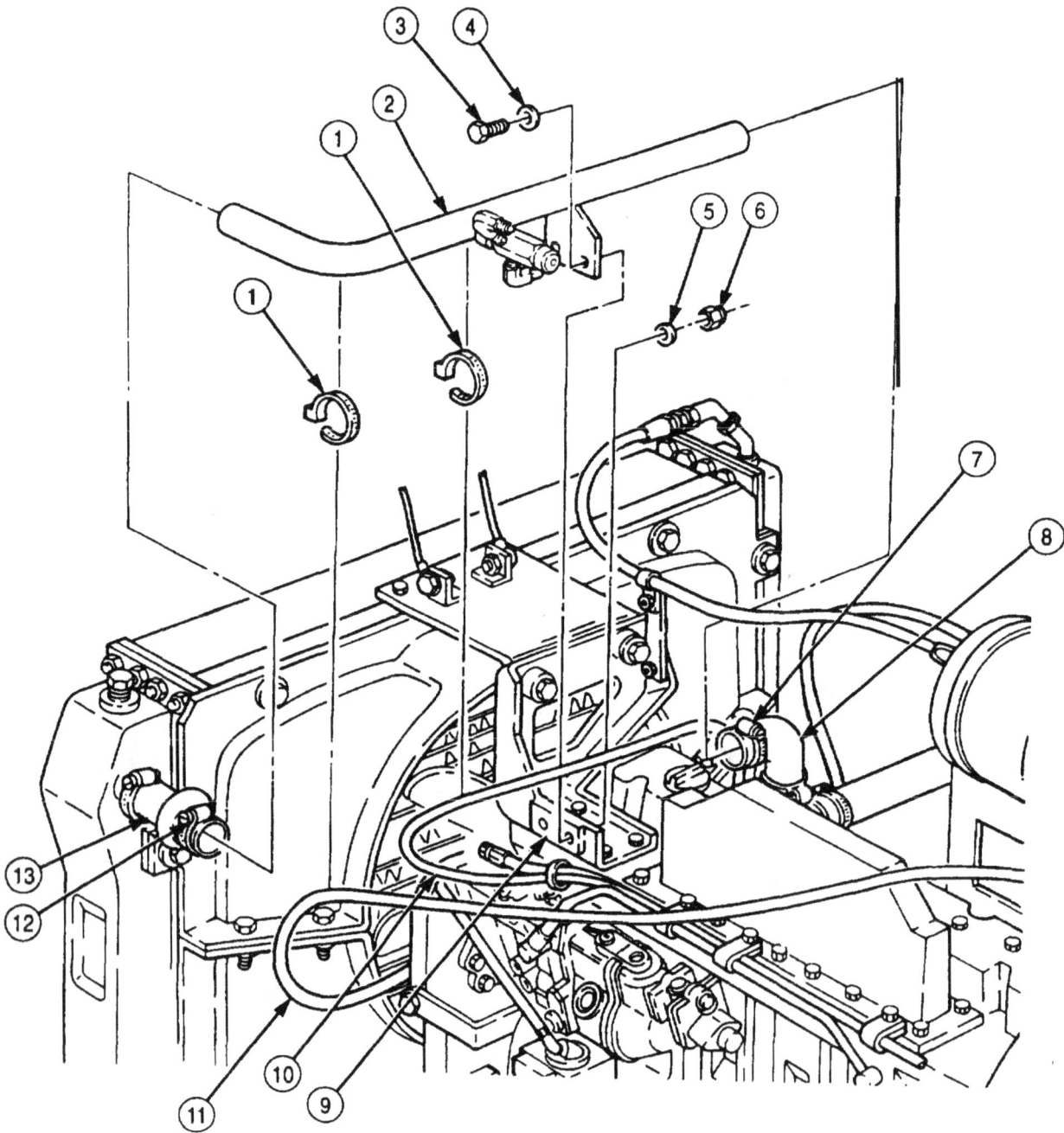

3-6. VALVE COVER (M939A2) MAINTENANCE (Contd)

10. Loosen clamps (19), (24), and (26) on hoses (18), (23), and (25).
11. Remove hoses (17), (18), (23), and (25) from surge tank assembly (15).
12. Remove nut (20) and screw (14) from surge tank assembly (15) and angle support bracket (21).
13. Remove two screws (16) and surge tank assembly (15) from exhaust manifold (22).

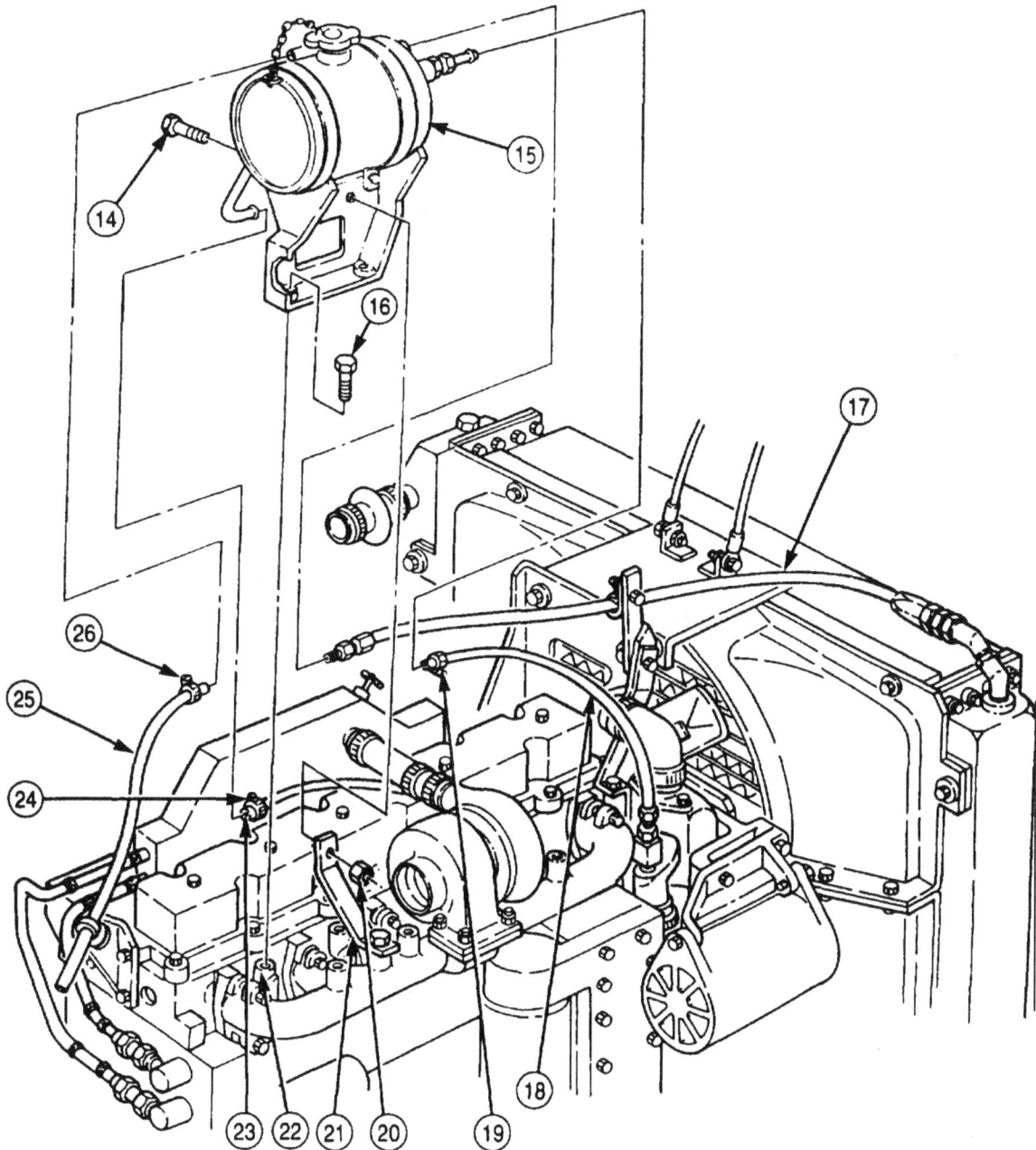

3-6. VALVE COVER (M939A2) MAINTENANCE (Contd)

14. Loosen four clamps (8) on crossover tube (9) and two hoses (7).

15. Slide two hoses (7) over crossover tube (9) and remove crossover tube (9) from turbocharger (13) and aftercooler (15).

16. Remove six screws (3), O-rings (2), and valve cover (1) from cylinder head (14). Discard O-rings (2).

17. Turn valve cover (1) upside down, depress locking tabs (6) on vent connector (5), and remove vent connector (5) and O-ring (4) from valve cover (1). Discard O-ring (4).

18. Remove filler cap (11) and seal (12) from valve cover (1). Discard seal (12).

b. Cleaning and Inspection

WARNING

Drycleaning solvent is flammable and toxic. Do not use near an open flame and always have a tire extinguisher nearby when solvent is used. Use only in well-ventilated places, wear protective clothing, and dispose of cleaning rags in approved container. Failure to do this may result in injury to personnel and/or damage to equipment.

1. Clean remains of valve cover gasket (16) and sealing compound from valve cover (1) and cylinder head (14).

2. Clean breather tube (18) with drycleaning solvent and dry with clean rag.

3. Inspect breather tube (18) for obstructions or collapsed and bent sections that could cause restrictions. Remove obstructions or replace breather tube (18) if bent.

4. Inspect hoses (19) and (20) for obstructions, tears, flat spots, or other deformities that could cause restrictions. Remove obstructions or replace hoses (19) and (20) if torn or deformed.

5. Inspect vent connector (5). Replace if nicked, cracked, or damaged.

6. Inspect valve cover (1) for distortion, cracks, or nicks in casting that would prevent a tight seal. Replace if damaged.

c. Installation

1. Install new seal (12) and filler cap (11) on valve cover (1).

CAUTION

Valve cover gasket must extend into overlap area. Trim excess overlap, and do not stretch valve cover gasket. Damage to gasket may result.

2. Turn valve cover (1) upside down and install new valve cover gasket (16) on valve cover (1), overlapping ends in the overlap area (17). Apply gasket compound to overlap area (17).

3. Compress locking tabs (6) of vent connector (5), and install new O-ring (4) and vent connector (5) on valve cover (1).

4. Install valve cover (1) on cylinder head (14) with six new O-rings (2) and screws (3). Tighten screws (3) 18 lb-ft (24 N•m).

CAUTION

Ensure covers are removed from turbocharger intake output port and aftercooler inlet port prior to installing crossover tube. Failure to do so may cause damage to equipment.

5. Install crossover tube (9) by sliding two hoses (7) onto turbocharger (13) and aftercooler (15) and tighten four clamps (8).

3-6. VALVE COVER (M939A2) MAINTENANCE (Contd)

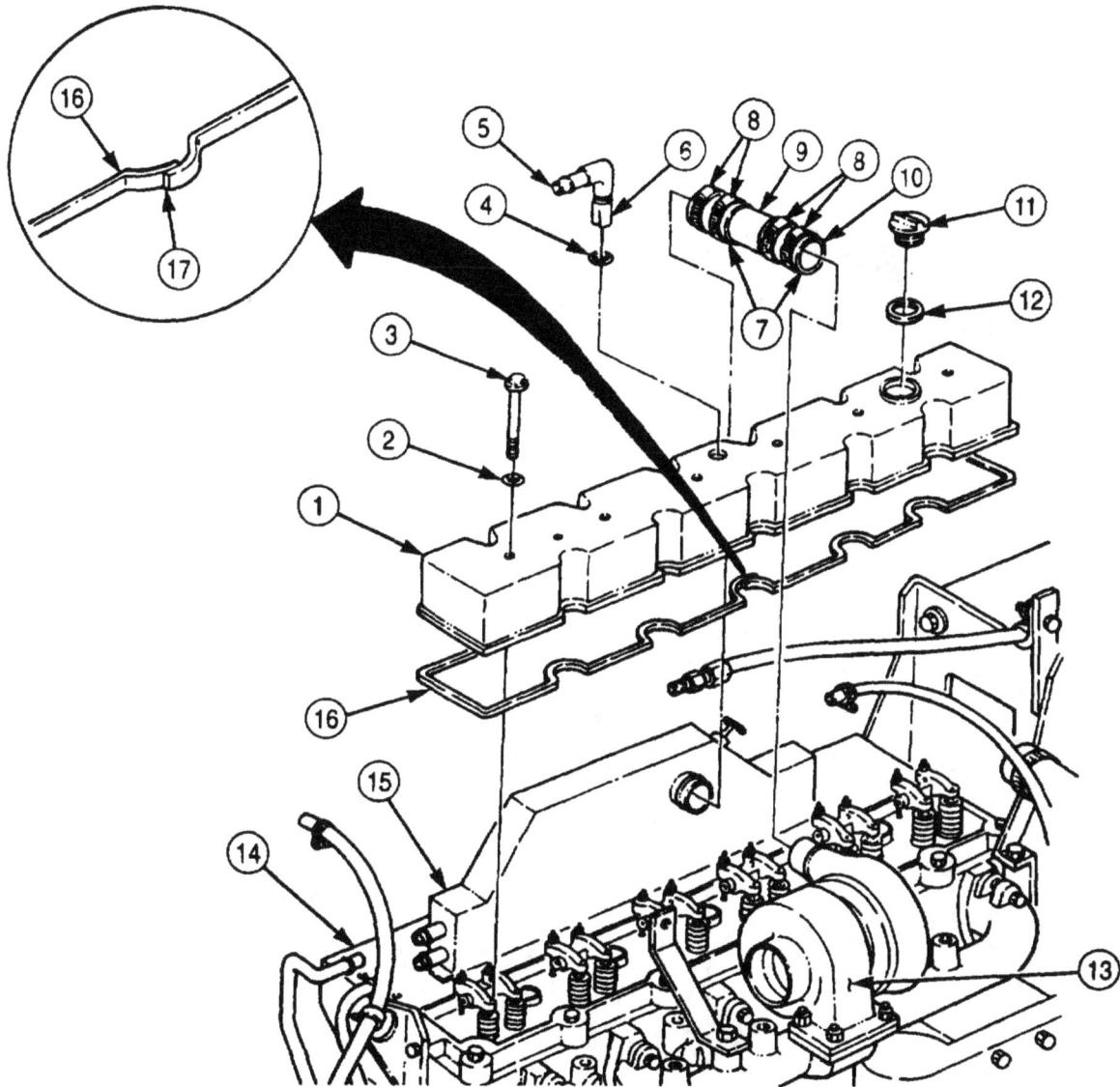

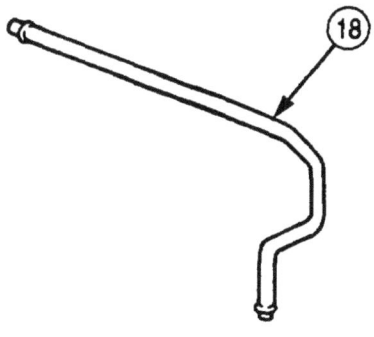

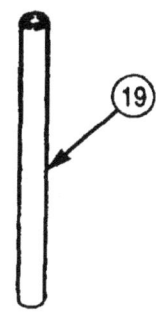

3-6. VALVE COVER (M939A2) MAINTENANCE (Contd)

6. Install surge tank assembly (2) on exhaust manifold (9) with two screws (3). Tighten screws (3) 50-55 lb-ft (68-75 N•m).

7. Install surge tank assembly (2) on angle support bracket (8) with screw (1) and nut (7). Tighten screw (1) 35-40 lb-ft (47-54 N•m).

8. Install hoses (4), (5), (10), (12), and (12) on surge tank assembly (2) and tighen clamps (6), (11), and (13).

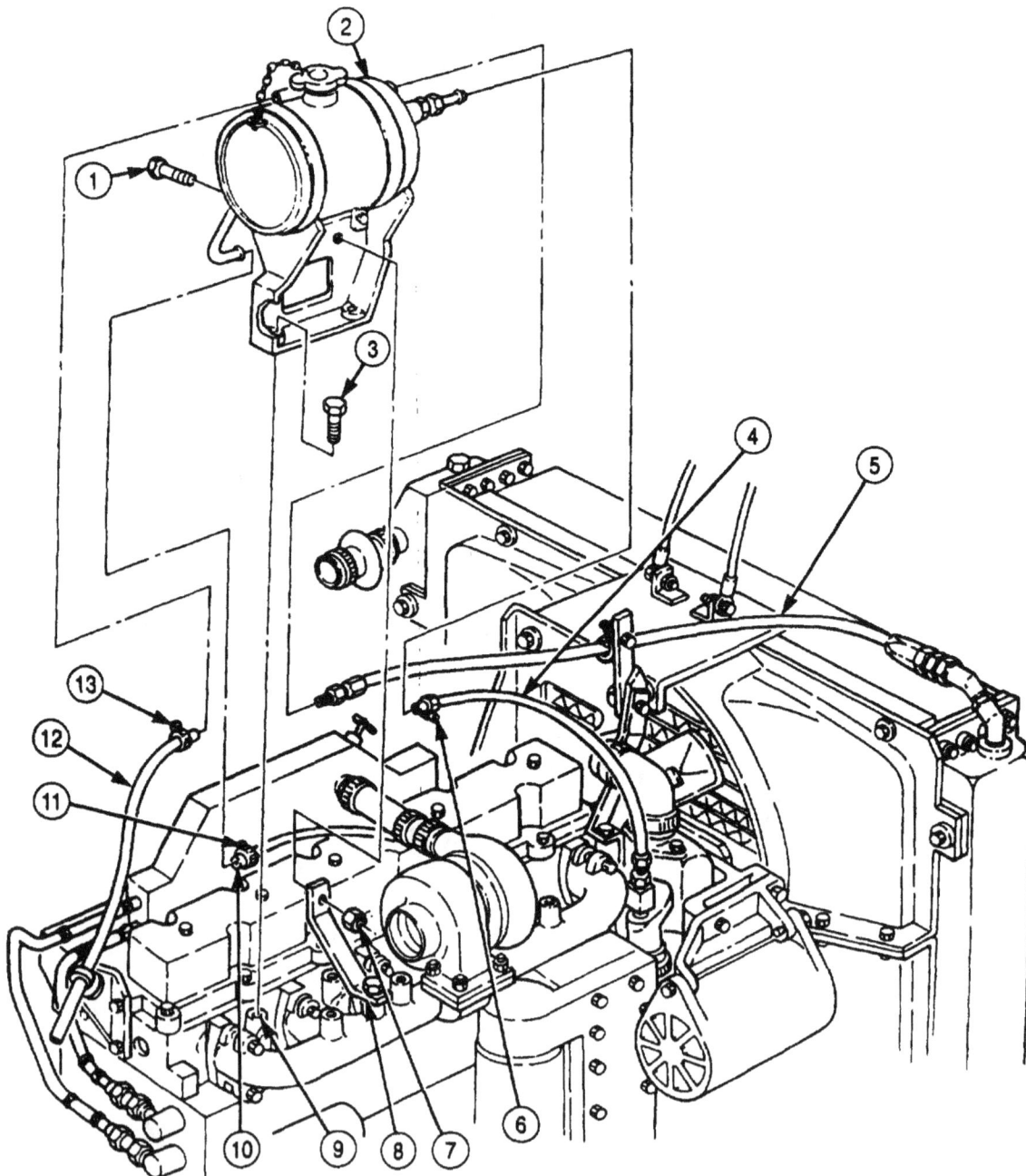

3-6. VALVE COVER (M939A2) MAINTENANCE (Contd)

9. Install radiator inlet tube (14) on hose (26) and elbow (21) and tighten clamps (25) and (20).

10. Install radiator inlet tube (14) on bracket (22) with two washers (17), screws (16), washers (18), and new locknuts (19).

11. Connect hose (23) and cable (24) to inlet tube (14) with tiedown straps (15), as required, to keep hose (23) and cable (24) clear of moving parts.

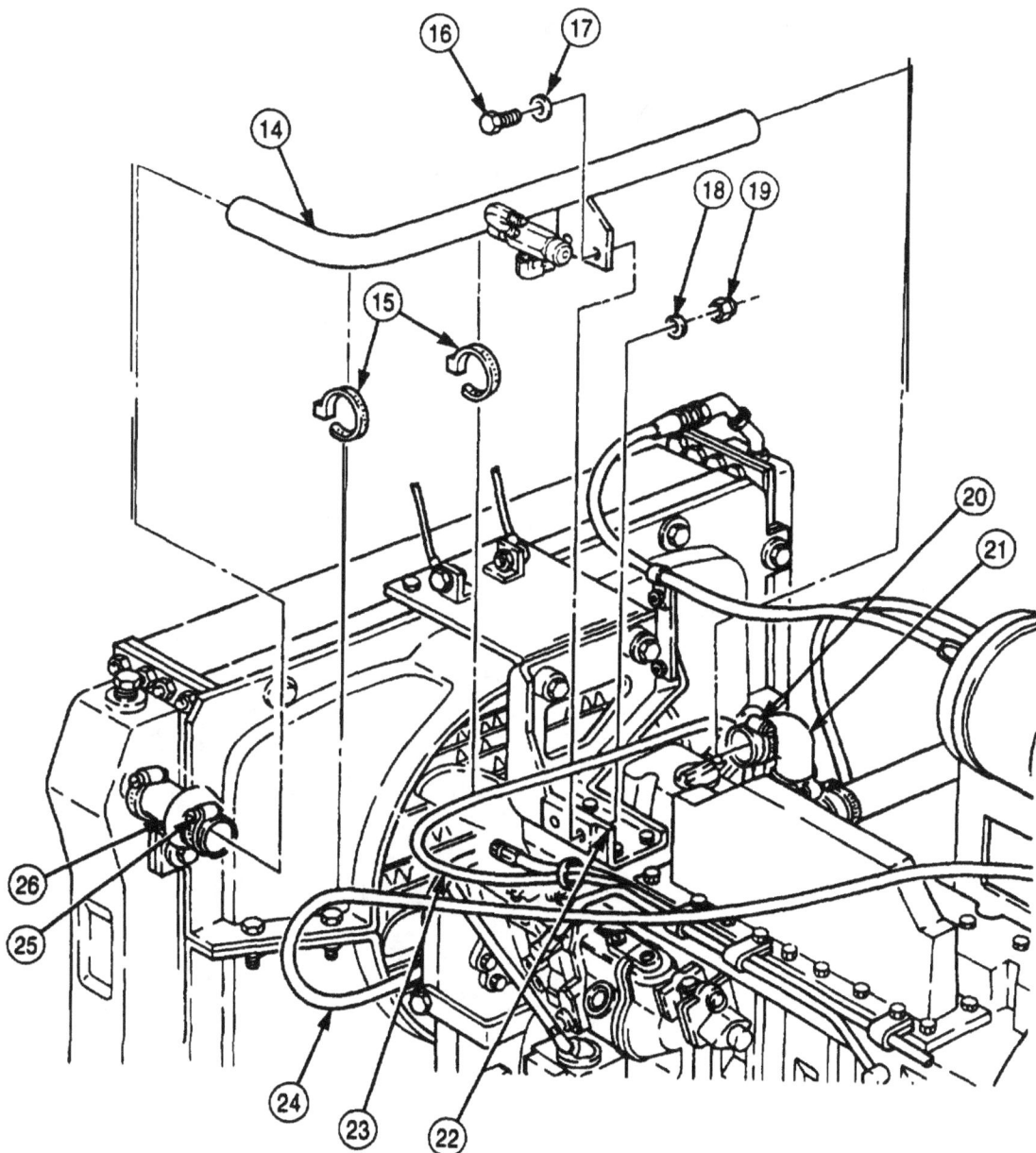

3-6. VALVE COVER (M939A2) MAINTENANCE (Contd)

12. Install breather tube (3) and hose (2) on vent connector (13) with two clamps (1).

13. Install hose (9) on breather tube (3) with clamp (8).

14. Install two clamps (5) on breather tube (3) and valve cover (6) with two screws (4).

NOTE

Remove plugs from all lines before making connections. Be prepared to catch any dripping of fluids when lines are uncapped.

15. Install two hoses (11) on aftercooler (12) and aftercooler tubes (7) with four clamps (10).

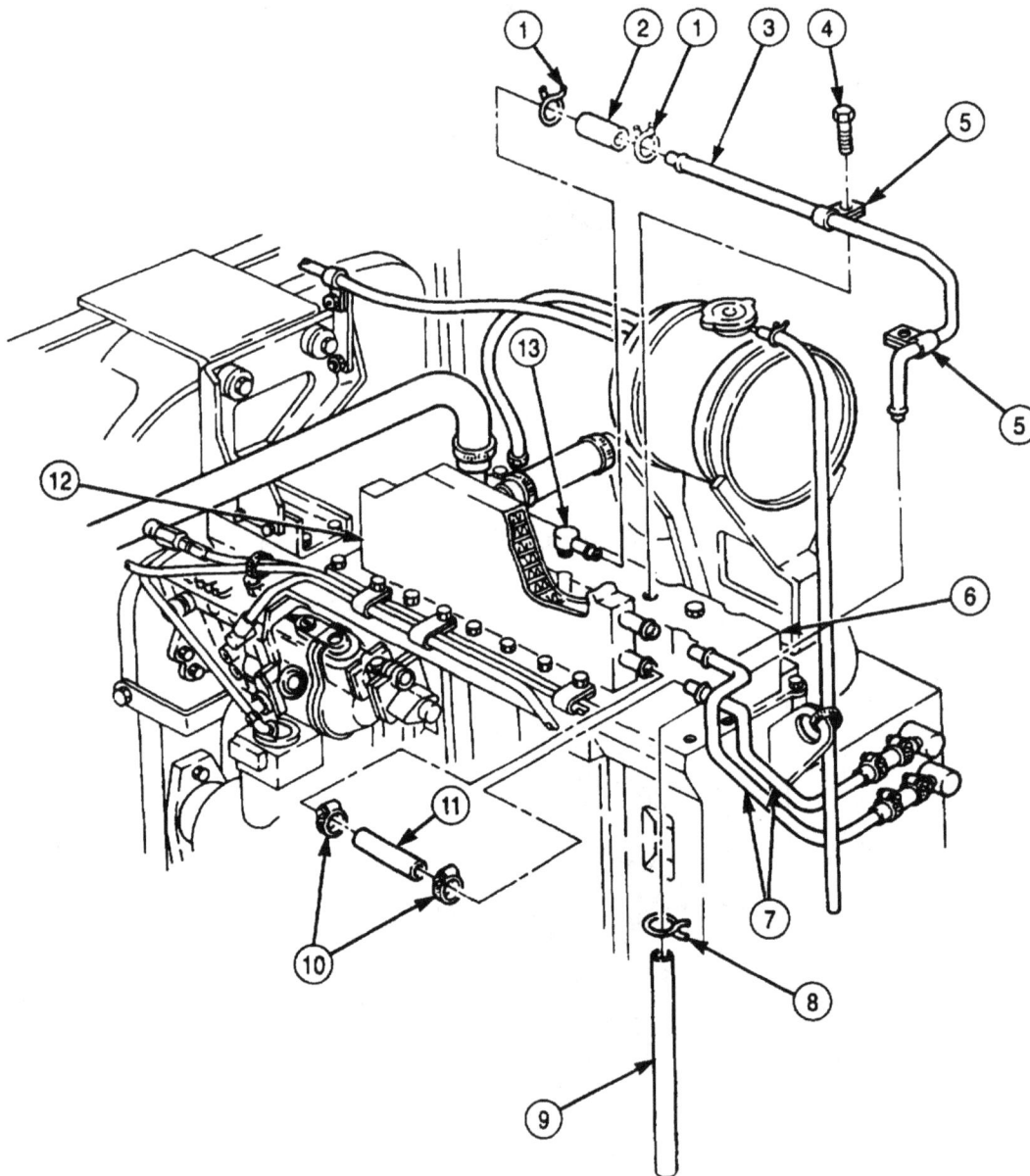

3-6. VALVE COVER (M939A2) MAINTENANCE (Contd)

16. Connect air lines (17) and (19) to elbows (16) and (18). Secure air lines (17) and (19) to radiator inlet tube (14) with tiedown straps (15), as required.

CAUTION

Ensure drainvalve on aftercooler is open when filling cooling system. Failure to do so may result in damage to equipment.

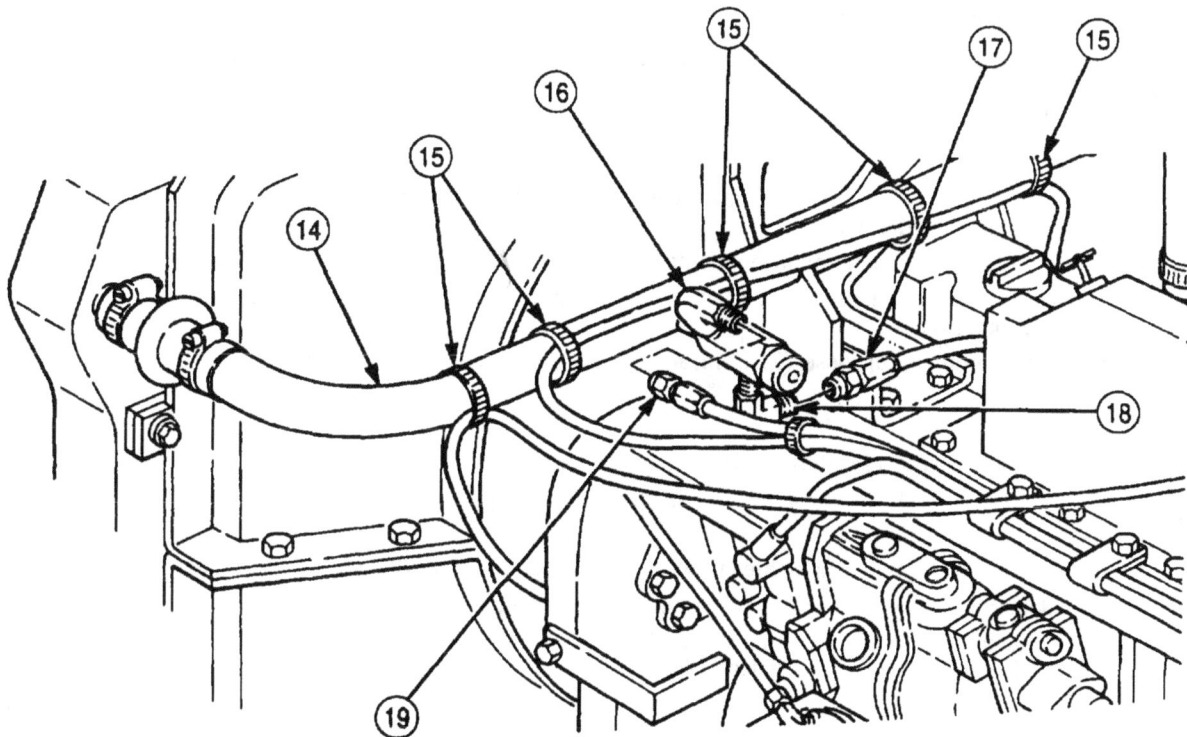

FOLLOW-ON TASKS: • Install air cleaner hose (para. 3-13).
• Fill coolant to proper level (para. 3-53).
• Start engine (TM 9-2320-272-10) and check for leaks.
• Check coolant level and add coolant as necessary (TM 9-2320-272-10).

3-7. OIL PUMP PICKUP HOSE MAINTENANCE

THIS TASK COVERS:

a. Removal

b. Inspection   c. Installation

INITIAL SETUP:

APPLICABLE MODELS

M939/A1

TOOLS

General mechanic's tool kit (Appendix E, Item 1)
Torque wrench (Appendix E, Item 146)

MATERIALS/PARTS

Antiseize tape (Appendix C, Item 72)

REFERENCES (TM)

LO 9-2320-272-12
TM 9-2320-272-10
TM 9-2320-272-24P

EQUIPMENT CONDITION

- Parking brake set (TM 9-2320-272-10).
- Hood raised and secured (TM 9-2320-272-10).
- Left splash shield removed (TM 9-2320-272-10).
- Engine oil drained (para. 3-5).

a. Removal

NOTE

- Engine oil pan is mounted with screw-assembled washers on late model engines.
- Have adequate drainage container ready to catch oil.

1. Remove two screws (6), washers (7), and clamps (5) from oil pump pickup and return hoses (2) and (8) and engine oil pan (3).

2. Disconnect oil pump pickup hose (2) from adapter flange (4) and oil pump (1).

b. Inspection

Inspect oil pump pickup hose (2). Replace if threads are stripped or hose (2) is cracked or frayed.

c. Installation

1. Apply antiseize tape to male pipe threads of oil pump pickup hose (2) and adapter flange (4).

2. Connect oil pump pickup hose (2) to oil pump (1) and adapter flange (4).

3. Install two clamps (5) on oil pump pickup and return hoses (2) and (8) and engine oil pan (3) with two washers (7) and screws (6). Tighten screws (6) 35-40 lb-ft (47-54 N•m).

3-7. OIL PUMP PICKUP HOSE MAINTENANCE (Contd)

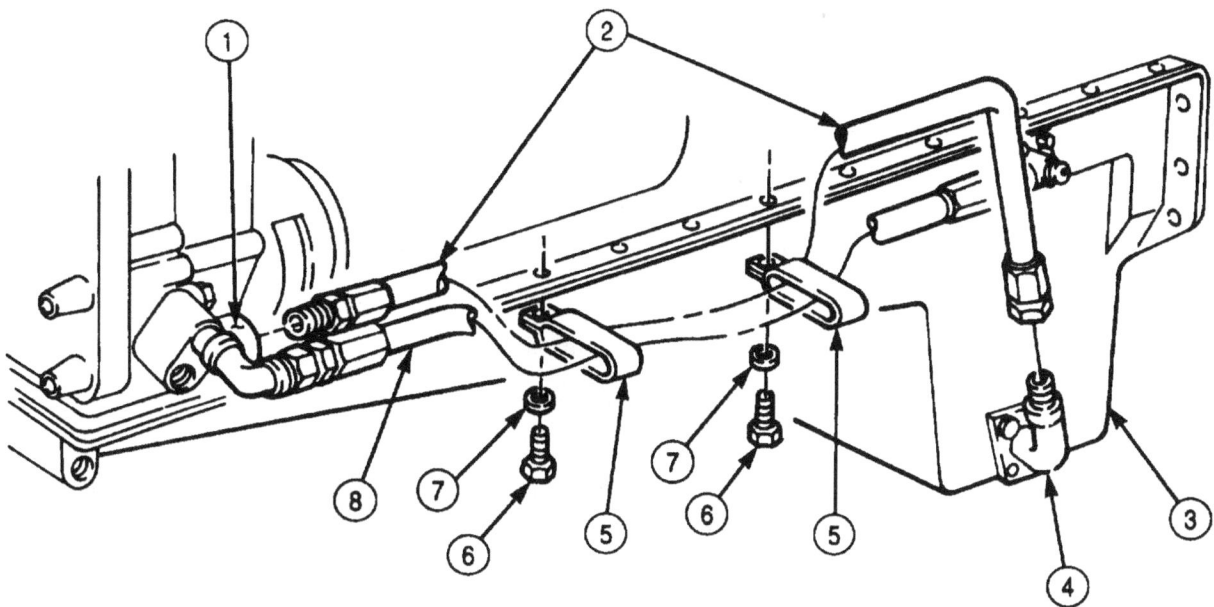

FOLLOW-ON TASKS:
- Fill engine with oil to proper level (LO 9-2320-272-12).
- Start engine (TM 9-2320-272-10) and check for leaks.
- Check oil level and add oil as necessary (TM 9-2320-272-10).
- Install left splash shield (TM 9-2320-272-10).

3-8. OIL PUMP RETURN HOSE MAINTENANCE

THIS TASK COVERS:

a. Removal
b. Inspection

c. Installation

<u>INITIAL SETUP:</u>

<u>APPLICABLE MODELS</u>
M939/A1

<u>TOOLS</u>
General mechanic's tool kit (Appendix E, Item 1)
Torque wrench (Appendix E, Item 146)

<u>MATERIALS/PARTS</u>
Antiseize tape (Appendix C, Item 72)

<u>REFERENCES</u> (TM)
LO 9-2320-272-12
TM 9-2320-272-10
TM 9-2320-272-24P

<u>EQUIPMENT CONDITION</u>
• Parking brake set (TM 9-2320-272-10).
• Hood raised and secured (TM 9-2320-272-10).
• Left splash shield removed (TM 9-2320-272-10).
• Engine oil drained (para. 3-5).

a. Removal

NOTE
Have adequate drainage container ready to catch oil.

1. Remove two screws (7), washers (8), and clamps (6) from oil pump pickup and return hoses (2) and (3) and engine oil pan (5).
2. Disconnect oil pump return hose (3) from adapter (9) and oil pan aerator (4).
3. Remove adapter (9), elbow (10), and nipple (11) from oil pump (1).

b. Inspection

1. Inspect oil pump return hose (3). Replace hose (3) if cracked, frayed, or split.
2. Inspect adapter (9), elbow (10), and nipple (11). Replace if adapter (9), elbow (10), or nipple (11) have cracks or threads are stripped or crossed.

c. Installation

1. Apply antiseize tape to male pipe threads of oil pump return hose (3), adapter (9), and nipple (11).
2. Install nipple (11), elbow (10), and adapter (9) on oil pump (1).
3. Connect oil pump return hose (3) to oil pan aerator (4) and adapter (91.
4. Install two clamps (6) and oil pump pickup and return hoses (2) and (3) on engine oil pan (5) with two washers (8) and screws (7). Tighten screws (7) 35-40 lb-ft (47-54 N•m).

3-8. OIL PUMP RETURN HOSE MAINTENANCE (Contd)

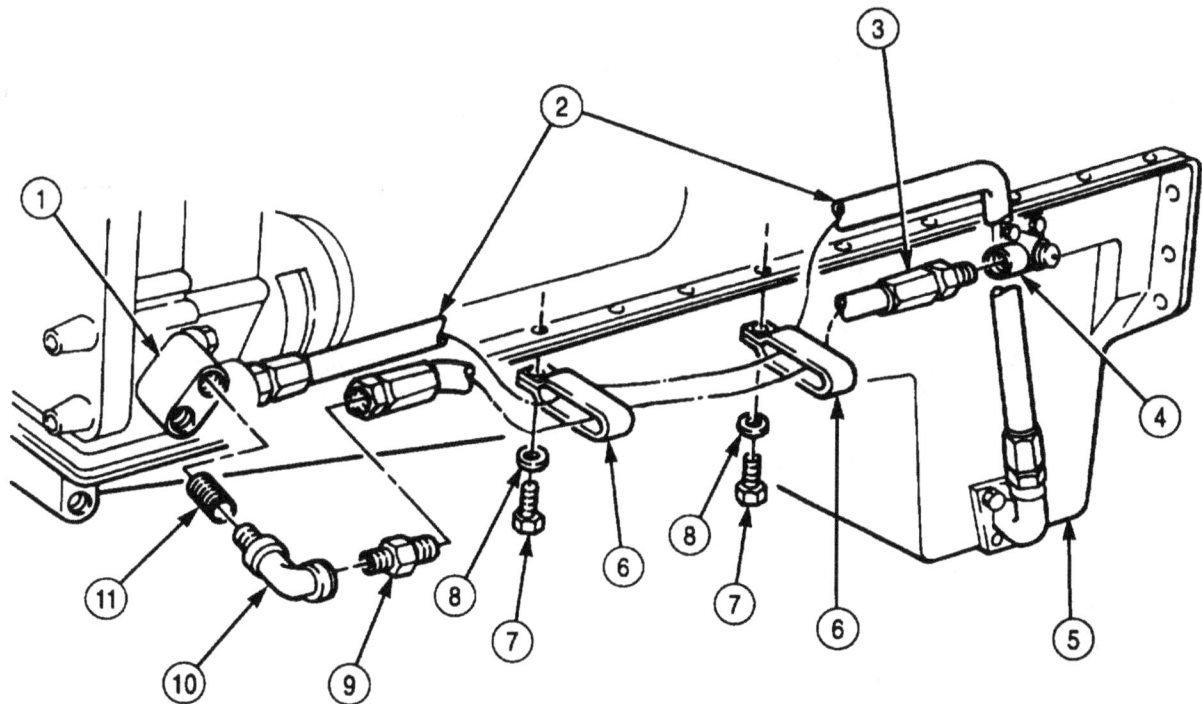

FOLLOW-ON TASKS: • Fill engine with oil to proper level (LO 9-2320-272-12).
• Start engine (TM 9-2320-272-10) and check for leaks.
• Check oil level and add oil as necessary (TM 9-2320-272-10).
• Install left splash shield (TM 9-2320-272-10).

3-9. FRONT SUMP TUBE MAINTENANCE

THIS TASK COVERS:

a. Removal   c. Installation
b. Inspection

INITIAL SETUP:

APPLICABLE MODELS
M939/A1

TOOLS
General mechanic's tool kit (Appendix E, Item 1)

MATERIALS/PARTS
Two bushings (Appendix D, Item 23)
Antiseize tape (Appendix C. Item 72)

REFERENCES (TM)
TM 9-2320-272-10
TM 9-2320-272-24P

EQUIPMENT CONDITION
• Parking brake set (TM 9-2320-272-10).
• Hood raised and secured (TM 9-2320-272-10).
• Left splash shield removed (TM 9-2320-272-10).

---

a. Removal

NOTE
Have adequate drainage container ready to catch oil.

1. Loosen two nuts (4) on sump tube (5), and slide nuts (4) toward center of front sump tube (5).

2. Tighten elbow (3) 1/4 turn to free sump tube (5). Remove sump tube (5) from adapter (6) and elbow (3).

3. Remove two bushings (7) from sump tube (5). Discard bushings (7).

NOTE
Mark position of elbow for installation.

4. Remove adapter (6) from oil pan (2), and elbow (3) from oil pump (1).

b. Inspection

1. Inspect two nuts (4), elbow (3), and adapter (6). Replace nuts (4), elbow (3), or adapter (6) if threads are damaged.

2. Inspect sump tube (5). Replace if cracked.

c. Installation

1. Apply antiseize tape to male pipe threads of adapter (6) and elbow (3).

NOTE
Ensure elbow is less than 1/8 turn from final position to allow
room for installation of front sump tube.

2. Install adapter (6) on oil pan (2), and elbow (3) on oil pump (1).

3. Install two nuts (4) and new bushings (7) on sump tube (5), and slide toward center of sump tube (5).

4. Position one end of sump tube (5) on adapter (6), and start elbow (3) in other end of sump tube (5). Turn elbow (3) to tighten into final position, pressing front sump tube (5) into elbow (3) and aligning front sump tube (5) with elbow (3).

5. Connect sump tube (5) to adapter (6) and elbow (3) by tightening two nuts (4).

3-9. FRONT SUMP TUBE MAINTENANCE (Contd)

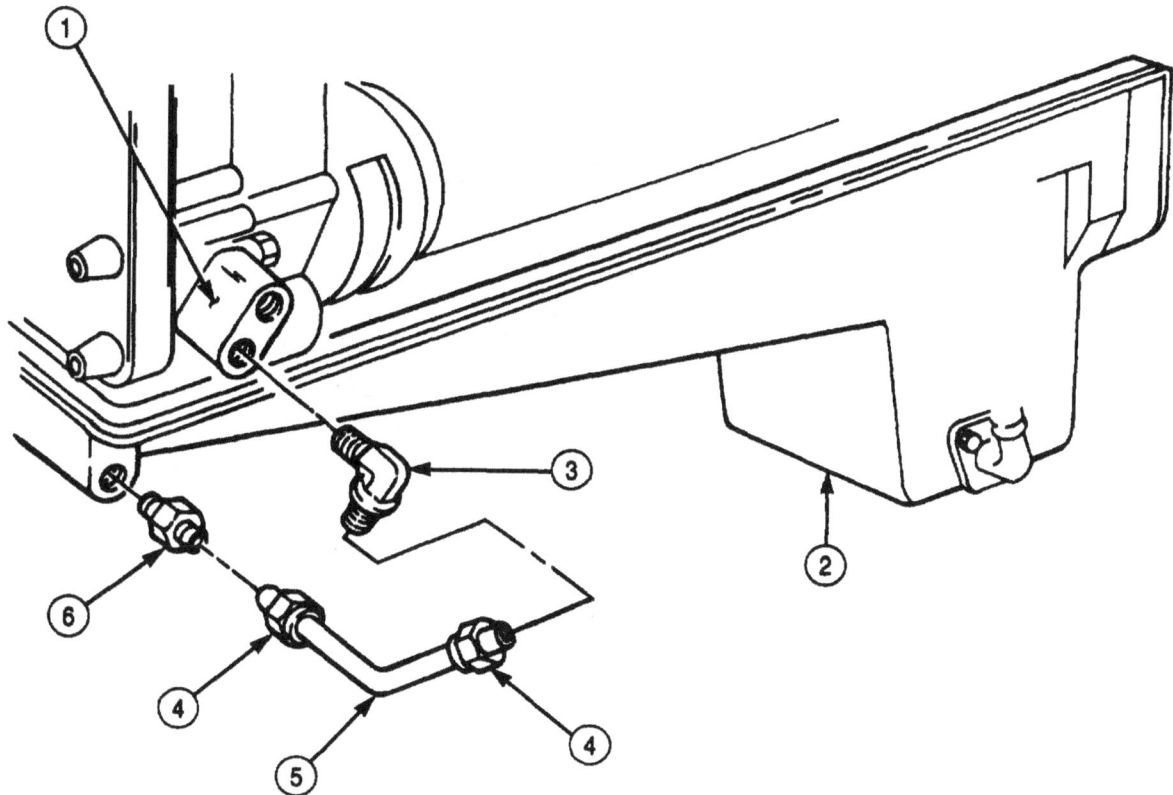

FOLLOW-ON TASKS:
- Start engine (TM 9-2320-272-10) and check for leaks.
- Check oil level and add oil as necessary (TM 9-2320-272-10).
- Install left splash shield (TM 9-2320-272-10).

3-10. ENGINE AOAP VALVE MAINTENANCE

THIS TASK COVERS:

a Oil Sampling
b. Removal

c. Installation

INITIAL SETUP:

APPLICABLE MODELS

All

TOOLS

General mechanic's tool kit (Appendix E, Item 1)

MATERIALS/PARTS

Bag, plastic (Appendix C, Item 11)
Bottle, oil sample (Appendix C, Item 12)
Hose (Appendix C, Item 34)
Sack, shipping (Appendix C, Item 60)
Gasket sealant (Appendix C, Item 30)

REFERENCES (TM)

LO 9-2320-272-12
TM 9-2320-272-10
TM 9-2320-272-24P

EQUIPMENT CONDITION

• Hood raised and secured (TM 9-2320-272-10).
• Parking brake set (TM 9-2320-272-10).
• Right splash shield removed (TM 9-2320-272-10).

GENERAL SAFETY INSTRUCTIONS

• Do not perform task when engine is hot, or drain oil when hot.
• Prevent spilling of oil. See Army POL (para. 1-8).

WARNING

Accidental or intentional introduction of liquid contaminants into the environment is in violation of state, federal, and military regulations. Refer to Army POL (para. 1-8) for information concerning storage, use, and disposal of these liquids. Failure to do so may result in injury or death.

a. Oil Sampling

NOTE

• Sampling procedures are similar for M939/A1/A2 series vehicles. This procedure is for M939A2 series vehicles.

• Engine should be at operating temperature to ensure circulating oil has reached a uniform consistency. Failure to allow engine to reach operating temperature may result in inaccurate AOAP analysis.

1. Start engine (TM 9-2320-272-10) and run until operating temperature is reached.

2. Remove capnut (2) from oil sampling valve (1).

3. Install rubber hose (3) on sampling valve (1).

NOTE

• Perform step 4 to flush engine oil from oil sampling valve.

• Have clean drainage container ready to catch oil that is flushed for return to the engine. Do not use sample bottle.

4. Open oil sampling valve (1) and drain 1 pt (0.473 L) of oil into clean drainage container.

5. Close oil sampling valve (1).

6. Return flushed oil to engine (LO 9-2320-272-12).

7. Place sample bottle (4) under oil sampling valve (1) and place rubber hose (3) in oil sample bottle (4).

8. Open sampling valve (1) and fill oil sampling bottle (4) to within 0.5 in. (1.3 cm) from top of sample bottle (4).

3-10. ENGINE AOAP VALVE MAINTENANCE (Contd)

9. Close sampling valve (1).

NOTE

Seal oil sample in plastic bag and shipping sack.

10. Remove rubber hose (3) from oil sample bottle (4) and oil sampling valve (1). Seal oil sample bottle (4).
11. Install capnut (2) on oil sampling valve (1).
12. Stop engine (TM 9-2320-272-10).
13. Notify supervisor for processing AOAP oil sample.

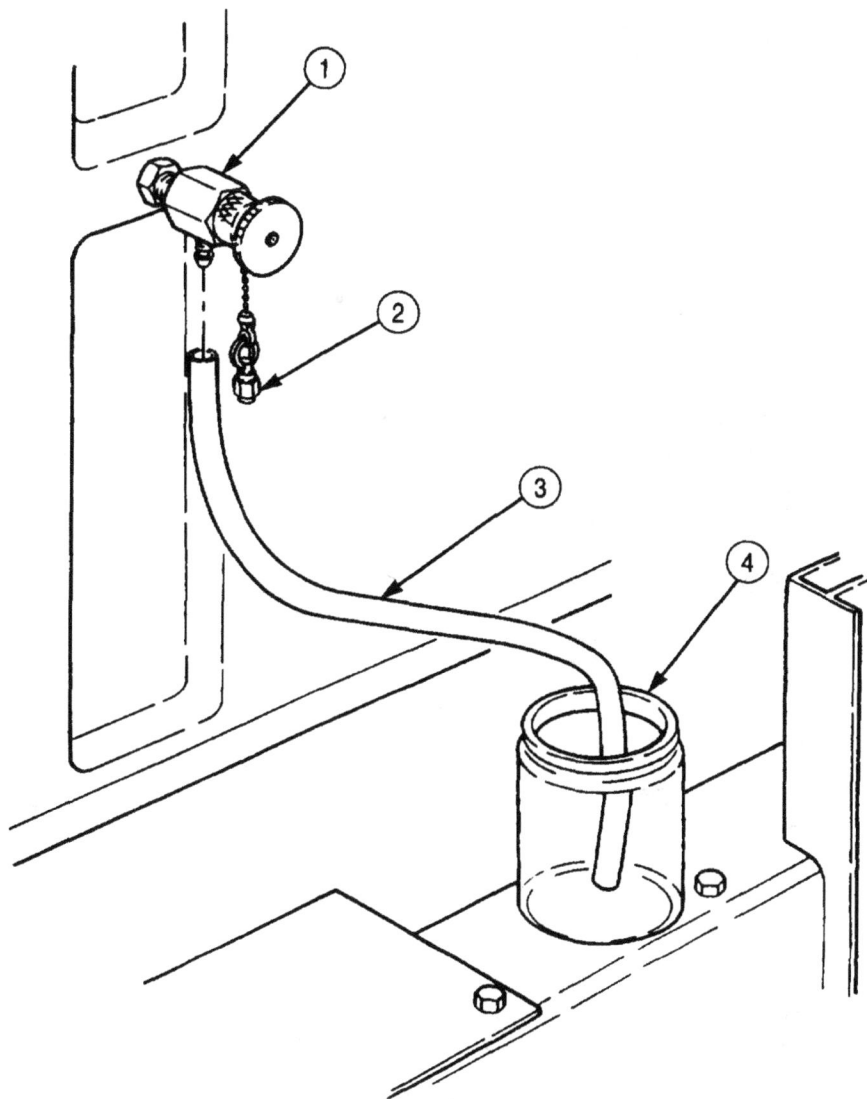

3-10. ENGINE AOAP VALVE MAINTENANCE (Contd)

b. Removal

WARNING

Do not perform this task when engine is hot. Doing so may result in injury to personnel.

NOTE

- Have adequate drainage container ready to catch oil.
- Perform step 1 for M939/A1 series vehicles and step 2 for M939A2 series vehicles.

1. Remove oil sampling valve (2) from engine oil cooler (1).
2. Remove oil sampling valve (2) from engine block (3).

c. Installation

1. Apply a thin coat of gasket sealant to male threads of oil sampling valve (2).

NOTE

Perform step 2 for M939/A1 series vehicles and step 3 for M939A2 series vehicles.

2. Install oil sampling valve (2) on engine oil cooler (1).
3. Install oil sampling valve (2) on engine block (3).

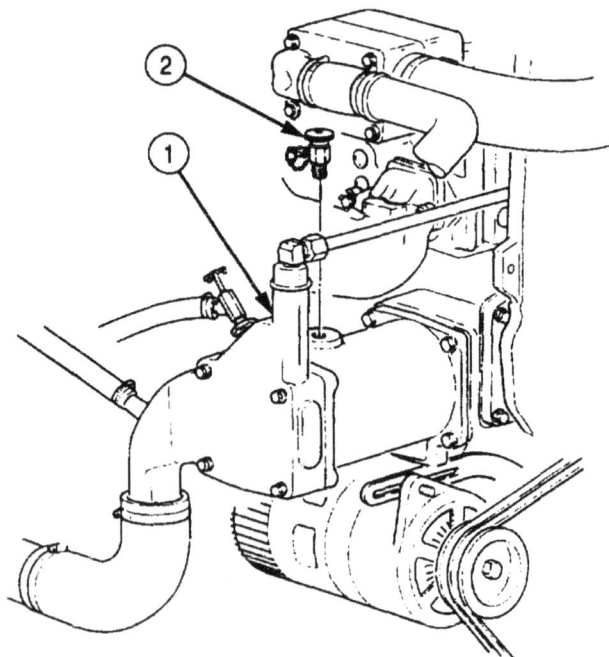

M939/A1

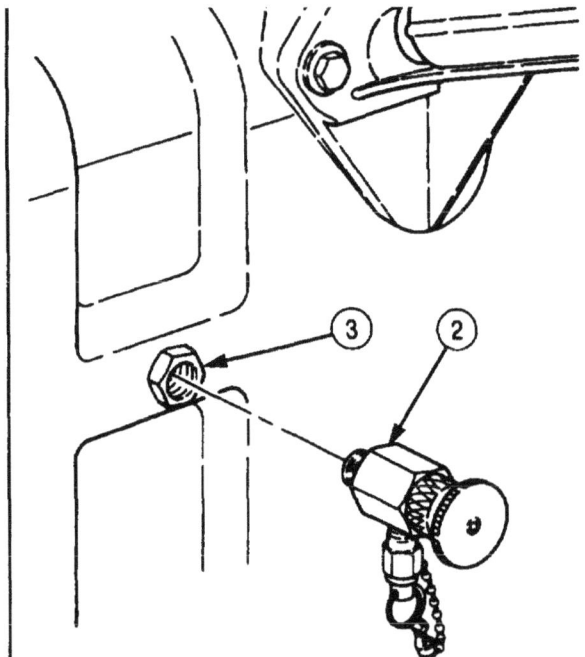

M939A2

FOLLOW-ON TASKS: • Start engine (TM 9-2320-272-10) and check for leaks.
  • Check oil level and add oil as necessary (TM 9-2320-272-10).
  • Install right splash shield (TM 9-2320-272-10).

Section II. AIR INTAKE SYSTEM MAINTENANCE INDEX

3-11. AIR INTAKE SYSTEM MAINTENANCE INDEX

3-12. AIR CLEANER INDICATOR AND TUBE MAINTENANCE

THIS TASK COVERS:

a. Testing

b. Removal

c. Cleaning and Inspection

d. Installation

INITIAL SETUP:

APPLICABLE MODELS

All

TOOLS

General mechanic's tool kit (Appendix E, Item 1)

MATERIALS/PARTS

Two locknuts (Appendix D, Item 313)

Antiseize tape (Appendix C, Item 72)

REFERENCES (TM)

TM 9-2320-272-10

TM 9-2320-272-24P

EQUIPMENT CONDITION

- Parking brake set (TM 9-2320-272-10).
- Hood raised and secured (TM 9-2320-272-10).
- Left splash shield removed (TM 9-2320-272-10).

GENERAL SAFETY INSTRUCTIONS

When cleaning with compressed air, wear eyeshields and ensure source pressure does not exceed 30 psi (207 kPa).

a. Testing

NOTE

Perform step 1 for M939/A1 series vehicles.

1. Remove nut (3), screw (6), and extension tube cap (2) from extension tube (5).

2. Start engine (TM 9-2320-272-10) and run at 1,200 rpm.

3. Cover 90 percent of intake extension tube (5) opening with cardboard (4).

4. Observe filter indicator (1) to ensure red band is visible. If red band is visible, indicator (1) works properly. If not, replace indicator (1).

5. Remove cardboard (4) and reset indicator (1) (TM 9-2320-272-10) if working properly.

NOTE

Perform step 6 for M939/A1 series vehicles.

6. Install extension tube cap (2) on extension tube (5) with screw (6) and nut (3).

3-12. AIR CLEANER INDICATOR AND TUBE MAINTENANCE (Contd)

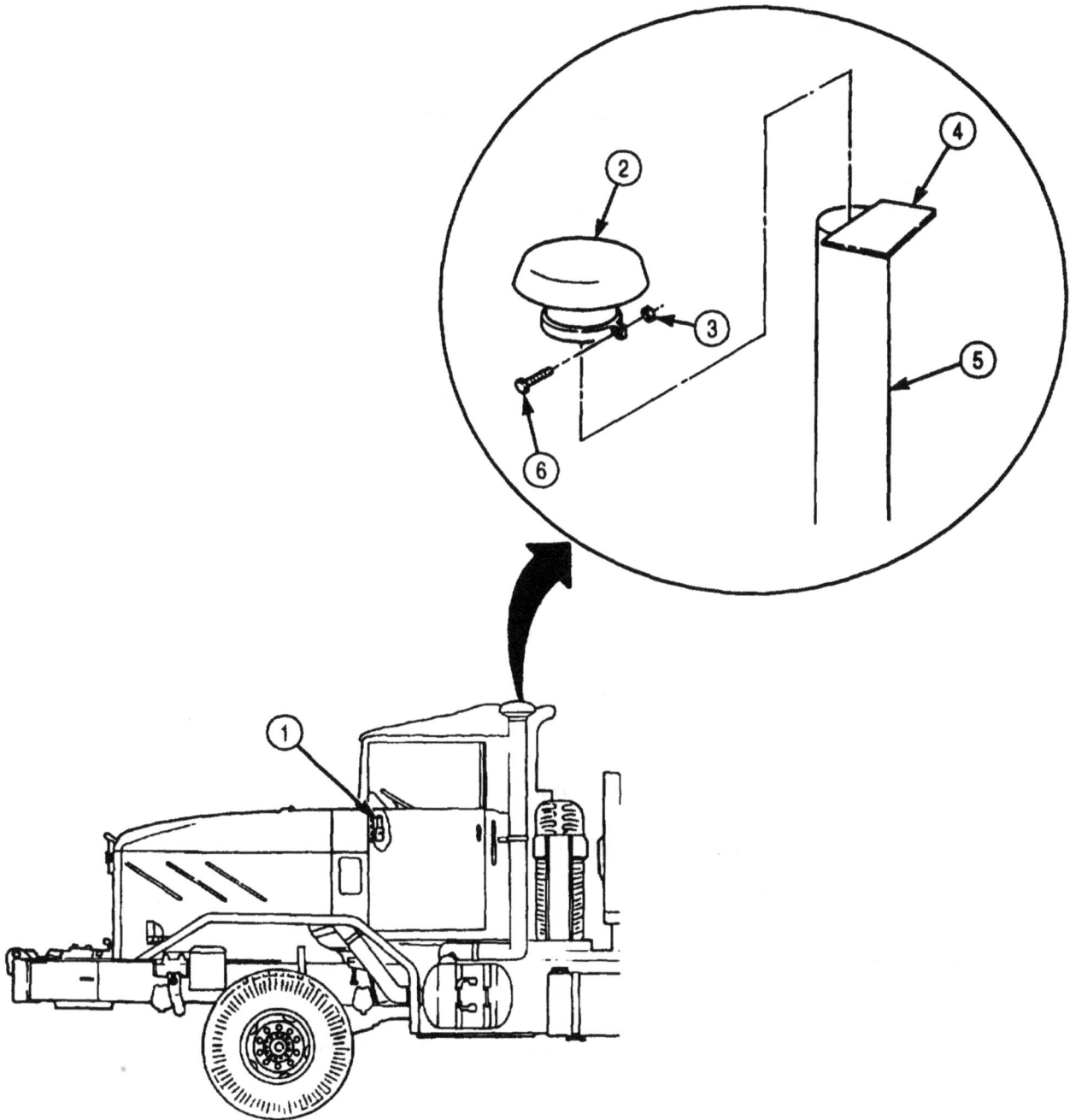

3-12. AIR CLEANER INDICATOR AND TUBE MAINTENANCE (Contd)

b. Removal

NOTE
Perform steps 1 through 7 for M939A2 series vehicles.

1. Remove clamp (14) and air cleaner indicator tube (1) from connector (13).
2. Remove connector (13) and filter (12) from air tube (11).
3. Remove four screws (5) and plate (4) from instrument panel (10).
4. Remove clamp (9) and indicator tube (1) from air cleaner indicator (6).
5. Remove two locknuts (8), screws (7), and air cleaner indicator (6) from plate (4). Discard locknuts (8).
6. Remove indicator tube (1) and grommet (2) from firewall (3).
7. Remove grommet (2) from indicator tube (1).

NOTE
Perform steps 8 through 15 for M939 and M939A1 series vehicles.

8. Remove four screws (26), plate (29), and air cleaner indicator (22) from instrument panel (25).
9. Disconnect indicator tube nut (20) from elbow (21).
10. Remove elbow (21) from adapter (30) and adapter (30) from air cleaner indicator (22).
11. Remove two locknuts (31), screws (27), plate (29), and gasket (28) from air cleaner indicator (22). Discard locknuts (31).
12. Disconnect indicator tube nut (16) from elbow (15).
13. Remove grommet (18) and indicator tube (17) from firewall (19).
14. Remove grommet (18) from indicator tube (17).
15. Remove elbow (15) and filter (24) from intake manifold (23).

c. Cleaning and Inspection

WARNING

Eyeshields must be worn when cleaning with compressed air. Compressed air source will not exceed 30 psi (207 kPa). Failure to do so may result in injury to personnel.

NOTE
Cleaning and inspection of air cleaner indicator, indicator tube, and filter are basically the same for M939, M939A1, and M939A2 series vehicles. This cleaning and inspection procedure covers M939A2 series vehicles.

1. Clean air cleaner indicator tube (1) with compressed air.
2. For general inspection instructions, refer to para. 2-15.
3. Inspect indicator tube (1) and filter (12) for clogs, kinks, cracks, and breaks. Replace indicator tube (1) and filter (12) if clogged, kinked, cracked, or broken.
4. Inspect air cleaner indicator (6) for cracks. Replace air cleaner indicator (6) if cracked.

3-12. AIR CLEANER INDICATOR AND TUBE MAINTENANCE (Contd)

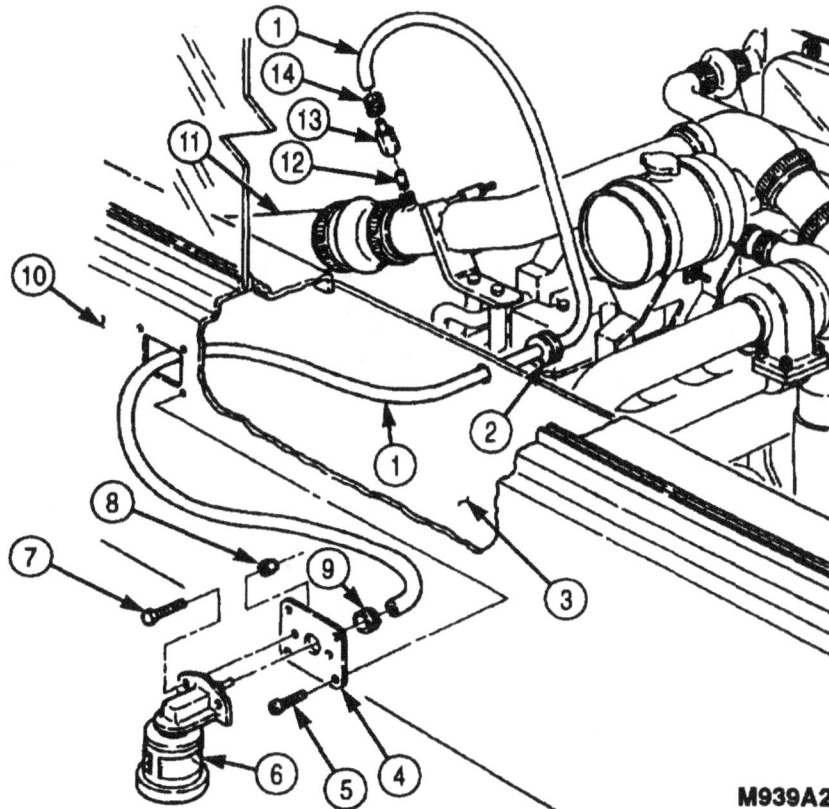

M939A2

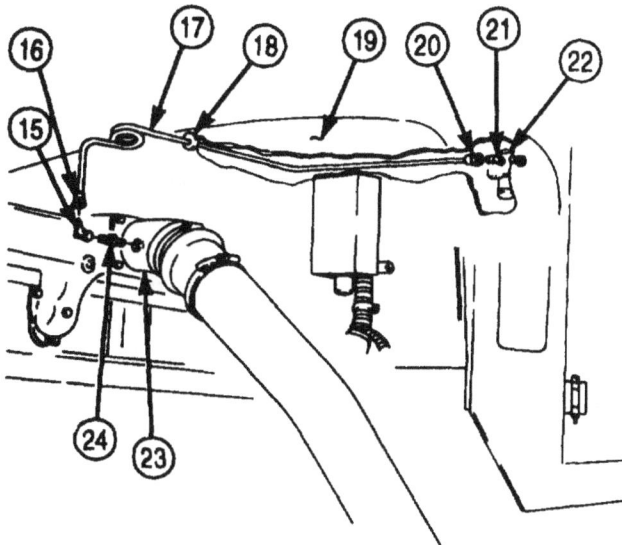

M939/A1

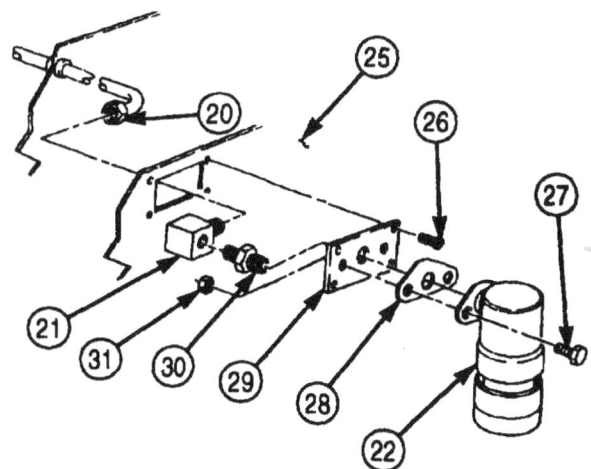

M939/A1

3-12. AIR CLEANER INDICATOR AND TUBE MAINTENANCE (Contd)

d. Installation

NOTE
- Perform steps 1 through 8 for M939A2 series vehicles.
- Male pipe threads must be wrapped with antiseize tape before installation.

1. Install filter (12) and connector (13) on air tube (11).
2. Install indicator tube (1) on connector (13) with clamp (14).
3. Insert indicator tube (1) through firewall (3).
4. Position grommet (2) on indicator tube (1) and install on firewall (3).
5. Insert indicator tube (1) through instrument panel (10).
6. Install air cleaner indicator (6) on plate (4) with two screws (7) and new locknuts (8).
7. Install indicator tube (1) on air cleaner indicator (6) with clamp (9).
8. Install plate (4) and air cleaner indicator (6) on instrument panel (10) with four screws (5).

NOTE
Perform steps 9 through 16 for M939 and M939A1 series vehicles.

9. Install filter (24) and elbow (15) on intake manifold (23).
10. Position grommet (18) on indicator tube (17), and install indicator tube (17) and grommet (18) through firewall (19).
11. Connect indicator tube nut (20) to elbow (21).
12. Install adapter (30) on elbow (21).
13. Install plate (29) and gasket (28) over adapter (30).
14. Install air cleaner indicator (22) on adapter (30).
15. Install air cleaner indicator (22) on gasket (28) and plate (29) with two screws (27) and new locknuts (31).
16. Connect indicator tube nut (16) to elbow (15) and install plate (29) on instrument panel (25) with four screws (26).

3-12. AIR CLEANER INDICATOR AND TUBE MAINTENANCE (Contd)

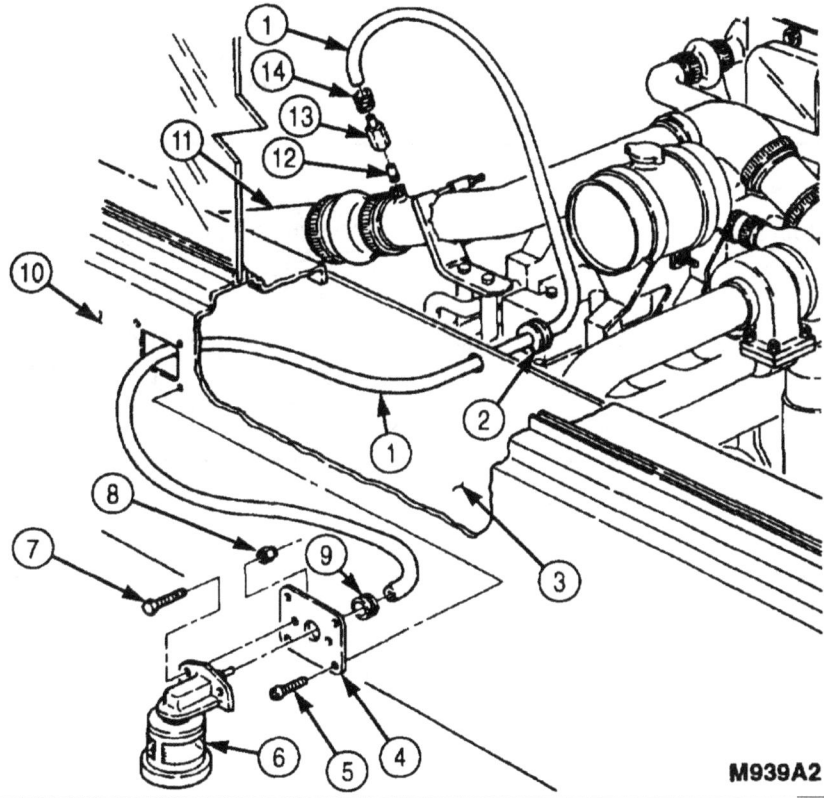

M939A2

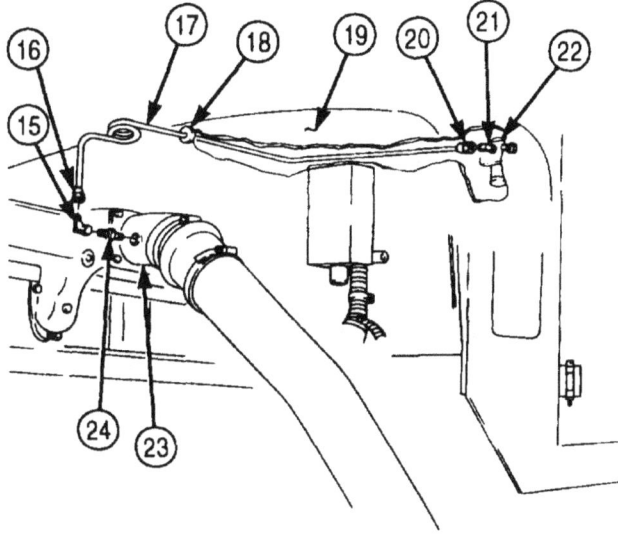

M939/A1

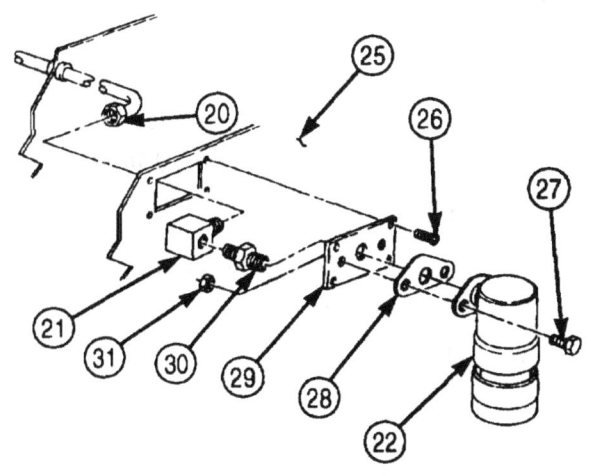

M939/A1

FOLLOW-ON TASK: Install left splash shield (TM 9-2320-272-10).

3-13. AIR CLEANER HOSE (M939A2) MAINTENANCE

THIS TASK COVERS:

a. Removal   d. Assembly
b. Disassembly   e. Installation
c. Inspection

INITIAL SETUP:

APPLICABLE MODELS
M939A2

TOOLS
General mechanic's tool kit (Appendix E, Item 1)

MATERIALS/PARTS
Tiedown strap (Appendix D, Item 688)

REFERENCES (TM)
TM 9-2320-272-10
TM 9-2320-272-24P

EQUIPMENT CONDITION
• Parking brake set (TM 9-2320-272-10).
• Hood raised and secured (TM 9-2320-272-10).

---

a. Removal

1. Remove clamp (3) and air indicator tube (1) from connector (4).
2. Remove connector (4) from nipple (2).
3. Remove tiedown strap (17) and tachometer drive cable (18) from air intake ducting (7). Discard tiedown strap (17).
4. Remove two nuts (6), bracket (5), and U-bolt (14) from air intake ducting (7) and support bracket (13).

NOTE
Tag spacers for installation.

5. Remove two screws (15), washers (16), support bracket (13), and two spacers (12) from cylinder head (10).
6. Remove two clamps (8) and air intake ducting (7) from turbocharger (9) and pipe (11).

b. Disassembly

1. Remove clamp (20) and hose (19) from tube (21).
2. Remove clamp (22) and tube (21) from elbow (23).
3. Remove clamp (24) and elbow (23) from tube (25).
4. Remove clamp (26) and tube (25) from elbow (27).

c. Inspection

1. Inspect tubes (21) and (25) for cracks and dents. Replace tube (21) or (25) if cracked or dented.
2. Inspect elbows (23) and (27) for cracks and dents. Replace elbow (23) or (27) if cracked or dented.

d. Assembly

1. Install tube (25) on elbow (27) with clamp (26). Do not tighten clamp (26).
2. Install elbow (23) on tube (25) with clamp (24). Do not tighten clamp (24).
3. Install tube (21) on elbow (23) with clamp (22). Do not tighten clamp (22).
4. Install hose (19) on tube (21) with clamp (20). Do not tighten clamp (20).

3-13. AIR CLEANER HOSE (M939A2) MAINTENANCE (Contd)

e. Installation

1. Install air intake ducting (7) on turbocharger (9) and pipe (11) with two clamps (8).
2. Install two spacers (12) and support bracket (13) on cylinder head (10) with two washers (16) and screws (15).
3. Install air intake ducting (7) on support bracket (13) with U-bolt (14), bracket (5), and two nuts (6).
4. Tighten clamps (20), (22), (24), and (26).
5. Install tachometer drive cable (18) on air intake ducting (7) with new tiedown strap (17).
6. Install connector (4) on nipple (2).
7. Install air indicator tube (1) on connector (4) with clamp (3).

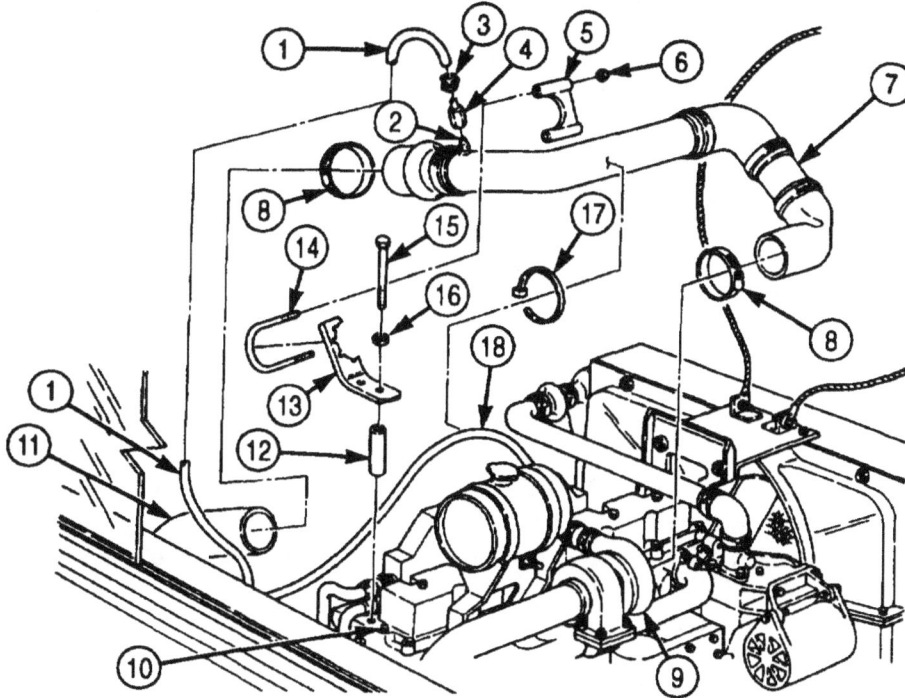

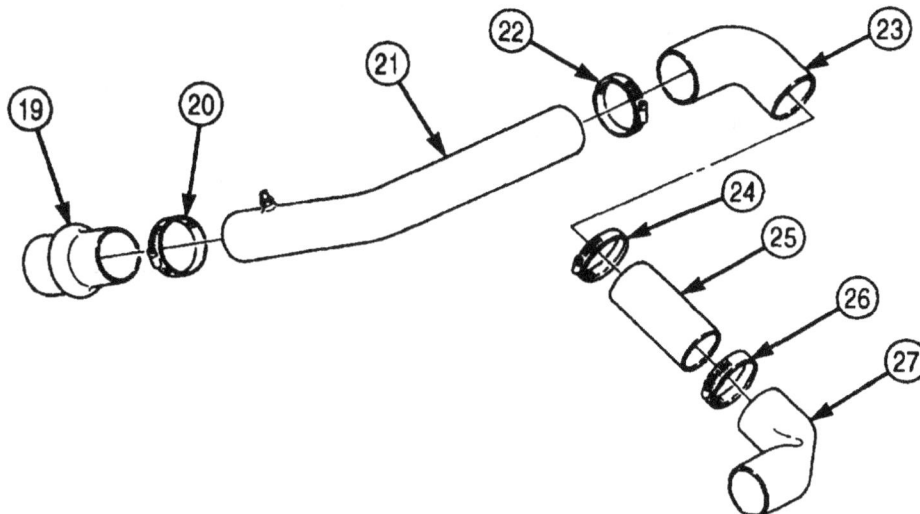

3-14. AIR CLEANER INTAKE PIPE (M939/A1) MAINTENANCE

THIS TASK COVERS:

a. Removal   c. Installation
b. Inspection

INITIAL SETUP:

APPLICABLE MODELS
M939/A1

TOOLS
General mechanic's tool kit (Appendix E, Item 1)

MATERIALS/PARTS
Three locknuts (Appendix D, Item 288)

REFERENCES (TM)
TM 9-2320-272-10
TM 9-2320-272-24P

EQUIPMENT CONDITION
- Parking brake set (TM 9-2320-272-10).
- Hood raised and secured (TM 9-2320-272-10).
- Left splash shield removed (TM 9-2320-272-10).
- Ether valve and bracket removed (para. 3-35).

---

a. Removal

1. Remove locknut (9), air intake clamp (15), screw (6), and harness cable clamp (7) from hanger strap (8). Discard locknut (9).
2. Loosen hose clamps (13) and (5) and remove air intake pipe (14) and hose clamps (13) and (5) from lower hump hose (12) and upper hump hose (4).
3. Loosen hose clamp (3) and remove upper hump hose (4) and hose clamp (3) from air intake manifold (2).
4. Loosen hose clamp (11) and remove lower hump hose (12) and hose clamp (11) from air cleaner (10).
5. Remove locknut (22), washer (23), hanger strap (8), and screw (19) from mounting bracket (20). Discard locknut (22).

NOTE
Assistant will help with step 6.

6. Pull back floor mat (17) in cab (1), and remove locknut (21), mounting bracket (20), and screw (16) from floor (18). Discard locknut (21).
7. Remove intake clamp (15) from intake pipe (14).

b. Inspection

Inspect upper and lower hump hoses (4) and (12) and intake pipe (14) for cracks. Replace if cracked.

c. Installation

1. Install hanger strap (8) on mounting bracket (20) with screw (19), washer (23), and new locknut (22).
2. Install mounting bracket (20) on floor (18) with screw (16) and new locknut (21). Install floor mat (17).

NOTE
Assistant will help with step 3.

3. Position hose clamp (11) on lower hump hose (12), install lower hump hose (12) on air cleaner (10), and tighten hose clamp (11).

3-14. AIR CLEANER INTAKE PIPE (M939/A1) MAINTENANCE (Contd)

4. Position hose clamp (3) on upper hump hose (4), install upper hump hose (4) on air intake manifold (2), and tighten hose clamp (3).

5. Position intake pipe clamp (15) over intake pipe (14).

NOTE

Assistant will help with step 6.

6. Install intake pipe (14) on lower hump hose (12) and upper hump hose (4) with hose clamps (13) and (5).

7. Install intake pipe clamp (15) and harness cable clamp (7) on hanger strap (8) with screw (6) and new locknut (9).

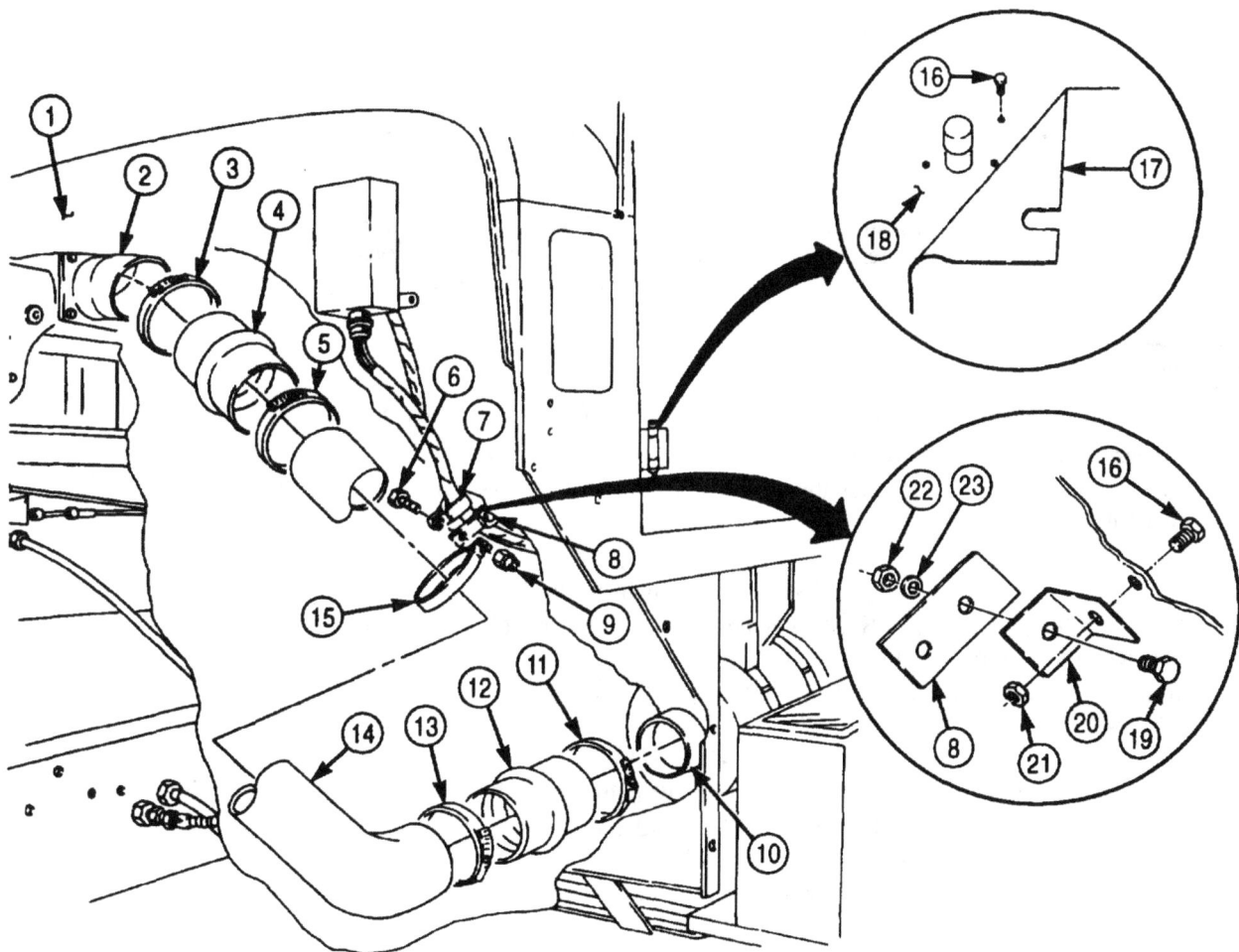

FOLLOW-ON TASKS: • Install ether valve and bracket (para. 3-35)
  • Start engine (TM 9-2320-272-10) and check for air leaks.

3-15. AIR CLEANER ASSEMBLY AND MOUNTING BRACKET MAINTENANCE

THIS TASK COVERS:

a. Removal   c. Installation
b. Inspection

INITIAL SETUP:

APPLICABLE MODELS
All

TOOLS
General mechanic's tool kit (Appendix E, Item 1)

MATERIALS/PARTS
Three locknuts (Appendix D, Item 288)
Five lockwashers (Appendix D, Item 354)

REFERENCES (TM)
TM 9-2320-272-10
TM 9-2320-272-24P

EQUIPMENT CONDITION
Parking brake set (TM 9-2320-272-10).

GENERAL SAFETY INSTRUCTIONS
If NBC exposure is suspected, all air filter media should be handled by personnel wearing protective equipment.

WARNING

If NBC exposure is suspected, all air filter media should be handled by personnel wearing protective equipment. Consult your unit NBC officer or NBC NCO for appropriate handling or disposal instructions.

a. Removal

1. Loosen hose clamps (6) and (12) on air cleaner tube hose (4) and cleaner-to-intake pipe hump hose (13).

2. Release support strap latch (8) from mounting bracket (5) and air cleaner assembly (9).

3. Remove two locknuts (1), mounting band (7), and two screws (2) from mounting bracket (3). Discard locknuts (1).

NOTE
Assistant will help with step 4.

4. Remove air cleaner assembly (9) and hose clamps (6) and (12) from air cleaner tube hose (4) and cleaner-to-intake pipe hump hose (13).

5. Remove locknut (10), screw (11), mounting band (7), and support strap (8) from air cleaner assembly (9). Discard locknut (10).

NOTE
Floor mat must be pulled back to gain access to one screw.

6. Remove two screws (17), lockwashers (15), and mounting bracket (3) from mounting bracket (5) and subfloor (18). Discard lockwashers (15).

7. Remove three screws (14), lockwashers (15), and mounting bracket (5) from cab floor (16). Discard lockwashers (15).

b. Inspection

Inspect air cleaner assembly (9) for cracks and splits that would allow unfiltered air to enter. Replace if cracked or split.

3-15. AIR CLEANER ASSEMBLY AND MOUNTING BRACKET MAINTENANCE (Contd)

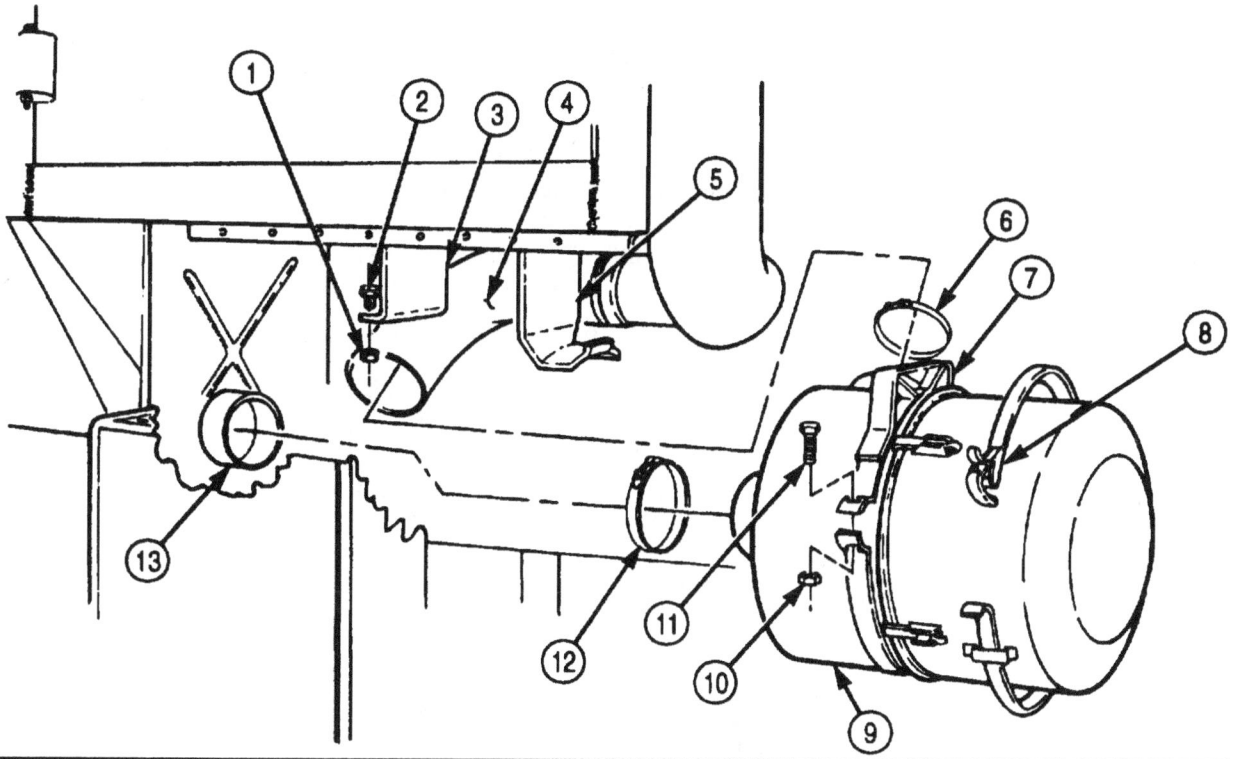

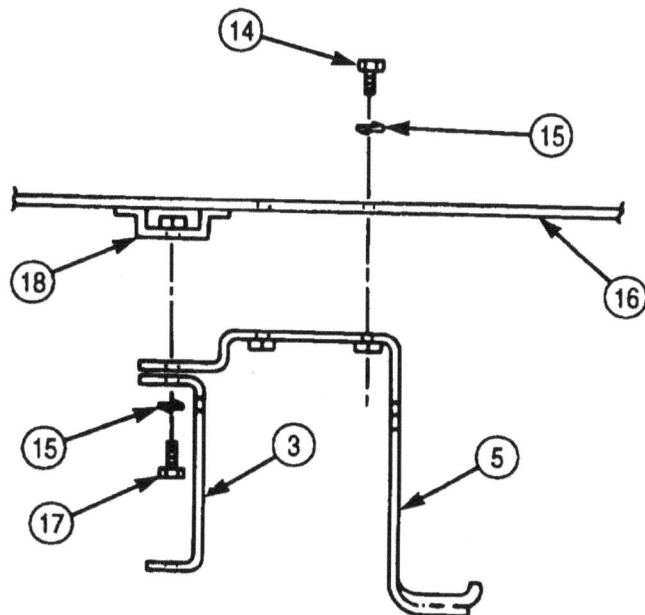

3-15. AIR CLEANER ASSEMBLY AND MOUNTING BRACKET MAINTENANCE (Contd)

c. Installation

NOTE

Floor mat must be pulled back to gain access to one screw hole in
steps 1 and 2.

1. Install mounting bracket (5) on cab floor (16) with three new lockwashers (15) and screws (14).

2. Install mounting bracket (3) on mounting bracket (5) and subfloor (18) with two new lockwashers (15) and screws (17).

3. Install mounting band (7) on air cleaner assembly (9) with screw (11) and new locknut (10).

4. Position support strap (8) on air cleaner assembly (9).

NOTE

Assistant will help with step 5.

5. Position hose clamps (6) and (12) on air cleaner tube hose (4) and cleaner-to-intake pipe hump hose (13), and install air cleaner assembly (9) on air cleaner tube hose (4) and cleaner-to-intake pipe hump hose (13) with hose clamps (6) and (12).

6. Install air cleaner assembly (9) on mounting brackets (5) and (3) with support strap (8), two screws (2), mounting band (7), and two new locknuts (1). Fasten support strap (8).

7. Tighten hose clamps (6) and (12).

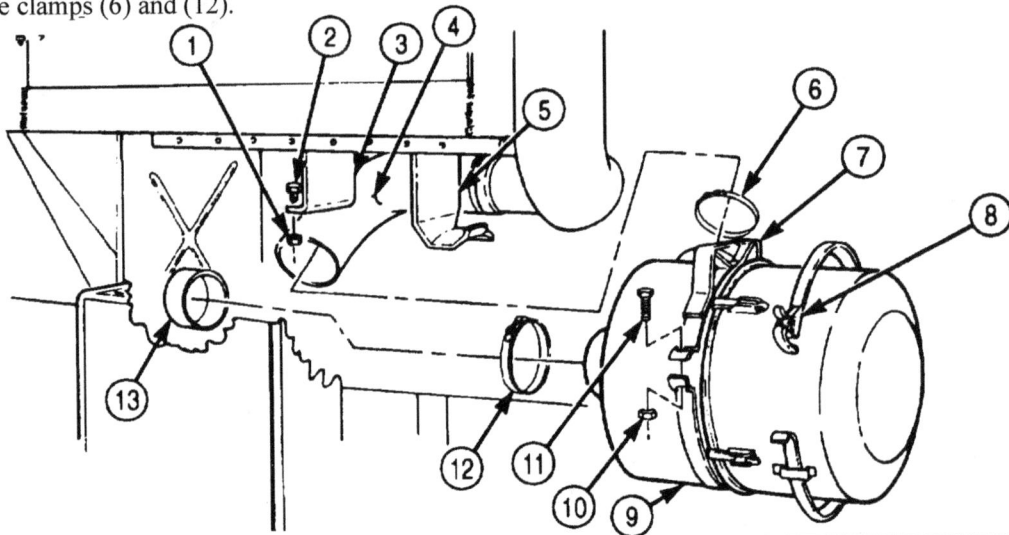

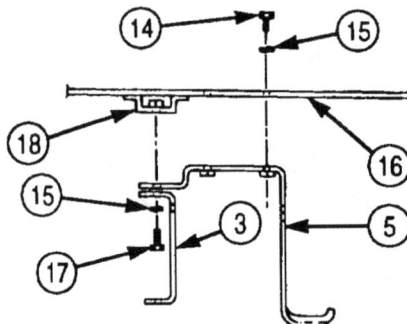

FOLLOW-ON TASKS: • Start engine (TM 9-2320-272-10) and check for air leaks.
• Ensure air filter indicator in cab indicates green (TM 9-2320-272-10).

Section III. FUEL SYSTEM MAINTENANCE

3-16. FUEL SYSTEM MAINTENANCE INDEX

3-17. FUEL DRAIN AND BYPASS TUBING (M939/A1) REPLACEMENT

THIS TASK COVERS:

a. Removal   b. Installation

INITIAL SETUP:

APPLICABLE MODELS
M939/A1

TOOLS
General mechanic's tool kit (Appendix E, Item 1)
Torque wrench (Appendix E, Item 144)

MATERIALS/PARTS
Five lockwashers (Appendix D, Item 348)
Lockwasher (Appendix D, Item 349)
Antiseize tape (Appendix C, Item 72)

REFERENCES (TM)
TM 9-2320-272-10
TM 9-2320-272-24P

EQUIPMENT CONDITION
- Parking brake set (TM 9-2320-272-10).
- Hood raised and secured (TM 9-2320-272-10).
- Water pump removed (para. 3-68).
- Air compressor-to-engine coolant return tube removed (para. 3-55).

---

a. Removal

NOTE
Have drainage container ready to catch fuel.

1. Disconnect fuel supply tube (7) from fuel pump shutoff valve (8).

2. Disconnect fuel return tube (10) from fuel pump elbow (23).

3. Remove screw (16), lockwasher (15), washer (14), clamp (13), and spacer (12) from left side of engine block (1) and bracket (11). Discard lockwasher (15).

4. Remove screw (17), lockwasher (18), and bracket (11) from left side of engine block (1). Discard lockwasher (18).

NOTE
Step 5 may not be required for all vehicles.

5. Remove screw (28), lockwasher (27), washer (26), two clamps (25), fuel supply tube (7), and fuel return tube (6) from intake manifold (24). Discard lockwasher (27).

6. Remove nut (2), lockwasher (3), screw (5), and clamp (4) from left side of engine block (1). Discard lockwasher (3).

7. Remove nut (21), lockwasher (22), screw (19), and clamp (20) from tubes (6) and (7) left side of engine block (1). Discard lockwasher (22).

8. Disconnect fuel return tubes (6) and (10) from tee (9).

3-17. FUEL DRAIN AND BYPASS TUBING (M939/A1) REPLACEMENT (Contd)

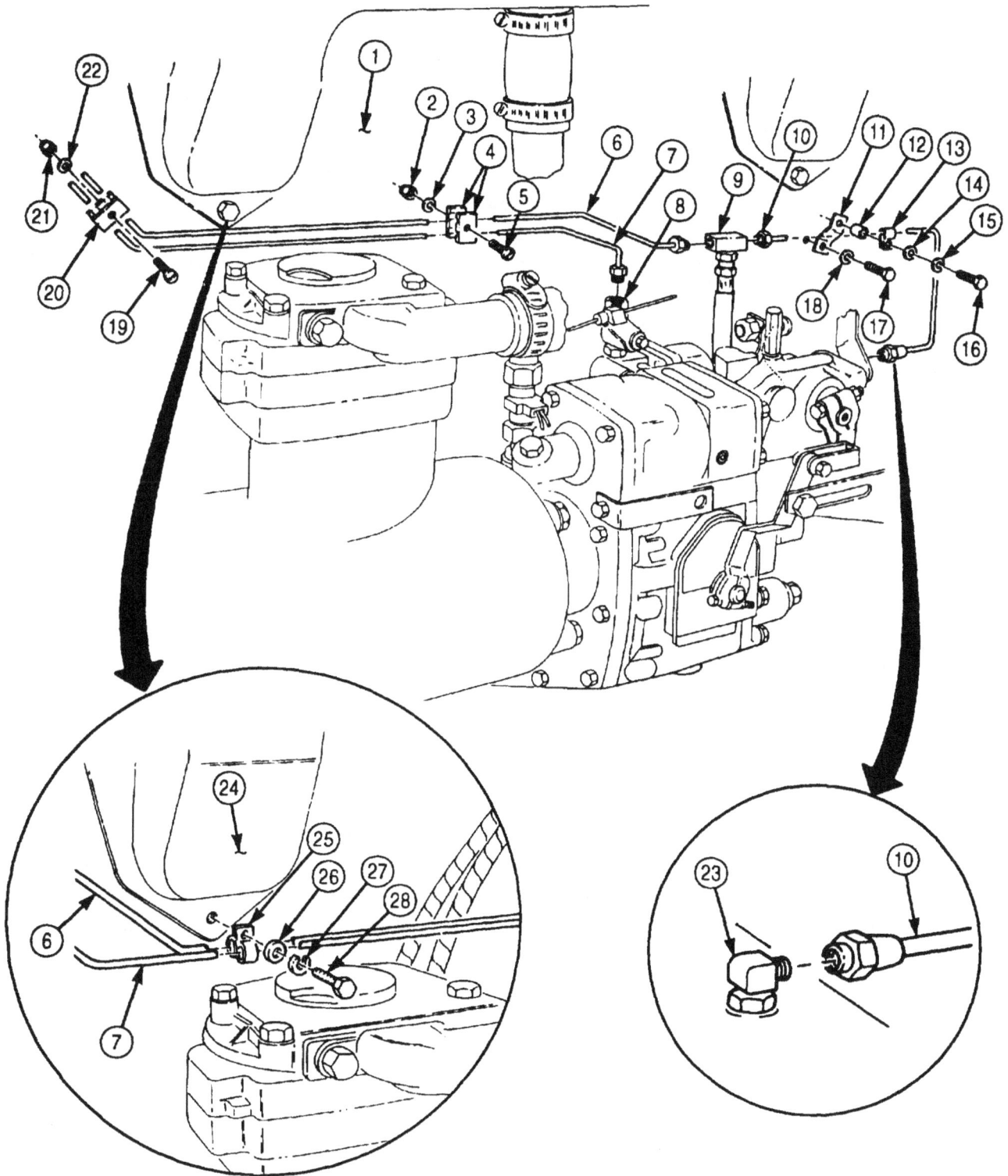

3-17. FUEL DRAIN AND BYPASS TUBING (M939/A1) REPLACEMENT (Contd)

9. Remove screw (10), lockwasher (11), washer (12), clamp (13) with fuel supply tube (4), support bracket (14), and spacer (15) from front of cylinder head (2). Discard lockwasher (11).

NOTE

Perform step 10 for vehicles with early model engines.

10. Remove fuel supply tube (4) from adapter (5).

NOTE

Perform step 11 for vehicles equipped with late model engines.

11. Remove fuel supply tube (4) and fuel pressure transducer (6) from tee (7).
12. Remove fuel return tube (1) from adapter (3).
13. Remove two screws (8) and upper radiator support bracket (9) from front of cylinder head (2).

NOTE

Perform step 14 for early model engines or step 15 for late model engines.

14. Remove two adapters (3) and (5) from front of cylinder head (2).
15. Remove tee (7) and adapter (3) from front of cylinder head (2).

b. Installation

NOTE

- Wrap all male pipe threads with antiseize tape before installation.
- Perform step 1 for early model engines or step 2 for late model engines.

1. Install two adapters (3) and (5) on front of cylinder head (2).
2. Install tee (7) and adapter (3) on front of cylinder head (2).
3. Install upper radiator support bracket (9) on front of cylinder head (2) with two screws (8). Tighten screws (8) 55-65 lb-ft (75-88 N.m).
4. Connect fuel return tube (1) to adapter (3).

NOTE

Perform step 5 for vehicles with early model engines.

5. Connect fuel supply tube (4) to adapter (5).

NOTE

Perform step 6 for vehicles with late model engines.

6. Connect fuel supply tube (4) and fuel pressure transducer (6) to tee (7).
7. Install spacer (15), support bracket (14), and clamp (13) with fuel supply tube (4) on front of cylinder head (2) with washer (12), new lockwasher (11), and screw (10).

3-17. FUEL DRAIN AND BYPASS TUBING (M939/A1) REPLACEMENT (Contd)

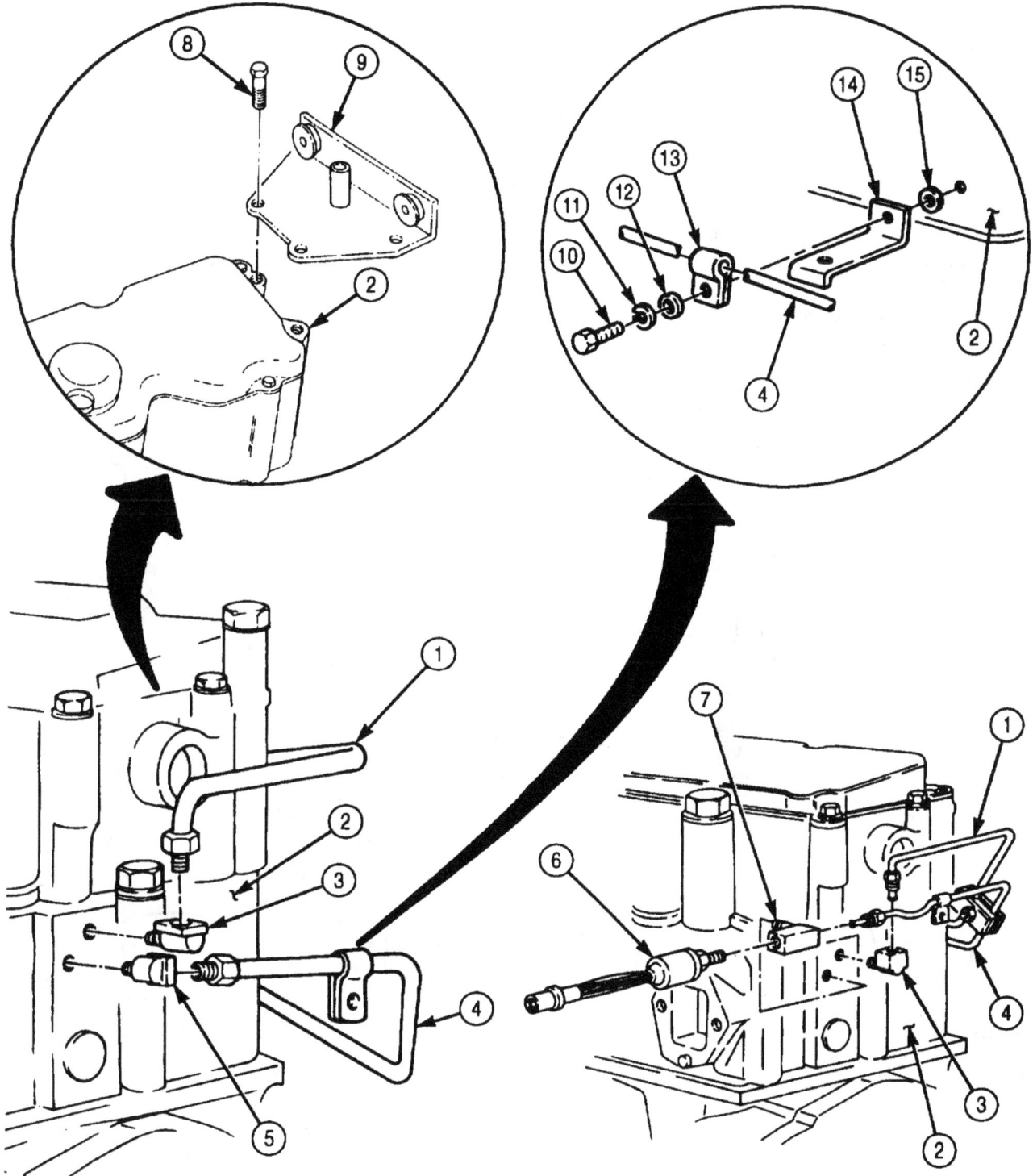

EARLY MODEL ENGINE

LATE MODEL ENGINE

3-17. FUEL DRAIN AND BYPASS TUBING (M939/A1) REPLACEMENT (Contd)

8. Install clamp (22) on fuel return tube (4) and fuel supply tube (9) with screw (21), new lockwasher (2), and nut (1).

9. Install clamp (7) on fuel return tube (4) and fuel supply tube (9) with screw (8), new lockwasher (6), and nut (5).

NOTE
Step 10 may not be required for all vehicles.

10. Install two clamps (25) and tubes (4) and (9) on manifold (24) with washer (26), new lockwasher (27), and screw (28).

11. Install bracket (11) on engine block (3) with new lockwasher (19) and screw (18).

12. Install clamp (13) and tube (17) on bracket (11) with spacer (12), washer (14), new luckwasher (15), and screw (16).

13. Connect fuel return tube (17) to fuel pump elbow (23).

14. Connect, fuel supply tube (9) to fuel pump shutoff valve (20).

15. Connect tubes (17) and (4) to tee (10).

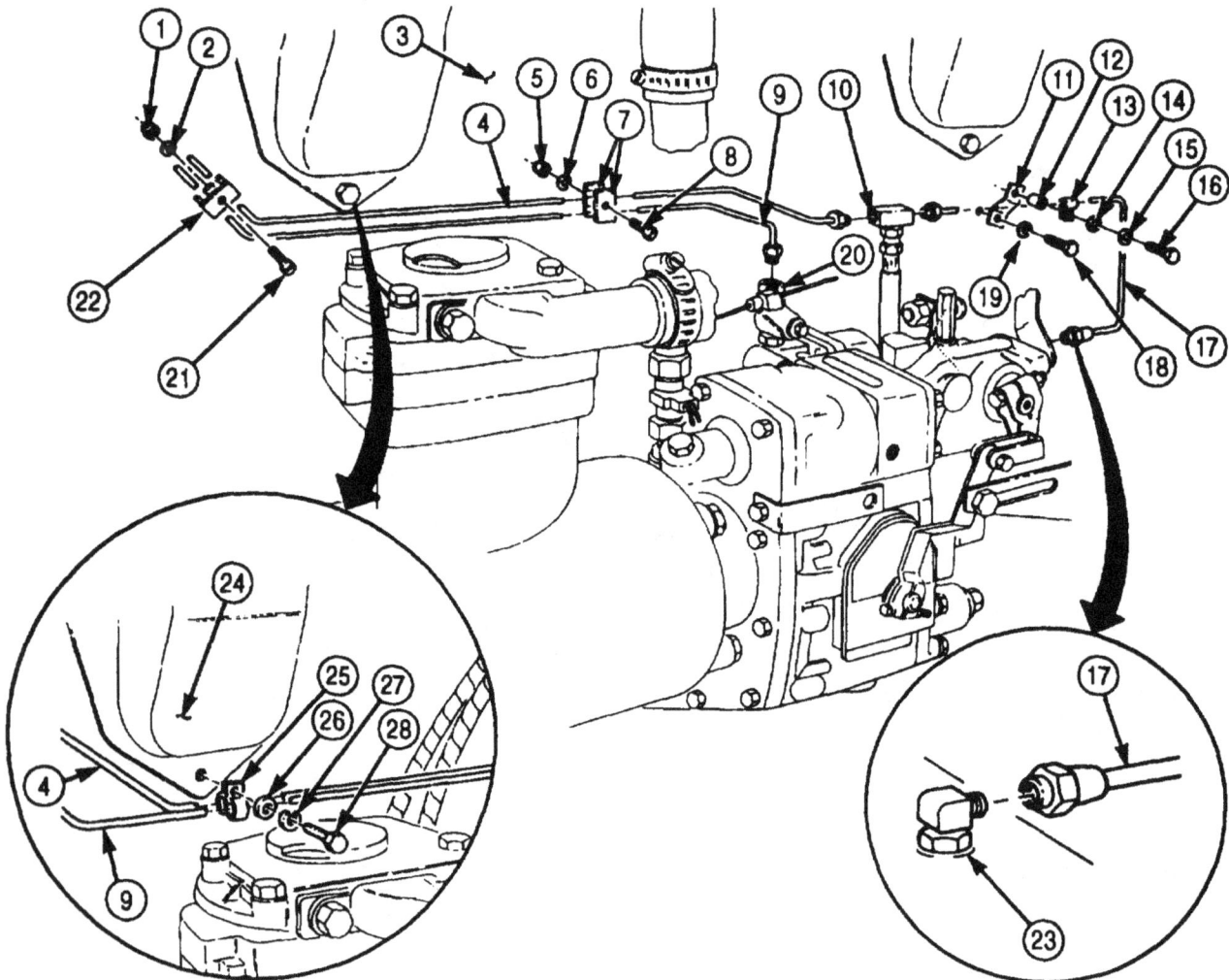

FOLLOW-ON TASKS:
 • Install air compressor-to-engine coolant return tube (para. 3-55).
 • Install water pump (para. 3-68).

3-18. AIR FUEL CONTROL (AFC) TUBE (M939A2) MAINTENANCE

THIS TASK COVERS:

a. Removal   c. Installation

b. Inspection

INITIAL SETUP:

APPLICABLE MODELS
M939A2

TOOLS
General mechanic's tool kit (Appendix E, Item 1)

MATERIALS/PARTS
Two seal washers (Appendix D, Item 644)

REFERENCES (TM)
TM 9-2320-272-10
TM 9-2320-272-24P

EQUIPMENT CONDITION
- Parking brake set (TM 9-2320-272-10).
- Hood raised and secured (TM 9-2320-272-10).

a. Removal

1. Disconnect AFC tube (2) from adaptor (1).
2. Remove screw (3), two seal washers (4), and AFC tube (2) from fuel pump (5). Discard seal washers (4).

b. Inspection

1. For general inspection instructions, refer to para. 2-15.
2. Replace all parts failing inspection.

c. Installation

1. Install AFC tube (2) on fuel pump (5) with new seal washers (4) and screw (3).
2. Connect AFC tube (2) to adapter (1).

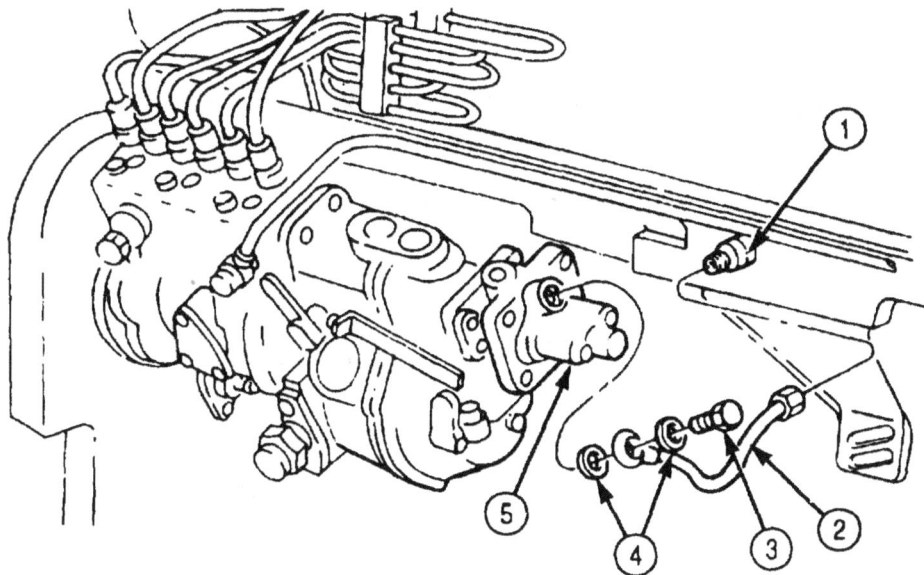

FOLLOW-ON TASK: Start engine (TM 9-2320-272-10) and check for leaks.

3-19. FUEL INJECTOR TUBE (M939A2) MAINTENANCE

THIS TASK COVERS:

a. Removal
b. Cleaning and Inspection

c. Installation

INITIAL SETUP:

APPLICABLE MODELS
M939A2

TOOLS
General mechanic's tool kit (Appendix E, Item 1)

MATERIALS/PARTS
Seven banjo seals (Appendix D, Item 9)
Cap and plug set (Appendix C, Item 14)

REFERENCES (TM)
TM 9-2320-272-10
TM 9-2320-272-24P

EQUIPMENT CONDITION
• Parking brake set (TM 9-2320-272-10)
• Hood raised and secured (TM 9-2320-272-10)

GENERAL SAFETY INSTRUCTIONS
Diesel fuel is flammable. Do not perform this task near open flames.

a. Removal

WARNING

Diesel fuel is highly flammable. Do not perform fuel system procedures near open flame. Injury to personnel may result.

CAUTION

Cap or plug all openings immediately after disconnecting fuel lines to prevent contamination. Failure to do so may result in damage to fuel pump.

1. Remove seven screws (3), banjo seals (4), and fuel manifold (5) from six fuel injectors (6) and fuel return tube (7). Discard banjo seals (4).

NOTE
Tag injector tubes for installation.

2. Loosen six fuel line nuts (10) and disconnect fuel injector tubes (11) from fuel injectors (6).

3. Loosen six fuel line nuts (10) and disconnect fuel injector tubes (11) from fuel injector pump (9).

4. Remove screw (1), washer (2), and fuel injector tubes (11) from cylinder head (8).

NOTE
Tag brace assemblies for installation.

5. Remove two screws (25), washers (28), tube braces (26), and isolators (27) from fuel injector tubes (11).

6. Remove screw (21), washer (24), two tube braces (22), and isolators (23) from fuel injector tubes (11).

7. Remove four screws (12), washers (15), two tubes braces (13), and isolators (14) from injector tubes (11).

8. Remove four screws (16), washers (17), two tube braces (18), brace (20), and four isolators (19) from injector tubes (11).

3-19. FUEL INJECTOR TUBE (M939A2) MAINTENANCE (Contd)

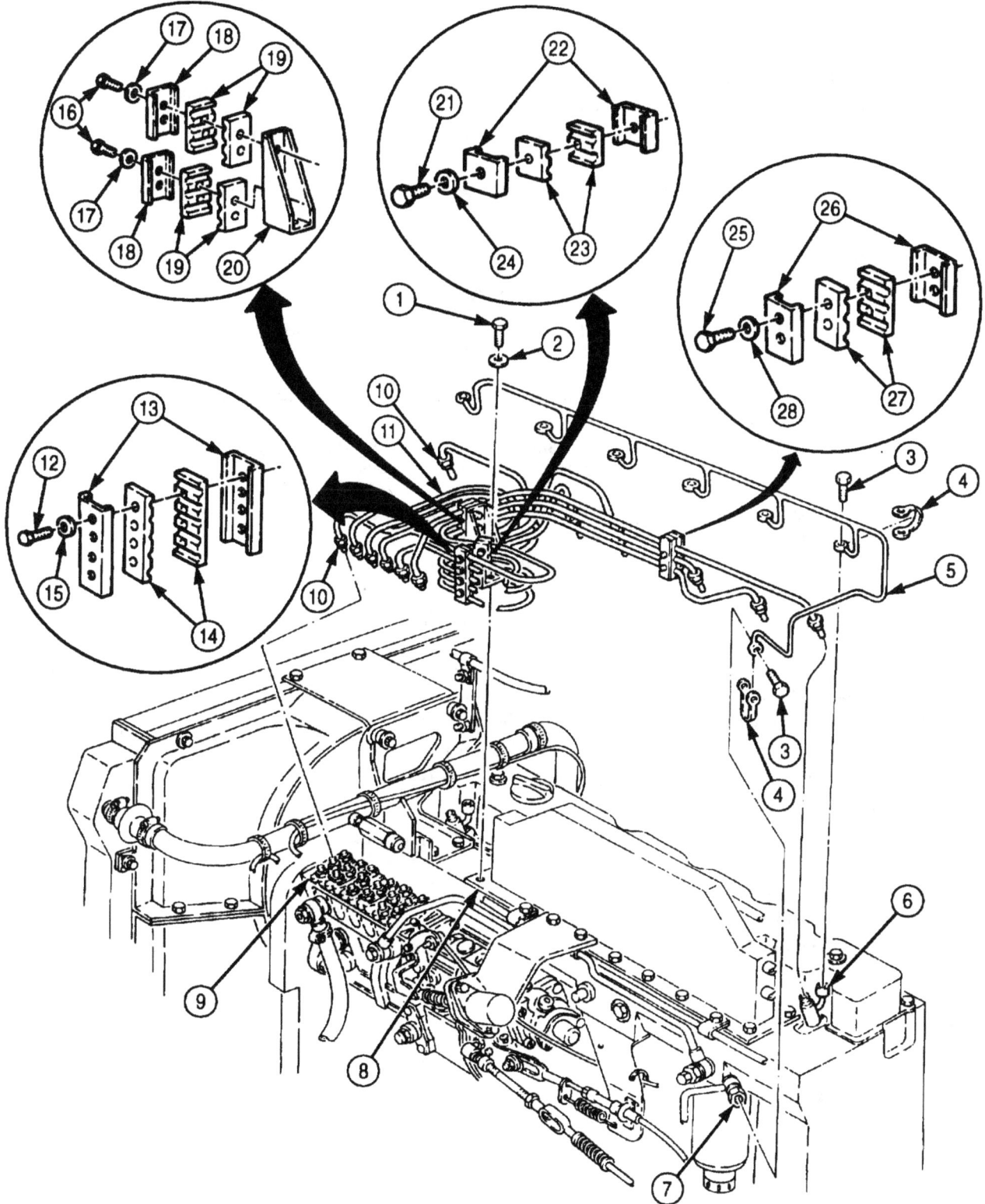

3-19. FUEL INJECTOR TUBE (M939A2) MAINTENANCE (Contd)

b. Cleaning and Inspection

1. For general cleaning instructions, refer to para. 2-14.
2. For general inspection instructions, refer to para. 2-15.
3. Inspect fuel injection tubes (11) for cracks or dents. If cracked or dented, replace fuel injection tubes (11).
4. Inspect fuel manifold (5) for cracks or dents. If cracked or dented, replace fuel manifold (5).
5. Replace all parts failing inspection.

c. Installation

CAUTION

- Injector tubes deliver fuel at high pressure causing them to expand and contract. Ensure injector tubes are securely clamped and routed so they do not come in contact with each other or any engine component. Failure to do so may result in premature injector tube failure.
- Injector tubes must be installed in correct positions as tagged. Failure to do so may result in damage to tubes and engine malfunction.

1. Install four isolators (19), brace (20), and two tube braces (18) on six fuel injector tubes (11) with four washers (17) and screws (16). Finger tighten screws (16).
2. Install two isolators (14) and tube braces (13) on fuel injector tubes (11) with four washers (15) and screws (12). Finger tighten screws (12).
3. Install two isolators (23) and tube braces (22) on fuel injector tubes (11) with washer (24) and screw (21). Finger tighten screw (21).
4. Install two isolators (27) and tube braces (26) on injector tubes (11) with two washers (28) and screws (25). Finger tighten screws (25).
5. Install six fuel injector tubes (11) on cylinder head (8) with washer (2) and screw (1).
6. Position six fuel injector tubes (11) on fuel injection pump (9) and fuel injectors (6).
7. Connect six fuel tube nuts (10) to fuel injection pump (9).
8. Connect six fuel tube nuts (10) to fuel injectors (6).
9. Tighten two screws (25), screw (21), four screws (12), and screws (16).

NOTE
Banjo seals must face toward fuel injectors to allow installation of valve cover.

10. Install fuel manifold (5) and seven new banjo seals (4) on six fuel injectors (6) and fuel supply tube (7) with seven screws (3).

3-19. FUEL INJECTOR TUBE (M939A2) MAINTENANCE (Contd)

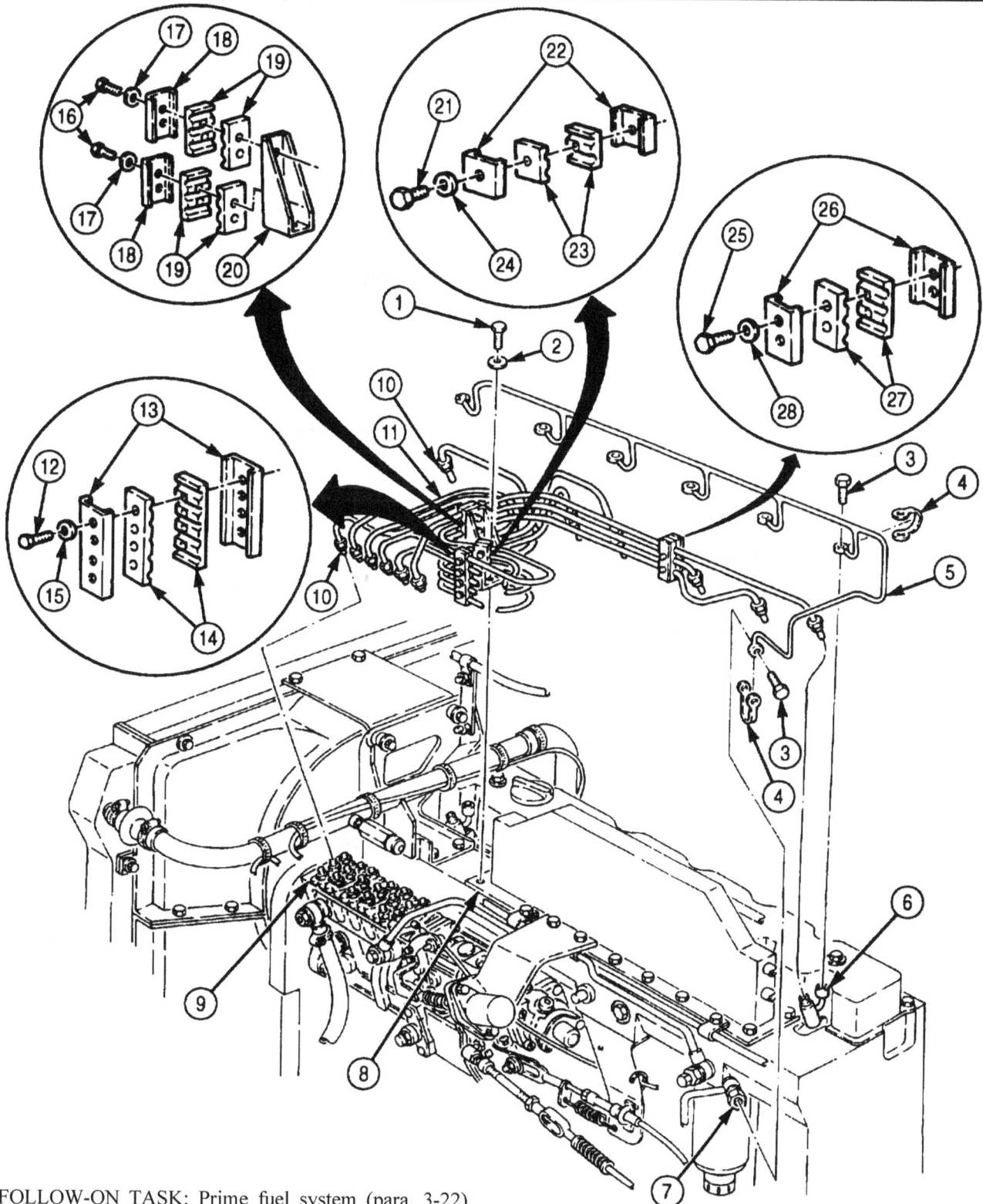

FOLLOW-ON TASK: Prime fuel system (para. 3-22).

3-20. FUEL TRANSFER PUMP AND SUPPLY LINES (M939A2) MAINTENANCE

THIS TASK COVERS:

a. Removal   c. Installation
b. Cleaning and Inspection

INITIAL SETUP:

APPLICABLE MODELS
M939A2

TOOLS
General mechanic's tool kit (Appendix E, Item 1)

MATERIALS/PARTS
Banjo seal (Appendix D, Item 9)
Two seal washers (Appendix D, Item 645)
Two seal washers (Appendix D, Item 646)
Gasket (Appendix D, Item 139)
Antiseize tape (Appendix C, Item 72)

REFERENCES (TM)
TM 9-2320-272-10
TM 9-2320-272-24P

EQUIPMENT CONDITION
• Parking brake set (TM 9-2320-272-10).
• Hood raised and secured (TM 9-2320-272-10).

GENERAL SAFETY INSTRUCTIONS
Diesel fuel is flammable. Do not perform this task near open flames.

---

a. Removal

> **WARNING**
>
> Diesel fuel is highly flammable. Do not perform fuel system
> procedures near open flames. Injury to personnel may result.

1. Disconnect fuel pressure transducer (1) from wiring harness (2).

2. Remove fuel pressure transducer (1) from adapter (10).

3. Remove adapter (10), two seal washers (11), and fuel supply tube (5) from fuel transfer pump (13). Discard seal washers (11).

4. Remove fuel supply line (16) from elbow (15) on fuel transfer pump (13).

5. Remove two screws (14), fuel transfer pump (13), and gasket (12) from engine block (17). Discard gasket (12).

6. Remove elbow (15) from fuel transfer pump (13).

> **NOTE**
> Perform steps 7 and 8 if fuel supply tube is damaged.

7. Remove screw (8), banjo seal (9), and fuel drain manifold (7) from adapter (6). Discard banjo seal (9).

8. Remove adapter (6), two seal washers (4), and fuel supply tube (5) from fuel filter head (3). Discard seal washers (4).

b. Cleaning and Inspection

1. For general cleaning instructions, refer to para. 2-14.

2. For general inspection instructions, refer to para. 2-15.

3. Replace all parts failing inspection.

3-20. FUEL TRANSFER PUMP AND SUPPLY LINES (M939A2) MAINTENANCE (Contd)

c. Installation

1. Apply antiseize tape to threads of elbow (15) and install elbow (15) on fuel transfer pump (13).

2. Install new gasket (12) and fuel transfer pump (13) on engine block (17) with two screws (14).

3. Connect fuel supply line (16) to elbow (15) on fuel transfer pump (13).

4. Install fuel supply tube (5) on fuel transfer pump (13) with two new seal washers (11) and adapter (10). Finger tighten adapter (10).

NOTE
Perform steps 5 through 7 if fuel supply tube was removed.

5. Install fuel supply tube (5) on fuel filter head (3) with two new seal washers (4) and adapter (6).

6. Tighten adapter (10).

7. Place new banjo seal (9) around fuel drain manifold (7) and install on adapter (6) with screw (8).

8. Apply antiseize tape to threads on fuel pressure transducer (1) and install fuel pressure transducer (1) on adapter (10).

9. Connect wiring harness (2) to fuel pressure transducer (1).

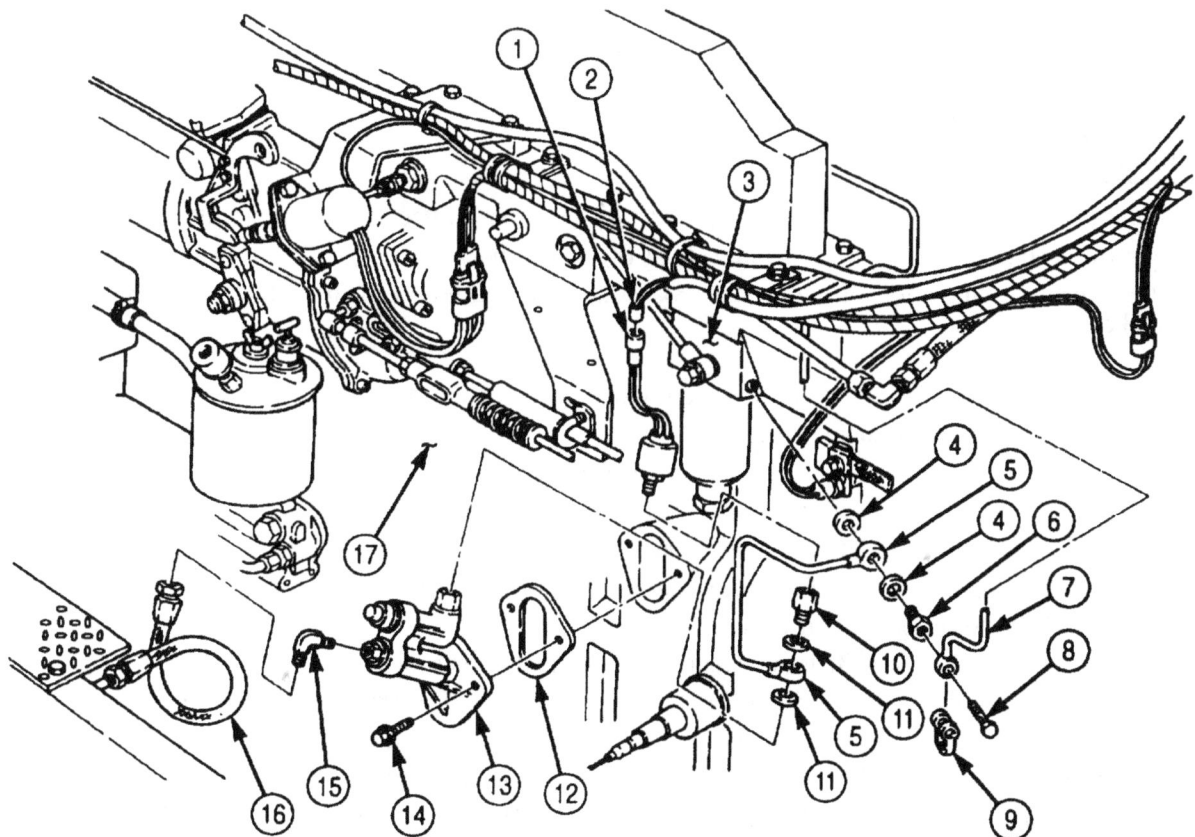

FOLLOW-ON TASK: Prime fuel system (para. 3-22).

3-21. TURBOCHARGER AND COOLANT LINES (M939A2) REPLACEMENT

THIS TASK COVERS:

a. Removal   b. Installation

INITIAL SETUP:

APPLICABLE MODELS
M939A2

TOOLS
General mechanic's tool kit (Appendix E, Item 1)
Torque wrench (Appendix E, Item 146)

MATERIALS/PARTS
O-ring (Appendix D, Item 476)
Gasket (Appendix D, Item 141)
Gasket (Appendix D, Item 140)
Antiseize compound (Appendix C, Item 10)
Antiseize tape (Appendix C, Item 72)
Lubricating oil (Appendix C, Item 50)

REFERENCES (TM)
LO 9-2320-272-12
TM 9-2320-272-10
TM 9-2320-272-24P

EQUIPMENT CONDITION
• Parking brake set (TM 9-2320-272-10).
• Hood raised and secured (TM 9-2320-272-10).
• Coolant drained (para. 3-53).

GENERAL SAFETY INSTRUCTIONS
Do not perform this task when exhaust system is hot.

WARNING

Do not touch hot exhaust system components with bare hands.
Injury to personnel may result.

a. Removal

1. Remove four clamps (1), two hoses (2), and tube (3) from turbocharger (4) and aftercooler (5).

2. Remove two screws (7) from oil return tube (8) and turbocharger (4).

3. Remove two clamps (9), oil return tube (8), gasket (6), and hose (10) from tube (11). Discard gasket (6).

NOTE
Perform step 4 if tube is damaged.

4. Remove tube (11) from engine block (12).

NOTE
• Perform steps 5 and 6 for early model engines.
• Perform steps 7 and 8 for late model engines.

5. Remove oil supply hose (13) from turbocharger (4) and connector (14).

6. Remove connector (14) from oil cooler (16).

7. Remove oil supply hose (13) from turbocharger (4) and connector (23).

8. Remove connector (23) and O-ring (24) from oil cooler (16). Discard O-ring (24).

9. Loosen clamp (15) and disconnect air intake tube (20) from turbocharger (4).

10. Loosen clamp (19) and disconnect exhaust pipe (18) from turbocharger (4).

11. Remove four nuts (22), turbocharger (4), and gasket (21) from exhaust manifold (17). Discard gasket (21).

3-21. TURBOCHARGER AND COOLANT LINES (M939A2) REPLACEMENT (Contd)

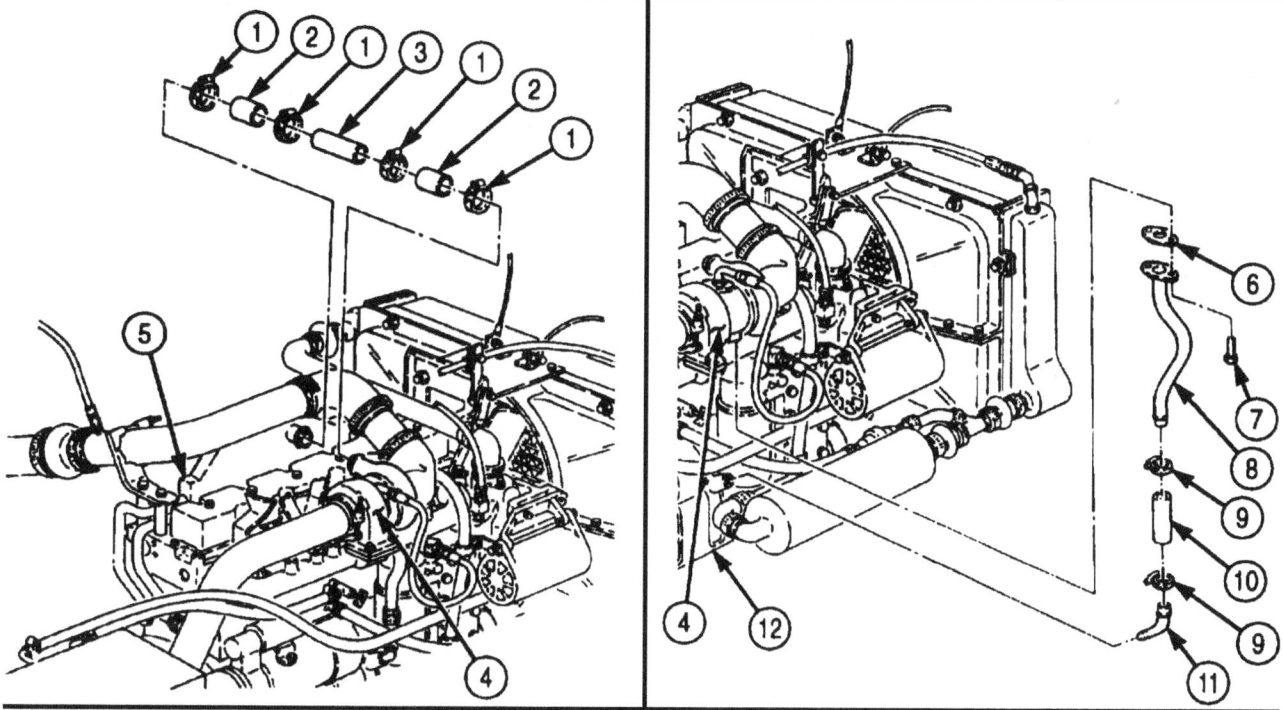

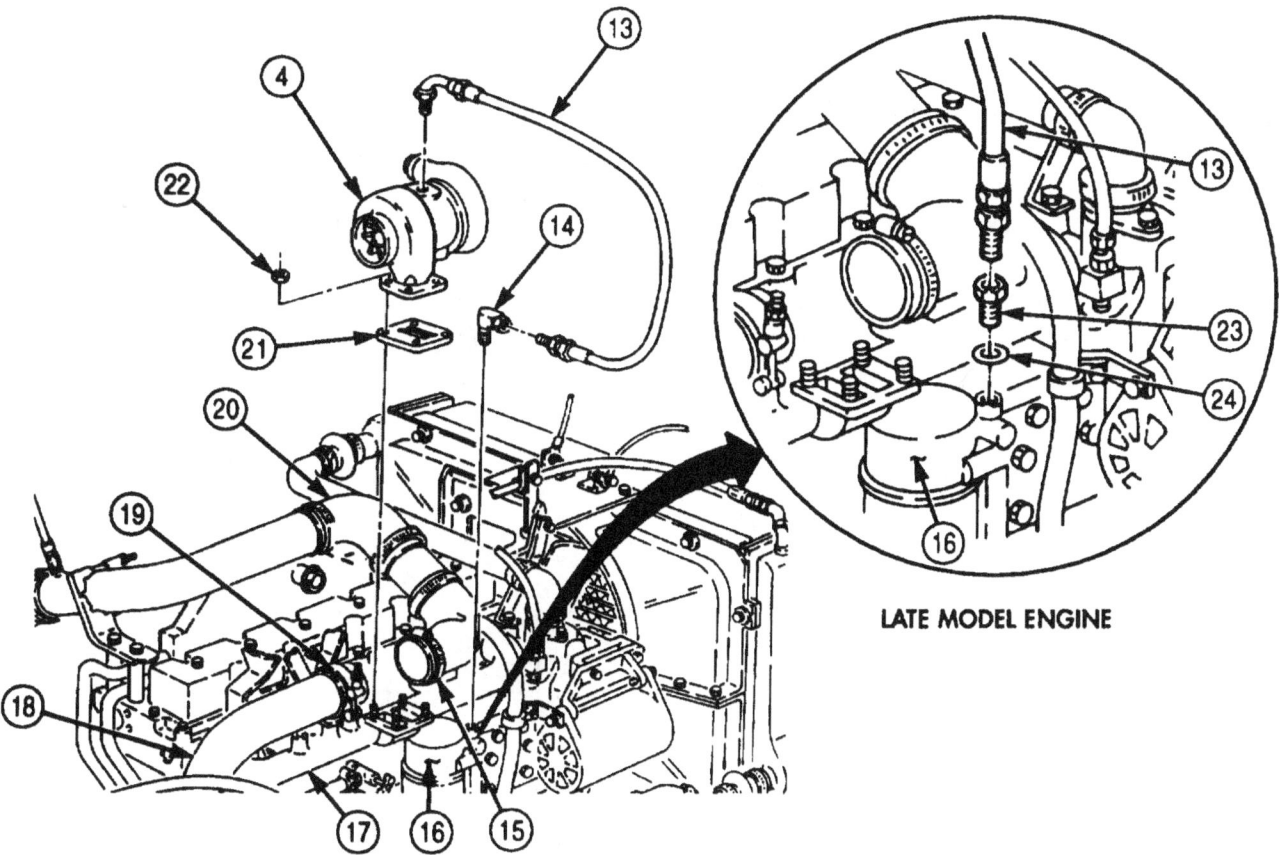

LATE MODEL ENGINE

3-21. TURBOCHARGER AND COOLANT LINES (M939A2) REPLACEMENT (Contd)

b. Installation

1. Apply antiseize compound to four mounting studs (6) and install new gasket (12) and turbocharger (1) on exhaust manifold (7) with four nuts (13). Tighten nuts (13) to 25 lb-ft (34 N.m).

CAUTION

Turbocharger must be lubricated with clean engine oil prior to starting engine, or bearing damage may result.

2. With a small funnel positioned in turbocharger oil supply hole (2), pour 2-3 ounces (57-85 grams) of clean engine oil into turbocharger (1) (LO 9-2320-272-12).

3. Spin impeller/turbine blades (14) by hand to coat bearings with oil (LO 9-2320-272-12).

4. Connect exhaust pipe (8) to turbocharger (1) and tighten clamp (9).

5. Connect air intake tube (11) to turbocharger (1) and tighten clamp (10).

NOTE

- Perform steps 6 and 7 for early model engines.

- Perform steps 8 and 9 for late model engines,

- Wrap all male pipe threads with antiseize tape before installation.

6. Install connector (4) on oil cooler (5).

7. Install oil supply hose (3) on turbocharger (1) and connector (4).

8. Install new O-ring (16) and connector (15) on oil cooler (5).

9. Install oil supply hose (3) on turbocharger (1) and connector (15).

NOTE

Perform step 10 if tube was removed.

10. Install tube (22) on engine block (23).

11. Install new gasket (17) and oil return tube (19) on turbocharger (1) with two screws (18).

12. Install hose (21) and oil return tube (19) on tube (22) with two clamps (20).

13. Install tube (26) and two hoses (25) on aftercooler (27) and turbocharger (1) with four clamps (24).

CAUTION

When filling cooling system, ensure drainvalve on aftercooler is open. Failure to do so may result in damage to equipment.

3-21. TURBOCHARGER AND COOLANT LINES (M939A2) REPLACEMENT (Contd)

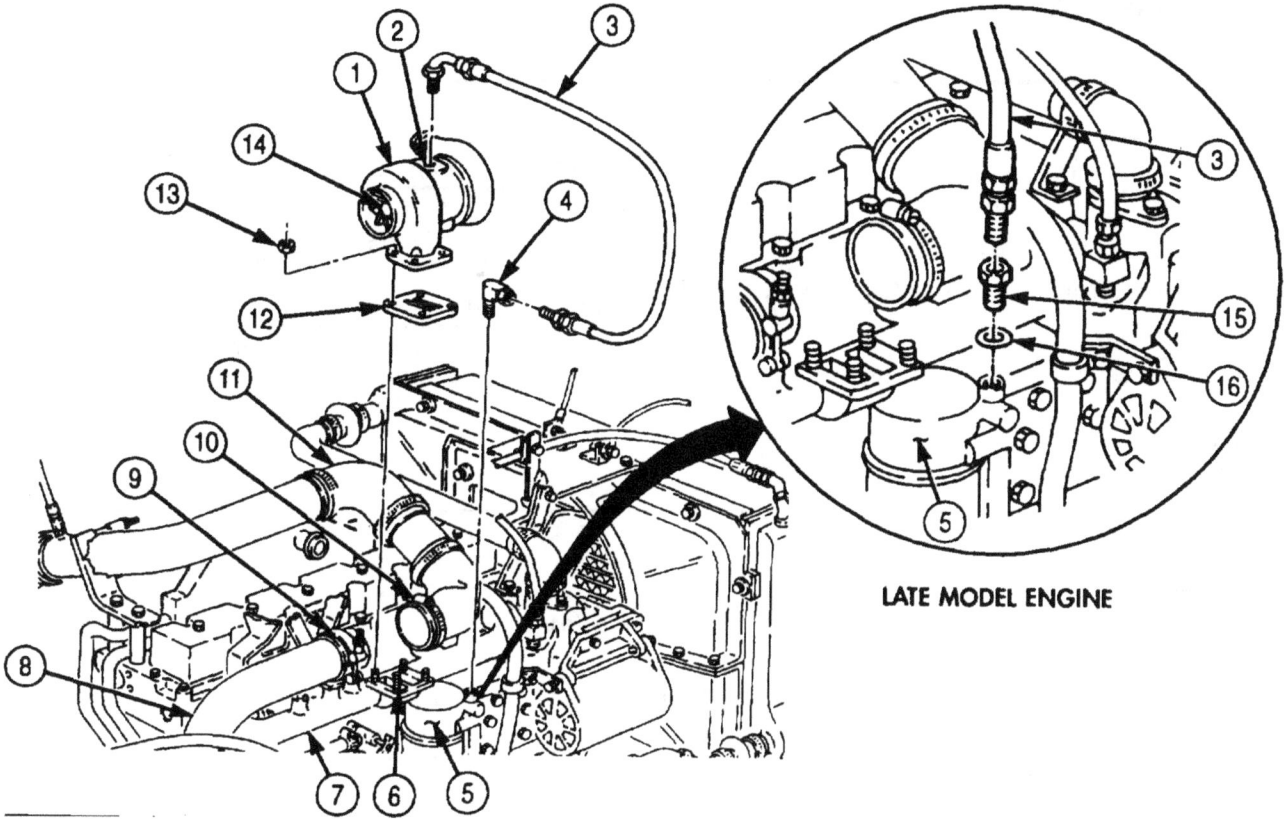

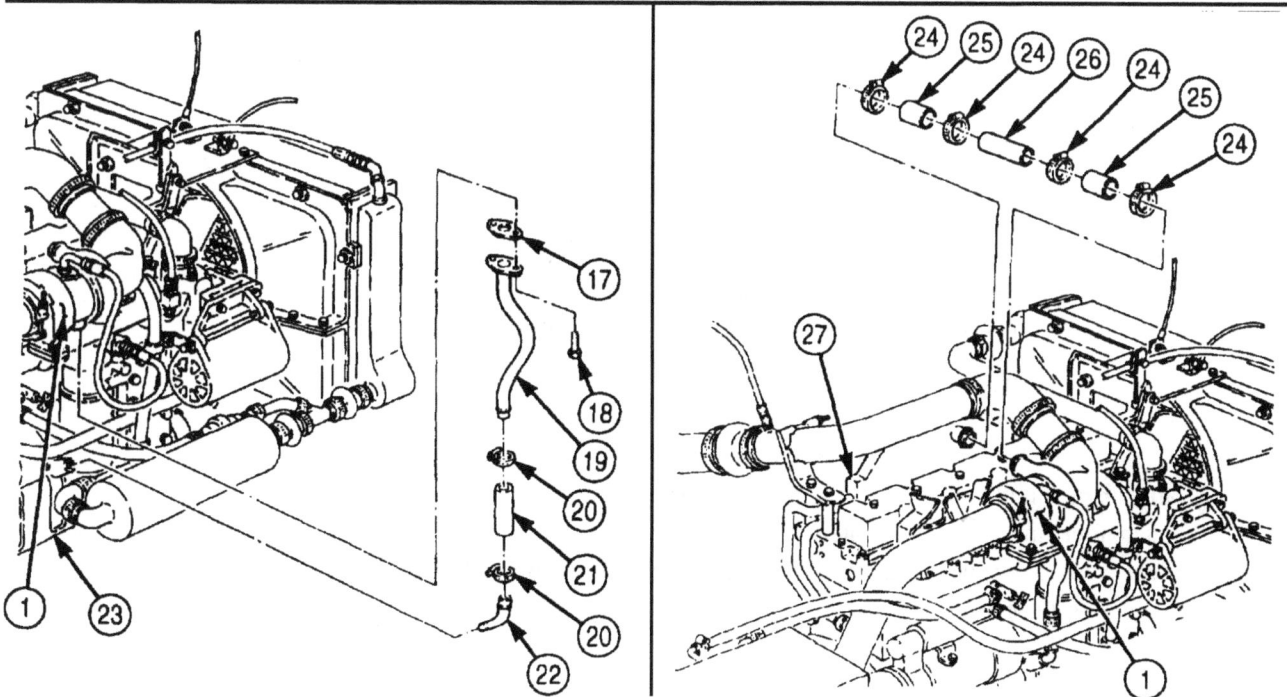

LATE MODEL ENGINE

FOLLOW-ON TASK: Fill coolant system (para. 3-53).

3-22. PRIMING FUEL SYSTEM (M939A2)

THIS TASK COVERS:
Priming

INITIAL SETUP:

APPLICABLE MODELS
M939A2

TOOLS
General mechanic's tool kit (Appendix E, Item 1)

MATERIALS/PARTS
Cleaning cloth (Appendix C, Item 21)

REFERENCES (TM)
TM 9-2320-272-10
TM 9-2320-272-24P

EQUIPMENT CONDITION
- Parking brake set (TM 9-2302-272-10).
- Hood raised and secured (TM 9-2320-272-10).
- Air cleaner hose removed (para. 3-13).

GENERAL SAFETY INSTRUCTIONS
- All personnel must stand clear during engine cranking operations.
- Use eyeshields and gloves when loosening fuel lines.

Priming

1. Loosen adapter (6) and vent screw (7) on fuel pump (5).

NOTE
Perform step 2 until all air is purged from fuel system.

2. Press plunger (3) on fuel transfer pump (4) until air is purged from fuel system.
3. Tighten vent screw (7) and adapter (6) on fuel pump (5).
4. Start engine (TM 9-2320-272-10).

NOTE
If engine fails to start, or runs rough, perform steps 5 through 10.

5. Loosen fuel line (2) from fuel injector (1).
6. Place cleaning cloth around fuel line (2) and fuel injector (1).

WARNING

- All personnel must stand clear during engine cranking operation. Failure to do so may result in injury to personnel.
- Fuel pressure is sufficient to penetrate skin. Wear hand protection and safety goggles at all times when removing injector tubes. Failure to do so may result in injury to personnel.

NOTE
Assistant will help with step 7.

7. Crank engine until all air is vented at fuel injector (1).
8. Tighten fuel line (2) at fuel injector (1).
9. Repeat steps 5 through 8 until all fuel injectors (1) are purged free of air.
10. Stop engine (TM 9-2320-272-10).

3-22. PRIMING FUEL SYSTEM (M939A2) (Contd)

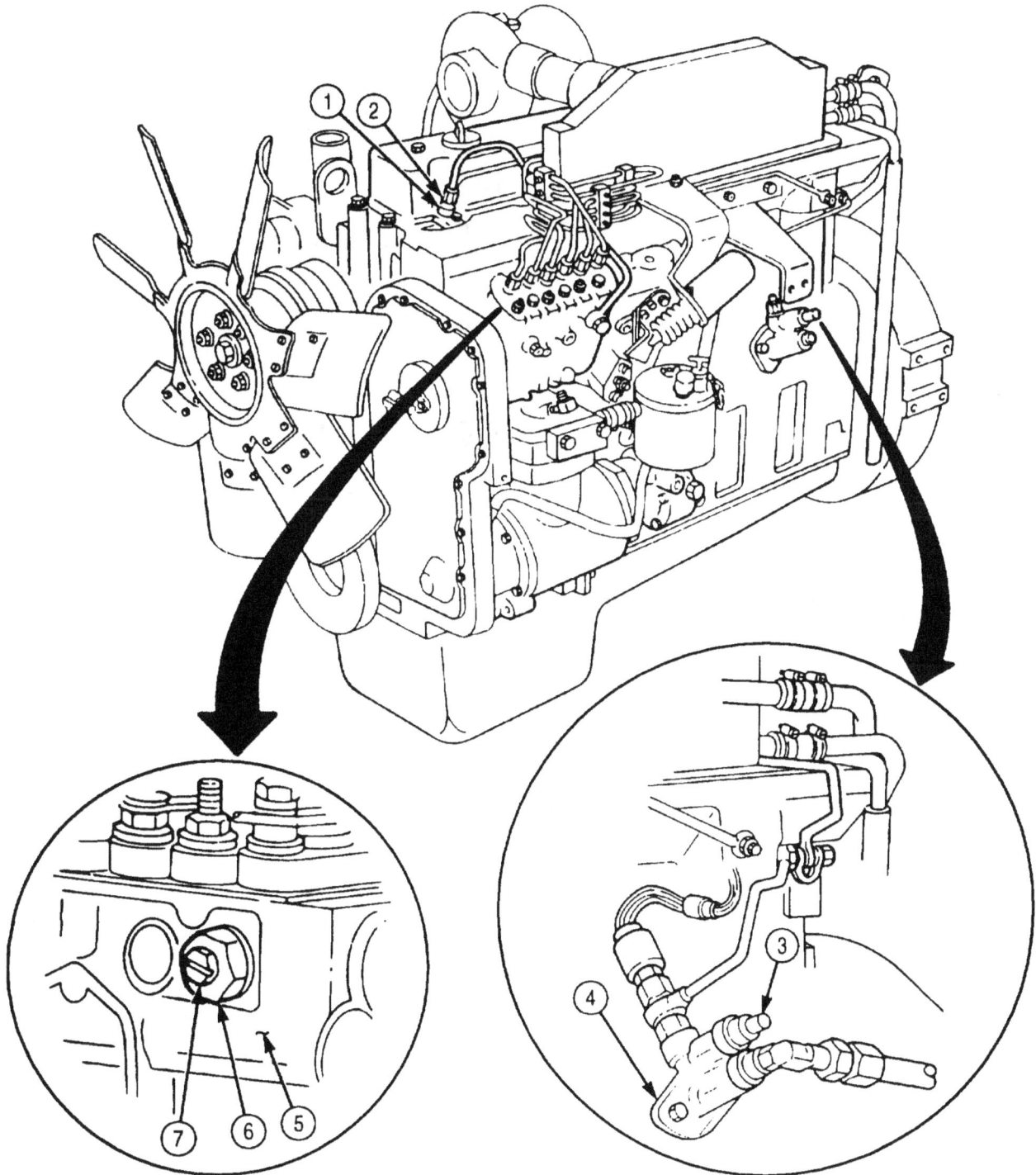

FOLLOW-ON TASKS: • Install air cleaner hose (para. 3-13).
• Start engine (TM 9-2320-272-10) and check for leaks.

3-23. FUEL SELECTOR VALVE FLEX HOSE REPLACEMENT

THIS TASK COVERS:

a. Removal   b. Installation

INITIAL SETUP:

APPLICABLE MODELS
M929/A1/A2, M930/A1/A2, M931/A1/A2,
M932/A1/A2, M936/A1/A2

TOOLS
General mechanic's tool kit (Appendix E, Item 1)

MATERIALS/PARTS
Cap and plug set (Appendix C, Item 14)
Antiseize tape (Appendix C, Item 72)

REFERENCES (TM)
TM 9-2320-272-10
TM 9-2320-272-24P

EQUIPMENT CONDITION
• Parking brake set (TM 9-2320-272-10).
• Hood raised and secured (TM 9-2320-272-10).

GENERAL SAFETY INSTRUCTIONS
Diesel fuel is flammable. Do not perform this
procedure near flames.

WARNING

Diesel fuel is highly flammable. Do not perform fuel system
procedures near flames. Injury to personnel may result.

a. Removal

NOTE
• All hoses are disconnected the same.
• All hoses must be disconnected from fuel lines first.
• Have drainage container ready to catch fuel.
• Plug all fuel lines using protective cap plugs.
• Tag lines and fittings for installation.

1. Disconnect six flex hoses (4) from four fuel lines (3) and two fuel line-to-hose adapters (5).
2. Disconnect six flex hoses (4) from two valve ports (6) and four elbows (2) on selector valve (1).

b. Installation

NOTE
• Male pipe threads must be wrapped with antiseize tape before
  installation.
• Remove protective cap plugs from fuel lines before installation.

1. Connect six flex hoses (4) to two valve ports (6) and four elbows (2) on selector valve (1).
2. Connect six flex hoses (4) to four fuel lines (3) and two fuel line-to-hose adapters (5).

3-23. FUEL SELECTOR VALVE FLEX HOSE REPLACEMENT (Contd)

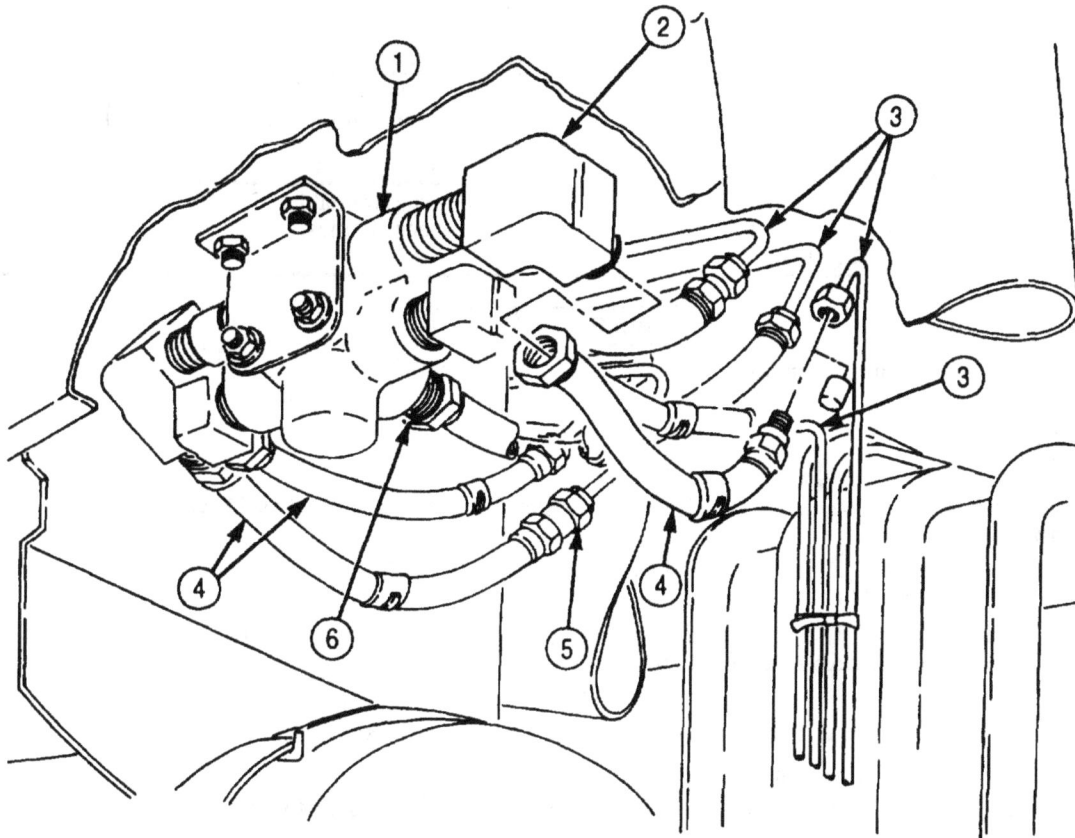

FOLLOW-ON TASK: Start engine (TM 9-2320-272-10) and test selector valve for fuel leaks and proper
operation.

3-24. FUEL SELECTOR VALVE, MOUNTING BRACKET, AND INDICATOR PLATE MAINTENANCE

THIS TASK COVERS:

a. Removal
b. Inspection

c. Installation

INITIAL SETUP:

APPLICABLE MODELS
M929/A1/A2, M930/A1/A2, M931/A1/A2, M932/A1/A2, M936/A1/A2

TOOLS
General mechanic's tool kit (Appendix E, Item 1)

MATERIALS/PARTS
One lockwashers (Appendix D, Item 352)
Two lockwashers (Appendix D, Item 354)
Two locknuts (Appendix D, Item 313)
Antiseize tape (Appendix C, Item 72)

REFERENCES (TM)
TM 9-2320-272-10
TM 9-2320-272-24P

EQUIPMENT CONDITION
• Parking brake set (TM 9-2320-272-10).
• Selector valve flex hoses removed (para. 3-23).

GENERAL SAFETY INSTRUCTIONS
Diesel fuel is flammable. Do not perform this procedure near flames.

WARNING

Diesel fuel is highly flammable. Do not perform fuel system procedures near open flame. Injury to personnel may result.

a. Removal

1. Remove screw (1), lockwasher (2), and lever (3) from left rear cab floor (5). Discard lockwasher (2).
2. Remove two screws (12), lockwashers (11), and selector valve (9) from mounting bracket (10). Discard lockwashers (11).
3. Remove two locknuts (13), screws (15), washers (14), indicator plate (4), and mounting bracket (10) from cab floor (5). Discard locknuts (13).

NOTE

• Use soft-nosed vise to hold selector valve.
• Tag fittings for installation.

4. Remove two elbows (7), pipe nipples (6), and elbows (8) from fuel selector valve (9).

b. Inspection

1. Install lever (3) on selector valve (9) with screw (1).
2. Turn lever (3) left, then right, while checking inside of selector valve (9) for burrs and nicks. Replace selector valve (9) if nicked or burred.
3. Remove screw (1) and lever (3) from selector valve (9).
4. Inspect two pipe nipples (6) and elbows (7) and (8) for stripped threads. Replace any part if threads are stripped.

3-24. FUEL SELECTOR VALVE MOUNTING BRACKET, AND INDICATOR PLATE MAINTENANCE (Contd)

c. Installation

NOTE

Locknuts are installed under cab floor.

1. Align indicator plate (4) and mounting bracket (10) with holes in cab floor (5) and install with two washers (14), screws (15), and new locknuts (13).

NOTE

Male pipe threads must be wrapped with antiseize tape before installation.

2. Install two elbows (8), pipe nipples (6), and elbows (7) on selector valve (9).

3. Install selector valve (9) on mounting bracket (10) with two new lockwashers (11) and screws (12).

4. Install lever (3) on selector valve (9) with new lockwasher (2) and screw (1). Ensure lever (3) pointer is positioned opposite notch on stem of selector valve (9).

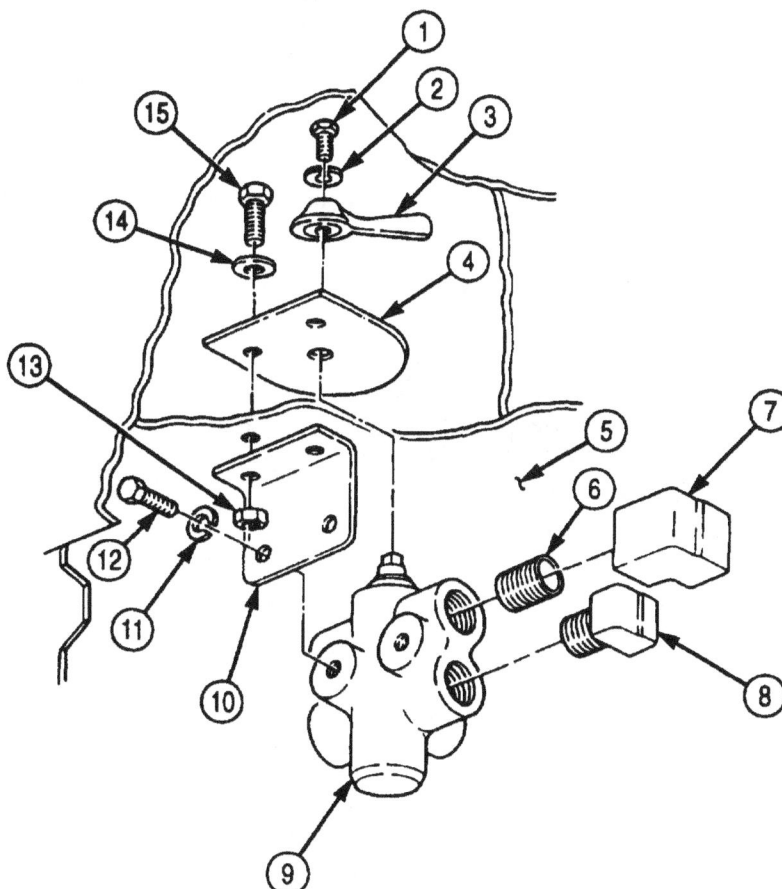

FOLLOW-ON TASKS: • Connect selector valve flex hoses (para. 3-23).
• Start engine (TM 9-2320-272-10) and check for fuel leaks at valve when switching selector valve to both tanks.

3-25. FUEL TANK FILLER CAP AND SPOUT REPLACEMENT

THIS TASK COVERS:

a. Removal   b. Installation

INITIAL SETUP:

APPLICABLE MODELS
All

TOOLS
General mechanic's tool kit (Appendix E, Item 1)

MATERIALS/PARTS
Two gaskets (Appendix D, Item 142)

REFERENCES (TM)
TM 9-2320-272-10
TM 9-2320-272-24P

EQUIPMENT CONDITION
• Parking brake set (TM 9-2320-272-10).
• Filler cap removed (TM 9-2320-272-10).

GENERAL SAFETY INSTRUCTIONS
Diesel fuel is flammable. Do not perform this procedure near flames.

WARNING
Diesel fuel is highly flammable. Do not perform fuel system procedures near open flames. Injury to personnel may result.

a. Removal

1. Remove gasket (2) from filler cap (1). Discard gasket (2).
2. Remove filler cap retaining chain (6) from filler spout assembly (3).
3. Turn filler spout assembly (3) counterclockwise and remove it from fuel tank (5).
4. Remove gasket (4) from filler spout assembly (3). Discard gasket (4).

b. Installation

NOTE
M939/A1 vehicles use same gaskets in steps 1 and 2. M939A2 vehicles use different gaskets for steps 1 and 2.

1. Install new gasket (2) on filler cap (1).
2. Install new gasket (4) on filler spout assembly (3).
3. Install filler spout assembly (3) in fuel tank (5).
4. Attach filler cap retaining chain (6) to filler spout assembly (3).

3-25. FUEL TANK FILLER CAP AND SPOUT REPLACEMENT (Contd)

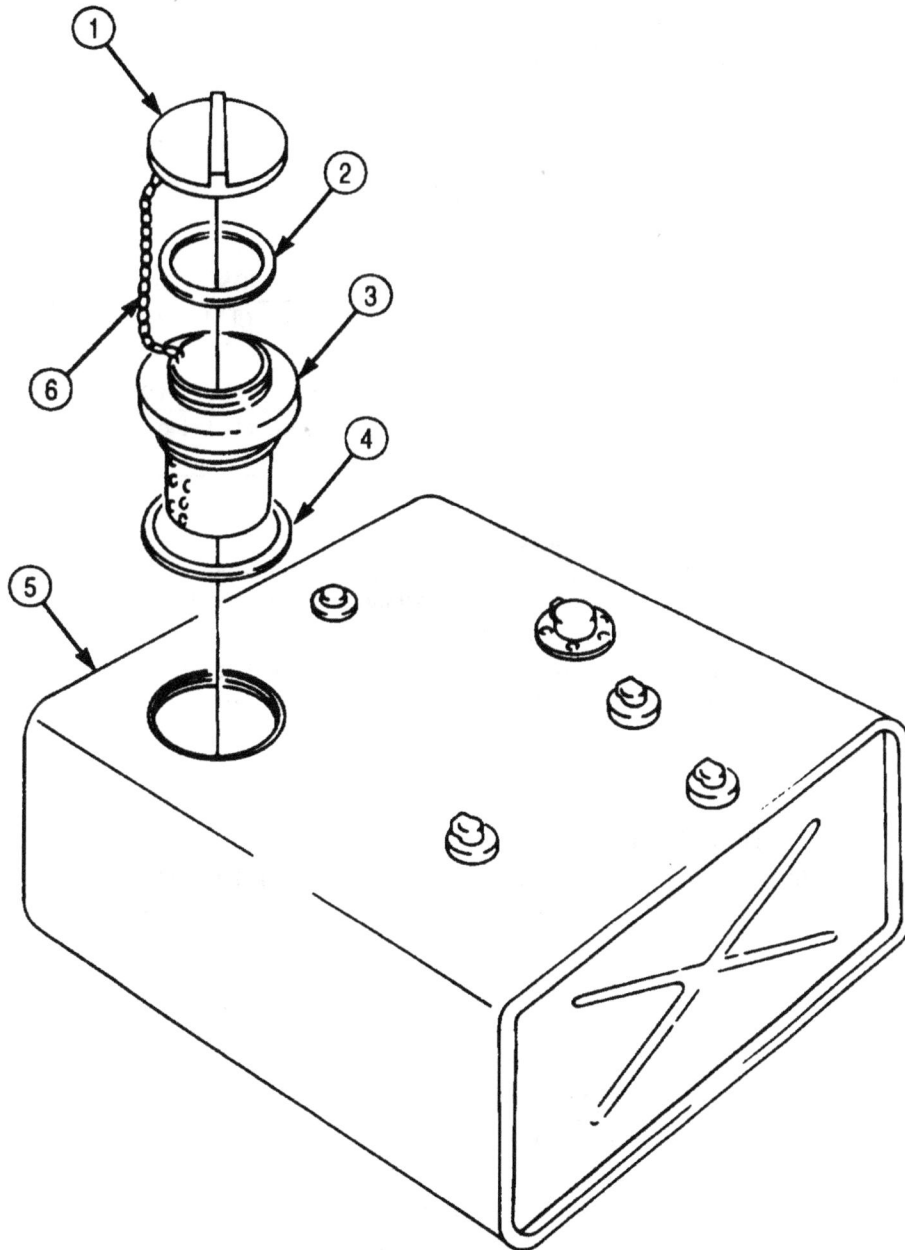

FOLLOW-ON TASK: Install filler cap (TM 9-2320-272-10).

3-26. FUEL TANK MAINTENANCE

THIS TASK COVERS:

a. Draining
b. Removal

c. Inspection
d. Installation

INITIAL SETUP:

APPLICABLE MODELS
All except M936/A1/A2

TOOLS
General mechanic's tool kit (Appendix E, Item 1)

MATERIALS/PARTS
Gasket (Appendix D, Item 144)
Gasket (Appendix D, Item 143)
Four locknuts (Appendix D, Item 288)
Antiseize tape (Appendix C, Item 72)

REFERENCES (TM)
FM 43-2
TM 9-2320-272-10
TM 9-2320-272-24P

EQUIPMENT CONDITION
• Parking brake set (TM 9-2320-272-10).
• Fuel tank filler cap and spout removed (para. 3-25).

GENERAL SAFETY INSTRUCTIONS
Diesel fuel is flammable. Do not perform this procedure near flames.

WARNING

Diesel fuel is highly flammable. Do not perform fuel system procedures near open flames. Injury to personnel may result.

NOTE

The replacement procedure for right and left single fuel tanks is basically the same. This procedure covers the left fuel tank.

a. Draining

NOTE
Have drainage container ready to catch fuel.

1. Remove drainplug (10) and gasket (9) from bottom of fuel tank (16). Discard gasket (9).
2. After draining is complete, install new gasket (9) and drainplug (10) on bottom of fuel tank (16).

b. Removal

NOTE
Tag lines for installation.

1. Disconnect two vent lines (6) from elbows (7).
2. Disconnect fuel return line (13) from elbow (12).
3. Disconnect fuel supply line (5) from elbow (14).
4. Disconnect fuel transmitter wire (2) from fuel transmitter unit (15).
5. Remove screw (3) and ground wire (4) from fuel tank (16).
6. Remove two locknuts (11) from hanger straps (8). Discard locknuts (11).
7. Remove two locknuts (22), screws (21), and hanger straps (8) from hangers (1). Discard locknuts (22).

3-26. FUEL TANK MAINTENANCE (Contd)

NOTE
Perform step 8 for M934/A1/A2 model vehicles.

8. Disconnect personnel heater fuel supply line (18), adapter (19), and fitting (20) from top of fuel tank (16).

NOTE
Assistant will help with step 9.

9. Remove fuel tank (16) from two hangers (1).

NOTE
Mark direction of elbows for installation before removing.

10. Remove fuel supply tube with elbow (14) from top of fuel tank (16).

11. Remove two vent line elbows (7) and return line elbow (12) from top of fuel tank (16).

12. Remove four screws (3), fuel transmitter unit (15), and gasket (17) from fuel tank (16). Discard gasket (17).

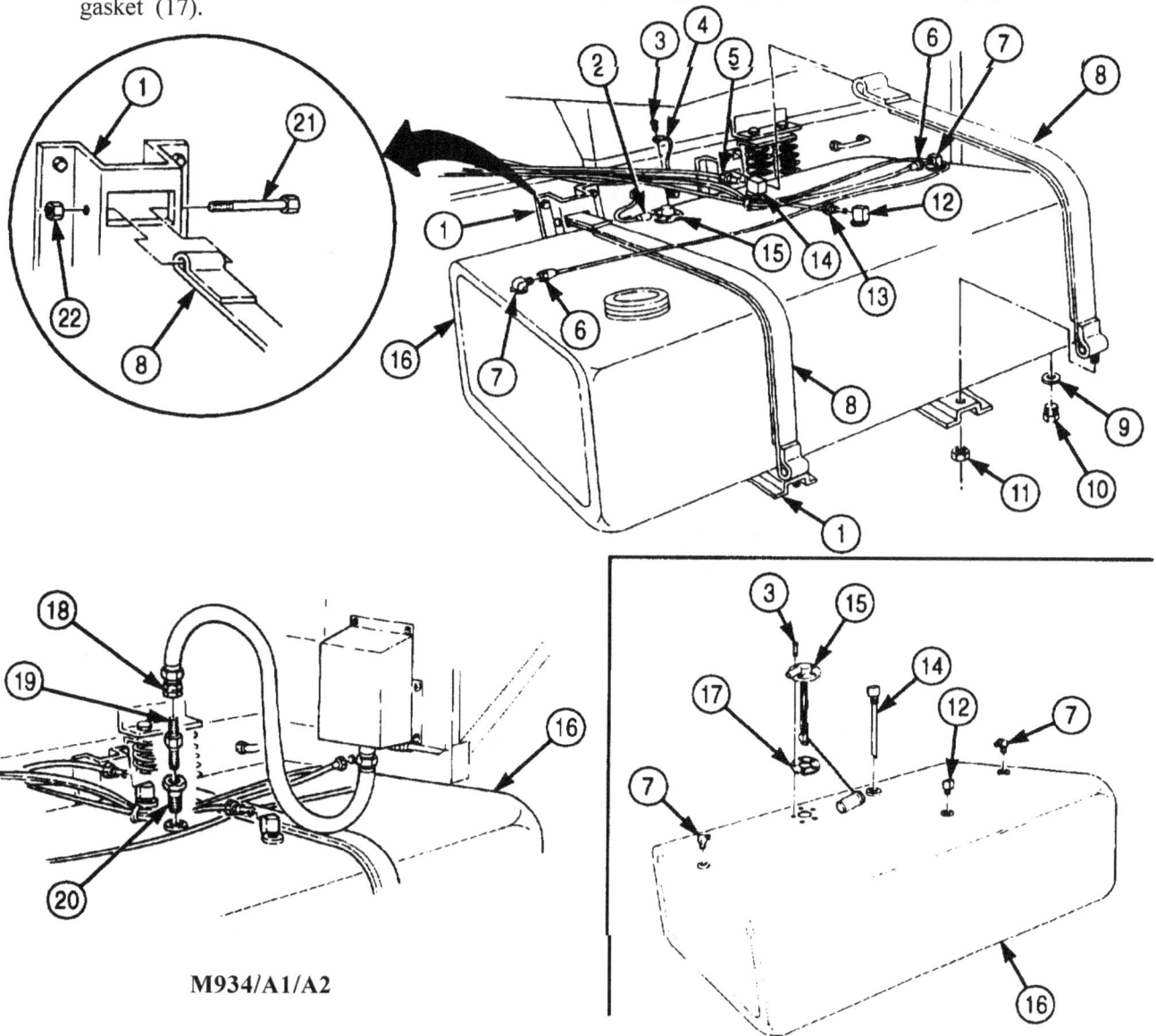

M934/A1/A2

3-26. FUEL TANK MAINTENANCE (Contd)

c. Inspection

Inspect fuel tank (6) for cracks, holes, and stripped threads (FM 43-2.)

d. Installation

NOTE
- Male pipe threads must be wrapped with antiseize tape before installation.
- When installing fuel transmitter unit, do not use screw hole closest to vehicle frame. Ground wire will be installed at that location.

1. Position new gasket (7) on fuel transmitter unit (2) and install unit (2) on fuel tank (6) with four screws (1).
2. Install return line elbow (4) and two vent line elbows (5) on top of fuel tank (6).
3. Install fuel supply tube with elbow (3) on top of tank (6).

NOTE
Assistant will help with step 4.

4. Place fuel tank (6) on two hangers (8).
5. Install two hangers straps (14) on hangers (8) with two screws (18) and new locknuts (19).
6. Install outer ends of two hanger straps (14) on hangers (8) with two new locknuts (15).

NOTE
Perform step 7 for M934/A1/A2 model vehicles.

7. Install fitting (22), adapter (21), and personnel heater fuel supply line (20) on fuel tank (6).
8. Install ground wire (11) on fuel transmitter unit (2) with screw (10).
9. Connect fuel transmitter wire (9) to fuel transmitter unit (2).
10. Connect fuel supply line (12) to elbow (3).
11. Connect fuel return line (16) to elbow (4).
12. Connect vent lines (13) and (17) to elbows (5).

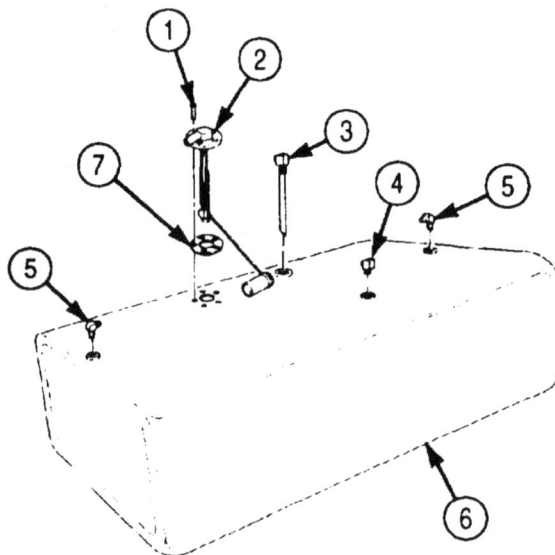

3-26. FUEL TANK MAINTENANCE (Contd)

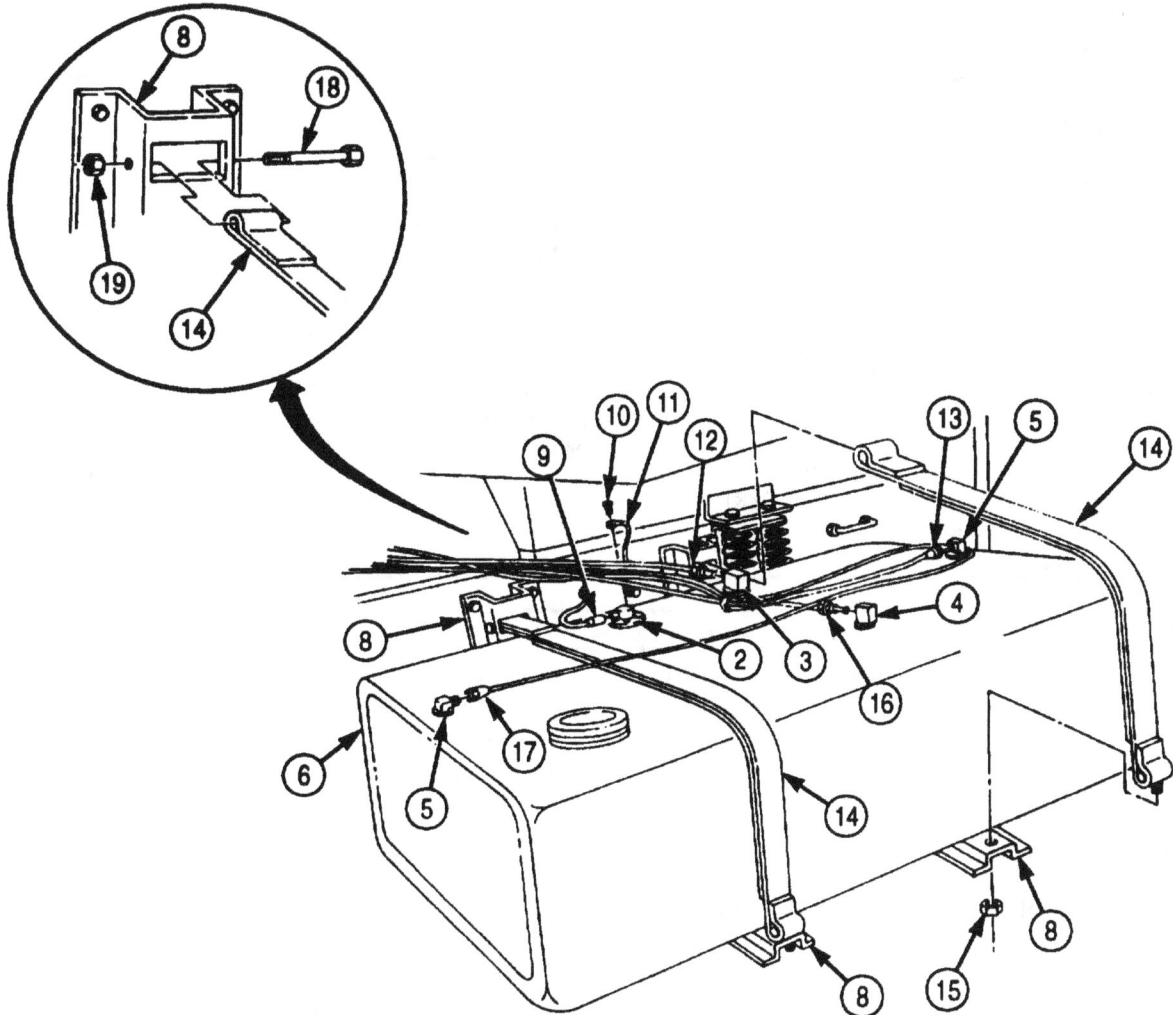

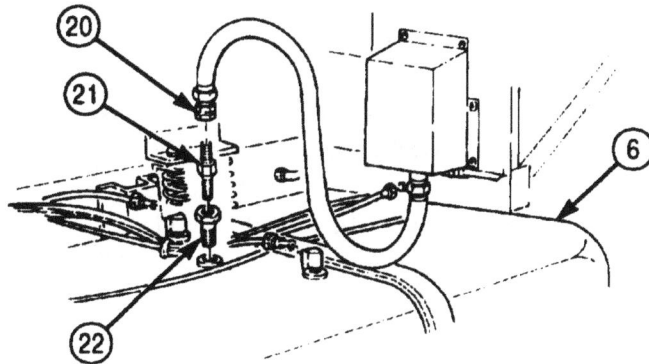

FOLLOW-ON TASK: Install fuel tank filler cap and spout (para. 3-25).

3-27. FUEL TANK (M936/A1/A2) MAINTENANCE

THIS TASK COVERS:

a. Draining   c. Inspection
b. Removal   d. Installation

INITIAL SETUP:

APPLICABLE MODELS
M936/A1/A2

TOOLS
General mechanic's tool kit (Appendix E, Item 1)

MATERIALS/PARTS
Two lockwashers (Appendix D, Item 358)
Gasket (Appendix D, Item 144)
Gasket (Appendix D, Item 143)
Gasket (Appendix D, Item 142)
Locknut (Appendix D, Item 299)
Three locknuts (Appendix D, Item 288)
Lockwasher (Appendix D, Item 379)
Antiseize tape (Appendix C, Item 72)

REFERENCES (TM)
FM 43-2
TM 9-2320-272-10
TM 9-2320-272-24P

EQUIPMENT CONDITION
Parking brake set (TM 9-2320-272-10).

GENERAL SAFETY INSTRUCTIONS
Diesel fuel is flammable. Do not perform this procedure near flames.

WARNING

Diesel fuel is highly flammable. Do not perform fuel system procedures near open flame. Injury to personnel may result.

NOTE

The replacement procedure for right and left fuel tanks is basically the same. This procedure covers the left fuel tank.

a. Draining

1. Loosen fuel filler cap (2) at top of tank (1).

NOTE
Have drainage container ready to catch fuel.

2. Remove drainplug (4) and gasket (3) from bottom of fuel tank (1). Discard gasket (3).
3. After draining is complete, install new gasket (3) and drainplug (4) in fuel tank (1).
4. Tighten fuel filler cap (2) on fuel tank (1).

b. Removal

1. Loosen two nuts on screws (6) on wrecker body (5), and slide stop plate (7) upward.
2. Remove outrigger (8) from body (5).
3. Remove locknut (14), screw (9), washer (10), ground wire (11), and lockwasher (12) from left side of frame (13) and double check valve (15). Discard locknut (14) and lockwasher (12).

3-27. FUEL TANK (M936/A1/A2) MAINTENANCE (Contd)

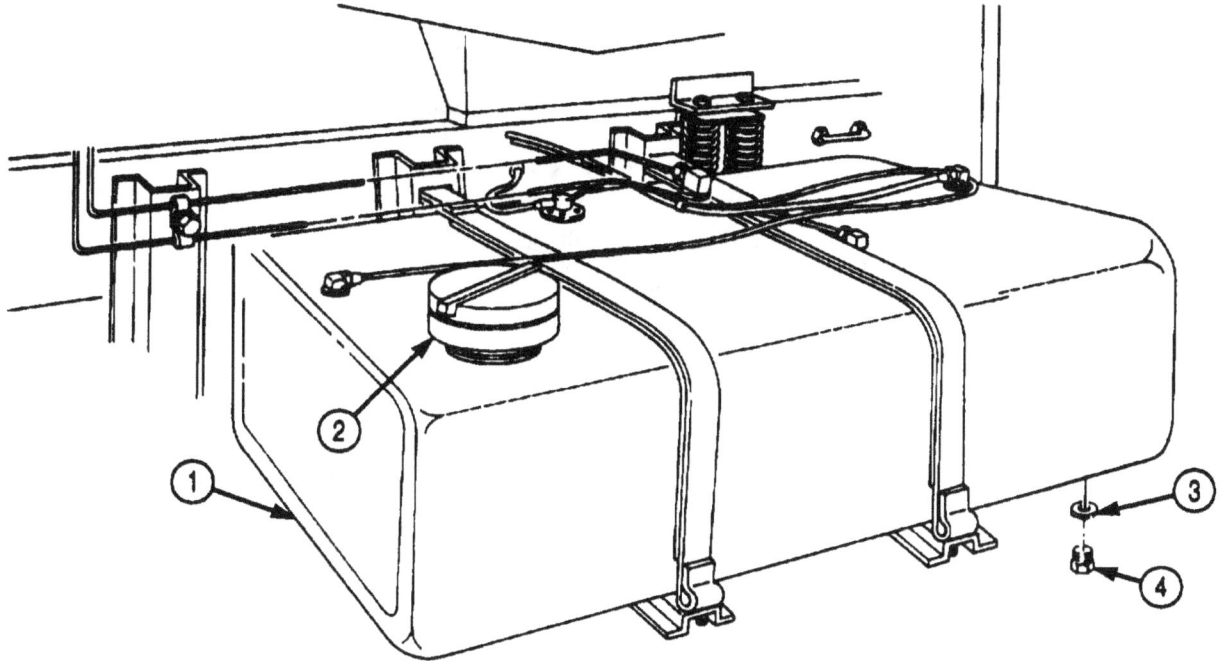

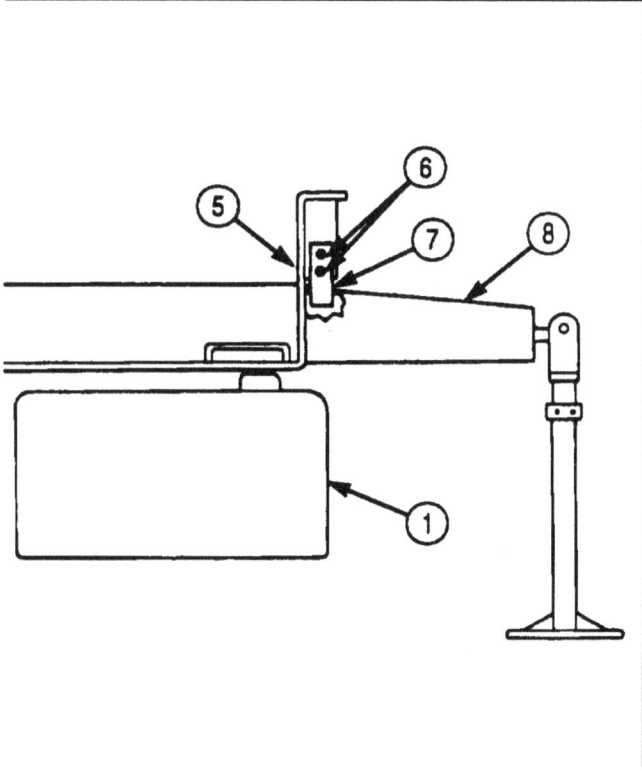

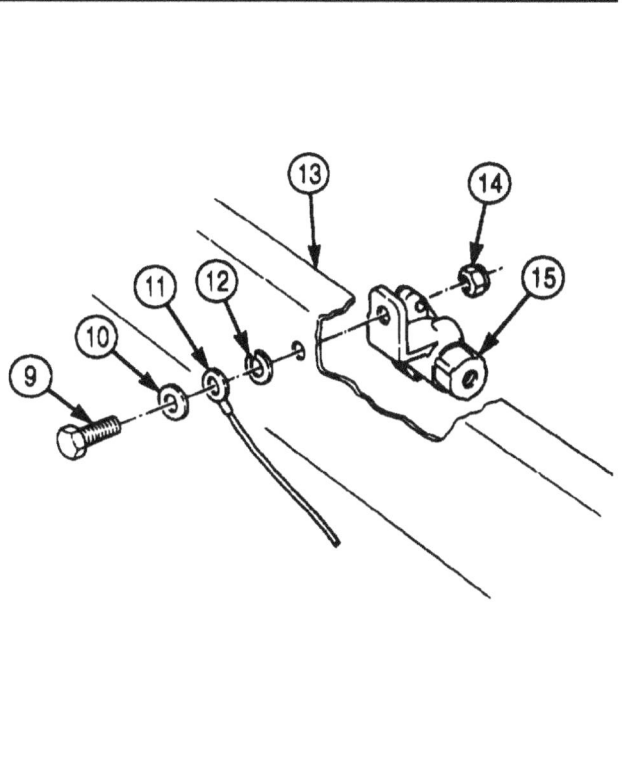

3-27. FUEL TANK (M936/A1/A2) MAINTENANCE (Contd)

4. Disconnect fuel transmitter wire (6) from fuel transmitter unit (16).

NOTE

Perform steps 5, 6, and 7 for left fuel tank only.

5. Disconnect rubber hose adapter (4) from fuel supply line (3).

6. Disconnect rubber hose adapter (1) from fuel return line (2).

7. Remove locknut (24), screw (21), and clamps (22) and (23) from left side of frame (5). Discard locknut (24).

NOTE

• Perform steps 8 and 9 for right side fuel tank only.

• Tag lines and fittings for installation.

8. Disconnect fuel supply line (7) from elbow (15).

9. Disconnect fuel return line (14) from elbow (11).

10. Remove vent lines (9) and (19) from elbows (10) and (20).

11. Disconnect wire connector (25) from right side marker light (26).

12. Disconnect wire connector (29) from left side marker light (32).

NOTE

• Assistant will help with steps 13 through 15.

• Screws for marker light are accessible through storage compartment 1A (TM 9-2320-272-10).

13. Remove two nuts (31), lockwashers (30), screws (27), and left side marker light (32) from body (28). Discard lockwashers (30).

14. Remove two locknuts (12) and hanger straps (8) and (17) from hangers (13). Discard locknuts (12).

15. Remove fuel tank (18) from hangers (13).

3-27. FUEL TANK (M936/A1/A2) MAINTENANCE (Contd)

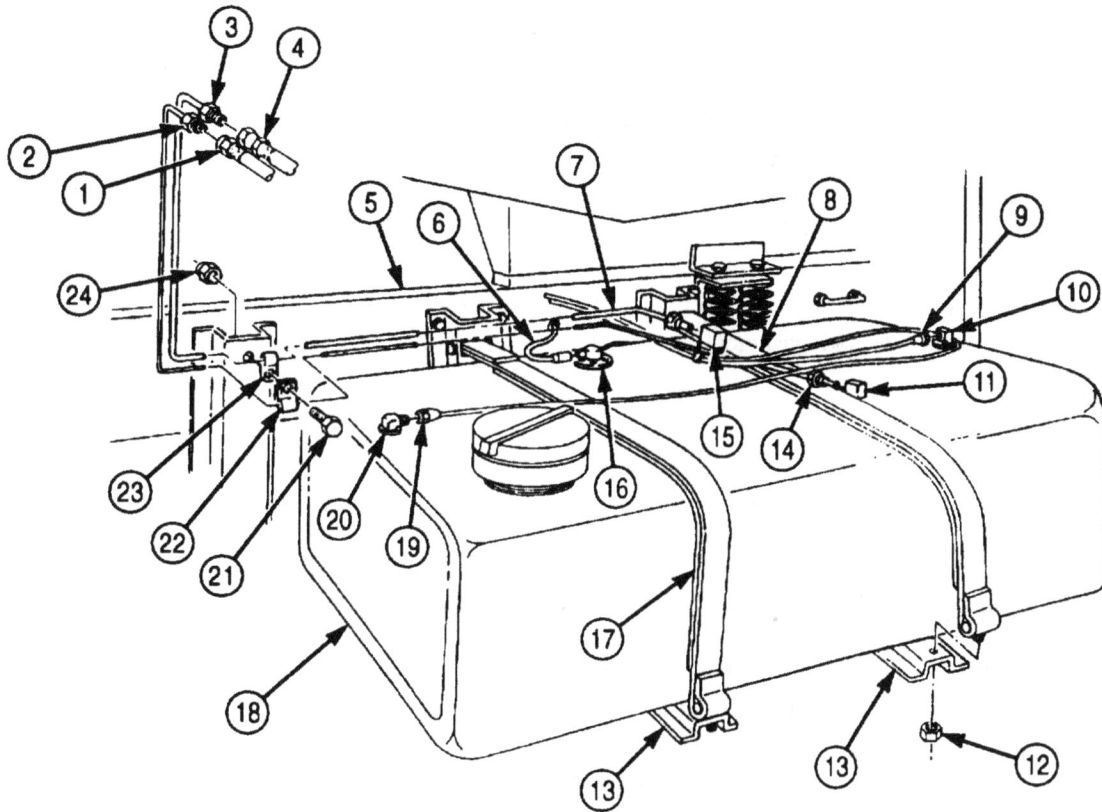

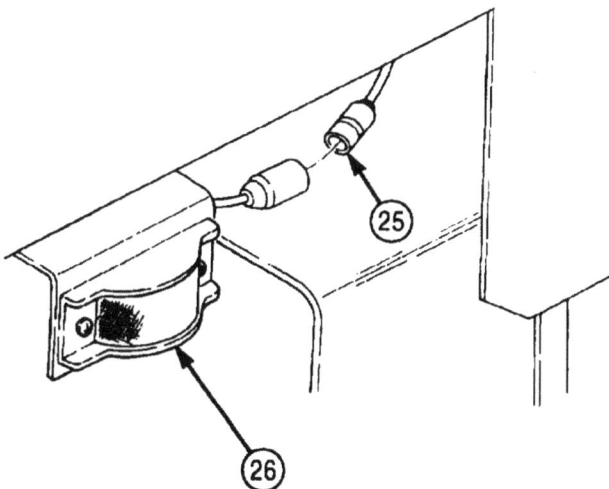

RIGHT SIDE

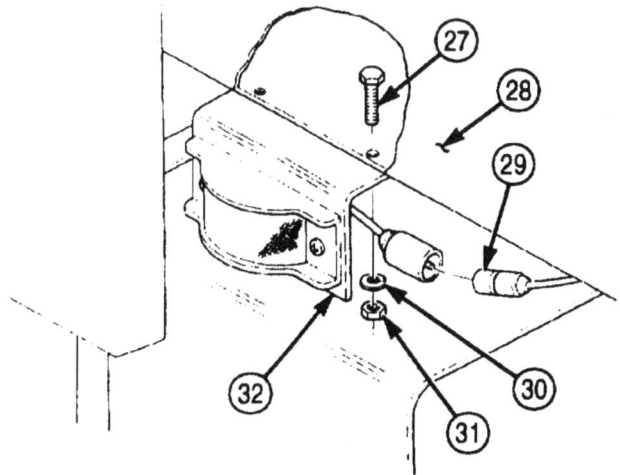

LEFT SIDE

3-27. FUEL TANK (M936/A1/A2) MAINTENANCE (Contd)

NOTE

Tag fittings and transmitter unit for direction during installation.

16. Remove fuel supply tube and elbow (5) from top of fuel tank (8).

17. Remove vent line elbows (7) and (9) from top of fuel tank (8).

18. Remove return line elbow (6) from top of fuel tank (8).

19. Remove screw (2) and ground wire (3) from fuel transmitter unit (4).

20. Remove four screws (2), fuel transmitter unit (4), and gasket (11) from fuel tank (8). Discard gasket (11).

21. Remove filler cap (1) and disconnect S-chain (13) from fuel strainer (12).

22. Remove fuel strainer (12) and gasket (10) from fuel tank (8). Discard gasket (10).

c. Inspection

Inspect fuel tank (8) for cracks, holes, and stripped threads (FM 43-2).

d. Installation

NOTE

Male pipe threads must be wrapped with antiseize tape before installation.

1. Install new gasket (10) and fuel strainer (12) in fuel tank (8).

2. Connect S-chain (13) to fuel strainer (12) and install filler cap (1).

NOTE

Do not us screw hole nearest to vehicle frame.

3. Install new gasket (11) and fuel transmitter unit (4) on fuel tank (8) with four screws (2).

4. Install ground wire (3) on fuel transmitter unit (4) with screw (2).

5. Install return line elbow (6) and vent line elbows (7) and (9) on top of fuel tank (8).

3-27. FUEL TANK (M936/A1/A2) MAINTENANCE (Contd)

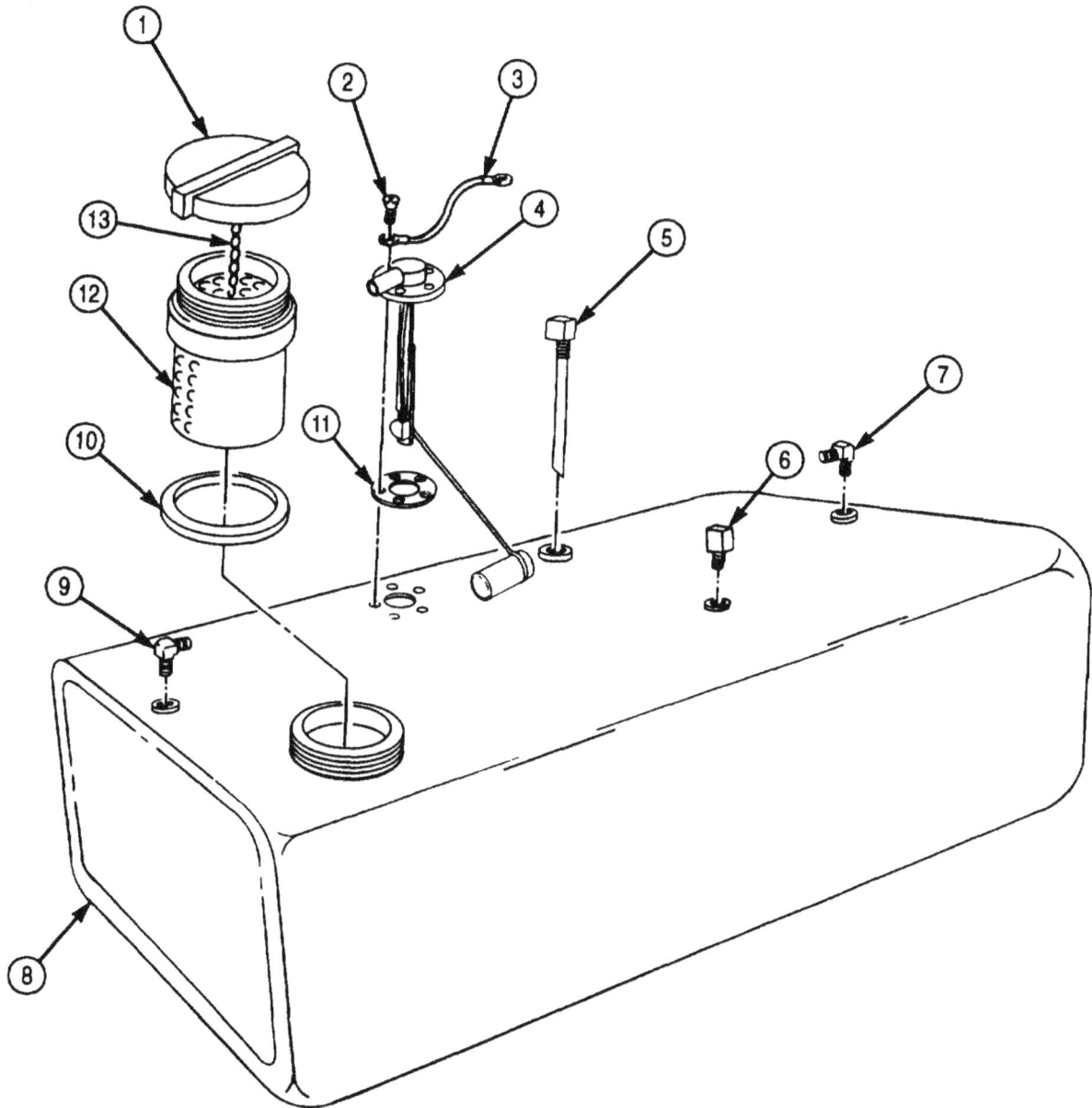

3-27. FUEL TANK (M936/A1/A2) MAINTENANCE (Contd)

NOTE
Assistant will help with steps 6 through 10.

6. Place fuel tank (18) on two hangers (13) far enough to support tank (18).
7. Connect vent lines (9) and (19) to elbows (10) and (20).
8. Connect fuel supply line (7) to elbow (15).
9. Connect fuel return line (14) to elbow (11).
10. Install hanger straps (8) and (17) on hangers (13) with two new locknuts (12).

NOTE
Slide fuel tank all the way in before performing steps 11 and 12.

11. Install left side marker light (30) on body (26) with two screws (25), new lockwashers (28), and nuts (29).
12. Connect left side marker light (30) to harness connector (27).
13. Connect right side marker light (32) to harness connector (31).

NOTE
Perform steps 14, 15, and 16 for left side tank only

14. Connect fuel supply line (3) to rubber hose adapter (4).
15. Connect fuel return line (2) to rubber hose adapter (1).
16. Install clamps (22) and (23) on frame (5) with screw (21) and new locknut (24).
17. Connect fuel transmitter wire (6) to fuel transmitter unit (16).

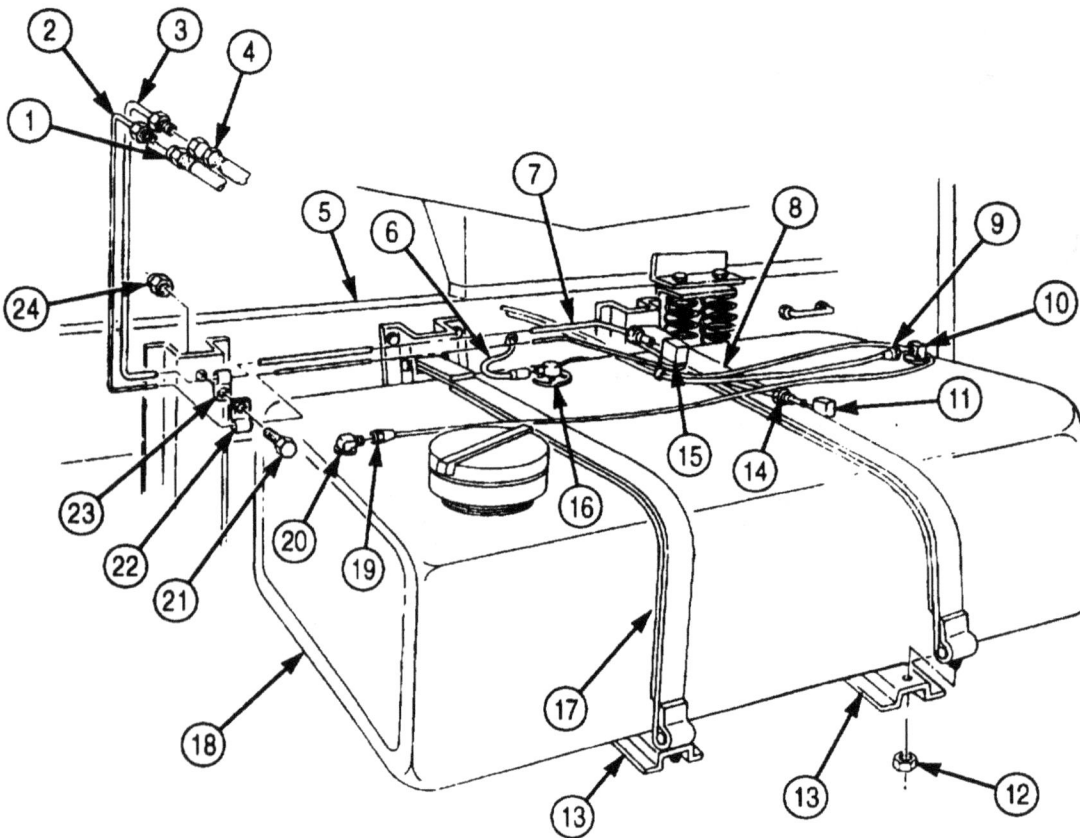

3-27. FUEL TANK (M936/A1/A2) MAINTENANCE (Contd)

18. Install ground wire (37) on frame (5) and double check valve (34) with new lockwasher (38), washer (36), screw (35), and new locknut (33).

19. Loosen two nuts on screws (40) and slide stop plate (41) upward.

20. Install outrigger (42) in wrecker body (39) and slide stop plate (41) down. Stop plate (41) must contact back of slot in top of outrigger (42).

21. Tighten two nuts on screws (40).

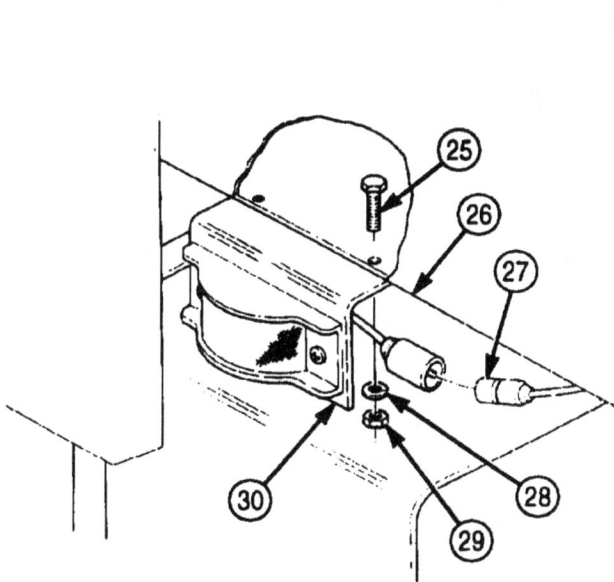

LEFT SIDE

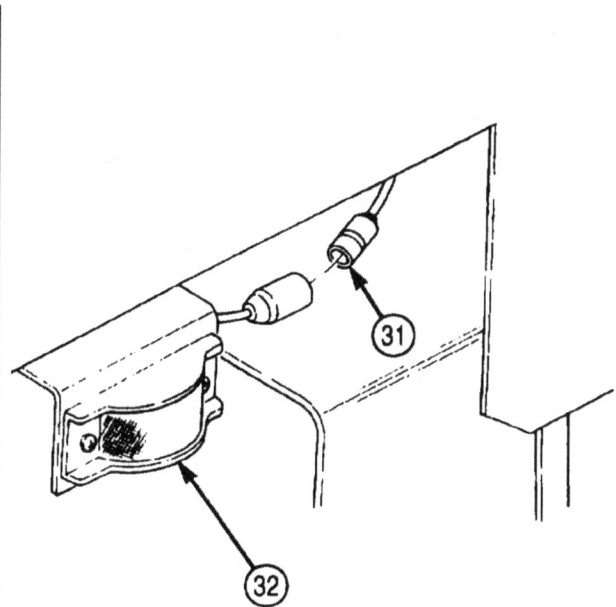

RIGHT SIDE

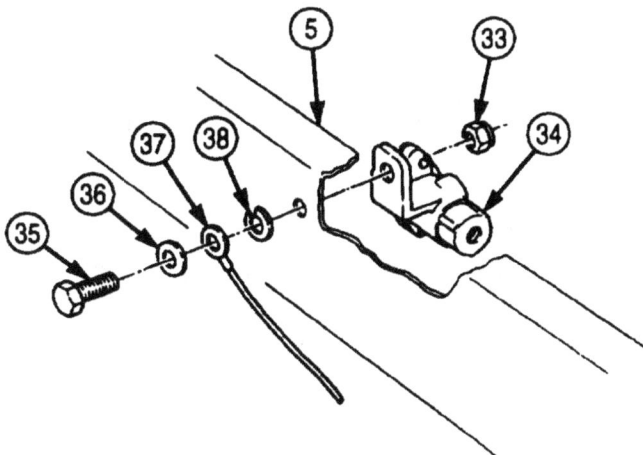

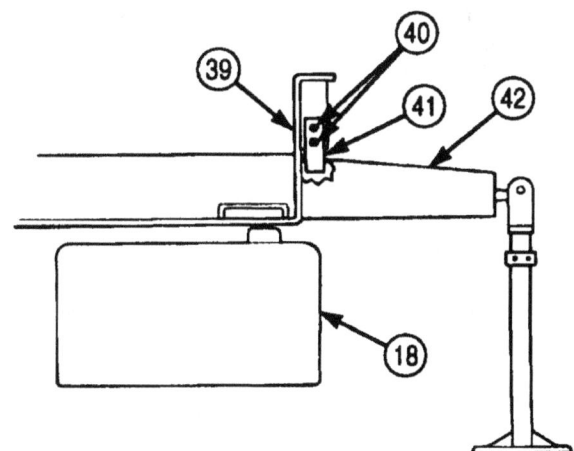

FOLLOW-ON TASKS: Start engine (TM 9-2320-272-10) and road test vehicle.

3-28. FUEL TANK HANGERS AND RETAINING STRAPS MAINTENANCE

THIS TASK COVERS:

a. Removal
b. Inspection

c. Installation

INITIAL SETUP:

APPLICABLE MODELS
All except M936/A1/A2

TOOLS
General mechanic's tool kit (Appendix E, Item 1)

MATERIALS/PARTS
Ten locknuts (Appendix D, Item 288)
Five locknuts (Appendix D, Item 291)
Two lockwashers (Appendix D, Item 379)
Eight locknuts (Appendix D, Item 309)
Adhesive sealant (Appendix C, Item 6)

REFERENCES (TM)
TM 9-2320-272-10
TM 9-2320-272-24P

EQUIPMENT CONDITION
- Parking brake set (TM 9-2320-272-10).
- Fuel tank(s) removed (para. 3-26).
- Dual fuel tank wet air reservoir removed (M929/A1/A2, M930/A1/A2, M931/A1/A2, and M932/A1/A2 model vehicles (right side fuel tank only) (para. 3-200).

NOTE
- On models with dual fuel tanks (M929/A1/A2, M930/A1/A2, M931/A1/A2, and M932/A1/A2), right side forward fuel tank hanger must be removed and installed at direct support maintenance.
- The number of locknuts and lockwashers varies between models. Refer to parts list for quantity for particular model.

a. Removal

1. Remove two locknuts (19), screws (8), retaining straps (22), and insulators (1) from two fuel tank hangers (14). Discard locknuts (19).

2. Remove eight locknuts (13), screws (15), and two hanger stabilizing straps (12) from hangers (14). Discard locknuts (13).

3. Remove four locknuts (9), screws (11), and two support brackets (10) from fuel tank hangers (14). Discard locknuts (9).

4. Remove locknut (6), bracket (5), lockwasher (4), washer (3), screw (21), lockwasher (4), washer (3), ground lead (20), and lockwasher (4) from hanger (14) and frame (7). Discard lockwashers (4) and locknut (6).

NOTE
Assistant will help with step 5.

6. Remove eight locknuts (2), screws (18), and two fuel tank hangers (14) from frame (7). Discard locknuts (2).

6. Remove two rubber sheets (16) and rubber sheets (17) from two fuel tank hangers (14).

3-28. FUEL TANK HANGERS AND RETAINING STRAPS MAINTENANCE (Contd)

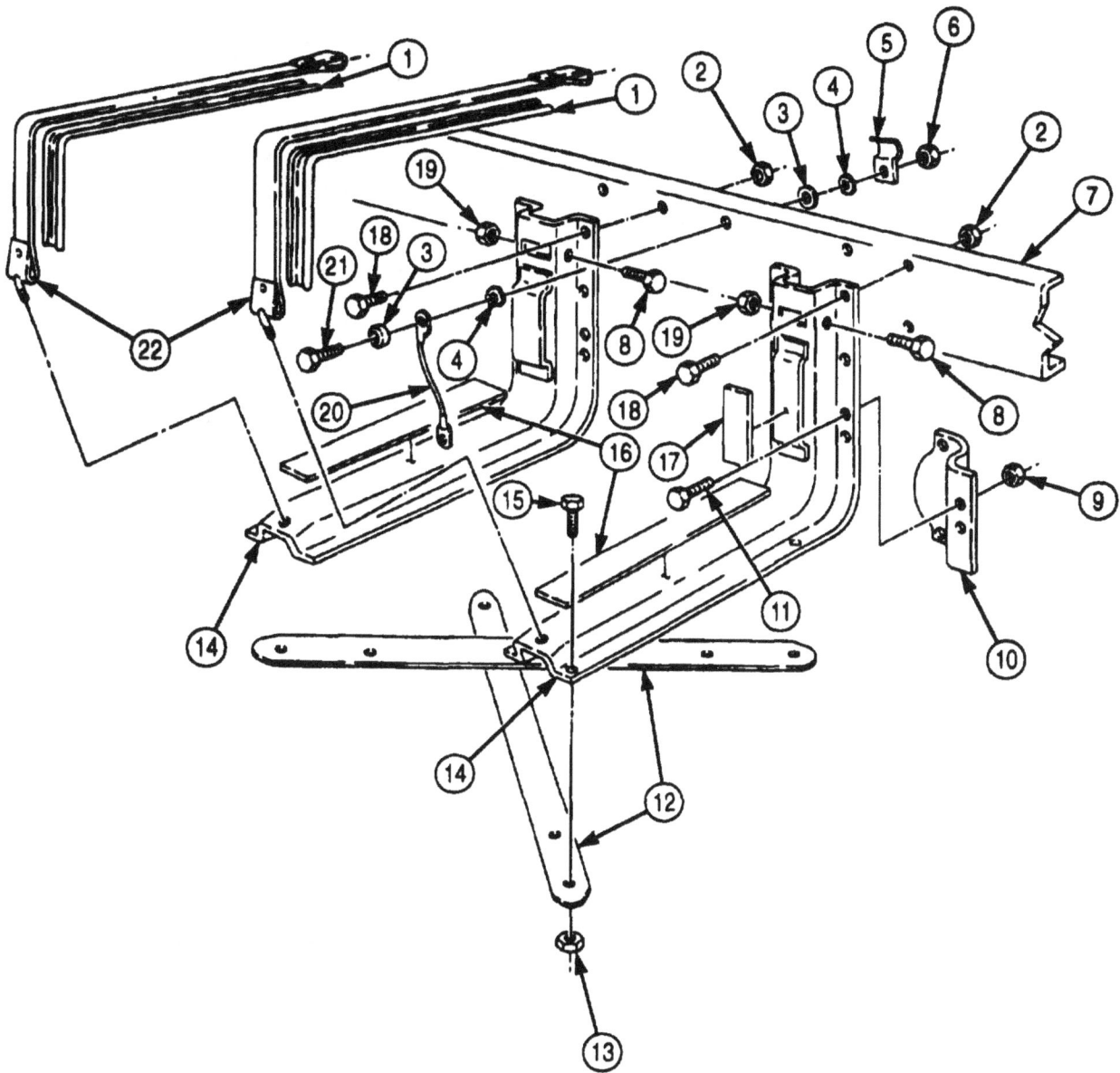

3-28. FUEL TANK HANGERS AND RETAINING STRAPS MAINTENANCE (Contd)

b. Inspection

1. Inspect two rubber sheets (16) and (17) for presence and damage. Replace if missing or damaged.
2. Inspect two insulators (1) for deterioration. Replace if missing or damaged.

c. Installation

1. Apply adhesive sealant to two insulators (1) and install as needed, on retaining straps (22).
2. Apply adhesive sealant to two rubber sheets (16) and rubber sheets (17) and install, as needed, on hangers (14).

NOTE

Assistant will help with step 2.

3. Install two fuel tank hangers (14) on frame (7) with eight screws (18) and new locknuts (2).
4. Place wiring harness bracket (5) over wiring harness and install new lockwasher (4), ground wire (20), washer (3), screw (21), washer (3), new lockwasher (4), wiring harness bracket (5), and new locknut (6) on frame (7).
5. Install two support brackets (10) on hangers (14) with four screws (11) and new locknuts (9).
6. Install two fuel hanger stabilizing straps (12) on hangers (14) with eight screws (15) and new locknuts (13).
7. Install two retaining straps (22) and insulators (1) on hangers (14) with two screws (8) and new locknuts (19).

3-28. FUEL TANK HANGERS AND RETAINING STRAPS MAINTENANCE (Contd)

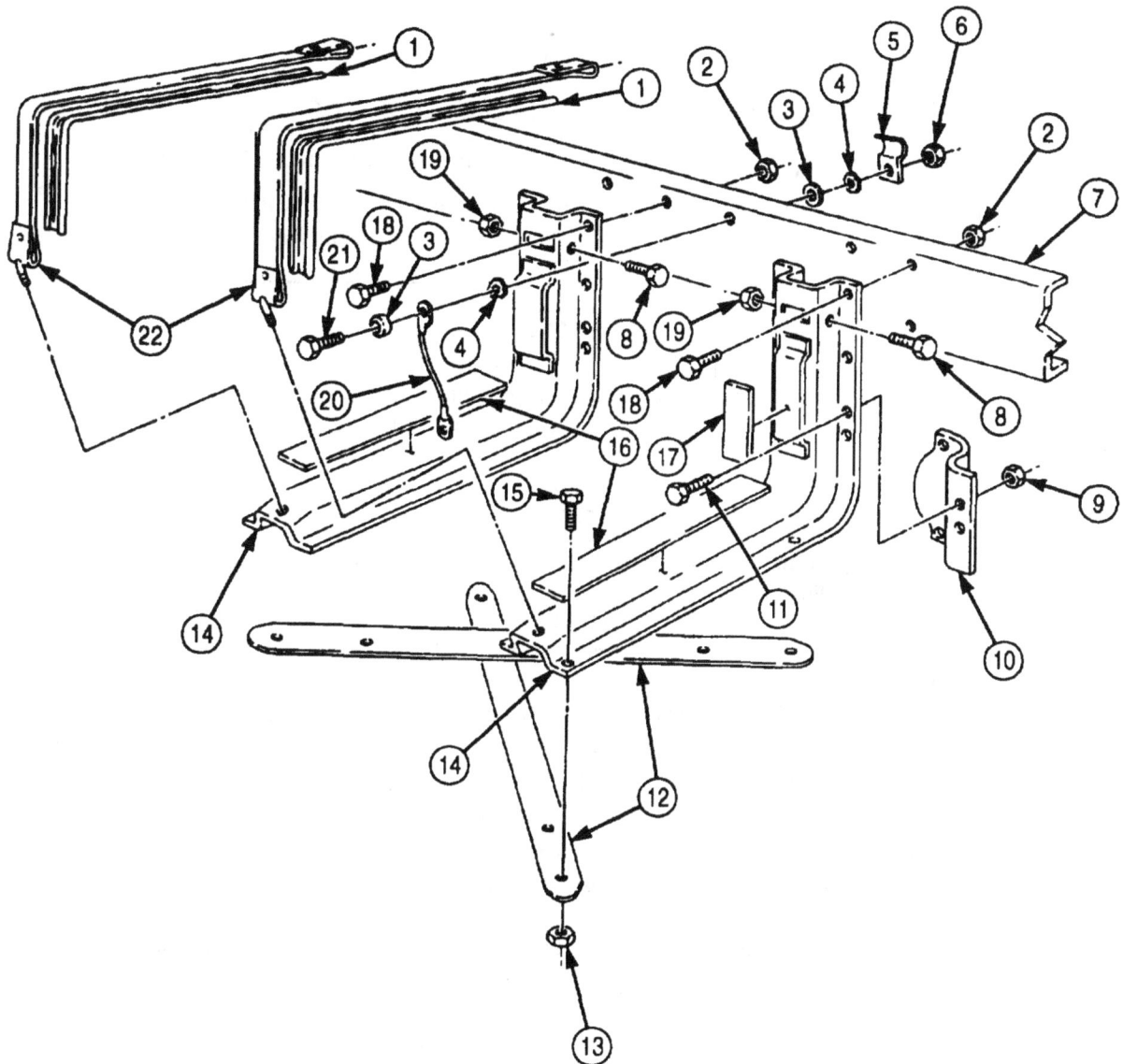

FOLLOW-ON TASKS: • Install fuel tank(s) (para. 3-26).
• Install dual fuel tank wet air reservoir (M929/A1/A2, M930/A1/A2, M931/A1/A2, and M932/A1/A2 model vehicles) (right side fuel tank only) (para. 3-200).

3-29. FUEL TANK LINES (M939A2) REPLACEMENT

THIS TASK COVERS:

a. Removal   b. Installation

INITIAL SETUP:

APPLICABLE MODELS
M939A2

TOOLS
General mechanic's tool kit (Appendix E, Item 11

MATERIALS/PARTS
Locknut (Appendix D, Item 300)
Tiedown straps (Appendix D, Item 690)
Cap and plug set (Appendix C, Item 14)

REFERENCES (TM)
TM 9-2320-272-10
TM 9-2320-272-24P

EQUIPMENT CONDITION
- Parking brake set (TM 9-2320-272-10).
- Hood raised and secured (TM 9-2320-272-10).

GENERAL SAFETY INSTRUCTIONS
Diesel fuel is flammable. Do not perform this task near open flames.

WARNING

Diesel fuel is highly flammable. Do not perform fuel system procedures near open flame. Injury to personnel may result.

a. Removal

CAUTION

Cap or plug all openings immediately after disconnecting lines and hoses to prevent contamination. Failure to do so may result in fuel system damage.

NOTE

Remove and discard tiedown straps as required and note locations for installation.

1. Disconnect two tubing nuts (3) and remove fuel supply hose (4) from elbow (2) and fuel supply line (7).

NOTE

- Perform step 2 if fuel filter is damaged.
- Note position and direction of arrow for installation.

2. Remove two clamps (5) and fuel filter (6) from fuel supply hose (4).
3. Remove locknut (12), screw (8), fuel return hose (14), and clamp (13) from bracket (11). Discard locknut (12).
4. Loosen two hose clamps (10) and remove fuel return hose (14) from fitting (1) and fuel return line (9).

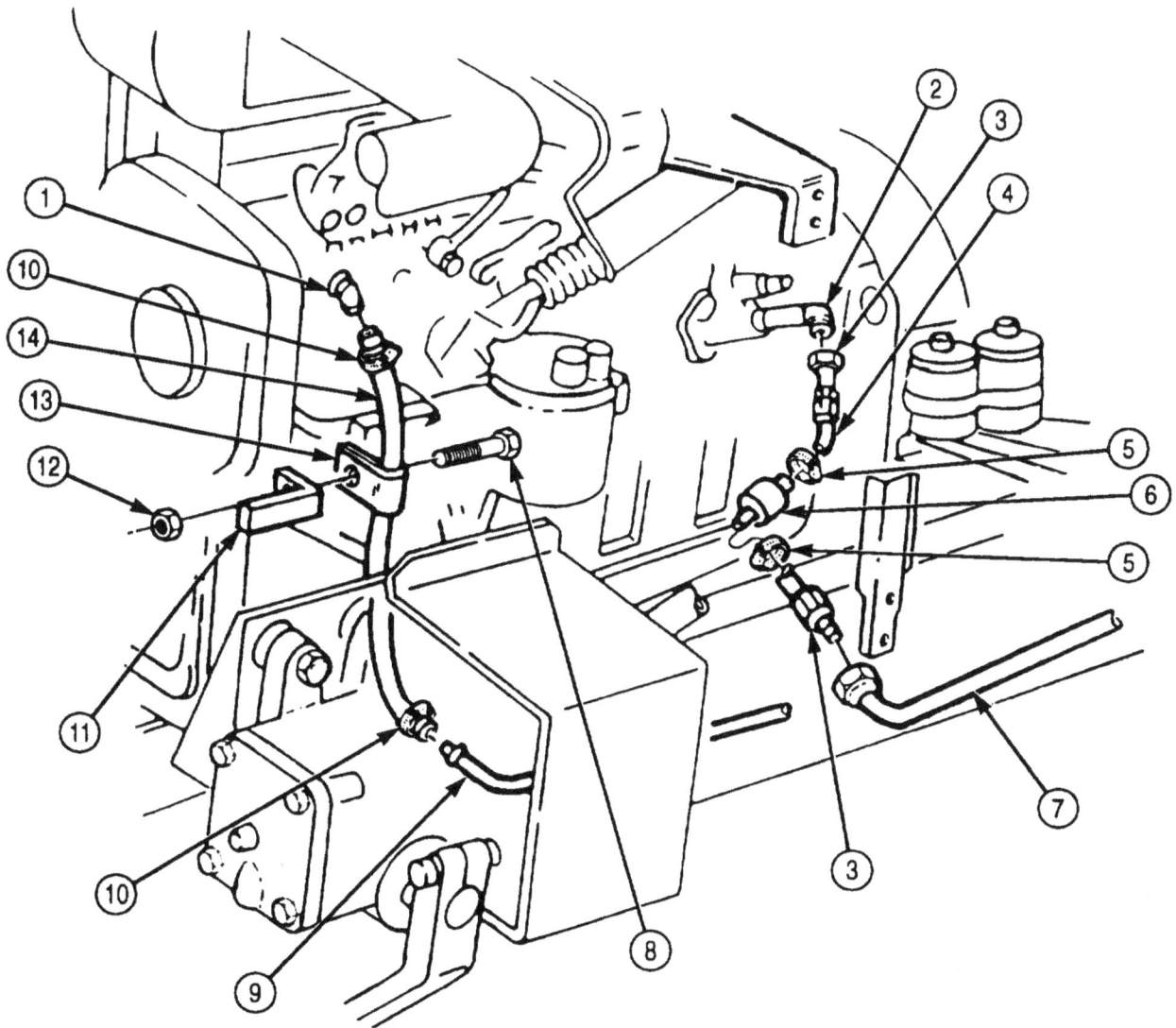

3-29. FUEL TANK LINES (M939A2) REPLACEMENT (Contd)

5. Remove nut (4), screw (1), and clamp (3) from fuel return line (2).
6. Remove nut (8), screw (5), and clamp (7) from fuel supply line (6).
7. Remove nut (11) from screw (9).
8. Remove two clamps (10) from fuel supply line (6) and fuel return line (2).
9. Remove nut (17), screw (15), and two clamps (16) from fuel supply line (6) and fuel return line (2).
10. Disconnect two tubing nuts (18) and remove tee fitting (19) and front half of fuel supply line (6) from rear half of fuel supply line (6).
11. Remove nut (14), screw (12), and clamp (13) from fuel return line (2).

3-29. FUEL TANK LINES (M939A2) REPLACEMENT (Contd)

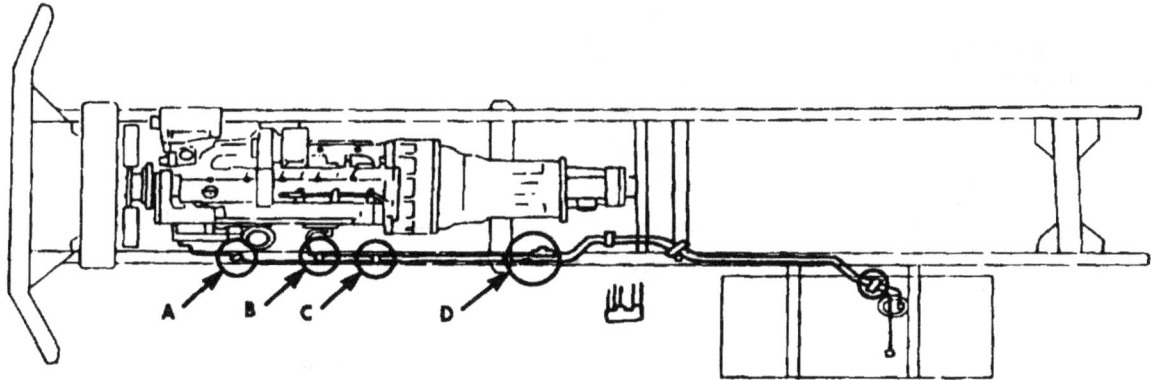

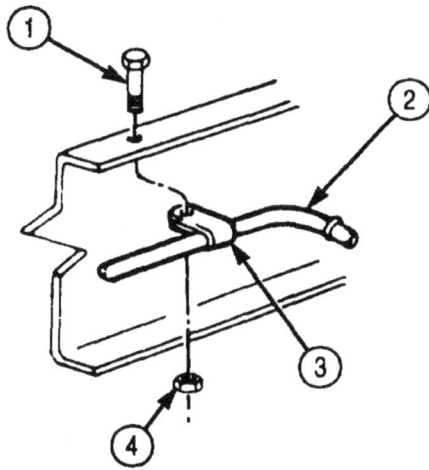

VIEW A

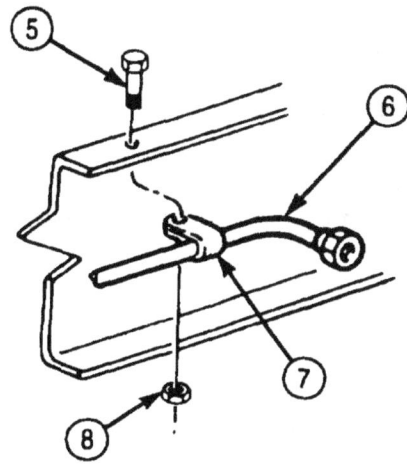

VIEW B

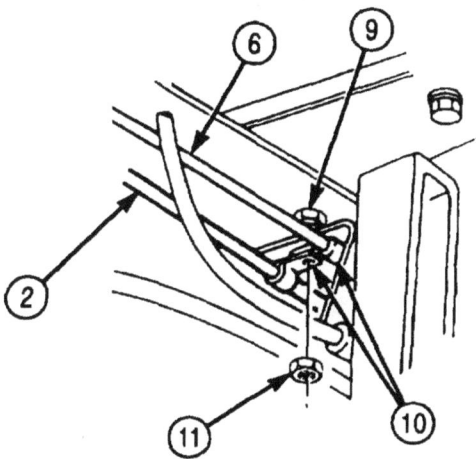

VIEW C

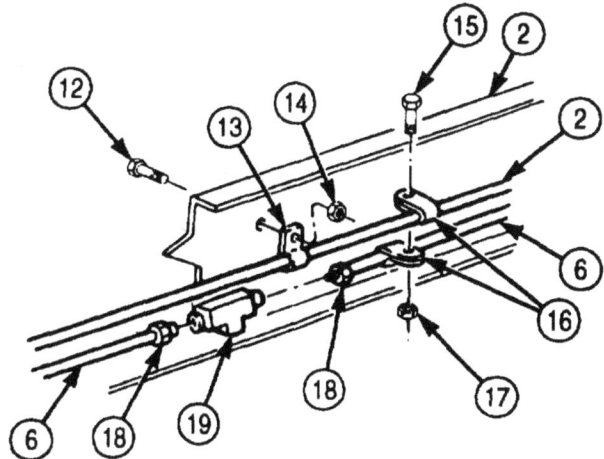

VIEW D

3-29. FUEL TANK LINES (M939A2) REPLACEMENT (Contd)

12. Remove nut (5), screw (1), and two clamps (4) from fuel supply line (2) and fuel return line (3).
13. Remove nut (11), screw (8), and two clamps (10) from fuel supply line (2), fuel return line (3), and bracket (7).

NOTE
Perform step 14 if bracket is damaged.

14. Remove nut (9), screw (6), and bracket (7) from frame rail (12).
16. Remove nut (15), screw (13), and two clamps (14) from fuel supply line (2) and fuel return line (3).
16. Disconnect two tubing nuts (17) from fittings (18), and remove fuel supply line (2) and fuel return line (3) from fuel tank (16).

NOTE
Perform steps 17 through 19 for vehicles with dual tanks.

17. Remove nut (22) and screw (24) from two clamps (23).
18. Disconnect two tubing nuts (21) from adapter fittings (20) on selector valve (19).
19. Remove fuel supply line (2) and fuel return line (3) from adapter fittings (20).

b. Installation

1. Connect rear half of fuel supply line (2) and fuel return line (3) to fittings (18) on fuel tank (16) with tubing nuts (17).
2. Install two clamps (14) on fuel supply line (2) and fuel return line (3) with screw (13) and nut (15).

NOTE
Perform step 3 if bracket was removed.

3. Install bracket (7) on frame rail (12) with screw (6) and nut (9).
4. Install two clamps (10) on fuel supply line (2), fuel return line (3), and bracket (7) with screw (8) and nut (11).
5. Install two clamps (4) on fuel supply line (2) and fuel return line (3) with screw (1) and nut (6).

NOTE
Perform steps 6 and 7 for vehicles with dual tanks.

6. Install fuel supply line (2) and fuel return line (3) on adapter fittings (20) of selector valve (19) with tubing nuts (21).
7. Install two clamps (23) on fuel supply line (2) and fuel return line (3) with screw (24) and nut (22).

3-29. FUEL TANK LINES (M939A2) REPLACEMENT (Contd)

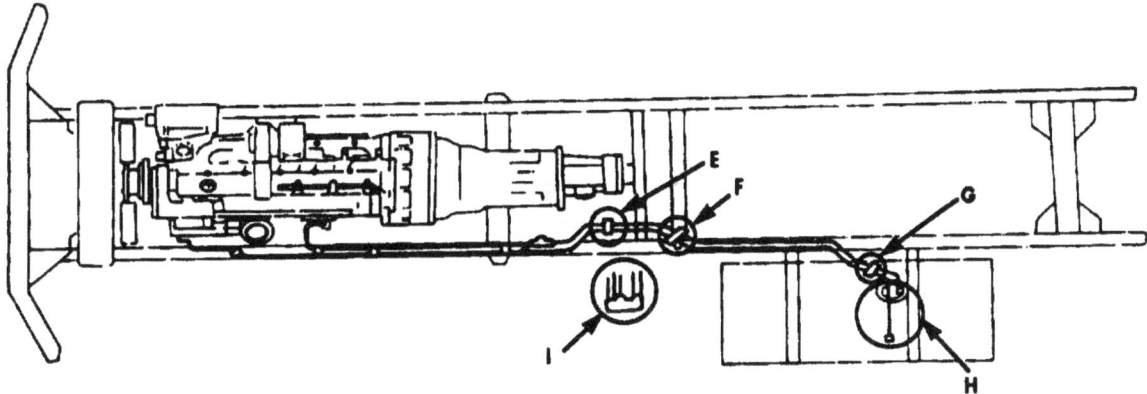

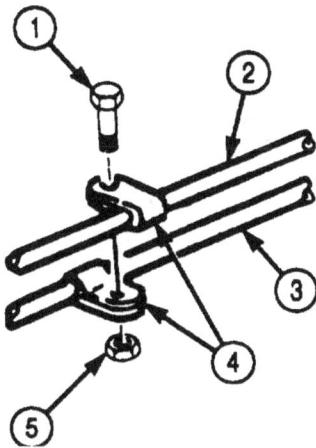

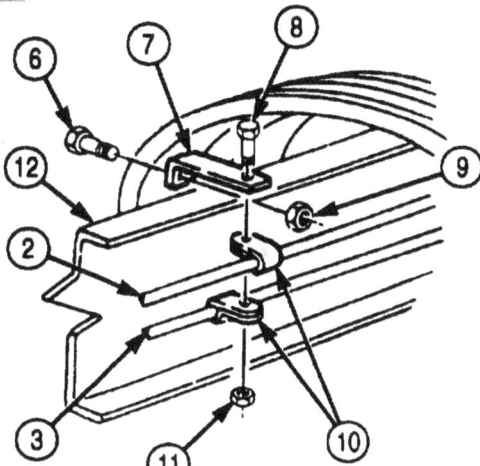

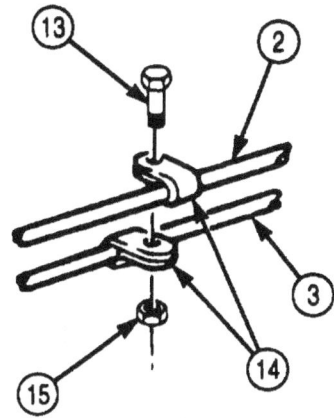

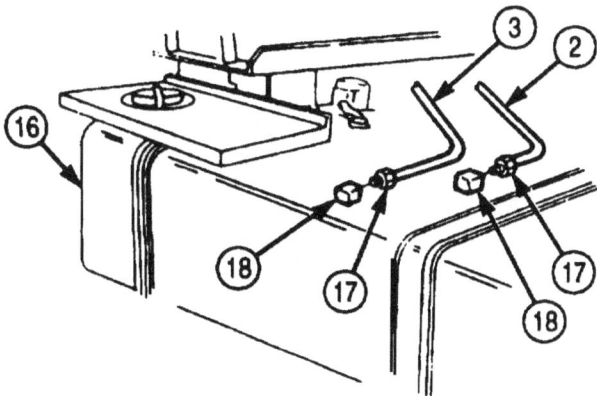

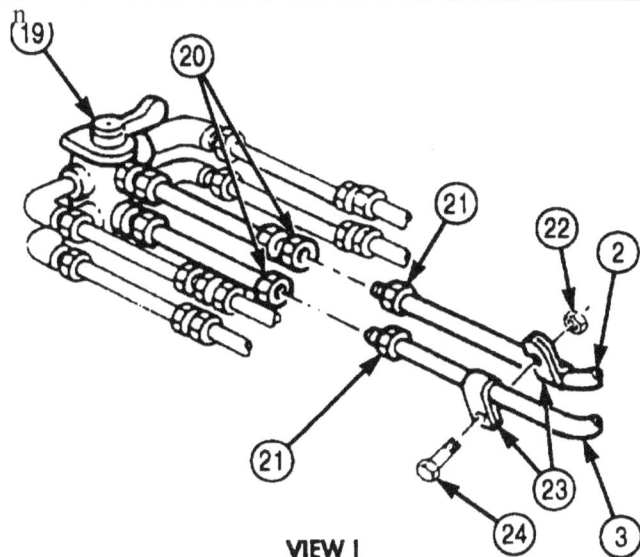

VIEW E
SINGLE TANK

VIEW F
SINGLE TANK

VIEW G
SINGLE TANK

VIEW H
SINGLE TANK

VIEW I
DUAL TANK

3-29. FUEL TANK LINES (M939A2) REPLACEMENT (Contd)

8. Install clamp (2) on fuel return line (5) with screw (1) and nut (3).
9. Install front half of fuel supply line (6) and rear half of fuel supply line (6) on tee (10) with tubing nuts (9).
10. Install two clamps (7) on fuel supply line (6) and fuel return line (5) with screw (4) and nut (8).
11. Install two clamps (12) on fuel supply line (6) and fuel return line (5) with screw (11) and nut (13).
12. Install clamp (15) on fuel supply line (6) with screw (14) and nut (16).
13. Install clamp (18) on fuel return line (5) with screw (17) and nut (19).
14. Install fuel return hose (28) on fitting (20) and fuel return line (5) with hose clamps (24).
15. Install fuel return hose (28) on bracket (25) with clamp (27), screw (23), and new locknut (26).

NOTE

Perform step 16 if fuel filter was removed.

16. Install fuel filter (25) on fuel supply hose (23) with two clamps (24).
17. Install fuel supply hose (23) on fuel supply line (6) and elbow (21) with with tubing nuts (22).
18. Install new tiedown straps as required.

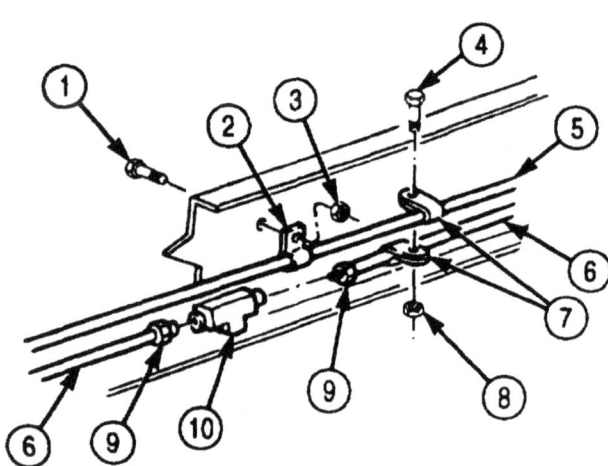

VIEW A

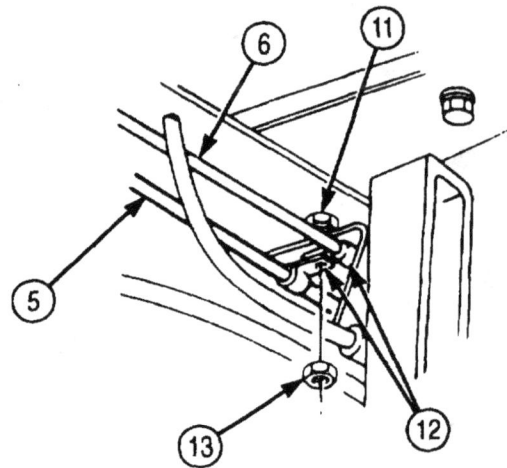

VIEW B

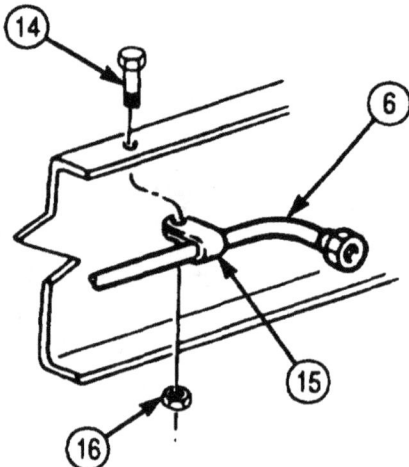

VIEW-C

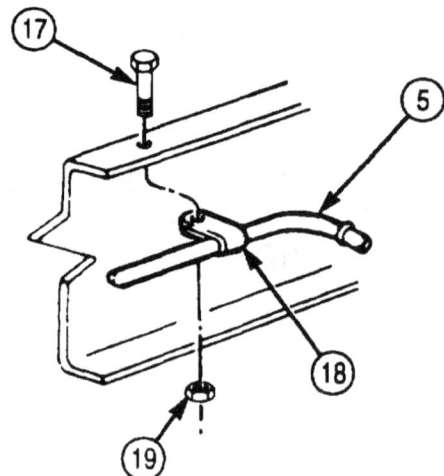

VIEW-D

3-29. FUEL TANK LINES (M939A2) REPLACEMENT (Contd)

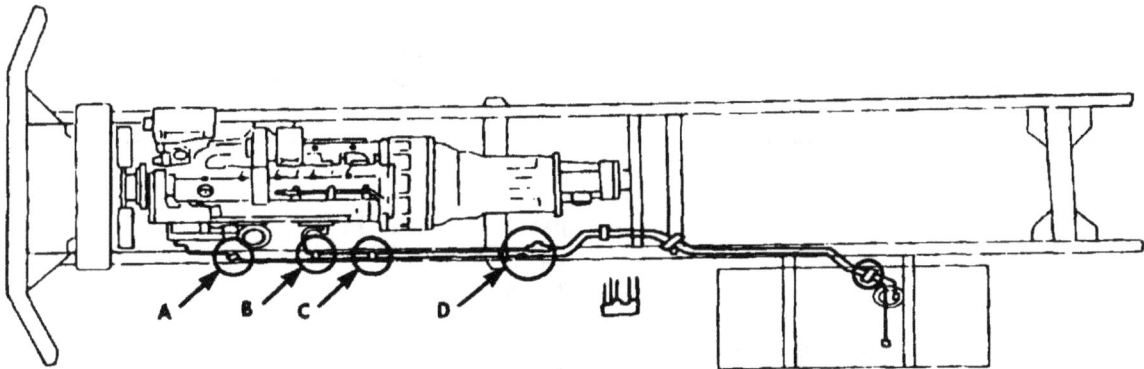

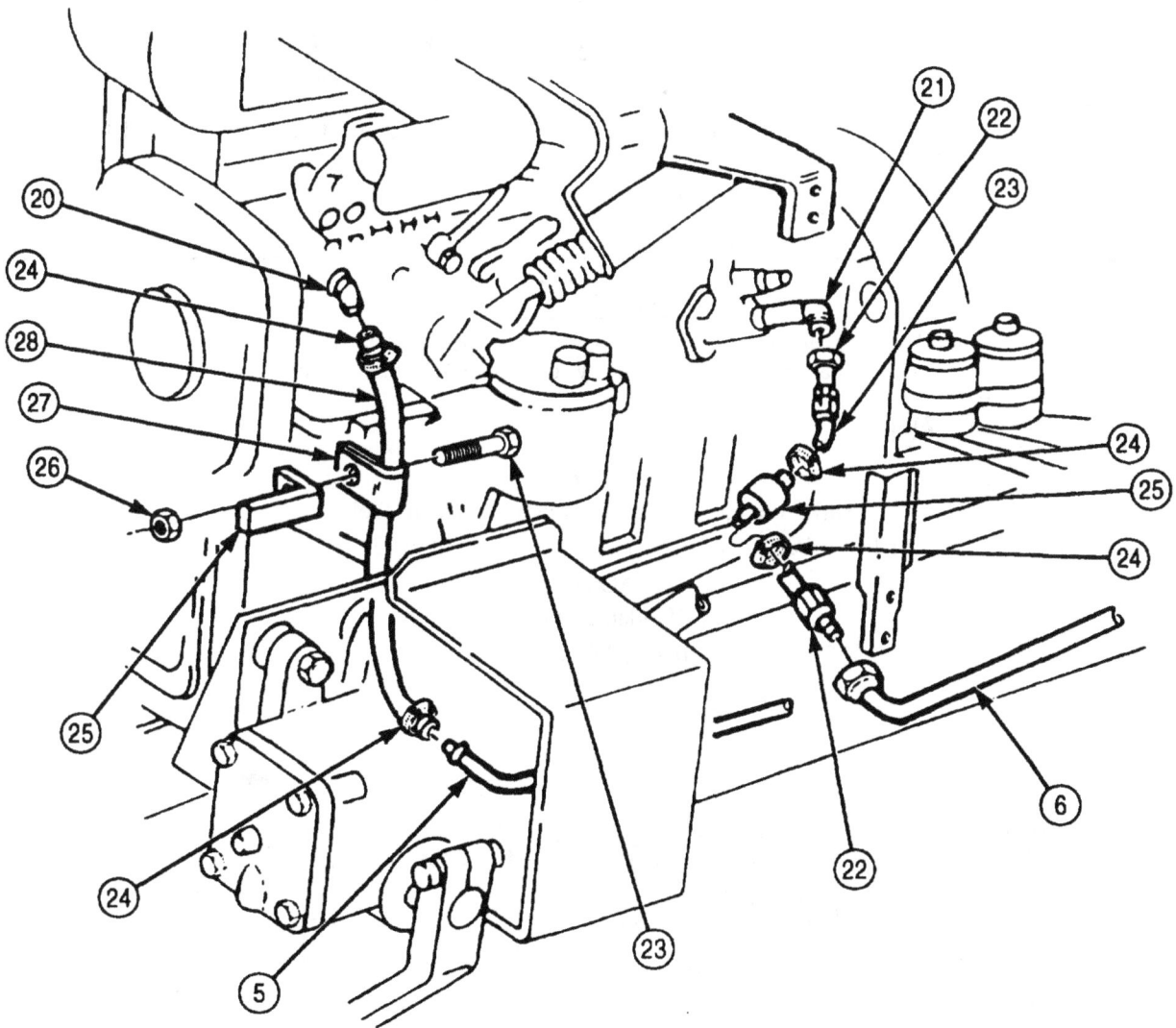

FOLLOW-ON TASKS: • Prime fuel system (para. 3-22).
                      • Start engine (TM 9-2320-272-10) and check for fuel leaks.

3-30. FUEL FILTER AND COVER MAINTENANCE

THIS TASK COVERS:

a. Removal (M939A2)
b. Removal (M939/A1)
c. Cleaning and Inspection

d. Installation (M939/A1)
e. Installation (M939A2)

INITIAL SETUP:

APPLICABLE MODELS

All

TOOLS

General mechanic's tool kit (Appendix E, Item 1)
Torque wrench (Appendix E, Item 146)

MATERIALS/PARTS

Adapter (M939A2) (Appendix D, Item 1)
Fluid pressure parts kit (M939/A1)
  (Appendix D, Item 129)
Fuel filter (M939A2) (Appendix D, Item 135)
Three locknuts (Appendix D, Item 288)
Lint-free cloth (Appendix C, Item 21)
Antiseize tape (Appendix C, Item 72)
Drycleaning solvent (Appendix C, Item 71)

REFERENCES (TM)

TM 9-2320-272-10
TM 9-2320-272-24P

EQUIPMENT CONDITION

- Parking brake set (TM 9-2320-272-10).
- Hood raised and secured (TM 9-2320-272-10).

GENERAL SAFETY INSTRUCTIONS

- Keep fire extinguisher nearby when using dry cleaning solvent.
- Dry cleaning solvent is flammable and toxic. Do not use near open flame.
- Diesel fuel is flammable. Do not perform this procedure near flames.

WARNING

Diesel fuel is highly flammable. Do not perform fuel system
procedures near open flames. Injury to personnel may result.

a. Removal (M939A2)

NOTE

Have drainage container ready to catch fuel.

1. Remove fuel filter (3) from fuel filter head (1). Discard fuel filter (3).

NOTE

Perform step 2 if adapter threads are stripped.

2. Remove adapter (2) from fuel filter head (1). Discard adapter (2).

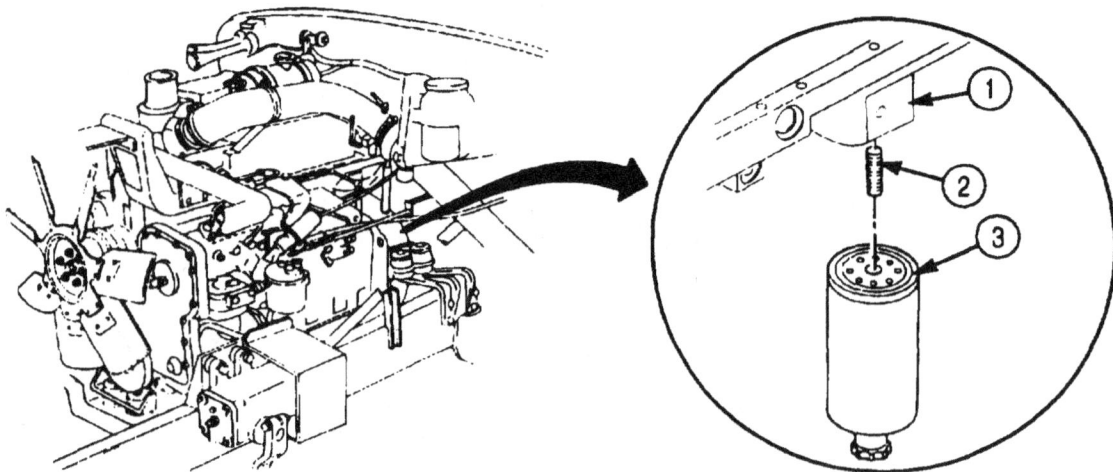

3-30. FUEL FILTER AND COVER MAINTENANCE (Contd)

b. Removal (M939/A1)

NOTE

Have drainage container ready to catch fuel.

1. Open filter drainvalve (9) and inlet drainvalve (7) located at the left front underside of cab.
2. Close drainvalves (9) and (7) when fuel drainage is complete.
3. Holding filter case (8), remove center bolt (4), square washer (5), and small O-ring (6) from top of filter cover (12). Discard O-ring (6).
4. Remove filter case (8), fuel filter element (10), and large gasket (11) from filter cover (12). Discard gasket (11) and fuel filter element (10).
5. Remove drainvalve (9) from filter case (10).

NOTE

If only the filter is to be replaced, perform task c, steps 2 through 5.

TM 9-2320-272-24-1

3-30. FUEL FILTER AND COVER MAINTENANCE (Contd)

NOTE

Perform steps 6 through 11 to repair or replace filter cover or any fittings.

6. Disconnect filter inlet line (8) from adapter (7).
7. Disconnect filter outlet line (1) from elbow (2).

NOTE

- Assistant will help with steps 8 through 10.
- Record the directions of elbow and tee before removal.

8. Remove three locknuts (5), screws (3), and filter cover (11) from mounting bracket (4). Discard locknuts (5).
9. Remove elbow (2) from filter cover (11).
10. Remove adapter (7), tee (6), and elbow (10) from filter cover (11).
11. Remove drainvalve (9) from tee (6).

c. Cleaning and Inspection

NOTE

Perform all steps of task c if fuel filter cover was removed from vehicle.

1. Inspect elbows (2) and (10), tee (6), adapter (7), and drainvalve (9) for stripped threads. Replace any damaged part(s).

WARNING

Drycleaning solvent is flammable and will not be used near open flame. Use only in well-ventilated places. Failure to do this may result in injury to personnel.

2. Clean filter cover (11) and filter case (15) with drycleaning solvent and dry with lint-free cloth.
3. Inspect fuel filter cover (11) for cracks or damage to gasket seating area. Replace if cracked or seating area is damaged.
4. Inspect fuel filter case (15) for cracks, stripped threads, or damage to gasket seating area. Replace if cracked, threads are stripped, or seating area is damaged.
5. Inspect drainvalve (16) for stripped threads. Replace if threads are stripped.

d. Installation (M939/A1)

NOTE

- All male pipe threads must be wrapped with antiseize tape before installation.
- If only the fuel filter element was changed, go to step 5.

1. Install adapter elbows (2) and (10) on filter cover (11) in directions noted during removal.

NOTE

Ensure drainvalve port on tee faces down.

2. Install tee (6) on elbow (10).
3. Install drainvalve (9) and adapter (7) on tee (6).
4. Connect filter outlet line (1) to elbow (2), and attach filter cover (11) to mounting bracket (4) with three screws (3) and new locknuts (5).

3-30. FUEL FILTER AND COVER MAINTENANCE (Contd)

5. Install drainvalve (16) in fuel filter case (15).

6. Position new gasket (19) in fuel filter cover (11).

7. Place new filter element (17) in filter case (15), with handle (18) up.

8. Position filter case (15) against filter cover (11) and install filter case (15) on filter cover (11) with new O-ring (14). washer (13), and screw (12). Tighten screw (12) 20-25 lb-ft (27-34 N•m).

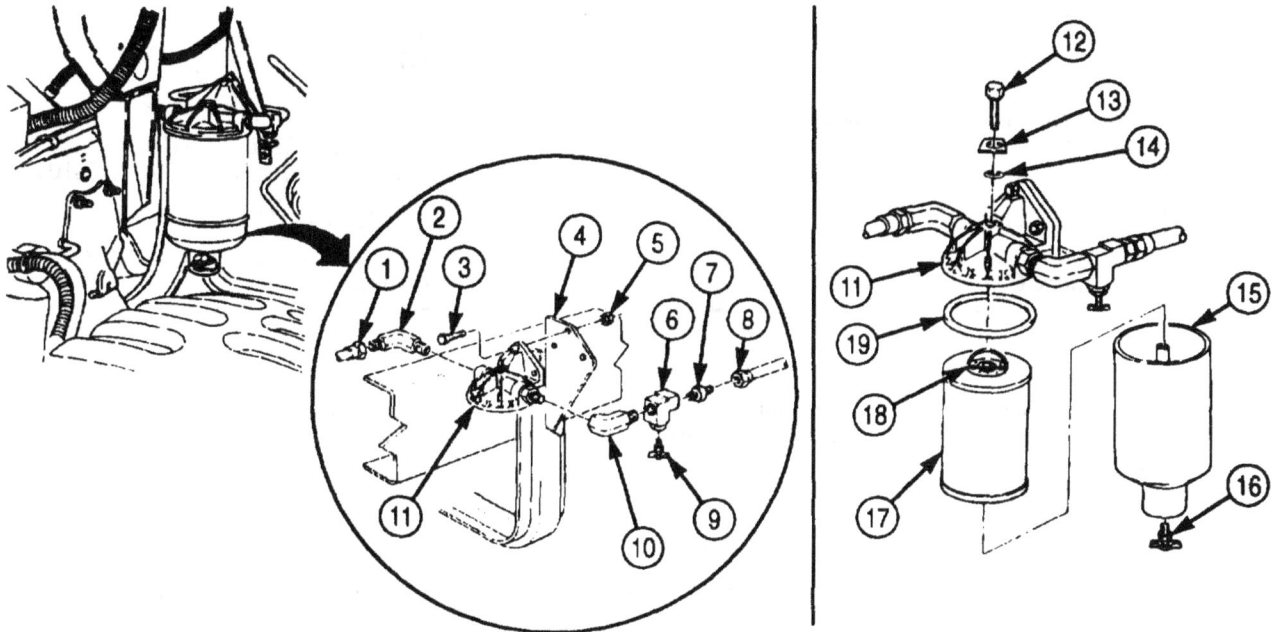

e. Installation (M939A2)

NOTE

Perform step 1 if adapter was removed.

1. Install new adapter (20) on fuel filter head (21).

2. Fill new fuel filter (23) with clean diesel fuel, Coat seal (22) with diesel fuel.

3. Install new fuel filter (23) on fuel filter head (21). Hand-tighten fuel filter (23), then tighten an additional 3/4 turn.

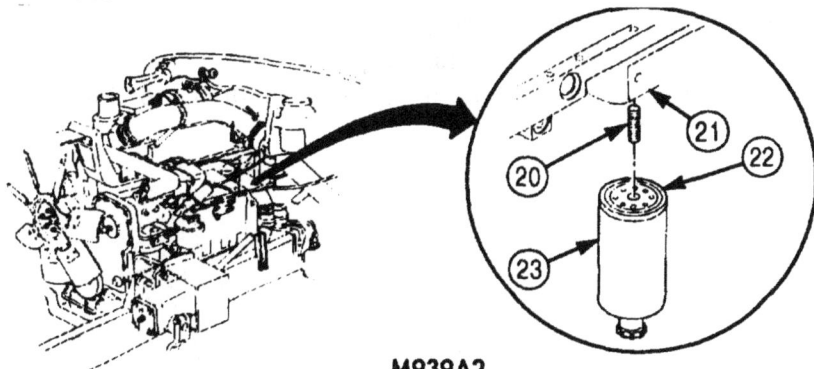

M939A2

FOLLOW-ON TASKS: • Prime fuel system (para. 3-22 for M939A2; TM 9-2320-272-10 for M939/A1).
• Start engine (TM 9-2320-272-10) and check for fuel leaks.

3-31. AFC FUEL PUMP FILTER REPLACEMENT

THIS TASK COVERS:

a. Removal   b. Installation

INITIAL SETUP:

APPLICABLE MODELS
All except M936/A1/A2

TOOLS
General mechanic's tool kit (Appendix E, Item 1)
Torque wrench (Appendix E, Item 146)

MATERIALS/PARTS
Filter (Appendix D, Item 121)
Gasket (Appendix D, Item 146)
Rages (Appendix C, Item 58)

REFERENCES (TM)
TM 9-2320-272-10
TM 9-2320-272-24P

EQUIPMENT CONDITION
- Parking brake set (TM 9-2320-272-10).
- Left splash shield removed (TM 9-2320-272-10).

GENERAL SAFETY INSTRUCTIONS
Diesel fuel is flammable. Do not perform this
procedure near flames.

WARNING

Diesel fuel is highly flammable. Do not perform fuel system
procedures near open flame. Injury to personnel may result.

a. Removal

CAUTION

Fuel pump exterior must be cleaned before filter cap is removed to
prevent foreign particles from entering fuel pump.

1. Using clean, dry cloth, clean exterior of fuel pump (5) located on left side of engine (6).
2. Remove filter cap (3) and gasket (4) from fuel pump (5). Discard gasket (4).
3. Remove filter spring (2) and filter (1) from fuel pump (5). Discard filter (1).

b. Installation

CAUTION

Ensure small end of filter spring is against filter to assure proper
fuel flow. Failure to do this will result in fuel pump damage.

1. Install new filter (1), open end first, in fuel pump (5).
2. Install new gasket (4) and filter spring (2) on filter cap (3).
3. Install filter cap (3) on fuel pump (5). Tighten cap (3) to 8-12 lb-ft (11-16 N-m).

3-31. AFC FUEL PUMP FILTER REPLACEMENT (Contd)

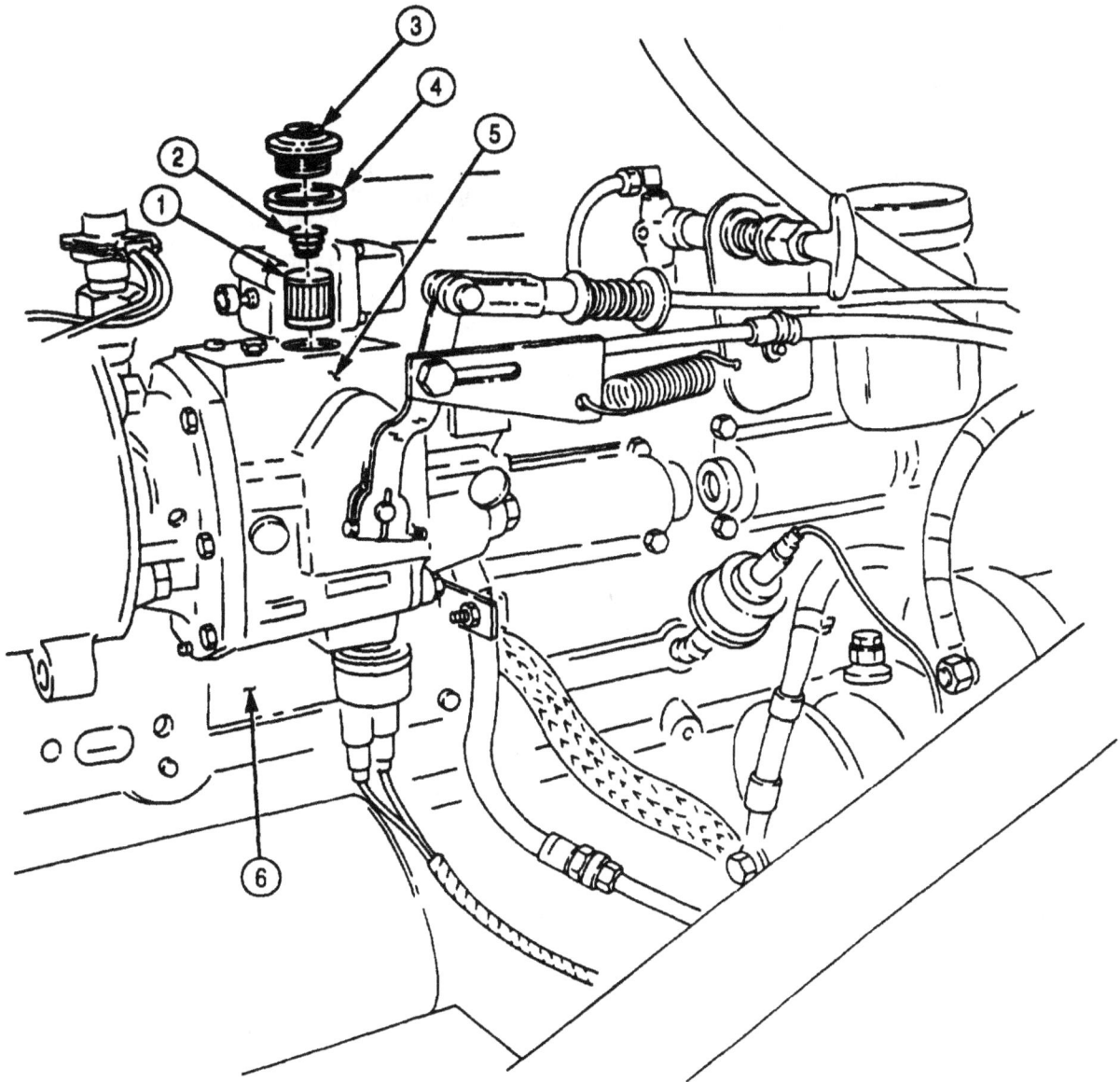

FOLLOW-ON TASKS: • Prime fuel system (para. 3-22 for M939A2; TM 9-2320-272-10 for M939/A1).
• Install left splash shield (TM 9-2320-272-10)
• Start engine (TM 9-2320-272-10) and check for fuel leaks.

3-32. FUEL PUMP WITH VS GOVERNOR FILTER REPLACEMENT

THIS TASK COVERS:

a. Removal   b. Installation

INITIAL SETUP:

APPLICABLE MODELS
M936/A1/A2

TOOLS
General mechanic's tool kit (Appendix E, Item 1)

MATERIALS/PARTS
O-ring (Appendix D, Item 424)
Filter (Appendix D, Item 121)
Cleaning cloth (Appendix C, Item 21)

REFERENCES (TM)
TM 9-2320-272-10
TM 9-2320-272-24P

EQUIPMENT CONDITION
- Parking brake set (TM 9-2320-272-10).
- Left splash shield removed (TM 9-2320-272-10).

GENERAL SAFETY INSTRUCTIONS
Diesel fuel is flammable. Do not perform this procedure near flames.

WARNING

Diesel fuel is highly flammable. Do not perform fuel system procedures near open flame. Injury to personnel may result.

a. Removal

CAUTION

Fuel pump exterior must be cleaned before filter cap is removed to prevent foreign particles from entering fuel pump.

1. Using a clean, dry cloth, clean fuel pump (7) located on left side of engine (6).

NOTE

Place hand under fuel filter to catch assembly when snapring is removed.

2. Remove snapring (5), retainer (3), filter spring (2), and filter (1) from fuel pump (7).

3. Remove O-ring (4) from retainer (3). Discard O-ring (4).

4. Inspect filter (1) for holes or embedded metal particles. If holes or metal particles are found, discard filter (1).

b. Installation

1. Install new O-ring (4) on retainer (3).

CAUTION

Small end of filter spring and hole in filter screen must be facing up toward pump to allow fuel flow. Failure to do this will result in fuel pump damage.

2. Install filter (1), filter spring (2), and retainer (3) in fuel pump (7) with snapring (5).

3-32. FUEL PUMP WITH VS GOVERNOR FILTER REPLACEMENT (Contd)

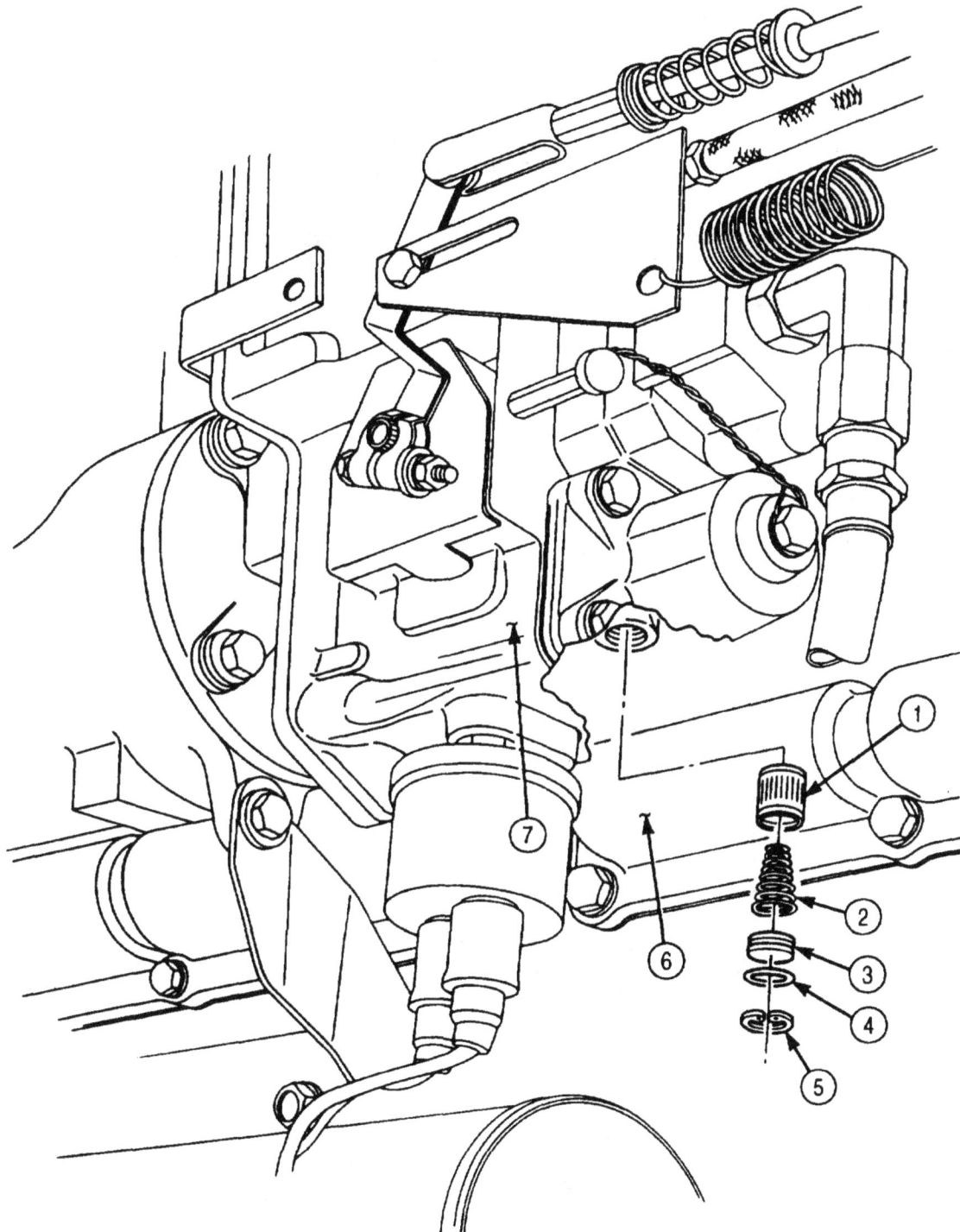

FOLLOW-ON TASKS: • Prime fuel system (para. 3-22 for M939A2; TM 9-2320-272-10 for M939/A1).
                   • Install left splash shield (TM 9-2320-272-10).
                   • Start engine (TM 9-2320-272-10) and check fuel pump for leaks and operation.

3-33. ETHER START SWITCH REPLACEMENT

THIS TASK COVERS:

a. Removal   b. Installation

INITIAL SETUP:

APPLICABLE MODELS
All

TOOLS
General mechanic's tool kit (Appendix E, Item 1)

REFERENCES (TM)
TM 9-2320-272-10
TM 9-2320-272-24P

EQUIPMENT CONDITION
- Parking brake set (TM 9-2320-272-10).
- Battery ground cables disconnected (para. 3-126).

a. Removal

NOTE
Prior to removal, tag wires for installation.

1. Disconnect two ether start switch wires (3) from harness wires (2) under left side of dash (1).
2. Remove nut (5) and ether start switch (4) from left side of dash (1).

b. Installation

1. Install ether start switch (4) on left side of dash (1) with nut (5).
2. Connect two ether start switch wires (3) to wiring harness wires (2) under left side of dash (1).

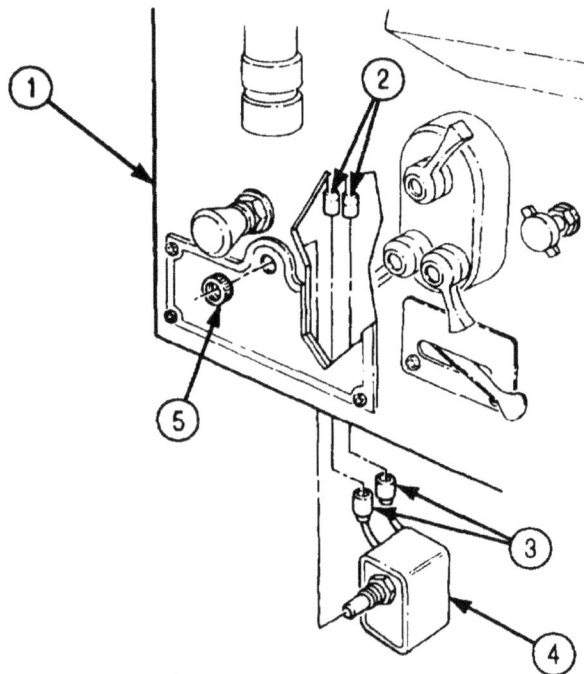

FOLLOW-ON TASK: Connect battery ground cables (para. 3-126).

3-34. ETHER START FUEL PRESSURE SWITCH (M939/A1) REPLACEMENT

THIS TASK COVERS:

a. Removal

b. Installation

INITIAL SETUP:

APPLICABLE MODELS
M9391/A1

TOOLS
General mechanic's tool kit (Appendix E, Item 1)

REFERENCES (TM)
TM 9-2320-272-10
TM 9-2320-272-24P

EQUIPMENT CONDITION
- Parking brake set (TM 9-2320-272-10).
- Left splash shield removed (TM 9-2320-272-10).
- Battery ground cables disconnected (para. 3-126).

---

a. Removal

NOTE
Have drainage container ready to catch fuel.

1. Disconnect two wires (2) from ether start pressure switch (3).
2. Remove ether start fuel pressure switch (3) from bottom of fuel pump (1).

b. Installation

1. Install ether start fuel pressure switch (3) on bottom of fuel pump (1).
2. Connect two wires (2) to ether start pressure switch (3).

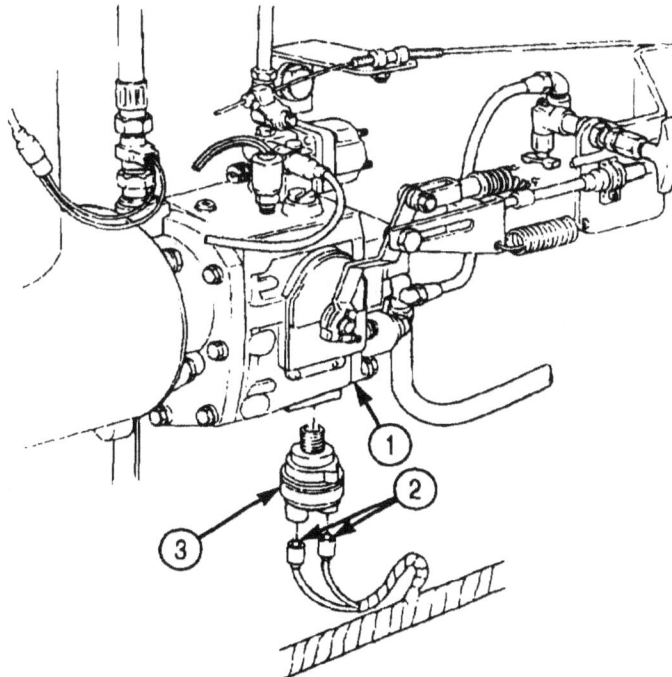

FOLLOW-ON TASKS: • Connect battery ground cables (para. 3-126).
  • Install left splash shield (TM 9-2320-272-10).

3-35. ETHER CYLINDER AND VALVE REPLACEMENT

THIS TASK COVERS:

a. Ether Cylinder Removal   c. Ether Valve Installation
b. Ether Valve Removal   d. Ether Cylinder Installation

INITIAL SETUP:

APPLICABLE MODELS
All

TOOLS
General mechanic's tool kit (Appendix E, Item 1)

MATERIALS/PARTS
Two locknuts (Appendix D, Item 299)

REFERENCES (TM)
TM 9-2320-272-10
TM 9-2320-272-24P

EQUIPMENT CONDITION
- Parking brake set (TM 9-2320-272-10).
- Hood raised and secured (TM 9-2320-272-10).
- Left splash shield removed (TM 9-2320-272-10).
- Battery ground cables disconnected (para. 3-126).

GENERAL SAFETY INSTRUCTIONS
Ether is flammable. Do not perform this task **near** open flames.

WARNING

Ether is extremely flammable. Do not perform ether system procedures near open flames. Injury to personnel may result.

a. Ether Cylinder Removal

1. Loosen ether cylinder clamp (12) on ether cylinder bracket (11).

NOTE

Ether cylinder must be removed quickly to allow ether cylinder check valve to close and prevent loss of ether.

2. Remove ether cylinder (1) from ether valve (10).

b. Ether Valve Removal

CAUTION

When ether cylinder is removed, cover must be installed to prevent dust or dirt from entering valve.

NOTE
Tag wires for installation.

1. Disconnect wire (7) from ether valve wire (6).
2. Disconnect ether supply tube (8) from ether valve adapter (9).
3. Remove two locknuts (5), screws (2), ground wire (3), ether valve bracket (4), and ether valve (10) from ether cylinder bracket (11). Discard locknuts (5).

3-35. ETHER CYLINDER AND VALVE REPLACEMENT (Contd)

c. Ether Valve Installation

1. Install ether valve bracket (4), ether valve (10), and ground wire (3) on ether cylinder bracket (11) with two screws (2) and new locknuts (5).
2. Connect wire (7) to ether valve wire (6).
3. Connect ether supply tube (8) to ether valve fitting (9).

d. Ether Cylinder Installation

Install ether cylinder (1) on ether valve (10) and ether cylinder bracket (11) and tighten clamp (12).

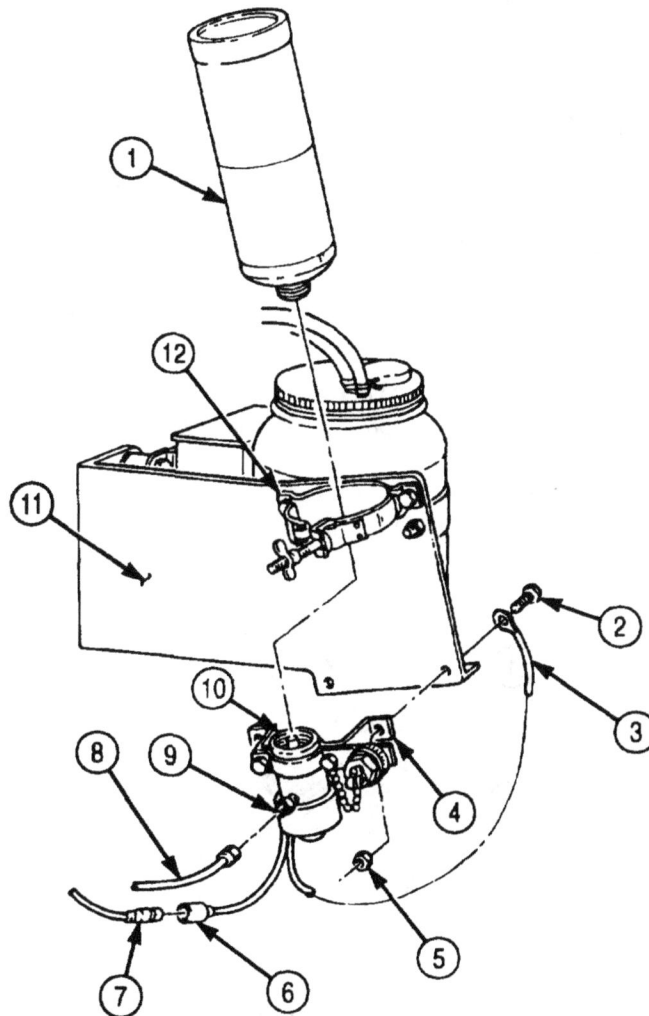

FOLLOW-ON TASKS: • Connect battery ground cables (para. 3-126).
• Install left splash shield (TM 9-2320-272-10).

3-36. ETHER THERMAL CLOSE VALVE AND BUSHING REPLACEMENT

THIS TASK COVERS:

a. Removal   b. Installation

INITIAL SETUP:

APPLICABLE MODELS
All

TOOLS
General mechanic's tool kit (Appendix E, Item 1)

MATERIALS/PARTS
Antiseize tape (Appendix C, Item 72)

REFERENCES (TM)
TM 9-2320-272-10
TM 9-2320-272-24P

EQUIPMENT CONDITION
- Parking brake set (TM 9-2320-272-10).
- Hood raised and secured (TM 9-2320-272-10).

---

a. Removal

NOTE
Tag tubes for installation.

1. Disconnect thermal close ether supply tube (3) and atomizer ether supply tube (1) from adapters (2) on thermal close valve (6).
2. Remove thermal close valve (6) from adapter bushing (5).
3. Remove adapter bushing (5) from water manifold (4).

b. Installation

NOTE
Male pipe threads must be wrapped with antiseize tape before installation.

1. Install adapter bushing (5) in water manifold (4).
2. Install thermal close valve (6) in bushing (5).
3. Connect atomizer ether supply tube (1) and thermal close ether supply tube (3) to adapters (2) on thermal close valve (6).

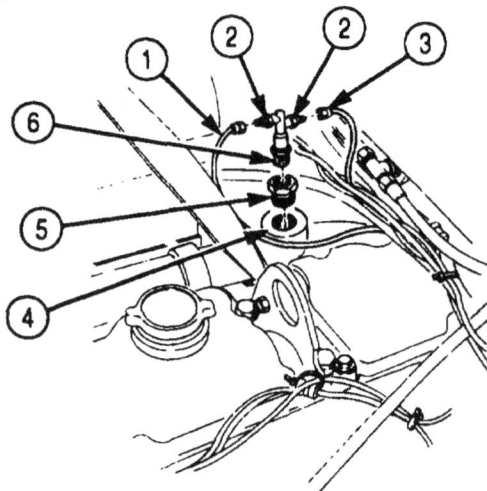

FOLLOW-ON TASK: Start engine (TM 9-2320-272-10) and check for coolant leaks around thermal close valve and adapter bushing.

3-116

3-37. ETHER ATOMIZER (M939/A1) REPLACEMENT

THIS TASK COVERS:

a. Removal   b. Installation

INITIAL SETUP:

APPLICABLE MODELS
M939/A1

TOOLS
General mechanic's tool kit (Appendix E, Item 1)

MATERIALS/PARTS
Antiseize tape (Appendix C, Item 72)

REFERENCES (TM)
TM 9-2320-272-10
TM 9-2320-272-24P

EQUIPMENT CONDITION
- Parking brake set (TM 9-2320-272-10).
- Hood raised and secured (TM 9-2320-272-10).

a. Removal

1. Disconnect ether atomizer tube (2) from ether atomizer (1).
2. Remove ether atomizer (1) from intake manifold (3).

b. Installation

NOTE

Male pipe threads must be wrapped with antiseize tape before installation.

1. Install atomizer (1) on intake manifold (3).
2. Connect ether atomizer supply tube (2) to atomizer (1).

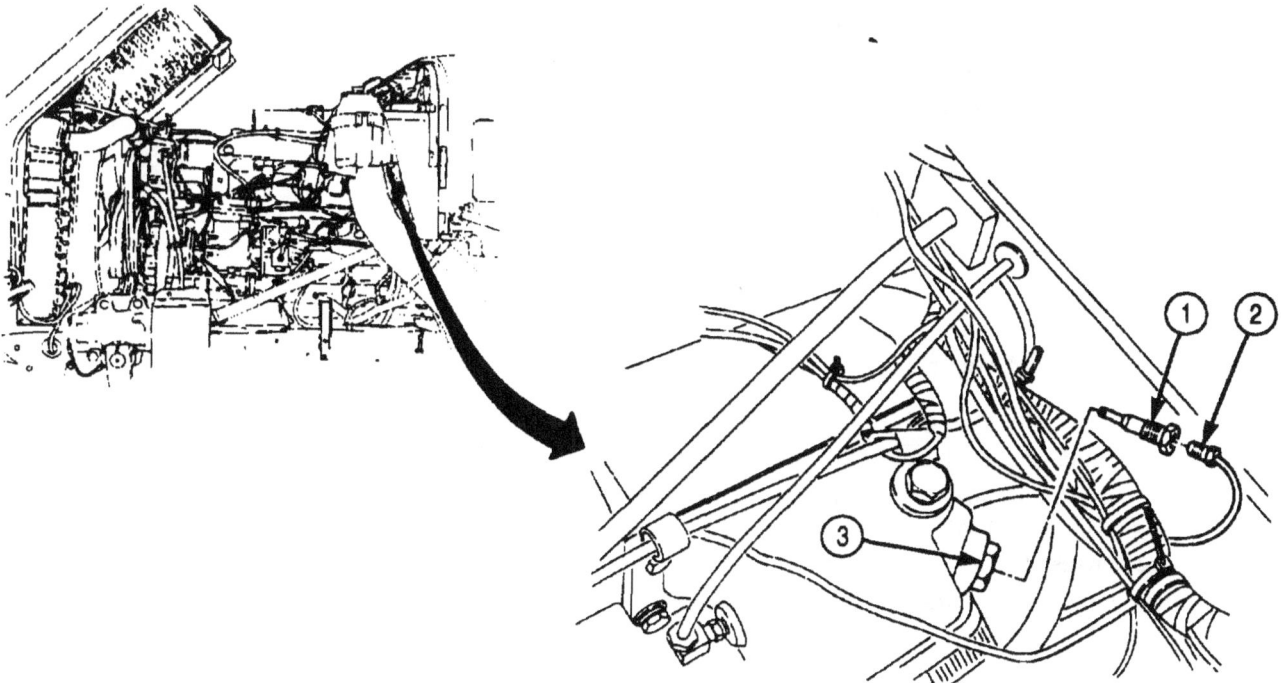

3-38. ETHER ATOMIZER AND TEMPERATURE SENSOR (M939A2) REPLACEMENT

THIS TASK COVERS:

a. Removal   b. Installation

INITIAL SETUP:

APPLICABLE MODELS
M939A2

TOOLS
General mechanic's tool kit (Appendix E, Item 1)

MATERIALS/PARTS
Lockwasher (Appendix D, Item 364)
Tiedown straps (Appendix D, Item 684)
Antiseize compound (Appendix C, Item 10)
Cap and plug set (Appendix C, Item 14)

REFERENCES (TM)
TM 9-2320-272-10
TM 9-2320-272-24P

EQUIPMENT CONDITION
- Parking brake set (TM 9-2320-272-10).
- Hood raised and secured (TM 9-2320-272-10).

GENERAL SAFETY INSTRUCTIONS
Ether is flammable. Do not perform this task near open flames.

WARNING

Ether is extremely flammable. Do not perform ether system procedures near open flame. Injury to personnel may result.

a. Removal

CAUTION

Cap or plug all openings immediately after disconnecting lines and hoses to prevent contamination. Failure to do so may result in damage to engine.

NOTE

Record locations of tiedown straps.

1. Remove tiedown straps (3) as required. Discard tiedown straps (3).
2. Disconnect wire connector (5) from wiring harness (4).
3. Remove screw (11), washer (10), ground strap (9), lockwasher (8), and temperature sensor (7) from engine block (6). Discard lockwasher (8).
4. Disconnect ether supply tube (1) from ether atomizer (2).
5. Remove ether atomizer (2) from adapter (12).

b. Installation

1. Coat threads of ether atomizer (2) with antiseize compound and install ether atomizer (2) in adapter (12).
2. Tighten ether atomizer (2) until indicator mark is in three or nine o'clock position.
3. Connect ether supply tube (1) to ether atomizer (2).
4. Install temperature sensor (7) and ground strap (9) on engine block (6) with new lockwasher (8), washer (10), and screw (11).
5. Connect wire (5) to wiring harness (4).
6. Install new tiedown straps (3) as required.

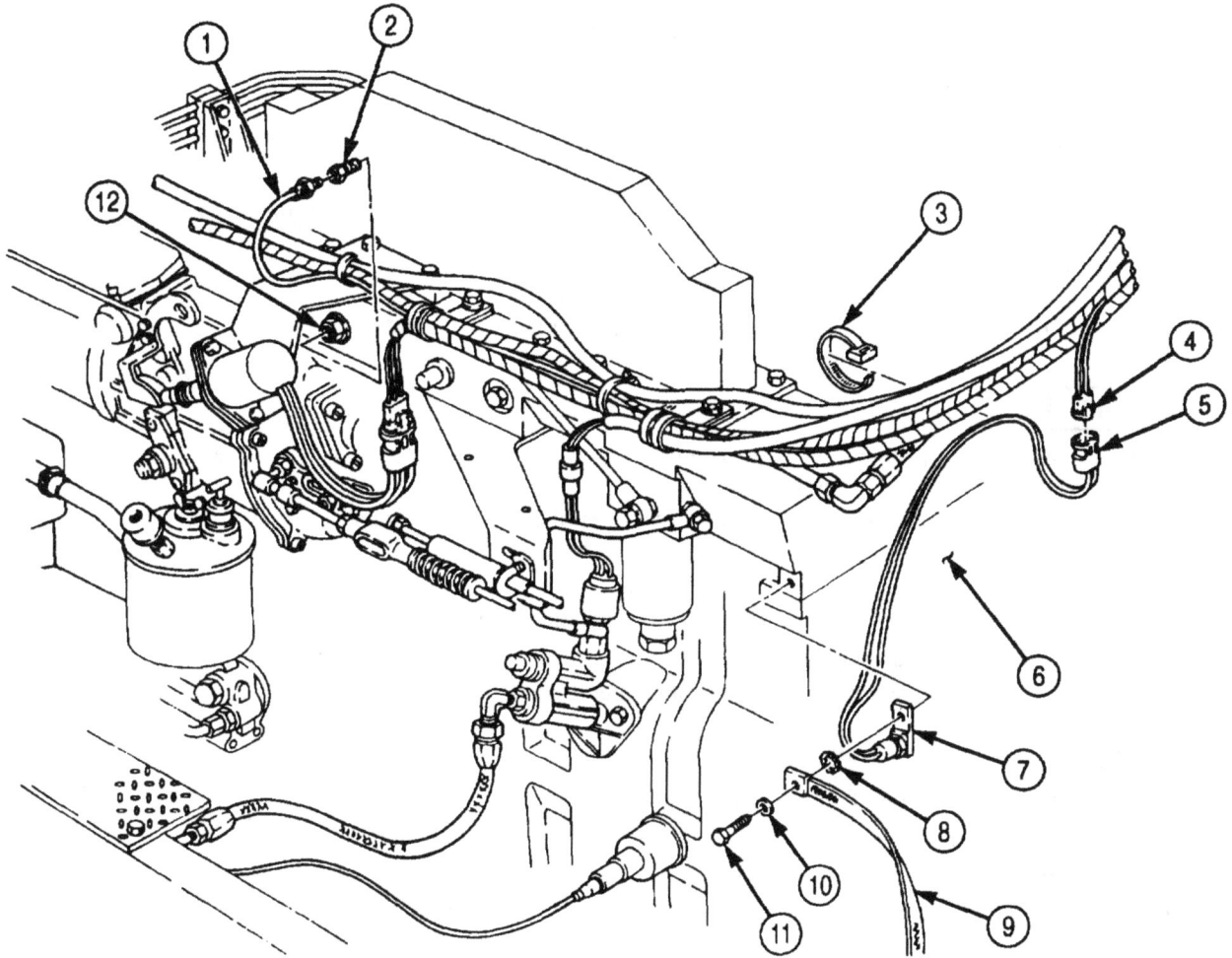

3-39. ETHER TUBING REPLACEMENT

THIS TASK COVERS:

a. Removal   b. Installation

INITIAL SETUP:

APPLICABLE MODELS
All

TOOLS
General mechanic's tool kit (Appendix E, Item 1)

MATERIALS/PARTS
Three tiedown straps (Appendix D, Item 684)

REFERENCES (TM)
TM 9-2320-272-10
TM 9-2320-272-24P

EQUIPMENT CONDITION
- Parking brake set (TM 9-2320-272-10).
- Left splash shield removed (TM 9-2320-272-10).

GENERAL SAFETY INSTRUCTIONS
Ether is flammable. Keep a fire extinguisher nearby.

WARNING

Ether is extremely flammable. Do not perform this procedure near
open flame. Injury to personnel may result.

a. Removal

1. Remove three tiedown straps (5) from atomizer ether supply tube (1) and thermal close ether supply tube (4). Discard tiedown straps (5).
2. Disconnect thermal close ether supply tube (4) from thermal close adapter (3) and ether valve adapter (9).

NOTE
Tag tubes for installation.

3. Disconnect atomizer ether supply tube (1) from thermal close adapter (2) and atomizer (6).
4. Disconnect ether cylinder relief tube (8) from ether cylinder relief inlet adapter (7).

b. Installation

1. Connect ether cylinder relief tube (8) to ether cylinder relief inlet adapter (7).
2. Connect atomizer ether supply tube (1) to thermal close adapter (2) and atomizer (6).
3. Connect thermal close ether supply tube (4) to thermal close adapter (3) and ether valve adapter (9).
4. Secure atomizer ether supply tube (1) and thermal close ether supply tube (4) with three new tiedown straps (5).

3-39. ETHER TUBING REPLACEMENT (Contd)

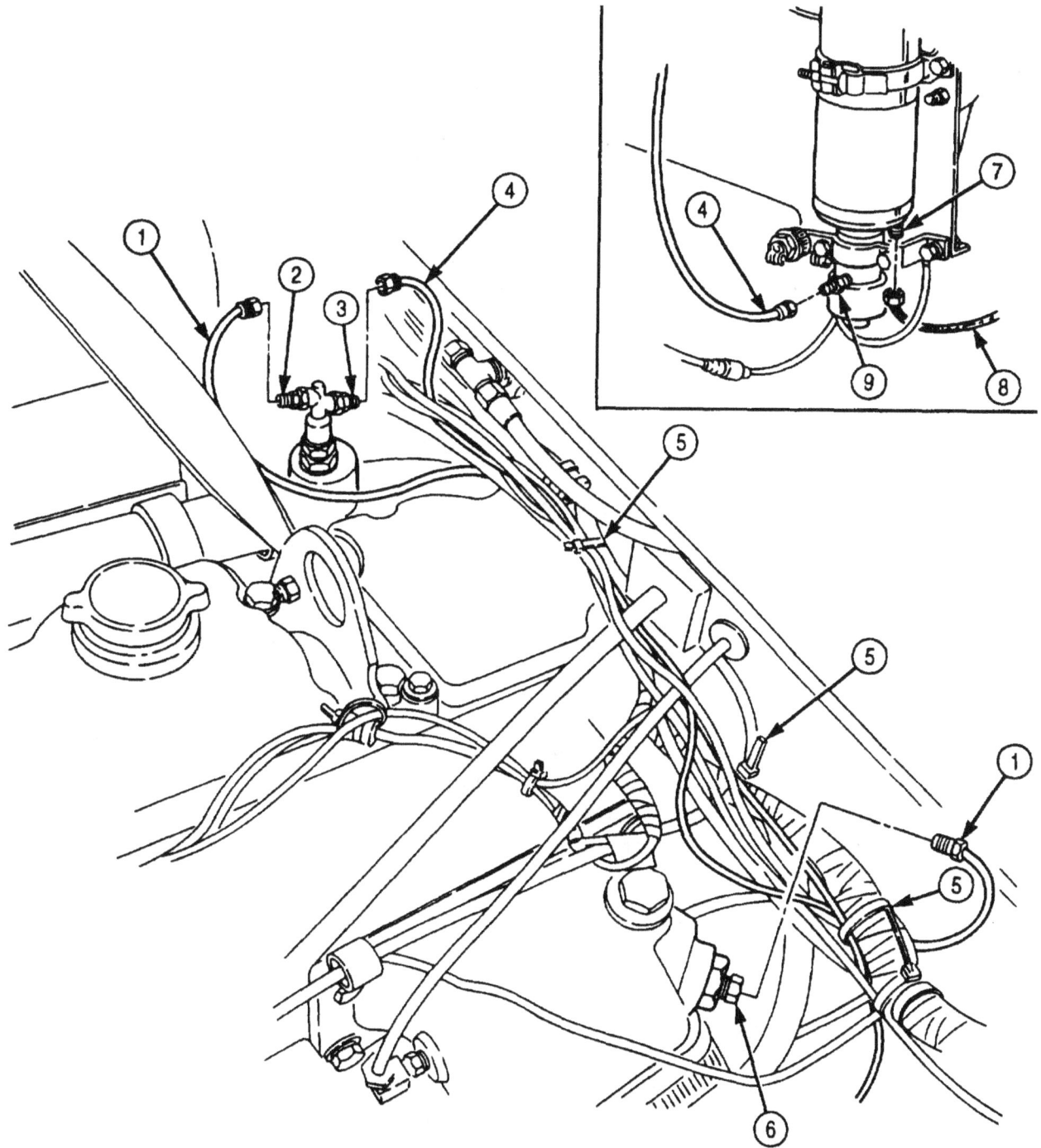

FOLLOW-ON TASK: Install left splash shield (TM 9-2320-272-10).

3-40. FUEL PRIMER PUMP (M939/A1) REPLACEMENT

THIS TASK COVERS:

a. Removal   b. Installation

INITIAL SETUP:

APPLICABLE MODELS
M939/A1

TOOLS
General mechanic's tool kit (Appendix E, Item 1)

REFERENCES (TM)
TM 9-2320-272-10
TM 9-2320-272-24P

EQUIPMENT CONDITION
- Parking brake set (TM 9-2320-272-10).
- Hood raised and secured (TM 9-2320-272-10).

a. Removal

1. Open drainvalve (7) on primer pump (8).
2. Disconnect fuel primer supply line (1) from fuel primer adapter (9).
3. Loosen plunger retainer locknut (6) from plunger retainer (5).
4. Loosen and remove plunger (4) and retainer (5) from primer pump (8).
5. Remove plunger retainer locknut (6) from primer pump (8).
6. Remove jamnut (3) and primer pump (8) from accelerator bracket (2).
7. Remove drainvalve (7) and fuel primer adapter (9) from primer pump (8).

b. Installation

1. Install drainvalve (7) and fuel primer adapter (9) on primer pump (8).
2. Install primer pump (8) on accelerator bracket (2) with jamnut (3). Tighten jamnut (3) against accelerator bracket (2).
3. Install plunger retainer locknut (6) on primer pump (8).
4. Install plunger (4) and plunger retainer (5) in primer pump (8).
5. Tighten plunger retainer (5) on pump (8).
6. Tighten locknut (6) against plunger retainer (5).
7. Connect fuel primer supply line (1) to fuel primer adapter (9).

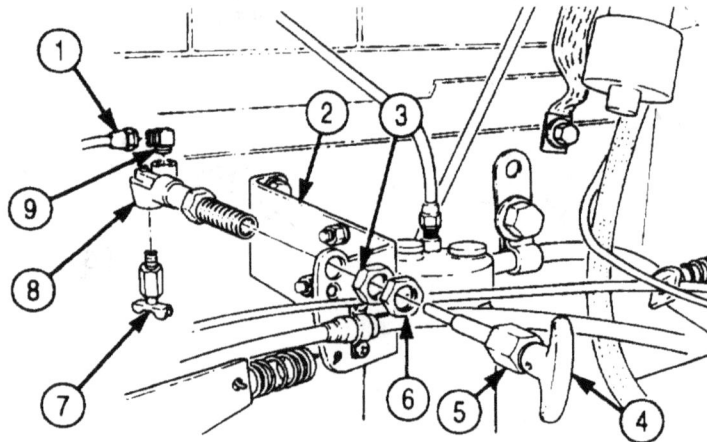

FOLLOW-ON TASK: Prime fuel system (para. 3-22 for M939A2; TM 9-2320-272-10 for M939/A1).

3-41. ACCELERATOR PEDAL, BRACKET, ROD, AND STOPSCREW REPLACEMENT

THIS TASK COVERS:
a. Removal   b. Installation

INITIAL SETUP:

APPLICABLE MODELS
All

TOOLS
General mechanic's tool kit (Appendix E, Item 1)

MATERIALS/PARTS
Two cotter pins (Appendix D, Item 66)

REFERENCES (TM)
TM 9-2320-272-10
TM 9-2320-272-24P

EQUIPMENT CONDITION
- Parking brake set (TM 9-2320-272-10).
- Left splash shield removed (TM 9-2320-272-10).

a. Removal

1. Remove cotter pin (2) and washer (3) from accelerator pedal push rod (1). Discard cotter pin (2).
2. Remove hinge pin (8) and accelerator pedal (4) from accelerator pedal bracket (6).
3. Remove two screws (5) and bracket (6) from cab floor (7).
4. Remove cotter pin (13), washer (11), and accelerator pedal rod (1) from link assembly (12). Discard cotter pin (13).
5. Loosen jamnut (10) above cab floor (7) and remove accelerator pedal stopscrew (9).
6. Remove jamnut (10) from pedal stopscrew (9).

b. Installation

1. Install jamnut (10) on pedal stopscrew (9) to limit of threads.
2. Install pedal stopscrew (9) on cab floor (7) and tighten jamnut (10) against cab floor (7).
3. Install accelerator pedal bracket (6) on cab floor (7) with two screws (5).
4. Install accelerator pedal (4) on bracket (6) with hinge pin (8).
5. Install accelerator pedal push rod (1) on accelerator pedal (4) and link assembly (12) with washers (3) and (11) and new cotter pins (13) and (2). Spread ends of cotter pins (2) and (13) after adjustment.

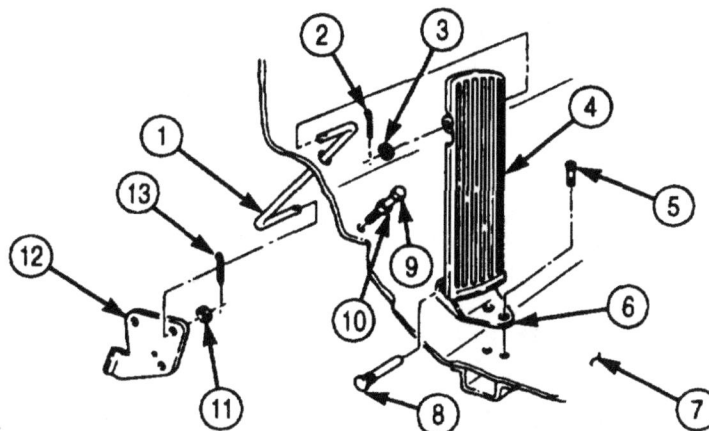

FOLLOW-ON TASKS: • Adjust accelerator linkage (para. 3-42 or 3-43).
• Install left splash shield (TM 9-2320-272-10).

3-42. ACCELERATOR LINKAGE (M939/A1) MAINTENANCE

THIS TASK COVERS:

a. Removal

b. Installation

c. Adjustment

INITIAL SETUP:

APPLICABLE MODELS
M939/A1

TOOLS
General mechanic's tool kit (Appendix E, Item 1)

MATERALS/PARTS
Locknut (Appendix D, Item 276)
Locknut (Appendix D, Item 277)
Screw (Appendix D, Item 569)
Spring pin (Appendix D, Item 674)

REFERENCES (TM)
TM 9-2320-272-10
TM 9-2320-272-24P

EQUIPMENT CONDITION
- Parking brake set (TM 9-2320-272-10).
- Hood raised and secured (TM 9-2329-272-10).
- Left splash shield removed (TM 9-2320-272-10).

---

a. Removal

1. Remove locknut (2), screw (6), and clevis rod (1) from throttle lever (7). Discard screw (6) and locknut (2).

2. Disconnect return spring (8) from accelerator rod (5) and link assembly (12).

3. While holding nut (11), remove locknut (13) and accelerator rod (5) from link assembly (12). Discard locknut (13).

NOTE

Record position of accelerator rod in ball joint for installation.

4. Loosen jamnut (9) and remove ball joint (10) from accelerator rod (5).

6. Remove spring pin (3), clevis rod (1), and spring (4) from accelerator rod (5). Discard spring pin (3).

b. Installation

1. Install spring (4) and clevis rod (1) on accelerator rod (5) and secure with new spring pin (3).

2. Install ball joint (10) on accelerator rod (5).

3. Position accelerator rod (5) and ball joint (10) on link assembly (12).

4. While holding nut (11), install and tighten new locknut (13) on ball joint (10).

NOTE

If hole in clevis and throttle lever do not align, go to adjustment, task c.

5. Connect clevis rod (1) to throttle lever (7) with new screw (6) and new locknut (2).

6. Connect return spring (8) to accelerator rod (5) and link assembly (12).

c. Adjustment

NOTE

Perform step 1 only if link assembly was removed.

1. Disconnect return spring (8) from link assembly (12) and accelerator rod (5).

2. Remove screw (6) and locknut (2). Discard locknut (2) and screw (6).

3-42. ACCELERATOR LINKAGE (M939/A1) MAINTENANCE (Contd)

NOTE

Assistant will help with steps 3 through 8.

3. Push throttle lever (7) forward to FULL THROTTLE position.

4. Pull clevis rod (1) forward as far as possible.

6. Loosen jamnut (9) and hand-turn accelerator rod (5) to shorten or lengthen as needed to align holes in clevis rod (1) and throttle lever (7).

6. Install clevis (1) on throttle lever (7) with new screw (6) and new locknut (2).

7. Tighten jamnut (9) against ball joint (10).

8. Connect return spring (8) to accelerator rod (6) and link assembly (12).

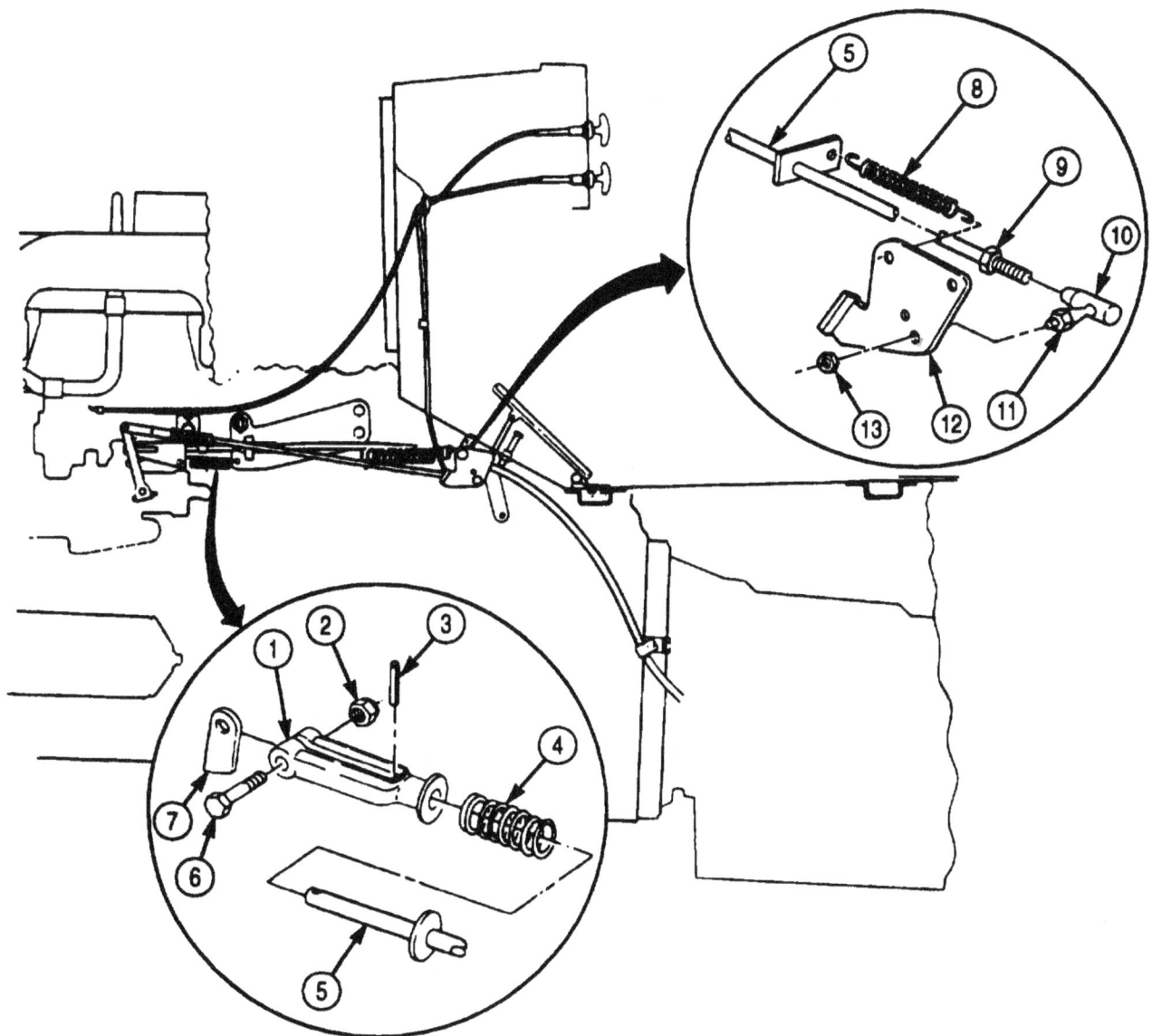

FOLLOW-ON TASK: Install left splash shield (TM 9-2320-272-10).

3-43. ACCELERATOR LINKAGE (M939A2) MAINTENANCE

THIS TASK COVERS:

a. Removal
b. Disassembly
c. Cleaning and Inspection

d. Assembly
e. Installation
f. Adjustment

INITIAL SETUP:

APPLICABLE MODELS
M939A2

TOOLS
General mechanic's tool kit (Appendix E, Item 1)

MATERIALS/PARTS
Locknut (Appendix D, Item 277)
Two cotter pins (Appendix D. Item 46)

REFERENCES (TM)
TM 9-2320-272-10
TM 9-2320-272-24P

EQUIPMENT CONDITION
- Parking brake set (TM 9-2320-272-10.
- Hood raised and secured (TM 9-2320-272-10).

a. Removal

1. Compress spring-loaded sleeve (22) and remove quick-disconnect socket (21) from ball joint (1) on control lever (2).
2. Remove throttle return spring (14) from accelerator rod (15) and throttle shaft (6).
3. Remove locknut (9) and ball joint (12) from bracket (8). Discard locknut (9).
4. Loosen screw (11) and remove connector (10) and cable (3) from bracket (8).
5. Remove two cotter pins (5), washers (7), throttle shaft (61, and bracket (8) from bracket (4). Discard cotter pins (5).

b. Disassembly

NOTE
Assistant will help with step 1.

1. Compress spring (17) and remove pin (16), spring (17), and yoke (18) from accelerator rod (15).
2. Loosen nut (19) and remove yoke (18) from rod (20).
3. Remove nut (19) from rod (20).
4. Remove rod (20) from quick-disconnect socket (211.

NOTE
Mark position of nut on accelerator rod for installation.

5. Loosen nut (13) and remove ball joint (12) from accelerator rod (15).

3-43. ACCELERATOR LINKAGE (M939A2) MAINTENANCE (Contd)

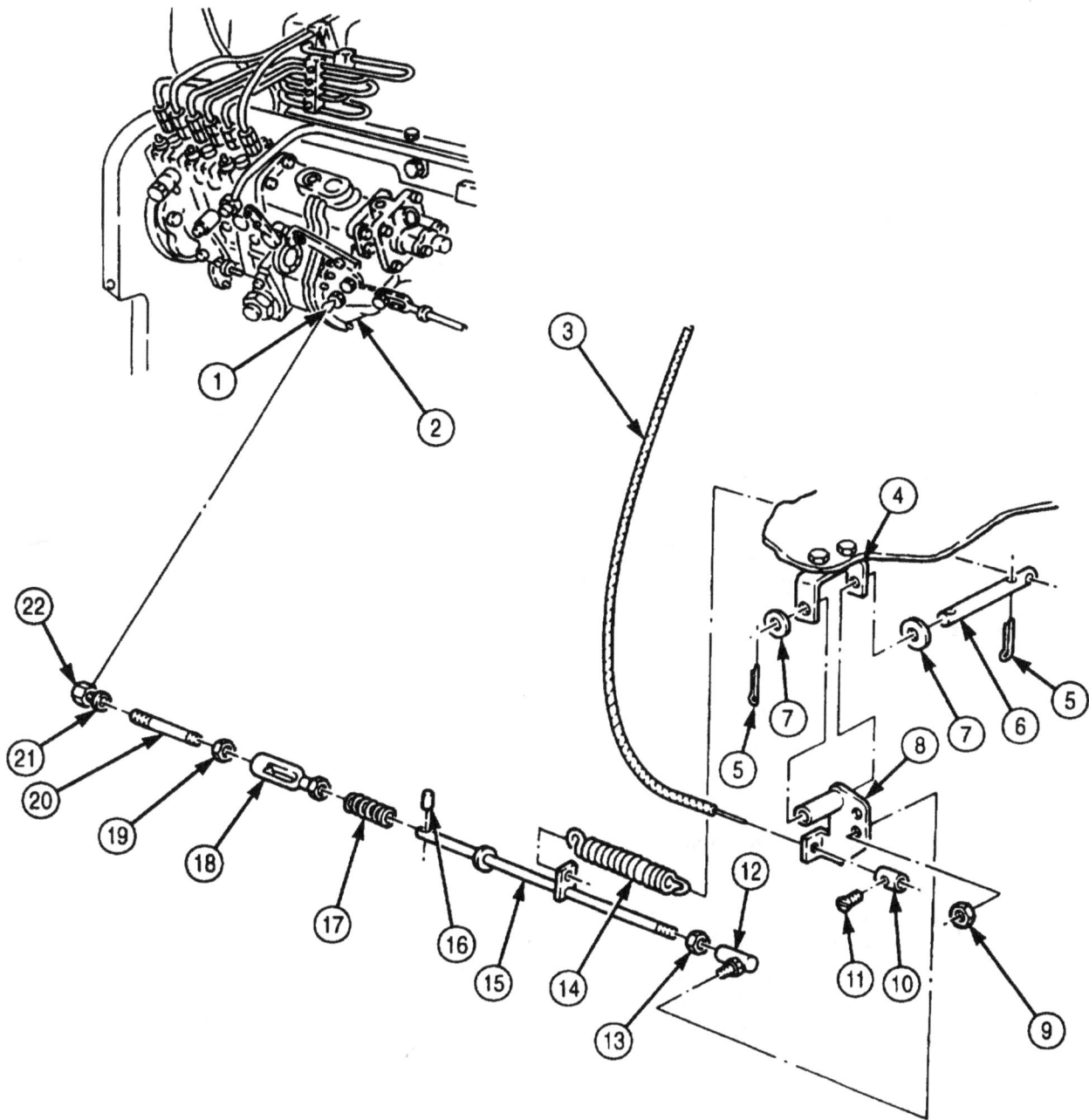

3-43. ACCELERATOR LINKAGE (M939A2) MAINTENANCE (Contd)

c. Cleaning and Inspection

1. For general cleaning instructions, refer to para. 2-14.
2. For general inspection instructions, refer to para. 2-15.
3. Replace all parts failing inspection.

d. Assembly

1. Position nut (13) to mark on accelerator rod (15).
2. Install ball joint (12) on accelerator rod (15).
3. Install rod (20) on quick-disconnect socket (21).
4. Install nut (19) on rod (20). Finger tighten nut (19).
5. Install yoke (18) on rod (20) so several threads are between yoke (18) and nut (19).
6. Install yoke (18) and spring (17) on accelerator rod (15).

NOTE
Assistant will help with step 7.

7. Compress spring (17) and install pin (16) in accelerator rod (15).

e. Installation

1. Install bracket (8) on bracket (4) with throttle shaft (6), two washers (7), and new cotter pins (5).
2. Install connector (10) and cable (3) on bracket (8) and tighten screw (11).
3. Install ball joint (12) on bracket (8) with new locknut (9).
4. Install throttle return spring (14) on accelerator rod (15) and throttle shaft (6).

f. Adjustment

NOTE

1. Position control lever (2) to idle position.
2. Turn rod (20) until control lever (2) moves forward 0.125 in. (3.2 mm).
3. Tighten nut (19) securely.
4. Depress spring-loaded sleeve (22) and install quick-disconnect socket (21) on ball joint (1).
5. Check for engine idle at 550-650 rpm.

NOTE
Perform steps 6, 7, and 8 as necessary to obtain correct idle speed.

6. Compress spring-loaded sleeve (22) and disconnect socket (21) from ball joint (1).
7. Loosen jamnut (19) and lengthen or shorten rod (20) by turning rod (20) into or out of yoke (18) to obtain correct movement for idle speed.
8. Depress spring-loaded sleeve (22) and install quick-disconnect socket (21) on ball joint (1).
9. Tighten jamnut (19) when correct idle speed is obtained.

3-43. ACCELERATOR LINKAGE (M939A2) MAINTENANCE (Contd)

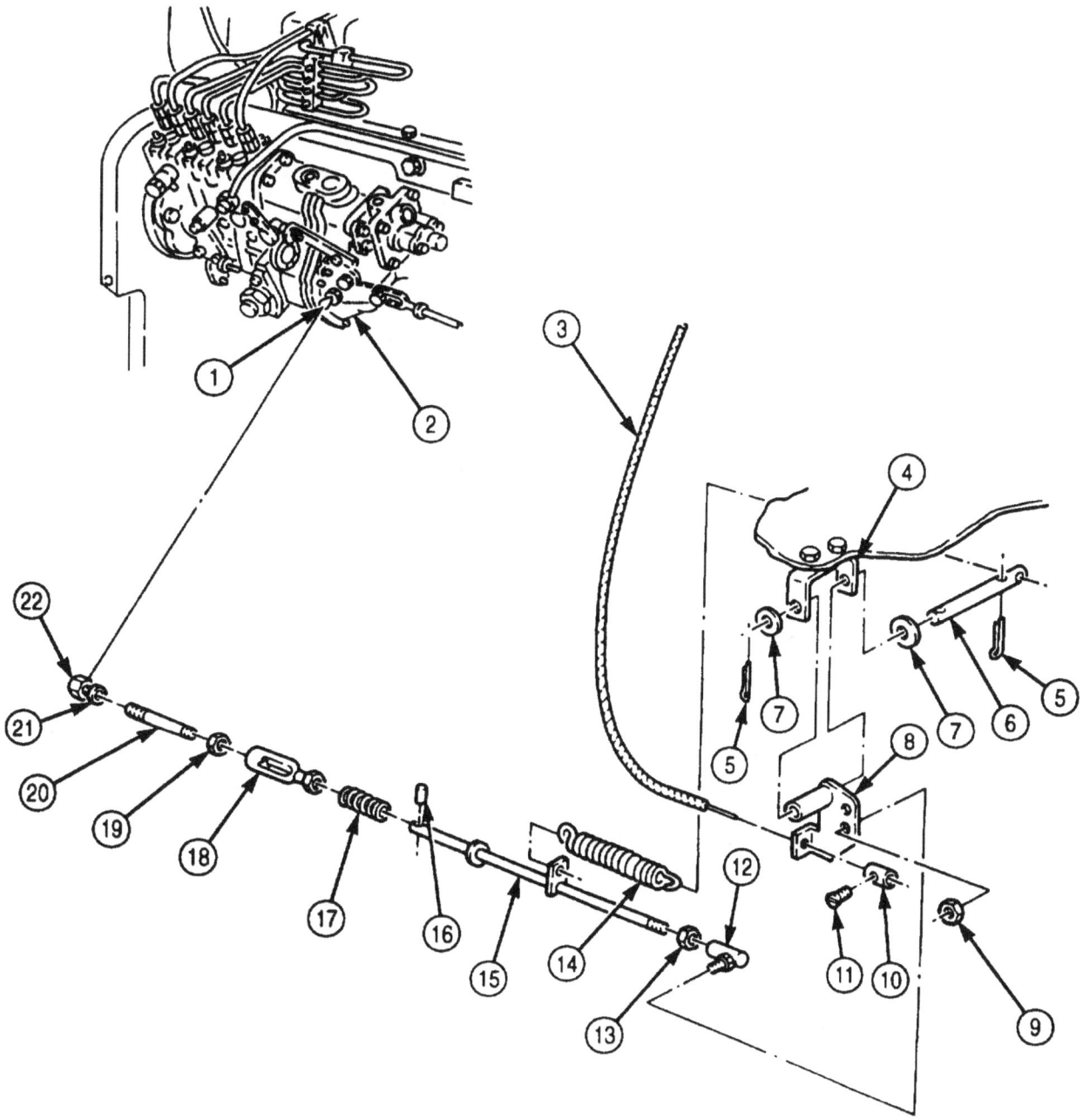

FOLLOW-ON TASK: Start engine (TM 9-2320-272-10) and check engine idle.

3-44. EMERGENCY STOP CONTROL CABLE MAINTENANCE

THIS TASK COVERS:

a. Removal
b. Inspection

c. Installation

INITIAL SETUP:

APPLICABLE MODELS

All

TOOLS

General mechanic's tool kit (Appendix E, Item 1)

MATERIALS/PARTS

Locknut (M939/A1) (Appendix D, Item 313)
Lockwasher (M939/A1) (Appendix D, Item 381)
Lockwasher (Appendix D, Item 354)
Lockpin (M939A2) (Appendix D, Item 271)
GAA grease (Appendix C, Item 28)

REFERENCES (TM)

TM 9-2320-272-10
TM 9-2320-272-24P

EQUIPMENT CONDITION

- Parking brake set (TM 9-2320-272-10).
- Hood raised and secured (TM 9-2320-272-10).
- Battery ground cables disconnected (para. 3-126).
- Left splash shield removed (TM 9-2320-272-10).

a. Removal

NOTE
Steps 1 through 4 are for M939A2 series vehicles.

1. Remove lockpin (10), washer (11), and cable pivot (4) from shutoff valve lever (1). Discard lockpin (10).
2. Remove screw (2), connector (3), and cable pivot (4) from cable (5).
3. Remove screw (9), clamp (7), and cable conduit (6) from fuel bracket (8).
4. Remove clamp (7) from conduit (6).

NOTE
Steps 5 through 9 are for M939/A1 series vehicles.

5. Remove connector screw (18) and connector (19) from shutoff valve control lever (28) and cable (5).
6. Remove connector (19) from stop control cable (5).
7. Remove locknut (24), washer (25), and screw (21) from clamp (20) on conduit (6) and conduit clamp bracket (26). Discard locknut (24).
8. Remove screw (23), lockwasher (22), and clamp bracket (26) from engine (27). Discard lockwasher (22).
9. Remove clamp (20) from conduit (6).

NOTE
Steps 10 through 13 applies to all vehicles.

10. From behind instrument panel (15), remove nut (12) and lockwasher (13) on emergency stop control (14). Discard lockwasher (13).
11. Pull emergency stop control (14), conduit (6), and cable (5) through firewall (17), grommet (16), and instrument panel (151.
12. Remove grommet (16) from tirewall (17).

NOTE
Perform step 13 only if replacing stop control cable.

13. Remove stop control cable (5) and handle of stop control (14) from conduit (6).

3-44. EMERGENCY STOP CONTROL CABLE MAINTENANCE (Contd)

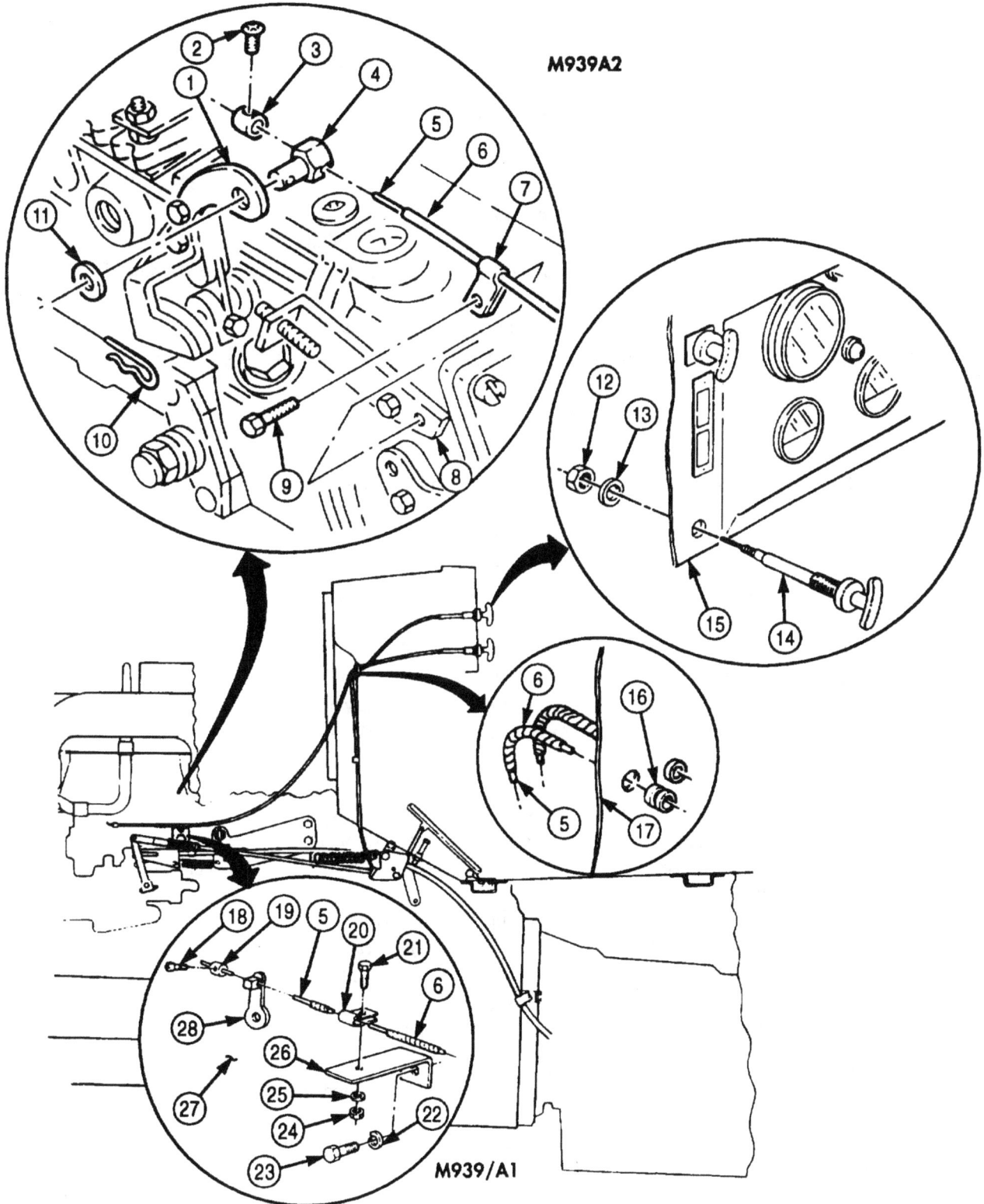

M939A2

M939/A1

3-44. EMERGENCY STOP CONTROL CABLE MAINTENANCE (Contd)

b. Inspection

1. Inspect stop control cable (8) for bends, breaks, or damage. Replace cable (8) if broken, bent, or damaged.
2. Inspect conduit (5) for breaks on kinks that could cause cable (8) to bind. Replace conduit (5) if broken or kinked.

c. Installation

NOTE

Perform step 1 only if replacing cable in conduit.

1. Coat stop control cable (8) with light film of GM grease and thread through cable conduit (5)

NOTE

Steps 2 through 5 applies to all vehicles.

2. Thread stop control conduit (5) and cable (8) through instrument panel (4) and seat stop control (3) in panel (4).
3. Place new lockwasher (2) and nut (1) over conduit (5) and install on stop control (3). Tighten nut (1) on stop control (3).
4. Install grommet (6) in firewall (7).
5. Thread cable (8) and conduit (5) through grommet (6) and firewall (7).

NOTE

Steps 6 through 11 apply to M939/A1 series vehicles.

6. Thread cable (8) through hole in shutoff valve control lever (18).
7. Push shutoff lever (18) all the way forward and slide connector (10) over cable (8) and against shutoff lever (18).
8. Install screw (9) in connector (10) and tighten screw (9) against cable (8).
9. Bend end of cable (8) projecting beyond connector (10) upward at a 90° angle.
10. Install clamp bracket (17) on engine (19) with screw (14) and new lockwasher (13).
11. Install stop control clamp (11) on conduit (5) and clamp bracket (17) with screw (12), washer (16), and new locknut (15).

NOTE

Steps 12 through 16 apply to M939A2 series vehicles.

12. Position clamp (24) over cable conduit (5) and install clamp (24) on fuel bracket (25) with screw (26).
13. Position cable pivot (23) on cable (8), put pivot (23) into shutoff lever (20), and secure with washer (28) and new lockpin (27).
14. Place connector (22) over cable (8) and snugged up to cable pivot (23).
15. Install screw (21) in connector (22) to secure connector (22) on cable (8).
16. Bend end of cable (8) projecting beyond connector (22) upward at 90° angle.

NOTE

Step 17 applies to all vehicles.

17. Start engine (TM 9-2320-272-10) and check for correct emergency stop operation.
18. For M939A2 series vehicles, loosen screw (21) in connector (22) and adjust as needed to achieve engine cutoff.
19. For M939/A1 series vehicles, loosen screw (9) in connector (10) and adjust as needed to achieve engine cutoff.

3-44. EMERGENCY STOP CONTROL CABLE MAINTENANCE (Contd)

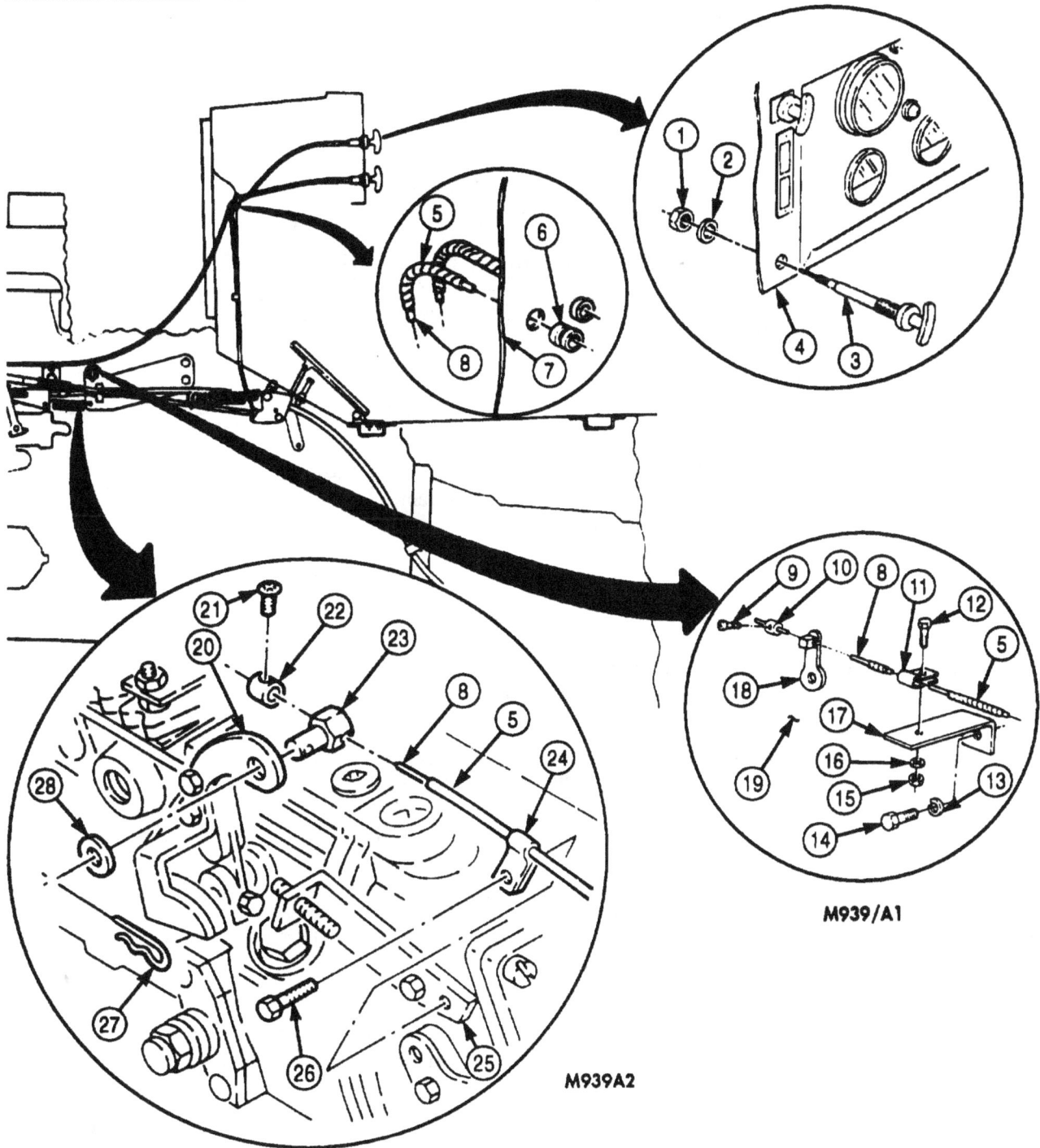

M939/A1

M939A2

FOLLOW-ON TASKS: • Connect battery ground cables (para. 3-126).
  • Install left splash shield (TM 9-2320-272-10).
  • Start engine (TM 9-2320-272-10) and test emergency stop control for proper operation.

3-45. THROTTLE CONTROL CABLE MAINTENANCE

THIS TASK COVERS:

a. Removal   c. Installation

b. Inspection

INITIAL SETUP:

APPLICABLE MODELS

All

TOOLS

General mechanic's tool kit (Appendix E, Item 1)

MATERIALS/PARTS

Lockwasher (Appendix D, Item 372)
Lockwasher (Appendix D, Item 381)
Locknut (Appendix D, Item 277)
GM grease (Appendix C, Item 28)

REFERENCES (TM)

TM 9-2320-272-10
TM 9-2320-272-24P

EQUIPMENT CONDITION
- Parking brake set (TM 9-2320-272-10).
- Hood raised and secured (TM 9-2320-272-10).
- Left splash shield removed (TM 9-2320-272-10).

NOTE

Throttle control cable maintenance for all vehicles is basically the same. This procedure covers M939/A1 series vehicles.

a. Removal

NOTE

Throttle cable end may have to be straightened before performing step 1.

1. Remove screw (18) and connector (17) from throttle cable (19).

2. Remove throttle cable (191 from throttle rod link (16).

NOTE

Perform step 3 if unmodified throttle routing (throttle cable on bracket). Perform step 4 if throttle cable clamp is mounted on firewall.

3. Remove locknut (12), washer (11), conduit clamp (14), and screw (15) from bracket (9) on firewall (10). Discard locknut (12).

4. Remove screw (20), conduit (13) with clamp (21), wire harness (22) with clamp (23), washer (24), ground strap (25), and lockwasher (26) from firewall (10). Discard lockwasher (26).

5. Pull conduit (13) and throttle cable (19) through grommet (1) in firewall (10) and into vehicle cab.

6. Remove grommet (1) from firewall (10).

7. Remove four screws (4) from throttle control plate (7) and instrument panel (2).

8. Remove nut (3), lockwasher (8), and plate (7) from throttle control (6). Discard lockwasher (8).

9. Remove throttle control (6), plate (7), and conduit (13) from instrument panel (2).

NOTE

Perform step 10 if throttle control cable is to be replaced.

10. Remove throttle control handle (5) and cable (19) from conduit (13).

3-45. THROTTLE CONTROL CABLE MAINTENANCE (Contd)

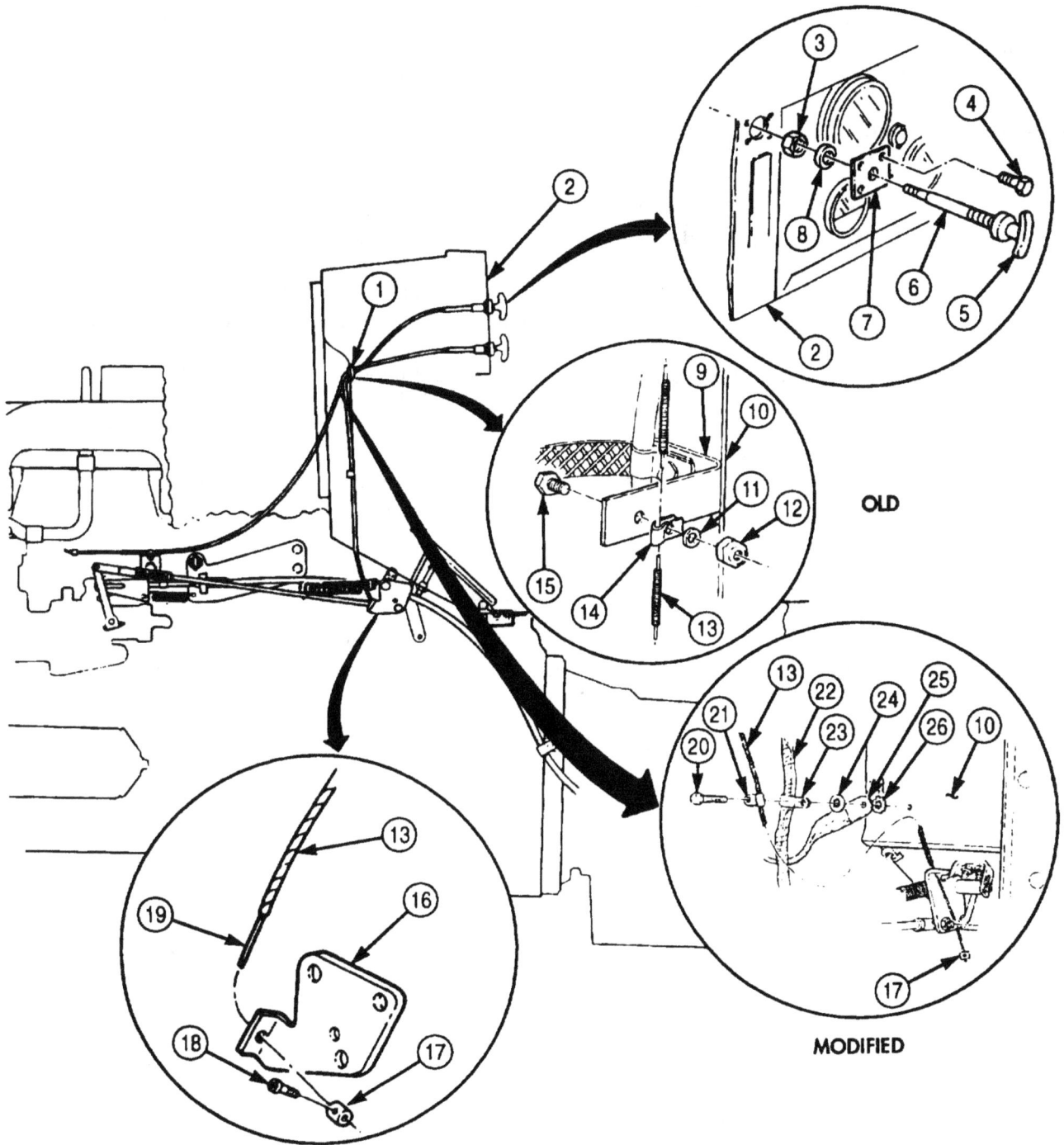

OLD

MODIFIED

3-45. THROTTLE CONTROL CABLE MAINTENANCE (Contd)

b. Inspection

1. Inspect throttle cable (14) for bends, breaks, and damage. Replace if bent, broken, or damaged,
2. Inspect conduit (10) for breaks and kinks that could bind on cable (14). Replace conduit (10) if broken or kinked.

c. Installation

NOTE

- If throttle control cable is being replaced, installed, or if throttle sticks in the DOWN position, follow this installation procedure.
- Perform step 1 if throttle cable is new or has been removed from conduit.

1. Coat throttle cable (14) with light film of GM grease and thread through cable conduit (10).
2. Install plate (7) over conduit (10) and secure to throttle control (6) with new lockwasher (8) and nut (3).
3. Coat control cable (14) with light coat of GAA grease and feed throttle control cable (14) and conduit (10) through instrument panel (2) and secure plate (7) to instrument panel (2) with four screws (4).
4. Install grommet (1) in firewall (22).
5. Thread control cable (14) and conduit (10) through grommet (1) and into engine side of firewall (22).
6. Position conduit clamp (16) over conduit (10) and feed control cable (14) through hole in throttle rod link (11).

NOTE

Throttle rod link and handle on end of control cable must be in closed throttle position before performing step 7.

7. Ensuring handle (5) is fully seated on control (6), install connector (12) over control cable (14), slide connector (12) up cable (14) into contact with link (11), and install screw (13). Bend projecting end of cable (14) up at connector (12).
8. Install conduit clamp (16) and wiring harness (17) with clamp (18), washer (19), strap (20). and new lockwasher (21) on firewall (22) with screw (15).

3-45. THROTTLE CONTROL CABLE MAINTENANCE (Contd)

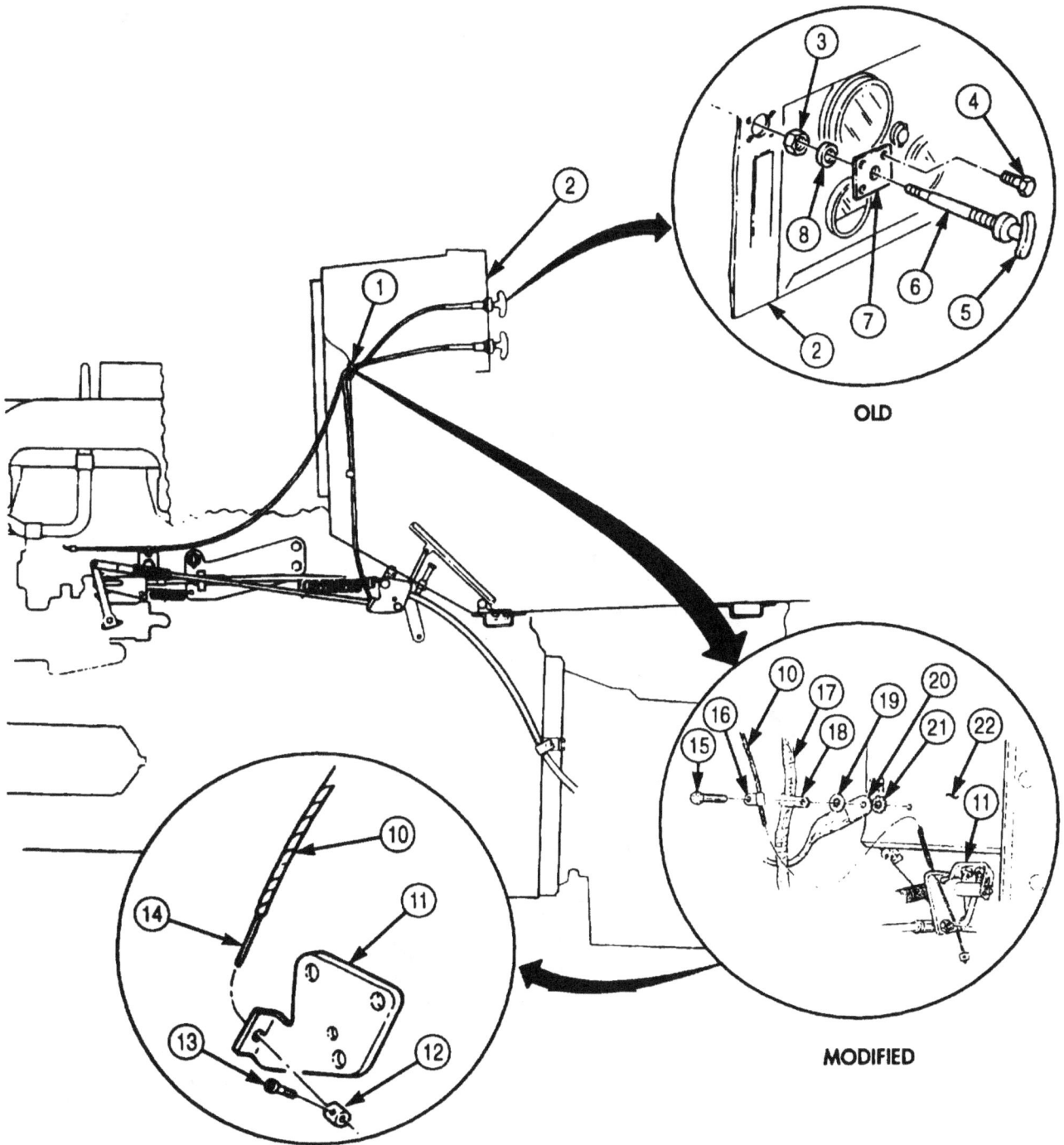

OLD

MODIFIED

FOLLOW-ON TASKS: • Install left splash shield (TM 9-2320-272-10).
• Start engine (TM 9-2320-272-10) and test throttle control for proper operation.

3-46. THROTTLE CONTROL SOLENOID (M939A2) REPLACEMENT

THIS TASK COVERS:

a. Removal   b. Installation

INITIAL SETUP:

APPLICABLE MODELS   **REFERENCES (TM)**
M939A2   TM 9-2320-272-10
                                              TM 9-2320-272-24P
TOOLS
General mechanic's tool kit (Appendix E, Item 1)   **EQUIPMENT CONDITION**
                                              • Parking brake set (TM 9-2320-272-10).
                                              • Hood raised and secured (TM 9-2320-272-10).
                                              • Battery ground cables disconnected (para. 3-126).

a. Removal

1. Disconnect plug (6) from wiring harness (7).
2. Remove two screws (4) and throttle control solenoid (5) from bracket (3).
3. Remove boot (8) from throttle control solenoid (5).
4. Remove two screws (2) and bracket (3) from engine (1).

b. Installation

1. Install bracket (3) on engine (1) with two screws (2).
2. Install boot (8) on throttle control solenoid (5).
3. Install throttle control solenoid (5) on bracket (3) with two screws (4).
4. Connect plug (6) to wiring harness (7).

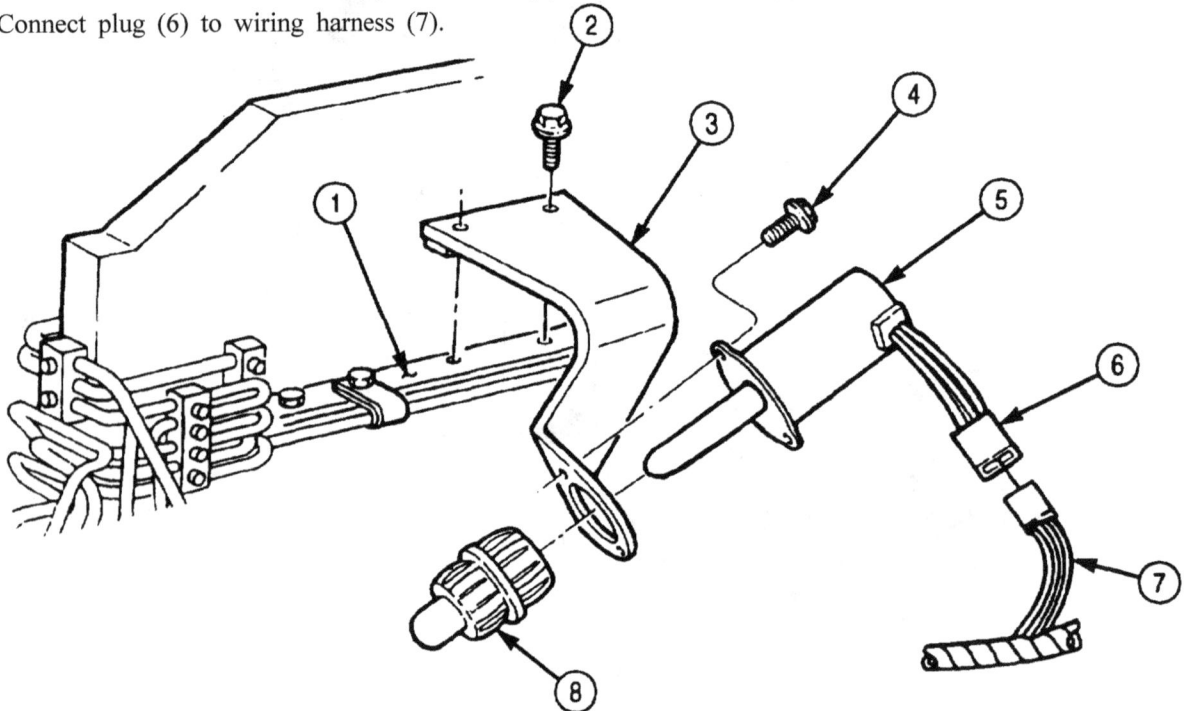

FOLLOW-ON TASK: Connect battery ground cables (para. 3-126).

Section IV. EXHAUST SYSTEM MAINTENANCE

3-47. EXHAUST SYSTEM MAINTENANCE INDEX

3-48. EXHAUST STACK REPLACEMENT

THIS TASK COVERS:

a. Removal   b. Installation

INITIAL SETUP:

APPLICABLE MODELS   REFERENCES (TM)
All   TM 9-2320-272-10
                                                            TM 9-2320-272-24P
TOOLS
General mechanic's tool kit (Appendix E, Item 1)   EQUIPMENT CONDITION
                                                            Parking brake set (TM 9-2320-272-10).
MATERIALS/PARTS
Gasket (Appendix D, Item 170)   GENERAL SAFETY INSTRUCTIONS
Locknut (Appendix D, Item 317)   Do not touch hot exhaust system components with
GM grease (Appendix C, Item 28)   bare hands.

a. Removal

WARNING

Do not touch hot exhaust system components with bare hands.
Injury to personnel may result.

1. Remove locknut (6) and screw (3) from clamp (2) at right rear of cab (7). Discard locknut (6).

NOTE
Assistant will help with step 2.

2. Remove clamp (2) from exhaust stack (1) and muffler (4).

3. Remove exhaust stack (1) and gasket (5) from muffler (4). Discard gasket (5) and clean gasket remains from mating surfaces.

b. Installation

1. Apply small amount of GM grease on muffler mating surface, and position new gasket (5) and exhaust stack (1) over muffler (4) so opening of exhaust stack (1) is directly away from cab (7).

NOTE
Assistant will help with step 2.

2. Secure exhaust stack (1) to muffler (4) with clamp (2), screw (3), and new locknut (6).

3-48. EXHAUST STACK REPLACEMENT (Contd)

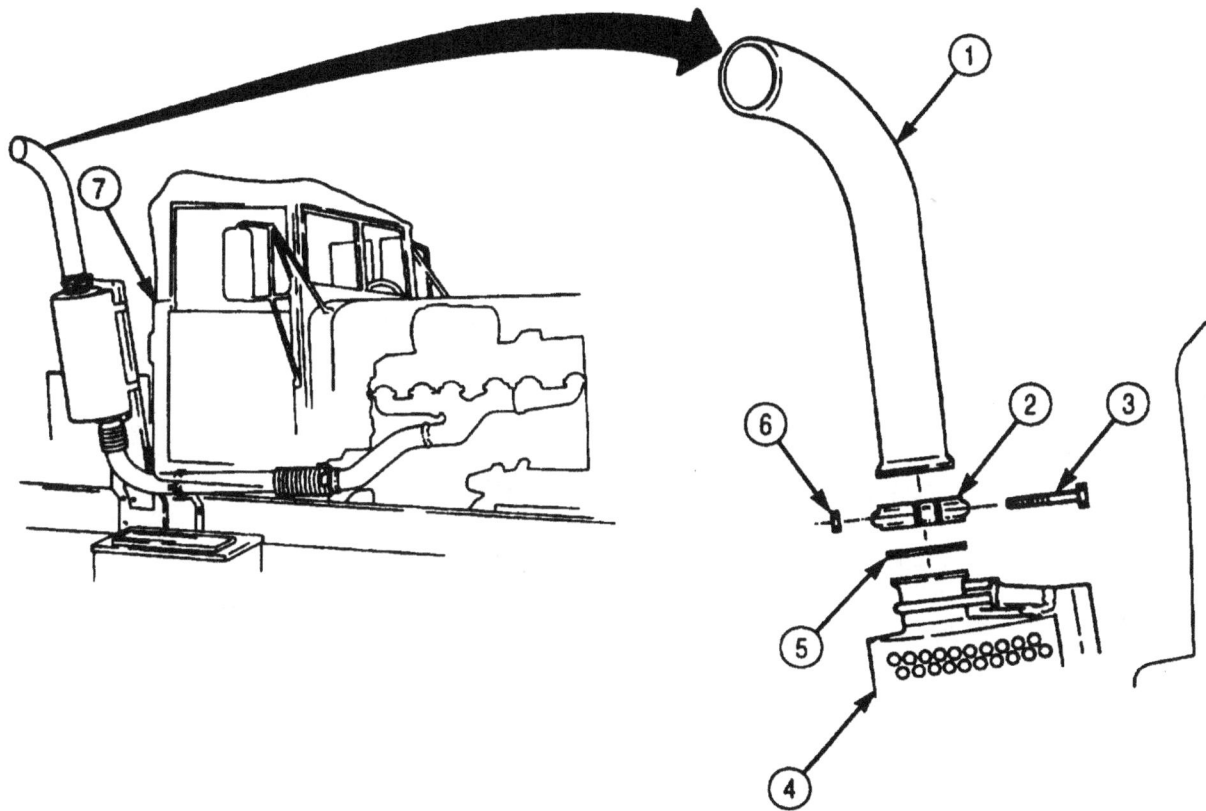

FOLLOW-ON TASK: Start engine (TM 9-2320-272-10) and check for exhaust leaks.

3-49. REAR EXHAUST PIPE, SUPPORT BRACKET, AND CAB HEAT SHIELD MAINTENANCE

THIS TASK COVERS:

a. Removal
b. Inspection

c. Installation

INITIAL SETUP:

APPLICABLE MODELS
All

TOOLS
General mechanic's tool kit (Appendix E, Item 1)

MATERIALS/PARTS
Two gaskets (Appendix D, Item 170)
Two locknuts (Appendix D, Item 317)
Two locknuts (Appendix D, Item 291)
Two locknuts (Appendix D, Item 294)
GAA grease (Appendix C, Item 28)

REFERENCES (TM)
TM 9-2320-272-10
TM 9-2320-272-24P

EQUIPMENT CONDITION
Parking brake set (TM 9-2320-272-10).

GENERAL SAFETY INSTRUCTIONS
Do not touch hot exhaust system components with bare hands.

---

a. Removal

> **WARNING**
>
> Do not touch hot exhaust system components with bare hands.
> Injury to personnel may result.

1. Remove two screws (14) and heat shield (13) from cab heat shield brackets (7).

2. Remove locknut (3), screw (20), exhaust pipe coupling clamp (4), and gasket (2) from muffler flange (1). Discard locknut (3) and gasket (2).

3. Remove locknut (12), screw (8), exhaust pipe coupling clamp (9), and gasket (10) from front exhaust pipe flange (11). Discard locknut (12) and gasket (10).

4. Remove two locknuts (15), screws (5), top support bracket (6), and rear exhaust pipe (19) from bottom support bracket (6), muffler flange (11), and front exhaust pipe flange (11). Discard locknuts (15).

5. Remove two locknuts (17), screws (16), and bottom support bracket (6) from crossmember (18). Discard locknuts (17).

6. Remove gasket remains from rear exhaust pipe (19) flange, front exhaust pipe flange (11), and muffler flange (1).

b. Inspection

Inspect rear exhaust pipe (19), front exhaust pipe flange (11), and muffler flange (1) for cracks. Replace if cracked.

c. Installation

1. Install bottom support bracket (6) on crossmember (18) with two screws (16) and new locknuts (17).

2. Apply small amount of GAA grease on new gaskets (2) and (10), and position gaskets (2) and (10) on muffler flange (1) and front exhaust pipe flange (11).

3. Install rear exhaust pipe (19) on bottom support bracket (6) with top support bracket (6), two screws (5), and new locknuts (15).

3-49. REAR EXHAUST PIPE, SUPPORT BRACKET, AND CAB HEAT SHIELD MAINTENANCE (Contd)

4. Install rear exhaust pipe (19) on muffler flange (1) and front exhaust pipe flange (11) with exhaust pipe coupling clamps (4) and (9), screws (20) and (8), and new locknuts (3) and (12).

5. Install heat shield (13) on two cab heat shield brackets (7) with screws (14).

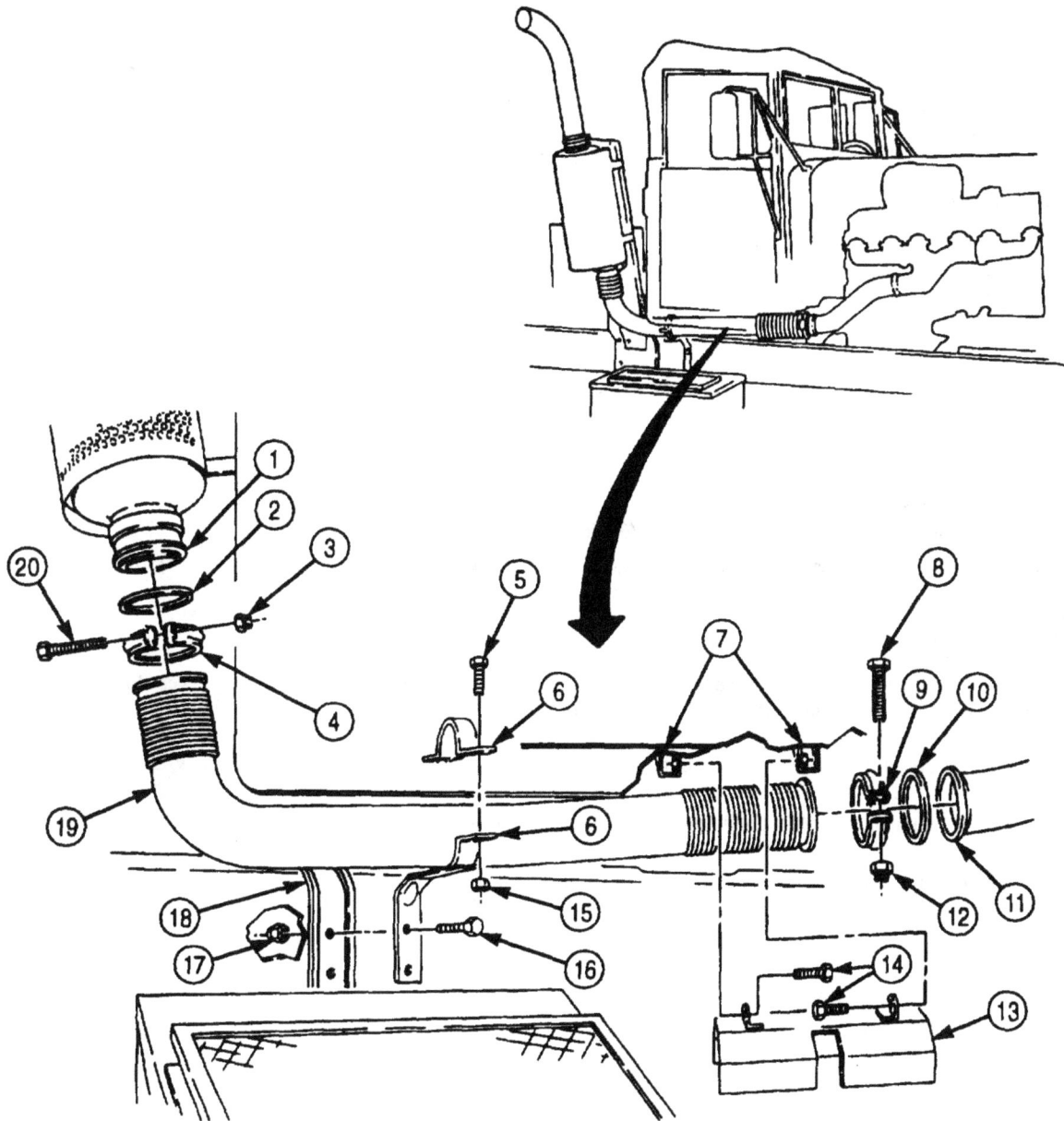

FOLLOW-ON TASK: Start engine (TM 9-2320-272-10) and check for exhaust leaks.

3-50. FRONT EXHAUST PIPE REPLACEMENT

THIS TASK COVERS:

a. Removal   b. Installation

INITIAL SETUP:

APPLICABLE/MODELS
All

TOOLS
General mechanic's tool kit (Appendix E, Item 1)

MATERIALS/PARTS
Locknut (Appendix D, Item 317)
Gasket (Appendix D, Item 170)
Gasket (Appendix D, Item 175)
GAA grease (Appendix C, Item 28)

REFERENCES (TM)
TM 9-2320-272-10
TM 9-2320-272-24P

EQUIPMENT CONDITION
• Parking brake set (TM 9-2320-272-10).
• Hood raised and secured (TM 9-2320-272-10).
• Right splash shield removed (TM 9-2320-272-10).

GENERAL SAFETY INSTRUCTIONS
Do not touch hot exhaust system components with bare hands.

---

a. Removal

> **WARNING**
>
> Do not touch hot exhaust system components with bare hands.
> Injury to personnel may result.

1. Remove two screws (12) and heat shield (13) from cab heat shield brackets (11).

2. Remove locknut (4) and manifold coupling clamp (1) from front exhaust pipe (5) and exhaust manifold (2). Retain locknut (4).

3. Remove front exhaust pipe (5) and gasket (3) from exhaust manifold (2). Discard gasket (3).

4. Remove locknut (8), screw (10), coupling clamp (7), front exhaust pipe (5), and gasket (6) from rear flex pipe (9). Discard locknut (8) and gasket (6).

b. Installation

1. Apply small amount of GAA grease on flange of front exhaust pipe (5), position new gasket (6) on front exhaust pipe (5), and install pipe (5) on flex pipe (9) with coupling clamp (7), screw (10), and new locknut (8).

2. Apply small amount of GM grease on flange of exhaust manifold (2), and position new gasket (3) and front exhaust pipe (5) on exhaust manifold (2).

3. Install front exhaust pipe (5) on flange of exhaust manifold (2) with manifold coupling clamp (1) and locknut (4).

4. Install heat shield (13) on two cab heat shield brackets (11) with screws (12).

3-50. FRONT EXHAUST PIPE REPLACEMENT (Contd)

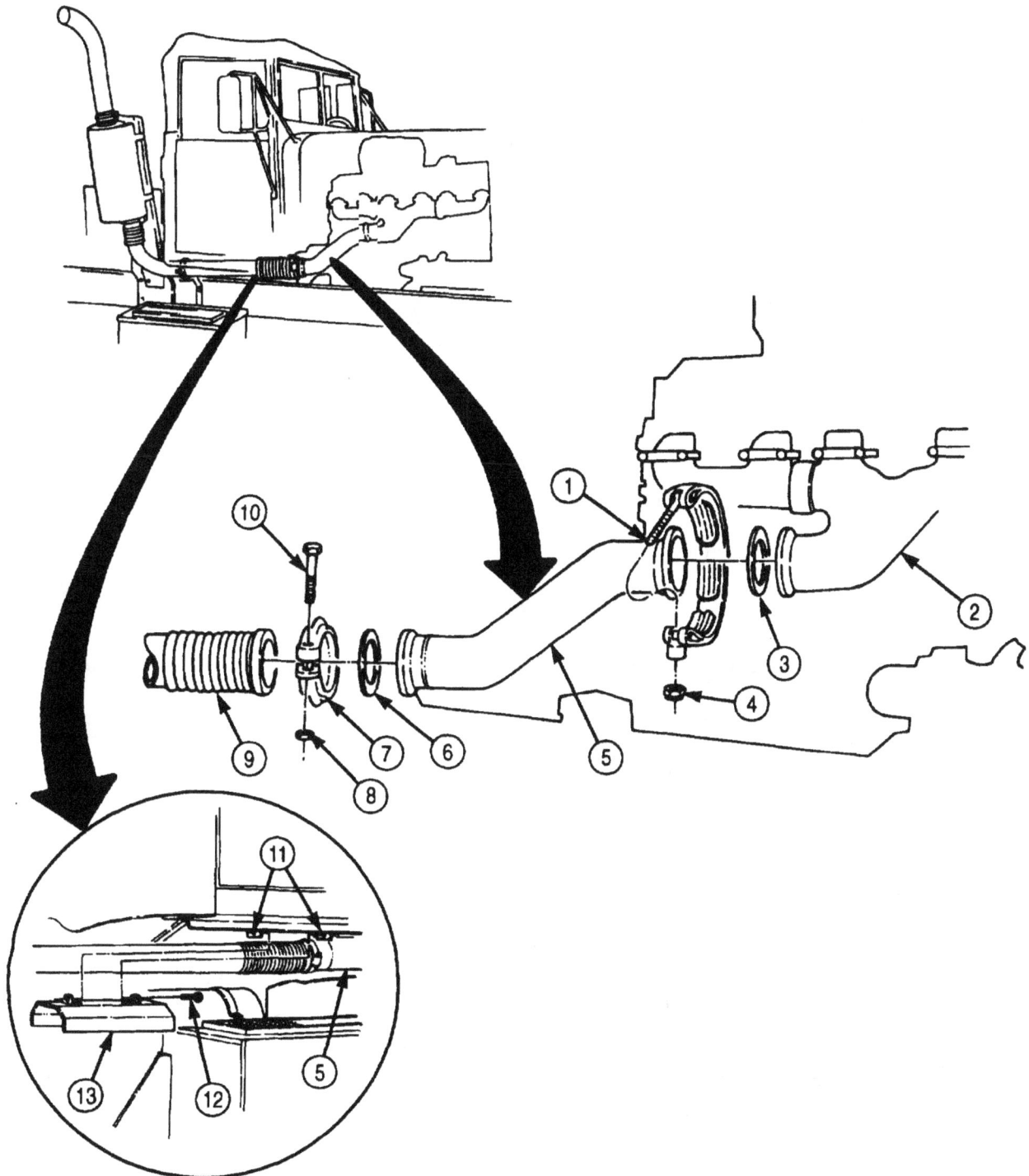

FOLLOW-ON TASKS: • Start engine (TM 9-2320-272-10) and check for exhaust leaks.
• Install right splash shield (TM 9-2320-272-10).

3-51. MUFFLER AND SHIELD MAINTENANCE

THIS TASK COVERS:

a. Removal   c. Installation
b. Inspection

INITIAL SETUP:

APPLICABLE MODELS
All

TOOLS
General mechanic's tool kit (Appendix E, Item 1)

MATERIALS/PARTS
Gasket (Appendix D, Item 170)
Four locknuts (Appendix D, Item 291)
Locknut (Appendix D, Item 317)
GAA Grease (Appendix C, Item 28)

REFERENCES (TM)
TM 9-2320-272-10
TM 9-2320-272-24P

EQUIPMENT CONDITION
• Parking brake set (TM 9-2320-272-10).
• Exhaust stack removed (para. 3-48).

GENERAL SAFETY INSTRUCTIONS
Do not touch hot exhaust system components with
bare hands.

a. Removal

WARNING

Do not touch hot exhaust system components with bare hands.
Injury to personnel may result.

1. Remove four locknuts (8), screws (14), and muffler shield (1) from muffler shield support (5).
   Discard locknuts (8).

NOTE
Assistant will help with steps 2 and 3.

2. Remove four nuts (7), lockwashers (6), two U-bolts (2), U-clamps (4), and muffler (3) from muffler
   shield support (5). Retain lockwashers (6).

3. Remove locknut (13), screw (9), exhaust pipe coupling clamp (12), muffler (3), and exhaust pipe
   gasket (11) from exhaust pipe (10). Discard locknut (13) and gasket (11).

b. Inspection

Inspect muffler (3) for cracks. Replace if cracked.

c. Installation

NOTE
Assistant will help with steps 1 and 2.

1. Apply small amount of GM grease on flange of exhaust pipe (10), position muffler (3) with narrow
   side facing away from cab (15), and install new gasket (11) and muffler (3) on exhaust pipe (10) with
   coupling clamp (12), screw (9), and new locknut (13).

2. Install muffler (3) on muffler shield support (5) with two U-clamps (4), U-bolts (2), four
   lockwashers (6), and nuts (7).

3. Install muffler shield (1) on muffler shield support (5) with four screws (14) and new locknuts (8).

3-51. MUFFLER AND SHIELD MAINTENANCE (Contd)

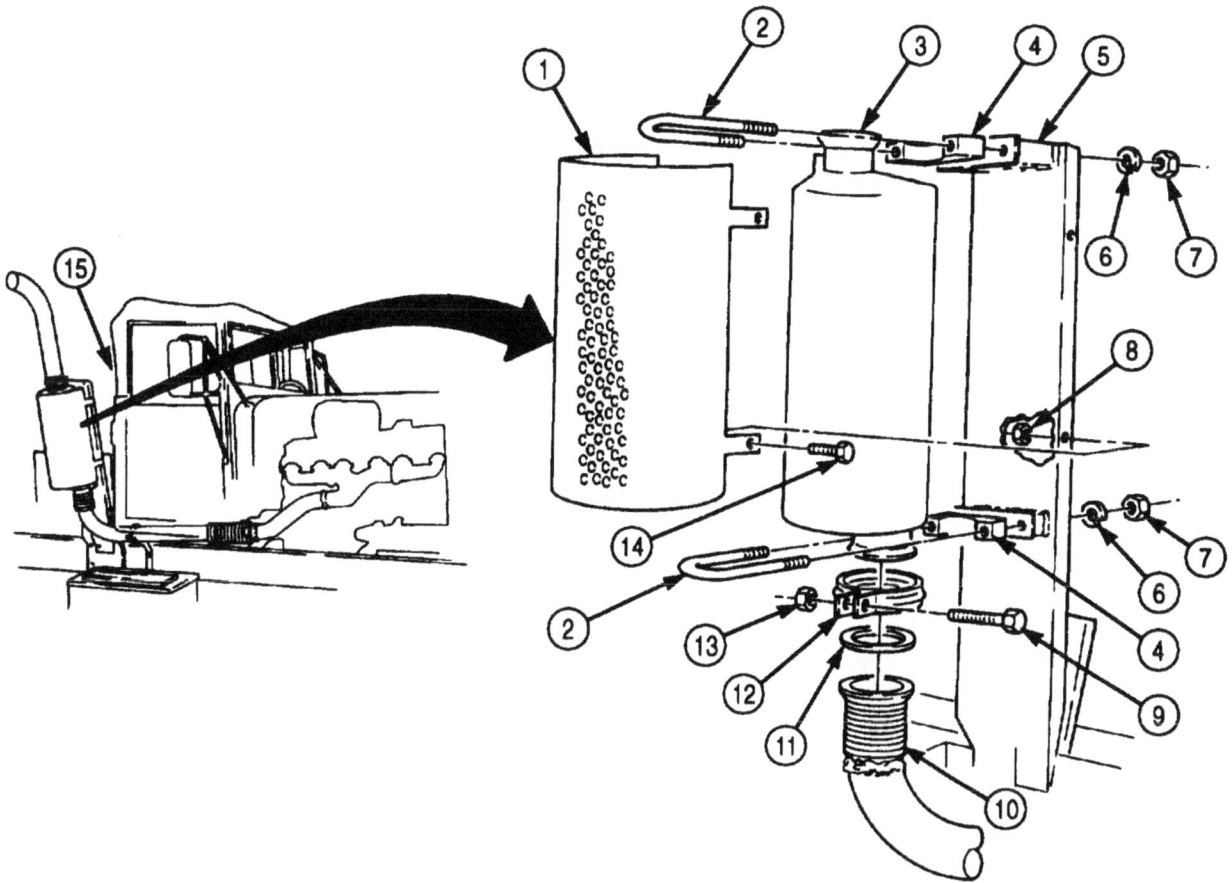

FOLLOW-ON TASK: Install exhaust stack (para. 3-48).

Section V. COOLING SYSTEM MAINTENANCE

3-52. COOLING SYSTEM MAINTENANCE INDEX

THIS TASK COVERS:

a. Depressurizing System   c. Cleaning and Flushing System
b. Draining System   d. Filling System

INITIAL SETUP:

APPLICABLE MODELS

All

TOOLS

General mechanic's tool kit (Appendix E, Item 1)

MATERIALS/PARTS

Cleaning compound kit (Appendix C, Item 19)
Antifreeze (Appendix C, Item 8 or 9)

REFERENCES (TM)

TM 9-2320-272-10
TM 9-2320-272-24P

EQUIPMENT CINDITION

• Parking brake set (TM 9-2320-272-10).
• Hood raised and secured (TM 9-2320-272-10).
• Right splash shield removed (TM 9-2320-272-10).

GENERAL SAFETY INSTRUCTIONS

Cooling system must be depressurized to remove surge tank cap when engine temperature is above 175°F (79°C).

---

a. Depressurizing System

WARNING

Care should be taken when removing surge tank filler cap. Steam or hot coolant under pressure may cause injury to personnel.

If engine is hot, place thick cloth over cap and turn cap to first stop. Wait until pressure has escaped then remove cap from surge tank.

b. Draining System

NOTE

Have drainage container ready to catch coolant.

1. Remove cap (1) from surge tank (2) and open drainvalve (4) on radiator (3) and allow system to drain.

NOTE

Open drainvalve on aftercooler to permit cooling system to fully drain.

2. Open aftercooler (6) drainvalve (5) on M939A2 series vehicles and allow cooling system to drain.

3. Inspect coolant for rust and foreign particles.

4. If drained coolant is heavily rusted, the partially clogged system must be cleaned and flushed.

c. Cleaning and Flushing System

Clean and flush radiator (3) and cooling system with cleaning compound kit. Follow instructions provided with kit.

3-53. COOLING SYSTEM SERVICING (Contd)

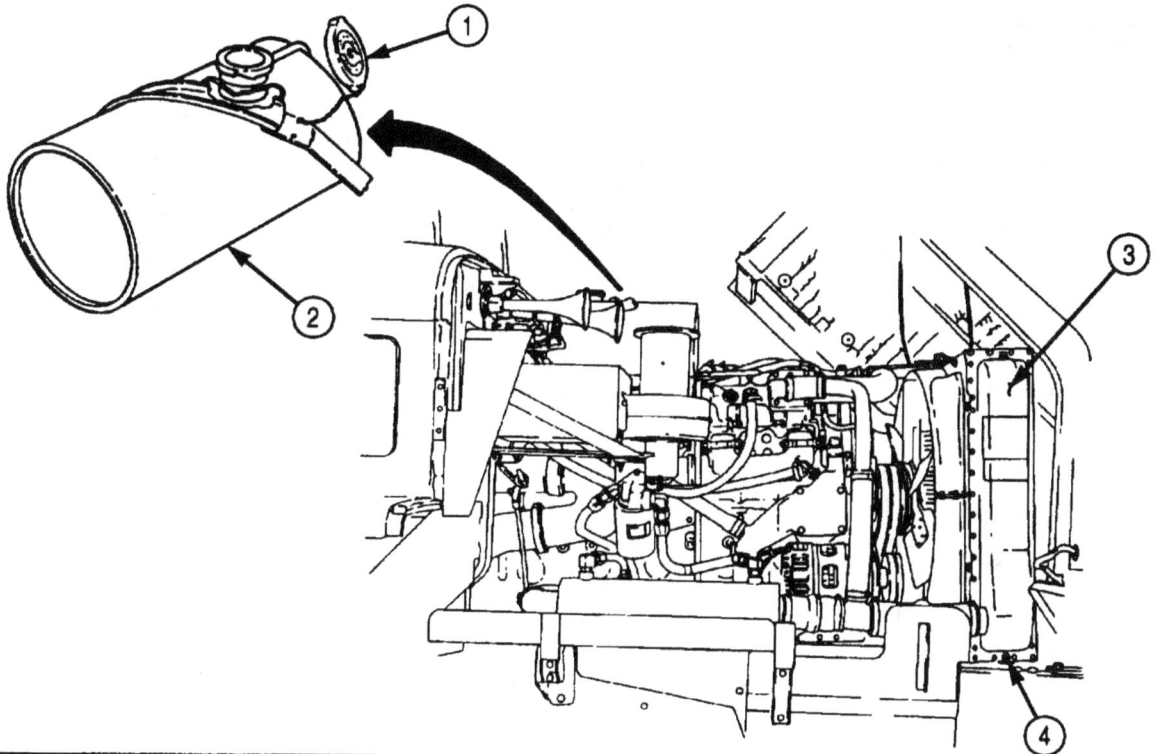

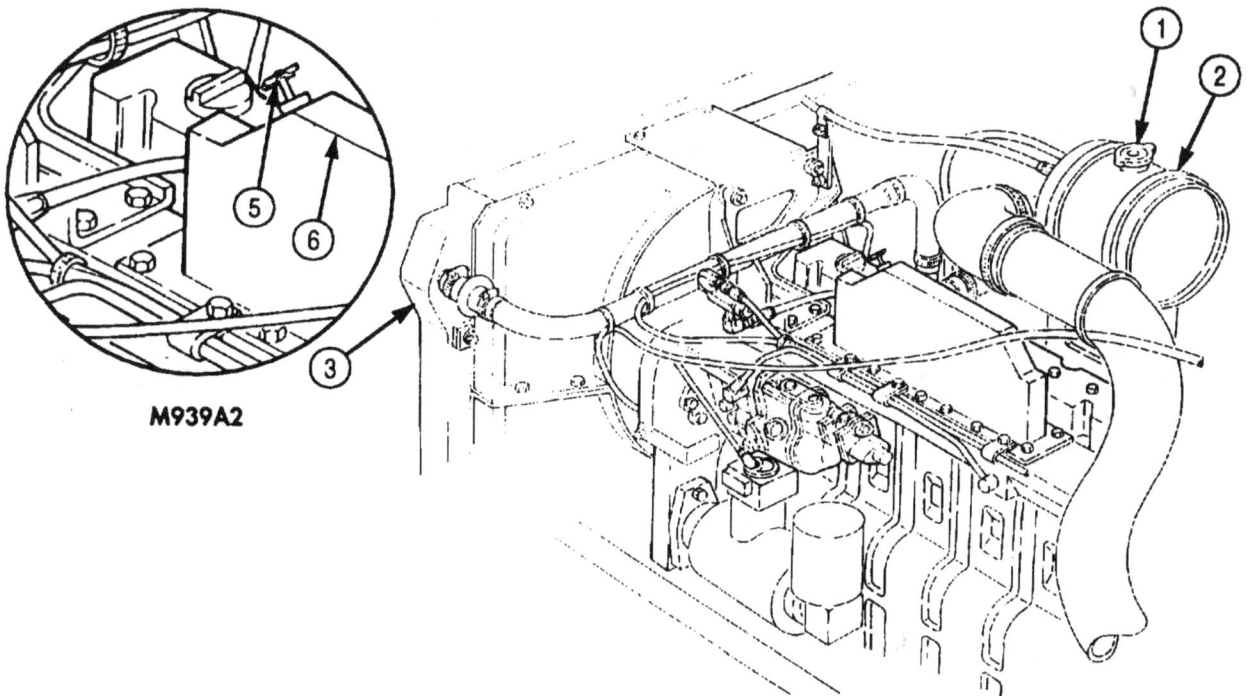

M939A2

3-53. COOLING SYSTEM SERVICING (Contd)

d. Filling System

Table 3-1. Guide for Preparation of Antifreeze Solutions.

ETHYLENE-GLYCOL -60°F (-51 °C) INHIBITED ML-A-46153		
LOWEST EXPECTED AMBIENT TEMPERATURE	QUARTS OF ANTIFREEZE REQUIRED	ARCTIC GRADE ANTIFREEZE -90°F (-68°C) ML-A-11755
F C		Freezing point of -90°F (-68°C). Issued ready for use and must not be mixed with any other liquid.
+20 -7	9	
+10 -12	11-3/4	
0 -18	16	
-10 -23	19	
-20 -29	20-1/2	
-30 -34	23-1/2	
-40 -40	25	
-50 -46	22-6-1/2	
-55 -48	28	
Below -60 Below -51	Use arctic grade antifreeze (-90°F)(-68°C)	

CAUTION

When filling cooling system on M939A2 series vehicles, ensure drainvalve on aftercooler is open. Failure to do so may result in damage to equipment.

NOTE

The cooling system for the vehicles covered in this manual has a 47 qt (44.51) capacity.

1. Fill cooling system with required amount of antifreeze and add water to full mark. Close aftercooler (6) drainvalve (5) when coolant starts running out of drainvalve (5) for on M939A2 series vehicles.

2. Install filler cap (1) on surge tank (2).

3. Start engine (TM 9-2320-272-10) and run at fast idle (800-1,000 rpm) until engine temperature reaches 185°F (85°C). At this temperature, thermostats should fully open.

4. Check antifreeze protection level using antifreeze tester. Follow surge tank filler cap (1) removal procedure of task a.

3-53. COOLING SYSTEM SERVICING (Contd)

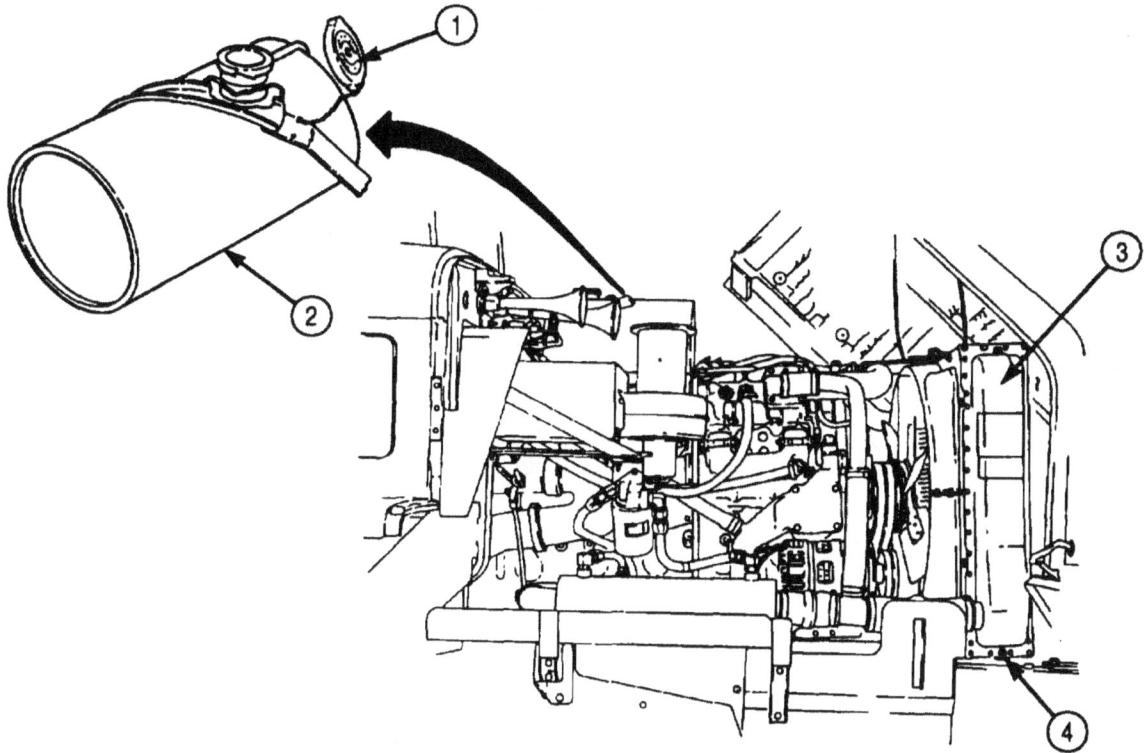

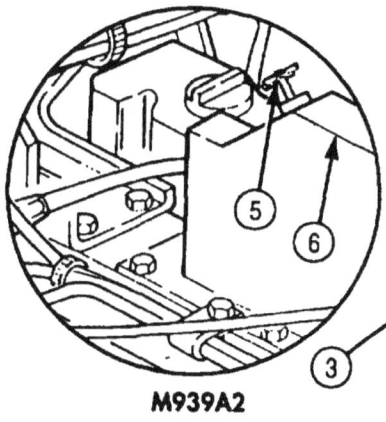

M939A2

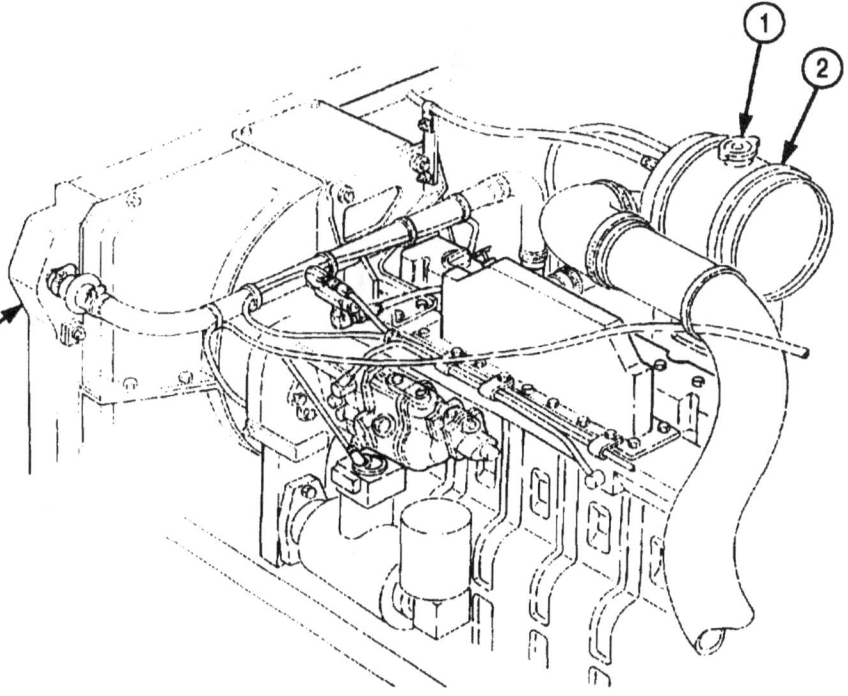

FOLLOW-ON TASKS: • Install right splash shield (TM 9-2320-272-10).
  • Start engine (TM 9-2320-272-10) and check for coolant leaks.

3-54. COOLANT HOSES AND TUBES (M939/A1) REPLACEMENT

THIS TASK COVERS:

a. Radiator Inlet Hoses and Tube Removal
b. Thermostat Housing Hose, Radiator Bypass Tube, and Hose Removal
c. Surge Tank Hose Removal
d. Radiator Outlet Hoses and Tee Removal
e. Transmission Oil Cooler Hoses and Tube Removal
f. Transmission Oil Cooler Hoses and Tubes Installation
g. Radiator Outlet Hoses and Tee Installation
h. Surge Tank Hose Installation
i. Thermostat Housing Hose, Radiator Bypass Tube, and Hose Installation
j. Radiator Inlet Hoses and Tube Installation

INITIAL SETUP:

APPLICABLE MODELS
M939/A2

TOOLS
General mechanic's tool kit (Appendix E, Item 1)

MATERIALS/PARTS
Two locknuts (Appendix D, Item 313)

REFERENCES (TM)
TM 9-2320-272-10
TM 9-2320-272-24P

EQUIPMENT CONDITION
• Parking brake set (TM 9-2320-272-10).
• Hood raised and secured TM 9-2320-272-10).
• Right splash shield removed (TM 9-2320-272-10).
• Coolant drained (para. 3-53).

a. Radiator Inlet Hoses and Tube Removal

1. Loosen two clamps (2) and remove radiator inlet hose (1) from radiator (3) and inlet tube (18).

2. Loosen two clamps (17) and remove hose (8) from thermostat housing (16) and inlet tube (18).

3. Remove two locknuts (4), screws (7), clamps (6), and inlet tube (18) from upper radiator bracket (5). Discard locknuts (4).

b. Thermostat Housing Hose, Radiator Bypass Tube, and Hose Removal

1. Loosen two clamps (14) and remove thermostat hose (15) from thermostat housing (16) and bypass tube (9).

2. Loosen clamp (10) and remove tube (9) from hose (11).

3. Loosen clamp (12) and remove hose (11) from tee (13).

3-54. COOLANT HOSES AND TUBES (M939/A1) REPLACEMENT (Contd)

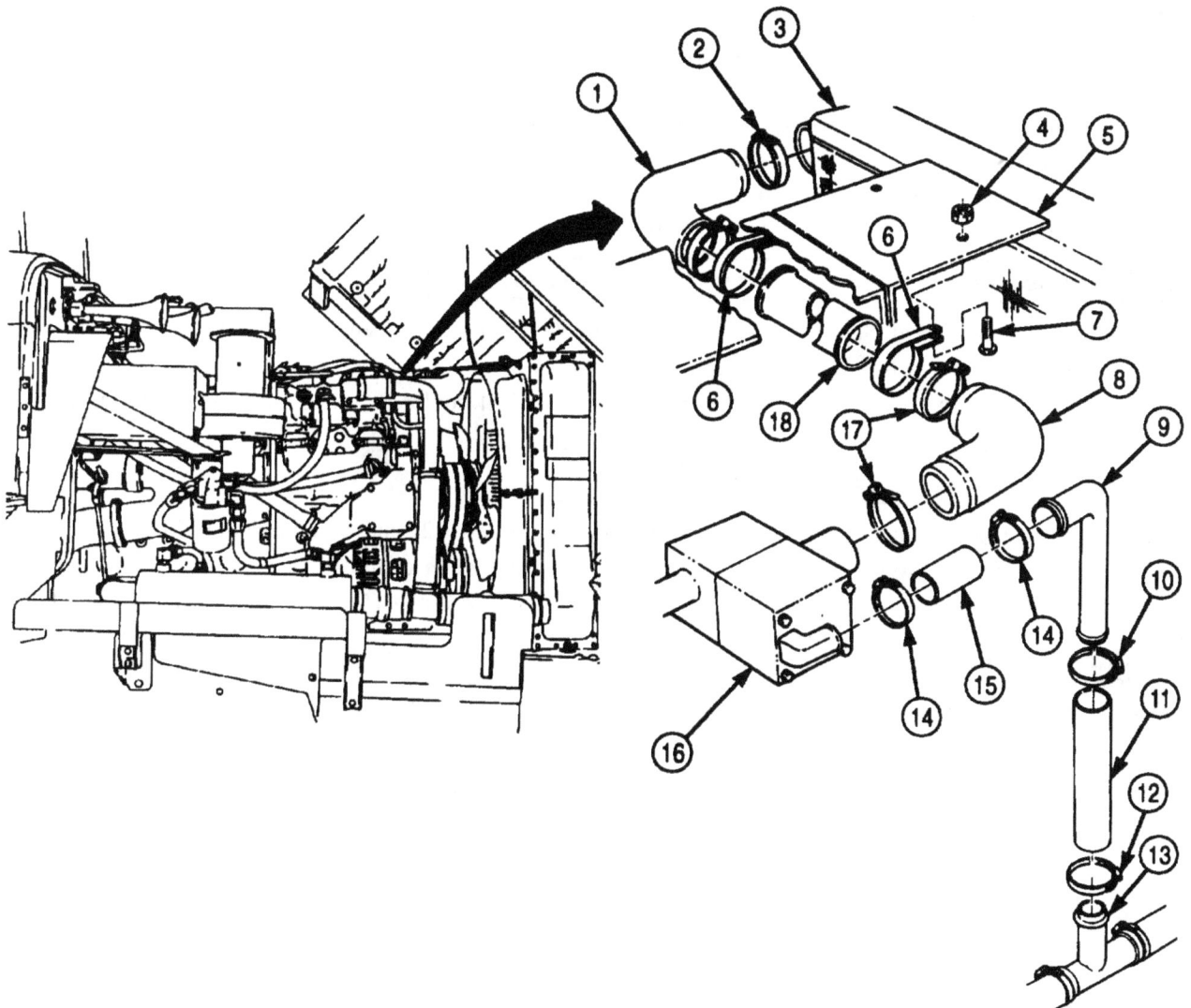

3-54. COOLANT HOSES AND TUBES (M939/A1) REPLACEMENT (Contd)

c. Surge Tank Hose Removal

Loosen two clamps (2) and remove surge tank hose (3) from surge tank (1) and engine oil cooler (4).

d. Radiator Outlet Hoses and Tee Removal

1. Loosen two clamps (6) and remove hose (7) from outlet (5) and tee (8).
2. Loosen clamp (9) and remove tee (8) from hose (10).
3. Loosen clamp (11) and remove hose (10) from transmission oil cooler (12).

e. Transmission Oil Cooler Hoses and Tube Removal

1. Loosen two clamps (13) and remove hose (14) from transmission oil cooler (12) and tube (15).
2. Loosen clamp (16) and remove tube (15) from hose (17).
3. Loosen clamp (18) and remove hose (17) from engine oil cooler (4).

f. Transmission Oil Cooler Hoses and Tubes Installation

1. Install hose (17) and clamp (18) on engine oil cooler (15) and tighten clamp (18).
2. Install tube (15) in hose (17) and tighten clamp (16).
3. Install two clamps (13) and hose (14) on transmission oil cooler (12) and tube (15) and tighten clamps (13).

g. Radiator Outlet Hoses and Tee Installation

1. Install clamp (11) and hose (10) on transmission oil cooler (12) and tighten clamp (11).
2. Install clamp (9) over hose (10) and install tee (8) and tighten clamp (9).
3. Install two clamps (6) and hose (7) on tee (8) and radiator outlet (5) and tighten clamps (6).

h. Surge Tank Hose Installation

Install surge tank hose (3) on surge tank (1) and engine oil cooler (4) and tighten two clamps (2).

3-54. COOLANT HOSES AND TUBES (M939/A1) REPLACEMENT (Contd)

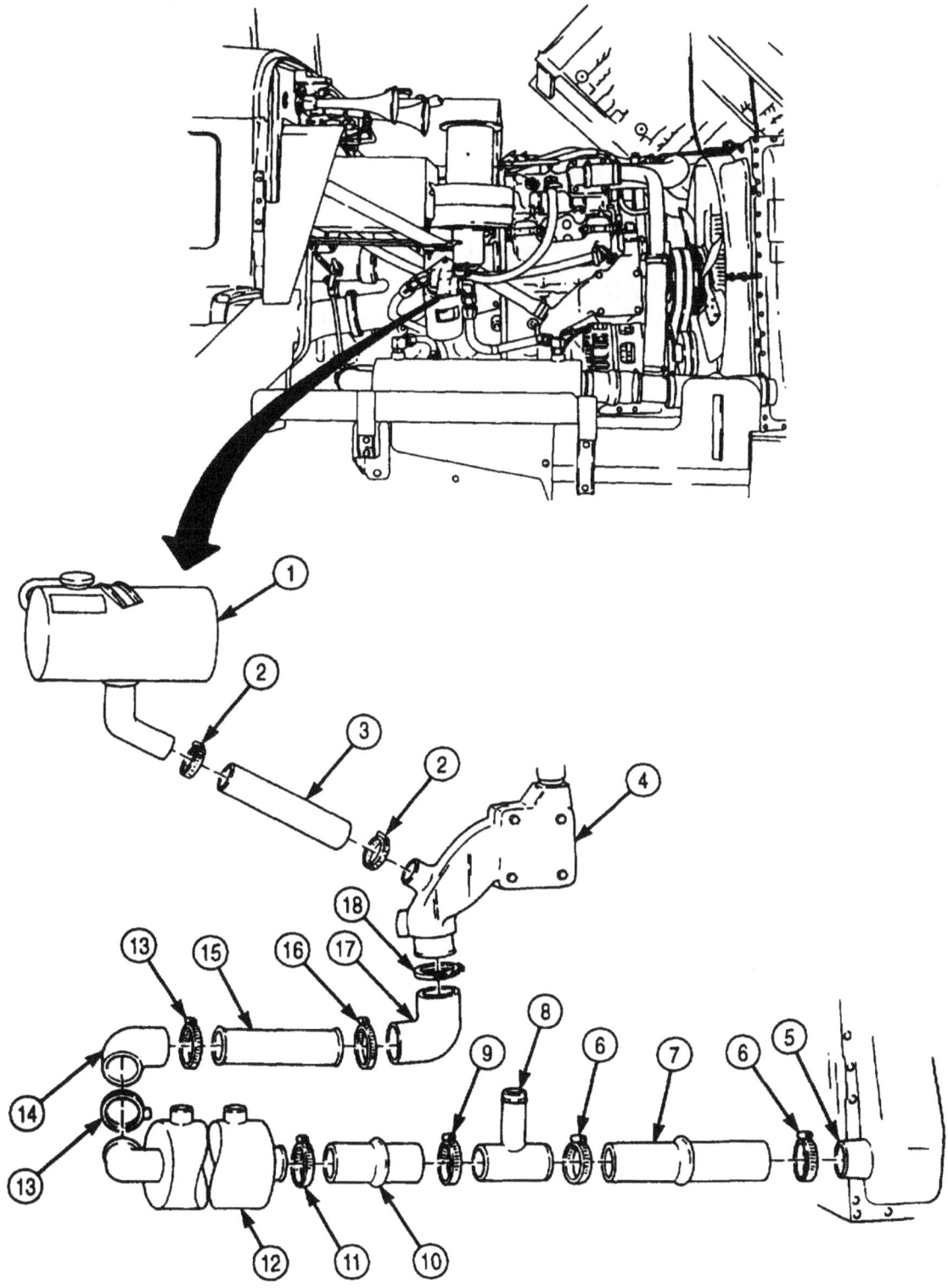

3-54. COOLANT HOSES AND TUBES (M939/A1) REPLACEMENT (Contd)

i. Thermostat Housing Hose, Radiator Bypass Tube, and Hose Installation

1. Install bypass hose (10) and clamp (11) on tee (12) and tighten clamp (11).
2. Install tube (8) and clamp (9) on hose (10) and tighten clamp (9).
3. Install thermostat hose (14) on tube (8) and thermostat housing (15) and tighten two clamps (13).

j. Radiator Inlet Hoses and Tube Installation

1. Install two clamps (19) on upper radiator support (5) with two screws (6) and new locknuts (3).
2. Install clamp (16) and hose (7) on thermostat housing (15). Do not tighten clamp (16) at this time.
3. Install clamp (17) on tube (18) and install tube (18) in hose (7). Tighten clamps (16) and (17).
4. Install inlet hose (1) on radiator (4) and tube (18) with two clamps (2).

3-54. COOLANT HOSES AND TUBES (M939/A1) REPLACEMENT (Contd)

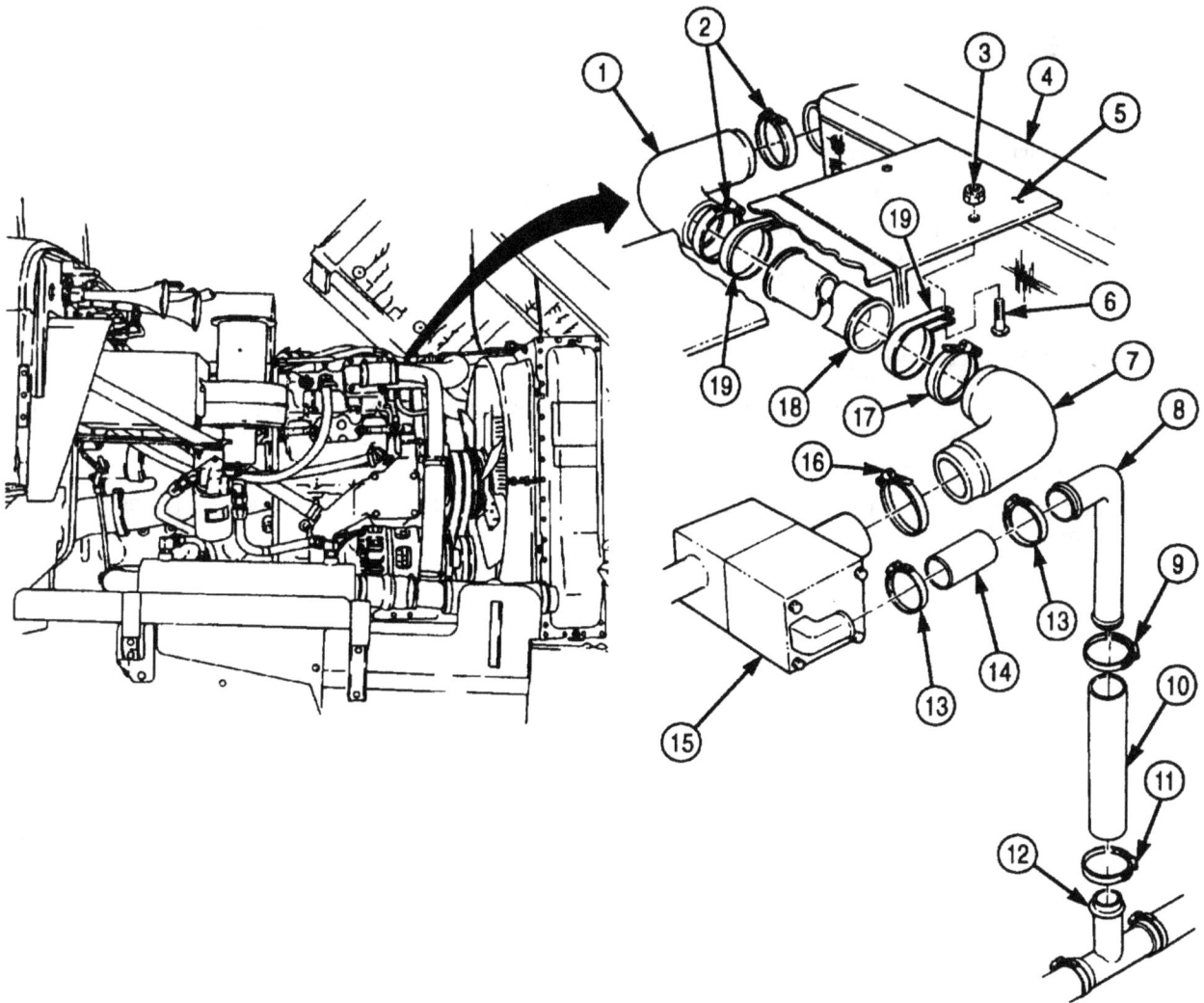

FOLLOW-ON TASKS: • Fill cooling system to proper level (para. 3-53).
• Install right splash shield (TM 9-2320-272-10).
• Start engine (TM 9-2320-272-10) and check for coolant leaks.

3-55. AIR COMPRESSOR COOLANT SUPPLY AND RETURN TUBES (M939/A1) MAINTENANCE

THIS TASK COVERS:

a. Supply Tube Removal
b. Return Tube Removal
c. Inspection

d. Supply Tube Installation
e. Return Tube Installation

INITIAL SETUP:

APPLICABLE MODELS

M939/A1

TOOLS

General mechanic's tool kit (Appendix E, Item 1)

MATERIALS/PARTS

Lockwasher (Appendix D, Item 349)
Four bushings (Appendix D, Item 21)
Antiseize tape (Appendix C, Item 72)

REFERENCES (TM)

TM 9-243
TM 9-2320-272-10
TM 9-2320-272-24P

EQUIPMENT CONDITION

• Parking brake set (TM 9-2320-272-10).
• Hood raised and secured (TM 9-2320-272-10).
• Cooling system drained (para. 3-63).

a. Supply Tube Removal

1. Loosen two flare nuts (4) and remove coolant supply tube (6) from water pump adapter (7) and air compressor elbow (2), and slide nuts (4) to center of tube (6).

2. Remove and discard two bushings (3) from tube (6).

3. Remove water pump adapter (7) and air compressor elbow (2) from engine oil cooler (1) and air compressor (5).

3-55. AIR COMPRESSOR COOLANT SUPPLY AND RETURN TUBES (M939/A1) MAINTENANCE (Contd)

b. Return Tube Removal

1. Loosen flare nut (16) and disconnect coolant return tube (12) from engine oil cooler elbow (18).

2. Remove nut (13), lockwasher (14), washer (15), screw (8), washer (9), and coolant return tube clamp (10) from bracket (11). Discard lockwasher (14).

3. Loosen flare nut (20) and remove coolant return tube (12) from elbow (22).

4. Remove bushings (21) and (17) from coolant return tube (12). Discard bushings (17) and (21).

5. Remove elbows (18) and (22) from engine oil cooler (19) and air compressor (23).

c. Inspection

1. Inspect flare nuts (20) and (16), adapter (7), elbow (2), and coolant supply tube (6) for cracks and stripped threads. Replace parts if cracked or are damaged threads. Refer to TM 9-243 for fabrication of supply tube (6).

2. Inspect tubing (12) for cracks and severe bends. Discard tubing (12) if cracked or badly bent. Refer to TM 9-243 for fabrication of tubing.

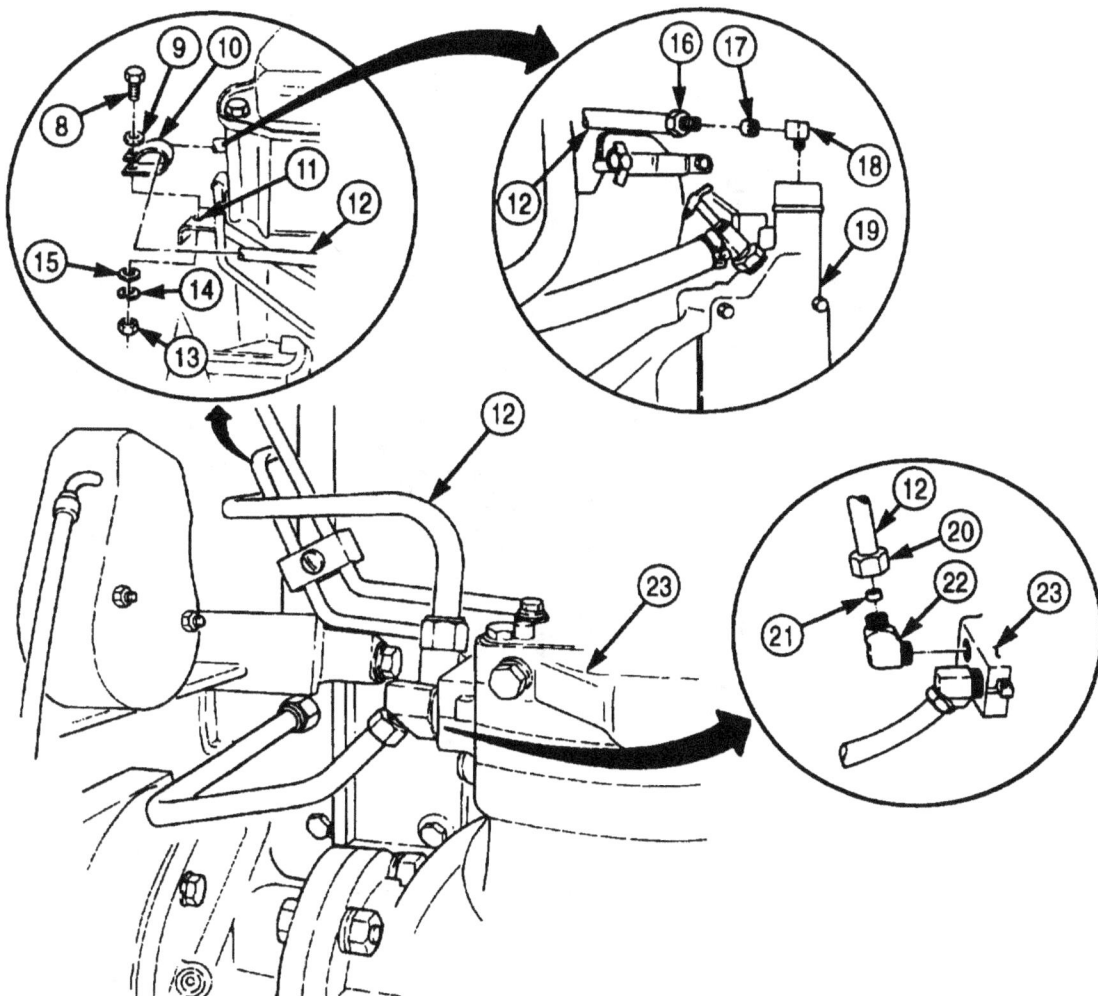

3-55. AIR COMPRESSOR COOLANT SUPPLY AND RETURN TUBES (M939/A1) MAINTENANCE (Contd)

d. Supply Tube Installation

NOTE

Male pipe threads must be wrapped with antiseize tape before installation.

1. Install water pump adapter adapter (7) in engine oil cooler (1).
2. Install air compressor elbow (2) on air compressor (3).
3. Install two new bushings (4) in coolant supply tube (6)
4. Install coolant supply tube (6) on air compressor elbow (2) and water pump adapter (7).

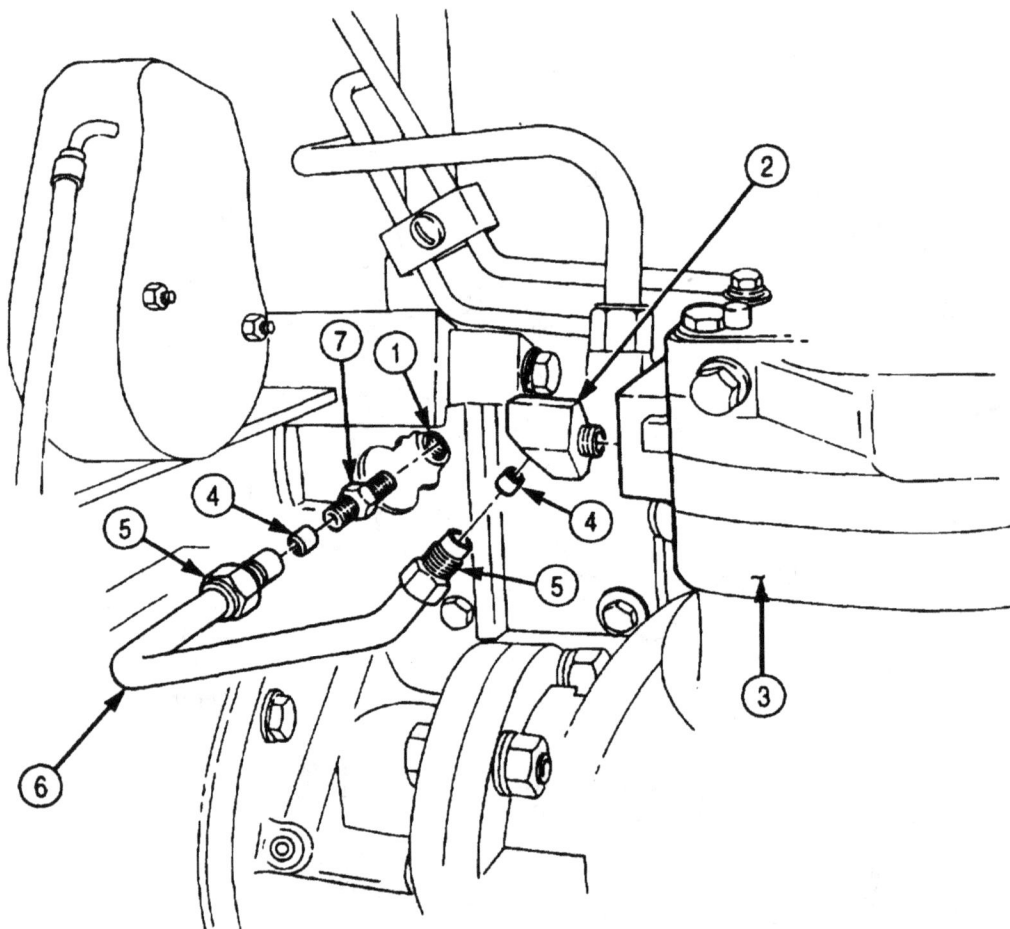

3-55. AIR COMPRESSOR COOLANT SUPPLY AND RETURN TUBES (M939/A1) MAINTENANCE (Contd)

e. Return Tube Installation

NOTE

Male pipe threads must be wrapped with antiseize tape before installation.

1. Install air compressor elbow (22) on air compressor (23).
2. Install engine oil cooler elbow (18) on engine oil cooler (19).
3. Install return tube (12) on compressor elbow (22) with new bushing (21) and flare nut (20). Do not tighten flare nut (20).
4. Connect return tube (12) on engine oil cooler elbow (18) with new bushing (17) and flare nut (16). Do not tighten flare nut (16).
5. Install coolant return tube (12) on support bracket (11) with clamp (10), washers (9) and (15), new lockwasher (14), and nut (13). Tighten nut (13).
6. Tighten flare nuts (16) and (20).

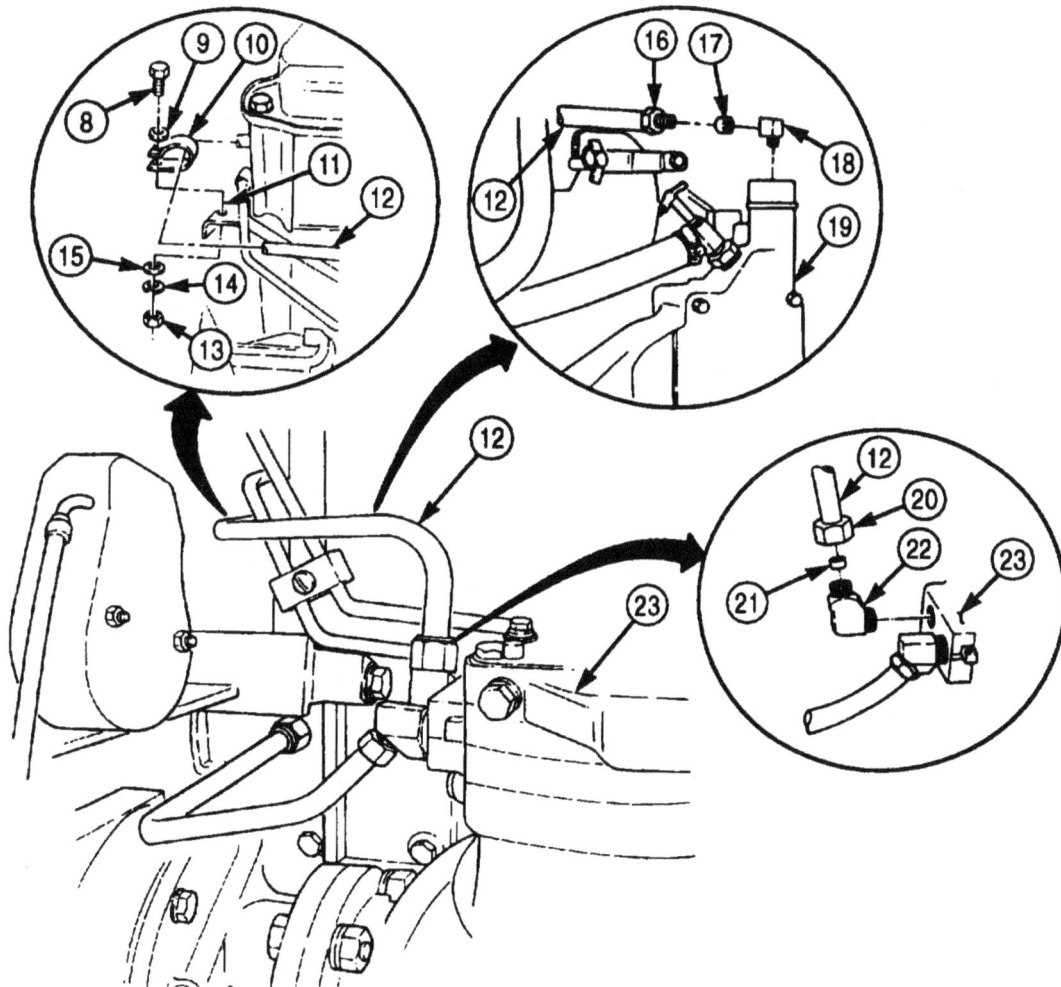

FOLLOW-ON TASKS: • Fill cooling system to proper level (para. 3-53).
• Start engine (TM 9-2320-272-10) and check for coolant leaks.

3-56. COOLANT HOSES AND LINES (M939A2) REPLACEMENT

THIS TASK COVERS:

a. Removal   b. Installation

INITIAL SETUP:

APPLICABLE MODELS
M939A2

TOOLS
General mechanic's tool kit (Appendix E, Item 1)

MATERIALS/PARTS
Locknut (Appendix D, Item 306)
Antiseize tape (Appendix C, Item 72)

REFERENCES (TM)
TM 9-2320-272-10
TM 9-2320-272-24P

EQUIPMENT CONDITION
• Hood raised and secured (TM 9-2320-272-10).
• Right splash shield removed (TM 9-2320-272-10).
• Cooling system drained (para. 3-53).

a. Removal

NOTE
If removal of tiedown straps is necessary, note location for installation.

1. Remove locknut (5), screw (2), clamp (3), and hose (1) from bracket (6). Discard locknut (5).
2. Disconnect hose (1) from adapter (4) and surge tank (18).
3. Remove two clamps (19) and hose (20) from surge tank (18) and thermostat housing (9).
4. Remove adapter (4) and two elbows (7) from radiator (8).
5. Remove two clamps (12) and remove hose (13) from heater valve (16) and fitting (10).
6. Remove two clamps (11) and remove hose (14) from heater valve (15) and fitting (17).
7. Remove nut (23), screw (28), clamp (27), and hose (26) from bracket (24).
8. Remove clamp (47) and hose (48) from surge tank fitting (21).
9. Remove clamp (29) and hose (26) from tube (40).

NOTE
Perform steps 10 and 11 for internal bypass systems.

10. Remove two clamps (36) and hose (37) from fitting (46) and thermostat canister (35).
11. Remove two clamps (38), hose (39), and thermostat canister (35) from tube (32).

NOTE
Perform step 12 for external bypass systems.

12. Remove two clamps (36) and hose (37) from fitting (49) and tube (32).
13. Remove two clamps (30) and hump hose (31) from outlet tube (25) and tube (32).
14. Remove two clamps (33) and tube (32) and hump hose (34) from transmission oil cooler (45).
15. Remove two clamps (41) and elbow hose (42) from outlet port (22) and tube (40).
16. Remove two clamps (43) and elbow hose (44) and tube (40) from transmission oil cooler (45).

3-56. COOLANT HOSES AND LINES (M939A2) REPLACEMENT (Contd)

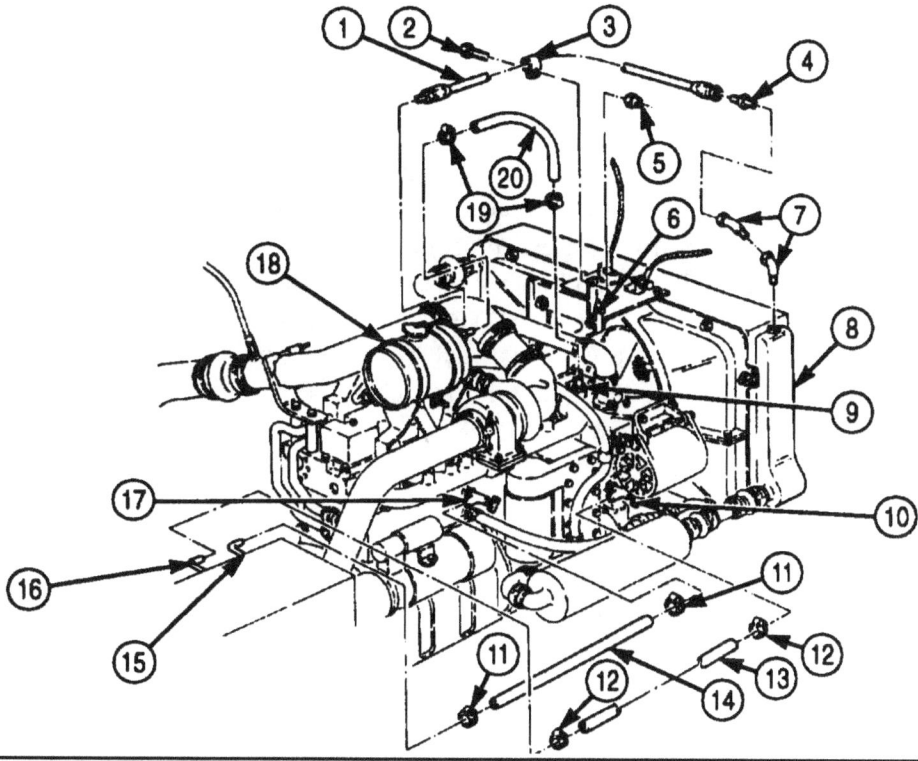

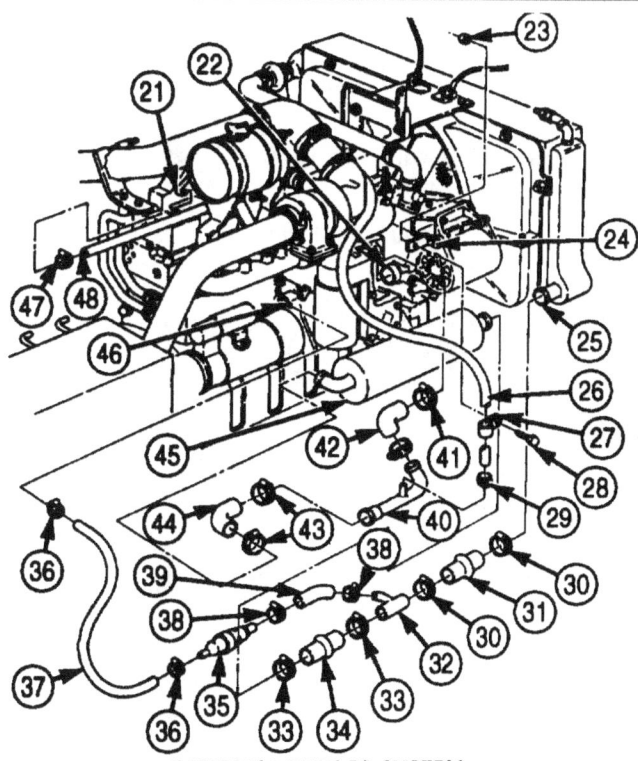

INTERNAL BYPASS SYSTEM

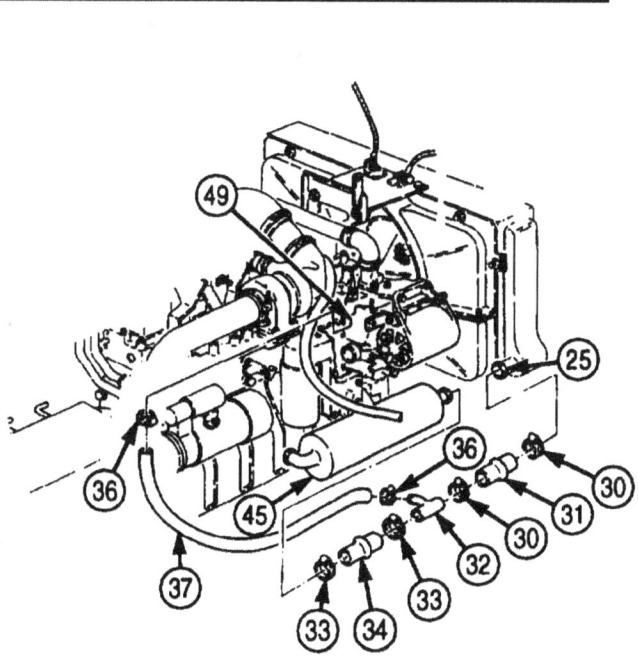

EXTERNAL BYPASS SYSTEM

3-56. COOLANT HOSES AND LINES (M939A2) REPLACEMENT (Contd)

b. Installation

1. Install elbow hose (24) on transmission oil cooler (25) with clamp (23).
2. Install tube (20) on elbow (24) with clamp (23).
3. Install elbow hose (22) on outlet port (2) and tube (20) with two clamps (21).
4. Install hump hose (16) on transmission oil cooler (25) with clamp (14).
5. Install tube (12) on hump hose (16) with clamp (14).
6. Install hump hose (11) on tube (12) and outlet tube (5) with two clamps (10).

NOTE

Perform step 7 for external bypass systems.

7. Install hose (19) on fitting (29) and tube (12) with two clamps (18).

NOTE

Perform steps 8, 9, and 10 for internal bypass systems.

8. Install hose (15) on tube (12) with clamp (13).
9. Install thermostat canister (17) on hose (15) with clamp (13).
10. Install hose (19) on fitting (26) and thermostat canister (17) with two clamps (18).
11. Install hose (6) on tube (20) with clamp (9).
12. Install hose (28) on surge tank fitting (1) with clamp (27).
13. Install hose (6) and clamp (7) on bracket (4) with screw (8) and nut (3).
14. Install hose (43) on heater valve (44) and fitting (46) with two clamps (42).
15. Install hose (41) on heater valve (45) and fitting (39) with two clamps (40).

NOTE

Wrap male pipe threads with antiseize tape before installation.

16. Install two elbows (36) and adapter (33) on radiator (37).
17. Install hose (49) on surge tank (47) and thermostat housing (38) with two clamps (48).
18. Install hose (30) on adapter (33) and surge tank (47).
19. Install clamp (32) and hose (30) on bracket (35) with screw (31) and new locknut (34).

CAUTION

When filling cooling system, ensure drainvalve on aftercooler is open. Failure to do so may result in damage to equipment.

3-56. COOLANT HOSES AND LINES (M939A2) REPLACEMENT (Contd)

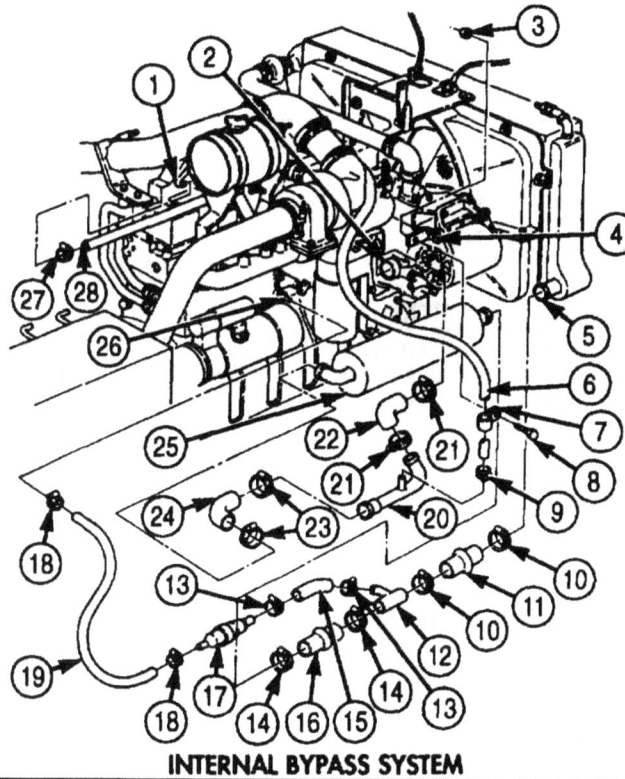

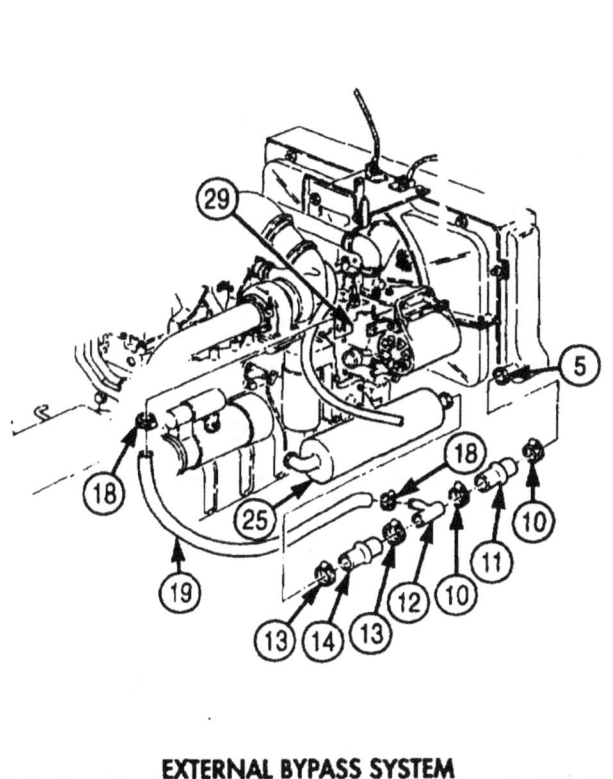

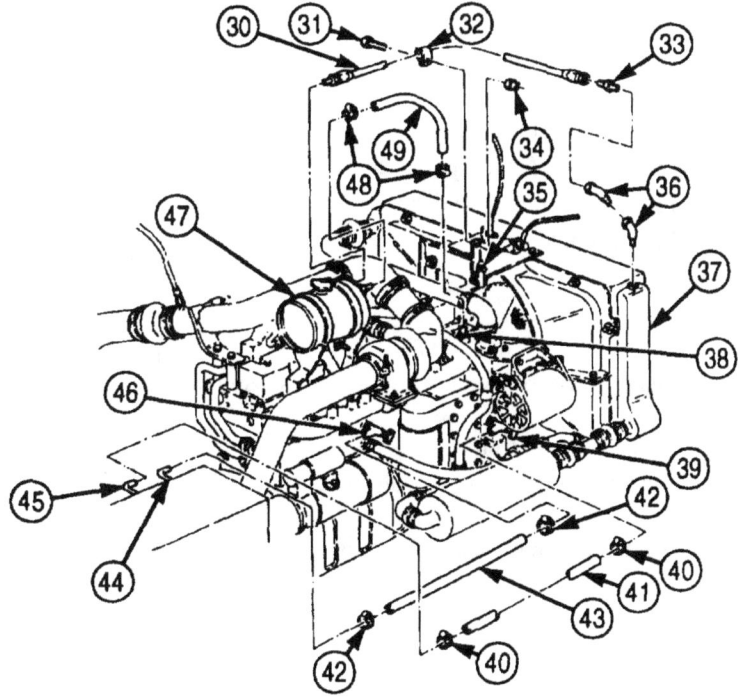

INTERNAL BYPASS SYSTEM

EXTERNAL BYPASS SYSTEM

FOLLOW-ON TASKS: • Fill cooling system to proper level (para. 3-53).
• Install right splash shield (TM 9-2320-272-10).
• Start engine (TM 9-2320-272-10) and check for coolant leaks. Tighten fittings and clamps as necessary.

3-57. AIR COMPRESSOR COOLANT LINES (M939A2) MAINTENANCE

THIS TASK COVERS:

a. Removal

b. Cleaning and Inspection

c. Installation

INITIAL SETUP:

APPLICABLE MODELS
M939A2

TOOLS
General mechanic's tool kit (Appendix E, Item 1)

MATERIALS/PARTS
Four seals (Appendix D, Item 625)
Cap and plug-set (Appendix C, Item 14)
Antiseize tape (Appendix C, Item 72)

REFERENCES (TM)
TM 9-2320-272-10
TM 9-2320-272-24P

EQUIPMENT CONDITION
• Hood raised and secured (TM 9-2320-272-10).
• Right splash shield removed (TM 9-2320-272-10).
• Cooling system drained (para. 3-53).

---

a. Removal

NOTE

When disconnecting water lines or hoses, plug ends and tag for identification during installation.

1. Disconnect two tubing nuts (10) and remove water outlet tube (7) and two seals (5) from air compressor (11) and elbow (2) in cylinder head (1). Discard seals (5).

2. Disconnect two tubing nuts (9) and remove water inlet tube (6) and two seals (4) from air compressor (11) and elbow (3) in engine block (8). Discard seals (4).

3. Remove elbow (2) from cylinder head (1) and elbow (3) from engine block (8).

b. Cleaning and Inspection

1. For general cleaning instructions, refer to para. 2-14.

2. For general inspection instructions, refer to para. 2-15.

3. Inspect tubes (6) and (7) for kinks, cracks, and bends.

4. Replace all parts failing inspection.

c. Installation

NOTE

Male pipe threads must be wrapped with antiseize tape before installation.

1. Install elbow (3) on engine block (8) and elbow (2) on cylinder head (1).

2. Install water inlet tube (6) and two new seals (4) on air compressor (11) and elbow (3) on engine block (8).

3. Install water outlet tube (7) and two new seals (5) on air compressor (11) and elbow (2) on cylinder head (1).

3-57. AIR COMPRESSOR COOLANT LINES (M939A2) MAINTENANCE (Contd)

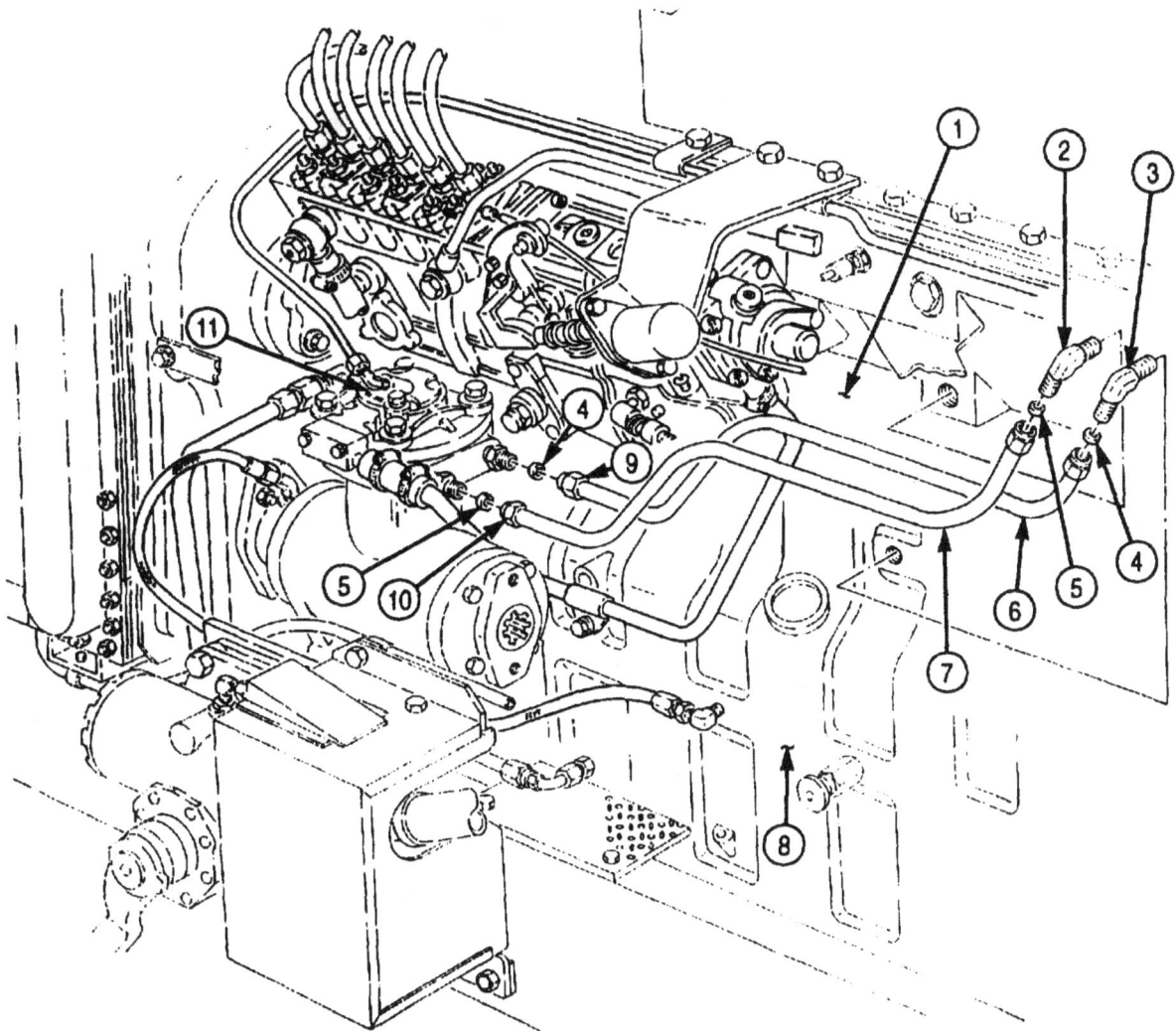

FOLLOW-ON TASKS: • Fill cooling system to proper level (para. 3-53).
                   • Install right splash shield (TM 9-2320-272-10).
                   • Start engine (TM 9-2320-272-10) and check for coolant leaks.

3-58. UPPER RADIATOR HOSES AND BRACKETS (M939A2) REPLACEMENT

THIS TASK COVERS:

a. Removal   b. Installation

INITIAL SETUP:

APPLICABLE MODELS
M939A2

TOOLS
General mechanic's tool kit (Appendix E, Item 1)
Torque wrench (Appendix E, Item 144)

MATERIALS/PARTS
Locknut (Appendix D, Item 306)
Two locknuts (Appendix D, Item 280)
Two lockwashers (Appendix D, Item 360)
Two locknuts (Appendix D, Item 278)
Four isolators (Appendix D, Item 263)
Tiedown straps (Appendix D, Item 685)

REFERENCES (TM)
TM 9-2320-272-10
TM 9-2320-272-24P

EQUIPMENT CONDITION
• Hood raised and secured (TM 9-2320-272-10).
• Right splash shield removed (TM 9-2320-272-10).
• Coolant drained as necessary (para. 3-53).
• Fan actuator disconnected (para. 3-74)

---

a. Removal

1. Remove tiedown straps (4), tachometer drive cable (11), and cable (12) from radiator inlet tube (3). Discard tiedown straps (4).
2. Remove two locknuts (8), washers (9), screws (5), and washers (6) from radiator inlet tube (3) and bracket (14). Discard locknuts (8).
3. Remove four clamps (1), hose (2), elbow (7), and radiator inlet tube (3) from radiator (13) and thermostat housing (10).
4. Remove two nuts (16), screws (19), washers (18), and hood stop cables (17) from bracket (40).
5. Straighten tabs of two lockwashers (41) and remove two screws (15) and lockwashers (41) from bracket (40) and radiator (13). Discard lockwashers (41).
6. Remove two locknuts (37), washers (36), isolators (38), screws (33), washers (34), and isolators (35) from bracket (14). Discard locknuts (37) and isolators (35) and (38).
7. Remove locknut (22), clamp (23), nut (20), washer (21), screws (24) and (26), and bracket (25) from brackets (27) and (14). Discard locknut (22).
8. Remove screw (28), washer (29), and bracket (27) from engine (30).
9. Remove four screws (32), washers (31), and bracket (14) from engine (30).

b. Installation

1. Install bracket (14) on engine (30) with four washers (31) and screws (32). Finger tighten screws (32).
2. Install bracket (27) on engine (30) with washer (29) and screw (28). Finger tighten screw (28).
3. Install brackets (25) and (27) on bracket (14) with screw (26), washer (21), and nut (20).
4. Install clamp (23) on bracket (25) with screw (24) and new locknut (22).
5. Install two new isolators (38), new isolators (35), and bracket (40) on bracket (14) with two washers (34), screws (33), washers (36), and new locknuts (37). Tighten screws (33) 30 lb-ft (41 N•m).
6. Install bracket (40) on radiator (13) with two washers (39), new lockwashers (41), and screws (15). Tighten screws (15) 40 lb-ft (54 N•m). Tighten screws (32) 45 lb-ft (61 N•m). Tighten screw (28).
7. Bend tabs of two lockwashers (41) against flats of screws (15).
8. Install two hood stop cables (17) on bracket (40) with two washers (18), screws (19), and nuts (16).
9. Install hose (2), elbow (7), and radiator inlet tube (3) on radiator (13) and thermostat housing (10) with four clamps (1).

3-58. UPPER RADIATOR HOSES AND BRACKETS (M939A2) REPLACEMENT (Contd)

10. Install radiator inlet tube (3) on bracket (14) with two washers (6), screws (5), washers (9), and new locknuts (8).

11. Secure tachometer drive cable (11) and cable (12) to radiator inlet tube (3) with new tiedown straps (4).

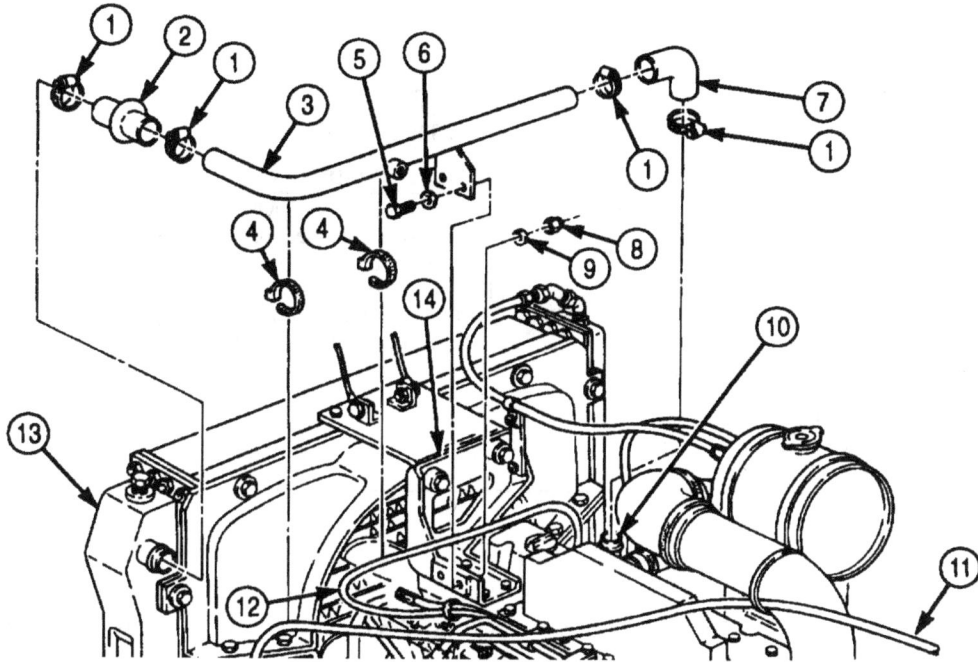

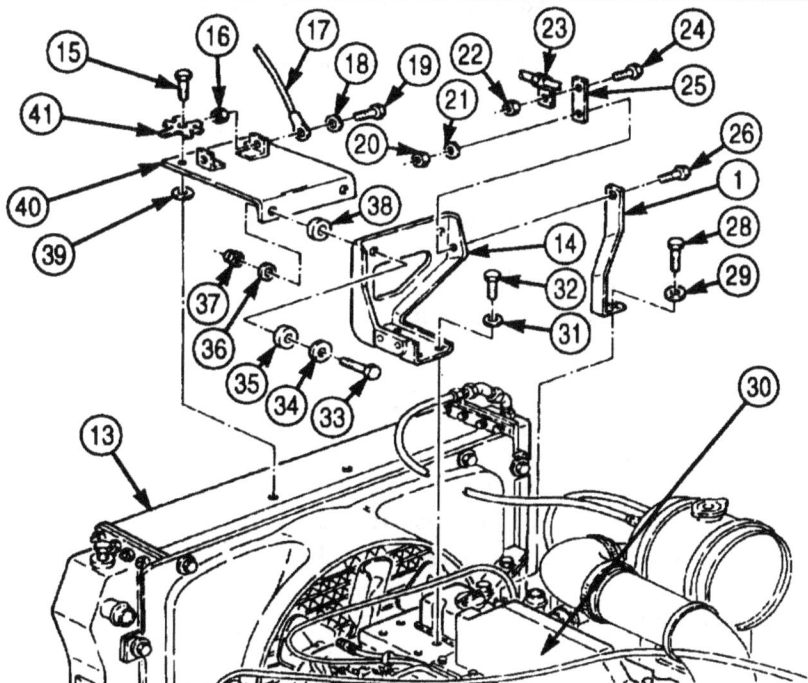

FOLLOW-ON TASKS: • Connect fan actuator (para. 3-74).
• Fill cooling system (para. 3-53).
• Install right splash shield (TM 9-2320-272-10).
• Start engine (TM 9-2320-2272-10) and check for leaks.

3-59. RADIATOR (M939/A1) REPLACEMENT

THIS TASK COVERS:

a. Removal   b. Installation

INITIAL SETUP:

APPLICABLE MODELS
M939/A1

TOOLS
General mechanic's tool kit (Appendix E, Item 1)

MATERIALS/PARTS
Locknut (Appendix D, Item 273)
Five locknuts (Appendix D, Item 291)
Two locknuts (Appendix D, Item 313)
Six locknuts (Appendix D, Item 294)
Two lockwashers (Appendix D, Item 360)
Two lockwashers (Appendix D, Item 382)
Six lockwashers (Appendix D, Item 350)
Three rubber mounts (Appendix D, Item 263)
Antiseize tape (Appendix C, Item 72)

REFERENCES (TM)
TM 9-2320-272-10
TM 9-2320-272~24P

EQUIPMENT CONDITION
- Parking brake set (TM 9-2320-272-10).
- Hood raised and secured (TM 9-2320-272-10).
- Right and left splash shields removed (TM 9-2320-272-1).
- Cooling system drained (para. 3-53).
- Fan shroud removed (para. 3-63).

---

a. Removal

1. Disconnect radiator vent hose (18) from radiator (19).
2. Remove two locknuts (6), washers (8), screws (9), and hood stop cables (7) from upper radiator bracket (1). Discard locknuts (6).
3. Remove two screws (4), lockwashers (3), and washers (2) from upper radiator bracket (1). Discard lockwashers (3).
4. Remove two locknuts (5), screws (15), and clamps (20) from upper radiator bracket (1). Discard locknuts (5).
5. Remove two locknuts (11), washers (12), screws (14), and upper radiator bracket (1) from engine bracket (10). Discard locknuts (11).

NOTE
Perform step 6 only if rubber mounts are damaged or new radiator
is being installed.

6. &move two rubber mounts (13) from engine bracket (10). Discard rubber mounts (13).
7. Loosen clamp (17) and disconnect radiator hose (16) from radiator (19).
8. Remove adapter (22) and two elbows (21) from radiator (19).
9. Remove two plugs (23) and drainvalve (24) from radiator (19).
10. Loosen clamp (25) and disconnect radiator outlet hose (26) from radiator (19).
11. Remove locknut (29), screw (37), and washer (36) from brackets (34) and (28). Discard locknut (29).

NOTE
Assistant will help with steps 12 and 13.

12. Remove four locknuts (30), two screws (33), and screws (38) from radiator support (34). Discard locknuts (30).
13. Remove radiator (19) from vehicle.

NOTE
Perform step 14 only if rubber mount is damaged or a new
radiator is being installed.

3-59. RADIATOR (M939/A1) REPLACEMENT (Contd)

14. Remove and discard rubber mount (45).

NOTE

Perform step 15 only if bracket is damaged or being replaced.

15. Remove two screws (31), lockwashers (32), washers (27), and radiator support bracket (28) from vehicle. Discard lockwashers (32).

16. Remove locknut (44), screw (48), and radiator support (34) from radiator (19). Discard locknut (44).

17. Remove two locknuts (40), screws (50), and control assembly (49) from mounting bracket (39). Discard locknuts (40).

18. Remove four screws (43), lockwashers (42), washers (41), and mounting bracket (39) from radiator (19). Discard lockwashers (42).

19. Remove two screws (46), lockwashers (47), and bracket (35) from radiator (19). Discard lockwashers (47).

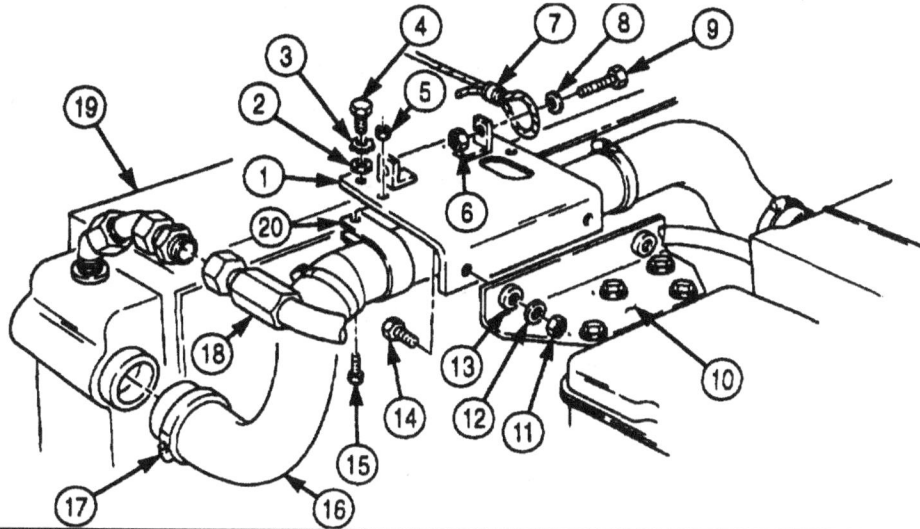

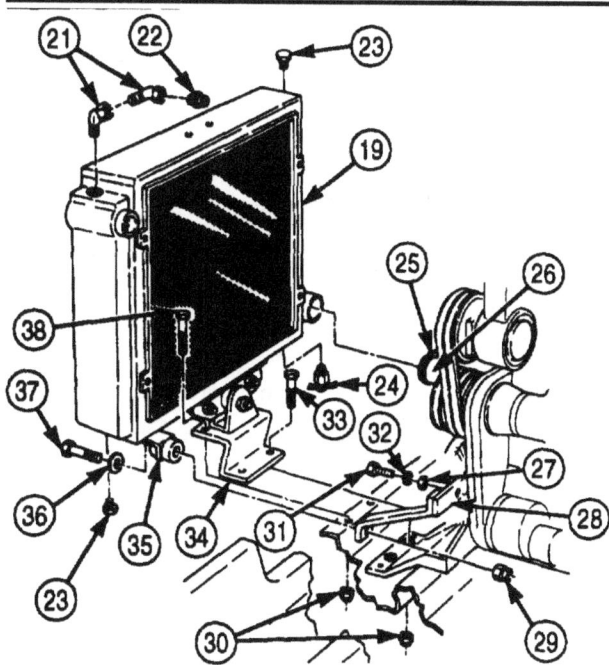

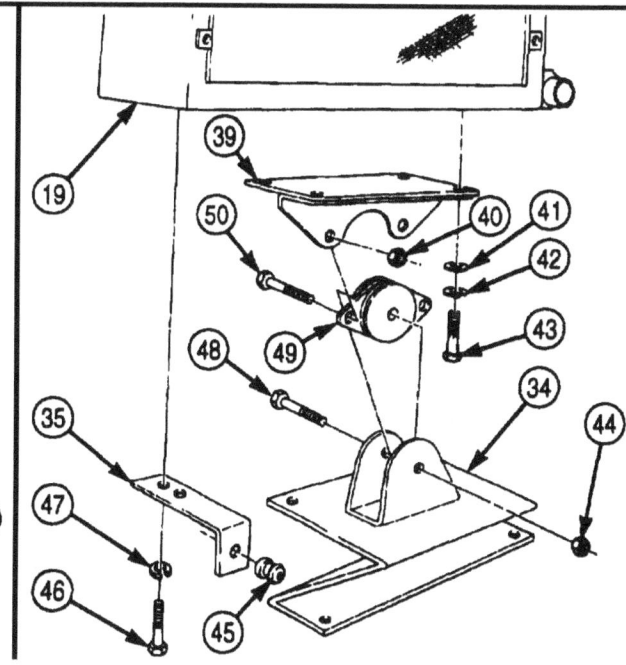

3-59. RADIATOR (M939/A1) REPLACEMENT (Contd)

b. Installation

NOTE

- If new radiator is being installed, use attaching parts and fittings from old radiator.
- Fittings must be inspected for cracks or stripped threads.
- Male pipe threads must be wrapped with antiseize tape before installation.

1. Install new rubber mount (9) in bracket (12).

2. Install bracket (12) on radiator (1) with two new lockwashers (11) and screws (10).

3. Install mounting bracket (2) on radiator (1) with four new lockwashers (5), washers (4), and screws (6).

4. Install control assembly (14) on bracket (2) with two screws (15) and new locknuts (3).

5. Install radiator support (7) on control assembly (14) with screw (13) and new locknut (8).

NOTE

Perform step 6 only if bracket was removed.

6. Install radiator support bracket (23) with two washers (22), new lockwashers (28), and screws (27).

NOTE

Assistant is needed for steps 7, 8, and 9.

7. Install radiator (1) on frame (26).

8. Install radiator support (7) on support bracket (23) with two screws (29), screws (32), and four new locknuts (25).

9. Install washer (30), screw (31), and new locknut (24) on brackets (12) and (23).

10. Connect radiator outlet hose (21) to radiator (1) and tighten clamp (20).

11. Install two plugs (18) and drainvalve (19) on radiator (1).

12. Install two elbows (16) and adapter (17) on radiator (1).

13. Connect radiator inlet hose (49) to radiator (1) and tighten clamp (50).

NOTE

Perform step 14 only if rubber mounts were removed.

14. Install two new rubber mounts (46) on engine bracket (43).

15. Install upper radiator bracket (34) on engine bracket (43) with two screws (47), washers (45), and new locknuts (44).

16. Install upper radiator bracket (34) on radiator (1) with two screws (37), new lockwashers (36), and washers (35).

17. Install two hood stop cables (39) on upper radiator bracket (34) with two screws (41), washers (40), and new locknuts (42).

18. Install two clamps (33) on upper radiator bracket (34) with two screws (48) and new locknuts (38).

19. Connect radiator vent hose (51) on adapter (17).

3-59. RADIATOR (M939/A1) REPLACEMENT (Contd)

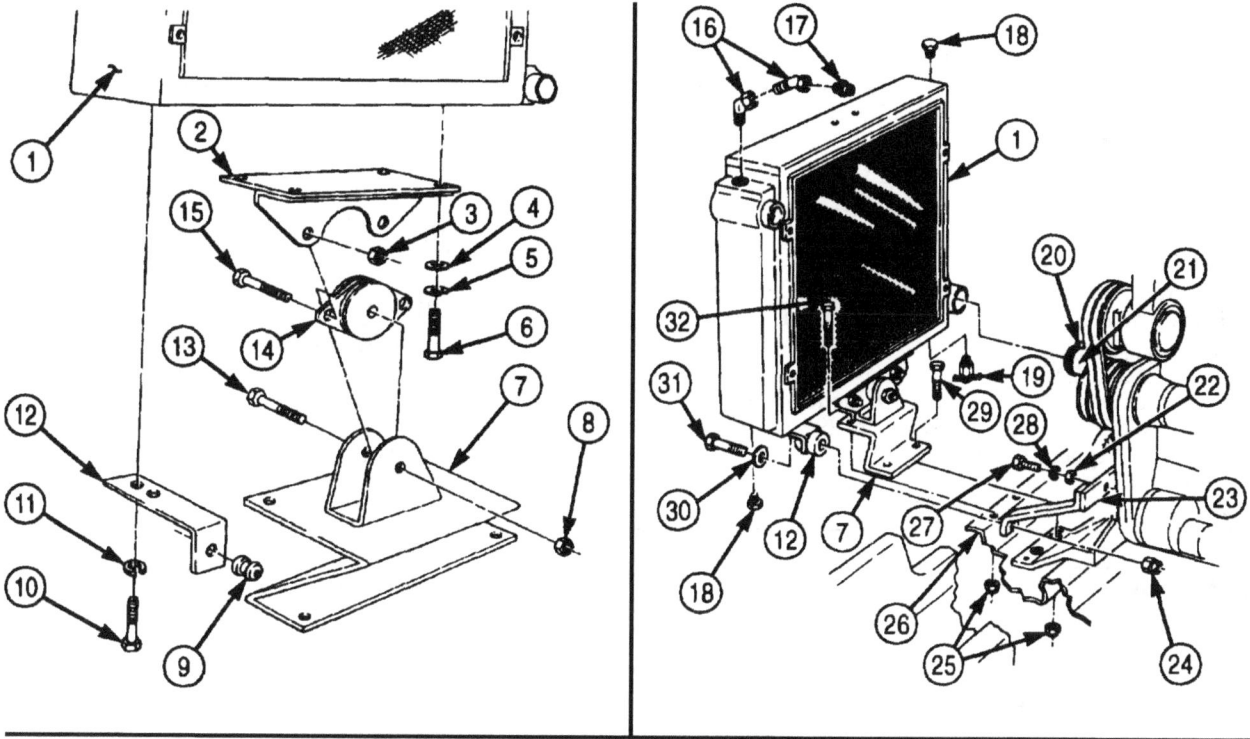

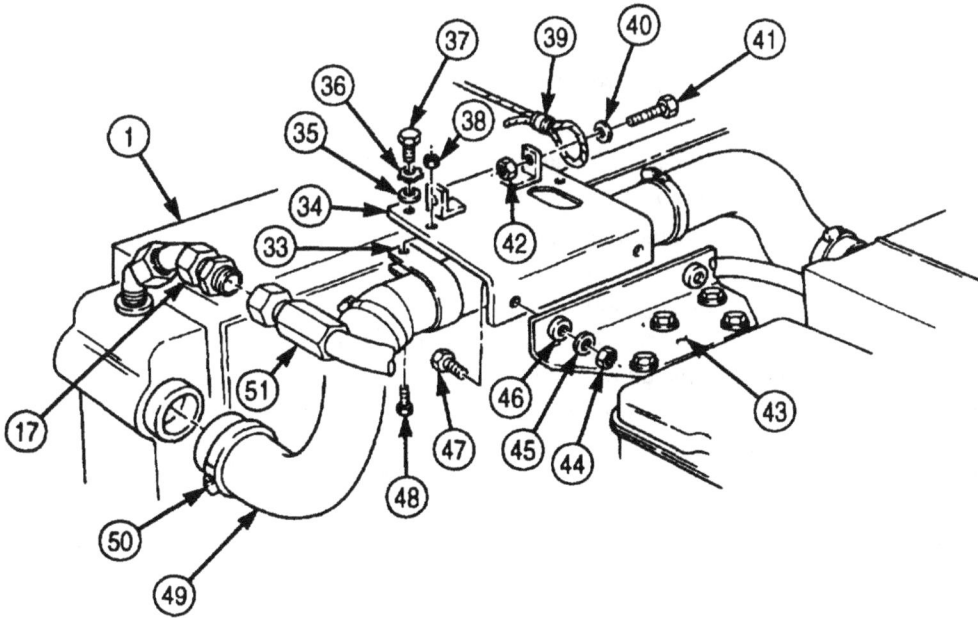

FOLLOW-ON TASKS: • Install fan shroud (para. 3-63).
• Fill cooling system to proper level (para. 3-53).
• Install left and right splash shields (TM 9-2320-272-10).
• Start engine (TM 9-2320-272-10) and check cooling system for leaks.

3-60. RADIATOR (M939A2) MAINTENANCE

THIS TASK COVERS:

a. Removal
b. Cleaning and Inspection

c. Installation

<u>INITIAL SETUP:</u>

<u>APPLICABLE MODELS</u>
M939A2

<u>TOOLS</u>
General mechanic's tool kit (Appendix E, Item 1)
Torque wrench (Appendix E, Item 146)

<u>MATERIALS/PARTS</u>
Two lockwashers (Appendix D, Item 360)
Two locknuts (Appendix D, Item 294)
Locknut (Appendix D, Item 273)
Four locknuts (Appendix D, Item 319)
Antiseize tape (Appendix C, Item 72)

<u>REFERENCES (TM)</u>
TM 750-254
TM 9-2320-272-10
TM 9-2320-272-243?

<u>EQUIPMENT CONDITION</u>
• Parking brake set (TM 9-2320-272-10).
• Right and left splash shields removed (TM 9-2320-272-10).
• Coolant drained (para. 3-53).
• Hood removed (para. 3-275).
• Oil cooler removed (para. 3-4).

a. Removal

1. Straighten tabs on two lockwashers (29) and remove screws (30), lockwashers (29), and washers (28) from bracket (27) and radiator (1). Discard lockwashers (29).

2. Disconnect hose (5) from adapter (2).

3. Remove adaptor (2), elbow (3), and elbow (4) from radiator (1).

4. Loosen clamp (25) and disconnect hump hose (26) from inlet of radiator (1).

5. Loosen clamp (7) and disconnect lower hump hose (6) from outlet of radiator (1).

6. Remove screws (10) and (11) and washers (9) from radiator (1).

NOTE
Assistant will help with steps 7 and 8.

7. Remove four screws (21) and washers (20) from mounting bracket (19).

8. Remove radiator (1) from engine compartment (12).

NOTE
• Replace mounting bracket only if damaged.
• Retain attaching hardware, fan shroud, and fittings for installation of new radiator.

9. Remove two locknuts (22), screws (18), washers (16), and radiator mounting bracket (19) from control assembly (23). Discard locknuts (22).

10. Remove locknut (24), screw (15). and control assembly (23) from radiator support (13). Discard locknut (24).

11. Remove four locknuts (8), screws (17), washers (14), and radiator support (13) from engine compartment (12). Discard locknuts (8).

3-60. RADIATOR (M939A2) MAINTENANCE (Contd)

b. Cleaning and Inspection

1. Refer to TM 750-254 for radiator cleaning, inspection, and repair instructions.
2. Inspect mounting and fitting parts (para. 2-15).
3. Replace all parts failing inspection.

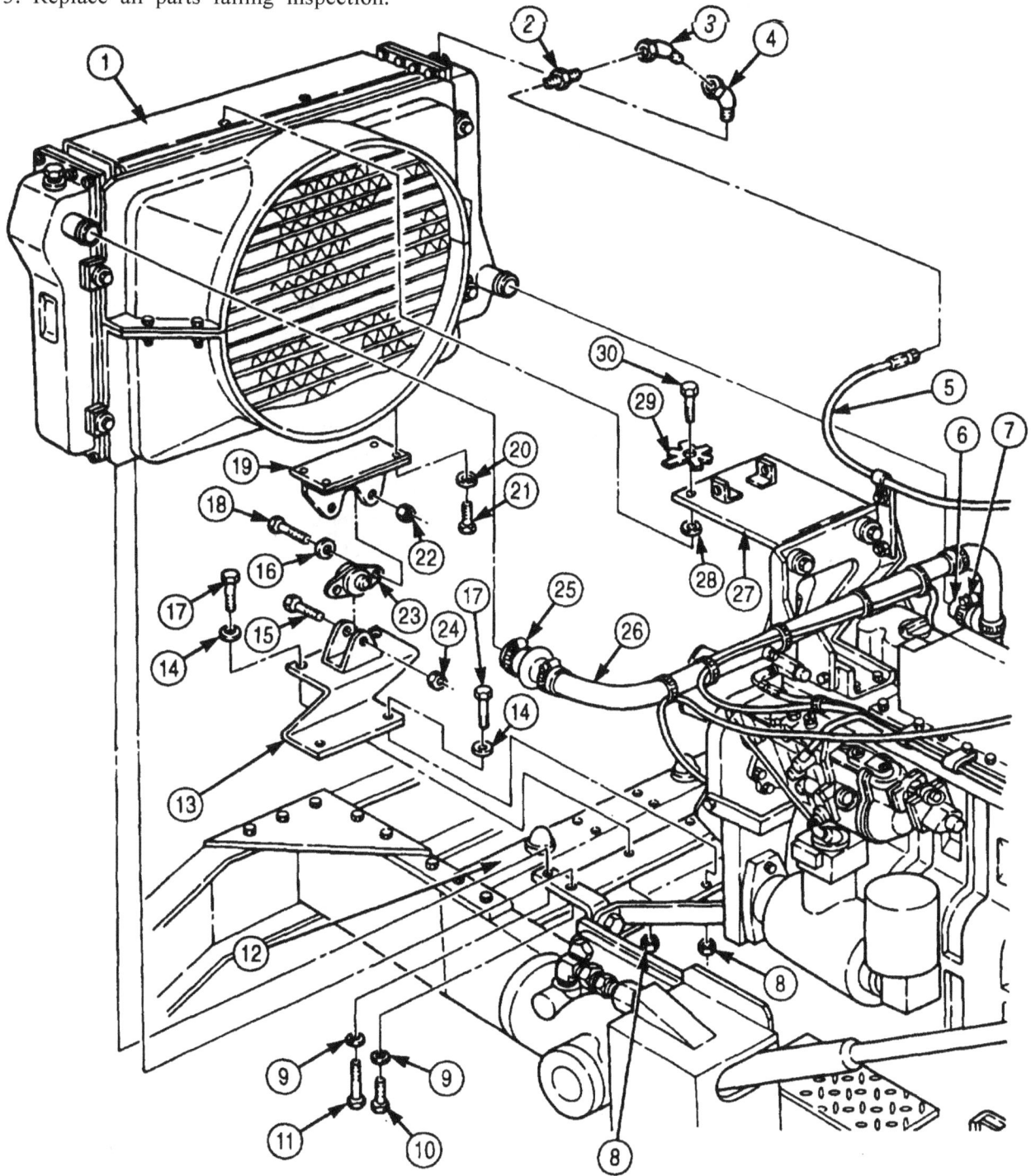

3-60. RADIATOR (M939A2) MAINTENANCE (Contd)

c. Installation

NOTE
- Refer to para. 3-64 for removing and installing fan and shroud.
- Perform steps 1 through 3 only if parts were removed.

1. Install radiator support (13) in engine compartment (12) with four washers (14), screws (17), and new locknuts (8).

2. Install control assembly (23) on radiator support (12) with screw (15) and new locknut (24).

3. Install radiator mounting bracket (19) on control assembly (23) with two screws (18), washers (16), and new locknuts (22).

NOTE
Assistant needed for steps 4 and 5.

4. Position radiator (1) in vehicle.

5. Attach radiator (1) to mounting bracket (19) with four screws (21) and washers (20). Hand tighten screws (21).

6. Install two washers (9) and screws (10) and (11) through bracket (19) and into bottom of radiator (1). Tighten screws (10), (11), and (21) to 40 lb-ft (55 N•m).

7. Connect lower hump hose (6) to outlet of radiator (1) and tighten clamp (7).

8. Connect upper hump hose (26) to inlet of radiator (1) and tighten clamp (25).

9. Wrap male threads of elbows (3) and (4) with antiseize tape and install in radiator (1).

10. Install adapter (2) in elbow (3) and connect hose (5) to adapter (2).

11. Install two washers (28), new lockwashers (29), and screws (30) through bracket (27) and into radiator (1). Tighten screws (30) to 40 lb-ft (55 N•m). Bend tabs of lockwashers (29) over flats of screws (30).

3-60. RADIATOR (M939A2) MAINTENANCE (Contd)

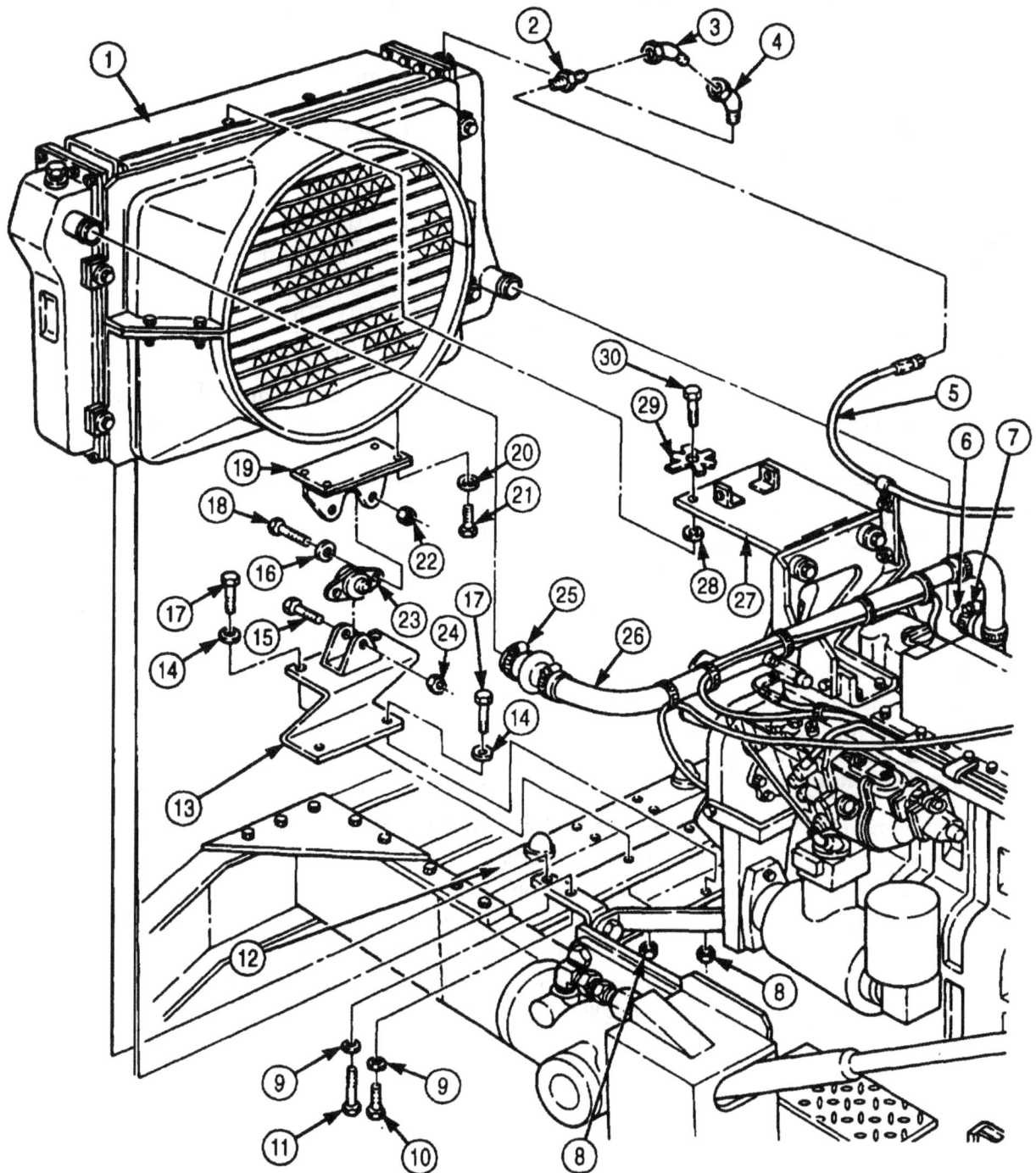

FOLLOW-ON TASKS: • Install oil cooler (para. 3-4).
• Fill cooling system to proper level (para. 3-53).
• Start engine (TM 9-2320-272-10) and check cooling system for leaks. Tighten fittings and clamps as necessary.
• Install right and left splash shields (TM 9-2320-272-10).
• Install hood (para. 3-275).

3-61. SURGE TANK, RADIATOR VENT HOSE, AND MANIFOLD RETURN HOSE (M939/A1) REPLACEMENT

THIS TASK COVERS:

a. Radiator Vent and Manifold Return Hoses Removal
b. Surge Tank Removal

c. Surge Tank Installation
d. Radiator Vent and Manifold Return Hoses Installation

INITIAL SETUP:

APPLICABLE MODELS

M939/A1

TOOLS

General mechanic's tool kit (Appendix E, Item 1)
Torque wrench (Appendix E, Item 144)

MATERIALS/PARTS

Four keywashers (Appendix D, Item 267)
Locknut (Appendix D, Item 288)
Antiseize tape (Appendix C, Item 72)

REFERENCES (TM)

TM 9-2320-272-10
TM 9-2320-272-24P

EQUIPMENT CONDITION

• Hood raised and secured (TM 9-2320-272-10).
• Parking brake set (para. 3-301).
• Right splash shield removed (TM 9-2320-272-10).
• Cooling system drained as required (para. 3-63).

---

a. Radiator Vent and Manifold Return Hoses Removal

1. Disconnect radiator vent hose (7) from adapter (13) on radiator (8).
2. Remove adapter (13) and elbows (12) and (11) from radiator (8).
3. Remove screw (5), washer (6), hose clamp (4), and spacer (10) from engine bracket (9).
4. Disconnect manifold return hose (2) from elbow (3) on water manifold (14).
5. Remove manifold return hose (2) and radiator vent hose (7) from surge tank (1).
6. Remove elbow (3) from water manifold (14).

b. Surge Tank Removal

1. Remove cap (15) from surge tank (1) and disconnect cap (15) from retaining pin (16).
2. Remove pin (19) and chain (17) from surge tank support (24).
3. Remove screw (18), washer (21), and surge tank support extension (28) from surge tank (1).
4. Remove locknut (23), screw (20), and surge tank support (24) from lifting eye (22). Discard locknut (23).
5. Loosen clamp (27) and disconnect engine oil cooler hose (26) from surge tank (1).
6. Unlock four keywashers (30) and remove four screws (29), keywashers (30), washers (31), and two exhaust manifold clamps (32) from cylinder head (25). Discard keywashers (30).
7. Remove surge tank (1) from vehicle.

3-61. SURGE TANK, RADIATOR VENT HOSE, AND MANIFOLD RETURN HOSE (M939/A1) REPLACEMENT (Contd)

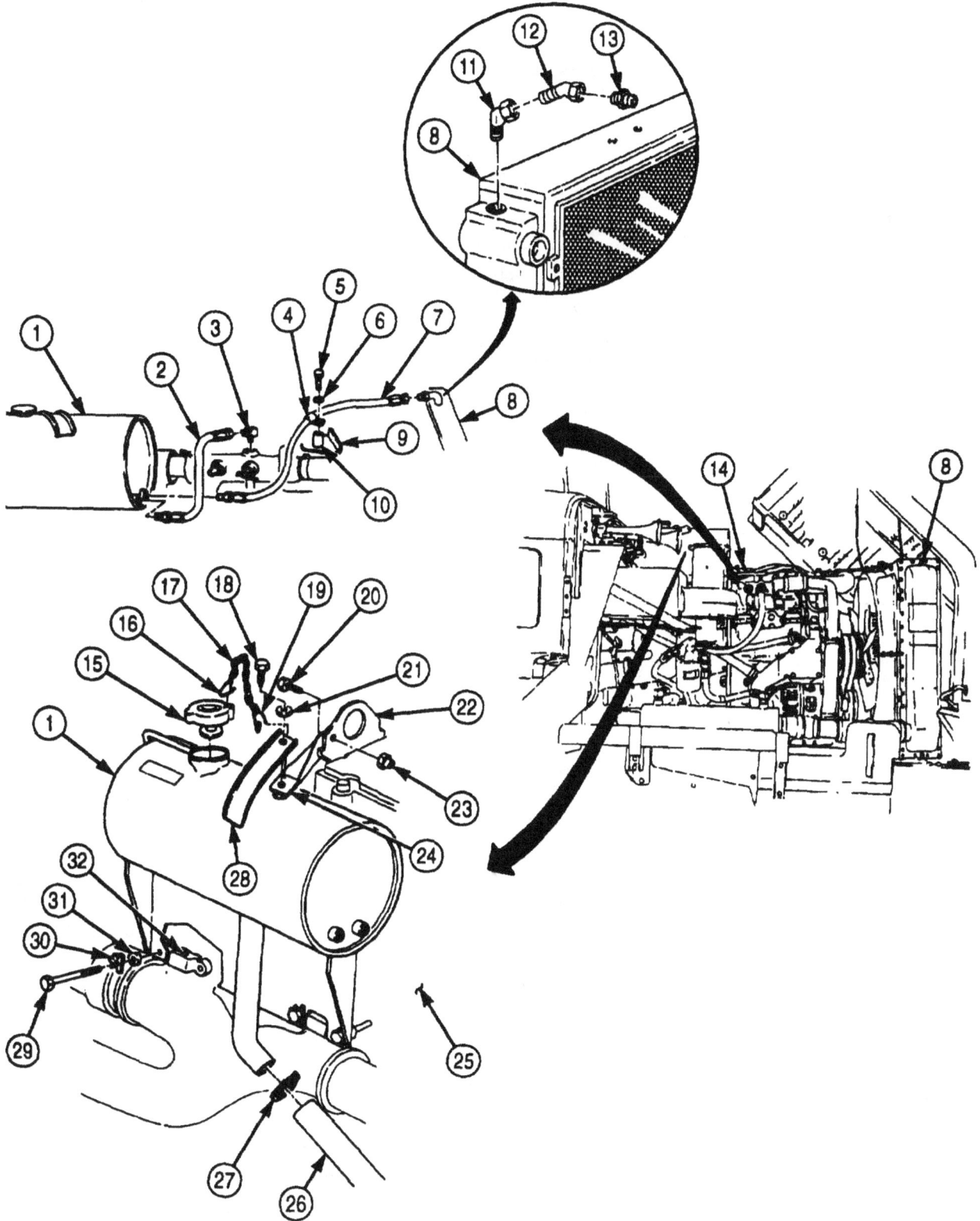

3-61. SURGE TANK, RADIATOR VENT HOSE, AND MANIFOLD RETURN HOSE (M939/A1) REPLACEMENT (Contd)

c. Surge Tank Installation

1. Install surge tank (3) on cylinder head (14) with two exhaust manifold clamps (21), four washers (20), new keywashers (19), and screws (18). Ensure clamps (21) are parallel to surface of cylinder head (14). Tighten screws (18) 15-20 lb-ft (20-27 N•m).

2. Retighten screws (18) 40-45 lb-ft (54-61 N-m) and bend tabs of keywashers (19) against flats of screws (18).

3. Connect engine oil cooler hose (15) to surge tank (3) and tighten clamp (16).

4. Install surge tank support (13) on lifting eye (11) with screw (9) and new locknut (12).

5. Install surge tank support extension (17) on support (13) with washer (10) and screw (7).

6. Connect chain (6) to cap (4) and surge tank support (13) with pin (8).

7. Install cap (4) on surge tank (3) with retaining pin (5).

d. Radiator Vent and Manifold Return Hoses Installation

NOTE

- Fittings must be cleaned and inspected for cracks and crossed or stripped threads.
- Male pipe threads must be wrapped with antiseize tape before installation.

1. Install manifold return hose (22) and radiator vent hose (27) on surge tank (3).

2. Install elbow (23) on water manifold (1).

3. Install elbows (30) and (31) and adapter (32) on radiator (2).

4. Connect manifold return hose (22) to elbow (23) on water manifold (1).

5. Connect radiator vent hose (27) to adapter (32) on radiator (2).

6. Install spacer (29) and hose clamp (24) on engine bracket (28) with washer (26) and screw (25). Tighten screw (25) 75-95 lb-in (8-10 N•m).

3-61. SURGE TANK, RADIATOR VENT HOSE, AND MANIFOLD RETURN HOSE (M939/A1) REPLACEMENT (Contd)

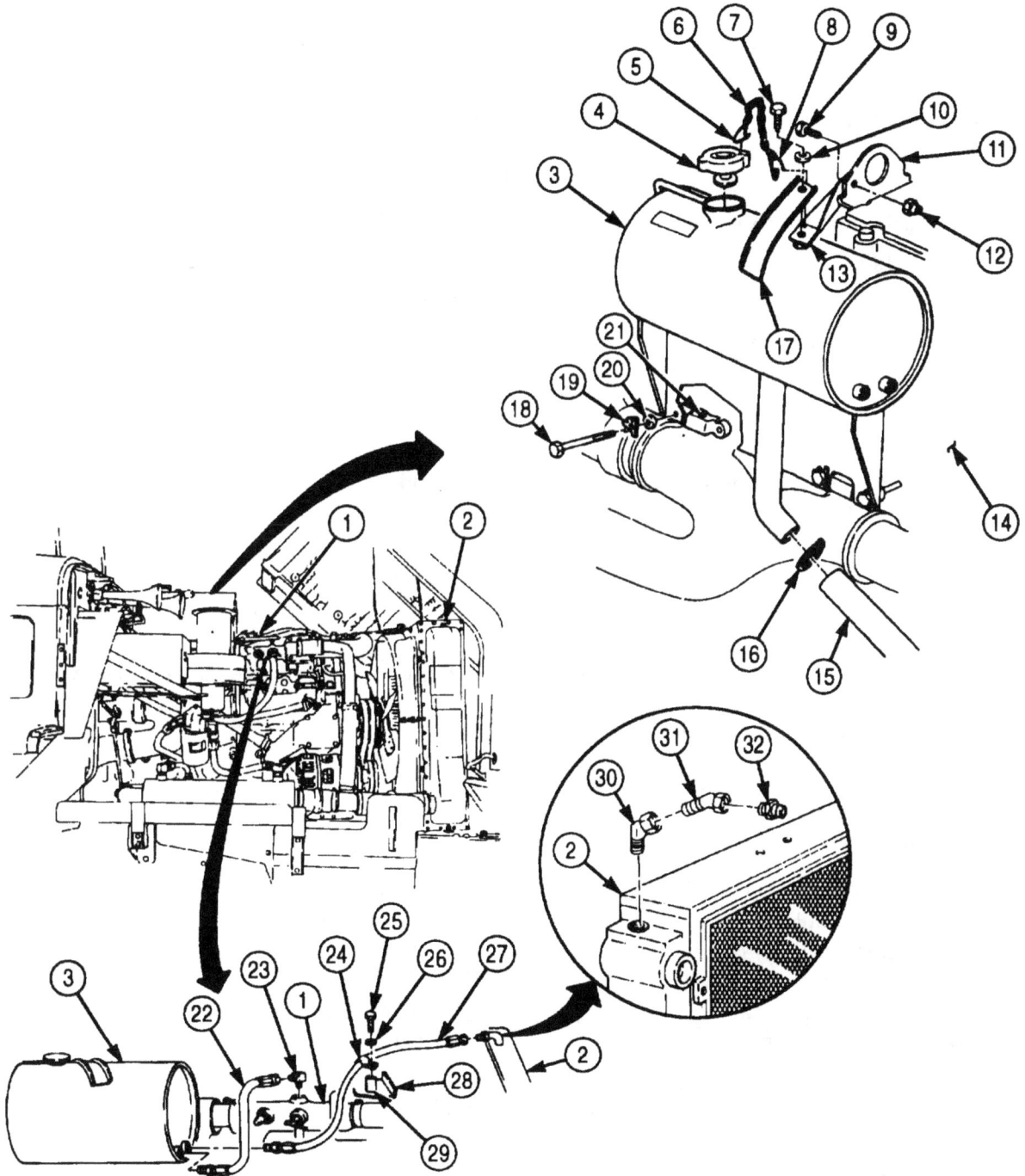

FOLLOW-ON TASKS: • Install right splash shield (TM 9-2320-272-10).
• Fill cooling system (para. 3-53).
• Start engine (TM 9-2320-272-10) and check cooling system for leaks.

3-62. SURGE TANK AND BRACKET (M939A2) REPLACEMENT

THIS TASK COVERS:

a. Removal   b. Installation

INITIAL SETUP:

APPLICABLE MODELS
M939A2

TOOLS
General mechanic's tool kit (Appendix E, Item 1)

MATERIALS/PARTS
Two locknuts (Appendix D, Item 279)
Tiedown straps (Appendix D, Item 685)
Antiseize tape (Appendix C, Item 72)

REFERENCES (TM)
TM 9-2320-272-10
TM 9-2320-272-24P

EQUIPMENT CONDITION
• Parking brake set (TM 9-2320-272-10).
• Hood raised and secured (TM 9-2320-272-10).
• Cooling system drained as required (para. 3-53).
• Exhaust pipe removed (para. 3-50).

---

a. Removal

NOTE
Remove tiedown straps as necessary. Note location for installation.

1. Loosen clamp (18) and disconnect hose (17) from surge tank (3).
2. Loosen clamp (16) and disconnect hose (14) from surge tank (3).
3. Loosen clamp (12) and disconnect hose (13) from surge tank (3).
4. Remove fitting (4) from surge tank (3).
5. Disconnect hose (11) from surge tank (3).
6. Remove two locknuts (6), washers (5), screws (24), nuts (7), clamps (23), surge tank (3), and two seats (22) from bracket (20). Discard locknuts (6).
7. Remove cap (2), chain (1), and S-hook (25) from surge tank (3).
8. Remove two screws (19), screw (10), and bracket (20) from engine (15).
9. Remove nut (9), screw (21), and bracket (8) from bracket (20).

b. Installation

1. Install S-hook (25), chain (1), and cap (2) on surge tank (3).
2. Install bracket (8) on bracket (20) with screw (21) and nut (9). Do not tighten nut (9).
3. Install bracket (8) and (20) on engine (15) with two screws (19) and screw (10).
4. Position two seats (22) on bracket (20).
5. Position two clamps (23) on surge tank (3) and install clamps (23) and surge tank (3) on bracket (20) with two screws (24), nuts (7), washers (5), and new locknuts (6).
6. Connect hose (14) on surge tank (3) and tighten clamp (16).
7. Connect hose (17) to surge tank (3) and tighten clamp (18).

NOTE
Male pipe threads must be wrapped with antiseize tape before installation.

3-62. SURGE TANK AND BRACKET (M939A2) REPLACEMENT (Contd)

7. Install fitting (4) on surge tank (3).
8. Connect hose (13) to fitting (4) and tighten clamp (12).
9. Connect hose (11) to surge tank (3).

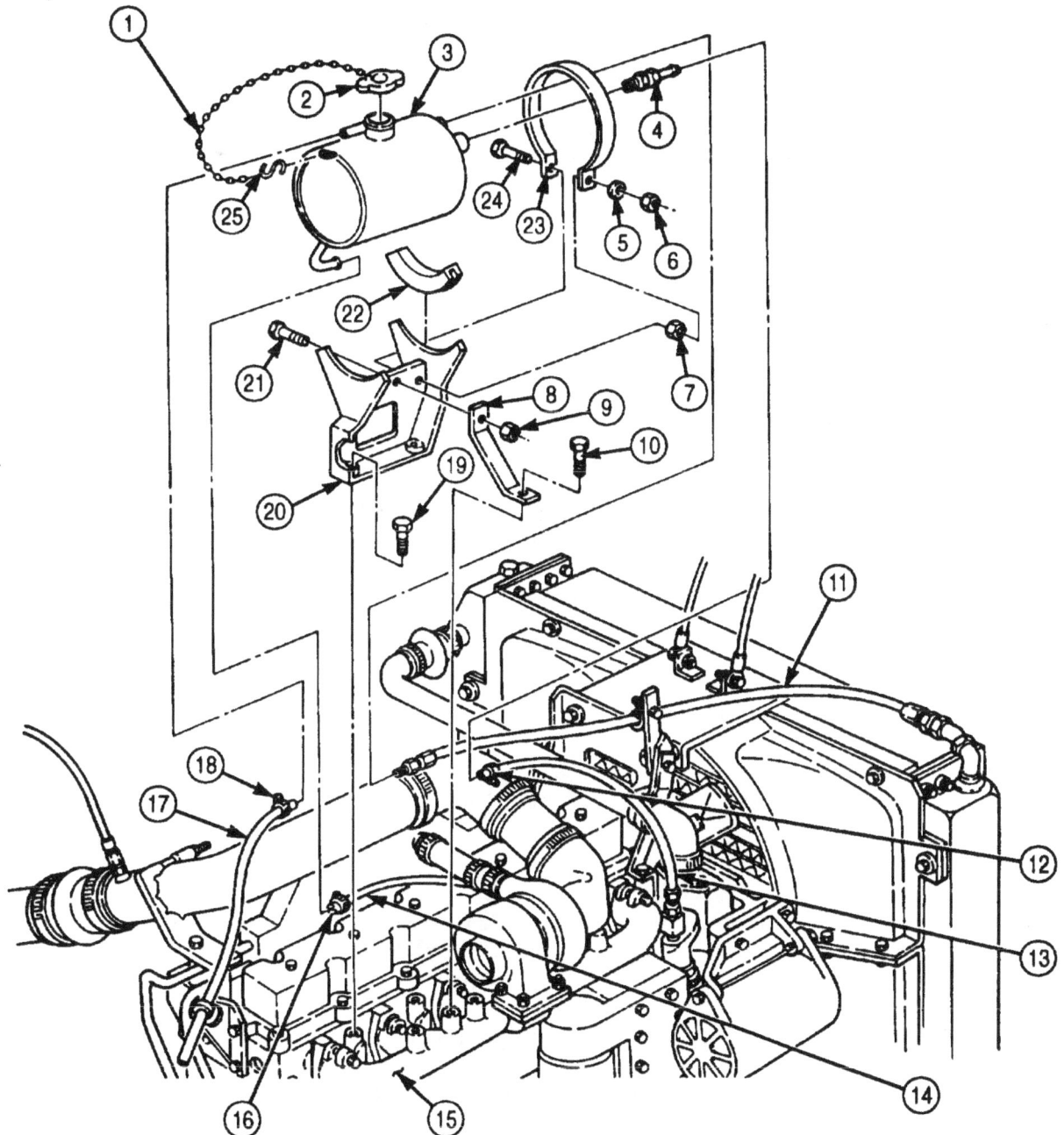

FOLLOW-ON TASKS: • Install exhaust pipe (para. 3-50).
                   • Fill cooling system to proper level (para. 3-53).
                   • Start engine (TM 9-2320-272-10) and check for coolant leaks.

3-63. RADIATOR FAN SHROUD (M939/A1) REPLACEMENT

THIS TASK COVERS

a. Removal   b. Installation

INITIAL SETUP:

APPLICABLE MODELS   REFERENCES (TM)
M939/A1   TM 9-2320-272-10
                                     TM 9-2320-272-24P
TOOLS
General mechanic's tool kit (Appendix E, Item 1)   EQUIPMENT CONDITION
Torque wrench (Appendix E, Item 146)   • Parking brake set (TM 9-2320-272-10).
                                                    • Hood raised and secured (TM 9-2320-272-10).
MATERIALS/PARTS   • Left and right splash shields removed
Six lockwashers (Appendix D, Item 364)   (TM 9-2320-272-10).
Four locknuts (Appendix D, Item 299)   • Cooling system drained as needed (para. 3-53).
Four locknuts (Appendix D, Item 313)
Eight locknuts (Appendix D, Item 304)

---

a. Removal

NOTE
Have drainage container ready to catch coolant.

1. Loosen hose clamp (2) and disconnect radiator inlet hose (1) from radiator pipe (3).

2. Remove screw (20), lockwasher (21), and washer (22) from comer shroud (23). Discard lockwasher (21).

3. Remove locknut (5), screw (18), and washers (6) and (19) from radiator (4) and comer shroud (23). Discard locknut (5).

4. Remove four locknuts (10), eight washers (9), four screws (17), and comer shroud (23) from radiator (4). Discard locknuts (10).

5. Remove three locknuts (7), washers (8), screws (13), and washers (12) from fan shroud (11) from radiator (4). Discard locknuts (7).

6. Remove five screws (16), washers (14), and lockwashers (15) from radiator (4). Discard lockwashers (15).

7. Remove fan shroud (11) from radiator (4).

8. Remove eight locknuts (27), sixteen washers (24), eight screws (25), and four brackets (26) from radiator (4). Discard locknuts (27).

b. Installation

1. Install four brackets (26) on radiator (4) with sixteen washers (24), eight screws (25), and new locknuts (27). Tighten screws (25) 10-14 lb-ft (14-19 N•m).

2. Install fan shroud (11) on radiator (4) with five washers (14), new lockwashers (15), and screws (16). Tighten screws (16) to 110-14 lb-ft (14-19 N•m).

3. Install fan shroud (11) on radiator (4) with three washers (12), screws (13), washers (8), and new locknuts (7).

4. Install corner shroud (23) on radiator (4) with four washers (9), screws (17), washers (9), and new locknuts (10). Tighten screws (17) 66-86 lb-in. (7-10 N•m).

5. Install comer shroud (23) on radiator (4) with washer (19), screw (18), washer (6), and new locknut (5). Tighten screw (18) 10-14 lb-ft (14-19 N•m).

6. Secure corner shroud (23) on radiator (4) with washer (22), new lockwasher (21), and screw (20). Tighten screw (20) 110-14 lb-ft (14-19 N•m).

3-63. RADIATOR FAN SHROUD (M939/A1) REPLACEMENT (Contd)

7.   Connect radiator hose (1) to radiator inlet (3) and tighten clamp (2).

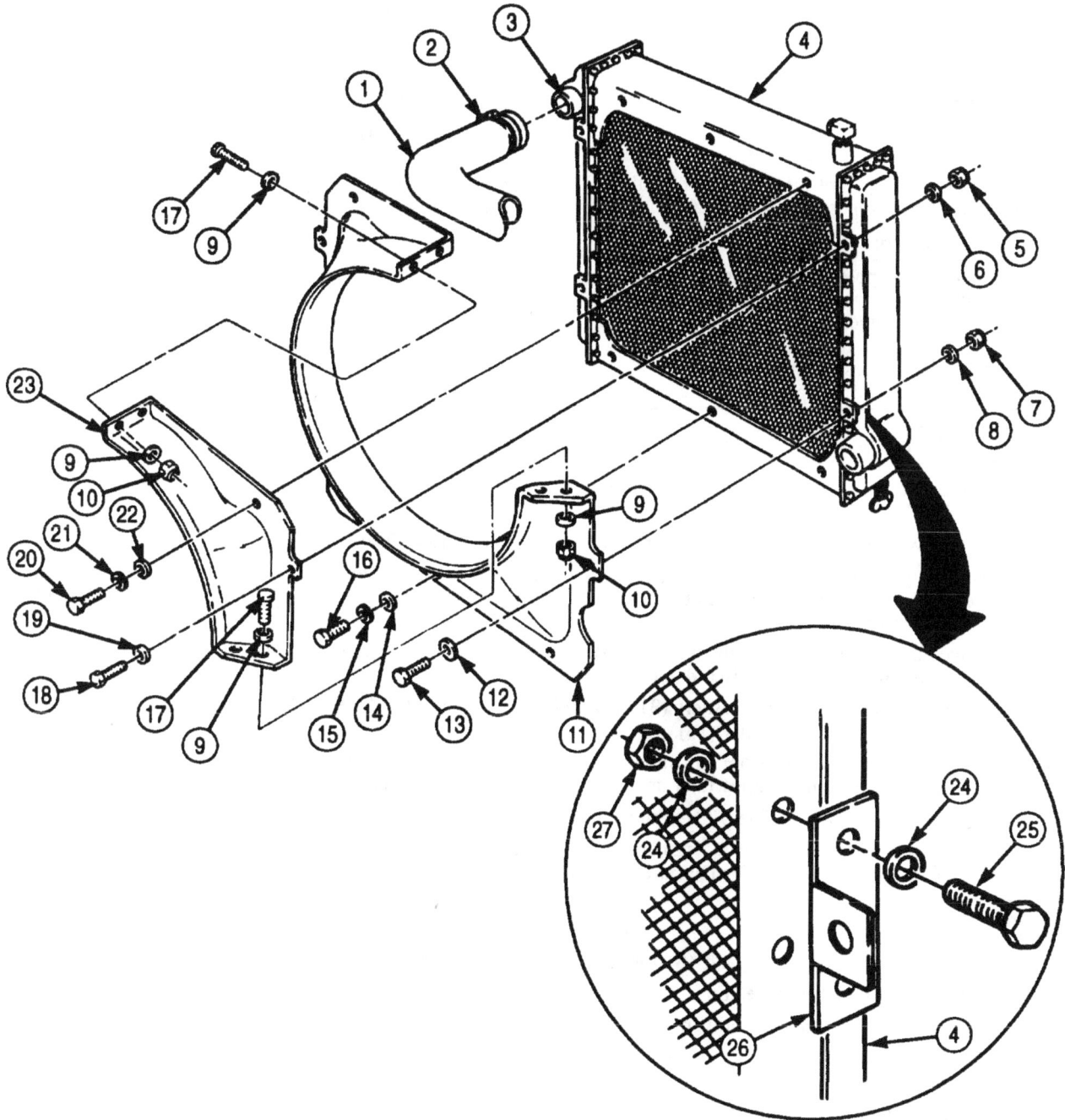

FOLLOW-ON TASKS: • Fill cooling system to proper level (para. 3-53).
                   • Start engine (TM 9-2320-272-10) and check for coolant leaks.
                   • Install left and right splash shields (TM 9-2320-272-10).

3-64. FAN AND FAN SHROUD (M939A2) MAINTENANCE

THIS TASK COVERS:

a. Removal   c. Installation
b. Inspection

INITIAL SETUP:

APPLICABLE MODELS
M939A2

TOOLS
General mechanic's tool kit (Appendix E, Item 1)
Torque wrench (Appendix E, Item 146)

MATERIALS/PARTS
Four locknuts (Appendix D, Item 299)
Four locknuts (Appendix D, Item 313)
Two locknuts (Appendix D, Item 278)
Two lockwashers (Appendix D, Item 360)
Two sets of isolators (Appendix D, Item 263)
Antiseize compound (Appendix C, Item 10)

REFERENCES (TM}
TM 9-2320-272-10
TM 9-2320-272-24P

EQUIPMENT CONDITION
• Parking brake set (TM 9-2320-272-10).
• Hood raised and secured (TM 9-2320-272-10).

---

a. Removal

1. Remove two locknuts (24), snubbing washers (25), screws (13), and washers (14) from brackets (18) and (23). Discard locknuts (24).

2. Remove two sets of isolators (12) from brackets (23) and (18). Discard isolators (12).

3. Straighten tabs on two lockwashers (20) and remove screws (21), lockwashers (20), bracket (23), and spacer (19) from radiator (17). Let bracket (23) hang from attaching cables (22) on vehicle hood (1). Discard lockwashers (20).

4. Remove four locknuts (5), eight washers (6), and four screws (7) securing top half of shroud (2) to bottom half of shroud (8). Discard locknuts (5).

6. Remove five screws (4), six washers (3), two locknuts (16), and top half of shroud (2) from radiator (17). Discard locknuts (16).

6. Remove two screws (27) from fan mounting bracket (26), position fan drive pulley (28) with fan clutch (29) so fan clutch holes (34) are aligned, and install two screws (27) in fan clutch holes (34).

7. Remove six nuts (30), washers (31), and fan blade (32) from fan clutch (29).

8. Remove five screws (10), seven washers (11), two locknuts (15), clip (9), and bottom half of shroud (8) from radiator (17). Discard locknuts (15).

b. Inspection

1. Inspect fan (32) shrouds (2) and (8) for cracks and missing pieces.

2. Lay fan (32) on flat surface. Ensure all blades of fan (32) contact the flat surface the same way.

3. Inspect fan (32) for cracks, broken pieces, and loose anchorage and studs (33) for cracks or stripped threads.

4. Inspect all other parts (para. 2-15).

5. Replace all parts failing inspection.

c. Installation

1. Install bottom half of shroud (8) on radiator (17) with clip (9), five screws (10), seven washers (11), and two new locknuts (15). Ensure clip (9) is mounted under correct screw (10).

3-64. FAN AND FAN SHROUD (M939A2) MAINTENANCE (Contd)

2. Align fan blade (32) with fan clutch holes (34) and install six washers (31) and nuts (30). Tighten nuts (30) 36 lb-ft (47 N•m).

3. Remove two screws (27) from fan clutch (29).

4. Coat screw (27) threads with antiseize compound and install screws (27) on fan mounting bracket (26).

5. Install top half of shroud (2) on radiator (17) with five screws (4), six washers (3), and two new locknuts (16). Finger tighten screws (4).

6. Secure halves of shroud (2) and (8) with eight washers (6), four screws (7), and new locknuts (6).

7. Tighten five screws (4) and two locknuts (16).

8. Install two new sets of isolators (12) in brackets (18) and (23).

9. Install spacer (19) and bracket (23) on bracket (18) and radiator (17) with two screws (13), washers (14), snubbing washers (26), and new locknuts (24). Finger tighten locknuts (24).

10. Install two new lockwashers (20) and screws (21) on bracket (23) and radiator (17). Tighten screws (21) 40 lb-ft (55 N•m). Bend tabs on lockwashers (20) against flats of screws (21).

11. Tighten locknuts (24) 35 lb-ft (50 N•m).

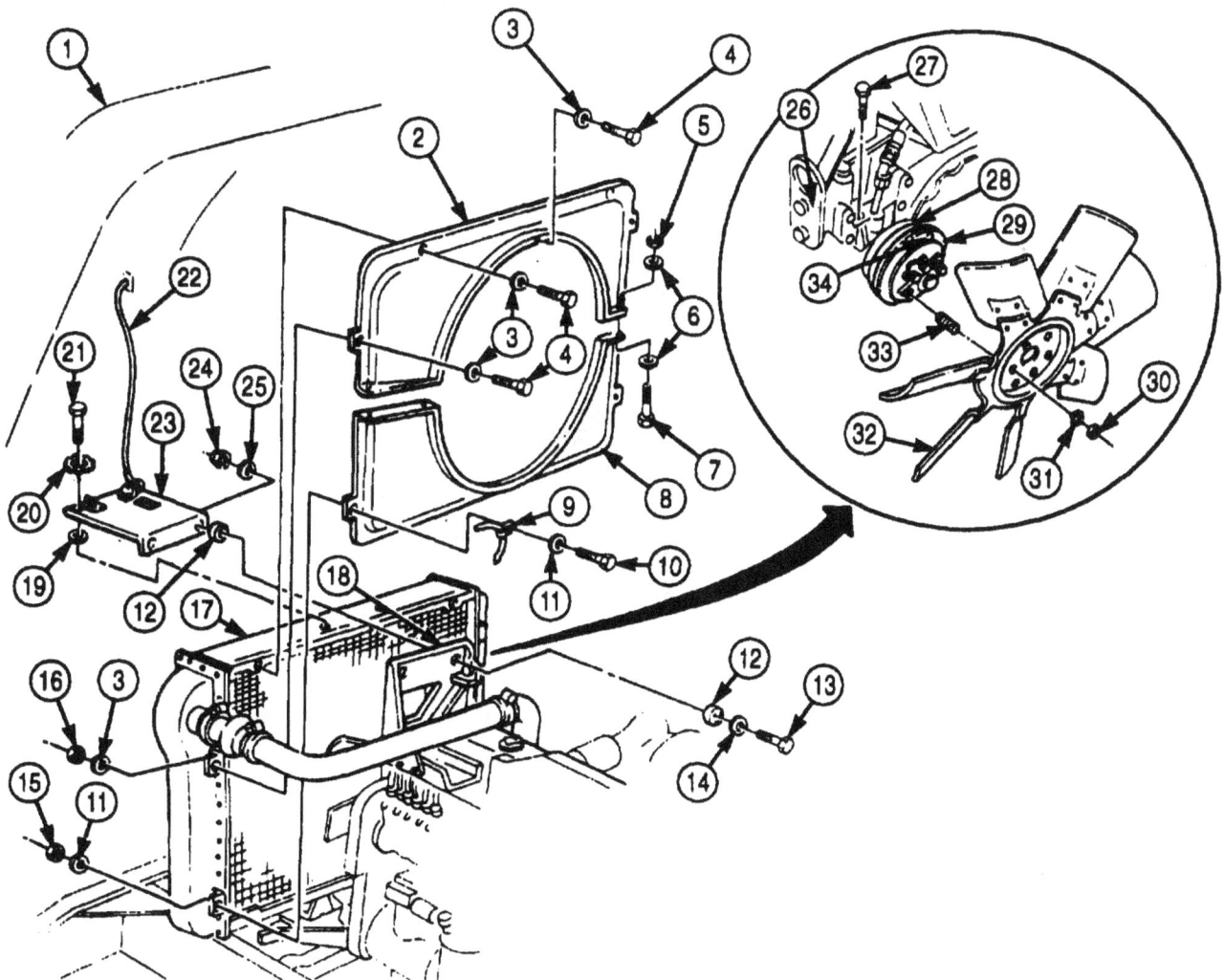

FOLLOW-ON TASK: Start engine (TM 9-2320-272-10) and run until fan clutch locks up. Check fan for alignment.

3-65. THERMOSTAT (M939/A1) MAINTENANCE

THIS TASK COVERS:

a. Removal   c. Installation
b. Testing

INITIAL SETUP:

APPLICABLE MODELS
M939/A1

TOOLS
General mechanic's tool kit (Appendix E, Item 1)
Hot water thermometer

MATERIALS/PARTS
Four screw-assembled lockwashers (Appendix D, Item 592)
Gasket (Appendix D, Item 147)
Thermostat seal (Appendix D, Item 597)
GAA grease (Appendix C, Item 28)

REFERENCES (TM)
TM 9-2320-272-10
TM 9-2320-272-24P

EQUIPMENT CONDITION
• Hood raised and secured (TM 9-2320-272-10).
• Parking brake set (TM 9-2320-272-10).
• Right splash shield removed (TM 9-2320-272-10).
• Cooling system drained as necessary (para. 3-53).

GENERAL SAFETY INSTRUCTIONS
Use caution when testing thermostat. Scalding hot water may cause severe burns.

---

a. Removal

1. Loosen clamp (6) and disconnect radiator inlet hose (7) from thermostat housing (5).

2. Loosen two clamps (9) and remove radiator bypass hose (10) from thermostat housing (5) and radiator bypass tube (8).

3. Remove four screw-assembled lockwashers (11) from thermostat housing (5) and water manifold header (1). Discard screw-assembled lockwashers (11).

4. Remove thermostat housing (5) and gasket (2) from water manifold header (1). Discard gasket (2) and clean gasket remains from mating surfaces.

5. Remove thermostat (3) from thermostat housing (5).

6. Remove seal (4) from thermostat housing (5). Discard seal (4).

b. Testing

WARNING
Use caution when testing thermostat. Hot water may cause injury to personnel.

1. Place thermostat (3) in a container of water at 197°F (92°C).

NOTE
Do not let thermostat touch container sides during testing.

2. Observe thermostat (3) to see if valve opens. Replace thermostat (3) if valve does not fully open.

c. Installation

1. Position new seal (4) in thermostat housing (5).

2. Install thermostat (3) in thermostat housing (5).

3. Coat one surface of gasket (2) with light coat of GAA grease and position new gasket (2) over four holes on water manifold header (1).

3-65. THERMOSTAT (M939/A1) MAINTENANCE (Contd)

4. Aligning holes in thermostat housing (5) and header (1), install housing (5) with four new screw-assembled lockwashers (11).

5. Connect radiator bypass hose (10) to thermostat housing (5) and bypass tube (8) and tighten two clamps (9).

6. Connect radiator inlet hose (7) to thermostat housing (5) and tighten clamp (6).

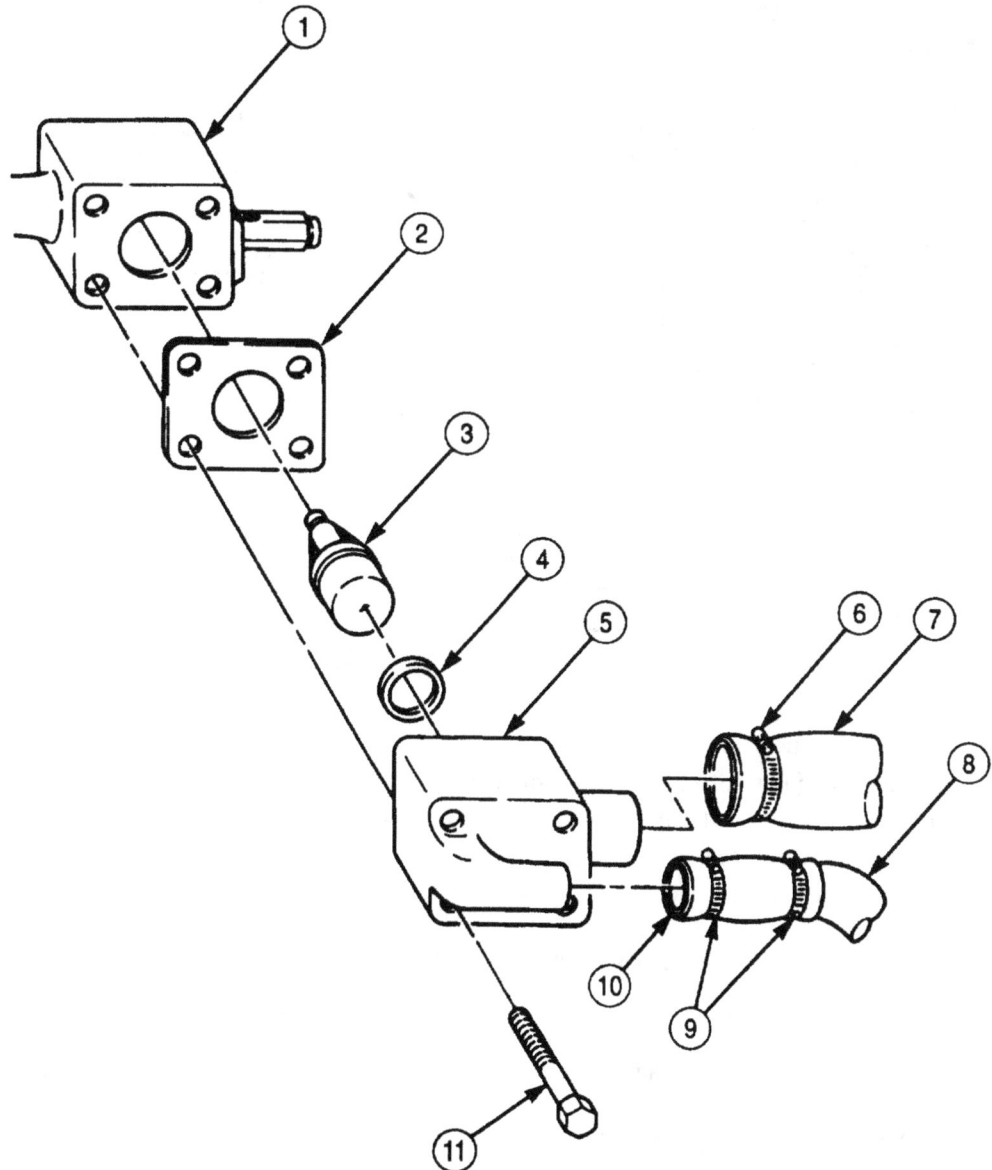

FOLLOW-ON TASKS: • Fill cooling system to proper level (para. 3-53).
                    • Start engine (TM 9-2320-272-10), check for coolant leaks, and check temperature gauge for normal reading of 175°F to 195°F (70°C to 97°C).
                    • Install right splash shield (TM 9-2320-272-10).

3-66. THERMOSTATS AND HOUSING (M939A2) MAINTENANCE

THIS TASK COVERS:

a. Removal (Internal Bypass)   d. Removal (External Bypass)
b. Cleaning and Inspection (Both Types)   e. Installation (External Bypass)
c. Installation (Internal Bypass)

INITIAL SETUP:

APPLICABLE MODELS
M939A2

TOOLS
General mechanic's too kit (Appendix E, Item 1)
Torque wrench (Appendix E, Item 146)

MATERIALS/PARTS
Gasket (Appendix D, Item 148)
Gasket (Appendix D, Item 149)
Gasket (Appendix D, Item 150)
Gasket (Appendix D, Item 151)
Gasket (Appendix D, Item 152)
Seal (Appendix D, Item 598)
Antiseize tape (Appendix C, Item 72)

REFERENCES (TM)
TM 9-2320-272-10
TM 9-2320-272-24P

EQUIPMENT CONDITION
• Parking brake set (TM 9-2320-272-10).
• Hood raised and secured (TM 9-2320-272-10).
• Drivebelt tensioner removed (para. 3-71)

GENERAL SAFETY INSTRUCTIONS
Do not drain coolant when engine is hot.

NOTE

There are two different bypass systems used for M939A2 engines. An identifiable difference is the thermostat housing for external bypass and internal bypass engines.

a. Removal (Internal Bypass)

WARNING

Do not drain coolant when engine is hot. Severe bums to personnel may result.

1. Drain three quarts of coolant from oil cooler drainvalve (11) into container.

2. Loosen clamp (3) and remove hose (2) from adapter (4).

3. Loosen alternator link screw (12).

4. While supporting alternator (15), remove two screws (13) and washers (14) from bracket (16). Pivot alternator (15) down and move bracket (16) aside.

5. Remove four screws (17) and bracket (18) from thermostat housing (7).

NOTE
Tag screws for identification in installation.

6. Remove two screws (1), adapter (4), and gasket (5) from thermostat housing (7). Discard gasket (5)

7. Remove two screws (6), thermostat housing (7), and gasket (8) from engine block (10). Discard gasket (8).

CAUTION

Cover coolant passages. Dirt in cooling system will cause overheating and engine failure.

8. Remove two thermostats (9) from engine block (10).

3-66. THERMOSTATS AND HOUSING (M939A2) MAINTENANCE (Contd)

b. Cleaning and Inspection (Both Types)

1. Clean all parts and mating surfaces (para. 2-14).
2. Inspect all other parts (para. 2-15).
3. Replace all parts failing inspection.

c. Installation (Internal Bypass)

1. Install two thermostats (9) in engine block (10).
2. Install new gasket (8) on engine block (10).
3. Install thermostat housing (7) on engine block (10) with two screws (6).
4. Install new gasket (5) and adapter (4) on thermostat housing (7) with two screws (1). Tighten screws (1) and (6)
5. Install bracket (18) on thermostat housing (7) with four screws (17).
6. Move alternator (15) up with bracket (16), and install two washers (14) and screws (13). Tighten screws (13) 35 lb-ft (45 N°m).
7. Tighten link screw (12).
8. Install hose (2) on adapter (4) with clamp (3).

NOTE
Go to follow-on tasks.

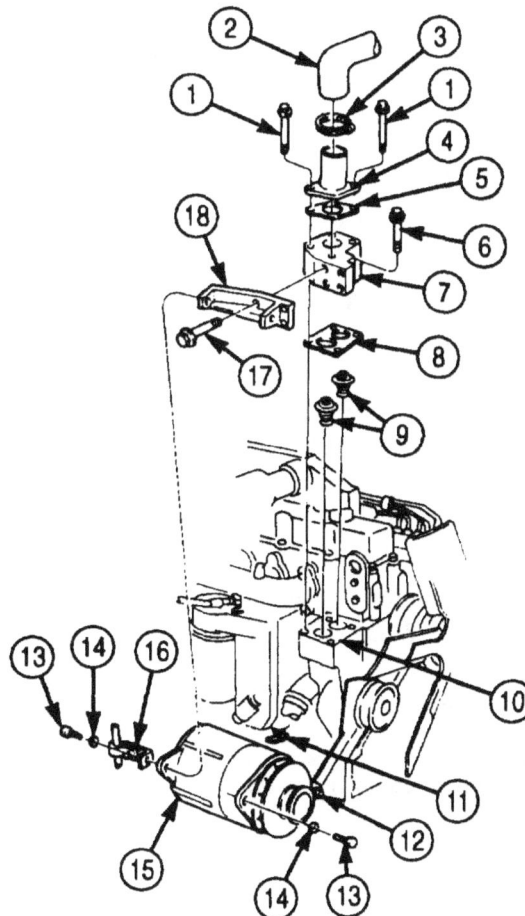

3-66. THERMOSTATS AND HOUSING (M939A2) MAINTENANCE (Contd)

d. Removal (External Bypass)

WARNING

Do not drain coolant when engine is hot. Severe bums may result.

1. Drain three quarts of coolant from oil cooler drain valve (17) into container.
2. Loosen clamp (5) and remove hose (4) from adapter (8).
3. Loosen alternator link screw (18).
4. While supporting alternator (21), remove two screws (19) and washers (20) from brackets (22) and (25) and move alternator (21) aside.
5. Remove four screws (23) and bracket (25) from support (14).
6. Loosen clamp (33) and remove hose (1) from valve (32).
7. Loosen tube (29) and fittings (2) and (30) from tee (31).
8. Remove tee (31) from thermostat housing (10).
9. Remove valve (32) from tee (31).

NOTE

Note the location of long and short screws.

10. Remove short screw (6), long screw (3), adapter (8), and gasket (9) from thermostat housing (10). Discard gasket (9).
11. Loosen clamp (28) and remove hose (26) from housing (10).
12. Remove two screws (7), housing (10), thermostat (12), and gasket (13) from support (14). Discard gasket (13).
13. Remove seal (11) from housing (10). Discard seal (11).
14. Remove two screws (27), support (14), and gasket (15) from engine block (16). Discard gasket (15).
15. Remove plug (24) from support (14).
16. Go to task b for inspection procedures.

e. Installation (External Bypass)

NOTE

If support was removed, perform steps 1 and 2.

1. Install plug (24) in support (14).
2. Install new gasket (15) and support (14) on engine block (16) with two screws (27).
3. With housing (10) upside down, install new seal (11), with cupped side down, until seal (11) is seated.

CAUTION

If seal is damaged, engine failure may result.

4. Install thermostat (12) into seal (11) until thermostat (12) seats against housing (10).
5. Install new gasket (13) and housing (10) on support (14) with two screws (7).
6. Install hose (26) on housing (10) and tighten clamp (28).
7. Install new gasket (9) and adapter (8) on housing (10) with long screw (3) and short screw (6).

Ensure that drainvalve on aftercooler is open or cooling system may overheat.

3-66. THERMOSTATS AND HOUSING (M939A2) MAINTENANCE (Contd)

8. Wrap threads of valve (32) with antiseize tape and install in tee (31), and install tee (31) on housing (10).

9. Install tube (29), fitting (2), and fitting (30) on tee (31).

10. Install hose (1) on valve (32) and tighten clamp (33).

11. Install bracket (25) on support (14) with four screws (23).

12. Move alternator (21) up and align bracket (22) to support (14) and install two washers (20) and screws (19). Tighten screws (19) 35 lb-ft (45 N•m).

13. Tighten link screw (18).

14. Install hose (4) on adapter (8) and tighten clamp (5).

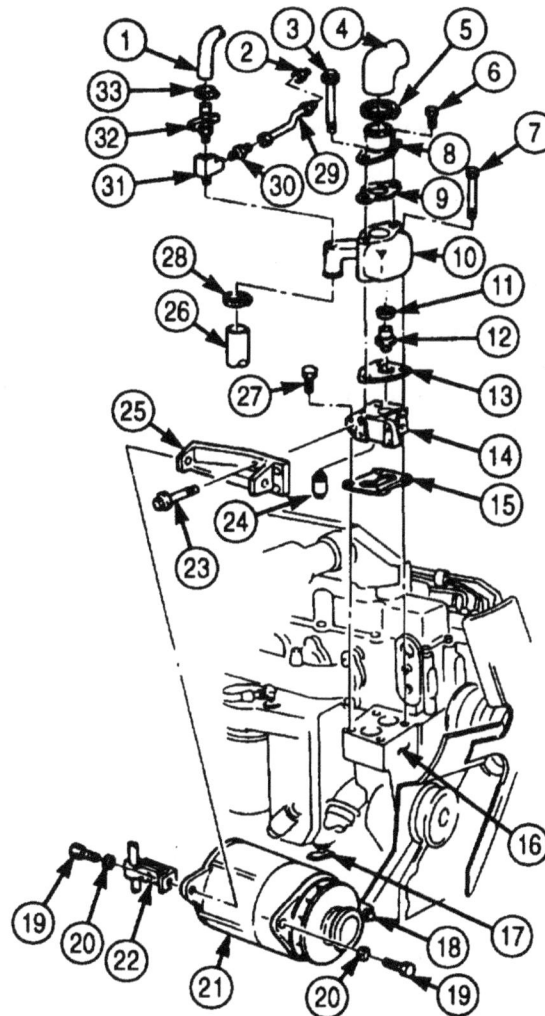

FOLLOW-ON TASKS: • Install drivebelt tensioner (3-71).
• Fill cooling system to proper level (para. 3-53).
• Start engine (TM 9-2320-272-10), check for leaks, and check temperature gauge for indication of 180° F to 205° F (85' C to 95° C).

3-67. WATER PUMP DRIVEBELT (M939/A1) MAINTENANCE

THIS TASK COVERS:

a. Adjustment
b. Removal

c. Inspection
d. Installation

INITIAL SETUP:

APPLICABLE MODELS
M939/A1

TOOLS
General mechanic's tool kit (Appendix E, Item 1)
Belt tension gauge (Appendix E, Item 16)

MATERIALS/PARTS
Three lockwashers (Appendix D, Item 350)

REFERENCES (TM)
TM 9-2320-2722-10
TM 9-2320-2722-24P

EQUIPMENT CONDITION

• Parking brake set (TM 9-2320-272-10).
• Hood raised and secured (TM 9-2320-272-10).
• Right splash shield removed (TM 9-2320-272-10).
• Fan drivebelts removed (para. 3-70).
• Power steering pump drivebelts removed (para. 3-230).

a. Adjustment

1. Check for proper belt tension by positioning belt tension gauge on drivebelt (1) between pump housing (6) and accessory drive pulley (2). New belt tension should be 95-105 lb-ft (423-467 N•m). Used belt tension should be 90-95 lb-ft (378-422 N•m).
2. Loosen six screws (4) securing support bracket (3) to engine (7).
3. Place brass drift punch against stud (5) on water pump housing (6).
4. Referring to values in step 1, move stud (5) clockwise facing pump housing (6) to tighten drivebelt (1) tension. Move stud (6) counterclockwise to loosen drivebelt (1) tension.
5. Tighten six screws (4) securing support bracket (3) to engine (7).
6. Recheck drivebelt (1) tension. If drivebelt (1) tension cannot be adjusted, replace drivebelt (1).

3-67. WATER PUMP DRIVEBELT (M939/A1) MAINTENANCE (Contd)

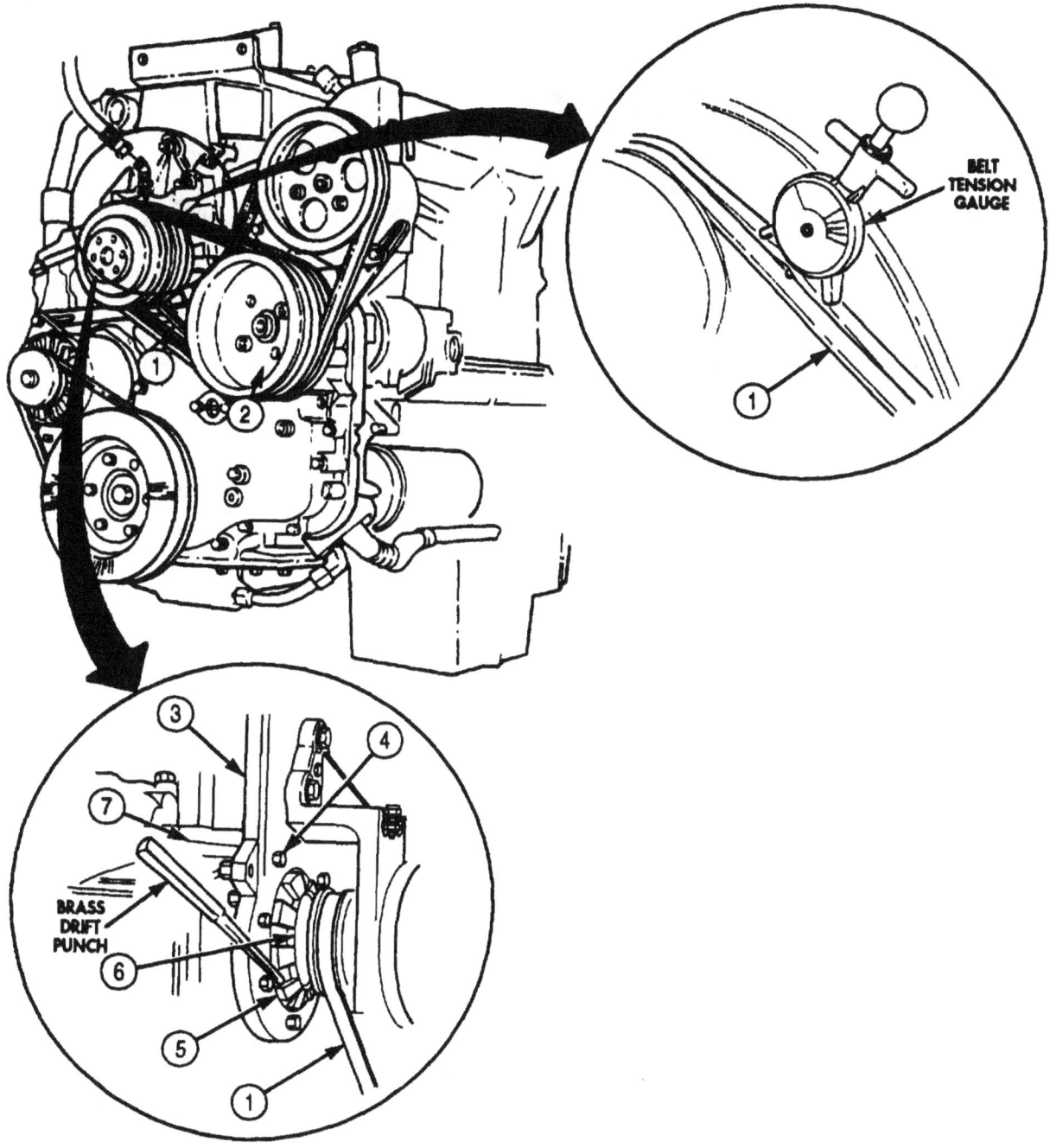

BELT TENSION GAUGE

BRASS DRIFT PUNCH

3-67. WATER PUMP DRIVEBELT (M939/A1) MAINTENANCE (Contd)

b. Removal

NOTE
Assistant will help with step 1.

1. Remove screw (1), lockwasher (2), two screws (11), washers (12), screws (10), lockwashers (9), and washers (8) from fan pulley bracket (3). Discard lockwashers (2) and (9).
2. Lower fan pulley bracket (3).
3. Loosen six screws (13) holding support bracket (4) to engine (16).
4. Place a brass drift punch against stud (14) on water pump housing (15) and turn stud (14) counterclockwise to release drivebelt (7) tension.
5. Remove drivebelt (7) from pump pulley (6) and accessory drive pulley (5).

c. Inspection

Inspect drivebelt for cracks, splits, and breaks. Replace drivebelt if cracked, split, or broken.

d. Installation

1. Install drivebelt (7) over pump pulley (6) and accessory drive pulley (5).
2. Using brass drift punch, turn stud (14) clockwise for preliminary belt (7) tightening.
3. Check drivebelt (7) tension. Refer to task a.
4. Tighten six screws (13).

NOTE
Assistant will help with step 5.

5. Raise fan pulley bracket (3) and install new lockwasher (2), screw (1), two washers (8), new lockwashers (9), screws (10), washers (12), and screws (11). Do not tighten screws (1), (11), and (10).
6. Recheck drivebelt tension. Refer to task a.
7. If drivebelt (7) tension is satisfactory, tighten screws (1), (11), and (10).

3-67. WATER PUMP DRIVEBELT (M939/A1) MAINTENANCE (Contd)

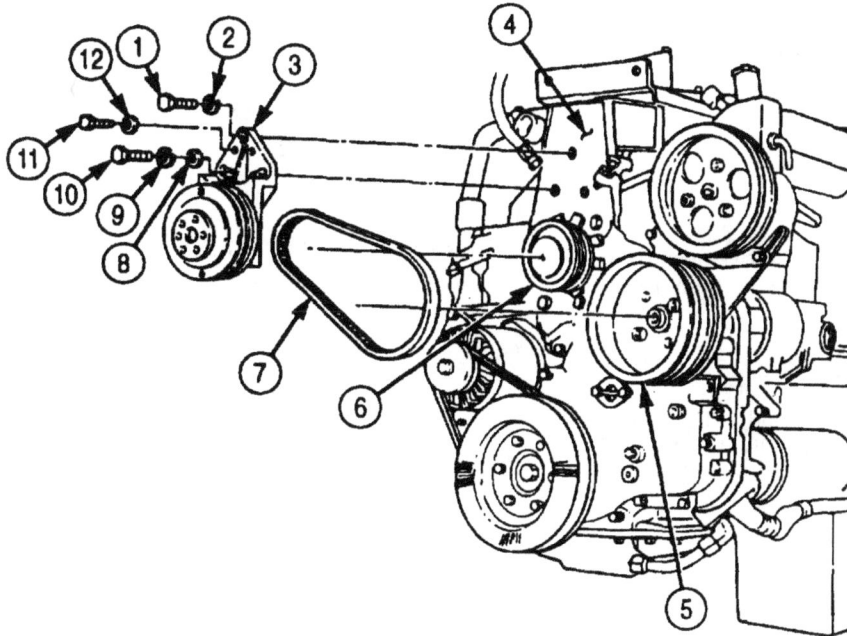

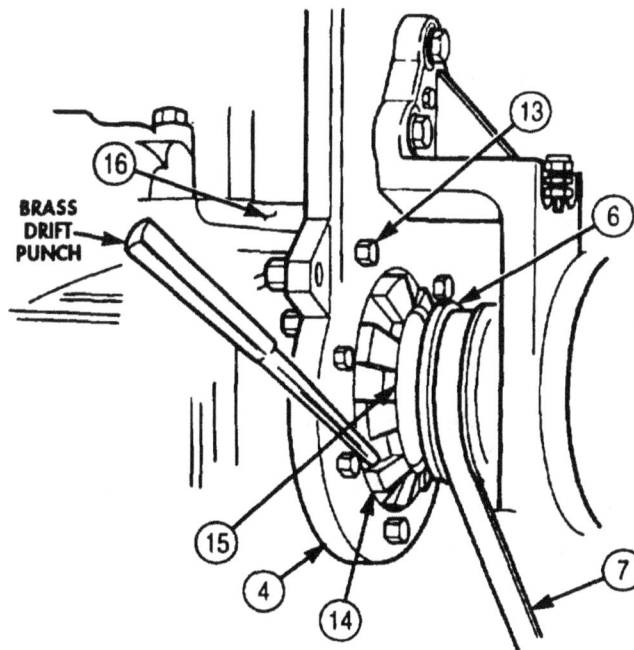

BRASS
DRIFT
PUNCH

FOLLOW-ON TASKS:
- Install power steering pump drivebelts (para. 3-230).
- Install fan drivebelts (para. 3-70).
- Start engine (TM 9-2320-272-10), idle engine for five minutes, and stop engine.
- Install right splash shield (TM 9-2320-272-10).

3-68. WATER PUMP (M939/A1) REPLACEMENT

THIS TASK COVERS:

a. Removal   b. Installation

INITIAL SETUP:

APPLICABLE MODELS
M939/A1

TOOLS
General mechanic's tool kit (Appendix E, Item 1)
Torque wrench (Appendix E, Item 146)

MATERIALS/PARTS
Gasket (Appendix D, Item 153)
O-ring (Appendix D, Item 425)
Eight lockwashers (Appendix D, Item 349)
Two lockwashers (Appendix D, Item 350)
GAA grease (Appendix C, Item 28)

REFERENCES (TM)
TM 9-2320-272-10
TM 9-2320-272-24P

EQUIPMENT CONDITION
• Parking brake set (TM 9-2320-272-10).
• Hood raised and secured (TM 9-2320-272-10).
• Cooling system drained (para. 3-53).
• Fan drive clutch removed (para. 3-75).
• Water pump drivebelt removed (para. 3-67).
• Alternator adjusting link removed (para. 3-79).

---

a. Removal

1. Remove screw (4), lockwasher (5), hose clamp (6), and spacer (7) from engine bracket (8). Discard lockwasher (5).
2. Remove screw (3), lockwasher (2), and washer (1) from engine bracket (8). Discard lockwasher (2).
3. Remove six screws (11), lockwashers (10), and support bracket (9) from engine (12). Discard lockwashers (10).
4. Remove water pump body (16) and O-ring (15) from water pump support (14). Discard O-ring (15).
5. Remove two screws (17) and lockwashers (18) from water pump support (14) and engine (12). Discard lockwasher (18).
6. Remove water pump support (14) and gasket (13) from engine (12). Discard gasket (13).

b. Installation

1. Coat both sides of new gasket (13) lightly with clean GAA grease and position on water pump support (14).
2. Position water pump support (14) with gasket (13) on engine (12).
3. Install water pump support (14) on engine (12) with two new lockwashers (18) and screws (17).
4. Put light coat of GAA grease on O-ring (15), mount O-ring (15) on water pump body (16), and install water pump body (16) into support (14).
5. Install support bracket (9) on engine (12) with six new lockwashers (10) and screws (11). Tighten screws (11) 30 lb-ft (41 N•m).
6. Install washer (1), new lockwasher (2), and screw (3) through engine bracket (8) and into support bracket (9).
7. Install hose clamp (6) and spacer (7) on engine bracket (8) with new lockwasher (5) and screw (4).

3-68. WATER PUMP (M939/A1) REPLACEMENT (Contd)

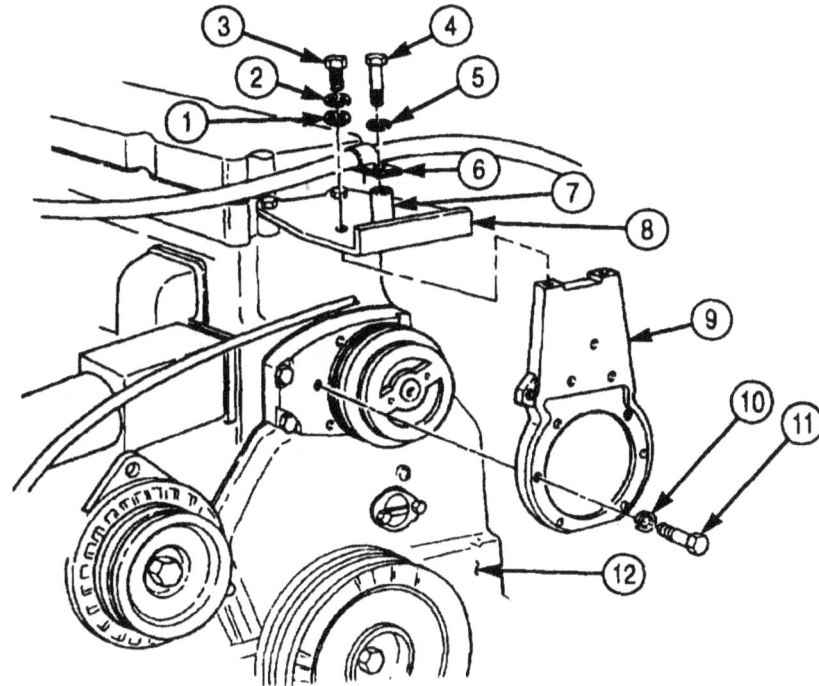

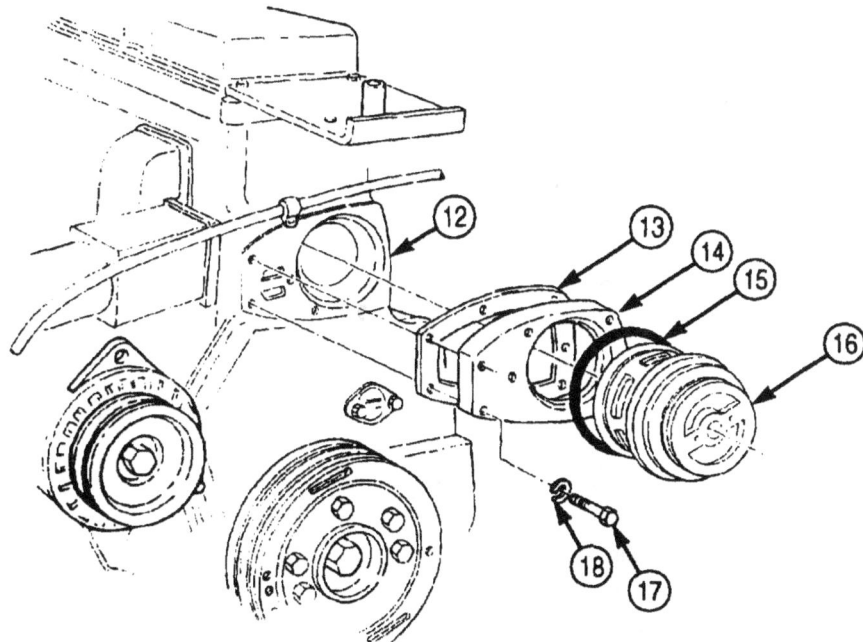

FOLLOW-ON TASKS:
* Install and adjust water pump drivebelt (para. 3-67).
* Install fan drive clutch (para. 3-75).
* Fill cooling system to proper level (para. 3-53).
* Install alternator adjusting link (para. 3-79).
* Start engine (TM 9-2320-272-10) and check for coolant leaks.

3-69. WATER PUMP (M939A2) MAINTENANCE

THIS TASK COVERS:

a. Removal   c. Installation
b. Cleaning and Inspection

INITIAL SETUP:

APPLICABLE MODELS
M939A2

TOOLS
General mechanic's tool kit (Appendix E, Item 1)

MATERIALS/PARTS
O-ring (Appendix D, Item 462)

REFERENCES (TM)
TM 9-2320-272-10
TM 9-2320-272-24P

EQUIPMENT CONDITION
• Parking brake set (TM 9-2320-272-16).
• Hood raised and secured (TM 9-2320-272-10).
• Engine coolant drained (para. 3-53).
• Fan drivebelt removed (para. 3-71).

a. Removal

1. Loosen alternator link screw (10) and remove long screw (8) and two short screws (2) from water pump (3). Note locations of long and short screws.

CAUTION

Do not strike or pry on water pump pulley or shaft. Coolant leaks may result.

2. Tap edge of water pump flange (4) to release water pump (3) from engine block (1).
3. Remove water pump (3) and O-ring (7) from engine block (1). Discard O-ring (7).

b. Cleaning and Inspection

1. Clean mating surfaces of water pump (3) and engine block (1).
2. Inspect water pump flange (4) for cracks and nicks.
3. Check for smooth shaft rotation.
4. Inspect impeller blades (6) for corrosion or wear.
5. Inspect weep hole (5) for evidence of leakage.
6. Replace pump (3) if any parts fail inspection.

c. Installation

CAUTION

Do not pinch, cut, or crimp O-ring while performing steps 1 and 2. Damaged O-ring may result in coolant leak.

1. Install new O-ring (7) on water pump (3).
2. Install water pump (3) engine block (1) with two short screws (2). Hand tighten screws (2).
3. Align alternator link (9) with water pump flange (4) and install long screw (8). Hand tighten screw (8).
4. Tighten alternator link screw (10) and short and long screws (2) and (8).

3-69. WATER PUMP (M939A2) MAINTENANCE (Contd)

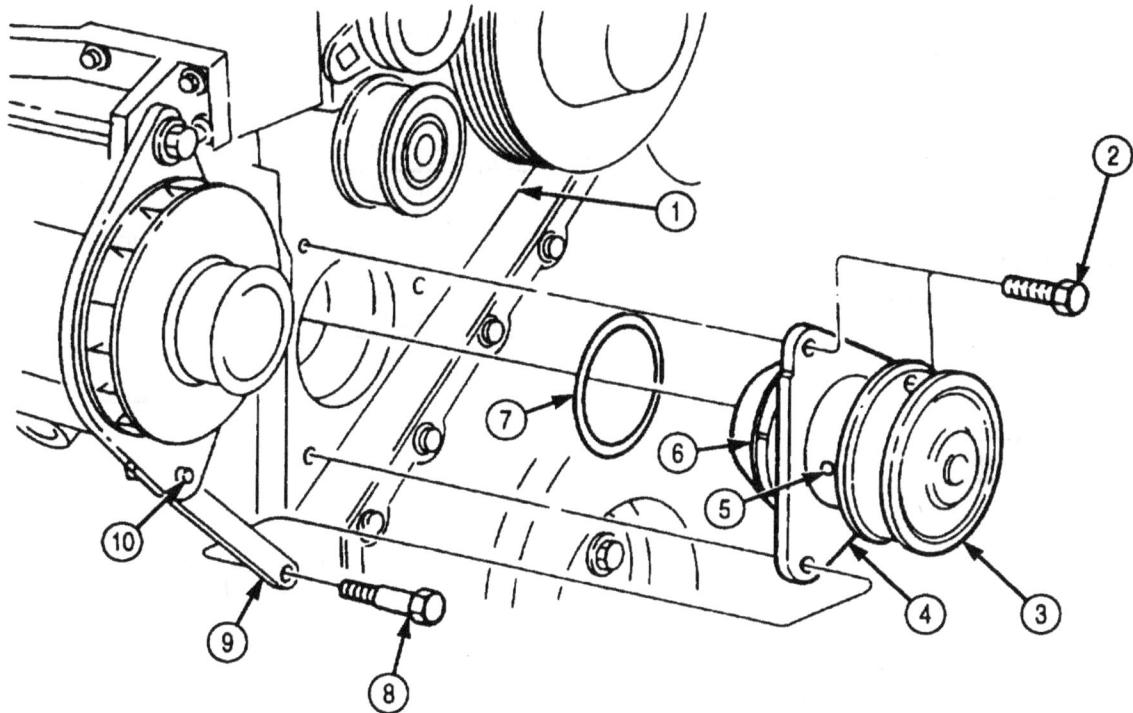

FOLLOW-ON TASKS: • Install fan drivebelt (para. 3-71).
• Fill cooling system to proper level (para. 3-53).
• Start engine (TM 9-2320-272-10), idle engine for five minutes, and check for coolant leaks when temperature gauge reads 180°-205° F (85°-95°C).

3-70. FAN DRIVEBELTS (M939/A1) MAINTENANCE

THIS TASK COVERS:

a. Adjustment
b. Removal

c. Inspection
d. Installation

INITIAL SETUP:

APPLICABLE MODELS
M939/A1

TOOLS
General mechanic's tool kit (Appendix E, Item 1)
Belt tension gauge (Appendix E, Item 16)
Torque wrench (Appendix E, Item 146)

MATERIALS/PARTS
Cap and plug set (Appendix C, Item 14)
Antiseize tape (Appendix C, Item 72)

REFERENCES (TM
TM 9-2320-272-10
TM 9-2320-272-24P

EQUIPMENT CONDITION
• Parking brake set (TM 9-2320-272-10).
• Hood raised and secured (TM 9-2320-272-10).
• Air reservoirs drained (TM 9-2320-272-10).
• Right splash shield removed (TM 9-2320-272-10).

GENERAL SAFETY INSTRUCTIONS
Do not disconnect air lines or hoses before draining air reservoirs.

a. Adjustment

WARNING

Do not disconnect air lines or hoses before draining air reservoirs.
Small parts under pressure may shoot out with high velocity,
causing injury to personnel.

1. Disconnect air line (7) from elbow (8).
2. Remove elbow (8) and adapter (9) from union (10).
3. Loosen three screws (1) on fan pulley bracket (6).
4. Loosen jamnut (2) on adjusting screw (3) on water pump clamp ring (4).
5. Using belt tension gauge on inner belt midway between pulleys, adjust tension of inner belt (5) by turning screw (3). Tension should be 95-105 lb-ft (423-467 N•m) for a new belt (5) and 85-95 lb-ft (378-422 N•m) for a used belt (5).
6. Repeat tension check of step 5 for outer belt (5). Recheck inner belt (5).

NOTE
Belts must be replaced as a matched set.

7. If either inner or outer belts (5) cannot be adjusted in proper range, replace both belts (5).
8. Tighten top screw (1) 25-35 lb-ft (34-47 N•m). Tighten two side screws (1) 25-35 lb-ft (34-47 N•m).

NOTE
Male pipe threads must be wrapped with antiseize tape before installation.

9. Install adapter (9) and elbow (8) on union (10).
10. Connect air line (7) to elbow (8).

b. Removal

1. Disconnect air line (7) from elbow (8).
2. Remove elbow (8) and adapter (9) from union (10).
3. Loosen three screws (1) on bracket (6).

3-70. FAN DRIVEBELTS (M939/A1) MAINTENANCE (Contd)

4. Loosen jamnut (2) on adjusting screw (3).

5. Push fan pulley bracket (6) toward accessory drive pulley (12) and remove two fan drivebelts (5).

c. Inspection

Inspect two fan drivebelts (6) for cracks, splits, and breaks. Replace both belts (5) if either one fails inspection.

d. Installation

CAUTION

Fan drivebelts must be replaced in matched sets. Failure to do so will result in premature belt wear or failure.

1. Position two fan drivebelts (5) over fan clutch pulley (11) and in first and second grooves of accessory drive pulley (12).

2. Complete installation with correct tensioning of drivebelts (5). Refer to task a.

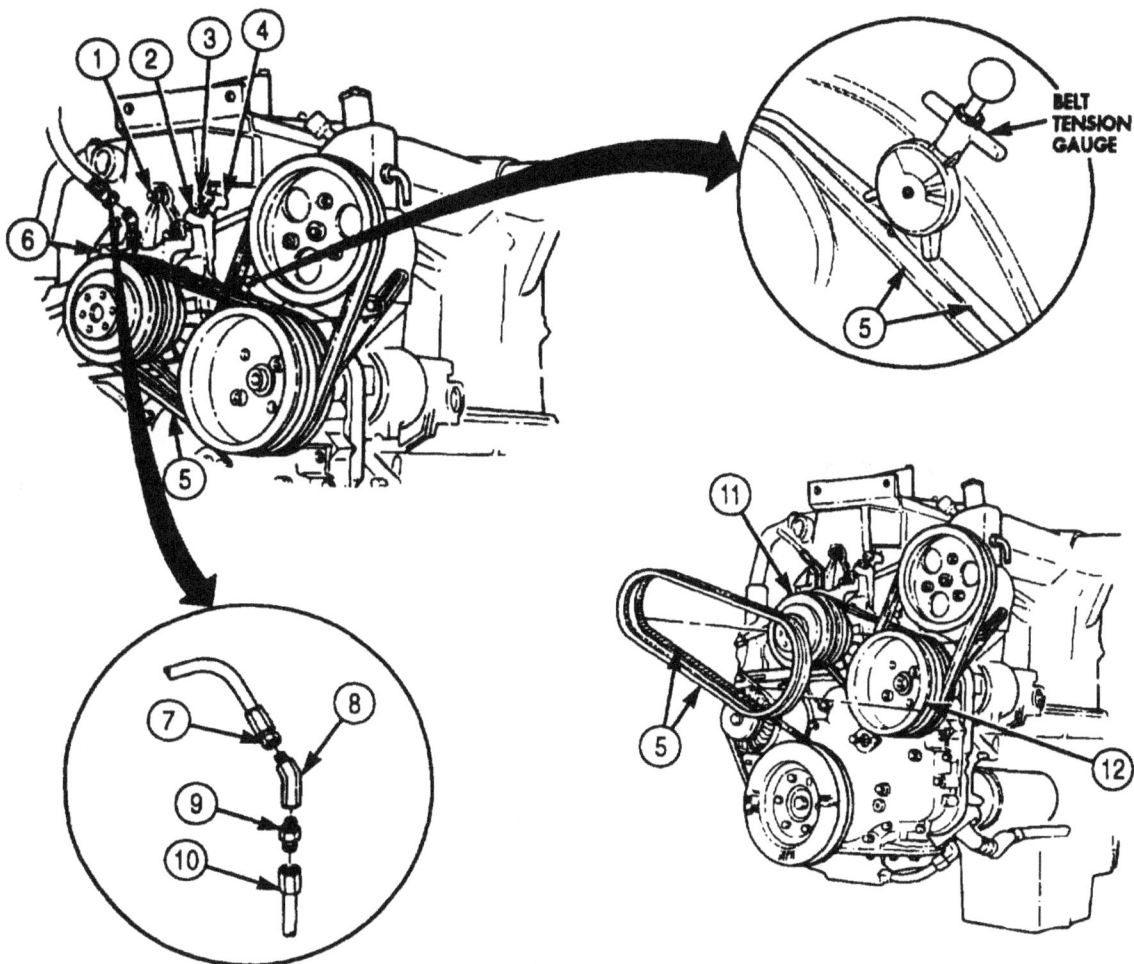

FOLLOW-ON TASKS: • Start engine (TM 9-2320-272-10) and idle for five minutes.
  • Install right splash shield (TM 9-2320-272-10).

3-71. FAN DRIVEBELT AND DRIVEBELT TENSIONER (M939A2) REPLACEMENT

THIS TASK COVERS:

a. Removal   b. Installation

INITIAL SETUP:

APPLICABLE MODELS
M939A2

TOOLS
General mechanic's tool kit (Appendix E, Item 1)
Torque wrench (Appendix E, Item 146)

REFERENCES (TM)
TM 9-2320-272-10
TM 9-2320-272-24P

EQUIPMENT CONDITION
• Parking brake set (TM 9-2320-2722-10).
• Hood raised and secured (TM 9-2320-272-10).
• Right splash shield removed (TM 9-2320-272-10).

a. Removal

1. Using a breaker bar in square drive hole (7) of belt tensioner (4), move belt tensioner (4) up.

NOTE
Assistant will help with step 2.

2. Remove drivebelt (8) from alternator pulley (11), and release belt tensioner (4).
3. Remove drivebelt (8) from vibration damper pulley (9), water pump pulley (10), and fan pulley (1).
4. Remove two screws (6) and belt tensioner bracket (2) from thermostat housing (12).
5. Remove screw (5) and belt tensioner (4) from bracket (2).

b. Installation

1. Install belt tensioner bracket (2) on thermostat housing (12) with two screws (6).
2. Align locating pin (3) on belt tensioner (4) with belt tensioner bracket (2) and install screw (5). Tighten screw (5) 30 lb-ft (45 N•m).
3. Install drivebelt (8) over fan pulley (1) and under vibration damper pulley (9).
4. Install drivebelt (8) between water pump pulley (10) and belt tensioner (4).

NOTE
Assistant will help with step 5.

5. Using breaker bar in square drive hole (7) of belt tensioner (4), move belt tensioner (4) up and install drivebelt (8) on alternator pulley (11). Release belt tensioner (4).

3-71. FAN DRIVEBELT AND DRIVEBELT TENSIONER (M939A2) REPLACEMENT (Contd)

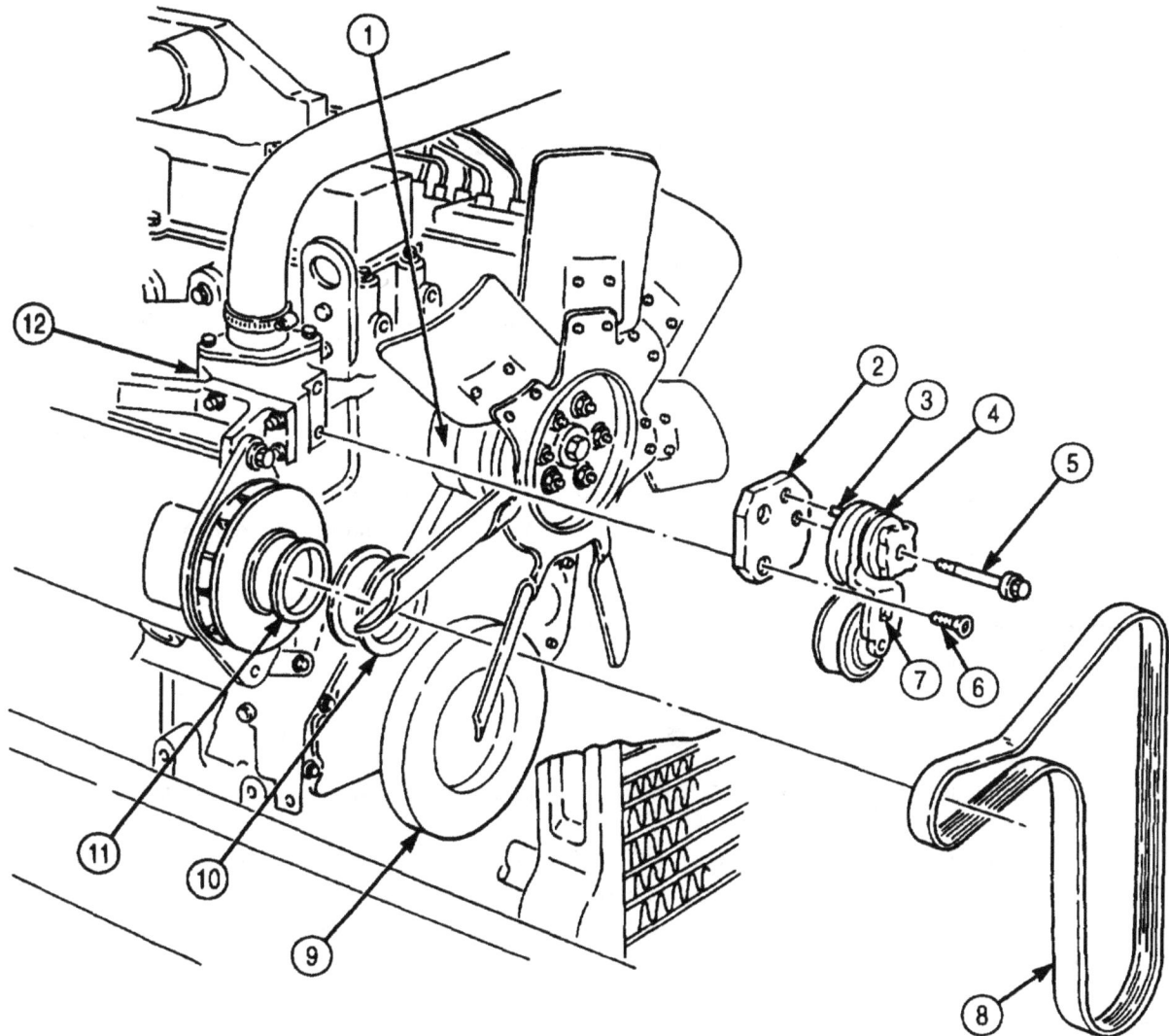

FOLLOW-ON TASKS: • Start engine (TM 9-2320-272-10) and check drivebelt.
   • Install right splash shield (TM 9-2320-272-10).

3-72. RADIATOR FAN BLADE (M939/A1) REPLACEMENT

THIS TASK COVERS:

a. Removal   b. Installation

INITIAL SETUP:

APPLICABLE MODELS
M939/A1

TOOLS
General mechanic's tool kit (Appendix E, Item 1)
Torque wrench (Appendix E, Item 146)

MATERIALS/PARTS
Four locknuts (Appendix D, Item 313)
Locknut (Appendix D, Item 299)
Six lockwashers (Appendix D, Item 354)

REFERENCES (TM)
TM 9-2320-272- 10
TM 9-2320-272-24P

EQUIPMENT CONDITION
• Parking brake set (TM 9-2320-272-10).
• Hood raised and secured (TM 9-3220-272-10).
• Right splash shield removed (TM 9-2320-272-10).

a. Removal

1. Remove four locknuts (1), eight washers (2), and four screws (3) from corner shroud (4). Discard locknuts (1).
2. Remove screw (5) and washer (6) from corner shroud (4) and radiator (7).
3. Remove locknut (11), washers (12) and (9), screw (8), and comer shroud (4) from bracket (10). Discard locknut (11).
4. Remove six screws (13), lockwashers (14), and fan (16) from fan hub (15). Discard lockwashers (14).

b. Installation

NOTE
Ensure fan installation does not block fan drive clutch lockup
holes.

1. Install fan blade (16) on fan hub (15) with six new lockwashers (14) and screws (13). Tighten screws (13) 25-31 lb-ft (34-42 N•m).
2. Install comer shroud (4) on bracket (10) with washer (9), screw (8), washer (12), and new locknut (11).
3. Install comer shroud (4) on radiator (7) with washer (6) and screw (5).
4. Secure comer shroud (4) on radiator (7) with four washers (2), screws (3), washers (2), and new locknuts (1).

3-72. RADIATOR FAN BLADE (M939/A1) REPLACEMENT (Contd)

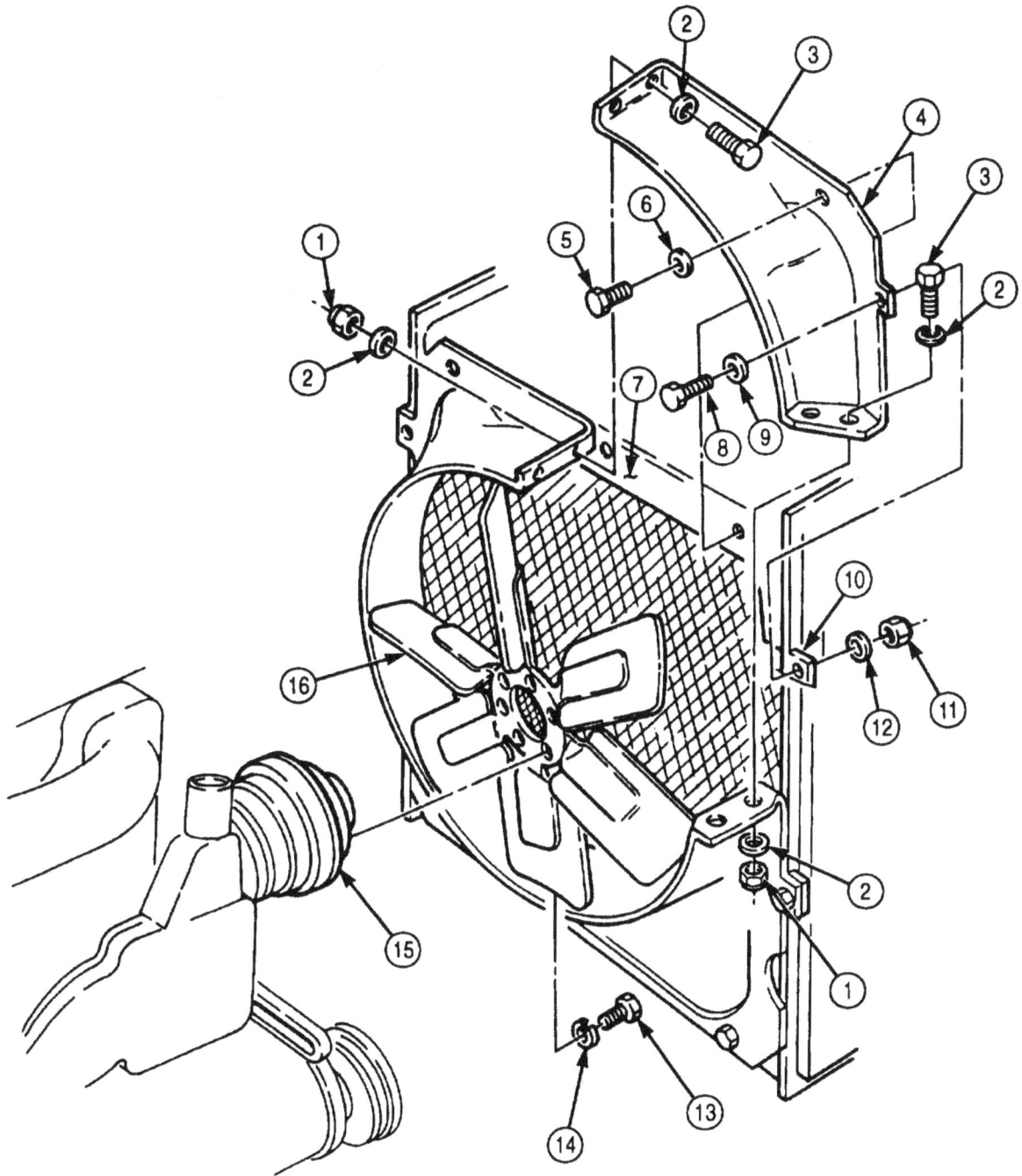

FOLLOW-ON TASK: Install right splash shield (TM 9-2320-272-10).

3-73. FAN DRIVE CLUTCH ACTUATOR (M939/A1) MAINTENANCE

THIS TASK COVERS:

a. Removal
b. Inspection

c. Installation

INITIAL SETUP:

APPLICABLE MODELS

M939/A1

TOOLS

General mechanic's tool kit (Appendix E, Item 1)

MATERIALS/PARTS

Cap and plug set (Appendix C, Item 14)
Antiseize tape (Appendix C, Item 72)

REFERENCES (TM)

TM 9-2320-272-10
TM 9-2320-272-24P

EQUIPMENT CONDITION

• Parking brake set (TM 9-2320-272-10).
• Primary and secondary air tanks drained (TM 9-2320-272-10).
• Hood raised and secured (TM 9-2320-272-10).
• Coolant drained below water manifold level (para. 3-53).

GENERAL SAFETY INSTRUCTIONS

Do not disconnect air lines or hoses before draining air reservoirs.

---

a. Removal

WARNING

Do not disconnect air lines or hoses before draining air reservoirs.
Small parts under pressure may shoot out with high velocity,
causing injury to personnel.

1. Disconnect air hose (3) from adapter (4) on fan drive clutch actuator (5).

2. Remove adapter (4) from fan drive actuator (5).

3. Remove actuator-to-fan hose (2) from adapter (6).

4. Remove adapter (6) from actuator (5).

5. Remove actuator (5) from water manifold (1).

b. Inspection

Inspect all connections for cracks and stripped threads. Replace any parts with cracks, or stripped or damage threads.

c. Installation

NOTE

Male pipe threads must be wrapped with antiseize tape before
installation.

1. Install actuator (5) on water manifold (1).

2. Install adapter (6) on actuator (5).

3. Connect actuator-to-fan hose (2) on adapter (6).

4. Install adapter (4) in actuator (5).

5. Connect air hose (3) to adapter (4).

3-73. FAN DRIVE CLUTCH ACTUATOR (M939/A1) MAINTENANCE (Contd)

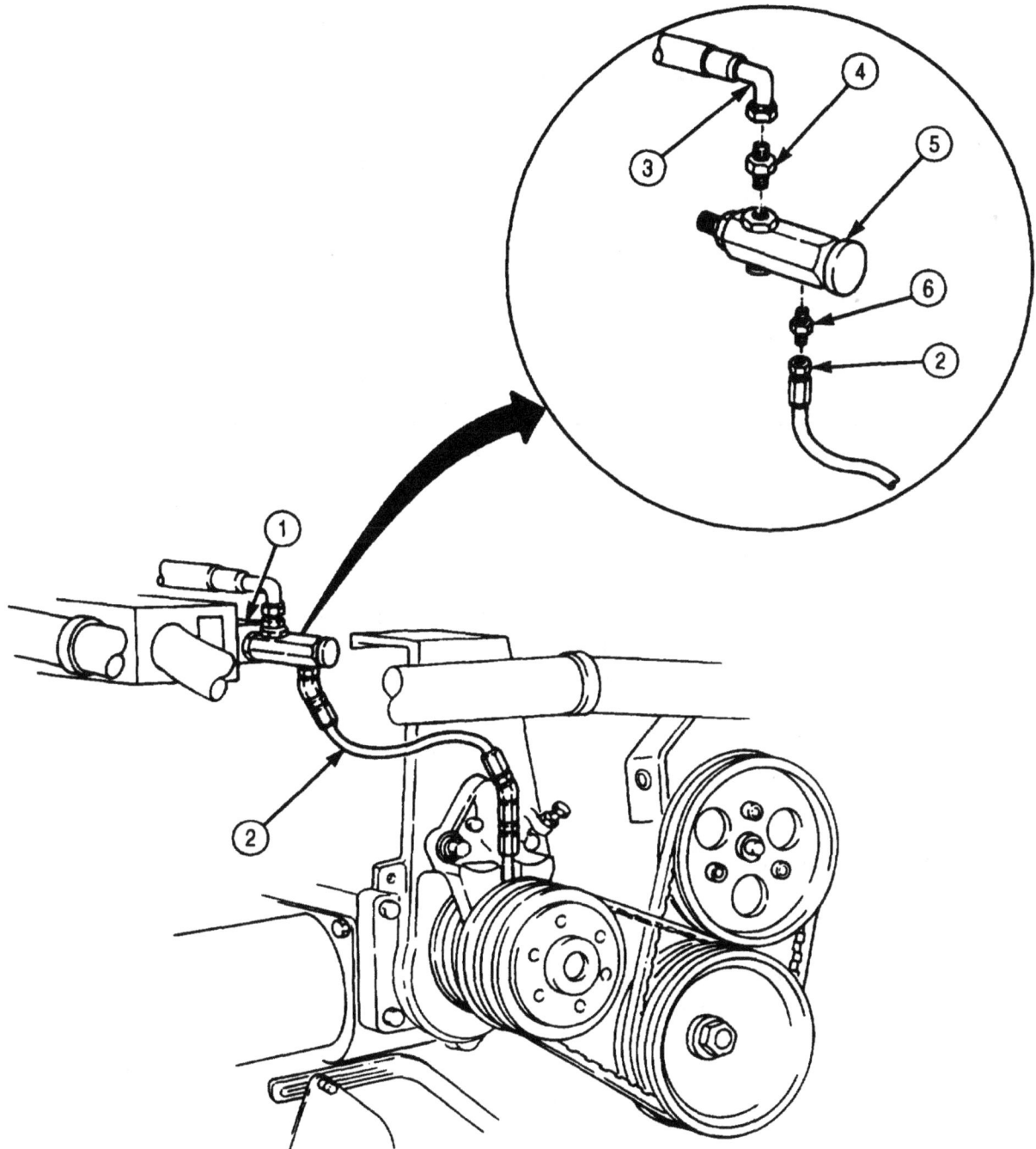

FOLLOW-ON TASKS: • Fill cooling system to proper level (para. 3-53).
• Start engine (TM 9-2320-272-10) and check for air and coolant leaks.
• Check for proper operation of actuator (TM 9-2320-272-10).

3-74. FAN CLUTCH AND FAN ACTUATOR (M939A2) MAINTENANCE

THIS TASK COVERS:

a. Removal   c. Installation
b. Cleaning and Inspection

INITIAL SETUP:

APPLICABLE MODELS
M939A2

TOOLS
General mechanic's tool kit (Appendix E, Item 1)

MATERIALS/PARTS
Cap and plug set (Appendix C, Item 14)
Antiseize tape (Appendix C, Item 72)

REFERENCES (TM)
TM 9-2320-272-10
TM 9-2320-272-24P

EQUIPMENT CONDITION
- Parking brake set (TM 9-2320-272-10).
- Hood raised and secured (TM 9-2320-272-10).
- Coolant drained below water manifold level (para. 3-53).
- Fan and fan shroud removed (para. 3-64).
- Drivebelt removed (para. 3-71).

a. Removal

1. Disconnect air line (6) and elbow (2) from actuator (3).
2. Disconnect air line (5) and elbow (4) from actuator (3).

NOTE
Note position of actuator for installation.

3. Remove actuator (3) from radiator inlet tube (1).
4. Disconnect air line (5) from elbow (7).

NOTE
Assistant will support fan clutch while removing screws.

5. Remove four screws (10), washers (9), and fan clutch (11) from engine (8).
6. Remove elbow (7) from fan clutch (11).

b Cleaning and Inspection

1. Clean all parts (para. 2-14).
2. Inspect air lines (5) and (6) for cracks, kinks, and wear.
3. Inspect other parts (para. 2-15).
4. Replace all parts failing inspection.

3-74. FAN CLUTCH AND FAN ACTUATOR (M939A2) MAINTENANCE (Contd)

c. Installation

NOTE
- All male threads must be wrapped with antiseize tape before installation.
- Assistant will support fan clutch during installation.

1. Install fan clutch (11) on engine (8) with four washers (9) and screws (10).
2. Install elbow (7) on fan clutch (11).
3. Connect air line (5) to elbow (7).
4. Wrap male pipe threads of actuator (3) with antiseize tape and install in radiator inlet tube (1).
5. Install air line (6) and elbow (2) on actuator (3).
6. Install air line (5) and elbow (4) on actuator (3).

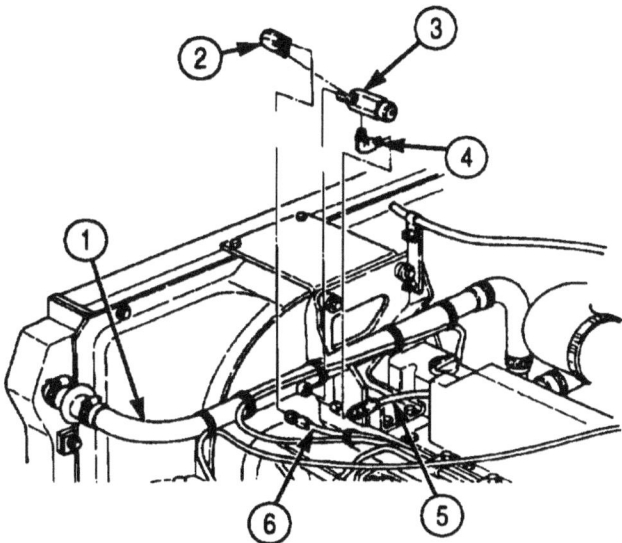

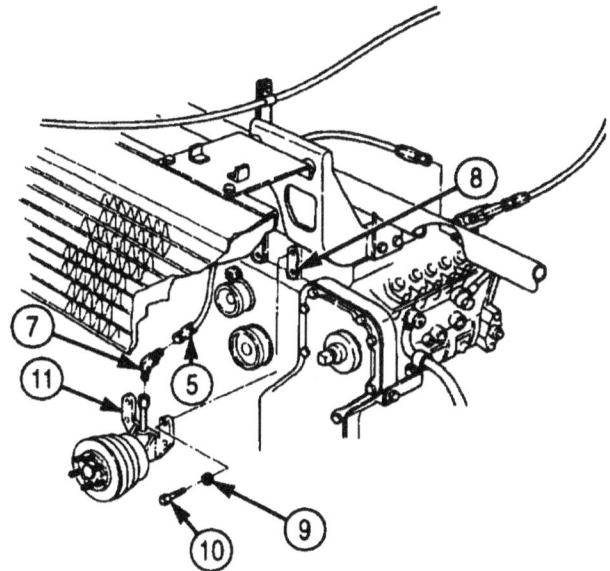

FOLLOW-ON TASKS:
- Fill cooling system to proper level (para. 3-53).
- Install drivebelt (para. 3-71).
- Install fan and fan shroud (para. 3-64).
- Start engine (TM 9-2320-272-10), check for leaks, and tighten any loose fittings.

3-213

3-75. FAN DRIVE CLUTCH (M939/A1) REPLACEMENT

THIS TASK COVERS:

a. Removal   b. Installation

INITIAL SETUP:

APPLICABLE MODELS
M939/A1

TOOLS
General mechanic's tool kit (Appendix E, Item 1)

MATERIALS/PARTS
Three lockwashers (Appendix D, Item 350)
Loctite (Appendix C, Item 63)
Cap and plug set (Appendix C, Item 14)
Antiseize tape (Appendix C, Item 72)

REFERENCES (TM)
TM 9-2320-272-10
TM 9-2320-272-24P

EQUIPMENT CONDITION
• Parking brake set (TM 9-2320-272-10).
• Hood raised and secured (TM 9-2320-272-10).
• Fan drivebelts removed (para. 3-70).
• Fan removed (para. 3-72).

a. Removal

1. Disconnect hose (2) from fan drive actuator (1) and elbow (4) on fan pulley bracket (3).

2. Remove elbow (4) and adapter (6) from union (5).

3. Remove two fan clutch override screws (7) and washers (8) from fan bracket (3) or fan drive clutch (13) holes (12).

4. Remove two screws (16), lockwashers (15), washers (14), screw (9), lockwasher (10), fan drive clutch (13), and fan pulley bracket (3) from support bracket (11). Discard lockwashers (15) and (10).

b. Installation

1. Install fan drive clutch (13) and fan pulley bracket (3) on support bracket (11) with new lockwasher (10), two new lockwashers (15), two washers (14), screws (16), and screw (9). Do not tighten screws (16) and (9) until drivebelts are installed.

2. Apply Loctite on threads of screws (7). Install two fan clutch override washers (8) and screws (7) in fan pulley bracket (3).

NOTE

Male pipe threads must be wrapped with antiseize tape before installation.

3. Install adapter (6) and elbow (4) on union (5).

4. Connect hose (2) to elbow (4) and fan drive actuator (1).

3-75. FAN DRIVE CLUTCH (M939/A1) REPLACEMENT (Contd)

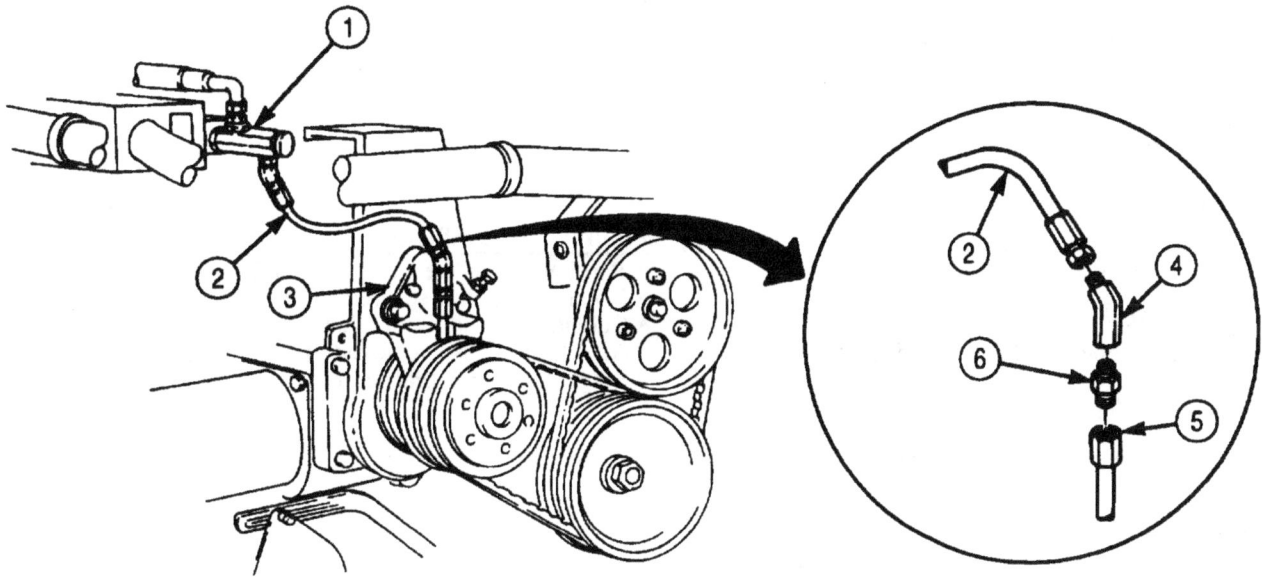

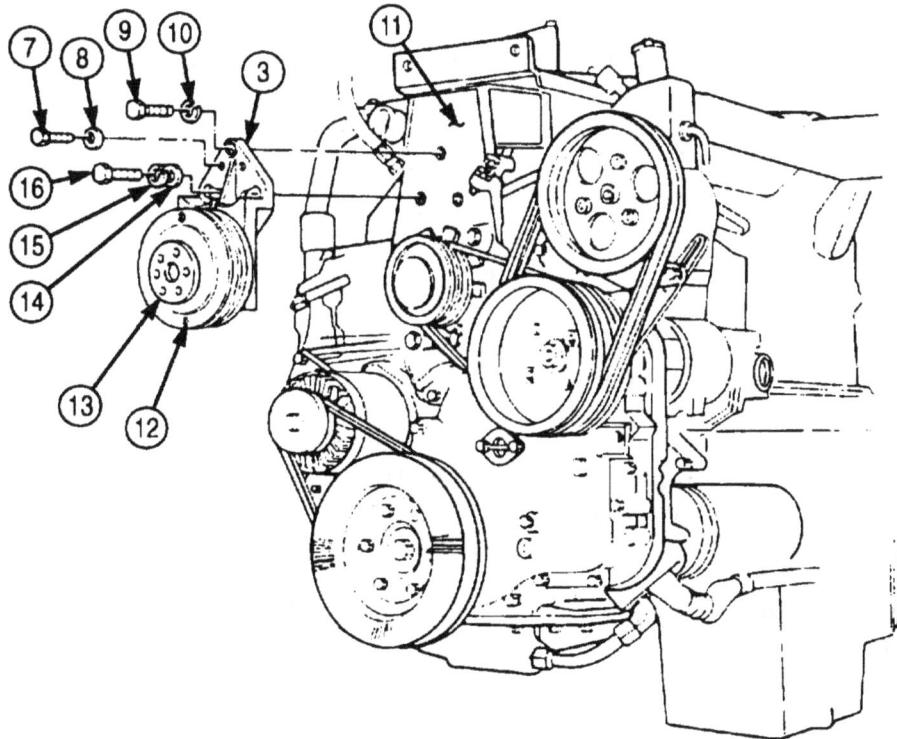

FOLLOW-ON TASKS: • Install drivebelts (para. 3-70).
                       • Install fan (para. 3-72).
                       • Start engine (TM 9-2320-272-10), check for leaks, and check for proper fan operation.

3-76. AFTERCOOLER AND TUBES (M939A2) REPLACEMENT

THIS TASK COVERS:

a. Removal   b. Installation

INITIAL SETUP:

APPLICABLE MODELS

M939A2

TOOLS

General mechanic's tool kit (Appendix E, Item 1)
Torque wrench (Appendix E, Item 146)

MATERIALS/PARTS

Gasket (Appendix D, Item 154)
Two seals (Appendix D, Item 599)
Antiseize tape (Appendix C, Item 72)
Cleaning cloth (Appendix C, Item 21)

REFERENCES (TM)

TM 9-2320-272-10
TM 9-2320-272-24P

EQUIPMENT CONDITION

• Parking brake set (TM 9-2320-272-10).
• Hood raised and secured (TM 9-2320-272-10).
• Coolant drained from block (para. 3-53).
• Air cleaner hose removed (para. 3-13).
• Surge tank and bracket removed (para. 3-62).

a. Removal

1. Remove screw (5) and tube brace (12) from aftercooler (6).

2. Remove screws (4) and (7) from aftercooler (6) and clamps (8) and rotate clamps (8) away from aftercooler (6).

3. Disconnect plug (1) from wiring harness plug (13).

4. Remove two screws (3) and throttle control solenoid bracket (2) from aftercooler (6). Move clamp (10) clear of aftercooler (6).

5. Remove two screws (9) and clamps (11) from aftercooler (6).

6. Loosen two clamps (16) and disconnect crossover tube (15) from aftercooler (6) and turbocharger (28).

7. Remove screw (24) and clamping plate (25) from engine block (26).

NOTE

Tag tubes for installation.

8. Loosen four hose clamps (16) and remove two hoses (18) and water return tube (20) from aftercooler (6) and elbow (23).

9. Loosen four hose clamps (19) and remove two hoses (17) and water supply tube (27) from aftercooler (6) and elbow (21).

10. Remove elbows (23) and (21) and two seals (22) from engine block (26). Discard seals (22).

CAUTION

Cover intake manifold with lint-free cloth to prevent contamination from entering engine. Failure to do so may result in damage to engine.

11. Remove thirteen screws (14), aftercooler (6), and gasket (29) from engine block (26). Discard gasket (29).

3-76. AFTERCOOLER AND TUBES (M939A2) REPLACEMENT (Contd)

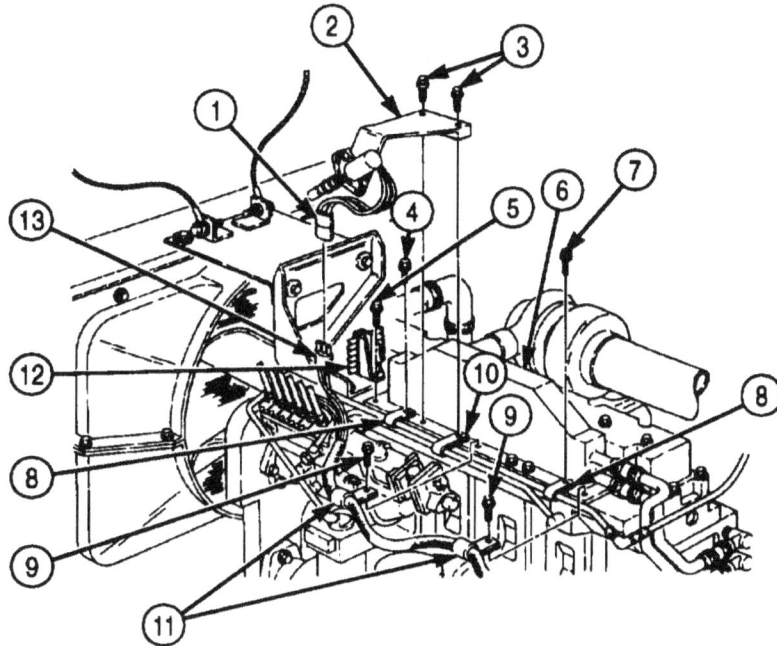

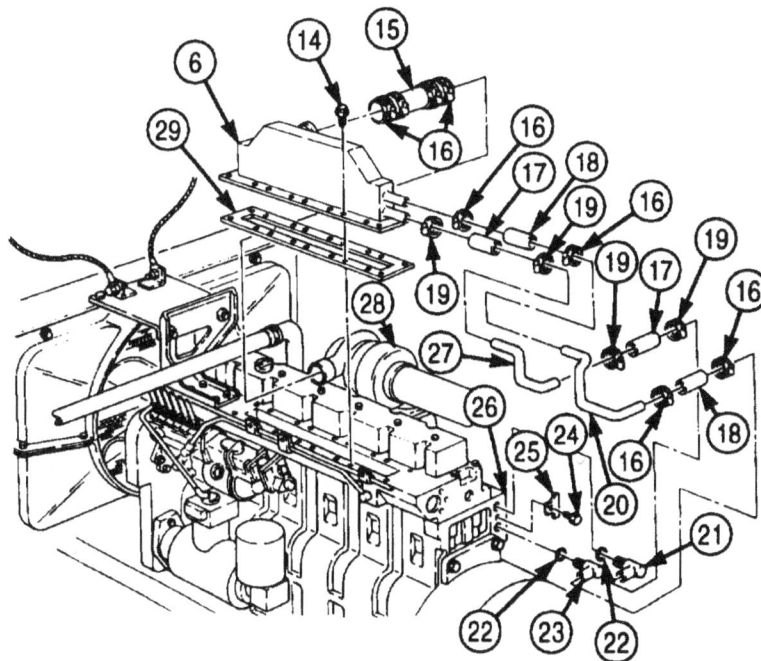

3-76. AFTERCOOLER AND TUBES (M939A2) REPLACEMENT (Contd)

b. Installation

NOTE

Male pipe threads must be wrapped with antiseize tape before installation.

1. Install new gasket (29) and aftercooler (6) on engine block (26) with thirteen screws (14). Tighten screws (14) 28 lb-ft (24 N•m).

2. Install two new seals (22) and elbows (21) and (23) on engine block (26).

3. Install two hoses (17) and water supply tube (27) on aftercooler (6) and elbow (21) with four hose clamps (19).

4. Install two hoses (18) and water return tube (20) on aftercooler (6) and elbow (23) with four hose clamps (16).

5. Install clamping plate (25) on engine block (26) with screw (24).

6. Install crossover tube (15) on aftercooler (6) and turbocharger (28) with two clamps (16).

7. Install two clamps (11) on aftercooler (6) with two screws (9).

8. Install clamp (10) and throttle control solenoid bracket (2) on aftercooler (6) with two screws (3).

9. Connect plug (1) to wiring harness plug (13).

10. Rotate two clamps (8) into position on aftercooler (6) and install with screws (7) and (4).

11. Install tube brace (12) on aftercooler (6) with screw (5).

3-76. AFTERCOOLER AND TUBES (M939A2) REPLACEMENT (Contd)

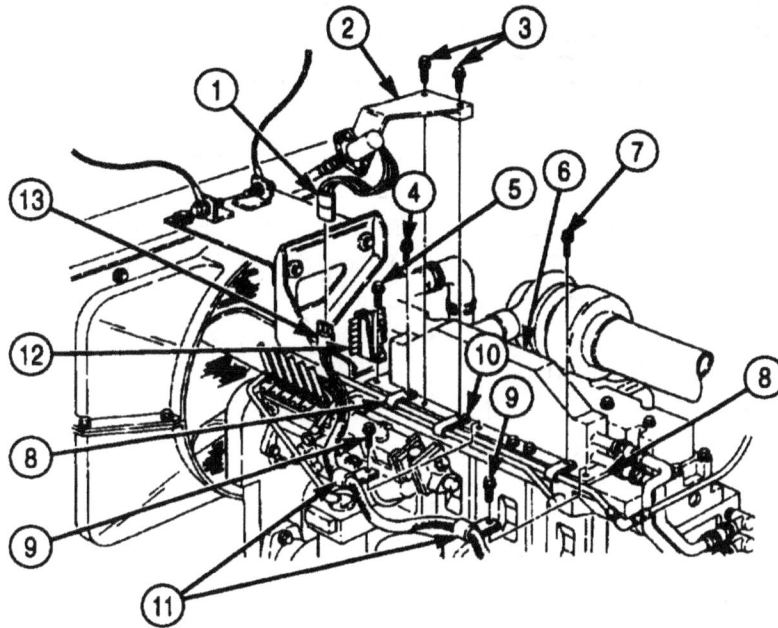

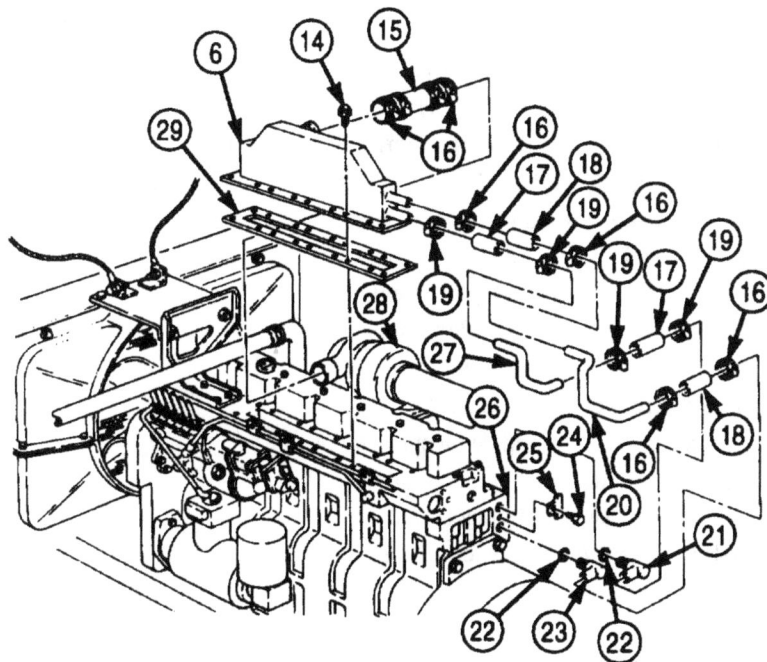

FOLLOW-ON TASKS: • Install surge tank and bracket (para. 3-62).
                        • Install air cleaner hose (para. 3-13).

TM 9-2320-272-24-1

Section VI. ELECTRICAL SYSTEM MAINTENANCE

3-77. ELECTRICAL SYSTEM MAINTENANCE INDEX

3-77. ELECTRICAL SYSTEM MAINTENANCE INDEX (Contd)

3-78. ALTERNATOR DRIVEBELTS (M939/A1) MAINTENANCE

THIS TASK COVERS:
a. Removal
b. Inspection

c. Installation and Adjustment

INITIAL SETUP:

APPLICABLE MODELS
M939/A1

TOOLS
General mechanic's tool kit (Appendix E, Item 1)
Belt tension gauge (Appendix E, Item 16)
Torque wrench (Appendix E, Item 144)
Prybar

REFERENCES (TM)
TM 9-2320-272-10
TM 9-2320-272-24P

EQUIPMENT CONDITION
• Parking brake set (TM 9-2320-272-10).
• Right splash shield removed (TM 9-2320-272-10).
• Hood raised and secured (TM 9-2320-272-10).

NOTE
For replacement of alternator/fan drivebelt on M939A2 series
vehicles, refer to paragraph 3-71.

a. Removal

1. Loosen screws (3) and (6) at adjusting link (2). Do not remove screws (3) and (6).
2. Loosen two screws (9) at alternator mounting bracket (10). Do not remove screws (9).
3. Push alternator (1) toward engine (5) until slack exists in two alternator belts (8). Remove two alternator belts (8) from alternator pulley (4) and vibration damper (7).

b. Inspection

Inspect two alternator belts (8) for breaks, cracks, or splits. Replace if broken, cracked, or split.

c. Installation and Adjustment

NOTE
Alternator belts must be replaced as matched sets.

1. Position two alternator belts (8) over vibration damper (7) and alternator pulley (4).
2. Insert prybar between engine (5) and alternator (1) and pull prybar down until alternator belts (8) appear tight.
3. Tighten screw (3) on adjusting link (2) 15-20 lb-ft (20-27 N•m).
4. Tighten screw (6) on adjusting link (2) 25-31 lb-ft (34-42 N•m).
5. Using belt tension gauge, check for proper belt tension. Tension of new belts (8) should be 95-105 lb (423-467 N). Tension of used belts (8) should be 85-95 lb (378-422 N). Replace belts (8) if they cannot be adjusted.
6. Tighten two screws (9) on mounting bracket (10) 39-49 lb-ft (53-66 N•m).

3-78. ALTERNATOR DRIVEBELTS (M939/A1) MAINTENANCE (Contd)

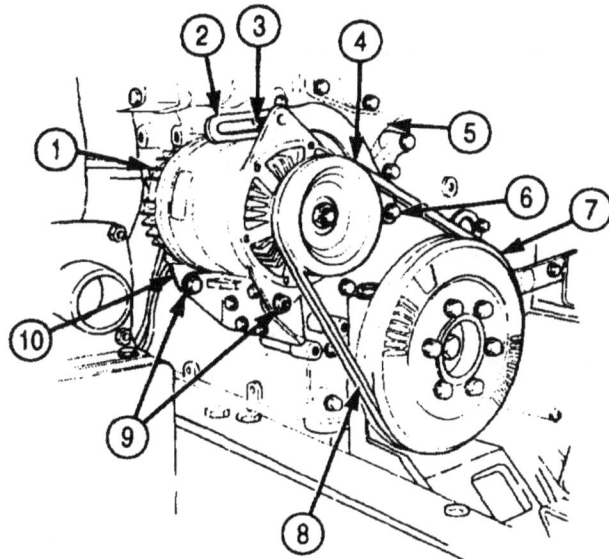

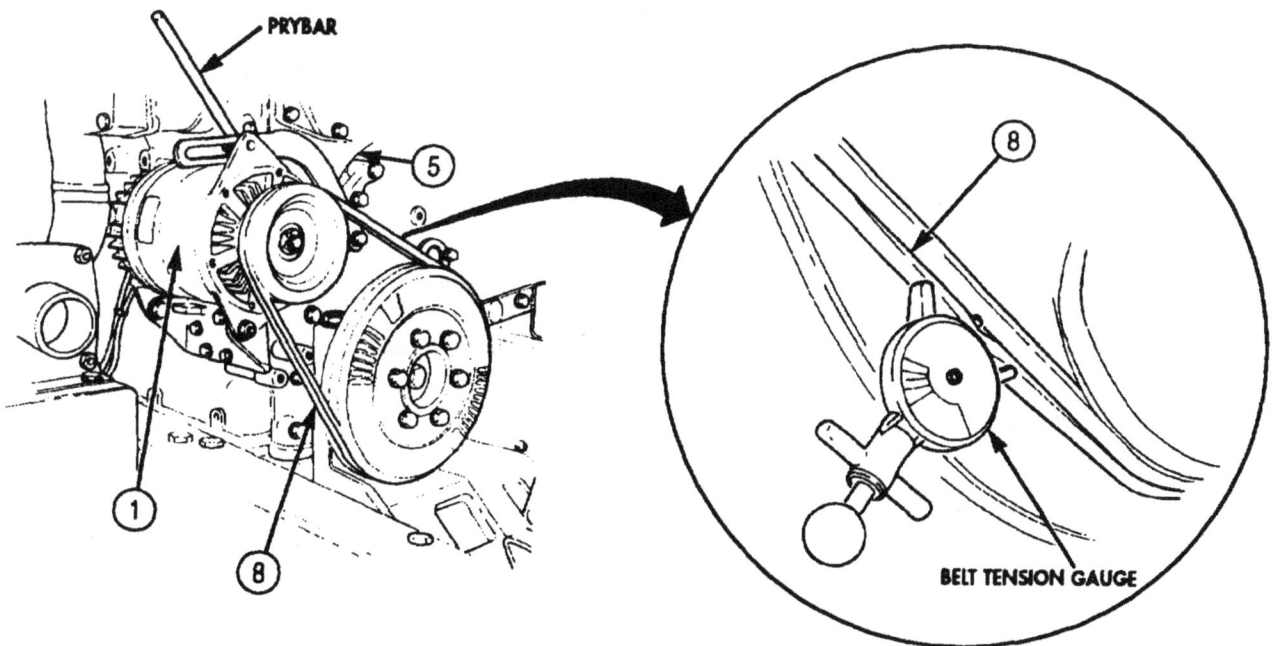

PRYBAR

BELT TENSION GAUGE

FOLLOW-ON TASKS: • Start engine (TM 9-2320-272-10) and check alternator operation.
                      • Install right splash shield (TM 9-2320-272-10).

3-79. ALTERNATOR AND MOUNTING BRACKET (M939/A1) REPLACEMENT

THIS TASK COVERS:

a. Removal   b. Installation

INITIAL SETUP:

APPLICABLE MODELS
M939/A1

TOOLS
General mechanic's tool kit (Appendix E, Item 1)
Torque wrench (Appendix E, Item 144)

MATERIALS/PARTS
Two screw-assembled lockwashers
  (Appendix D, Item 587)
Five lockwashers (Appendix D, Item 354)
Two lockwashers (Appendix D, Item 376)
Two locknuts (Appendix D, Item 281)
Lockwasher (Appendix D, Item 401)
Lockwasher (Appendix D, Item 403)
Tiedown strap (Appendix D, Item 684)
Lockwasher (Appendix D, Item 364)
Gasket sealant (Appendix C, Item 30)

REFERENCES (TM)
TM 9-2320-272-10
TM 9-2320-272-24P

EQUIPMENT DESCRIPTION

• Parking brake set (TM 9-2320-272-10).
• Battery ground cables disconnected (para. 3-126).
• Alternator drivebelts removed (para. 3-78).

GENERAL SAFETY INSTRUCTIONS
Alternator is heavy. Use assistance when replacing alternator.

a. Removal

1. Remove two screw-assembled lockwashers (5) and terminal cover (4) from alternator (10). Discard screw-assembled lockwashers (5).

2. Remove two screws (3), lockwashers (2), and wire retaining strap (1) from alternator (10). Discard lockwashers (2).

NOTE
• Sealant must be removed before removing wires.
• Tag all wires for installation.

3. Remove screw (15), lockwasher (14), and wire (13) from alternator (10). Discard lockwasher (14).

4. Remove nut (19), lockwasher (18), washer (17), and wire (16) from alternator (10). Discard lockwasher (18).

5. Remove nut (6), lockwasher (7), washer (8), and wire (9) from alternator (10). Discard lockwasher (7).

6. Disconnect connector (11) from alternator (10).

7. Remove and discard tiedown strap (12).

WARNING
Alternator is heavy. Assistant will help with alternator removal.
Failure to do so may result in injury to personnel.

8. Remove screw (32), lockwasher (20), washers (21) and (22), screw-assembled washer (23), and adjusting link (31) from alternator (10). Discard lockwasher (20).

9. Remove two locknuts (24), washers (25), screws (30), and alternator (10) from mounting bracket (26). Discard locknuts (24).

10. Remove four screws (29), lockwashers (28), and mounting bracket (26) from engine (27). Discard lockwashers (28).

3-79. ALTERNATOR AND MOUNTING BRACKET (M939/A1) REPLACEMENT (Contd)

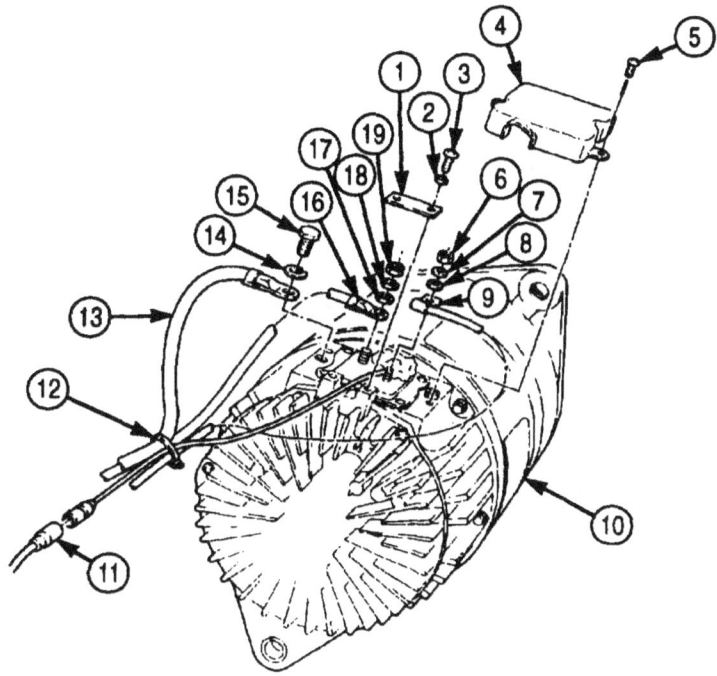

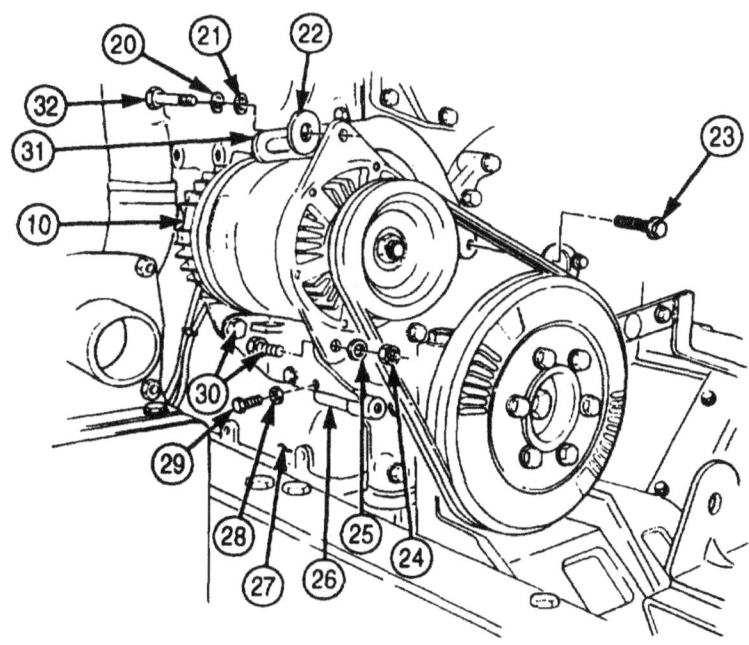

3-79. ALTERNATOR AND MOUNTING BRACKET (M939/A1) REPLACEMENT (Contd)

WARNING

Alternator is heavy. Assistant will help with alternator installation. Failure to do so may result in injury to personnel.

b. Installation

1. Install mounting bracket (7) on engine (8) with four new lockwashers (9) and screws (10).
2. Install alternator (12) on mounting bracket (7) with two screws (11), washers (6), and new locknuts (5). Finger tighten locknuts (5).

NOTE
Ensure wire connecting points are thoroughly cleaned before connections are made.

3. Install wire (29) on alternator (12) with washer (30), new lockwasher (31), and nut (32). Tighten nut (32) 45-55 bin. (5-6 N•m).
4. Install wire (23) on alternator (12) with washer (22), new lockwasher (21), and nut (20). Tighten nut (20) 20-25 lb-in. (2-3 N•m).
5. Connect connector (24) to alternator (12).
6. Install wire (26) on alternator (12) with new lockwasher (27) and screw (28). Tighten screw (28) 82-102 lb-in. (9-12 N•m).

NOTE
Wires are held in place with strap.

7. Install wire retaining strap (15) on alternator (12) with two new lockwashers (16) and screws (17).
8. Install washer (3) and adjusting link (13) on alternator (12) with screw-assembled washer (4), washer (2), new lockwasher (1), and screw (14).
9. Seal wires (29) and (23) and connectors with gasket sealant.
10. Install terminal cover (18) on alternator (12) with two new screw-assembled lockwashers (19).
11. Install new tiedown strap (25) on alternator wires.

3-79. ALTERNATOR AND MOUNTING BRACKET (M939/A1) REPLACEMENT (Contd)

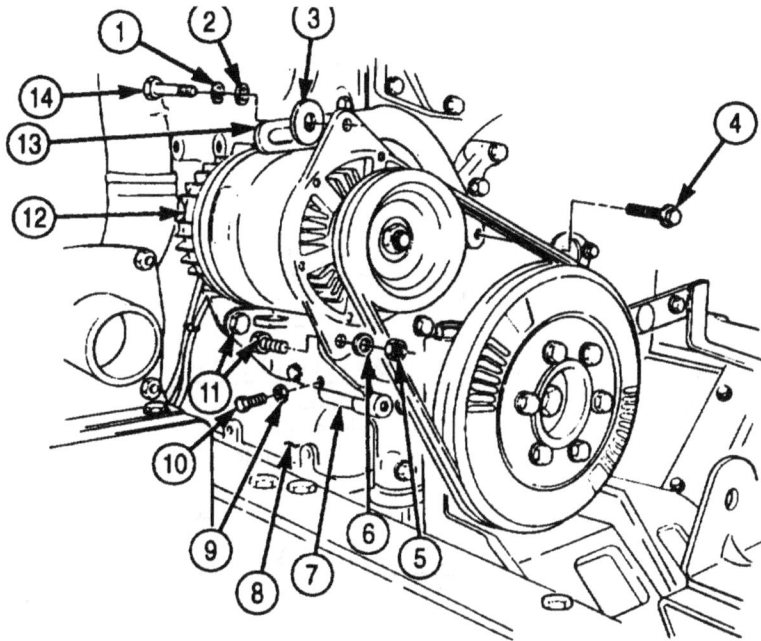

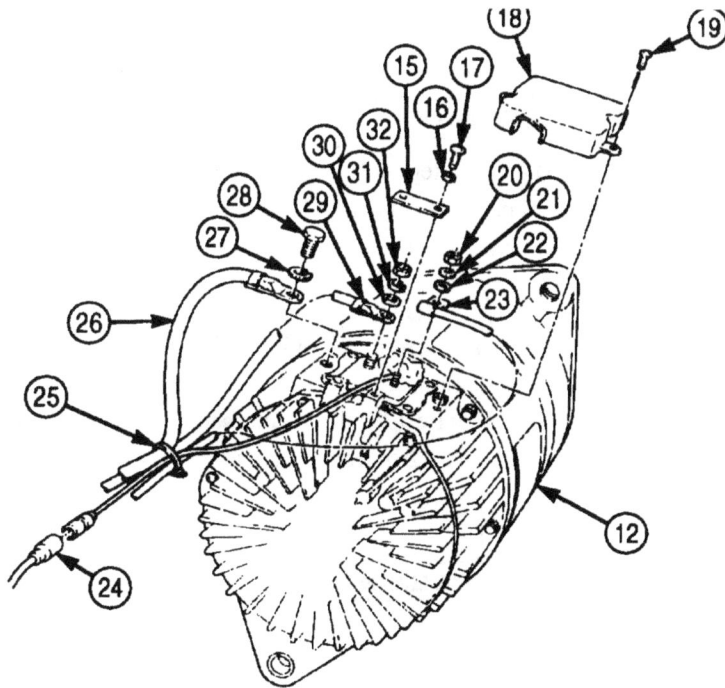

FOLLOW-ON TASKS: • Install alternator belts (para. 3-78).
                 • Connect battery ground cable (para. 3-126).
                 • Start engine (TM 9-2320-272-10) and check alternator operation.

3-80. ALTERNATOR, MOUNTING BRACKET, AND PULLEY (M939A2) MAINTENANCE

THIS TASK COVERS:

a. Alternator and Mounting Bracket
   Removal
b. Pulley Removal
c. Pulley Installation

d. Alternator and Mounting Bracket
   Installation
e. Adjustment

INITIAL SETUP:

APPLICABLE MODELS
M939A2

TOOLS
General mechanic's tool kit (Appendix E, Item 1)
Torque wrench (Appendix E, Item 146)
Universal puller (Appendix E, Item 103)
Multimeter (Appendix E, Item 86)
Vise

MATERIALS/PARTS
Two screw-aesembled lockwashers
   (Appendix D, Item 587)
Lockwasher (Appendix D, Item 364)
Locknut (Appendix D, Item 282)
Woodruff Key (Appendix D, Item 726)
Lockwasher (Appendix D, Item 403)
Lockwasher (Appendix D, Item 401)
Gasket sealant (Appendix C, Item 30)

REFERENCES (TM)
TM 9-2320-272-10
TM 9-2320-272-24P

EQUIPMENT CONDITION
• Parking brake set (TM 9-2320-272-10).
• Hood raised and secured (TM 9-2320-272-10).
• Right splash shield removed (TM 9-2320-272-10).
• Battery ground cables disconnected (para. 3-126).
• Fan drivebelt removed (para. 3-71).

GENERAL SAFETY INSTRUCTIONS
Alternator is heavy. Use assistant when replacing
alternator.

a. Alternator and Mounting Bracket Removal

1. Remove screw (16) and washer (15) from alternator link (14).

2. Loosen two screws (2) and pivot alternator (1) upward to access terminal cover (23).

3. Remove two screw-assembled lockwashers (24) and terminal cover (23) from alternator (1). Discard
   screw-assembled lockwashers (24).

NOTE
- Sealant must be removed before removing wires.
- Tag all leads for installation.

4. Remove screw (22), lockwasher (21), and wire (27) from alternator (1). Discard lockwasher (21).

5. Remove nut (20), lockwasher (19), and wire (26) from alternator (1). Discard lockwasher (19).

6. Remove nut (18), lockwasher (17), and wire (25) from alternator (1). Discard lockwasher (17).

7. Disconnect wire (28) from connector (29).

WARNING

Alternator is heavy. Assistant will help with alternator removal.
Failure to do so may result in injury to personnel.

8. Remove two screws (2), washers (3), and alternator (1) from brackets (4) and (6). Move bracket (4) to
   one side.

9. Remove four screws (7) and bracket (6) from engine (5).

10. Remove screw (12), washer (13), and alternator link (14) from engine (5).

3-80. ALTERNATOR, MOUNTING BRACKET, AND PULLEY (M939A2) MAINTENANCE (Contd)

b. Pulley Removal

1. Place alternator pulley (11) in vise.
2. Remove locknut (10) and washer (9) from alternator (1). Diecard locknut (10).
3. Remove alternator pulley (11) from vise.
4. Using universal puller, remove alternator pulley (11) and woodruff key (8) from alternator (1). Discard woodruff key (8).

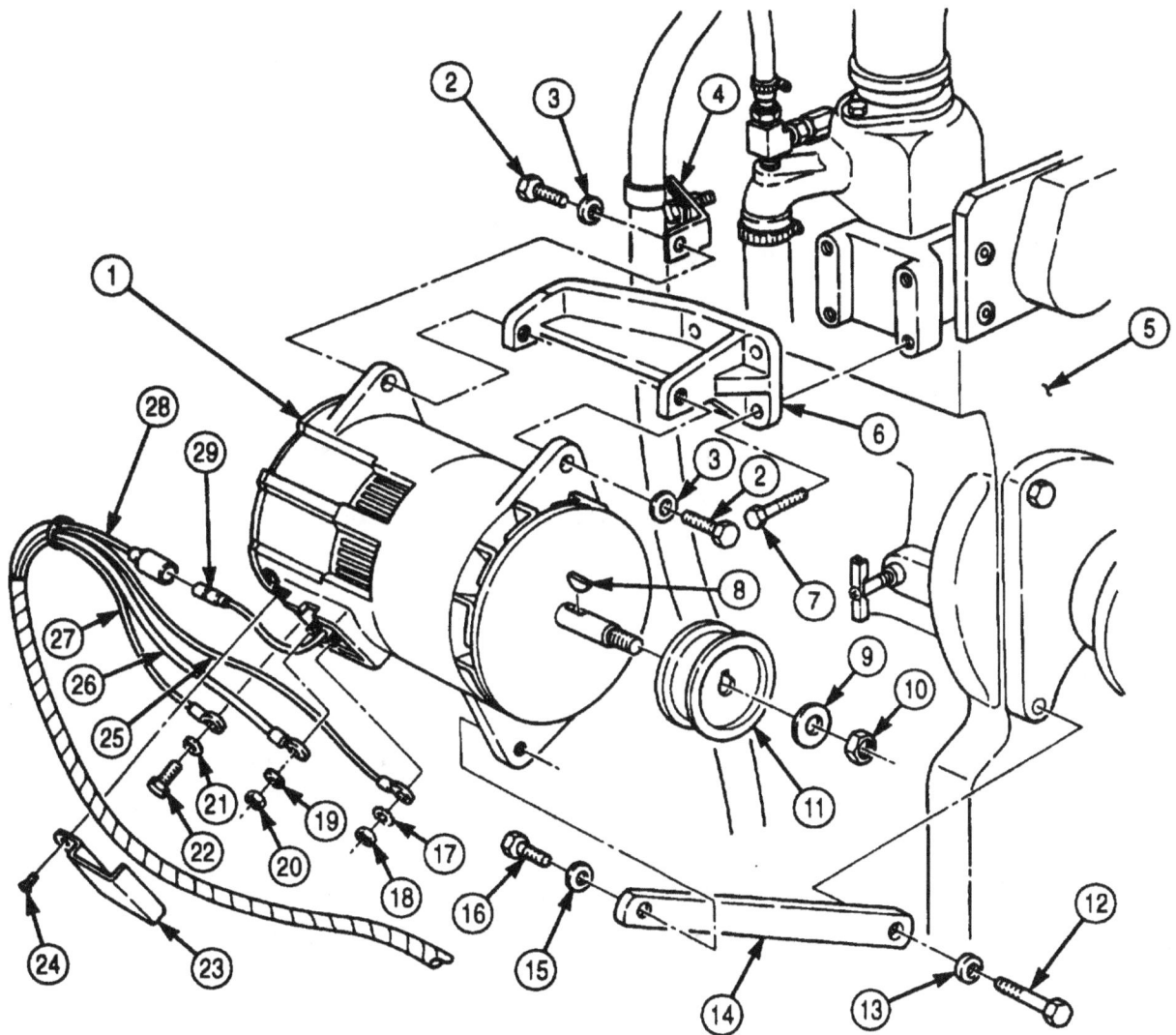

3-80. ALTERNATOR, MOUNTING BRACKET, AND PULLEY (M939A2) MAINTENANCE (Contd)

c. Pulley Installation

1. Install new woodruff key (8) and alternator pulley (9) on alternator (1).
2. Place alternator pulley (9) in vise.
3. Install washer (11) and new locknut (10) on alternator (1). Tighten locknut (10) 90-100 lb-ft (122-136 N•m).
4. Remove alternator pulley (9) from vise.

d. Alternator and Mounting Bracket Installation

WARNING

Alternator is heavy. Assistant will help with alternator installation. Failure to do so may result in injury to personnel.

NOTE

Ensure wire connecting points are thoroughly cleaned before connections are made.

1. Install alternator link (14) on engine (5) with washer (13) and screw (12).
2. Install mounting bracket (6) on engine (5) with four screws (7).
3. Install alternator (1) on mounting bracket (6) and bracket (4) with two washers (3) and screws (2). Finger-tighten screws (2).
4. Connect wire (28) to connector (29).
5. Install wire (25) on alternator (1) with new lockwasher (17) and nut (18). Tighten nut (18) 45-55 lb-in (5-6 N•m).
6. Install wire (26) on alternator (1) with new lockwasher (19) and nut (20). Tighten nut (20) 20-25 lb-in (2-3 N•m).
7. Install wire (27) on alternator (1) with new lockwasher (21) and screw (22). Tighten screw (22) 82-102 lb-in (9-12 N•m).
8. Pivot alternator (1) to align with alternator link (14).
9. Install alternator link (14) on alternator (1) with washer (15) and screw (16).
10. Install fan drivebelt (para. 3-71).
11. Connect battery ground cables (para. 3-126).
12. Adjust alternator (1) (task e.).

e. Adjustment

1. Start engine (TM 9-2320-272-10).
2. Set engine speed to 1200 rpm (TM 9-2320-272-10)
3. Turn on headlights (TM 9-2320-272-10) to place load on alternator (1).
4. Using multimeter, check alternator (1) output voltage. Connect black lead to ground cable (27) and touch red lead to wire (28). Output voltage should be 27.8-28.2 VDC. If adjustment is required, continue with next step. If no adjustment is required, proceed to step 8.
5. Using hex-head driver, remove pipe plug (30) from alternator (1).
6. Turn adjusting screw counterclockwise to increase or clockwise to decrease voltage.

3-80. ALTERNATOR, MOUNTING BRACKET, AND PULLEY (M939A2) MAINTENANCE (Contd)

7. Apply gasket sealant to pipe plug (30) threads. Using hex-head driver, install pipe plug (30) and tighten 24-36 lb-in. (3-4 N•m).
8. Turn off headlights (TM 9-2320-272-10).
9. Stop engine (TM 9-2320-272-10).
10. Seal wires (28), (27), (26), and (25) and connector (29) completely with gasket sealant.
11. Install terminal cover (23) on alternator (1) with two new screw-assembled lockwashers (24).

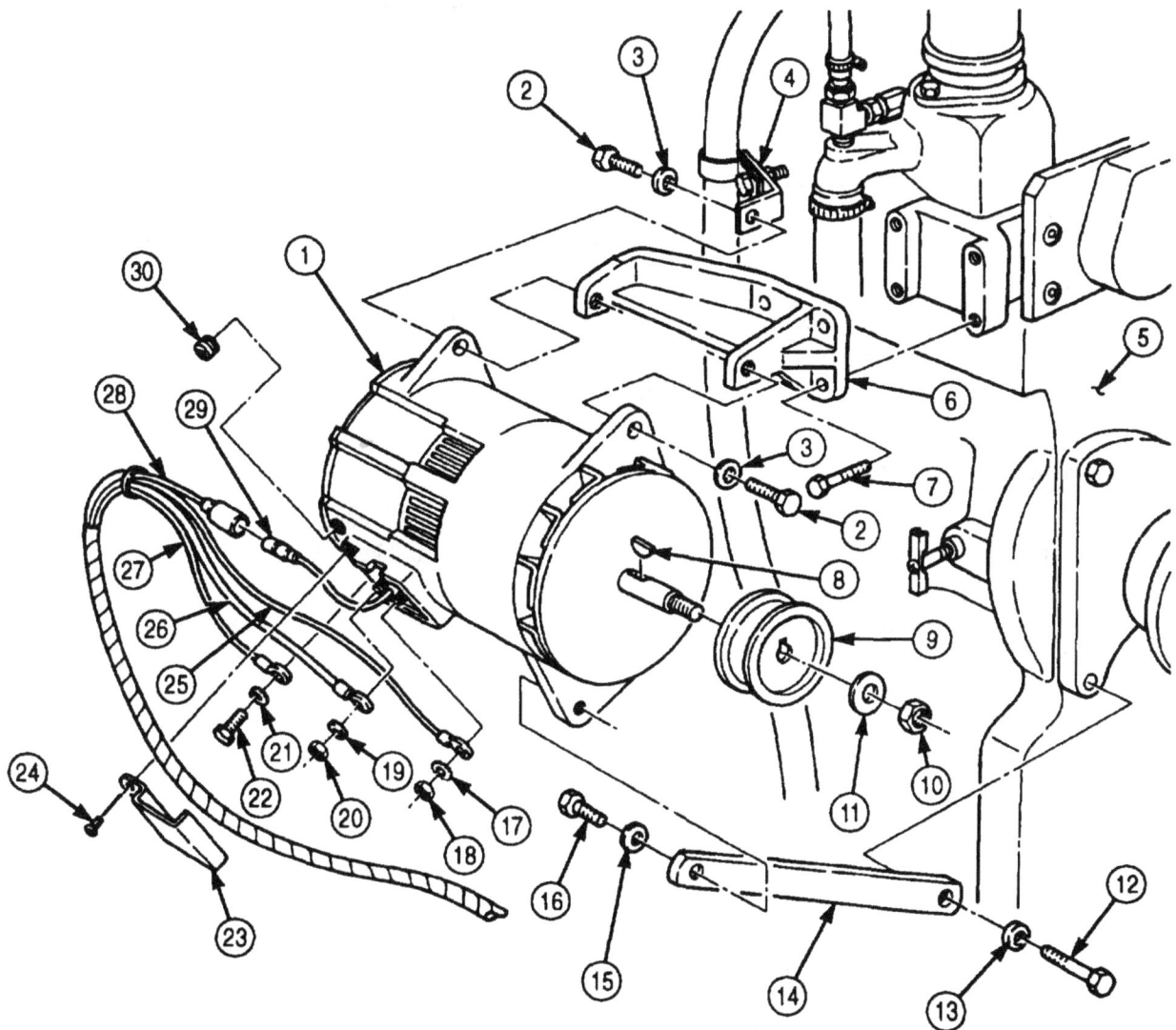

FOLLOW-ON TASK: Install right splash shield (TM 9-2320-272-10).

3-81. ALTERNATOR PULLEY (M939/A1) REPLACEMENT

THIS TASK COVERS

a. Removal   b. Installation

INITIAL SETUP:

APPLICABLE MODELS

M939/A1

TOOLS

General mechanic's tool kit (Appendix E, Item 1)
Torque wrench (Appendix E, Item 144)
Universal puller (Appendix E, Item 103)
Vise

MATERIALS/PARTS

Locknut (Appendix D, Item 282)
Woodruff key (Appendix D, Item 726)

REFERENCES (TM)

TM 9-2320-272-10
TM 9-2320-272-24P

EQUIPMENT CONDITION

• Parking brake set (TM 9-2320-272-10).
• Hood raised and secured (TM 9-2320-272-10).
• Alternator removed (para. 3-79).

a. Removal

1. Position alternator pulley (1) in vise.
2. Remove locknut (3) and washer (2) from pulley shaft (5). Discard locknut (3).
3. Remove alternator pulley (1) from vise.
4. Using universal puller, remove alternator pulley (1) and woodruff key (4) from pulley shaft (5). Discard woodruff key (4).

b. Installation

1. Install new woodruff key (4) in groove of pulley shaft (5) with flat edge up.
2. Install pulley (1) on pulley shaft (5) with keyway (6) on woodruff key (4).
3. Place alternator pulley (2) in vise.
4. Install washer (2) and new locknut (3) on pulley shaft (5). Tighten locknut (3) 90-95 lb-ft (122- 128 N•m).

3-81. ALTERNATOR PULLEY (M939/A1) REPLACEMENT (Contd)

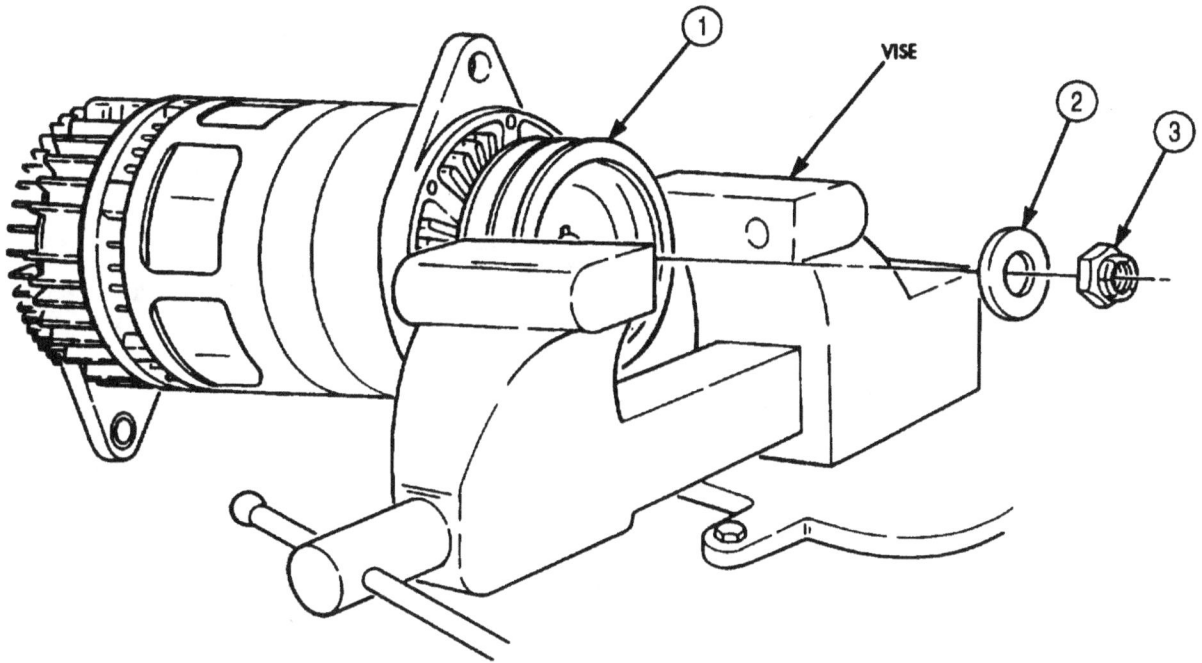

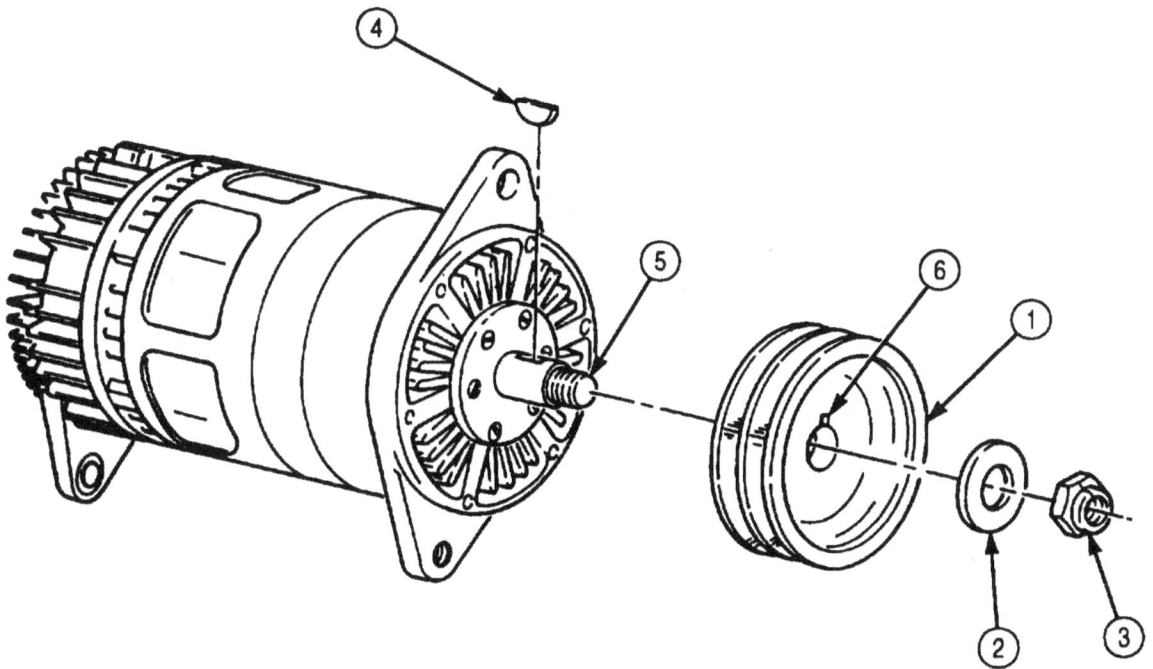

FOLLOW-ON TASK: Install alternator (para. 3-79).

3-82. STARTER REPLACEMENT

THIS TASK COVERS:

a. Removal   b. Installation

INITIAL SETUP:

APPLICABLE MODELS
All

TOOLS
General mechanic's tool kit (Appendix E, Item 1)

MATERIALS/PARTS
Two lockwashers (M939/A1) (Appendix D, Item 416)
Four lockwashers (M939A2) (Appendix D, Item 416)
Gasket (M939/A1) (Appendix D, Item 155)
Gasket (M939/A1) (Appendix D, Item 156)
Gasket (M939A2) (Appendix D, Item 157)
Gasket (M939A2) (Appendix D, Item 158)

REFERENCES (TM)
TM 9-2320-272-10
TM 9-2320-272-20P

EQUIPMENT CONDITION
-]Parking brake set (TM 9-2320-272-10).
-]Hood raised and secured (TM 9-2320-272-10).
-]Battery ground cables disconnected (para. 3-126).

a. Removal

NOTE
- Tag wires for installation.
- Step 1 is for M939A2 vehicles only.

1. Remove nut (24), lockwasher (23), and wire (22) from solenoid (2). Discard lockwasher (23).

2. Remove nut (16), lockwasher (17), and wires (18) and (15) from solenoid (2). Discard lockwasher (17).

NOTE
Ground strap will be present on M939/A1 vehicles only.

3. Remove nut (12), lockwasher (13), wires (14) and (11), and ground strap (10) from starter (9). Discard lockwasher (13).

NOTE
Step 4 is for M939A2 vehicles only.

4. Remove nut (25), lockwasher (26), and wire (27) from starter (9). Discard lockwasher (26).

NOTE
Step 5 is for M939/A1 vehicles only.

5. Remove screw (21), clip (20), and wire (19) from solenoid (2).

NOTE
- Steps 6 and 7 are for M939A2 vehicles only.
- Assistant will help with step 6.

6. Remove three screws (8), washers (7), starter (9), gasket (6), spacer (5), and gasket (4) from flywheel housing (3). Discard gaskets (6) and (4).

7. Remove three sleeves (28) from starter (9).

NOTE
- Steps 8 and 9 are for M939/A1 only.
- Assistant will help with step 9.

8. Remove screw-assembled washer (1) from starter (9).

3-82. STARTER REPLACEMENT (Contd)

9. Remove two screws (8), washers (7), starter (9), gasket (6), spacer (5), and gasket (4) from flywheel housing (3). Discard gaskets (6) and (4).

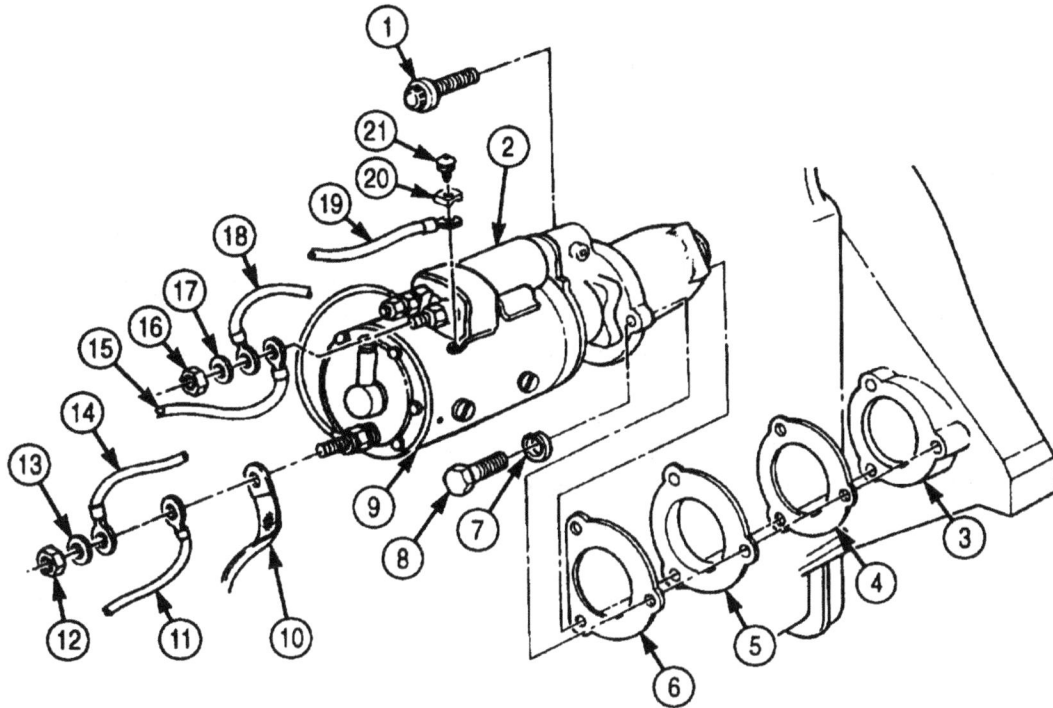

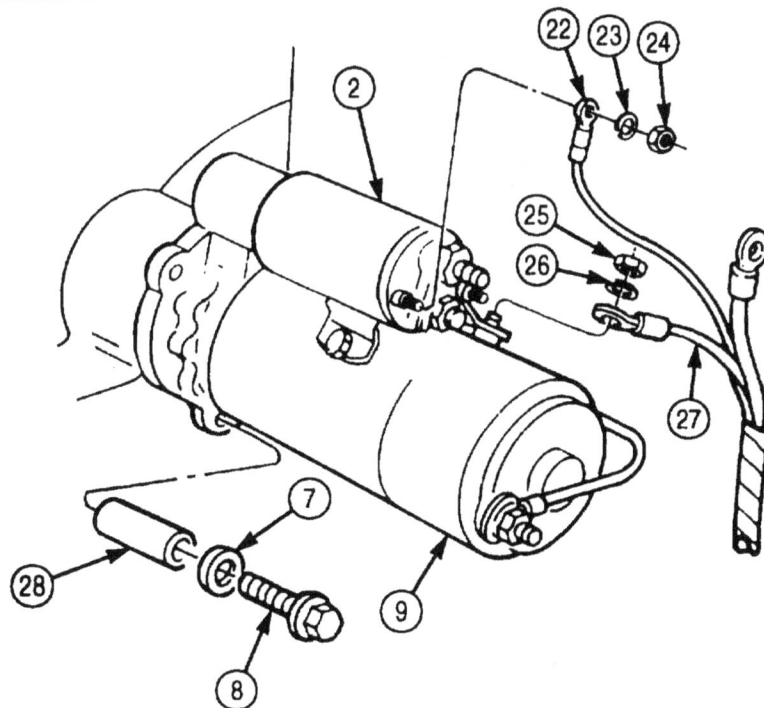

3-82. STARTER REPLACEMENT (Contd)

b. Installation

NOTE

- Steps 1 and 2 are for M939/A1 vehicles only.
- Assistant will help with step 1.

1. Install new gasket (4), spacer (5), new gasket (6), and starter (9) on flywheel housing (3) with two washers (7) and screws (8). Tighten screws (8) 55 lb-ft (76 N•m).

2. Install screw-assembled washer (1) in starter (9).

NOTE

- Steps 3 and 4 are for M939A2 vehicles only.
- Assistant will help with step 4.

3. Install three sleeves (28) on starter (9).

4. Install new gasket (4), spacer (6), new gasket (6), and starter (9) on flywheel housing (3) with three washers (7) and screws (8). Tighten screws (8) 55 lb-ft (75 N•m).

NOTE

Step 5 is for M939/A1 vehicles only.

5. Install wire (19) on solenoid (2) with clip (20) and screw (21).

NOTE

Step 6 is for M939A2 vehicles only.

6. Install wire (27) on starter (9) with new lockwasher (26) and nut (25).

NOTE

Ground strap will be present on M939/A1 vehicles only.

7. Install ground strap (10) and wires (11) and (14) on starter (9) with new lockwasher (13) and nut (12).

8. Install wires (15) and (18) on solenoid (2) with new lockwasher (17) and nut (16).

NOTE

Step 6 is for M939A2 vehicles only.

9. Install wire (22) on solenoid (2) with new lockwasher (23) and nut (24).

3-82. STARTER REPLACEMENT (Contd)

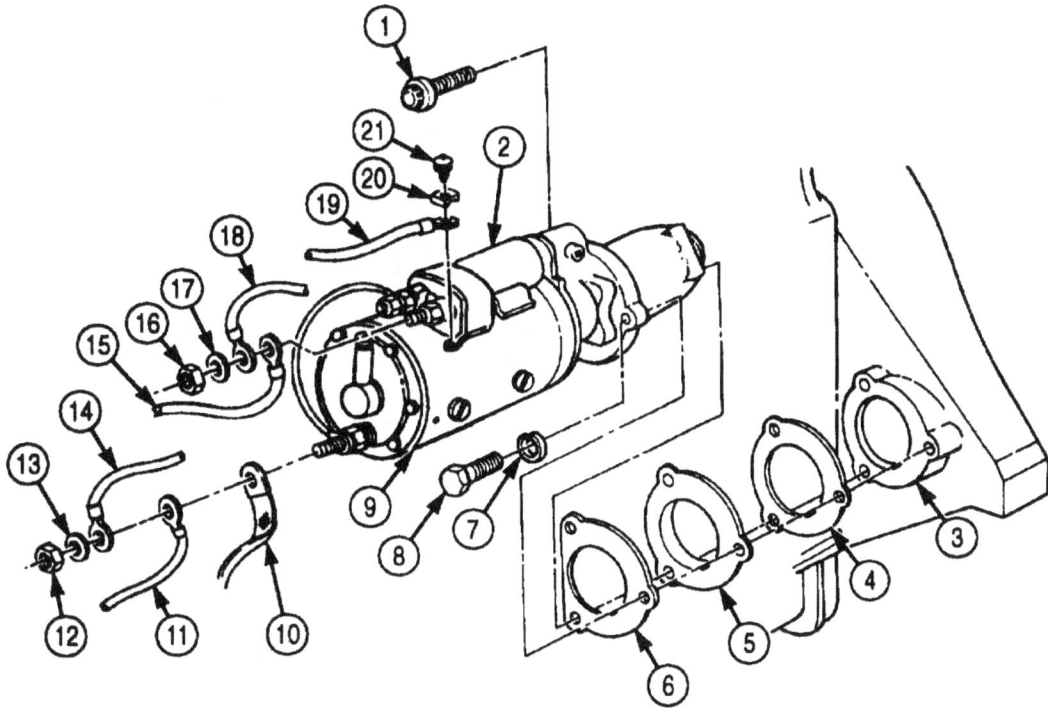

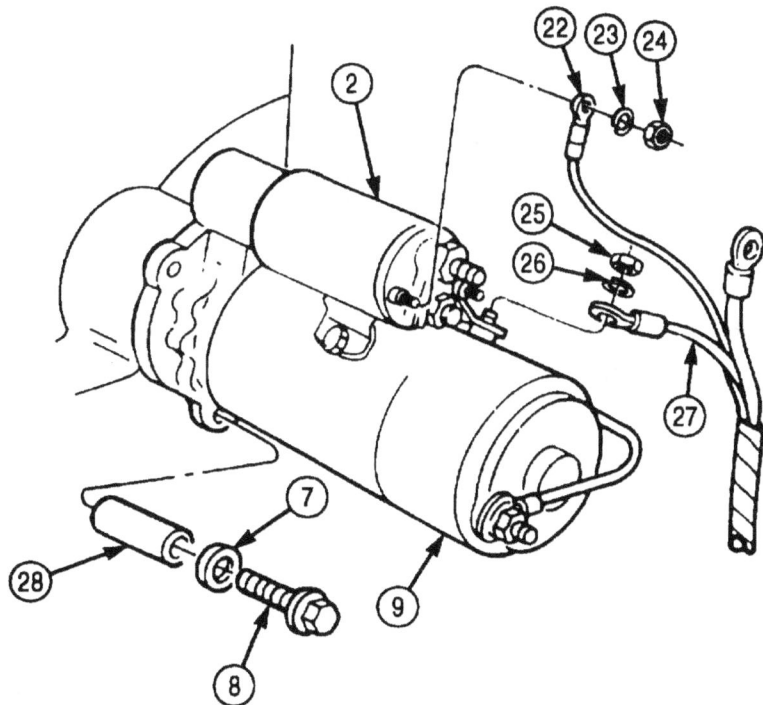

FOLLOW-ON TASK: Connect battery ground cables (para. 3-126).

3-83. INSTRUMENT CLUSTER MAINTENANCE

THIS TASK COVERS:

a. Removal
b. Disassembly

c. Assembly
d. Installation

INITIAL SETUP:

APPLICABLE MODELS

All

TOOLS

General mechanic's tool kit (Appendix E, Item 1)

MATERIALS/PARTS

Three spring nuts (Appendix D, Item 673)
Three cotter pins (Appendix D, Item 53)

REFERENCES (TM)

TM 9-2320-272-10
TM 9-2320-272-20P

EQUIPMENT CONDITION
• Parking brake set (TM 9-2320-272-10).
• Air reservoirs drained (TM 9-2320-272-10).
• Hood raised and secured (TM 9-2320-272-10).
• Battery ground cables disconnected (para. 3-126).
• Disconnect CTIS electrical components (M939A2) (para. 3-470).

GENERAL SAFETY INSTRUCTIONS
Do not disconnect air lines before draining air reservoirs. Small parts under pressure may shoot out with high velocity, causing injury to personnel.

a. Removal

1. Remove screw (15), clamp (3), cable (1), and retaining clip (4) from diverter bracket (6).
2. Remove cotter pin (16) from control rod (5). Discard cotter pin (16).

NOTE
Tag cables for installation.

3. Remove cable (1) and spring nut (2) from control rod (5). Discard spring nut (2).
4. Remove screw (14), clamp (10), cable (13), and retaining clip (8) from diverter bracket (7).
5. Remove cotter pin (12) from control rod (9). Discard cotter pin (12).
6. Remove heat control cable (13) and spring nut (11) from control rod (9). Discard spring nut (11).
7. Remove screw (20), retaining nut (22), and clamp (21) from heater assembly (17).
8. Remove cotter pin (23) from heater assembly (17). Discard cotter pin (23).
9. Remove fresh air control cable (19) and spring nut (18) from heater assembly (17). Discard spring

10. Remove eight screws (26) from instrument cluster (25).
11. Separate instrument cluster (25) from instrument panel (24).

3-83. INSTRUMENT CLUSTER MAINTENANCE (Contd)

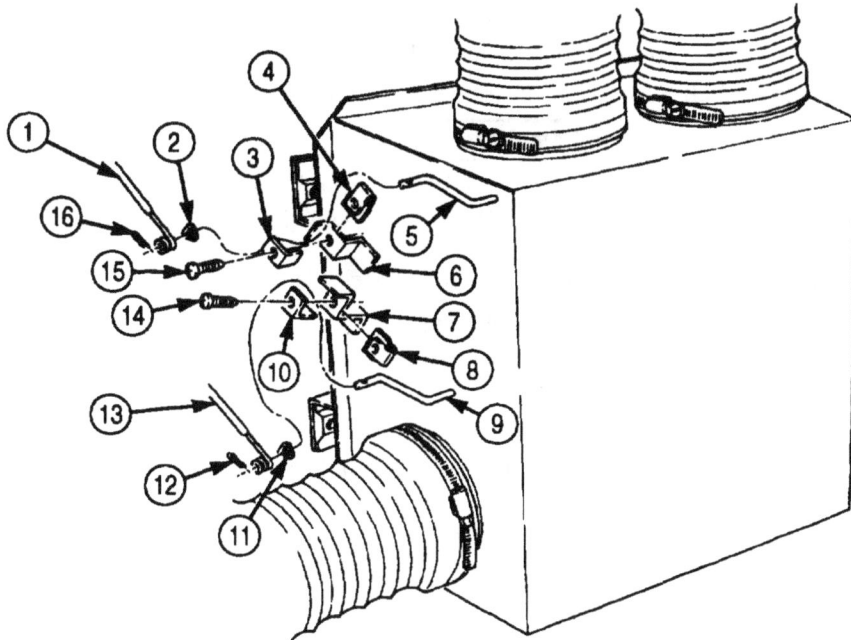

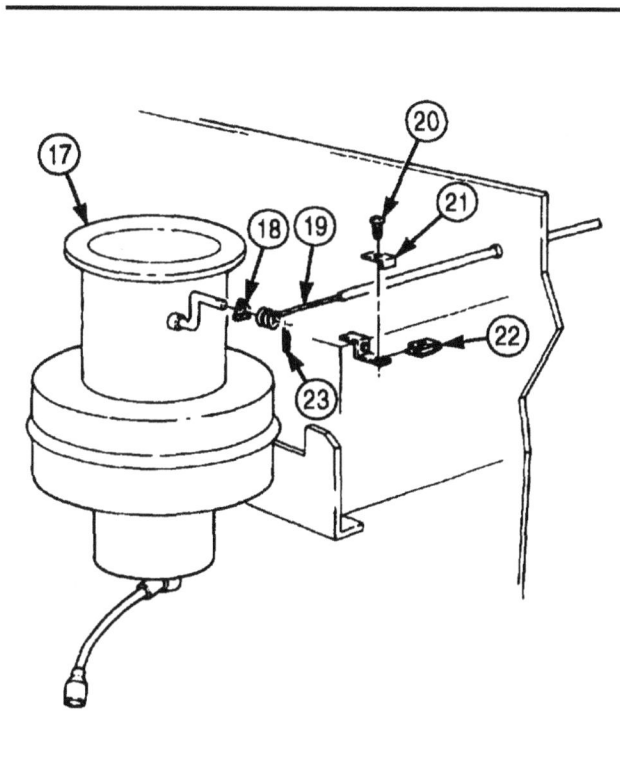

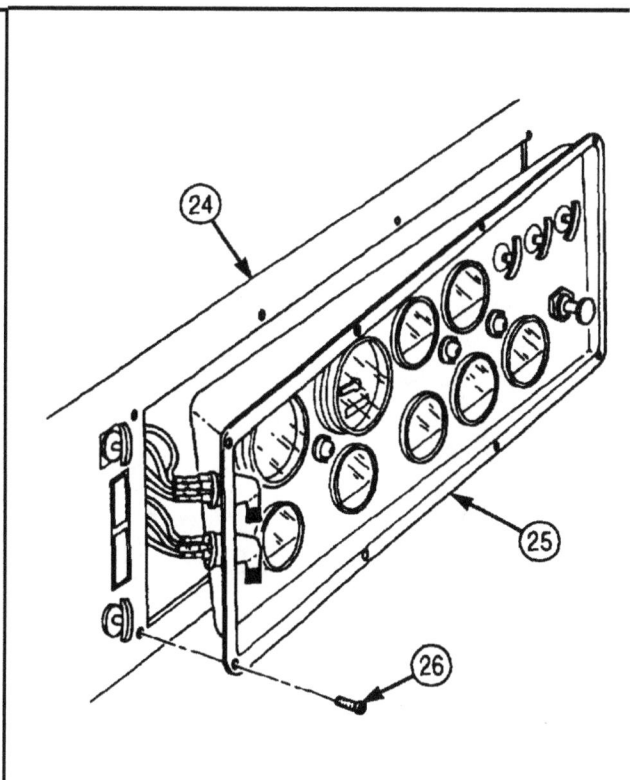

3-83. INSTRUMENT CLUSTER MAINTENANCE (Contd)

NOTE

Tag wires and drive shaft for installation.

12. Disconnect tachometer driveshaft (9) from instrument cluster (4).

13. Disconnect speedometer driveshaft (14) from instrument cluster (4).

WARNING

Do not disconnect air lines before draining air reservoirs. Small parts under pressure may shoot out with high velocity, causing injury to personnel.

14. Disconnect air lines (5), (19), and (21) from instrument cluster (4).

15. Disconnect four wires (10) from instrument cluster (4).

16. Disconnect three wires (11) from instrument cluster (4).

17. Disconnect wire (15) from wire (17).

NOTE

Wires in steps 20 through 28 are located behind instrument cluster.

18. Disconnect five wires (8) from instrument cluster (4).

19. Disconnect wires (12) and (13) from instrument cluster (4).

20. Disconnect wires (16) and (18) from instrument cluster (4).

21. Disconnect wires (24) and (23) from instrument cluster (4).

22. Disconnect wires (6) and (7) from instrument cluster (4).

23. Disconnect wire (20) from instrument cluster (4).

24. Disconnect two wires (22) from instrument cluster (4).

25. Disconnect cables (1), (2), and (3) from instrument cluster (4).

3-83. INSTRUMENT CLUSTER MAINTENANCE (Contd)

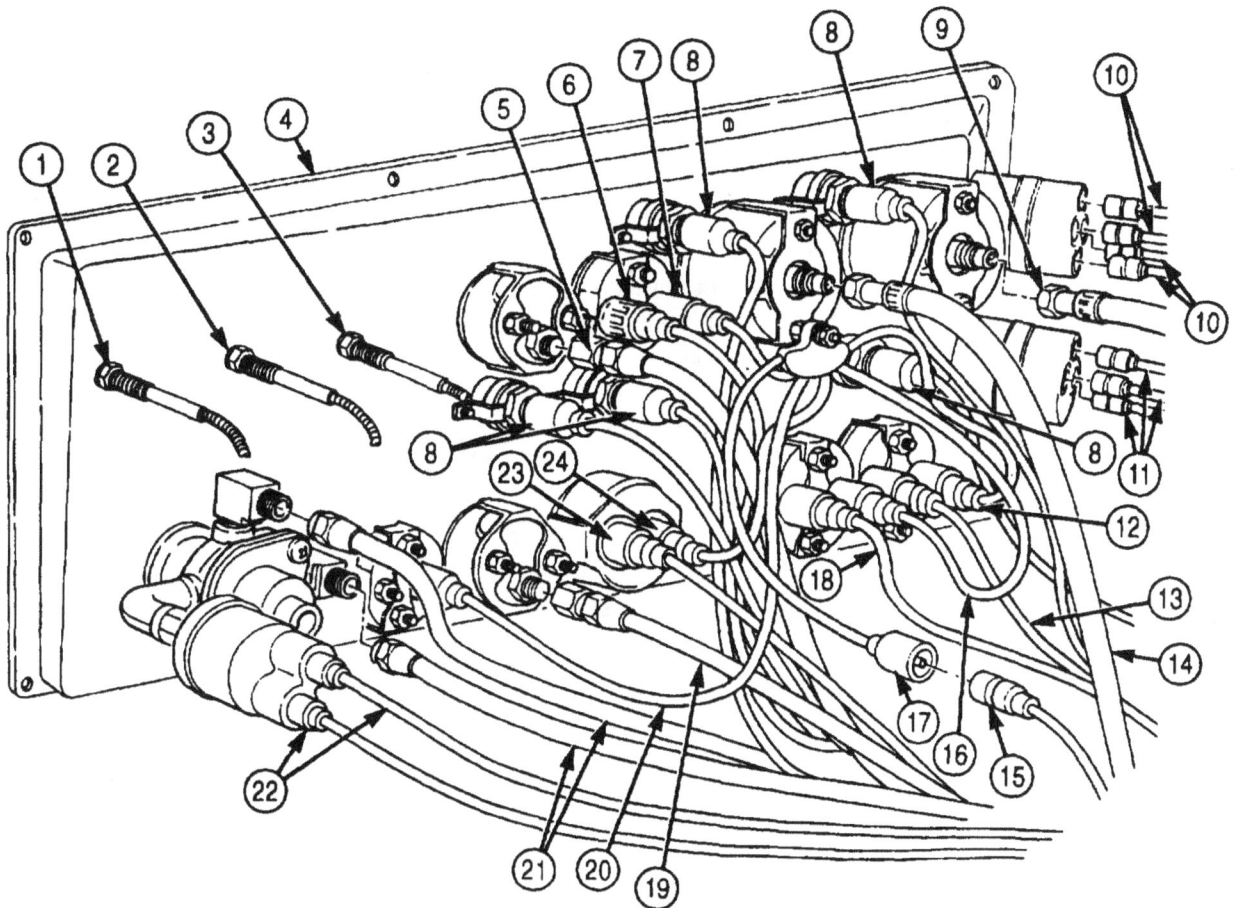

3-83. INSTRUMENT CLUSTER MAINTENANCE (Contd)

b. Disassembly

1. Remove battery switch lever (1) from instrument cluster (13) (para. 3-107).
2. Remove starter switch lever (12) from instrument cluster (13) (para. 3-107).
3. Remove five indicator panel lights (10) from instrument cluster (13) (para. 3-84).
4. Remove tachometer (2) from instrument cluster (13) (para. 3-88).
5. Remove speedometer (3) from instrument cluster (13) (para. 3-88).
6. Remove five electrical gauges (4) from instrument cluster (13) (para. 3-86).
7. Remove primary air gauge (5) from instrument cluster (13) (para. 3-87).
8. Remove secondary air gauge (11) from instrument cluster (13) (para. 3-87).
9. Remove fresh air vent control (8) from instrument cluster (13) (para. 3-293).
10. Remove defroster control (6) from instrument cluster (13) (para. 3-294).
11. Remove heater control (7) from instrument cluster (13) (para. 3-294).
12. Remove spring brake pressure switch (9) from instrument cluster (13) (para. 3-100).

c. Assembly

1. Install spring brake pressure switch (9) on instrument cluster (13) (para. 3-100).
2. Install heater control (7) on instrument cluster (13) (para. 3-294).
3. Install defroster control (6) on instrument cluster (13) (para. 3-294).
4. Install fresh air vent control (8) on instrument cluster (13) (para. 3-293).
5. Install secondary air gauge (11) on instrument cluster (13) (para. 3-87).
6. Install primary air gauge (5) on instrument cluster (13) (para. 3-87).
7. Install five electrical gauges (4) on instrument cluster (13) (para. 3-86).
8. Install speedometer (3) on instrument cluster (13) (para. 3-88).
9. Install tachometer (2) on instrument cluster (13) (para. 3-88).
10. Install five indicator panel lights (10) on instrument cluster (13) (para. 3-84).
11. Install starter switch lever (12) on instrument cluster (13) (para. 3-107).
12. Install battery switch lever (1) on instrument cluster (13) (para. 3-107).

3-83. INSTRUMENT CLUSTER MAINTENANCE (Contd)

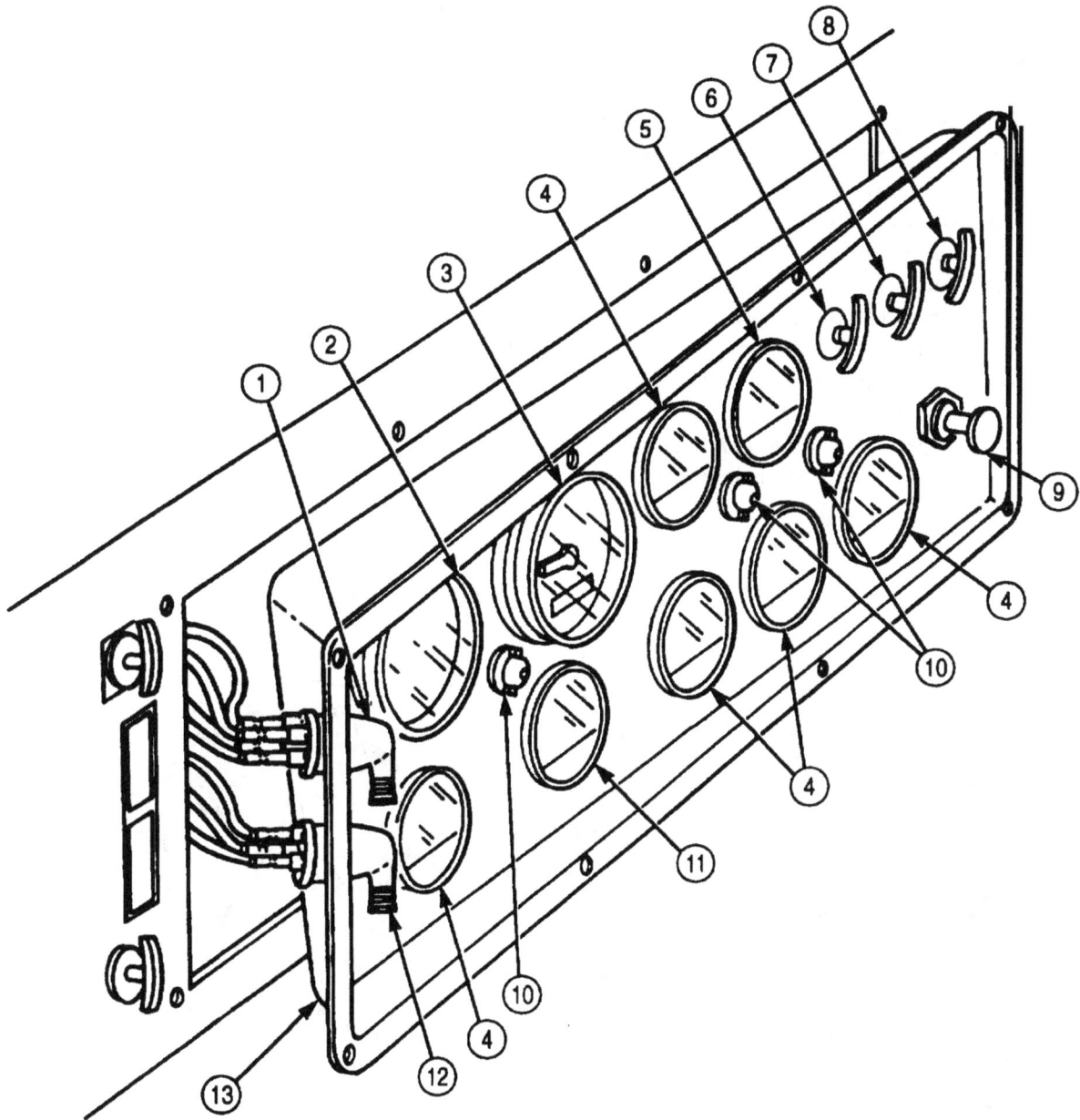

3-83. INSTRUMENT CLUSTER MAINTENANCE (Contd)

d. Installation

1. Connect cables (3), (2), and (1) to instrument cluster (4).
2. Connect two wires (22) to instrument cluster (4).
3. Connect wire (20) to instrument cluster (4).
4. Connect wires (7) and (6) to instrument cluster (4).
5. Connect wires (23) and (24) to battery instrument cluster (4).
6. Connect wires (18) and (16) to instrument cluster (4).
7. Connect wires (13) and (12) to instrument cluster (4).
8. Connect five wires (8) to instrument cluster (4).
9. Connect wire (15) to wire (17).
10. Connect three wires (11) to instrument cluster (4).
11. Connect four wires (10) to instrument cluster (4).
12. Connect air lines (21), (19), and (5) to instrument cluster (4).
13. Connect speedometer driveshaft (14) to instrument cluster (4).
14. Connect tachometer driveshaft (9) to instrument cluster (4).
15. Install fresh air control cable (27) and new spring nut (26) on heater assembly (25).
16. Install new cotter pin (31) on heater assembly (25).
17. Install clamp (29), retaining nut (30), and screw (28) on heater assembly (25).

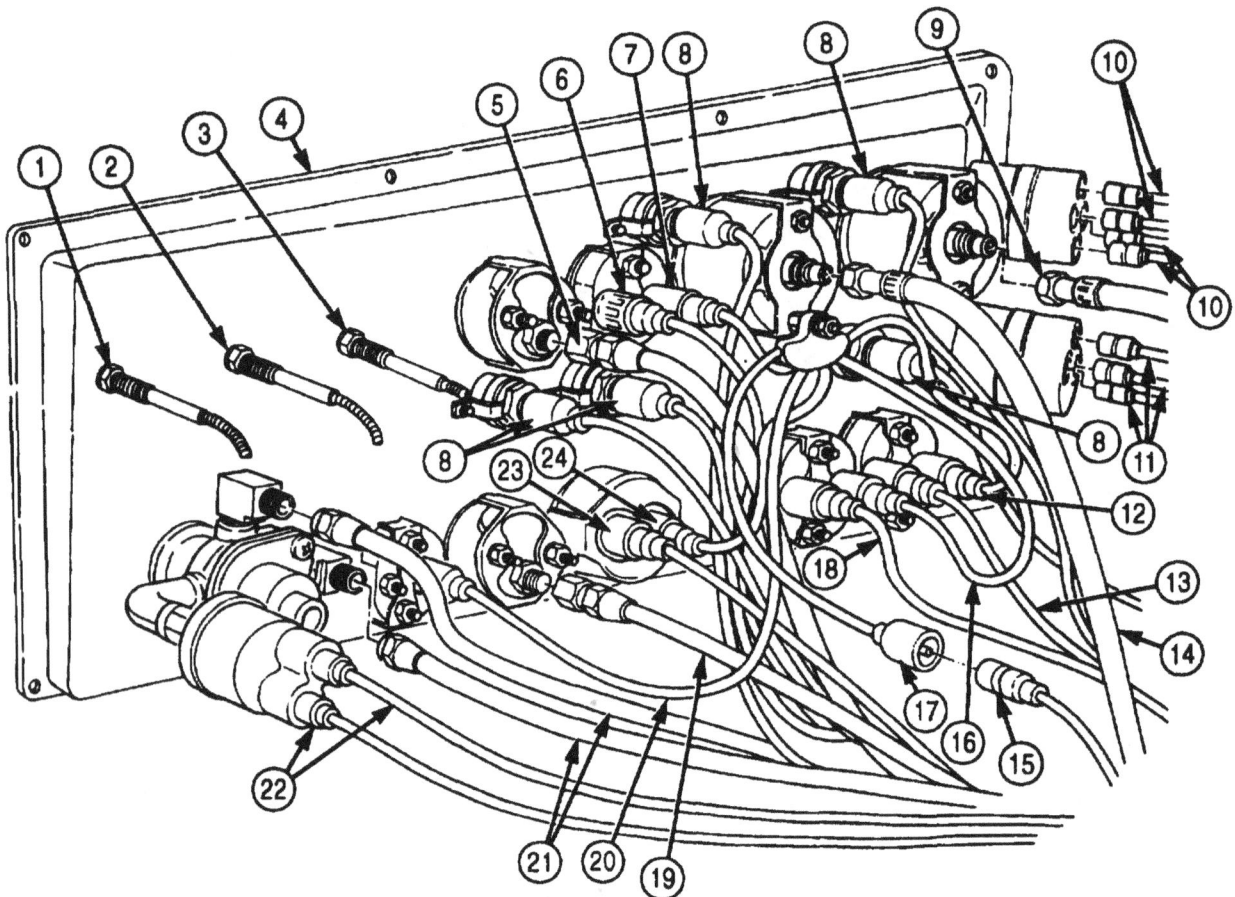

3-83. INSTRUMENT CLUSTER MAINTENANCE (Contd)

18. Route heat control cable (43) and defroster control cable (32) through diverter brackets (37) and (38).
19. Install heat control cable (43), defroster control cable (32), and new spring nuts (33) and (41) on control panel rods (40) and (36).
20. Install new cotter pins (42) and (47) on control rods (40) and (36).
21. Install retaining clips (35) and (39), cables (32) and (43), and clamps (34) and (44) on diverter brackets (37) and (38) with screws (46) and (45).
22. Install instrument cluster (4) on instrument panel (48) with eight screws (49).

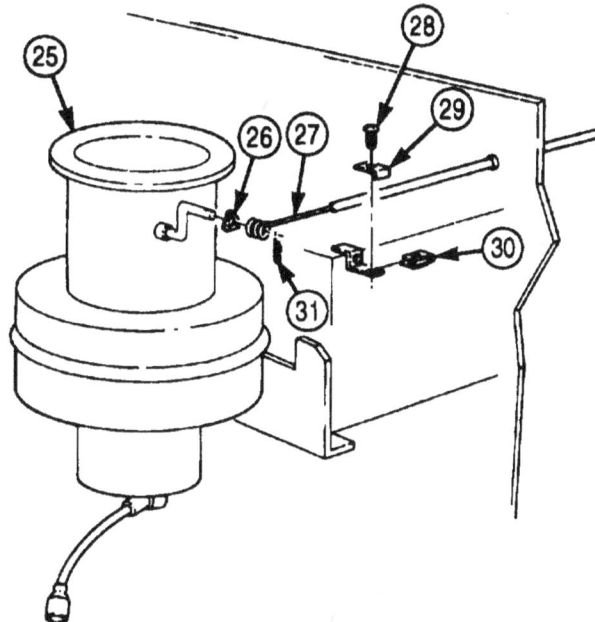

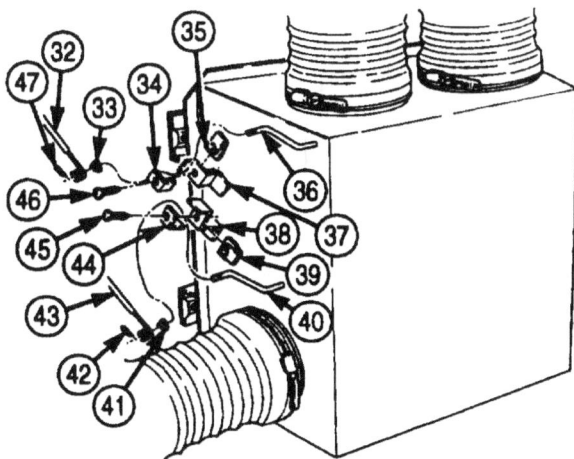

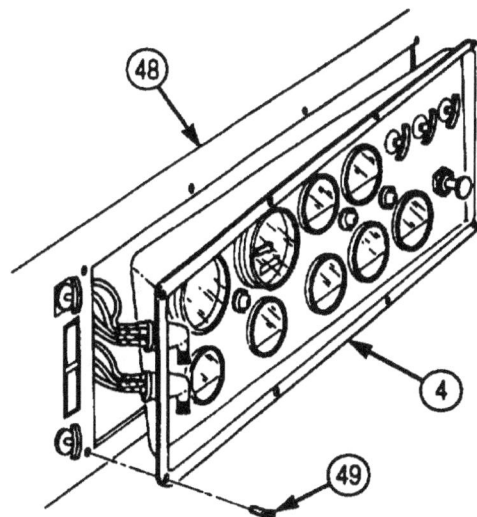

FOLLOW-ON TASKS:
- Connect CTIS electrical components (M939A2) (para. 3-470).
- Connect battery ground cables (para. 3-126).
- Start engine (TM 9-2320-272-10) and check gauges for proper operation. Check for air leaks at air gauges.

3-84. INDICATOR PANEL LIGHT AND LAMP REPLACEMENT

THIS TASK COVERS:

a. Light Assembly Removal
b. Light Assembly Installation

c. Lamp Removal
d. Lamp Installation

INITIAL SETUP:

APPLICABLE MODELS
All

TOOLS
General mechanic's tool kit (Appendix E, Item 1)

MATERIALS/PARTS
Two lockwashers (Appendix D. Item 351)

REFERENCES (TM)
TM 9-2320-272-10
TM 9-2320-272-24P

EQUIPMENT CONDITION
• Parking brake set (TM 9-2326-272-10).
• Battery ground cables disconnected (para. 3-126).

---

a. Light Assembly Removal

1. Remove four screws (7) from warning light panel (1) and pull warning light panel (1) away from instrument panel (2).
2. Remove lamp lens (10) from lamp holder (5).
3. Remove two screws (9) and lockwashers (8) from warning light panel assembly (1) and lamp holder bracket (6). Discard lockwashers (8).

NOTE

Tag wire for installation.

4. Disconnect wire (3) from lamp holder connector (4).
5. Remove lamp holder (5) and bracket (6) from warning light panel assembly (1).

b. Light Assembly Installation

1. Install lamp holder (5) and bracket (6) on warning light panel assembly (1) with two new lockwashers (8) and screws (9).
2. Connect wire (3) to lamp holder connector (4).
3. Install lamp lens (10) on lamp holder (5).
4. Install warning light panel (1) on instrument panel (2) with four screws (7).

c. Lamp Removal

NOTE

All panel light lamps are removed and installed the same.

1. Remove lamp lens (10) from lamp holder (5).
2. Push lamp (11) inward and turn counterclockwise to remove.

d. Lamp Installation

1. Push lamp (11) inward and turn clockwise to install.
2. Install lamp lens (10) on lamp holder (5).

3-84. INDICATOR PANEL LIGHT AND LAMP REPLACEMENT (Contd)

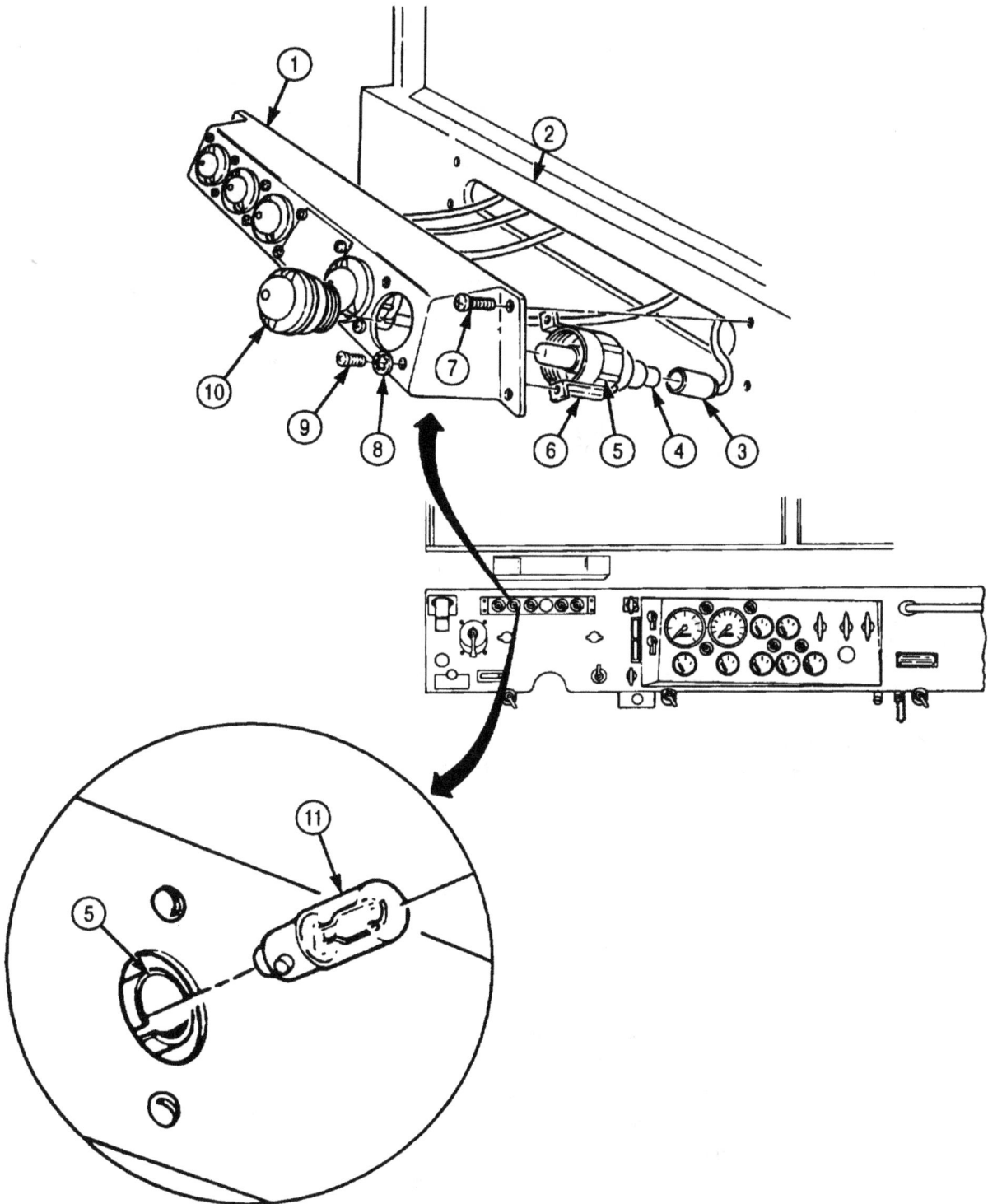

FOLLOW-ON TASK: Connect battery ground cables (para. 3-126).

3-85. INSTRUMENT PANEL REPLACEMENT

THIS TASK COVERS:

a. Removal   b. Installation

INITIAL SETUP:

APPLICABLE MODELS

All

TOOLS

General mechanic's tool kit (Appendix E, Item 1)

REFERENCES (TM)

TM 9-2320-272-10
TM 9-2320-272-24P

EQUIPMENT CONDITION

• Parking brake set (TM 9-2320-272-10).
• Battery ground cables disconnected (para. 3-126).
• Instrument cluster removed (para. 3-83).
• Electrical switches removed (para. 3-107).
• Fuel selector valve switch removed (para. 3-109).

EQUIPMENT CONDITION (Contd)

• Floodlight control switch removed (para. 3-120).
• Pressure gauge removed (para. 3-86).
• Steering wheel removed (para. 3-226).
• Cold-start indicator and lamp removed (para. 3-84).
• Engine stop and throttle control cable removed (para. 3-45).
• Turn signal flasher removed (para. 3-114).
• Instrument panel circuit breaker removed (para. 3-105).
• Windshield wiper pump removed (para. 3-281).
• Windshield wiper hoses removed (para. 3-282).
• Heater control box removed (para. 3-295).
• Personnel heater control cables removed (para. 3-294).

NOTE

It may be necessary to remove the upper steering column. If so, notify DS maintenance.

a. Removal

1. Remove two screws (5) from instrument panel (4) and bracket (7).
2. Remove five screws (3) from instrument panel (4).
3. Remove four screws (2) from instrument panel (4).
4. Remove two screw-assembled washers (1) from instrument panel (4).
5. Remove screw (6) from instrument panel (4).

NOTE

Assistant will help with step 5.

6. Remove instrument panel (4) from cab (8) and brackets (7) and (9).

b. Installation

NOTE

Assistant will help with step 1.

1. Position instrument panel (4) on cab (8) and brackets (7) and (9).
2. Install screw (6) on instrument panel (4).
3. Install two screw-assembled washers (1) on instrument panel (4).
4. Install four screws (2) on instrument panel (4).
5. Install five screws (3) on instrument panel (4).
6. Install two screws (5) on instrument panel (4) and bracket (7).

3-85. INSTRUMENT PANEL REPLACEMENT (Contd)

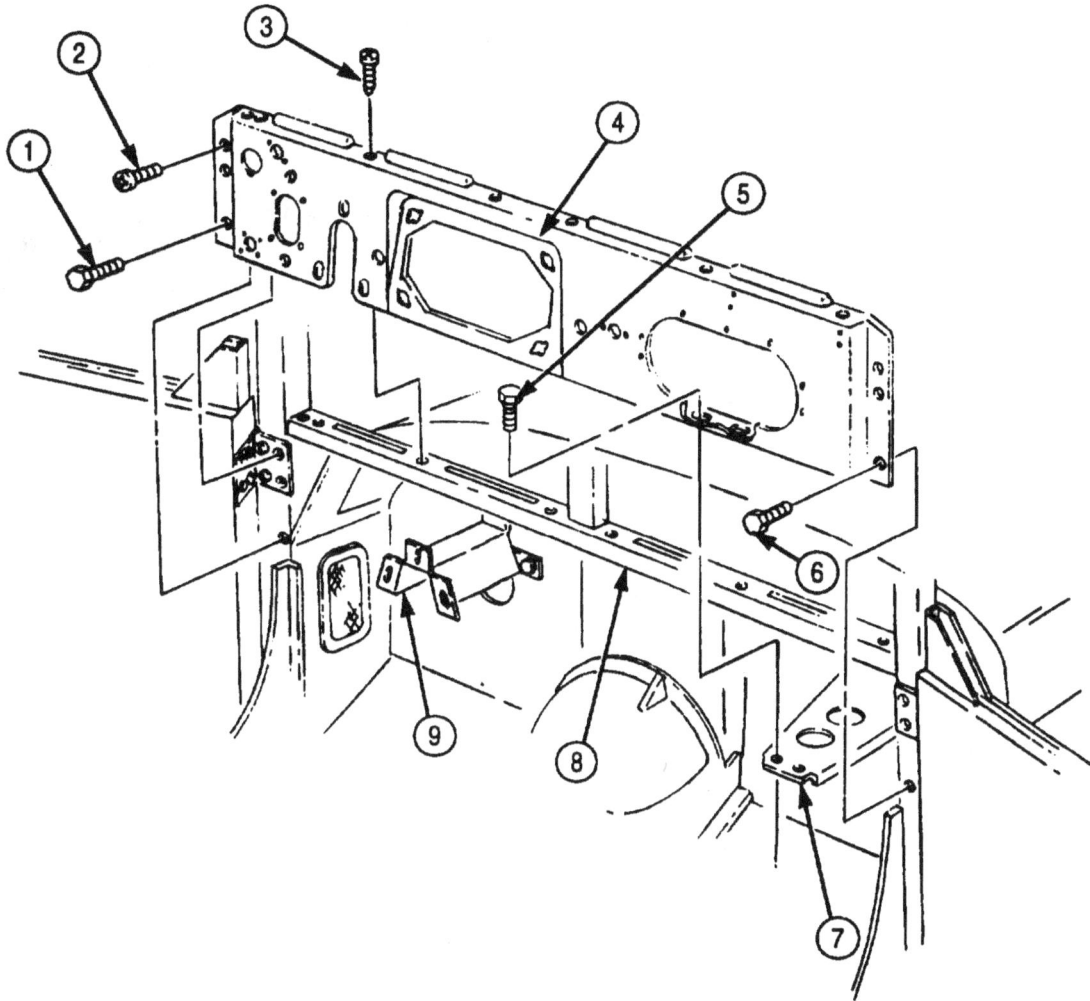

FOLLOW-ON TASKS:
- Install turn signal flasher (para. 3-114).
- Install instrument cluster (para. 3-83).
- Install electrical switches (para. 3-107).
- Install fuel selector valve switch (para. 3-109).
- Install floodlight control switch (para. 3-120).
- Install pressure gauge switch (para. 3-86).
- Install steering wheel (para. 3-226).
- Install cold-start indicator and lamp (para. 3-84).
- Install engine stop and throttle control cables (para. 3-45).
- Install instrument panel circuit breaker (para. 3-105).
- Install personnel heater control cables (para. 3-294).
- Install heater control box (para. 3-295).
- Install windshield wiper hoses (para. 3-282).
- Install windshield wiper pump (para. 3-281).
- Connect battery ground cables (para. 3-126).

3-86. ELECTRICAL GAUGES REPLACEMENT

THIS TASK COVERS:

a. Removal   b. Installation

INITIAL SETUP:

APPLICABLE MODELS
All

TOOLS
General mechanic's tool kit (Appendix E, Item 1)

MATERIALS/PARTS
Two lockwashers (Appendix D, Item 352)

REFERENCES (TM)
TM 9-2320-272-10
TM 9-2320-272-24P

EQUIPMENT CONDITION
• Parking brake set (TM 9-2320-272-10).
• Battery ground cables disconnected (para. 3-126).

NOTE
Engine coolant temperature, transmission oil temperature, engine oil pressure, battery/alternator, and fuel level gauges are removed and installed the same.

a. Removal

1. Remove eight screws (7) from instrument panel (4).
2. Pull instrument cluster (5) away from instrument panel (4).

NOTE
• Battery indicator gauge has only one wire to disconnect.
• Tag wires for installation.

3. Disconnect wires (3) and (9) from fuel level gauge (6).
4. Remove two nuts (1), lockwashers (2), and gauge mounting bracket (8) from fuel level gauge (6). Discard lockwashers (2).
5. Remove fuel level gauge (6) from instrument cluster (5).

b. Installation

1. Install fuel level gauge (6) on instrument cluster (5).
2. Install gauge mounting bracket (8) on fuel level gauge (6) with two new lockwashers (2) and nuts (1).
3. Connect wires (3) and (9) to fuel level gauge (6).
4. Install instrument cluster (5) on instrument panel (4) with eight screws (7).

3-86. ELECTRICAL GAUGES REPLACEMENT (Contd)

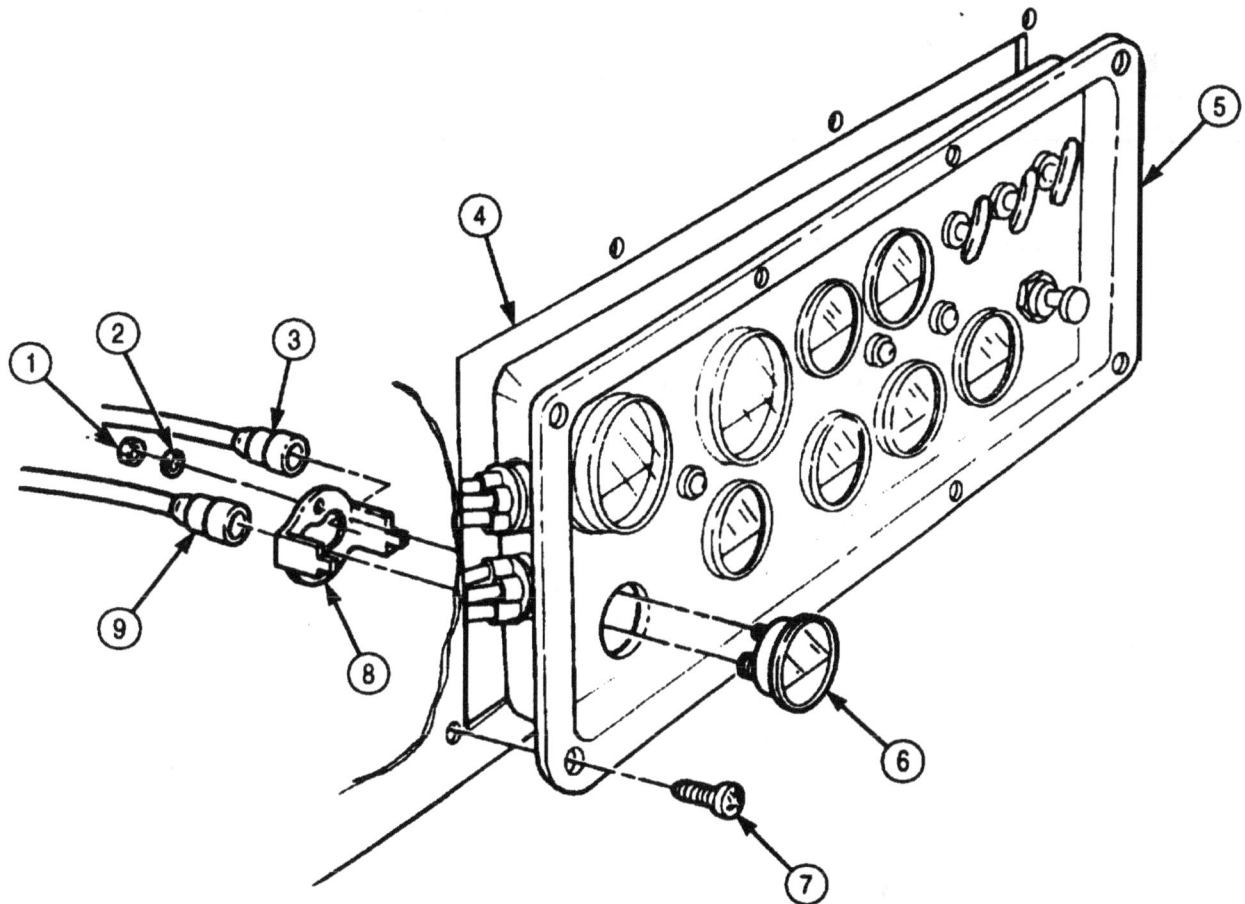

FOLLOW-ON TASKS: • Connect battery ground cables (para. 3-126).
                 • Start engine (TM 9-2320-272-10) and check gauge for proper operation.

3-87. AIR GAUGE REPLACEMENT

THIS TASK COVERS:

a. Removal   b. Installation

INITIAL SETUP:

APPLICABLE MODELS
All

TOOLS
General mechanic's tool kit (Appendix E, Item 1)

MATERIALS/PARTS
Two lockwashers (Appendix D, Item 352)
Antiseize tape (Appendix C. Item 72)

REFERENCES (TM)
TM 9-2320-272-10
TM 9-2320-272-24P

EQUIPMENT CONDITION
• Parking brake set (TM 9-2320-272-10).
• Air reservoirs drained (TM 9-2320-272-10).
• Battery ground cables disconnected (para. 3-126).

NOTE
The primary and secondary air pressure gauges are removed and installed in the same manner.

a. Removal

1. Remove eight screws (7) from instrument cluster (6).
2. Pull instrument cluster (6) away from instrument panel (8).

NOTE
Tag air line for installation.

3. Disconnect air line (1) from primary air gauge (5).
4. Remove two nuts (2), lockwashers (3), and gauge mounting bracket (4) from primary air gauge (5). Discard lockwashers (3).
5. Remove primary air gauge (5) from instrument cluster (6).

b. Installation

1. Install primary air gauge (5) on instrument cluster (6).
2. Install gauge mounting bracket (4) on primary air gauge (5) with two new lockwashers (3) and nuts (2).

NOTE
Male pipe threads must be wrapped with antiseize tape before installation.

3. Connect air line (1) to primary air gauge (5).
4. Install instrument cluster (6) on instrument panel (8) with eight screws (7).

3-87. AIR GAUGE REPLACEMENT (Contd)

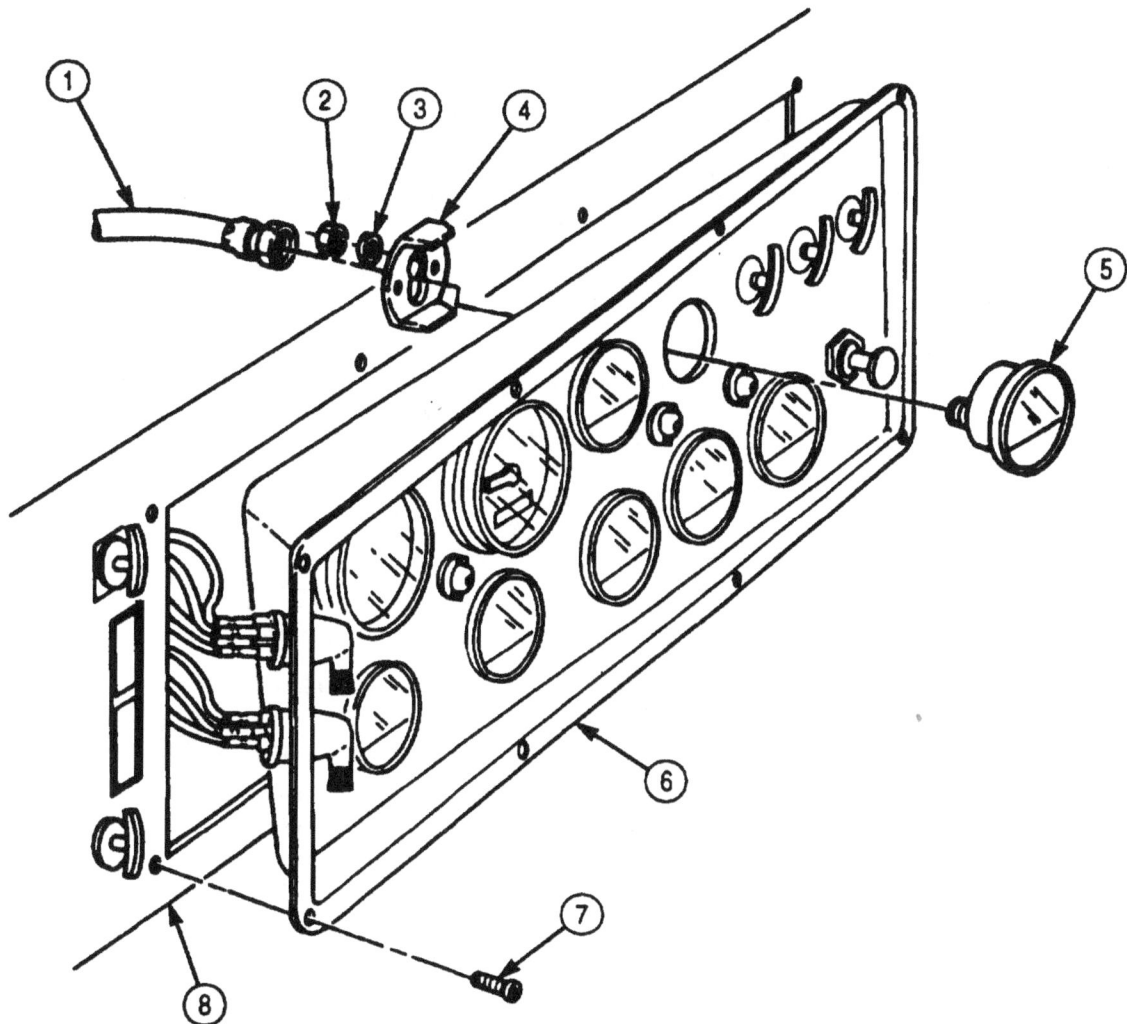

FOLLOW-ON TASKS: • Connect battery ground cables (para. 3-126).
• Start engine (TM 9-2320-272-10), allow air pressure to build up to normal operating range, and check for air leaks at gauge.
• Check if gauge is indicating air pressure (TM 9-2320-272-10).

3-88. SPEEDOMETER AND TACHOMETER REPLACEMENT

THIS TASK COVERS:

a. Removal   b. Installation

INITIAL SETUP:

APPLICABLE MODELS
All

TOOLS
General mechanic's tool kit (Appendix E, Item 1)

MATERIALS/PARTS
Two lockwashers (Appendix D, Item 352)

REFERENCES (TM)
TM 9-2320-272-10
TM 9-2320-272-24P

EQUIPMENT CONDITION
- Parking brake set (TM 9-2320-272-10).
- Battery ground cables disconnected (para. 3-126).

NOTE

Speedometer and tachometer are replaced the same. This procedure covers speedometer.

a. Removal

1. Remove eight screws (7) from instrument cluster (8).

2. Pull instrument cluster (8) away from instrument panel (9).

3. Loosen nut (2) and disconnect speedometer driveshaft (1) from speedometer (6).

4. Remove two nuts (3), cable assembly (11), extension stud (10), washer (12), and two lockwashers (4) from speedometer mounting bracket (5). Discard lockwashers (4).

5. Remove mounting bracket (5) and speedometer (6) from instrument cluster (8).

b. Installation

1. Install speedometer (6) and mounting bracket (5) on instrument cluster (8) with two new lockwashers (4), extension stud (10), and nut (3).

2. Install cable assembly (11) on extension stud (10) with washer (12) and nut (3).

3. Install speedometer driveshaft (1) on speedometer (6) and tighten nut. (2).

4. Install instrument cluster (8) on instrument panel (9) with eight screws (7).

3-88. SPEEDOMETER AND TACHOMETER REPLACEMENT (Contd)

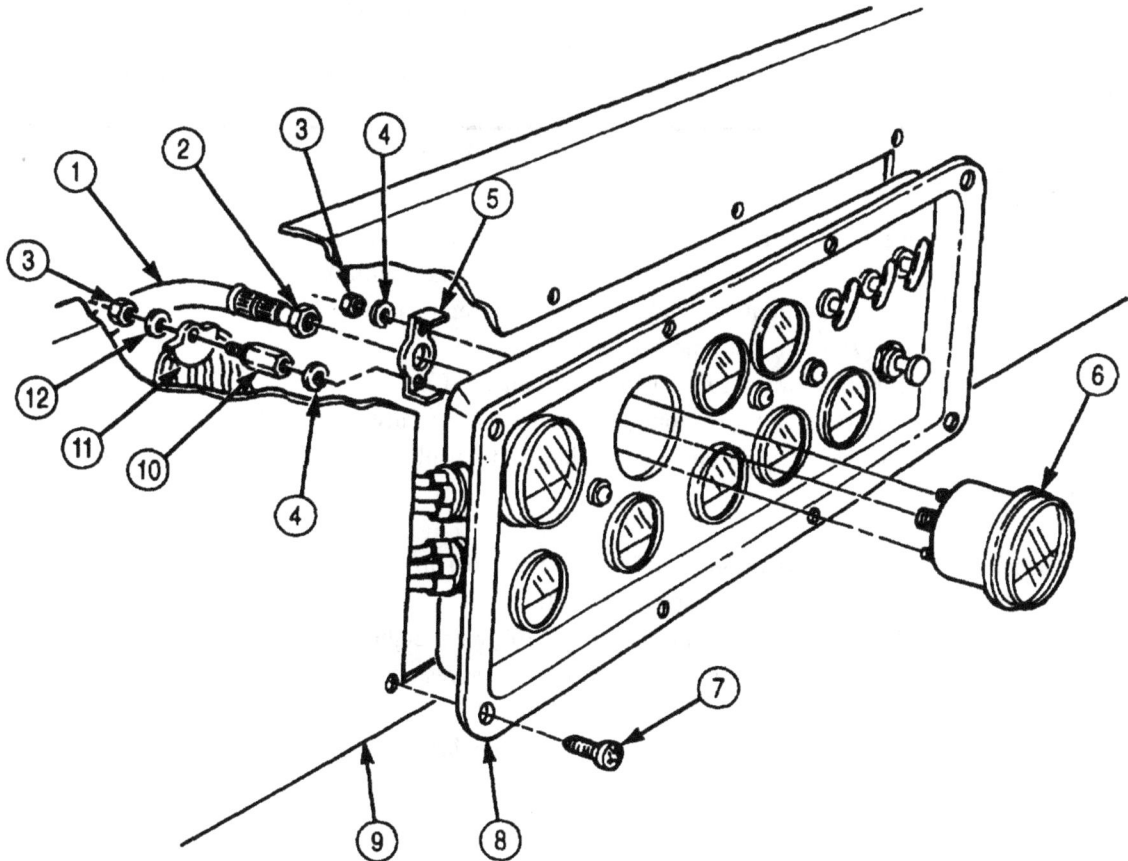

FOLLOW-ON TASKS: • Connect battery ground cables (para. 3-126).
                    • Start engine (TM 9-2320-272-10) and road test to check speedometer and tachometer for proper operation.

3-89. TACHOMETER DRIVESHAFT (M939/A1) REPLACEMENT

THIS TASK COVERS:

a. Removal   b. Installation

INITIAL SETUP:

APPLICABLE MODELS
M939/A1

TOOLS
General mechanic's tool kit (Appendix E, Item 1)

MATERIALS/PARTS
Rubber grommet (Appendix D, Item 565)

REFERENCES (TM)
TM 9-2320-272-10
TM 9-2320-27224P

EQUIPMENT CONDITION
• Parking brake set (TM 9-2320-272-10).
• Left splash shield removed (TM 9-2320-272-10).

---

a. Removal

1. Disconnect tachometer driveshaft (5) from tachometer pulse sender unit (12).
2. Remove screw (1), washer (2), and loop clamp (3) from intake manifold (11).
3. Remove rubber grommet (4) from firewall (6). Discard grommet (4).
4. Remove eight screws (9) from instrument cluster (8).
5. Pull instrument cluster (8) away from instrument panel (10).
6. Disconnect tachometer driveshaft (5) from tachometer (7) on instrument cluster (8).
7. Remove tachometer driveshaft (5) from engine side of firewall (6). Replace if bent or cracked.

b. Installation

1. Push tachometer driveshaft (5) through hole in firewall (6) to tachometer (7).
2. Install tachometer driveshaft (5) on tachometer (7).
3. Install instrument cluster (8) on instrument panel (10) with eight screws (9).
4. Slide new rubber grommet (4) over disconnected end of driveshaft (5) and install in firewall (6).
5. Install driveshaft (5) on tachometer pulse sender unit (12).
6. Position loop clamp (3) over driveshaft (5) and install on intake manifold (11) with washer (2) and screw (1).

3-89. TACHOMETER DRIVESHAFT (M939/A1) REPLACEMENT (Contd)

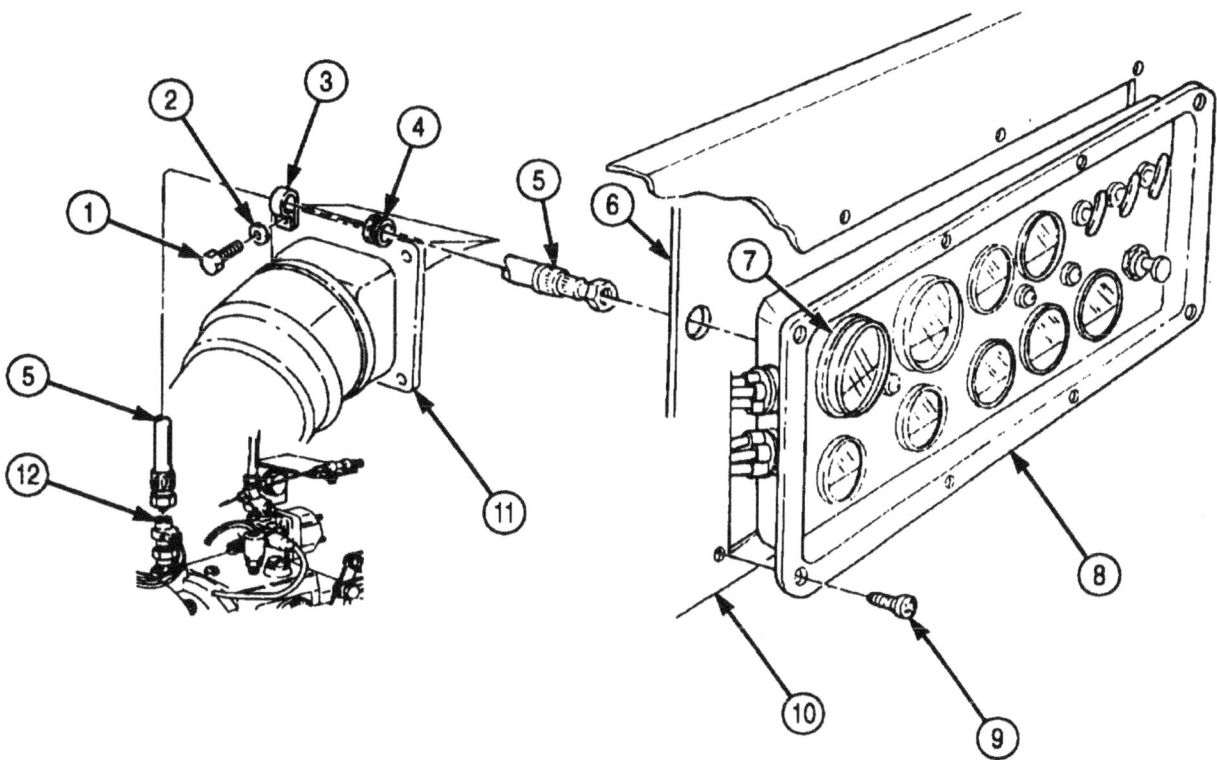

FOLLOW-ON TASKS: • Install left splash shield (TM 9-2320-272-10).
• Start engine (TM 9-2320-272-10) and check tachometer for proper operation.

3-90. SPEEDOMETER DRIVESHAFT MAINTENANCE

THIS TASK COVERS:

a. Removal

b. Inspection

c. Installation

INITIAL SETUP:

APPLICABLE MODELS

All

TOOLS

General mechanic's tool kit (Appendix E, Item 1)
Torque wrench (Appendix E, Item 144)

MATERIALS/PARTS

Five tiedown straps (Appendix D, Item 690)
GAA grease (Appendix C, Item 28)

REFERENCES (TM)

TM 9-2320-272-10
TM 9-2320-272-24P

EQUIPMENT CONDITION

Parking brake set (TM 9-2320-272-10).

---

a. Removal

1. Remove screw (5), washer (4), two clamps (2), speedometer driveshaft (6), and clamp (3) from air intake manifold (1).

2. Remove eight screws (11) from instrument cluster (10) and pull away from instrument panel (9).

3. Loosen shaft nut (7) and disconnect speedometer driveshaft (6) from speedometer (8).

4. Remove nut (15), screw (20), and clamps (19) and (14) from bracket (16) on engine bell housing (18).

5. Remove screw (17) and bracket (16) from engine bell housing (18).

6. Remove clamp (19) from speedometer driveshaft (6).

7. Remove screw (23) and clamp (22) from transfer case (21).

8. Remove clamp (22) from speedometer driveshaft (6).

9. Disconnect shaft nut (24) from transfer case input gear cover (25).

10. Remove five tiedown straps (26) from speedometer driveshaft (6). Discard tiedown straps (26).

11. Remove speedometer driveshaft (6) and grommet (13) from firewall (12).

12. Remove grommet (13) from speedometer driveshaft (6).

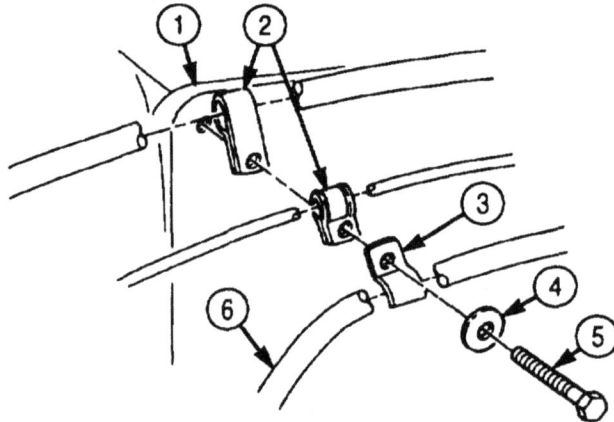

3-90. SPEEDOMETER DRIVESHAFT MAINTENANCE (Contd)

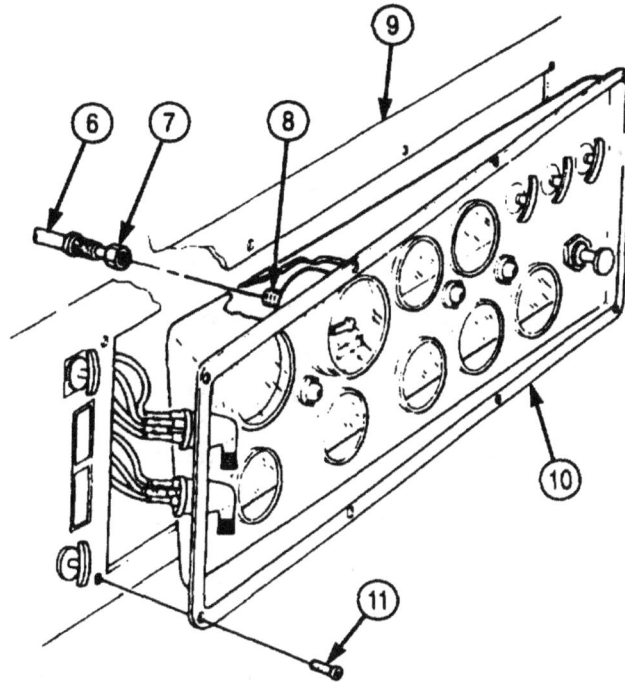

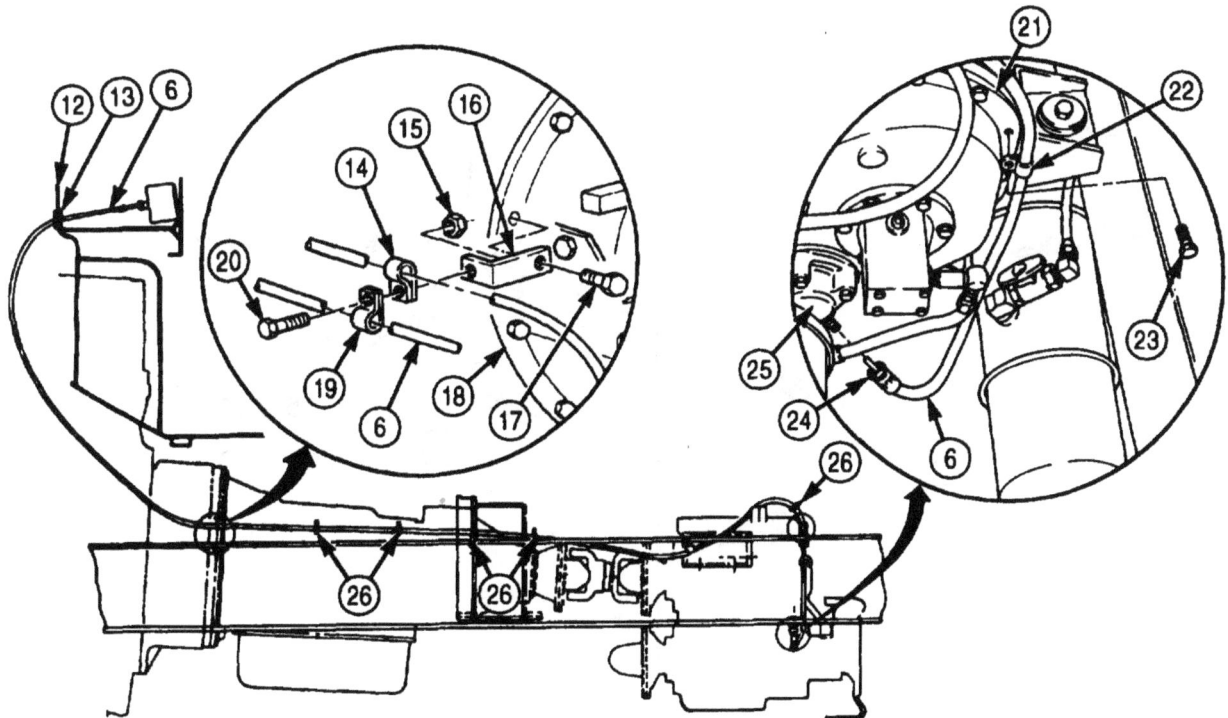

3-90. SPEEDOMETER DRIVESHAFT MAINTENANCE (Contd)

NOTE

Pull driveshaft core out from driveshaft conduit 1-1/2 in. (38 mm)
to gain access to retaining washer.

13. Remove retaining washer (3) and driveshaft core (1) from driveshaft (2).

b. Inspection

1. Inspect driveshaft (2) for cracks. Replace if cracked.

2. Inspect driveshaft core (1) for breaks. Replace if broken.

c. Installation

NOTE

Apply thin coat of GM grease to driveshaft core.

1. Install driveshaft core (1) and retaining washer (3) on driveshaft (2).

2. Install grommet (10) with speedometer driveshaft (2) on firewall (9).

3. Connect shaft nut (21) to transfer case input gear cover (22).

4. Install speedometer driveshaft (2) and clamp (19) on transfer case (18) with screw (20). Tighten screw (20) 40-65 lb-ft (54-88 N•m).

5. Install bracket (13) on engine bell housing (15) with screw (14). Tighten screw (14) 25-31 lb-ft (34-42 N•m).

6. Install speedometer driveshaft (2), clamps (11) and (16) on bracket (13) with screw (17) and nut (12).

7. Install five new tiedown straps (23) on speedometer driveshaft (2).

8. Connect shaft nut (24) to speedometer (25).

9. Install instrument cluster (27) on instrument panel (26) with eight screws (28).

10. Install two clamps (5), speedometer driveshaft (2), clamp (6), washer (7), and screw (8) on air intake manifold (4).

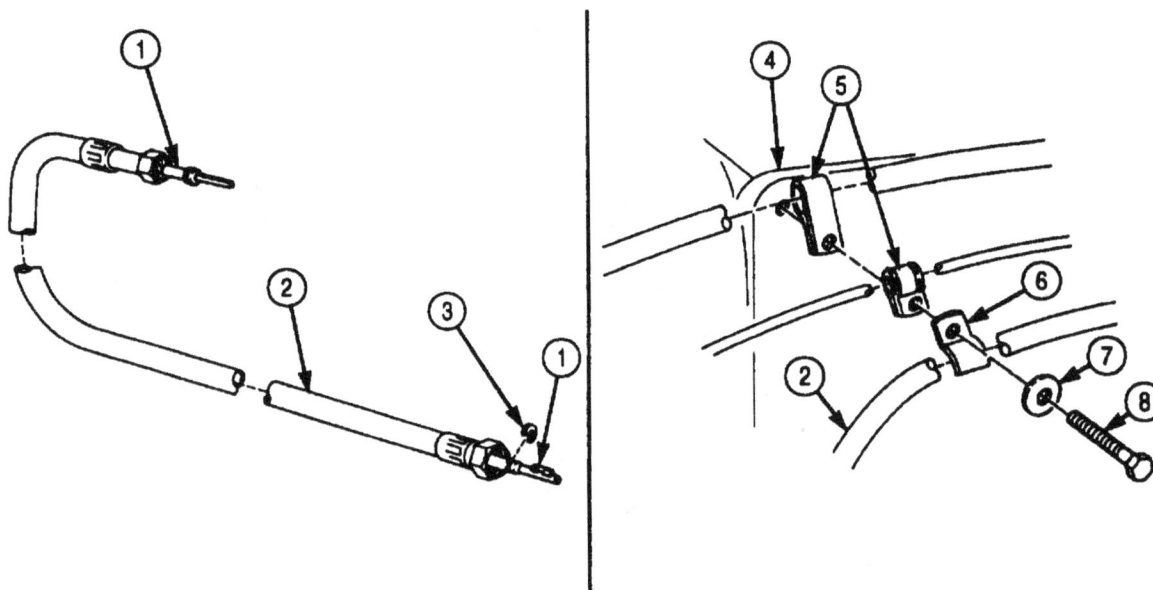

3-90. SPEEDOMETER DRIVESHAFT MAINTENANCE (Contd)

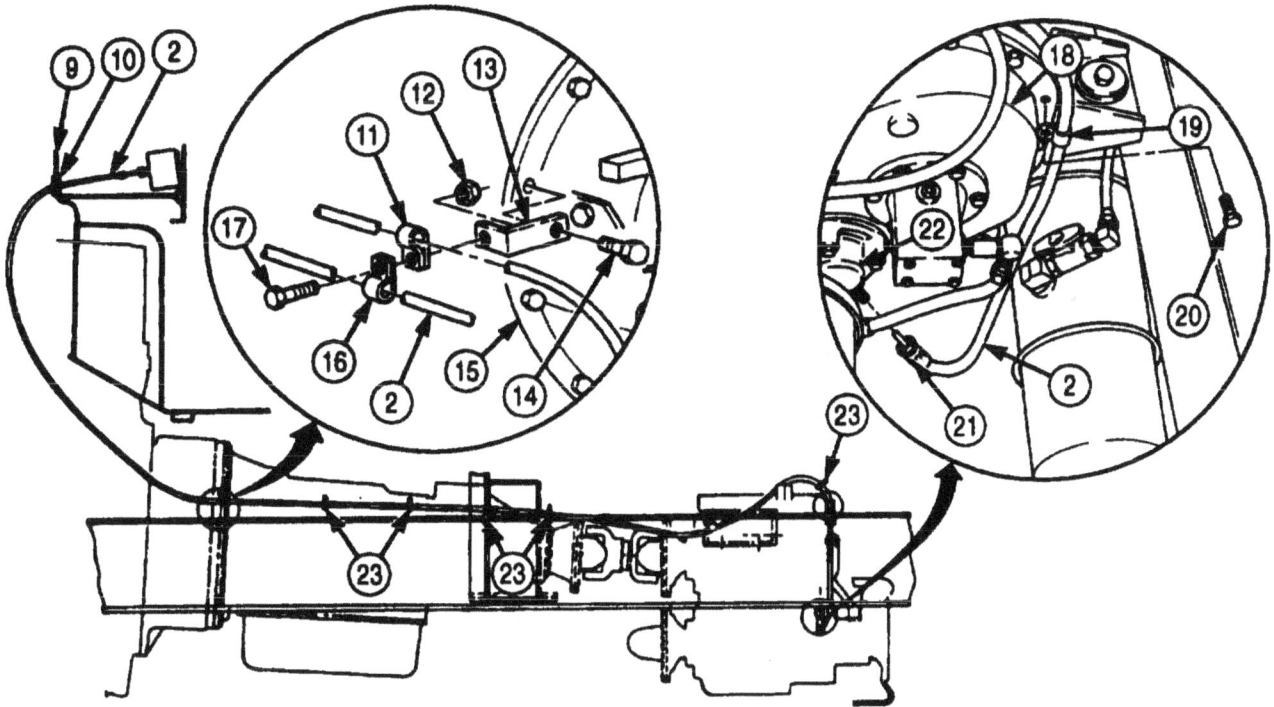

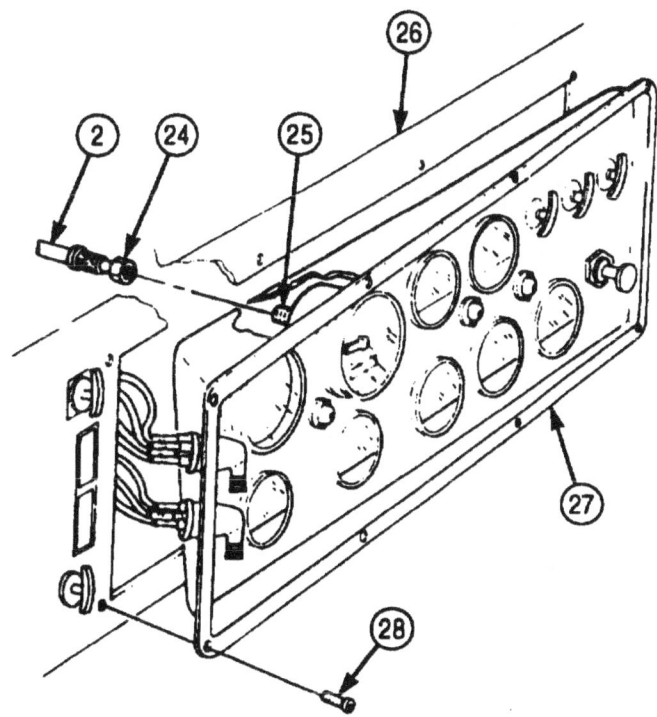

FOLLOW-ON TASK: Start engine (TM 9-2320-272-10) and road test for proper speedometer operation.

3-91. OIL PRESSURE SENDING UNIT REPLACEMENT

THIS TASK COVERS:

a. Removal   b. Installation

INITIAL SETUP:

APPLICABLE MODELS	**REFERENCES** (TM)
All	TM 9-2320-272-10
	TM 9-2320-272-24P
TOOLS	
General mechanic's tool kit (Appendix E, Item 1)	**EQUIPMENT CONDITION**
	Parking brake set (TM 9-2320-272-10).
MATERIALS/PARTS	Left splash shield removed (TM 9-2320-272-10).
Antiseize tape (Appendix C, Item 72)	Battery ground cables disconnected (para. 3-126).

---

a. Removal

1. Disconnect wire (1) from oil pressure sending unit (2).
2. Remove oil sending unit (2) from adapter fitting (5).

NOTE

Step 3 is for M939/A1 models only

3. Remove adapter fitting (5) and elbow (4) from left side of engine (3).

b. Installation

NOTE

- Male pipe threads must be wrapped with antiseize tape before installation.
- Steps 1 and 2 are for M939/A1 models only.

1. Install elbow (4) on left side of engine (3).
2. Install adapter fitting (5) on elbow (4).
3. Install oil sending unit (2) on adapter fitting (5).
4. Connect wire (1) to oil pressure sending unit (2).

3-91. OIL PRESSURE SENDING UNIT REPLACEMENT (Contd)

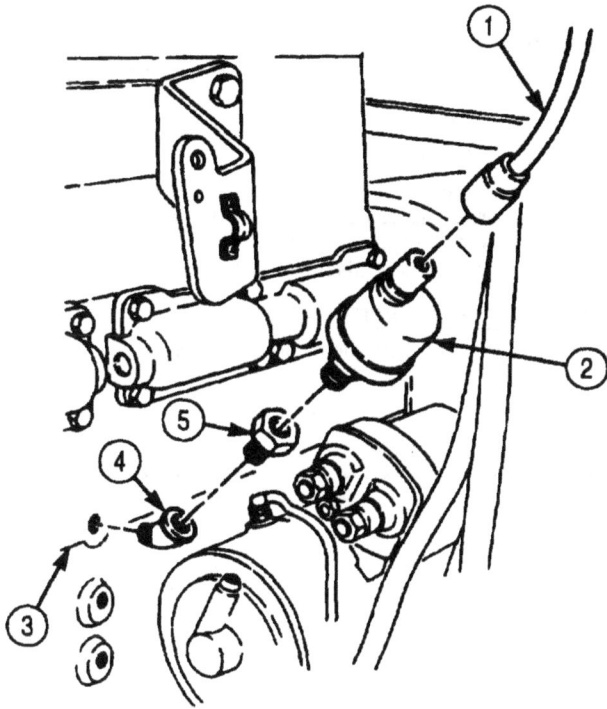

M939/A1 VEHICLES

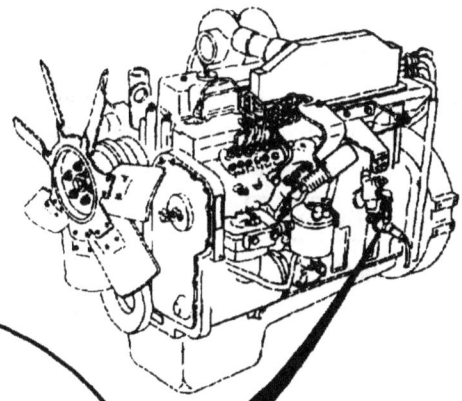

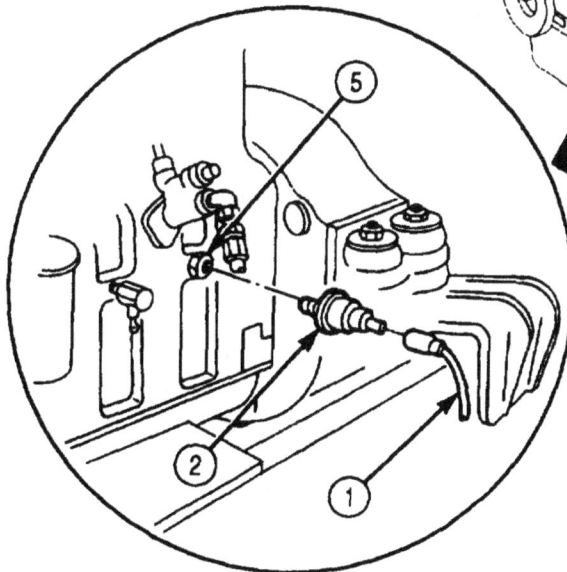

M939A2 VEHICLES

FOLLOW-ON TASKS: • Install left splash shield (TM 9-2320-272-10).
                 • Connect battery ground cables (para. 3-126).
                 • Start engine (TM 9-2320-272-10) and check oil pressure sending unit operation.

3-92. FUEL LEVEL SENDING UNIT REPLACEMENT

THIS TASK COVERS:

a. Removal   b. Installation

INITIAL SETUP:

APPLICABLE MODELS
All

TOOLS
General mechanic's tool kit (Appendix E, Item 1)

MATERIALS/PARTS
Casket (Appendix D, Item 143)

REFERENCES (TM)
TM 9-2320-272-10
TM 9-2320-272-24P

EQUIPMENT CONDITION
• Parking brake set (TM 9-2320-272-10).
• Spare tire carrier access step removed (M931/A1/A2 and M932/A1/A2) (para. 3-260).
• Battery ground cables disconnected (para. 3-126).

GENERAL SAFETY INSTRUCTIONS
Do not perform this task near open flame.

WARNING

Diesel fuel is highly flammable. Do not perform this procedure near open flame. Injury to personnel may result.

a. Removal

1. Disconnect wire (2) from sending unit (3).
2. Remove five screws (1) and ground wire (6) from sending unit (3) and fuel tank (5).
3. Remove sending unit (3) and gasket (4) from fuel tank (5). Discard gasket (4).

b. Installation

1. Install new gasket (4), sending unit (3), and ground wire (6) on fuel tank (5) with five screws (1).
2. Connect wire (2) to sending unit (3).

3-92. FUEL LEVEL SENDING UNIT REPLACEMENT (Contd)

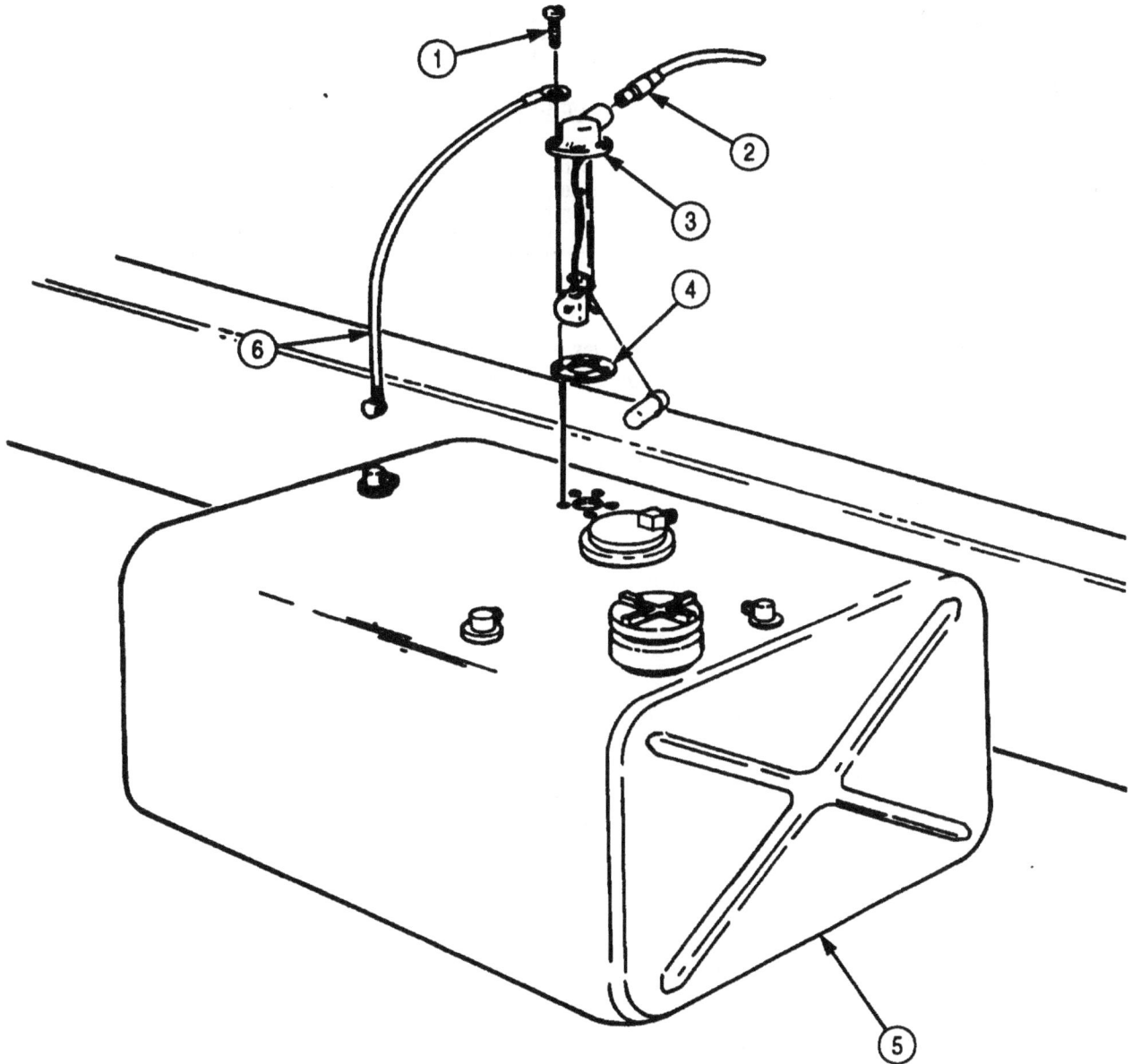

FOLLOW-ON TASKS: • Install spare tire carrier access step (M931/A1/A2 and M932/A1/A2) (para. 3-260).
• Connect battery ground cables (para. 3-126).
• Start engine (TM 9-2320-272-10) and check fuel gauge.

3-93. WATER TEMPERATURE SENDING UNIT REPLACEMENT

THIS TASK COVERS:

a. Removal   b. Installation

INITIAL SETUP:

APPLICABLE MODELS
All

TOOLS
General mechanic's tool kit (Appendix E, Item 1)

MATERIALS/PARTS
Antiseize tape (Appendix C, Item 72)

REFERENCES (TM)
TM 9-2320-272-10
TM 9-2320-272-24P

EQUIPMENT CONDITION
• Parking brake set (TM 9-2320-272-10).
• Battery ground cables disconnected (para. 3-126).
• Right splash shield removed (TM 9-2320-272-10).
• Surge tank cap removed (TM 9-2320-272-10).
• Engine coolant drained (para. 3-53).

a. Removal

1. Disconnect wire (3) from water temperature sending unit (2).
2. Remove water temperature sending unit (2) from water manifold (1).

b. Installation

NOTE

Male pipe thread must be wrapped with antiseize tape before installation.

1. Install water temperature sending unit (2) on water manifold (1).
2. Connect wire (3) to water temperature sending unit (2).

3-93. WATER TEMPERATURE SENDING UNIT REPLACEMENT (Contd)

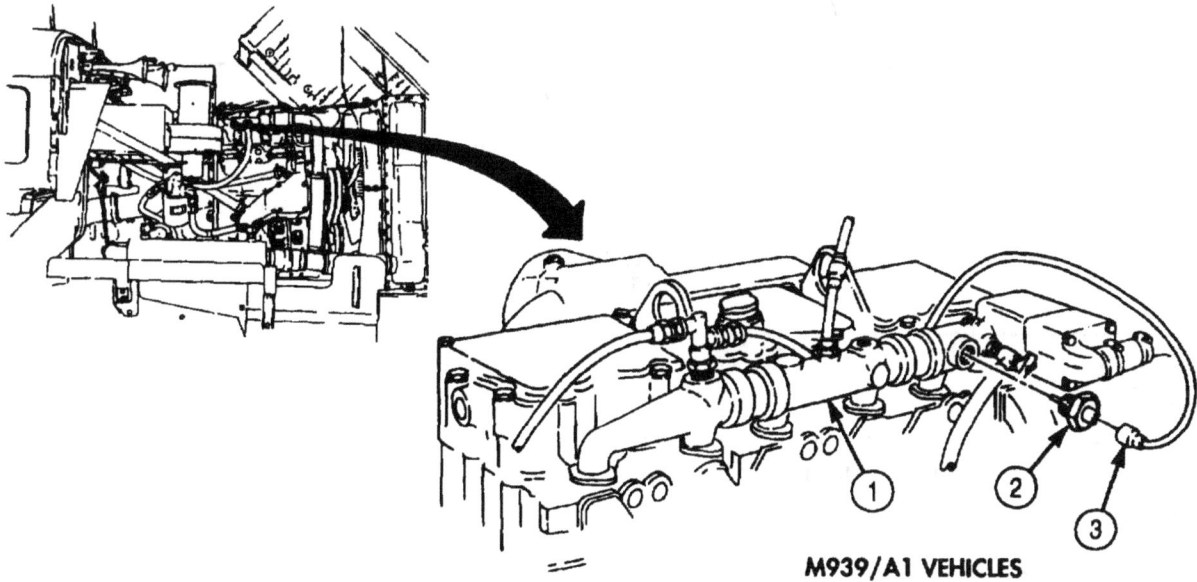

M939/A1 VEHICLES

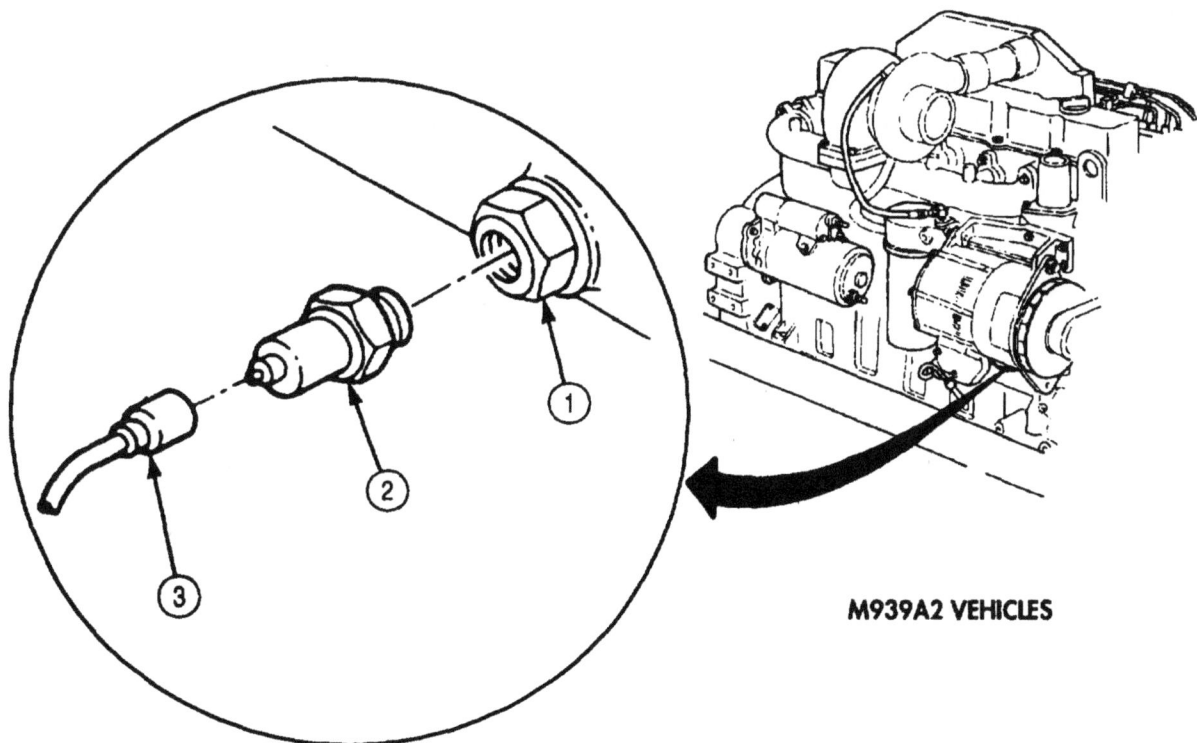

M939A2 VEHICLES

FOLLOW-ON TASKS: • Connect battery ground cables (para. 3-126).
• Fill coolant (para. 3-53).
• Install surge tank cap (TM 9-2320-272-10).
• Start engine (TM 9-2320-272-10) and check for leaks at water manifold. Check temperature gauge for proper operation.
• Install right splash shield (TM 9-2320-272-10).

3-94. TACHOMETER PULSE SENDER (M939/A1) REPLACEMENT

THIS TASK COVERS:

a. Removal   b. Installation

INITIAL SETUP:

APPLICABLE MODELS
M939/A1

TOOLS
General mechanic's tool kit (Appendix E, Item 1)

REFERENCES (TM)
TM 9-2320-272-10
TM 9-2320-272-24P

EQUIPMENT CONDITION
• Parking brake set (TM 9-2320-272-10).
• Battery ground cables disconnected (para. 3-126).

a. Removal

1. Disconnect tachometer pulse sender connector (1) from pulse sender receptacle (6).
2. Remove tachometer cable (2) from tachometer pulse sender (3).
3. Remove tachometer pulse sender (3) from adapter fitting (5).
4. Remove drive tip (4) from adapter fitting (5).

b. Installation

1. Install drive tip (4) on adapter fitting (5).
2. Install tachometer pulse sender (3) on adapter fitting (5).
3. Install tachometer cable (2) on tachometer pulse sender (3).
4. Connect tachometer pulse sender connector (1) to pulse sender receptacle (6).

3-94. TACHOMETER PULSE SENDER (M939/A1) REPLACEMENT (Contd)

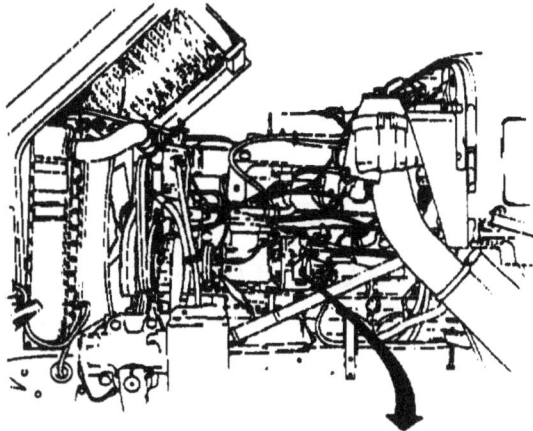

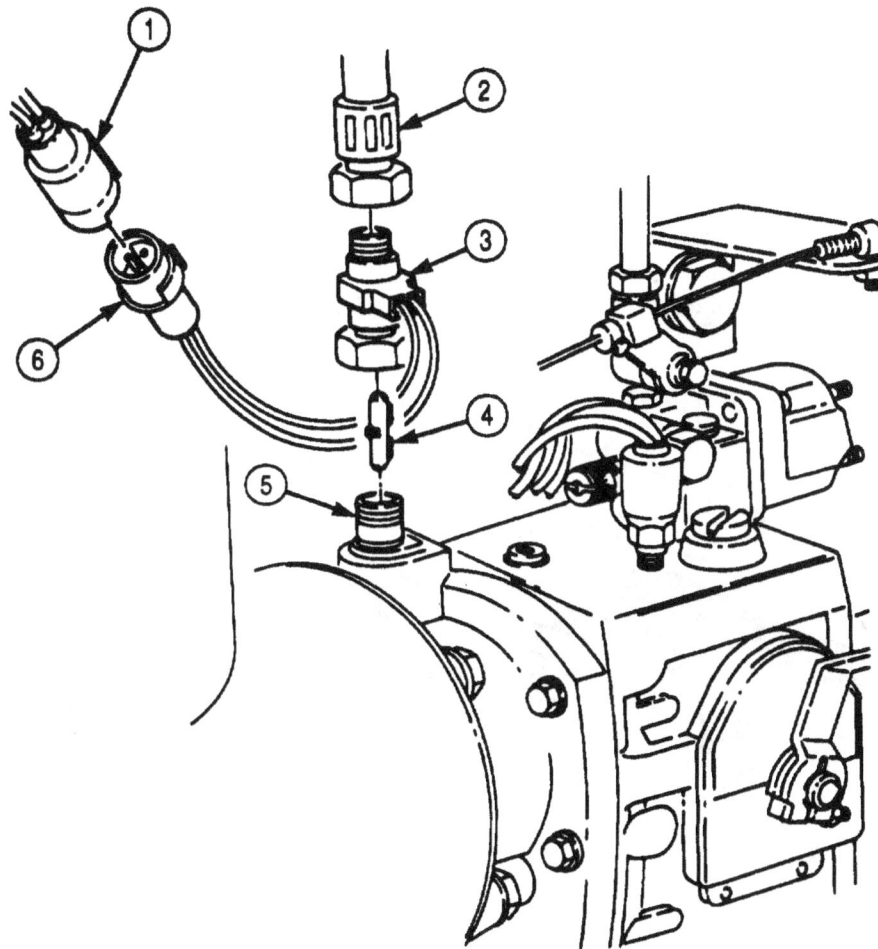

FOLLOW-ON TASKS: • Connect battery ground cables (para. 3-126).
                    • Start engine (TM 9-2320-272-10) and test tachometer for proper operation.

3-95. TACHOMETER PULSE SENDER (M939A2) REPLACEMENT

THIS TASK COVERS:

a. Removal   b. Installation

INITIAL SETUP:

APPLICABLE MODELS
M939A2

TOOLS
General mechanic's tool kit (Appendix E, Item 1)

REFERENCES (TM)
TM 9-2320-272-10
TM 9-2320-272-24P

EQUIPMENT CONDITION
• Parking brake set (TM 9-2320-272-10).
• Hood raised and secured (TM 9-2320-272-10).
• Battery ground cables disconnected (para. 3-126).

a. Removal

1. Remove eight screws (3) and pull instrument cluster (2) from instrument panel (1).
2. Disconnect tachometer drive cable (8) from tachometer pulse sensor (7).
3. Disconnect plug (4) from connector (5).
4. Remove tachometer pulse sensor (7) from tachometer gauge (6).

b. Installation

1. Install tachometer pulse sensor (7) on tachometer gauge (6).
2. Connect plug (4) to connector (5).
3. Connect tachometer drive cable (8) to tachometer pulse sensor (7).
4. Install instrument cluster (2) on instrument panel (1) with eight screws (3).

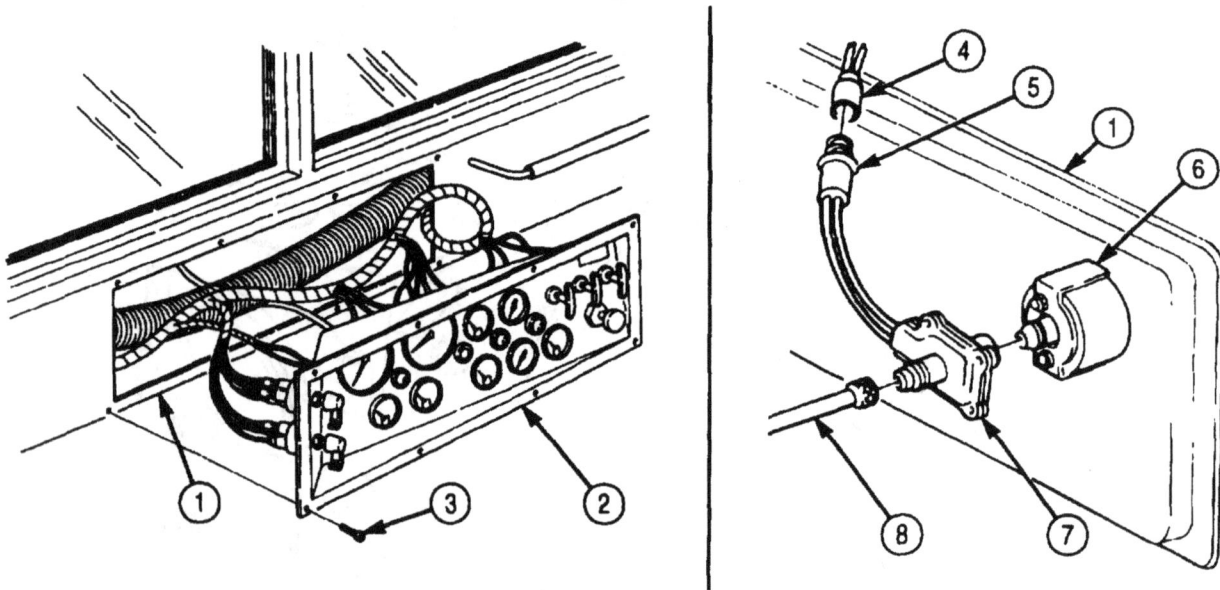

FOLLOW-ON TASKS: • Connect battery ground cables (para. 3-126).
 • Start engine (TM 9-2320-272-10) and test tachometer for proper operation.

3-96. FUEL PRESSURE TRANSDUCER (M939/A1) REPLACEMENT

THIS TASK COVERS:

a. Removal   b. Installation

INITIAL SETUP:

APPLICABLE MODELS
All

TOOLS
General mechanic's tool kit (Appendix E, Item 1)

MATERIALS/PARTS
Antiseize tape (Appendix D, Item 72)

REFERENCES (TM)
TM 9-2320-272-10
TM 9-2320-272-24P

EQUIPMENT CONDITION
• Parking brake set (TM 9-2320-272-10).
• Left splash shield removed (TM 9-2320-272-10).
• Battery ground cables disconnected (para. 3-126).

---

a. Removal

1. Disconnect fuel pressure transducer connector (4) from harness wire (1).
2. Remove fuel pressure transducer (2) from fuel pump (3).

b. Installation

NOTE

Male pipe threads must be wrapped with antiseize tape before installation.

1. Install fuel pressure transducer (2) on fuel pump (3).
2. Connect fuel pressure transducer connector (4) to harness wire (1).

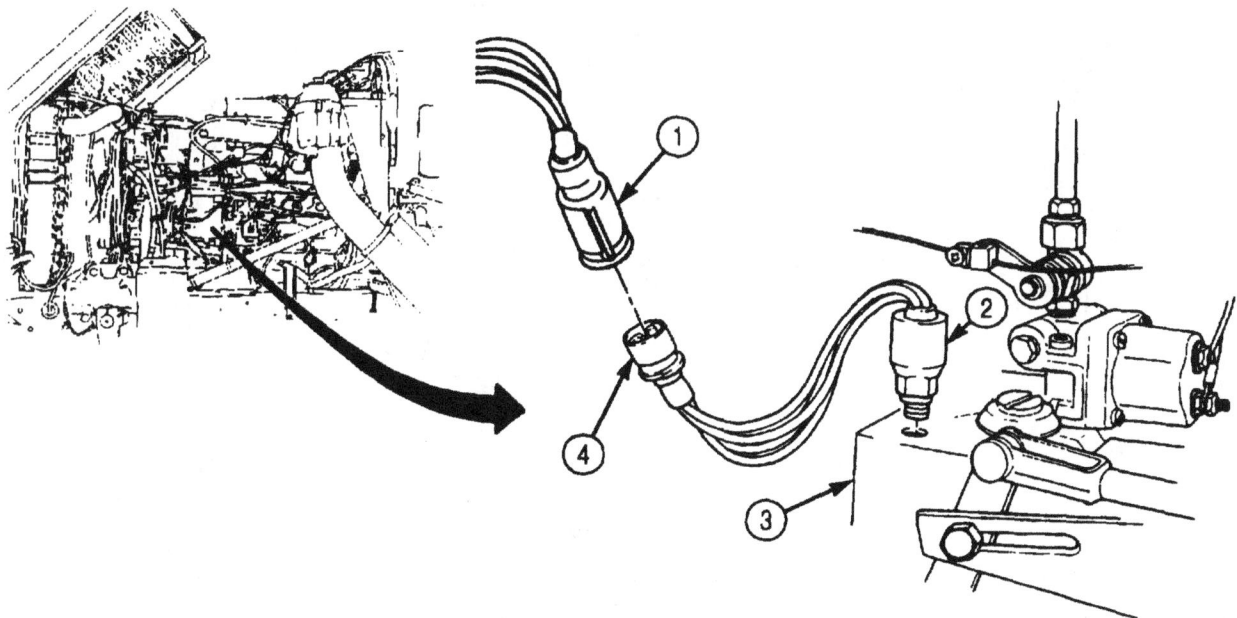

FOLLOW-ON TASKS: • Connect battery ground cables (para. 3-126).
• Start engine (TM 9-2320-272-10) and check fuel pressure gauge for proper operation.
• Install left splash shield (TM 9-2320-272-10).

3-271

3-97. TRANSMISSION TEMPERATURE TRANSMITTER REPLACEMENT

THIS TASK COVERS:

a. Removal   b. Installation

INITIAL SETUP:

APPLICABLE MODELS
All

TOOLS
General mechanic's tool kit (Appendix E, Item 1)

REFERENCES (TM)
TM 9-2320-272-10
TM 9-2320-272-24P

EQUIPMENT CONDITION
• Parking brake set (TM 9-2320-272-10).
• Battery ground cables disconnected (para. 3-126).

NOTE
Access to transmitter can be gained through access plate on cab floor.

a. Removal

1. Disconnect wire (1) from temperature transmitter (2).
2. Remove temperature transmitter (2) from transmission adapter fitting (3).

b. Installation

1. Install temperature transmitter (2) on transmission adapter fitting (3).
2. Connect wire (1) to temperature transmitter (2).

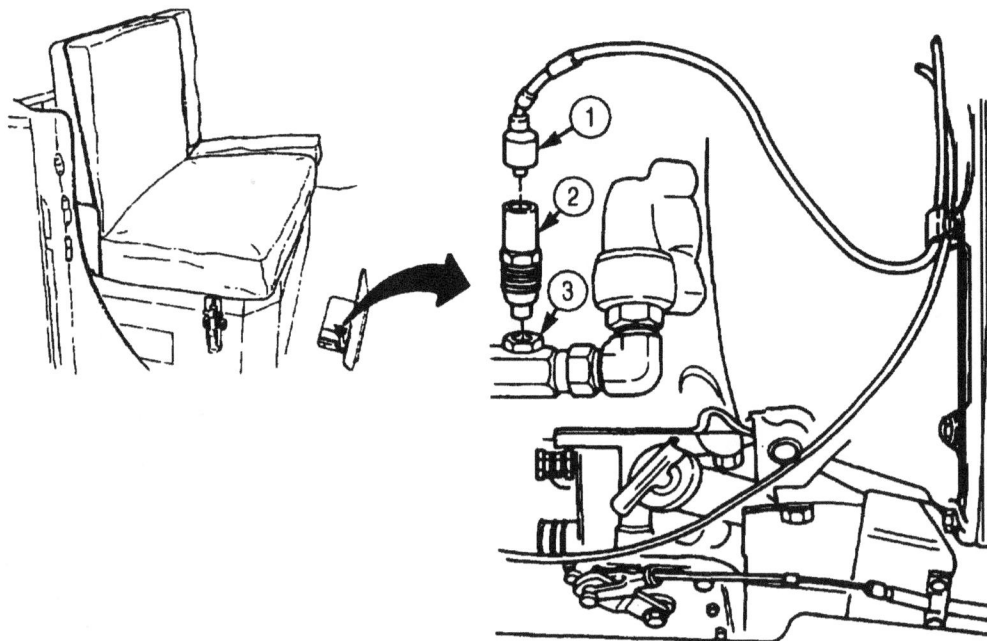

FOLLOW-ON TASKS: • Connect battery ground cables (para. 3-126).
• Start engine (TM 9-2320-272-10) and check transmission temperature gauge for operation.

3-98. TRANSMISSION NEUTRAL START SWITCH REPLACEMENT

THIS TASK COVERS:

a. Removal   b. Installation

<u>INITIAL SETUP:</u>

<u>APPLICABLE MODELS</u> All	<u>REFERENCES</u> (TM) TM 9-2320-272-10 TM 9-2320-272-24P
<u>TOOLS</u> General mechanic's tool kit (Appendix E, Item 1)	<u>EQUIPMENT CONDITION</u> • Parking brake set (TM 9-2320-272-10). • Battery ground cables disconnected (para. 3-126).

---

a. Removal

1. Remove nut (2) and retainer shift linkage (3) from left side of transmission (1).

NOTE

Tag wires for installation.

2. Disconnect wires (5) and (6) from neutral start switch (4).
3. Remove neutral start switch (4) from left side of transmission (1).

1. Install neutral start switch (4) on left side of transmission (1). Do not overtighten.
2. Connect wires (6) and (5) to neutral start switch (4).
3. Install retainer shift linkage (3) on left side of transmission (1) with nut (2).

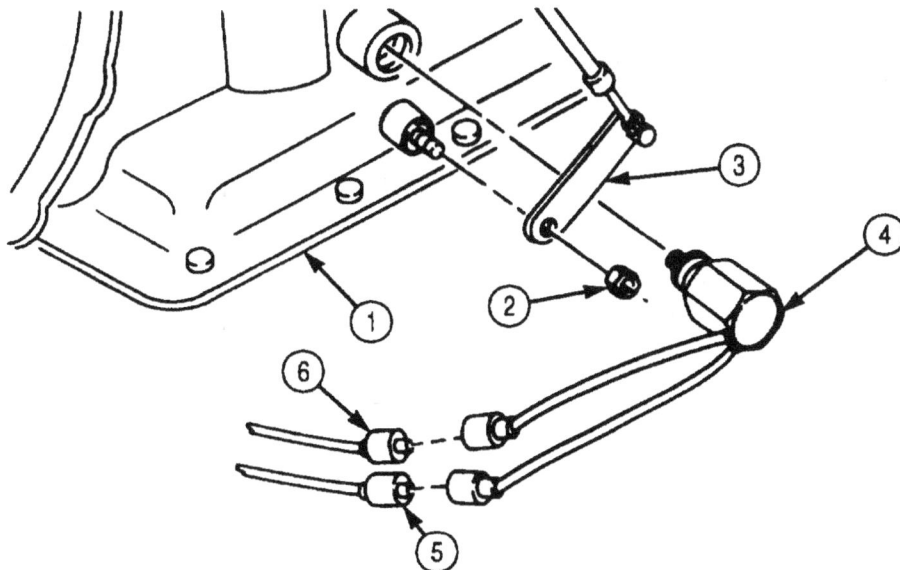

FOLLOW-ON TASKS: • Connect battery ground cables (para. 3-126).
• Check operation of neutral start switch (TM 9-2320-272-10).

3-99. PRIMARY AND SECONDARY LOW AIR PRESSURE SWITCH REPLACEMENT

THIS TASK COVERS:

a. Primary Switch Removal
b. Primary Switch Installation

c. Secondary Switch Removal
d. Secondary Switch Installation

INITIAL SETUP:

APPLICABLE MODELS

All

TOOLS

General mechanic's tool kit (Appendix E, Item 1)

MATERIALS/PARTS

Antiseize tape (Appendix C, Item 72)

REFERENCES (TM)

TM 9-2320-272-10
TM 9-2320-272-24P

EQUIPMENT CONDITION
• Parking brake set (TM 9-2320-272-10).
• Air reservoirs drained (TM 9-2320-272-10).
• Battery ground cables disconnected (para. 3-126).

GENERAL SAFETY INSTRUCTIONS
• Do not disconnect air lines before draining reservoirs.
• Exhaust system components are hot.

WARNING

- Do not disconnect air lines before draining air reservoirs. Small parts under pressure may shoot out with high velocity, causing injury to personnel.
- Do not touch hot exhaust system components with bare hands. Injury to personnel may result.

a. Primary Switch Removal

NOTE
Tag wires for installation.

1. Disconnect wires (1) and (2) from primary low air pressure switch (5).
2. Remove low air pressure switch (5) and adapter fitting (4) from adapter elbow (3).

b. Primary Switch Installation

NOTE
Male pipe threads must be wrapped with antiseize tape before installation

1. Install adapter fitting (4) and low air pressure switch (5) on adapter elbow (3).
2. Connect wires (2) and (1) to primary low air pressure switch (5).

3-99. PRIMARY AND SECONDARY LOW AIR PRESSURE SWITCH REPLACEMENT (Contd)

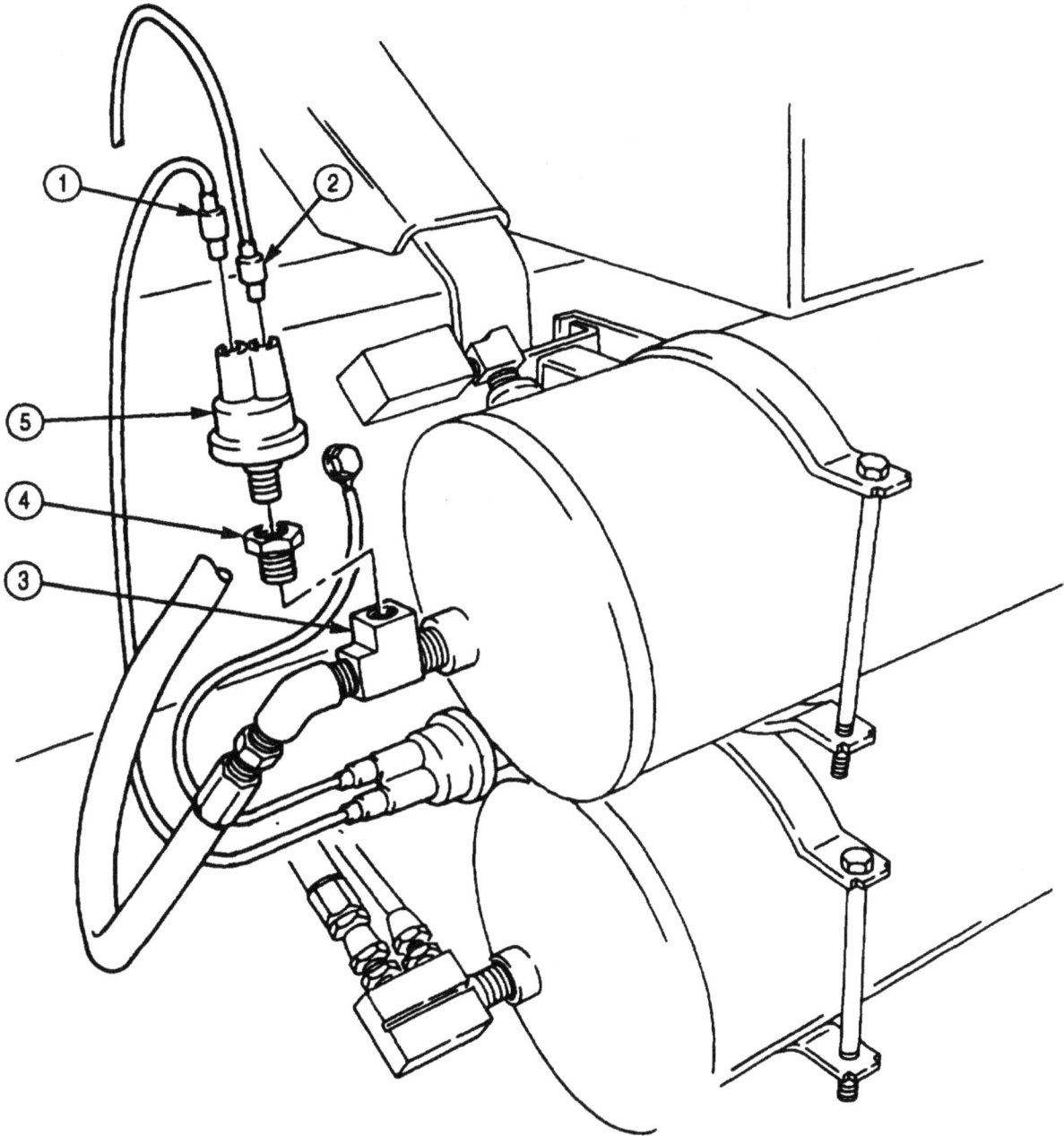

3-99. PRIMARY AND SECONDARY LOW AIR PRESSURE SWITCH REPLACEMENT (Contd)

c. Secondary Switch Removal

NOTE

Tag wires for installation.

1. Disconnect wire (8) and ground wire (1) from secondary low air pressure switch (7).

NOTE

Assistant will help with steps 2 and 3.

2. Remove two nuts (4) and screws (3) from secondary air reservoir (6).

3. Push and rotate secondary air reservoir (5) toward rear of vehicle.

4. Remove low air pressure switch (7) and adapter fitting (6) from adapter elbow (2).

d. Secondary Switch Installation

NOTE

Male pipe threads must be wrapped with antiseize tape before installation.

1. Install adapter fitting (6) and secondary low air pressure switch (7) on adapter elbow (2).

NOTE

Assistant will help with step 2.

2. Install two screws (3) and nuts (4) on secondary air reservoir (6).

3. Connect wire (8) and ground wire (1) to secondary low air pressure switch (7).

3-99. PRIMARY AND SECONDARY LOW AIR PRESSURE SWITCH REPLACEMENT (Contd)

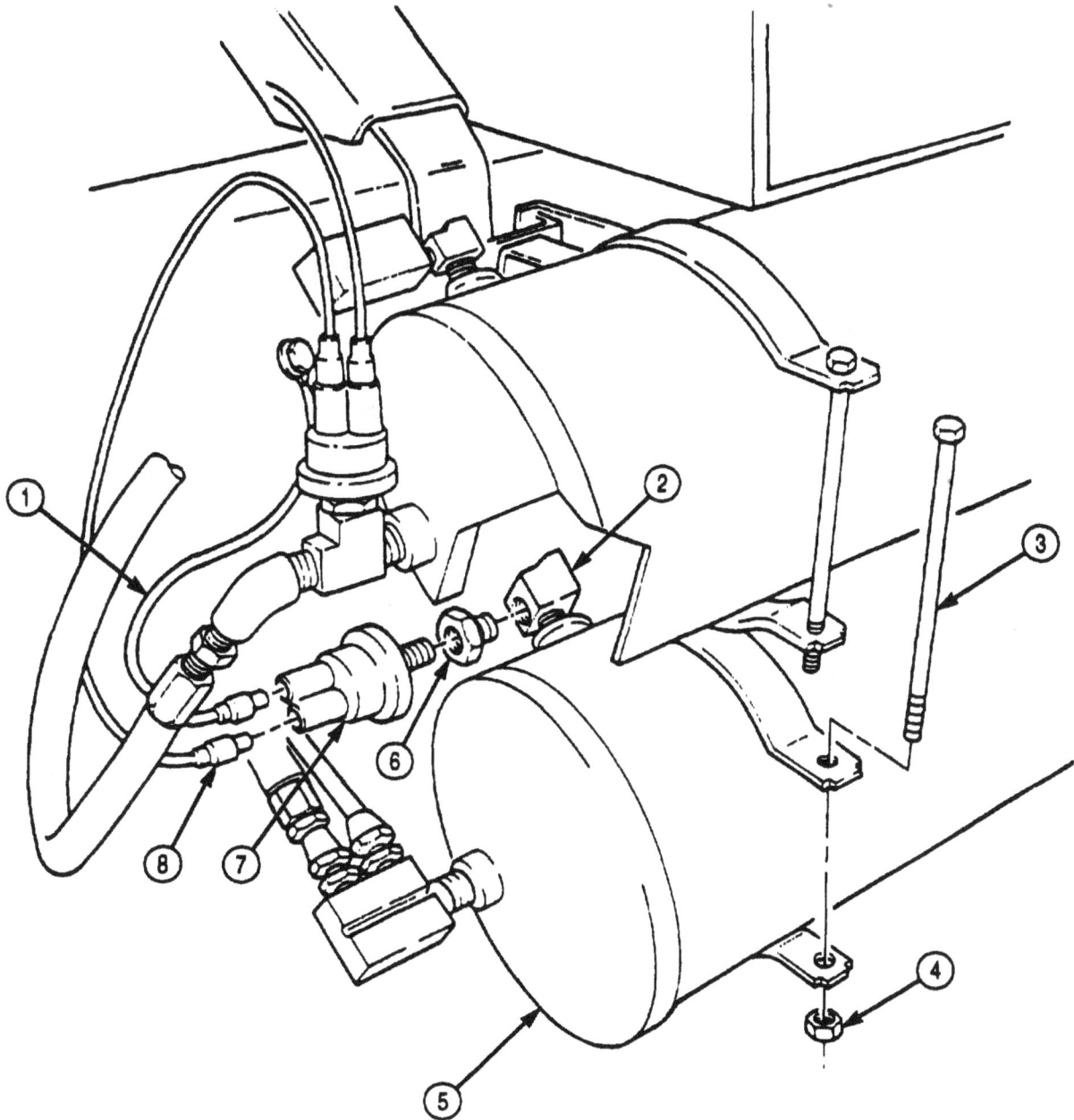

FOLLOW-ON TASKS: • Connect battery ground cables (para. 3-126).
 • Start engine (TM 9-2320-272-10) and allow air pressure to build to normal
 operating range, Check for air leaks at switch. Check if air pressure warning light
 and buzzer stop operating when air pressure has built to 60 psi (413 kPa).

3-100. SPRING BRAKE PRESSURE SWITCH REPLACEMENT

THIS TASK COVERS:

a. Removal   b. Installation

INITIAL SETUP:

APPLICABLE MODELS
All

TOOLS
General mechanic's tool kit (Appendix E, Item 1)

MATERIALS/PARTS
Antiseize tape (Appendix C, Item 72)

REFERENCES (TM)
TM 9-2320-272-10
TM 9-2320-272-24P

EQUIPMENT CONDITION
- Parking brake set (TM 9-2320-272-10).
- Air reservoirs drained (TM 9-2320-272-10).
- Battery ground cables disconnected (para. 3-126).

GENERAL SAFETY INSTRUCTIONS
Do not disconnect air lines before draining air reservoirs.

WARNING

Do not disconnect air lines before draining air reservoirs. Small
parts under pressure may shoot out with high velocity, causing
injury to personnel.

a. Removal

1. Remove eight screws (3) from instrument cluster (2).
2. Pull instrument cluster (2) away from instrument panel (1).
3. Disconnect two wires (8) from spring brake pressure switch (7).
4. Remove spring brake pressure switch (7), fitting (6), and elbow (5) from spring brake release control valve (4).

b. Installation

NOTE
Male pipe threads must be wrapped with antiseize tape before
installation.

1. Install elbow (5), fitting (6), and spring brake pressure switch (7) on spring brake release control valve (4).
2. Connect two wires (8) to spring brake pressure switch (7).
3. Install instrument cluster (2) on instrument panel (1) with eight screws (3).

3-100. SPRING BRAKE PRESSURE SWITCH REPLACEMENT (Contd)

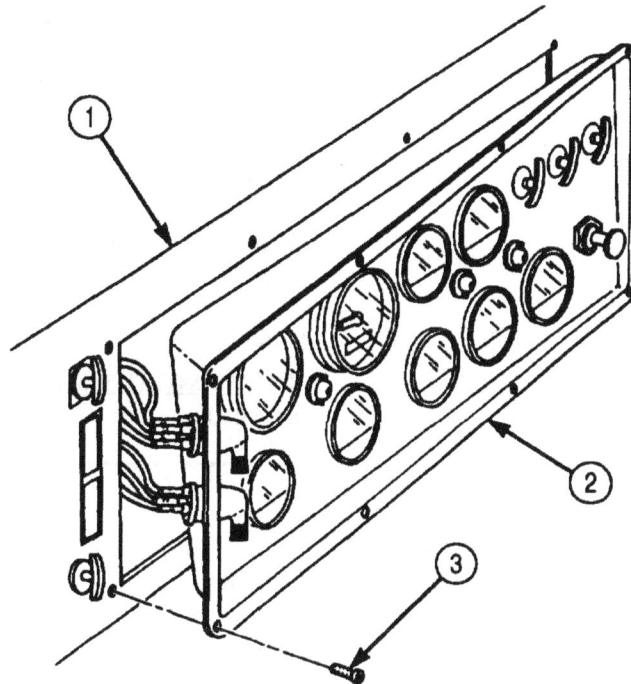

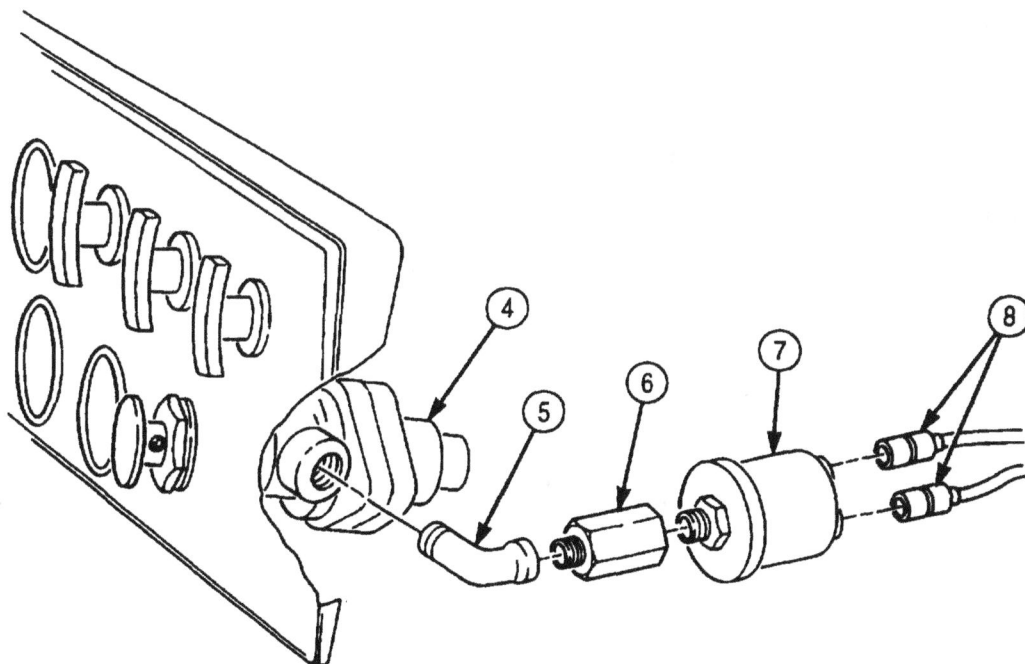

FOLLOW-ON TASKS: • Connect battery ground cables (para. 3-126).
• Start engine (TM 9-2320-272-10) and allow air pressure to build up to normal operating range. Check for air leaks at point switch is attached to valve. Stop engine and engage spring brakes. Spring brake warning light should glow.

3-101. FRONT WHEEL DRIVE LOCK-IN SWITCH REPLACEMENTT

THIS TASK COVERS:

a. Removal   b. Installation

INITIAL SETUP:

APPLICABLE MODELS
All

TOOLS
General mechanic's tool kit (Appendix E, Item 1)

MATERIALS/PARTS
Antiseize tape (Appendix C, Item 72)

REFERENCES (TM)
TM 9-2320-272-10
TM 9-2320-272-24P

EQUIPMENT CONDITION
• Parking brake set (TM 9-2320-272-10).
• Air reservoirs drained (TM 9-2320-272-10).
• Battery ground cables disconnected (para. 3-126).

GENERAL SAFETY INSTRUCTIONS
Do not disconnect air lines before draining air reservoirs.

WARNING

Do not disconnect air lines before draining air reservoirs. Small parts under pressure may shoot out with high velocity, causing injury to personnel.

a. Removal

1. Disconnect wire connector (3) from front wheel drive lock-in switch (2).
2. Remove lock-in switch (2) from front wheel drive lock-in valve elbow (1).

b. Installation

NOTE
Male pipe threads must be wrapped with antiseize tape before installation.

1. Install lock-in switch (2) on front wheel drive lock-in valve elbow (1).
2. Connect wire connector (3) to front wheel drive lock-in switch (2).

3-101. FRONT WHEEL DRIVE LOCK-IN SWITCH REPLACEMENT (Contd)

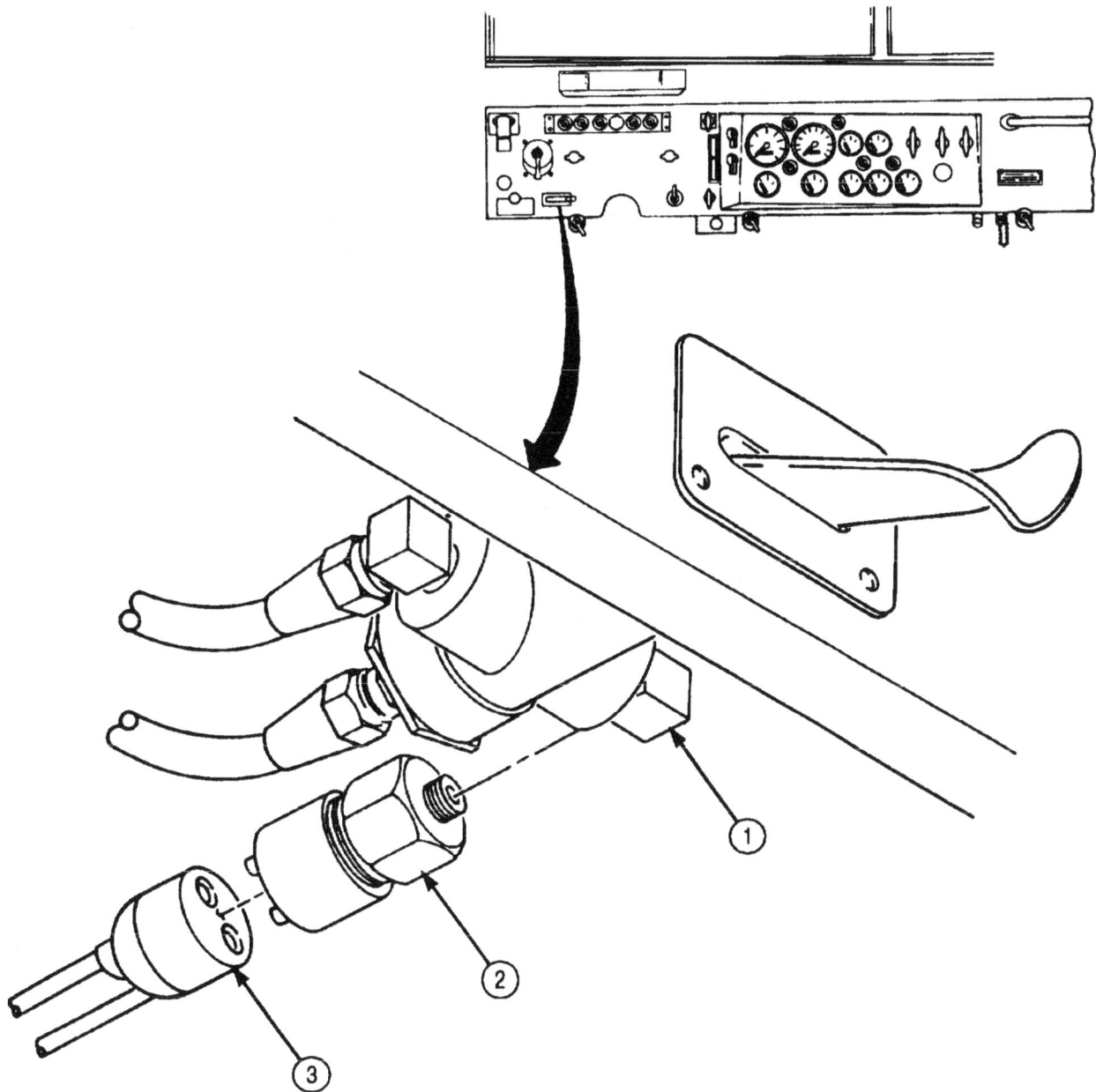

FOLLOW-ON TASKS: • Connect battery ground cables (para. 3-126).
• Start engine (TM 9-2320-272-10) and allow air pressure to build up to normal operating range. Check for leaks at switch.
• Engage front wheel drive and check if axle lock-in indicator light is illuminated.

3-102. HORN CONTACT BRUSH REPLACEMENT

THIS TASK COVERS:
a. Removal   b. Installation

INITIAL SETUP:

APPLICABLE MODELS
All

TOOLS
General mechanic's tool kit (Appendix E, Item 1)

REFERENCES (TM)
TM 9-2320-272-10
TM 9-2320-272-24P

EQUIPMENT CONDITION
- Parking brake set (TM 9-2320-272-10).
- Left splash shield removed (TM 9-2320-272-10).
- Battery ground cables disconnected (para. 3-126).

---

a. Removal

1. Loosen two screws (26) and nuts (28) on outer cab firewall (27).

NOTE
Step 2 applies to M931 and M932 models only.

2. Loosen two screws (22) on trailer brake control valve bracket (25).
3. Loosen two screws (23) on steering column bracket (24).
4. Turn steering column (14) until contact brush cover (20) is free from firewall (13).
5. Separate floormat (11) from steering column (14) and cab floor (12).
6. Disconnect connector (1) and boot (21) from brush cover (20).
7. Remove four screws (4), wire (5), contact brush cover (20), and gasket (8) from steering column (14).
8. Remove screw (19) and locktab (18) from contact brush (9). Disconnect wire (17) and capacitor (16).
9. Remove two screws (15), capacitor (16), contact brush (9), and pad (10) from steering column (14).
10. Remove nut (2), two washers (3), washer (6), wire (17), screw (7), and boot (21) from contact brush cover (20).

b. Installation

1. Install pad (10), contact brush (9). and capacitor (16) on steering column (14) with two screws (15).
2. Install boot (21) and wire (17) on contact brush cover (20) with screw (7), two washers (3), washer (6), and nut (2).
3. Install wire (17) and capacitor (16) on contact brush (9) with locktab (18) and screw (19).
4. Install gasket (8), contact brush cover (20), and ground wire (5) on steering column (14) with four screws (4).
5. Connect connector (1) and boot (21) to brush cover (20).
6. Turn steering column (14) until contact brush cover (20) is toward firewall (13).
7. Position floormat (11) against steering column (14) on cab floor (12).
8. Tighten two screws (23) on steering column bracket (24).

NOTE
Step 9 applies to M931 and M932 vehicles only

9. Tighten two screws (22) on trailer brake control valve bracket (25).
10. Tighten two screws (26) and nuts (28) on outer cab firewall (27).

3-102. HORN CONTACT BRUSH REPLACEMENT (Contd)

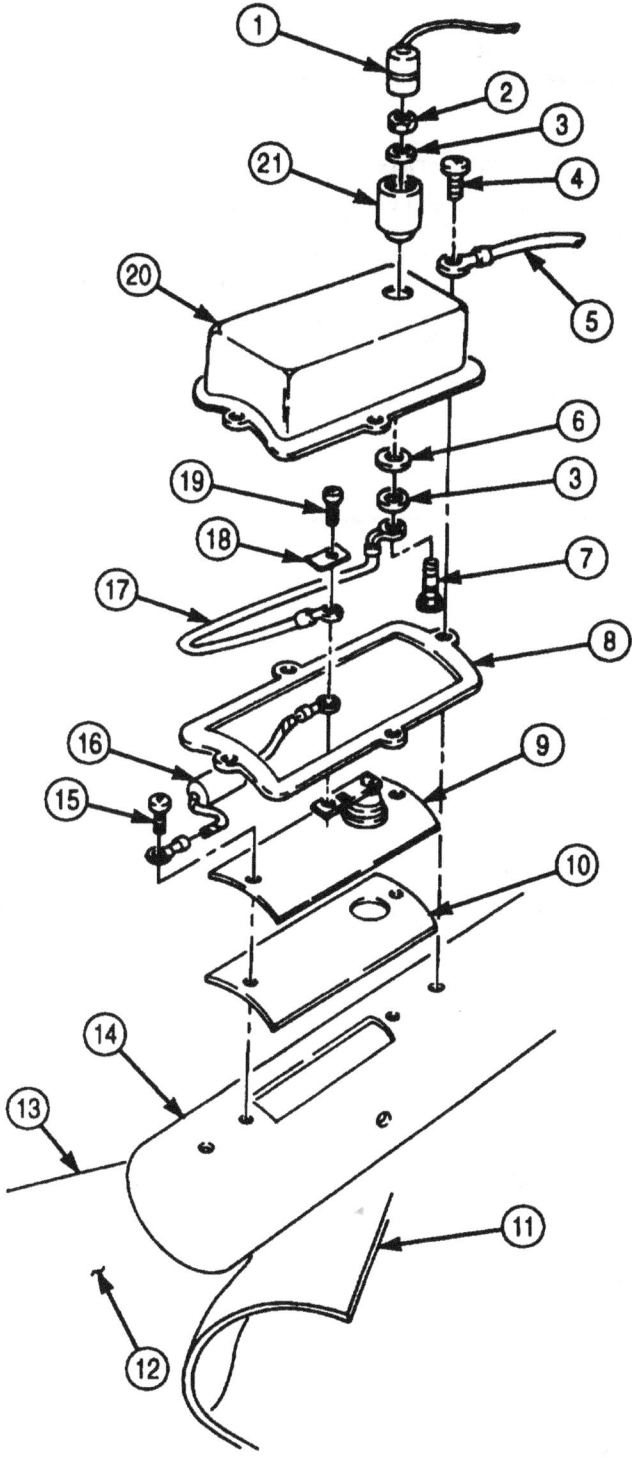

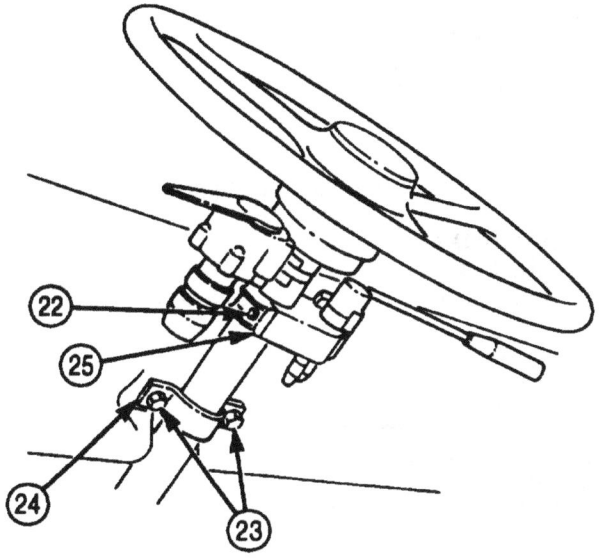

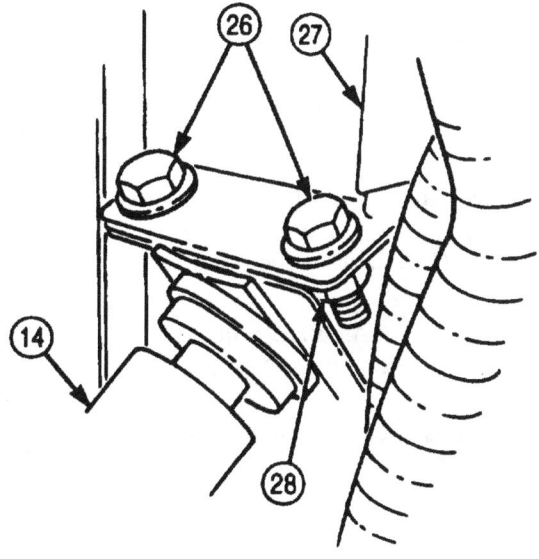

FOLLOW-ON TASKS: • Connect battery ground cables (para. 3-126).
                • Check horn for proper operation (TM 9-2320-272-10).
                • Install left splash shield (TM 9-2320-272-10).

3-103. HORN, SOLENOID, AND BRACKET REPLACEMENT

THIS TASK COVERS:

a. Removal   b. Installation

INITIAL SETUP:

APPLICABLE MODELS
All

TOOLS
General mechanic's tool kit (Appendix E, Item 1)

MATERIALS/PARTS
Two lockwashers (Appendix D, Item 371)
Antiseize tape (Appendix C, Item 72)

REFERENCES (TM)
TM 9-2320-272-10
TM 9-2320-272-24P

EQUIPMENT CONDITION
• Parking brake ret (TM 9-2320-272-10).
• Air reservoirs drained (TM 9-2320-272-10).
• Hood raised and secured (TM 9-2320-272-10).
• Battery ground cables disconnected (para. 3-126).

GENERAL SAFETY INSTRUCTIONS
Do not disconnect air lines before draining air reservoirs.

a. Removal

1. Disconnect wires (6) and (7) from horn solenoid (6).

WARNING

Do not disconnect air lines before draining air reservoirs. Small parts under pressure may shoot out with high velocity, causing injury to personnel.

2. Disconnect air line (4) from elbow (3).

3. Remove elbow (3) from horn solenoid (6).

NOTE

Horn solenoid fitting may become disconnected from horn solenoid during removal. Horn solenoid and fitting are replaced as an assembly.

4. Remove horn solenoid (6) from horn (2).

5. Remove two nuts (10), wire (9), two lockwashers (11), screws (1), and horn (2) from bracket (13). Discard lockwashers (11).

NOTE

• Assistant will help with step 6.
• Pull insulation away from inside cab firewall to allow access to nut.

6. Remove screw-assembled washer (8), nut (15), washer (14), screw (12), and bracket (13) from firewall (16).

3-103. HORN, SOLENOID, AND BRACKET REPLACEMENT (Contd)

b. Installation

NOTE

- When new solenoid is installed, fitting from old solenoid may be used. Fitting must be cleaned and inspected for cracks or stripped threads.
- Male pipe threads must be wrapped with antiseize tape before installation.
- Assistant will help with step 1.

1. Install bracket (13) on firewall (16) with washer (14), screw (12), nut (15), and screw-assembled washer (8).
2. Install horn (2) on bracket (13) with two screws (1), new lockwashers (11), wire (9), end two nuts (10).
3. Install horn solenoid (6) on horn (2).
4. Install elbow (3) on horn solenoid (6).
5. Connect air line (4) to elbow (3).
6. Connect wires (6) and (7) to horn solenoid (5).

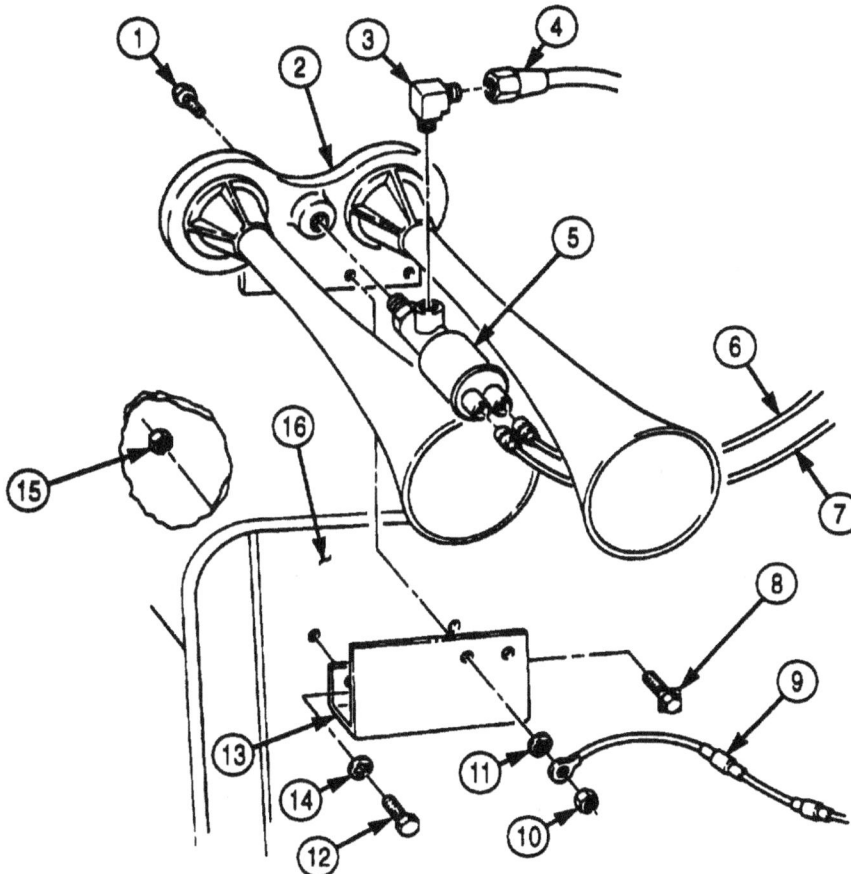

FOLLOW-ON TASKS: • Connect battery ground cables (para. 3-126).
- Start engine (TM 9-2320-272-10) and allow air pressure to build up to normal operating range. Check for air leaks at horn solenoid.
- Check horn for proper operation (TM 9-2320-272-10).

3-104. HORN SWITCH REPLACEMENT

THIS TASK COVERS:

a. Removal   b. Installation

INITIAL SETUP:

APPLICABLE MODELS
All

TOOLS
General mechanic's tool kit (Appendix E, Item 1)

MATRIALS/PARTS
O-ring (Appendix D, Item 428)

REFERENCES (TM)
TM 9-2320-272-10
TM 9-2320-272-24P

EQUIPMENT CONDITION
- Parking brake set (TM 9-2320-272-10).
- Battery ground cables disconnect (para. 3-126).

a. Removal

1. Remove three screws (6) and horn switch assembly (2) from steering column (1).
2. Disconnect horn switch connector (9) from diode lead (3).
3. Remove spring (11) and seat (4) from diode lead (3).
4. Remove retaining ring (8), horn switch (7), and O-ring (10) from adapter (5). Discard O-ring (10).

b. Installation

1. Install new O-ring (10), horn switch (7), and retaining ring (8) on adapter (5).
2. Install seat (4) and spring (11) over diode lead (3).
3. Connect horn switch connector (9) to diode lead (3).
4. Install horn switch assembly (2) on steering column (1) with three screws (6).

3-104. HORN SWITCH REPLACEMENT (Contd)

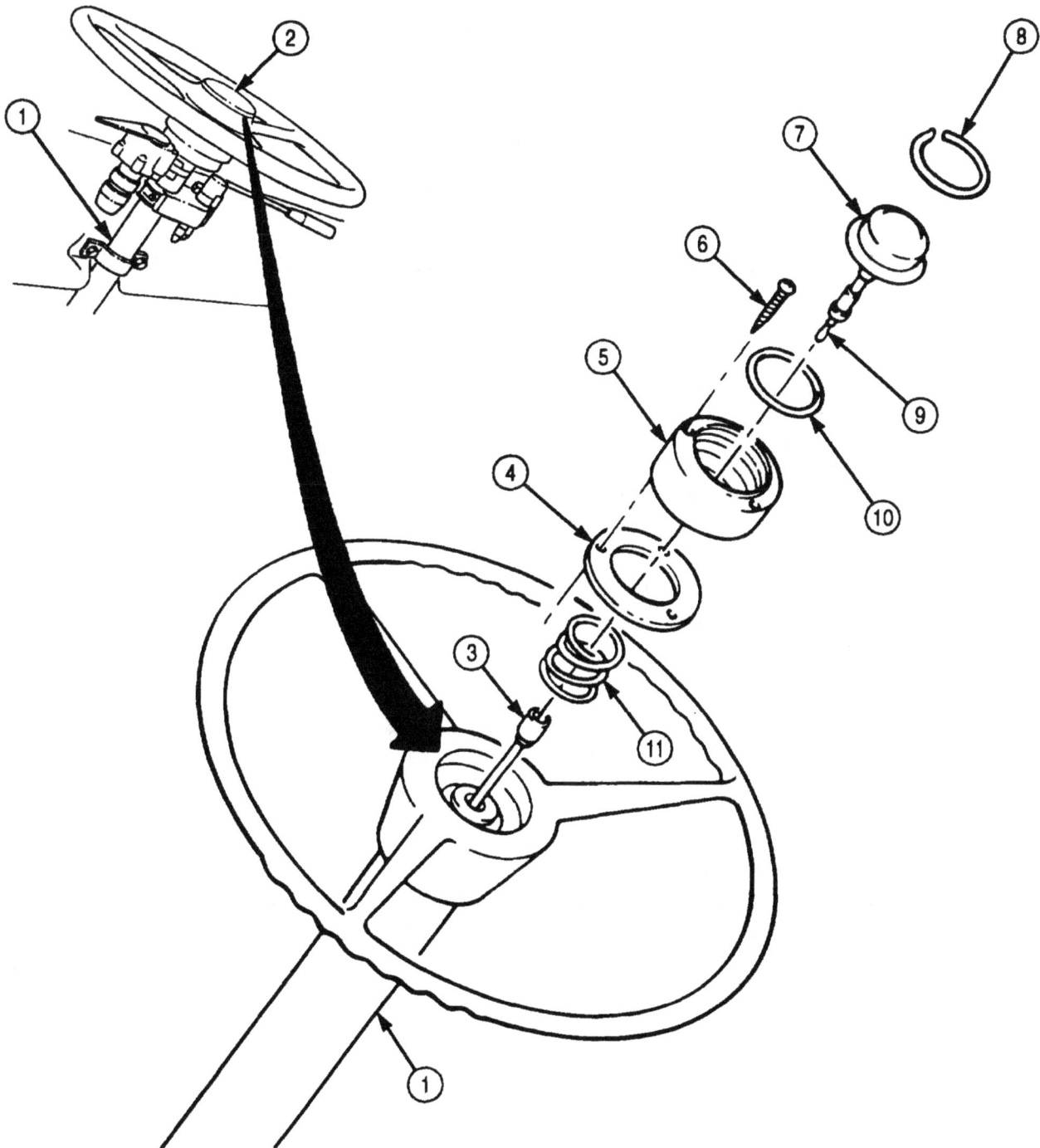

FOLLOW-ON TASKS: • Connect battery ground cables (para. 3-126).
                    • Check horn for proper operation (TM 9-2320-272-10).

3-105. CIRCUIT BREAKER REPLACEMENT

THIS TASK COVERS:

a. Removal   b. Installation

INITIAL SETUP:

APPLICABLE MODELS
All

TOOLS
General mechanic's tool kit (Appendix E, Item 1)

REFERENCES (TM)
TM 9-2320-272-10
TM 9-2320-272-24P

EQUIPMENT CONDITION
• Parking brake set (TM 9-2320-272-10).
• Hood raised and secured (TM 9-2320-272-10).
• Battery ground cables disconnected (para. 3-126).

---

a. Removal

NOTE

- Perform steps 1 and 2 to remove circuit breakers located on engine firewall.
- Perform steps 3 and 4 to remove circuit breakers located behind instrument panel and above steering column.
- Tag wires for installation

1. Disconnect wires (4) and (5) from circuit breaker (2).
2. Remove two screws (3) and circuit breaker (2) from firewall (1).
3. Disconnect wires (6) and (7) from circuit breaker (9).
4. Remove two screws (10) and circuit breaker (9) from instrument panel brace (8).

b. Installation

NOTE

- Perform steps 1 and 2 to install circuit breakers on engine firewall.
- Perform steps 3 and 4 to install circuit breakers behind instrument panel.

1. Install circuit breaker (2) on firewall (1) with two screws (3).
2. Connect wires (5) and (4) to circuit breaker (2).
3. Install circuit breaker (9) on instrument panel brace (8) with two screws (10).
4. Connect wires (7) and (6) to circuit breaker (9).

3-105. CIRCUIT BREAKER REPLACEMENT (Contd)

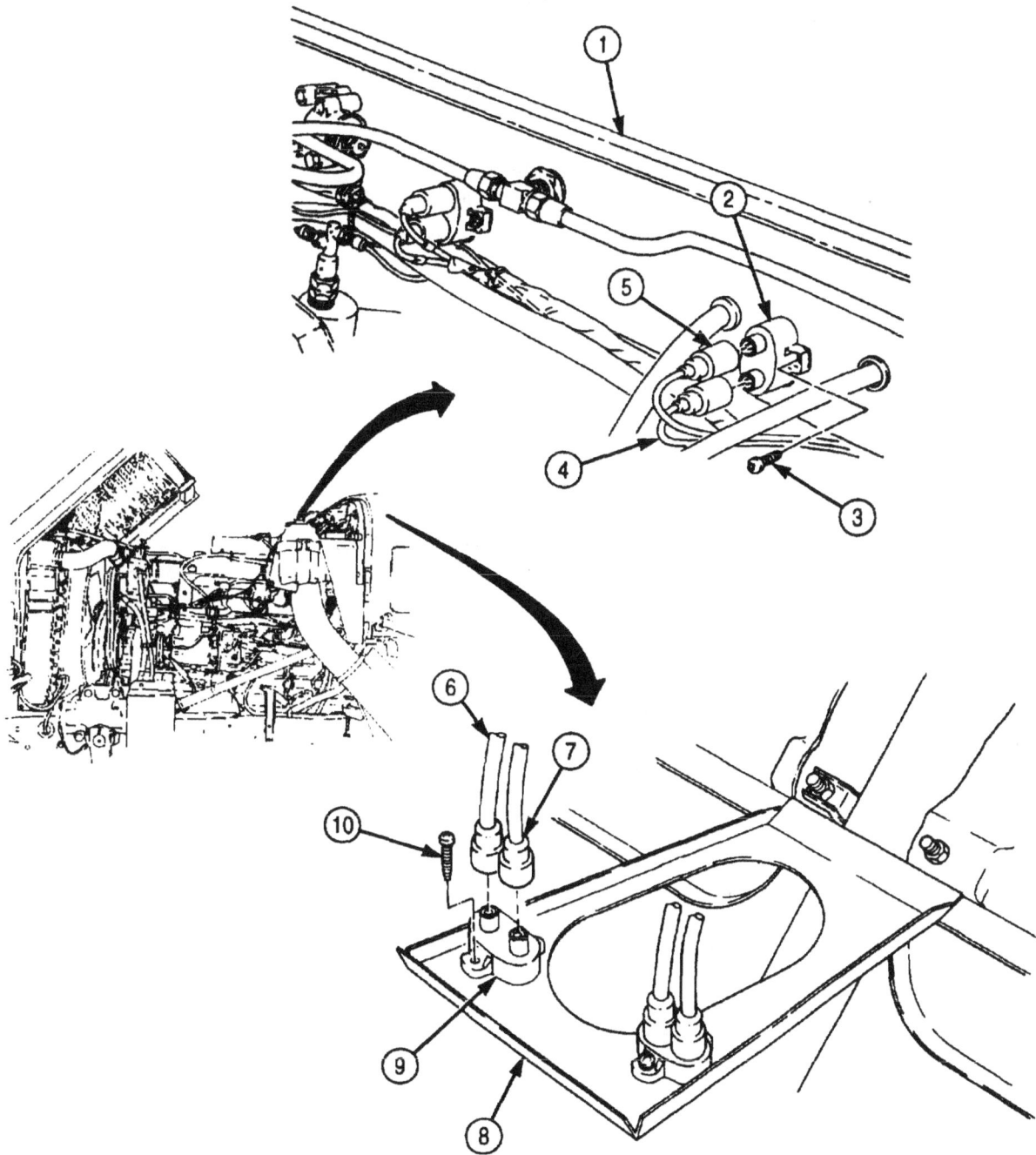

FOLLOW-ON TASKS: Connect battery ground cables (para. 3-126).

3-106. FAILSAFE WARNING MODULE REPLACEMENT

THIS TASK COVERS:

a. Removal

b. Installation

INITIAL SETUP:

APPLICABLE MODELS
All

TOOLS
General mechanic's tool kit (Appendix E, Item 1)

REFERENCES (TM)
TM 9-2320-272-10
TM 9-2320-272-24P

EQUIPMENT CONDITION
- Parking brake set (TM 9-2320-272-10).
- Battery ground cables disconnected (para. 3-126).

---

a. Removal

1. Disconnect harness connector (2) from failsafe warning module (6).
2. Remove screw (5), ground wire (4), and washer (3) from failsafe warning module (6).
3. Remove screw (7), washer (8), and failsafe warning module (6) from cowl (1).

b. Installation

1. Install failsafe warning module (6) on cowl (1) with washer (8) and screw (7).
2. Install washer (3) and ground wire (4) on failsafe warning module (6) with screw (5).
3. Connect harness connector (2) to failsafe warning module (6).

3-106. FAILSAFE WARNING MODULE REPLACEMENT (Contd)

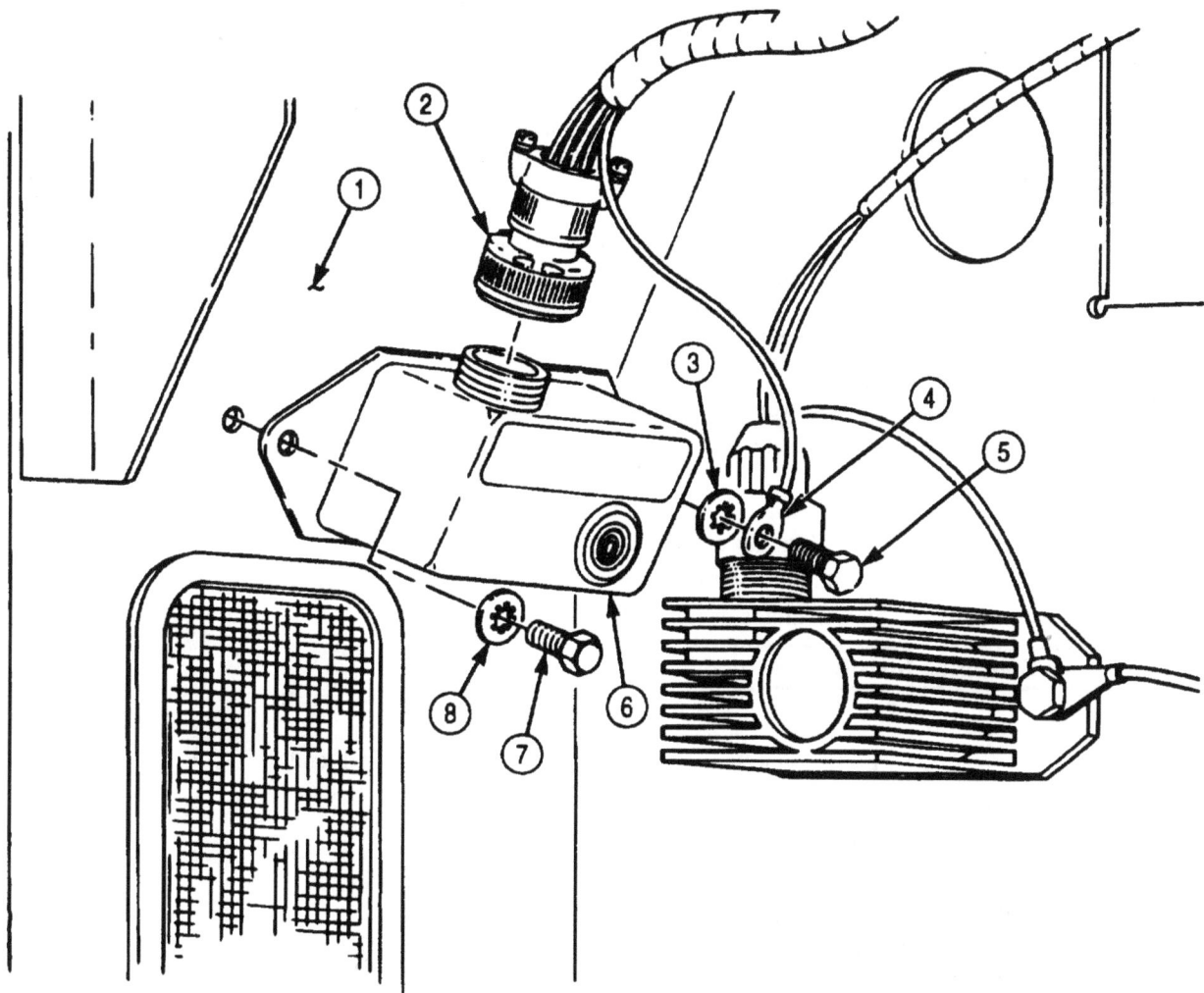

FOLLOW-ON TASKS: • Connect battery ground cables (para. 3-126).
  • Start engine (TM 9-2320-272-10) and check failsafe warning module for proper operation.

3-107. ELECTRICAL SWITCHES REPLACEMENT

THIS TASK COVERS:

a. Accessory Switcher Removal   c. Battery and Starter Switch Removal
b. Accessory Switcher Installation   d. Battery and Starter Switch Installation

---

INITIAL SETUP:

APPLICABLE MODELS
All

Tools
General mechanic's tool kit (Appendix E, Item 1)

MATERIALS/PARTS
Three lockwashers (Appendix D, Item 355)

REFERENCES (TM)
TM 9-2320-272-10
TM 9-2320-272-24P

EQUIPMENT CONDITION
- Parking brake set (TM 9-2320-272-10).
- Blower motor in OFF position (TM 9-2320-272-10).
- Battery ground cables disconnected (para. 9-126).

NOTE
Heater switch, floodlight switch, and warning switches are all
replaced the same way, Tasks a. and b. covers heater switch.

a. Accessory Switches Removal

1. Remove screw (1) and lever (2) from blower motor switch (9).
2. Remove nut (3), lockwasher (4), and blower motor switch (9) from instrument panel (10). Discard lockwasher (4).

NOTE
Tag all connectors for installation.

3. Disconnect connectors (6), (7), and (8) from blower motor switch (9).
4. Remove plug (16) from blower motor switch (9).

b. Accessory Switches Installation

CAUTION
Ensure each connector is inserted into proper terminal end.

1. Install plug (16) into terminal (13).
2. Install connector (7) into terminal (11).
3. Install connector (8) into terminal (12).
4. Install connector (6) into terminal (14).
6. Install blower motor switch (9) on instrument panel (10), with key (6) into keyway behind instrument panel (10), with new lockwasher (4) and nut (3).
6. Position lever (2) on blower motor switch (9) with pointing edge placed in OFF position and install with screw (1).

NOTE
Battery and starter switches are replaced basically the same.
Tasks c. and d. cover battery switch.

c. Battery and Starter Switch Removal

1. Remove screw (24), lockwasher (23), switch lever (16), felt washer (17), and washer (18) from battery switch (20). Discard lockwasher (23).

3-107. ELECTRICAL SWITCHES REPLACEMENT (Contd)

2. Remove nut (22) and lockwasher (19) from instrument cluster (21) and pull switch (20) from behind instrument cluster (21).

NOTE

Tag wires for installation.

3. Disconnect wires (25), (26), (27), and (28) and remove battery switch (20) from instrument cluster (21).

d. Battery and Starter Switch Installation

NOTE

If new switch is being installed, use mounting hardware supplied with switch, Starter switch has only three wires.

1. Connect wires (28),(27), (26), and (25) to battery switch (20).

2. Position battery switch (20) through instrument cluster (21) from the rear and install with new lockwasher (19) and nut (22).

3. Position washer (18), felt washer (17), and switch lever (16) on battery switch (20) and install with new lockwasher (23) and screw (24).

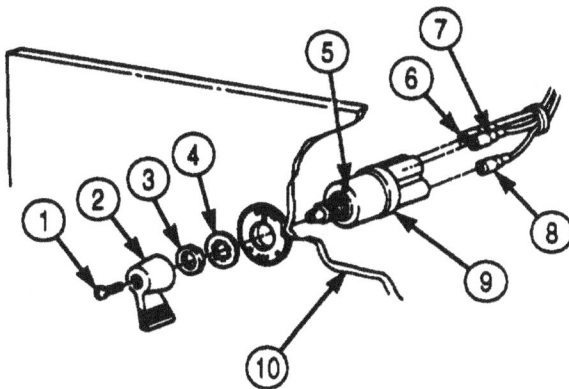

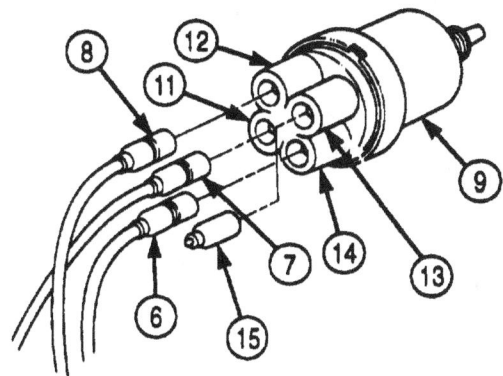

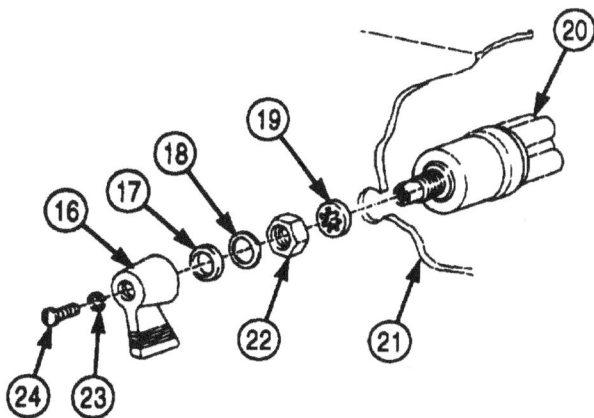

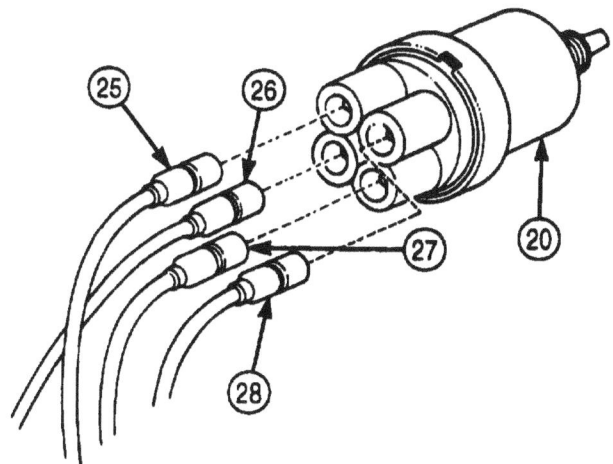

FOLLOW-ON TASK: Connect battery ground cables (para. 3-126).

3-108. MAIN LIGHT SWITCH REPLACEMENT

THIS TASK COVERS:

a. Removal   b. Installation

INITIAL SETUP:

APPLICABLE MODELS	REFERENCES (TM)
All	TM 9-2320-272-10
	TM 9-2320-272-24P

TOOLS

General mechanic's tool kit (Appendix E, Item 1)

EQUIPMENT CONDITION
- Parking brake set (TM 9-2320-272-10).
- Battery ground cables disconnected (para. 3-126).

NOTE
Two different types of light switches are used.

a. Removal

1. Remove three screws (8), washers (7), switch levers (6), and washers (5) from main light switch (1).

NOTE
Switch is removed from behind instrument panel.

2. Remove four screws (4) and light switch (1) from instrument panel (3).
3. Disconnect harness connector (2) from main light switch (1).

b. Installation

1. Connect harness connector (2) to main light switch (1).
2. Install main light switch (1) on instrument panel (3) with four screws (4).
3. Install three washers (5) and switch levers (6) on main light switch (1) with three washers (7) and screws (8).

3-108. MAIN LIGHT SWITCH REPLACEMENT (Contd)

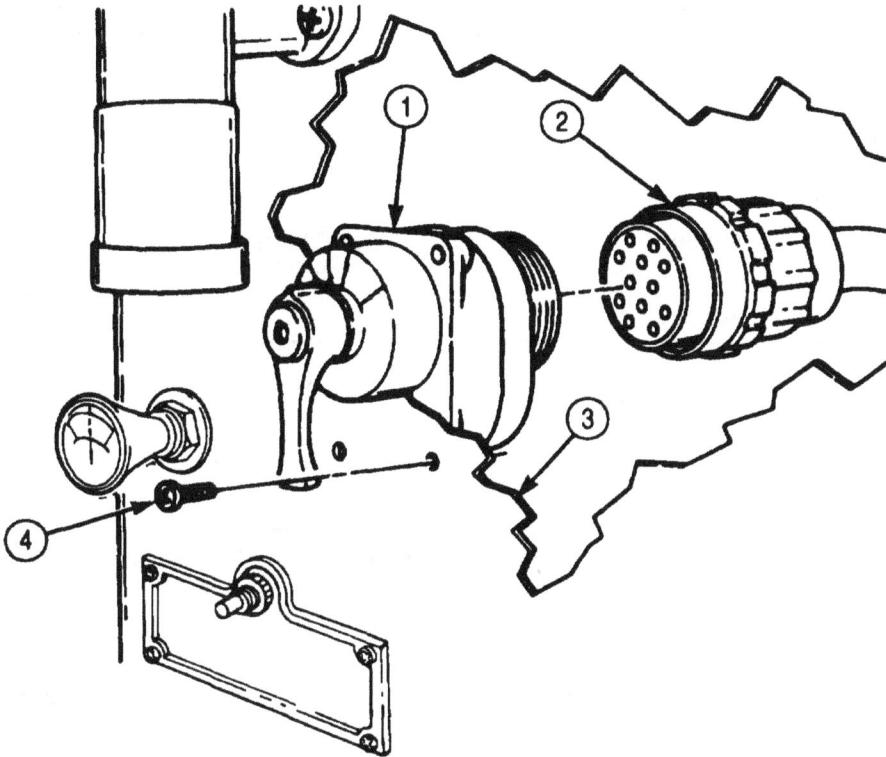

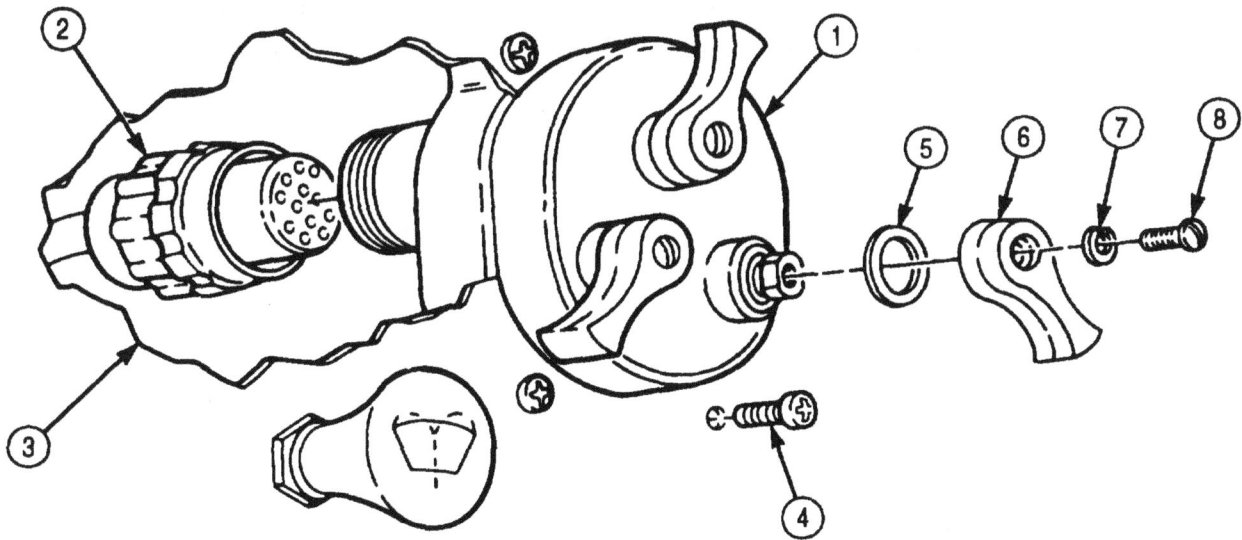

FOLLOW-ON TASKS: • Connect battery ground cables (para. 3-126).
• Check lights for proper operation (TM 9-2320-272-10).

3-109. FUEL SELECTOR VALVE SWITCH REPLACEMENT

THIS TASK COVERS:

a. Removal   b. Installation

INITIAL SETUP:

APPLICABLE MODELS   **REFERENCES (TM)**
All   TM 9-2320-272-10
                                         TM 9-2320-272-24P
TOOLS
General mechanic's tool kit (Appendix E, Item 1)   **EQUIPMENT CONDITION**
                                         • Parking brake set (TM 9-2320-272-10).
MATERIALS/PARTS   • Battery ground cables disconnected (para. 3-126).
Two locknuts (Appendix D. Item 283)

---

a. Removal

NOTE
Tag wires for installation.

1. Disconnect two wires (2) and wire (14) from fuel selector valve switch (3).
2. Remove screw (9), washer (10), lever (8), felt washer (11), and washer (7) from fuel selector valve switch (3).
3. Remove nut (12), washer (6), and plate (5) from fuel selector valve switch (3).
4. Remove fuel selector valve switch (3) from bracket (4).
6. Remove two locknuts (1), screws (13), and bracket (4) from instrument panel (15). Discard locknuts (1).

b. Installation

1. Install bracket (4) on instrument panel (15) with two screws (13) and new locknuts (1).
2. Position fuel selector valve switch (3) on bracket (4) and install plate (5), washer (6), and nut (12).
3. Install lever (8) on fuel selector valve switch (3) with washer (7), felt washer (11), washer (10), and screw (9).
4. Connect two wires (2) and wire (14) to fuel selector valve switch (3).

3-109. FUEL SELECTOR VALVE SWITCH REPLACEMENT (Contd)

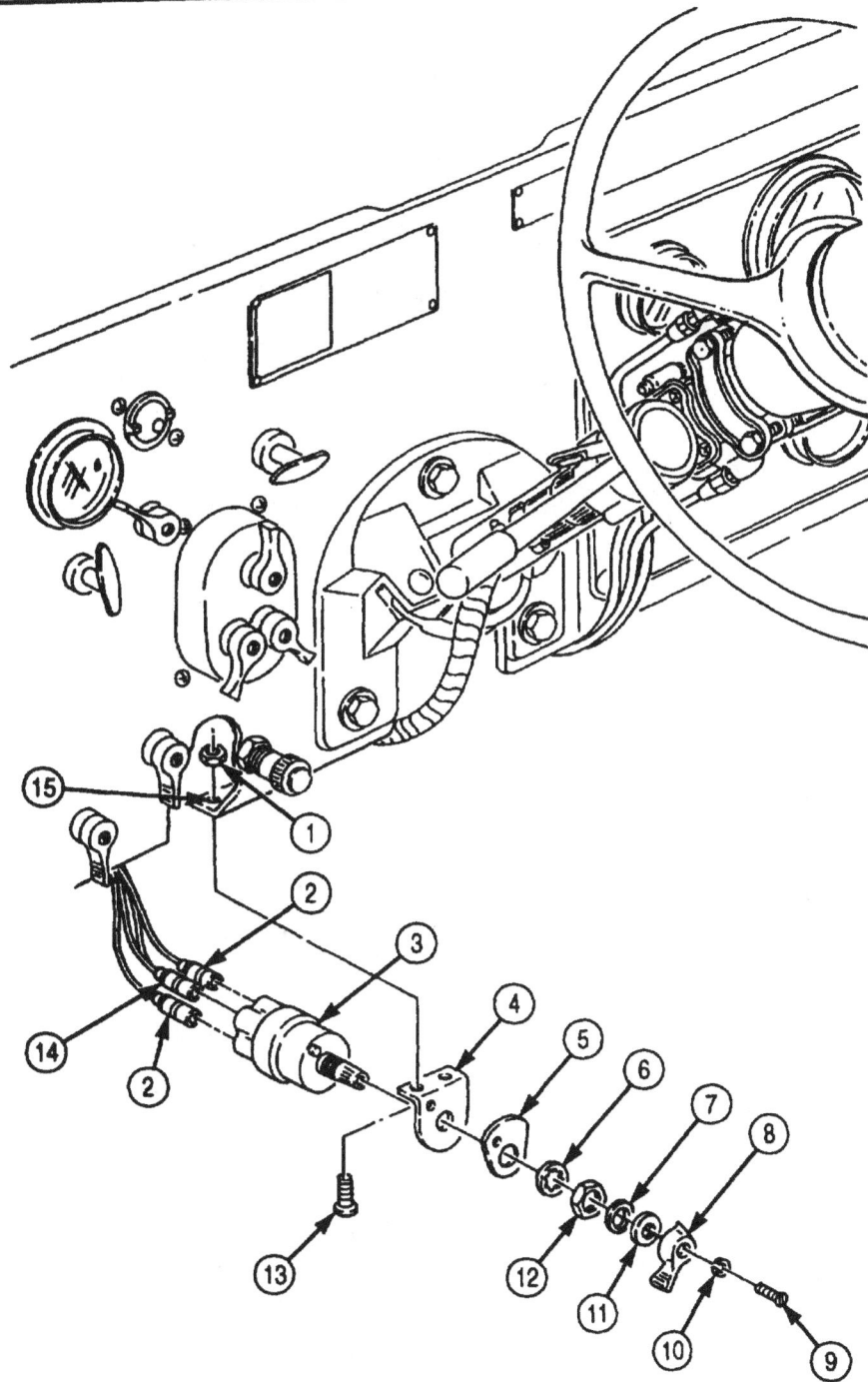

FOLLOW-ON TASKS:
- Connect battery ground cables (para. 3-126).
- Check fuel selector valve switch for proper operation (TM 9-2320-272-10).

3-110. AUXILIARY OUTLET SOCKET AND RECEPTACLE (M936/A1/A2) REPLACEMENT

THIS TASK COVERS:

a. Removal   b. Installation

INITIAL SETUP:

APPLICABLE MODELS
M936/A1/A2

TOOLS
General mechanic's tool kit (Appendix E, Item 1)

REFERENCES (TM)
TM 9-2320-272-10
TM 9-2320-272-24P

EQUIPMENT CONDITION
• Parking brake set (TM 9-2320-272-10).
• Battery ground cables disconnected (para. 3-126).

---

a. Removal

1. Remove screw (13) and outlet cable (11) from auxiliary outlet socket (14).
2. Remove screw (10), clamp (2), and auxiliary outlet socket (14) from instrument panel (1).
3. Remove clip (12) from instrument panel (1).
4. Disconnect wire (5) from wire (4).
5. Remove four nuts (3), screws (9), cover (8), and auxiliary outlet receptacle (7) from instrument panel (1).

b. Installation

1. Insert wire (4) through hole (6) on instrument panel (1) and connect to wire (5).
2. Install auxiliary outlet receptacle (7) and cover (8) on instrument panel (1) with four screws (9) and nuts (3).
3. Install clip (12) on instrument panel (1).
4. Install auxiliary outlet socket (14) on instrument panel (1) with clamp (2) and screw (10).
5. Install outlet cable (11) on auxiliary outlet socket (14) with screw (13).
6. Install outlet cable (11) on clip (12).

3-110. AUXILIARY OUTLET SOCKET AND RECEPTACLE (M936/A1/A2) REPLACEMENT (Contd)

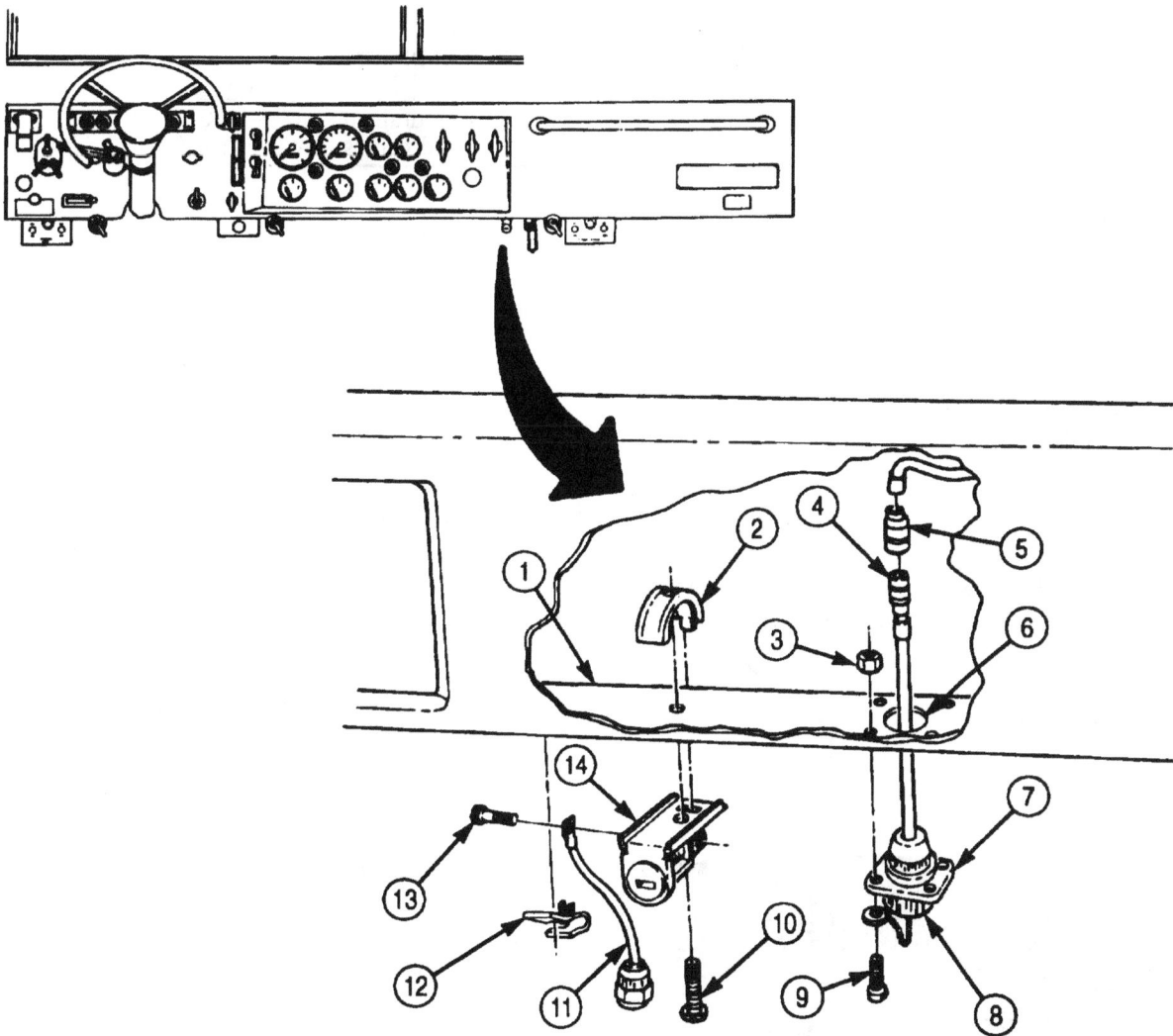

FOLLOW-ON TASK: Connect battery ground cables (para. 3-126).

3-111. STOPLIGHT SWITCH REPLACEMENT

THIS TASK COVERS:

a. Removal   b. Installation

INITIAL SETUP:

APPLICABLE MODELS
All

TOOLS
General mechanic's tool kit (Appendix E, Item 1)

MATERIALS/PARTS
Antiseize tape (Appendix C, Item 72)

REFERENCES (TM)
TM 9-2320-272-10
TM 9-2320-272-24P

EQUIPMENT CONDITION
- Parking brake set (TM 9-2320-272-10).
- Air reservoirs drained (TM 9-2320-272-10).
- Battery ground cables disconnected (para. 3-126).

a. Removal

NOTE
Tag wires for installation.

1. Disconnect two wires (1) from stoplight switch (2).
2. Remove stoplight switch (2) from double check valve (3).

b. Installation

NOTE
Male pipe threads must be wrapped with antiseize tape before installation.

1. Install stoplight switch (2) on double check valve (3).
2. Connect two wires (1) to stoplight switch (2).

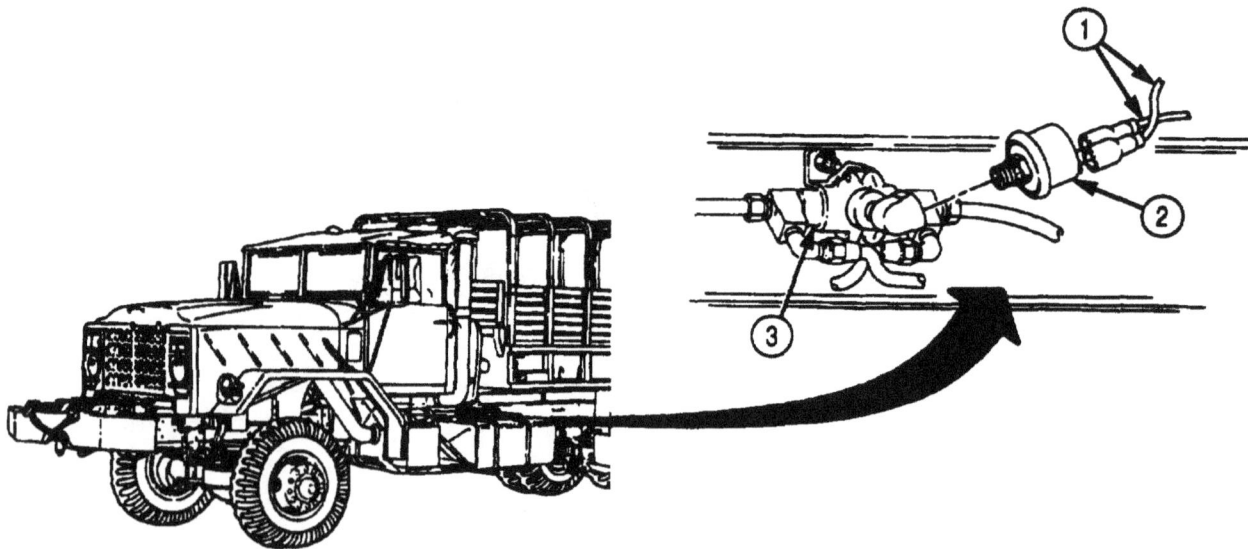

FOLLOW-ON TASKS: • Connect battery ground cables (para. 3-126).
- Start engine (TM 9-2320-272-10) and allow air pressure to build up to normal operating range. Check for air leaks at switch.
- Check stoplights for proper operation (TM 9-2320-272-10).

3-112. HIGH BEAM SELECTOR SWITCH REPLACEMENT

THIS TASK COVERS:
a. Removal   b. Installation

___

INITIAL SETUP:

APPLICABLE MODELS
ALL

TOOLS
General mechanic's tool kit (Appendix E, Item 1)

REFERENCES (TM)
TM 9-2320-272-10
TM 9-2320-272-24P

EQUIPMENT CONDITION
• Parking brake set (TM 9-2320-272-10).
• Battery ground cables disconnected (para. 3-126).

___

a. Removal

NOTE
Tag all wires for installation.

1. Disconnect electrical wires (3), (4), and (5) from selector switch (6) under cab floor (2).

NOTE
Assistant will help with step 2.

2. Remove two screws (1) and selector switch (6) from cab floor (2).

b. Installation

NOTE
Assistant will help with step 1.

1. Install selector switch (6) on cab floor (2) with two screws (1).
2. Connect electrical wires (5), (4), and (3) to selector switch (6).

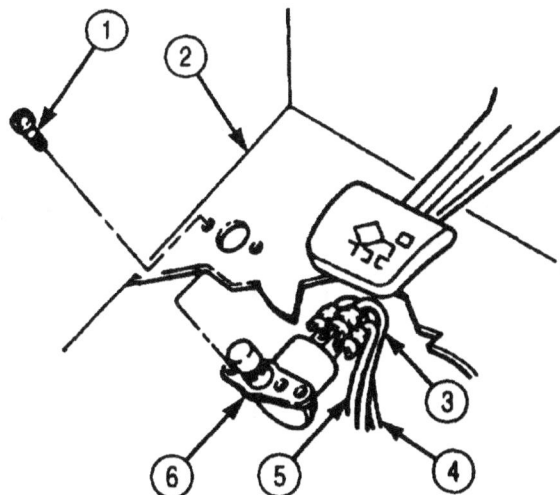

FOLLOW-ON TASKS: • Connect battery ground cables (para. 3-126).
• Check headlight beam selector for proper operation (TM 9-2320-272-10).

3-301

3-113. TURN SIGNAL CONTROL AND INDICATOR LAMP REPLACEMENT

THIS TASK COVERS:

a. Removal   **b. Installation**

INITIAL SETUP:

APPLICABLE MODELS   **REFERENCES (TM)**
All   TM 9-2320-272-10
                                               TM 9-2320-272-24P
TOOLS
General mechanic's tool kit (Appendix E, Item 1)   **EQUIPMENT CONDITION**
                                               • Parking brake set (TM 9-2320-272-10).
                                               • Battery ground cables disconnected (para. 3-126).

a. Removal

1. Disconnect connector (9) from turn signal control (1).
2. Remove clamp (7), signal self-canceller (8), and turn signal control (1) from steering column (6).
3. Remove lamp lens (3) and lamp (2) from lamp socket (4).

b. Installation

1. Install lamp (2) on lamp socket (4).
2. Install lamp lens (3) on turn signal control (1).
3. Align signal self-canceller (8) with cancelling pin (5), and install turn signal control (1) on steering column (6) with clamp (7).
4. Connect connector (9) to turn signal control (1).

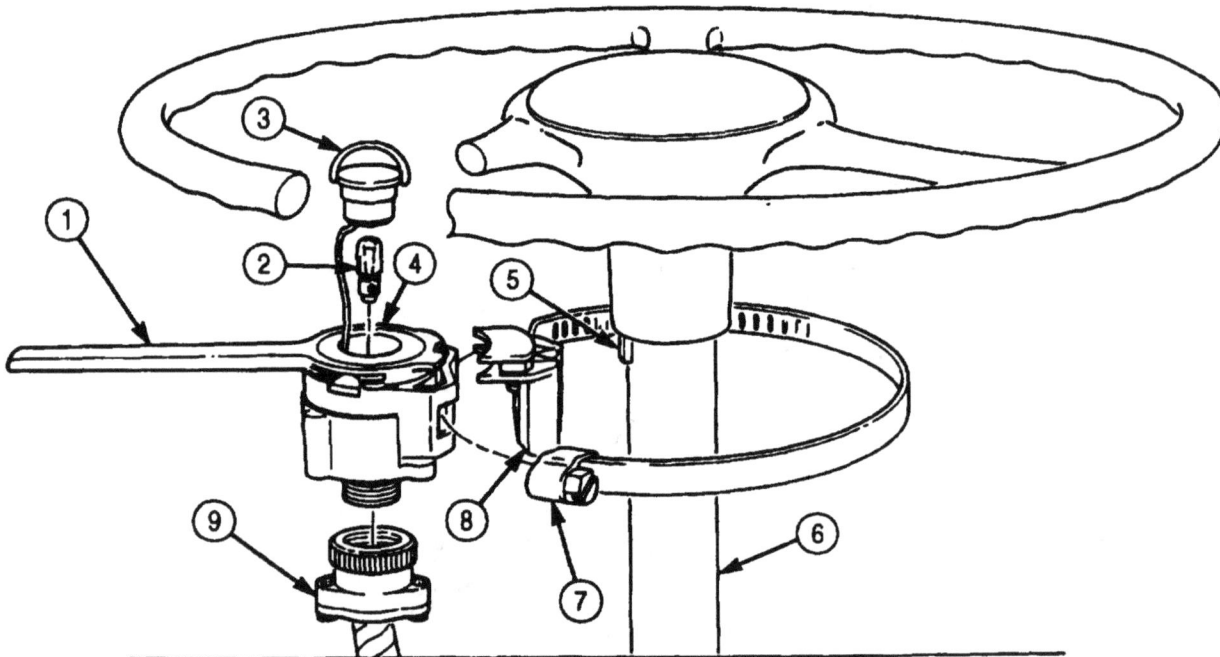

FOLLOW-ON TASKS: • Connect battery ground cables (para. 3-126).
                   • Check turn signal control for proper operation (TM 9-2320-272-10).

3-114. TURN SIGNAL FLASHER REPLACEMENT

THIS TASK COVERS:

a. Removal   b. Installation

INITIAL SETUP:

APPLICABLE MODELS
All

TOOLS
General mechanic's tool kit (Appendix E, Item 1)

REFERENCES (TM)
TM 9-2320-272-10
TM 9-2320-272-24P

EQUIPMENT CONDITION
• Parking brake set (TM 9-2320-272-10).
• Battery ground cables disconnected (para. 3-126).

a. Removal

1. Remove screws (7) and (8), washers (6) and (9), wire (5), horn ground wire (4), and flasher (3) from firewall (1).
2. Remove harness connector (2) from flasher (3)

b. Installation

1. Install harness connector (2) on flasher (3).
2. Install flasher (3), horn ground wire (4), and wire (5) on firewall (1) with washers (6) and (9) and screws (7) and (8).

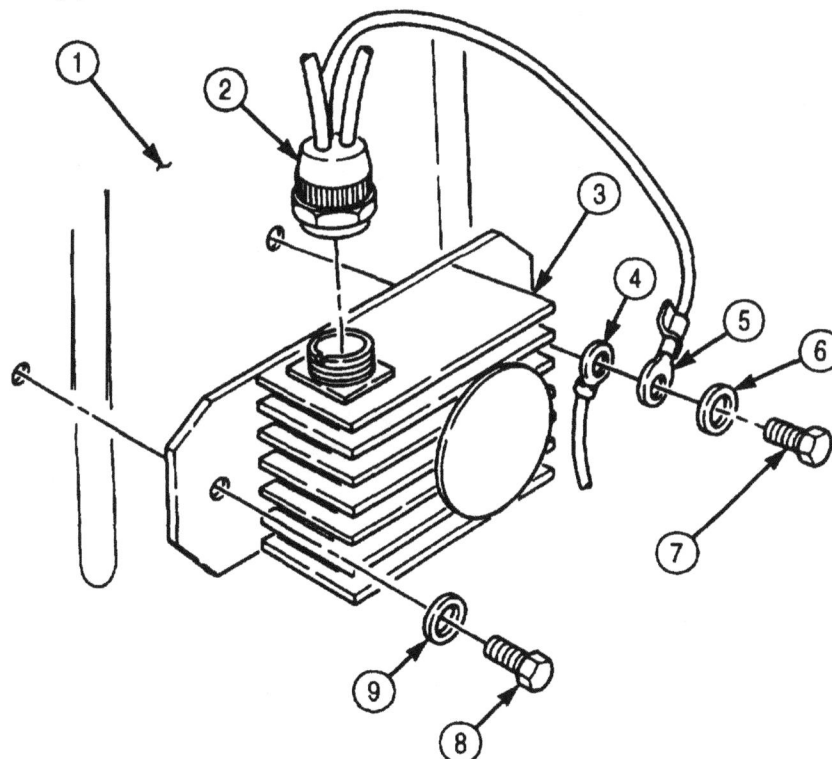

FOLLOW-ON TASKS: • Connect battery ground cables (para. 3-126).
• Check turn signal flasher for proper operation (TM 9-2320-272-10).

3-303

3-115. PROTECTIVE CONTROL BOX REPLACEMENT

THIS TASK COVERS:

a. Removal   b. Installation

INITIAL SETUP:

APPLICABLE MODELS
All

TOOLS
General mechanic's tool kit (Appendix E, Item 1)
Torque wrench (Appendix E, Item 146)

MATERIALS/PARTS
Six lockwashers (Appendix D, Item 355)

REFERENCES (TM)
TM 9-2320-272-10
TM 9-2320-272-24P

EQUIPMENT CONDITION
- Parking brake set (TM 9-2320-272-10).
- Battery ground cables disconnected (para. 3-126).
- Ether cylinder and valve removed (para. 3-35).

a. Removal

1. Remove windshield washer bottle lid (2) and washer bottle (3) from windshield washer bottle bracket (4). Do not disconnect windshield washer hoses (1).
2. Disconnect harness connector (12) from protective control box (13).
3. Remove four screws (11), lockwashers (10), ground wire (9), protective control box (13), two lockwashers (8), spacers (7), spacers (5), and washer bottle bracket (4) from firewall (6). Discard lockwashers (10) and (8).

b. Installation

1. Install washer bottle bracket (4) on firewall (6) with two spacers (5), new lockwashers (8), spacers (7). protective control box (13). ground wire (9), four new lockwashers (10), and screws (11).
2. Connect harness connector (12) on protective control box (13). Tighten connector (12) 10 lb-ft (14 N•m).
3. Install washer bottle (3) and windshield washer bottle lid (2) on windshield washer bottle bracket (4).

3-115. PROTECTIVE CONTROL BOX REPLACEMENT (Contd)

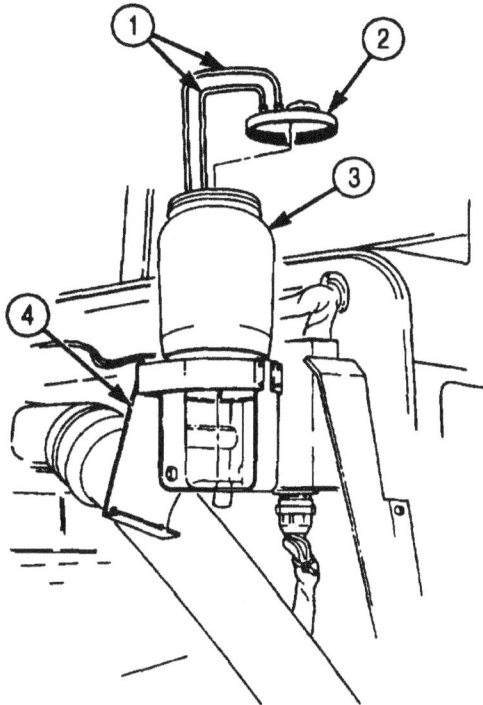

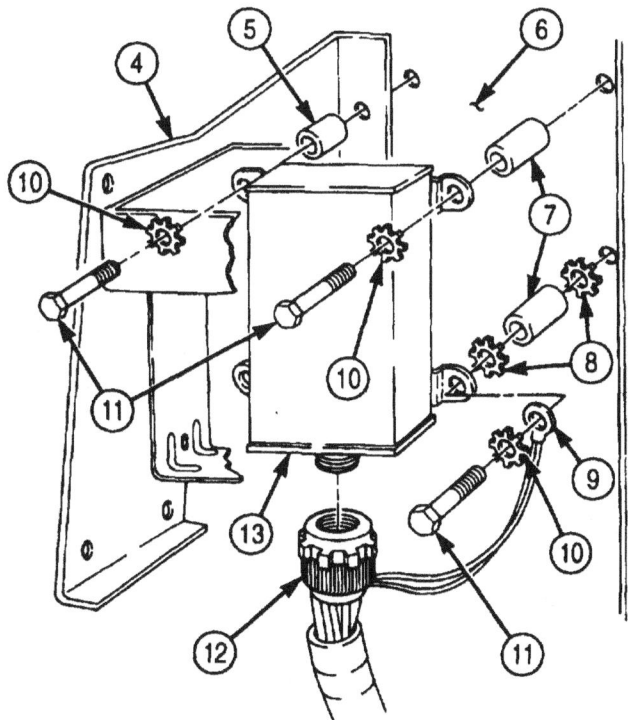

FOLLOW-ON TASKS: • Install ether cylinder and valve (para. 3-35).
                  • Connect battery ground cables (para. 3-126).

3-116. HEADLIGHT MAINTENANCE

THIS TASK COVERS:

a. Removal   c. Alignment
b. Installation

INITIAL SETUP:

APPLICABLE MODELS
All

TOOLS
General mechanic's tool kit (Appendix E, Item 1)

MATERIALS/PARTS
Three lockwashers (Appendix D, Item 345)

REFERENCES (TM)
TM 9-2320-272-10
TM 9-2320-272-24P

EQUIPMENT CONDITION
- Parking brake set (TM 9-2320-272-10).
- Hood raised and secured (TM 9-2320-272-10).
- Battery ground cables disconnected (para. 3-126).

---

a. Removal

NOTE
Tag all connectors for installation.

1. Remove three screws (3) and retaining ring (2) from headlamp housing (1).
2. Pull headlamp (4) out from headlamp housing (1).
3. Disconnect three connectors (5) from headlamp housing (1).
4. Remove three nuts (10), lockwashers (8), and headlamp housing (1) from hood (9). Discard lockwashers (8).
5. Disconnect three connectors (11) from headlamp housing (1).
6. Remove three headlamp connectors (7) and grommets (6) from headlamp housing (1).

b. Installation

1. Install three grommets (6) and headlamp connectors (7) on headlamp housing (1).
2. Connect three connectors (11) to headlamp housing (1).
3. Install headlamp housing (1) on hood (9) with three new lockwashers (8) and nuts (10).
4. Connect three connectors (5) to headlamp housing (1).
5. Position headlamp (4) in headlamp housing (1) with the words SEALED BEAM positioned at bottom of headlamp (4).
6. Install retaining ring (2) on headlamp housing (1) with three screws (3).
7. Connect battery ground cables (para. 3-126).

c. Alignment

1. Draw a horizontal line (12) on a wall the height of center of headlights (15).
2. Park truck facing wall so headlights (15) are 25 ft (7.6 m) from wall.
3. Draw a vertical line (14) through horizontal line (12) so it is in line with center of headlight (15).
4. Turn headlights on low beam (TM 9-2320-260-102
5. Adjust headlight horizontal direction with adjusting screw (16) until left edge of bright light area (13) on wall is 2-6 in. (5.08-15.24 cm) right of vertical line (14).
6. Adjust headlight vertical direction with adjusting screw (17) until top edge of bright light area (13) on wall is touching lower side of horizontal line (12).
7. Adjust other headlight using same procedure.

3-116. HEADLIGHT MAINTENANCE (Contd)

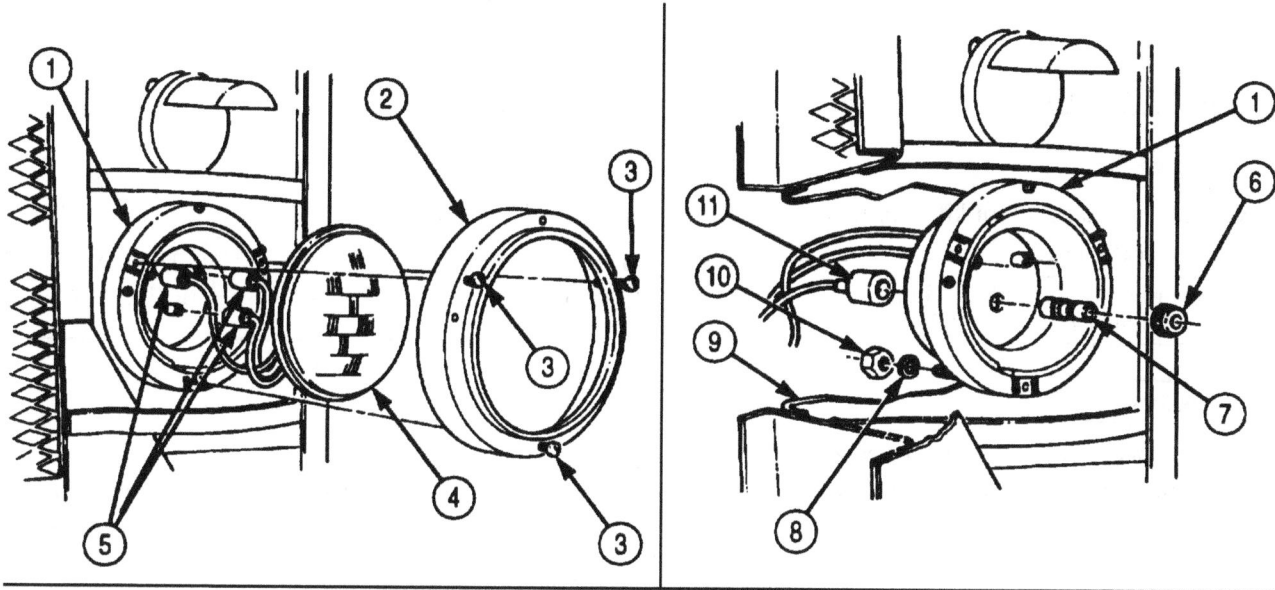

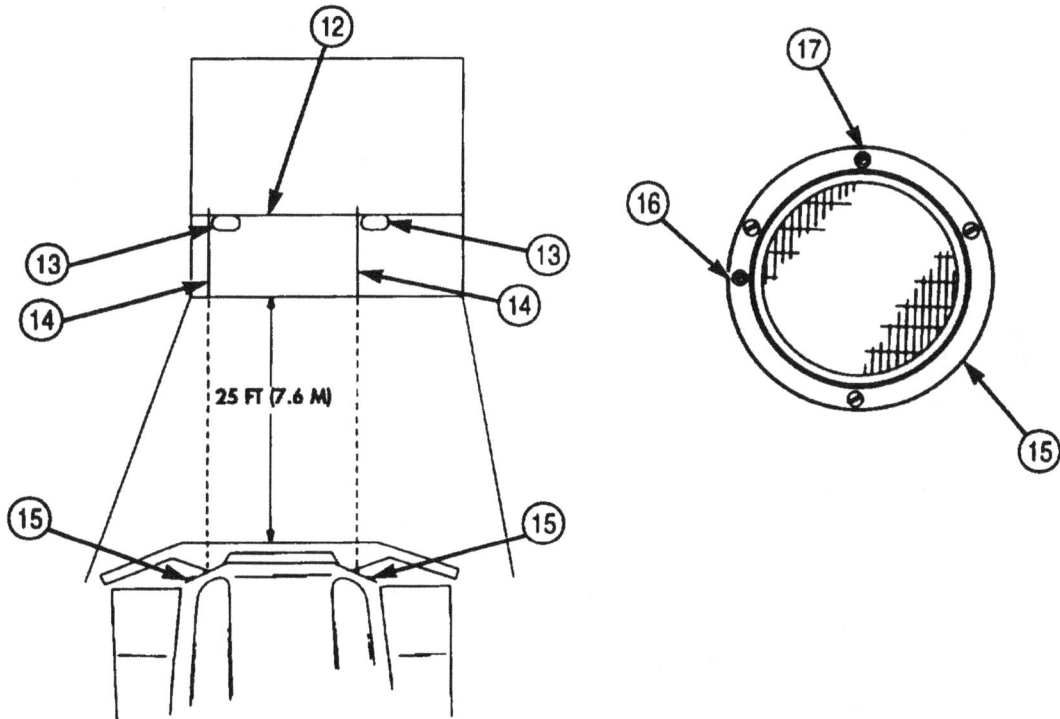

FOLLOW-ON TASK: Check headlamps for proper operation (TM 9-2320-272-10).

3-117. BLACKOUT LIGHT MAINTENANCE

THIS TASK COVERS:

a. Removal   c. Assembly
b. Disassembly   d. Installation

INITIAL SETUP:

APPLICABLE MODELS
All

TOOLS
General mechanic's tool kit (Appendix E, Item 1)

MATERIALS/PARTS
Lockwasher (Appendix D, Item 354)
Four locknuts (Appendix D, Item 313)

REFERENCES (TM)
TM 9-2320-272-10
TM 9-2320-272-24P

EQUIPMENT CONDITION
- Parking brake set (TM 9-2320-272-10).
- Hood raised and secured (TM 9-2320-272-10).
- Battery ground cables disconnected (para. 3-126).

---

a. Removal

1. Disconnect wire (8) from blackout light connector (12).
2. Remove nut (7), lockwasher (9), and adjustment washer (10) from mounting bracket (3). Discard lockwasher (9).

NOTE
Assistant will help with step 3.

3. Remove four locknuts (2), screws (5) and (6), and mounting bracket (3) from hood (4). Discard locknuts (2).
4. Remove blackout light (1) and spacer (11) from mounting bracket (3).

b. Disassembly

1. Loosen three screws (17) on door (13).
2. Remove door (13) from light housing (14).
3. Remove gasket (16) from door (13).
4. Remove lamp (15) from light housing (14).

c. Assembly

1. Install lamp (15) on lamp housing (14).
2. Install gasket (16) on door (13).
3. Install door (13) on light housing (14) and tighten three screws (17).

d. Installation

1. Install blackout light (I) and spacer (11) on mounting bracket (3).

NOTE
Assistant will help with step 2.

2. Install mounting bracket (3) on hood (4) with two screws (5) and (6), and four new locknuts (2).
3. Install adjustment washer (10), new lockwasher (9), and nut (7) on mounting bracket (3).
4. Connect wire (8) to blackout light connector (12).

3-117. BLACKOUT LIGHT MAINTENANCE (Contd)

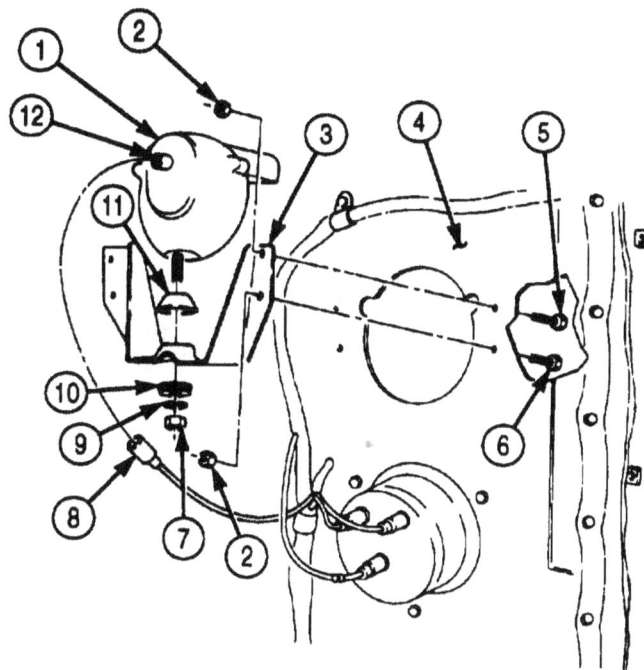

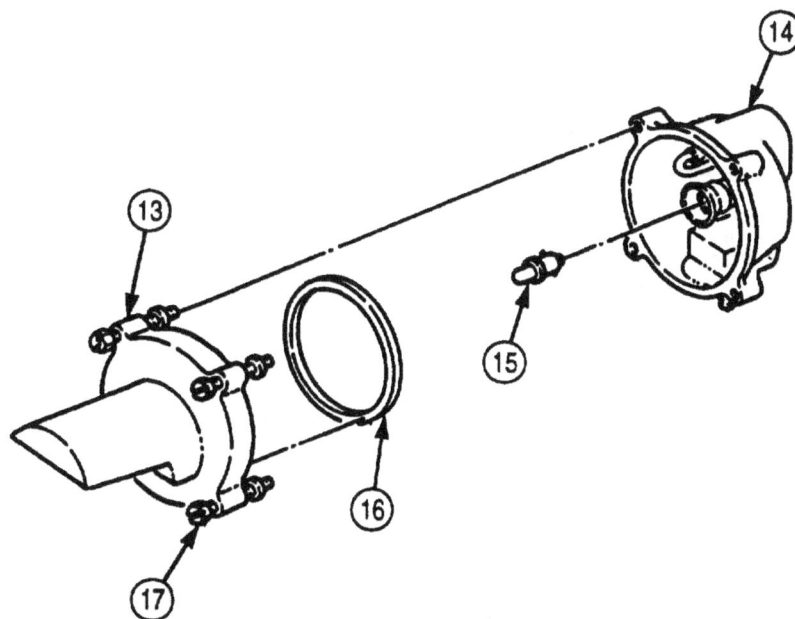

FOLLOW-ON TASKS: • Connect battery ground cables (para. 3-126).
 • Check blackout light for proper operation (TM 9-2320-272-10).

3-118. FRONT AND REAR COMPOSITE LIGHT AND BRACKET REPLACEMENT

THIS TASK COVERS:

a. Front Composite Light Removal
b. Front Composite Light Installation
c. Removal (M923/A1/A2)
d. Installation (M923/A1/A2)
e. Removal (M936/A1/A2)
f. Installation (M936/A1/A2)
g. Removal (M929/A1/A2)

h. Installation (M929/A1/A2)
i. Removal (M931/A1/A2)
j. Installation (M931/A1/A2)
k. Removal (M934/A1/A2)
L. Installation (M934/A1/A2)
m. Composite Light Lamp Removal
n. Composite Light Lamp Installation

INITIAL SETUP:

APPLICABLE MODELS
All

TOOLS
General mechanic's tool kit (Appendix E, Item 1)

MATERIALS/PARTS
Two lockwashers (Appendix D, Item 356)
Four locknuts (Appendix D, Item 299)
Four lockwashers (M936/A1/A2)
 (Appendix D, Item 345)
Two locknuts (M936/A1/A2) (Appendix D,
 Item 291)
Three locknuts (M929/A1/A2) (Appendix D,
 Item 313)
Two locknuts (M931/A1/A2) (Appendix D,
 Item 322)
Locknut (M931/A1/A2) (Appendix D, Item 288)
Four locknuts (M931/A1/A2)
 (Appendix D, Item 313)
Two locknuts (M934/A1/A2) (Appendix D,
 Item 288)
Two screw-assembled lockwashers
 (Appendix D, Item 583)

REFERENCES (TM)
TM 9-2320-272-10
TM 9-2320-272-24P

EQUIPMENT CONDITION
- Parking brake set (TM 9-2320-272-10).
- Battery ground cables disconnected (para. 3-126).

a. Front Composite Light Removal

1. Remove four locknuts (13) and screws (5) from fender (3). Discard locknuts (13).

NOTE
Tag wires for installation.

2. Disconnect wires (7), (8), (9), and (10).
3. Remove wiring cover (11) from cable (12).
4. Remove grommet (14) from fender (3).
5. Remove two screws (1), lockwasher (2), and front composite light (6) from fender (3). Discard lockwasher (2).
6. Remove locknut (19), two washers (17) and lockwashers (15), screw (16), and mounting bracket (4) from fender (3). Discard lockwashers (15) and locknut (19).

3-118. FRONT AND REAR COMPOSITE LIGHT AND BRACKET REPLACEMENT (Contd)

b. Front Composite Light Installation

1. Install mounting bracket (4) on fender (3) with screw (16), two washers (17), new lockwashers (15), and new locknut (19).
2. Insert wires (7), (8), (9), and (10) through hole in fender (3).
3. Install grommet (14) on fender (3).
4. Install composite light (6) on fender (3) with new lockwasher (2) and two screws (1).
5. Insert wires (7), (8), (9), and (10) through hole in wiring cover (11) and connect.
6. Install wiring cover (11) with four screws (5) and new locknuts (13).

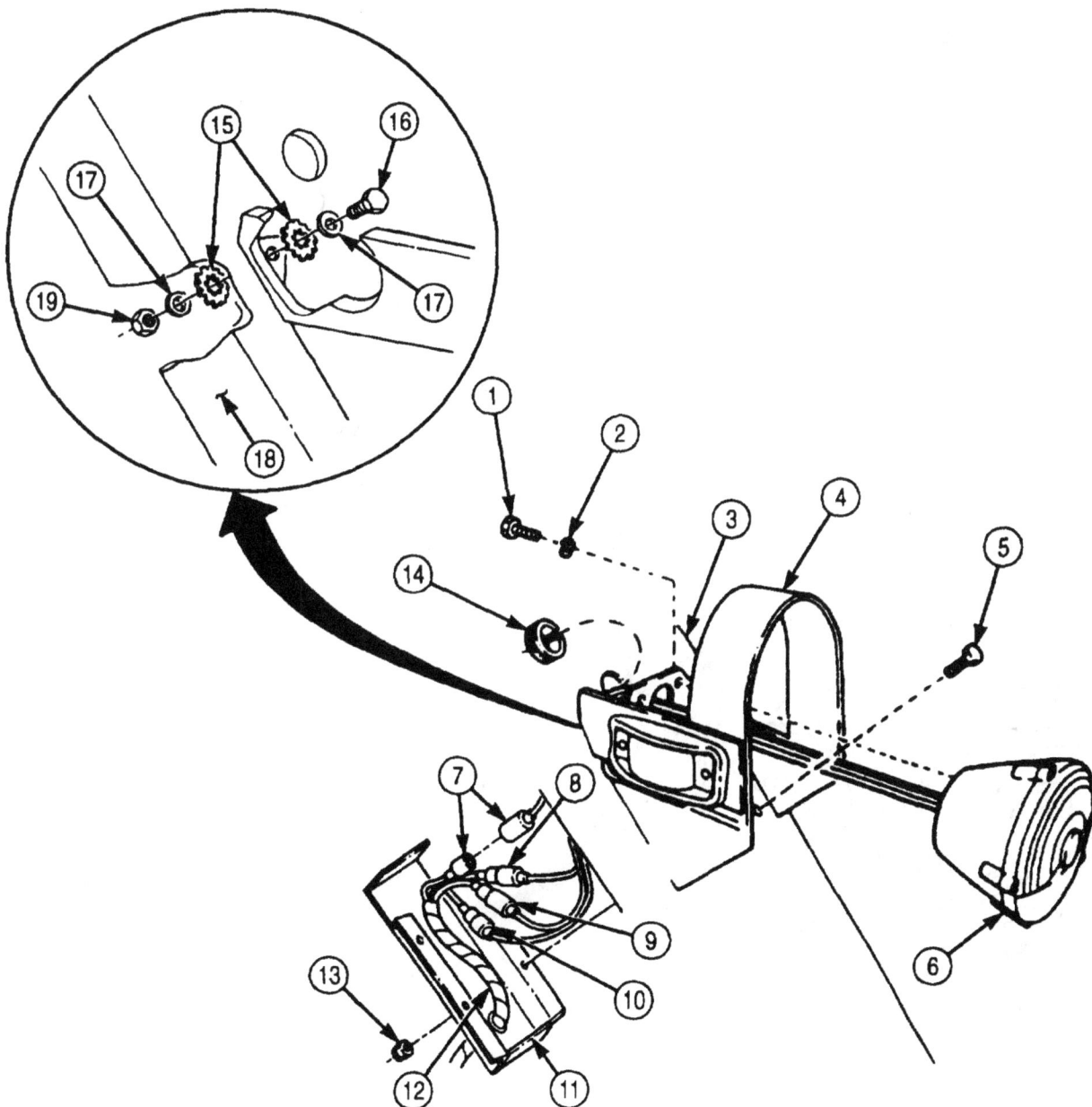

3-118. FRONT AND REAR COMPOSITE LIGHT AND BRACKET REPLACEMENT (Contd)

c. Removal (M923/A1/A2)

NOTE

Left and right composite lights are replaced the same way.

1. Remove two screw-assembled lockwashers (2), clamp (3), and rear composite light (6) from bracket (1). Discard screw-assembled lockwashers (2).

NOTE

Tag connectors for installation.

2. Disconnect four connectors (4) from wires (5).

d. Installation (M923/A1/A2

1. Connect four connectors (4) to wires (5).

2. Install rear composite light (6) and clamp (3) on bracket (1) with two new screw-assembled lockwashers (2).

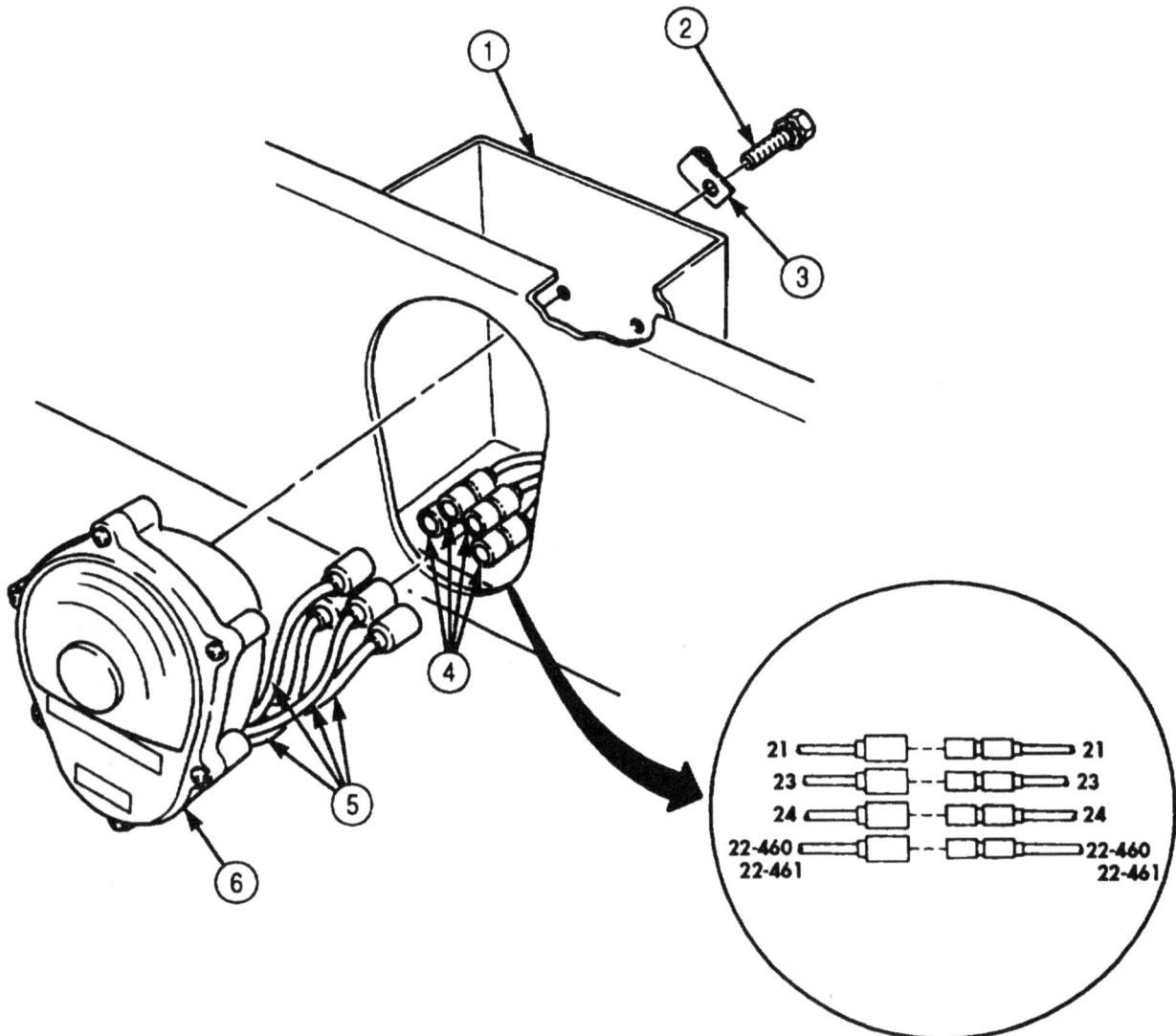

3-118. FRONT AND REAR COMPOSITE LIGHT AND BRACKET REPLACEMENT (Contd)

e. Removal (M936/A1/A2)

NOTE
Left and right composite lights and brackets are replaced the same way.

1. Remove two locknuts (14), screws (16), and reflector (15) from taillight guard (11). Discard locknuts (14).
2. Remove four screws (13), lockwashers (12), and taillight guard (11) from bracket (20). Discard lockwasher (12).

NOTE
Tag wires for installation.

3. Disconnect four connectors (17) from wires (18).
4. Remove two screw-assembled lockwashers (10) and rear composite light (19) from bracket (20). Discard screw-assembled lockwashers (10).
5. Remove two locknuts (9), screws (8), and bracket (20) from ladder bracket (7). Discard locknuts (9).

f. Installation (M936/A1/A2)

1. Install bracket (20) on ladder bracket (7) with two screws (8) and new locknuts (9).
2. Install rear composite light (19) on bracket (20) with two new screw-assembled lockwashers (10).
3. Connect four connectors (17) to wires (18).
4. Install taillight guard (11) on bracket (20) with four new lockwashers (12) and screws (13).
5. Install reflector (15) on taillight guard (12) with two screws (16) and new locknuts (14).

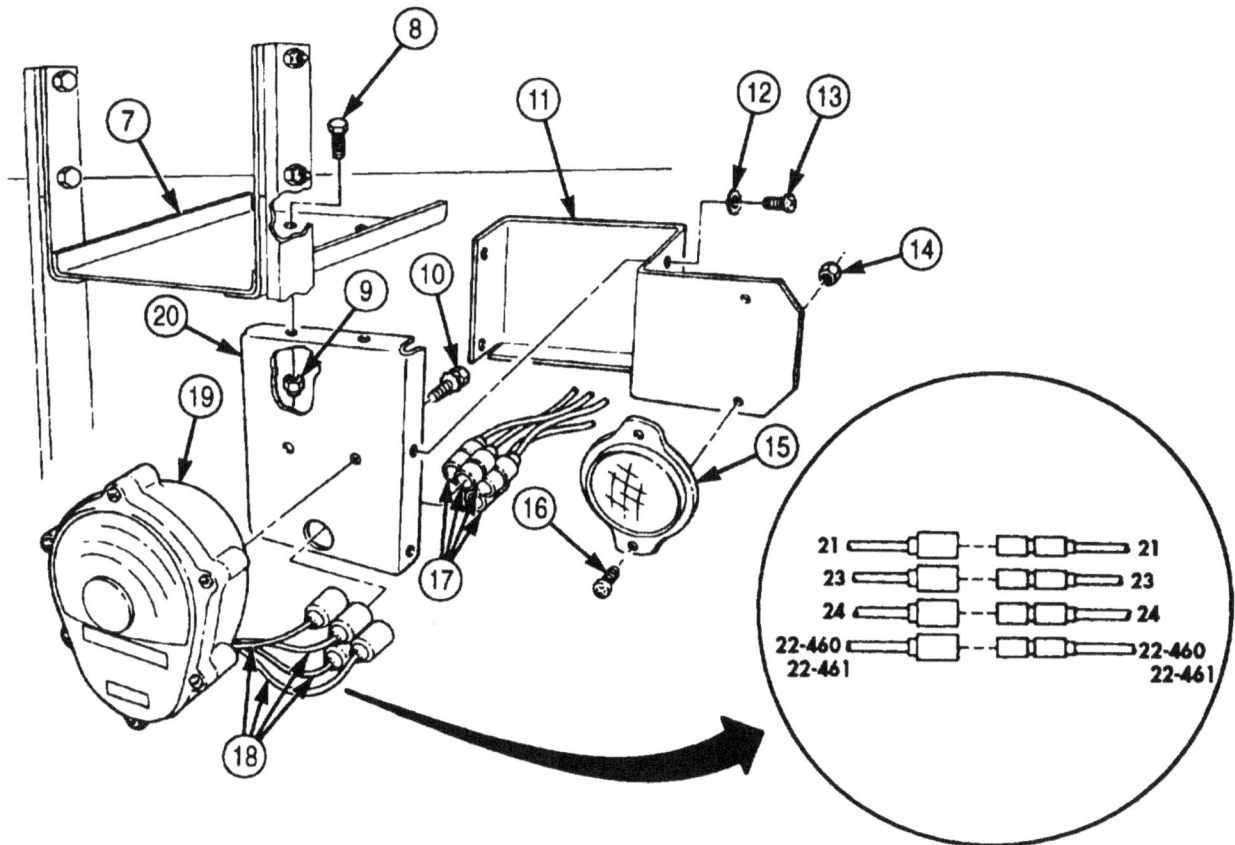

3-118. FRONT AND REAR COMPOSITE LIGHT AND BRACKET REPLACEMENT (Contd)

g. Removal (M929/A1/A2)

NOTE

Left and right composite lights and covers are replaced the same way.

1. Remove three locknuts (4), screws (5), and cover (3) from dump bed (1). Discard locknuts (4).

NOTE

Tag wires for installation.

2. Disconnect four connectors (6) from wires (7).

3. Remove two screw-assembled lockwashers (2) and rear composite light (8) from dump bed (1). Discard screw-assembled lockwashers (2).

h. Installation (M929/A1/A2)

1. Install rear composite light (8) on dump bed (1) with two new screw-assembled lockwashers (2).

2. Connect four connectors (6) to wires (7).

3. Install cover (3) on dump bed (1) with three screws (5) and new locknuts (4).

3-118. FRONT AND REAR COMPOSITE LIGHT AND BRACKET REPLACEMENT (Contd)

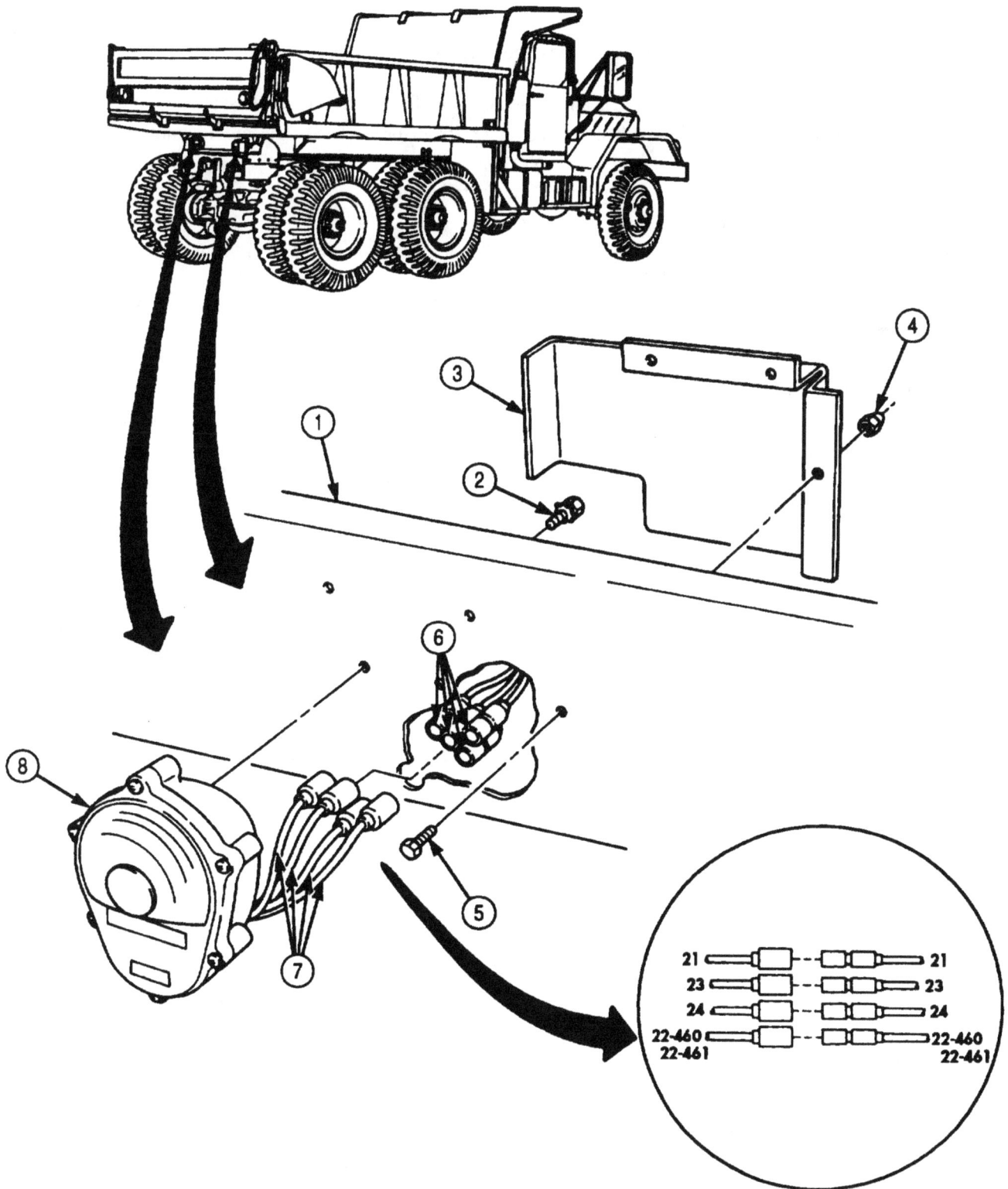

3-118. FRONT AND REAR COMPOSITE LIGHT AND BRACKET REPLACEMENT (Contd)

NOTE
Left and right composite lights, covers, and brackets are replaced
the same way.

1. Remove two locknuts (6), screws (12), and reflector (11) from bracket (5). Discard locknuts (6).

2. Remove four locknuts (9), screws (13), and cover (8) from bracket (5). Discard locknuts (9).

NOTE
Tag wires for installation.

3. Disconnect four connectors (10) from wires (14).

4. Remove two screw-assembled lockwashers (7) and rear composite light (15) from bracket (5). Dircard screw-assembled lockwashers (7).

5. Remove nut (2), locknut (3), two screws (4), and bracket (5) from frame (1). Discard locknut (3).

j. Installation (M931/A1/A2)

1. Install bracket (5) on frame (1) with two screws (4), new locknut (3), and nut (2).

2. Install rear composite light (15) on bracket (5) with two new screw-assembled lockwashers (7).

3. Connect four connectors (10) to wires (14).

4. Install cover (8) on bracket (5) with four screws (13) and new locknuts (9).

5. Install reflector (11) on bracket (5) with two screws (12) and new locknuts (6).

3-118. FRONT AND REAR COMPOSITE LIGHT AND BRACKET REPLACEMENT (Contd)

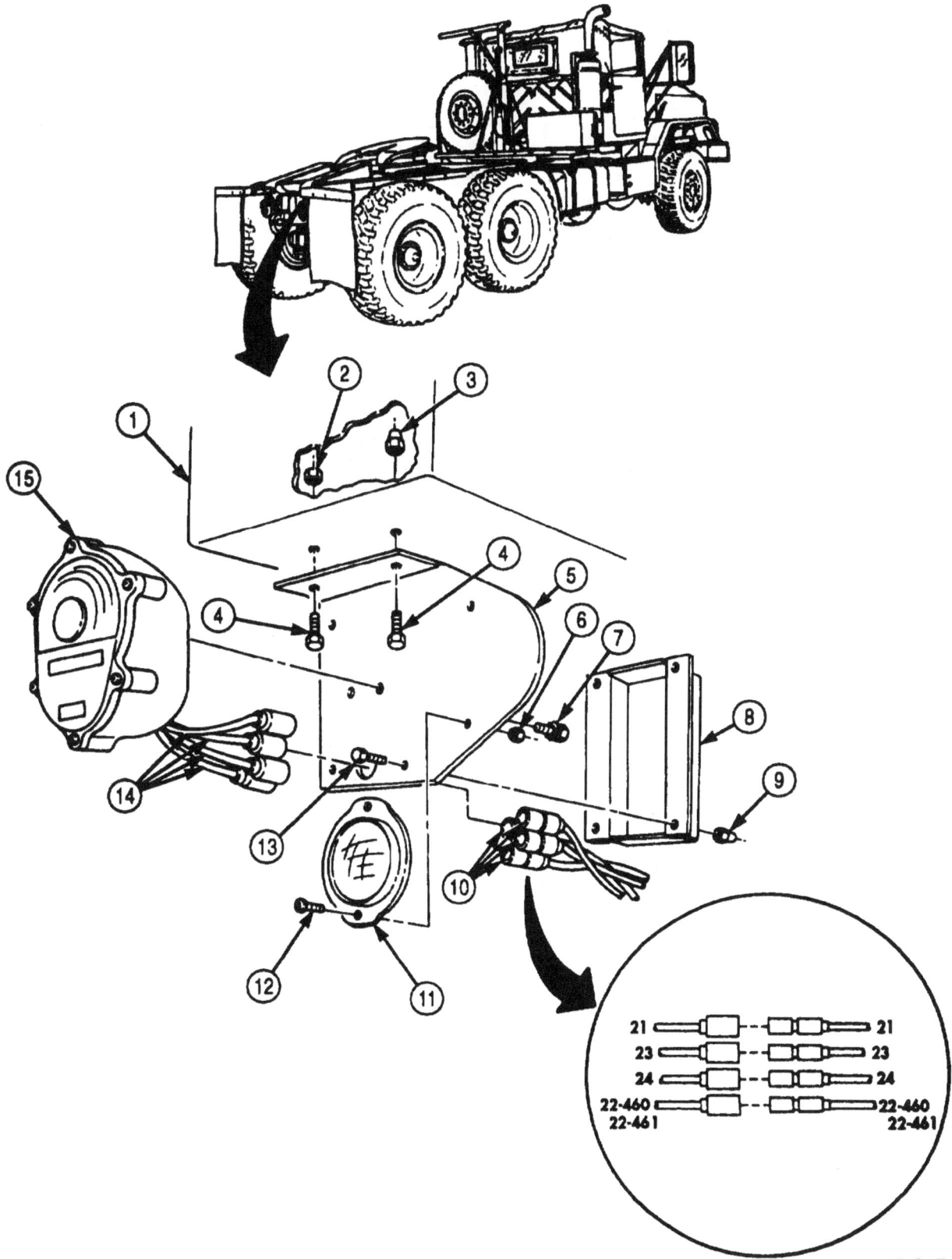

21 ⊏━━━ - - - ⊐━━ 21
23 ⊏━━━ - - - ⊐━━ 23
24 ⊏━━━ - - - ⊐━━ 24
22-460 ⊏━━ - - - ⊐━ 22-460
22-461 22-461

3-118. FRONT AND REAR COMPOSITE LIGHT AND BRACKET REPLACEMENT (Contd)

k. Removal (M934/A1/A2)

NOTE

- Left and right composite lights, brackets, covers, and braces are replaced the same way.
- Tag wires for installation.

1. Remove four screws (6) and cover (5) from bracket (9)
2. Disconnect four connectors (8) from wires (10).
3. Remove two screw-assembled lockwashers (4) and rear composite light (11) from bracket (9). Discard screw-assembled lockwashers (4).
4. Remove locknuts (14) and (7), two screws (3), and bracket (9) from brace (2). Discard locknuts (14) and (7).
5. Remove two nuts (12), screws (13), and brace (2) from frame rail (1).

l. Installation (M934/A1/A2)

1. Install brace (2) on frame rail (1) with two screws (13) and nuts (12).
2. Install bracket (9) on brace (2) with two screws (3) and new locknuts (14) and (7).
3. Install rear composite light (11) on bracket (9) with two new screw-assembled lockwashers (4).
4. Connect four connectors (8) to wires (10).
5. Install cover (5) on bracket (9) with four screws (6).

3-118. FRONT AND REAR COMPOSITE LIGHT AND BRACKET REPLACEMENT (Contd)

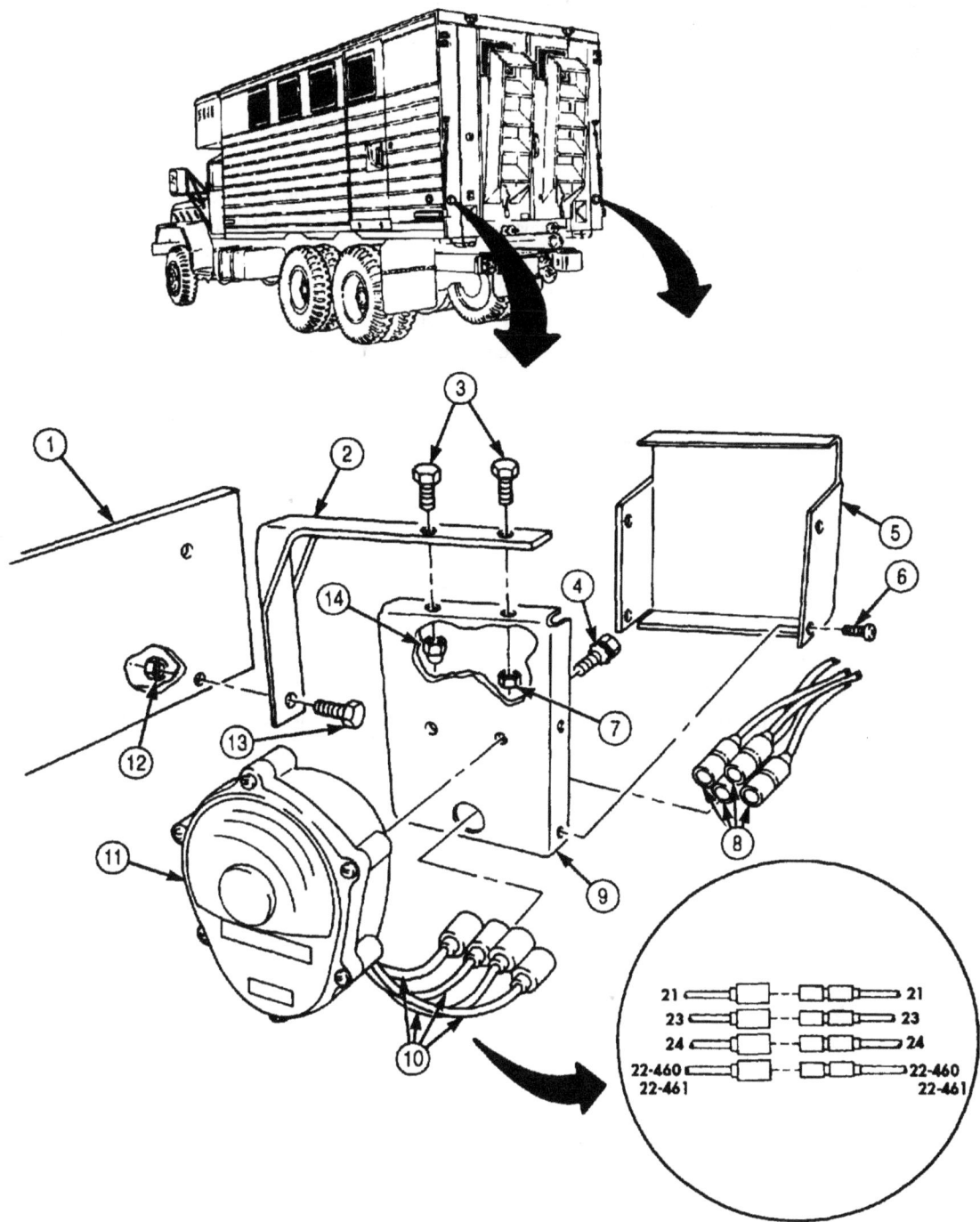

21	21
23	23
24	24
22-460	22-460
22-461	22-461

TM 9-2320-272-24-1

3-118. FRONT AND REAR COMPOSITE LIGHT AND BRACKET REPLACEMENT (Contd)

m. Composite Light Lamp Removal

NOTE

Composite light is off vehicle for this task.

1. Loosen six screws (8) on composite light body (4).

2. Remove composite light door (1) and seal (7) from composite light body (4).

3. Remove stoplight lamp (5), blackout marker lamp (6), turn signal lamp (2), and parking lamp (3) from composite light body (4).

NOTE

Perform step 4 for diode configured rear composite lights only.

4. Remove stoplight lamp (5). blackout marker diode (6), turn signal lamp (2), and parking lamp (3) from composite light body (4).

n. Composite Light Lamp Installation

NOTE

Composite light is off vehicle for this task.

1. Install stoplight lamp (5), blackout marker lamp (6), turn signal lamp (2), and parking lamp (3) on composite light body (4).

NOTE

Perform step 2 for diode configured rear composite lights only.

2. Install stoplight lamp (5), blackout marker diode (6), turn signal lamp (2), and parking lamp (3) on composite light body (4).

3. Install seal (7) and composite light door (1) on composite light body (4) and tighten six screws (8).

3-320

3-118. FRONT AND REAR COMPOSITE LIGHT AND BRACKET REPLACEMENT (Contd)

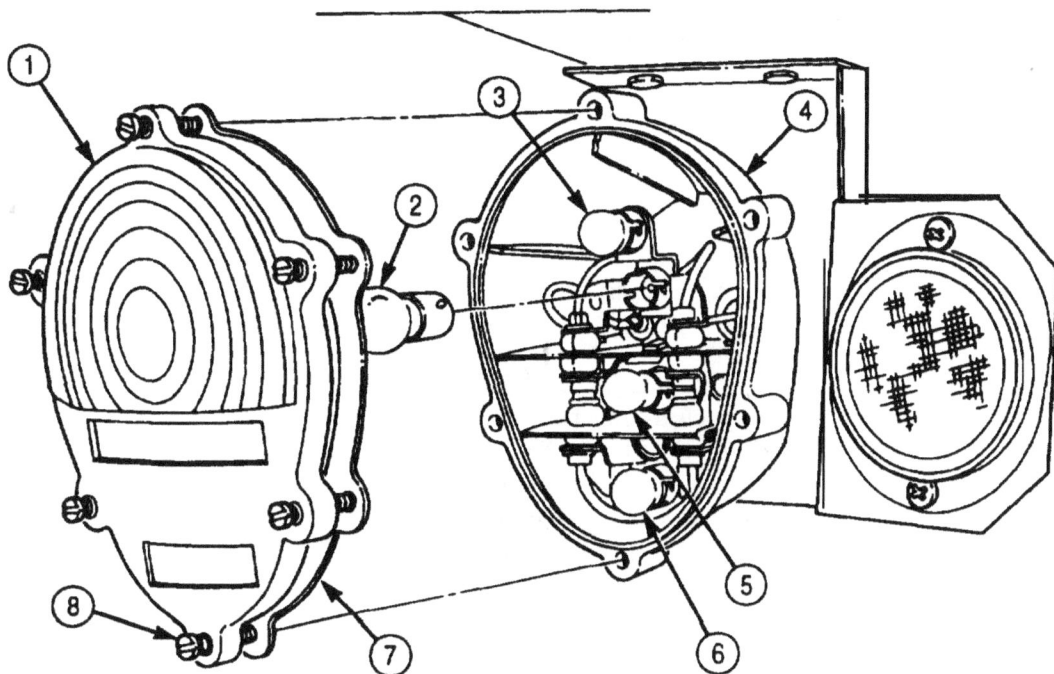

FOLLOW-ON TASKS: • Connect battery ground cables (para. 3-126).
• Check rear composite light for proper operation (TM 9-2320-272-10).

3-119. FLOODLIGHT SEALED BEAM LAMP AND DOOR (M936/A1/A2) REPLACEMENT

THIS TASK COVERS:

a. Removal   b. Installation

INITIAL SETUP:

APPLICABLE MODELS
M936/A1/A2

TOOLS
General mechanic's tool kit (Appendix E, Item 1)

REFERENCES (TM)
TM 9-2320-272-10
TM 9-2320-272-24P

EQUIPMENT CONDITION
- Parking brake set (TM 9-2320-272-10).
- Battery ground cables disconnected (para. 3-126).

GENERAL SAFETY INSTRUCTIONS
Lamp door retaining clips are under tension.

NOTE
All floodlight sealed beam lamps and doors are replaced the same way.

a. Removal

1. Remove three screws (1) and retaining rings (2) from lamp door (10).
2. Separate lamp door (10) from lamp housing (6).
3. Remove two screws (5) and wires (8) from sealed beam lamp (4).

WARNING

Lamp door retaining clips are under tension and must be removed with firm grip or injury to personnel may result.

NOTE
Note position of clips for installation.

4. Remove four retaining clips (3) from lamp door (10).
5. Remove sealed beam lamp (4) from lamp door (10).

b. Installation

1. Install sealed beam lamp (4) on lamp door (10). Align lamp (4) with notch (11) on lamp door (10).
2. Install four retaining clips (3) on lamp door (10).
3. Install two wires (8) on lamp (4) with two screws (5).
4. Align drain hole (9) in door (10) with notch (7) in lamp housing (6) and install lamp door (10) on lamp housing (6) with three screws (1) and retaining rings (2).

3-119. FLOODLIGHT SEALED BEAM LAMP AND DOOR (M936/A1/A2) REPLACEMENT (Contd)

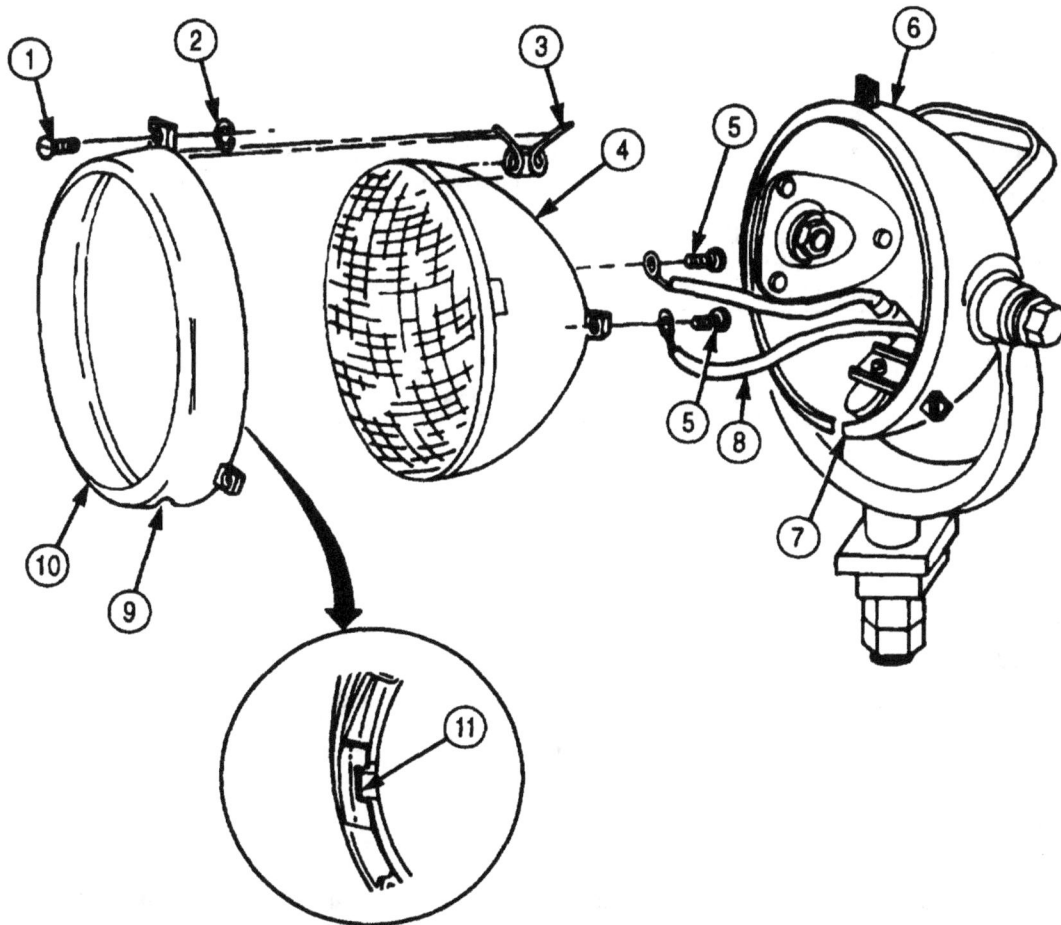

FOLLOW-ON TASKS: • Connect battery ground cables (para. 3-126).
                • Check floodlight for proper operation (TM 9-2320-272-10).

3-120. FLOODLIGHT CONTROL SWITCH MAINTENANCE

THIS TASK COVERS:

a. Removal   d. Assembly
b. Disassembly   e. Installation
c. Inspection

INITIAL SETUP:

APPLICABLE MODELS
M936/A1/A2

TOOLS
General mechanic's tool kit (Appendix E, Item 1)

MATERIALS/PARTS
Two locknuts (Appendix D, Item 283)
Lockwasher (Appendix D, Item 355)
Lockwasher (Appendix D, Item 351)

REFERENCES (TM)
TM 9-2320-272-10
TM 9-2320-272-24P

EQUIPMENT CONDITION
- Parking brake set (TM 9-2320-272-10).
- Battery ground cables disconnected (para. 3-126).

a. Removal

1. Remove two locknuts (1) and mounting screws (11) from angle bracket (2) and instrument panel (16). Discard locknuts (1).

2. Lower floodlight control switch (12).

NOTE
Tag connectors for installation.

3. Disconnect four connectors (15) from floodlight control switch (12).

b. Disassembly

1. Remove screw (7), lockwasher (6), switch lever (8), washer (5), and felt washer (9) from floodlight control switch (12). Discard lockwasher (6).

2. Remove nut (4), lockwasher (10), identification plate (3), angle bracket (21, plug (14), and shell (13) from floodlight control switch (12). Discard lockwasher (10).

c. Inspection

1. Inspect floodlight control switch (12) for breaks and cracks in housing. Replace if housing is broken or cracked.

2. Inspect switch lever (8) for breaks and cracks. Replace if broken or cracked.

d. Assembly

1. Install shell (13), plug (14), angle bracket (2), and identification plate (3) on floodlight control switch (12) with new lockwasher (10) and nut (4).

2. Install felt washer (9), washer (5), and switch lever (8) on floodlight control switch (12) with new lockwasher (6) and screw (7).

3-120. FLOODLIGHT CONTROL SWITCH MAINTENANCE (Contd)

e. Installation

1. Install floodlight control switch (12) and angle bracket (2) on instrument panel (16) with two screws (11) and new locknuts (1).

2. Connect four connectors (15) to floodlight control switch (12).

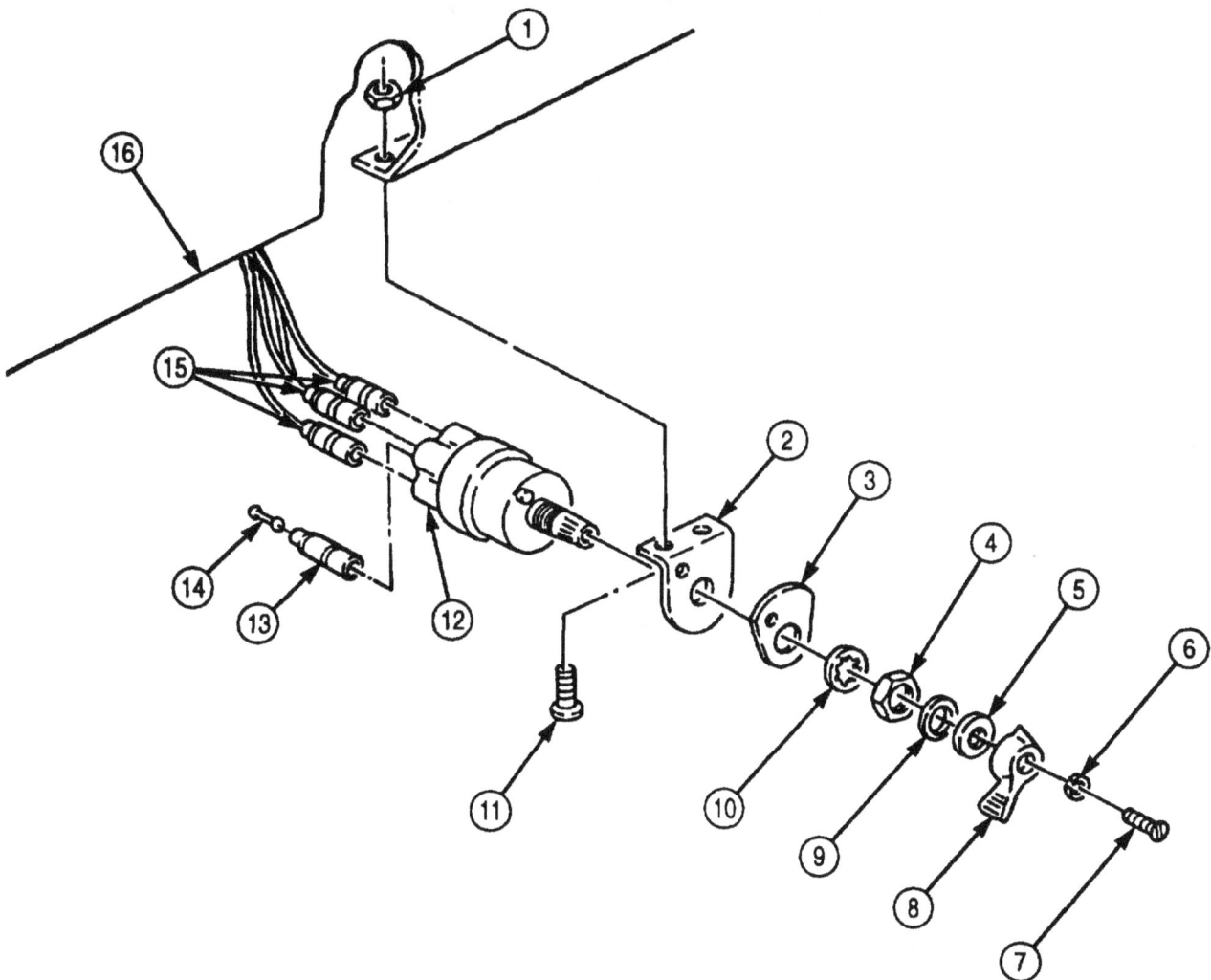

FOLLOW-ON TASKS: • Connect battery ground cables (para. 3-126).
                 • Check operation of floodlight control switch (TM 9-2320-272-10).

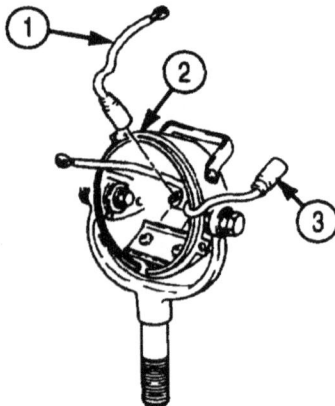

3-121. FLOODLIGHT MAINTENANCE

THIS TASK COVERS:

a. Removal
b. Disassembly

c. Assembly
d. Installation

INITIAL SETUP:

APPLICABLE MODELS
M936/A1/A2

TOOLS
General mechanic's tool kit (Appendix E, Item 1)

MATERIALS/PARTS
Two lockwashers (Appendix D, Item 351)
Two lockwashers (Appendix D, Item 354)

REFERENCES (TM)
TM 9-2320-272-10
TM 9-2320-272-24P

EQUIPMENT CONDITION
• Parking brake set (TM 9-2320-272-10).
• Floodlight sealed beam lamp and door removed (para. 3-119).

NOTE
All floodlights are maintained the same way.

a. Removal

NOTE
Tag wires for installation.

1. Disconnect two wires (6) from connectors (4).
2. Remove two nuts (9), washer (7), floodlight housing (2), and washer (7), from floodlight bracket (8).

b. Disassembly

NOTE
Tag wires for installation.

1. Disconnect switch wire (3) and lamp wire (1) from floodlight housing (2).
2. Remove two connectors (4) and grommets (5) from floodlight housing (2).
3. Remove two screws (10), retainer (11), and pressure switch (12) from floodlight housing (2).
4. Remove two screws (15), lockwashers (14), and switch housing (13) from floodlight housing (2). Discard lockwashers (14).

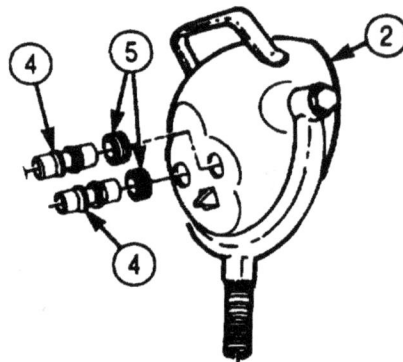

3-121. FLOODLIGHT MAINTENANCE (Contd)

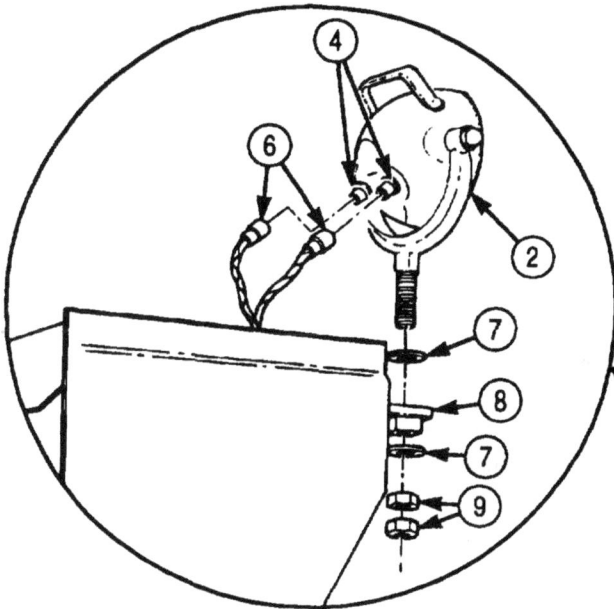

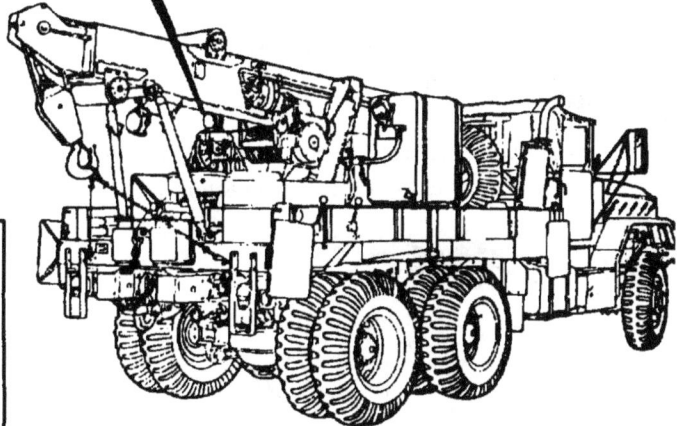

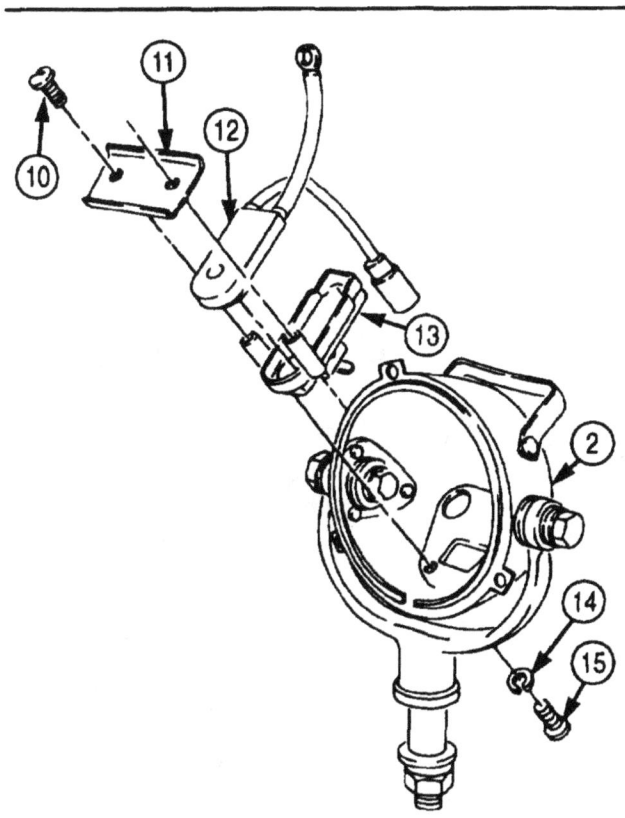

3-121. FLOODLIGHT MAINTENANCE (Contd)

5. Remove two nuts (1), lockwashers (10), and washers (2) from floodlight housing (3) and screws (5). Discard lockwashers (10).

6. Remove two screws (5), washers (4), four spring washers (6), two washers (4), floodlight housing (3), and two washers (4) from mounting bracket (9).

7. Remove pin (8) and stud (7) from mounting bracket (9).

8. Remove two spacers (18) and grommets (17) from floodlight assembly (3).

c. Assembly

1. Install two grommets (17) and spacers (18) on floodlight housing (3).

2. Install stud (7) on mounting bracket (9) with pin (8).

3. Install floodlight housing (3) and two washers (4) on mounting bracket (9) with four spring washers (6), two washers (4), and screws (5).

4. Install two washers (2), new lockwashers (10), and nuts (1) on floodlight housing (3) and screws (5).

5. Install switch housing (14) on floodlight housing (3) with two new lockwashers (15) and screws (16).

6. Install pressure switch (13) on floodlight housing (3) with retainer (12) and two screws (11).

7. Install two grommets (20) and connectors (19) on floodlight housing (3).

8. Connect lamp wire (21) and switch wire (22) to floodlight housing (3).

d. Installation

1. Install washer (24) and floodlight housing (3) on floodlight bracket (25) with washer (24) and two nuts (26).

2. Connect two wires (23) to connectors (19).

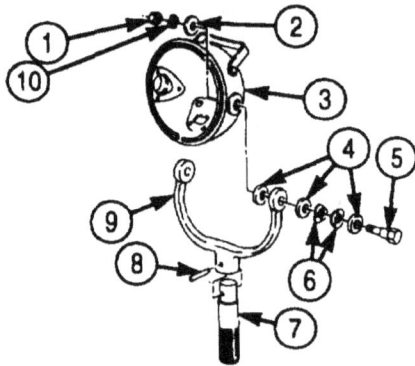

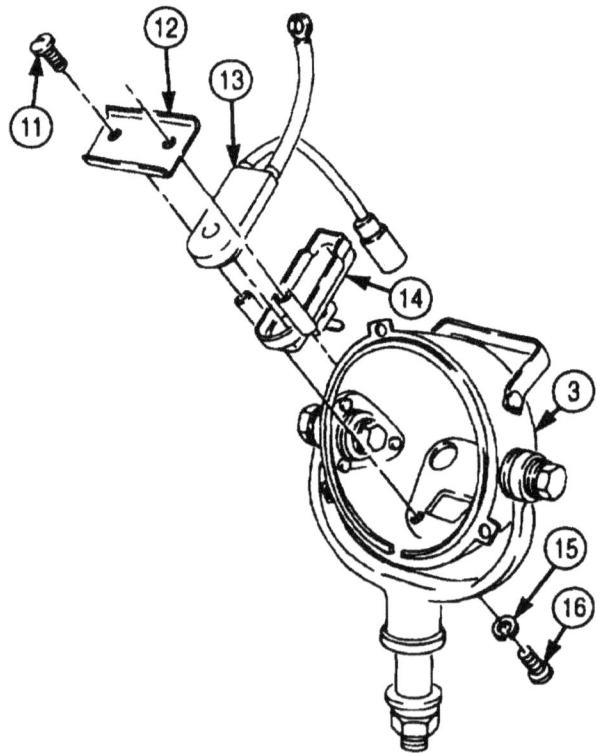

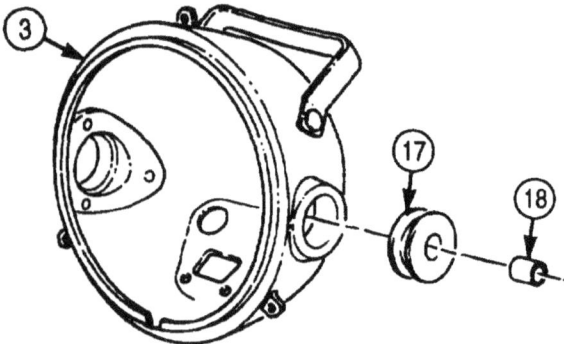

3-121. FLOODLIGHT MAINTENANCE (Contd)

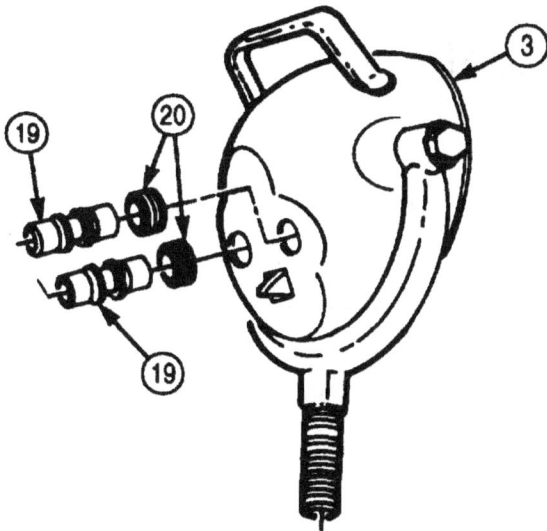

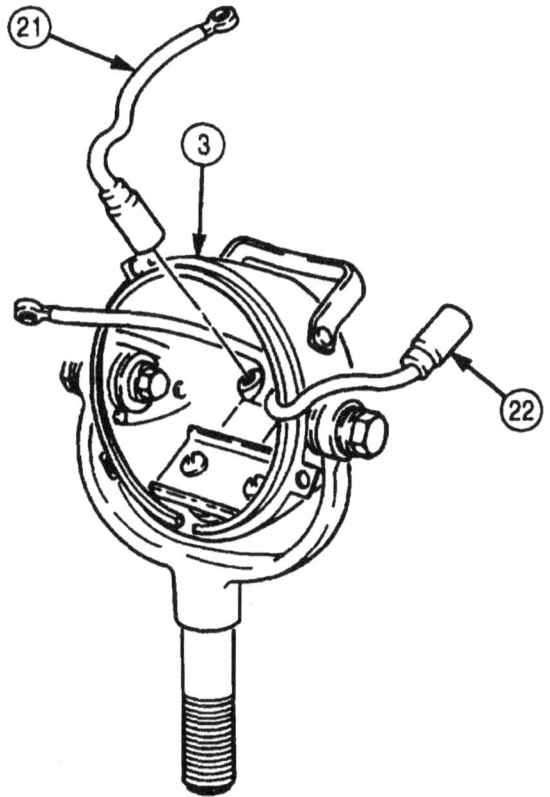

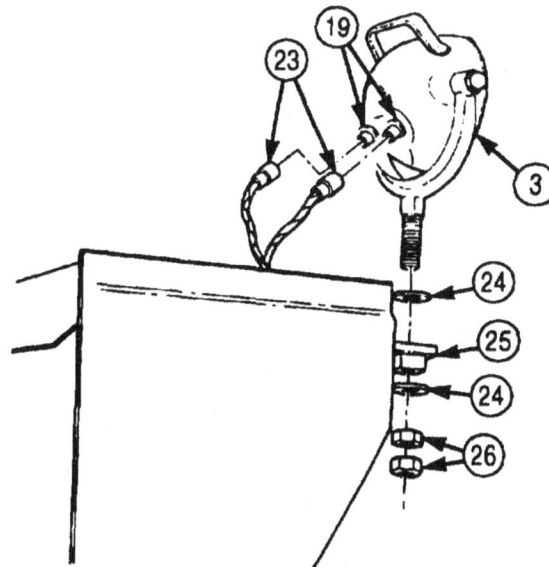

FOLLOW-ON TASKS: • Install floodlight sealed beam lamp and door (para. 3-119).
• Check floodlight for proper operation (TM 9-2320-272-10).

3-122. SIDE MARKER LIGHTS AND BRACKET REPLACEMENT

THIS TASK COVERS:

a. Lamp Removal
b. Light Removal

c. Light Installation
d. Lamp Installation

INITIAL SETUP:

APPLICABLE MODELS
All

TOOLS
General mechanic's tool kit (Appendix E, Item 1)

MATERIALS/PARTS
Four lockwashers (Appendix D, Item 356)
Four lockwashers (Appendix D, Item 357)
Two locknuts (Appendix D, Item 288)

REFERENCES (TM)
TM 9-2320-272-10
TM 9-2320-272-24P

EQUIPMENT CONDITION
- Parking brake set (TM 9-2320-272-10)
- Battery ground cables disconnected (para. 3-126).

NOTE
- Clearance lights have the same design as side marker lights. Only their location on the vehicle is different.
- Front and rear side markers are replaced basically the same. This procedure is for rear side marker light.

a. Lamp Removal

1. Remove two screws (4), cover (3), and lens cover (2) from side marker light (5).
2. Remove lamp (1) from side marker light (5).

b. Light Removal

1. Disconnect side marker light plug (15) from wiring harness lead (16).
2. Remove four nuts (14), lockwashers (13), screws (12), and side marker light (5) from bracket (9). Discard lockwashers (13).
3. Remove two locknuts (6), screws (11), four lockwashers (7), clamp (8), and bracket (9) from rail (10). Discard locknuts (6) and lockwashers (7).
4. Remove clamp (8) from wiring harness lead (16).

c. Light Installation

1. Position clamp (8) on wiring harness lead (16).
2. Install bracket (9) and clamp (8) on rail (10) with four new lockwashers (7). two screws (11), and new locknuts (6).
3. Install side marker light (5) on bracket (9) with four screws (12), new lockwashers (13), and nuts (14).
4. Connect side marker light plug (15) to wiring harness lead (16).

d. Lamp Installation

1. Install lamp (1) on side marker light (5).
2. Install lens cover (2) and cover (3) on side marker light (5) with two screws (4).

3-122. SIDE MARKER LIGHTS AND BRACKET REPLACEMENT (Contd)

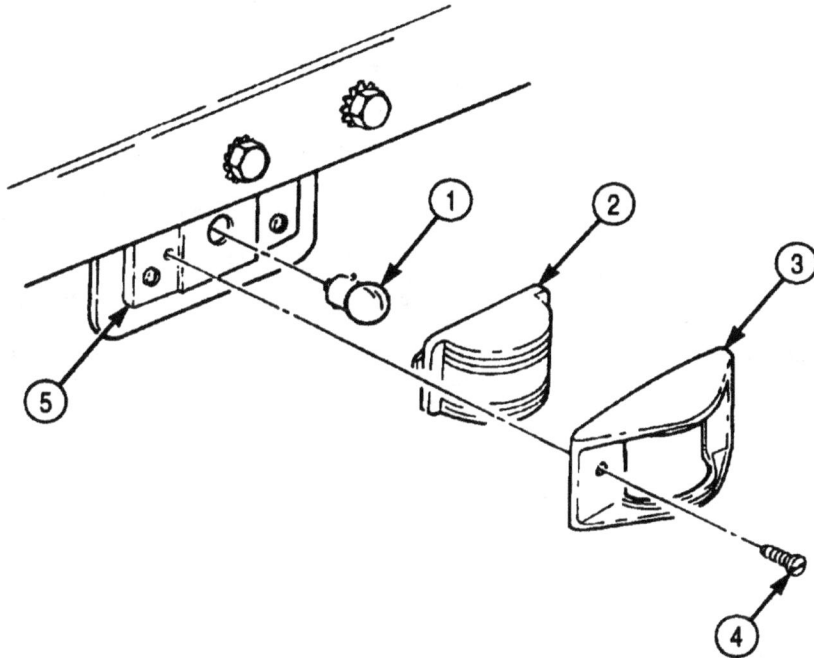

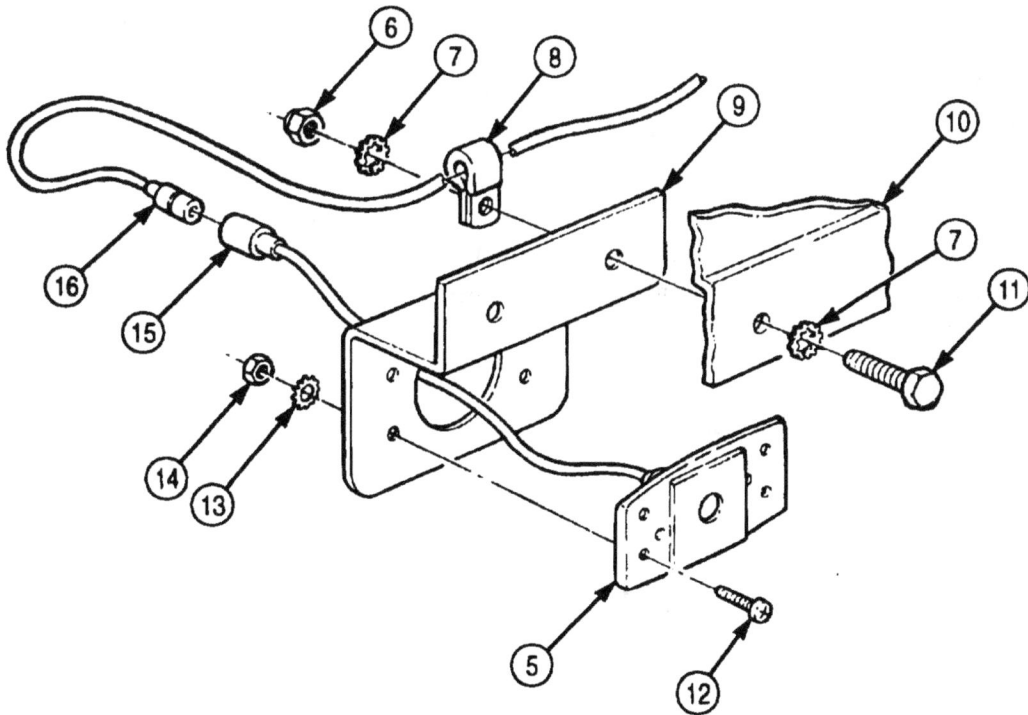

FOLLOW-ON TASKS • Connect battery ground cables (para. 3-126).
• Check operation of clearance and side marker lights (TM 9-2320-272-10).

3-123. BATTERY BOX COVER REPLACEMENT

THIS TASK COVERS:

a. Removal   **b. Installation**

INITIAL SETUP:

APPLICABLE MODELS
All

TOOLS
General mechanic's tool kit (Appendix E, Item 1)

MATERIALS/PARTS
Six locknuts (Appendix D, Item 283)

REFERENCES (TM)
TM 9-2320-272-10
TM 9-2320-272-24P

EQUIPMENT CONDITION
- Parking brake set (TM 9-2320-272-10).
- Companion seat cushion removed (para. 3-286).

GENERAL SAFETY INSTRUCTIONS
- Remove all jewelry before beginning procedure.
- Ensure batteries are clamped down, rubber boots are installed, clamps are well down on battery posts, and all batery cables lie flat against top of batteries.

WARNING

- When performing battery maintenance, ensure batteries are seated and clamped down, all rubber boots are installed, clamps are well down on battery posts, and all battery cables lie flat against top of batteries. Failure to do so may result in injury to personnel or damage to equipment.

- Remove all jewelry. If jewelry or disconnecter battery ground cable contacts battery terminal, a direct short will result and may cause injury to personnel.

a. Removal

NOTE

Assistant will hold battery box cover open while mechanic removes support rod.

1. Release two latches (2) and raise battery box cover (1).

2. Remove six locknuts (5), screws (3), and battery box cover (1) from battery box cover hinge (4). Discard locknuts (5).

b. Installation

1. Install battery box cover (1) on battery box cover hinge (4) with six screws (3) and new locknuts (5).

2. Lower battery box cover (1) and secure with two latches (2).

3-123. BATTERY BOX COVER REPLACEMENT (Contd)

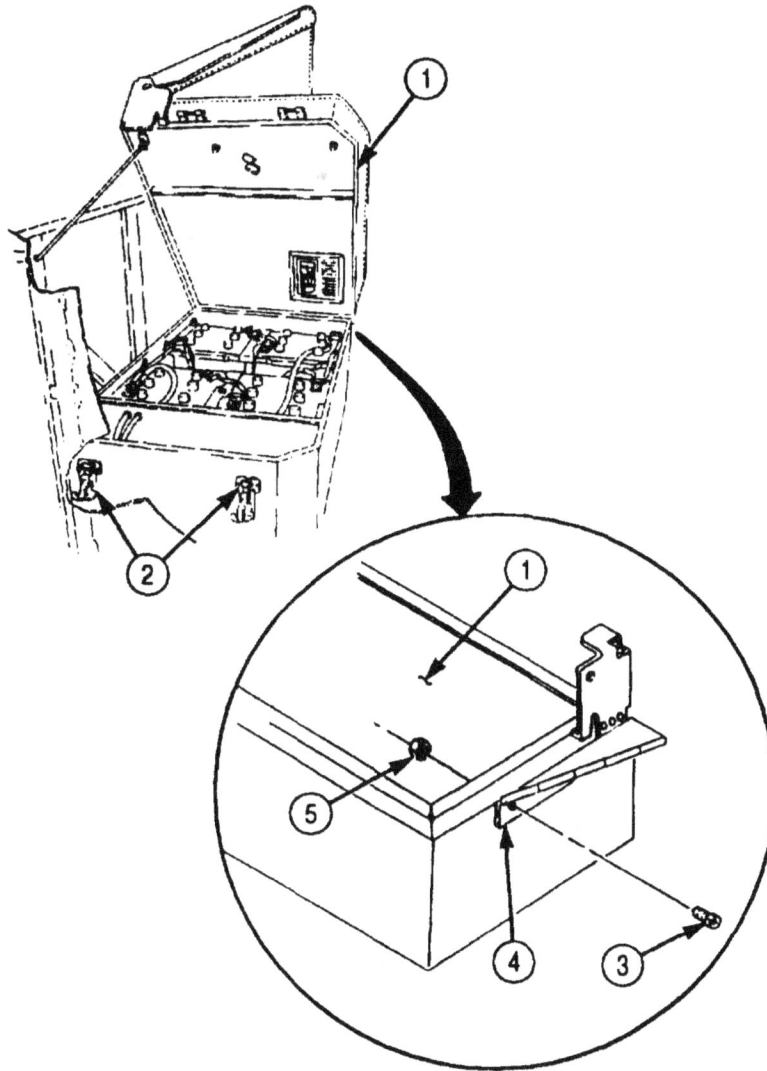

FOLLOW-ON TASK: Install companion seat cushion (para. 3-286).

3-124. BATTERY CABLE AND TERMINAL ADAPTER REPLACEMENT

THIS TASK COVERS:

a. Removal   b. Installation

INITIAL SETUP:

APPLICABLE MODELS
All

TOOLS
General mechanic's tool kit (Appendix E, Item 1)

REFERENCES (TM)
TM 9-2320-272-10
TM 9-2320-272-24P
TM 9-6140-200-14

EQUIPMENT CONDITION
- Parking brake set (TM 9-2320-272-10).
- Battery ground cables disconnected (para. 3-126).

GENERAL SAFETY INSTRUCTIONS
- Remove all jewelry before performing procedure.
- Ensure batteries are clamped down, rubber boots are installed, clamps are well down on battery posts, and all battery cables lie flat against top of batteries.

WARNING

Remove all jewelry. If jewelry or disconnected battery ground cable contacts battery terminal, a direct short will result and may cause injury to personnel.

NOTE

- All battery cables and terminal adapters are replaced the same way.
- Tag cables for installation.

a. Removal

1. Remove nut (1) and screw (6) from terminal adapter (4).
2. Remove terminal adapter (4) and rubber boot (8) from battery post (7).
3. Remove nut (2), screw (5), and two battery cables (3) from terminal adapter (4).

NOTE

Refer to TM 9-6140-200-14 for inspection and service of battery cables and adapters.

b. Installation

WARNING

When performing battery maintenance, ensure batteries are seated and clamped down, all rubber boots are installed, clamps are well down on battery posts, and all battery cables lie flat against top of batteries. Failure to do so may result in injury to personnel or damage to equipment.

NOTE

When installing one cable to an adapter, place the cable under bolt head. When installing two cables, place one cable on each side of adapter.

3-124. BATTERY CABLE AND TERMINAL ADAPTER REPLACEMENT (Contd)

1. Install rubber boot (8) on battery post (7).
2. Install two cables (3) on terminal adapter (4) with screw (5) and nut (2).
3. Install terminal adapter (4) on battery post (7) and with screw (6) and nut (1).

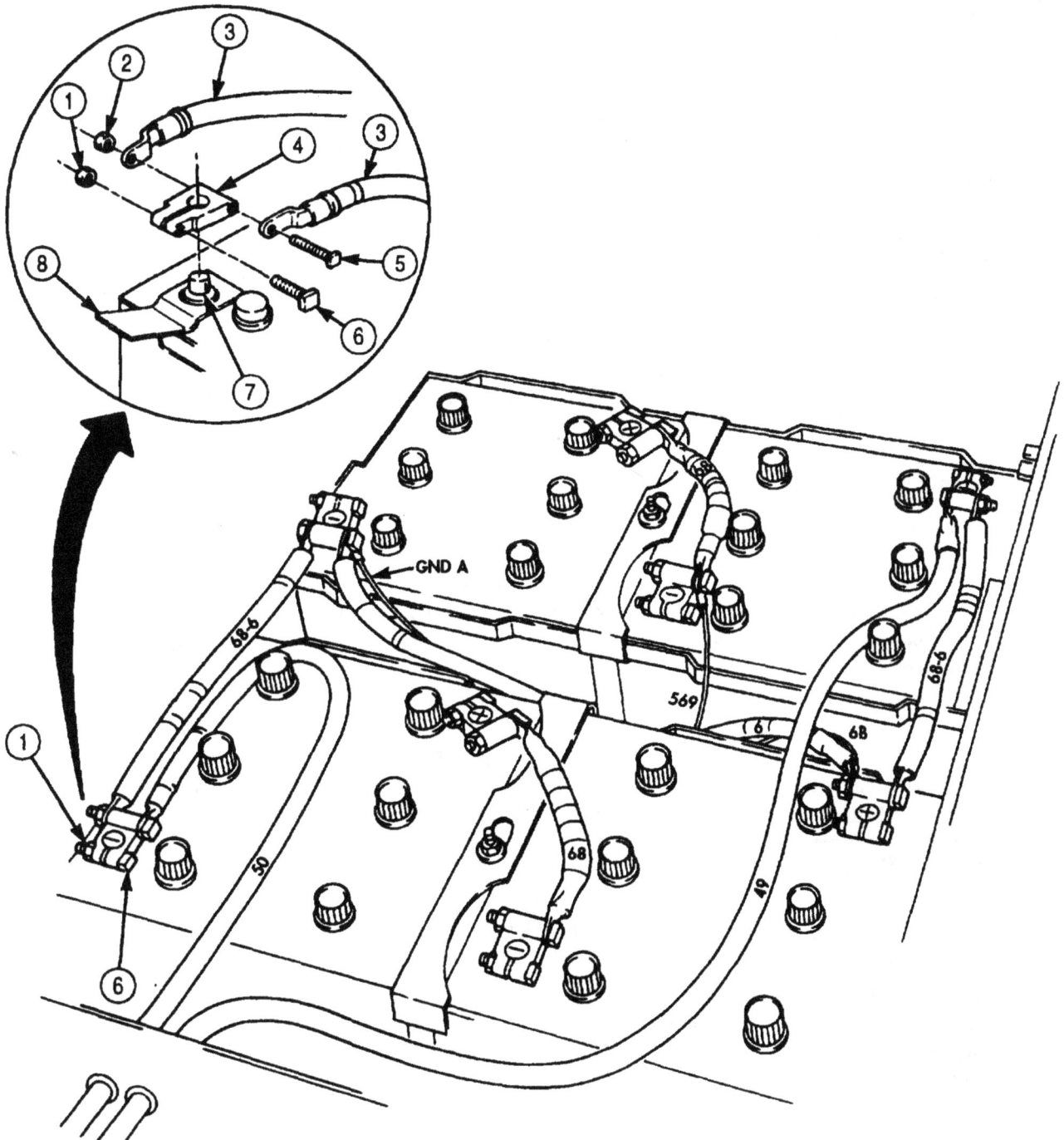

FOLLOW-ON TASK: Connect battery ground cables (para. 3-126).

3-125. BATTERY MAINTENANCE

THIS TASK COVERS:

a. Removal
b. Inspection and Cleaning

c. Installation

INITIAL SETUP:

APPLICABLE MODELS
All

TOOLS
General mechanic's tool kit (Appendix E, Item 1)

MATERIALS/PARTS
Ten lockwashers (Appendix D, Item 354)
Rags (Appendix C, Item 58)
Sodium bicarbonate (Appendix C, Item 69)

REFERENCES (TM)
TM 9-2320-272-10
TM 9-2320-272-24P
TM 9-6140-200-14

EQUIPMENT CONDITION
- Parking brake set (TM 9-2320-272-10).
- Battery ground cables disconnected (para. 3-126).

GENERAL SAFETY INSTRUCTIONS
Remove all jewelry before performing procedure.

WARNING

Remove all jewelry when working on electrical circuits. Jewelry coming in contact with electrical circuits may produce a short circuit, causing extreme heat, explosions, and fling particles of metal. Failure to comply may result in injury to personnel.

a. Removal

1. Loosen six nuts (4) and screws (5) on terminal adapters (2).

NOTE
Tag cables for installation.

2. Remove six terminal adapters (2) and eight rubber boots (3) from four batteries (1).
3. Remove ten nuts (6), lockwashers (7), and washers (8) from battery tiedown bolts (10). Discard lockwashers (7).
4. Remove two battery tiedowns (11) from battery box (9).

NOTE
Assistant will help with step 5.

5. Remove four batteries (1) from battery box (9).
6. Remove ten battery tiedown bolts (10) from battery box (9).

b. Inspection and Cleaning

NOTE
Refer to TM 9-6140-200-14 for battery inspection and service.

Inspect battery box (9) for corrosion and acid deposits. If found:

a. Apply sodium bicarbonate and water solution to inside of battery box.
b. Let solution set for five minutes.
c. Rinse with clean water.
d. Wipe dry with clean rag.

3-125. BATTERY MAINTENANCE (Contd)

c. Installation

1. Install ten battery tiedown bolts (10) in battery box (9).

NOTE

Assistant will help with step 2.

2. Lower four batteries (1) into battery box (9).
3. Install two battery tiedowns (11) over ten bolts (10) with washers (8), new lockwashers (7), and nuts (6).
4. Install eight rubber boots (3) on batteries (1).
5. Install six terminal adapters (2) on batteries (1) and tighten screws (5) and nuts (4).

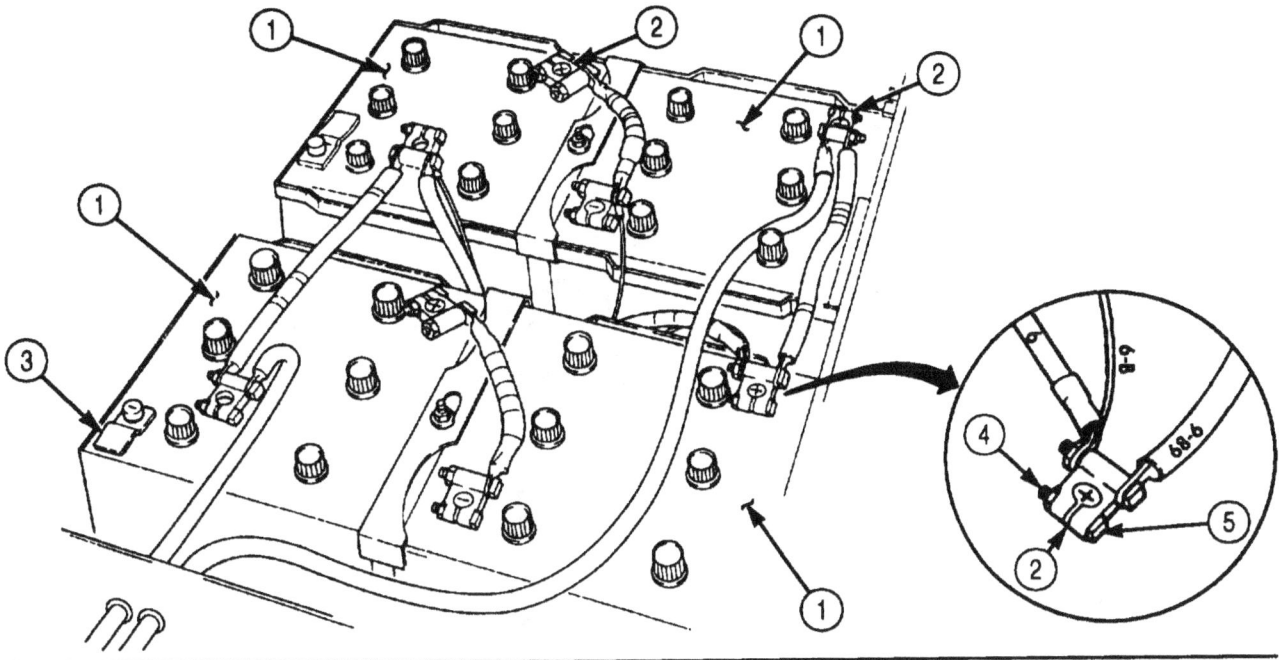

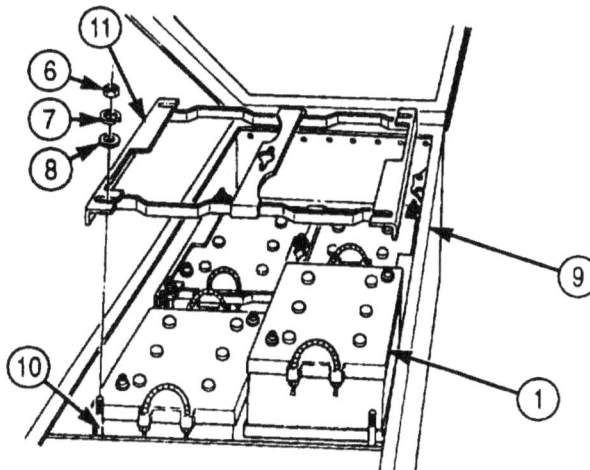

FOLLOW-ON TASK: Connect battery ground cables (para. 3-126).

3-126. BATTERY GROUND CABLE MAINTENANCE

THIS TASK COVERS:

a. Disconnection   b. Connection

INITIAL SETUP:

APPLICABLE MODELS

All

TOOLS

General mechanic's tool kit (Appendix E, Item 1)

REFERENCES (TM)

TM 9-2320-272-10
TM 9-2320-272-24P

EQUIPMENT CONDITION

1 Parking brake set (TM 9-2320-272-10).
1 Battery box cover raised and secured
   (TM 9-2320-272-10).

GENERAL SAFETY INSTRUCTIONS

- Remove all jewelry before performing procedure.
- Ensure batteries are seated and clamped down, rubber boots are installed, clamps are well down on battery posts and all battery cables be flat on top of batteries.

WARNING

- Remove all jewelry. If jewelry or disconnected battery ground cable contacts battery terminal, a direct short will result and may cause injury to personnel.
- When performing battery maintenance, ensure batteries are seated and clamped down, all rubber boots are installed, clamps are well down on battery posts, and all battery cables lie flat against top of batteries. Failure to do so may result in injury to personnel or damage to equipment.

a. Disconnection

1. Loosen nut (5) and screw (3) on terminal adapter (6).

NOTE

Tag all terminal adapters and rubber boots for installation.

2. Remove terminal adapter (6) and rubber boot (4) from battery (7).
3. Loosen nut (10) and screw (2) on terminal adapter (1).
4. Remove terminal adapter (1) and rubber boot (9) from battery (8).

b. Connection

1. Install rubber boots (4) and (9) on batteries (7) and (8).
2. Install terminal adapter (1) on battery (8) and tighten screw (2) and nut (10).
3. Install terminal adapter (6) on battery (7) and tighten screw (3) and nut (5).

3-126. BATTERY GROUND CABLE MAINTENANCE (Contd)

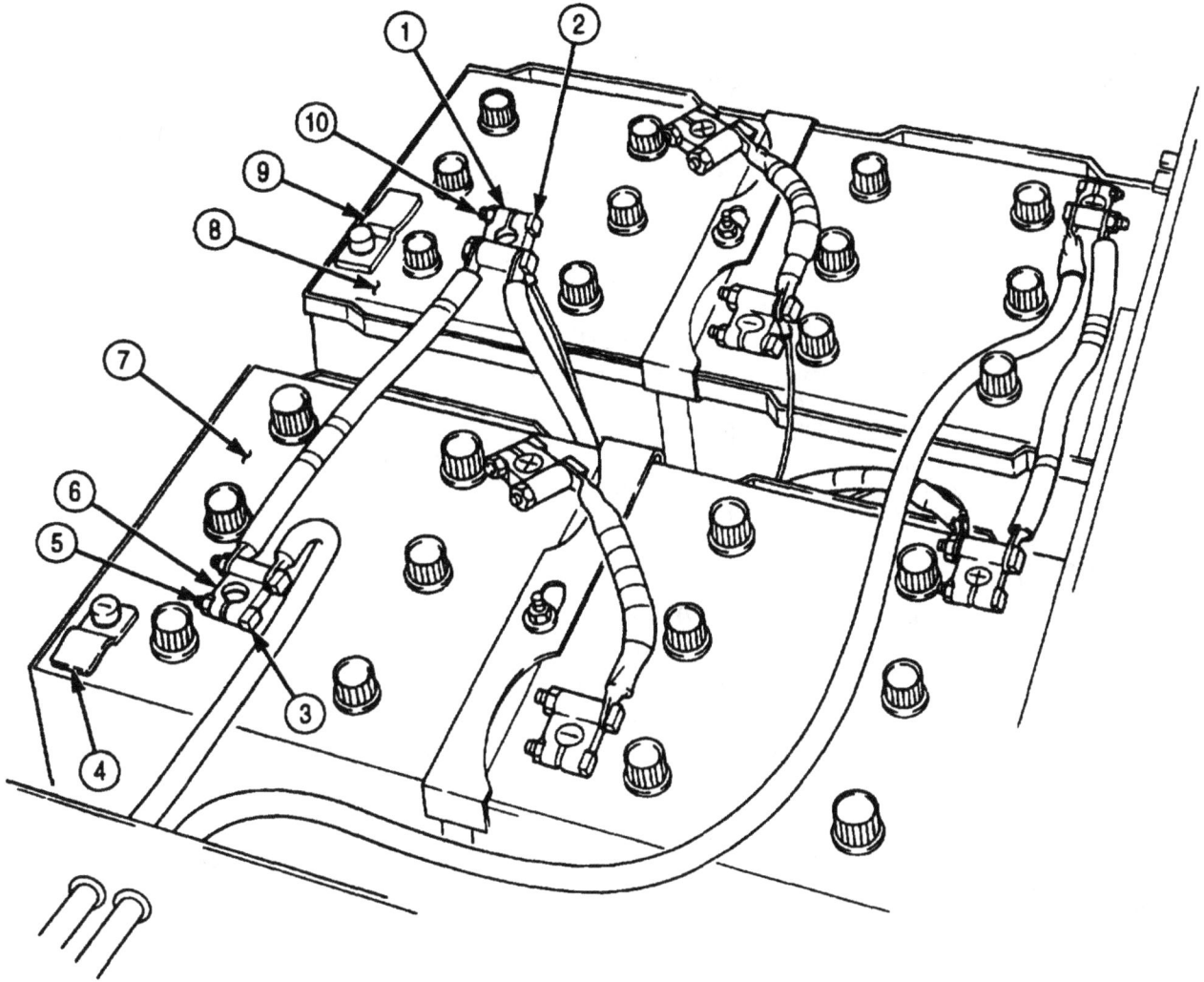

FOLLOW-ON TASK: Lower battery box cover and secure (TM 9-2320-272-10).

3-127. SLAVE RECEPTACLE REPLACEMENT

THIS TASK COVERS:

a. Removal   b. Installation

INITIAL SETUP:

APPLICABLE MODELS
All

TOOLS
General mechanic's tool kit (Appendix E, Item 1)

MATERIALS/PARTS
Two lockwashers (Appendix D, Item 354)

REFERENCES (TM)
TM 9-23201272-10
TM 9-2320-272-24P

EQUIPMENT CONDITION
- Parking brake set (TM 9-2320-272-10).
- Battery ground cables disconnected (para. 3-126).

GENERAL SAFETY INSTRUCTIONS
- Remove all jewelry before performing procedure.
- Disconnect battery cables before removing slave receptacles.

WARNING

- Remove all jewelry. If jewelry or disconnected battery ground cable contacts battery terminal, a direct short will result and may cause injury to personnel.
- Do not remove slave receptacle before disconnecting battery ground cables. If energized battery cable contacts cab, a direct short will result and may cause injury to personnel.

a. Removal

1. Remove four nuts (7), screws (10), rope (9), and cover (11) from cab (6).

NOTE

Insulating hose remains inside cab.

2. Pull slave receptacle (12) until battery cables (3) and (8) are exposed.
3. Remove two screws (1), lockwashers (2), and battery cables (3) and (8) from slave receptacle (12). Discard lockwashers (2).
4. Remove insulator (5) and gasket (4) from slave receptacle (12).
5. Pull cables (3) and (8) clear of insulating hose (13) and remove from battery box (14).

b. Installation

1. Route cables (3) and (8) through insulating hose (13) and battery box (14).
2. Pull cables (3) and (8) through insulating hose (13) until exposed outside cab (6).
3. Install gasket (4) and insulator (5) on slave receptacle (12).
4. Install battery cables (3) and (8) on slave receptacle (12) with two new lockwashers (2) and screws (1).
5. Insert slave receptacle (12) in cab (6) and position insulating hose (13) over screws (1). Install slave receptacle (12) with three screws (10) and nuts (7).
6. Install cover (11) and rope (9) on slave receptacle (12) with screw (10) and nut (7).

3-127. SLAVE RECEPTACLE REPLACEMENT (Contd)

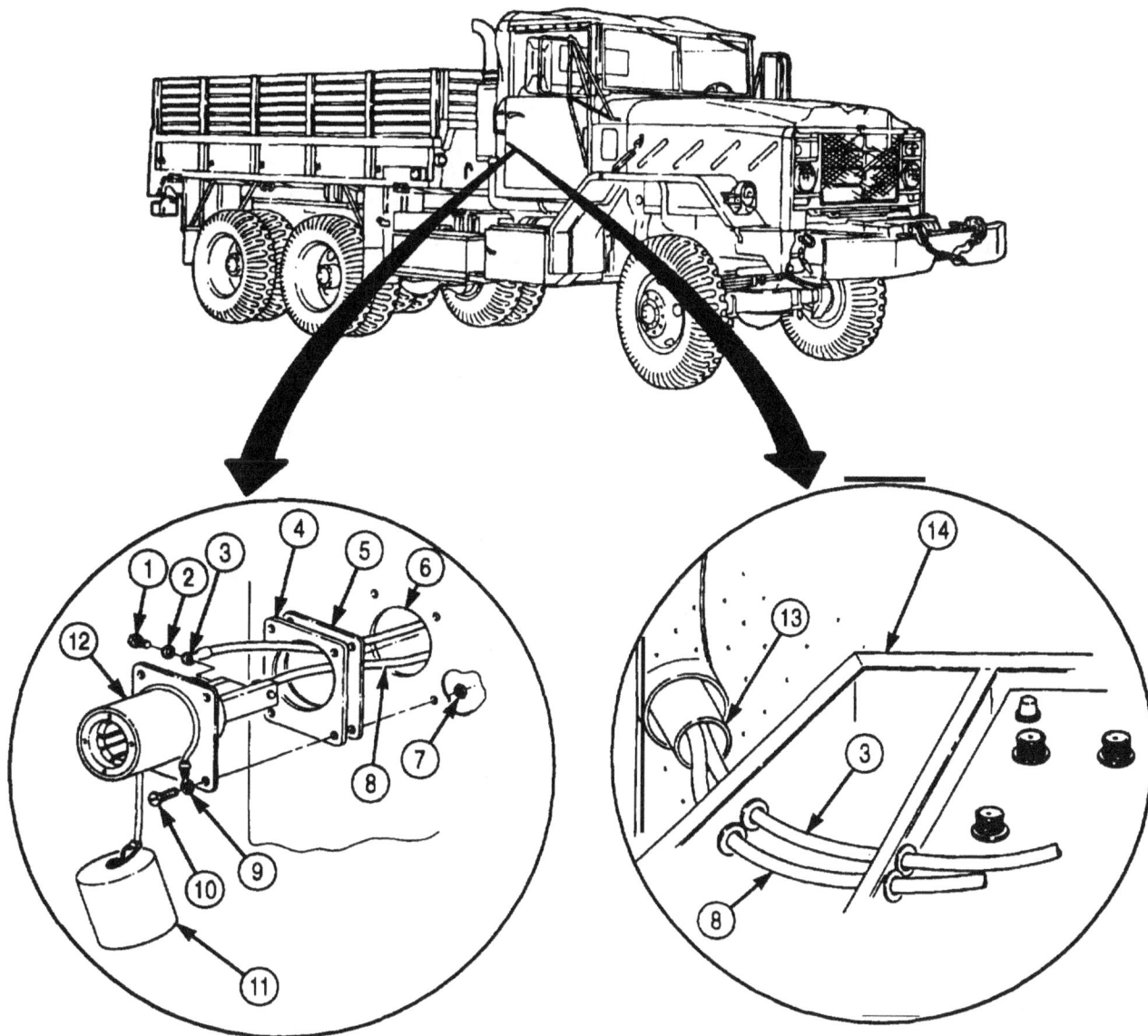

FOLLOW-ON TASK: Install battery ground cables (para. 3-126).

3-128. BATTERY BOX REPLACEMENT

THIS TASK COVER:

a. Removal   **b. Installation**

INITIAL SETUP:

APPLICABLE MODELS All	**REFERENCES (TM)** TM 9-2320-272-10 TM 9-2320-272-24P
TOOLS General mechanic's tool kit (Appendix E, Item 1) Torque wrench (Appendix D, Item 146)	**EQUIPMENT CONDITION** • Parking brake set (TM 9-2320-272-10). • Battery box cover removed (para. 3-123).
MATERIALS/PARTS Four lockwashers (Appendix D, Item 358)	• Battery cable terminal adapters removed (para. 3-124). • Batteries removed (para. 3-125).

a. Removal

1. Pull slave receptacle cables (12) and (13) through grommets (14) and cab (11).
2. Loosen two clamps (5) on hose (2).
3. Remove hose (2) from battery box (6) and cab (11).
4. Remove wood battery support blocks (1) from battery box (6).
5. Remove four screws (3) and lockwashers (4) from battery box (6). Discard lockwashers (4).
6. Push battery cables (7) through grommets (8). Remove battery box (6) from cab (11).
7. Remove four grommets (14) and two grommets (8) from battery box (6).

b. Installation

1. Install four grommets (14) and two grommets (8) on battery box (6).
2. Align vent tube (9) with hole in cab floor (10) and install battery box (6) on cab (11) with cable grommets (8) aligned with cables (7).
3. Pull battery cables (7) through grommets (8).
4. Secure battery box (6) to cab (11) with four screws (3) and new lockwashers (4). Tighten screws (3) to 25 lb-ft (34 N•m).
5. Place wood battery support blocks (1) in battery box (6).
6. Install hose (2) on battery box (6) and cab (11) with two clamps (5).
7. Pull slave receptacle cables (13) and (12) through grommets (14).

3-128. BATTERY BOX REPLACEMENT (Contd)

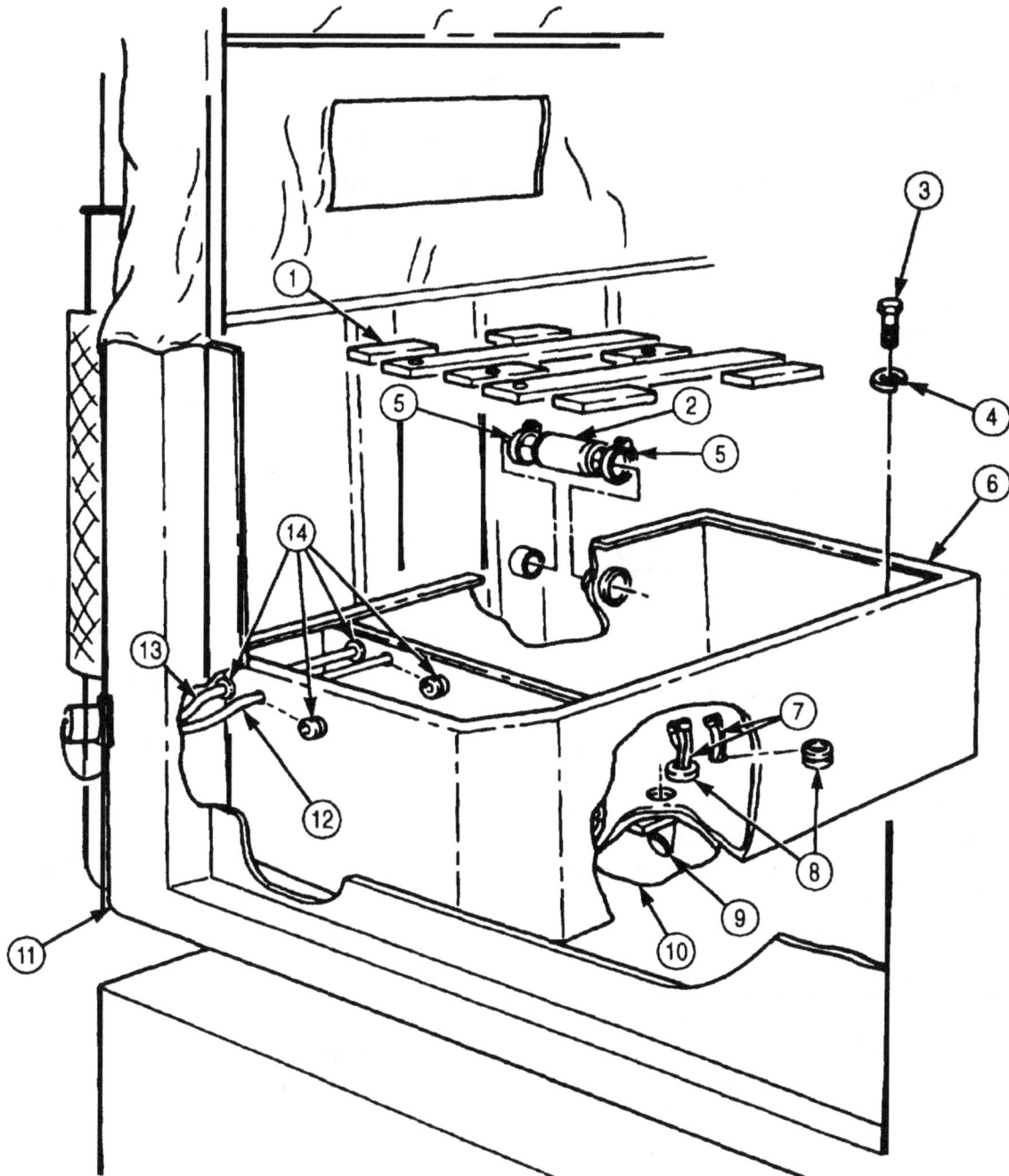

FOLLOW-ON TASKS: • Install batteries (para. 3-125).
                  • Install battery cable terminal adapters (para. 3-124).
                  • Install battery box cover (para. 3-123).

3-129. GROUND STRAP REPLACEMENT

THIS TASK COVERS:

a. Ground Strap (M939/A1) Removal
b. Ground Strap (M939/A1) Installation

c. Ground Strap (M939A2) Removal
d. Ground Strap (M939A2) Installation

INITIAL SETUP:

APPLICABLE MODELS
All

TOOLS
General mechanic's tool kit (Appendix E, Item 1)

MATERIALS/PARTS
Three lockwashers (M939/A1)
  (Appendix D, Item 379)
Lockwasher (M939/A1) (Appendix D, Item 372)
Lockwasher (M939/A1) (Appendix D, Item 382)
Locknut (M939/A1) (Appendix D, Item 291)
Three lockwashers (M939A2)
  (Appendix D, Item 356)
Four lockwashers (M939A2)
  (Appendix D, Item 372)
Two lockwashers (M939A2)
  (Appendix D, Item 360)

REFERENCES (TM)
TM 9-2320-272-10
TM 9-2320-272-24P

EQUIPMENT CONDITION
- Parking brake set (TM 9-2320-272-10).
- Hood raised and secured (TM 9-2320-272-10).

a. Ground Strap (M939/A1) Removal

1. Remove screw (1), washer (2), ground strap (6), lockwasher (4), and clamp (3) from firewall (5). Discard lockwasher (4).
2. Remove screw (7), wire (8), ground strap (6), and lockwasher (9) from engine (10). Discard lockwasher (9).
3. Remove screw (11), two washers (12), ground strap (15), and lockwasher (14) from air compressor (13). Discard lockwasher (14).
4. Remove locknut (18), clamp (19), screw (16), lockwasher (20), and ground strap (15) from frame rail (17). Discard locknut (18) and lockwasher (20).
5. Remove screw (24), washer (23), ground strap (22), and lockwasher (21) from engine (10). Discard lockwasher (21).
6. Remove nut (28), washer (27), wires (29) and (26), and ground strap (22) from engine starter (25).

b. Ground Strap (M939/A1) Installation

1. Install ground strap (22) on engine starter (25) with wires (26) and (29), washer (27), and nut (28).
2. Install new lockwasher (21), ground strap (22), washer (23), and screw (24) on engine (10).
3. Install screw (16), new lockwasher (20), ground strap (15), clamp (19), and new locknut (18) on frame rail (17).
4. Install new lockwasher (14), ground strap (15), two washers (12), and screw (11) on air compressor (13).
5. Install new lockwasher (9), ground strap (6), wire (8), and screw (7) on engine (10).
6. Install new lockwasher (4), ground strap (6), clamp (3), washer (2), and screw (1) on firewall (5).

3-129. GROUND STRAP REPLACEMENT (Contd)

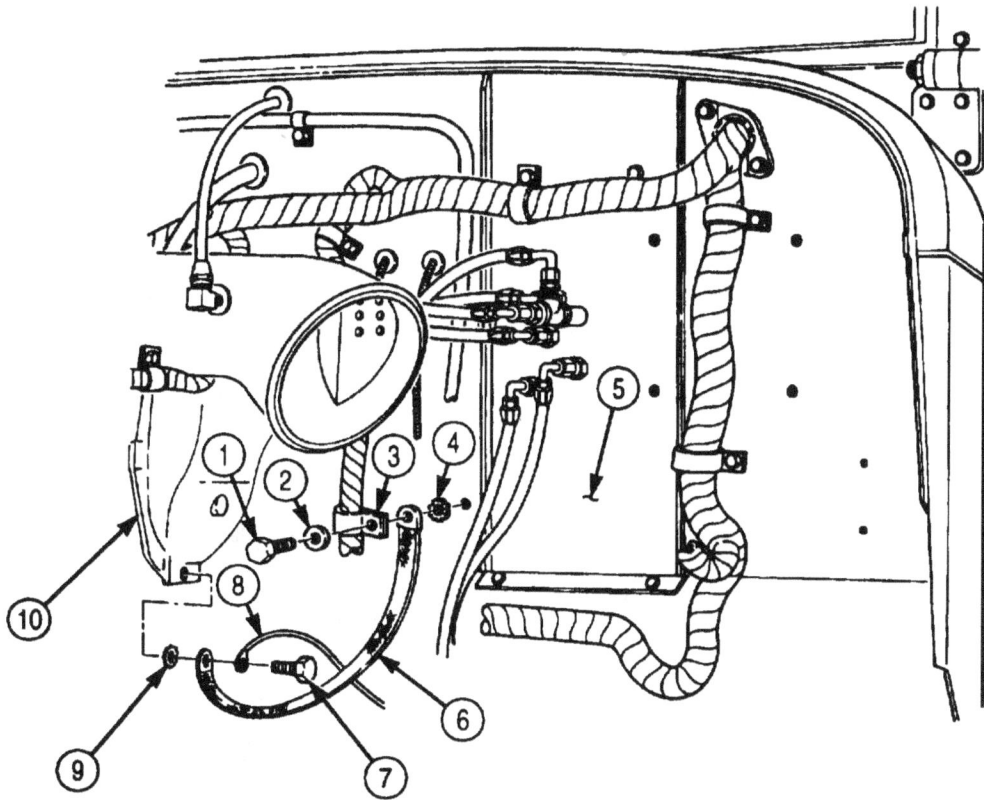

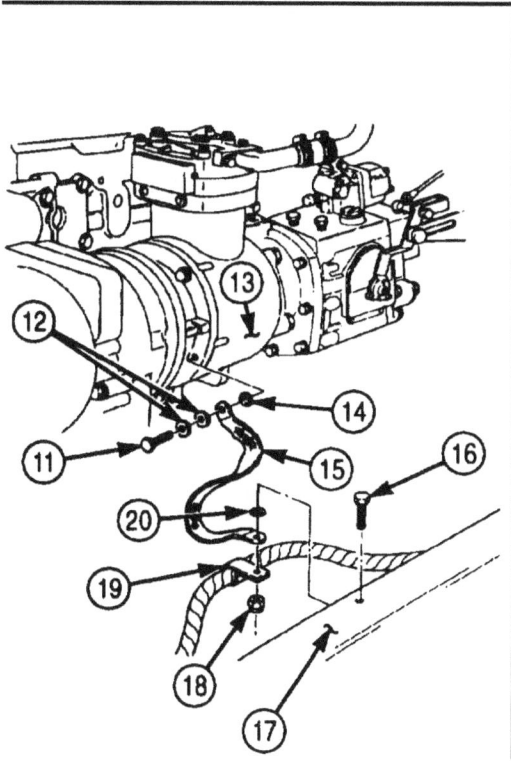

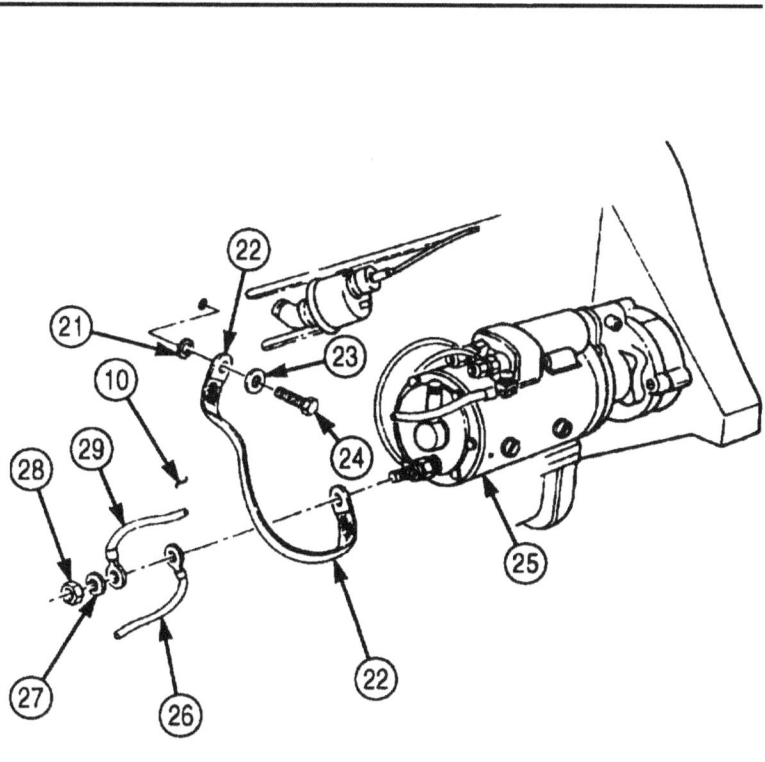

3-129. GROUND STRAP REPLACEMENT (Contd)

c. Ground Strap (M939A2) Removal

1. Remove nut (1), washer (2), lockwasher (12), ground strap (5), lockwasher (11), and screw (9) from frame rail (10). Discard lockwashers (12) and (11).
2. Remove screw (6), washer (7), ground strap (5), three wires (8), and lockwasher (4) from engine block (3). Discard lockwasher (4).
3. Remove nut (20), lockwasher (21), screw (25), washer (24), ground strap (17), and lockwasher (23) from frame rail (22). Discard lockwashers (21) and (23).
4. Remove nut (14), lockwasher (15), screw (19), washer (18), ground strap (17), and lockwasher (16) from hood (13) Discard lockwashers (15) and (16).
5. Remove screw (35), lockwasher (33), clamp (36), ground strap (29), and bracket (34) from firewall (32). Discard lockwasher (33).
6. Remove screw (31), washer (30), ground strap (29), lockwasher (28), temperature sensor (27), and washer (26) from engine block (3). Discard lockwasher (28).

d. Ground Strap (M939A2) Installation

1. Install washer (26), temperature sensor (27), new lockwasher (28), ground strap (29), washer (30), and screw (31) on engine block (3).
2. Install bracket (34), ground strap (29), clamp (36), new lockwasher (33), and screw (35) on firewall (32).
3. Install new lockwasher (16), ground strap (17), washer (18), screw (19), new lockwasher (15), and nut (14) on hood (13).
4. Install new lockwasher (23), ground strap (17), washer (24), screw (25), lockwasher (21), and nut (20) on frame rail (22).
5. Install new lockwasher (4), three wires (8), ground strap (5), washer (7), and screw (6) on engine block (3).
6. Install new lockwasher (11), ground strap (5), new lockwasher (12), screw (9), washer (2), and nut (1) on frame rail (10).

3-129. GROUND STRAP REPLACEMENT (Contd)

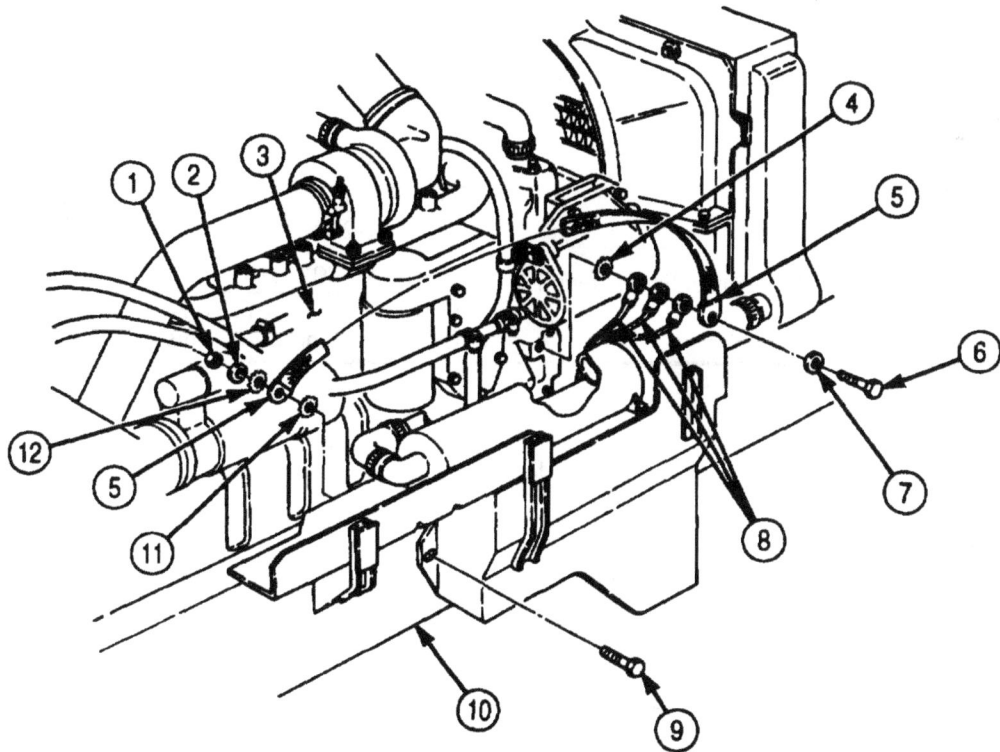

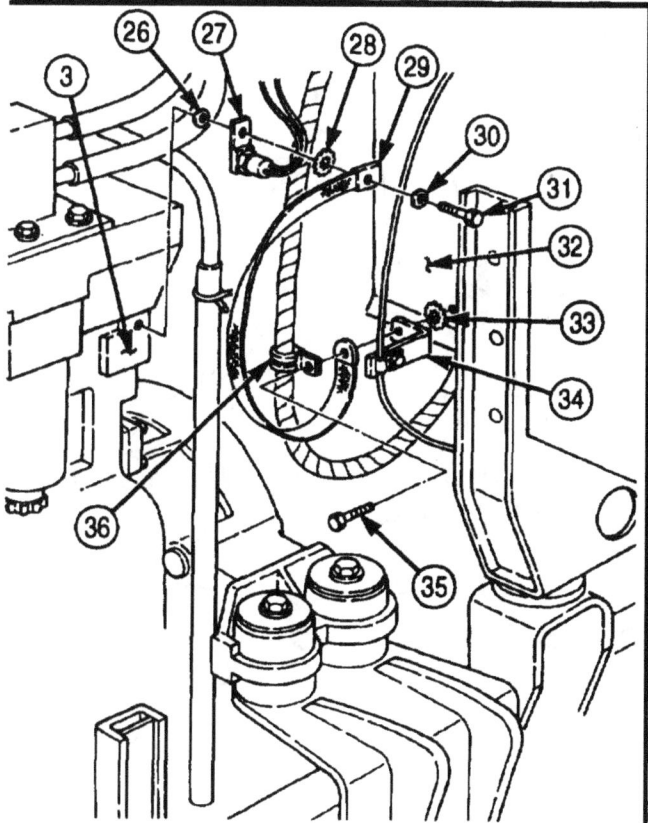

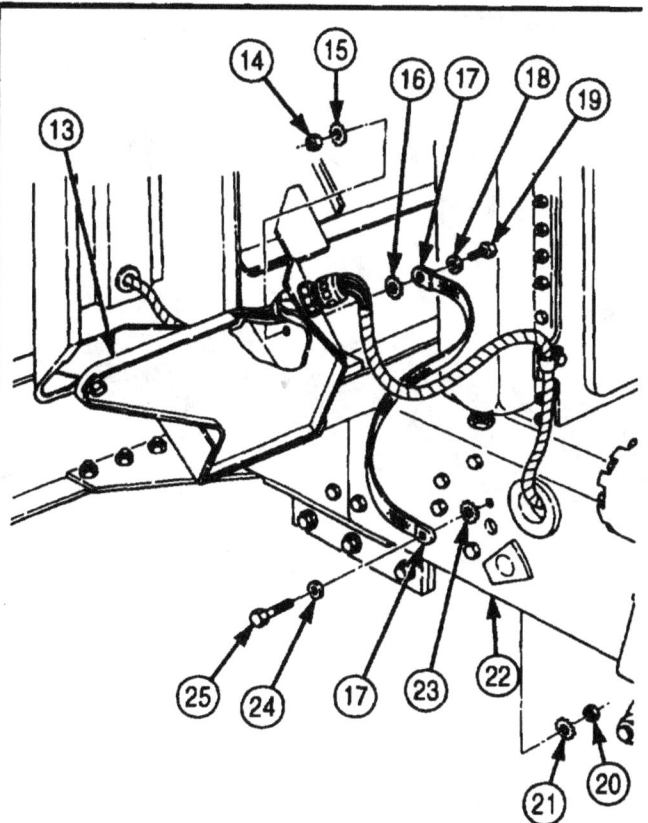

3-130. INSTRUMENT CLUSTER HARNESS REPLACEMENT

THIS TASK COVERS:

a. Removal   b. Installation

INITIAL SETUP:

APPLICABLE MODELS
All

Tools
General mechanic's tool kit (Appendix E, Item 1)

REFERENCES (TM)
TM 9-2320-272-10
TM 9-2320-272-24P

EQUIPMENT CONDITION
- Parking brake set (TM 9-2320-272-10).
- Battery ground cables disconnected (para. 3-126).

---

a. Removal

Disconnect five leads (5), nut (3), washer (2), cable assembly (4), and washer (1) from speedometer stud (6).

b. Installation

Install cable assembly (4) on speedometer stud (6) with washer (1), washer (2), and nut (3) and connect five leads (5).

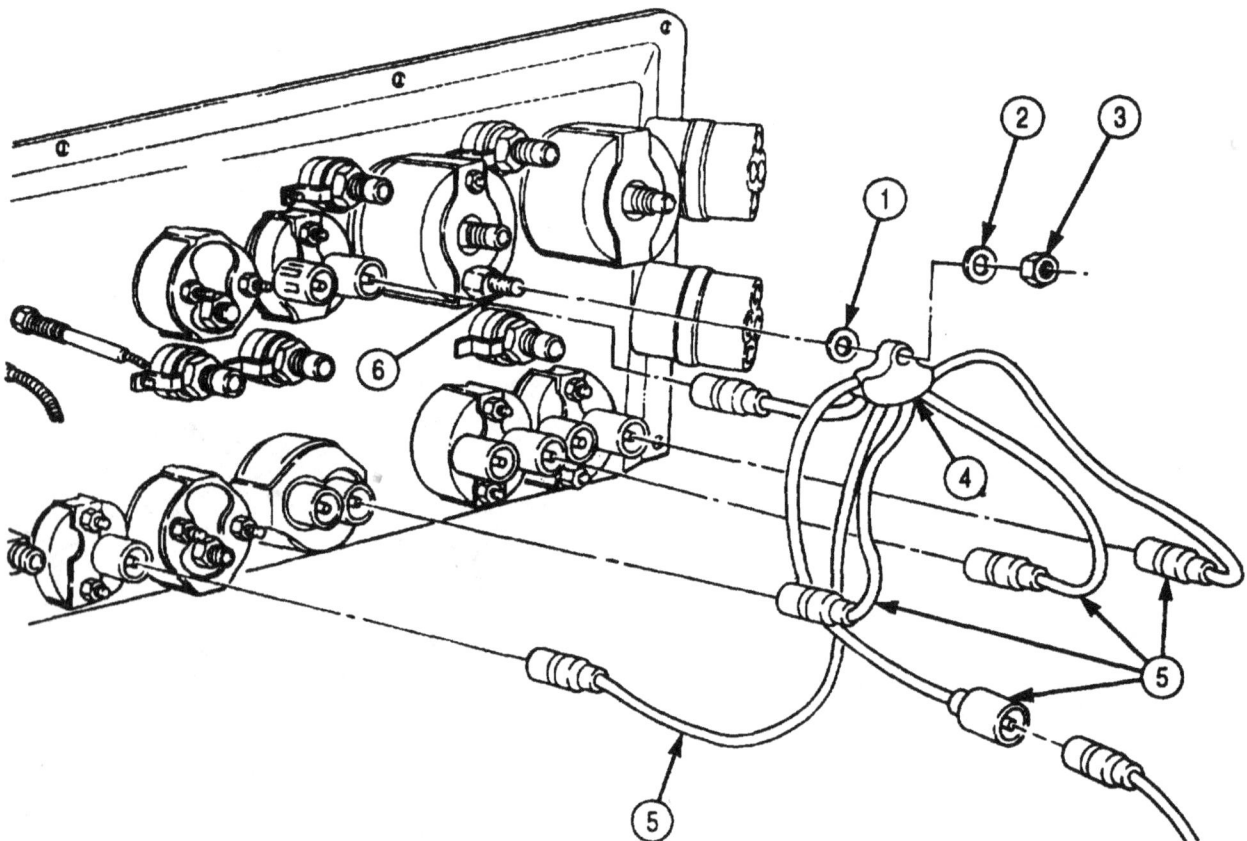

FOLLOW-ON TASK: Connect battery ground cables (para. 3-126).

3-348

3-131. WIRING HARNESS REPAIR

THIS TASK COVERS:

a. Terminal-Type Cable Connector
b. Male Cable Connector
c. Female Connector (With Sleeve)

d. Plug Assembly
e. Receptacle Assembly

INITIAL SETUP:

APPLICABLE MODELS

All

TOOLS

General mechanic's tool kit (Appendix E, Item 1)
Electrical tool kit (Appendix E, Item 40)
Heat gun (Appendix E, Item 62)
Soldering torch kit (Appendix E, Item 126)

MATERIALS/PARTS

Solder (Appendix C, Item 70)

REFERENCES (TM)

TB SIG-222
TM 9-2320-272-10
TM 9-232-272-24P

EQUIPMENT CONDITION

- Parking brake set (TM 9-2320-272-10).
- Battery ground cables disconnected (para. 3-126).

GENERAL SAFETY INSTRUCTIONS

Do not wear jewelry when repairing harnesses.

WARNING

Do not wear jewelry when repairing harnesses. Injury to personnel may result if circuit is suddenly energized.

NOTE

If a wiring harness is damaged beyond repair, notify Direct Support maintenance for replacement.

a. Terminal-Type Cable Connector

1. Strip cable insulation (1) from cable (2) equal to depth of terminal well (4).
2. Slide insulator (3) over cable (2).
3. Insert cable (2) into terminal well (4) and crimp.
4. Slide insulator (3) over crimped end of terminal (5).

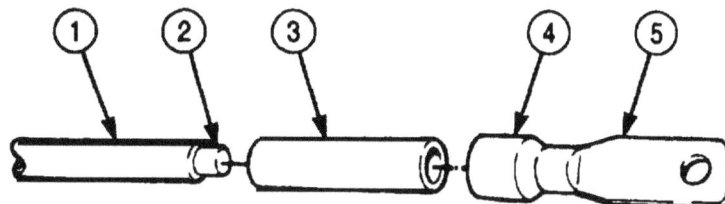

3-131. WIRING HARNESS REPAIR (Contd)

b. Male Cable Connector

1. Strip cable insulation (1) from cable (2) equal to depth of ferrule well (5).
2. Slide shell (3) over cable (2).
3. Insert cable (2) into ferrule well (5) and crimp.
4. Place C-washer (4) over crimped junction at terminal (6).
6. Slide shell (3) over C-washer (4) and terminal (6).

c. Female Cable Connector (With Sleeve)

1. Strip cable insulation (7) from cable (8) equal to depth of terminal well (11).
2. Slide shell (9) and sleeve (10) over cable (8).
3. Insert cable (8) into terminal well (11) and crimp.
4. Slide shell (9) and sleeve (10) over terminal (12).

d. Plug Assembly

NOTE
Refer to TB SIG-222 for soldering instructions.

1. Strip cable insulation (13) from cable (15) equal to depth of solder wells (16) of inserts (17).
2. Pass cable ends (15) through grommet retaining nut (14), grommet (18), and coupling nut (20).
3. Insert cable ends (15) into solder wells (16) of inserts (17) and solder.
4. Slide grommet (18) over inserts (17) and press into shell assembly (19) until seated.
5. Slide retaining nut (14) up cable (15) and install on shell assembly (19).

e. Receptacle Assembly

NOTE
Refer to TB SIG-222 for soldering instructions.

1. Strip cable insulation (21) from cable (23) equal to depth of solder wells (24) of inserts (25).
2. Pass cable ends (23) through grommet retaining nut (22) and grommet (26).
3. Insert cable ends (23) into solder wells (24) of insert (25) and solder.
4. Slide grommet (26) over inserts (25) and press into receptacle assembly (27) until seated.
5. Thread grommet retaining nut (22) into receptacle (27) until seated.

3-131. WIRING HARNESS REPAIR (Contd)

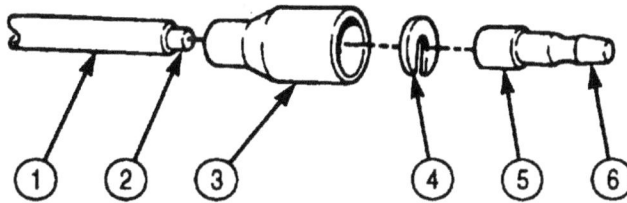

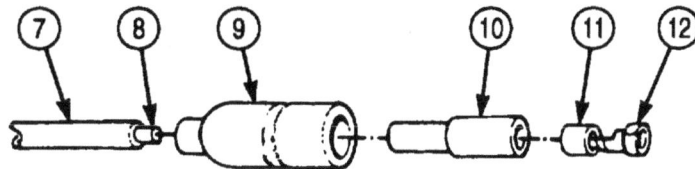

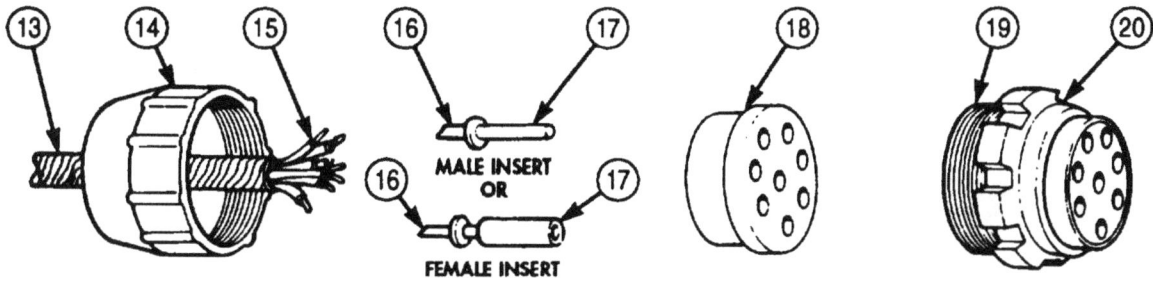

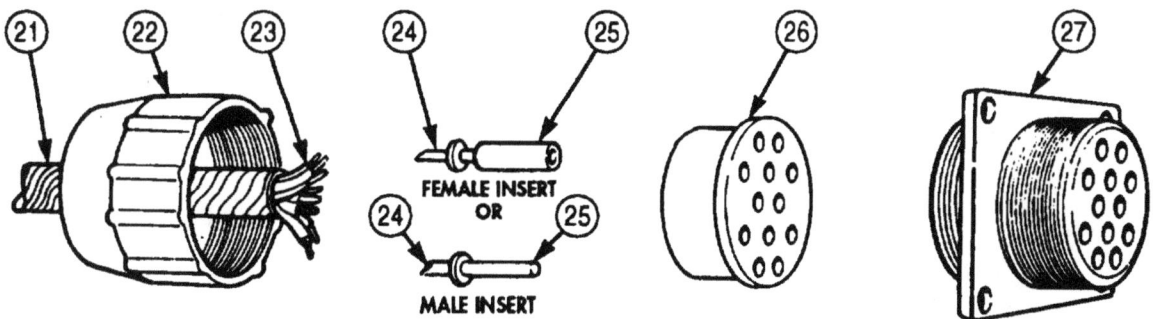

FOLLOW-ON TASK: Connect battery ground cables (para. 3-126).

Section VII. TRANSMISSION AND TRANSFER CASE MAINTENANCE

3-132. TRANSMISSION MAINTENANCE INDEX

3-133. TRANSMISSION OIL SERVICE INSTRUCTIONS

THIS TASK COVERS:

a. Draining Oil
b. Oil Filter Removal
c. Oil Filter Installation

d. Governor Filter Removal
e. Governor Filter Installation
f. Replenishing Oil

INITIAL SETUP:

APPLICABLE MODELS
All

TOOLS
General mechanic's tool kit (Appendix E, Item 1)
Torque wrench Appendix E, Item 144)

MATERIALS/PARTS
Governor filter (Appendix D, Item 249)
Governor filter O-ring (Appendix D, Item 250)
Rubber O-ring (Appendix D, Item 429)
Transmission oil filter (Appendix D, Item 699)
Transmission oil pan gasket (Appendix D, Item 159)
Tiedown strap (Appendix D, Item 690)
Fiber washer (Appendix D, Item 119)
Lubricating oil (Appendix C, Item 49)
Drycleaning solvent (Appendix C, Item 71)

REFERENCES (TM)
LO 9-2320-272-12
TM 9-2320-272-10
TM 9-2320-27224P

EQUIPMENT CONDITION
Parking brake set (TM 9-2320-272-10).

GENERAL SAFETY INSTRUCTIONS
- Exhaust gases can kill. Operate vehicle only in well-ventilated area.
- Store or dispose of used oil properly.
- Keep fire extinguisher nearby when using drycleaning solvent.

WARNING

- Exhaust gases can kill. Operate vehicle only in a well-ventilated area. Failure to do this may result in injury to personnel.
- Accidental or intentional introduction of liquid contaminants into the environment is in violation of state, federal, and military regulations. Refer to Army POL (para. 1-8) for information concerning storage, use, and disposal of these liquids. Failure to do so may result in injury or death.

a. Draining Oil

NOTE
Do not shift transmission through driving gear ranges when warming transmission oil. This is a procedure used only when replenishing transmission oil.

1. Start engine (TM 9-2320-272-10), operate at 700-750 rpm until transmission oil reaches normal operating temperature of 120°-220°F (49°-105°C), then shut off engine.

NOTE
Have drainage container ready to catch oil.

2. Remove drainplug (3) and fiber washer (2) from right rear of transmission oil pan (1) and drain oil. Discard fiber washer (2).

NOTE
Inspect oil for grit, foaminess, and/or milkiness. If present, notify your supervisor.

3. Install new fiber washer (2) and drainplug (3) in transmission oil pan (1).

3-133. TRANSMISSION OIL SERVICE INSTRUCTIONS (Contd)

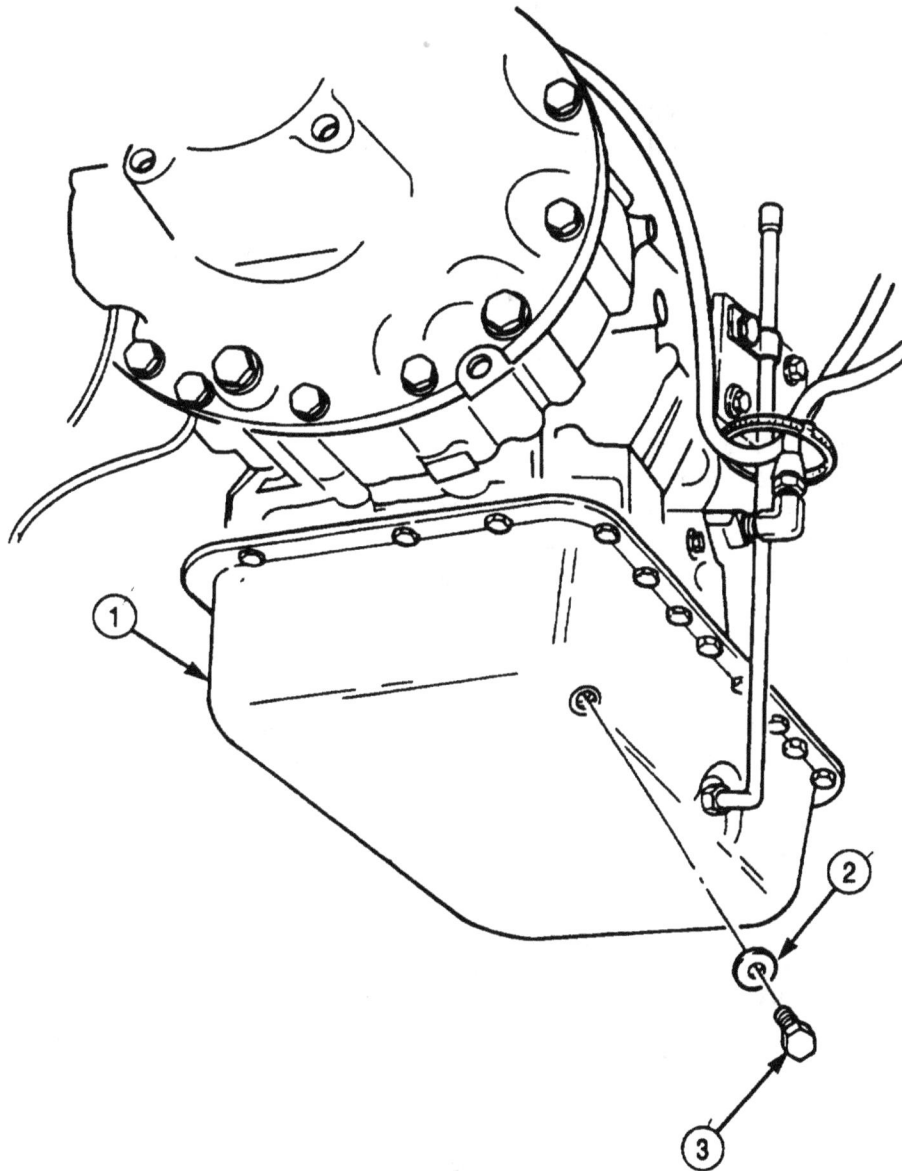

3-133. TRANSMISSION OIL SERVICE INSTRUCTIONS (Contd)

b. Oil Filter Removal

NOTE

Removal of transmission oil filter is basically the same for M939, M939A1, and M939A2 series vehicles. This procedure covers M939 and M939A1 series vehicles only.

1. Open access door (1) in cab floor.

CAUTION

Wipe area around dipstick tube before removal to prevent entry of dirt Damage may occur if dirt or dust enters transmission.

2. Remove screw (5) and washer (4) from transmission (2).
3. Remove tiedown strap (7) from dipstick tube (3) and tube (6). Discard tiedown strap (7).
4. Loosen flare nut (8) and remove dipstick tube (3) from transmission oil pan (14).
5. Remove twenty-one screws (15) from transmission oil pan (14).
6. Remove oil pan (14) and oil pan gasket (13) from transmission (2). Discard gasket (13).
7. Remove screw (16) and oil filter assembly (17) from transmission (2).
8. Remove oil filter assembly (17) from suction tube (12). Discard oil filter assembly (17).

WARNING

Drycleaning solvent is flammable and will not be used near open flame. Use only in well-ventilated areas. Failure to do so may result in injury to personnel.

9. Remove suction tube (12) and rubber O-ring (11) from transmission (2). Discard O-ring (11) and clean suction tube (12) thoroughly with drycleaning solvent.
10. Remove transmission oil cooler filter (para. 3-139).
11. Disconnect transmission oil cooler hoses (18) and (20) from transmission oil filter base (19).

CAUTION

Compressed air source will not exceed 60 psi (41 kPa). Doing so may cause damage to internal components of the transmission.

12. Using compressed air, drain remaining oil from transmission oil cooler hoses (18) and (20).
13. Connect transmission oil cooler hoses (18) and (20) to transmission oil filter base (19).
14. Install transmission oil cooler filter (para. 3-139).

c. Oil Filter Installation

NOTE

Installation of transmission oil filter is basically the same for M939, M939A1, and M939A2 series vehicles. This procedure covers M939 and M939A1 series vehicles only.

1. Position new rubber O-ring (11) on suction tube (12) and slide downward until O-ring (11) contacts suction tube lip (21).
2. Install suction tube (12) in transmission oil input port (10), pressing until O-ring (11) seats on oil input port (10).
3. Install new oil filter assembly (17) on suction tube (12), pressing until oil filter assembly intake grommet (22) contacts suction tube lip (23).
4. Install oil filter assembly (17) on transmission mounting boss (9) with screw (16). Tighten screw (16) 10-15 lb-ft (14-20 N•m).

3-133. TRANSMISSION OIL SERVICE INSTRUCTIONS (Contd)

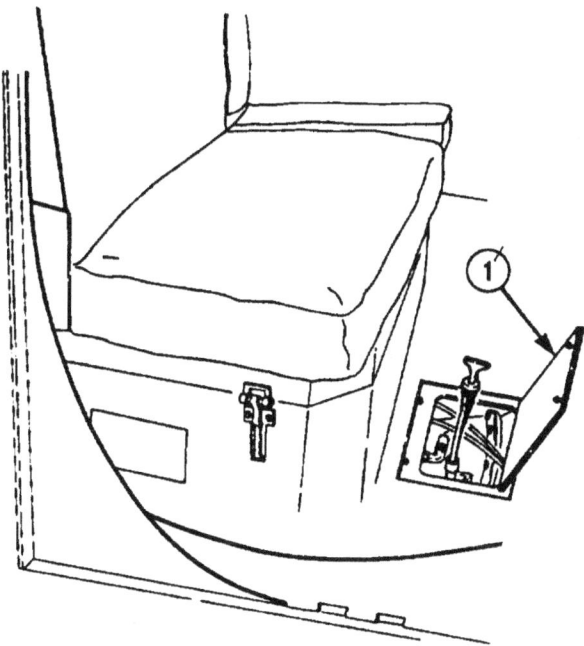

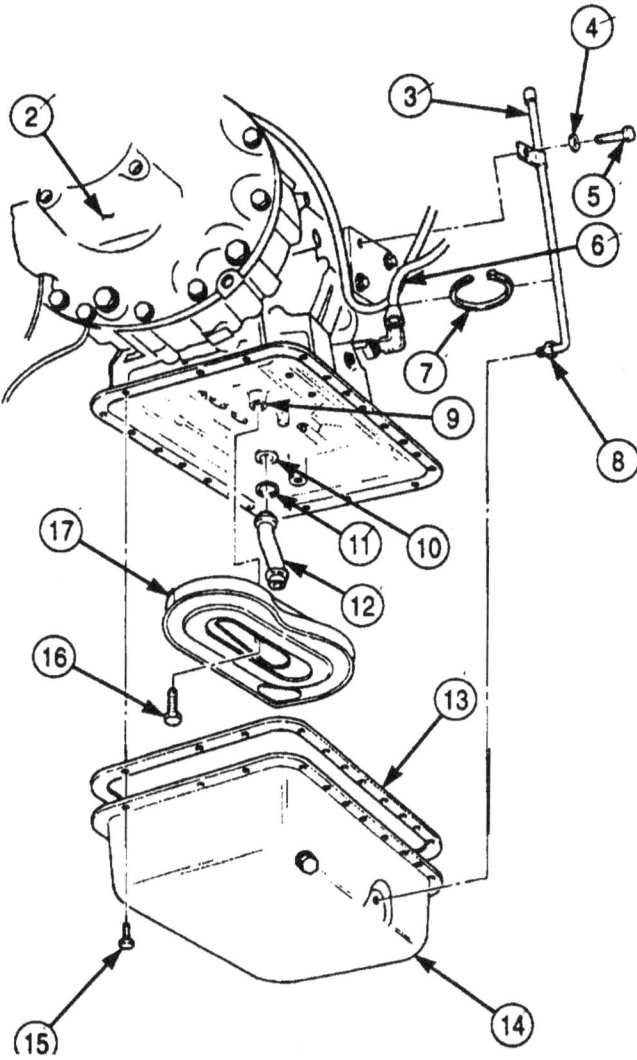

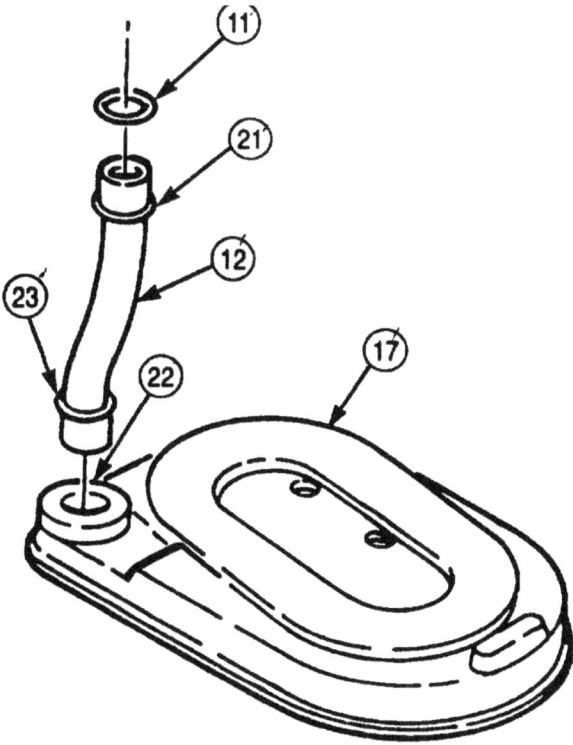

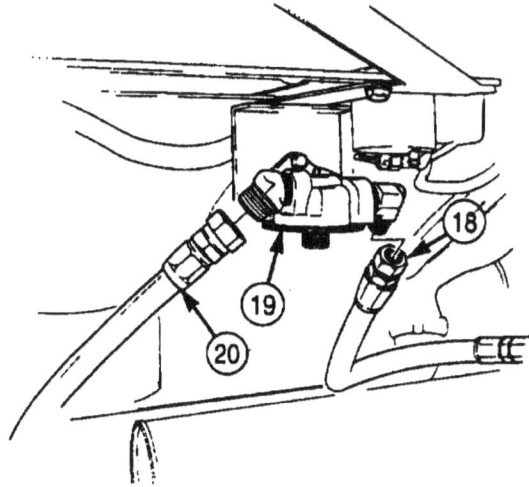

3-133. TRANSMISSION OIL SERVICE INSTRUCTIONS (Contd)

CAUTION

Do not use gasket sealing compound when installing oil pan gasket; oil leakage will result. If necessary, oil or light grease coating may be used to hold oil pan gasket in position during installation.

5. Install new oil pan gasket (2) and oil pan (4) on transmission (1) with twenty-one screws (3).
6. Tighten twenty-one screws (3) 10-15 lb-ft (14-20 N•m) in sequence shown.

NOTE

Due to gasket compression, torque values will be lost and screws must be retightened.

7. After oil pan gasket (2) is seated, loosen and retighten screws (3) 5 lb-ft (7 N-m) in sequence shown.
8. Install dipstick tube (5) on transmission (1) with washer (8), clamp (6), and screw (7).
9. Install dipstick tube (5) in oil pan port (12) with flare nut (11). Tighten flare nut (11) 10-25 lb-ft (14-34 N•m).
10. Install new tiedown strap (9) on dipstick tube (5) and tube (10).

d. Governor Filter Removal

WARNING

Drycleaning solvent is flammable and will not be used near open flame. Use only in well-ventilated areas. Failure to do so may result in injury to personnel.

Remove governor filter plug (13) and O-ring (14) from transmission (1). Discard O-ring (14) and clean plug (13) thoroughly with drycleaning solvent.

CAUTION

Do not pry governor tilter from governor filter housing. Use thin, pliable wire as a hook to reach into the housing to slide out filter. Failure to do this will result in governor filter housing damage.

2. Remove governor filter (16) from governor fiber housing (15). Discard filter (16).

e. Governor Filter Installation

NOTE

Failure to properly install governor filter into governor filter housing will greatly reduce its effectiveness.

1. Install governor filter (16), with open end inserted first, into governor filter housing (16).
2. Install O-ring (14) on governor filter plug (13).
3. Install governor filter plug (13) in filter housing (15). Tighten governor filter plug (13) 15 lb-ft (20 N•m).

3-133. TRANSMISSION OIL SERVICE INSTRUCTIONS (Contd)

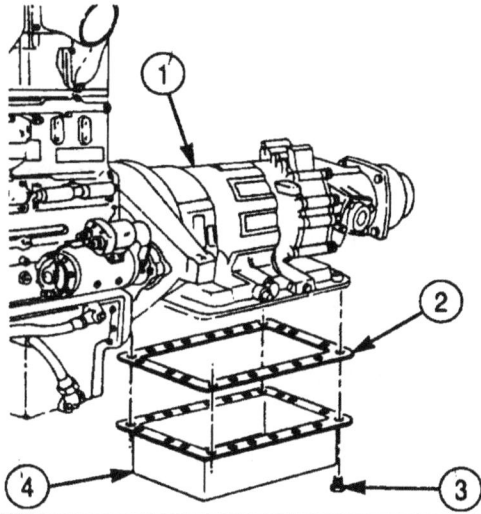

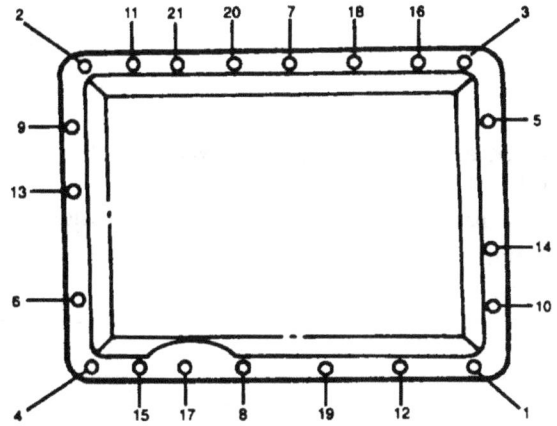

TIGHTENING SEQUENCE

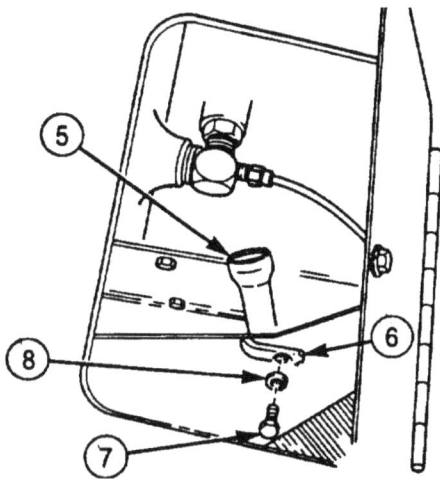

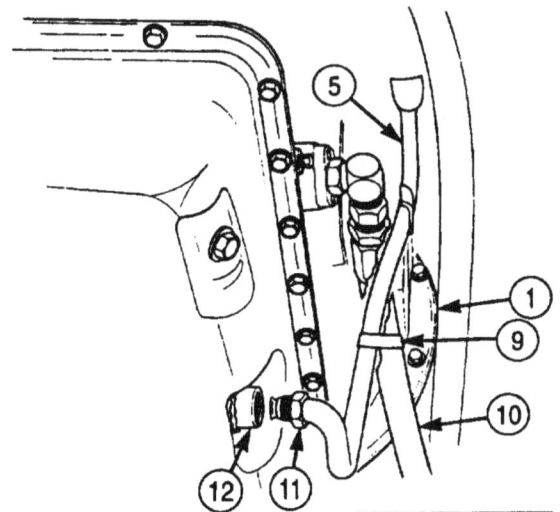

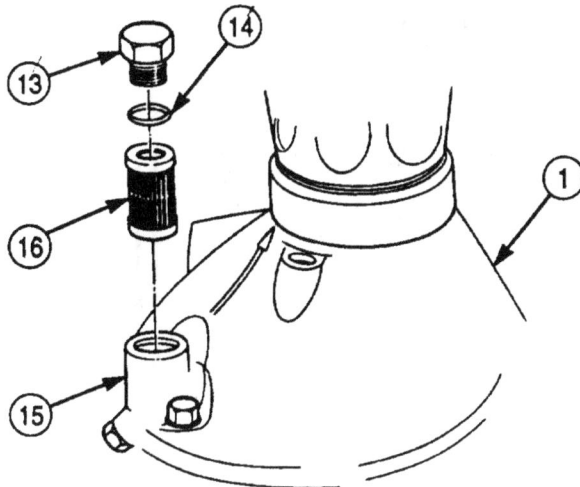

3-133. TRANSMISSION OIL SERVICE INSTRUCTIONS (Contd)

f. Replenishing Oil

NOTE

Refer to LO 9-2320-272-12 for drain and refill capacity and recommended grade of transmission oil.

1. Open access door (2) inside vehicle cab and remove transmission oil dipstick (1) from transmission oil dipstick tube (3).

2. Add recommended quantity of transmission oil in transmission oil dipstick tube (3).

3. Install transmission oil dipstick (1) in dipstick tube (3).

WARNING

Exhaust gases can kill. Operate vehicle only in a well-ventilated area. Failure to do so may result in injury to personnel.

4. Start engine (TM 9-2320-272-10), operate engine at 700-750 rpm until transmission oil reaches normal operating temperature of 120°-220°F (49°-105°C), then shut off engine.

NOTE

Begin shifting transmission through driving ranges only after transmission oil has reached normal operating temperature. Perform shifting procedure for approximately two minutes before returning transmission to neutral and shutting off engine.

5. Shift transmission (4) through driving ranges to allow oil to circulate throughout transmission (4).

6. Check transmission oil and fill to proper level (LO 9-2320-272-12).

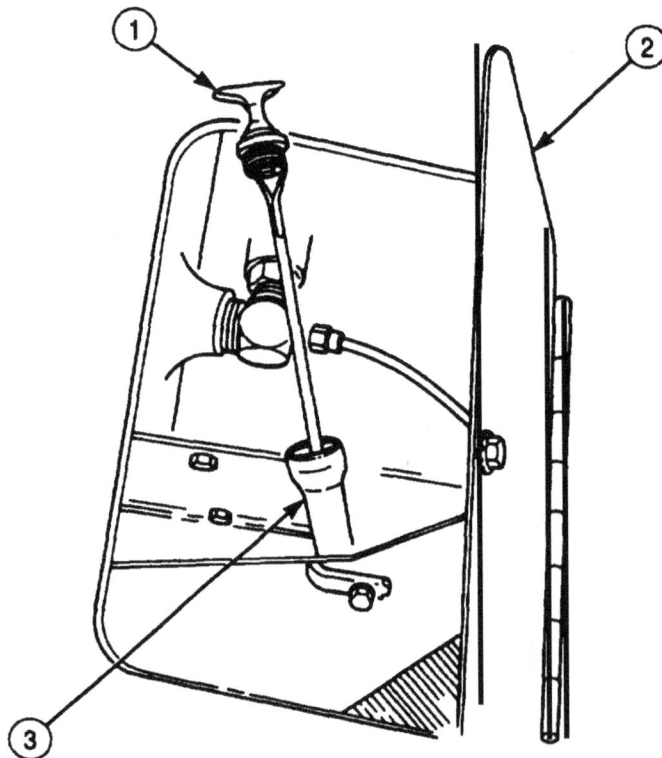

3-133. TRANSMISSION OIL SERVICE INSTRUCTIONS (Contd)

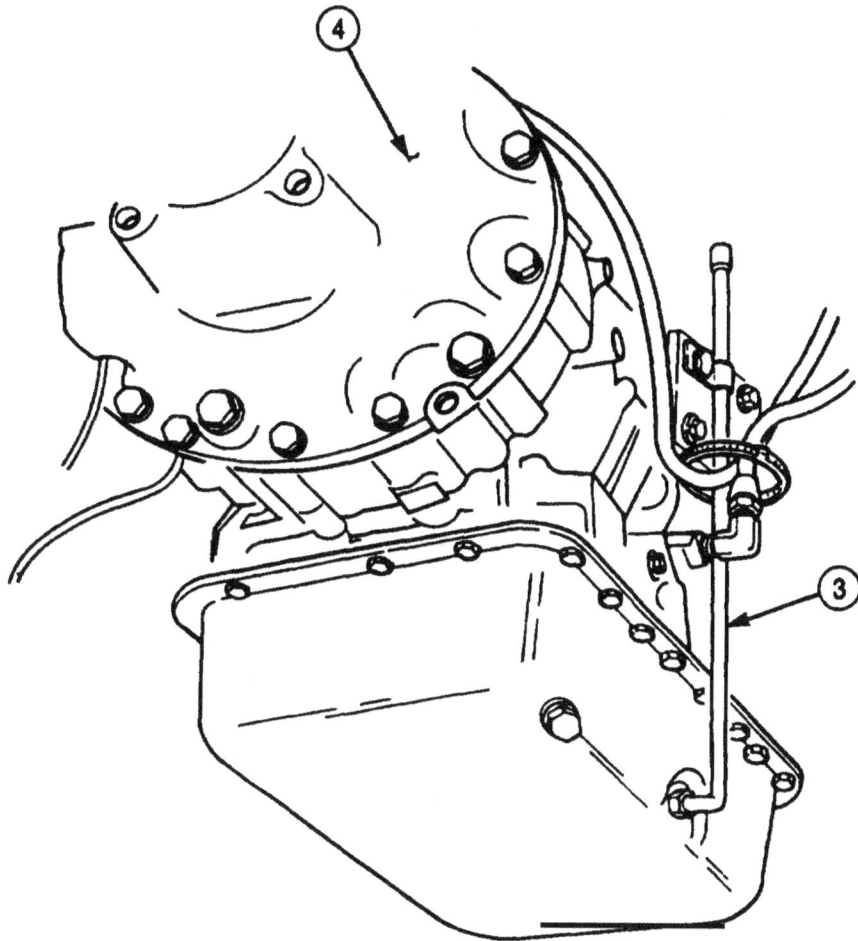

FOLLOW-ON TASK: Start engine (TM 9-2320-272-10), check for leaks, and road test vehicle.

3-134. TRANSMISSION DIPSTICK TUBE AND DIPSTICK MAINTENANCE

THIS TASK COVERS:

a. Removal

c. Installation

b. Inspection

INITIAL SETUP:

APPLICABLE MODELS:

All

TOOLS

General mechanic's tool kit (Appendix E, Item 1)
Torque wrench (Appendix E, Item 144)

MATERIALS/PARTS

Tiedown strap (Appendix D, Item 690)
Rags (Appendix C, Item 58)

REFERENCES (TM)

LO 9-2320-272-12
TM 9-2320-272-10
TM 9-2320-272-24P

EQUIPMENT CONDITION

- Parking brake set (TM 9-2320-272-10).
- Hood raised and secured (TM 9-2320-272-10).
- Transmission oil drained (para. 3-133).

a. Removal

CAUTION

Wipe area around dipstick tube before removal to prevent entry of
dirt. Damage may occur if dirt or dust enters transmission.

NOTE

Perform steps 1 through 3 for M939/A1 model vehicles only.

1. Open access door (3) in cab floor and remove transmission oil dipstick (1) from transmission oil dipstick tube (2).
2. Remove screw (7), washer (6), clip (5), and dipstick tube (2) from flywheel housing (4).
3. Remove tiedown strap (9) from dipstick tube (2) and tube (8). Discard tiedown strap (9).

NOTE

Perform steps 4 and 5 for M939A2 model vehicles only.

4. Remove transmission oil dipstick (1) from transmission oil dipstick tube (2).
5. Remove screw (7), clip (5), and dipstick tube (2) from flywheel housing (4).
6. Loosen flare nut (10) and remove dipstick tube (2) from transmission oil pan (12).

b. Inspection

1. Inspect dipstick tube (2) for cracks, blockage, weld damage at clip (5), and damage to flared end, If blocked, remove obstruction. If damaged, replace.
2. Inspect flare nut (10) for cross-threading and burrs. If cross-threaded or burred, replace dipstick tube (2).
3. Inspect oil pan dipstick tube port (11) for cracked or broken welds, burrs, and cross-threading. If cracked or broken, replace oil pan (12). If burred or cross-threaded, repair threads.

c. Intstallation

1. Install dipstick tube (2) on transmission oil pan (12) with flare nut (10). Tighten flare nut (10) 10-25 lb-ft (14-34 N•m).

NOTE

Perform steps 2 and 3 for M939A2 model vehicles only.

2. Install dipstick tube (2) on flywheel housing (4) with clip (5) and screw (7).

3-134. TRANSMISSION DIPSTICK TUBE AND DIPSTICK MAINTENANCE (Contd)

3. Install dipstick (1) in dipstick tube (2).

NOTE

Perform steps 4 through 6 for M939/A1 model vehicles only.

4. Install dipstick tube (2) and clip (5) on flywheel housing (4) with washer (6) and screw (7).
5. Install dipstick tube (2) on tube (8) with new tiedown strap (9).
6. Install dipstick (1) in dipstick tube (2).

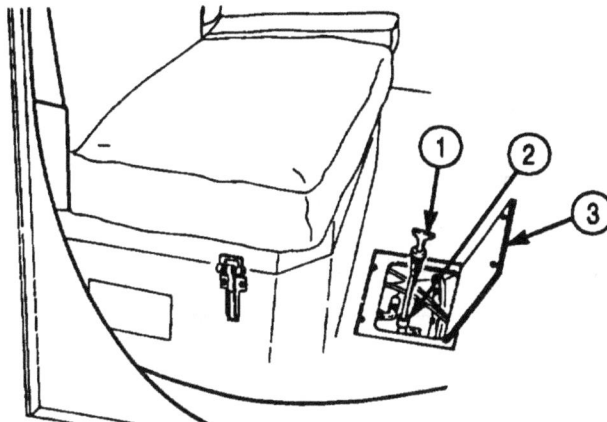

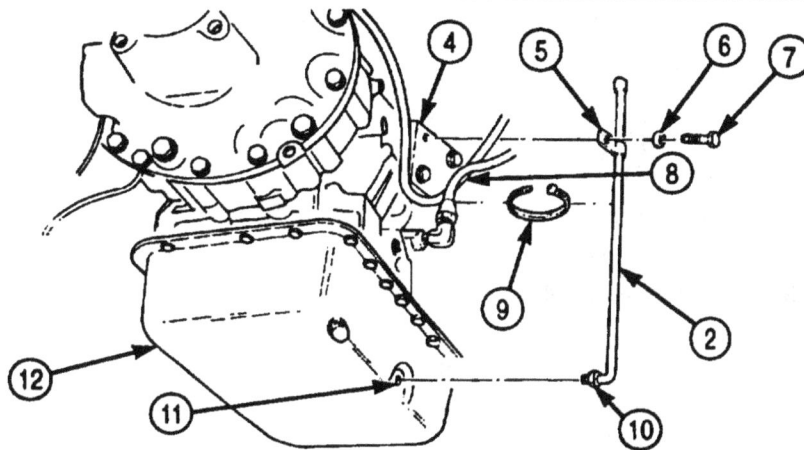

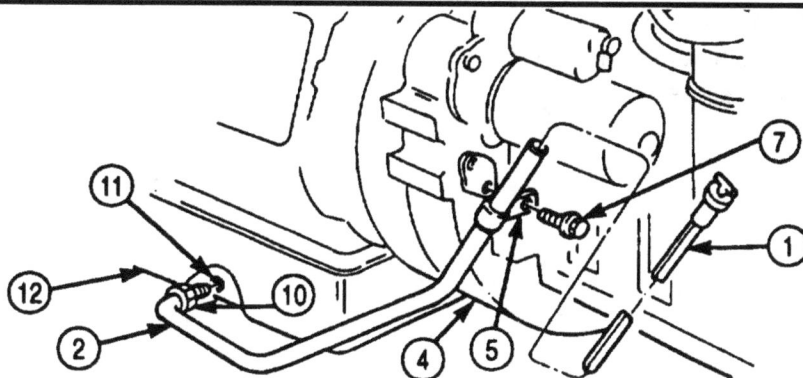

FOLLOW-ON TASK: Fill transmission (LO 9-2320-272-12).

3-135. GOVERNOR PIPING AND CAPACITOR MAINTENANCE

THIS TASK COVERS:

a. Removal   c. Installation

b. Inspection

INITIAL SETUP:

APPLICABLE MODELS
M936/A1/A2

TOOLS
a. General mechanic's tool kit (Appendix E, Item 1)

MATERIALS/PARTS
Two locknuts (Appendix D, Item 313)
Lockwasher (Appendix D, Item 350)
Two cotter pins (Appendix D, Item 66)
Three tiedown straps (Appendix D, Item 684)
Antiseize tape (Appendix C, Item 72)

REFERENCES (TM)
TM 9-2320-272-24-1
TM 9-2320-272-24P

EQUIPMENT CONDITION
Parking brake set (TM 9-2320-272-10).

---

a. Removal

NOTE

- Perform steps 1 and 2 for M936/A1 model vehicles only.
- Have drainage container ready to catch oil.

1. Remove hose (13) from elbow (14) and tee (2).
2. Remove elbow (14) from governor (15).
3. Remove cap (1) from tee (2).
4. Remove hose (10) from tee (2).
5. Remove tee (2) from nipple (3).
6. Remove nipple (3) from adapter (4).

NOTE

Note location of tiedown strap for installation.

7. Remove tiedown strap (12) from hoses (10) and (11). Discard tiedown strap (12).
8. Remove hose (10) from adapter (9).
9. Remove adapter (9) from tee (5).
10. Remove hose (7) from elbows (8) and (6).
11. Remove elbow (8) from tee (5).

3-135. GOVERNOR PIPING AND CAPACITOR MAINTENANCE (Contd)

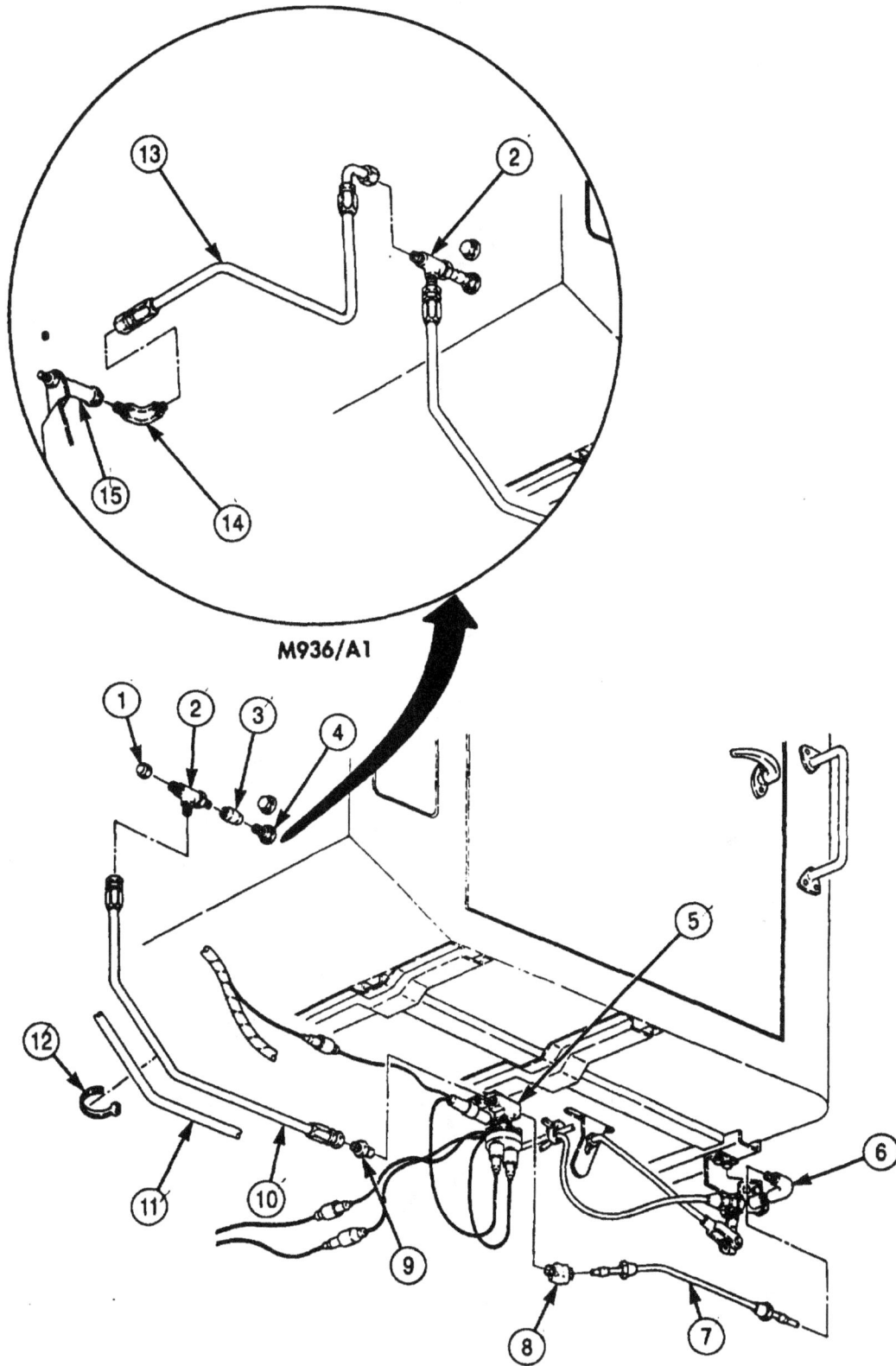

M936/A1

3-135. GOVERNOR PIPING AND CAPACITOR MAINTENANCE (Contd)

NOTE

Tag all leads for installation.

12. Disconnect capacitor leads (8) and (9) from transmission 5th gear lock-up pressure switch (7).
13. Remove pressure switch (7) from tee (6).
14. Disconnect capacitor leads (2), (12), and (15) from leads (1), (13), and (14).
15. Remove screw (10), lockwasher (11), tee (6), and capacitor (5) from body (3). Discard lockwasher (11).
16. Remove clamp (4) from capacitor (5).

NOTE

Note location of tiedown straps for installation.

17. Remove two tiedown straps (31) from hoses (32) and (35). Discard tiedown straps (31).
18. Remove hose (32) from adapters (30) and (37).
19. Remove adapters (30) and (37) from valves (19) and (38).
20. Remove cotter pin (33), washer (34), and rod (17) from lever (16). Discard cotter pin (33).
21. Remove cotter pin (27), washer (26), pin (24), and clevis (25) from valve (19). Discard cotter pin (27).
22. Remove clevis (25) from rod (17).
23. Remove breather (22) from adapter (23).
24. Remove elbow (21) and adapter (23) from valve (19).
25. Remove two locknuts (28), screws (20), and valve (19) From bracket (18). Discard locknuts (28).

NOTE

Perform step 26 for M936/Al series vehicles only

26. Remove two screws (36) and bracket (18) from body (3).
27. Remove two screws (29) and bracket (18) from body (3).

b. Inspection

Inspect hoses (39), (32), and (40). Replace hoses (39), (32), or (40) if damaged.

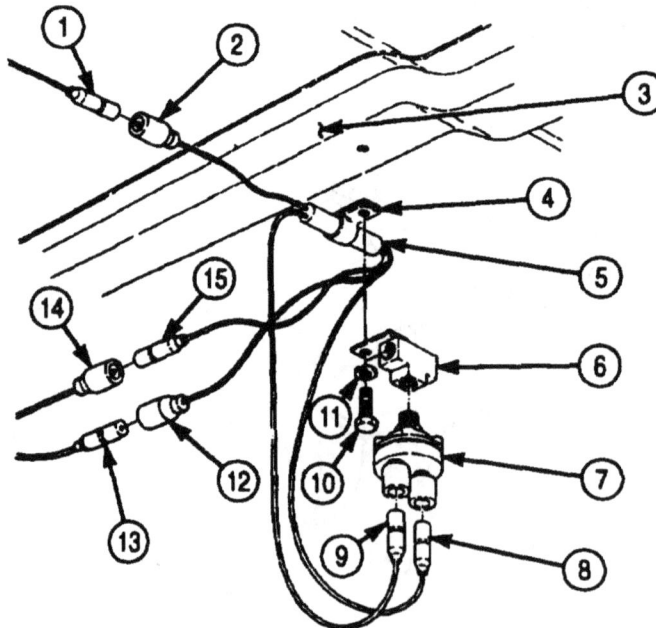

3-135. GOVERNOR PIPING AND CAPACITOR MAINTENANCE (Contd)

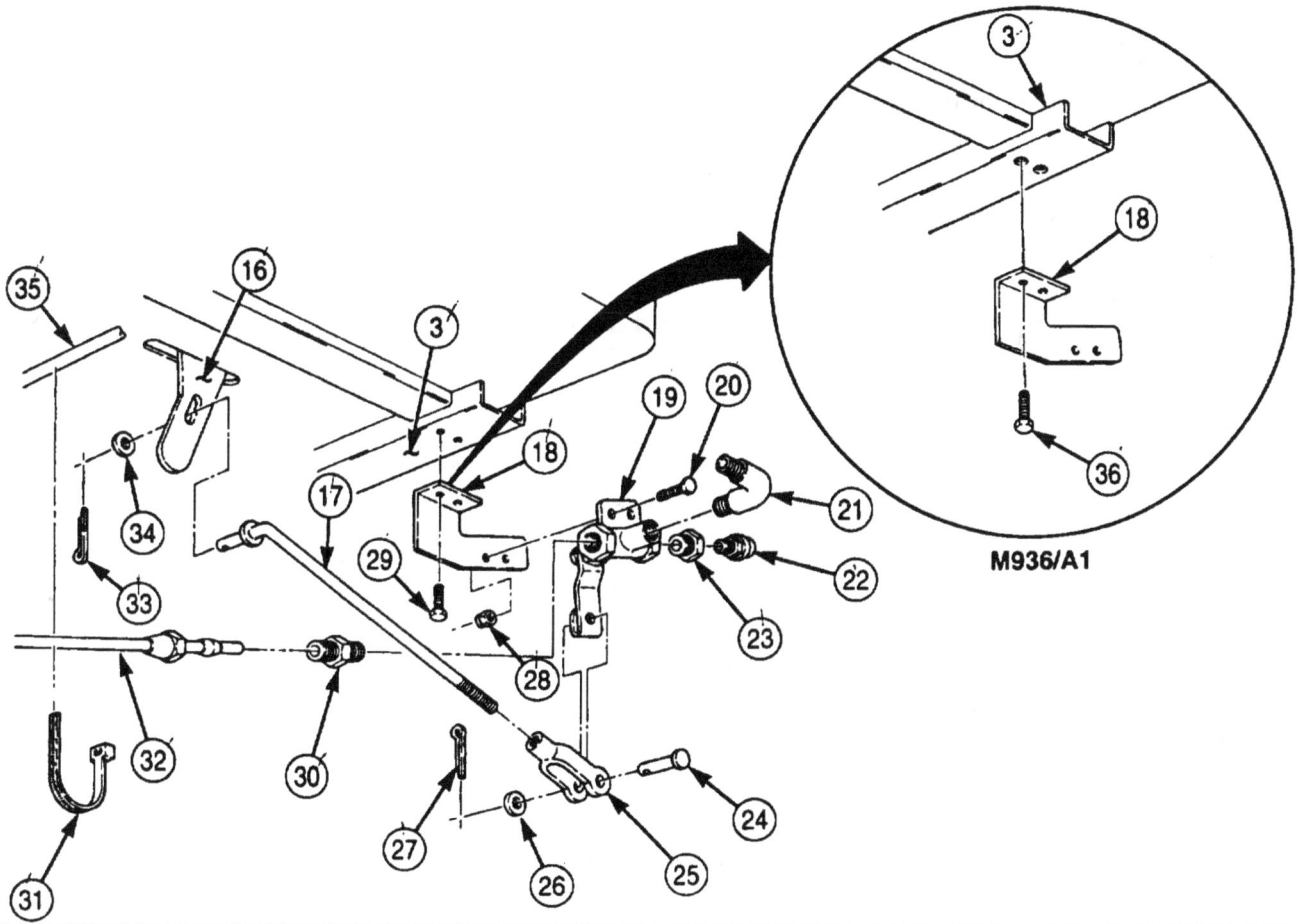

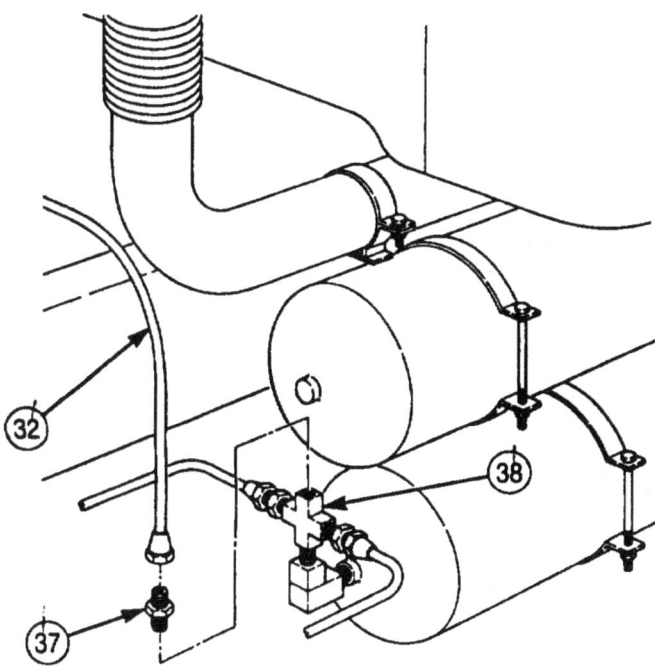

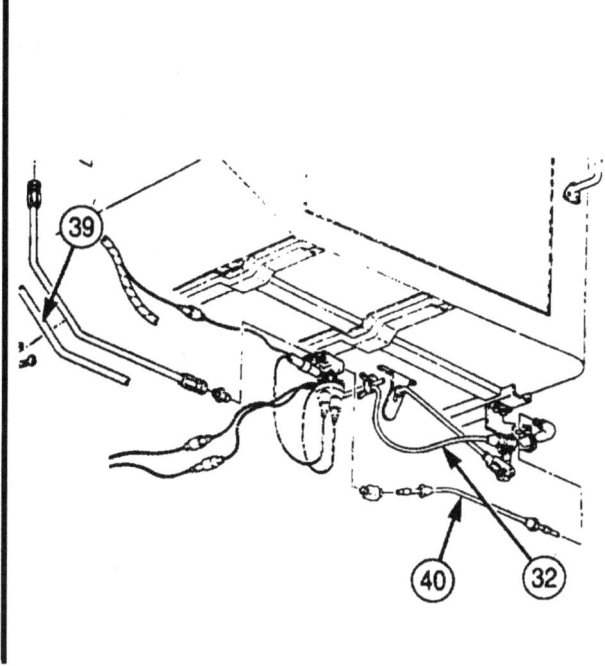

M936/A1

3-135. GOVERNOR PIPING AND CAPACITOR MAINTENANCE (Contd)

c. Installation

1. Install bracket (18) on body (3) with two screws (29).

NOTE

Perform step 2 for M936/A1 series vehicles only.

2. Install bracket (18) on body (3) with two screws (36).
3. Install valve (19) on bracket (18) with two screws (20) and new locknuts (28).

NOTE

Male pipe threads must be wrapped with antiseize tape before installation.

4. Install elbow (21) and adapter (23) on valve (19).
5. Install breather (22) on adapter (23).
6. Install clevis (25) on rod (17).
7. Install clevis (25) on valve (19) with pin (24), washer (26), and new cotter pin (27).
8. Install rod (17) on lever (16) with washer (34) and new cotter pin (33).
9. Install adapters (30) and (38) on valves (19) and (37).
10. Install hose (32) on adapters (30) and (38).
11. Install two new tiedown straps (31) on hoses (32) and (35).
12. Install clamp (4) on capacitor (5).
13. Install capacitor (5) and tee (6) on body (3) with new lockwasher (11) and screw (10).
14. Connect capacitor leads (2), (12), and (15) to leads (1), (13), and (14).
15. Install transmission 5th gear lock-up pressure switch (7) on tee (6).
16. Connect capacitor leads (8) and (9) to pressure switch (7).

3-135. GOVERNOR PIPING AND CAPACITOR MAINTENANCE (Contd)

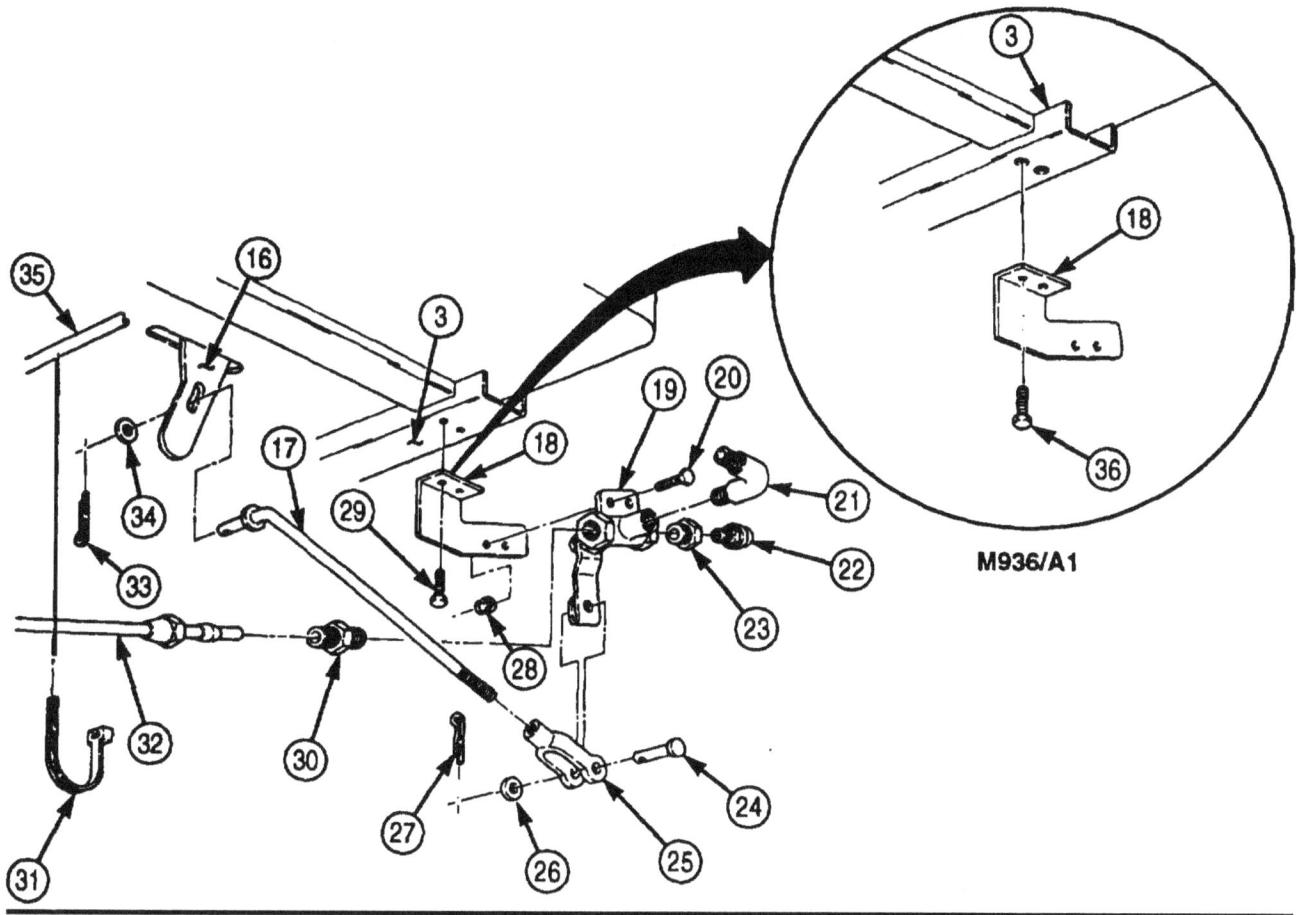

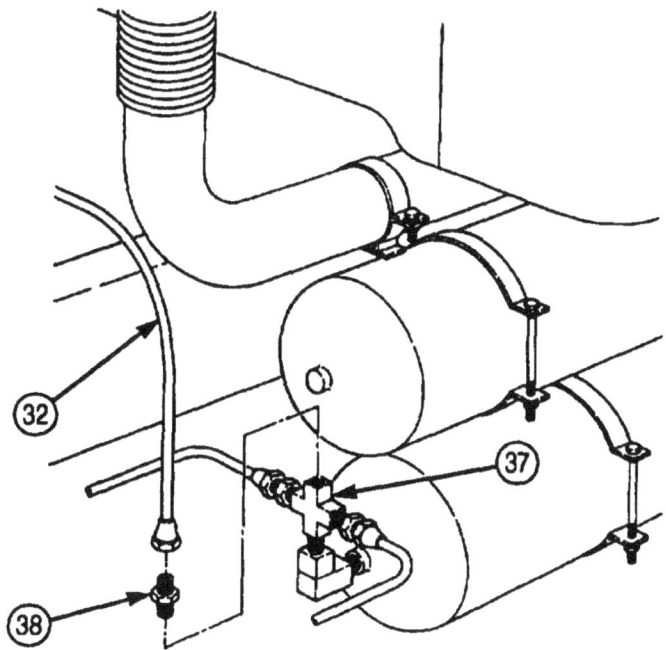

M936/A1

3-135. GOVERNOR PIPING AND CAPACITOR MAINTENANCE (Contd)

17.   Install elbow (11) on tee (8).

18.   Install hose (10) on elbows (11) and (9).

19.   Install adapter (12) on tee (8).

20.   Install hose (13) on adapter (12).

21.   Install new tiedown strap (15) on hoses (13) and (14).

22.   Install nipple (6) on adapter (7).

23.   Install tee (2) on nipple (6).

24.   Install hose (13) on tee (2).

NOTE

Perform steps 25 and 26 for M936/A1 series vehicles only.

25.   Install elbow (3) on governor (4).

26.   Install hose (1) on elbow (3) and tee (2).

27.   Install cap (5) on tee (2).

3-135. GOVERNOR PIPING AND CAPACITOR MAINTENANCE (Contd)

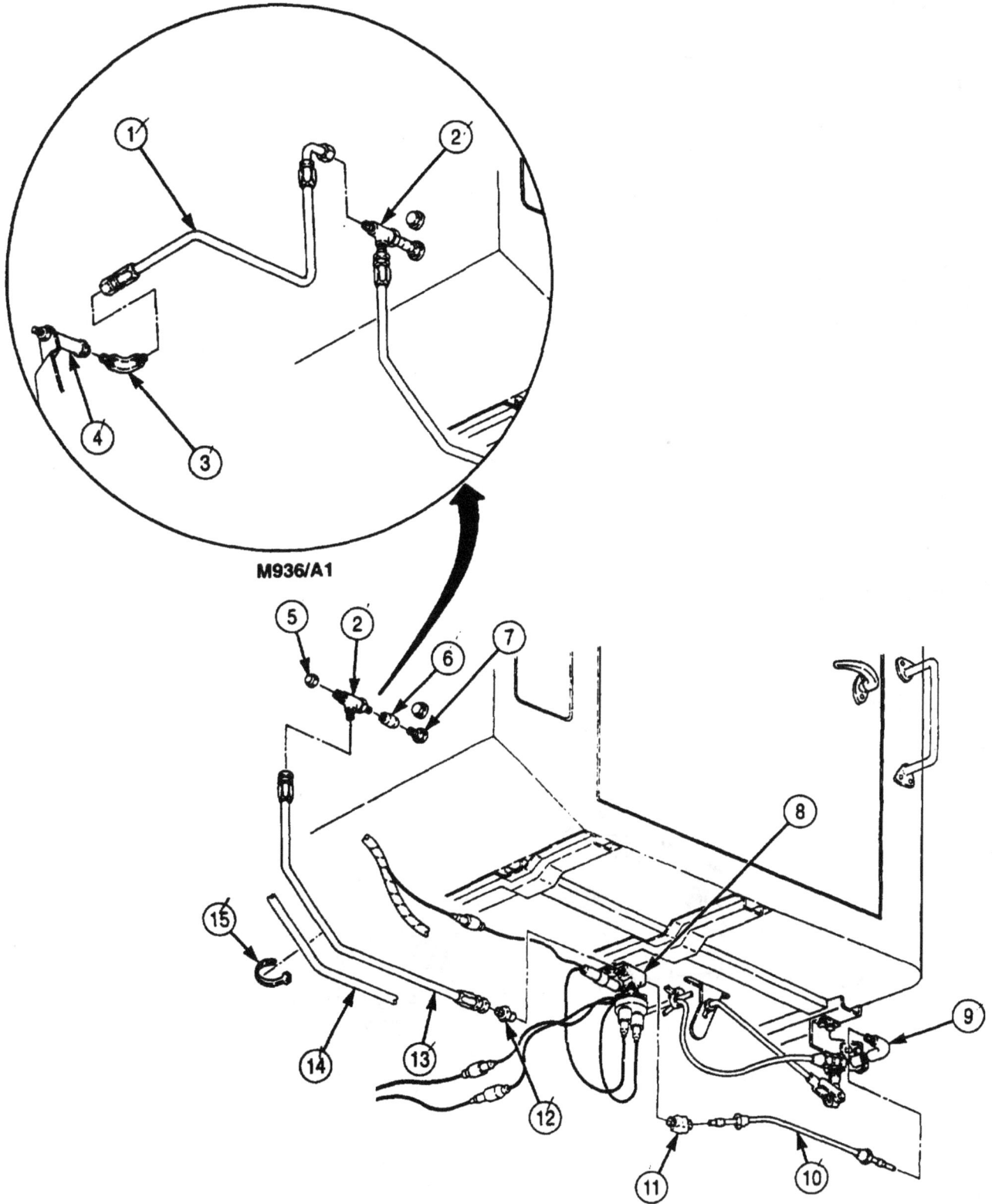

M936/A1

3-136. TRANSMISSION BREATHER REPLACEMENT

THIS TASK COVERS:

a. Removal   b. Installation

INITIAL SETUP:

APPLICABLE MODELS
All

TOOLS
General mechanic's tool kit (Appendix E, Item 1)

MATERIALS/PARTS
Antiseize tape (Appendix C, Item 72)
Rag (Appendix C, Item 58)

REFERENCES (TM)
TM 9-2320-272-10
TM 9-2320-272-24P

EQUIPMENT CONDITION
Parking brake set (TM 9-2320-272-10).

a. Removal

CAUTION

Clean area around breather before removal to prevent entry of dirt. Damage may occur if dirt or dust enters the transmission.

1. Disconnect vent line (2) from transmission breather (1).
2. Remove breather (1) and adapter (4) from transmission (3). Discard breather (1) and adapter (4) if threads are stripped.

b. Installation

NOTE

Male pipe threads must be wrapped with antiseize tape before installation.

1. Install adapter (4) and breather (1) on transmission (3).
2. Connect vent line (2) to transmission breather (1).

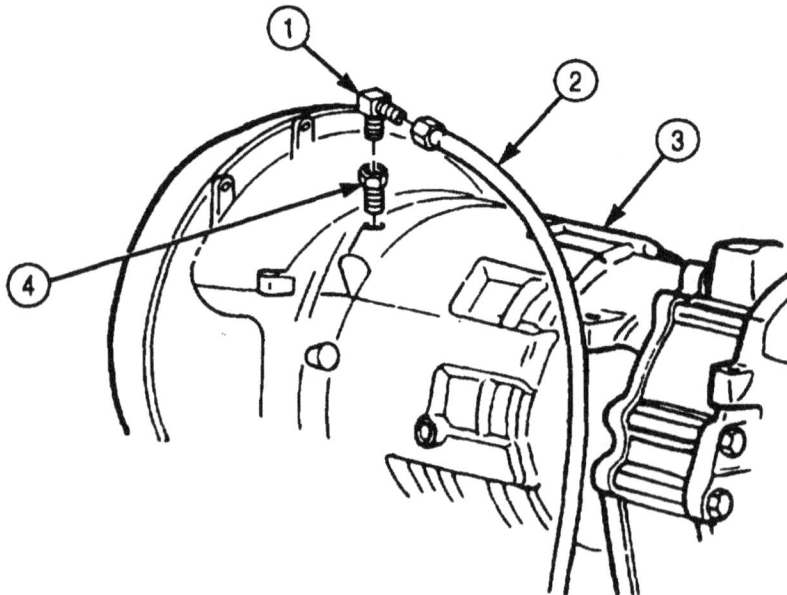

3-137. TRANSMISSION MOUNTING BRACKET AND ISOLATOR (M939A2) REPLACEMENT

THIS TASK COVERS:

a. Removal   b. Installation

INITIAL SETUP:

APPLICABLE MODELS
M939A2

TOOLS
General mechanic's tool kit (Appendix E, Item 1)
Transmission jack (Appendix E, Item 147)

MATERIALS/PARTS
Two lockwashers (Appendix D, Item 350)
Two lockwashers (Appendix D, Item 392)

REFERENCES (TM)
TM 9-2320-272-10
TM 9-2320-272-24P

EQUIPMENT CONDITION
Parking brake set (TM 9-2320-272-10).

GENERAL SAFETY INSTRUCTIONS
Keep hands clear of transmission during bracket
and isolator replacement.

a. Removal

WARNING

Keep hands clear of transmission when removing bracket and
isolator. A slipped jack may result in injury to personnel.

1. Using a transmission jack, raise transmission (7) and remove two screws (5) and lockwashers (6) from transmission mounting bracket (8). Discard lockwashers (6).

2. Remove two screws (2), lockwashers (3), isolator (4), and transmission mounting bracket (8) from bracket (1). Discard lockwashers (3).

b. Installation

NOTE

It will be necessary to raise and lower transmission slightly to
obtain proper alignment of screws.

1. Install transmission mounting bracket (8) on bracket (1) with isolator (4), two new lockwashers (3), and screws (2).

2. Using jack, position transmission (7) on transmission mounting bracket (8) and install with two new lockwashers (6) and screws (5).

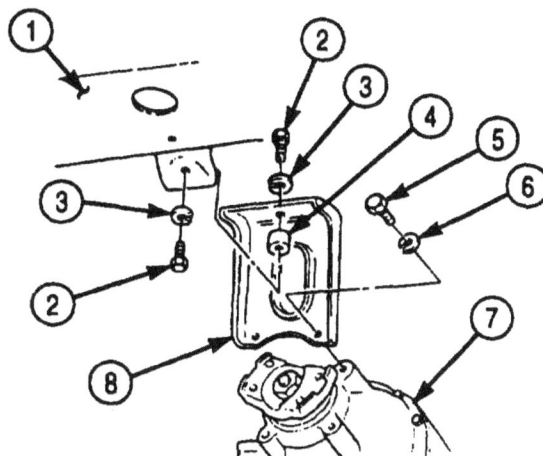

3-138. TRANSMISSION 5TH GEAR LOCK-UP PRESSURE SWITCH REPLACEMENT

THIS TASK COVERS:

a. Removal   b. Installation

INITIAL SETUP:

APPLICABLE MODELS
All

TOOLS
General mechanic's tool kit (Appendix E, Item 1)

REFERENCES (TM)
TM 9-2320-272-10
TM 9-2320-272-24P

EQUIPMENT CONDITION
- Parking brake set (TM 9-2320-272-10).
- Wheels chocked (TM 9-2320-272-10).
- Battery ground cable disconnected (para. 3-126).

a. Removal

1. Disconnect two capacitor leads (3) from pressure switch (2).
2. Remove pressure switch (2) from tee (1).

b. Installation

1. Install pressure switch (2) on tee (1).
2. Connect two capacitor leads (3) to pressure switch (2).

3-138. TRANSMISSION 5TH GEAR LOCK-UP PRESSURE SWITCH REPLACEMENT (Contd)

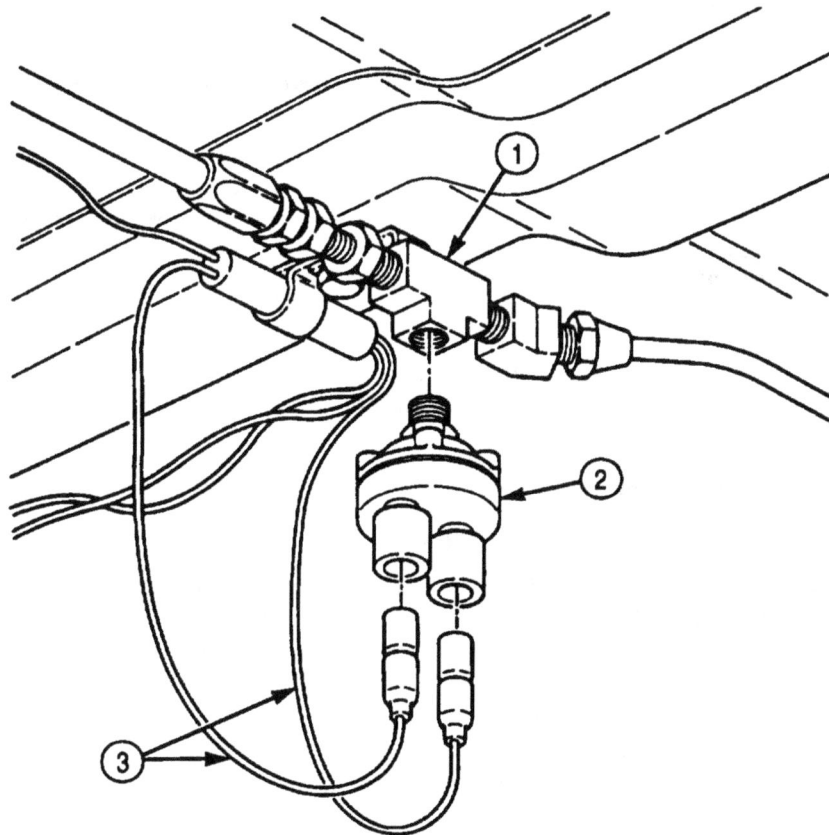

FOLLOW-ON TASK: Connect battery ground cable (para. 3-126).

3-139. TRANSMISSION OIL COOLER FILTER AND HEAD REPLACEMENT

THIS TASK COVERS:

a. Oil Cooler Filter Removal
b. Oil Cooler Filter Head Removal
c. Oil Cooler Filter Head Installation
d. Oil Cooler Filter Installation

INITIAL SETUP:

APPLICABLE MODELS
All

TOOLS
General mechanic's tool kit (Appendix E, Item 1)

MATERIALS/PARTS
Oil filter (Appendix D, Item 491)
Three locknuts (Appendix D, Item 299)
Lubricating oil (Appendix C, Item 48)
Rag (Appendix C, Item 58)
Cap and plug set (Appendix C, Item 14)

REFERENCES (TM)
LO 9-2320-272-12
TM 9-2320-272-10
TM 9-2320-272-24P

EQUIPMENT CONDITION
- Parking brake set (TM 9-2320-272-10).
- Hood raised and secured (TM 9-2320-272-10).

a. Oil Cooler Filter Removal

NOTE
Have drainage container ready to catch oil.

Remove oil filter (3) and oil filter seal (2) from oil filter head (1). Drain oil and discard oil filter (3) and oil filter seal (2).

b. Oil Cooler Filter Head Removal

CAUTION
- Clean area around hoses before removal to prevent entry of dirt. Damage may occur if dirt or dust enters the transmission.
- Cover or plug all open hoses and connections immediately after disconnection to prevent contamination. Failure to do this may result in transmission damage.

NOTE
- Have drainage container ready to catch oil.
- Tag hoses for installation.

1. Remove oil filter (3) as shown in task a.
2. Remove oil filter supply hose (8) from adapter elbow (7).
3. Remove supply hose (9) from adapter elbow (10).
4. Remove three locknuts (4), screws (6), and oil filter head (1) from oil filter bracket (5). Discard locknuts (4).

3-139. TRANSMISSION OIL COOLER FILTER AND HEAD REPLACEMENT (Contd)

c. Oil Cooler Filter Head Installation

CAUTION

Ensure plugs or covers are completely removed from all hoses and connections. Failure to do this may result in transmission damage.

NOTE

If new filter head is to be installed, use fittings from old filter head.

1. Install oil filter head (1) on oil filter bracket (5) with three screws (6) and new locknuts (4).
2. Install supply hose (9) on adapter elbow (10).
3. Install oil filter supply hose (8) on adapter elbow (7).
4. Install oil filter (3) as shown in task d.

d. Oil Cooler Filter Installation

1. Lightly coat new oil filter seal (2) with lubricating oil and position on oil filter head (1).
2. Install new oil filter (3).

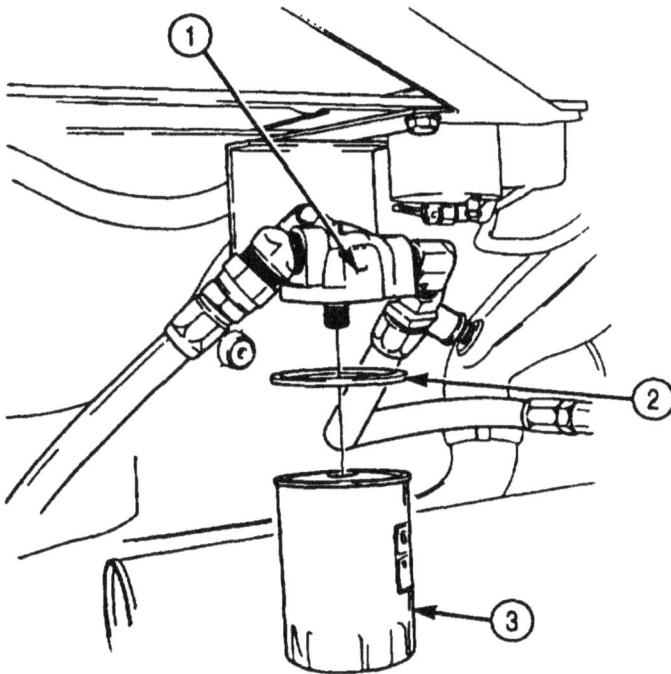

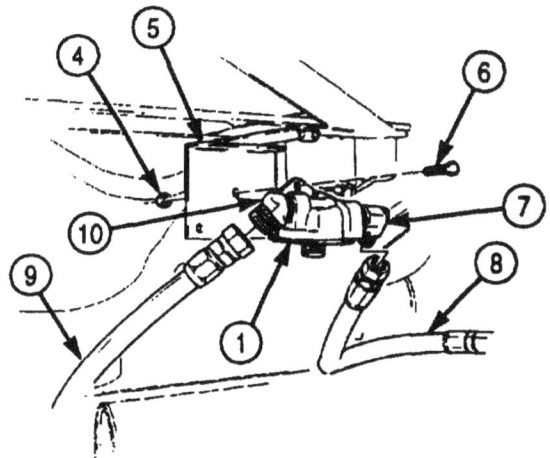

FOLLOW-ON TASKS: • Fill transmission oil to proper level (LO 9-2320-272-12).
 • Start engine (TM 9-2320-272-10), check for leaks, and road test vehicle.

3-140. TRANSMISSION OIL COOLER AND MOUNT (M939/A1) REPLACEMENT

THIS TASK COVERS:
a. Oil Cooler Removal
b. Oil Cooler Mount Removal
c. Oil Cooler Mount Installation
d. Oil Cooler Installation

INITIAL SETUP:

APPLICABLE MODELS
M939, M939A1

TOOLS
General mechanic's tool kit (Appendix E, Item 1)

MATERIALS/PARTS
Four locknuts (Appendix D, Item 291)
Four locknuts (Appendix D, Item 309)
Cap and plug set (Appendix C, Item 14)
Rag (Appendix C, Item 58)

REFERENCES (TM)
LO 9-2320-272-12
TM 9-2320-272-10
TM 9-2320-272-24P

EQUIPMENT CONDITION
- Parking brake set (TM 9-2320-272-10).
- Right splash shield removed (TM 9-2320-272-10).
- Cooling system drained (para. 3-53).

a. Oil Cooler Removal

CAUTION
- Clean area around hoses before removal to prevent entry of dirt. Damage may occur if dirt or dust enters the transmission.
- Cover or plug all open hoses and connections immediately after disconnection to prevent contamination. Failure to do this may result in transmission damage.

NOTE
- Have drainage container ready to catch oil and coolant.
- Tag hoses for installation.

1. Loosen hose clamp (7) and remove coolant supply hose (8) from coolant inlet flange (6) and clamp (7) from hose (8).
2. Loosen hose clamp (15) and remove coolant return hose (1) from coolant outlet flange (14) and clamp (15) from hose (1).
3. Disconnect oil cooler-to-cooler filter supply hose (4) from elbow (5).
4. Disconnect transmission-to-oil cooler supply hose (3) from elbow (2).
5. Remove four locknuts (11), screws (13), washers (12), and oil cooler (10) from two mounts (9). Discard locknuts (11) and empty coolant from oil cooler (10).

b. Oil Cooler Mount Removal

Remove four locknuts (16), screws (18), and two mounts (9) from frame rail (17). Discard locknuts (16).

c. Oil Cooler Mount Installation

Install two mounts (9) on frame rail (17) with four screws (18) and new locknuts (16).

3-140. TRANSMISSION OIL COOLER AND MOUNT (M939/A1) REPLACEMENT (Contd)

d. Oil Cooler Installation

CAUTION

Ensure plugs or covers are completely removed from all hoses and connections. Failure to do this may result in transmission and oil cooler damage.

NOTE

If new oil cooler is to be installed, use fittings from old oil cooler.

1. Install transmission oil cooler (10) on two mounts (9) with four washers (12), screws (13), and new locknuts (11).

2. Connect transmission-to-oil cooler supply hose (3) on elbow (2).

3. Connect oil cooler-to-cooler filter supply hose (4) on elbow (5).

4. Position hose clamp (15) on hose (1) and install hose (1) on flange (14) and tighten clamp (15).

5. Position clamp (7) on hose (8) and install hose (8) on flange (6) and tighten clamp (7).

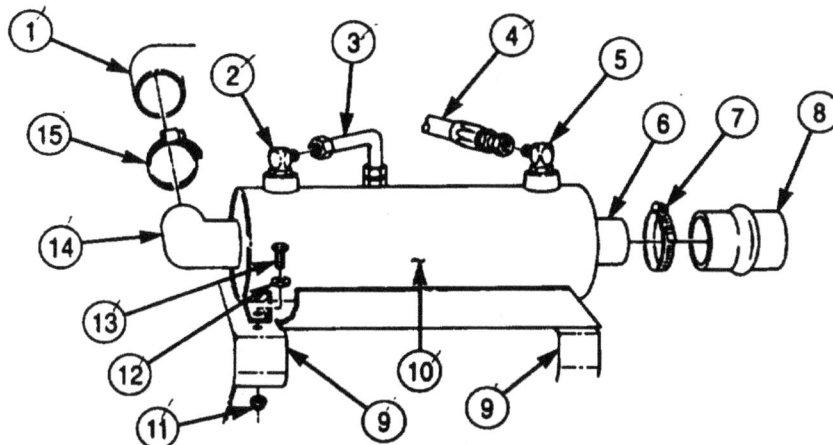

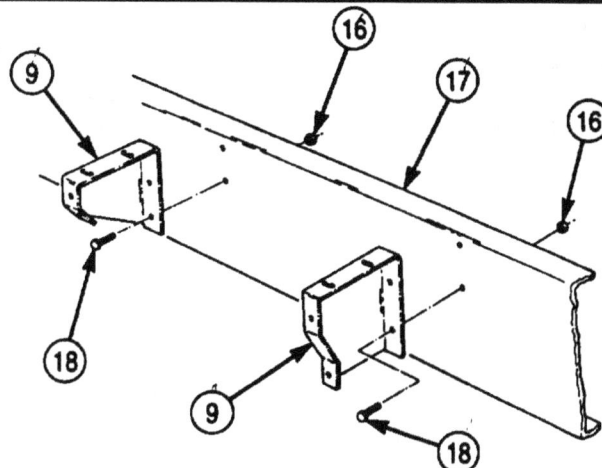

FOLLOW-ON TASKS: • Fill vehicle cooling system to proper operating level (para. 3-53).
• Install right splash shield (TM 9-2320-272-10).
• Fill transmission oil reservoir to proper level (LO 9-2320-272-12).
• Start engine (TM 9-2320-272-10), check for leaks, and road test vehicle.

3-141. TRANSMISSION OIL COOLER AND MOUNT (M939A2) REPLACEMENT

THIS TASK COVERS:

a. Removal   b. Installation

INITIAL SETUP:

APPLICABLE MODELS
M939A2

TOOLS
General mechanic's tool kit (Appendix E, Item 1)

MATERIALS/PARTS
Four locknuts (Appendix D, Item 291)
Four lockwashers (Appendix D, Item 354)
Two O-rings (Appendix D, Item 430)
Antiseize tape (Appendix C, Item 72)
Cap and plug ret (Appendix C, Item 14)

REFERENCES (TM)
TM 9-2320-272-10
TM 9-2320-272-24P

EQUIPMENT CONDITION
• Parking brake ret (TM 9-2320-272-10).
• Cooling system drained (para. 3-63).

---

a. Removal

CAUTION

• Clean area around hoses before removal to prevent entry of dirt Damage may occur if dirt or dust enters the transmission.

• Cover or plug all open hoses and connections immediately after disconnection to prevent contamination. Failure to do this may result in transmission damage.

NOTE

Have drainage container ready to catch oil.

1. Disconnect transmission oil lines (4) and (18) from oil sampling valve (5) and elbow (15).

2. Loosen clamps (1) and (17) and disconnect hoses (2) and (16) from oil cooler (11).

3. Remove four locknuts (20), washers (10), screws (9), washers (10), and oil cooler (11) from bracket (3) and splash panel extension (22). Discard locknuts (20).

4. Remove oil sampling valve (5), nut (6), washer (7), and O-ring (8) from oil cooler (11). Discard O-ring (8).

5. Remove elbow (16), nut (14), washer (13), and O-ring (12) from oil cooler (11). Discard O-ring (12).

6. Remove four locknuts (23), lockwashers (24), screws (21), and splash panel extension (22) from bracket (25) and frame rail (19). Discard locknuts (23) and lockwashers (24).

3-141. TRANSMISSION OIL COOLER AND MOUNT (M939A2) REPLACEMENT (Contd)

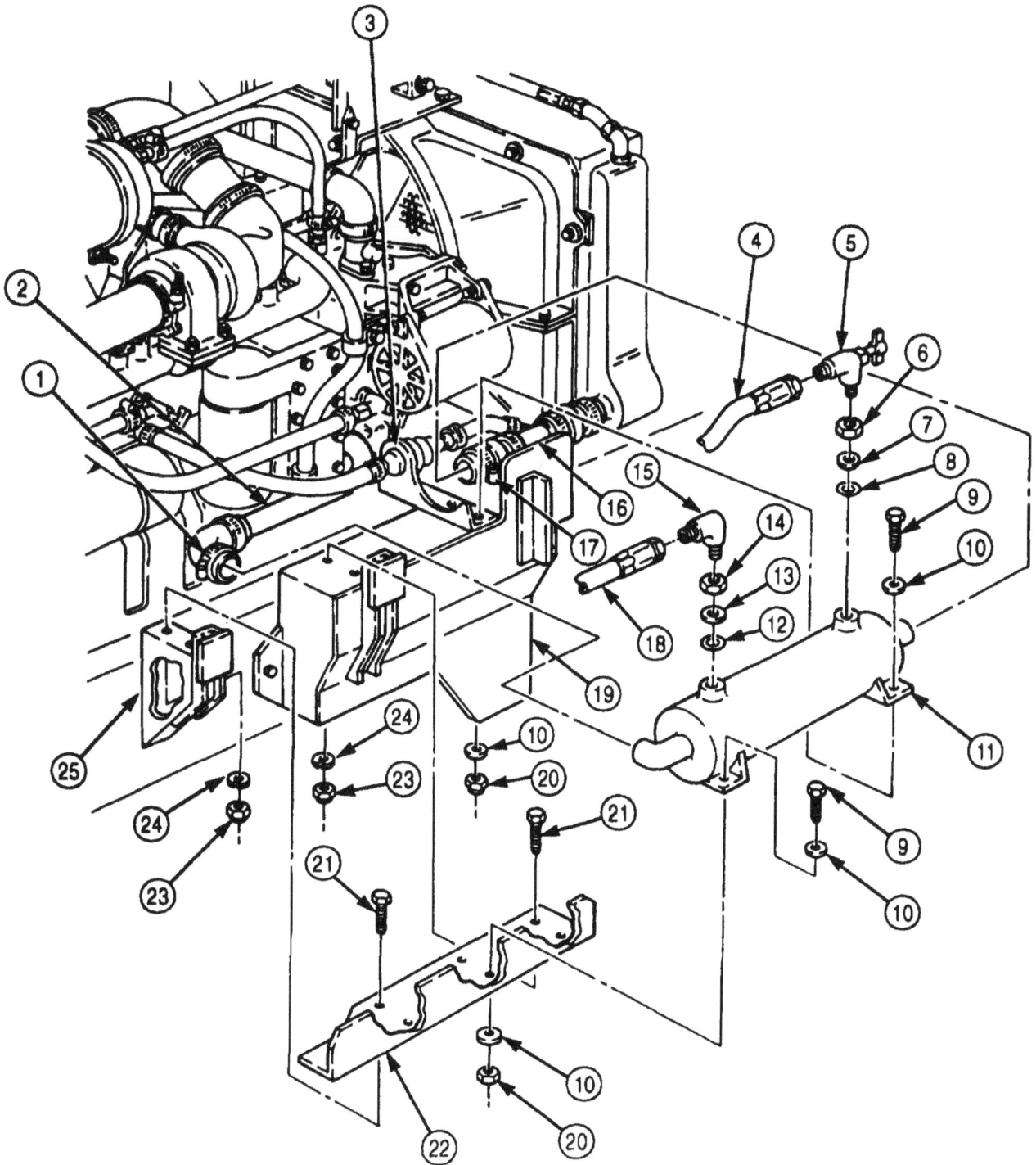

3-141. TRANSMISSION OIL COOLER AND MOUNT (M939A2) REPLACEMENT (Contd)

b. Installation

1. Install splash panel extension (22) on bracket (25) and frame rail (19) with four screws (21), new lockwashers (24), and new locknuts (23).

NOTE

Male pipe threads must be wrapped with antiseize tape before installation.

2. Install new O-ring (12), washer (13), nut (14), and elbow (15) on oil cooler (11).

3. Install new O-ring (8), washer (7), nut (6), and oil sampling valve (5) on oil cooler (11).

4. Install oil cooler (11) on bracket (3) and splash panel extension (22) with four washers (10), screws (9), washers (10), and new locknuts (20).

5. Connect hoses (2) and (16) to oil cooler (11) and tighten clamps (1) and (17).

6. Connect transmission oil lines (4) and (18) to oil sampling valve (5) and elbow (15).

CAUTION

When filling cooling system, ensure drain valve on aftercooler is open. Failure to do so may result in damage to equipment.

3-141. TRANSMISSION OIL COOLER AND MOUNT (M939A2) REPLACEMENT (Contd)

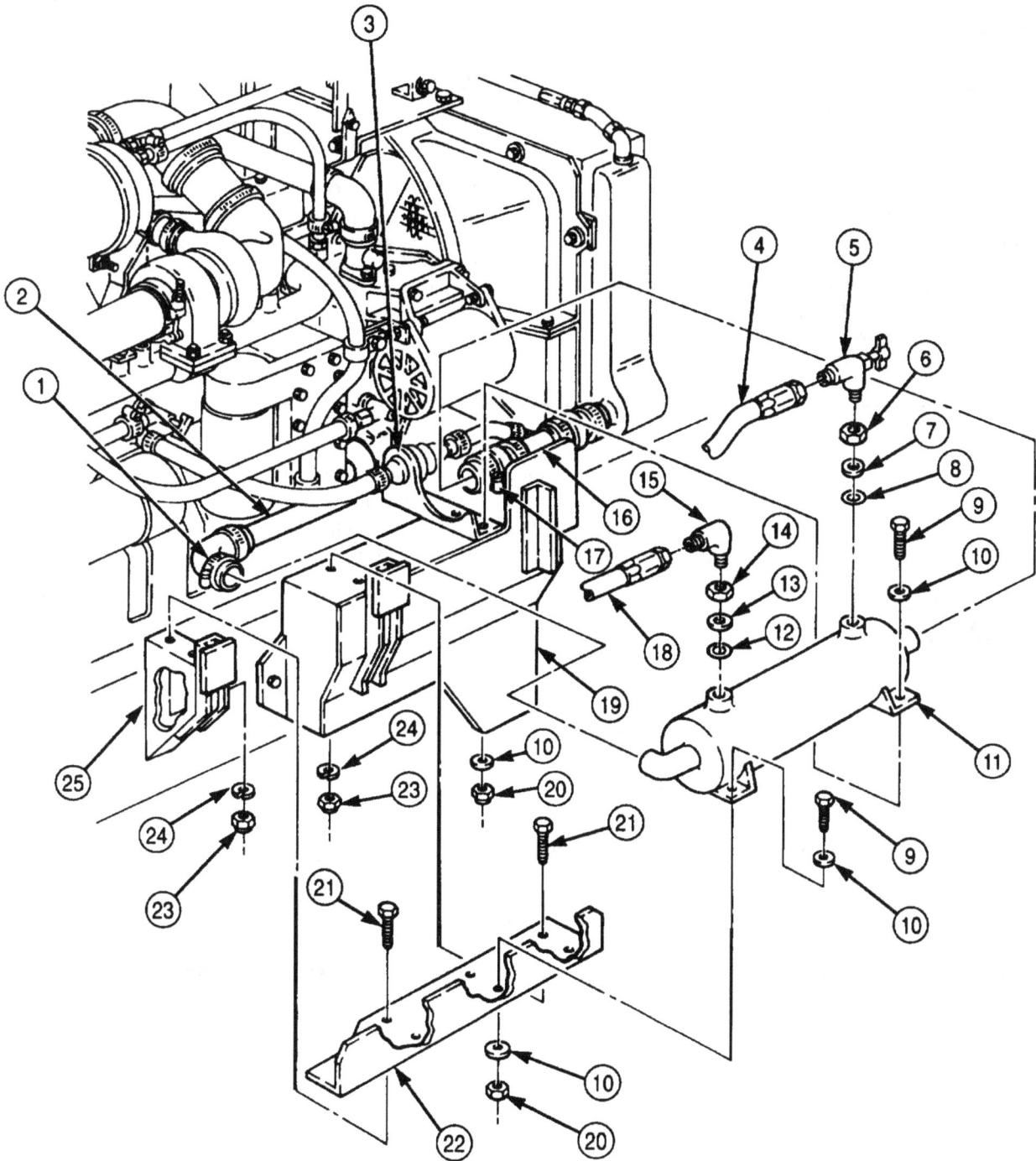

FOLLOW-ON TASKS: • Fill cooling system (para. 3-53).
• Check transmission oil level (TM 9-2320-272-10).
• Start engine (TM 9-2320-272-10) and check for leaks.

3-142. TRANSMISSION OIL COOLER HOSES REPLACEMENT

THIS TASK COVERS:

a. Removal   **b. Installation**

INITIAL SETUP:

APPLICABLE MODELS
All

TOOLS
General mechanic's tool kit (Appendix E, Item 1)

MATERIALS/PARTS
Two O-rings (Appendix D, Item 430)
Two O-rings (Appendix D, Item 470)
Two lockwashers (Appendix D, Item 416)
Locknut (Appendix D, Item 306)
Two tiedown straps (Appendix D, Item 697)
Cap and plug set (Appendix C, Item 14)
Antiseize tape (Appendix C, Item 72)

REFERENCES (TM)
LO 9-2320-272-12
TM 9-2320-272-10
TM 9-2320-272-24P

EQUIPMENT CONDITION
- Parking brake Bet (TM 9-2320-272-10).
- Eight splash shield removed (TM 9-2320-272-10).
- Transmission temperature transmitter unit removed (para. 3-97).

CAUTION
- Clean area around hoses before removal to prevent entry of dirt. Damage may occur if dirt or dust enters the transmission.
- Cover or plug all open hoses and connections immediately after disconnect to prevent contamination. Failure to do this may result in transmission damage.

a. Removal

NOTE
- Have drainage container ready to catch oil.
- Access to hose connection in step 1 is gained through access door in cab floor.
- Tag hoses for installation.

1. Disconnect supply hose (1) from adapter (6).
2. Remove adapter (6) and adapter (5) from elbow (4).
3. Remove elbow (4) and O-ring (3) from transmission (2). Discard O-ring (3).

NOTE
Perform step 4 for vehicles equipped with front winch.

4. Remove locknut (7) and screw (11) from hanger strap (8), and move winch hydraulic hoses (9) and (10) and two clamps (12) aside. Discard locknut (7).

3-142. TRANSMISSION OIL COOLER HOSES REPLACEMENT (Contd)

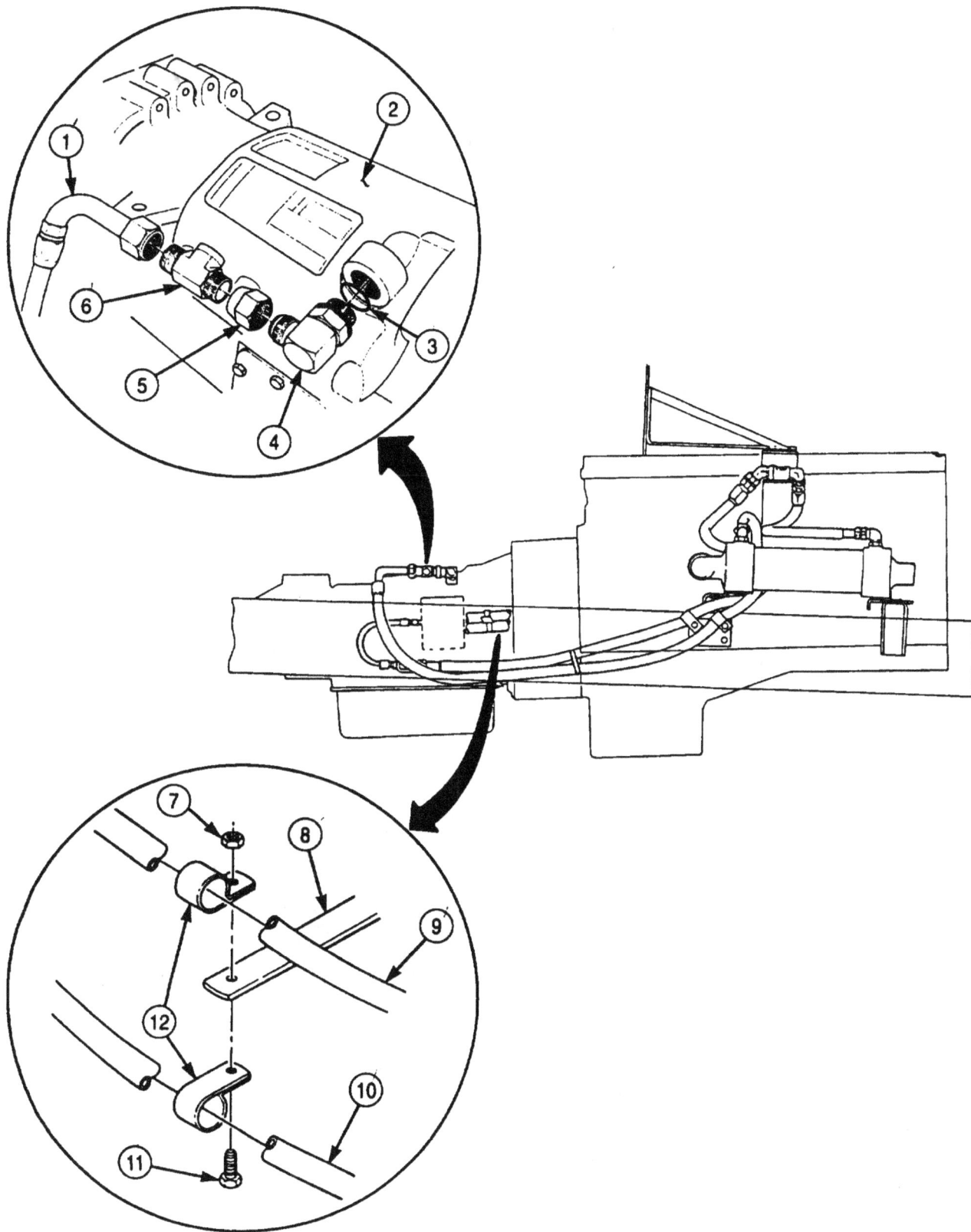

3-142. TRANSMISSION OIL COOLER HOSES REPLACEMENT (Contd)

NOTE
Perform steps 5 through 10 for vehicles equipped with transmission Power Takeoff (PTO).

5. Disconnect hose (11) from hose adapter (1).
6. Remove adapter (1) from adapter (2).
7. Remove adapter (2) from adapter (3).
8. Disconnect hose (11) from hose adapter (10).
9. Remove adapter (10) from adapter (9).
10. Remove adapter (9) from PTO (6).
11. Disconnect hose (4) from adapter (3).
12. Remove adapter (3) and O-ring (8) from valve housing (7). Discard O-ring (8).
13. Remove two tiedown straps (5) from hoses (4) and (17). Discard tiedown straps (5).
14. Remove two screws (13), washers (14), lockwashers (15), clamps (16), and hoses (4) and (17) from engine access cover (12). Discard lockwashers (15).

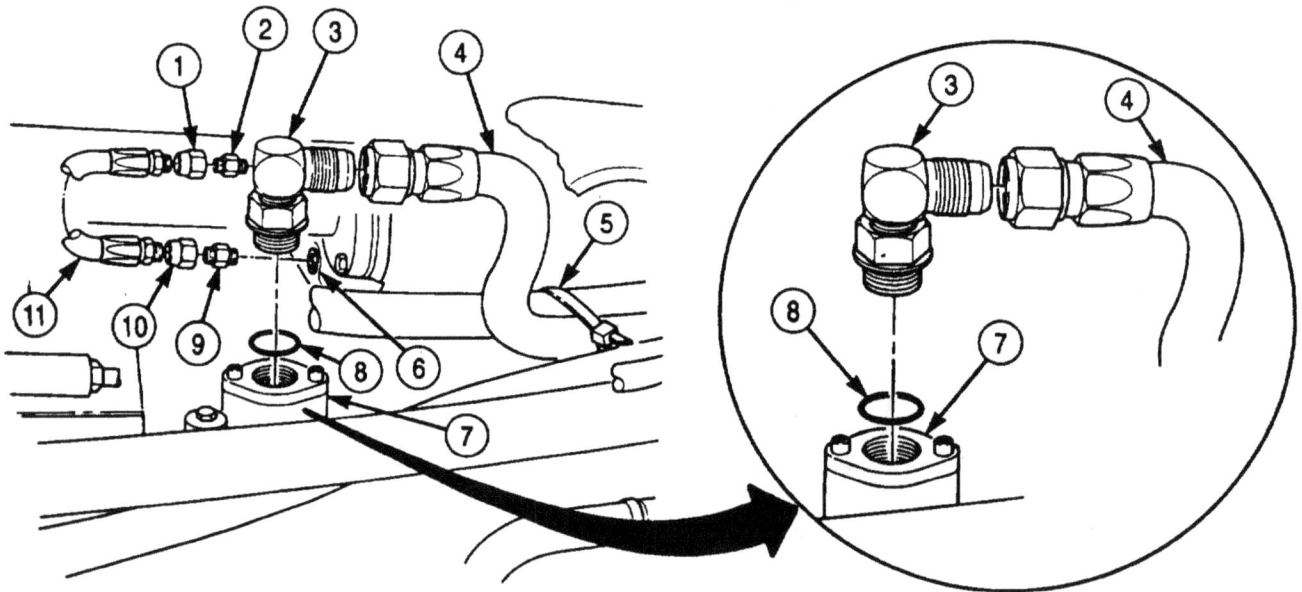

WITH PTO   WITHOUT PTO

3-142. TRANSMISSION OIL COOLER HOSES REPLACEMENT (Contd)

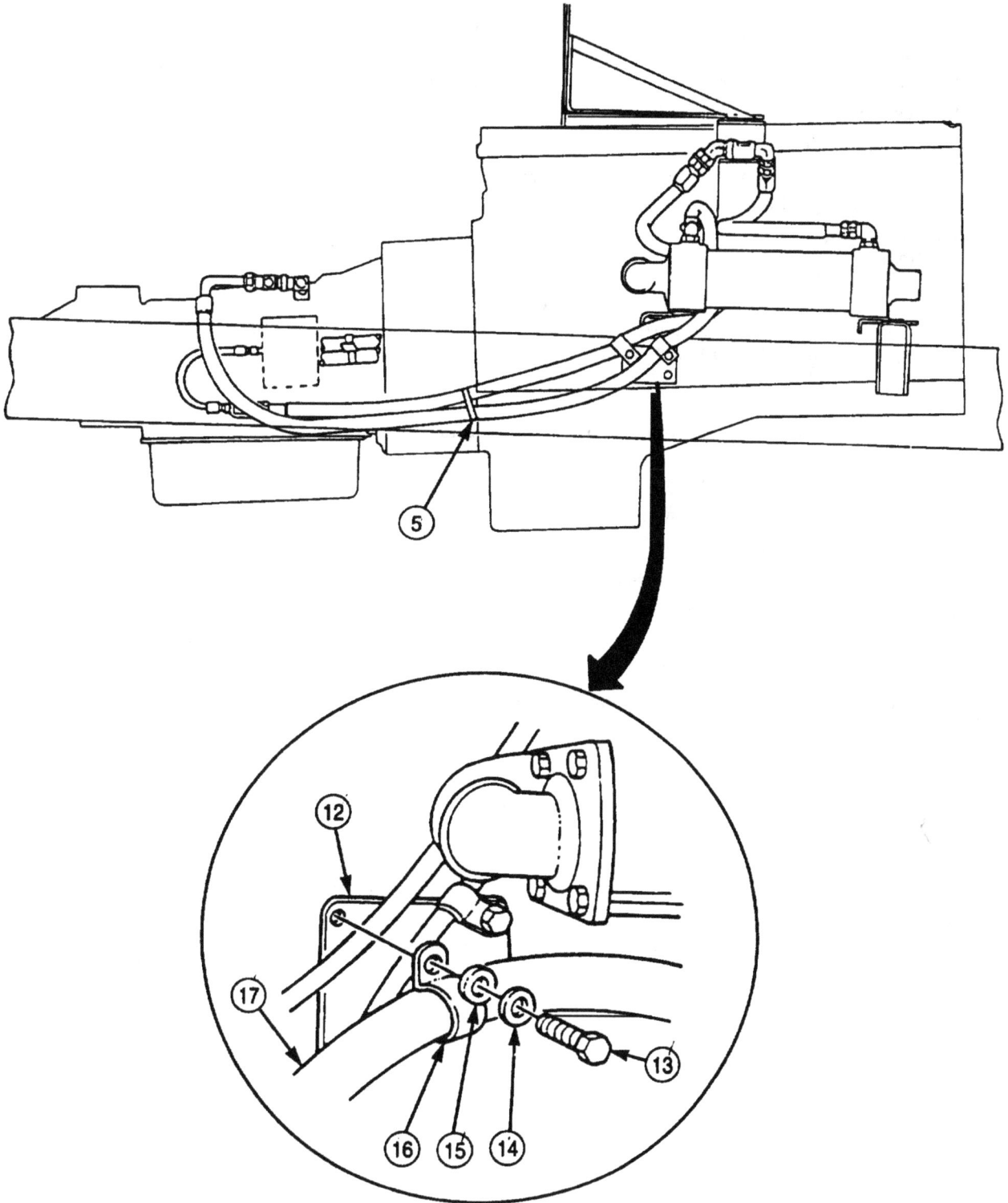

3-142. TRANSMISSION OIL COOLER HOSES REPLACEMENT (Contd)

16. Disconnect oil cooler supply hose (8) from elbow (11).
16. Remove elbow (11) and O-ring (10) from oil cooler (9). Discard O-ring (10).
17. Disconnect supply hose (6) from adapter elbow (6).
18. Remove elbow (6) and O-ring (7) from oil cooler (9). Discard O-ring (7).
19. Remove supply hose (6) from adapter elbow (4).
20. Remove adapter elbow (4) from oil filter housing (3).
21. Disconnect return hose (1) from elbow (2).
22. Remove elbow (2) from oil filter housing (3).

b. Installation

CAUTION

Ensure plugs or covers are completely removed from all hoses and connections. Failure to do this may result in transmission and oil cooler damage.

NOTE

Male pipe threads must he wrapped with antiseize tape before installation.

1. Install elbow (2) on housing (3).
2. Connect return hose (1) to elbow (2).
3. Install elbow (4) on housing (3).
4. Connect supply hose (6) to elbow (4).
6. Install new O-ring (7) and elbow (6) in oil cooler (9).
6. Connect supply hose (5) to elbow (6).
7. Install new O-ring (10) and elbow (11) in oil cooler (9).
8. Connect oil cooler supply hose (8) to elbow (11).

3-142. TRANSMISSION OIL COOLER HOSES REPLACEMENT (Contd)

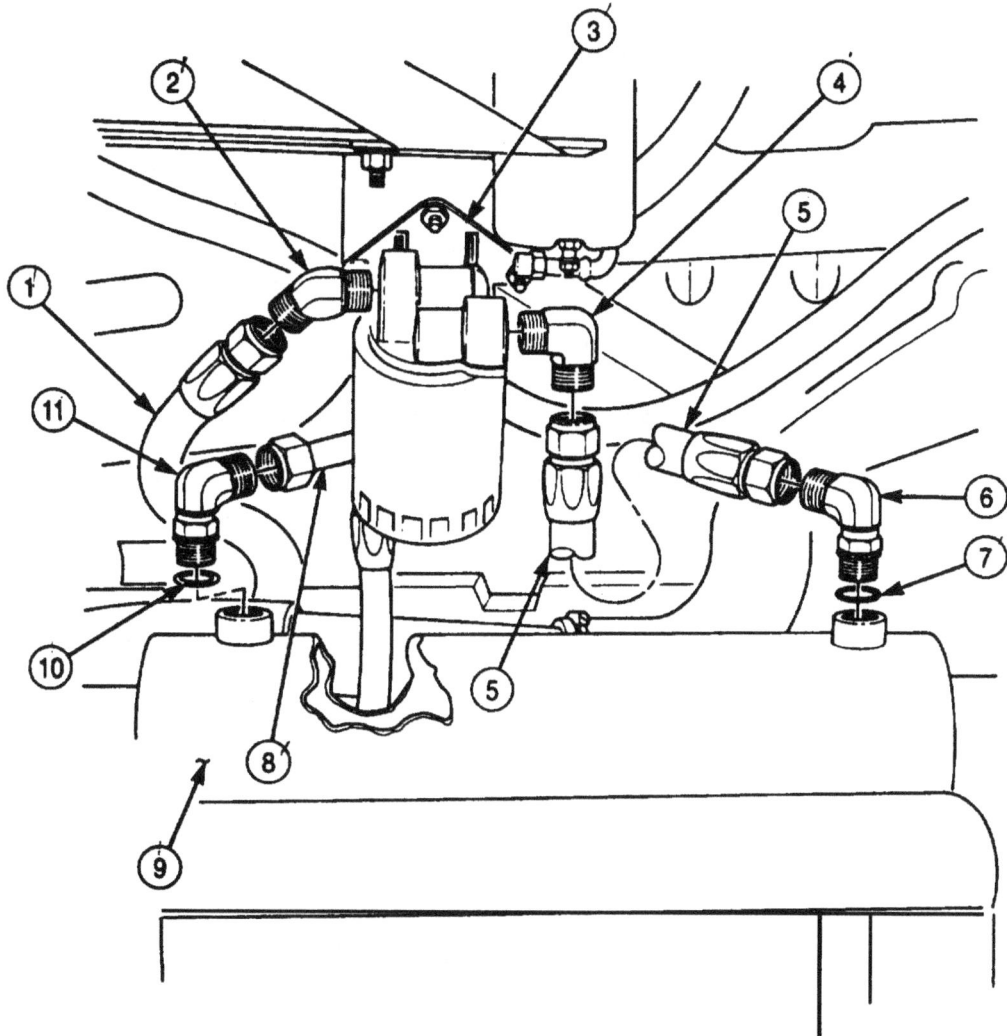

3-142. TRANSMISSION OIL COOLER HOSES REPLACEMENT (Contd)

9. Install two clamps (17) and hoses (4) and (18) on engine access cover (13) with two washers (15), new lockwashers (16), and screws (14).

10. Install two new tiedown straps (5) on hoses (4) and (18).

11. Install new O-ring (8) and adapter (3) in valve housing (7).

12. Connect hose (4) to adapter (3).

NOTE

Perform step 13 through 18 for vehicles equipped with transmission PTO.

13. Install adapter (9) on PTO (6).

14. Install adapter (10) on adapter (9).

16. Connect hose (11) to adapter (10).

16. Install adapter (2) on adapter (3).

17. Install adapter (1) on adapter (2).

18. Connect hose (11) to adapter (1).

NOTE

Perform step 19 for vehicles equipped with front winch.

19. Install hydraulic hoses (21) and (22) on hanger strap (20) with two clamps (24), screw (23), and new locknut (19).

20. Install new O-ring (26) and elbow (27) on transmission (12).

21. Install adapter (28) on elbow (27).

22. Install adapter (29) on adapter (28).

23. Connect supply hose (25) to adapter (29).

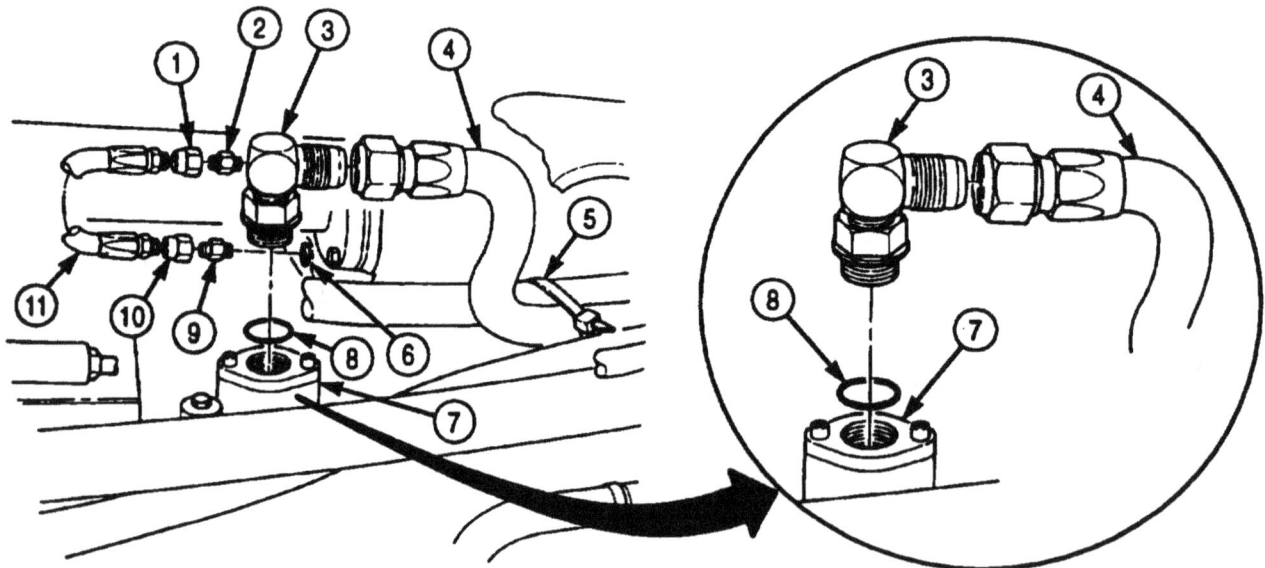

WITH PTO   WITHOUT PTO

3-142. TRANSMISSION OIL COOLER HOSES REPLACEMENT (Contd)

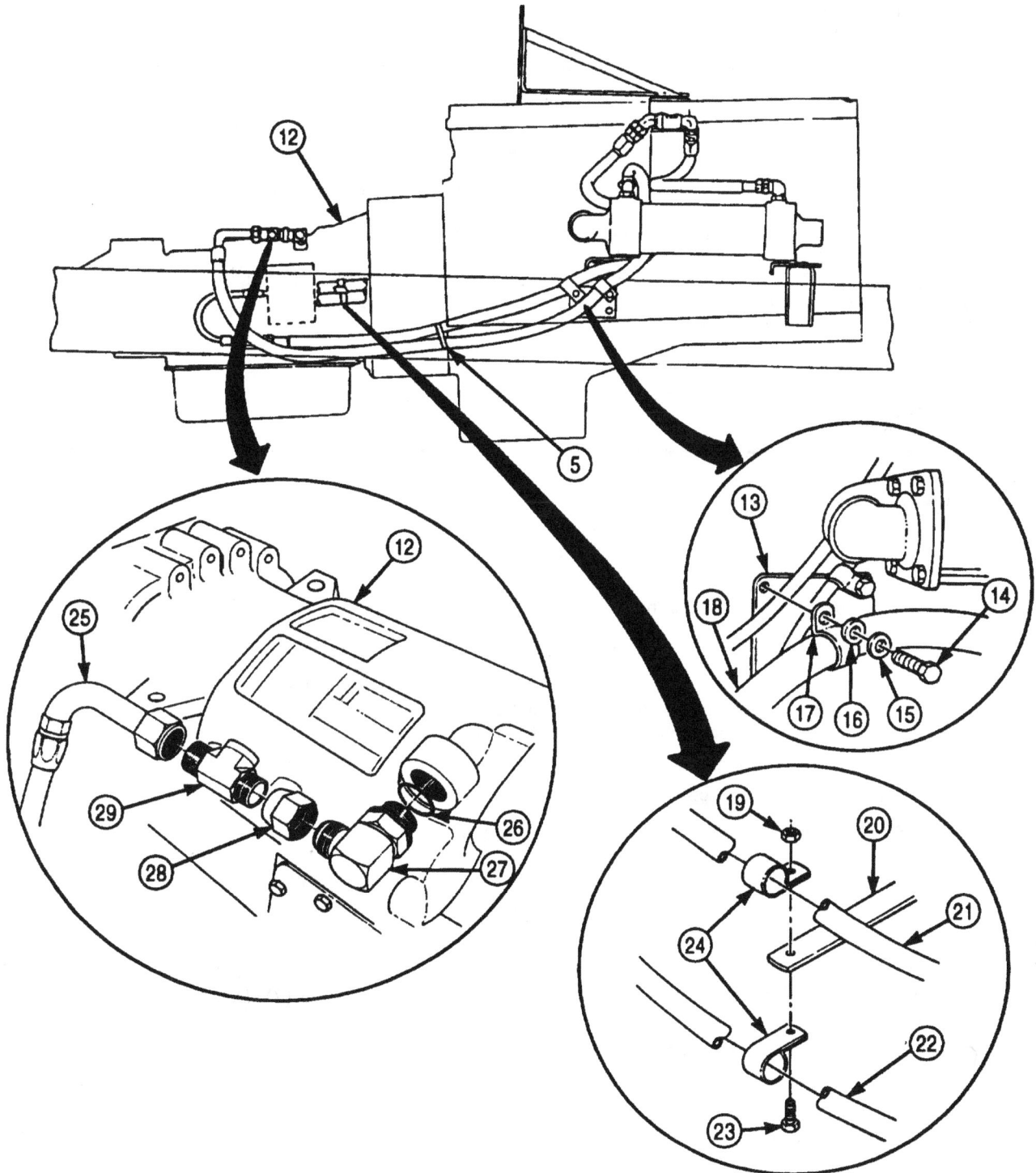

FOLLOW-ON TASKS:
- Install transmission temperature transmitter unit (para. 3-97).
- Fill transmission oil reservoir to proper level (LO 9-2320-272-12).
- Install right splash shield (TM 9-2320-272-10).
- Start engine (TM 9-2320-272-10), check for leaks, and road test vehicle.

3-143. TRANSFER CASE INTERLOCK VALVE REPLACEMENT

THIS TASK COVERS:

a. Removal

b. Installation

INITIAL SETUP:

APPLICABLE MODELS
All (Except M936/A1/A2)

TOOLS
General mechanic's tool kit (Appendix E, Item 1)

MATERIALS/PARTS
Lockwasher (Appendix D, Item 404)
Two locknuts (Appendix D, Item 306)
Antiseize tape (Appendix C, Item 72)

REFERENCES (TM)
TM 9-2320-272-10
TM 9-2320-272-24P

EQUIPMENT CONDITION
- Parking brake set (TM 9-2320-272-10).
- Air reservoirs drained (TM 9-2320-272-10).
- Dump spare tire carrier removed (M929/A1/A2, M930/A1/A2) (para. 3-257).

GENERAL SAFETY INSTRUCTIONS
Do not disconnect air lines before draining air reservoirs.

a. Removal

WARNING

Do not disconnect air lines before draining air reservoirs. Small parts under pressure may shoot out with high velocity, causing injury to personnel.

NOTE
- Replacement of the interlock valve on M936/A1/A2 series vehicles is not performed at unit level because the transfer case must be removed. Notify DS maintenance.

- Tag air lines and wires for installation.

1. Disconnect interlock valve supply line (6) from elbow (5).
2. Disconnect air cylinder supply line (1) from elbow (2).
3. Disconnect vent line (12) from elbow (11).
4. Disconnect wire (8) from connector (9).

NOTE
Assistant will help with step 5.

5. Remove locknut (17), washer (18), diode ground wire (19), locknut (20), interlock valve ground wire (16), lockwasher (21), cable clamp (15), and screw (10) from frame (7). Discard lockwasher (21) and locknuts (17) and (20).
6. Remove two screws (14) and interlock valve (3) from bracket (13).
7. Remove elbows (2), (5), and (11) and adapter (4) from interlock valve (3).

3-143. TRANSFER CASE INTERLOCK VALVE REPLACEMENT (Contd)

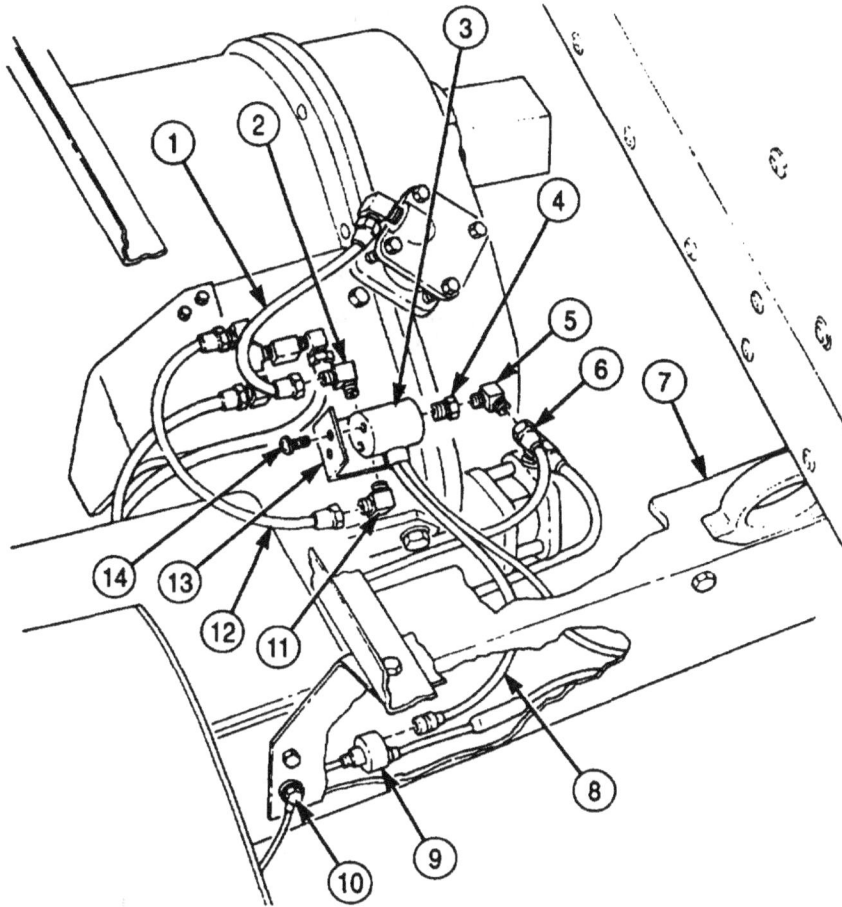

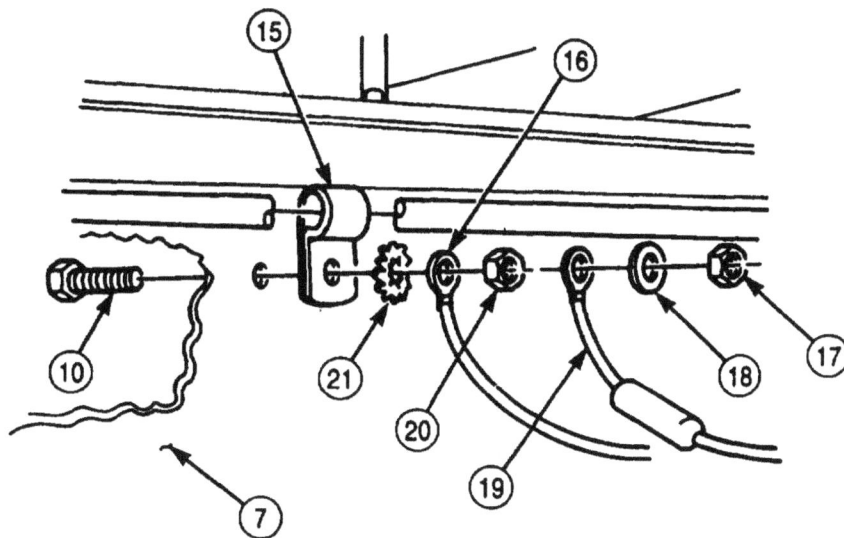

3-143. TRANSFER CASE INTERLOCK VALVE REPLACEMENT (Contd)

b. Installation

NOTE

Male pipe threads must be wrapped with antiseize tape before installation.

1. Install elbow (2), adapter (4), and two elbows (5) and (11) on interlock valve (3).
2. Install interlock valve (3) on bracket (13) with two screws (14).

NOTE

Assistant will help with step 3.

3. Install screw (10) and cable clamp (15) on Frame (7) with new lockwasher (21), interlock valve ground wire (16), new locknut (20), diode ground wire (19), washer (18), and new locknut (17).
4. Connect wire (8) to connector (9).
5. Connect vent line (12) to elbow (11).
6. Connect air cylinder supply line (1) to elbow (2).
7. Connect interlock valve supply line (6) to elbow (5).

3-143. TRANSFER CASE INTERLOCK VALVE REPLACEMENT (Contd)

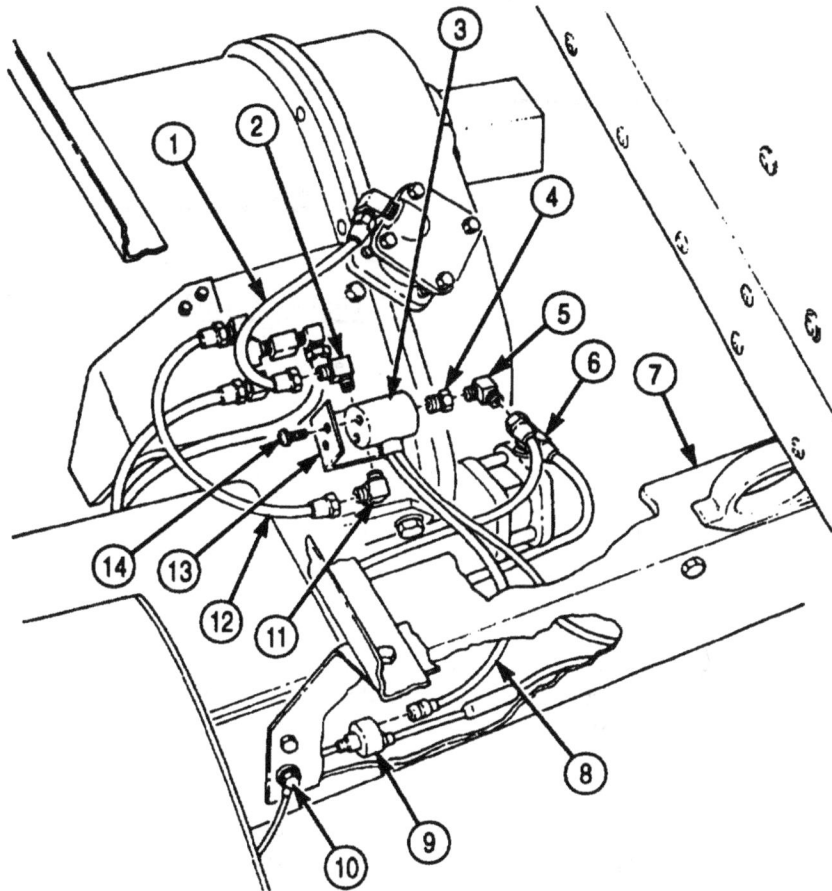

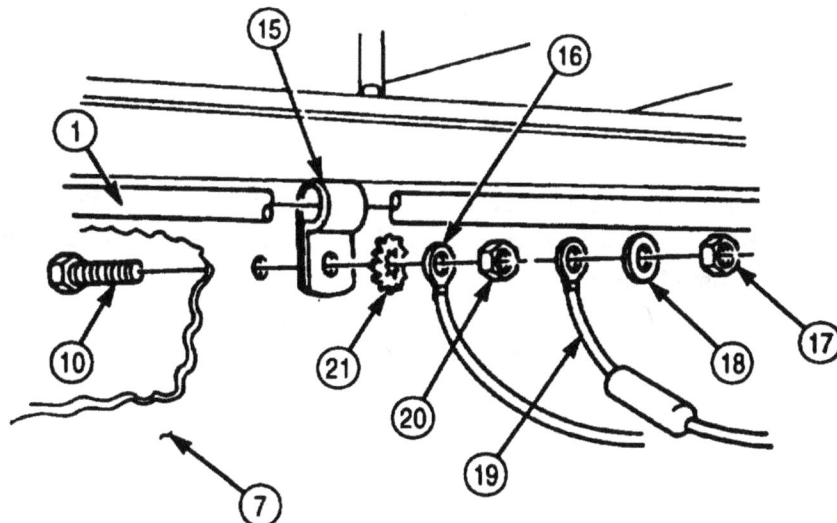

FOLLOW-ON TASKS: • Install dump spare tire carrier (M929/A1/A2, M930/A1/A2) (para. 3-257).
• Start engine (TM 9-2320-272-10), allow air pressure to build to normal operating range, and check interlock valve for leaks. Road test vehicle.

3-144. TRANSFER CASE FRONT AXLE ENGAGEMENT CONTROL VALVE REPLACEMENT

THIS TASK COVERS:

a Removal   b. Installation

INITIAL SETUP:

APPLICABLE MODELS
All

TOOLS
General mechanic's tool kit (Appendix E, 1)

MATERIALS/PARTS
Antiseize tape (Appendix D, Item 72)

REFERENCES (TM)
TM 9-2320-272-10
TM 9-2320-272-24P

EQUIPMENT CONDITION
- Parking brake set (TM 9-2320-272-10).
- Air reservoirs drained (TM 9-2320-272-10).

GENERAL SAFETY INSTRUCTIONS
Do not disconnect air lines before draining air reservoirs.

---

a. Removal

WARNING

Do not disconnect air lines before draining air reservoirs. Small parts under pressure may shoot out with high velocity, causing injury to personnel.

NOTE
Tag air lines for installation.

1. Disconnect three air lines (4) from elbows (3), (6), and (7).
2. Remove elbows (6) and (7) and pipes (5) and (8) from control valve (1).
3. Remove elbow (3) from control valve (1).
4. Remove two nuts (10), screws (2), and control valve (1) from front transfer case bracket (9).

b. Installation

NOTE
Wrap all male pipe threads with antiseize tape before installation.

1. Install central valve (1) on front transfer case bracket (9) with two screws (2) and nuts (10).
2. Install elbow (3) on control valve (1).
3. Install pipes (8) and (5) and elbows (7) and (6) on control valve (1).
4. Connect three air lines (4) to elbows (7), (6), and (3).

**3-144. TRANSFER CASE FRONT AXLE ENGAGEMENT CONTROL VALVE
REPLACEMENT (Contd)**

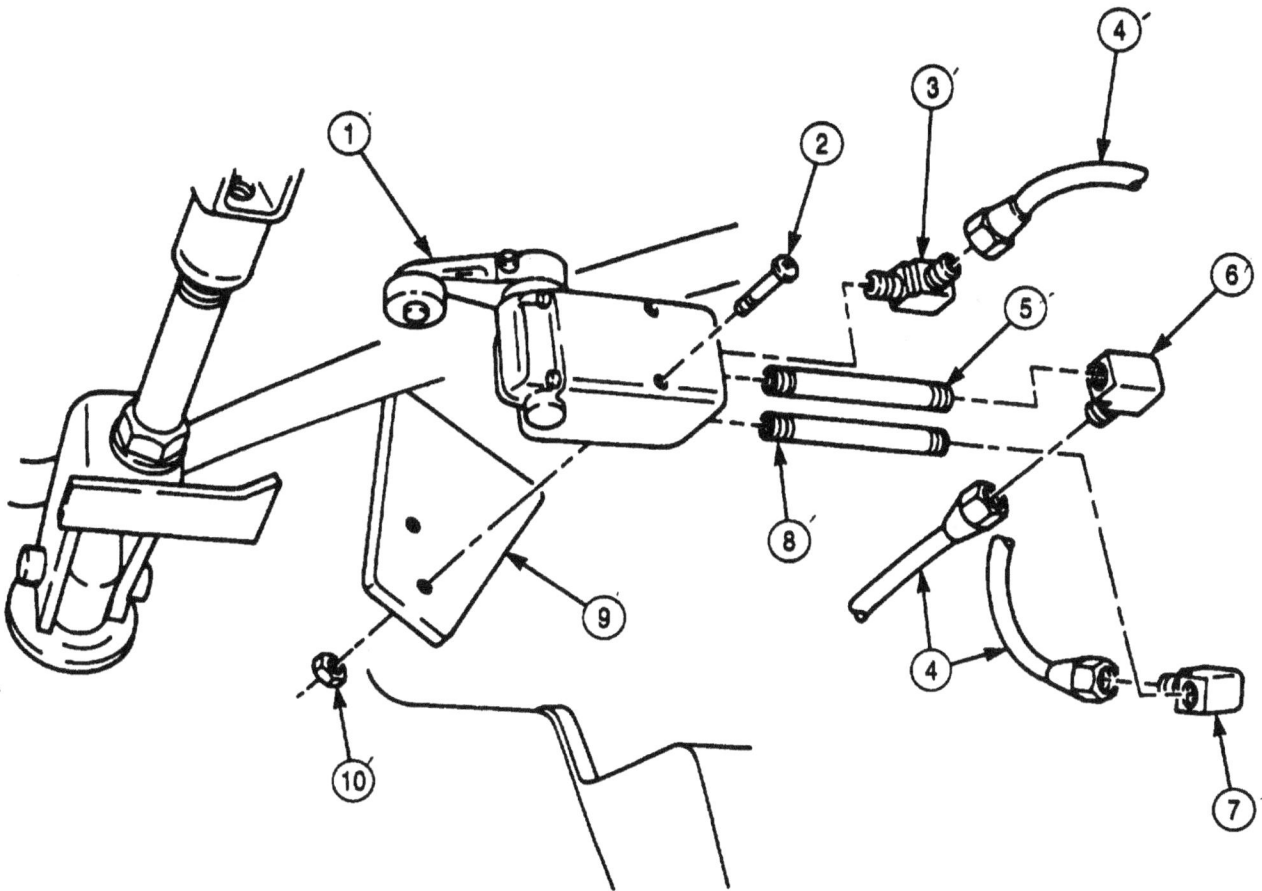

FOLLOW-ON TASK: Start engine (TM 9-2320-272-10), allow air pressure to build to normal operating
range, and check for air leaks and proper front axle engagement. Road test vehicle.

3-145. TRANSMISSION MODULATOR AND CABLE MAINTENANCE

THIS TASK COVERS:

a Removal   c. Adjustment
b. Installation

INITIAL SETUP:

APPLICABLE MODELS
All

TOOLS
General mechanic's tool kit (Appendix E, Item 1)
Torque wrench (Appendix E, Item 144)

MATERIALS/PARTS
O-ring (Appendix D, Item 454)
Locknut (Appendix D, Item 276)
Two locknuts (Appendix D, Item 283)
Locknut (Appendix D, Item 284)

REFERENCES (TM)
LO 9-2320-272-12
TM 9-2320-272-10
TM 9-2320-272-24P

EQUIPMENT CONDITION
- Parking brake set (TM 9-2320-272-10).
- Left splash shield removed (TM 9-2320-272-10).

a. Removal

NOTE

Have drainage container ready to catch oil.

1. Remove screw (15), retaining bracket (14), modulator (17), and O-ring (16) from transmission (13). Discard O-ring (16).

2. Remove locknut (22), screw (19), speedometer cable clamp (18), and modulator cable clamp (20) from transmission bracket (21). Discard locknut (22).

3. Remove modulator return spring (8) from modulator link (9) and bracket (2).

4. Remove two locknuts (3), screws (7), clamp (6), and shim (5) from bracket (2). Discard locknuts (3).

5. Remove locknut (1), screw (10), modulator link (9), and modulator cable (4) from throttle lever (11) on fuel pump (12). Discard locknut (1).

6. Remove modulator link (9) from modulator cable (4).

3-145. TRANSMISSION MODULATOR AND CABLE MAINTENANCE (Contd)

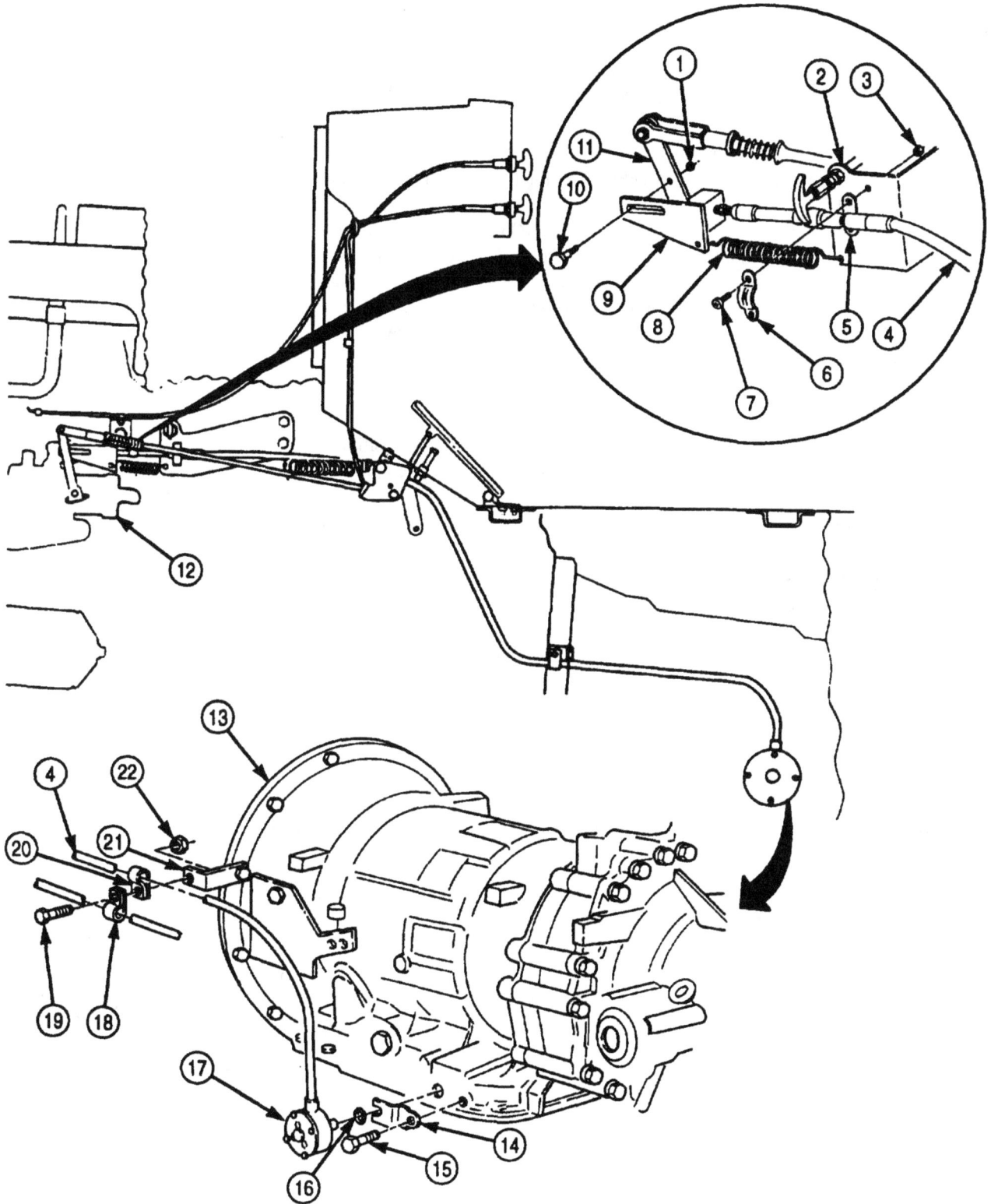

3-145. TRANSMISSION MODULATOR AND CABLE MAINTENANCE (Contd)

b. Installation

1. Install new O-ring (22) and modulator (23) on transmission (19) with retaining bracket (20) and screw (21). Tighten screw (21) 16-20 lb-ft (22-27 N.m).

NOTE

Clamp and shim must align on modulator cable.

2. Install shim (13) and modulator cable (14) on bracket (11) with clamp (15), two screws (16), and new locknuts (12).

NOTE

Locate modulator cable away from sharp edges and avoid sharp bends.

3. Install modulator cable (14) on transmission bracket (27) with modulator cable clamp (26), speedometer cable clamp (24), screw (25), and new locknut (28).

c. Adjustment

NOTE

Assistant will help with steps 1 through 4.

1. Move throttle lever (8) on fuel pump (6) to FULLY OPEN position and hold.
2. Pull threaded end of modulator cable (14) out to STOP position.
3. Loosen jamnut (10) and thread modulator link (18) onto modulator cable (14) until front of slot aligns with hole in throttle lever (8). Continue to loosen jamnut (10) as needed to position modulator link (18).
4. Back off modulator link (18) two turns to provide free pin clearance and tighten jamnut (10).
5. Install modulator link (18) on throttle lever (8) with screw (7) and new locknut (9).

NOTE

- Assistant will help with steps 6 and 7.
- Ensure throttle lever is still in FULLY OPEN position.

6. Loosen throttle stopscrew (3) and jamnut (4) on cab floor (1). Loosen nut (5) as needed.
7. Adjust stop screw (3) to barely touch cab floor (1) side of accelerator pedal (2). Tighten jamnut (4) and nut (5).
8. Release throttle lever (8) on fuel pump (6).
9. Install modulator return spring (17) on modulator link (18) and bracket (11).

3-145. TRANSMISSION MODULATOR AND CABLE MAINTENANCE (Contd)

FOLLOW-ON TASKS: • Fill transmission to proper level (LO 9-2320-272-12).
• Install left splash shield (TM 9-2320-272-10).
• Start engine (TM 9-2320-272-10) and road test vehicle.

Section VIII. PROPELLER SHAFTS, AXLES, AND SUSPENSION SYSTEM MAINTENANCE

3-146. PROPELLER SHAFTS, AXLES, AND SUSPENSION SYSTEM MAINTENANCE INDEX

3-147. TRANSFER CASE FRONT AXLE LOCK-IN CONTROL VALVE REPLACEMENT

THIS TASK COVERS:

a. Removal   b. Installation

INITIAL SETUP:

APPLICABLE MODELS
All

TOOLS
General mechanic's tool kit (Appendix E, Item 1)

MATERIALS/PARTS
Antiseize tape (Appendix C, Item 72)

REFERENCES (TM)
TM 9-2320-272-10
TM 9-2320-272-24P

EQUIPMENT CONDITION
- Parking brake set (TM 9-2320-272-10).
- Air reservoirs drained (TM 9-2320-272-10).
- Front axle lock-in switch removed (para 3-101).
- Fuel selector valve removed (if equipped) (para. 3-24).

GENERAL SAFETY INSTRUCTIONS
Do not disconnect air lines before draining air reservoirs.

WARNING

Do not disconnect air lines before draining air reservoirs. Small parts under pressure may shoot out with high velocity, causing injury to personnel.

a. Removal

NOTE
Tag air lines for installation.

1. Disconnect control line (8) from front axle control valve elbow (7).
2. Disconnect manifold tee supply line (1) from adapter fitting (2).
3. Remove two screws (4), instruction plate (5), and control valve (3) from instrument panel (6).

b. Installation

NOTE
If new control valve is being installed, use fitting from old valve.
Clean all male pipe threads and wrap with antiseize tape before installation.

1. Install control valve (3) and instruction plate (5) on instrument panel (6) with two screws (4).
2. Connect manifold tee supply line (1) to adapter fitting (2).
3. Connect control line (8) to front axle control valve elbow (7).

3-147. TRANSFER CASE FRONT AXLE LOCK-IN CONTROL VALVE REPLACEMENT (Contd)

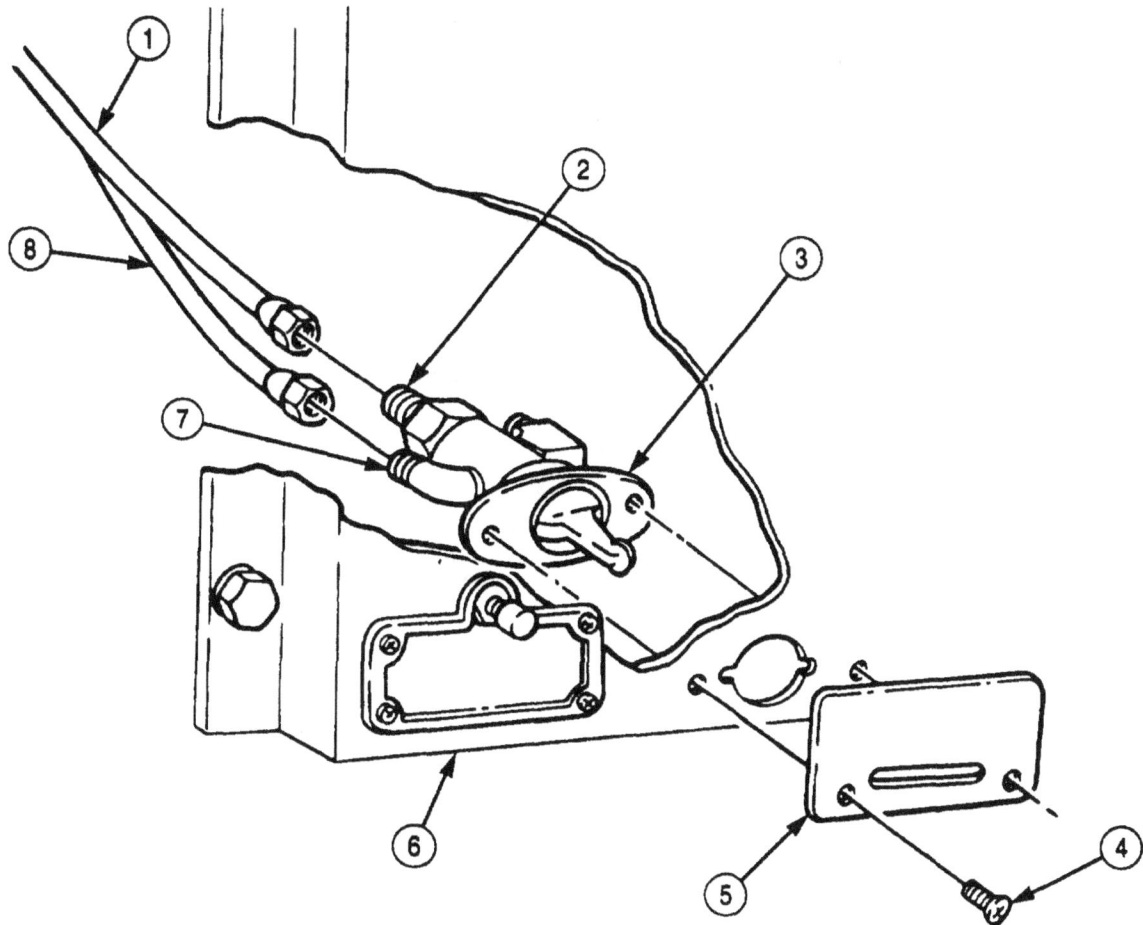

FOLLOW-ON TASKS: • Install front axle lock-in switch (para. 3-101).
• Install fuel selector (para. 3-24).
• Start engine (TM 9-2320-272-10) and allow air pressure to build up to normal operating pressure. Check for air leaks at axle lock-in valve. Road test vehicle.
• Check front axle lock-in valve for proper operation.

3-148. TRANSMISSION TO TRANSFER CASE PROPELLER SHAFT MAINTENANCE

THIS TASK COVERS:

a. Removal   c. Assembly
b. Disassembly   d. Installation

INITIAL SETUP:

APPLICABLE MODELS
All

TOOLS
General mechanic's tool kit (Appendix E, Item 1)
Torque wrench (Appendix E, Item 144)

MATERIALS/PARTS
Twelve lockwashers (Appendix D, Item 361)
Universal joint kit (Appendix D, Item 702)

REFERENCES (TM)
LO 9-2320-272-12
TM 9-2320-272-10
TM 9-2320-272-24P

EQUIPMENT CONDITION
- Parking brake set (TM 9-2320-272-10).
- Wheels chocked (TM 9-2320-272-10).

---

a. Removal

1. Place transfer case lever (1) in high position.
2. Remove four screws (5) and lockwashers (4) from transmission yoke (2) and transfer case yoke (3). Discard lockwashers (4).
3. Place transfer case lever (1) in neutral position.
4. Turn propeller shaft (6) until remaining screws (5) can be seen.
5. Place transfer case lever (1) to high position.
6. Remove four remaining screws (5), lockwashers (4), and propeller shaft (6) from transmission yoke (2) and transfer case yoke (3). Discard lockwashers (4).

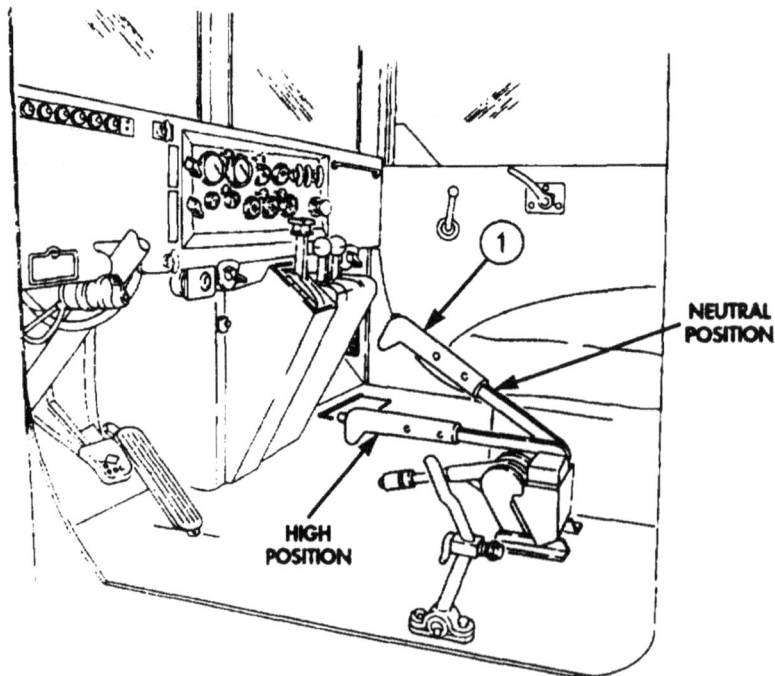

3-148. TRANSMISSION TO TRANSFER CASE PROPELLER SHAFT MAINTENANCE (Contd)

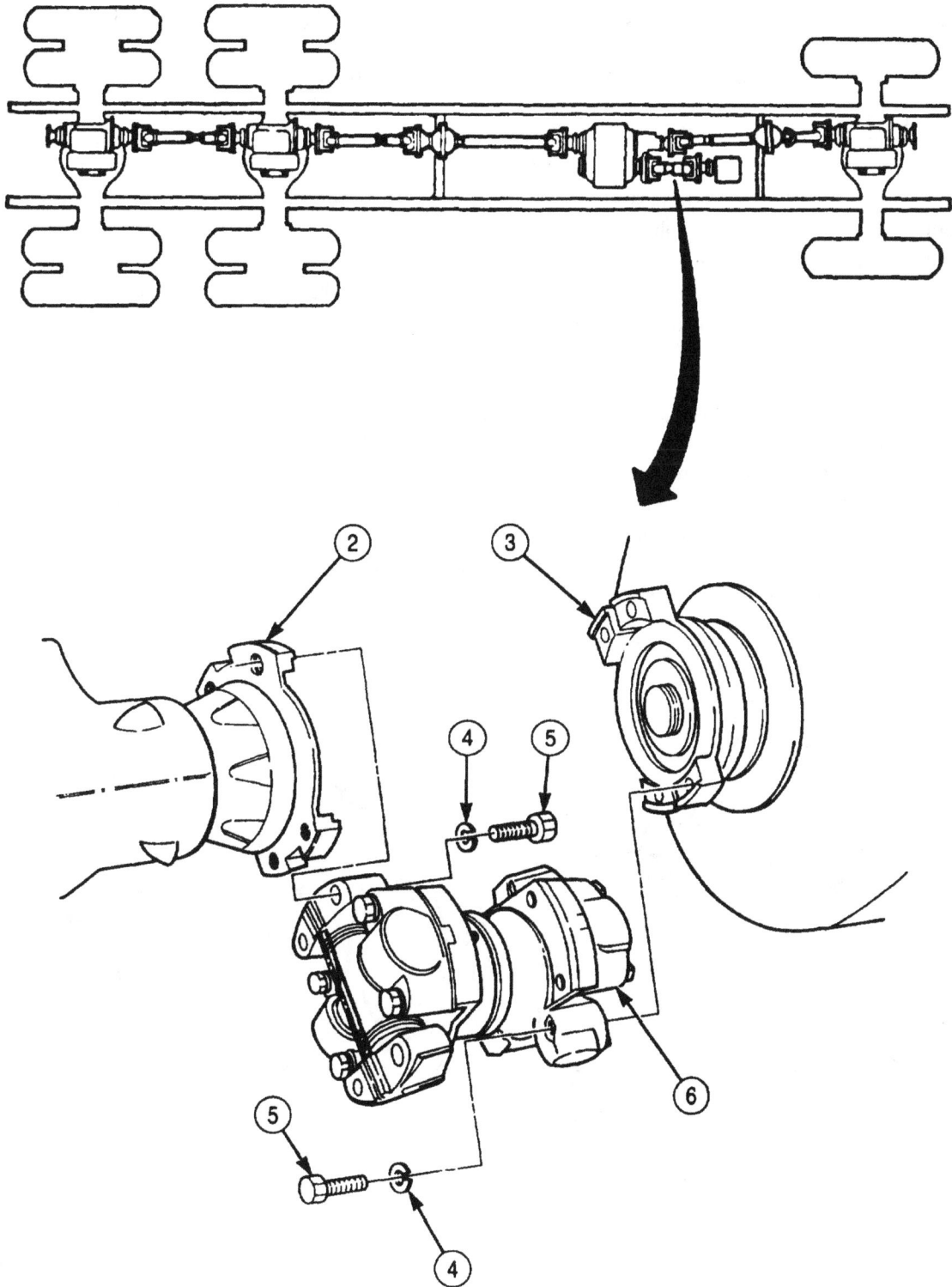

3-148. TRANSMISSION TO TRANSFER CASE PROPELLER SHAFT MAINTENANCE (Cod)

b. Disassembly

1. Remove grease fitting (5) from input yoke (4).
2. Remove four screws (1), lockwashers (2), and cross assembly (3) from input yoke (4). Discard lockwashers (2).
3. Separate input yoke (4) from output yoke (7). Replace oil seal (6) if damaged.

c. Assembly

NOTE
Perform step 1 only if seal was removed.

1. Install seal assembly (6) over splined output yoke (7).
2. Install output yoke (7) into input yoke (4).
3. Install cross assembly (3) on input yoke (4) with four new lockwashers (2) and screws (1).
4. Install grease fitting (5) on input yoke (4).

d. Installation

1. Install propeller shaft (13) between transmission yoke (9) and transfer case yoke (12) with four new lockwashers (10) and screws (11). Tighten screws (11) 90-110 lb-ft (122-149 N.m).
2. Place transfer case lever (8) in neutral position.
3. Turn propeller shaft (13) until remaining screws (11) can be installed.
4. Place transfer case lever (8) in high position.
5. Install propeller shaft (13) between transmission yoke (9) and transfer case yoke (12) with four remaining new lockwashers (10) and screws (11). Tighten screws (11) 90-110 lb-ft (122-149 N.m).

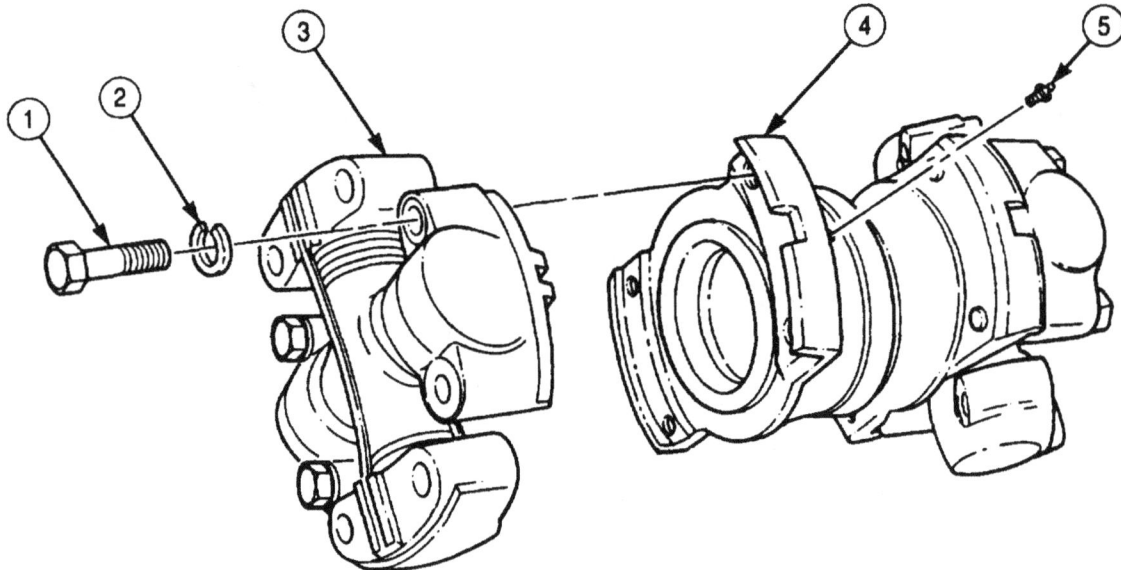

3-148. TRANSMISSION TO TRANSFER CASE PROPELLER SHAFT MAINTENANCE (Contd)

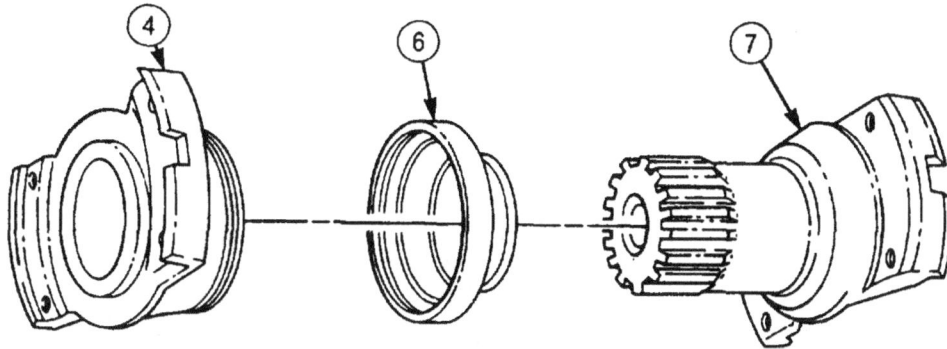

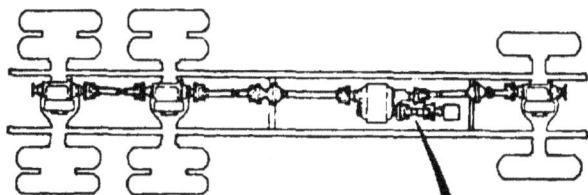

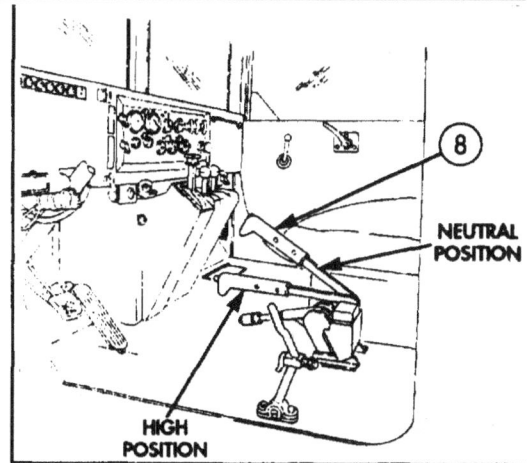

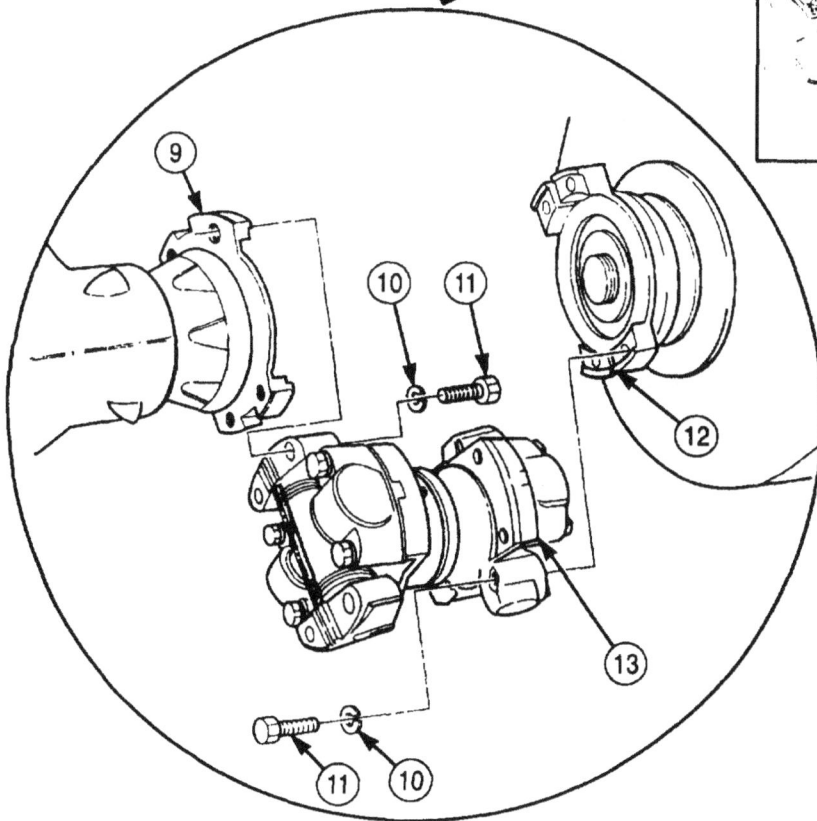

FOLLOW-ON TASK: Lubricate propeller shaft universal joints (LO 9-2320-272-12).

3-149. TRANSFER CASE TO FRONT AXLE PROPELLER SHAFT REPLACEMENT

THIS TASK COVERS:

a. Removal   b. Installation

INITIAL SETUP:

APPLICABLE MODELS
All

TOOLS
General mechanic's tool kit (Appendix E, Item 1)
Torque wrench (Appendix E, Item 146)

MATERIALS/PARTS
Sixteen locknuts (Appendix D, Item 291)
Two lockwashers (Appendix D, Item 382)

REFERENCES (TM)
LO 9-2320-272-12
TM 9-2320-272-10
TM 9-2320-272-24P

EQUIPMENT CONDITION
- Parking brake set (TM 9-2320-272-10).
- Wheels chocked (TM 9-2320-272-10).

---

a. Removal

NOTE
Step 1 refers to M939/A1 vehicles only

1. Remove two screws (4), lockwashers (5), and center bearing bracket (6) from bracket (7). Discard lockwashers (5).
2. Remove eight locknuts (10) and screws (8) from propeller shaft flange (11) and differential flange (9). Discard locknuts (10).
3. Remove eight locknuts (1) and screws (2) from propeller shaft flange (12) and transfer case flange (13). Discard locknuts (1).
4. Remove transfer case to front propeller shaft assembly (3).

b. Installation

1. Install transfer case to front axle propeller shaft assembly (3) on transfer case flange (13) with eight screws (2) and new locknuts (1). Tighten locknuts (1) 32-40 lb-ft (43-54 N.m).
2. Install transfer case to front axle propeller shaft assembly (3) on differential flange (9) with eight screws (8) and new locknuts (10). Tighten locknuts (10) 32-40 lb-ft (43-54 N.m).

NOTE
Step 4 refers to M939/A1 vehicles only.

3. Install center bearing bracket (6) on bracket (7) with two new lockwashers (5) and screws (4).

3-149. TRANSFER CASE TO FRONT AXLE PROPELLER SHAFT REPLACEMENT (Contd)

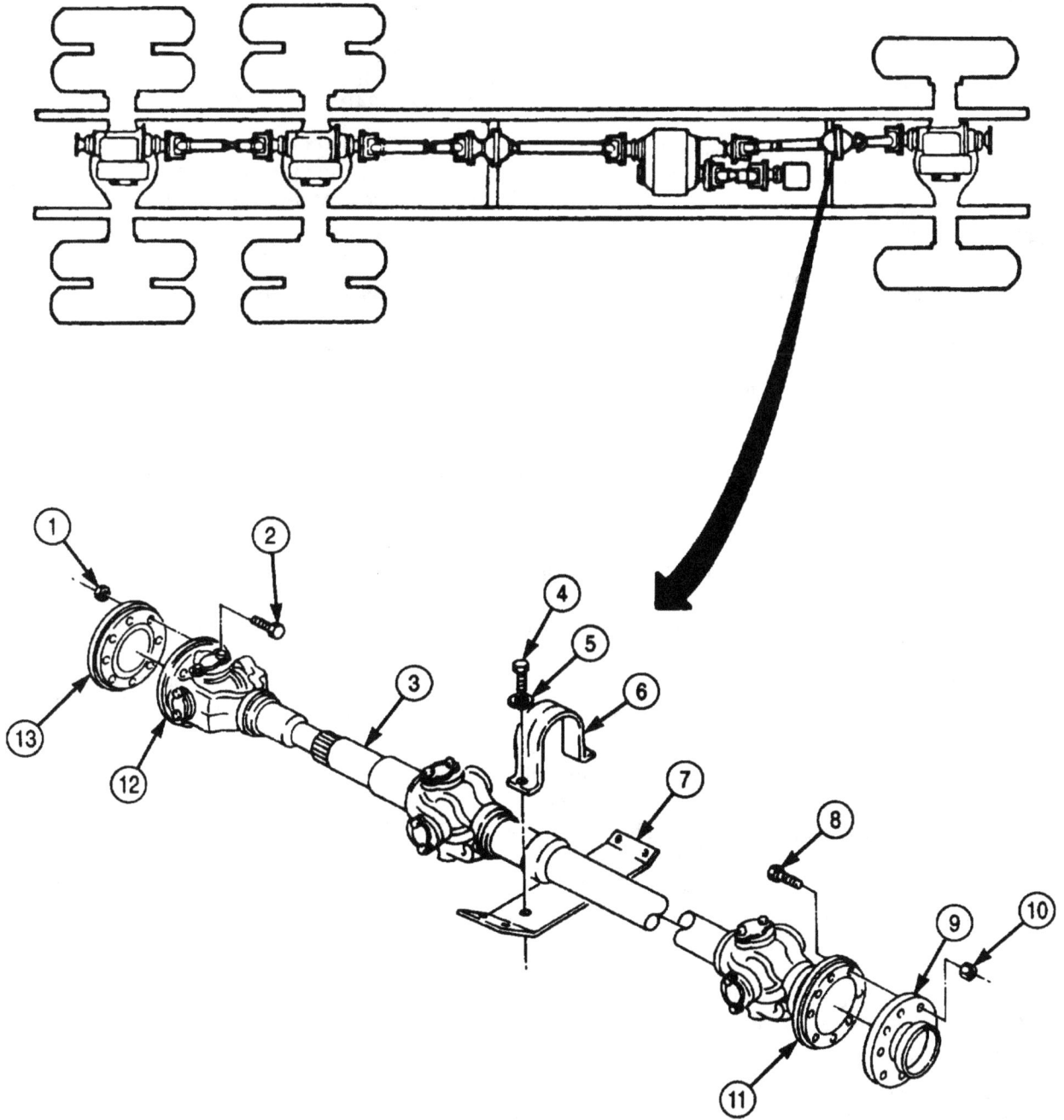

FOLLOW-ON TASK: Lubricate propeller shaft universal joints (LO 9-2320-272-12).

3-150. TRANSFER CASE TO FORWARD-REAR AXLE PROPELLER SHAFT MAINTENANCE

THIS TASK COVERS:

a. Removal
b. Disassembly
c. Cleaning and Inspection

d. Assembly
e. Installation

INITIAL SETUP:

APPLICABLE MODELS
M927/A1/A2, M928/A1/A2, M934/A1/A2, M935/A1/A2

TOOLS
General mechanic's tool kit (Appendix E, Item 1)
Torque wrench (Appendix E, Item 146)

MATERIALS/PARTS
Twenty-two locknuts (M939A2 old
    configuration) (Appendix D, Item 291)
Sixteen locknuts (Appendix D, Item 291)
Two universal plates (Appendix D, Item 704)
Cotter pin (M939/A1 old configuration)
    (Appendix D, Item 80)
Twelve locknuts (M939/A1 old configuration)
    (Appendix D, Item 291)
Drycleaning solvent (Appendix C, Item 71)
GAA Grease (Appendix C, Item 28)

REFERENCES (TM)
LO 9-2320-272-12
TM 9-2320-272-10
TM 9-2320-272-24P

EQUIPMENT CONDITION
- Parking brake set (TM 9-2320-272-10).
- Wheels chocked (TM 9-2320-272-10).

GENERAL SAFETY INSTRUCTIONS
- Drycleaning solvent is hazardous and flammable. Do not use near open flame.
- Keep fire extinguisher nearby when drycleaning solvent is used.

a. Removal

1. Remove eight locknuts (3) from transfer case brake drum (1) and transfer case shaft yoke flange (2). Discard locknuts (3).

2. Remove eight screws (13), locknuts (12), and rear propeller shaft yoke flange (14) from forward-rear axle yoke flange (11). Discard locknuts (12).

3. Remove forward propeller shaft front section (4) from forward propeller shaft aft section (24).

4. Remove dustcover (5) and spacer (6) from forward propeller shaft front section (4).

NOTE
- Perform steps 5 through 14 only for vehicles equipped with old forward-rear axle propeller shaft.
- Perform steps 5 through 9 for models M927/A1, M928/A1, M934/A1, and M935/A1 models only.

5. Remove eight screws (17) and locknuts (21) from rear propeller shaft companion flange (16) and center bearing companion flange (20). Discard locknuts (21).

6. Remove forward-rear axle propeller shaft (15) and center bearing flange (20) from center bearing bracket (10).

7. Remove four screws (8) and locknuts (7) from center bearing mounting bracket (10) and cross-member (9). Discard locknuts (7).

8. Remove cotter pin (18) and nut (19) from forward propeller shaft (4). Discard cotter pin (18).

9. Remove center bearing companion flange (20), dustcover (22), spacer (23), center bearing mounting bracket (10), spacer (6), and dustcover (5) from forward propeller shaft (4).

3-150. TRANSFER CASE TO FORWARD-REAR AXLE PROPELLER SHAFT MAINTENANCE (Contd)

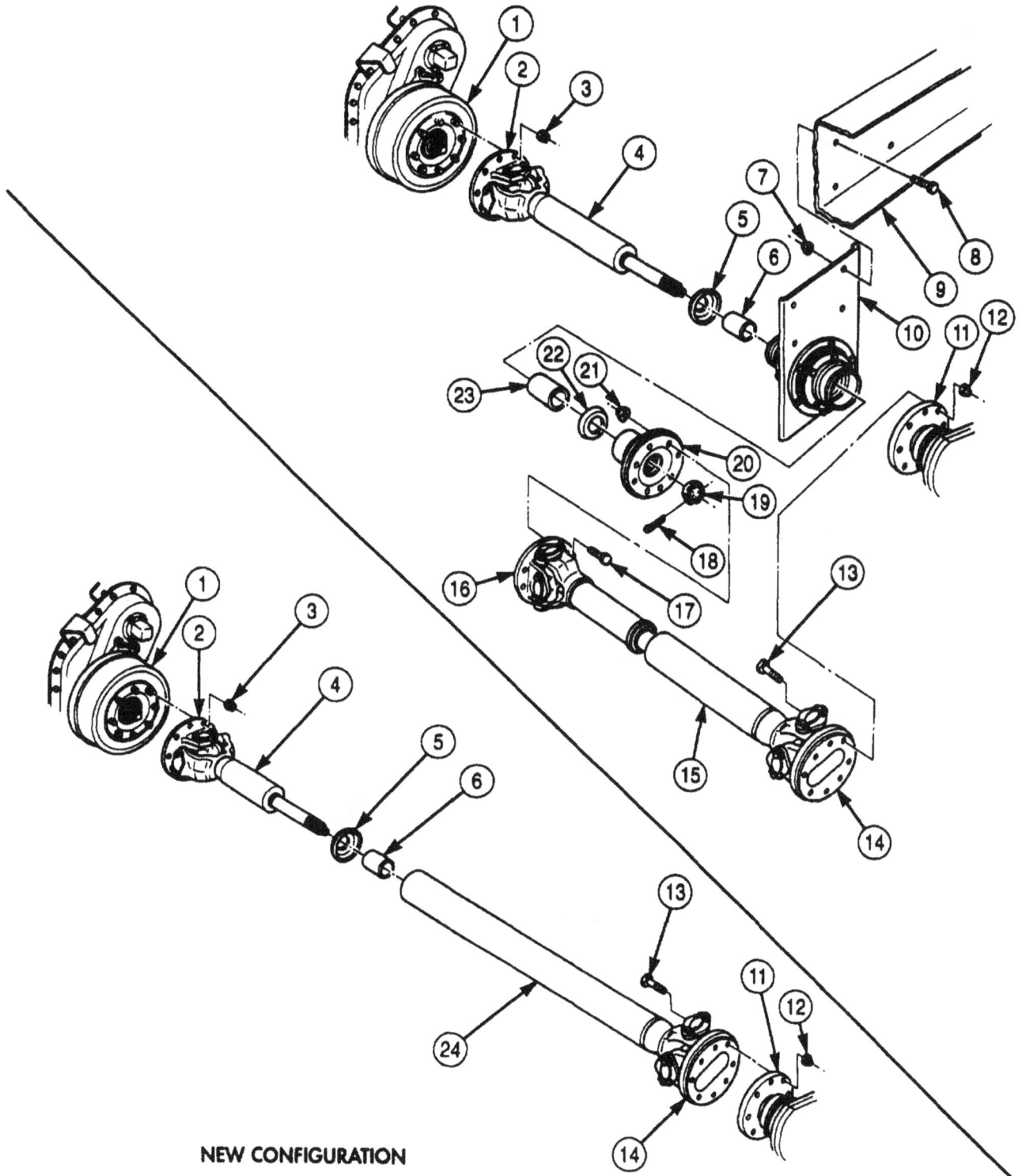

NEW CONFIGURATION

3-150. TRANSFER CASE TO FORWARD-REAR AXLE PROPELLER SHAFT MAINTENANCE (Contd)

NOTE

Perform steps 10 through 14 for M927A2, M928A2, M934A2, and M935A2 models only.

10. Remove eight locknuts (9), screws (7), and rear axle flange (8) from center bearing flange (6). Discard locknuts (9).

11. Remove eight locknuts (13) and flange (14) from transfer case brakedrum studs (1). Discard locknuts (13).

12. Remove two locknuts (12), screws (4), washers (3), center bearing (11), forward-rear axle propeller shaft (10), and vibration damper (5) from bracket (2). Discard locknuts (12).

13. Remove locknut (15), washer (16), clamp (17), spacer (18), screw (20), and clamp (19) from bracket (2) and crossmember (22). Discard locknut (15).

14. Remove three locknuts (24), washers (23), screws (21), and bracket (2) from crossmember (22). Discard locknuts (24).

15. Straighten tabs on two universal plates (26).

16. Remove four screws (25), two universal plates (26), bearing caps (34), and universal joint (28) from center bearing propeller shaft (27). Discard universal plates (26).

b. Disassembly

NOTE

Perform disassembly for M927A2, M928A2, M934A2, and M935A2 models only.

1. Loosen dust seal (31) on shaft (30).

2. Remove shaft (30) from shaft (32).

3. Remove dust seal (31) from shaft (32).

4. Remove grease fitting (33) from yoke (29).

c. Cleaning and Inspection

WARNING

Drycleaning solvent is flammable and toxic. Do not use near open flame and always have fire extinguisher nearby when solvents are used. Use only in well-ventilated places, wear protective clothing, and dispose of cleaning rags in approved container. Failure to do this may result in injury or death to personnel and/or damage to equipment.

1. Clean all parts in drycleaning solvent and allow to dry.

2. Inspect all parts for cracks, fitting, or scoring. Replace parts as necessary if cracked, pitted, or scored.

3-150. TRANSFER CASE TO FORWARD-REAR AXLE PROPELLER SHAFT MAINTENANCE (Contd)

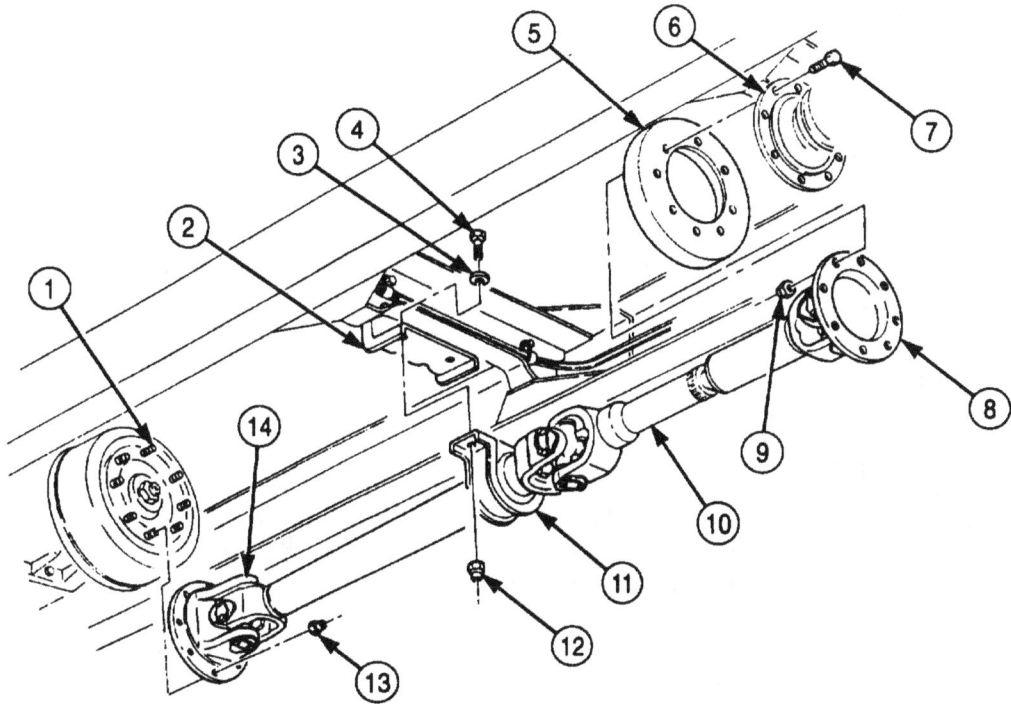

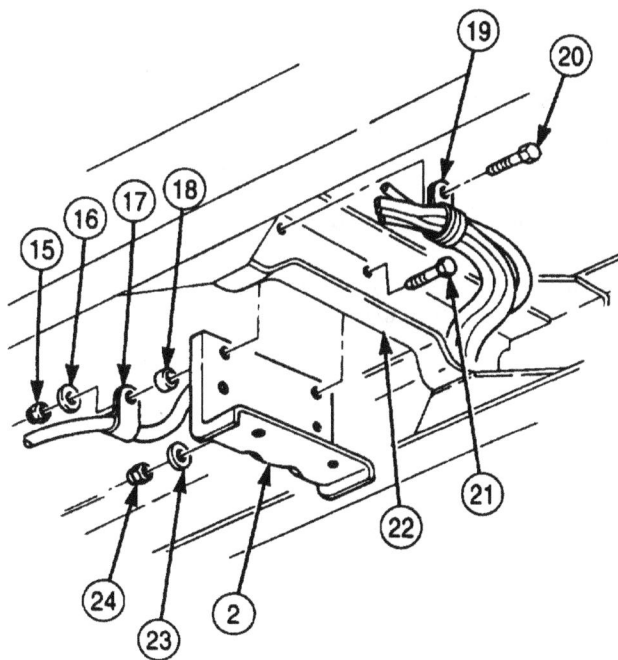

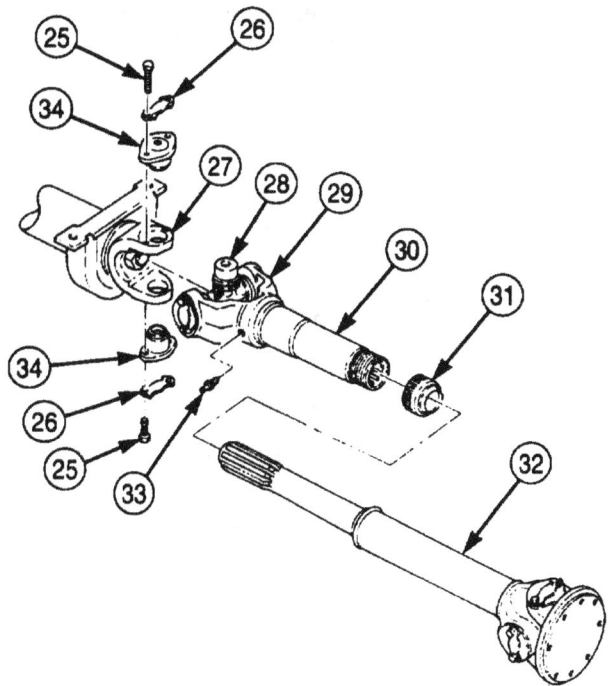

3-150. TRANSFER CASE TO FORWARD-REAR AXLE PROPELLER SHAFT MAINTENANCE (Contd)

d. Assembly

NOTE

Perform assembly for M927A2, M928A2, M934A2, and M935A2 models only.

1. Install grease fitting (10) in yoke (6).

NOTE

Apply light coat of GAA grease to dust seal and splines of shaft before installation.

2. Install dust seal (8) on splines of shaft (9).
3. Install shaft (7) on shaft (9).
4. Tighten dust seal (8) on shaft (7).

e. Installation

1. Install universal joint (5) on center bearing propeller shaft (4) with two bearing caps (1), new universal plates (3), and four screws (2). Tighten screw (2) 32-40 lb-ft (43-54 N.m).
2. Bend tabs on universal plates (3) against screws (2).

NOTE

Perform steps 3 through 7 for M927A2, M928A2, M934A2, and M935A2 models only.

3. Install bracket (19) on crossmember (18) with three screws (17), washers (20), and new locknuts (21).
4. Install clamps (15) and (13) on bracket (19) and crossmember (18) with screw (16), spacer (14), washer (121, and new locknut (11).

NOTE

Assistant will help with steps 5 through 7.

5. Holding forward-rear axle propeller shaft (27) on brakedrum studs (32) and forward-rear axle flange (24), install center bearing (28) on bracket (19) with two washers (33), screws (34), and new locknuts (29).
6. Install flange (31) on eight transfer case brakedrum studs (32) with new locknuts (30). Tighten locknuts (30) 32-40 lb-ft (43-54 N.m) in alternate sequence.
7. Install vibration dampener (22) and flange (25) on forward-rear flange (24) with eight screws (23) and new locknuts (26). Tighten locknuts (26) 30-40 lb-ft. (43-54 N.m).

3-150. TRANSFER CASE TO FORWARD-REAR AXLE PROPELLER SHAFT MAINTENANCE (Contd)

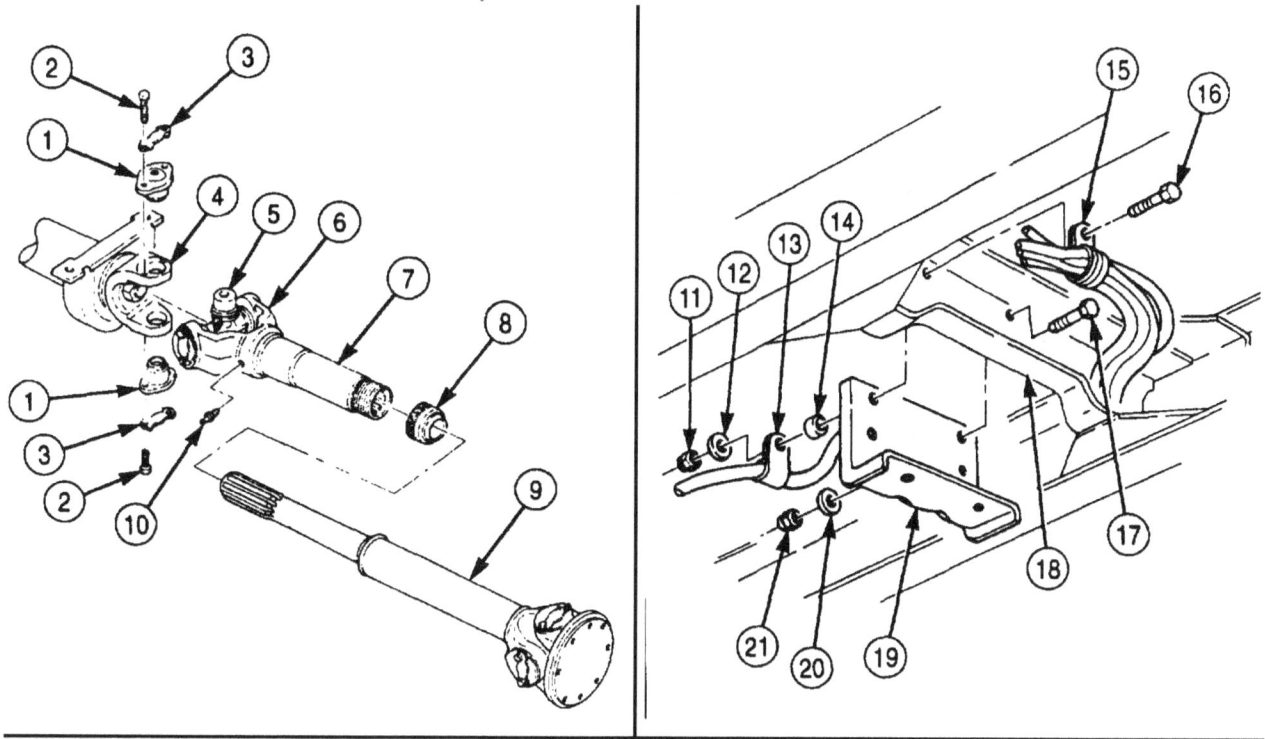

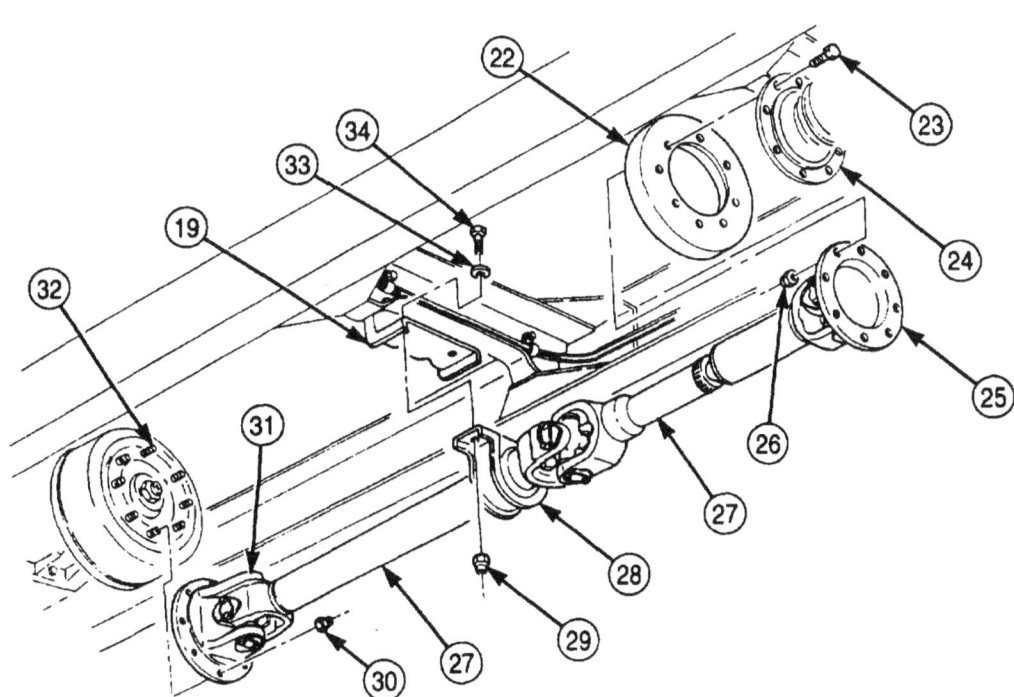

3-150. TRANSFER CASE TO FORWARD-REAR AXLE PROPELLER SHAFT MAINTENANCE (Contd)

NOTE

Perform steps 8 through 12 only for vehicles equipped with old propeller shaft.

8. Install dustcover (5), spacer (6), center bearing mounting bracket (10), spacer (23), dustcover (22), and center bearing companion flange (20) on forward propeller shaft (4) with nut (19). Tighten nut (19) 100-115 lb-ft (136-156 N.m). If needed, turn nut (19) clockwise to align nearest slot with hole in shaft (4) to install new cotter pin (18).

9. Install new cotter pin (18) on nut (19) and forward propeller shaft (4).

NOTE

Assistant will help with steps 10 through 14.

10. Install center bearing mounting bracket (10) on crossmember (9) with four screws (8) and new locknuts (7).

NOTE

Position rear-rear propeller shaft so that grease fitting on U-joint faces downward. Refer to new configuration for steps 13 and 14.

11. Install rear propeller shaft (15) companion flange (16) on center bearing companion flange (20) with eight screws (17) and new locknuts (21). Tighten locknuts (21) 30-40 lb-ft (41-54 N.m).

12. Install spacer (6) and dustcover (5) on forward propeller shaft front section (4).

13. Install forward propeller shaft front section (4) on forward propeller shaft aft section (24).

14. Install rear propeller shaft yoke flange (14) on forward-rear axle yoke flange (11) with eight screws (13) and new locknuts (12).

15. Install transfer case shaft yoke flange (2) on transfer case brake drum (1) with eight new locknuts (3).

3-150. TRANSFER CASE TO FORWARD-REAR AXLE PROPELLER SHAFT MAINTENANCE (Contd)

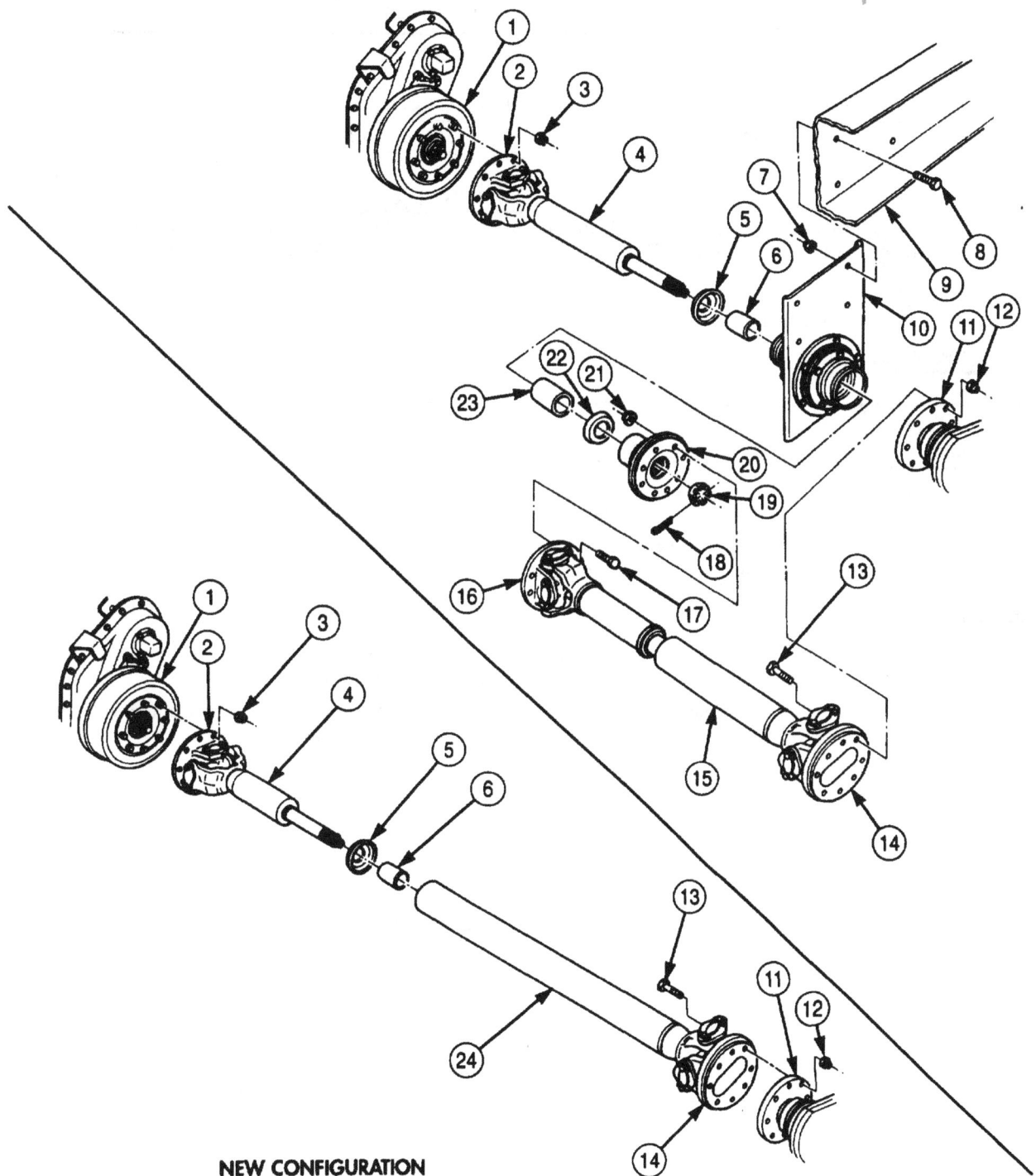

NEW CONFIGURATION

FOLLOW-ON TASK: Lubricate propeller shaft universal joints (LO 9-2320-272-12).

3-151. FORWARD-REAR TO REAR-REAR PROPELLER SHAFT REPLACEMENT

THIS TASK COVERS:

a. Removal
b. Installation

INITIAL SETUP:

APPLICABLE MODELS
All except M927/A1, M928/A1/A2, M934/A1/A2, M935/A1/A2

REFERENCES (TM)
TM 9-2320-272-12
TM 9-2320-272-10

TOOLS
General mechanic's tool kit (Appendix E, Item 1)
Torque wrench (Appendix E, Item 148)

EQUIPMENT CONDITION
• Parking brake set (TM 9-2320-272-10).
• Wheels chocked (TM 9-2320-272-10).

MATERIALS/PARTS
Sixteen locknuts (Appendix D, Item 291)

PERSONNEL REQUIRED
Two

a. Removal

1. Remove eight locknuts (4), screws (3), and forward-rear to rear-rear axle propeller shaft (2) from transfer case input flange (5). Discard locknuts (4).
2. Remove eight locknuts (7), screws (1), and forward-rear to rear-rear axle propeller shaft (2) from transmission output flange (6). Discard locknuts (7).

b. Installation

1. Install forward-rear to rear-rear axle propeller shaft (2) on transmission output flange (6) with eight screws (1) and new locknuts (7). Tighten locknuts (7) 32-40 lb-ft. (43-54 N.m).
2. Install forward-rear to rear-rear axle propeller shaft (2) on transfer case input flange (5) with eight screws (3) and new locknuts (4). Tighten locknuts (4) 32-40 lb-ft (43-54 N.m).

3-151. FORWARD-REAR TO REAR-REAR PROPELLER SHAFT REPLACEMENT (Contd)

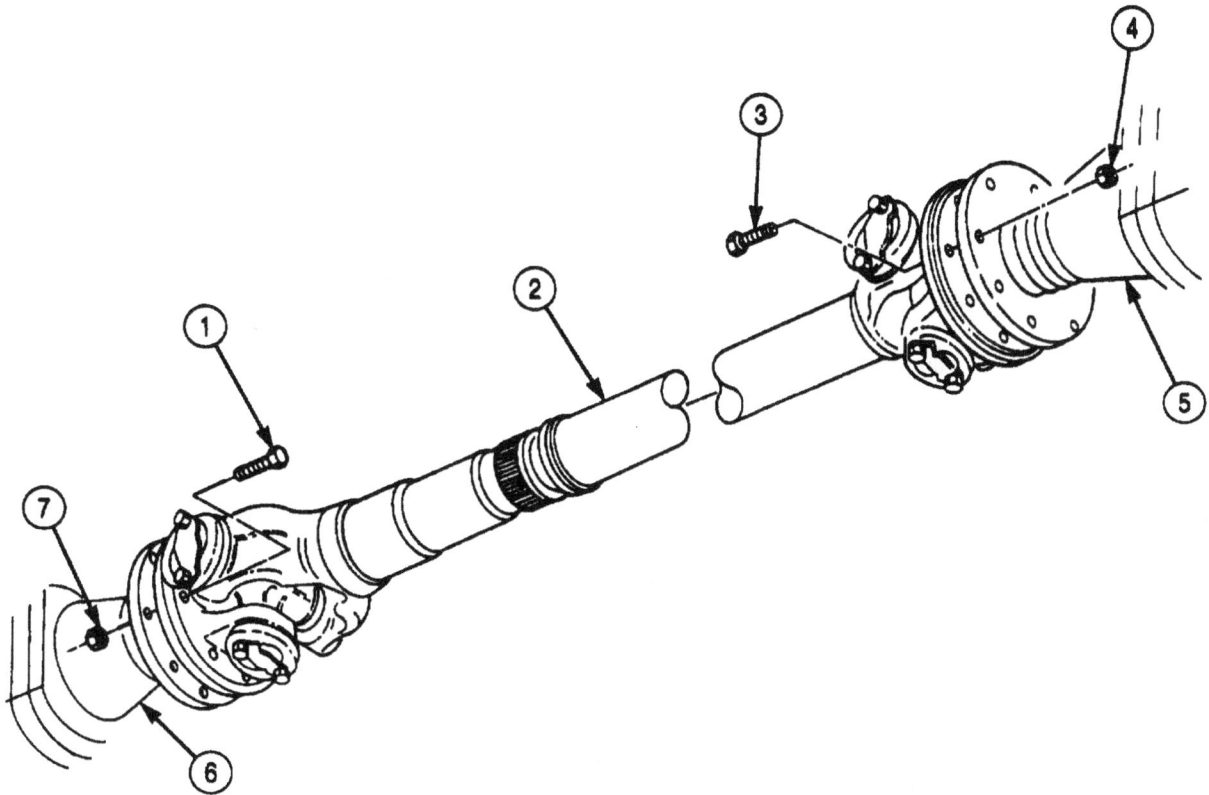

3-152. TOE-IN CHECK AND ADJUSTMENT

THIS TASK COVERS:

a. Toe-in Check   b. Toe-in Adjustment

INITIAL SETUP:

APPLICABLE MODELS
All

TOOLS
General mechanic's tool kit (Appendix E, Item 1)
Toe-in gauge (Appendix E, Item 56)
Torque wrench (Appendix E, Item 144)

REFERENCES (TM)
TM 9-2320-272-10
TM 9-2320-272-24P

EQUIPMENT CONDITION
- Vehicle parked on level ground (TM 9-2320-272-10).
- Engine off (TM 9-2320-272-10).
- Parking brake set (TM 9-2320-272-10).
- Tires properly inflated (TM 9-2320-272-10).
- Wheel bearings adjusted (para. 3-225).

a. Toe-in Check

1. Place front wheels (1) in straight-ahead position.
2. Place toe-in gauge between front wheels (1) at middle of tires (3) as far in front of axle (2) as possible.
3. Move toe-in gauge until chains just touch the ground.
4. Move scale so pointer reads 0.

NOTE
Assistant will help with steps 5 through 7.

5. Start engine (TM 9-2320-272-10).
6. With toe-in gauge in place, move vehicle forward until toe-in gauge is in back of axle (2) and chains just touch ground.
7. Stop engine (TM 9-2320-272-10).

b. Toe-in Adjustment

1. Read position of pointer on scale. Pointer should read 1/16-3/16-inch toe-in.
2. If reading is not within limits given, leave toe-in gauge in place and proceed to step 3.
3. Loosen two nuts (5) on tie rod (4).
4. Turn tie rod (4) until pointer reads 1/16-3/16-inch toe-in on scale.
5. Tighten nuts (5) 60-80 lb-ft (81-109 N.m).
6. Remove toe-in gauge.

3-152. TOE-IN CHECK AND ADJUSTMENT (Contd)

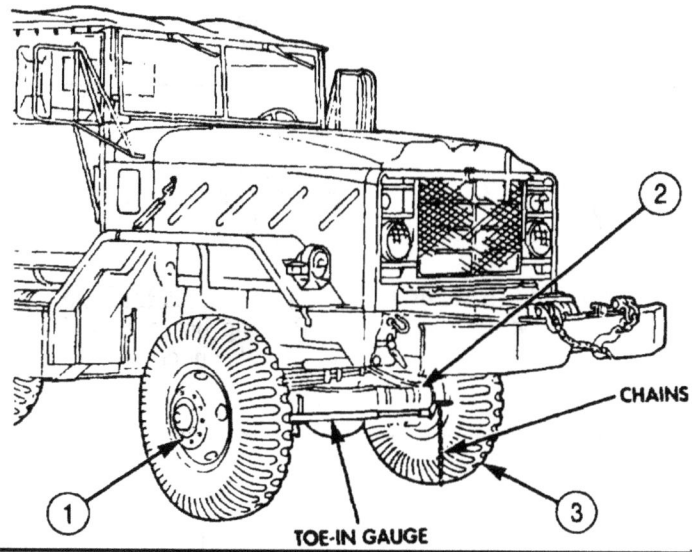

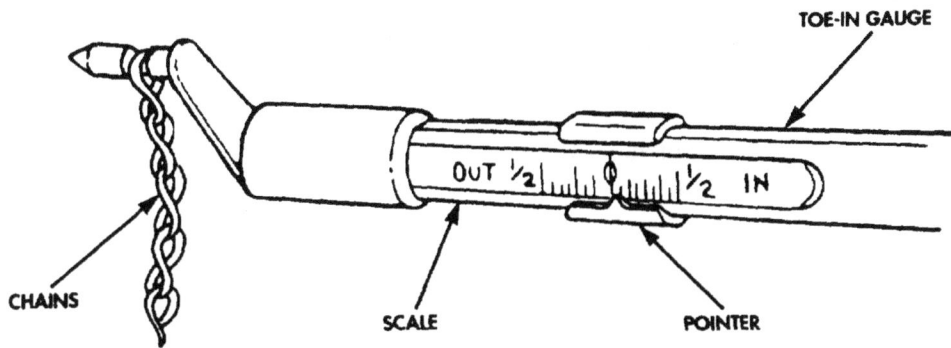

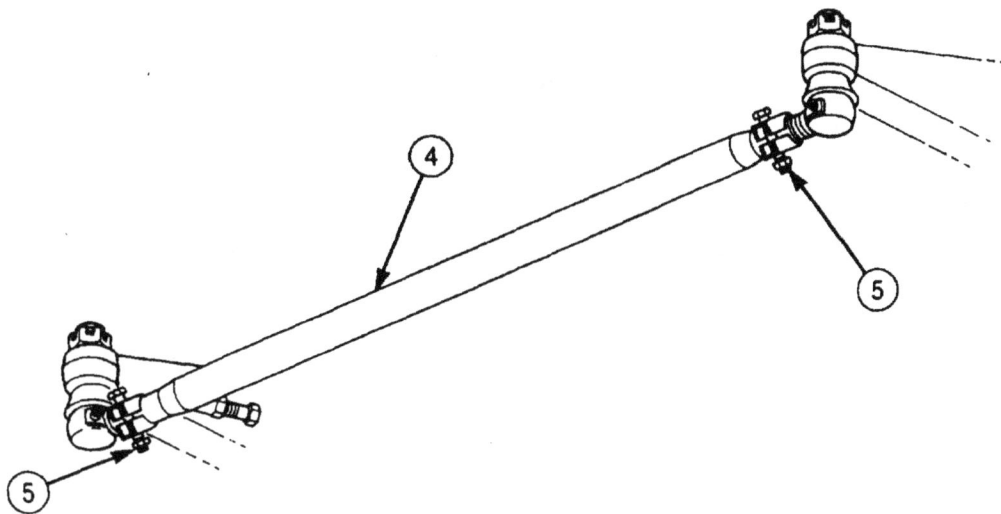

CHAINS

TOE-IN GAUGE

TOE-IN GAUGE

OUT ½   ½ IN

CHAINS

SCALE

POINTER

3-153. TIE ROD MAINTENANCE

THIS TASK COVERS:

a. Removal   c. Installation
b. Inspection

INITIAL SETUP:

APPLICABLE MODELS
All

TOOLS
General mechanic's tool kit (Appendix E, Item 1)
Torque wrench (Appendix E, Item 145)

MATERIALS/PARTS
Two cotter pins (Appendix D, Item 85)

REFERENCES (TM)
TM 9-2320-272-10
TM 9-2320-272-24P

EQUIPMENT CONDITION
- Parking brake set (TM 9-2320-272-10).
- Wheels chocked (TM 9-2320-272-10).

a. Removal

1. Remove two cotter pins (3) from tie rod ends (5) and nuts (1). Discard cotter pins (3).
2. Remove two nuts (1) from tie rod ends (5). Mark tie rod ends left and right.
3. Remove tie rod (10) from left and right steering knuckle arms (2).
4. Loosen two screws (8), nuts (6), and clamps (7) on tie rod (10).

NOTE
Record number of turns required to remove tie rod ends.

5. Remove two tie rod ends (5) from tie rod (10).
6. Remove two tie rod boots (4) and grease fittings (9) from tie rod ends (5).

b. Inspection

Inspect tie rod (10) and tie rod ends (5) for cracks, bends, and stripped threads. Replace tie rod (10) and/or tie rod ends (5) if bent, cracked, or stripped.

c. Installation

1. Install two tie rod boots (4) and grease fittings (9) on tie rod ends (5).

NOTE
Use same number of turns recorded during removal of tie rod ends.

2. Install two tie rod ends (5) on tie rod (10).
3. Tighten two screws (8) and nuts (6) on clamps (7). Tighten nuts (6) 60-80 lb-ft (81-109 N.m).
4. Install tie rod ends (5) on left and right steering knuckle arms (2) with two nuts (1). Tighten two nuts (1) to 160-180 lb-ft (217-244 N.m).
5. Install two new cotter pins (3) on tie rod ends (5) and nuts (1).

3-153. TIE ROD MAINTENANCE (Contd)

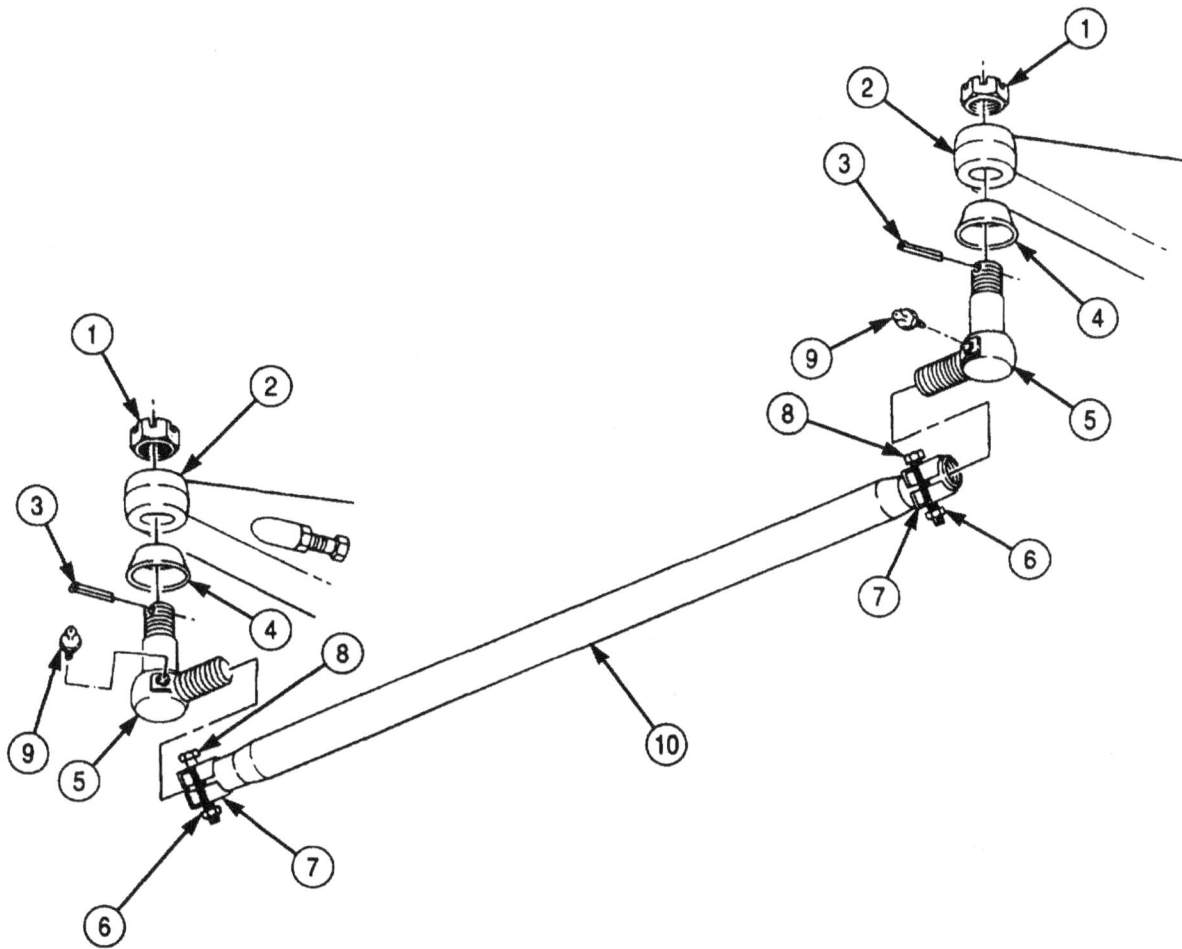

FOLLOW-ON TASK: Check and adjust toe-in (para. 3-152).

3-154. FRONT AXLE SHAFT AND UNIVERSAL JOINT MAINTENANCE

THIS TASK COVERS:

a. Removal
b. Cleaning and Inspection
c. Universal Joint Disassembly

d. Universal Joint Assembly
e. Installation

INITIAL SETUP:

APPLICABLE MODELS
All

TOOLS
General mechanic's tool kit (Appendix E, Item 1)
Vise

MATERIALS/PARTS
Cap and plug set (Appendix C, Item 14)
GAA grease (Appendix C, Item 28)
Sealing compound (Appendix C, Item 62)
Drycleaning solvent (Appendix C, Item 71)
Antiseize tape (Appendix C, Item 72)

REFERENCES (TM)
TM 9-2320-272-10
TM 9-2320-272-24P

EQUIPMENT CONDITION
- Parking brake set (TM 9-2320-272-10).
- Air reservoir drained (TM 9-2320-272-10).
- Front hub and drum removed (para. 3-223).
- Front hub and drum removed (M939A2) (para. 3-460).

GENERAL SAFETY INSTRUCTIONS
- Do not disconnect air lines before draining air reservoirs.
- Keep fire extinguisher nearby when drycleaning solvent is used.
- Drycleaning solvent is flammable and toxic. Do not use near open flame.

WARNING
Do not disconnect air lines before draining air reservoirs. Small parts under pressure may shoot out with high velocity, causing injury to personnel.

a. Removal

CAUTION
Cap or plug all openings immediately after disconnecting lines and hoses to prevent contamination. Failure to do so may result in brake system damage.

NOTE
Tag all lines and hoses for installation.

1. Disconnect vent air line (2) and service brake air line (1) from front brake chamber (4).
2. Remove ten nuts (7), washers (8), and brake spider (6) from steering knuckle housing (11).
3. Remove brake shoe spider and chamber assembly (5) from steering knuckle housing (11).
4. Remove spindle (3) from steering knuckle housing (11).
5. Remove axle shaft (10) from axle housing (12).
6. Remove washer (9) from axle shaft (10).

b. Cleaning and Inspection

WARNING
Drycleaning solvent is flammable and toxic. Do not use near open flame and always fave a fire extinguisher nearby when solvents are used. Use only in well-ventilated places, wear protective clothing, and dispose of cleaning rags in approved container. Failure to do this may result in injury or death to personnel and or damage to equipment.

1. Clean steering knuckle housing (11) with drycleaning solvent.

3-154. FRONT AXLE SHAFT AND UNIVERSAL JOINT MAINTENANCE (Contd)

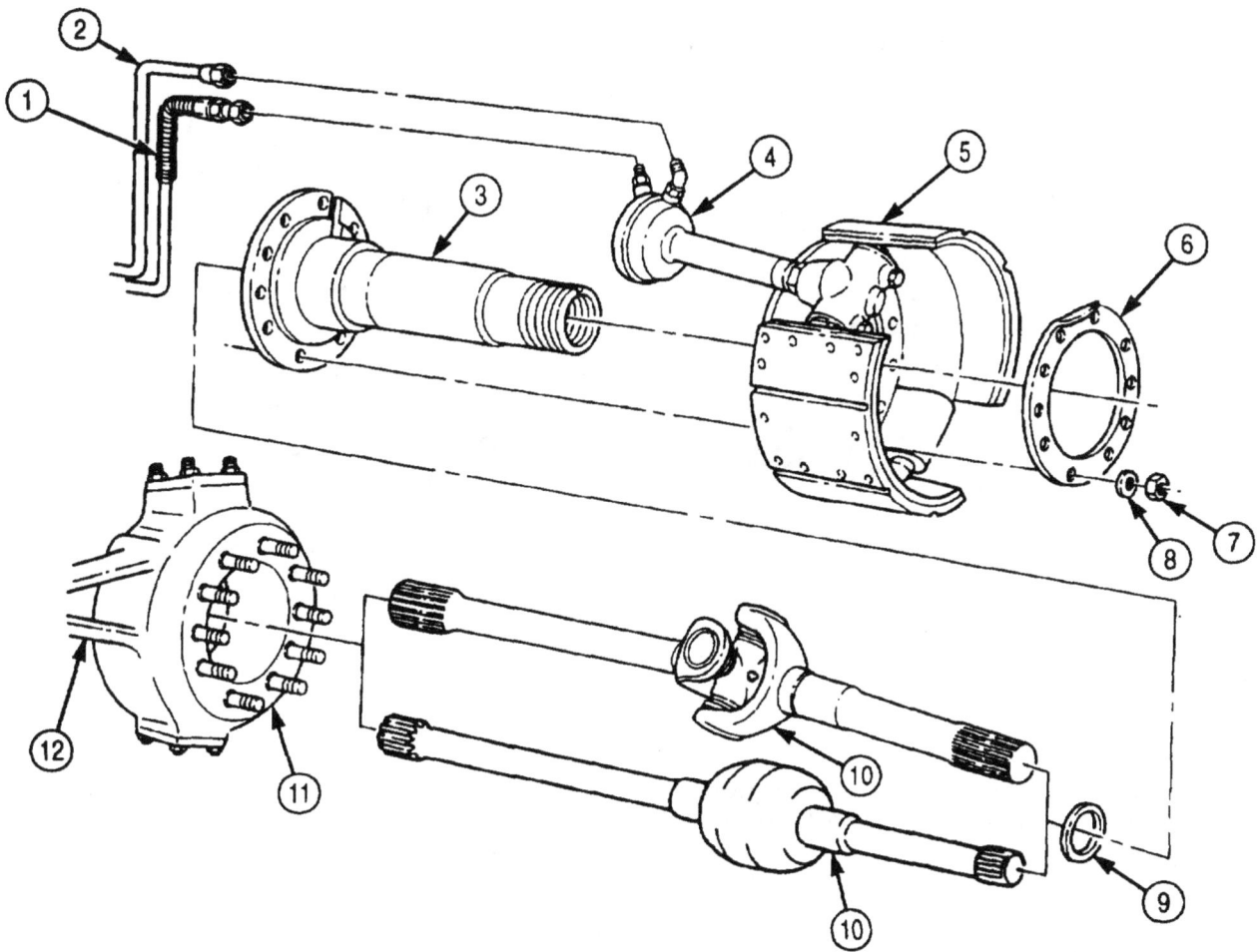

3-154. FRONT AXLE SHAFT AND UNIVERSAL JOINT MAINTENANCE (Contd)

NOTE

New configuration axle shafts have a CV joint in place of the universal joint. Axle shafts are no longer repairable and must be replaced if damaged.

2. Inspect axle shaft (5) for cracks and nicks at shaft splines (1). If shaft splines are cracked or nicked, replace axle shaft (5).

3. Place short end (4) of axle shaft (5) in soft-jawed vise. Pull and push up and down on inner shaft (2) and twist inner shaft (2). Replace axle shaft, (5) if any free play in universal or CV joint (3) is observed.

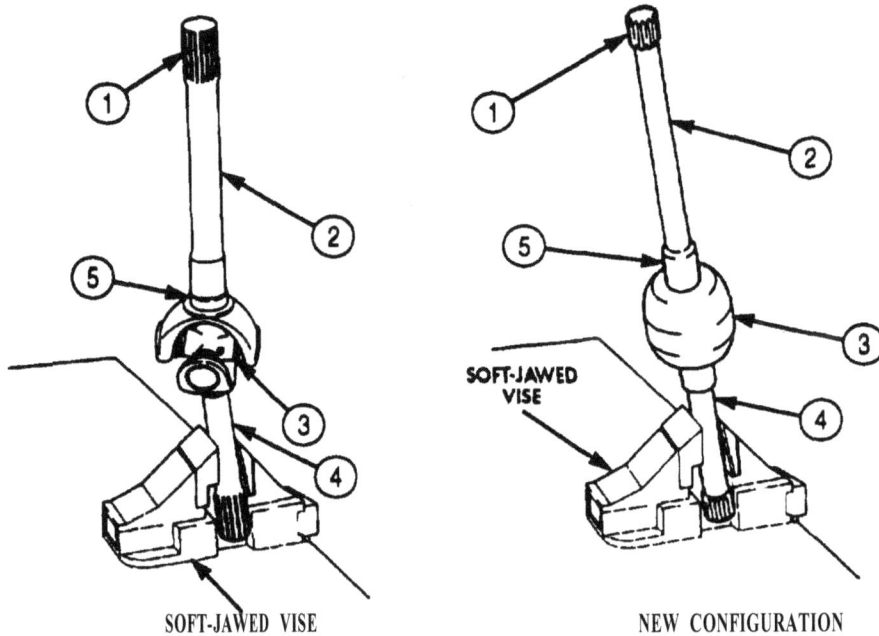

SOFT-JAWED VISE   NEW CONFIGURATION

e. Installation

1. Install washer (14) on axle shaft (15).

2. Lubricate universal joint (16) (if equipped), grease all bearing surfaces on axle shaft (15) and fill steering knuckle cavity (19) with GAA grease.

3. Support axle shaft (15) and install in steering knuckle housing (18).

4. Repack steering knuckle cavity (19) with GAA grease.

5. Apply sealing compound to spindle (8) and align with mounting studs (17) on steering knuckle housing (18). Ensure spindle (3) is slot end up.

6. Install brakeshoe spider and chamber assembly (10) over spindle (8) on steering knuckle housing (18).

7. Install brake spider (11) on steering knuckle housing (18) with ten washers (13) and nuts (12).

NOTE

Male pipe threads must be wrapped with antiseize tape before installation.

8. Connect vent air line (7) and service brake air line (6) to front brake chamber (9).

3-154. FRONT AXLE SHAFT AND UNIVERSAL JOINT MAINTENANCE (Contd)

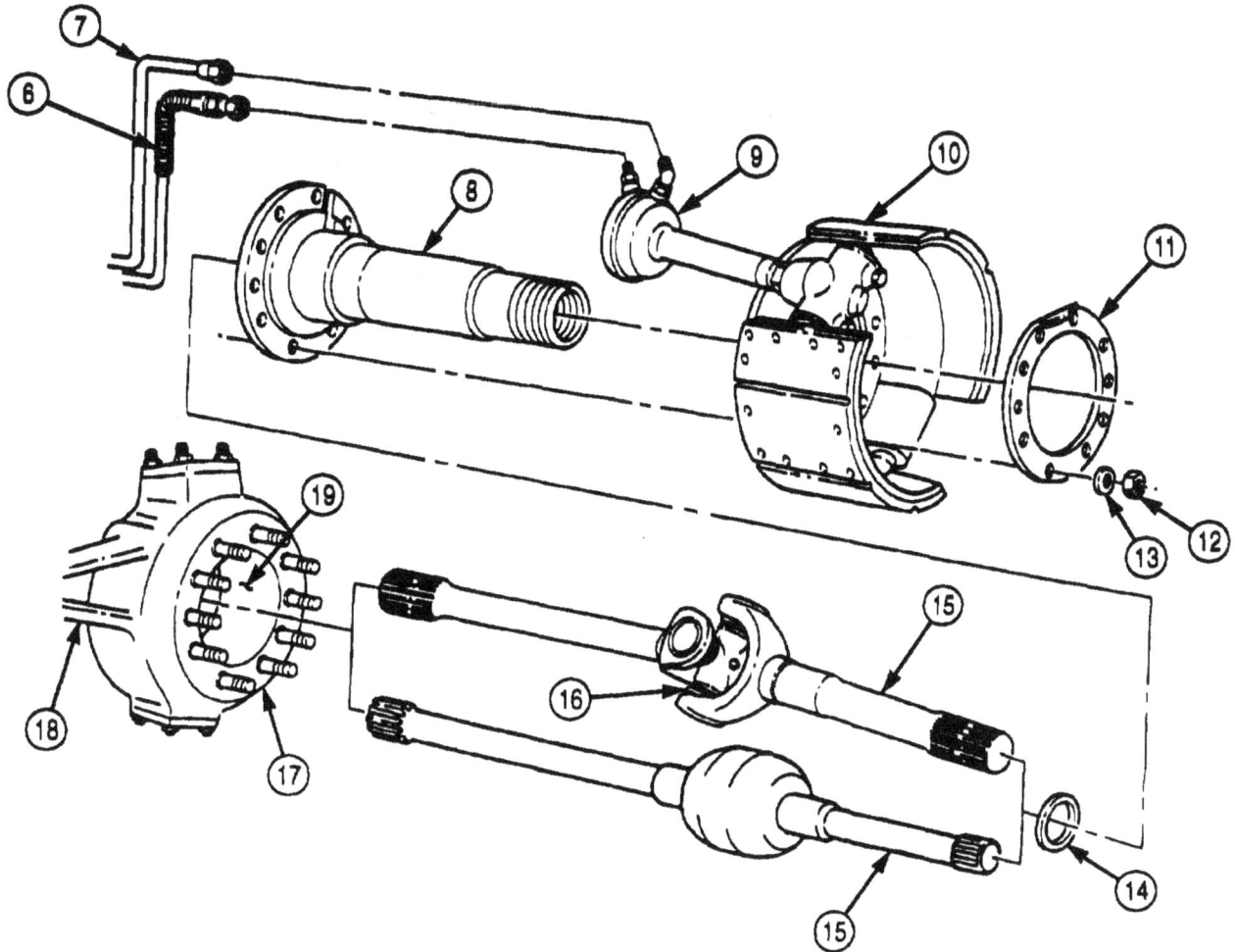

FOLLOW-ON TASKS: • Install front hub and drum (M939A2) (para. 3-460).
  • Install front hub and drum (para. 3-223).
  • Start engine (TM 9-2320-272-10) and check for leaks at air lines and brake chambers.

3-155. STEERING KNUCKLE BOOT MAINTENANCE

THIS TASK COVERS:

a. Removal   c. Installation
b. Cleaning and Inspection

INITIAL SETUP:

APPLICABLE MODELS

All

TOOLS

General mechanic's tool kit (Appendix E, Item 1)
Torque wrench (Appendix E, Item 144)

MATERIALS/PARTS

Steering knuckle boot (seal) replacement kit
  (Appendix D, Item 676)
Four lockwashers (Appendix D, Item 377)
Lockwire (Appendix D, Item 419)
Cleaning cloth (Appendix C, Item 21)
Adhesive (Appendix C, Item 5)

REFERENCES (TM)

LO 9-2320-272-12
TM 9-2320-272-10
TM 9-2320-272-24P

EQUIPMENT CONDITION

Parking brake set (TM 9-2320-272-10).

---

a. Removal

1. Remove four screws (4), lockwashers (3), and seal guard (5) from steering knuckle (2). Discard lockwashers (3).

2. Remove twelve screws (7), safety wire (1), and retaining plate (6) from steering knuckle (2). Discard safety wire (1).

3. Pull one side of steering knuckle seal (12) aside to expose inner retaining clamp screw (10).

4. Remove inner retaining clamp screw (10) and spacer (9) from inner retaining clamp (11).

5. Remove seal (12) and inner retaining clamp (11) from steering knuckle (2).

NOTE
Some seals may not have a zipper.

6. Open zipper (8) or cut to remove seal (12) from axle housing (13).

b. Cleaning and Inspection

1. Clean surface of steering knuckle (2), retaining plate (6), and steering knuckle seal (12) with cleaning cloth. Ensure there is no dirt or dust is inside steering knuckle (2).

2. Inspect steering knuckle seal (12) for cracks, tears, and damaged zipper (8). Replace if cracked, torn, or zipper (8) is damaged.

c. Installation

1. Place steering knuckle seal (12) on axle housing (13), with fabric side of zipper (8) facing steering knuckle (2).

2. Close zipper (8) and apply large amount of adhesive to locks and fabric of zipper (8). Allow adhesive to sit overnight.

3. Force inner lip of steering knuckle seal (12) into groove on axle housing (13). Ensure steering knuckle seal (12) is aligned to holes in steering knuckle (2).

4. Position inner retaining clamp (11) to lip of seal (12) and install clamp screw (10) and spacer (9).

3-155. STEERING KNUCKLE BOOT MAINTENANCE (Contd)

5. Lace zipper (8) locks with fine wire near edge of steering knuckle seal (13) and twist together.

6. Cut off excess zipper (8) and apply adhesive to exposed zipper (8) and fabric. Allow adhesive to sit overnight.

7. Align seal retaining plate (6) to holes in steering knuckle (2) and install twelve screws (7). Ensure notches in screw (7) heads are aligned so safety wire (1) can be installed.

8. Thread new safety wire (1) around each screw (7) and tie Off.

9. Install seal guard (5) on steering knuckle (2) with four new lockwashers (3) and screws (4). Tighten screws (4) 130-170 lb-ft (176-231 N.m).

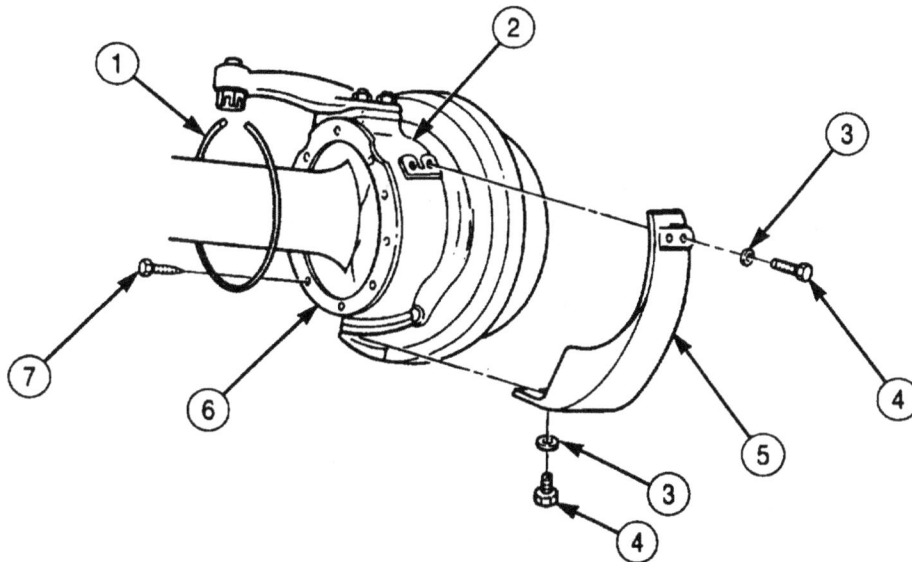

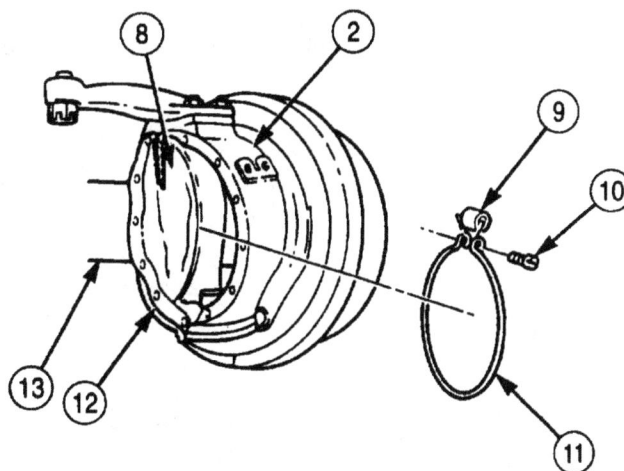

FOLLOW-ON TASK: Lubricate steering knuckle (LO 9-2320-272-12).

3-156. FRONT AXLE DRIVE FLANGE MAINTENANCE

THIS TASK COVERS:

a. Removal
b. Inspection

c. Installation

INITIAL SETUP:

APPLICABLE MODELS
All

TOOLS
General mechanic's tool kit (Appendix E, Item 1)
Torque wrench (Appendix E, Item 144)

MATERIALS/PARTS
Gasket sealant (Appendix C, Item 30)

REFERENCES (TM)
LO 9-2320-272-12
TM 9-2320-272-10
TM 9-2320-272-24P

EQUIPMENT CONDITION
Parking brake set (TM 9-2320-272-10).

a. Removal

1. Remove ten screws (1) and washers (2) from axle drive flange (3).
2. Install two screws (1) in threaded holes (7) and turn screws (1) evenly until flange (3) separates from hub (5).
3. Remove two screws (1) from axle drive flange (3).
4. Remove gasket (4) if present. Discard gasket (4) and/or clean gasket sealant or sealant remains from mating surfaces.

b. Inspection

1. Inspect mating surfaces of axle drive flange (3) and hub (5) for burrs, cracks, and gouges. File down surfaces having burrs. Replace axle drive flange (3) or hub (5) if cracked or gouged.
2. Inspect axle shaft (6) for damaged splines. Notify your supervisor if splines are damaged.

c. Installation

1. Coat surface of axle drive flange (3) with sealant and align with holes on hub (5).
2. Install axle drive flange (3) on hub (5) with ten washers (2) and screws (1). Tighten screws (1) 85-100 lb-ft (115-136 N.m).

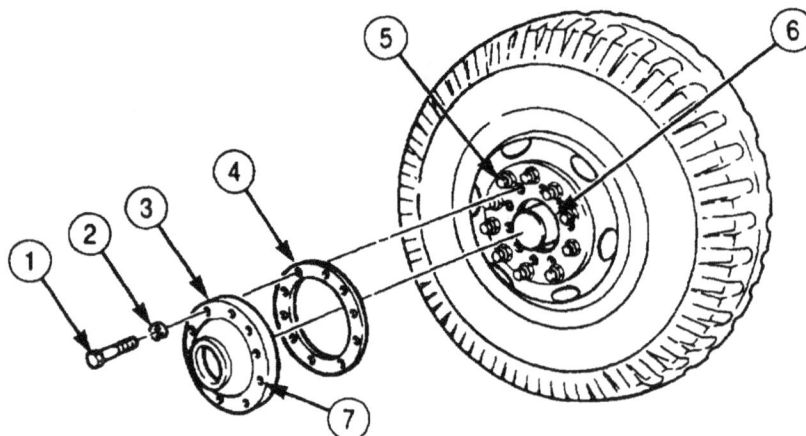

FOLLOW-ON TASK: Lubricate axle assembly (LO 9-2320-272-12).

3-157. REAR AXLE SHAFT MAINTENANCE

THIS TASK COVERS:

a. Removal   c. Installation
b. Inspection

<u>INITIAL SETUP:</u>

<u>APPLICABLE MODELS</u>
All

<u>TOOLS</u>
General mechanic's tool kit (Appendix E, Item 1)
Torque wrench (Appendix E, Item 144)

<u>MATERIALS/PARTS</u>
Gasket sealant (Appendix C, Item 30)

<u>REFERENCES (TM)</u>
LO 9-2320-272-12
TM 9-2320-272-10
TM 9-2320-272-24P

<u>EQUIPMENT CONDITION</u>
Parking brake set (TM 9-2320-272-10).

---

a. Removal

1. Remove ten screws (1) and washers (2) from axle shaft flange (3).
2. Tap shaft flange (3) and remove axle shaft flange (3) and axle shaft (4) from axle housing (6).
3. Remove gasket (5) if present, from axle shaft flange (3). Clean gasket or sealant from mating surfaces. Discard gasket (5).

b. Inspection

Inspect axle shaft (4) and axle housing (6) for cracks, burrs, and damaged splines on shaft. Replace axle shaft (4) or axle housing (6) if cracked, burred, or shaft has damaged splines.

b. Installation

1. Coat surface of axle shaft flange (3) with sealant.
2. Position splined end of axle shaft (4) into axle housing (6). Ensure axle shaft (4) slides into splined differential gear (7).
3. Align axle shaft flange (3) to hole in axle housing (6).
4. Install axle shaft flange (3) on axle housing (6) with ten washers (2) and screws (1). Tighten screws (1) 80-105 lb-ft (108-142 N.m).

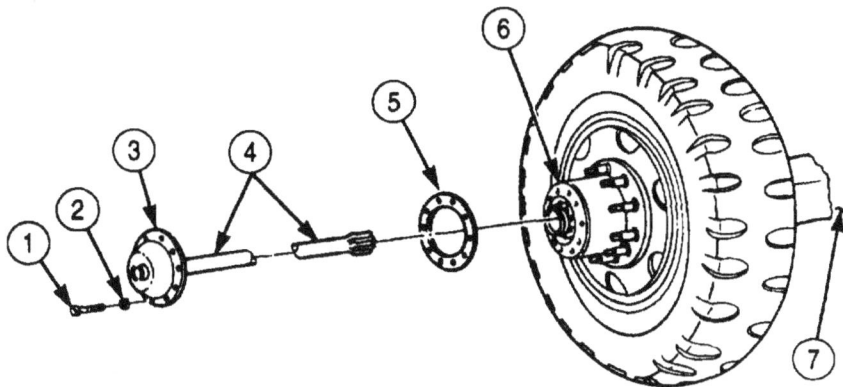

FOLLOW-ON TASK: Lubricate axle assembly (LO 9-2320-272-12).

3-158. CARRIER DIFFERENTIAL TOP COVER GASKET AND SIDE COVER GASKET REPLACEMENT

THIS TASK COVERS:

a. Top Cover Gasket Removal
b. Top Cover Gasket Installation

c. Side Cover Gasket Removal
d. Side Cover Gasket Installation

INITIAL SETUP:

APPLICABLE MODELS

All

TOOLS

General mechanic's tool kit (Appendix E, Item 1)
Torque wrench (Appendix E, Item 146)

MATERIALS/PARTS

Top cover gasket (Appendix D, Item 698)
Side cover gasket (Appendix D, Item 653)
Gasket sealant (Appendix C, Item 30)
Antiseize tape (Appendix C, Item 72)

REFERENCES (TM)

TM 9-2320-272-10
TM 9-2320-272-24P

EQUIPMENT CONDITION

- Parking brake set (TM 9-2320-272-10).
- Air reservoirs drained (TM 9-2320-272-10).
- Rear wheels chocked (TM 9-2320-272-10).

GENERAL SAFETY INSTRUCTIONS

Do not disconnect air lines before draining air reservoirs.

WARNING

Do not disconnect air lines before draining air reservoirs. Small parts under pressure may shoot out with high velocity, causing injury to personnel.

---

a. Top Cover Gasket Removal

NOTE

- Tag all air lines for installation.
- Perform steps 1 through 4 for forward-rear and rear-rear axle carrier differential top cover.

1. Remove four air lines (3) from adapters (2).
2. Remove two adapters (2) from manifolds (1).
3. Remove two screws (7), washers (6), and bracket (5) from carrier differential top cover (4).
4. Remove eight screws (7), washer (6), top cover (4), and gasket (8) from carrier differential housing (9). Clean gasket remains from mating surfaces. Discard gasket (8).

NOTE

Perform step 5 for front axle carrier differential top cover.

5. Remove ten screws (7), washers (6), top cover (4), and gasket (8) from carrier differential housing (9). Discard gasket (8).

3-158. CARRIER DIFFERENTIAL TOP COVER GASKET AND SIDE COVER GASKET REPLACEMENT (Contd)

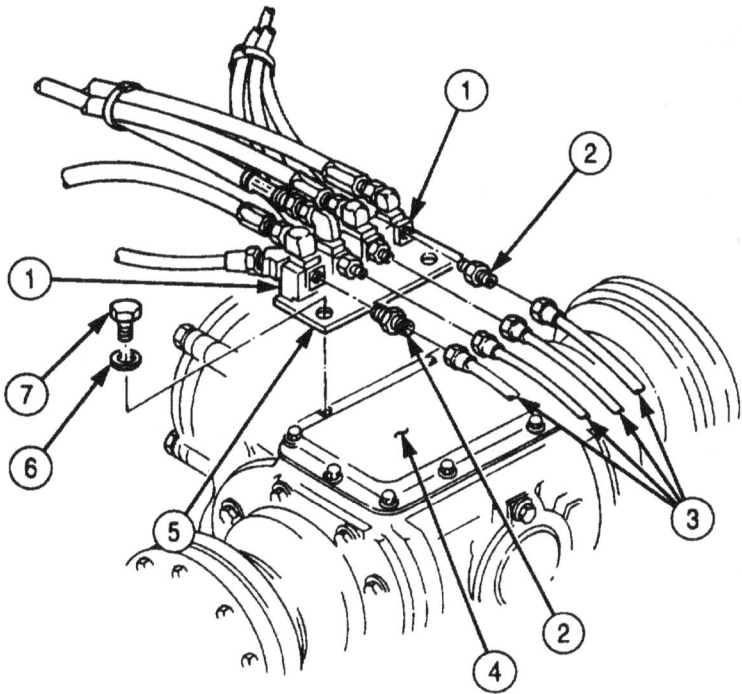

REAR AXLE

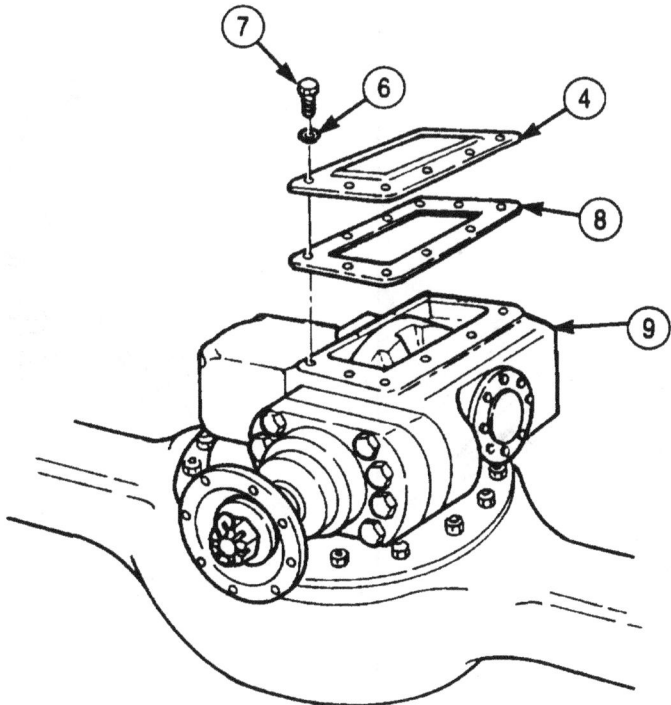

FRONT AND REAR AXLE

3-158. CARRIER DIFFERENTIAL TOP COVER GASKET AND SIDE COVER GASKET REPLACEMENT (Contd)

b. Top Cover Gasket Installation

1. Apply gasket sealant to mating surfaces of top cover (4).

NOTE

Perform step 2 for front axle carrier differential top cover,

2. Install new gasket (8) and top cover (4) on front axle carrier differential housing (9) with ten washers (6) and screws (7). Tighten screws (7) 27-34 lb-ft (37-48 N.m).

NOTE

- Perform steps 3 through 6 for forward-rear and rear-rear axle carrier differential top cover,
- Male pipe threads must be wrapped with antiseize tape prior to installation.

3. Install new gasket (8) and top cover (4) on carrier differential housing (9) with eight washer (6) and screws (7). Tighten screws (7) 27-34 lb-ft (37-48 N.m).

4. Install bracket (5) on carrier differential top cover (4) with two washers (6) and screws (7). Tighten screws (7) 27-34 lb-ft (37-48 N.m)

5. Install two adapters (2) on manifolds (1).

6. Install four air lines (8) on adapters (2).

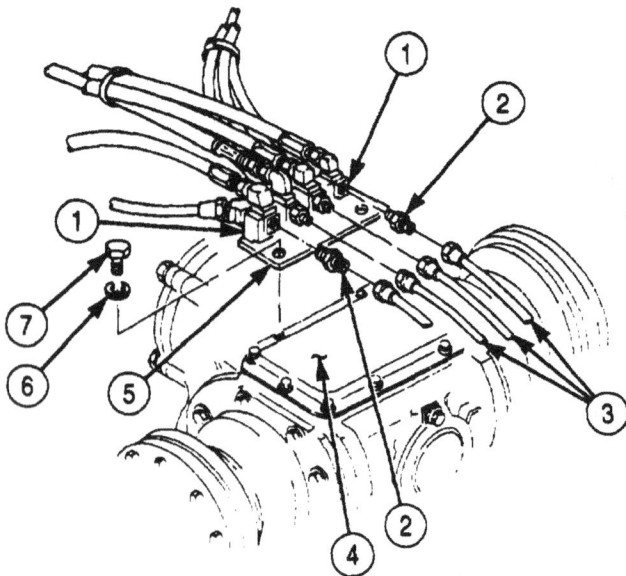

REAR AXLE

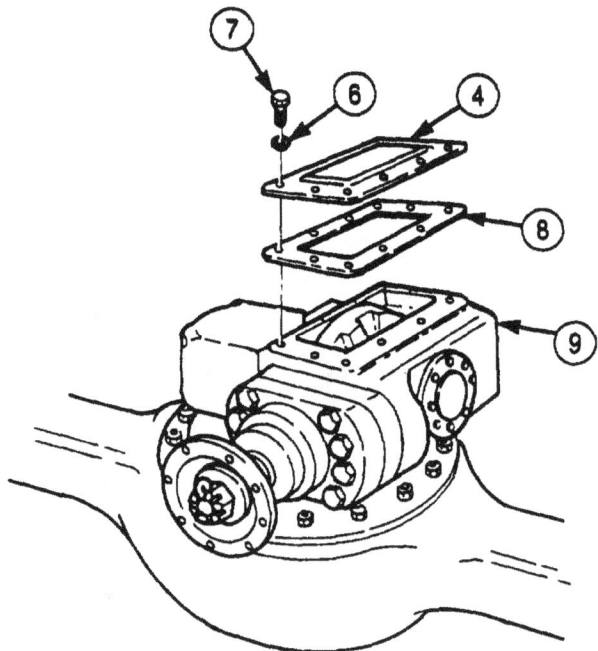

FRONT AND REAR AXLE

3-158. CARRIER DIFFERENTIAL TOP COVER GASKET AND SIDE COVER GASKET REPLACEMENT (Contd)

c. Side Cover Gasket Removal

1. Remove eight screws (13), washers (12), and carrier differential housing cover (11) from differential housing (9).
2. Remove gasket (10) from differential housing (9). Discard gasket (10). Clean gasket remains from carrier differential housing cover (11) and differential housing (9).

d. Side Cover Gasket Installation

1. Apply gasket sealant to mating surface of side cover (11) and differential housing (9).
2. Install new side cover gasket (10) and side cover (11) on differential housing (9) with eight washers (12) and screws (13). Tighten screws (13) 27-34 lb-ft (37-48) N.m).

3-159. FRONT AND REAR AXLE BREATHER REPLACEMENT

THIS TASK COVERS:

a. Removal

b. Installation

INITIAL SETUP:

APPLICABLE MODELS
All

TOOLS
General mechanic's tool kit (Appendix E, Item 1)

REFERENCES (TM)
TM 9-2320-272-10
TM 9-2320-272-24P

EQUIPMENT CONDITION
Parking brake set (TM 9-2320-272-10).

a. Removal

Remove breather assembly (1) from axle housing (2).

b. Installation

Install breather (1) on axle housing (2).

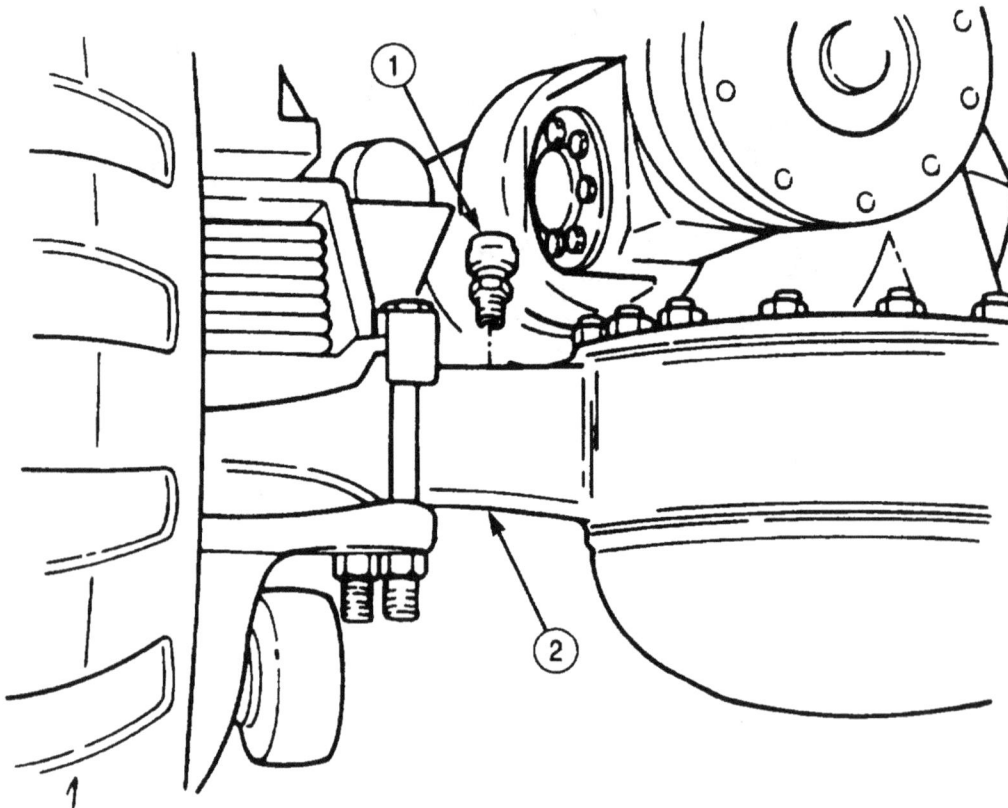

3-160. FRONT AXLE SEAL REPLACEMENT

THIS TASK COVERS:

a. Removal   b. Installation

INITIAL SETUP:

APPLICABLE MODELS
All

SPECIAL TOOLS
Oil seal installer tool (Appendix E, Item 91)

TOOLS
General mechanic's tool kit (Appendix E, Item 1)

MATERIALS/PARTS
Oil seal (M939/A1) (Appendix D, Item 496)
Oil seal (M939A2) (Appendix D, Item 495)

REFERENCES (TM)
TM 9-2320-272-10
TM 9-2320-272-24P

EQUIPMENT CONDITION
• Parking brake set (TM 9-2320-272-10).
• Front axle shaft and universal joint removed (para. 3-154).

a. Removal

Remove oil seal (1) from steering knuckle housing (2). Discard oil seal (1).

b. Installation

Using oil seal installer, install new oil seal (1) in steering knuckle housing (2).

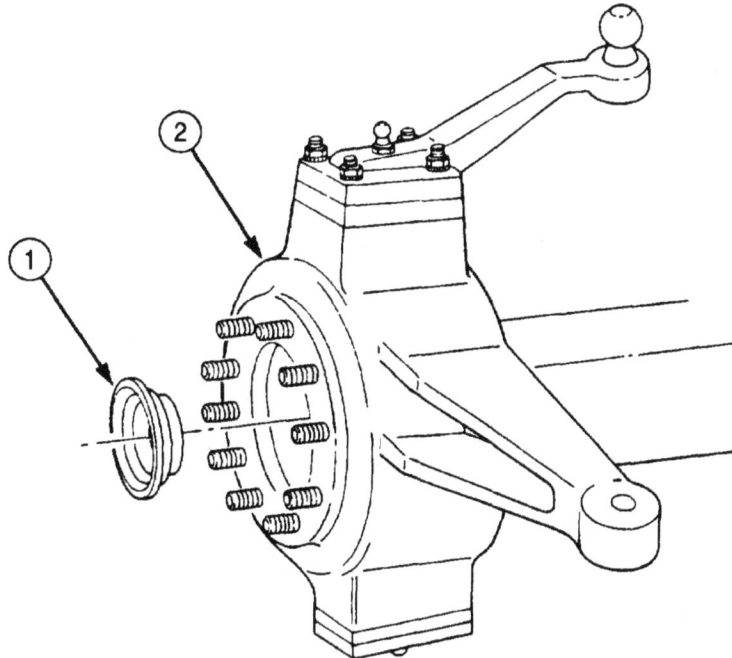

FOLLOW-ON TASK: Install front axle shaft and universal joint (para. 3-154).

3-161. FRONT SPRING AND MAIN LEAF REPLACEMENT

THIS TASK COVERS:

a. Front Spring Removal   c. Front Spring Main Leaf Installation
b. Front Spring Main Leaf Removal   d. Front Spring Installation

INITIAL SETUP:

APPLICABLE MODELS
All

TOOLS
General mechanic's tool kit (Appendix E, Item 1)
Torque wrench (Appendix E, Item 145)
C-clamps

MATERIALS/PARTS
Cotter pin (Appendix D, Item 49)
Two locknuts (Appendix D, Item 285)
Four lockwashers (Appendix D, Item 362)
Spring center bolt (Appendix D, Item 670)
Spring center nut (Appendix D, Item 669)
Antiseize tape (Appendix C, Item 72)
Cap and plug set (Appendix D, Item 14)
Powdered graphite (Appendix C, Item 31)

REFERENCES (TM)
TM 9-2320-272-10
TM 9-2320-272-24P

EQUIPMENT CONDITION
- Parking brake set (TM 9-2320-272-10).
- Air reservoirs drained (TM 9-2320-272 10).
- Right splash shield removed (TM 9-2320-272-10).
- Shock absorber removed (para. 3-166).
- Front wheels removed (TM 9-2320-272-10).

a. Front Spring Removal

NOTE
Steps 1 through 5 apply to right side of spring only.

1. Bend tabs (4) of steering cylinder dustcover (3) clear of steering cylinder (7) and right side spring hanger (1).

2. Remove cotter pin (6) from steering cylinder (7). Discard cotter pin (6).

3. Back out cylinder adjusting plug (5).

4. Remove steering cylinder (7) from ball stud (2).

5. Remove steering cylinder dustcover (3) from ball stud (2).

CAUTION
Cap or plug all openings immediately after disconnecting lines or hoses to prevent contamination. Failure to do so may result in brake system damage.

6. Disconnect two air lines (15) from service brake chamber (14).

7. Place jack stands under frame (17) at rear of hanger (10).

NOTE
Support axle when removing U-bolts.

8. Remove four nuts (13), lockwashers (12), and shock absorber mounting plate (11) from two U-bolts (9). Discard lockwashers (12).

NOTE
- Frame may have to be raised further to provide clearance for U-bolts removal.
- If U-bolts must be forced, place a piece of wood between hammer and bolt to prevent damage.

3-161. FRONT SPRING AND MAIN LEAF REPLACEMENT (Contd)

9. Remove two U-bolts (9) and upper spring seat (8) from springs (16).

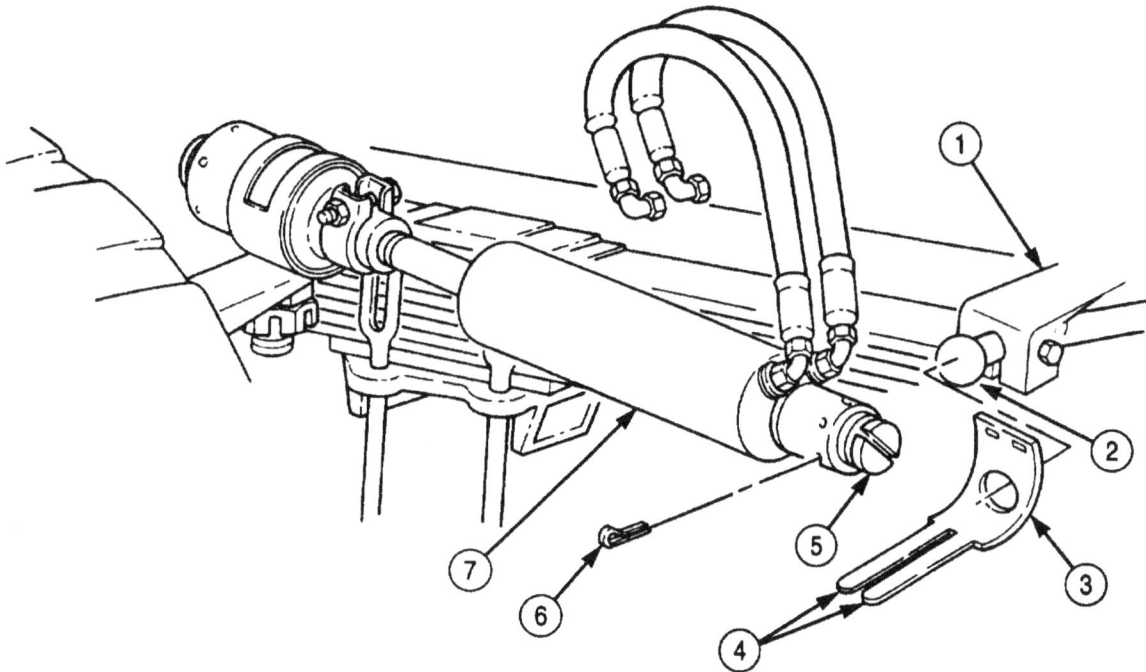

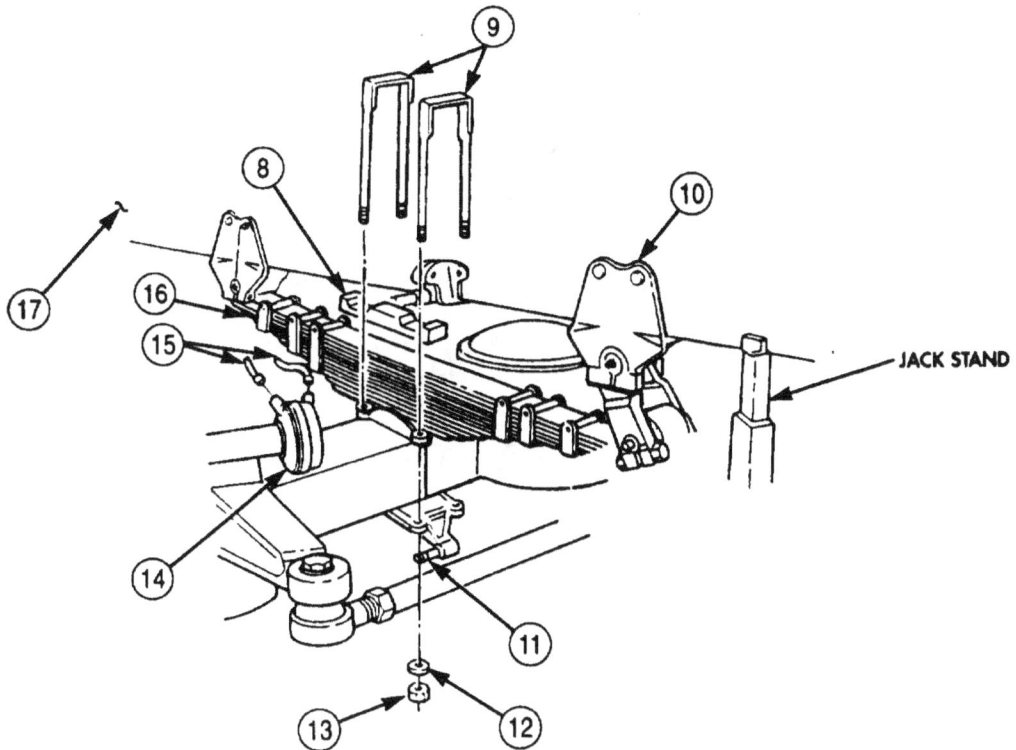

JACK STAND

3-161. FRONT SPRING AND MAIN LEAF REPLACEMENT (Contd)

NOTE

The spring pins in right spring hanger and left spring shackle
must be driven out from the inside. The remaining two pins can be
driven out from either side.

10. Remove pin retaining locknut (10) and screw (7) from spring hanger (1). Discard locknut (10).

11. Drive pin (9) out of spring hanger (1) and leaf spring eye (8).

12. Remove pin retaining locknut (13) and screw (11) from spring shackle (2). Discard locknut (13).

13. Drive pin (14) out of spring shackle (2) and leaf spring eye (12).

14. Lift spring assembly (6) from lower spring seat (3) and carefully move over rear axle (5) and tie rod (4) to the ground.

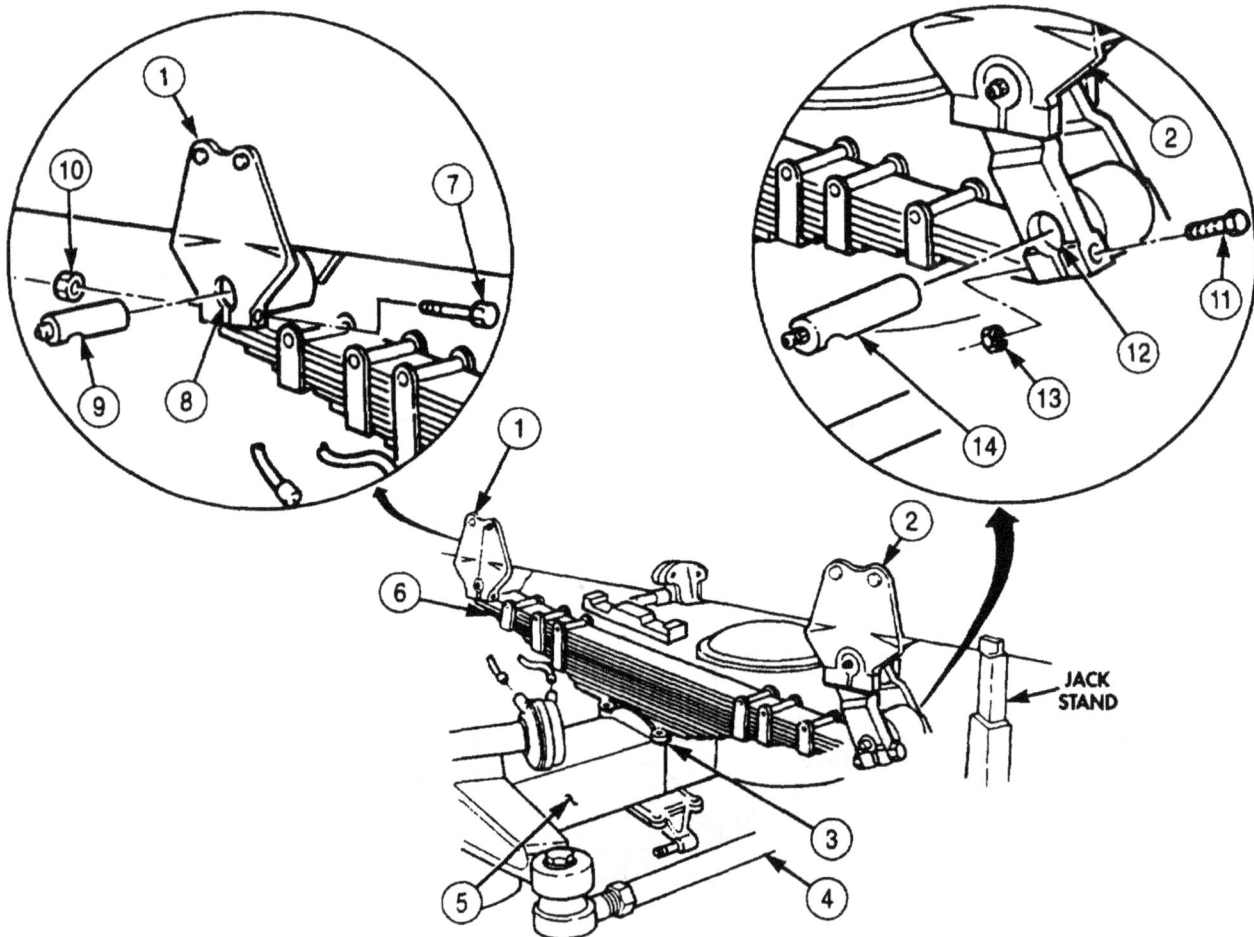

3-161. FRONT SPRING AND MAIN LEAF REPLACEMENT (Contd)

b. Front Spring Main Leaf Removal

1. Compress spring assembly (6) with C-clamps and remove six nuts (17), screws (20), and spacers (18) from six clips (19).

2. Remove nut (16) and spring center bolt (21) from spring assembly (6). Discard bolt (21) and nut (16).

3. Remove C-clamps from spring assembly (6).

4. Remove rebound leaf (15) from spring assembly (6), upper main leaf (23), and lower main leaf (22) from spring assembly (6). Separate upper main leaf (23) and lower main leaf (22).

c. Front Spring Main Leaf Installation

1. Position upper main leaf (23) into lower main leaf (22) and position both on spring assembly (6). Ensure rounded spring eye ends face upward.

2. Position rebound leaf (15) over upper main leaf (23).

3. Compress spring assembly (6) with C-clamps and align center bolt holes on spring assembly (6) leaves.

4. Install new spring center bolt (21) and new nut (16) through center holes in leaves and peen end of new bolt (21) over nut (16).

5. Install six spacers (18) between ends of spring leaf clips (19) with six screws (20) and nuts (17).

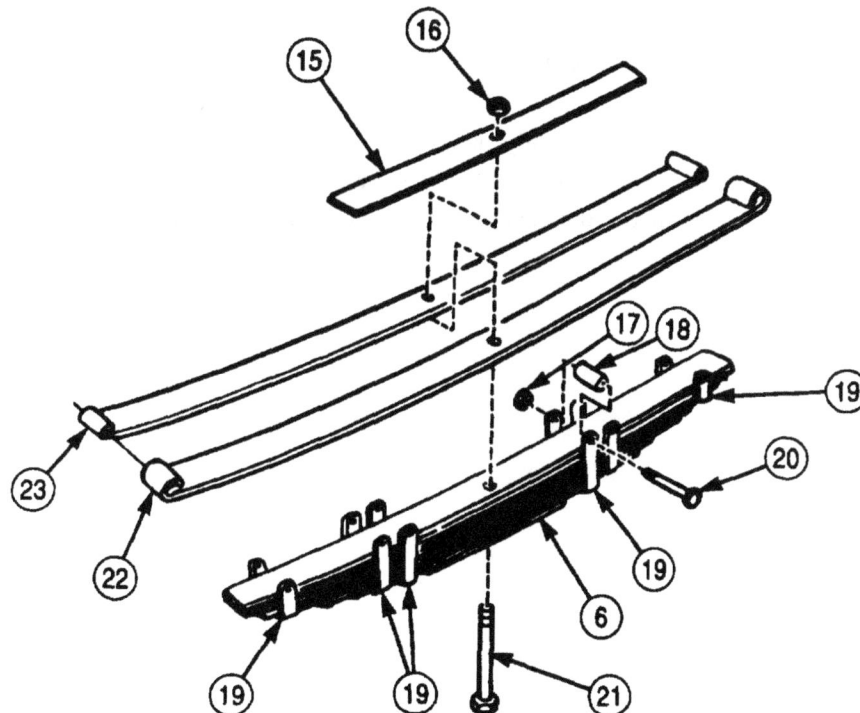

3-161. FRONT SPRING AND MAIN LEAF REPLACEMENT (Contd)

1. Raise one end of spring (11) over tie rod (4) and axle (8).

2. Edge spring (11) forward until spring center bolt (23) enters hole in center of lower spring seat (3).

3. Position upper spring seat (13) over top of spring (11).

4. Place two U-bolts (1) over upper spring seat (13) and through holes in lower spring seat (3), with long end toward wheel.

5. Install two U-bolts (1) on shock absorber mounting plate (5) with four new lockwashers (6) and nuts (7). Tighten nuts (7) 350-400 lb-ft (475-542 N•m).

6. Raise spring (11) to frame (12) until spring eyes (20) and (17) align with holes in hanger (16) and shackle (2).

7. Install pin (18) through front spring eye (17) and hanger (16) with screw (16) and new locknut (14). Groove in pin (18) must be in position to angle of screw holes before screw (16) can be installed,

8. Install pin (22) through spring eye (20) and shackle (2) with screw (19) and new locknut (21). Groove in pin (22) must be in position to angle of screw holes before screw (19) can be installed.

NOTE

All male pipe threads must be wrapped with antiseize tape prior to installation.

9. Connect two air lines (10) to service brake chamber (9).

10. Remove jack stands.

NOTE

Steps 11 through 16 apply to right side of spring only.

11. Install steering cylinder dustcover (26) on ball stud (25).

12. Position steering cylinder (30) over ball stud (25).

13. Tighten cylinder adjusting plug (28) until cylinder (30) does not move on ball stud (25).

14. Tighten cylinder adjusting plug (28) slightly more if necessary to align slot with hole in cylinder (30).

15. Install steering cylinder (30) on ball stud (25) with new cotter pin (29).

16. Bend tabs (27) of steering cylinder dustcover (26) around steering cylinder (30) and spring hanger (24).

3-161. FRONT SPRING AND MAIN LEAF REPLACEMENT (Contd)

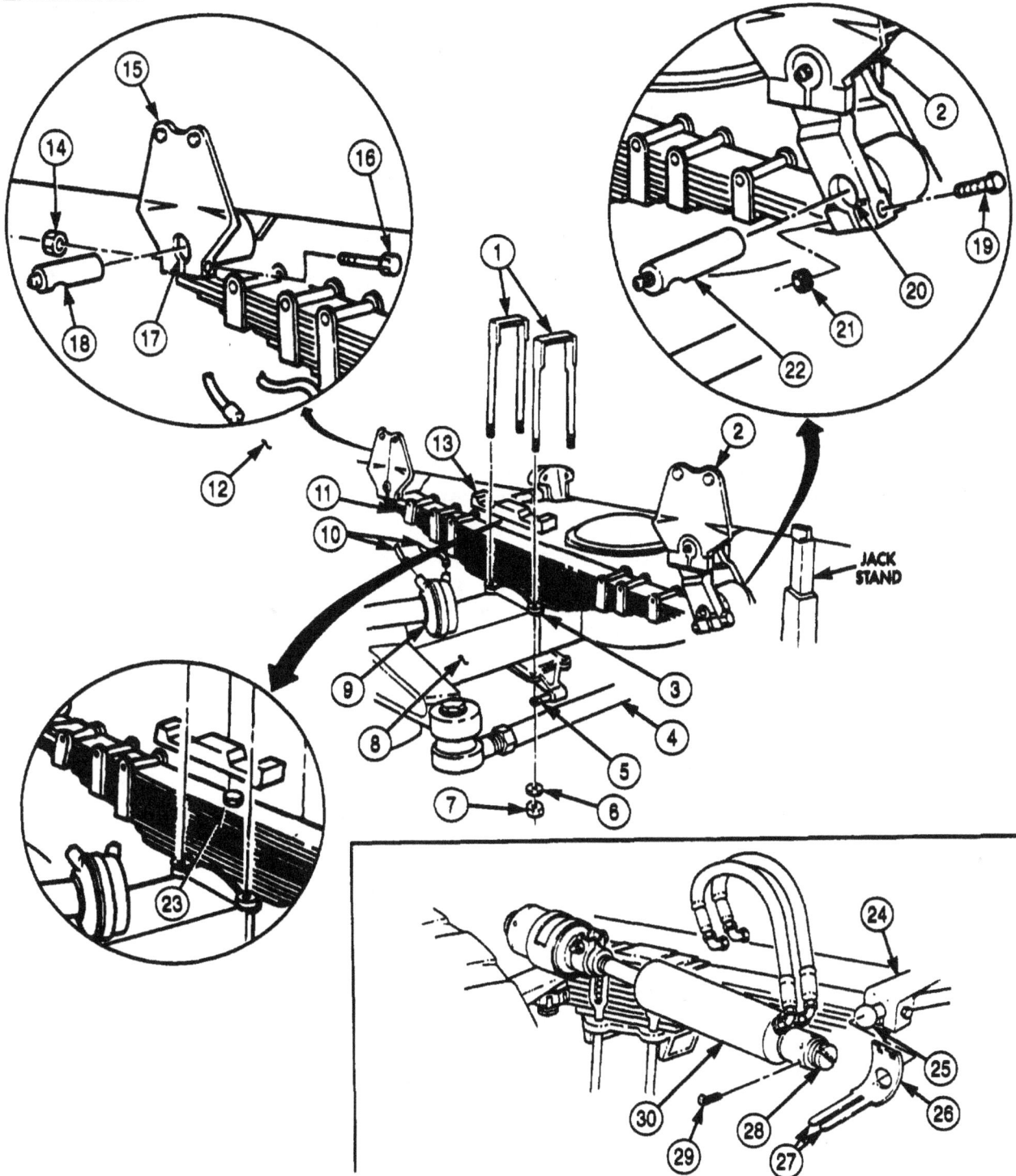

FOLLOW-ON TASKS: • Install shock absorber (para. 3-166).
                   • Install front wheels (TM 9-2320-272-10).
                   • Install right splash shield (TM 9-2320-272-10).

3-162. FRONT SPRING SHACKLE REPLACEMENT

THIS TASK COVERS:

a. Removal   b. Installation

INITIAL SETUP:

APPLICABLE MODELS
All

TOOLS
General mechanic's tool kit (Appendix E, Item 1)

MATERIALS/PARTS
Two locknuts (Appendix D, Item 285)
Screw (Appendix D. Item 570)

REFERENCES (TM)
LO 9-2320-272-12
TM 9-2320-272-10
TM 9-2320-272-24P

EQUIPMENT CONDITION
• Parking brake set (TM 9-2320-272-10).
• Front wheels removed (para. 3-220 or 3-221).

---

a. Removal

1. Raise vehicle to remove load from spring (12). Place jack stands under frame (16).
2. Remove locknut (1) and screw (3) from hanger (2). Discard locknut (1).
3. Remove locknut (6) from shackle (4). Discard locknut (6).
4. Loosen screw (7) in shackle (4).
5. Remove grease fitting (10) and lower shackle pin (9) from shackle (4) and spring eye (11). Pin (9) must be pushed through toward underside of vehicle.
6. Remove grease fitting (15) and upper shackle pin (14) from shackle (4) and hanger (2). Pin (14) must be pushed through to underside of vehicle.
7. Remove shackle (4) from hanger (2).
8. Remove screw (7) from hanger (2). Discard screw (7).
9. Remove bushing (5) from shackle (4).

b. Installation

1. Install bushing (5) in shackle (4).
2. Loosely thread new screw (7) on shackle (4).
3. Install shackle (4) on hanger (2).
4. Install shackle pin (14) in shackle (4) and hanger (2). Ensure pin slot (13) is positioned downward to allow installation of screw (3).
5. Install grease fitting (15) in shackle pin (14).
6. Install screw (3) and new locknut (1) on shackle (4) and hanger (2).
7. Lower vehicle until spring eye (11) aligns with holes in shackle (4).
8. Install shackle pin (9) in shackle (4) and spring eye (11). Make sure pin slot (8) is positioned downward to allow installation of screw (7).
9. Install grease fitting (10) on shackle pin (9) and hanger (2).
10. Tighten new screw (7) and install new locknut (6) on shackle (4).
11. Remove jack stands .

3-162. FRONT SPRING SHACKLE REPLACEMENT (Contd)

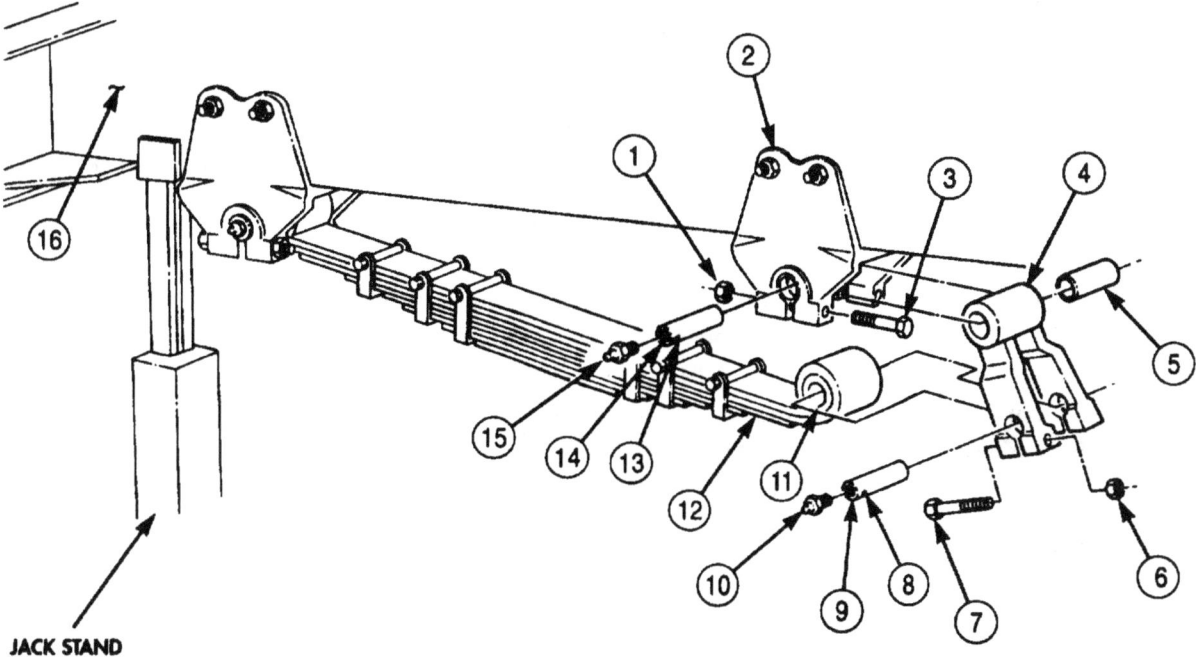

JACK STAND

FOLLOW-ON TASKS: • Lubricate shackle (LO 9-2320-272-12).
• Install front wheel (para. 3-220 or 3-221).

3-163. FRONT SPRING BUSHING MAINTENANCE

THIS TASK COVERS:

a. Inspection   c. Installation
b. Removal

INITIAL SETUP:

APPLICABLE MODELS
All

TOOLS
General mechanic's tool kit (Appendix E, Item 1)

MATERIALS/PARTS
Locknut (Appendix D, Item 285)
Bushing (Appendix D, Item 22)
Pin (Appendix D, Item 514)

REFERENCES (TM)
LO 9-2320-272-12
TM 9-2320-272-10
TM 9-2320-272-24P

EQUIPMENT CONDITION
• Parking brake set (TM 9-2320-272-10).
• Front wheels removed (para. 3-220 or 3-221).

---

a. Inspection

1. Insert prybar between spring hanger (1) and front spring shackle (3).
2. Press down on front spring shackle (3). If movement is evident between front spring shackle (3) and pin (2), perform task b.

b. Removal

1. Raise frame (4) to remove load from spring (12) and support with jack stands.
2. Remove locknut (7) and screw (8) from shackle (5). Discard locknut (7).
3. Remove grease fitting (11) from pin (10).
4. Remove pin (10) from shackle (5) and spring eye (13). Discard pin (10).
5. Remove spring (12) from shackle (5).
6. Remove bushing (6) from shackle (5). Discard bushing (6).

c. Installation

1. Install new bushing (6) on shackle (5).
2. Lower frame (4) until spring eye (13) is aligned with holes in shackle (5).
3. Install new pin (10) in shackle (5) and spring eye (13). Make sure pin slot (9) is positioned downward to allow installation of screw (8).
4. Install grease fitting (11) in pin (10).
5. Install screw (8) and new locknut (7) in shackle (5).

3-163. FRONT SPRING BUSHING MAINTENANCE (Contd)

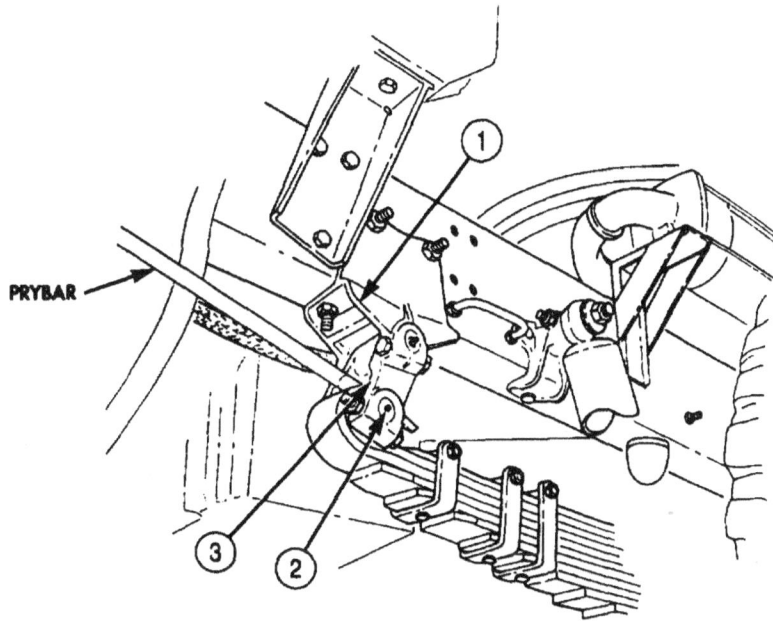

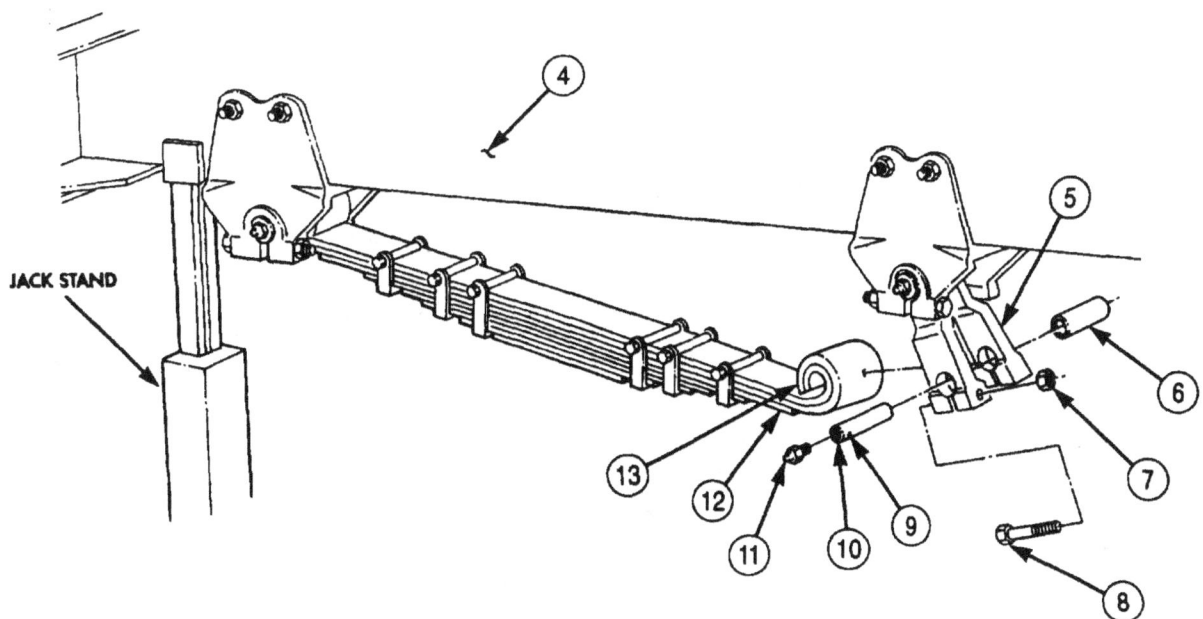

PRYBAR

JACK STAND

FOLLOW-ON TASKS: • Lubricate shackle (LO 9-2320-272-12).
• Install front wheels (para. 3-220 or 3-221).

3-164. FRONT SPRING BUMPER REPLACEMENT

THIS TASK COVERS:

a. Removal   b. Installation

INITIAL SETUP:

APPLICABLE MODELS
All

TOOLS
General mechanic's tool kit (Appendix E, Item 1)

MATERIALS/PARTS
Two locknuts (Appendix D, Item 280)

REFERENCES (TM)
TM 9-2320-272-10
TM 9-2320-272-24P

EQUIPMENT CONDITION
Parking brake set (TM 9-2320-272-10).

a. Removal

Remove two locknuts (1) and front spring bumper (2) from frame rail (3). Discard locknuts (1).

b. Installation

Install front spring bumper (2) on frame rail (3) with two new locknuts (1).

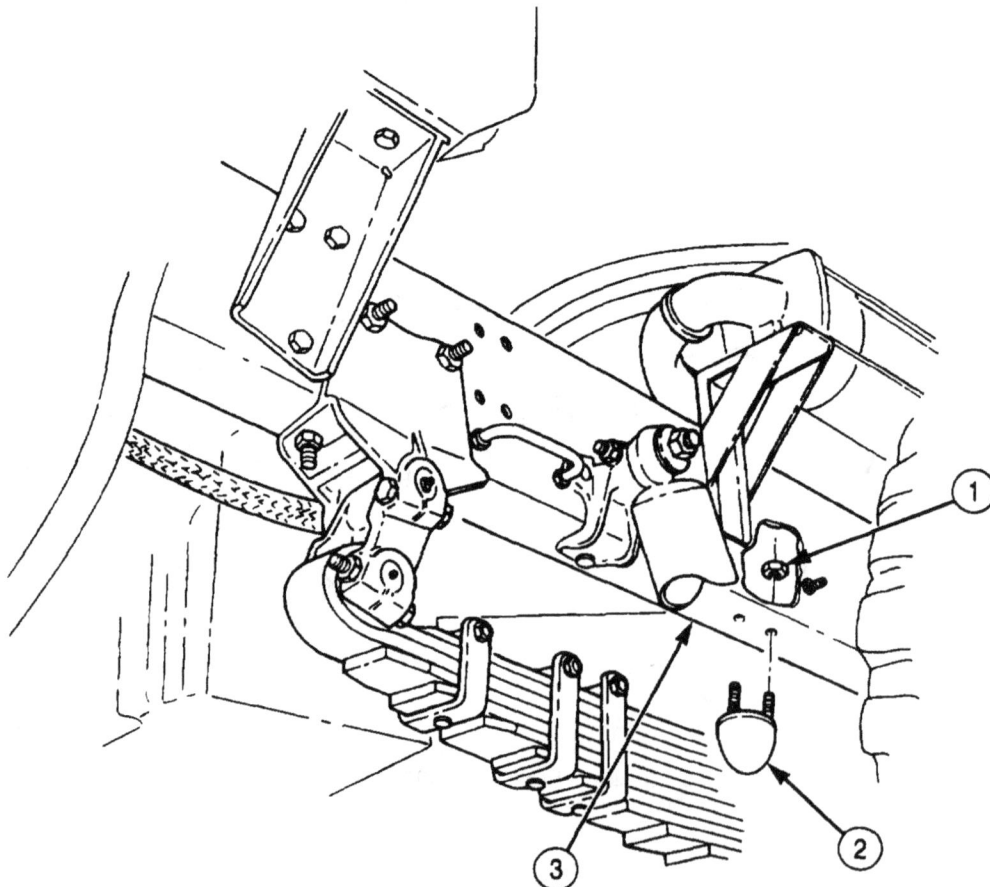

3-165. REAR SPRING BUMPER REPLACEMENT

THIS TASK COVERS:

a. Removal

b. Installation

INITIAL SETUP:

APPLICABLE MODELS
All

TOOLS
General mechanic's tool kit (Appendix E, Item 1)

MATERIALS/PARTS
Two locknuts (Appendix D, Item 280)

REFERENCES (TM)
TM 9-2320-272-10
TM 9-2320-272-24P

EQUIPMENT CONDITION
Parking brake set (TM 9-2320-272-10).

a. Removal

Remove two locknuts (1) and bumper (2) from bumper bracket (3). Discard locknuts (1).

b. Installation

Install bumper (2) on bumper bracket (3) with two new locknuts (1).

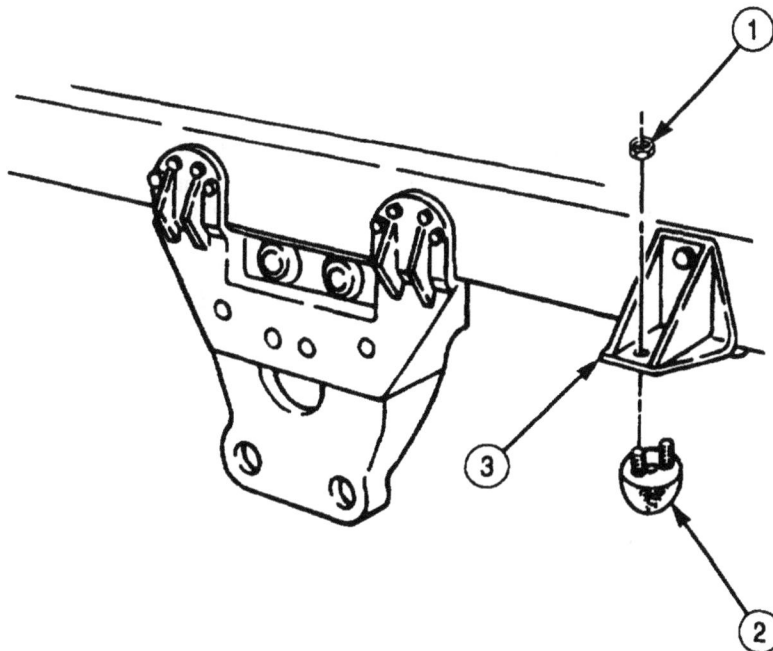

3-166. SHOCK ABSORBER AND MOUNTING PINS REPLACEMENT

THIS TASK COVERS:

a. Removal   b. Installation

INITIAL SETUP:

APPLICABLE MODELS
All

TOOLS
General mechanic's tool kit (Appendix E, Item 1)

MATERIALS/PARTS
Two locknuts (Appendix D, Item 286)
Two locknuts (Appendix D, Item 294)
Four rubber bushings (Appendix D, Item 564)

REFERENCES (TM)
TM 9-2320-272-10
TM 9-2320-272-24P

EQUIPMENT CONDITION
Parking brake set (TM 9-2320-272-10).

---

a. Removal

1. Remove two locknuts (1) and washers (2) from mounting pins (4). Discard locknuts (1).

2. Remove two rubber bushings (3), shock absorber (5), and two rubber bushings (3) from mounting studs (4). Discard rubber bushings (3).

3. Remove two locknuts (8) and screws (6) from mounting brackets (7). Discard locknuts (8).

4. Spread two mounting brackets (7) and remove mounting pins (4).

b. Installation

NOTE
Ensure pin slots are positioned to allow installation of screw.

1. Install two mounting pins (4) on mounting brackets (7) with two screws (6) and new locknuts (8).

2. Position four rubber bushings (3) and shock absorber (3) on two mounting pins (4).

3. Install four rubber bushings (3) and shock absorber (5) on two mounting pins (4) with washers (2) and new locknuts (1). Tighten locknuts (1) until seated and rubber bushings (3) are compressed.

3-166. SHOCK ABSORBER AND MOUNTING PINS REPLACEMENT (Contd)

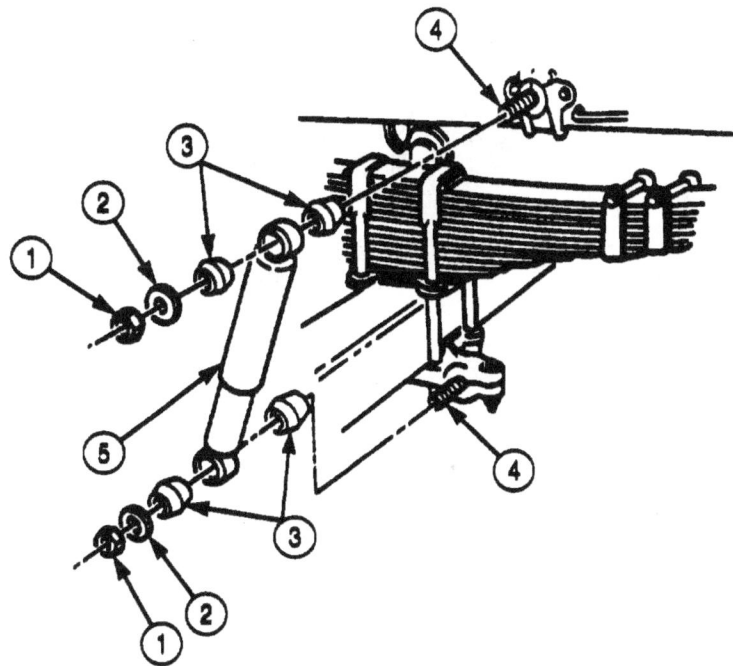

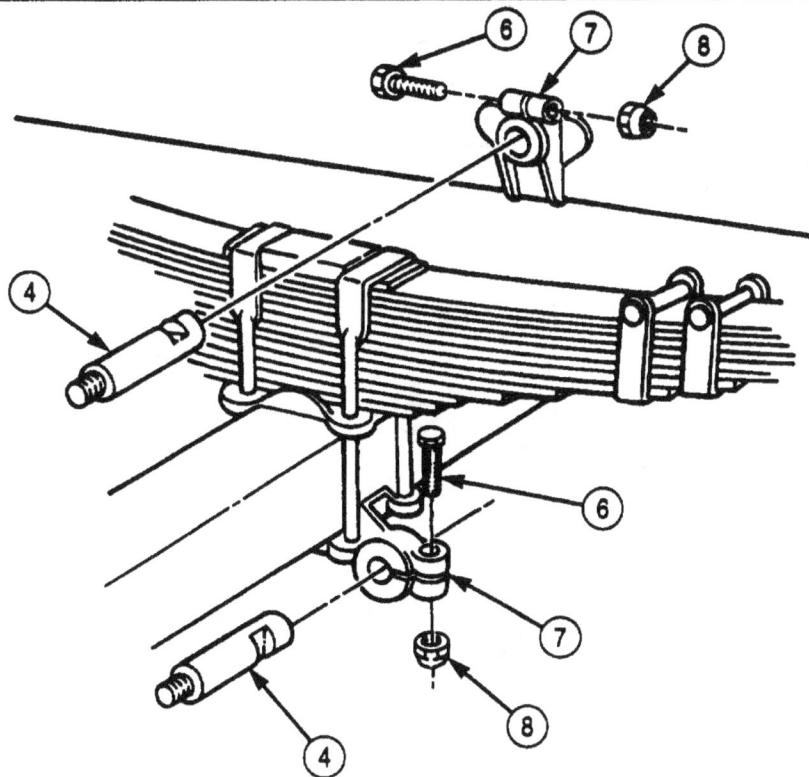

3-167. REAR SPRING REPLACEMENT

THIS TASK COVERS:

a. Removal   b. Installation

INITIAL SETUP:

APPLICABLE MODELS   REFERENCES (TM)
All   TM 9-2320-272-10
                                           TM 9-2320-272-24P
TOOLS
General mechanic's tool kit (Appendix E, Item 1)   EQUIPMENT CONDITION
Torque wrench (Appendix E, Item 145)   • Parking brake set (TM 9-2320-272-10).
                                           • Front wheels chocked (TM 9-2320-272-10).
MATERIALS/PARTS   • Rear wheels removed (para. 3-220 or 3-221).
Eight lockwashers (Appendix D, Item 363)

---

a. Removal

1. Raise vehicle and place two jack stands under cross tube (1), two jack stands under both forward-rear hubs (5), and two jack stands under both rear-rear hubs (10). Position tops of jack stands under cross tube (1) four inches above tops of jack stands to relieve pressure on rear springs (3).

2. Loosen two bolt clamps (9) on spring seat (8).

3. Remove four nuts (6) and lockwashers (7) from two U-bolts (2). Discard lockwashers (7).

NOTE

Step 4 applies to all models except M936 wrecker.

4. Remove two U-bolts (2) and upper spring saddle (4) from rear springs (3). It may be necessary to raise frame to remove U-bolts (2).

NOTE

Step 5 applies to M936 wrecker only

5. Remove two U-bolts (2) and stabilizer beam (11) from rear springs (3).

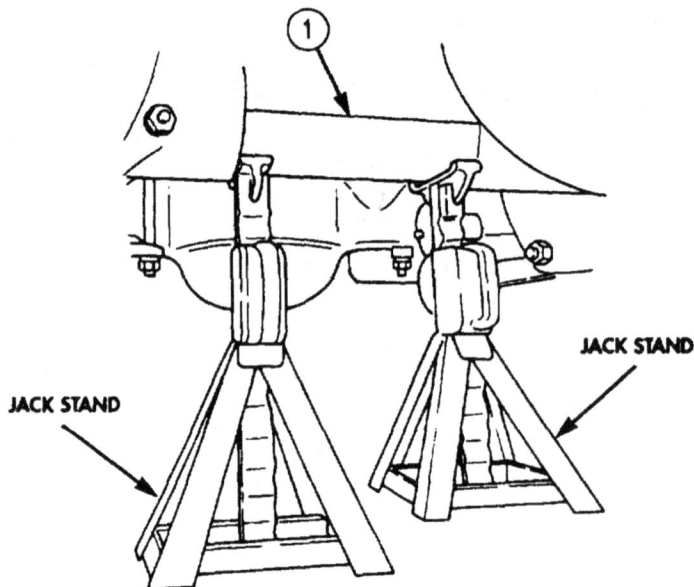

3-167. REAR SPRING REPLACEMENT (Contd)

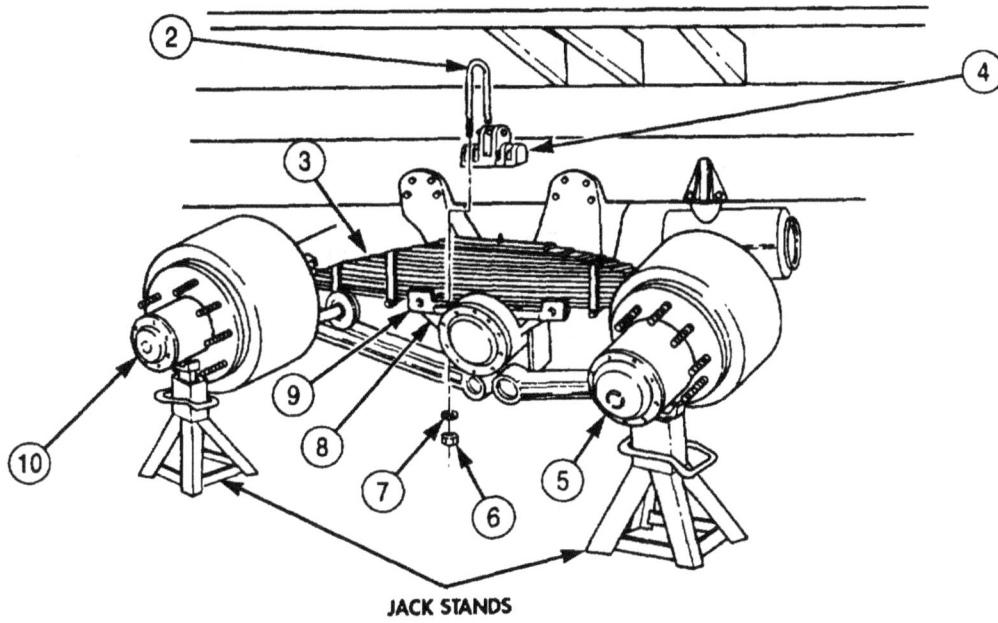

JACK STANDS

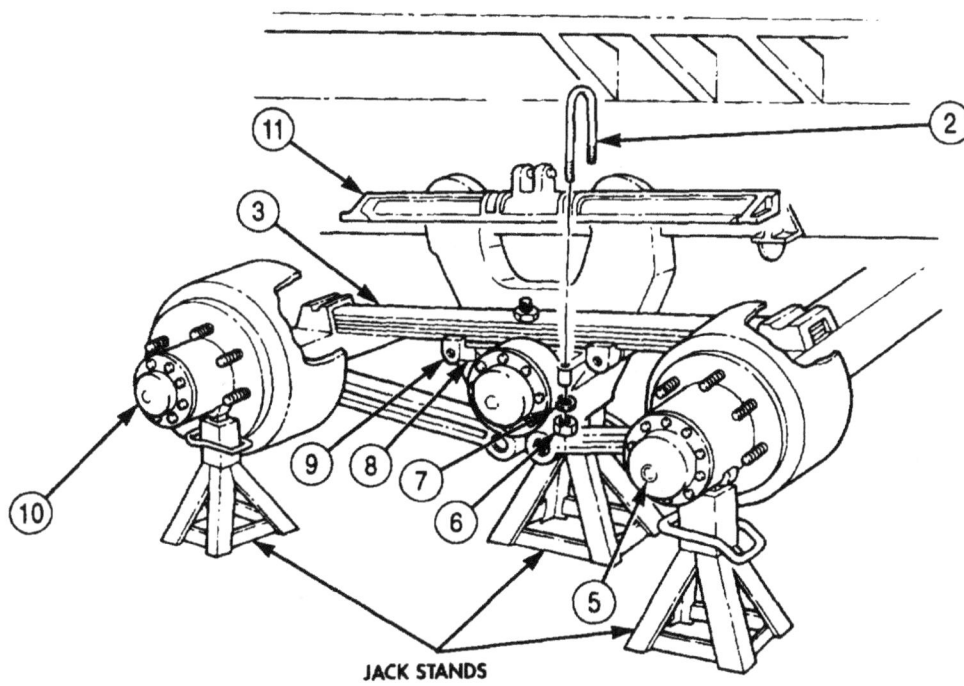

JACK STANDS

M936 WRECKER ONLY

3-167. REAR SPRING REPLACEMENT (Contd)

6. Remove two nuts (15), washers (14), U-bolt (5), and U-bolt bracket (8) from spring brake chamber (13).

NOTE

Assistant will help with steps 7 through 10.

7. Remove nut (9), lockwasher (10), washer (11), washer (2), screw (1), and spring brake chamber bracket (12) from upper spring bracket and wearpad (4). Discard lockwasher (10).

8. Remove two nuts (7), lockwashers (6), and screws (3) from upper spring bracket and wearpad (4). Discard lockwashers (6).

9. Remove two nuts (25), washers (24), U-bolt (18), and U-bolt bracket (19) from service brake chamber (17).

10. Remove nut (20), lockwasher (21), two washers (22), screw (16), and service brake chamber bracket (23) from upper spring bracket and wear pad (4). Discard lockwasher (21).

11. Remove upper spring bracket and wear pad (4) from axle housing (26).

NOTE

Assistants will help with steps 12 and 13.

12. Remove spring assembly (27) from spring seat (28).

NOTE

Step 13 applies to M936 wrecker only.

13. Remove spring assembly (29) from spring seat (28).

3-167. REAR SPRING REPLACEMENT (Contd)

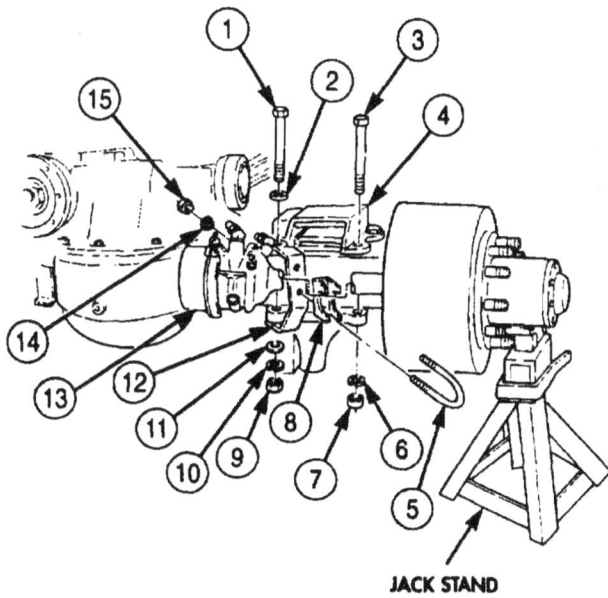

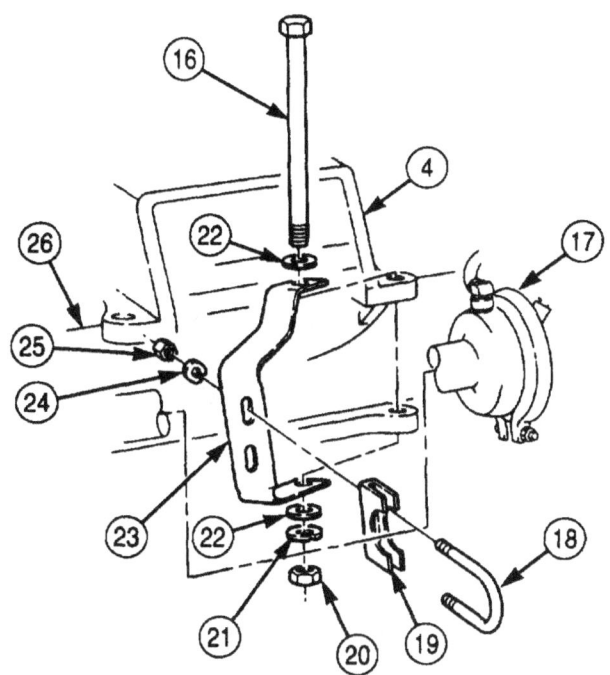

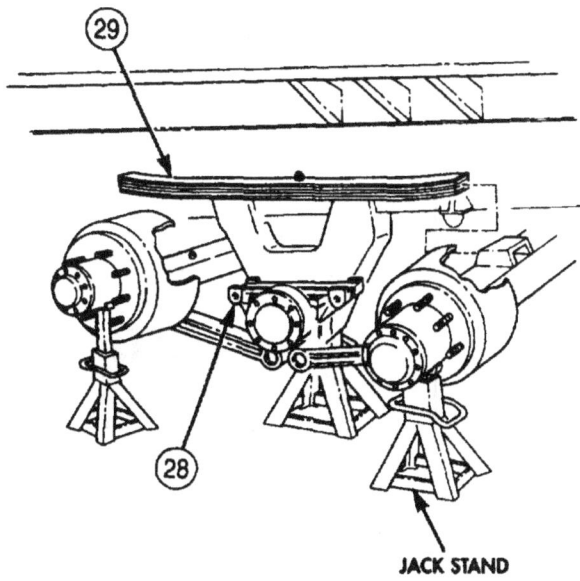

JACK STAND

JACK STAND

M936 WRECKER ONLY

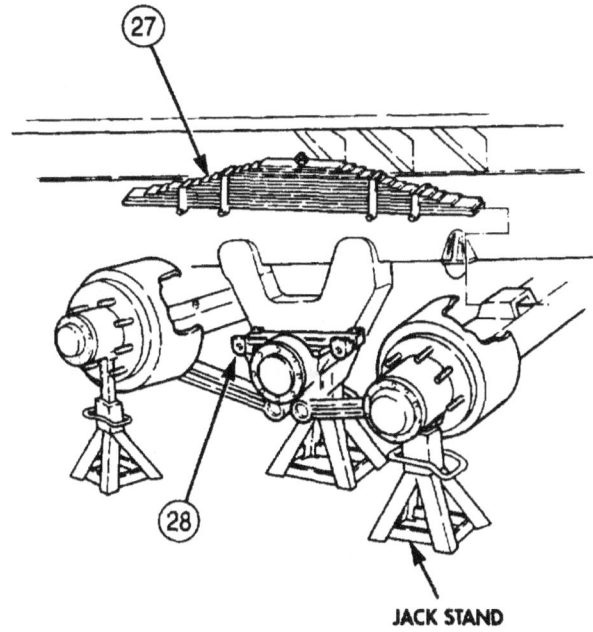

JACK STAND

3-167. REAR SPRING REPLACEMENT (Contd)

b. Installation

NOTE
- Assistants will help with steps 1 and 2.
- Step 1 applies to M936 wrecker only.

1. Position spring assembly (3) on spring seat (4).
2. Position spring assembly (1) on spring seat (2).
3. Slide upper spring bracket and wearpad (5) on main leaves (7) and position on axle housing (6) over dowel pin (8).
4. Install service brake chamber bracket (15) on upper spring bracket and wearpad (5) with screw (9), two washers (10), new lockwasher (14), and nut (13). Do not tighten nut (13).
5. Install U-bolt bracket (12) on service brake chamber bracket (15) with U-bolt (11), two washers (16), and nuts (17). Tighten nuts (17) 120-160 lb-ft (163-217 N•m).
6. Install spring brake chamber bracket (27) on upper spring bracket and wearpad (5) with screw (18), two washers (19), new lockwasher (26), and nut (25). Do not tighten nut (25).
7. Install U-bolt bracket (24) on spring brake chamber (28) with U-bolt (21), two washers (29), and nuts (30). Tighten nuts (30) 120-160 lb-ft (163-217 N•m).

NOTE
Assistant will help with steps 8 and 9.

8. Install two screws (20), new lockwashers (22), and nuts (23) on upper spring bracket and wearpad (5). Tighten nuts (13), (23), and (25) 280-360 lb-ft (380-488 N•m).
9. Install upper spring saddle (32) on spring seat (2) with two U-bolts (31), four new lockwashers (34), and nuts (33). Tighten nuts (33) 300-400 lb-ft (407-542 N•m).

NOTE
Step 10 applies to M936 wrecker only.

10. Install stabilizer beam (36) on spring seat (4) with two U-bolts (37), four new lockwashers (39), and nuts (38). Tighten nuts (38) 300-400 lb-ft (407-542 N•m).
11. Tighten two bolt clamps (35) on rear spring seat (2) or (4).

M936 WRECKER ONLY

3-167. REAR SPRING REPLACEMENT (Contd)

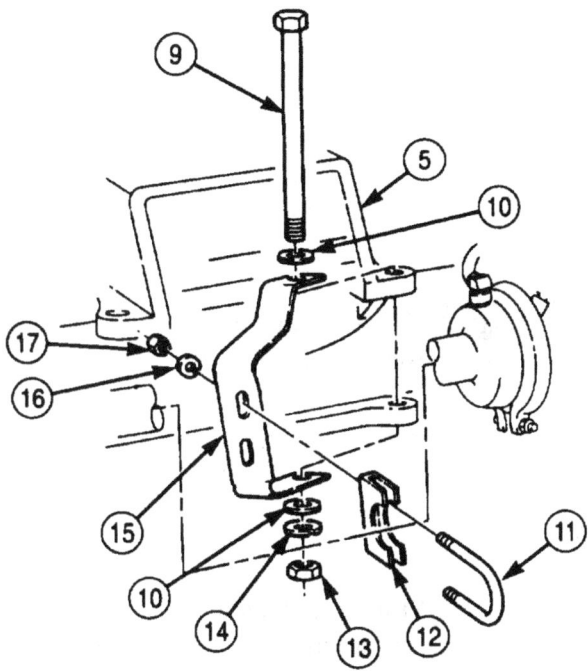

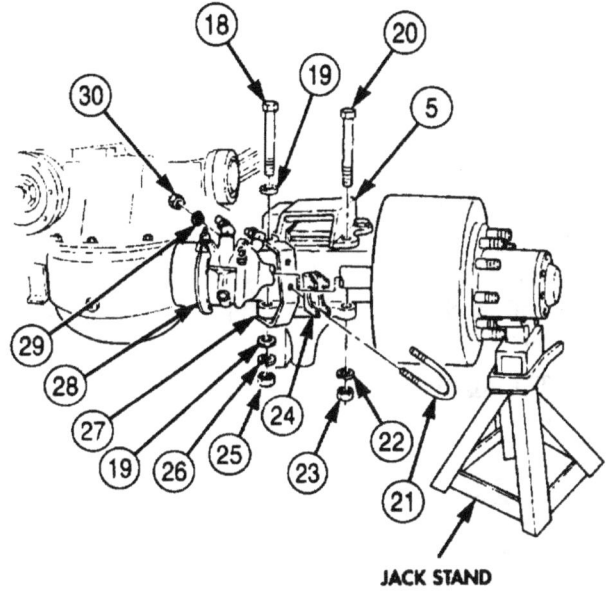

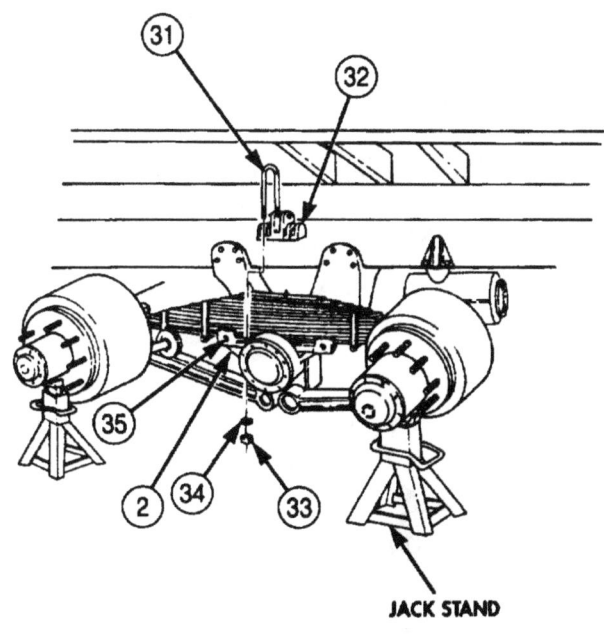

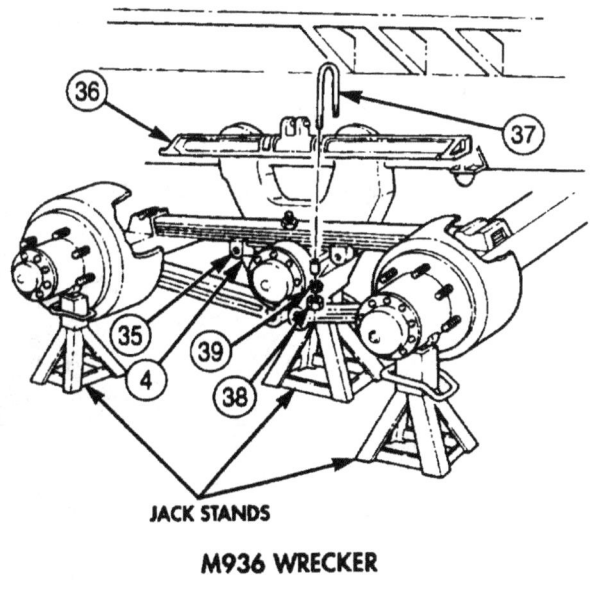

JACK STAND

JACK STAND

JACK STANDS

M936 WRECKER

FOLLOW-ON TASKS: • Install rear wheels (para. 3-220 or 3-221).
   • Start engine and road test vehicle (TM 9-2320-272-10).
   • Check U-bolts for tightness after road test.

3-168. REAR AXLES SPRING SEAT WEAR PADS AND UPPER BRACKET MAINTENANCE

THIS TASK COVERS:

a. Removal
b. Inspection

c. Installation

INITIAL SETUP:

APPLICABLE MODELS

All

Tools

General mechanic's tool kit (Appendix E, Item 1)
Torque wrench (Appendix E, Item 145)

MATERIAL/PARTS

Four lockwashers (Appendix D, Item 363)

REFERENCES (TM)

TM 9-2320-272-10
TM 9-2320-272-24P

EQUIPMENT CONDITION
• Parking brake ret (TM 9-2320-272-10)
• Rear wheels removed (para. 3-220 or 3-221).

a. Removal

1. Using hydraulic jack, raise vehicle at rear-rear axle differential housing (14) and place jack stands under wheel hubs (5).

2. Remove two nuts (16), washers (15), U-bolt (6), and U-bolt bracket (9) from spring brake chamber (13).

3. Remove nut (10), lockwasher (11), two washers (2), screw (1), and spring brake chamber bracket (12) from upper spring bracket (4). Discard lockwasher (11).

4. Remove two nuts (26), washers (25), U-bolt (21), and U-bolt bracket (22) from service brake chamber (20).

5. Remove nut (23), lockwasher (24), two washers (18), screw (17), and service brake chamber bracket (19) from upper spring bracket (4). Discard lockwasher (24).

6. Remove two nuts (8), lockwashers (7), and screws (3) from upper spring bracket (4). Discard lockwashers (7).

7. Place hydraulic jack under spring seat (31) and raise leaf spring assembly (32) until upper spring bracket (4) is clear of axle housing (28) and dowel pin (30).

8. Slide upper spring bracket (4) from bottom leaf (29) and remove from leaf spring assembly (32).

9. Remove spring seat wearpad (27) from upper spring bracket (4).

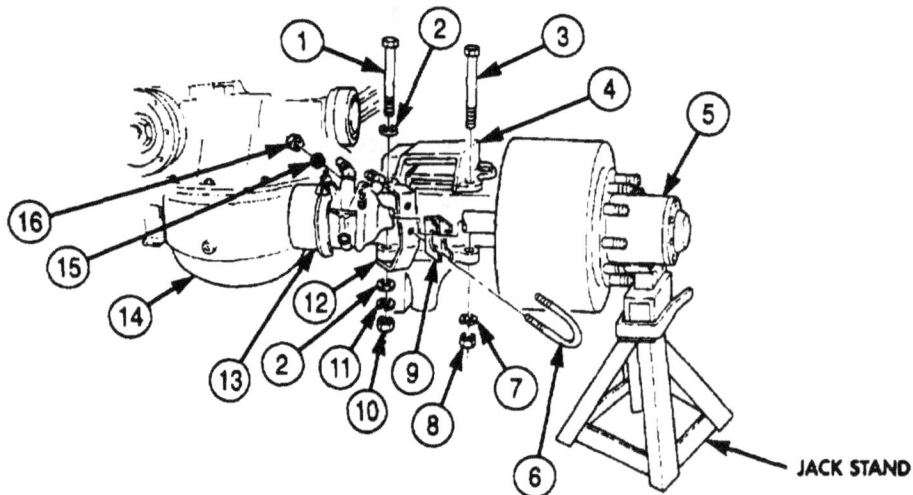

3-168. REAR AXLES SPRING SEAT WEAR PADS AND UPPER BRACKET MAINTENANCE (Contd)

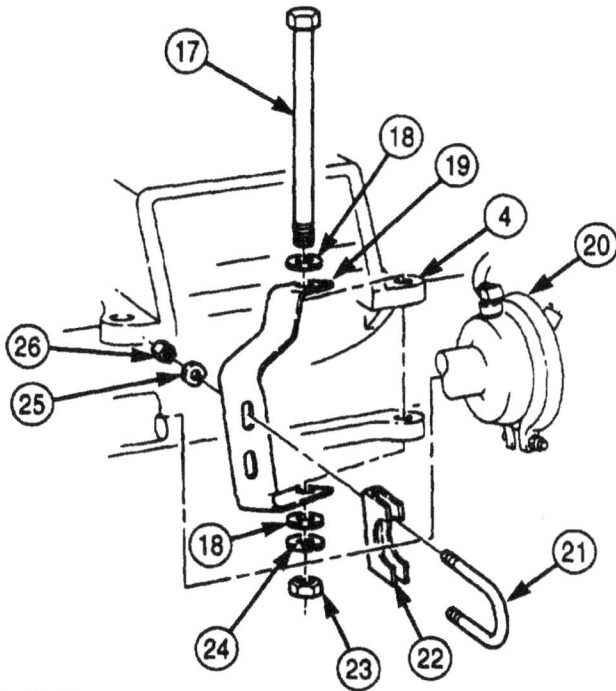

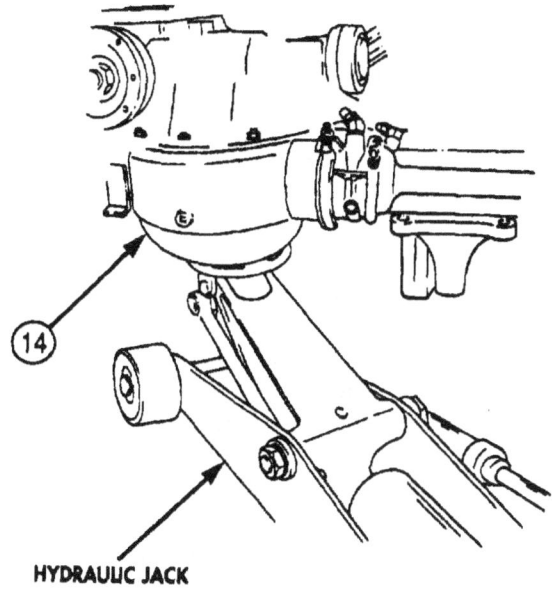

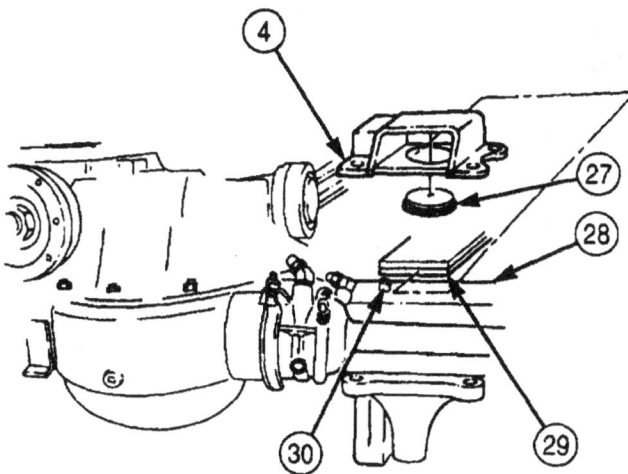

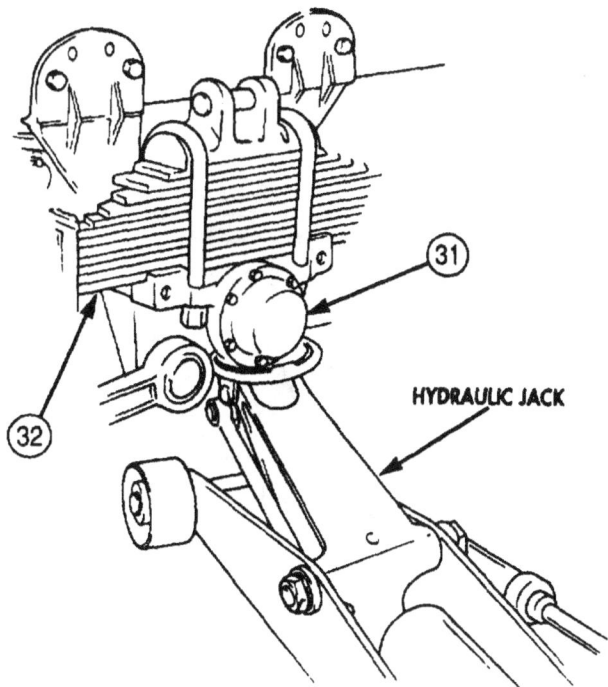

HYDRAULIC JACK

HYDRAULIC JACK

3-168. REAR AXLES SPRING SEAT WEAR PADS AND UPPER BRACKET MAINTENANCE (Contd)

b. Inspection

1. Inspect spring seat wearpad (2) for cracks. Replace spring seat wearpad (2) if cracked.
2. Inspect upper spring bracket (1) for cracks. Replace upper spring bracket (1) if cracked.

c. Installation

1. Install spring seat wearpad (2) in upper spring bracket (1).
2. Slide upper spring bracket (1) on bottom leaf (4) and install on axle housing (3) over dowel pin (5).
3. Lower hydraulic jack until leaf spring assembly (7) seats in upper spring bracket (1) and remove hydraulic jack from spring seat (6).
4. Position spring brake chamber bracket (17) over upper spring bracket (1) and install with screw (8), two washers (9), new lockwasher (16), and nut (15). Do not tighten nut (15).
5. Install service brake U-bolt bracket (14) between brake chamber (18) and brake chamber bracket (17) with U-bolt (11), two washers (19), and nuts (20). Tighten nuts (20) 120-160 lb-ft (163-217 N•m).
6. Position service brake chamber bracket (23) over upper spring bracket (1) and install with screw (21), two washers (22), new lockwasher (28), and nut (27). Do not tighten nut (27).
7. Install service brake U-bolt bracket (26) between service brake chamber (24) and brake chamber bracket (23) with U-bolt (25), two washers (29), and nuts (30). Tighten nuts (30) 120-160 lb-ft (163-217 N•m).
8. Install upper spring bracket (1) on spring brake chamber bracket (17) with two screws (10), new lockwashers (12), and nuts (13). Tighten nuts (13) 280-360 lb-ft (380-488 N•m).
9. Tighten nuts (15) and (27) 280-360 lb-ft, (380-488 N•m).

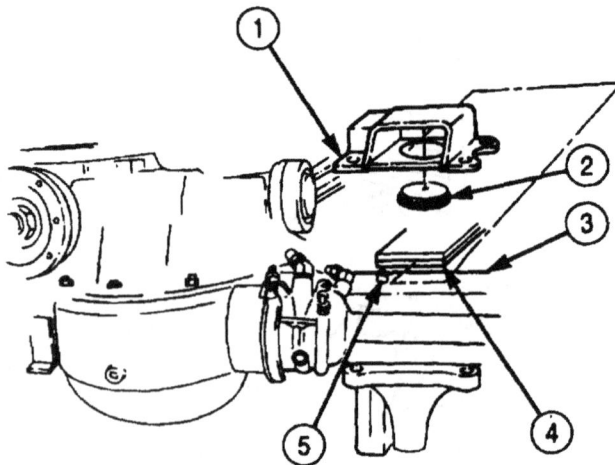

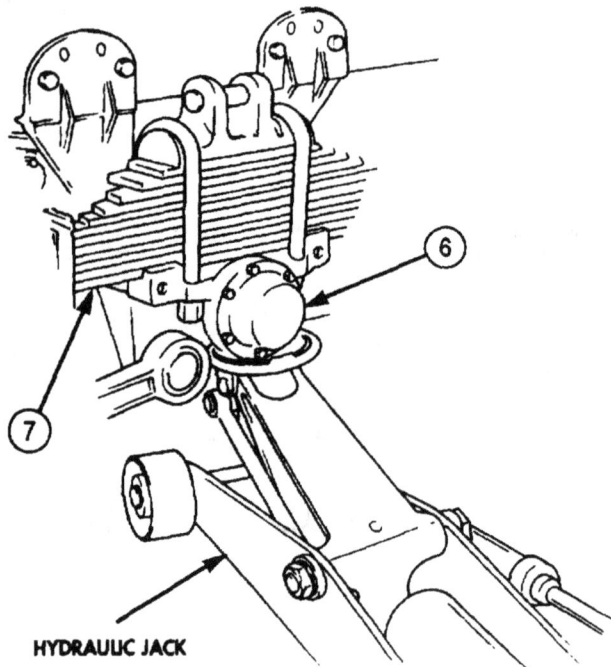

HYDRAULIC JACK

3-168. REAR AXLES SPRING SEAT WEAR PADS AND UPPER BRACKET MAINTENANCE (Contd)

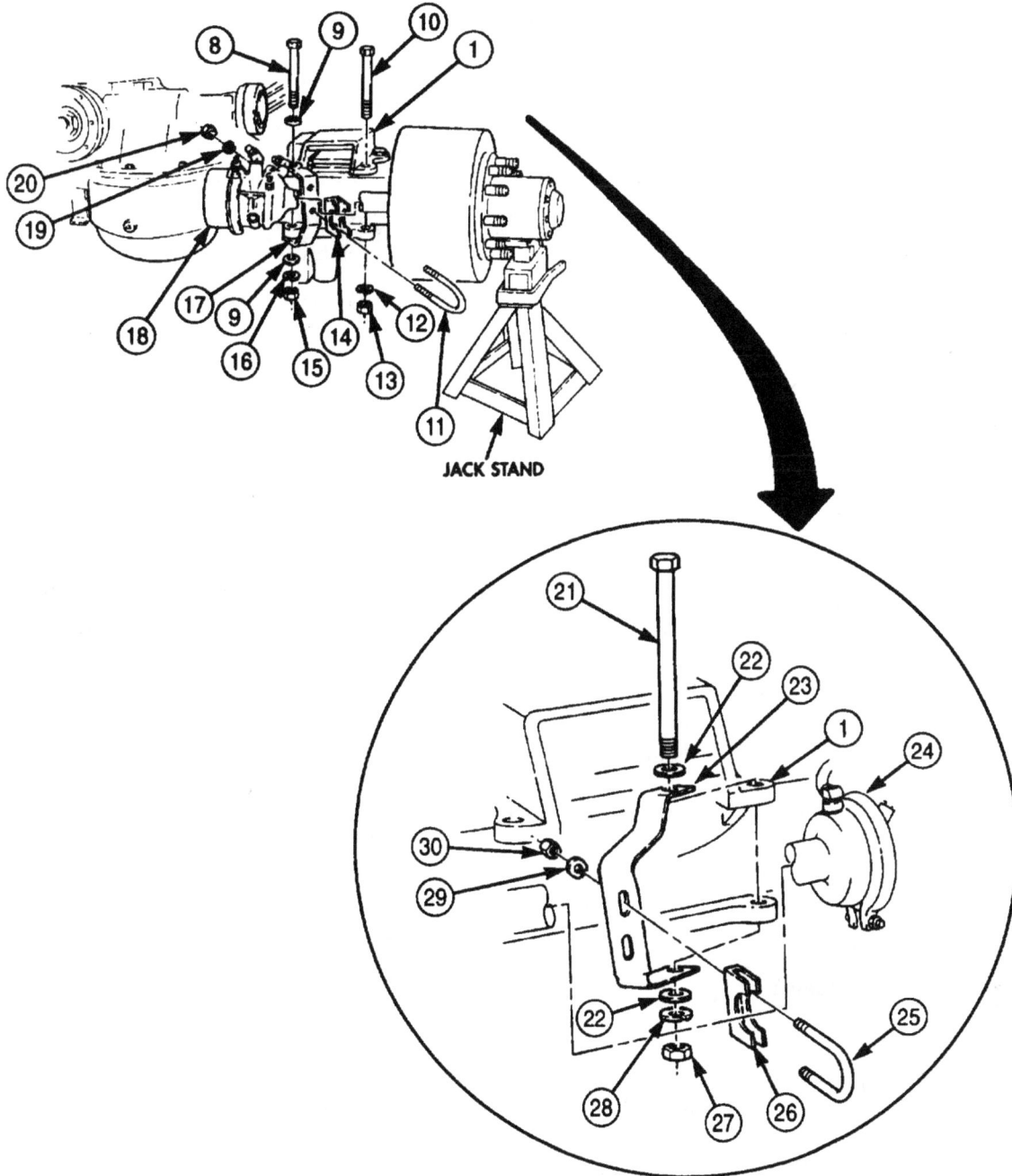

JACK STAND

FOLLOW-ON TASK: Install rear wheels (para. 3-220 or 3-221).

3-169. REAR SPRING SEAT MAINTENANCE

THIS TASK COVERS:

a. Removal
b. Cleaning and Inspection
c. Lubrication

d. Installation
e. Adjustment

INITIAL SETUP:

APPLICABLE MODELS

All

TOOLS

General mechanic's tool kit (Appendix E, Item 1)
Torque wrench (Appendix E, Item 144)
Hook tester scale (Appendix E, Item 55)

MATERIALS/PARTS

Six lockwashers (Appendix D, Item 364)
Two felt washers (Appendix D, Item 111)
Seal assembly (Appendix D, Item 634)
Gasket (Appendix D, Item 161)
GAA grease (Appendix C, Item 28)
Lubricating oil (Appendix C, Item 50)
Drycleaning solvent (Appendix C, Item 71)

REFERENCES (TM)

LO 9-2320-272-12
TM 9-2320-272-10
TM 9-2320-272-24P

EQUIPMENT CONDITION

Rear spring assembly removed (para. 3-167).

GENERAL SAFETY INSTRUCTIONS

- Fire extinguisher will be kept nearby when drycleaning solvent is used.
- Drycleaning solvent is flammable and toxic. Do not use near open flame.

---

a. Removal

1. Remove six screws (9) and lockwashers (8) from spring seat cap (10). Discard lockwashers (8).

2. Remove spring seat cap (10) and gasket (11) from spring seat (1). Discard gasket (11). Clean gasket (11) remains from mating surfaces.

3. Remove locking nut (12) and key washer (13) from crosstube (2).

4. Remove bearing adjusting nut (14) and outer bearing (15) from crosstube (2).

5. Remove spring seat (1) from crosstube (2).

6. Remove wiper ring (3), washer (4), two felt washers (5), packing retainer (6), grease seal assembly (7), and inner bearing (16) from spring seat (1). Discard felt washers (5) and grease seal assembly (7).

NOTE

Perform step 7 only when bearing races must be removed because of cracked, pitted, or scored condition, or if bearings are to be installed.

7. Remove bearing races (17) and (18) from spring seat (1).

b. Cleaning and Inspection

WARNING

Drycleaning solvent is flammable and toxic. Do not use near open flame and always have a fire extinguisher nearby when solvents are used. Use only in well-ventilated places, wear protective clothing, and dispose of cleaning rags in approved container. Failure to do this may result in injury or death to personnel and/or damage to equipment.

1. Clean all parts in drycleaning solvent and allow to dry

2. Inspect all parts for cracks, pitting, or scoring. Replace parts as necessary if cracked, pitted, or scored.

3-169. REAR SPRING SEAT MAINTENANCE (Contd)

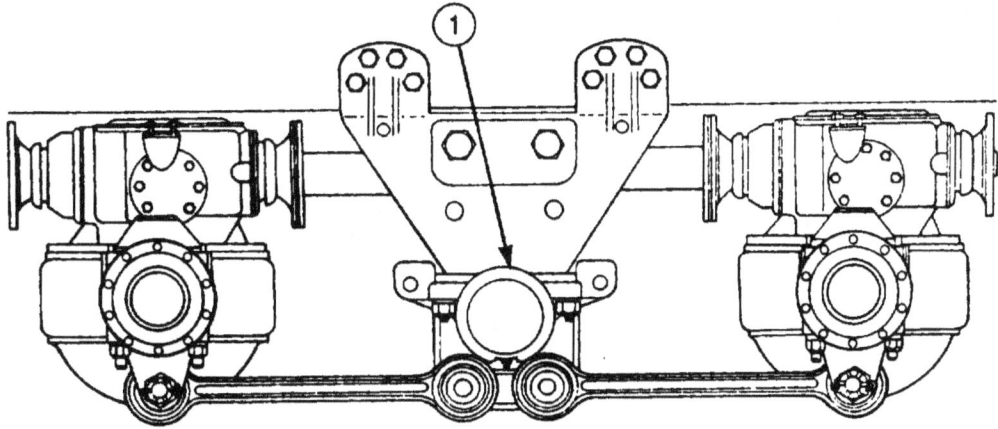

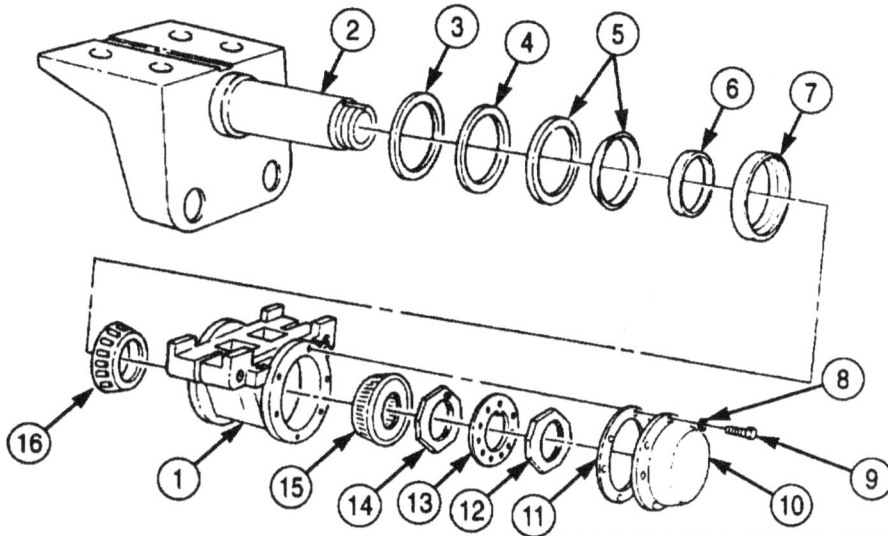

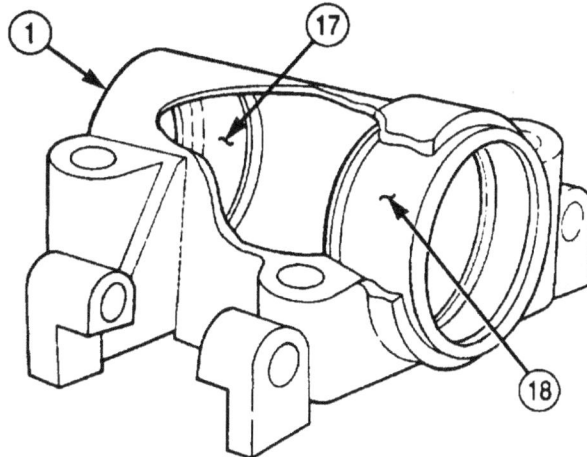

3-169. REAR SPRING SEAT MAINTENANCE (Contd)

c. Lubrication

Pack grease on inside of spring seat (1) and repack inner bearings (4) and outer bearing (5) with GAA grease (LO 9-2320-272-12).

d. Installation

NOTE

Perform step 1 if bearing races were removed.

1. Install bearing races (2) and (3) on spring seat (1).

NOTE

Soak felt washers with oil before installation.

2. Install wiper ring (7), washer (8), two new felt washers (9), and packing retainer (10) on crosstube (6).
3. Install inner bearing (4) in spring seat (1)
4. Install new grease seal assembly (11) over inner bearing (4) side of spring seat (1).
5. Install spring seat (1) on crosstube (6).
6. Install outer bearing (5) in spring seat (1).
7. Install bearing adjusting nut (18) on crosstube (6).

e. Adjustment

1. Connect hook tester scale in bolt hole (19).
2. Tighten adjusting nut (18) and pull tester scale downward.

NOTE

Bearings are correctly adjusted when pull on scale required to rotate seat is 25-33 lb (11-15 kg) This is equal to 15-20 lb-ft (20-27 N•m) preload on bearings.

3. Note pull required to rotate spring seat (1) around crosstube (6).
4. Install key washer (17) and locking nut (16). Repeat steps 2 and 3 to ensure bearing adjustment does not change. Tighten locking nut (16) 150-160 lb-ft (203-217 N•m).
5. Install new gasket (15) and spring seat cap (14) on spring seat (1) with six new lockwashers (12) and screws (13). Tighten screw (13) 16-20 lb-ft (22-27 N•m).

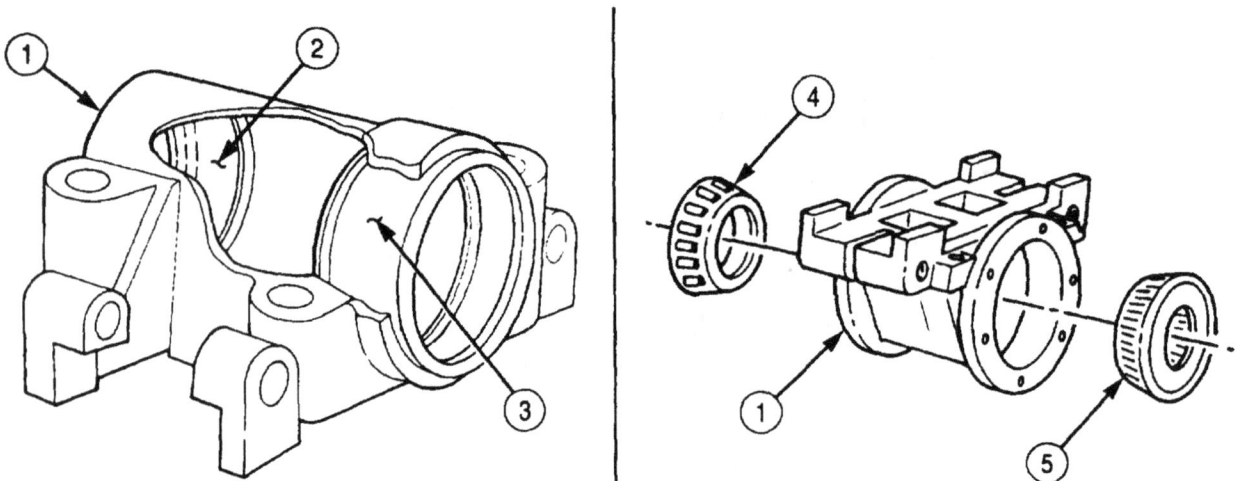

3-169. REAR SPRING SEAT MAINTENANCE (Contd)

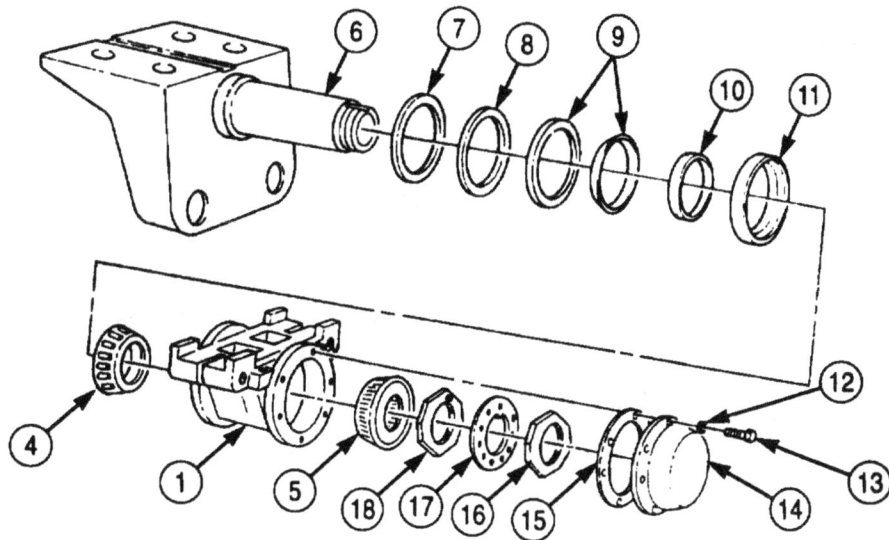

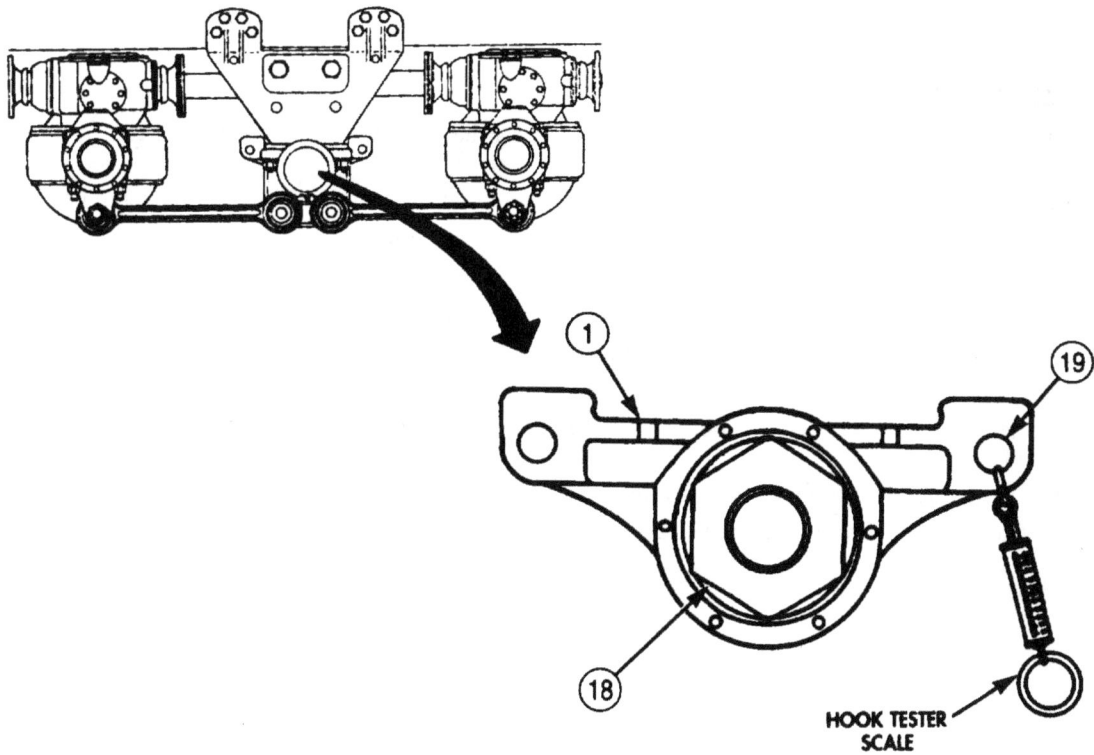

FOLLOW-ON TASK: • Install rear spring assembly (para 3-167).
• Lubricate spring seat (LO 9-2320-272-12).

3-170. UPPER AND LOWER TORQUE ROD MAINTENANCE

THIS TASK COVERS:

a. Upper Torque Rod Removal
b. Lower Torque Rod Removal
c. Cleaning and Inspection
d. Disassembly

e. Assembly
f. Setting Preload for Upper Torque Rod
g. Upper Torque Rod Installation
h. Lower Torque Rod Installation

INITIAL SETUP:

APPLICABLE MODELS

All

TOOLS

General mechanic's tool kit (Appendix E, Item 1)
Torque wrench (Appendix E, Item 145)
Arbor press
Vise

MATERIALS/PARTS

Four cotter pins (Appendix D. Item 47)
Four slotted nuts (Appendix D), Item 658)
Eight lockwashers (Appendix D, Item 362)

REFERENCES (TM)

TM 9-2320-272-10
TM 9-2320-272-24P
TM 9-237

EQUIPMENT CONDITION

• Parking brake set (TM 9-2320-272-10).
• Right rear spring assembly removed (for upper torque rods only) (para. 3-167).
• Two front-rear wheels removed (for lower torque rods only) (para. 3-220 or 3-221).

GENERAL SAFETY INSTRUCTIONS

Eye shields must be worn when cleaning with a wire brush.

a. Upper Torque Rod Removal

1. Remove four nuts (5), lockwashers (4), screws (3), and torque rod plate (6) from torque rod bracket (1). Discard lockwashers (4).

2. Remove cotter pin (8) from upper torque rod (2) and spring seat bracket (9). Discard cotter pin (8).

3. Back off slotted nut (7) until even with ball shaft (10) and use slotted nut (7) as a striking surface.

4. Drive torque rod ball shaft (10) out of spring seat bracket (9).

5. Remove slotted nut (7) from ball shaft (10). Discard slotted nut (7).

6. Remove upper torque rod (2) with upper torque rod bracket (1) from spring seat bracket (9).

7. Place upper torque rod (2) and bracket (1) in vise.

8. Remove cotter pin (11) and back off slotted nut (13) until even with end of ball shaft (12). Discard cotter pin (11).

9. Using nut (13) as striking surface, drive torque rod ball shaft (12) out of upper bracket (1).

10. Remove slotted nut (13) from ball shaft (12). Discard slotted nut (13).

3-170. UPPER AND LOWER TORQUE ROD MAINTENANCE (Contd)

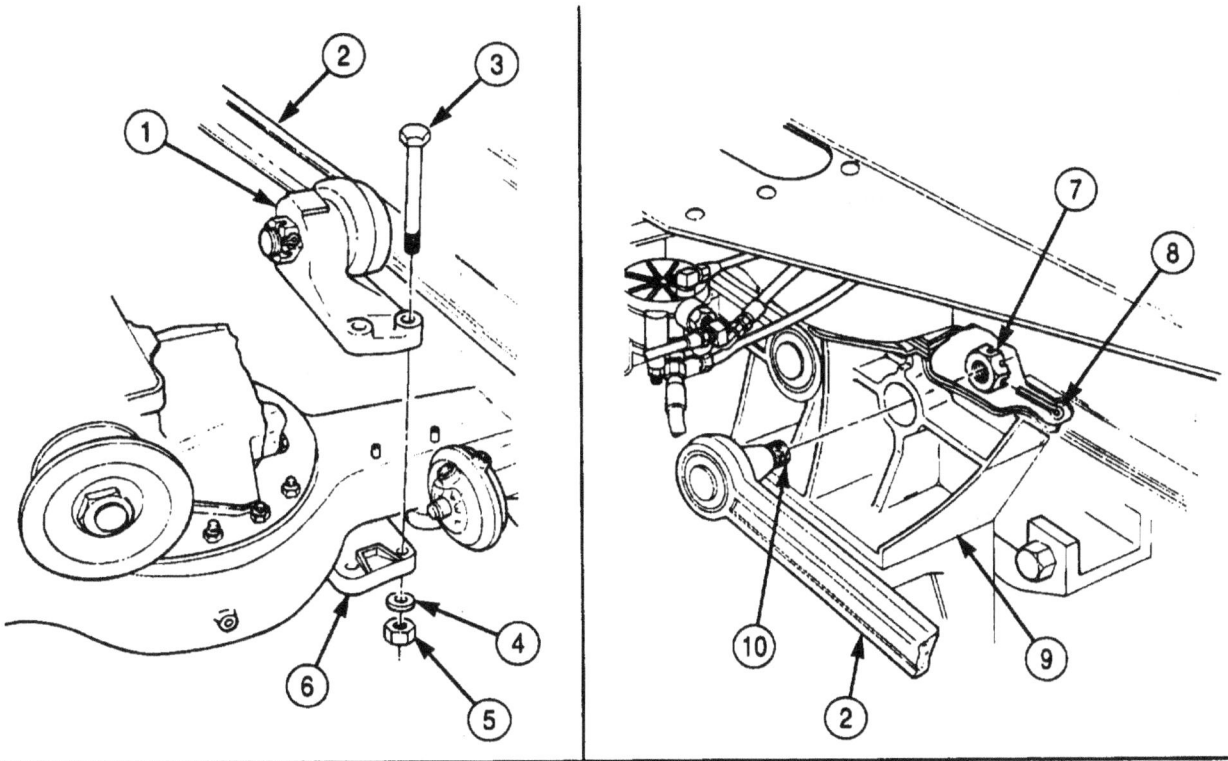

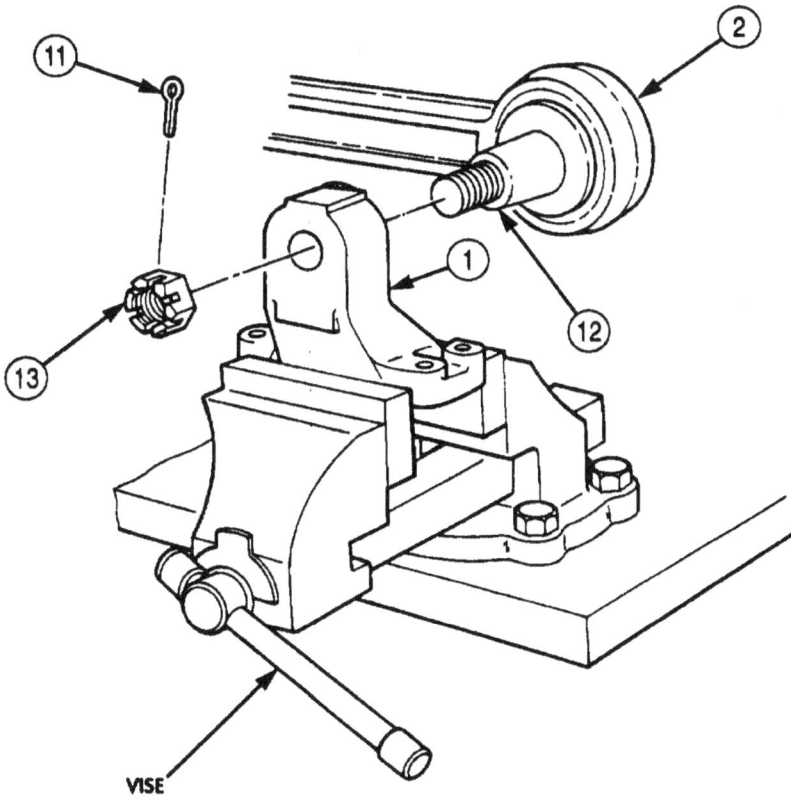

VISE

3-170. UPPER AND LOWER TORQUE ROD MAINTENANCE (Contd)

b. Lower Torque Rod Removal

1. Remove four nuts (8), lockwashers (9), screws (1), and upper spring bracket (14) from lower torque rod bracket (10). Discard lockwashers (9).
2. Remove cotter pin (5) from lower torque rod (7) and slotted nut (4). Discard cotter pin (5).
3. Back off slotted nut (4) until even with end of torque rod ball shaft (6).
4. Using slotted nut (4) as striking surface, drive torque rod ball shaft (6) out of spring seat bracket (3).
5. Remove slotted nut (4) and lower torque rod (7) from spring seat bracket (3) and rear axle (2). Discard slotted nut (4).
6. Place lower torque rod bracket (7) in vise.
7. Remove cotter pin (13) from lower torque rod (7) and slotted nut (12). Discard cotter pin (13).
8. Back off slotted nut (12) until even with end of torque rod ball shaft (11).
9. Using slotted nut (12) as striking surface, drive torque rod ball shaft (11) out of lower torque rod bracket (10).
10. Remove slotted nut (12) from lower torque rod bracket (10). Discard slotted nut (12).

c. Cleaning and Inspection

WARNING

Eyeshields must be worn when cleaning with a wire brush. Flying rust and metal particles may cause injury to personnel.

NOTE

Perform steps 1 through 8 for upper torque rod.

1. Clean rust from upper torque rod bracket (15) axle mating surface and dowel holes (16).
2. Inspect upper torque rod bracket (15) for breaks and cracks. Replace if broken or cracked.
3. Inspect upper torque rod plate (19) for breaks and cracks. Replace if broken or cracked.
4. Clean rear axle (2) and upper torque rod dowel (18) and upper spring bracket dowel (17).
5. Inspect upper torque rod dowel (18) and upper spring bracket dowel (17) for flat, broken, or out-of-round condition. Repeat if flat, broken, or out-of-round.

NOTE

Perform steps 6 and 7 only if dowels are to be replaced.

6. File weld securing upper torque rod bracket and upper spring bracket dowels (18) and (17) to rear axle (2) and remove dowels (18) and (17).
7. Tap in new upper torque rod bracket dowel (18) and upper spring bracket dowel (17) in rear axle (2) and spot-weld (TM 9-237). Dowel height must be 3/8 in. (9.53 mm).
8. Inspect upper and lower torque rods (7) around ball shafts (20) for breaks, cracks, and separation of rubber from torque rods (2).
9. Clean and inspect lower torque rod bracket (10) for breaks and cracks. Replace if broken or cracked.

d. Disassembly

Using arbor press, remove two balls and bushings (20) from ends of torque rod (7).

3-170. UPPER AND LOWER TORQUE ROD MAINTENANCE (Contd)

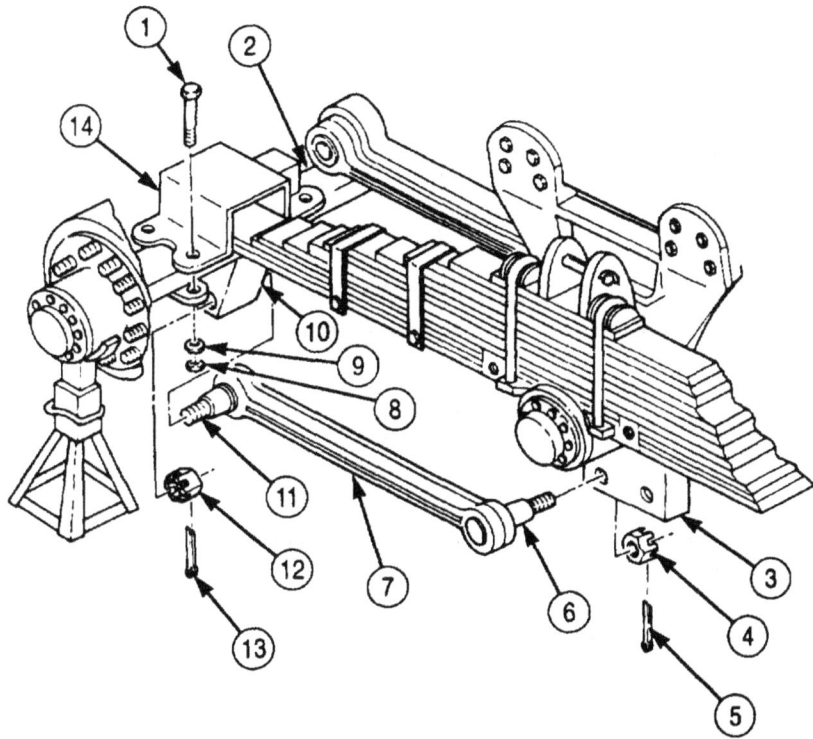

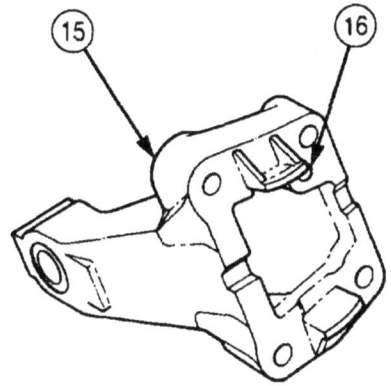

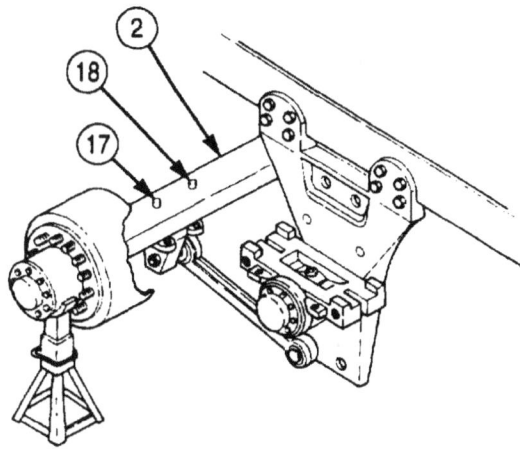

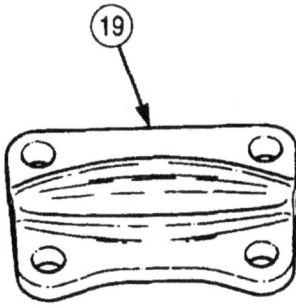

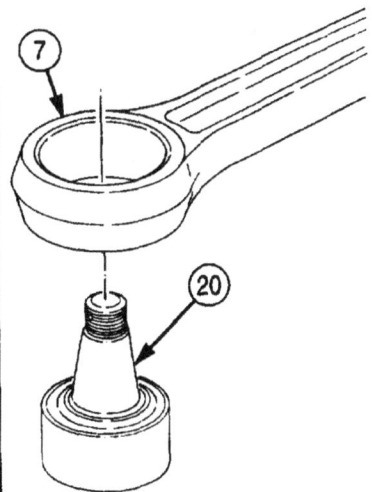

3-170. UPPER AND LOWER TORQUE ROD MAINTENANCE (Contd)

e. Assembly

NOTE

Ball and bushing must be pressed in on chamber side only.

Using arbor press, press balls and bushings (1) into torque rod (2) at chamber side (3).

f. Setting Preload for Upper Torque Rod

1. Position upper spring bracket (8) over upper spring bracket dowel (9).

NOTE

Perform step 2 for all models except M936/A1/A2 model vehicles.

2. Raise rear axle (10) until 6 in. (15.24 cm) is obtained between bottom of frame rail (7) and top of spring bracket (8).

NOTE

Perform step 3 for M936/A1/A2 model vehicles only.

3. Raise rear axle (10) until 7-1/4 in. (18.4 cm) is obtained between bottom of frame rail (7) and top of spring bracket (8).

4. Place ball shaft (4) on upper torque rod (5) in upper torque rod bracket (6) and install with new slotted nut (12). Finger-tighten slotted nut (12).

5. Place upper torque rod (5) and upper torque rod bracket (6) over dowel (11) on rear axle (10) and position torque rod ball shaft (4) in spring bracket (8).

NOTE

Ensure upper torque rod bracket is seated over dowel pin and ball shaft is in spring seat bracket before alignment is made.

6. Scribe an alignment mark from upper torque rod bracket (6) to torque rod (5).

7. Remove upper torque rod (5) and upper torque rod bracket (6) from rear axle (10) and spring bracket (8).

8. Place upper torque rod bracket (6) in vise with alignment marks aligned.

9. Tighten slotted nut (12) 350-400 lb-ft (475-542 N•m) and install new cotter pin (13).

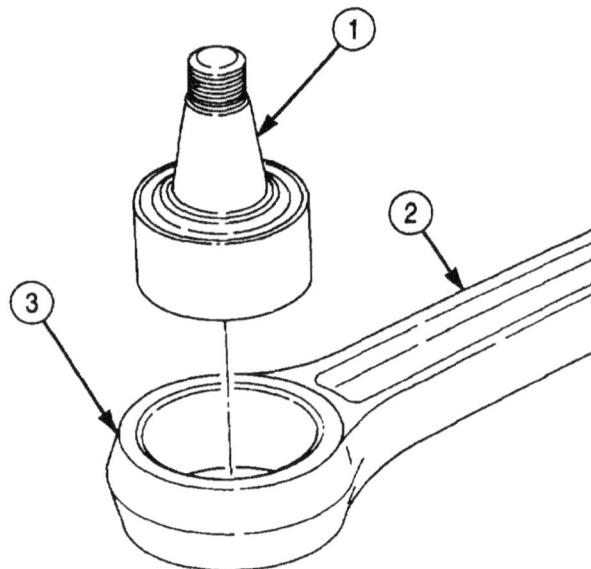

3-170. UPPER AND LOWER TORQUE ROD MAINTENANCE (Contd)

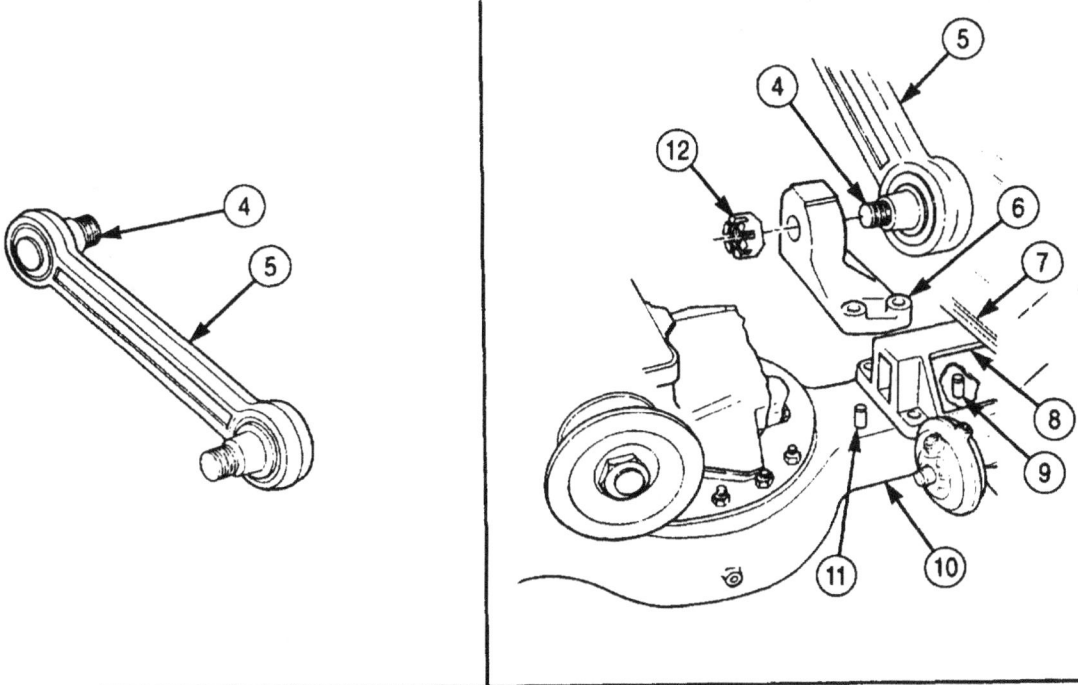

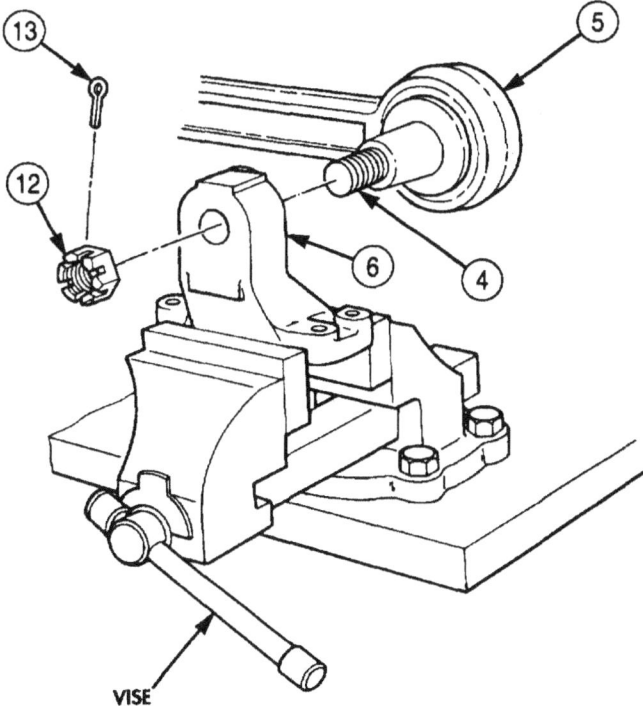

VISE

3-170. UPPER AND LOWER TORQUE ROD MAINTENANCE (Contd)

g. Upper Torque Rod Installation

1. Place upper torque rod (4) and upper torque rod bracket (12) over inner dowel pin (11) on rear axle (7) and torque rod ball shaft (5) in spring seat bracket (3).

CAUTION

Ensure upper bracket is seated over dowel pin before torque rod plate is installed. If not, dowel pin will be damaged.

2. Install upper torque rod plate (10) and bracket (12) on axle (7) with four screws (6), new lockwashers (8), and nuts (9). Tighten nuts (9) 280-360 lb-ft (380-488 N•m).

3. Install upper torque rod ball shaft (5) on spring seat bracket (3) with new slotted nut (1). Tighten slotted nut (1) 350-400 lb-ft (475-542 N•m) and install new cotter pin (2).

h. Lower Torque Rod Installation

CAUTION

Ensure 6 in. (15.24 cm) clearance remains between top of upper spring bracket and bottom of frame rail for all vehicles except M936/A1/A2 model vehicles. For the M936/A1/A2 model vehicles, this clearance is 7-1/4 in. (18.4 cm). If clearance is not proper, torque rods will be damaged.

1. Install lower torque rod (19) ball shaft (23) on lower torque rod bracket (20) with new slotted nut (24).

2. Install torque rod ball shaft (18) on spring seat bracket (15) with new slotted nut (16).

3. Tighten slotted nuts (16) and (24) to 350-400 lb-ft (475-542 N•m) and install new cotter pins (25) and (17).

4. Install lower torque rod bracket (20) and upper spring bracket (26) on rear axle (14) with four screws (13), new lockwashers (21), and nuts (22). Tighten nuts (22) 280-360 lb-ft (380-488 N•m). For M936/A1/A2 model vehicles tighten nuts (22) 320-425 (434-576 N•m).

3-170. UPPER AND LOWER TORQUE ROD MAINTENANCE (Contd)

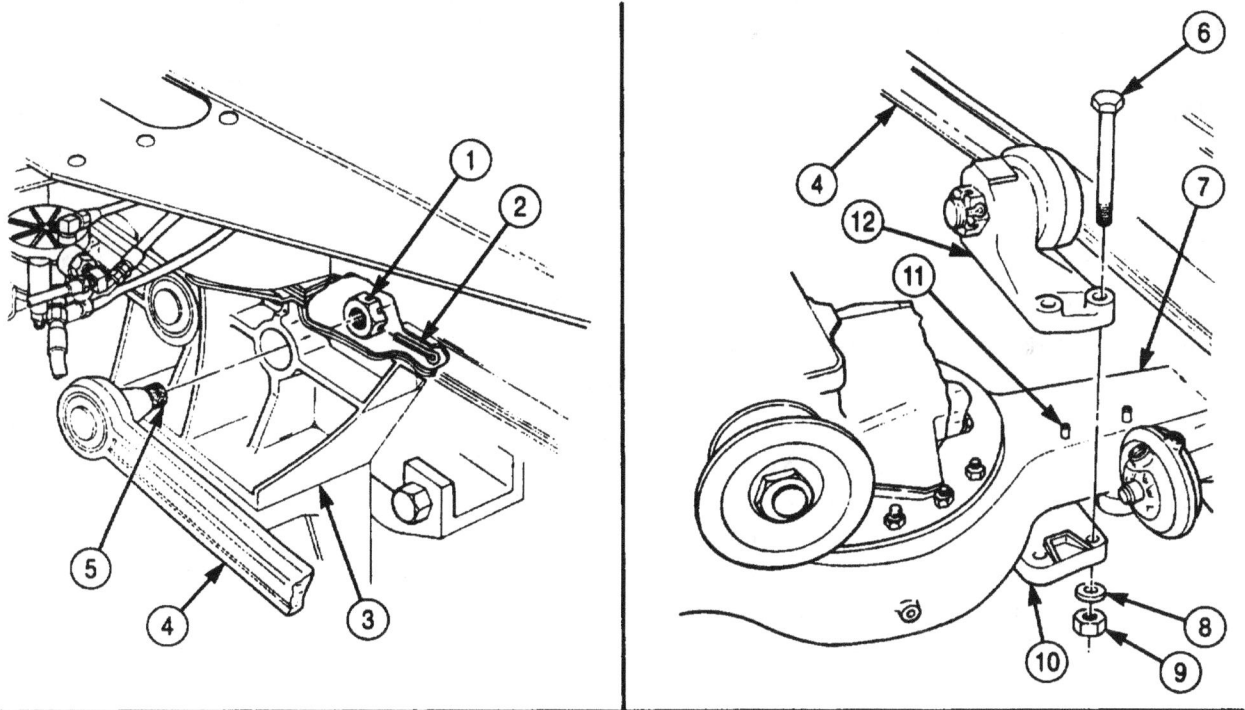

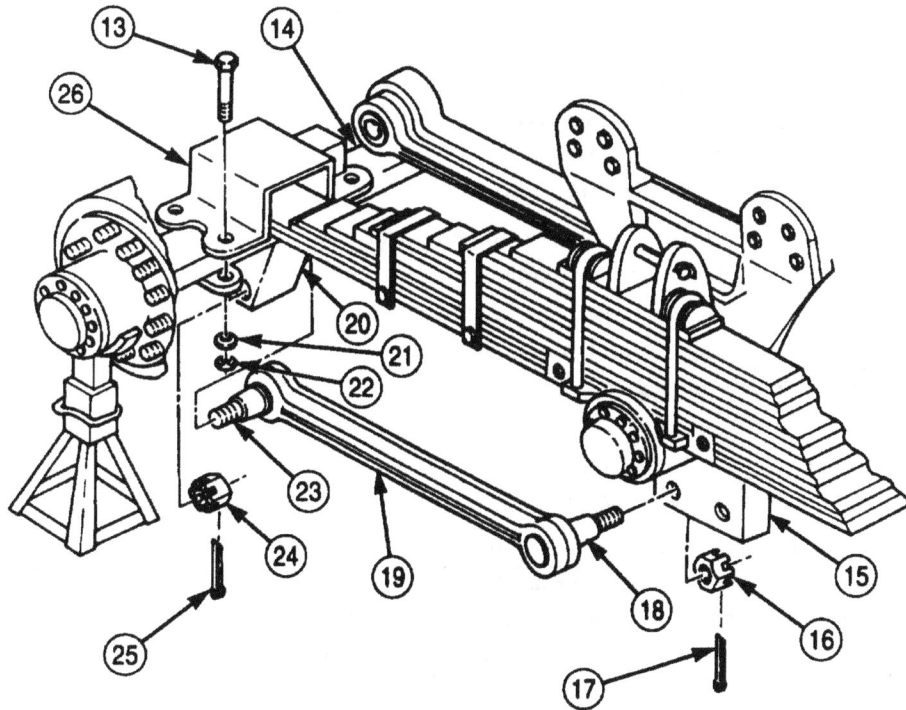

FOLLOW-ON TASKS: • If upper torque rod was replaced, install right rear spring assembly (para. 3-167).
• If lower torque rod was replaced, install two front-rear tires (para. 3-220 or 3-221).

Section IX. COMPRESSED AIR AND BRAKE SYSTEM MAINTENANCE

3-171. COMPRESSED AIR AND BRAKE SYSTEM MAINTENANCE INDEX

| 3-171. COMPRESSED AIR AND BRAKE SYSTEM MAINTENANCE INDEX (Contd) |

3-172. PARKING BRAKE ADJUSTMENT

THIS TASK COVERS:

a. Test

b. Minor Adjustment

c. Major Adjustment

d. Lever Arm Adjustment

INITIAL SETUP

APPLICABLE MODELS

All

TOOLS

General mechanic's tool kit (Appendix E, Item 1)

REFERENCES (TM)

TM 9-2320-272-10

EQUIPMENT CONDITION

• Wheels chocked (task c. only) (TM 9-2320-272-10).
• Parking brake disengaged (TM 9-2320-272-10).
• Transfer case shift lever in neutral (TM 9-2320-272-10).
• Transmission in neutral (TM 9-2320-272-10).
• Transfer case-to-forward rear axle propeller shaft removed (task c. only) (para. 3-150 or 3-151).

a. Test

1. Start engine (TM 9-2320-272-10) and observe air pressure gauge (5) located inside vehicle cab until air pressure builds to 90 psi (621 kPa).
2. Pull parking brake lever (3) up to engage.
3. Push in spring brake release control (6) on instrument panel (1) to release spring brakes.
4. Place transmission selector lever (4) in 1-5 (drive). If vehicle moves, perform task c.

b. Minor Adjustment

Turn parking brake adjusting cap (2) clockwise to increase; counterclockwise to decrease braking action.

3-172. PARKING BRAKE ADJUSTMENT (Contd)

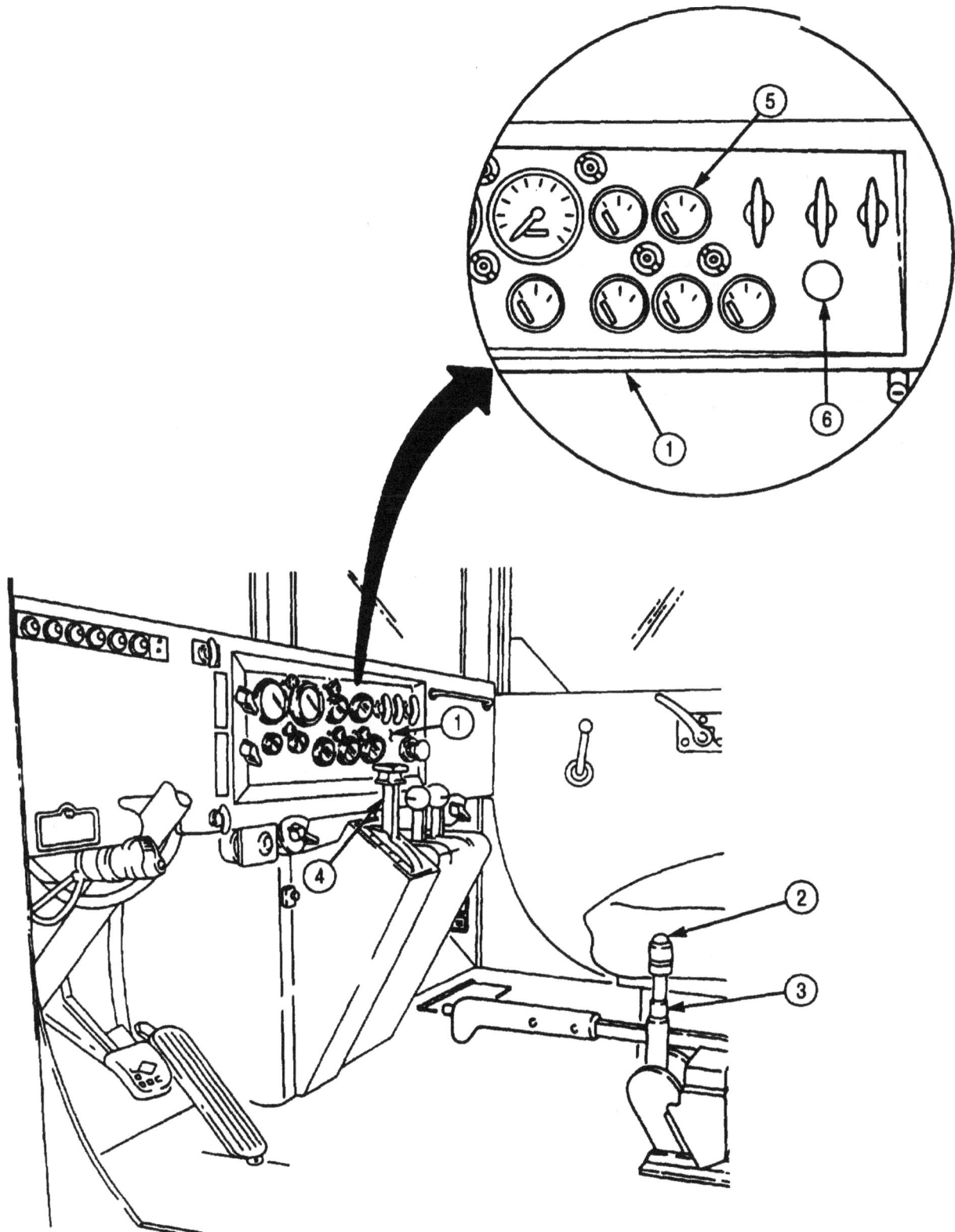

3-172. PARKING BRAKE ADJUSTMENT (Contd)

c. Major Adjustment

1. Loosen adjusting nut (4) on parking brake cable (1) until parking brake lever (3) has free travel.

2. Push parking brake lever (3) clockwise with one hand while turning drum (2) clockwise with other hand. If drum (2) turns freely when parking brake lever (3) is pushed clockwise, replace parking brakeshoes (para. 3-176). If drum (2) stops, go to step 3.

NOTE

Assistant will help with step 3.

3. Pull parking brake lever inside vehicle cab up to engage (TM 9-2320-272-10).

4. Tighten adjusting nut (4) on parking brake cable (1) against parking brake lever (3) until drum (2) does not move, then back off counterclockwise 1/2 turn.

5. Repeat step 1 until slight drag of brakeshoes against drum (2) is observed.

d. Lever Arm Adjustment

1. Position lever arm (7) horizontal to valve body (12).

2. Remove two locknuts (8), washers (9), mounting screws (5), and valve body (12) from bracket (6). Discard locknuts (8).

3. Loosen setscrew (10), adjust valve lever (11), and tighten setscrew (10).

4. Install valve body (12) on bracket (6) with two mounting screws (5), washers (9), and new lockwashers (8).

3-172. PARKING BRAKE ADJUSTMENT (Contd)

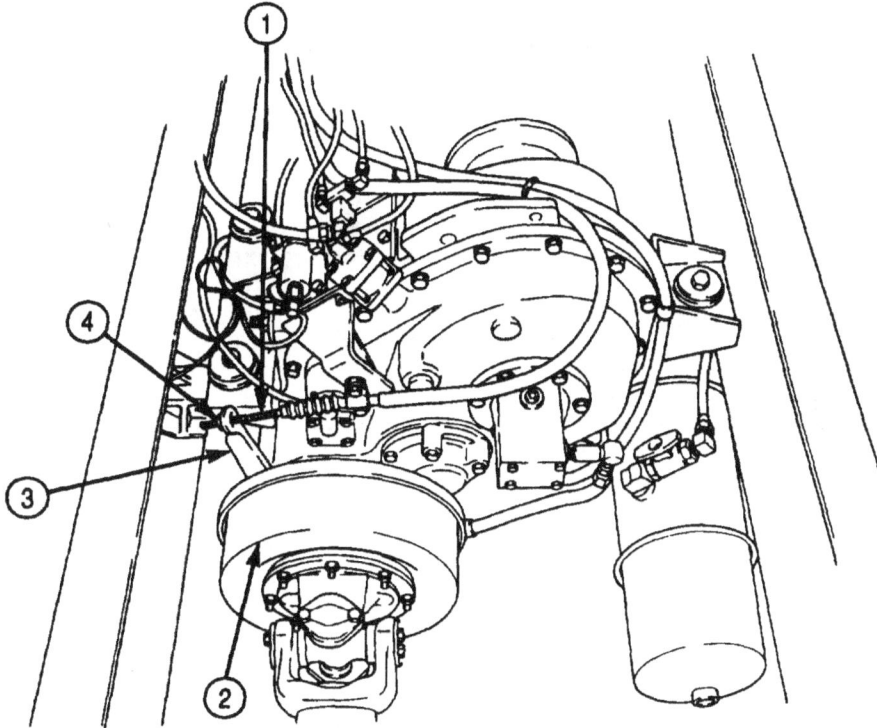

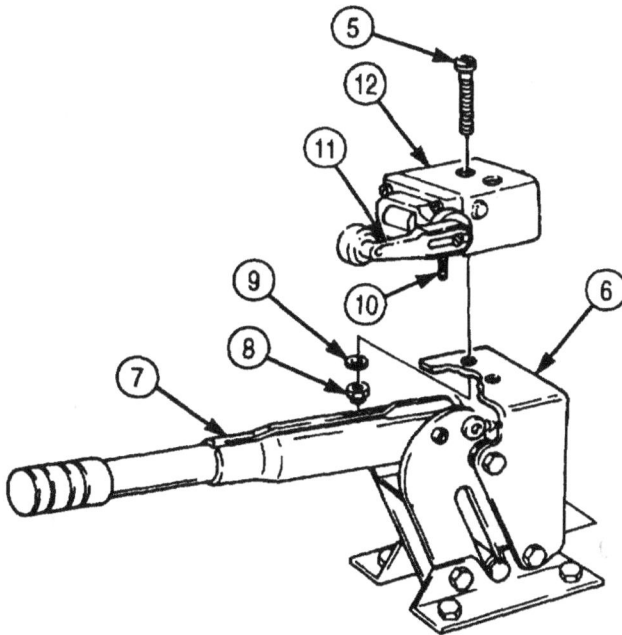

FOLLOW-ON TASKS:
- Install transfer case-to-forward rear axle propeller shaft (para. 3-150 or 3-151).
- Remove chocks and road test vehicle (TM 9-2320-272-10).

3-173. SPRING BRAKE VALVE MAINTENANCE

THIS TASK COVERS:

a. Removal   c. Installation
b. Adjustment

INITIAL SETUP:

APPLICABLE MODELS
All

TOOLS
General mechanic's tool kit (Appendix E, Item 1)

MATERIALS/PARTS
Two locknuts (Appendix D, Item 283)
Gasket sealant (Appendix C, Item 30)

REFERENCES (TM)
TM 9-2320-272-10
TM 9-2320-272-24P

EQUIPMENT CONDITION
• Wheels chocked (TM 9-2320-272-10).
• Drain air reservoirs (TM 9-2320-272-10).

GENERAL SAFETY INSTRUCTIONS
Do not disconnect air lines before draining air reservoirs.

---

1. Release parking brake lever (8) (TM 9-2320-272-10).

WARNING

Do not disconnect air lines before draining air reservoirs. Small parts under pressure may shoot out with high velocity, causing injury to personnel.

NOTE

Tag air lines for installation.

2. Disconnect three air lines (4) from adapter assemblies (3), (6), and (5).

3. Remove adapter assemblies (3), (6), and (5) from valve body (1).

4. Remove two locknuts (9), washers (10), screws (2), and valve body (1) from parking brake bracket (7). Discard locknuts (9).

1. Loosen setscrew (11) on valve lever (13) and position valve lever (13) parallel with valve body (1). Tighten setscrew (11).

2. Position valve body (1) on parking brake bracket (7). Valve lever (13) must be parallel with cab floor. If not, repeat step 1.

1. Install valve body (1) on parking brake bracket (7) with two screws (2), washers (10), and new locknuts (9). Valve lever roller (12) must be aligned with parking brake lever (8).

2. Lift up parking brake lever (8). Ensure parking brake lever (8) contacts valve lever roller (12) properly and release parking brake lever (8).

3-173. SPRING BRAKE VALVE MAINTENANCE (Contd)

CAUTION

Do not twist air lines. Twisted air lines will restrict air flow.

NOTE

Male pipe threads must be coated with gasket sealant before installation.

3. Install adapter assemblies (6), (5), and (3) on valve body (1).

4. Connect three air lines (4) to adapter assemblies (3), (6), and (5).

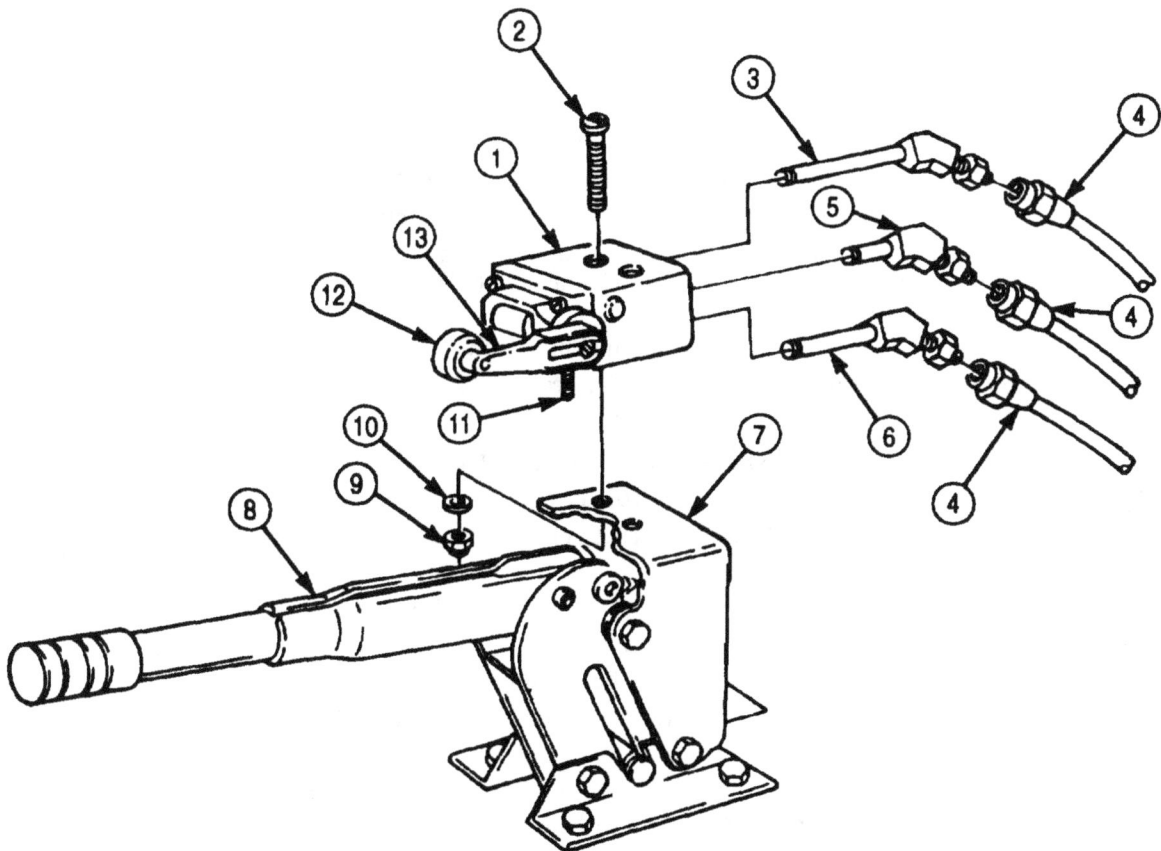

FOLLOW-ON TASKS: • Start engine (TM 9-2320-272-10) and allow air pressure to build to normal operating range. Check for air leaks at spring brake valve.
  • Set parking brake (TM 9-2320-272-10) and ensure spring brakes engage properly.
  • Remove chocks and road test vehicle (TM 9-2320-272-10)

3-174. PARKING BRAKE LEVER AND SWITCH REPLACEMENT

THIS TASK COVERS:

a. Removal   b. Installation

INITIAL SETUP

APPLICABLE MODELS
All

TOOLS
General mechanic's tool kit (Appendix E, Item 1)

MATERIALS/PARTS
Cotter pin (Appendix D, Item 46)
Four locknuts (Appendix D, Item 291)

REFERENCES (TM)
TM 9-2320-272-10
TM 9-2320-272-24P

EQUIPMENT CONDITION
• Wheels chocked (TM 9-2320-272-10).
• Battery ground cable disconnected (para. 3-126).

---

a. Removal

1. Release parking brake lever (2) (TM 9-2320-272-101.
2. Turn adjusting cap (1) on parking brake lever (2) completely out.
3. Remove cotter pin (5), washer (6), and clevis pin (7) from cable clevis (14). Discard cotter pin (6).
4. Remove three locknuts (3), screws (9), spacer washer (10), and spring parking brake valve and bracket (8) from parking brake housing (4) and brackets (13). Carefully set valve and bracket (8) aside. Discard locknuts (3).
5. Remove parking brake housing (4) from parking brake brackets (13).
6. Disconnect wire (11) from wire (12) under vehicle cab.

NOTE
Assistant will help with step 7.

7. Remove locknut (18), wire (17), and screw (16) from cab floor (19). Discard locknut (18).
8. Remove parking brake switch (15) from brackets (13).

b. Installation

NOTE
Assistant will help with step 1.

1. Install wire (17) on cab floor (19) with screw (16) and new locknut (18).
2. Connect wire (11) to wire (12).
3. Install parking brake housing (4), spring parking brake valve and bracket (8), and parking brake switch (16) on brackets (13) with three screws (9), spacer washer (10), and three new locknuts (3).
4. Install clevis pin (7) on cable clevis (14) with washer (6) and new cotter pin (5).

3-174. PARKING BRAKE LEVER AND SWITCH REPLACEMENT (Contd)

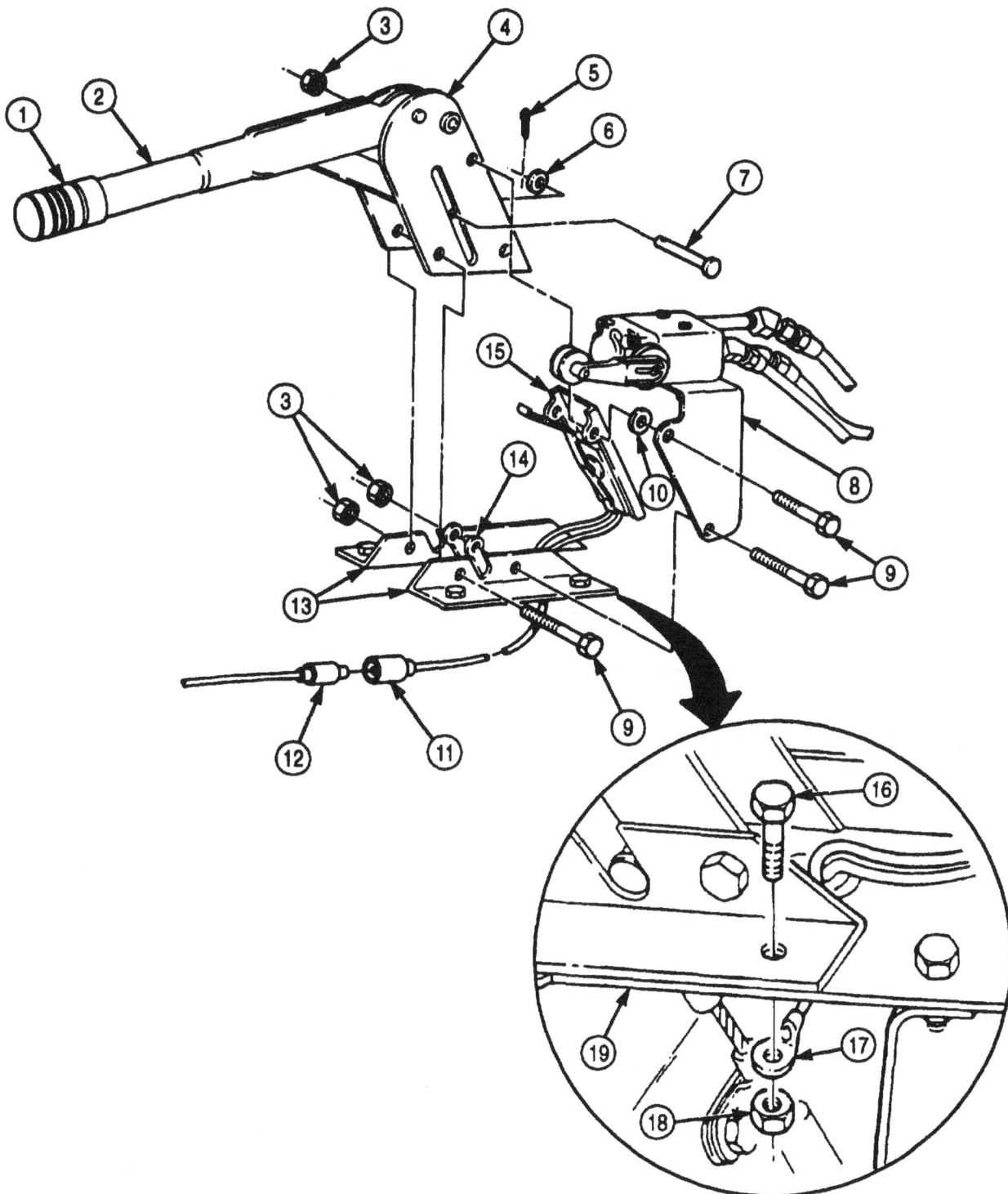

FOLLOW-ON TASKS: • Adjust parking brake (para. 3-172).
                 • Connect battery ground cable (para. 3-126).
                 • Remove chocks and road test vehicle (TM 9-2320-272-10).

3-175. PARKING BRAKE CABLE AND BRACKET REPLACEMENT

THIS TASK COVERS:

a. Removal   b. Installation

INITIAL SETUP

APPLICABLE MODELS
All

TOOLS
General mechanic's tool kit (Appendix E, Item 1)

MATERIALS/PARTS
Cotter pin (Appendix D, Item 46)
Eight locknuts (Appendix D, Item 291)
Four tiedown straps (Appendix D, Item 696)

REFERENCES (TM)
TM 9-2320-272-10
TM 9-2320-272-24P

EQUIPMENT CONDITION
Wheels chocked (TM 9-2320-272-10).

a. Removal

1. Remove cotter pin (1), washer (2), clevis pin (3), and parking brake cable (6) from parking brake lever (11). Discard cotter pin (1).

NOTE

Assistant will help with step 2.

2. Remove four locknuts (7), cable clamp bracket (5), four screws (10), and two parking lever brackets (9) from cab floor (8). Discard locknuts (7).

3. Remove grommet (4) and parking brake cable (6) from cab floor (8).

4. Remove two locknuts (15), screws (13), cable clamp (12), and spacer (14) from cable clamp bracket (5). Discard locknuts (15).

5. Remove four tiedown straps (18), screw (16), and two clamps (28) from parking brake cable (6). Discard tiedown straps (18).

6. Remove two locknuts (22), screws (19), cable clamp (20), and spacer (21) from transfer case bracket (26) and parking brake cable (6). Discard locknuts (22).

7. Remove nut (23) and parking brake cable (6) from brakedrum lever (24).

8. Remove two screws (25), washers (27), and transfer case bracket (26) from transfer case (17).

3-175. PARKING BRAKE CABLE AND BRACKET REPLACEMENT (Contd)

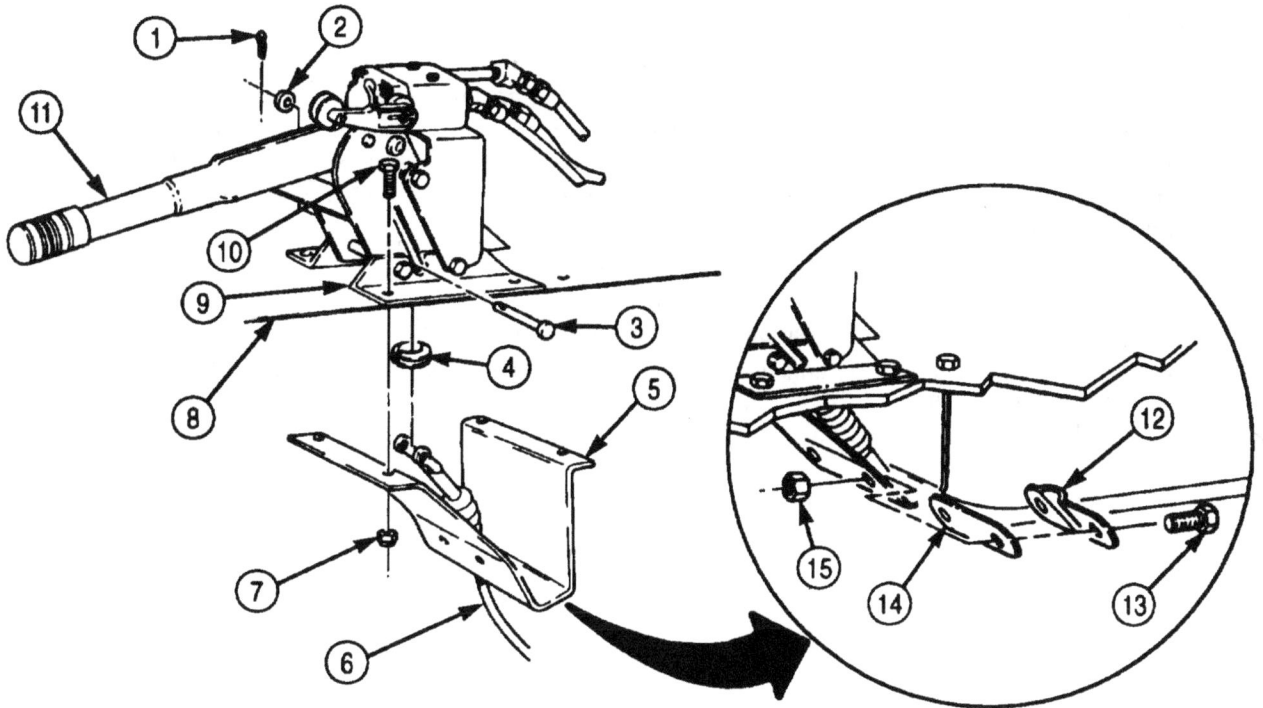

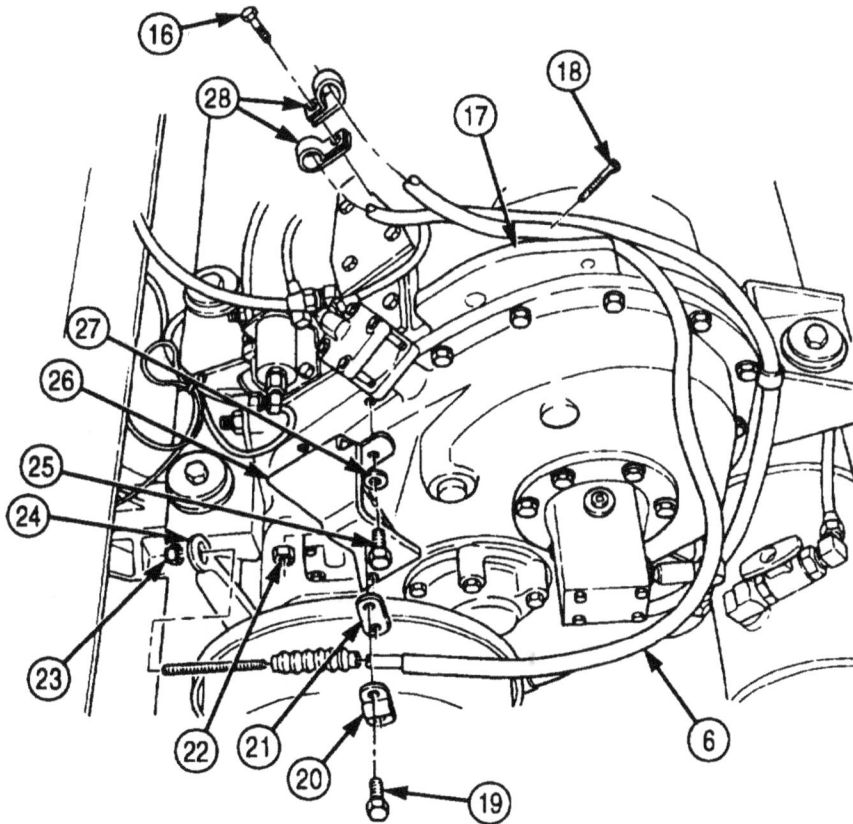

3-175. PARKING BRAKE CABLE AND BRACKET REPLACEMENT (Contd)

b. Installation

1. Install transfer case bracket (13) on transfer case (2) with two washers (5) and screws (6).

2. Install threaded end of parking brake cable (4) on brakedrum lever (12) with nut (11).

3. Position spacer (9) between transfer case bracket (13) and clamp (8), and install parking brake cable (4) on transfer case bracket (13) with spacer (9), clamp (8), two screws (7), and new locknuts (10).

4. Install parking brake cable (4) on cable clamp bracket (21) with clamp (26), spacer (27), two screws (25), and new locknuts (28).

5. Install grommet (20) around cable (4) and in cab floor (23).

NOTE

Assistant will help with step 6.

6. Install cable clamp bracket (21) and two parking lever brackets (18) on cab floor (23) with four screws (17) and new locknuts (22).

7. Install parking brake cable (4) on parking brake lever (24) with clevis pin (19), washer (16), and new cotter pin (15).

8. Install four new tiedown straps (3) on parking brake cable (4).

9. Install two clamps (14) on transfer case (2) with screw (1).

3-175. PARKING BRAKE CABLE AND BRACKET REPLACEMENT (Contd)

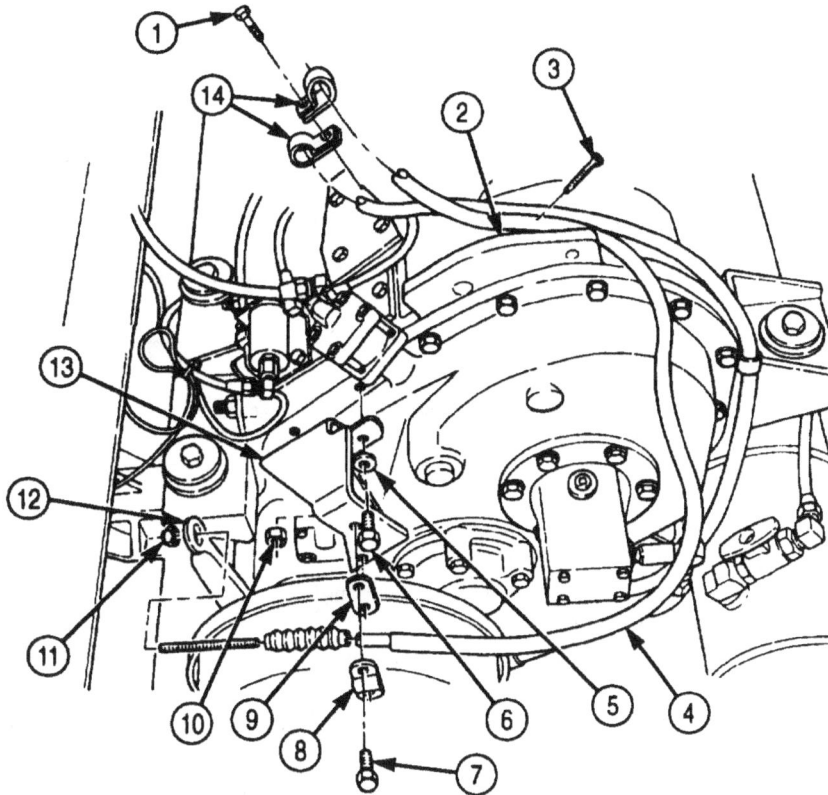

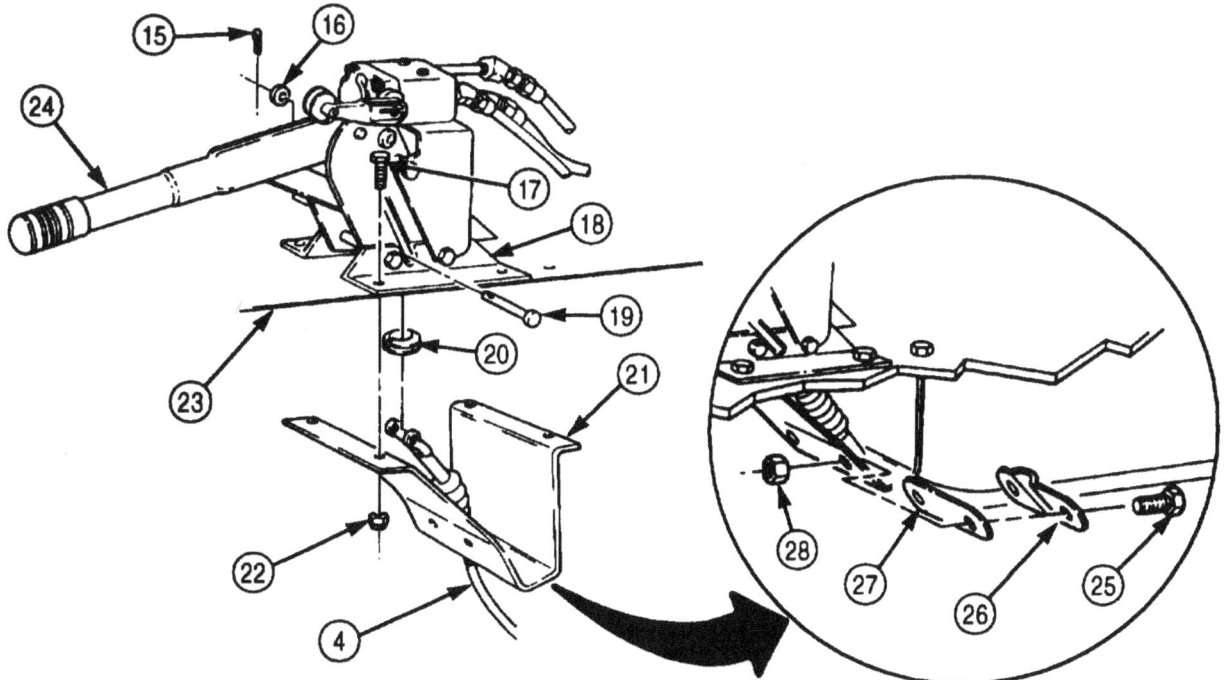

FOLLOW-ON TASK: Adjust parking brake (para. 3-172).

3-176. PARKING BRAKESHOES REPLACEMENT

THIS TASK COVERS:

a. Removal   b. Installation

INITIAL SETUP

APPLICABLE MODELS
All

SPECIAL TOOLS
Brake repair pliers (Appendix E, Item 20)

TOOLS
General mechanic's tool kit (Appendix E, Item 1)
Torque wrench (Appendix E, Item 144)

REFERENCES (TM)
TM 9-2320-272-10
TM 9-2320-272-24P

EQUIPMENT CONDITION
• Wheels chocked (TM 9-2320-272-10).
• Transfer case-to-forward rear axle propeller shaft.
  removed (para. 3-150 or 3-151).

NOTE
Parking brakeshoes should be replaced when parking brake lever
full travel is over 2.0 in. (5.1 cm).

a. Removal

1. Remove thrust nut (6) and washer (5) from transfer output shaft (1).
2. Remove parking brakedrum (3) and transfer output shaft flange (4) from transfer output shaft (1).
3. Remove transfer output shaft flange (4) from parking brakedrum (3).
4. Remove eight studs (2) from parking brakedrum (3).
5. Using brake repair pliers, remove two brakeshoe return springs (8) from parking brakeshoes (9).
6. Remove two parking brakeshoes (9) and actuating plate (10) from brakeshoe backing plate (7).

3-176. PARKING BRAKESHOES REPLACEMENT (Contd)

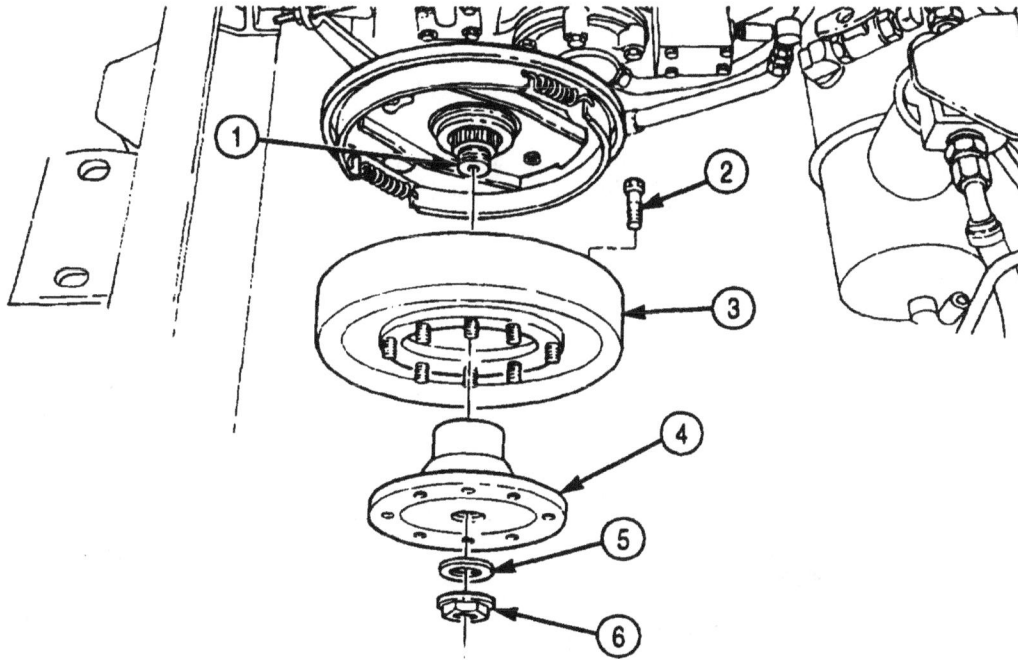

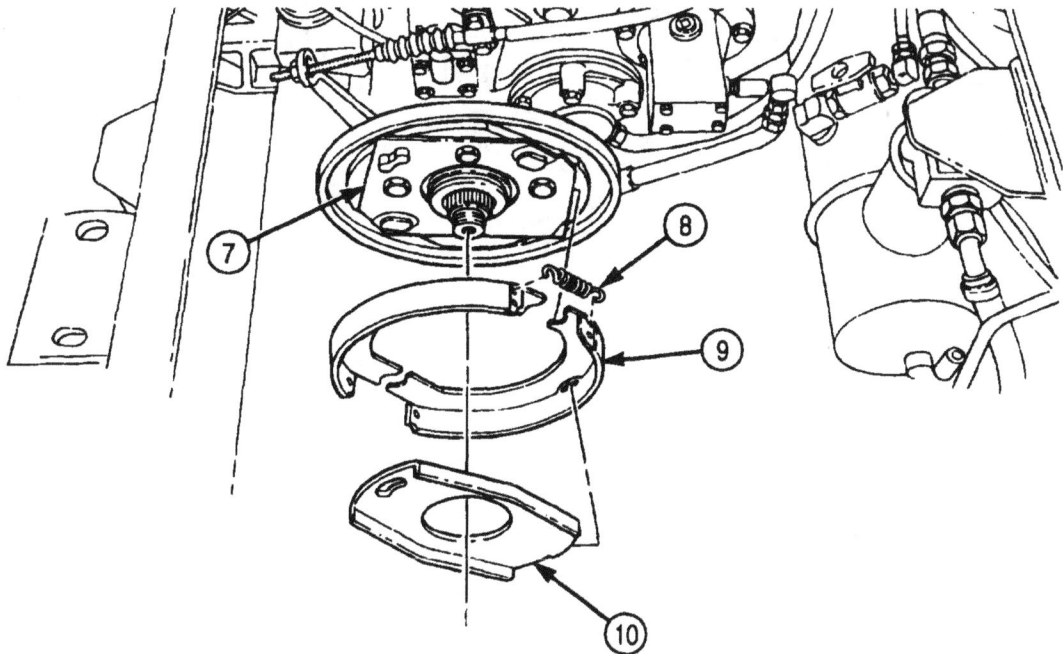

3-176. PARKING BRAKESHOES REPLACEMENT (Contd)

b. Installation

1. Install actuating plate (5) against brakeshoe backing plate (8) so retainer opening (6) fits over retaining stud (1) of backing plate (8).

2. Install two brakeshoes (4) on backing plate shoe studs (7) with two shoe return springs (2). Ensure springs (2) are attached to inside holes (3) of brakeshoes (4).

3. Install eight stude (10) on parking brakedrum (11).

4. Install parking brakedrum (11) and transfer output shaft flange (12) on transfer output shaft (9) with washer (13) and thrust nut (14). Tighten thrust nut (14) 450-600 lb-ft (610-814

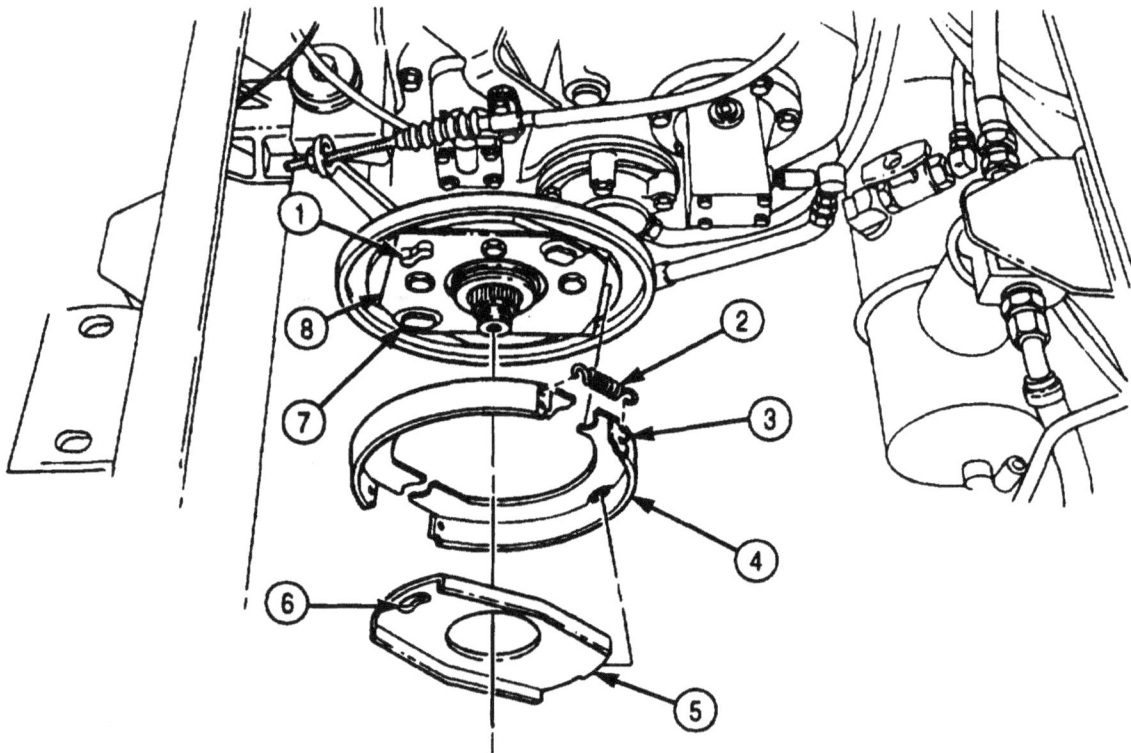

3-176. PARKING BRAKESHOES REPLACEMENT (Contd)

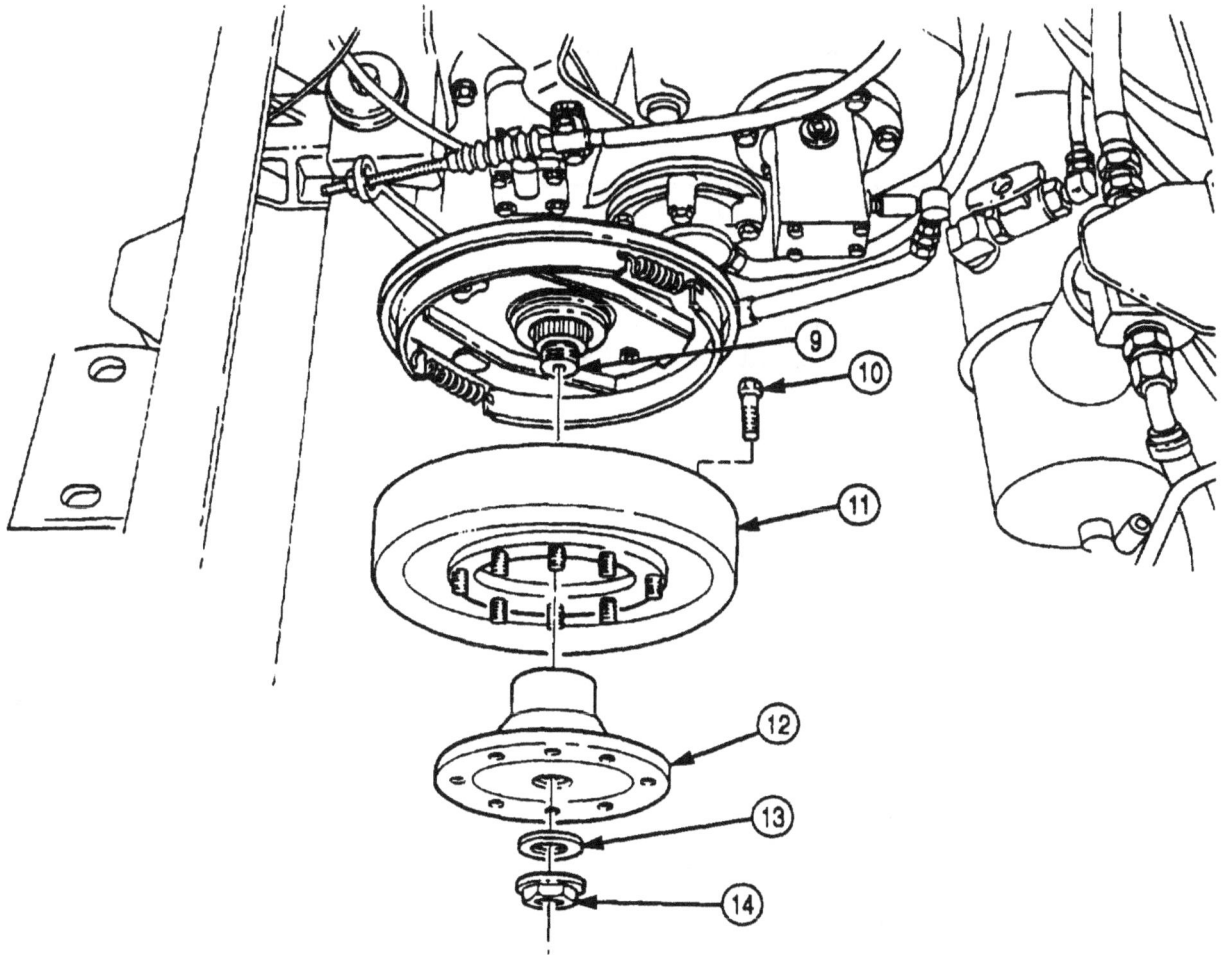

3-177. PARKING BRAKEDRUM DUSTCOVER REPLACEMENT

THIS TASK COVERS:

a. Removal   b. Installation

INITIAL SETUP:

APPLICABLE MODELS
All

TOOLS
General mechanic's tool kit (Appendix E, Item 1)
Torque wrench (Appendix E, Item 145)

MATERIALS/PARTS
Locknut (Appendix D, Item 287)

REFERENCES (TM)
TM 9-2320-272-10
TM 9-2320-272-24P

EQUIPMENT CONDITION
• Wheels chocked (TM 9-2320-272-10).
• Parking brakeshoes removed (para. 3-176).

a. Removal

1. Remove locknut (1), washer (2), brake lever (8), and brake lever cam (7) from parking brakedrum dustcover (4). Discard locknut (1).
2. Remove four screws (6), backing plate (5), and parking brakedrum dustcover (4) from backing plate companion flange (3).

b. Installation

1. Install parking brakedrum dustcover (4) and backing plate (5) on backing plate companion flange (3) with four screws (6). Tighten screws (6) 180-230 lb-ft (244-312 N-m).
2. Install brake lever (8) on parking brakedrum dustcover (4) with brake lever cam (7), washer (2), and new locknut (1).

3-177. PARKING BRAKEDRUM DUSTCOVER REPLACEMENT (Contd)

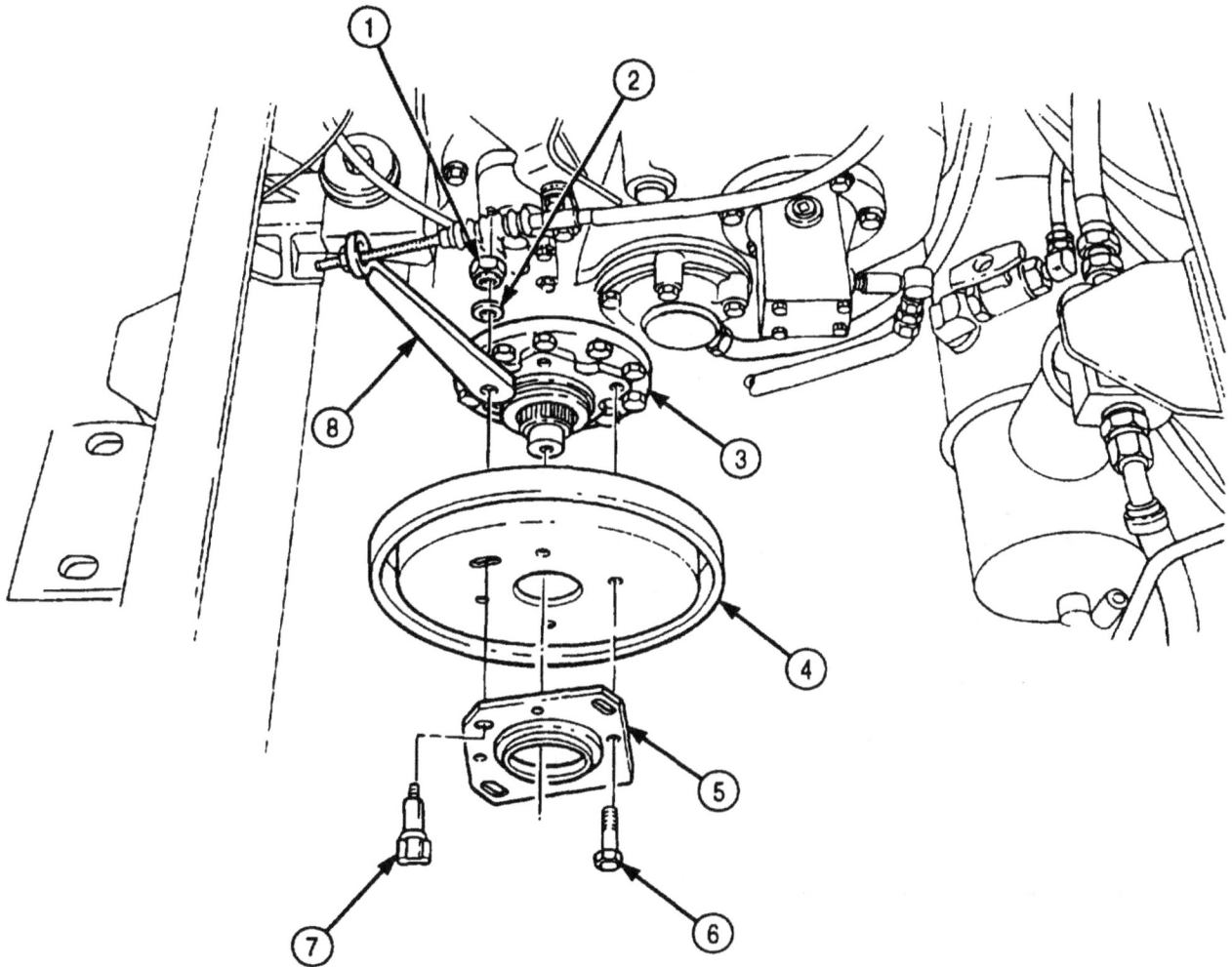

FOLLOW-ON TASK: Install parking brakeshoes (para. 3-176).

3-178. WHEEL BRAKEDRUM DUSTCOVERS REPLACEMENT

THIS TASK COVERS:
a. Front Wheel Dustcovers Removal
b. Rear Wheel Dustcovers Removal
c. Front Wheel Dustcovers Installation
d. Rear Wheel Dustcovers Installation

INITIAL SETUP:

APPLICABLE MODELS
All

TOOLS
General mechanic's tool kit (Appendix E, Item 1)

MATERIALS/PARTS
Eight lockwashers (Appendix D, Item 365)

REFERENCES (TM)
TM 9-2320-272-10
TM 9-2320-272-24P

EQUIPMENT CONDITION
Parking brake set (TM 9-2320-272-10).

a. Front Wheel Dustcovers Removal

NOTE

Left and right front brakedrum dustcovers are replaced basically the same. This procedure covers left front and rear brakedrum dustcovers.

1. Remove four screws (1), lockwashers (2), and dustcovers (3) and (5) from left front brake assembly (4). Discard lockwashers (2).
2. Remove four rubber plugs (6) from dustcovers (3) and (5).

b. Rear Wheel Dustcovers Removal

1. Remove four screws (12), lockwashers (10), and two dustcovers (7) from left rear brake assembly (11). Discard lockwashers (10).
2. Remove six rubber plugs (8) and two plastic plugs (9) from two dustcovers (7).

c. Front Wheel Dustcovers Installation

1. Install four rubber plugs (6) on dustcovers (3) and (5).
2. Install dustcovers (3) and (5) on left front brake assembly (4) with four new lockwashers (2) and screws (1).

d. Rear Wheel Dustcovers Installation

1. Install six rubber plugs (8) and two plastic plugs (9) on two dustcovers (7).
2. Install two dustcovers (7) on left rear brake assembly (11) with four new lockwashers (10) and screws (12).

3-178. WHEEL BRAKEDRUM DUSTCOVERS REPLACEMENT (Contd)

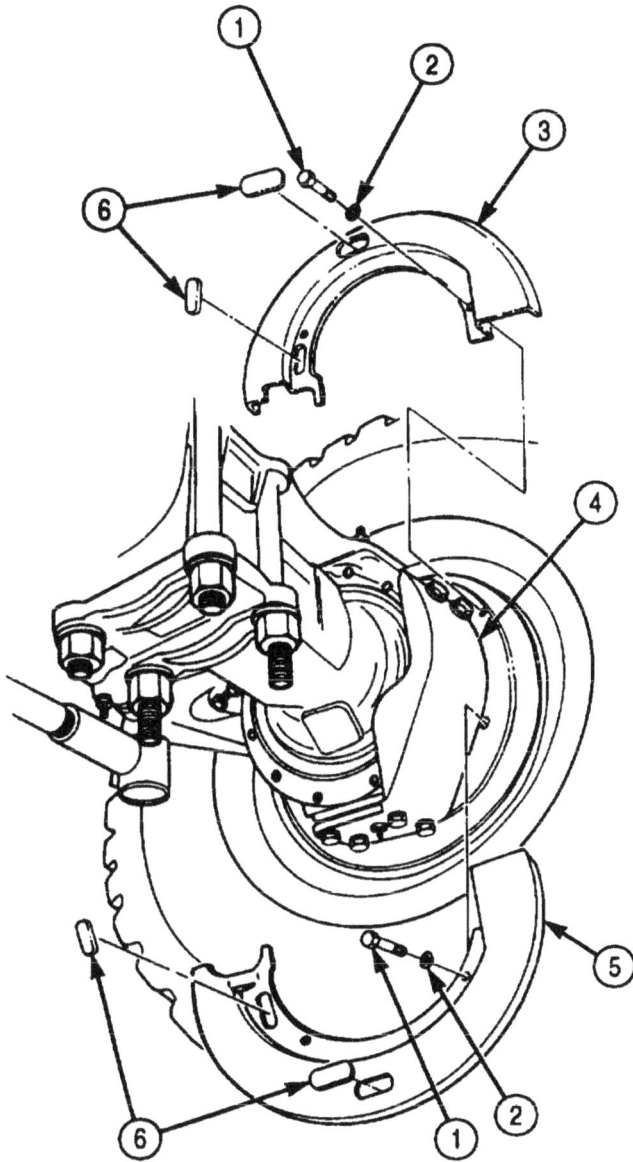

FRONT BRAKE DRUM

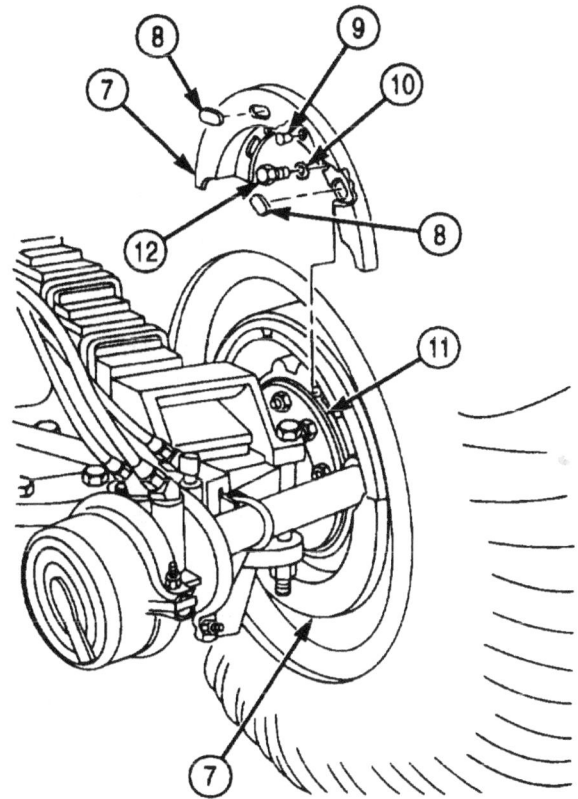

REAR BRAKE DRUM

3-179. BRAKE MECHANISM CHECKS AND ADJUSTMENTS (FRONT AND REAR)

THIS TASK COVERS:

a. Brakeshoe Check and Adjustment   c. Mechanism Inspection
b. Checking Brakeshoe Wear

INITIAL SETUP:

APPLICABLE MODELS
All

SPECIAL TOOLS
Brakeshoe adjusting tool (Appendix E, Item 21)

TOOLS
General mechanic's tool kit (Appendix E, Item 1)

PERSONNEL REQUIRED
TWO

REFERENCES (TM)
TM 9-2320-272-10
TM 9-2320-272-24P

EQUIPMENT CONDITION
• Parking brake set (TM 9-2320-272-10).
• Spring (emergency) brake caged (TM 9-2320-272-10)
• Front hub and drum removed (task c. only)
  (para. 3-223 or 3-460).
• Rear hub and drum removed (task c. only)
  (para. 3-224 or 3-461).

a. Brakeshoe Check and Adjustment

1. Remove two rubber inspection hole covers (1) from brakedrum dustcovers (4).

2. Check brakeshoe lining (7) to brakedrum (6) clearance through inspection hole (5). Clearance should be 0.020-0.040 in. (0.508-1.016 mm).

3. Remove two rubber adjusting hole covers (2) from brakedrum dustcovers (4).

4. Using brakeshoe adjusting tool through adjusting holes (3), rotate star wheel (8) until proper clearance is obtained.

5. Install two rubber inspection hole covers (1) and two rubber adjusting hole covers (2) on brakedrum dustcovers (4).

3-179. BRAKE MECHANISM CHECKS AND ADJUSTMENTS (FRONT AND REAR) (Contd)

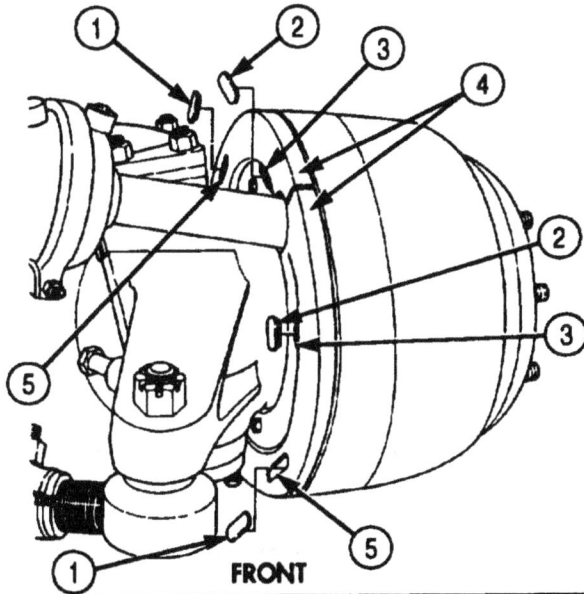

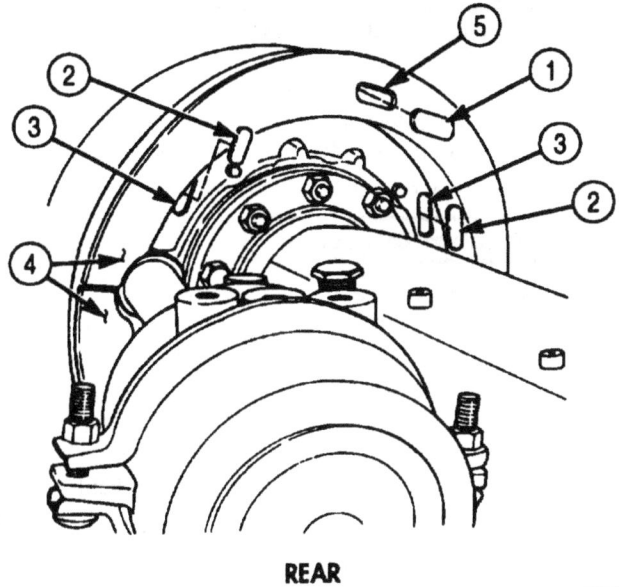

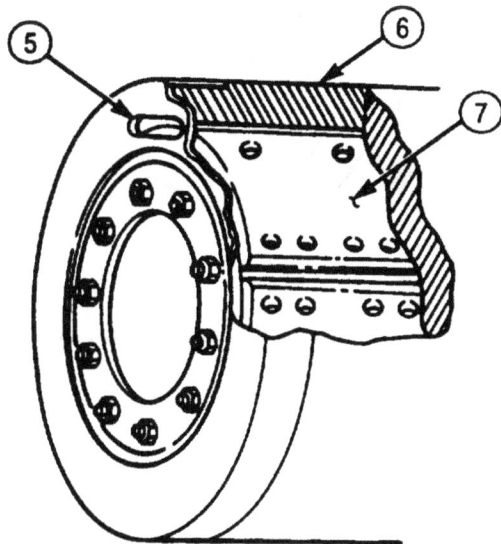

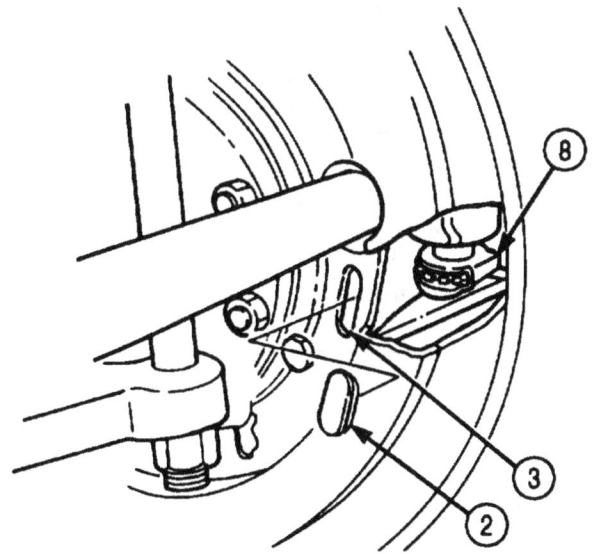

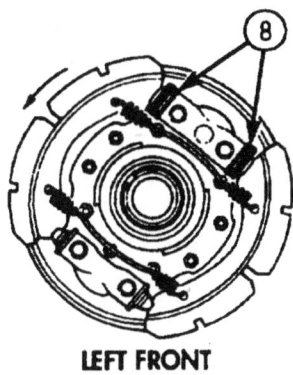

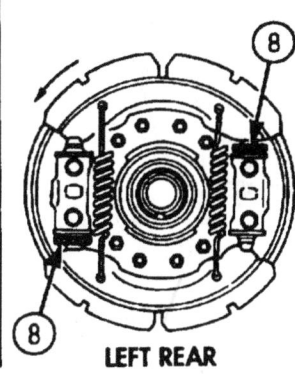

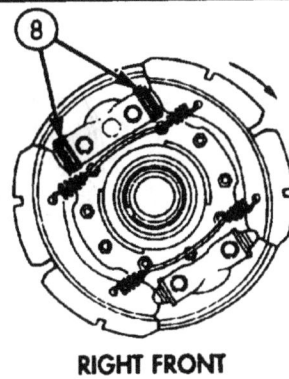

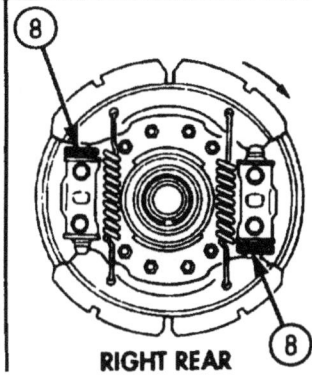

FRONT

REAR

LEFT FRONT

LEFT REAR

RIGHT FRONT

RIGHT REAR

TM 9-2320-272-24-1

3-179. BRAKE MECHANISM CHECKS AND ADJUSTMENTS (FRONT AND REAR) (Contd)

b. Checking Brakeshoe Wear

1. Remove two rubber inspection hole covers (1) from brakedrum dustcover (3).
2. Inspect chamfer (4) on brakeshoe lining (5) through inspection hole (2). If brakeshoe lining (5) is worn to depth of chamfer (4), replace brakeshoes (8) (para. 3-180).
3. Install two rubber inspection hole covers (1) on brakedrum dustcover (3).

c. Mechanism Inspection

1. Inspect brakeshoe linings (5) for cracks, chips, and oil contamination. If cracked, chipped, or contaminated, replace brakeshoes (8) (pare. 3-180).
2. Inspect plunger seals (10). Notify DS maintenance if rotted or tom.
3. Inspect shoe return springs (9) for stretching, bluing, cracks, and uneven coils. Replace return springs (9) showing any of these defects.
4. Inspect brake chambers (6) for cracks and bends at point where chamber (6) enters plunger (7). Replace chamber(s) (6) if cracked or bent.

FOLLOW-ON TASKS: • Install front hub and drum (task c. only) (para. 3-223 or 3-460).
• Install rear hub and drum (task c. only) (para. 3-224 or 3-461).

3-500

3-180. BRAKESHOE REPLACEMENT

THIS TASK COVERS:
a. Removal   b. Installation

INITIAL SETUP:

APPLICABLE MODELS
All

SPECIAL TOOLS
Brake repair pliers (Appendix E, Item 20)

TOOLS
General mechanic's tool kit (Appendix E, Item 1)

MATERIALS/PARTS
Brakeshoe and lining assembly kit
  (Appendix D, Item 17)

PERSONNEL REQUIRED
TWO

REFERENCES (TM)
TM 9-2320-272-10
TM 9-2320-272-24P

EQUIPMENT CONDITION
• Spring (emergency) brake caged (TM 9-2320-272-10).
• Front hub and drum removed (task c. only)
  (para. 3-223 or 3-460).
• Rear hub and drum removed (task c. only)
  (para. 3-224 or 3-461).

a. Removal

1. Using brake repair pliers, remove two return springs (3) from brakeshoes (1).
2. Remove two brakeshoes (1) from each anchor plunger (4), adjustable plunger (5), and brakeshoe retaining clips (2).

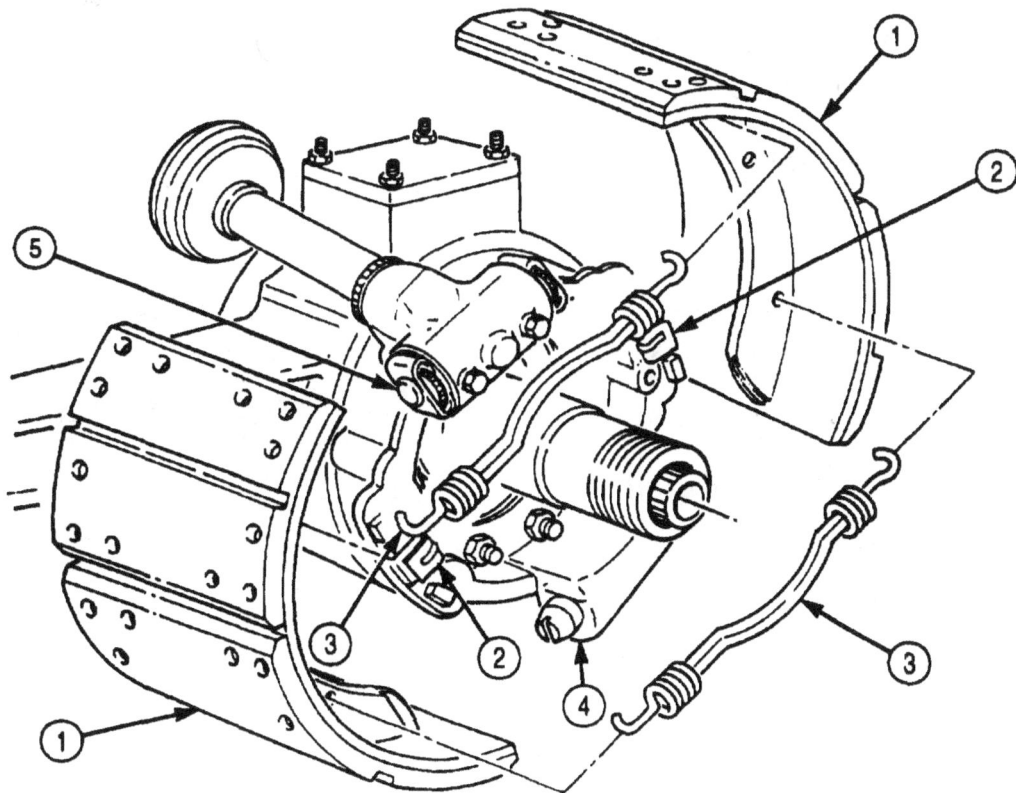

3-501

3-180. BRAKESHOE REPLACEMENT (Contd)

b. Installation

NOTE

- Ensure arrow stamped on shoe web points to anchor plunger.
- Brakeshoe and lining assemblies must be replaced in sets on both sides of axle with linings coded to the color and manufacturer.
- Bear brake mechanisms have one adjusting plunger in each plunger housing. Front brake mechanisms have two adjusting plungers in rear of plunger housing.

Position two brakeshoes (1) into each anchor plunger (4), adjustable plunger (5), and brakeshoe retaining clips (2) and secure with two return springs (3).

WHEEL ROTATION   **WHEEL ROTATION**

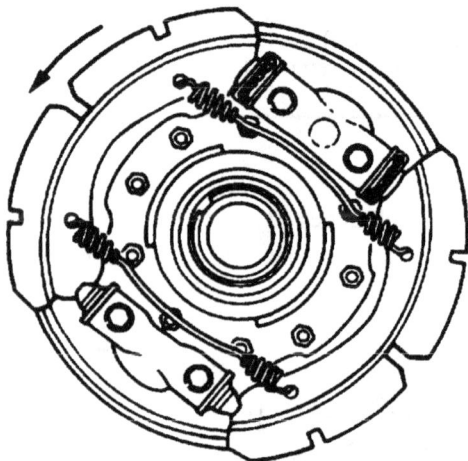

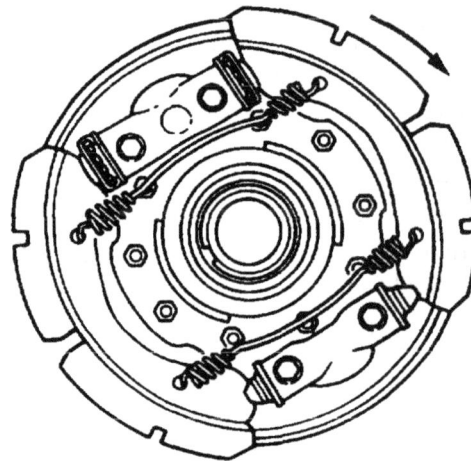

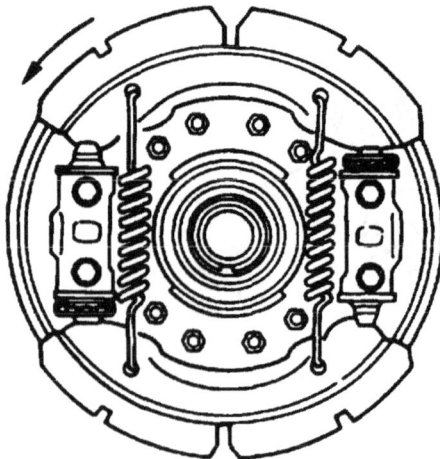

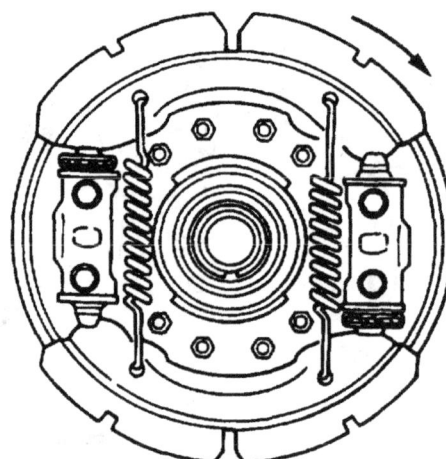

LEFT FRONT   **RIGHT FRONT**

WHEEL ROTATION   **WHEEL ROTATION**

LEFT REAR   **RIGHT REAR**

3-180. BRAKESHOE REPLACEMENT (Contd)

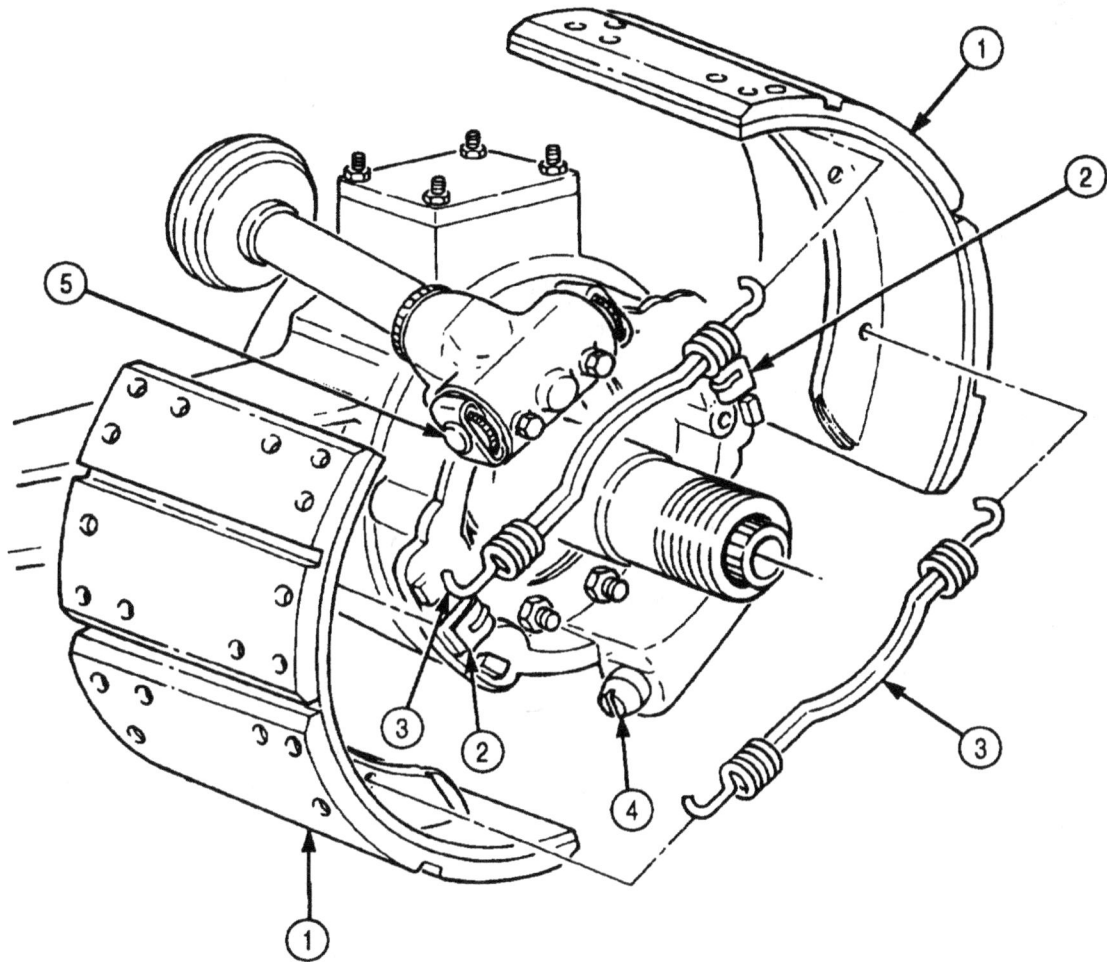

FOLLOW-ON TASKS: • Install front hub and drum (task c. only) (para. 3-223 or 3-460).
• Install rear hub and drum (task c. only) (para. 3-224 or 3-461).
• Check brakeshoe-to-drum clearance and adjust if necessary (para. 3-179).

3-181. SERVICE BRAKE CHAMBER REPLACEMENT

THIS TASK COVERS:

a. Removal   **b. Installation**

INITIAL SETUP:

APPLICABLE MODELS
All

TOOLS
General mechanic's tool kit (Appendix E, Item 1)

MATERIALS/PARTS
Antiseize tape (Appendix C, Item 72)

REFERENCES (TM)
TM 9-2320-272-10
TM 9-2320-272-24P

EQUIPMENT CONDITION
• Parking brake set (TM 9-2320-272-10).
• Air reservoirs drained (TM 9-2320-272-10).
• Wheel brakedrum dustcovers removed (para. 3-178).

GENERAL SAFETY INSTRUCTIONS
Do not disconnect air lines before draining air reservoirs.

WARNING

Do not disconnect air lines before draining air reservoirs. Small parts under pressure may shoot out with high velocity, causing injury to personnel.

a. Removal

NOTE
Tag all lines for installation.

1. Disconnect service brake control line (6) from brake adapter (7).
2. Remove brake adapter (7) from service brake chamber (3).
3. Disconnect vent line (5) from elbow (4).
4. Remove elbow (4) from service brake chamber (3).

NOTE
Perform step 5 for rear service brake only.

5. Remove two nuts (9), washers (10), U-bolt (12), and clamp (11) from service brake chamber (3).
6. Loosen collet nut (2) and remove service brake chamber (3) from actuator housing (8). Record position for installation.
7. Remove collet nut (2) from service brake chamber (3).

b. Installation

NOTE
Wrap all male pipe threads with antiseize tape before installation.

1. Install collet nut (2) on brake chamber (3).
2. Ensuring collet nut (2) is loose, install brake chamber (3) over wedge assembly (1) and into actuator housing (8), with brake chamber (3) positioned for air line connection. Thread collet nut (2) to bottom of brake chamber (3) and tighten 3/16 in. (4.8 mm) or 1-1/2 teeth.

3-181. SERVICE BRAKE CHAMBER REPLACEMENT (Contd)

NOTE

Perform step 3 for rear service brake only.

3. Install U-bolt (12) and clamp (11) on brake chamber (3) with two washers (10) and nuts (9).
4. Install elbow (4) on brake chamber (3).
5. Connect vent line (5) to elbow (4).
6. Install brake adapter (7) on brake chamber (3).
7. Connect service brake control line (6) to adapter (7).

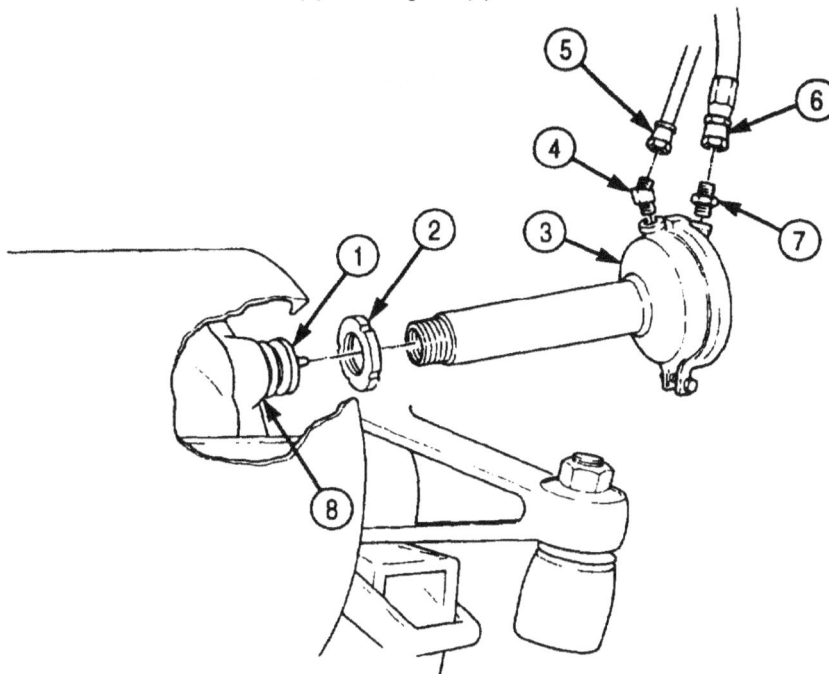

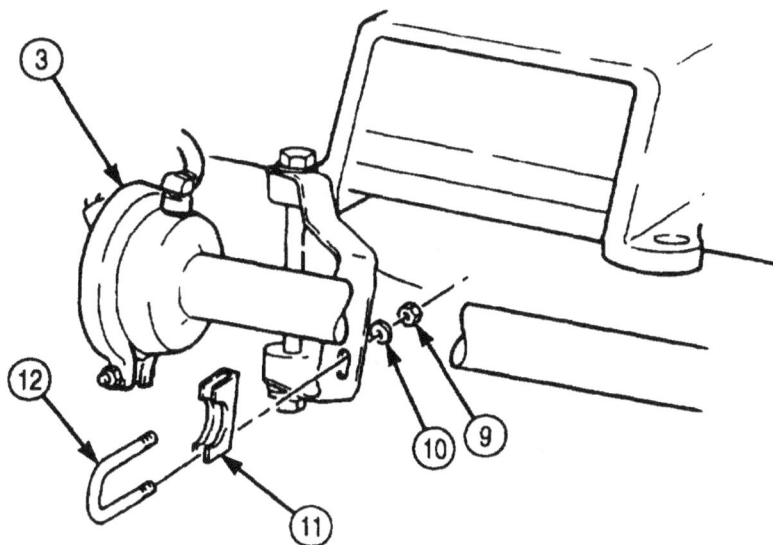

FOLLOW-ON TASK: Install wheel brakedrum dustcovers (para. 3-178).

3-182. COMBINATION SPRING (EMERGENCY) AND SERVICE BRAKE CHAMBER REPLACEMENT

THIS TASK COVERS:

a. Removal   b. Installation

INITIAL SETUP:

APPLICABLE MODELS
All

TOOLS
General mechanic's tool kit (Appendix E, Item 1)

MATERIALS/PARTS
Antiseize tape (Appendix C, Item 72)

REFERENCES (TM)
TM 9-2320-272-10
TM 9-2320-272-24P

EQUIPMENT CONDITION
• Air reservoirs drained (TM 9-2320-272-10).
• Spring (emergency) brake caged (TM 9-2320-272-10).

GENERAL SAFETY INSTRUCTIONS
Do not disconnect air lines before draining air reservoirs.

WARNING

Do not disconnect air lines before draining air reservoirs. Small parts under pressure may shoot out with high velocity, causing injury to personnel.

a. Removal

NOTE
Tag lines for installation.

1. Disconnect two vent lines (4) from tee (5).
2. Disconnect supply line (1) from elbow (3).
3. Disconnect control line (2) from elbow (13).
4. Loosen collet nut (10) and remove two nuts (6), washers (7), U-bolt (11), and bracket (8) from combination spring and service brake chamber (12).

CAUTION
Ensure wedge assembly does not fall out of plunger housing.

5. Remove combination spring and service brake chamber (12) from plunger housing (9) and pull straight away from plunger.

b. Installation

NOTE
Wrap all male pipe threads with antiseize tape before installation.

1. Ensuring collet nut (10) is loose, install combination spring and service brake chamber (12) in plunger housing (9) until secure, with service brake chamber (12) positioned for air line connections. Thread collet nut (10) to bottom of service brake chamber (12) and tighten 3/16 in. (4.8 mm) or 1-1/2 teeth.
2. Connect control line (2) to elbow (13).
3. Connect supply line (1) to elbow (3).
4. Connect two vent lines (4) to tee (5).
5. Install U-bolt (11) and bracket (8) on service brake chamber (12) with two washers (7) and nuts (6).

3-182. COMBINATION SPRING (EMERGENCY) AND SERVICE BRAKE CHAMBER REPLACEMENT (Contd)

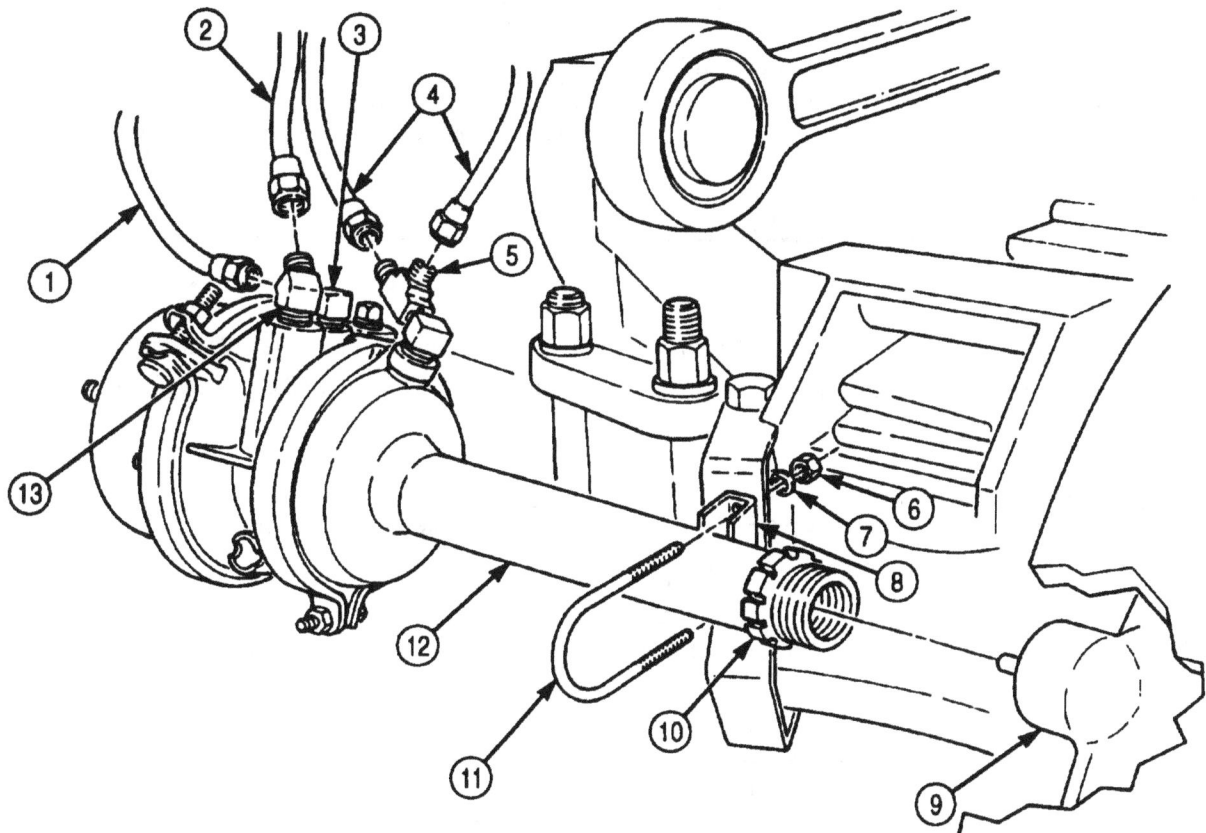

FOLLOW-ON TASKS: • Uncage spring brake chamber (TM 9-2320-272-10).
• Start engine (TM 9-2320-272-10) and allow air pressure to build to normal operating range. Check for air leaks at combination spring brake chamber. Road test vehicle.

3-l83. FRONT BRAKE SPIDER REPLACEMENT

THIS TASK COVERS:

a. Removal   b. Installation

INITIAL SETUP:

APPLICABLE MODELS
All

TOOLS
General mechanic's tool kit (Appendix E, Item 1)
Torque wrench (Appendix E, Item 144)

MATERIALS/PARTS
Two lockwashers (Appendix D, Item 365)
Gasket sealant (Appendix C, Item 30)

REFERENCES (TM)
TM 9-2320-272-10
TM 9-2320-272-24P

EQUIPMENT CONDITON
• Brakeshoes removed (para. 3-180).
• Service brake chamber removed (para. 3-181).

---

a. Removal

1. Remove nine nuts (6), washers (6), brake spider slinger (4), and brake spider (2) from studs (12).
2. Remove two nuts (10), lockwashers (9), screws (7), and clips (8) from brake spider (2). Discard lockwashers (9).

b. Installation

1. Install two clips (8) on brake spider (2) with two screws (7), new lockwashers (9), and nuts (10).
2. Apply gasket sealant to spindle (11) and brake spider (2) mating surfaces.

CAUTION
Failure to tighten nuts in proper sequence can crack brake spider.

NOTE
Tighten front brake spider after installing front wheel dustcovers.

3. Install brake spider (2) and brake spider slinger (4) on studs (12) with anchor plunger (3) at 1 o'clock position (left side) and anchor plunger (1) at 11 o'clock position (right side) with nine washers (5) and nuts (6). Tighten nuts (6) 110-145 lb-ft (149-196 N•m) in sequence shown.

3-183. FRONT BRAKE SPIDER REPLACEMENT (Contd)

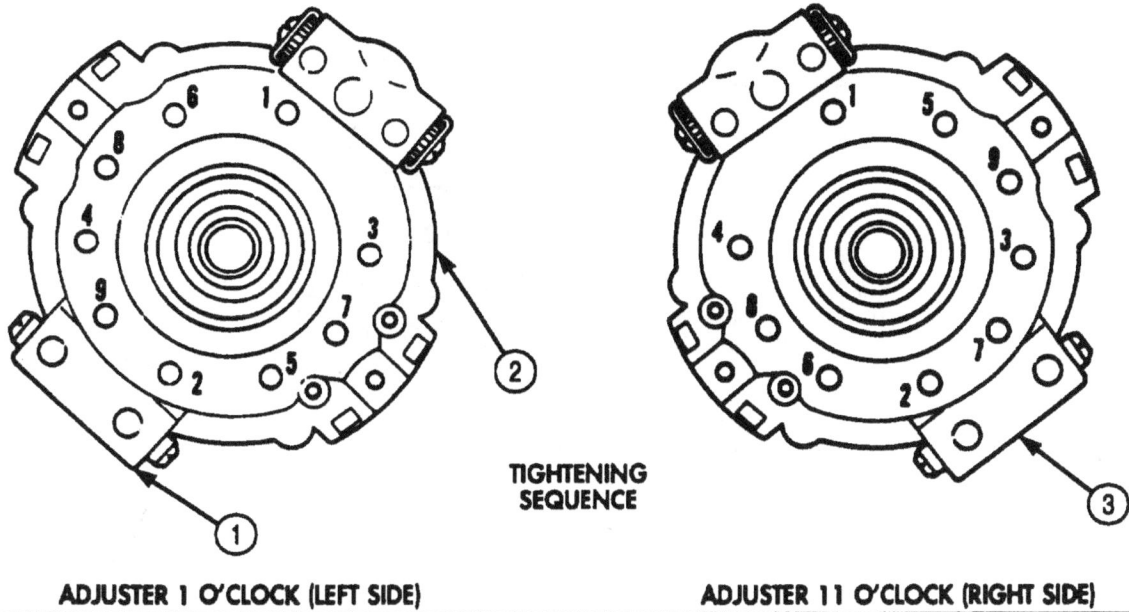

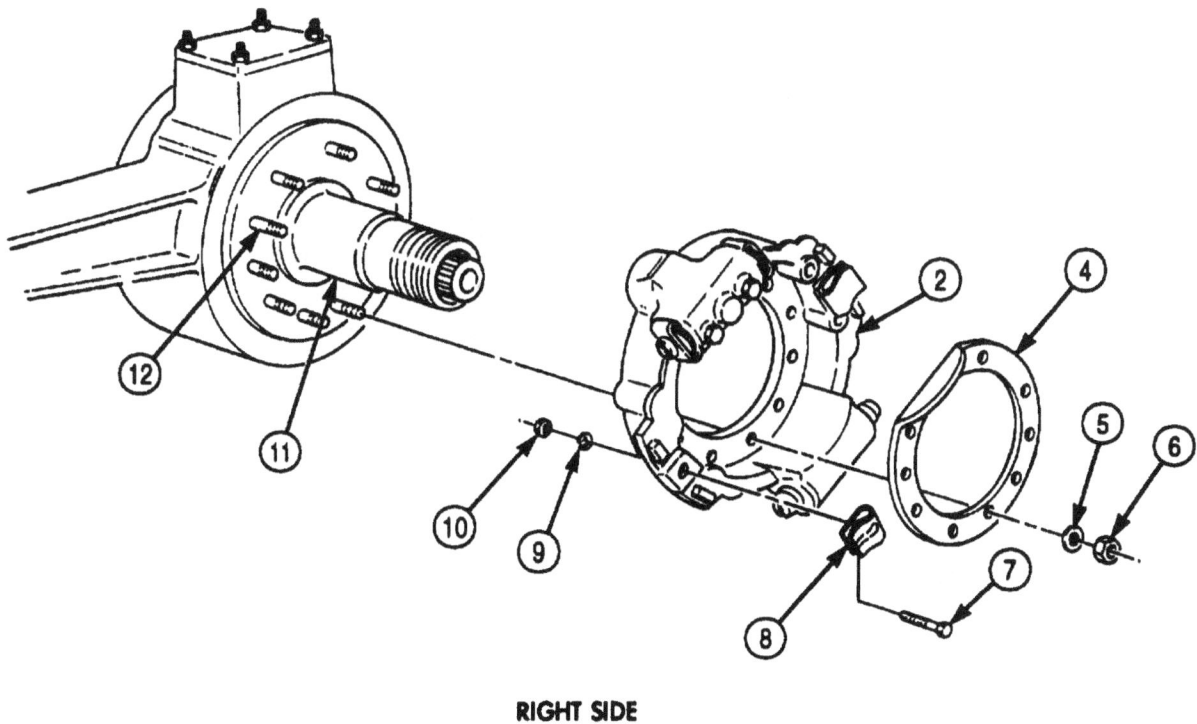

TIGHTENING
SEQUENCE

ADJUSTER 1 O'CLOCK (LEFT SIDE)   **ADJUSTER 11 O'CLOCK (RIGHT SIDE)**

RIGHT SIDE

FOLLOW-ON TASKS: • Install service brake chamber (para. 3-181).
   • Install brakeshoes (para. 3-180).
   • Start engine (TM 9-2320-272-10) and allow air pressure to build to normal
     operating range. Check brake system for proper operation. Road test vehicle.

3-184. REAR BRAKE SPIDER REPLACEMENT

THIS TASK COVERS:

a. Removal   b. Installation

INITIAL SETUP:

APPLICABLE MODELS
All

TOOLS
General mechanic's tool kit (Appendix E, Item 1)
Torque wrench (Appendix E, Item 144)

MATERIAL/PARTS
Two lockwashers (Appendix D, Item 365)
Gasket sealant (Appendix C, Item 30)

REFERENCES (TM)
TM 9-2320-272-10
TM 9-2320-272-24P

EQUIPMENT CONDITION
• Brakeshoes removed (para. 3-180).
• Service brake chamber removed (para. 3-181).
• Combination spring (emergency) and service brake chamber removed (para. 3-182).

NOTE

All rear brake spiders are replaced basically the same. This procedure covers right forward-rear axle.

a. Removal

1. Remove eight nuts (1), washers (2), screws (5), and washers (2) from brake spider (4).
2. Remove two screws (3) and washers (2) from inner side of rear axle housing flange (11).
3. Remove brake spider (4) from axle housing assembly (10).
4. Remove two nuts (9), lockwashers (8), screws (6), and clips (7) from brake spider (4). Discard lockwashers (8).

b. Installation

CAUTION

Failure to tighten nuts in proper sequence can crack brake spider.

1. Install two clips (7) on brake spider (4) with screws (6), new lockwashers (8), and nuts (9).
2. Apply gasket sealant on rear axle housing (10) and brake spider (4) mating surfaces.

NOTE

Assistant will help with step 3.

3. Position threaded holes 3 and 4 of brake spider (4) on axle housing mating surface at 3 and 9 o'clock positions, and install two washers (2) and screws (3) through back of flange (11) into threaded holes 3 and 4 of brake spider (4).
4. Install eight washers (2), screws (5), washers (2), and nuts (1) through brake spider (4) and flange (11). Tighten nuts (1) 110-145 lb-ft (149-196 N•m) in sequence shown.

3-184. REAR BRAKE SPIDER REPLACEMENT (Contd)

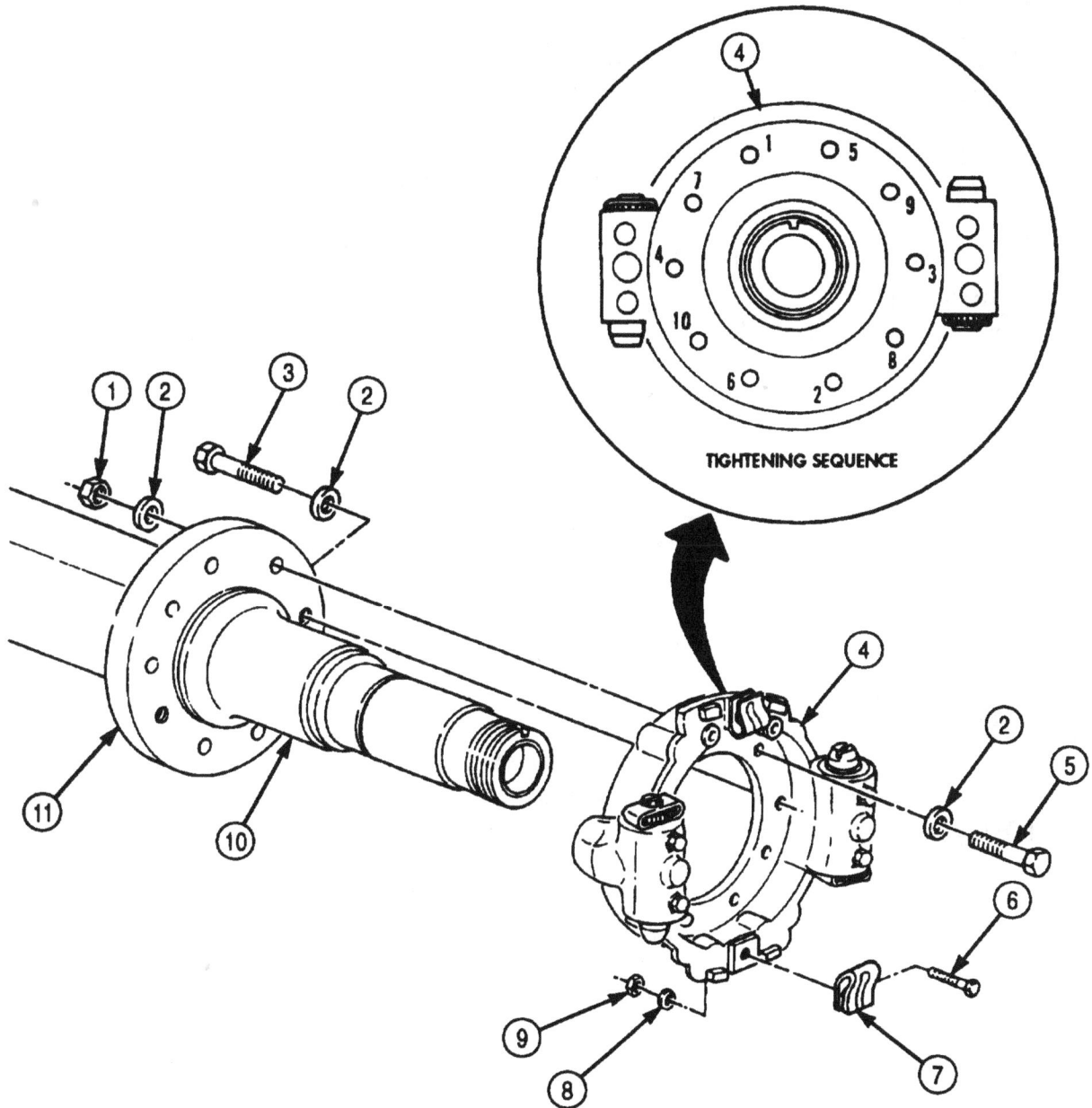

TIGHTENING SEQUENCE

RIGHT-HAND FORWARD-REAR AXLE

FOLLOW-ON TASKS: • Install combination spring (emergency) and service brake chamber (para. 3-182).
  • Install service brake chamber (para. 3-181).
  • Install brakeshoes (para. 3-180).
  • Check and adjust brakes (para 3-179).
  • Start engine (TM 9-2320-272-10) and allow air pressure to build to normal operating range. Check brake system for proper operation. Road test vehicle.

3-185. WET RESERVOIR (SUPPLY TANK) SAFETY VALVE REPLACEMENT

THIS TASK COVERS:

a. Removal   b. Installation

INITIAL SETUP:

APPLICABLE MODELS
All

TOOLS
General mechanic's tool kit (Appendix E, Item 1)

MATERIALS/PARTS
Antiseize tape (Appendix C, Item 72)

REFERENCES (TM)
TM 9-2320-272-10
TM 9-2320-272-24P

EQUIPMENT CONDITION
• Parking brake set (TM 9-2320-272-10).
• Air reservoirs drained (TM 9-2320-272-10).

GENERAL SAFETY INSTRUCTIONS
Do not remove safety valve before draining air reservoirs.

WARNING

Do not remove safety valve before draining air reservoirs. Small parts under pressure may shoot out with high velocity, causing injury to personnel.

a. Removal

Remove safety valve (1) from bushing (2) at inlet side of reservoir (3).

b. Installation

NOTE

Wrap male pipe threads with antiseize tape before installation.

Install safety valve (1) in bushing (2) at inlet side of reservoir (3).

3-185. WET RESERVOIR (SUPPLY TANK) SAFETY VALVE REPLACEMENT (Contd)

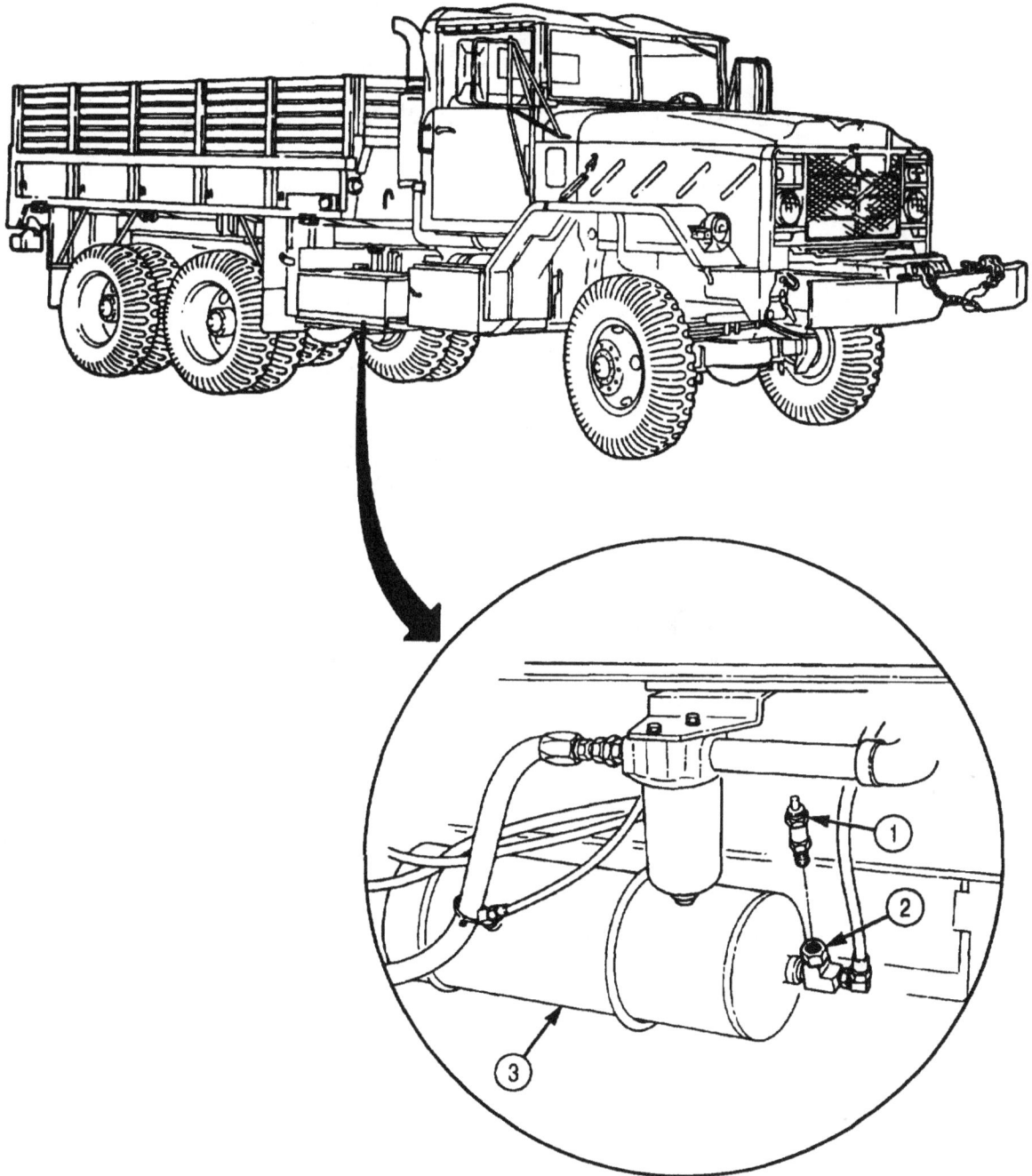

FOLLOW-ON TASK: Start engine (TM 9-2320-272-10) and allow air pressure to build to normal operating range. Check for air leaks at safety valve. Road test vehicle.

3-186. AIR RESERVOIR ONE-WAY CHECK VALVE REPLACEMENT

THIS TASK COVERS:

a. Removal   b. Installation

INITIAL SETUP:

APPLICABLE MODELS
All

TOOLS
General mechanic's tool kit (Appendix E, Item 1)

MATERIALS/PARTS
Antiseize tape (Appendix C, Item 72)

REFERENCES (TM)
TM 9-2320-272-10
TM 9-2320-272-24P

EQUIPMENT CONDITION
• Parking brake set (TM 9-2320-272-10).
• Air reservoirs drained (TM 9-2320-272-10).

GENERAL SAFETY INSTRUCTIONS
Do not disconnect air lines before draining air reservoirs.

WARNING

Do not disconnect air lines before draining air reservoirs. Small parts under pressure may shoot out with high velocity, causing injury to personnel.

NOTE

There is a one-way check valve mounted at inlet side of each reservoir. All three check valves are replaced basically the same. This procedure covers the spring brake one-way check.

1. Disconnect air line (1) from check valve adapter (2).

2. Remove check valve adapter (2) from check valve (3).

3. Remove check valve (3) from tee (4).

b. Installation

NOTE

Wrap all male pipe threads with antiseize tape before installation.

1. Install check valve (3) on tee (4).

2. Install check valve adapter (2) on check valve (3).

3. Connect air line (1) to check valve adapter (2).

3-186. AIR RESERVOIR ONE-WAY CHECK VALVE REPLACEMENT (Contd)

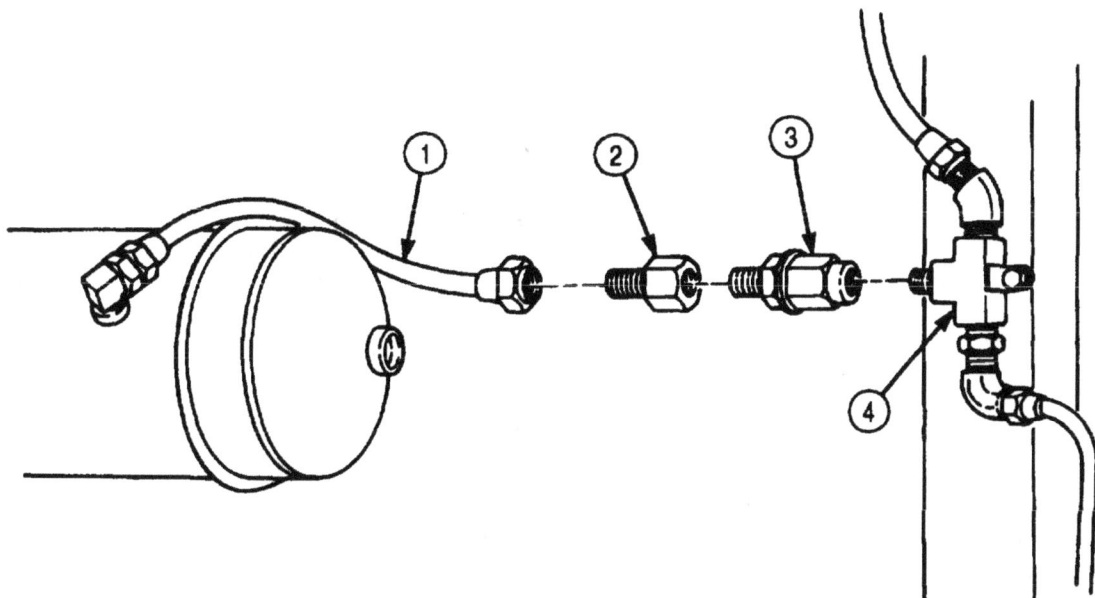

FOLLOW-ON TASK: Start engine (TM 9-2320-272-10) and allow air pressure to build to normal operating range. Check for air leaks at check valve. Road test vehicle.

3-187. FRONT RELAY VALVE REPLACEMENT

THIS TASK COVERS:

a. Removal   b. Installation

INITIAL SETUP:

APPLICABLE MODELS
All

TOOLS
General mechanic's tool kit (Appendix E, Item 1)

MATERIALS/PARTS
Two locknuts (Appendix D, Item 291)
Antiseize tape (Appendix C, Item 72)

REFERENCES (TM)
TM 9-2320-272-10
TM 9-2320-272-24P

EQUIPMENT CONDITION
• Parking brake set (TM 9-2320-272-10).
• Air reservoirs drained (TM 9-2320-272-10).

GENERAL SAFETY INSTRUCTIONS
Do not disconnect air lines before draining air reservoirs.

---

a. Removal

WARNING

Do not disconnect air lines before draining air reservoirs. Small parts under pressure may shoot out with high velocity, causing injury to personnel.

NOTE
• Tag lines and fittings for installation.
• Scribe fitting directions for installation.

1. Disconnect vent lines (4), (9), and (11) from elbow (3) and tee (10).
2. Disconnect delivery lines (8) and (14) from elbows (7) and (13).
3. Disconnect control line (12) from elbow (6).
4. Disconnect supply line (16) from adapter (17).
5. Remove two locknuts (18), screws (15), and front relay valve (2) from bracket (1). Discard locknuts (18).
6. Remove elbows (3), (6), (7), and (13), tee (10), and adapters (6) and (17) from front relay valve (2).

b. Installation

NOTE
• Use fittings from old valve.
• Wrap all male pipe threads with antiseize tape before installation.

1. Install elbows (3), (6), (7), and (13), tee (10), and adapters (6) and (17) on front relay valve (2).
2. Install front relay valve (2) on bracket (1) with two screws (15) and new locknuts (18).
3. Connect supply line (16) to adapter (17).
4. Connect control line (12) to elbow (6).
6. Connect delivery lines (8) and (14) to elbows (7) and (13).
6. Connect vent lines (4), (9), and (11) to elbow (3) and tee (10).

3-187. FRONT RELAY VALVE REPLACEMENT (Contd)

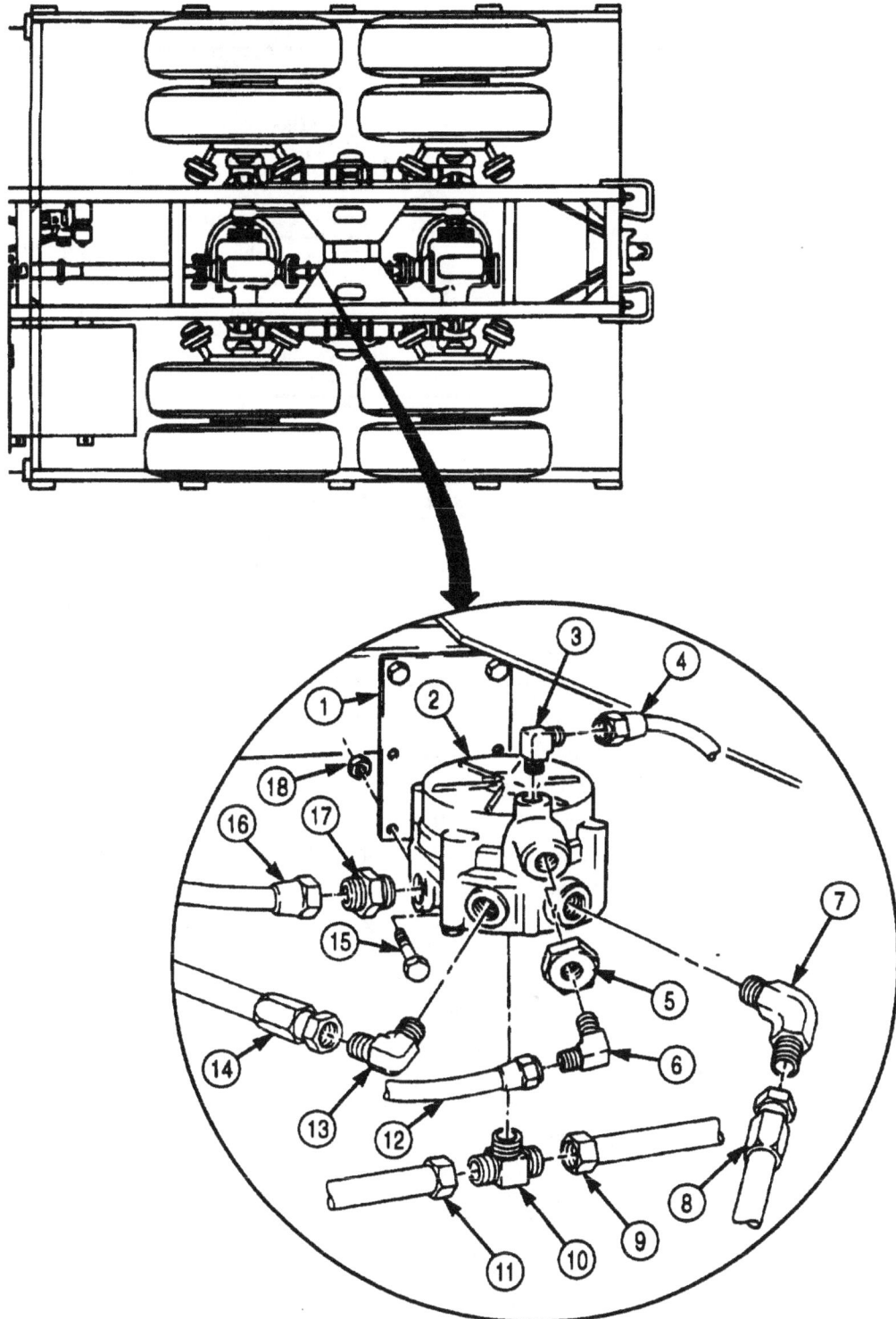

FOLLOW-ON TASK: Start engine (TM 9-2320-272-10) and allow air pressure to build to normal operating range. Check for air leaks at front relay valve. Road test vehicle.

3-188. REAR RELAY VALVE REPLACEMENT

THIS TASK COVERS:
a. Removal   b. Installation

INITIAL SETUP:

APPLICABLE MODELS
All

TOOLS
General mechanic's tool kit (Appendix E, Item 1)

MATERIALS/PARTS
Two locknuts (Appendix D, Item 291)
Antiseize tape (Appendix C, Item 72)

REFERENCES (TM)
TM 9-2320-272-10
TM 9-2320-272-24P

EQUIPMENT CONDITION
• Parking brake set (TM 9-2320-272-10).
• Air reservoirs drained (TM 9-2320-272-10).

GENERAL SAFETY INSTRUCTIONS
Do not disconnect air lines before draining air reservoirs.

---

a. Removal

WARNING
Do not disconnect air lines before draining air reservoirs. Small parts under pressure may shoot out with high velocity, causing injury to personnel.

NOTE
• Tag lines and fittings for installation.
• Scribe fitting directions for installation.

1. Disconnect vent lines (3) and (11) from elbows (2) and (12).
2. Disconnect delivery lines (8) and (15) from elbows (7).
3. Disconnect supply line (6) from adapter (5).
4. Disconnect control line (13) from elbow (10).
5. Remove two locknuts (17), screws (16), and rear relay valve (4) from bracket (1). Discard locknuts (17).
6. Remove elbows (2), (7), (10), and (12), and adapters (5), (9), and (14) from rear relay valve (4).

b. Installation

NOTE
• Use fittings from old valve.
• Wrap all male pipe threads with antiseize tape before installation.

1. Install elbows (2), (7), (10), and (12), and adapters (5), (9), and (14) on rear relay valve (4).
2. Install front relay valve (4) on bracket (1) with two screws (16) and new locknuts (17).
3. Connect supply line (6) to adapter (5).
4. Connect control line (13) to elbow (10).
5. Connect delivery lines (8) and (15) to elbows (7).
6. Connect vent lines (3) and (11) to elbows (2) and (12).

3-188. REAR RELAY VALVE REPLACEMENT (Contd)

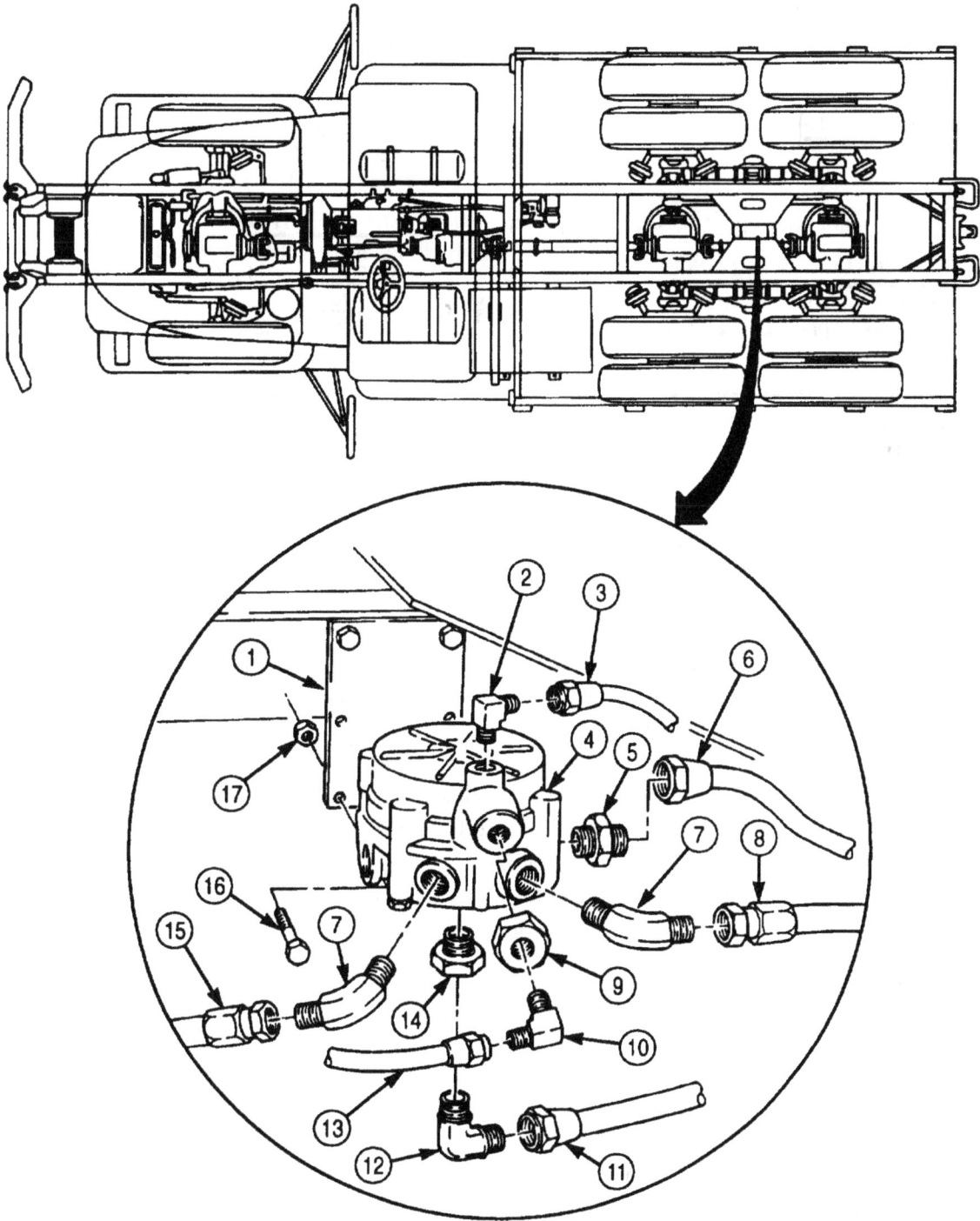

FOLLOW-ON TASK: Start engine (TM 9-2320-272-10) and allow air pressure to build to normal operating range. Check for air leaks at rear relay valve. Road test vehicle.

3-189. BRAKE CHAMBER AIR MANIFOLD TEE REPLACEMENT

THIS TASK COVERS:

a. Removal   b. Installation

INITIAL SETUP:

APPLICABLE MODELS
All

TOOLS
General mechanic's tool kit (Appendix E, Item 1)

MATERIAL/PARTS
Lockwasher (Appendix D, Item 354)
Antiseize tape (Appendix C, Item 72)

REFERENCES (TM)
TM 9-2320-272-10
TM 9-2320-272-24P

EQUIPMENT CONDITION
• Parking brake set (TM 9-2320-272-10).
• Air reservoirs drained (TM 9-2320-272-10).

GENERAL SAFETY INSTRUCTIONS
Do not disconnect air lines before draining air reservoirs.

a. Removal

WARNING

Do not disconnect air lines before draining air reservoirs. Small parts under pressure may shoot out with high velocity, causing injury to personnel.

NOTE

• All air manifold tees are replaced basically the same. This procedure covers the rear primary relay tee. Notice that only the two primary relay tees use a 45-degree elbow in addition to a 90-degree elbow.

• Tag air lines for installation.

• Scribe fitting directions for installation.

1. Disconnect right service brake chamber air line (10) from elbow (9).

2. Disconnect relay valve line (1) from adapter (2).

3. Disconnect left service brake chamber air line (4) from adapter (3).

4. Remove screw (6), lockwasher (7), and air manifold tee (8) from mounting bracket (5). Discard lockwasher (7).

b. Installation

NOTE

• If new tee is being installed, use fittings from old tee.

• Wrap all male pipe threads with antiseize tape before installation.

1. Install air manifold tee (8) on mounting bracket (5) with new lockwasher (7) and screw (6).

2. Connect left service brake chamber air line (4) to adapter (3).

3. Connect relay valve line (1) to adapter (2).

4. Connect right service brake chamber air line (10) to elbow (9).

3-189. BRAKE CHAMBER AIR MANIFOLD TEE REPLACEMENT (Contd)

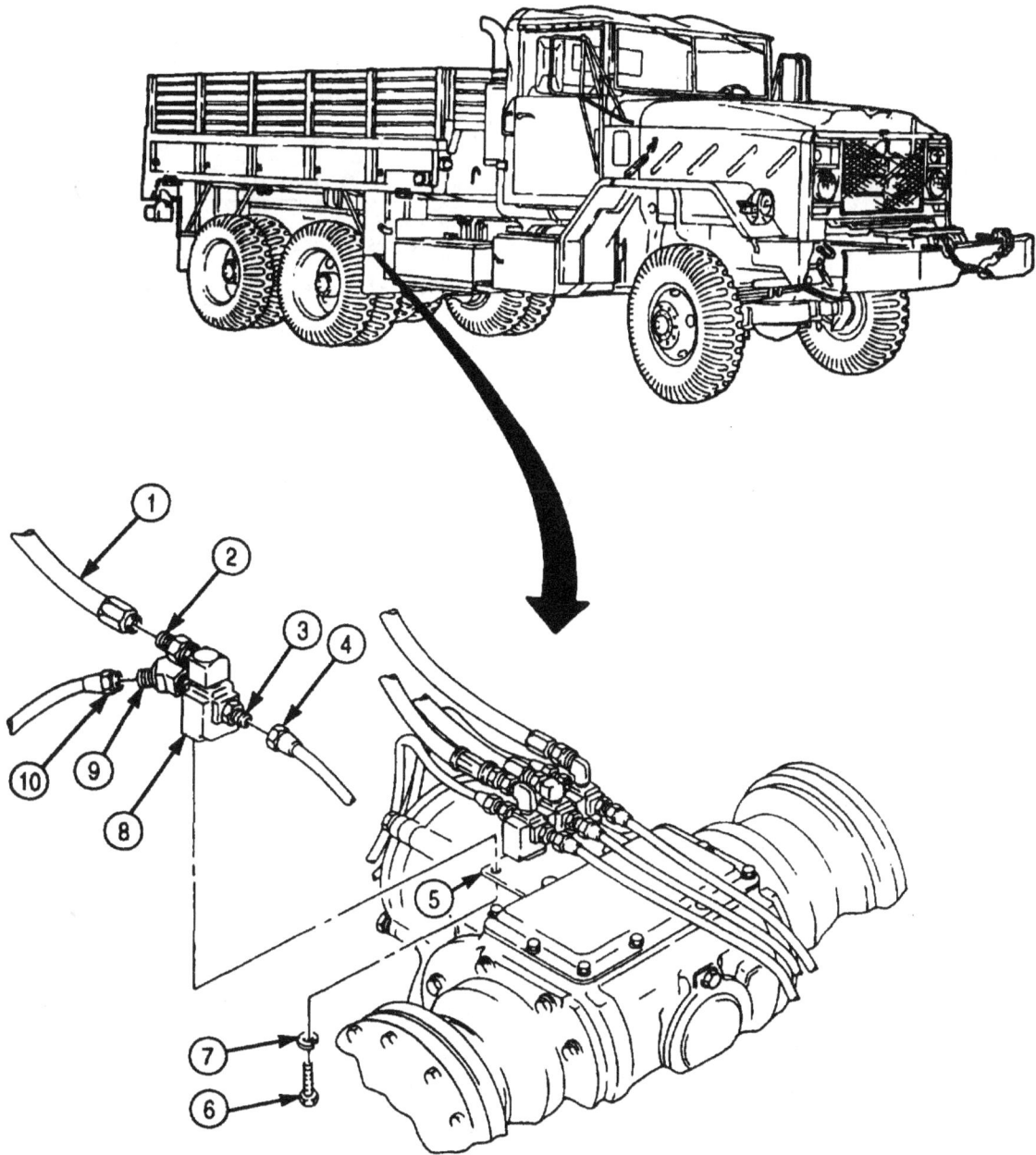

FOLLOW-ON TASK: Start engine (TM 9-2320-272-10) and allow air pressure to build to normal operating range. Check for air leaks at tee. Road test vehicle.

3-190. FRONT LIMITING VALVE REPLACEMENT

THIS TASK COVERS:
a. Removal   b. Installation

INITIAL SETUP:

APPLICABLE MODELS
All

TOOLS
General mechanic's tool kit (Appendix E, Item 1)

MATERIALS/PARTS
Two locknuts (Appendix D, Item 299)
Antiseize tape (Appendix C, Item 72)

REFERENCES (TM)
TM 9-2320-272-10
TM 9-2320-272-24P

EQUIPMENT CONDITION
• Parking brake set (TM 9-2320-272-10).
• Air reservoirs drained (TM 9-2320-272-10).

GENERAL SAFETY INSTRUCTIONS
Do not disconnect air lines before draining air reservoirs.

WARNING

Do not disconnect air lines before draining air reservoirs. Small parts under pressure may shoot out with high velocity, causing injury to personnel.

a. Removal

NOTE
• Tag air lines for installation.
• Scribe fitting directions for installation.

1. Disconnect two front service brake control lines (4) from tee (6).
2. Disconnect doublecheck valve No. 1 control line (1) from adapter (2).
3. Disconnect vent line (11) from elbow (6).
4. Remove two locknuts (9), screws (7), and limiting valve (10) from mounting bracket (8). Discard locknuts (9).
5. Remove tee (5), elbow (6), and adapters (2) and (3) from limiting valve (10).

b. Installation

NOTE
Wrap all male pipe threads with antiseize tape before installation.

1. Install adapters (3) and (2), tee (5), and elbow (6) on limiting valve (10).
2. Install limiting valve (10) on mounting bracket (8) with two screws (7) and new locknuts (9).
3. Connect vent line (11) to elbow (6).
4. Connect doublecheck valve No. 1 control line (1) to adapter (2).
5. Connect two front service brake control lines (4) to tee (5).

3-190. FRONT LIMITING VALVE REPLACEMENT (Contd)

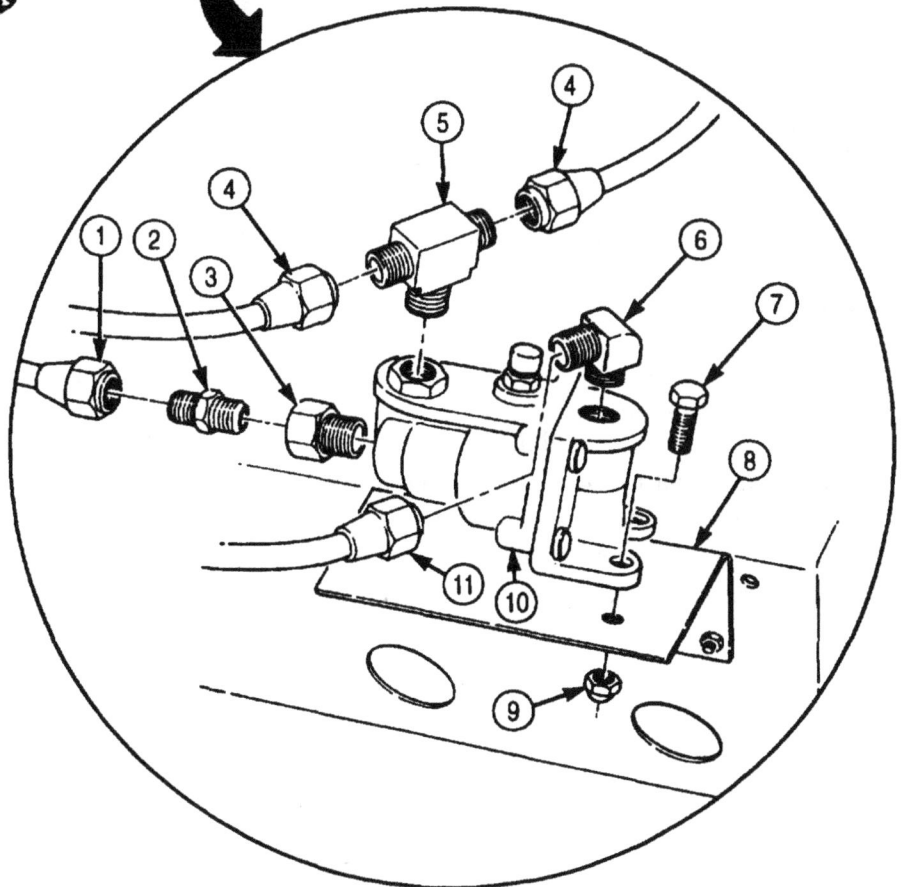

FOLLOW-ON TASK: Start engine (TM 9-2320-272-10) and allow air pressure to build to normal operating range. Check for air leaks at front limiting valve. Road test vehicle.

3-191. WASHER CONTROL VALVE REPLACEMENT

THIS TASK COVERS

a. Removal   b. Installation

INITIAL SETUP:

APPLICABLE MODELS

All

TOOLS
General mechanic"s tool kit (Appendix E, Item 1)

MATERIALS/PARTS
Antiseize tape (Appendix C, Item 72)

REFERENCES (TM)
TM 9-2320-272-10
TM 9-2320-272-24P

EQUIPMENT CONDITION
Air reservoirs drained (TM 9-2320-272-10),

GENERAL SAFETY INSTRUCTIONS
Do not dirconnect air liner before draining air
reservoirs.

WARNING

Do not disconnect air lines before draining air reservoirs. Small
parts under pressure may shoot out with high velocity, causing
injury to personnel.

a. Removal

1. Remove setscrew (6) and knob (7) from shaft (10).
2. Remove retaining ring (8), nut (5), washer (4), and washer control valve (3) from instrument panel (9).
3. Remove clamp (11) and washer bottle delivery line (12) from washer control valve (3).
4. Remove manifold tee supply line (1) from elbow (2).

b. Installation

NOTE

- If new valve is being installed, use fittings from old valve.
- Wrap all male pipe threads with antiseize tape before
  installation.

1. Install manifold tee supply line (1) on elbow (2).
2. Install delivery line (12) on washer control valve (3) with clamp (11).
3. Install washer control valve (3) on instrument panel (9) with washer (4) and nut (5).
4. Install retaining ring (8) and knob (7) on shaft (10) with setscrew (6).

3-191. WASHER CONTROL VALVE REPLACEMENT (Contd)

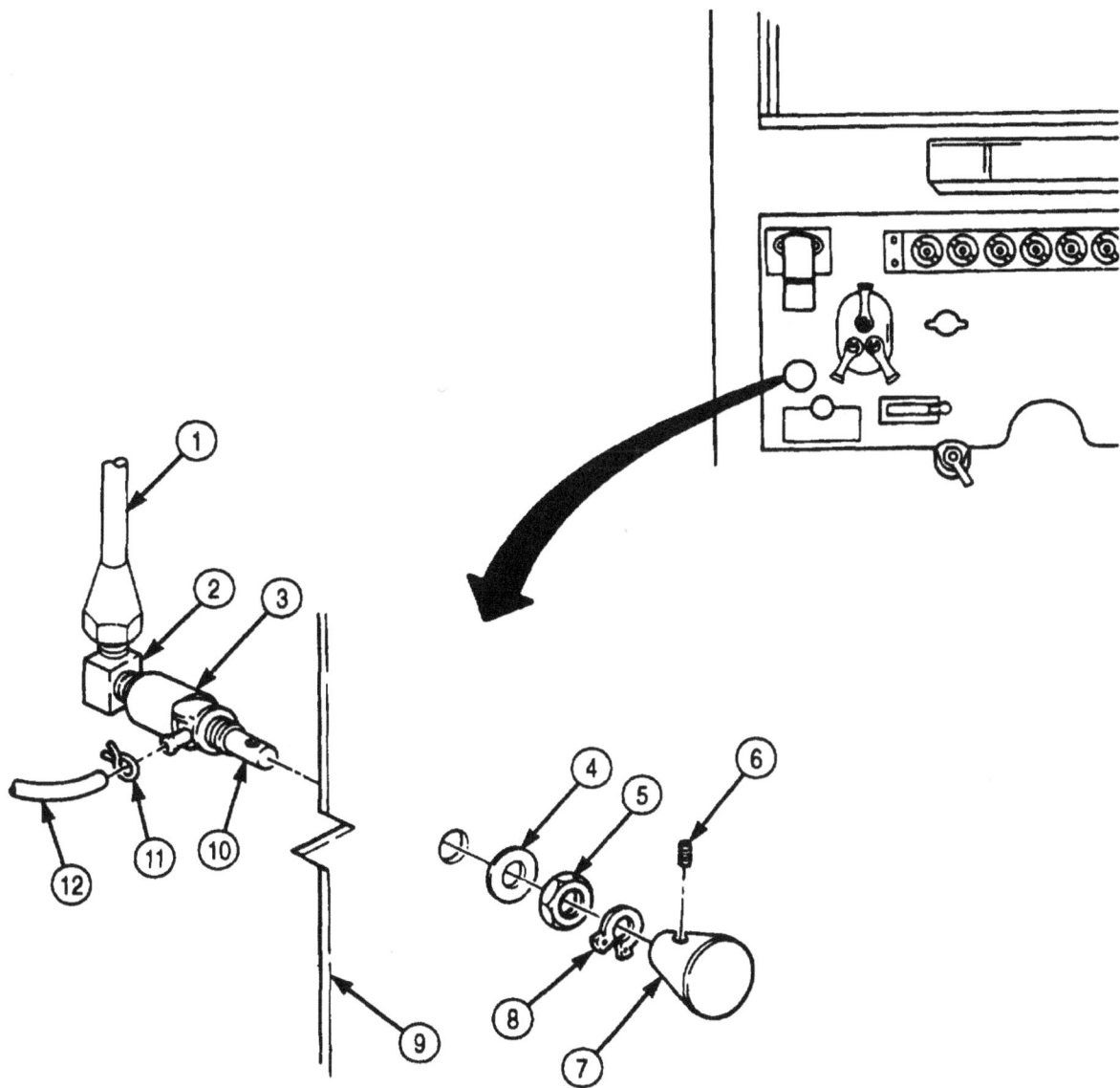

FOLLOW-ON TASK: Start engine (TM 9-2320-272-10) and check for leaks and proper washer control valve operation.

3-192. WINDSHIELD WIPER CONTROL VALVE REPLACEMENT

THIS TASK COVERS:
a. Removal   b. Installation

INITIAL SETUP:

APPLICABLE MODELS
All

TOOLS
General mechanic's tool kit (Appendix E, Item 1)

MATERIALS/PARTS
Lockwasher (Appendix D, Item 370)
Antiseize tape (Appendix C, Item 72)

REFERENCES (TM)
TM 9-2320-272-10
TM 9-2320-272-24P

EQUIPMENT CONDITION
• Parking brake set (TM 9-2320-272-10).
• Air reservoirs drained (TM 9-2320-272-10).

GENERAL SAFETY INSTRUCTIONS
Do not disconnect air lines before draining air reservoirs.

WARNING

Do not disconnect air lines before draining air reservoirs. Small parts under pressure may shoot out with high velocity, causing injury to personnel.

a. Removal

NOTE
Tag air lines for installation.

1. Remove setscrew (6) and remove knob (5) from shaft (4).
2. Remove nut (7), lockwasher (8), and control valve (10) from instrument panel (9). Discard lockwasher (8).
3. Remove clamp (2) and disconnect wiper motor run delivery line (1) from adapter (3).
4. Disconnect manifold tee supply line (14) from adapter (15).
5. Remove clamp (12) and disconnect motor park delivery line (13) from adapter (11).
6. Remove adapters (3), (15), and (11) from control valve (10).

b. Installation

NOTE
• If new wiper control valve is being installed, use fittings from old valve.
• Clean all male pipe threads and wrap with antiseize tape before installation.

1. Install adapters (3), (15), and (11) on control valve (10).
2. Connect motor park delivery line (13) to adapter (11) with clamp (12).
3. Connect manifold tee supply line (14) to adapter (15).
4. Connect wiper motor run delivery line (1) to adapter (3) with clamp (2).
5. Install control valve (10) on instrument panel (9) with new lockwasher (8) and nut (7).
6. Install knob (5) on shaft (4) with setscrew (6).

3-192. WINDSHIELD WIPER CONTROL VALVE REPLACEMENT (Contd)

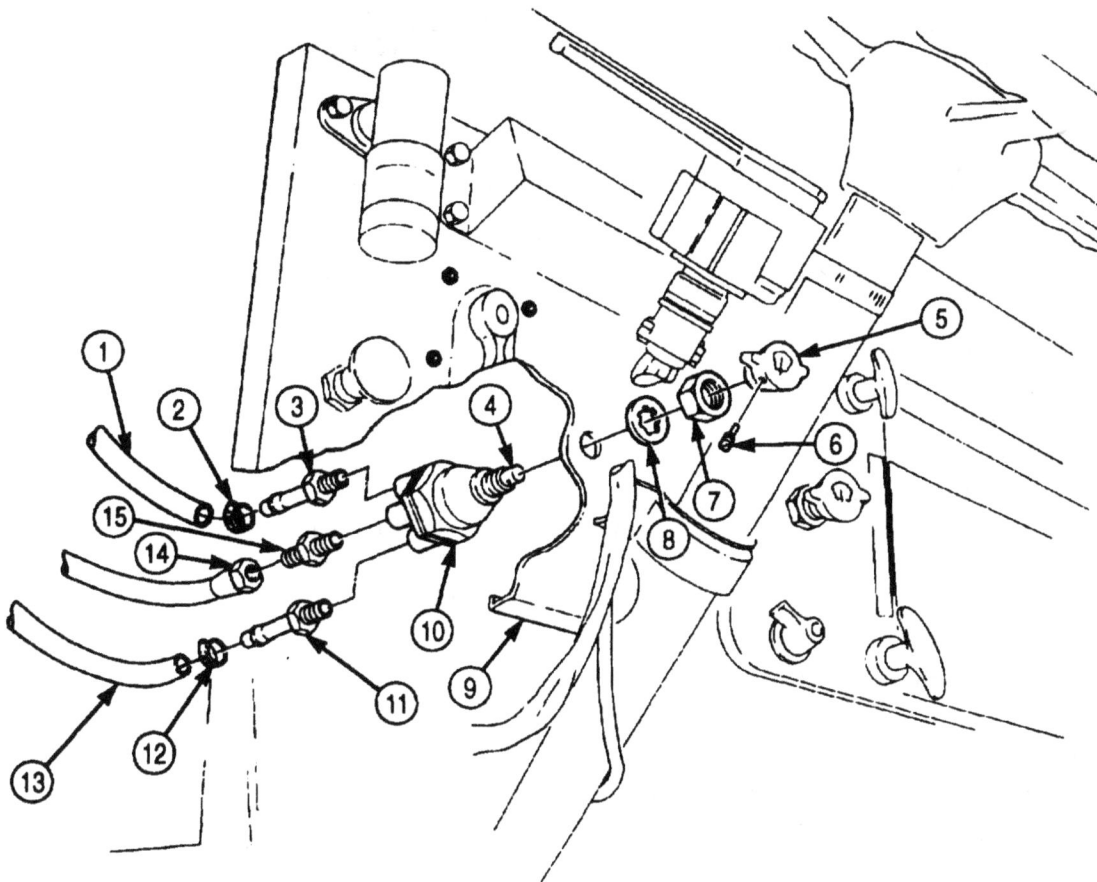

FOLLOW-ON TASK: Start engine (TM 9-2320-272-10) and allow air pressure to build to normal operating range. Check for air leaks at wiper control valve.

3-193. SPRING (EMERGENCY) BRAKE CHAMBER MAINTENANCE

THIS TASK COVERS:

a. Disassembly
b. Inspection

c. Assembly

INITIAL SETUP:

APPLICABLE MODELS
All

TOOLS
General mechanic's tool kit (Appendix E, Item 1)
Torque wrench (Appendix E, Item 146)

REFERENCES (TM)
TM 9-2320-272-10
TM 9-2320-272-24P

EQUIPMENT CONDITION
• Parking brake set (TM 9-2320-272-10)
• Air reservoirs drained (TM 9-2320-272-10).

GENERAL SAFETY INSTRUCTIONS
Do not disconnect air lines before draining air reservoirs.

WARNING

Do not disconnect air lines before draining air reservoirs. Small parts under pressure may shoot out with high velocity, causing injury to personnel.

a. Disassembly

1. Remove tube (1) from elbow (2).
2. Remove tube (10) from elbow (9).
3. Remove nut (7), screw (6), clamp (3), and housing assembly (8) from nonpressure housing (5).
4. Remove elbows (2) and (9) from housing assembly (8).

b. Inspection

Inspect spring brake chamber diaphragm (4). Replace if cracked, torn, or split.

c. Assembly

1. Install elbows (2) and (9) on housing assembly (8).
2. Install housing assembly (8) on nonpressure housing (5) with clamp (3), screw (6), and nut (7). Tighten nut (7) 18-25 lb-ft (24-34 N•m).
3. Install tube (10) on elbow (9).
4. Install tube (1) on elbow (2).

3-193. SPRING (EMERGENCY) BRAKE CHAMBER MAINTENANCE (Contd)

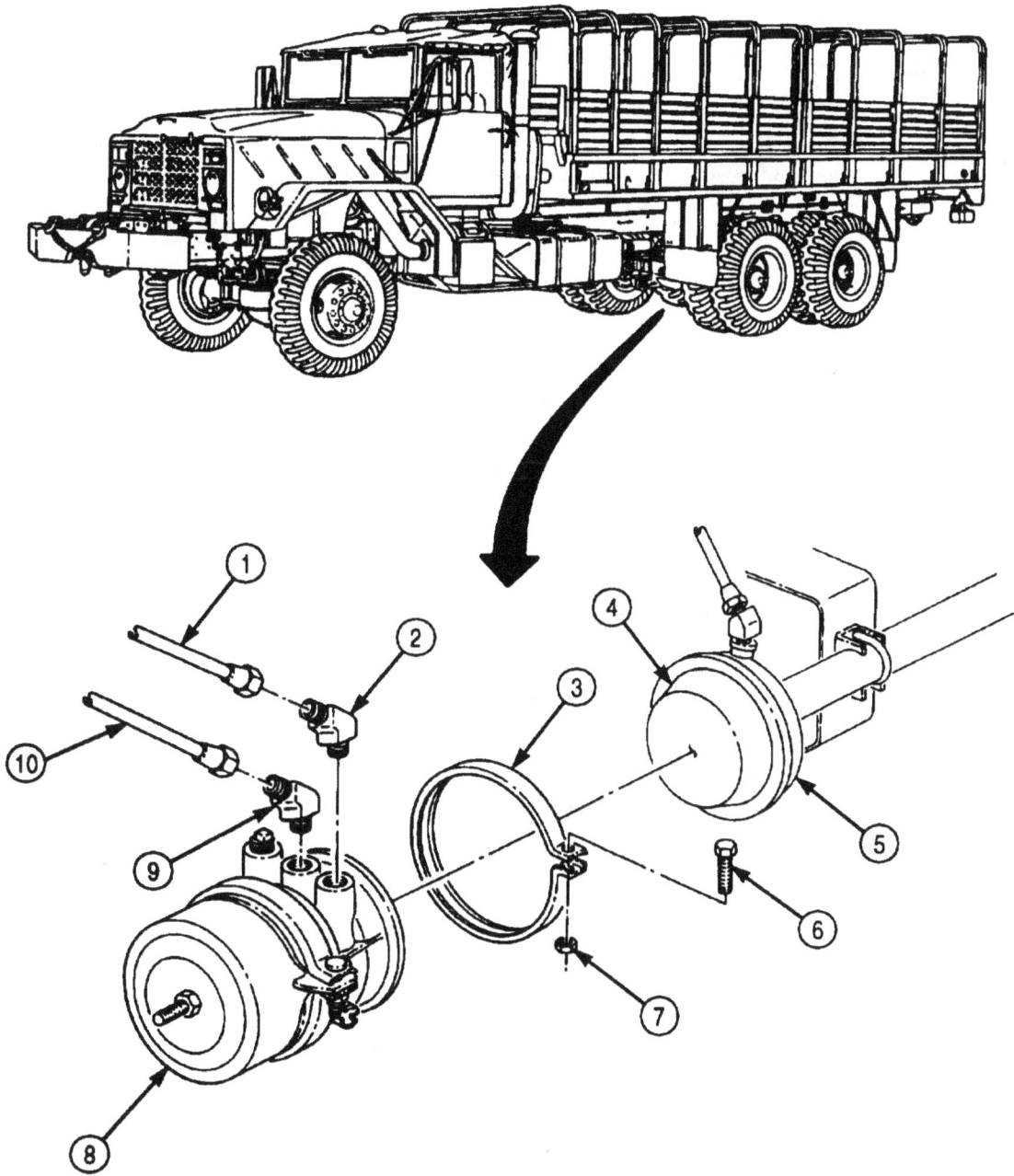

FOLLOW-ON TASK: Start engine (TM 9-2320-272-10) and allow air pressure to build to normal operating range. Check for air leaks at spring brake chamber. Road test vehicle.

3-194. SPRING PARKING BRAKE VALVE REPLACEMENT

THIS TASK COVERS:

a. Removal   b. Installation

INITIAL SETUP:

APPLICABLE MODELS
All

TOOLS
General mechanic's tool kit (Appendix E, Item 1)

MATERIALS/PARTS
Antiseize tape (Appendix C, Item 72)

REFERENCES (TM)
TM 9-2320-272-10
TM 9-2320-272-24P

EQUIPMENT CONDITION
• Parking brake set (TM 9-2320-272-10).
• Air reservoirs drained (TM 9-2320-272-10).

GENERAL SAFETY INSTRUCTIONS
Do not disconnect air lines before draining air reservoirs.

WARNING

Do not disconnect air lines before draining air reservoirs. Small parts under pressure may shoot out with high velocity, causing injury to personnel.

a. Removal

CAUTION

Use care to prevent excessive twisting when removing air lines.

NOTE

Tag air lines and fittings for installation.

1. Disconnect transfer case vent line (6) from adapter (3).
2. Disconnect reservoir supply line (5) from adapter (4).
3. Disconnect doublecheck control line (7) from elbow (8).
4. Remove two screws (2) and valve (1) from mounting bracket (9).

b. Installation

NOTE

• If new spring parking brake valve is being installed, use fittings from old valve.
• Wrap all male pipe threads with antiseize tape before installation.

1. Install valve (1) on mounting bracket (9) with two screws (2).
2. Connect doublecheck control line (7) to elbow (8).
3. Connect reservoir supply line (5) to adapter (4).
4. Connect transfer case vent line (6) to adapter (3).

3-194. SPRING PARKING BRAKE VALVE REPLACEMENT (Contd)

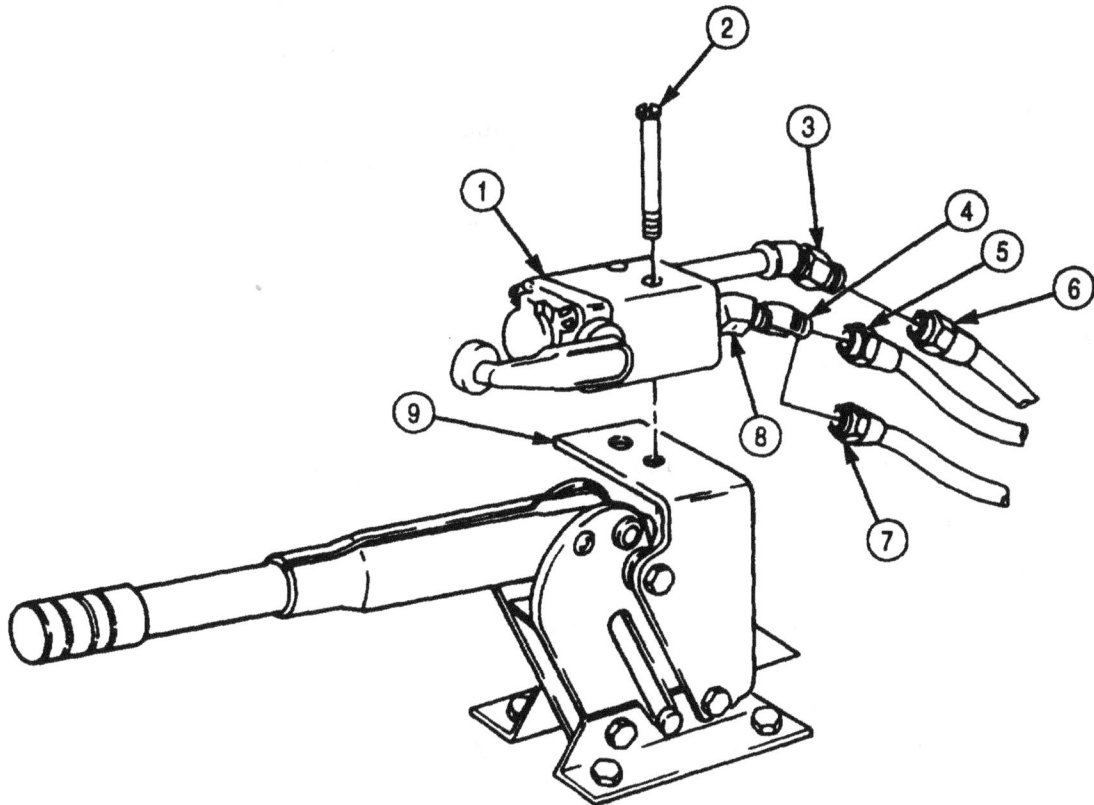

FOLLOW-ON TASK: Start engine (TM 9-2320-272-10) and allow air pressure to build to normal operating range. Ensure spring brake chamber releases spring when parking brake is applied. Road test vehicle.

3-l95. SPRING BRAKE DASH CONTROL VALVE REPLACEMENT

THIS TASK COVERS:

a. Removal   b. Installation

INITIAL SETUP:

APPLICABLE MODELS
All

TOOLS
General mechanic's tool kit (Appendix E, Item 1)
Toque wrench (Appendix E, Item 146)

MATERIALS/PARTS
Antiseize tape (Appendix C, Item 72)

REFERENCES (TM)
TM 9-2320-272-10
TM 9-2320-272-24P

EQUIPMENT CONDITION
• Parking brake set (TM 9-2320-272-10).
• Spring brake pressure switch removed (para. 3-100).

GENERAL SAFETY INSTRUCTIONS
Do not disconnect air lines before draining air reservoirs.

WARNING

Do not disconnect air lines before draining air reservoirs. Small parts under pressure may shoot out with high velocity, causing injury to personnel.

a. Removal

NOTE
Tag air lines for installation,

1. Disconnect doublecheck valve air line (5) from elbow (4).
2. Disconnect air line (3) from elbow (2).
3. Remove pin (7) and valve knob (9) from dash control valve assembly (6).
4. Remove nut (8) and dash control valve assembly (6) from instrument cluster (1).
6. Remove elbows (2) and (4) from dash control valve assembly (6).

b. Installation

NOTE
Wrap all male pipe threads with aniseize tape before installation,

1. Install elbows (2) and (4) on dash control valve assembly (6).
2. Install dash control valve assembly (6) on instrument cluster (1) with nut (8). Tighten nut (8) 150-300 lb-in. (17-34 N•m).
3. Install valve knob (9) on dash control valve assembly (6) with pin (7).
4. Connect air line (3) to elbow (2).
6. Connect doublecheck valve air line (5) to elbow (4).

3-195. SPRING BRAKE DASH CONTROL VALVE REPLACEMENT (Contd)

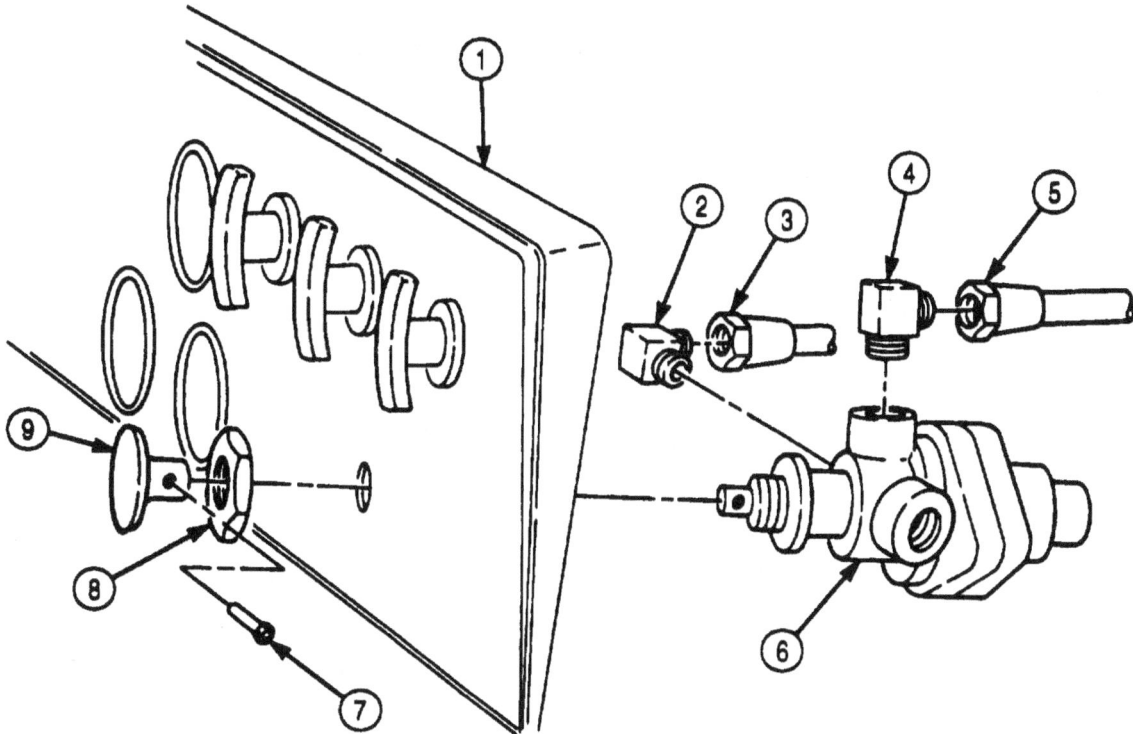

FOLLOW-ON TASKS:• Install spring brake pressure switch (para. 3-100).
       • Start engine (TM 9-2320-272-10) and allow air pressure to build to normal
        operating range. Check for air leaks at dash control valve. Check valve for proper
        operation. Road test vehicle.

3-196. BRAKE PEDAL MAINTENANCE

THIS TASK COVERS:

a. Removal   c. Adjustment
b. Installation

INITIAL SETUP:

APPLICABLE MODELS
All

TOOLS
General mechanic's tool kit (Appendix E, Item 1)

MATERIALS/PARTS
Two cotter pins (Appendix D, Item 48)

REFERENCES (TM)
TM 9-2320-272-10
TM 9-2320-272-24P

EQUIPMENT CONDITION
Parking brake set (TM 9-2320-272-10).

---

a. Removal

1. Remove return spring (5) from brake pedal (9) and cab (2).
2. Remove cotter pin (11), fulcrum pin (13), and brake pedal (9) from brake pedal valve (1). Discard cotter pin (11).
3. Remove jamnut (3) and pedal stop (4) from brake pedal valve (1).
4. Remove cotter pin (7), roller pin (10), and two rollers (6) from brake pedal (9). Discard cotter pin (7).
5. Remove rubber pad (8) from brake pedal (9).

b. Installation

1. Install rubber pad (8) on brake pedal (9).
2. Install two rollers (6) on brake pedal (9) with roller pin (10) and cotter pin (7).
3. Install pedal stop (4) on brake pedal valve (1) with jamnut (3). Do not tighten jamnut (3).
4. Install brake pedal (9) on brake pedal valve (1) with fulcrum pin (13) and cotter pin (11).
5. Install return spring (5) on brake pedal (9) and cab (2).

c. Adjustment

Use hand pressure to depress brake pedal (9) only until rollers (6) contact plunger (12), and position pedal stop (4) against brake pedal (9). Tighten jamnut (3).

3-196. BRAKE PEDAL MAINTENANCE (Contd)

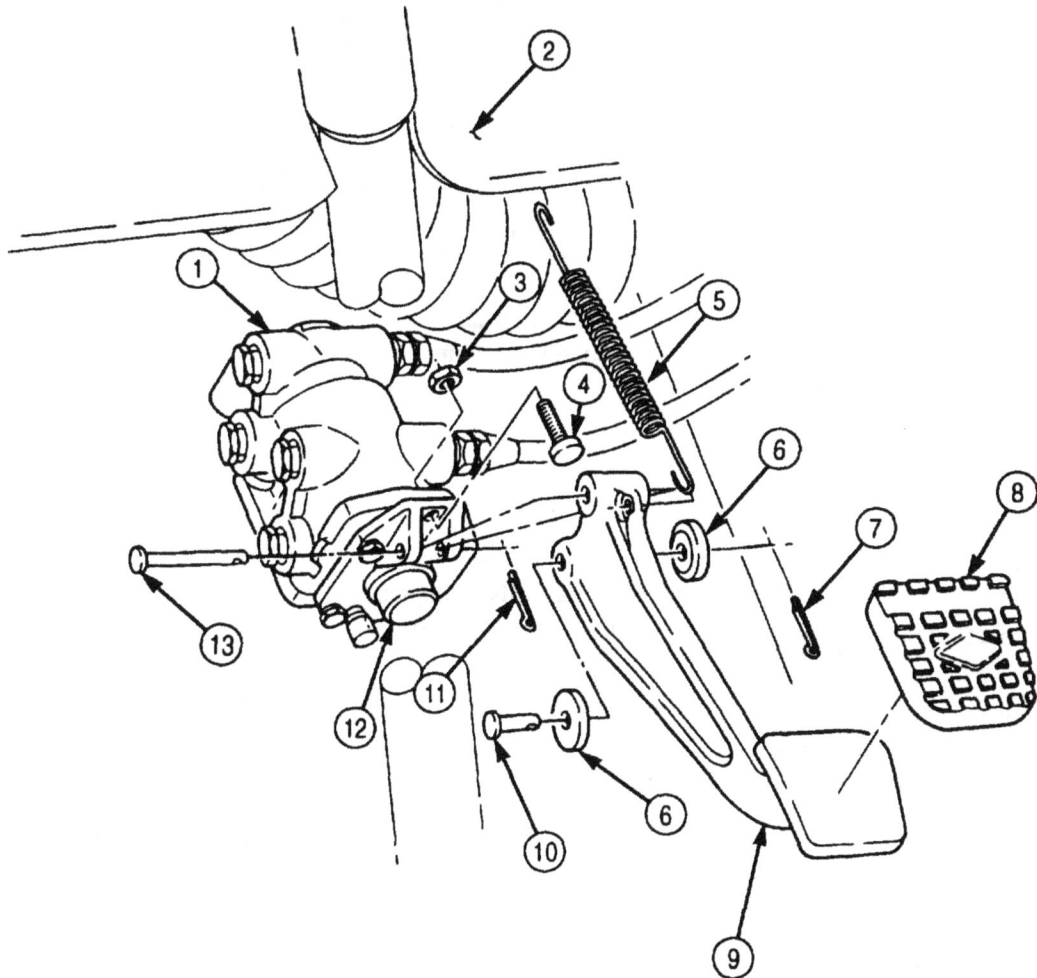

FOLLOW-ON TASK: Road test vehicle (TM 9-2320-272-10).

3-197. BRAKE PEDAL (TREADLE) VALVE REPLACEMENT

THIS TASK COVERS:

a. Removal   b. Installation

INITIAL SETUP:

APPLICABLE MODELS
All except M936/A1/A2

TOOLS
General mechanic's tool kit (Appendix E, Item 1)

MATERIALS/PARTS
Three locknuts (Appendix D, Item 288)
Antiseize tape (Appendix D, Item 72)

REFERENCES (TM)
TM 9-2320-272-10
TM 9-2320-272-24P

EQUIPMENT CONDITION
• Air reservoirs drained (TM 9-2320-272-10).
• Brake pedal removed (para. 3-196).
• Protective control box removed (para. 3-115).

GENERAL SAFETY INSTRUCTIONS
Do not disconnect air lines before draining air reservoirs.

WARNING

Do not disconnect air lines before draining air reservoirs. Small parts under pressure may shoot out with high velocity, causing injury to personnel.

a. Removal

NOTE
Tag air lines for installation.

1. Disconnect two air lines (2) from brake pedal valve adapters (1).
2. Disconnect air hoses (4) and (8) from adapters (7) and (9).
3. Remove adapters (7) and (9) from tee (10).
4. Disconnect air hose (6) from adapter (5).
5. Remove adapter (5) from firewall (3).
6. Remove tee (10) and pipe (11) from firewall (3).

3-197. BRAKE PEDAL (TREADLE) VALVE REPLACEMENT (Contd)

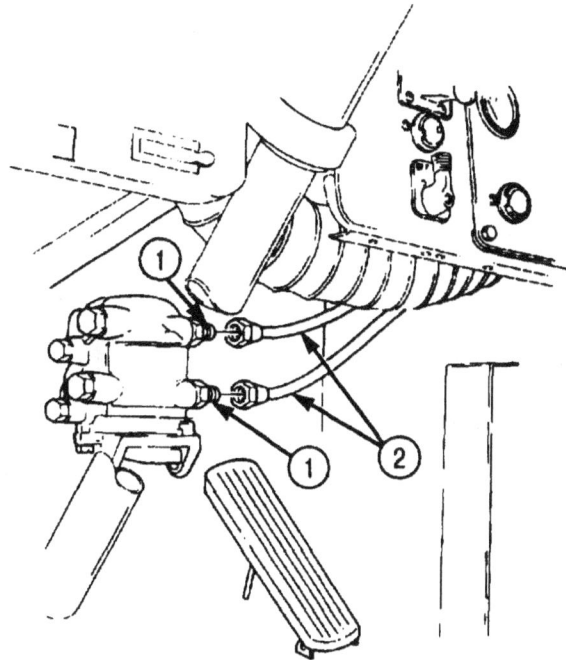

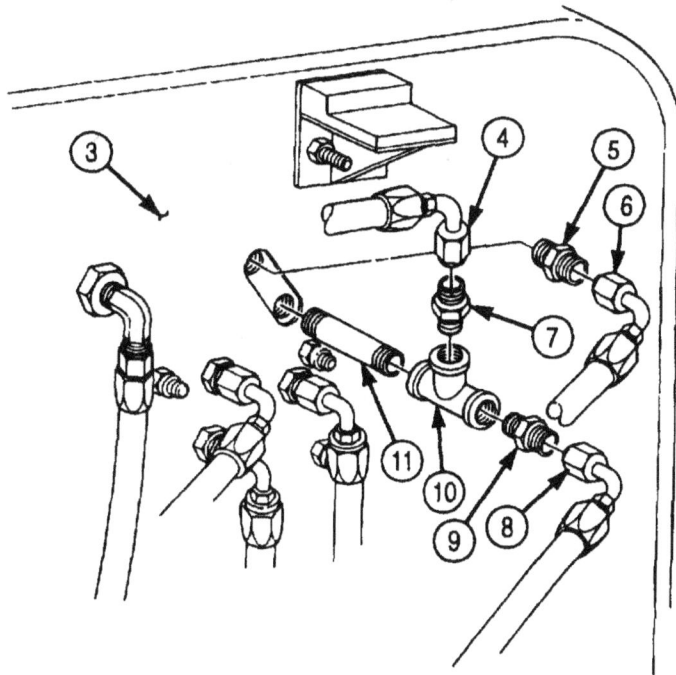

3-197. BRAKE PEDAL (TREADLE) VALVE REPLACEMENT (Contd)

7. Remove air hoses (3), (4), and (6) from adapters (2), (5), and (7).
8. Remove adapters (2), (5), and (7) from firewall (8).
9. Remove three locknuts (1) and brake pedal valve (9) from firewall (8). Discard locknuts (1).
10. Remove three studs (10) from brake pedal valve (9).
11. Remove two brake pedal valve adapters (12) from brake pedal valve (9).
12. Remove pressure relief valve (11) from brake pedal valve (9).

b. Installation

NOTE

- If new brake pedal valve is being installed, use attaching parts and fittings from old brake pedal valve.
- Fittings must be cleaned and inspected for cracks and stripped threads.
- Wrap all male pipe threads with antiseize tape before installation.

1. Install pressure relief valve (11) in brake pedal valve (9).
2. Install two brake pedal valve adapters (12) in brake pedal valve (9).
3. Install three studs (10) in brake pedal valve (9).
4. Install brake pedal valve (9) on firewall (8) with three new locknuts (1).
5. Install adapters (2), (5), and (7) on firewall (8).
6. Install air hoses (3), (4), and (6) on adapters (2), (5), and (7).
7. Install pipe (20) and tee (19) on firewall (8).
8. Install adapter (15) on firewall (8).
9. Connect air hose (16) to adapter (15).
10. Install adapters (14) and (18) on tee (19).
11. Connect air hoses (13) and (17) to adapters (14) and (18).
12. Connect two air lines (21) to brake pedal valve adapters (22).

3-197. BRAKE PEDAL (TREADLE) VALVE REPLACEMENT (Contd)

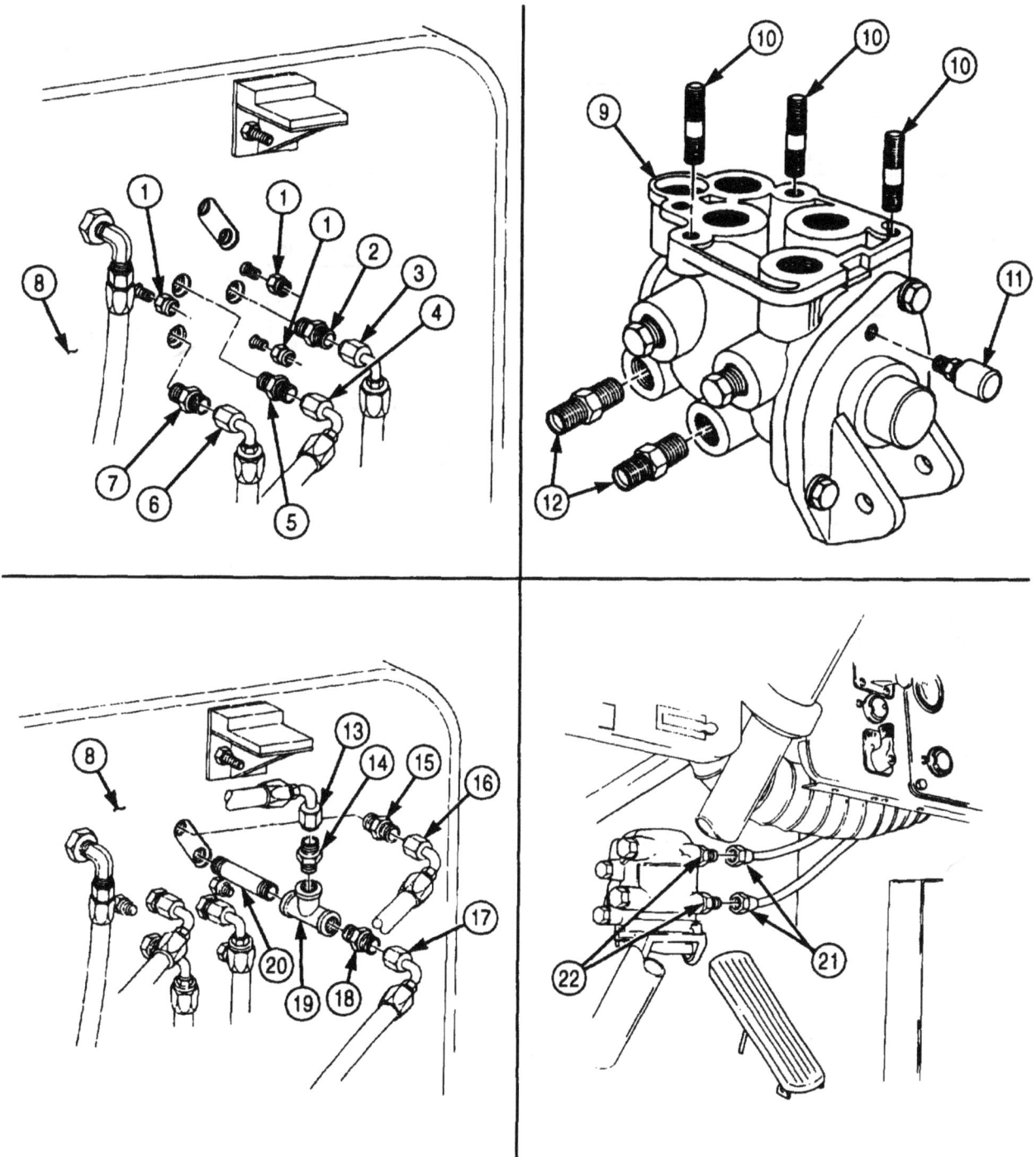

FOLLOW-ON TASKS: • Install protective control box (para. 3-115).
• Install brake pedal (para. 3-196).
• Start engine (TM 9-2320-272-10) and allow air pressure to build to normal operating range. Check for air leaks and proper brake operation. Road test vehicle.

3-198. BRAKE PEDAL (TREADLE) VALVE (M936/A1/A2) REPLACEMENT

THIS TASK COVERS:

a. Removal   b. Installation

INITIAL SETUP:

APPLICABLE MODELS
M986/A1/A2

TOOLS
General mechanic's tool kit (Appendix E, Item 1)

MATERIALS/PARTS
Three locknuts (Appendix D, Item 288)
Antiseize tape (Appendix C, Item 72)

REFERENCES (TM)
TM 9-2320-272-10
TM 9-2320-272-24P

EQUIPMENT CONDITION
• Parking brake set (TM 9-2320-272-10).
• Air reservoirs drained (TM 9-2320-272-10).
• Brake pedal removed (para. 3-196).
• Protective control box removed (para. 3-115).

GENERAL SAFETY INSTRUCTIONS
Do not disconnect air lines before draining air reservoirs.

WARNING

Do not disconnect air lines before draining air reservoirs. Small parts under pressure may shoot out with high velocity, causing injury to personnel.

a. Removal

1. Disconnect two air lines (2) from brake pedal valve adapters (1).
2. Disconnect air hoses (4) and (10) from adapters (7) and (9).
3. Remove adapters (7) and (9) from tee (8).
4. Remove air hose (6) and adapter (6) from firewall (3).
5. Remove tee (8) and pipe (13) from firewall (3).
6. Remove air hose (11) and adapter (12) from firewall (3).

3-198. BRAKE PEDAL (TREADLE) VALVE (M936/A1/A2) REPLACEMENT (Contd)

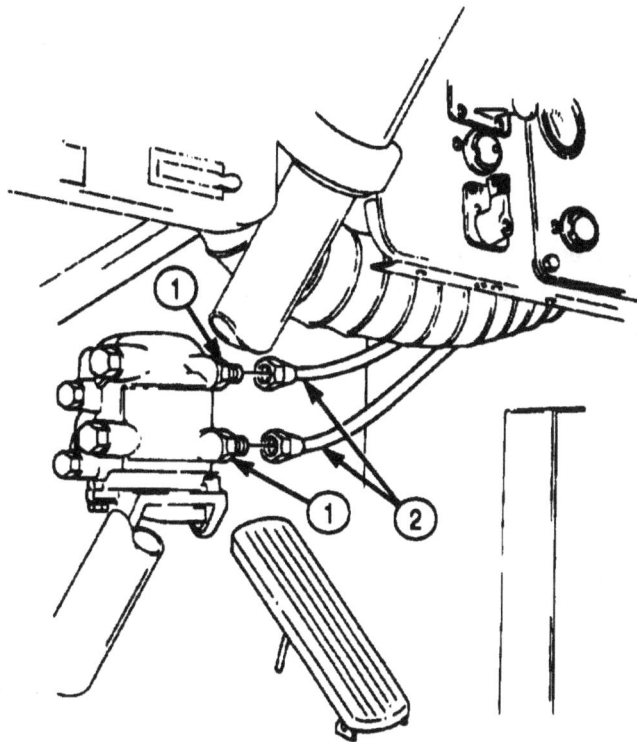

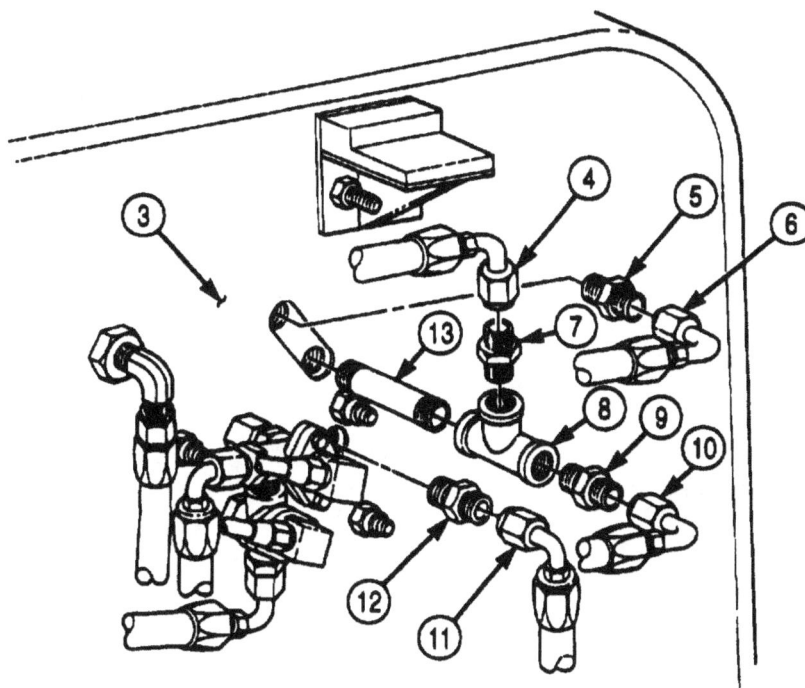

3-198. BRAKE PEDAL (TREADLE) VALVE (M936/A1/A2) REPLACEMENT (Contd)

7. Disconnect two air hoses (5) from elbows (4).

8. Remove two elbows (4) from control valves (3).

9. Disconnect air hoses (7) and (8) from adapters (6).

10. Remove two adapters (6) from control valves (3).

11. Remove two control valves (3) and adapter (2) from firewall (9).

12. Remove three locknuts (1) and brake pedal valve (10) from firewall (9). Discard locknuts (1).

13. Remove three studs (11) from brake pedal valve (10).

14. Remove two brake pedal valve adapters (13) from brake pedal valve (10).

15. Remove pressure relief valve (12) from brake pedal valve (10).

b. Installation

NOTE

- If new brake pedal valve is being installed, use attaching parts and fittings from old brake pedal valve.

- Fittings must be cleaned and inspected for cracks and stripped threads.

- Wrap all male pipe threads with antiseize tape before installation.

1. Install pressure relief valve (12) on brake pedal valve (10).

2. Install two brake pedal valve adapters (13) on brake pedal valve (10).

3. Install three studs (11) on brake pedal valve (10).

4. Install brake pedal valve (10) on firewall (9) with three new locknuts (1).

5. Install adapter (2) and two control valves (3) on firewall (9).

6. Install two adapters (6) on control valves (3).

7. Connect air hoses (7) and (8) to adapters (6).

8. Install two elbows (4) on control valves (3).

9. Connect two air hoses (5) to elbows (4).

3-198. BRAKE PEDAL (TREADLE) VALVE (M936/A1/A2) REPLACEMENT (Contd)

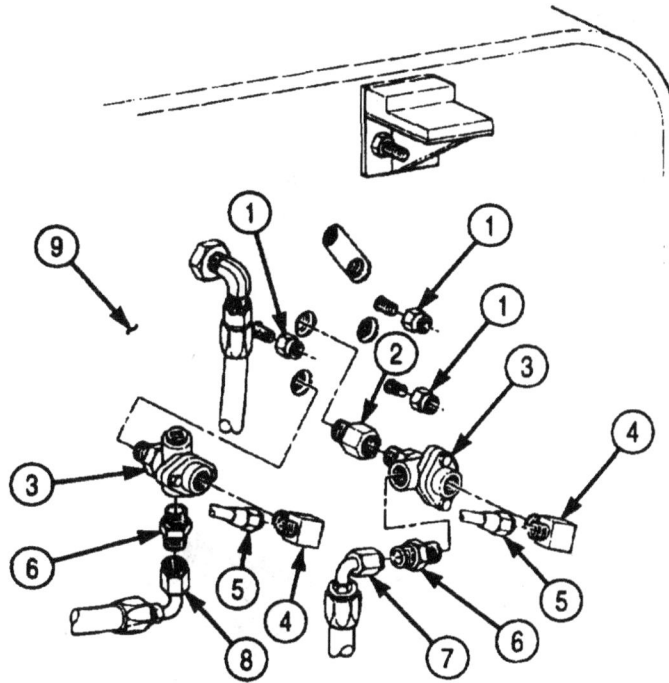

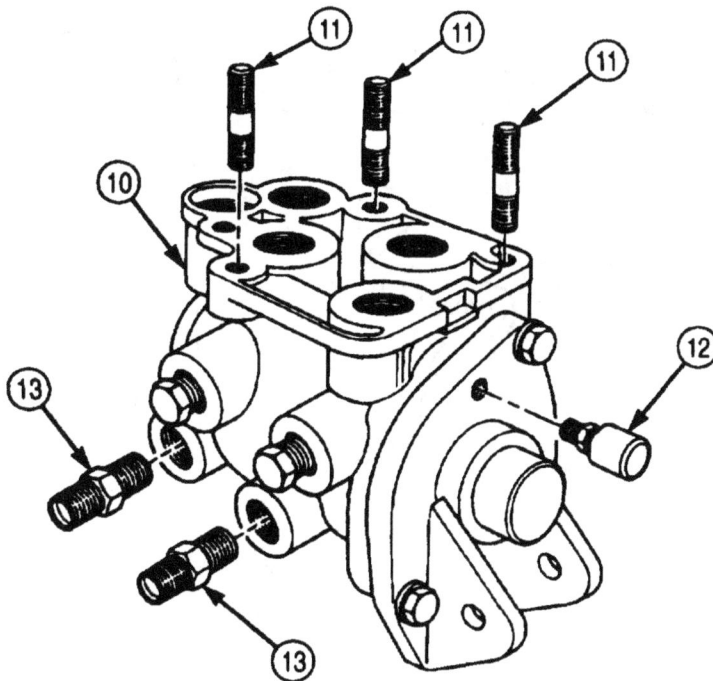

3-198. BRAKE PEDAL (TREADLE) VALVE (M936/A1/A2) REPLACEMENT (Contd)

10. Install adapter (12) and air hose (11) on firewall (3).
11. Install pipe (13) and tee (8) on firewall (3).
12. Install adapter (5) and air hose (6) on firewall (3).
13. Install adapters (7) and (9) on tee (8).
14. Connect air hoses (4) and (10) to adapters (7) and (9).
15. Connect two air lines (2) to brake pedal valve adapters (1).

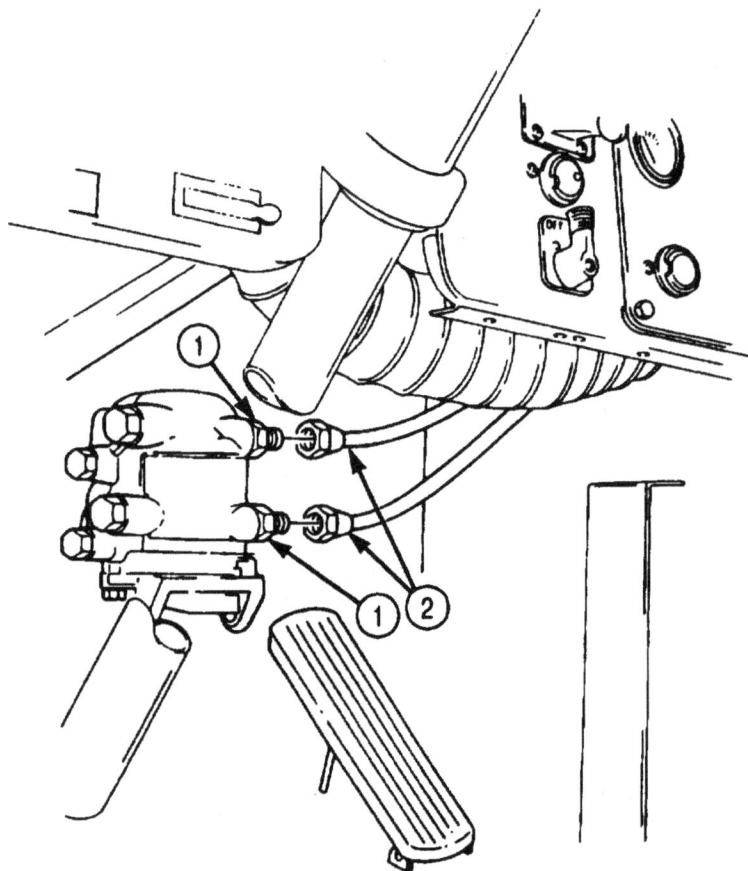

3-198. BRAKE PEDAL (TREADLE) VALVE (M936/A1/A2) REPLACEMENT (Contd)

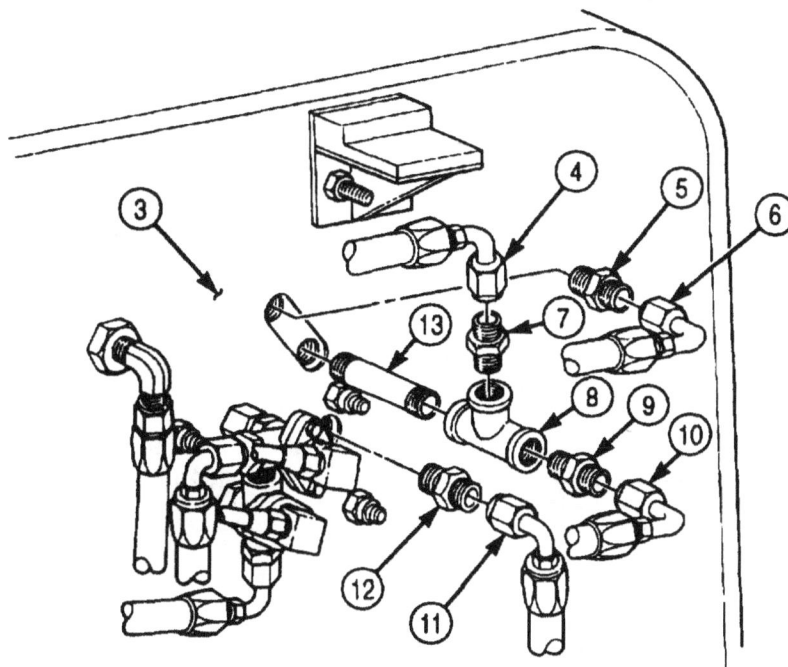

FOLLOW-ON TASKS: • Install protective control box (para. 3-115).
• Install brake pedal (para. 3-196).
• Start engine (TM 9-2320-272-10) and allow air pressure to build to normal operating range. Check for air leaks and proper brake operation. Road test vehicle.

3-199. BRAKE PROPORTIONING VALVE REPLACEMENT

THIS TASK COVERS:

a. Removal   b. Installation

INITIAL SETUP:

APPLICABLE MODELS
All

TOOLS
General mechanic's tool kit (Appendix E, Item 1)

MATERIALS/PARTS
Two locknuts (Appendix D, Item 299)
Two tiedown straps (Appendix D, Item 684)
Adhesive sealant (Appendix C, Item 4)

REFERENCES (TM)
TM 9-2320-272-10
TM 9-2320-272-24P

EQUIPMENT CONDITION
- Parking brake set (TM 9-2320-272-10).
- Air reservoirs drained (TM 9-2320-272-10).

GENERAL SAFETY INSTRUCTIONS
Do not disconnect air lines before draining air reservoirs.

---

a. Removal

WARNING

Do not disconnect air lines before draining air reservoirs. Small parts under pressure may shoot out with high velocity, causing injury to personnel.

NOTE

Tag air lines and fittings for installation.

1. Remove two tiedown straps (8) from air lines (1), (7), and (9). Discard tiedown straps (8).

2. Disconnect air lines (1), (3), (7), and (9) from elbow (10), adapter (4), elbow (5), and adapter (6).

3. Remove elbows (10) and (5) and adapters (4) and (6) from brake proportioning valve (2).

4. Remove two locknuts (11), screws (12), and brake proportioning valve (2) from bracket (13). Discard locknuts (11).

b. Installation

NOTE

- If new brake proportioning valve is being installed, use attaching parts and fittings from old brake proportioning valve.
- Fittings must be cleaned and inspected for cracks and stripped threads.
- Apply adhesive sealant to all male pipe threads before installation.

1. Install brake proportioning valve (2) on bracket (13) with two screws (12) and new locknuts (11).

2. Install elbows (10) and (5) and adapters (4) and (6) on brake proportioning valve (2).

3. Connect air lines (1), (3), (7), and (9) to elbow (10), adapter (4), elbow (5), and adapter (6).

4. Install two new tiedown straps (8) on air lines (1), (7), and (9).

3-199. BRAKE PROPORTIONING VALVE REPLACEMENT (Contd)

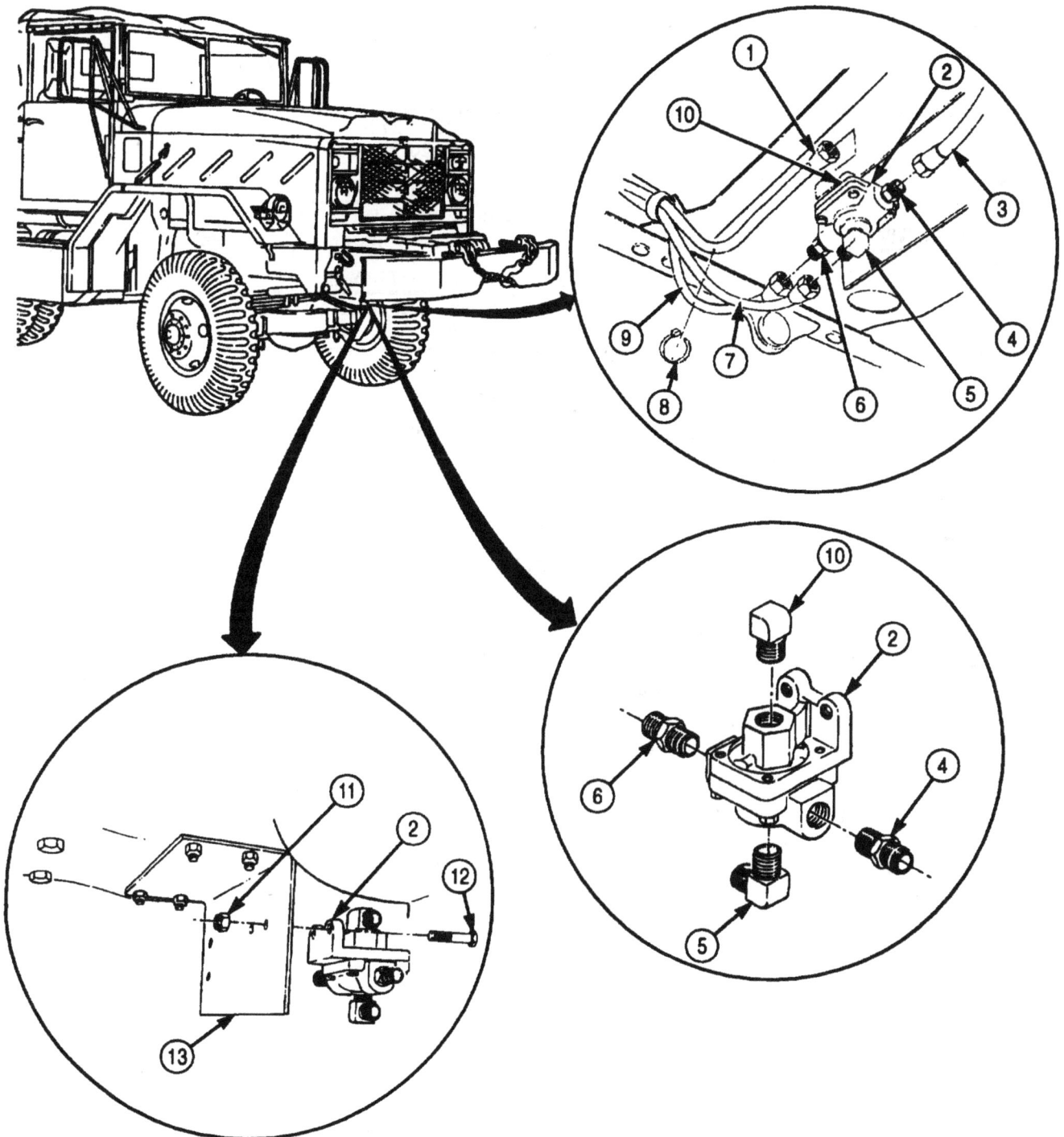

FOLLOW-ON TASK: Start engine (TM 9-2320-272-10) and allow air pressure to build to normal operating range. Check for air leaks and proper brake operation. Road test vehicle.

3-200. WET AIR RESERVOIR (SUPPLY TANK) AND BRACKET REPLACEMENT

THIS TASK COVERS:

a. Removal   b. Installation

INITIAL SETUP:

APPLICABLE MODELS
All except M936/A1/A2

TOOLS
General mechanic's tool kit (Appendix E, Item 1)

MATERIALS/PARTS
Eight locknuts (Appendix D, Item 291)
Antiseize tape (Appendix C, Item 72)

REFERENCES (TM)
TM 9-2320-272-10
TM 9-2320-272-24P

EQUIPMENT CONDITION
- Parking brake ret (TM 9-2320-272-10).
- Air reservoirs drained (TM 9-2320-272-10).

GENERAL SAFETY INSTRUCTIONS
Do not disconnect air lines before draining air reservoirs.

WARNING

Do not disconnect air lines before draining air reservoirs. Small parts under pressure may shoot out with high velocity, causing injury to personnel.

a. Removal

NOTE
Tag air liner for installation.

1. Disconnect drain line (12) from adapter (11).
2. Disconnect primary tank air line (14) from elbow (13).
3. Disconnect auxiliary air line (1) from elbow (2).
4. Disconnect trailer emergency air line (6) from elbow (3).
5. Disconnect secondary tank air line (4) from adapter (5).
6. Disconnect supply tank input air line (15) from elbow (16).
7. Remove four locknuts (8), two U-bolts (7), and supply tank (9) from two support brackets (10). Discard locknuts (8).
8. Remove four locknuts (18), screws (17), and two support brackets (10) from frame brace (19). Discard locknuts (18).

3-200. WET AIR RESERVOIR (SUPPLY TANK) AND BRACKET REPLACEMENT (Contd)

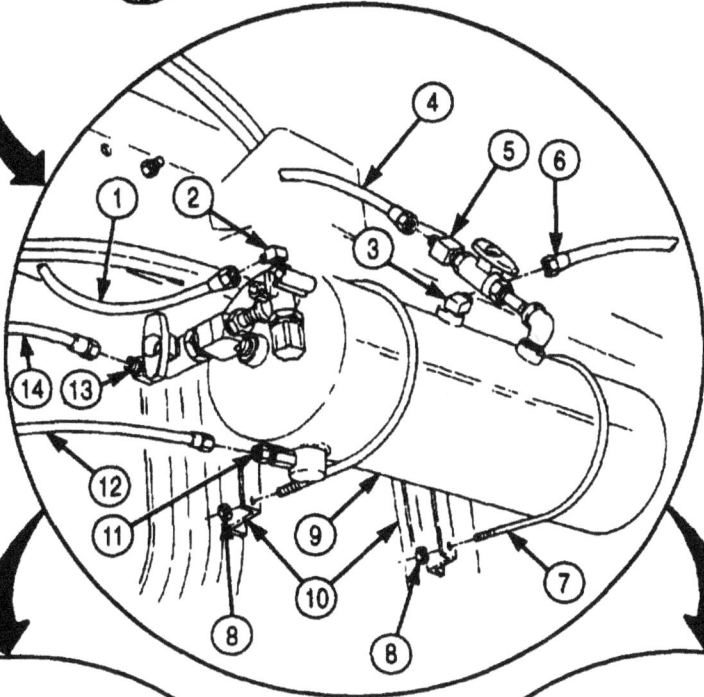

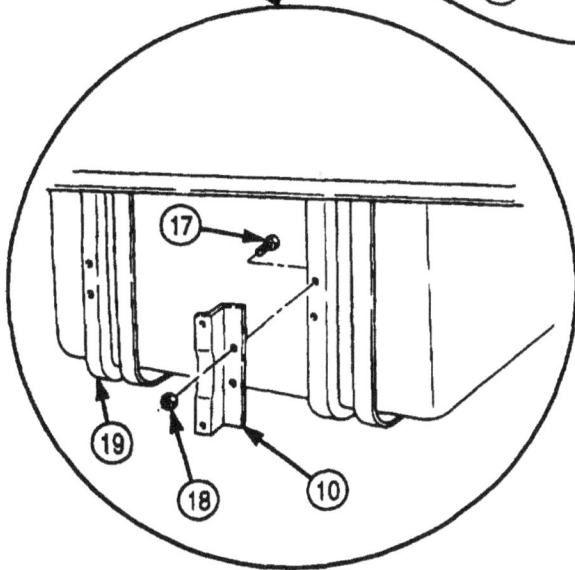

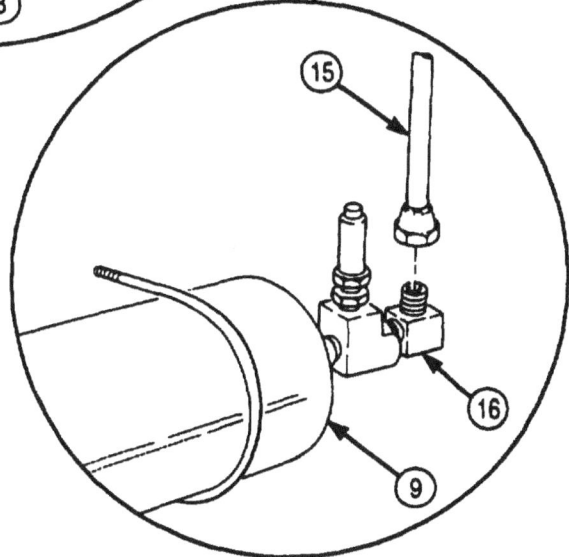

3-200. WET AIR RESERVOIR (SUPPLY TANK) AND BRACKET REPLACEMENT (Contd)

NOTE
Perform steps 9 through 14 for M934/A1/A2 vehicles.

9. Disconnect secondary tank air line (1) from adapter (2).

10. Disconnect trailer emergency air line (5) from elbow (6).

11. Remove four screws (11) and support plate (9) from frame (12).

12. Remove four locknuts (10), two U-bolts (8), and supply tank (7) from support plate (9). Discard locknuts (10).

CAUTION

Open-end wrench must be used to anchor tank boss fittings when connecting or disconnecting associated fittings. Damage to tank bosses may result if open-end wrench is not used.

13. Remove adapter (2), shutoff valve (3), and elbow (4) from supply tank (7).

14. Remove elbow (6) from supply tank (7).

15. Remove adapters (26) and (25) from supply tank (7).

16. Remove elbow (18), relief valve (19), nipple (17), and reducer (16) from tee (15).

17. Remove elbow (13), shutoff valve (14), tee (15), and nipple (27) from supply tank (7).

18. Remove adapter (21), shutoff valve (22), nipple (23), and elbow (24) from supply tank (7).

19. Remove elbow (20) from supply tank (7).

20. Remove elbow (30), safety valve (28), reducer (29), and tee (31) from supply tank (7).

3-200. WET AIR RESERVOIR (SUPPLY TANK) AND BRACKET REPLACEMENT (Cod)

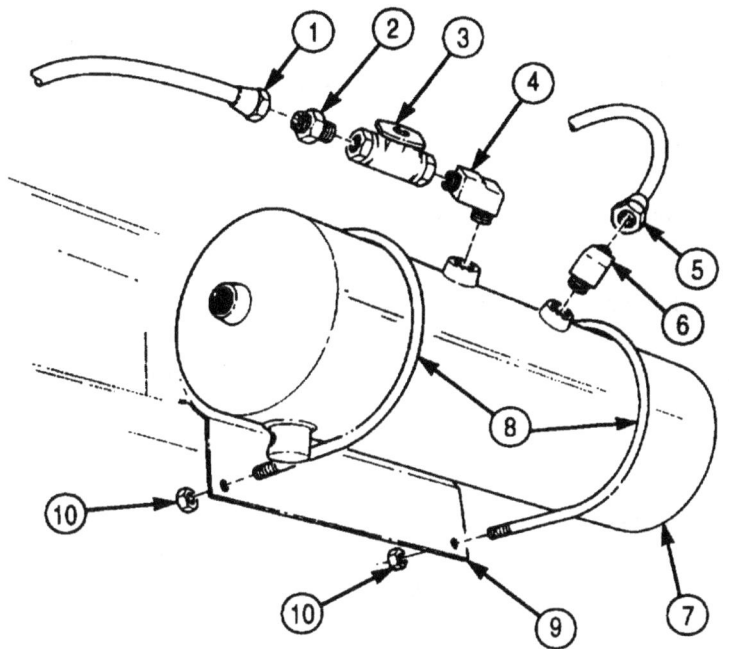

M934/A1/A2 VEHICLE

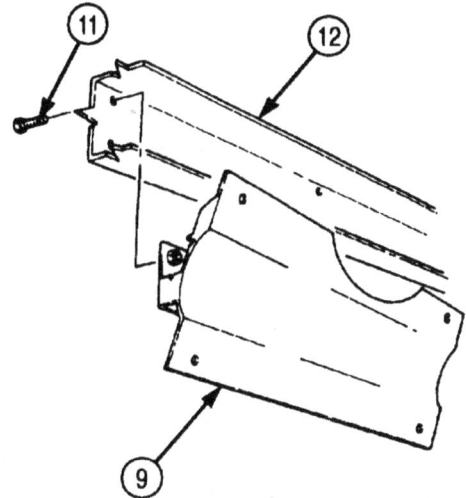

VAN MODEL M934/A1/A2

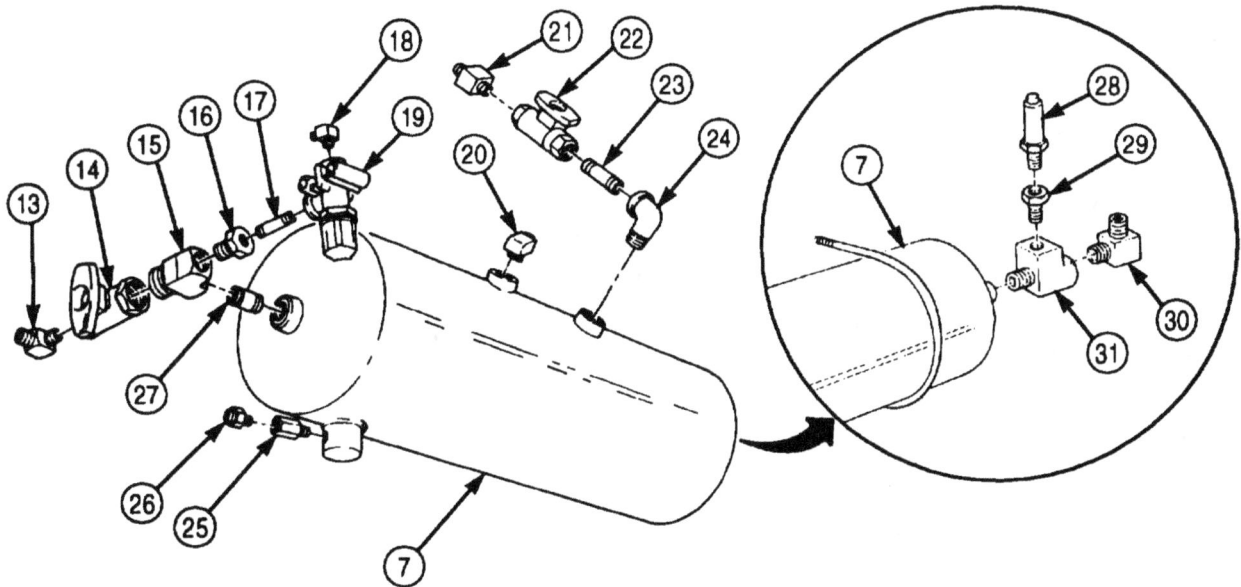

3-200. WET AIR RESERVOIR (SUPPLY TANK) AND BRACKET REPLACEMENT (Contd)

b. Installation

CAUTION

Open-end wrench must be used to anchor tank boss fittings when connecting or disconnecting associated fittings. Damage to tank bosses may result if open-end wrench is not used.

NOTE

- If new reservoir is being installed, fittings from old reservoir may be used. Fittings must be cleaned and inspected for cracks and stripped threads.
- Wrap male pipe threads with antiseize tape before installation.

1. Install adapters (18) and (19) on supply tank (17).
2. Install nipple (20), tee (7), reducer (8), nipple (9), relief valve (11), elbow (10), shutoff valve (6), and elbow (5) on supply tank (17).
3. Install elbow (16), nipple (15), shutoff valve (14), and adapter (13) on supply tank (17).
4. Install elbow (12) on supply tank (17).
5. Install tee (26), reducer (24), safety valve (23), and elbow (25) on supply tank (17).
6. Install two support brackets (4) on frame braces (1) with four screws (2) and new locknuts (3).
7. Install supply tank (17) on two support brackets (4) with two U-bolts (22) and four new locknuts (21).

NOTE

Perform steps 8 through 11 for M934/A1/A2 vehicles.

8. Install elbow (30) on supply tank (17).
9. Install elbow (29), shutoff valve (28), and adapter (27) on supply tank (17).
10. Install support plate (31) on frame (35) with four screws (34).
11. Install supply tank (17) on support plate (31) with two U-bolts (33) and four new locknuts (32).

3-200. WET AIR RESERVOIR (SUPPLY TANK) AND BRACKET REPLACEMENT (Contd)

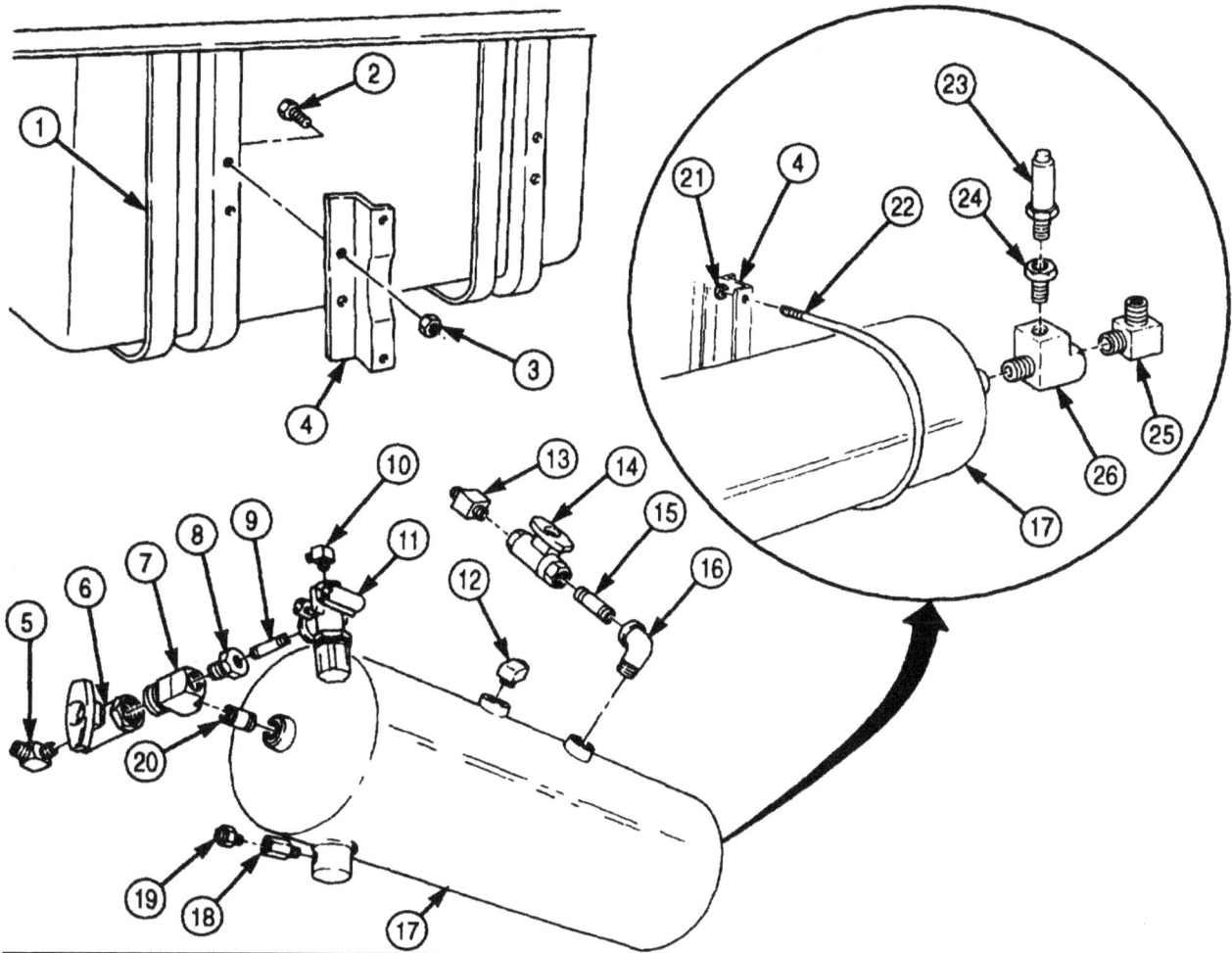

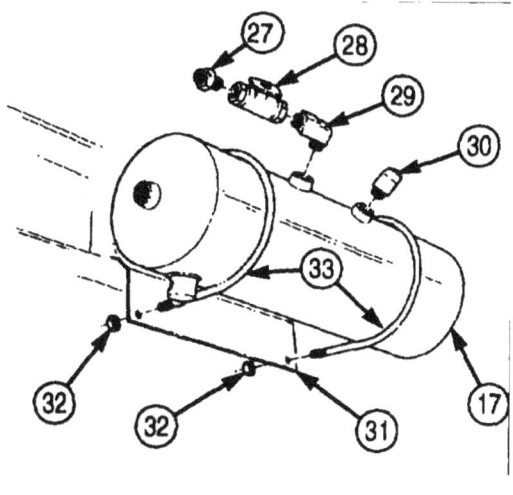

M934/A1/A2 VEHICLE

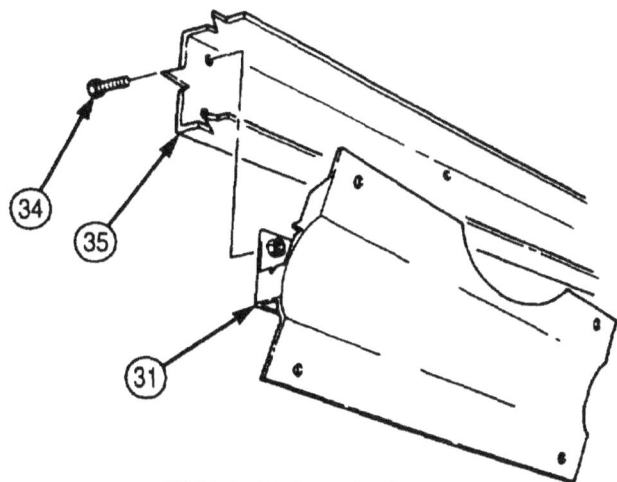

VAN MODEL M934/A1/A2

3-200. WET AIR RESERVOIR (SUPPLY TANK) AND BRACKET REPLACEMENT (Contd)

12. Connect supply tank input air line (11) to elbow (12).
13. Connect secondary tank air line (3) to adapter (5).
14. Connect trailer emergency air line (6) to elbow (4).
15. Connect auxiliary air line (1) to elbow (2).
16. Connect primary tank air line (10) to elbow (9).
17. Connect drain line (8) to adapter (7).
18. Connect trailer emergency air line (16) to elbow (15).
19. Connect secondary tank air line (13) to adapter (14).

3-200. WET AIR RESERVOIR (SUPPLY TANK) AND BRACKET REPLACEMENT (Contd)

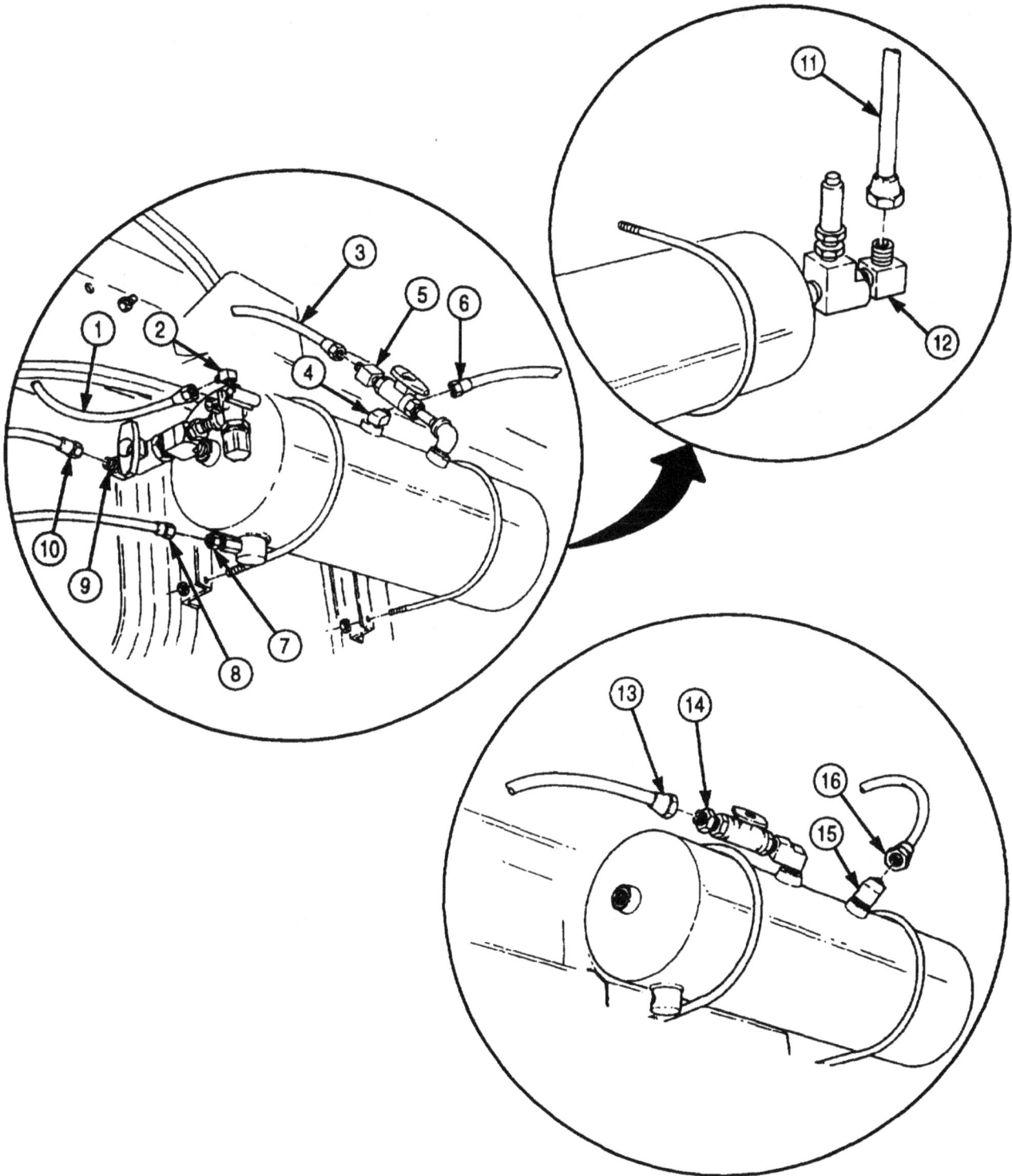

FOLLOW-ON TASK: Start engine (TM 9-2320-272-10) and allow air pressure to build to normal operating range. Check for air leaks. Road test vehicle.

3-201. WET AIR RESERVOIR (SUPPLY TANK) AND MOUNTING PLATE (M936/A1/A2) REPLACEMENT

THIS TASK COVERS:

a. Removal   b. Installation

INITIAL SETUP:

APPLICABLE MODELS
M936/A1/A2

TOOLS
General mechanic's tool kit (Appendix E, Item 1)

MATERIALS/PARTS
Four locknuts (Appendix D, Item 291)
Antiseize tape (Appendix C, Item 72)

REFERENCES (TM)
TM 9-2320-272-10
TM 9-2320-272-24P

EQUIPMENT CONDITION
- Parking brake set (TM 9-2320-272-10).
- Air reservoirs drained (TM 9-2320-272-10).

GENERAL SAFETY INSTRUCTIONS
Do not disconnect air lines before draining air reservoirs.

WARNING

Do not disconnect air lines before draining air reservoirs. Small parts under pressure may shoot out with high velocity, causing injury to personnel.

CAUTION

Anchor tank boss fittings with an open-end wrench before removing connecting fittings. Failure to do so may result in damage to equipment.

a. Removal

1. Disconnect primary tank air line (1) from elbow (2).
2. Disconnect trailer emergency air line (5) from elbow (6).
3. Disconnect auxiliary air line (3) from elbow (4).
4. Disconnect supply tank input air line (16) from adapter (15).
5. Disconnect secondary tank air line (14) from adapter (13).
6. Disconnect drain line (12) from adapter (11).
7. Remove four locknuts (9), two U-bolts (8), and supply tank (7) from support plate (10). Discard locknuts (9).

3-201. WET AIR RESERVOIR (SUPPLY TANK) AND MOUNTING PLATE (M936/A1/A2) REPLACEMENT (Contd)

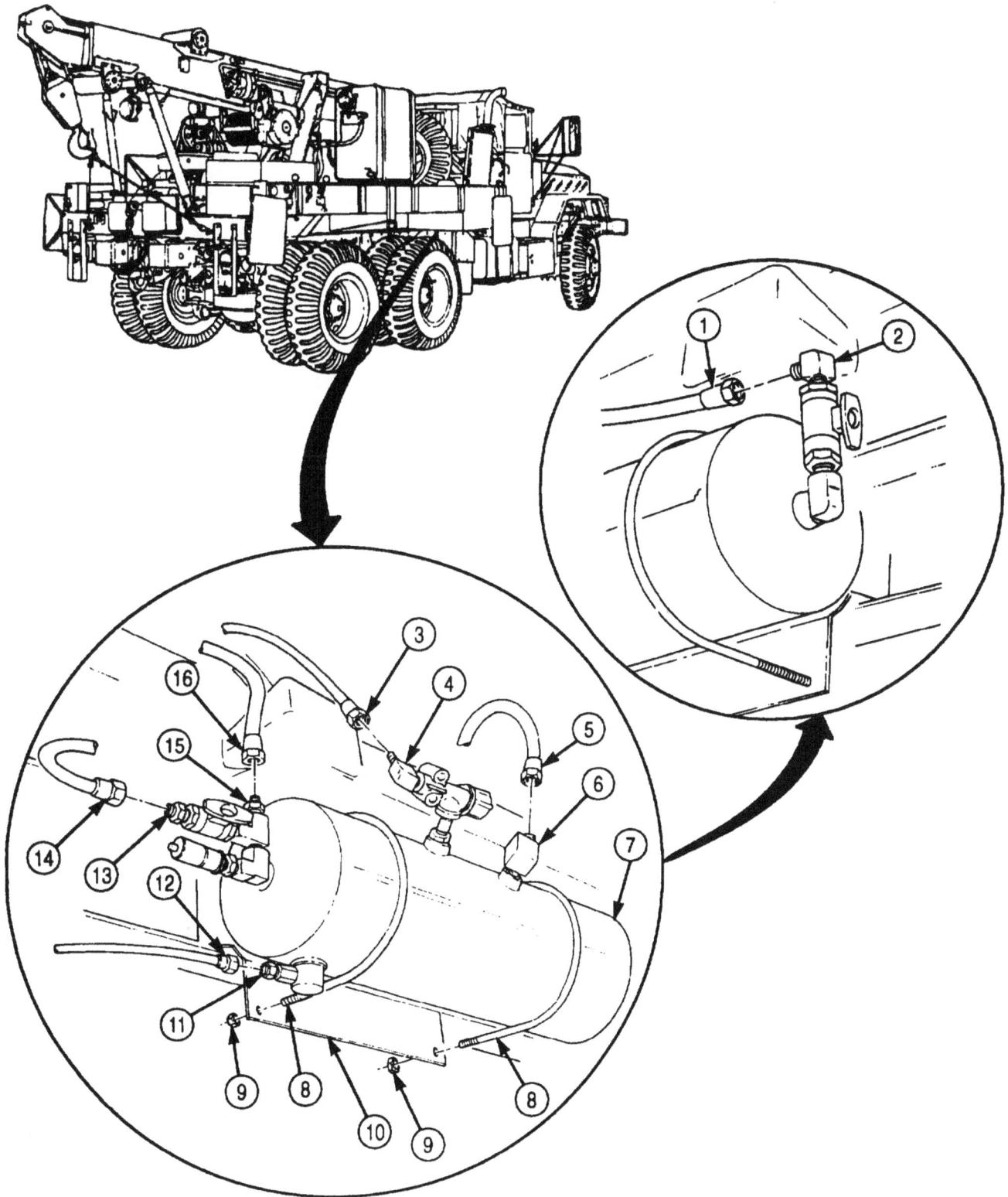

3-201. WET AIR RESERVOIR (SUPPLY TANK) AND MOUNTING PLATE (M936/A1/A2) REPLACEMENT (Contd)

8. Remove four screws (1) and support plate (3) from frame (2).
9. Remove elbow (20), shutoff valve (21), and elbow (22) from supply tank (14).
10. Remove elbow (13) from supply tank (14).
11. Remove elbow (9), relief valve (10), nipple (11), and reducer (12) from supply tank (14).
12. Remove adapters (16) and (15) from supply tank (14).
13. Remove adapters (7) and (4), safety valve (19), reducer (18), shutoff valve (5), tees (8) and (17), and nipple (6) from supply tank (14).

b. Installation

NOTE
- If new reservoir is being installed, fittings from old reservoir may be used. Fittings must be cleaned and inspected for cracks and stripped threads.
- Wrap male pipe threads with antiseize tape before installation.

1. Install nipple (6), tees (8) and (17), shutoff valve (5), reducer (18), safety valve (19), and adapters (7) and (4) on supply tank (14).
2. Install adapters (15) and (16) on supply tank (14).
3. Install reducer (12), nipple (11), relief valve (10), and elbow (9) on supply tank (14).
4. Install elbow (13) on supply tank (14).
5. Install elbow (22), shutoff valve (21), and elbow (20) on supply tank (14).
6. Install support plate (3) on frame (2) with four screws (1).

3-201. WET AIR RESERVOIR (SUPPLY TANK) AND MOUNTING PLATE (M936/A1/A2) REPLACEMENT (Contd)

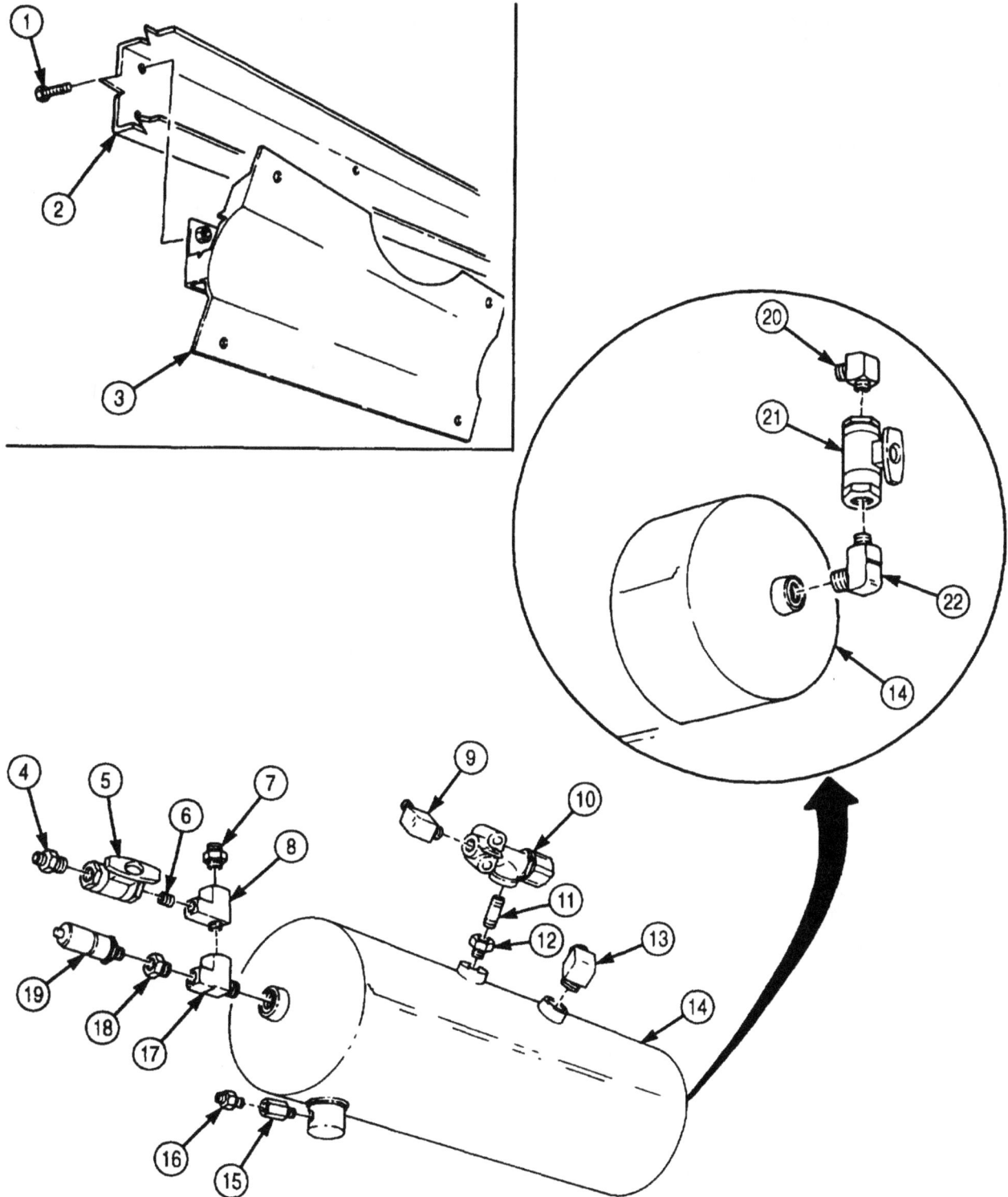

3-201. WET AIR RESERVOIR (SUPPLY TANK) AND MOUNTING PLATE (M936/A1/A2) REPLACEMENT (Contd)

7. Install supply tank (6) on support plate (9) with two U-bolts (7) and four new locknuts (8).

8. Connect drain line (11) to adapter (10).

9. Connect secondary tank air line (13) to adapter (12).

10. Connect supply tank input air line (1) to adapter (14).

11. Connect auxiliary air line (2) to elbow (3).

12. Connect trailer emergency air line (4) to elbow (5).

13. Connect primary tank air line (15) to elbow (16).

3-201. WET AIR RESERVOIR (SUPPLY TANK) AND MOUNTING PLATE (M936/A1/A2) REPLACEMENT (Contd)

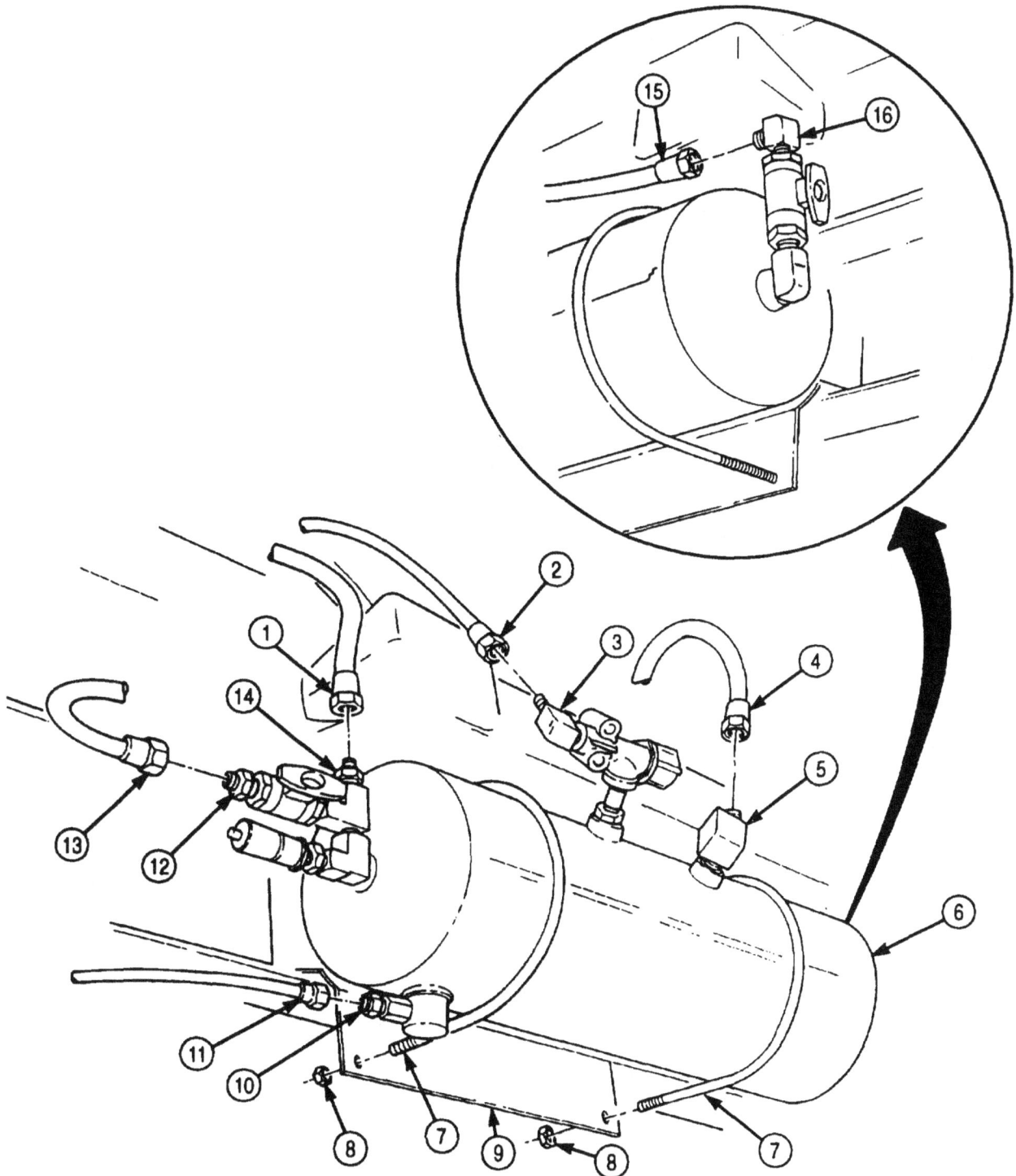

FOLLOW-ON TASK: Start engine (TM 9-2320-272-10) and allow air pressure to build to normal operating range. Check for air leaks. Road test vehicle.

3-202. PRIMARY AIR RESERVOIR (SUPPLY TANK) REPLACEMENT

THIS TASK COVERS:

a. Removal   b. Installation

INITIAL SETUP:

APPLICABLE MODELS
All except M936/A1/A2

TOOLS
General mechanic's tool kit (Appendix E, Item 1)

MATERIALS/PARTS
Four locknuts (Appendix D, Item 291)
Antiseize tape (Appendix C, Item 72)

REFERENCES (TM)
TM 9-2320-272-10
TM 9-2320-272-24P

EQUIPMENT CONDITION
- Parking brake set (TM 9-2320-272-10).
- Air reservoirs drained (TM 9-2320-272-10).
- Toolbox and step removed (para. 3-302).
- Primary low air pressure switch removed (para. 3-99).

GENERAL SAFETY INSTRUCTIONS
- Do not disconnect air lines before draining air reservoirs.
- Do not touch hot exhaust system components with bare hands.

WARNING

- Do not disconnect air lines before draining air reservoirs. Small parts under pressure may shoot out with high velocity, causing injury to personnel.
- Do not touch hot exhaust system components with bare hands. Injury to personnel may result.

a. Removal

NOTE
- Primary reservoir is located below the cab on right side frame rail above secondary reservoir.
- Tag air lines for installation.

1. Disconnect wet reservoir supply line (6) from elbow (7).
2. Disconnect drain line (8) from elbow (9).
3. Disconnect primary relay valve supply line (1) from elbow (10).
4. Disconnect supply line (2) from adapter (4).
5. Disconnect treadle valve supply line (3) from adapter (4).

NOTE
Assistant will help with steps 6 and 7.

6. Remove two locknuts (11) and screws (13) from clamps (12). Discard locknuts (11).
7. Remove two locknuts (11), screws (13), clamps (12), and reservoir (5) from clamps (14). Discard locknuts (11).

3-202. PRIMARY AIR RESERVOIR (SUPPLY TANK) REPLACEMENT (Contd)

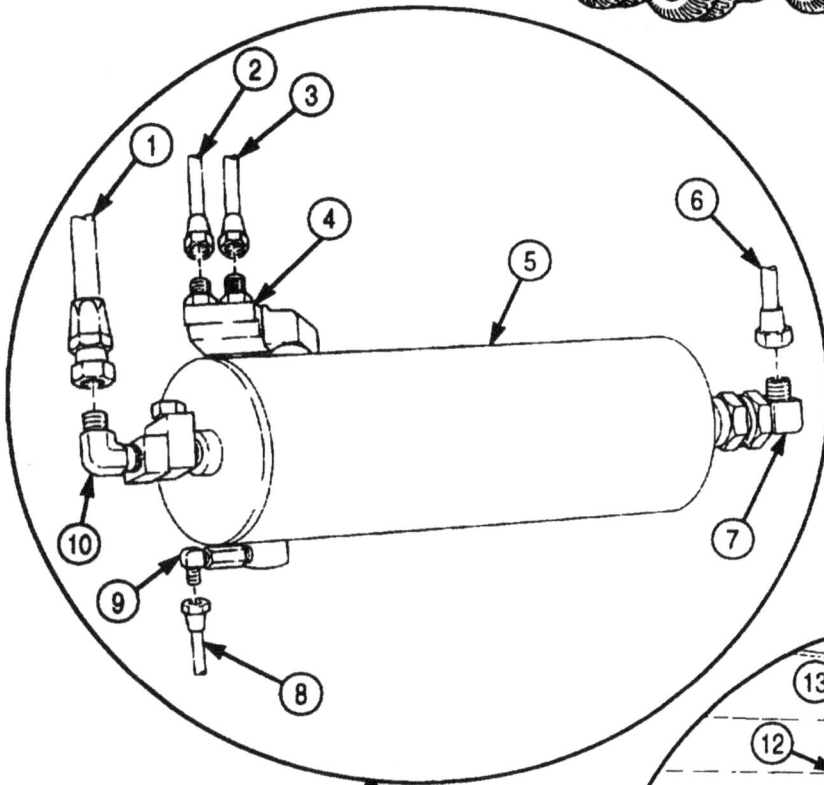

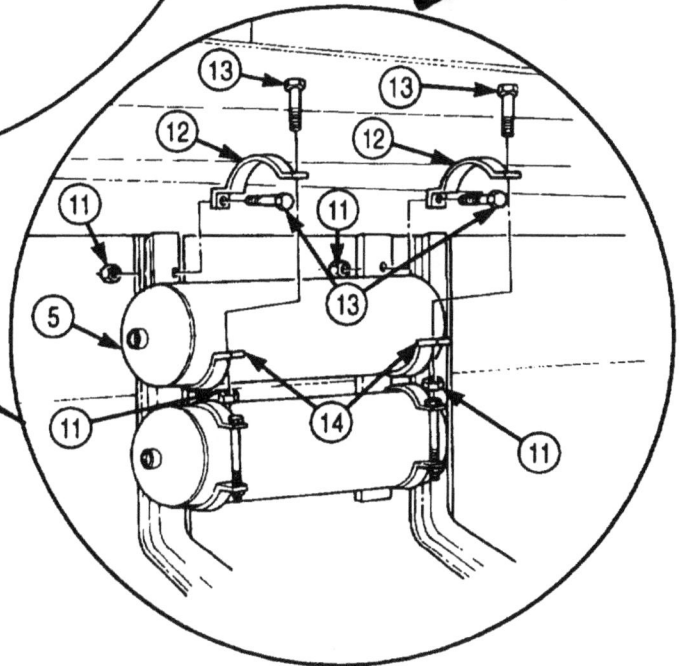

3-202. PRIMARY AIR RESERVOIR (SUPPLY TANK) REPLACEMENT (Contd)

b. Installation

NOTE

- If installing new reservoir, use old reservoir fittings.
- Wrap all male pipe threads with antiseize tape before installation.

1. Install reservoir (5) on two clamps (14) with clamps (12), screws (11), and new locknuts (13).
2. Install two screws (11) and new locknuts (13) on clamps (12).
3. Connect treadle valve supply line (3) to adapter (4).
4. Connect supply line (2) to adapter (4).
5. Connect primary relay valve supply line (1) to elbow (10).
6. Connect drain line (8) to elbow (9).
7. Connect wet reservoir supply line (6) to elbow (7).

3-202. PRIMARY AIR RESERVOIR (SUPPLY TANK) REPLACEMENT (Contd)

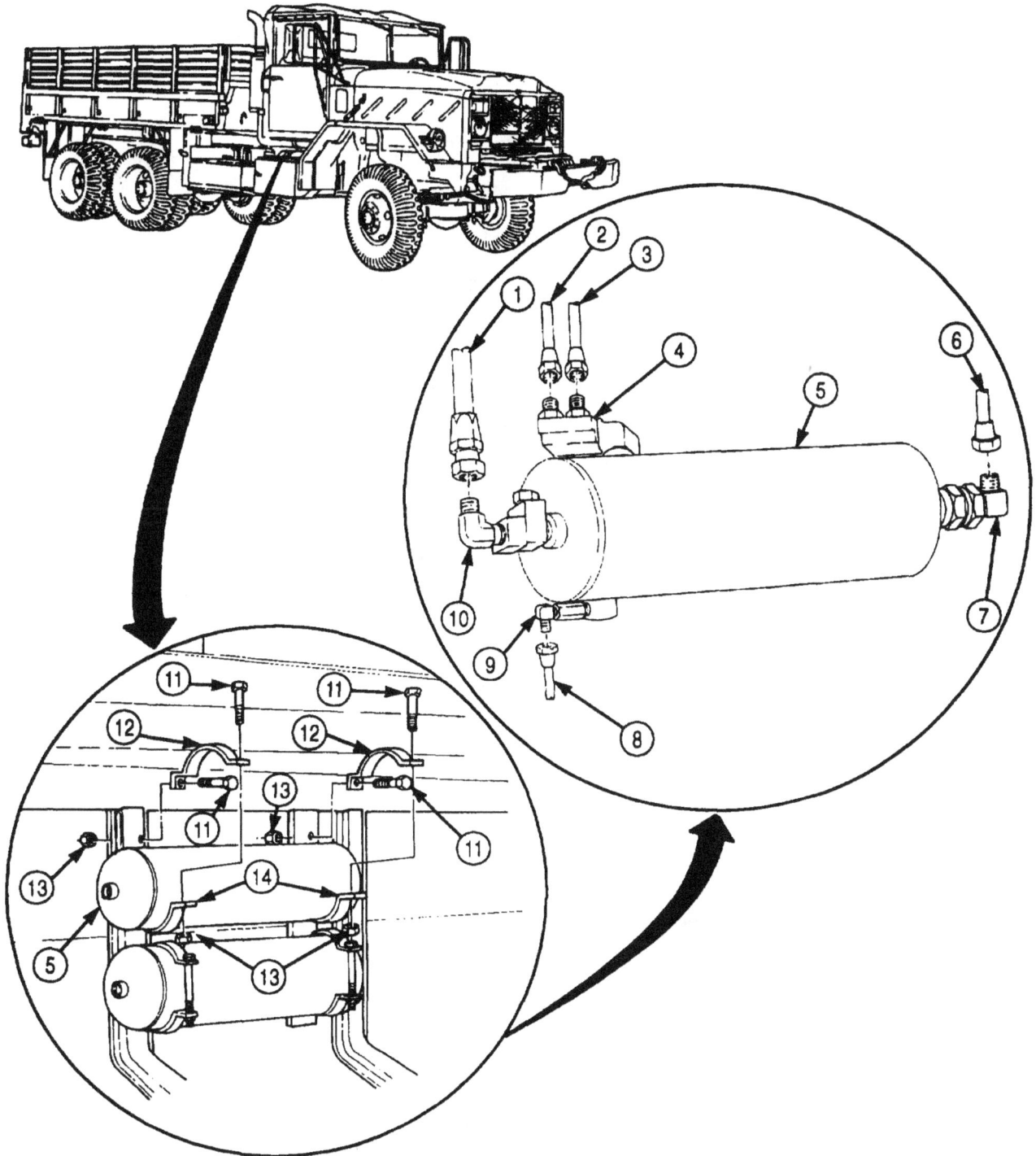

FOLLOW-ON TASKS:
- Install primary low air pressure switch (para. 3-99).
- Install toolbox and step (para. 3-302).
- Start engine (TM 9-2320-272-10) and allow air pressure to build to normal operating range. Check for air leaks at primary air reservoir. Road test vehicle.

3-203. SECONDARY AIR RESERVOIR (SUPPLY TANK) REPLACEMENT

THIS TASK COVERS:
a. Removal   b. Installation

INITIAL SETUP:

APPLICABLE MODELS
All except M936/A1/A2

TOOLS
General mechanic's tool kit (Appendix E, Item 1)
Soft-jawed vise

MATERIALS/PARTS
Four locknuts (Appendix D, Item 291)
Antiseize tape (Appendix C, Item 72)

REFERENCES (TM)
TM 9-2320-272-10
TM 9-2320-272-24P

EQUIPMENT CONDITION
- Parking brake set (TM 9-2320-272-10).
- Air reservoirs drained (TM 9-2320-272-10).
- Toolbox and step removed (para. 3-302).

GENERAL SAFETY INSTRUCTIONS
- Do not disconnect air lines before draining air reservoirs.
- Do not touch hot exhaust system components with bare hands.

WARNING

- Do not disconnect air lines before draining air reservoirs. Small parts under pressure may shoot out with high velocity, causing injury to personnel.
- Do not touch hot exhaust system components with bare hands. Injury to personnel may result.

a. Removal

NOTE
- Secondary reservoir is located below the cab on the right side frame rail below primary reservoir.
- Tag air lines for installation.

1. Disconnect wet reservoir supply line (8) from elbow (7).
2. Disconnect drainvalve line (11) from adapter (10).
3. Disconnect supply line (14) from elbow (13).
4. Disconnect supply line (15) from adapter (16).
5. Disconnect two wires (1) from secondary low-air pressure switch (2).

NOTE
Assistant will support reservoir during steps 6 and 7.

6. Remove four locknuts (20), screws (18), and two support clamps (19) from hangers (17). Discard locknuts (20).
7. Remove secondary reservoir (5) from two hangers (17) and place in soft-jawed vise.
8. Remove secondary low-air pressure switch (2), adapter (3), and fitting (4) from secondary reservoir (5).
9. Remove elbow (7) and check valve (6) from secondary reservoir (5).
10. Remove adapters (10) and (9) from secondary reservoir (5).
11. Remove elbow (13), adapter (16), and fitting (12) from secondary reservoir (5).
12. Remove secondary reservoir (5) from vise.

3-203. SECONDARY AIR RESERVOIR (SUPPLY TANK) REPLACEMENT (Contd)

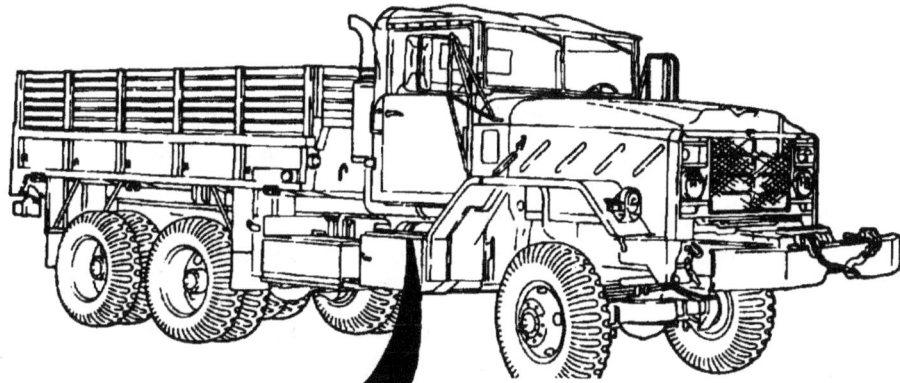

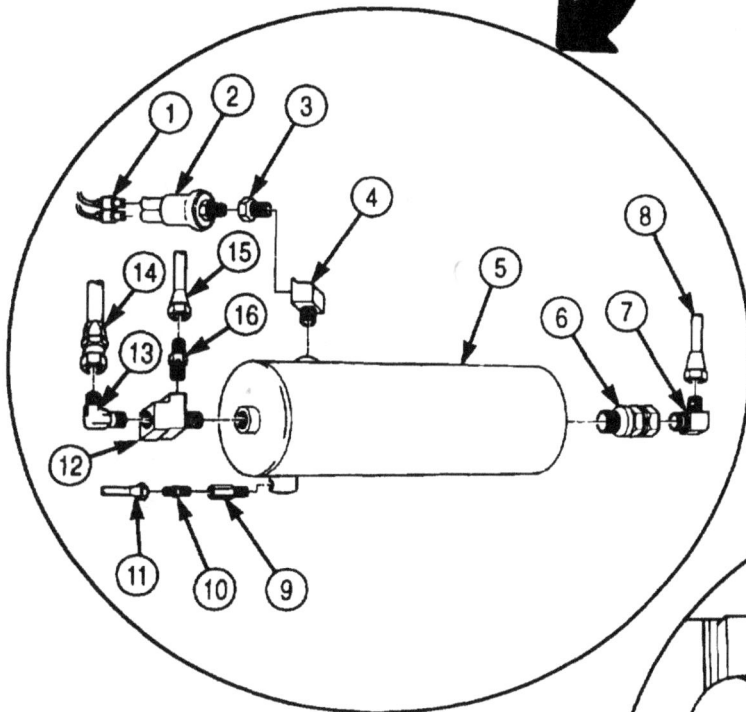

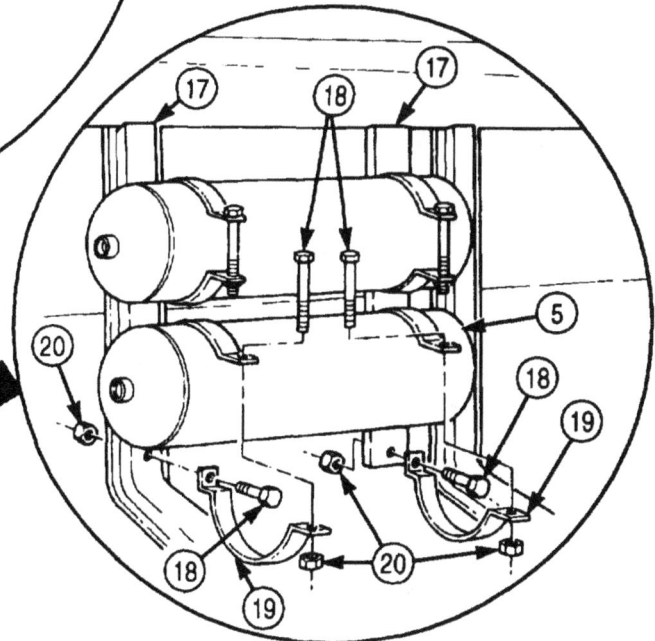

3-203. SECONDARY AIR RESERVOIR (SUPPLY TANK) REPLACEMENT (Contd)

b. Installation

NOTE
- If installing new reservoir, use old reservoir fittings.
- Wrap all male pipe threads with antiseize tape before installation.

1. Place secondary reservoir (5) in soft-jawed vise.
2. Install fitting (12), adapter (16), and elbow (13) on secondary reservoir (5).
3. Install adapters (9) and (10) on secondary reservoir (5).
4. Install check valve (6) and elbow (8) on secondary reservoir (5).
5. Install fitting (4), adapter (3), and secondary low-air pressure switch (2) on secondary reservoir (5).
6. Remove secondary reservoir (5) from soft-jawed vise.

NOTE
Assistant will support reservoir during step 7.

7. Install secondary reservoir (5) on two hangers (17) with support clamps (19), four screws (18), and new locknuts (20).
8. Connect two wires (1) to secondary low-air pressure switch (2).
9. Connect supply line (15) to adapter (16).
10. Connect supply line (14) to elbow (13).
11. Connect drainvalve line (11) to adapter (10).
12. Connect wet reservoir supply line (7) to elbow (8).

3-203. SECONDARY AIR RESERVOIR (SUPPLY TANK) REPLACEMENT (Contd)

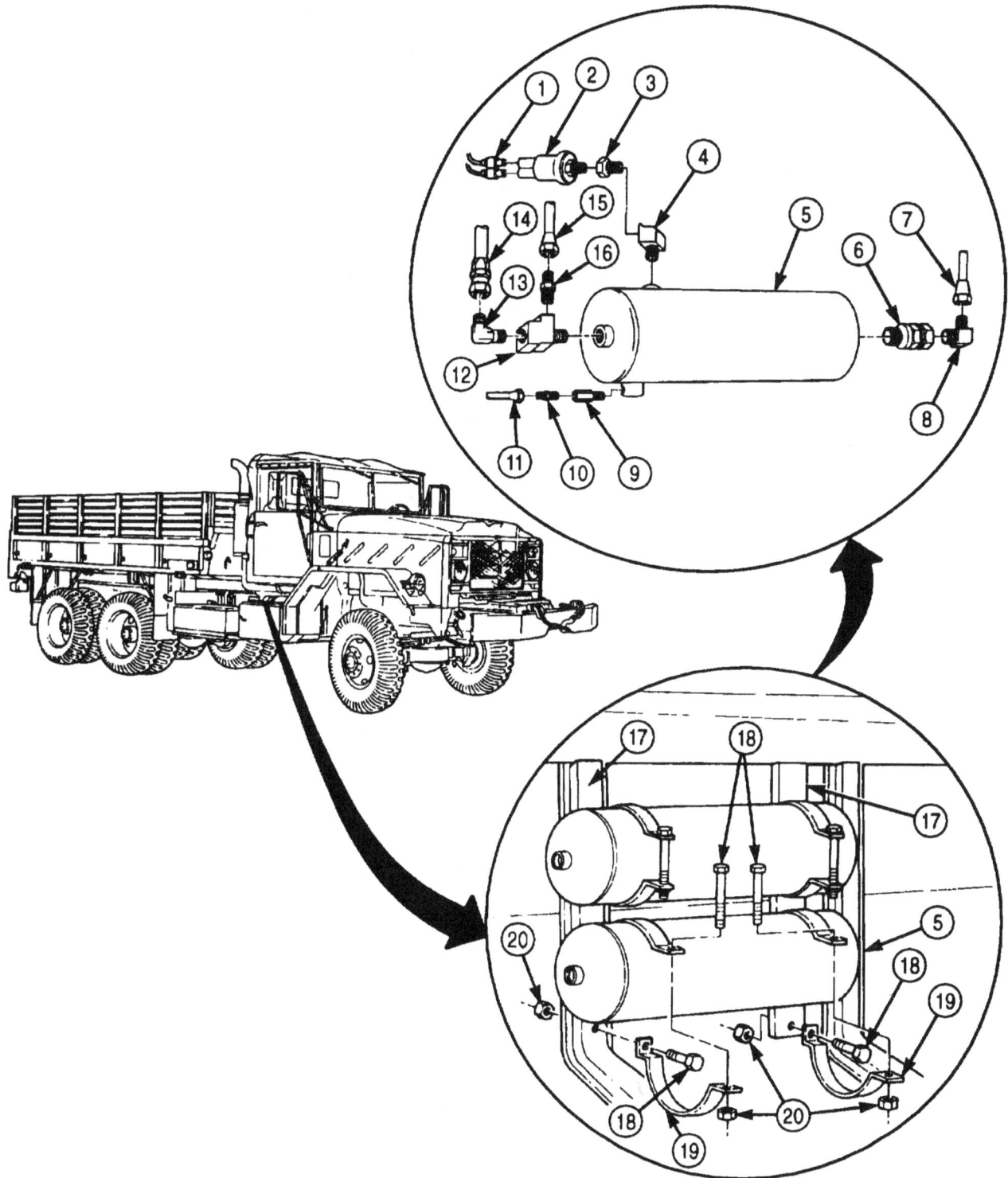

FOLLOW-ON TASKS: • Install toolbox and step (para. 3-302).
• Start engine (TM 9-2320-272-10) and allow air pressure to build to normal operating range. Check for air leaks at secondary air reservoir. Road test vehicle.

3-204. EMERGENCY SPRING BRAKE (SUPPLY TANK) AIR RESERVOIR REPLACEMENT

THIS TASK COVERS:

a. Removal   b. Installation

INITIAL SETUP:

APPLICABLE MODELS
All

TOOLS
General mechanic's tool kit (Appendix E, Item 1)

MATERIALS/PARTS
Four locknuts (Appendix D, Item 276)
Antiseize tape (Appendix C, Item 72)

REFERENCES (TM)
TM 9-2320-272-10
TM 9-2320-272-24P

EQUIPMENT CONDITION
- Parking brake set (TM 9-2320-272-10).
- Air reservoirs drained (TM 9-2320-272-10).
- Toolbox and step removed (para. 3-302).

GENERAL SAFETY INSTRUCTIONS
- Do not disconnect air lines before draining air reservoirs.
- Do not touch hot exhaust system components with bare hands.

WARNING

- Do not disconnect air lines before draining air reservoirs. Small parts under pressure may shoot out with high velocity, causing injury to personnel.
- Do not touch hot exhaust system components with bare hands. Injury to personnel may result.

a. Removal

NOTE
- Spring brake reservoir is located on the left frame rail next to air cleaner.
- Tag air lines for installation.

1. Disconnect air lines (5), (6), and (7) from elbows (10), (8), and (9).
2. Remove four locknuts (4), washers (3), two U-bolts (11), and emergency reservoir (1) from two support brackets (2). Discard locknuts (4).

b. Installation

NOTE
- If installing new reservoir, use old reservoir fittings. Wrap all male pipe threads with antiseize tape before installation.
- When installing emergency tank, ensure emergency tank drainvalve port faces downward and toward front of vehicle.

1. Install emergency reservoir (1) on two support brackets (2) with two U-bolts (11), four washers (3), and new locknuts (4).
2. Connect air lines (5), (6), and (7) to elbows (10), (8), and (9).

3-204. EMERGENCY SPRING BRAKE (SUPPLY TANK) AIR RESERVOIR REPLACEMENT (Contd)

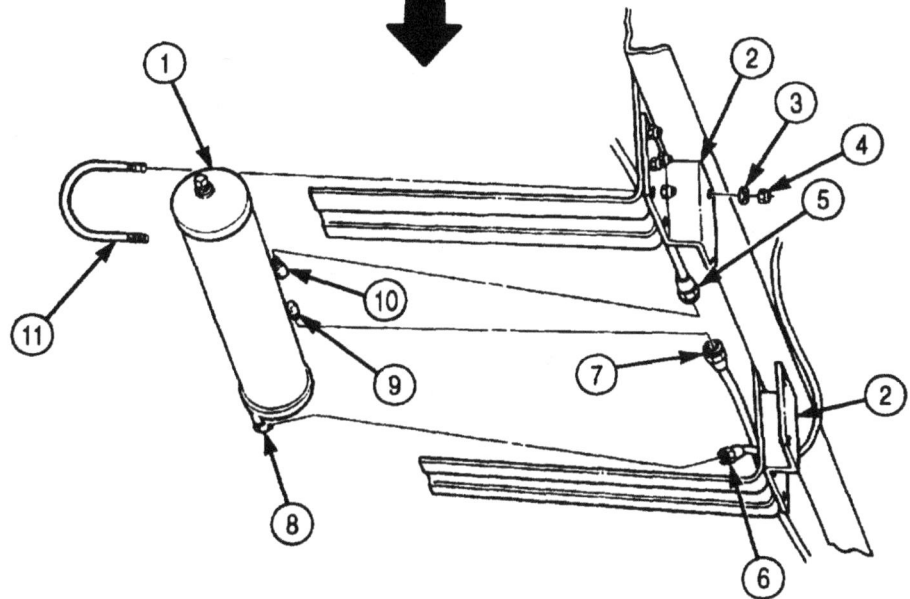

FOLLOW-ON TASKS:
- Install toolbox and step (para. 3-302).
- Start engine (TM 9-2320-272-10) and allow air pressure to build to normal operating range. Check for air leaks. Road test vehicle.

3-205. AIR RESERVOIR DRAINVALVES REPLACEMENT

THIS TASK COVERS:

a. Removal   b. Installation

INITIAL SETUP:

APPLICABLE MODELS
All

TOOLS
General mechanic's tool kit (Appendix E, Item 1)

MATERIALS/PARTS
Two locknuts (Appendix D, Item 274)
Antiseize tape (Appendix C, Item 72)

REFERENCES (TM)
TM 9-2320-272-10
TM 9-2320-272-24P

EQUIPMENT CONDITION
- Parking brake set (TM 9-2320-272-10).
- Air reservoirs drained (TM 9-2320-272-10).

GENERAL SAFETY INSTRUCTIONS
Do not disconnect air lines before draining air reservoirs.

WARNING

Do not disconnect air lines before draining air reservoirs. Small parts under pressure may shoot out with high velocity, causing injury to personnel.

a. Removal

NOTE
If removing more than one drainvalve, tag air lines for installation.

1. Disconnect air drain line (8) from coupling adapter (7).
2. Remove two locknuts (2), U-bolt (6), and drainvalve coupling (5) from bracket (1) at rear of right cab access step (4). Discard locknuts (2).
3. Remove coupling adapter (7) and drainvalve (3) from drainvalve coupling (5).

b. Installation

NOTE
Wrap all male pipe threads with antiseize tape before installation.

1. Install drainvalve (3) and coupling adapter (7) on drainvalve coupling (5).
2. Install drainvalve coupling (5) on bracket (1) of right cab access step (4) with U-bolt (6) and two new locknuts (2).
3. Connect air drain line (8) to coupling adapter (7).

3-205. AIR RESERVOIR DRAINVALVES REPLACEMENT (Contd)

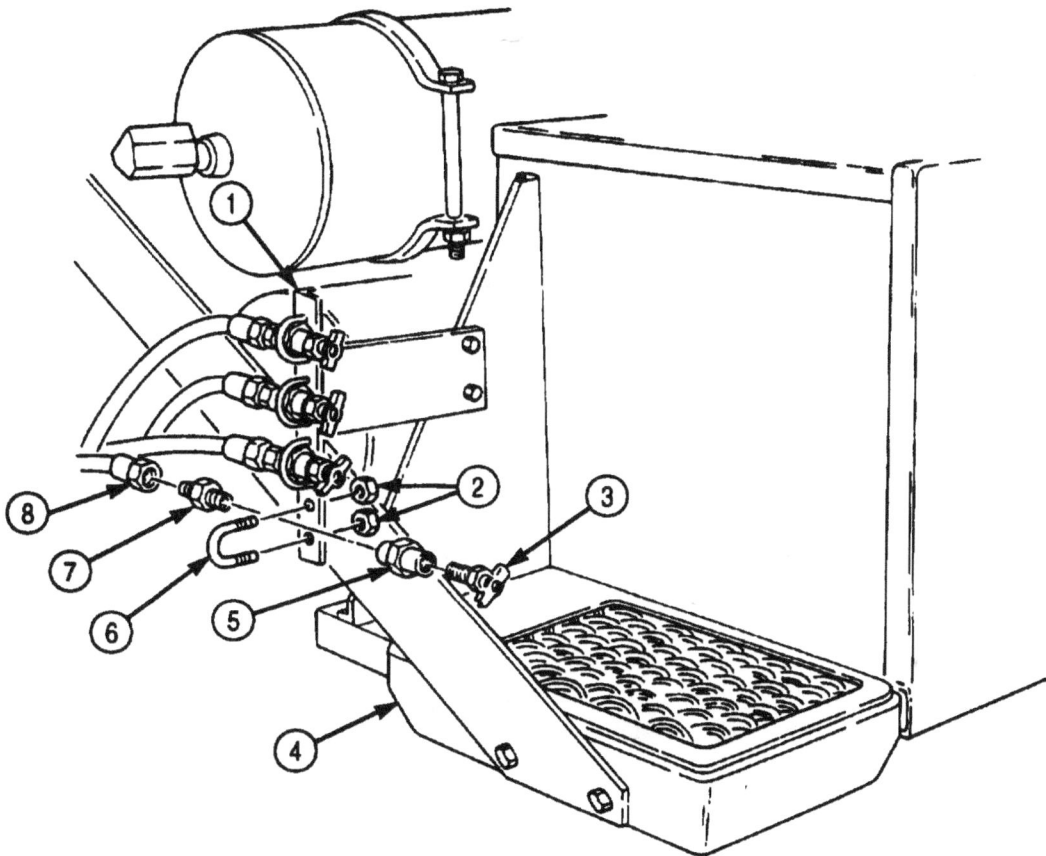

FOLLOW-ON TASK: Start engine (TM 9-2320-272-10) and allow air pressure to build to normal operating range. Check for air leaks at drainvalve. Road test vehicle.

3-206. AIR COMPRESSOR AND LINES (M939A2) REPLACEMENT

THIS TASK COVERS:

a. Removal   b. Installation

INITIAL SETUP:

APPLICABLE MODELS
M939A2

TOOLS
General mechanic's tool kit (Appendix E, Item 1)
Torque wrench (Appendix E, Item 144)

MATERIALS/PARTS
Four bushings (Appendix D, Item 21)
Gasket (Appendix D, Item 162)
Two locknuts (Appendix D, Item 289)
Gasket (Appendix D, Item 163)
Antiseize tape (Appendix C, Item 72)

PERSONNEL REQUIRED
TWO

REFERENCES (TM)
TM 9-2320-272-10
TM 9-2320-272-24P

EQUIPMENT CONDITION
- Air reservoirs drained (TM 9-2320-272-10).
- Coolant drained (para. 3-53).

a. Removal

1. Remove clamp (5) and hose (6) from power steering pump (3).

2. Disconnect supply line (4) from power steering pump (3).

3. Remove two screws (2), washers (1), power steering pump (3), and gasket (8) from air compressor (7). Discard gasket (8).

4. Disconnect air outlet tube (26) from elbow (27).

5. Remove air governor tube (9) from elbow (28).

6. Remove water outlet tube (19) and two bushings (14) from elbow (15) and air compressor (7). Discard bushings (14).

7. Remove water inlet tube (18) and two bushings (17) from elbow (16) and air compressor (7). Discard bushings (17).

8. Remove elbows (15) and (16) from engine block (20).

9. Remove screw (21) and air inlet tube (13) from engine block (20).

10. Loosen two hose clamps (10) and remove air inlet tube (13) from air inlet connector (24).

11. Loosen clamp (12) and remove air inlet tube (13) from adapter (11).

12. Remove adapter (11) from engine block (20).

13. Remove oil supply hose (23) from fitting (25) and elbow (22).

14. Remove elbow (22) from engine block (20).

3-206. AIR COMPRESSOR AND LINES (M939A2) REPLACEMENT (Contd)

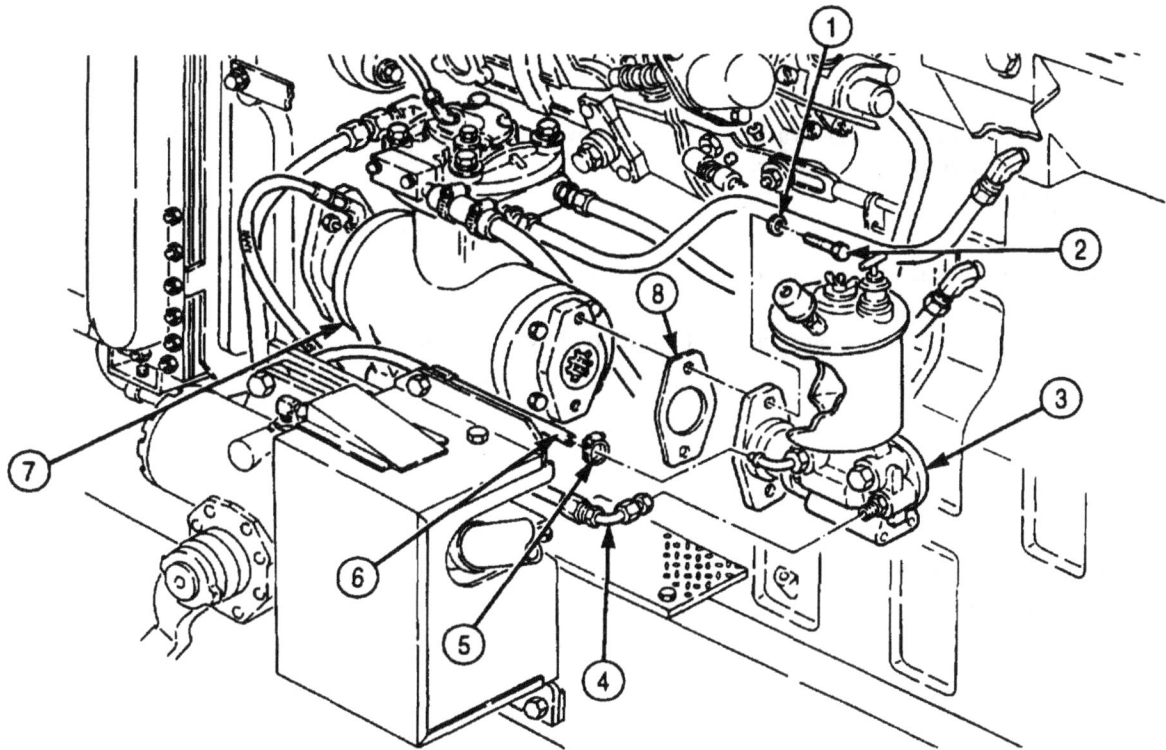

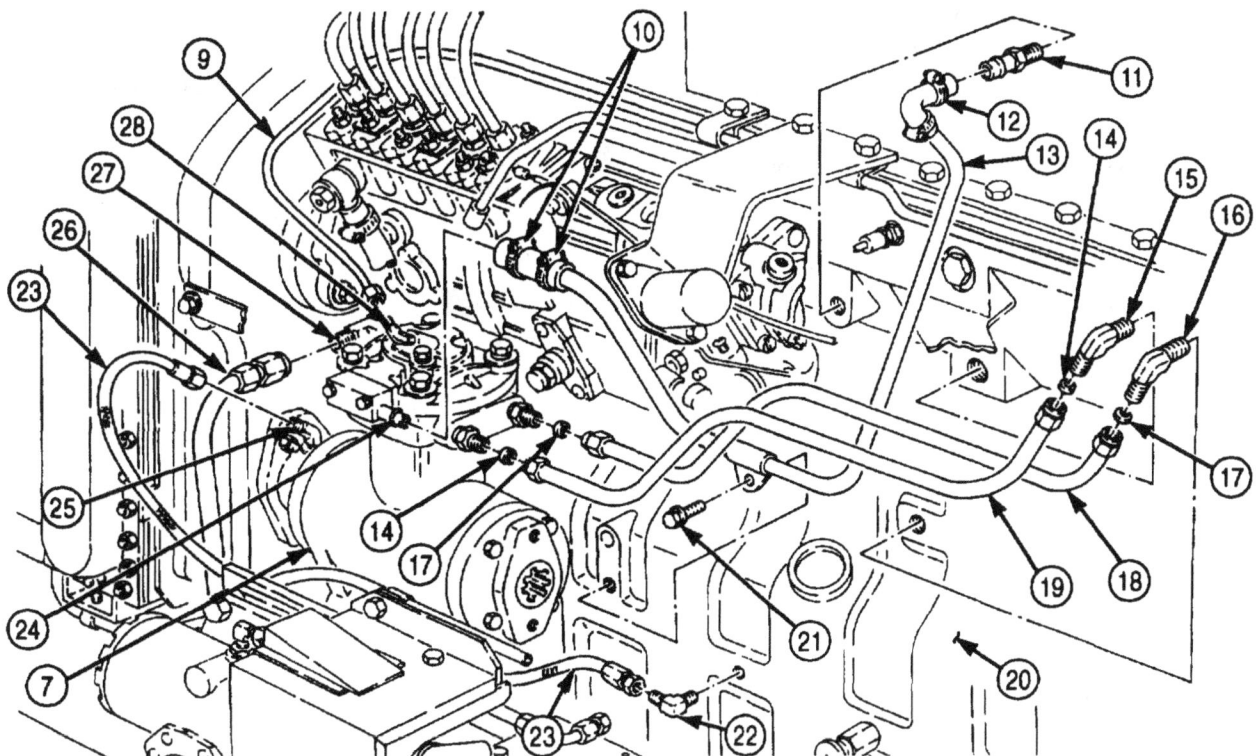

3-206. AIR COMPRESSOR AND LINES (M939A2) REPLACEMENT (Contd)

15. Remove two screws (12), clamp (11), and air outlet tube (10) from spacer (8).

16. Remove two screws (9) and spacer (8) from engine block (7) and brace (5).

17. Remove two screws (6) and brace (5) from air compressor (4).

18. Remove two locknuts (3), washers (2), air compressor (4), and gasket (1) from gear housing (13). Discard locknuts (3) and gasket (1).

b. Installation

1. Install new gasket (1) and air compressor (4) on gear housing (13) with two washers (2) and new locknuts (3). Tighten locknuts (3) 55 lb-ft (75 N.m).

2. Install brace (5) on air compressor (4) with two screws (6).

3. Install spacer (8) on engine block (7) and brace (5) with two screws (9).

4. Install air outlet tube (10) on spacer (8) with clamp (11) and two screws (12).

NOTE
Wrap all male pipe threads with antiseize tape before installation.

5. Install elbow (26) on engine block (7).

6. Install oil supply hose (27) on fitting (29) and elbow (26).

7. Install adapter (16) on engine block (7).

8. Install air inlet tube (18) on adapter (16) with clamp (17). Tighten clamp (17).

9. Install air inlet tube (18) on air inlet connector (28) with two hose clamps (15). Tighten hose clamps (15).

10. Install air inlet tube (18) on engine block (7) with screw (25).

11. Install elbows (20) and (21) on engine block (7).

12. Install two new bushings (22) and water inlet tube (23) on elbow (21) and air compressor (4).

13. Install two new bushings (19) and water outlet tube (24) on elbow (20) and air compressor (4).

14. Install air governor tube (14) on elbow (32).

15. Connect air outlet tube (30) to elbow (31).

3-206. AIR COMPRESSOR AND LINES (M939A2) REPLACEMENT (Contd)

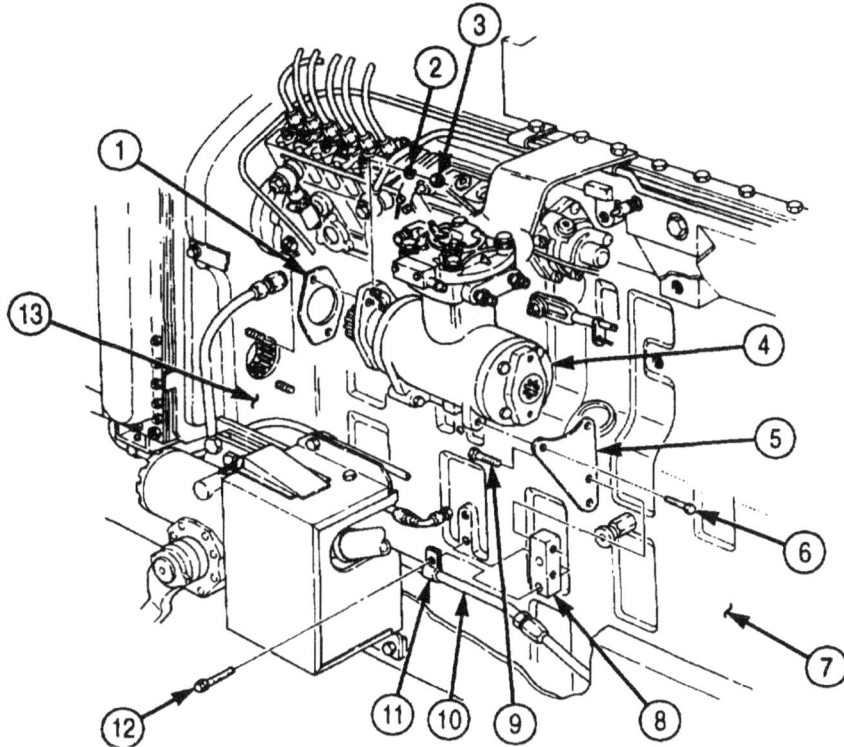

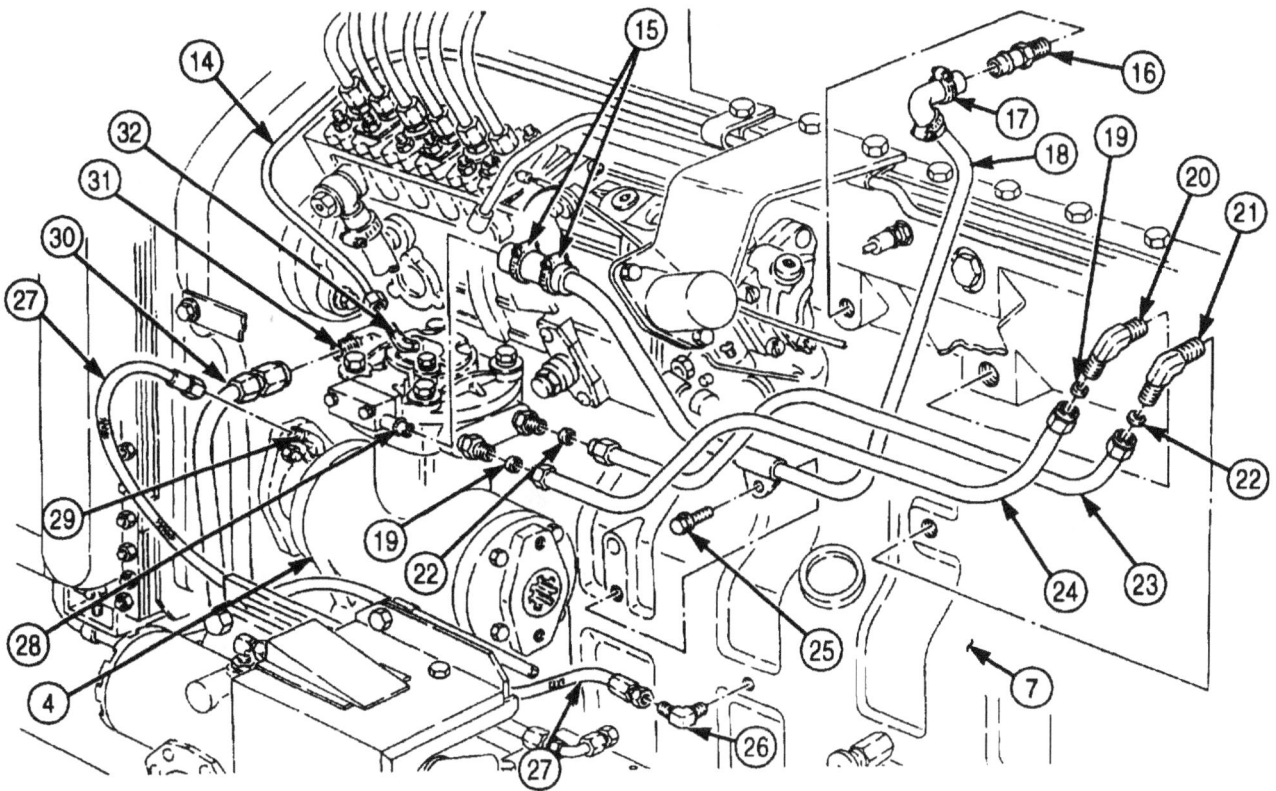

3-206. AIR COMPRESSOR AND LINES (M939A2) REPLACEMENT (Contd)

16. Install new gasket (8) and power steering pump (3) on air compressor (7) with two washers (1) and screws (2).

17. Connect supply line (4) to power steering pump (3).

18. Install hose (6) on power steering pump (3) with clamp (5).

CAUTION

When filling cooling system, ensure drainvalve on aftercooler is open. Failure to do so may result in damage to equipment.

3-206. AIR COMPRESSOR AND LINES (M939A2) REPLACEMENT (Contd)

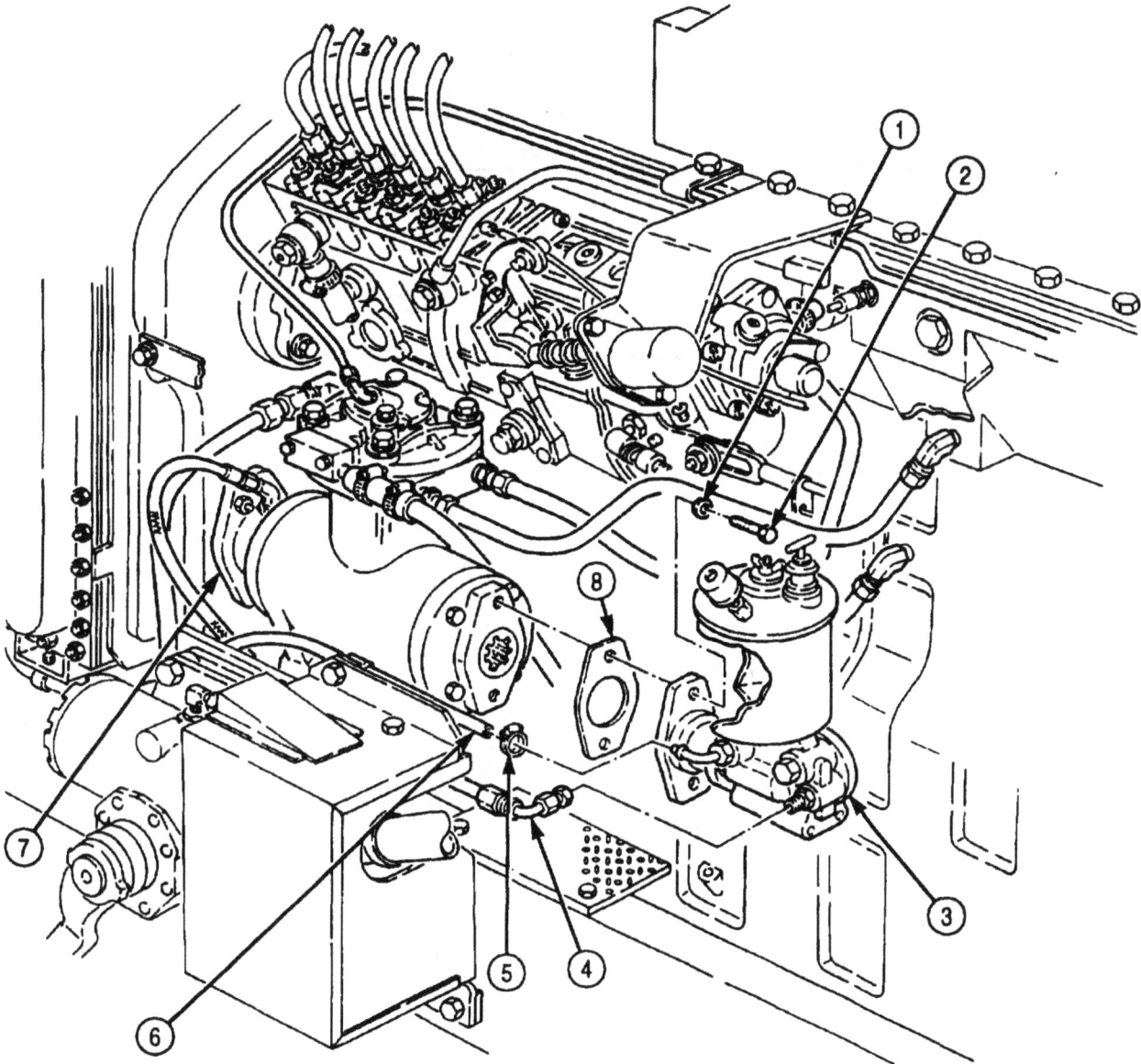

FOLLOW-ON TASK: Fill cooling system (para. 3-53).

3-207. AIR GOVERNOR MAINTENANCE

THIS TASK COVERS:

a. Removal
b. Installation

c. Testing
d. Adjustment

INITIAL SETUP:

APPLICABLE MODELS
All

TOOLS
General mechanic's tool kit (Appendix E, Item 1)

MATERIALS/PARTS
Antiseize tape (Appendix C, Item 72)

REFERENCES (TM)
TM 9-2320-272-10
TM 9-2320-272-24P

EQUIPMENT CONDITION
- Parking brake set (TM 9-2320-272-10).
- Air reservoirs drained (TM 9-2320-272-10).
- Right splash shield removed (TM 9-2320-272-10).

GENERAL SAFETY INSTRUCTIONS
Do not disconnect air lines before draining air reservoirs.

WARNING

Do not disconnect air lines before draining air reservoirs. Small parts under pressure may shoot out with high velocity, causing injury to personnel.

a. Removal

1. Disconnect governor supply air line (2) and governor-to-horn supply line (8) from tee (7).
2. Disconnect governor-to-compressor unloader line (4) from adapter (3).
3. Remove two screws (6), washers (5), and air governor (1) from cab cowl (9).

b. Installation

NOTE
- If installing new governor, use fitting from old governor.
- Wrap all male pipe threads with antiseize tape before installation.

1. Install air governor (1) on cab cowl (9) with two washers (5) and screws (6).
2. Connect governor-to-compressor unloader line (4) to adapter (3).
3. Connect governor supply air line (2) and governor-to-horn supply line (8) to tee (7).

3-207. AIR GOVERNOR MAINTENANCE (Contd)

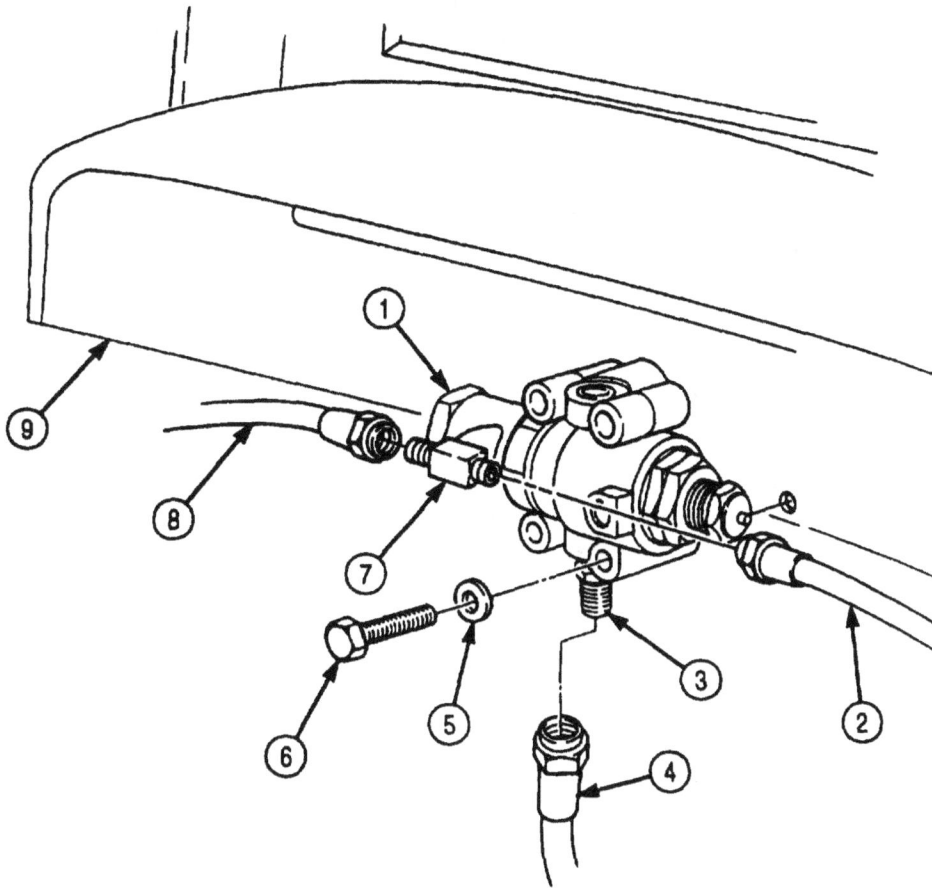

3-207. AIR GOVERNOR MAINTENANCE (Contd)

c. Testing

NOTE

Whenever governor is tested, instrument panel primary air gauge is used.

1. Start engine (TM 9-2320-272-10) and allow air pressure to build to normal operating range.

NOTE

Engine speed must be adjusted to 1,275 rpm before performing step 2.

2. Check air pressure gauge (4) for air compressor cut-out pressure. Air governor (1) should stop pressure buildup at 130 psi (896 kPa). If not, perform adjustment.

d. Adjustment

NOTE

Whenever governor is adjusted, instrument panel primary air gauge is used.

1. Loosen locknut (2) and turn adapter (3) clockwise to raise pressure; counterclockwise to lower pressure.

2. Depress brake pedal until pressure drops and air compression starts to build up pressure. Air governor (1) should stop pressure buildup at 130 psi (896 kPa). If not, repeat step 1.

3. Tighten locknut (2) when correct air pressure is reached.

NOTE

After air compression cut-out pressure is adjusted, perform step 4.

4. Check air pressure gauge (4) for air compressor cut-in pressure by depressing brake pedal repeatedly. Air governor (1) should stop pressure buildup above 80 psi (552 kPa). If not, replace air governor (1).

3-207. AIR GOVERNOR MAINTENANCE (Contd)

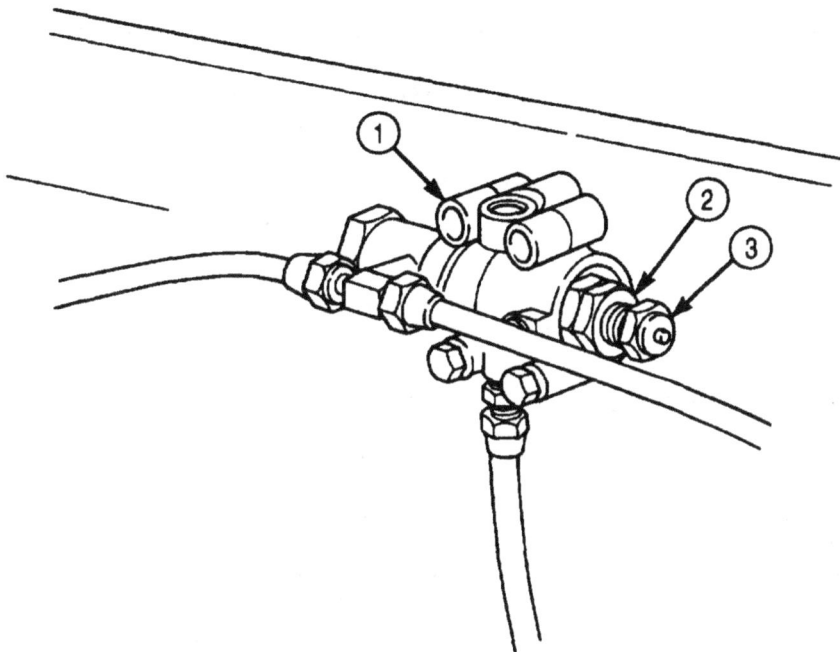

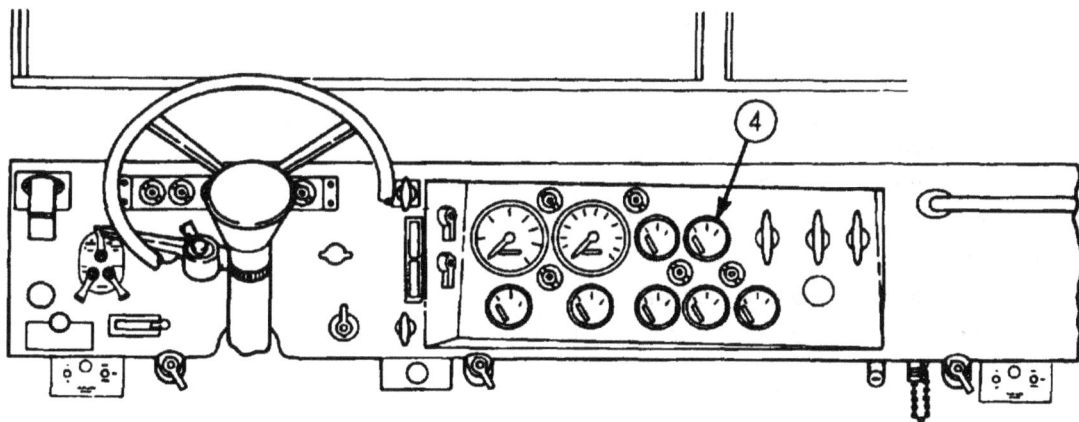

FOLLOW-ON TASKS: • Install right splash shield (TM 9-2320-272-10).
                   • Road test vehicle (TM 9-2320-272-10).

3-208. AIR COUPLINGS REPLACEMENT

THIS TASK COVERS:

a. Front Air Couplings Removal
b. Rear Air Couplings Removal
(M923/A1/A2)
c. Rear Air Couplings Removal
(M929/A1/A2, M931/A1/A2, M934/A1/A2
M936/A1/A2)

d. Front Air Couplings Installation
e. Rear Air Couplings Installation
(M923/A1/A2)
f. Rear Air Couplings Installation
(M929/A1/A2, M931/A1/A2, M934/A1/A2,
M936/A1/A2)

INITIAL SETUP:

APPLICABLE MODELS
All

TOOLS
General mechanic's tool kit (Appendix E, Item 1)
Soft-jawed vise

MATERIALS/PARTS
Six lockwashers (Appendix D, Item 366)
Twelve locknuts (Appendix D, Item 313)
Antiseize tape (Appendix D, Item 72)

REFERENCES (TM)
TM 9-2320-272-10
TM 9-2320-272-24P

EQUIPMENT CONDITION
- Parking brake set (TM 9-2320-272-10).
- Air reservoirs drained (TM 9-2320-272-10).
- Right and left splash shields removed
  (TM 9-2320-272-10).

GENERAL SAFETY INSTRUCTIONS
Do not disconnect air lines before draining air
reservoirs.

WARNING

Do not disconnect air lines before draining air reservoirs. Small
parts under pressure may shoot out with high velocity, causing
injury to personnel.

a. Front Air Couplings Removal

1. Remove dummy couplings (1) and (15) and S-hook (12) from air coupling (2) and bracket (5).
   Dummy coupling (1) has a built-in check valve.

   NOTE
   Perform steps 2 and 3 for emergency coupling on left side of
   vehicle.

2. Remove air coupling (2) from pipe nipple (14).
3. Remove pipe nipple (14) from valve (13).
4. Disconnect air line (11) from elbow (10).
5. Disconnect elbow (10) from adapter (4).
6. Remove nut (9), lockwasher (8), adapter (4), identification plate (6), and bracket (5) from frame rail (7).
   Discard lockwasher (8).
7. Remove adapter (4) from elbow (3).

   NOTE
   Perform step 8 for service coupling on right side of vehicle.

8. Remove elbow (3) from air coupling (2).

   NOTE
   Perform step 9 for emergency coupling on left side of vehicle.

9. Remove elbow (3) from valve (13).

3-208. AIR COUPLINGS REPLACEMENT (Contd)

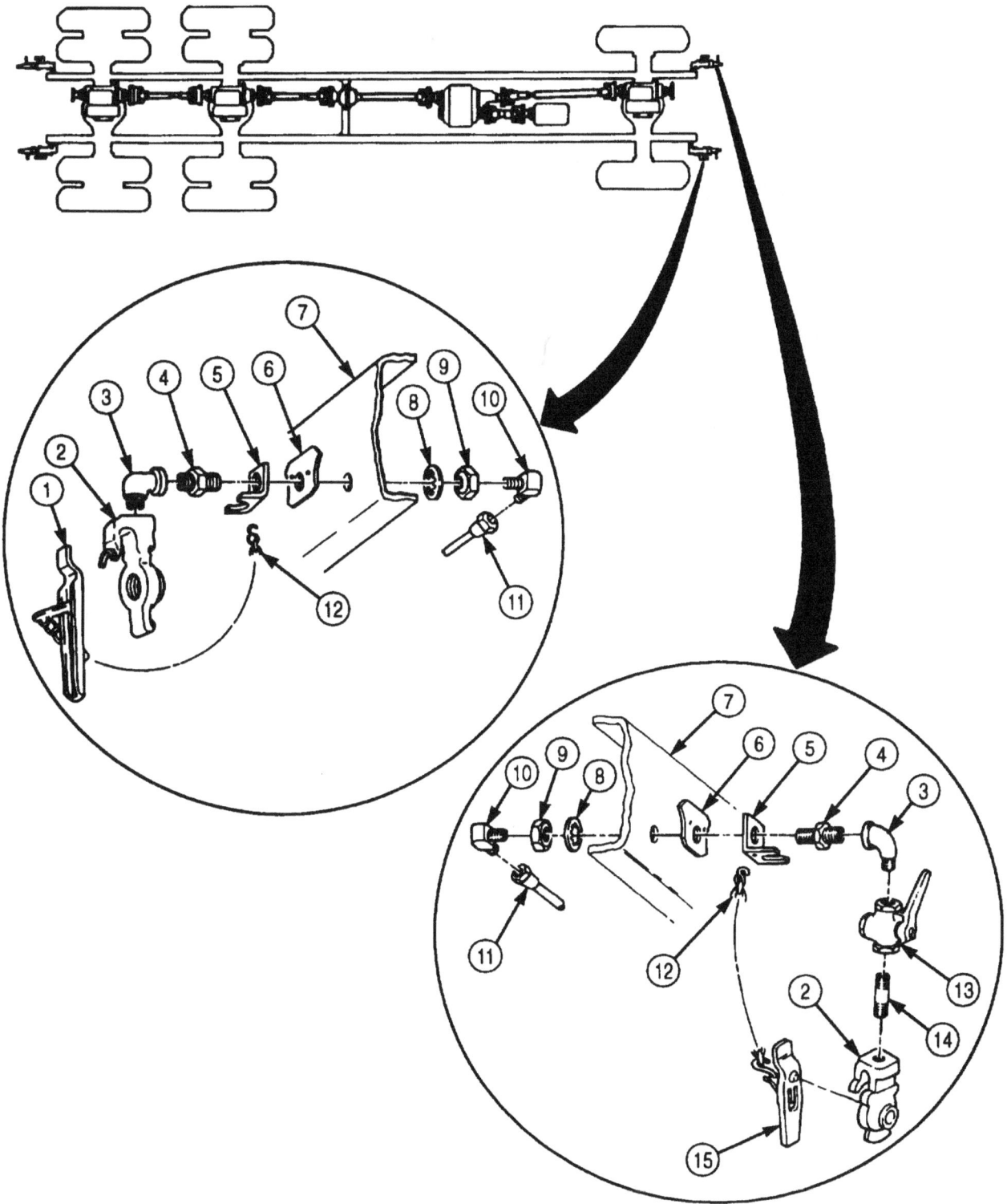

3-208. AIR COUPLINGS REPLACEMENT (Contd)

b. Rear Air Couplings Removal (M923/A1/A2)

1. Remove dummy coupling (9) and S-hook (8) from air coupling (10) and bracket (11).
2. Remove air coupling (10) and pipe nipple (7) from valve (12).
3. Disconnect air line (1) from elbow (2).
4. Remove elbow (2) from pipe coupling (3).
5. Remove pipe coupling (3) from adapter (4).
6. Remove four locknuts (15). washers (14), screws (6), bracket (11), and plate (13) from frame rail (5). Discard locknuts (15).

NOTE

Perform step 7 for emergency coupling on right side of vehicle.

7. Loosen nut (18) from adapter (4).
8. Remove valve (12) from elbow (16).

NOTE

Perform step 9 for service coupling on left side of vehicle.

9. Remove elbow (16) and pipe coupling (17) from adapter (4).
10. Remove elbow (16) from adapter (4).
11. Remove nut (18), lockwasher (19), identification plate (20), and adapter (4) from frame rail (5). Discard lockwasher (19).

3-208. AIR COUPLINGS REPLACEMENT (Contd)

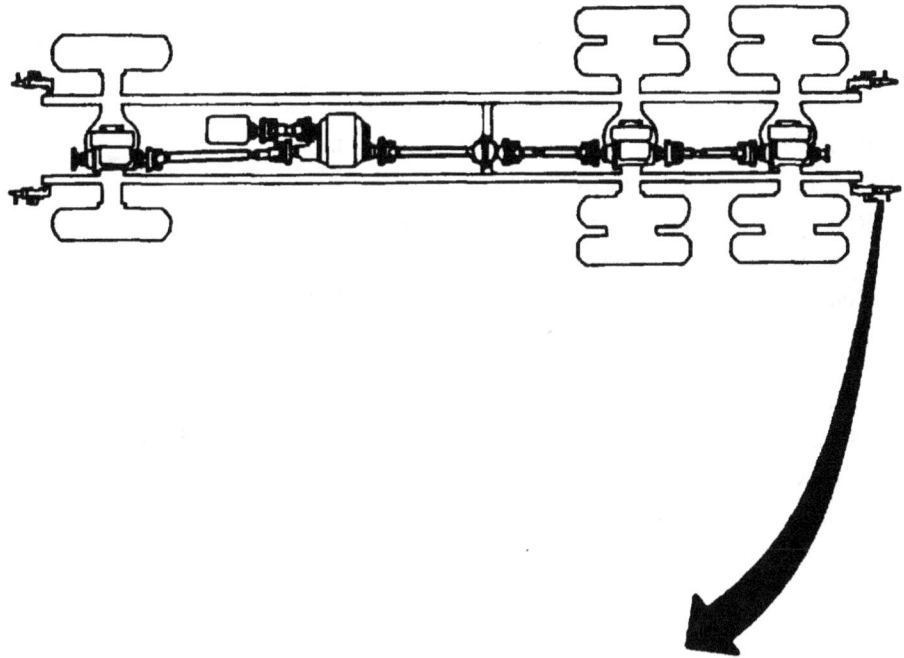

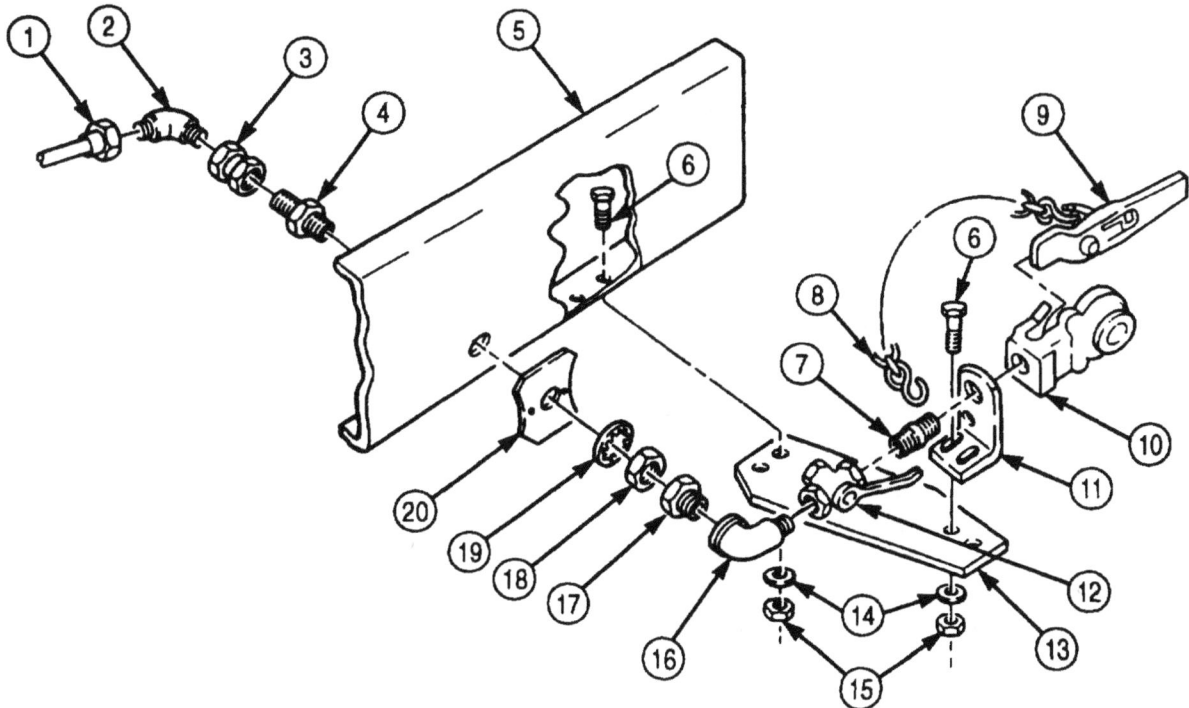

3-208. AIR COUPLINGS REPLACEMENT (Contd)

c. Rear Air Couplings Removal (M929/A1/A2, M931/A1/A2, M934/A1/A2, M936/A1/A2)

NOTE

Left and right (service and emergency) rear couplings are removed the same way.

1. Remove dummy coupling (10) and S-hook (13) from air coupling (9) and bracket (14).
2. Remove air coupling (9) from pipe nipple (11).
3. Disconnect air line (1) from adapter (2).
4. Remove adapter (2) from connector (21).
5. Remove nut (3) and lockwasher (4) from connector (21). Discard lockwasher (4).
6. Remove two screws (8) and identification plate (12) from bracket (14).
7. Remove two locknuts (17), washers (16), screws (5), plate (6), and bracket (14) from frame rail (7). Discard locknuts (17).
8. Remove pipe nipple (11) from elbow (15).
9. Place valve (18) in soft-jawed vise and remove elbow (15) from valve (18).
10. Remove connector (21) from elbow (20).
11. Remove elbow (20) from pipe nipple (19).
12. Remove pipe nipple (19) from valve (18).
13. Remove valve (18) from soft-jawed vise.

3-208. AIR COUPLINGS REPLACEMENT (Contd)

3-208. AIR COUPLINGS REPLACEMENT (Contd)

d. Front Air Couplings Installation

NOTE

- Fittings must be cleaned and inspected for cracks and stripped threads.
- Wrap all male pipe threads with antiseize tape before installation.
- Perform step 1 for emergency coupling on left side of vehicle.

1. Install elbow (3) in valve (13).

NOTE

Perform step 2 for service coupling on right side of vehicle.

2. Install elbow (3) in air coupling (2) and place air coupling (2) in soft-jawed vise.
3. Install adapter (4) in elbow (3) and remove air coupling (2) from vise.
4. Install identification plate (6), bracket (5), and adapter (4) in frame rail (7) with new lockwasher (8) and nut (9).
5. Install elbow (10) on adapter (4).
6. Connect air line (11) to elbow (10).

NOTE

Perform steps 7 and 8 for emergency coupling on left side of vehicle.

7. Install pipe nipple (14) in valve (13).
8. Install air coupling (2) on pipe nipple (14).
9. Install dummy couplings (15) and (1) and S-hook (12) on air coupling (2) and bracket (5).

3-208. AIR COUPLINGS REPLACEMENT (Contd)

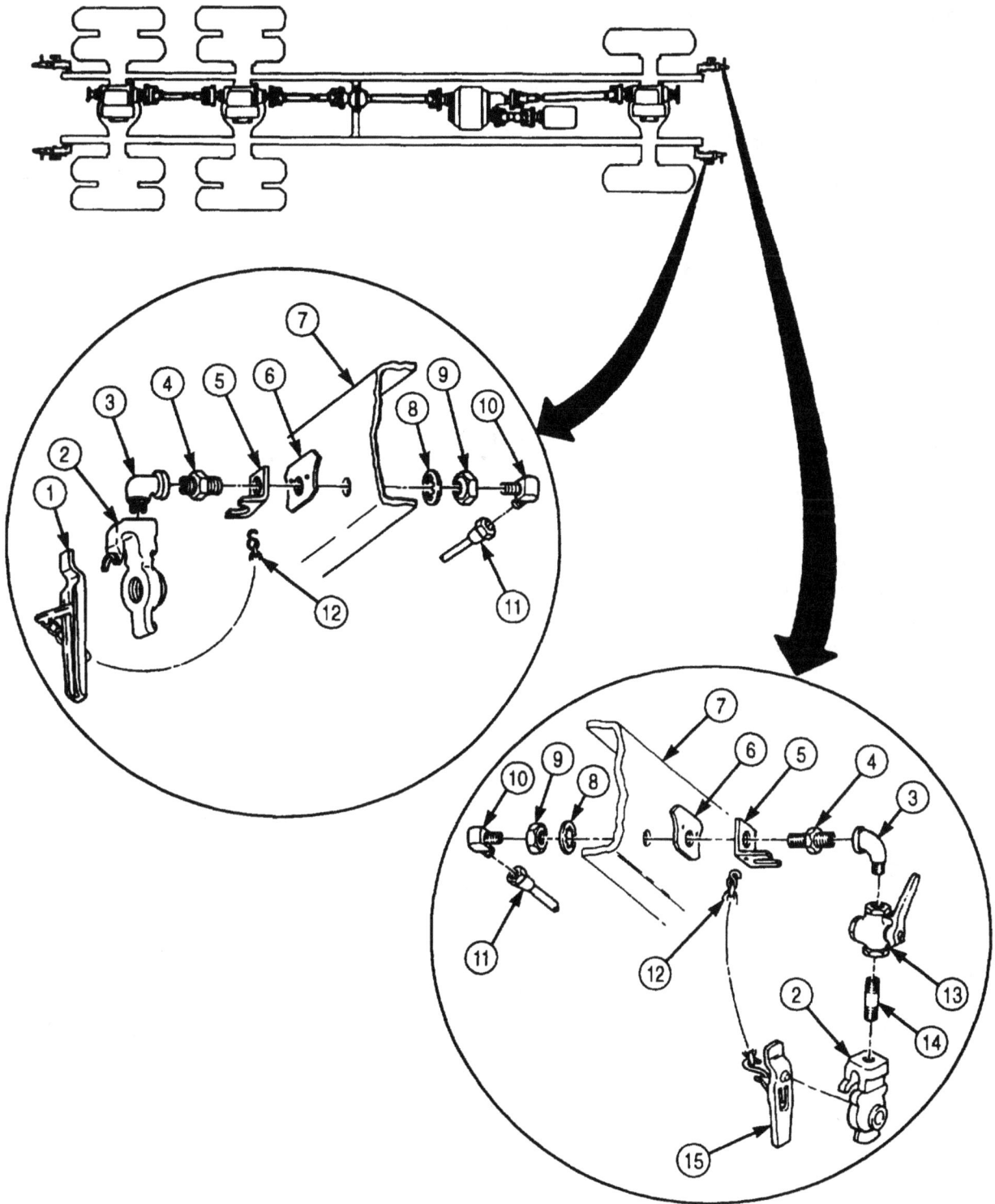

3-208. AIR COUPLINGS REPLACEMENT (Contd)

e. Rear Air Couplings Installation (M923/A1/A2)

NOTE

- Fittings must be cleaned and inspected for cracks and stripped threads.
- Wrap all male pipe threads with antiseize tape before installation.

1. Install adapter (4) and identification plate (20) in frame rail (5) with new lockwasher (19) and nut (18). Do not tighten nut (18).

NOTE

Perform steps 2 and 3 for service coupling on left side of vehicle.

2. Install pipe coupling (17) on adapter (4).

3. Install elbow (16) on pipe coupling (17).

4. Install valve (12) and pipe nipple (7) on elbow (16).

5. Install plate (13) and bracket (11) on frame rail (5) with four screws (6), washers (14), and new locknuts (15).

NOTE

Perform step 6 for emergency coupling on right side of vehicle.

6. Tighten nut (18).

7. Install pipe coupling (3) on adapter (4).

8. Install elbow (2) in pipe coupling (3).

9. Connect air line (1) to elbow (2).

10. Install air coupling (10) on pipe nipple (7).

11. Install dummy coupling (9) and S-hook (8) on air coupling (10) and bracket (11).

3-208. AIR COUPLINGS REPLACEMENT (Contd)

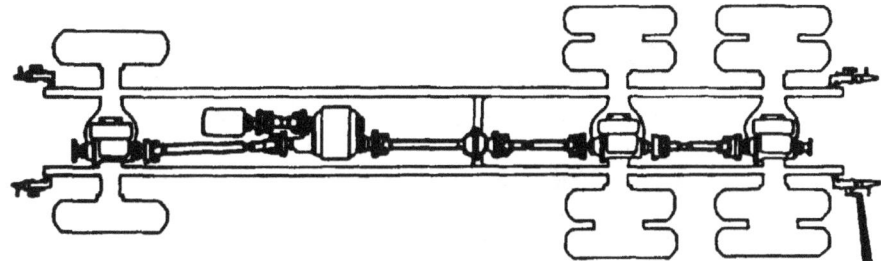

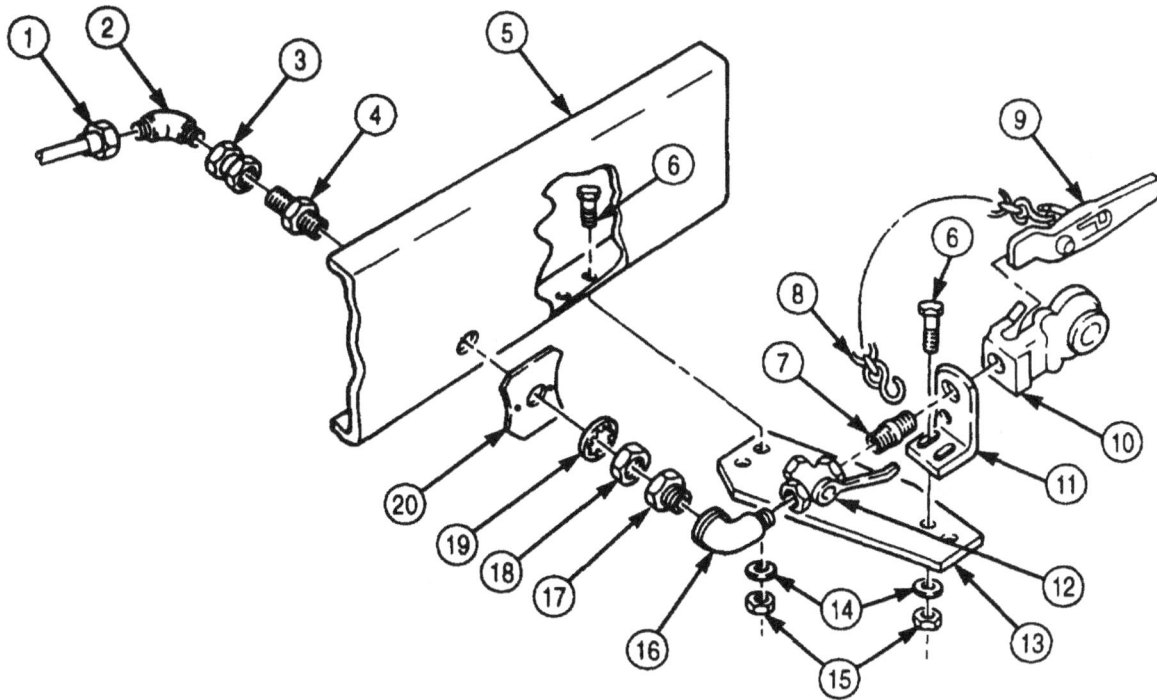

3-208. AIR COUPLINGS REPLACEMENT (Contd)

f. Rear Air Couplings Installation (M929/A1/A2, M931/A1/A2, M934/A1/A2, M936/A1/A2)

NOTE

- Left and right (service and emergency) rear couplings are installed the same way.
- Fittings must be cleaned and inspected for cracks and stripped threads.
- Wrap all male pipe threads with antiseize tape before installation.

1. Install bracket (14) and plate (6) on frame rail (7) with two screws (5), washers (16), and new locknuts (17).

2. Install identification plate (12) on bracket (14) with two screws (8).

3. Install valve (18), pipe nipple (19), elbow (20), connector (21), elbow (15), and pipe nipple (11) on bracket (14) and plate (6) with new lockwasher (4) and nut (3).

4. Install adapter (2) in connector (21).

5. Connect air line (1) to adapter (2).

6. Install air coupling (9) on pipe nipple (11).

7. Install dummy coupling (10) and S-hook (13) on air coupling (9) and bracket (14).

3-208. AIR COUPLINGS REPLACEMENT (Contd)

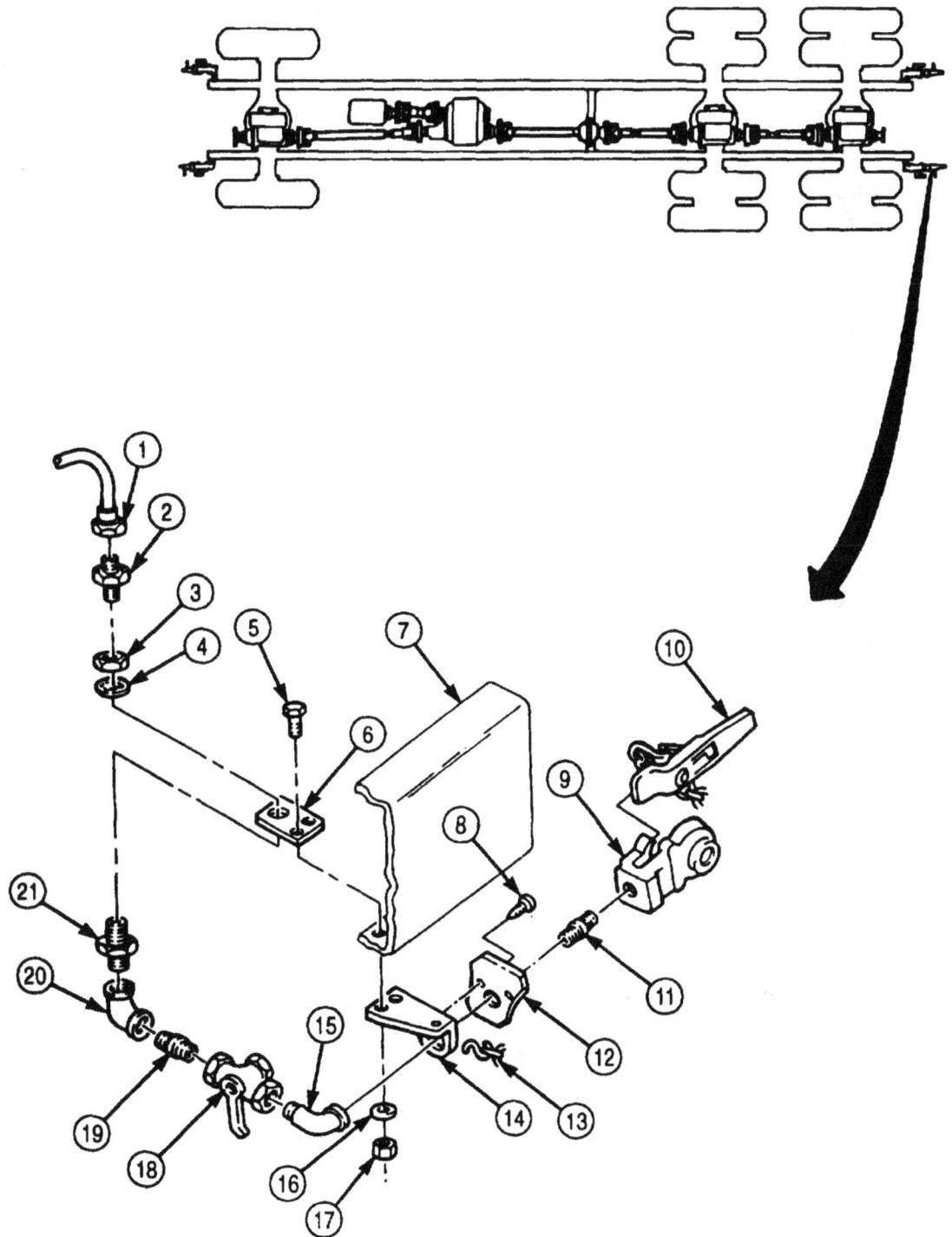

FOLLOW-ON TASKS: • Install right and left splash shields (TM 9-2320-272-10).
• Start engine (TM 9-2320-272-10) and allow air pressure to build to normal operating range. Check for air leaks and road test vehicle.

3-209. EMERGENCY AND TRAILER COUPLING HOSES (M931/A1/A2, M932/A1/A2) REPLACEMENT

THIS TASK COVERS:

a. Removal   b. Installation

INITIAL SETUP:

APPLICABLE MODELS
M931/A1/A2, M932/A1/A2

Tools
General mechanic's tool kit (Appendix E, Item 1)

MATERIALS/PARTS
Lockwasher (Appendix D, Item 366)
Antiseize tape (Appendix C, Item 72)

REFERENCES (TM)
TM 9-2320-272-10
TM 9-2320-272-24P

EQUIPMENT CONDITION
- Parking brake ret (TM 9-2320-272-10).
- Air reservoirs drained (TM 9-2320-272-10).

GENERAL SAFETY INSTRUCTIONS
Do not disconnect air lines before draining air reservoirs.

WARNING

Do not disconnect air lines before draining air reservoirs. Small parts under pressure may shoot out with high velocity, causing injury to personnel.

a. Removal

1. Remove coupling (1) from adapter (2).
2. Remove adapter (2) from coupling hose (3).
3. Remove coupling hose (3) from adapter (12).
4. Remove adapter (12) from cutoff valve (4).
6. Remove cutoff valve (4) from adapter (11).
6. Remove air line (9) from elbow (8).
7. Remove elbow (8) from adapter (11).
8. Remove nut (7), lockwasher (6), adapter (11), and identification plate (5) from frame hole (10). Discard lockwasher (6).

b. Installation

NOTE
Wrap all male pipe threads with antiseize tape before installation.

1. Insert adapter (11) through identification plate (5) and frame hole (10) and install with new lockwasher (6) and nut (7).
2. Install elbow (8) on adapter (11).
3. Install air line (9) on elbow (8).
4. Install cutoff valve (4) on adapter (11).
5. Install adapter (12) on cutoff valve (4).
6. Install coupling hose (3) on adapter (12).
7. Install adapter (2) on coupling hose (3).
8. Install coupling (1) on adapter (2).

3-209. EMERGENCY AND TRAILER COUPLING HOSES (M931/A1/A2, M932/A1/A2) REPLACEMENT

FOLLOW-ON TASK: Start engine (TM 9-2320-272-10) and allow air pressure to build to normal operating range. Check for air leaks and road test vehicle.

3-210. TRAILER AIRBRAKE HAND CONTROL VALVE (M931/A1/A2, M932/A1/A2) REPLACEMENT

THIS TASK COVERS:

a. Removal   b. Installation

INITIAL SETUP:

APPLICABLE MODELS
M931/A1/A2, M932/A1/A2

TOOLS
General mechanic's tool kit (Appendix E, Item 1)

MATERIALS/PARTS
Antiseize tape (Appendix D, Item 72)

REFERENCES (TM)
TM 9-2320-272-10
TM 9-2320-272-24P

EQUIPMENT CONDITION
• Parking brake set (TM 9-2320-272-10).
• Air reservoirs drained (TM 9-2320-272-10).

GENERAL SAFETY INSTRUCTIONS
Do not disconnect air lines before draining air reservoirs.

WARNING

Do not disconnect air lines before draining air reservoirs. Small parts under pressure may shoot out with high velocity, causing injury to personnel.

a. Removal

NOTE
- The airbrake hand control valve is mounted on the upper steering column opposite the turn signal control.
- Tag air lines for installation.

1. Disconnect vent line (10) from adapter fitting (9).
2. Disconnect delivery line (11) from adapter fitting (12).
3. Disconnect line (8) from adapter fitting (7).
4. Remove two screws (4), washers (3), tab (2), clamp (5), and control valve (1) from upper steering column (6).
5. Remove three adapter fittings (7), (9), and (12) from control valve (1).

b. Installation

NOTE
If new valve is being installed, use fittings from old valve.

Wrap all male pipe threads with antiseize tape before installation.

1. Install three adapter fittings (7), (9), and (12) on control valve (1).
2. Install control valve (1) on upper steering column (6) with clamp (5), tab (2), two washers (3), and screws (4).
3. Connect line (8) to adapter fitting (7).
4. Connect delivery line (11) to adapter fitting (12).
5. Connect vent line (10) to adapter fitting (9).

3-210. TRAILER AIRBRAKE HAND CONTROL VALVE (M931A1/A2, M932A1/A2) REPLACEMENT

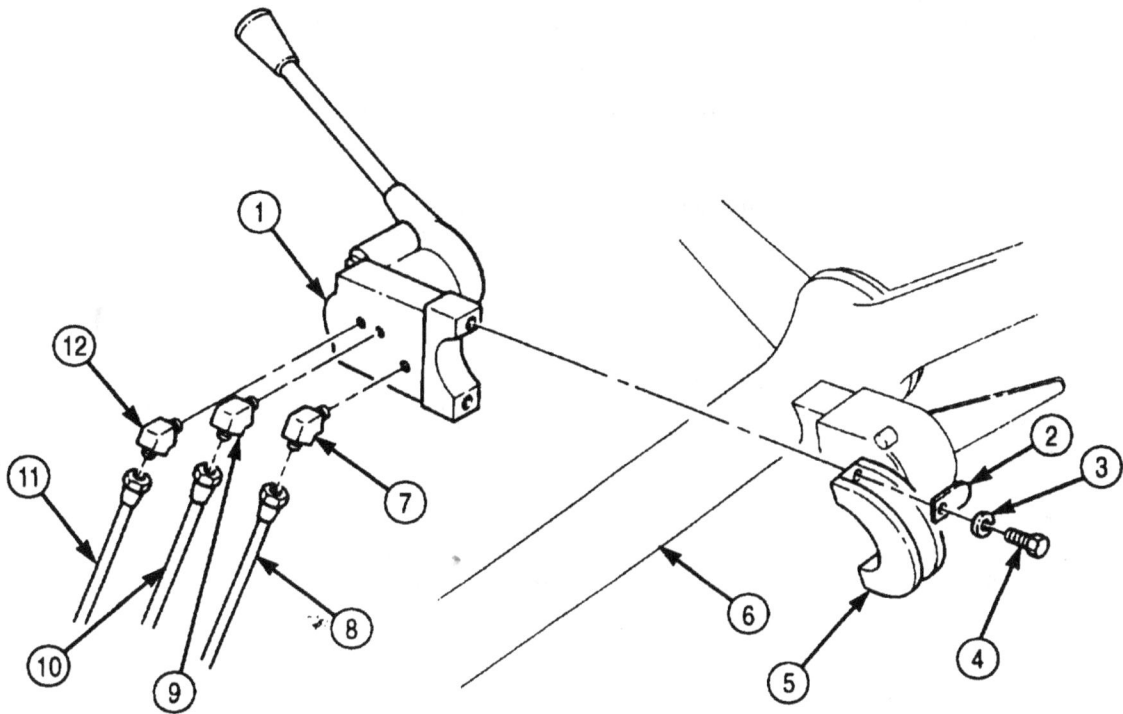

FOLLOW-ON TASK: Start engine (TM 9-2320-272-10) and allow air pressure to build to normal operating range. Check for air leaks at hand control valve. Road test vehicle.

3-211. ALCOHOL EVAPORATOR REPLACEMENT

THIS TASK COVERS:

a. Removal   b. Installation

INITIAL SETUP:

APPLICABLE MODELS
All

TOOLS
General mechanic's tool kit (Appendix E, Item 1)

MATERIALS/PARTS
Three locknuts (Appendix D, Item 313)

REFERENCES (TM)
TM 9-2320-272-10
TM 9-2320-272-24P

EQUIPMENT CONDITION
• Parking brake set (TM 9-2320-272-10).
• Hood raised and secured (TM 9-2320-272-10).

GENERAL SAFETY INSTRUCTIONS
Do not smoke during removal of alcohol evaporator container.

WARNING

Do not smoke when removing alcohol evaporator container. Injury
may result from improper handling of alcohol evaporator.

a. Removal

1. Remove alcohol evaporator container (7) from left side of engine.
2. Hold tube adapter (6) firmly in place and disconnect fitting (51 from adapter (6).
3. Remove three locknuts (11, screws (4), and cap (3) from engine-mounted bracket (2). Discard locknuts (1).

b. Installation

1. Install cap (3) on engine-mounted bracket (2) with three screws (4) and new locknuts (1).
2. Hold adapter (6) firmly in place and connect fitting (5) to adapter (6).
3. Install alcohol evaporator container (7) on left side of engine.

3-211. ALCOHOL EVAPORATOR REPLACEMENT (Contd)

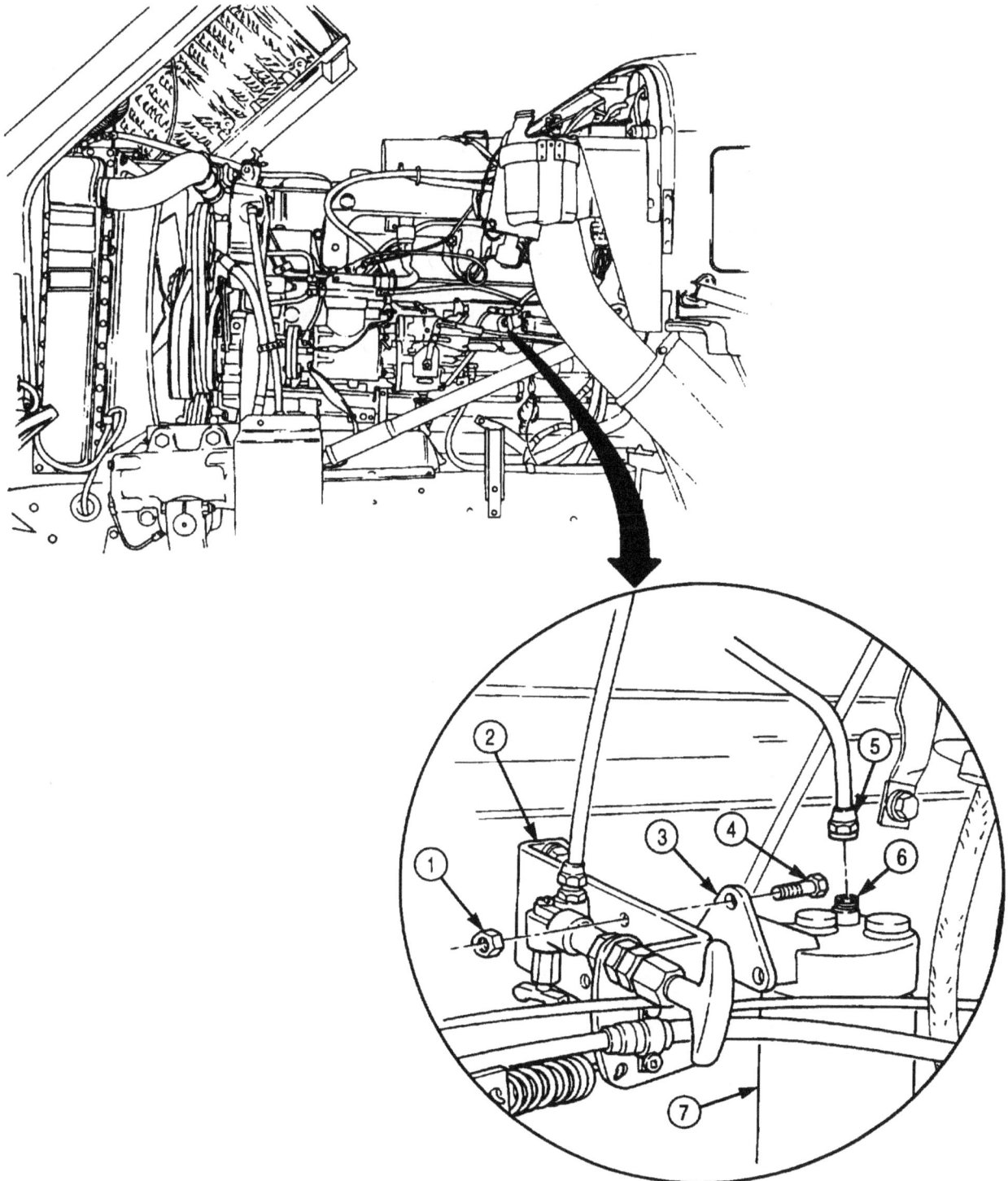

FOLLOW-ON TASK: Fill alcohol evaporator container as required (TM 9-2320-272-10).

3-212. DOUBLECHECK VALVE NO. 1 REPLACEMENT

THIS TASK COVERS:

a. Removal   b. Installation

INITIAL SETUP:

APPLICABLE MODELS
All

TOOLS
General mechanic's tool kit (Appendix E, Item 1)

MATERIALS/PARTS
Locknut (Appendix D, Item 299)
Antiseize tape (Appendix C, Item 72)

REFERENCES (TM)
TM 9-2320-272-10
TM 9-2320-272-24P

EQUIPMENT CONDITION
• Parking brake set (TM 9-2320-272-10).
• Air reservoirs drained (TM 9-2320-272-10).

GENERAL SAFETY INSTRUCTIONS
Do not disconnect air lines before draining air reservoirs.

WARNING

Do not disconnect air lines before draining air reservoirs. Small parts under pressure may shoot out with high velocity, causing injury to personnel.

a. Removal

NOTE
• Doublecheck valve No. 1 is located inside the front left frame rail and left of the front axle differential.
• Tag air lines for installation.

1. Disconnect line (1) from doublecheck valve No. 1 adapter (2).
2. Disconnect line (12) from valve adapter (11).
3. Disconnect line (9) from tee (10).
4. Remove tee (10) from doublecheck valve No. (3).
5. Disconnect line (5) from valve adapter (7).
6. Remove locknut (8), screw (4), and doublecheck valve No. 1 (3) from frame rail (6). Discard locknut (8).

b. Installation

NOTE
• If new doublecheck valve is being installed, use fittings from old doublecheck valve.
• Wrap all male pipe threads with antiseize tape before installation.

1. Position doublecheck valve No. 1 (3) against frame rail (6) and install with screw (4) and new locknut (8).
2. Connect line (5) to valve adapter (7).
3. Install tee (10) on doublecheck valve No. 1 (3).
4. Connect line (9) to tee (10).
5. Connect line (12) to valve adapter (11).
6. Connect line (1) to doublecheck valve No. 1 adapter (2).

3-212. DOUBLECHECK VALVE NO. 1 REPLACEMENT (Contd)

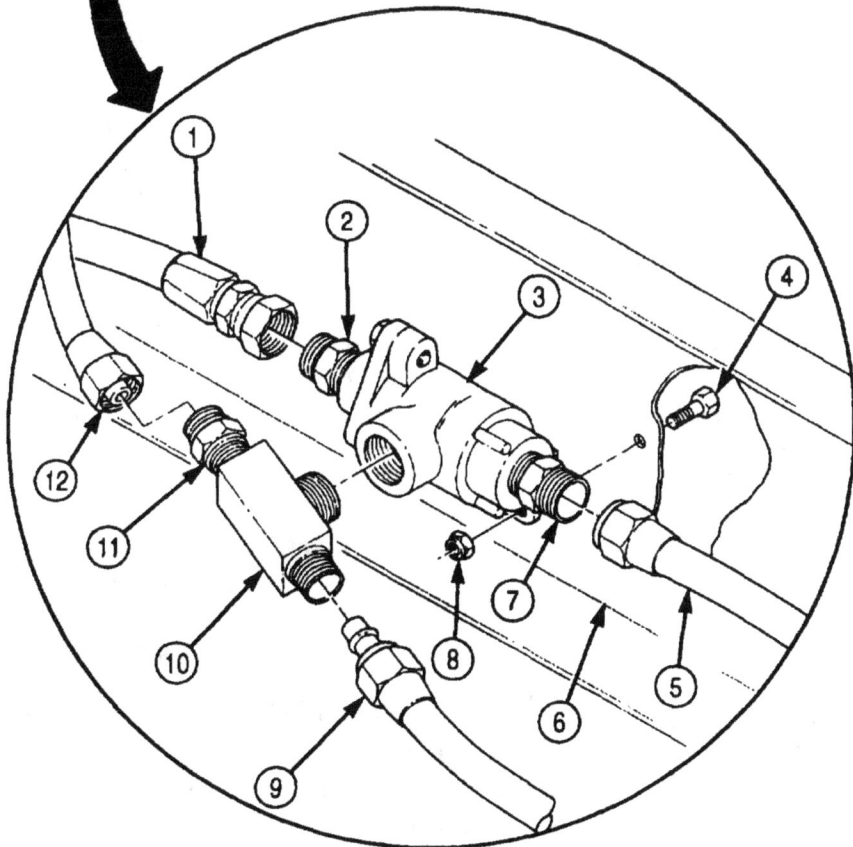

FOLLOW-ON TASK: Start engine (TM 9-2320-272-10) and allow air pressure to build to normal operating range. Check for air leaks at doublecheck valve No. 1. Road test vehicle.

3-213. DOUBLECHECK VALVE NO. 2 REPLACEMENT

THIS TASK COVERS:

a. Removal

b. Installation

INITIAL SETUP:

APPLICABLE MODELS
All except M931/A1/A2 and M932/A1/A2

TOOLS
General mechanic's tool kit (Appendix E, Item 1)

MATERIALS/PARTS
Locknut (Appendix D, Item 299)
Antiseize tape (Appendix C, Item 72)

REFERENCES (TM)
TM 9-2320-272-10
TM 9-2320-272-24P

EQUIPMENT CONDITION
Parking brake set (TM 9-2320-272-10).
Air reservoirs drained (TM 9-2320-272-10).

GENERAL SAFETY INSTRUCTIONS
Do not disconnect air lines before draining air reservoirs.

WARNING

Do not disconnect air lines before draining air reservoirs. Small parts under pressure may shoot out with high velocity, causing injury to personnel.

a. Removal

NOTE
• Doublecheck valve No. 2 is located left of the transfer case parking brakedrum inside left frame rail.
• Tag air lines for installation.

1. Disconnect two wires (6) from stoplight switch (5).
2. Disconnect treadle valve control line (7) from doublecheck valve No. 2 (3).
3. Disconnect secondary relay valve control line (9) from elbow (8).
4. Disconnect doublecheck valve No. 1 output line (10) from elbow (11).
5. Disconnect primary relay valve control line (13) from valve adapter tee (12).
6. Remove locknut (2), screw (1), and doublecheck valve No. 2 (3) from left frame rail (4). Discard locknut (2).

b. Installation

NOTE
• If new doublecheck valve is being installed, use fittings from old check valve.
• Wrap all male pipe threads with antiseize tape before installation.

1. Position doublecheck valve No. 2 (3) against frame rail (4) and install with screw (1) and new locknut (2).
2. Connect treadle valve control line (7) to doublecheck valve No. 2 (3).
3. Connect secondary relay valve control line (9) to elbow (8).
4. Connect doublecheck valve No. 1 output line (10) to elbow (11).
5. Connect primary relay valve control line (13) to valve adapter tee (12).
6. Connect two wires (6) to stoplight switch (5).

3-213. DOUBLECHECK VALVE NO. 2 REPLACEMENT (Contd)

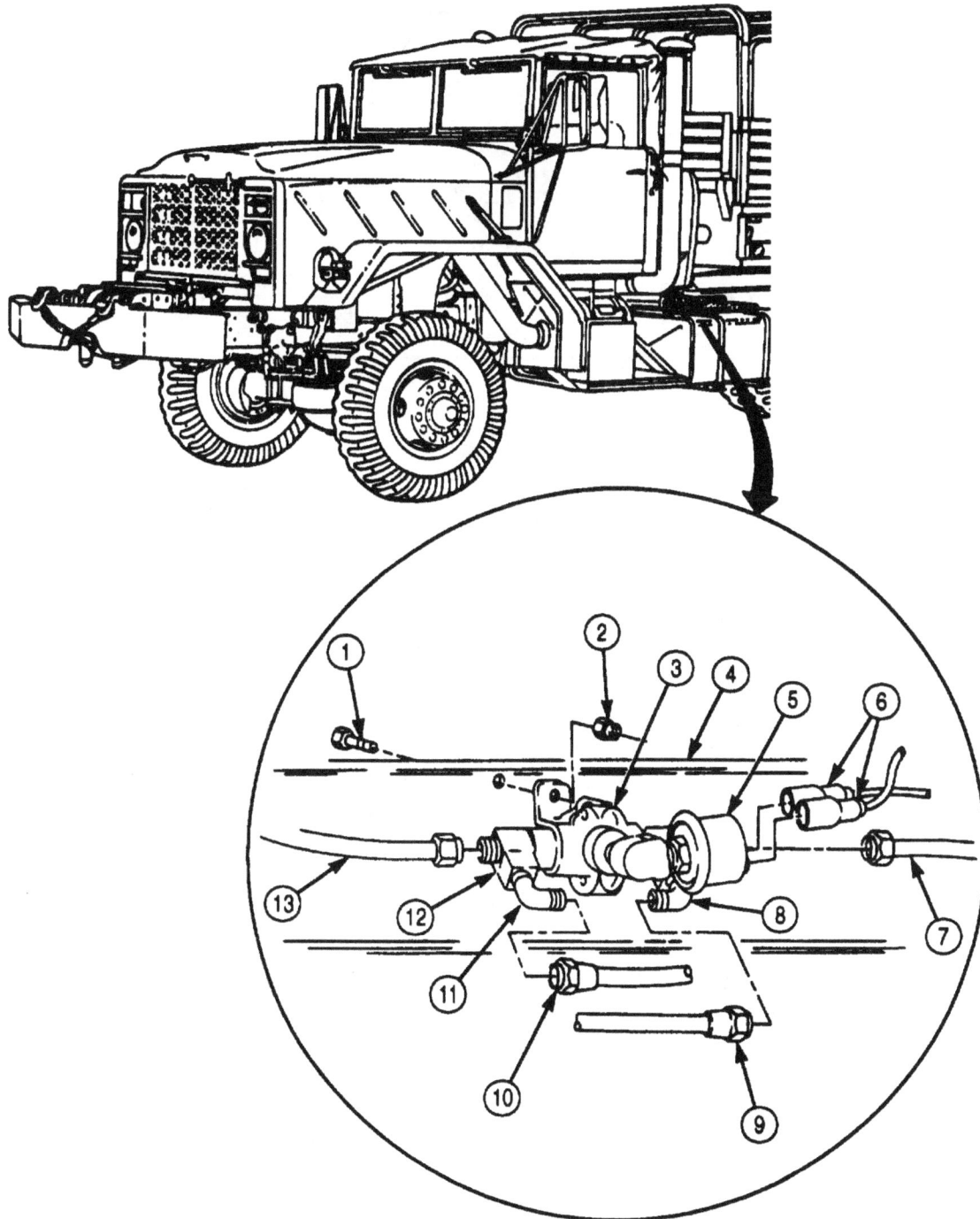

FOLLOW-ON TASK: Start engine (TM 9-2320-272-10) and allow air pressure to build to normal operating range. Check for air leaks at doublecheck valve No. 2. Road test vehicle.

3-214. DOUBLECHECK VALVE NO. 2 (M931/A1/A2, M932/A1/A2) REPLACEMENT

THIS TASK COVERS:

a. Removal   b. Installation

INITIAL SETUP:

APPLICABLE MODELS
M931/A1/A2, M932/A1/A2

TOOLS
General mechanic's tool kit (Appendix E, Item 1)

MATERIALS/PARTS
Locknut (Appendix D, Item 299)
Antiseize tape (Appendix C, Item 72)

REFERENCES (TM)
TM 9-2320-272-10
TM 9-2320-272-24P

EQUIPMENT CONDITION
• Parking brake set (TM 9-2320-272-10).
• Air reservoirs drained (TM 9-2320-272-10).

GENERAL SAFETY INSTRUCTIONS
Do not disconnect air lines before draining air reservoirs.

---

a. Removal

WARNING

Do not disconnect air lines before draining air reservoirs, Small parts under pressure may shoot out with high velocity, causing injury to personnel.

NOTE

• Doublecheck valve No. 5 is used with airbrake kits on M931/A1/A2 and M932/A1/A2 model vehicles. It is located inside the left frame rail in back of the stoplight switch.

• Tag air lines for installation.

1. Disconnect two wires (6) from stoplight switch (5).
2. Disconnect doublecheck valve No. 5 control line (8) from adapter tee (9).
3. Disconnect protection valve control line (10) and trailer handbrake control line (7) from doublecheck valve No. 2 (3).
4. Remove locknut (2), screw (1), and doublecheck valve No. 2 (3) from left frame rail (4). Discard locknut (2).

b. Installation

NOTE

If new doublecheck valve is being installed, use fittings from old check valve.

Wrap all male pipe threads with antiseize tape before installation.

1. Position doublecheck valve No. 2 (3) against left frame rail (4) and install with screw (1) and new locknut (2).
2. Connect protection valve control line (10) and trailer handbrake control line (7) to doublecheck valve No. 2 (3).
3. Connect doublecheck valve No. 5 control line (8) to adapter tee (9).
4. Connect two wires (6) to stoplight switch (5).

3-214. DOUBLECHECK VALVE NO. 2 (M931/A1/A2, M932/A1/A2) REPLACEMENT (Contd)

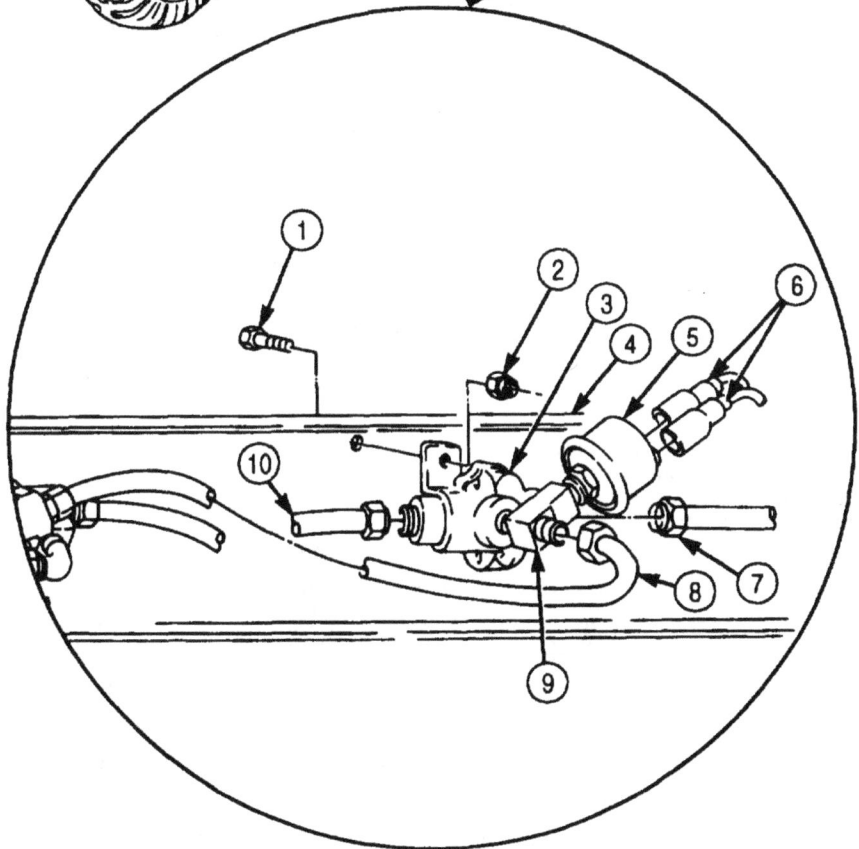

FOLLOW-ON TASK Start engine (TM 9-2320-272-10) and allow air pressure to build to normal operating range. Check for air leaks at doublecheck valve No. 2. Road test vehicle.

3-215. DOUBLECHECK VALVE NO. 5 (M931/A1/A2, M932/A1/A2) REPLACEMENT

THIS TASK COVERS:

a. Removal   b. Installation

INITIAL SETUP:

APPLICABLE MODELS
M931/A1/A2 and M932/A1/A2

TOOLS
General mechanic's tool kit (Appendix E, Item 1)

MATERIALS/PARTS
Locknut (Appendix D, Item 291)
Antiseize tape (Appendix C, Item 72)

REFERENCES (TM)
TM 9-2320-272-10
TM 9-2320-272-24P

EQUIPMENT CONDITION
Air reservoirs drained (TM 9-2320-272-10).

GENERAL SAFETY INSTRUCTIONS
Do not disconnect air lines before draining air reservoirs.

---

a. Removal

WARNING

Do not disconnect air lines before draining air reservoirs. Small parts under pressure may shoot out with high velocity, causing injury to personnel.

NOTE

Doublecheck valve No. 5 is used with airbrake kits on M931/A1/A2 and M932/A1/A2 model vehicles. It is located inside the left frame rail in back of the stoplight switch.

Tag air lines for installation.

1. Disconnect primary relay valve control line (14) from valve adapter tee (13).
2. Disconnect doublecheck valve No. 1 output line (11) from valve elbow (12).
3. Disconnect secondary relay valve control line (10) from valve elbow (9).
4. Disconnect doublecheck valve No. 2 control line (6) from bushing (5).
5. Disconnect treadle valve control line (7) from valve adapter tee (8).
6. Remove locknut (3), screw (1), and doublecheck valve No. 5 (4) from frame rail (2). Discard locknut (3).

b. Installation

NOTE

- If new valve is being installed, use fittings from old valve.
- Wrap all male pipe threads and wrap with antiseize tape before installation.

1. Position doublecheck valve No. 5 (4) against frame rail (2) and install with screw (1) and new locknut (3).
2. Connect treadle valve control line (7) to valve adapter tee (8).
3. Connect doublecheck valve No. 2 control line (6) to bushing (5).
4. Connect secondary relay valve control line (10) to valve elbow (9).
5. Connect doublecheck valve No. 1 output line (11) to valve elbow (12).
6. Connect primary relay valve control line (14) to valve adapter tee (13).

3-215. DOUBLECHECK VALVE NO. 5 (M931/A1/A2, M932/A1/A2) REPLACEMENT (Contd)

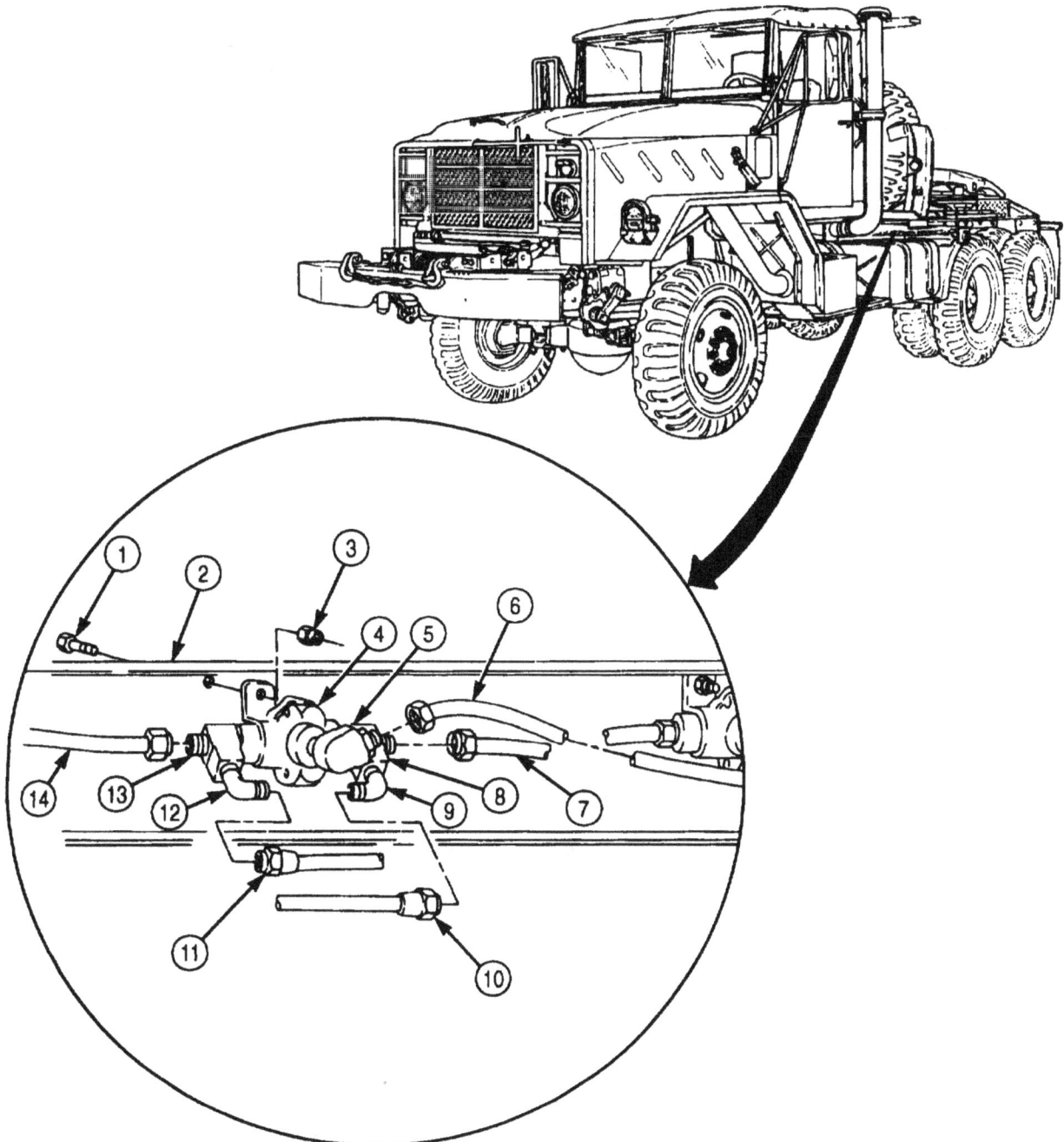

FOLLOW-ON TASK: Start engine (TM 9-2320-272-10) and allow air pressure to build to normal operating range. Check for air leaks at doublecheck valve No. 5. Road test vehicle.

3-216. DOUBLECHECK VALVE NO. 3, NO. 4, AND QUICK-RELEASE VALVE REPLACEMENT

THIS TASK COVERS:

a. Removal   b. Installation

INITIAL SETUP:

APPLICABLE MODELS
All

TOOLS
General mechanic's tool kit (Appendix E, Item 1)

MATERIALS/PARTS
Five locknuts (Appendix D, Item 299)
Eight lockwashers (Appendix D, Item 354)
Tiedown strap (Appendix D, Item 694)
Antiseize tape (Appendix C, Item 72)

REFERENCES (TM)
TM 9-2320-272-10
TM 9-2320-272-24P

EQUIPMENT CONDITION
• Parking brake set (TM 9-2320-272-10).
• Air reservoirs drained (TM 9-2320-272-10).

GENERAL SAFETY INSTRUCTIONS
Do not disconnect air lines before draining air reservoirs.

a. Removal

WARNING

Do not disconnect air lines before draining air reservoirs. Small parts under pressure may shoot out with high velocity, causing injury to personnel.

NOTE
Tag air lines for installation.

1. Remove locknut (4), screw (3), and clamp (5) from front step brace (1). Discard locknut (4)

2. Remove and discard tiedown strap (2).

3. Remove eight screws (6), lockwashers (24), and access box cover (22) from step box access (17). Discard lockwashers (24).

4. Disconnect air lines (14), (15), and (16) from doublecheck valve No. 4 (18).

5. Disconnect air lines (9), (13), and (14) from doublecheck valve No. 3 (21).

6. Disconnect air lines (10), (12), and (13) from quick-release valve (7).

7. Remove locknut (19), washer (20), screw (23), and doublecheck valve No. 4 (18) from access box cover (22). Discard locknut (19).

8. Remove locknut (19), washer (20), screw (23), and doublecheck valve No. 3 (21) from access box cover (22). Discard locknut (19).

9. Remove two locknuts (19), screws (23), and quick-release valve (7) from access box cover (22). Discard locknuts (19).

NOTE
Position valves in soft-jawed vise to remove fittings.

10. Remove two fittings (8) and fitting (5) from doublecheck valve No. 4 (18).

11. Remove two fittings (8) and fitting (11) from doublecheck valve No. 3 (21).

12. Remove three fittings (8) from quick-release valve (7).

3-216. DOUBLECHECK VALVE NO. 3, NO. 4, AND QUICK-RELEASE VALVE REPLACEMENT (Contd)

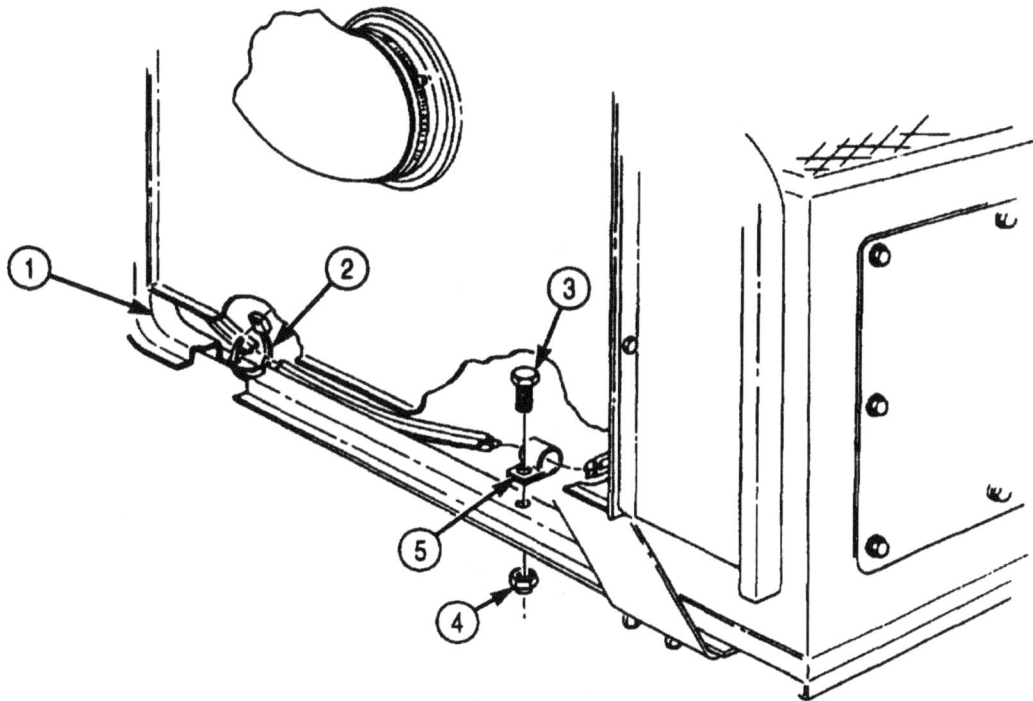

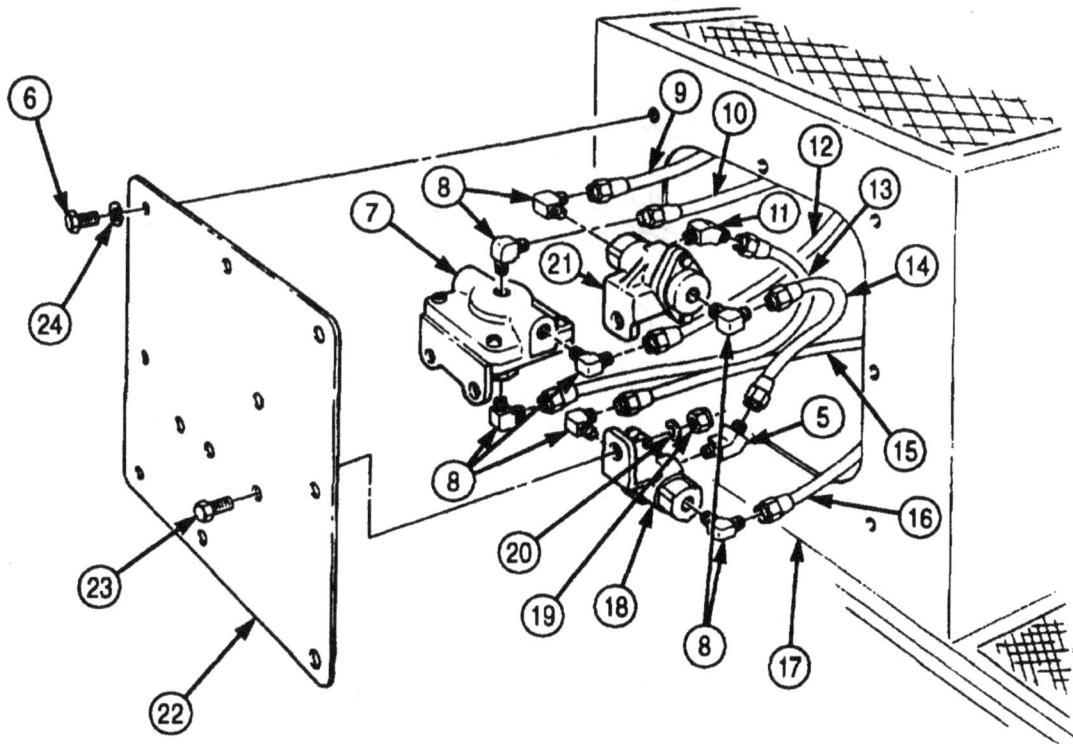

b. Installation

1. Install three fittings (2) on quick-release valve (1).

2. Install two fittings (2) and fitting (6) on doublecheck valve No. 3 (16).

3. Install two fittings (2) and fitting (5) on doublecheck valve No. 4 (12).

4. Install quick-release valve (1) on access box cover (16) with two screws (17), washers (14), and new locknuts (13).

5. Install doublecheck valve No. 3 (16) on access box cover (16) with screw (17), washer (14), and new locknut (13).

6. Install doublecheck valve No. 4 (12) on access box cover (16) with screw (17), washer (14), and new locknut (13).

7. Connect air lines (4), (6), and (7) to quick-release valve (1).

8. Connect air lines (3), (7), and (8) to doublecheck valve No. 3 (16).

9. Connect air lines (8), (9), and (10) to doublecheck valve No. 4 (12).

10. Install access box cover (16) on step box access (11) with eight new lockwashers (18) and screws (19).

11. Install new tiedown strap (21).

12. Install clamp (24) on front step brace (20) with screw (22) and new locknut (23).

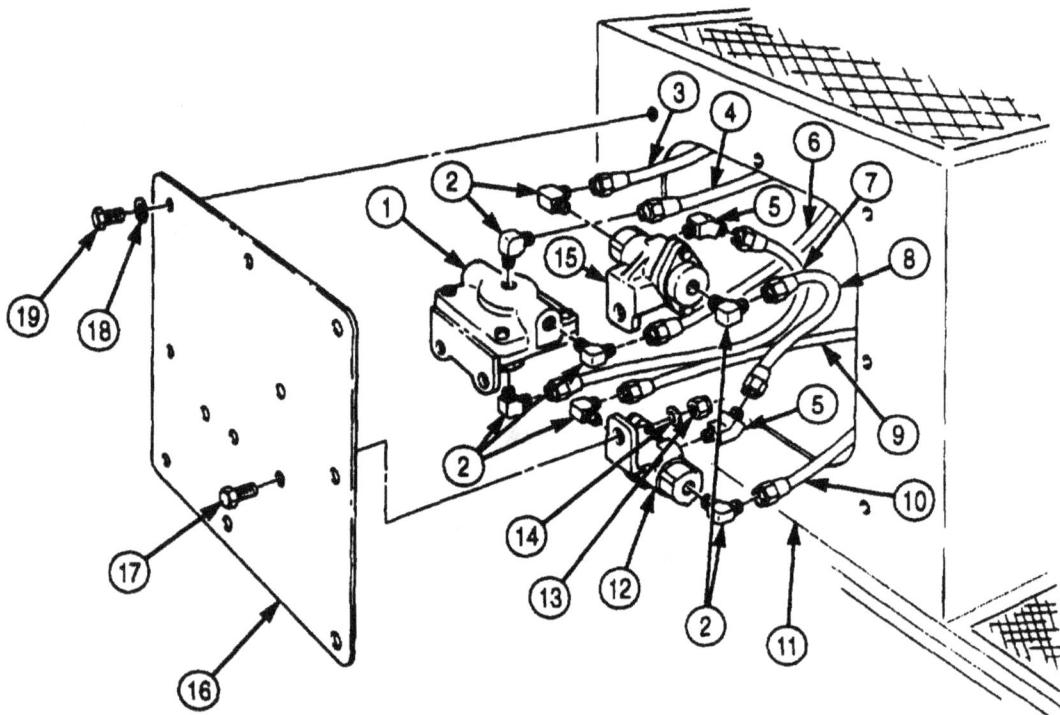

3-216. DOUBLECHECK VALVE NO. 3, NO. 4, AND QUICK-RELEASE VALVE REPLACEMENT (Contd)

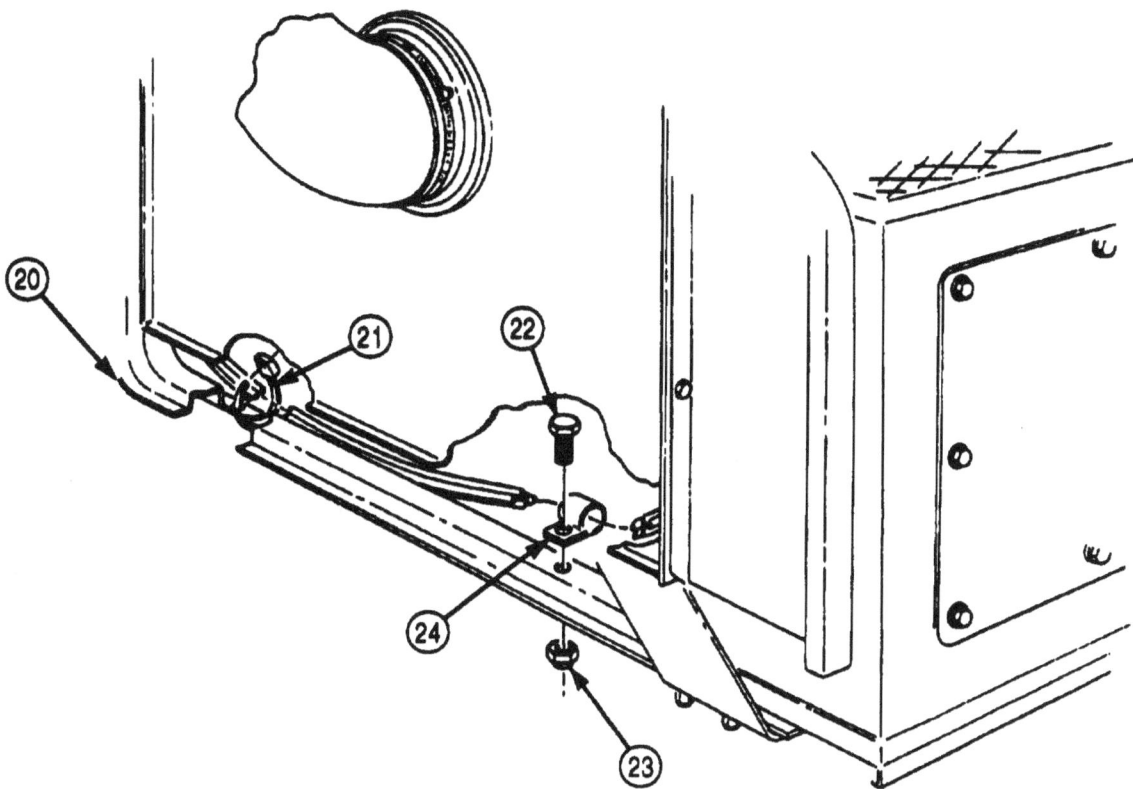

FOLLOW-ON TASK: Start engine (TM 9-2320-272-10) and allow air pressure to build to normal operating range. Check for air leaks at doublecheck valve No. 3, No. 4, and quick-release valve. Road test vehicle.

INDEX

INDEX (Contd)

INDEX (Contd)

INDEX (Contd)

INDEX (Contd)

INDEX (Contd)

INDEX (Contd)

INDEX (Contd)

INDEX (Contd)

INDEX (Contd)

INDEX (Contd)

INDEX (Contd)

INDEX (Contd)

INDEX (Contd)

INDEX (Contd)

INDEX (Contd)

INDEX (Contd)

INDEX (Contd)

INDEX (Contd)

INDEX (Contd)

INDEX (Contd)

INDEX (Contd)

INDEX (Contd)

INDEX (Contd)

THE METRIC SYSTEM AND EQUIVALENTS

LINEAR MEASURE
1 Centimeter = 10 Millimeters = 0.01 Meters = 0.3937 Inches
1 Meter = 100 Centimeters = 1,000 Millimeters = 39.37 Inches
1 Kilometer = 1,000 Meters = 0.621 Miles

SQUARE MEASURE
1 Sq Centimeter = 100 Sq Millimeters = 0.155 Sq Inches
1 Sq Meter = 10,000 Sq Centimeters = 10.76 Sq Feet
1 Sq Kilometer = 1,000,000 Sq Meters = 0.386 Sq Miles

CUBIC MEASURE
1 Cu Centimeter = 1,000 Cu Millimeters = 0.06 Cu Inches
1 Cu Meter = 1,000,000 Cu Centimeters = 35.31 Cu Feet

LIQUID MEASURE
1 Milliliter = 0.001 Liters = 0.0338 Fluid Ounces
1 Liter = 1,000 Milliliters = 33.82 Fluid Ounces

TEMPERATURE
$5/9 \, (°F - 32) = °C$
212° Fahrenheit is equivalent to 100° Celsius
90° Fahrenheit is equivalent to 32.2° Celsius
32° Fahrenheit is equivalent to 0° Celsius
$9/5 \, °C + 32 = °F$

WEIGHTS
1 Gram = 0.001 Kilograms = 1,000 Milligrams = 0.035 Ounces
1 Kilogram = 1,000 Grams = 2.2 Lb
1 Metric Ton = 1,000 Kilograms = 1 Megagram = 1.1 Short Tons

APPROXIMATE CONVERSION FACTORS

TO CHANGE	TO	MULTIPLY BY
Inches	Centimeters	2.540
Feet	Meters	0.305
Yards	Meters	0.914
Miles	Kilometers	1.609
Square Inches	Square Centimeters	6.451
Square Feet	Square Meters	0.093
Square Yards	Square Meters	0.836
Square Miles	Square Kilometers	2.590
Acres	Square Hectometers	0.405
Cubic Feet	Cubic Meters	0.028
Cubic Yards	Cubic Meters	0.765
Fluid Ounces	Milliliters	29.573
Pints	Liters	0.473
Quarts	Liters	0.946
Gallons	Liters	3.785
Ounces	Grams	28.349
Pounds	Kilograms	0.454
Short Tons	Metric Tons	0.907
Pound-Feet	Newton-Meters	1.356
Pounds Per Square Inch	Kilopascals	6.895
Miles Per Gallon	Kilometers Per Liter	0.425
Miles Per Hour	Kilometers Per Hour	1.609

TO CHANGE	TO	MULTIPLY BY
Centimeters	Inches	0.394
Meters	Feet	3.280
Meters	Yards	1.094
Kilometers	Miles	0.621
Square Centimeters	Square Inches	0.155
Square Meters	Square Feet	10.764
Square Meters	Square Yards	1.196
Square Kilometers	Square Miles	0.386
Square Hectometers	Acres	2.471
Cubic Meters	Cubic Feet	35.315
Cubic Meters	Cubic Yards	1.308
Milliliters	Fluid Ounces	0.034
Liters	Pints	2.113
Liters	Quarts	1.057
Liters	Gallons	0.264
Grams	Ounces	0.035
Kilograms	Pounds	2.205
Metric Tons	Short Tons	1.102
Newton-Meters	Pound-Feet	0.738
Kilopascals	Pounds Per Square Inch	0.145
Kilometers Per Liter	Miles Per Gallon	2.354
Kilometers Per Hour	Miles Per Hour	0.621

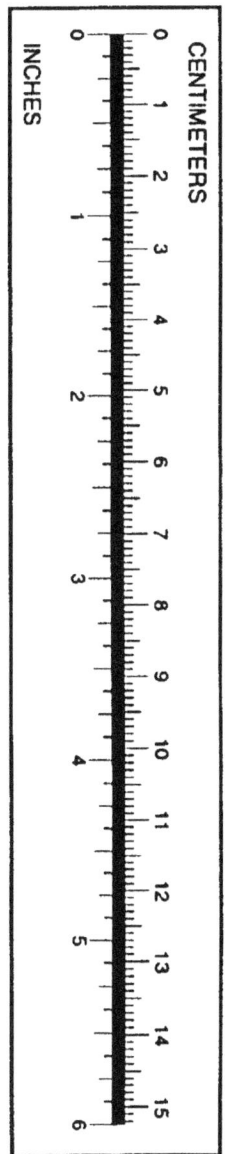

www.ingramcontent.com/pod-product-compliance
Lightning Source LLC
Chambersburg PA
CBHW081035050426
42335CB00052B/2426